What is Algebra?

The word "algebra" comes from a ninth-century text entitled *Hisâb al–jabr w'al–muqabâla*, which can be loosely translated as the "science of equations." However, algebra was used as early as 2000 B.C. in Babylonia and slightly later in Egypt. These ancients solved linear and some quadratic equations that arose in everyday activities. Their equations were written out in words with few symbols, and their number systems contained only positive integers and fractions. By about 200 B.C., mathematicians in China were working with systems of linear equations, and they made limited use of negative numbers.

In Greece starting around 500 B.C., geometric constructions were used to solve quadratic and cubic equations. The Greeks used letters to represent numbers, and their system was not suited to computations. However, they were the first mathematicians to conceive of irrational magnitudes. Also, around 250 A.D. the Greek Diophantus was the first mathematician to adopt a simplified algebraic notation consisting of special symbols and abbreviations. He is sometimes called the father of algebra.

Hindu mathematicians in the seventh century solved quadratic equations for both roots, including negative ones, and stated the rules for operations with signed quantities. However, negative numbers were not fully understood or accepted at this time. Our decimal system is based on the Hindu positional base 10 number system, which may have originated in China. The numerals 0, 1, 2, ..., 9 were introduced into western Europe by Arabs and are now called the Hindu-Arabic numerals.

Although the ancient Babylonians solved quadratic equations, it was not until the sixteenth century that algebraic formulas for solving cubic and quartic equations were developed. These formulas forced mathematicans to study the meaning of square roots of negative numbers. For example, the equation $x^3 = 15x + 4$ has 4 as a root, but the cubic formula gave 4 in the form

$$\sqrt[3]{2 + 11\sqrt{-1}} + \sqrt[3]{2 - 11\sqrt{-1}},$$

which would be written today as

$$2 + i + 2 - i,$$

where $i = \sqrt{-1}$. After solving the cubic and quartic equations, mathematicians began a search for formulas to solve equations of degree 5 and higher that was to last for 300 years.

In the seventeenth century the theory of equations was developed, and mathematical symbolism, including the use of x to denote the unknown, came close to its present form. Negative roots of equations were called "false roots," and $\sqrt{-1}$ was referred to as an "imaginary number."

Classical algebra culminated in the proof of two major theorems. The first of these, proven in 1799, states that in the complex number system, every polynomial equation of degree n has n roots. The second, proven in stages from about 1824 to 1830, states that general polynomial equations of degree 5 and higher cannot be solved by algebra. With these two results, algebra ceased to be the science of equations. Also, a graphical representation of complex numbers and their operations in the plane led to their general acceptance early in the nineteenth century, and by the middle of the century the defining properties of the real and complex number systems were determined. Other algebraic systems were investigated, and the modern era of abstract algebra began.

Algebraic Formulas

Exponents and Radicals

$x^m \cdot x^n = x^{m+n}$

$\dfrac{x^m}{x^n} = x^{m-n}$

$(xy)^n = x^n y^n$

$\left(\dfrac{x}{y}\right)^n = \dfrac{x^n}{y^n}$

$(x^m)^n = x^{mn}$

$x^{-n} = \dfrac{1}{x^n}$

$x^0 = 1, \quad x \neq 0$

$y = \sqrt[n]{x} \Leftrightarrow y^n = x,$
where $y > 0$ if $x > 0$

$x^{m/n} = \sqrt[n]{x^m} = (\sqrt[n]{x})^m$

$\sqrt[n]{xy} = \sqrt[n]{x}\,\sqrt[n]{y}$

$\sqrt[n]{\dfrac{x}{y}} = \dfrac{\sqrt[n]{x}}{\sqrt[n]{y}}$

Products

$(a + b)(a - b) = a^2 - b^2$
$(a + b)^2 = a^2 + 2ab + b^2$
$(a + b)^3 = a^3 + 3a^2b + 3ab^2 + b^3$

Binomial Theorem

$(a + b)^n = a^n + \cdots + {}_nC_r a^{n-r} b^r + \cdots + b^n,$

where ${}_nC_r = \dfrac{n!}{(n - r)!r!}$

Factoring

$a^2x^2 - b^2 = (ax + b)(ax - b)$
$a^2x^2 + 2abx + b^2 = (ax + b)^2$
$x^3 + a^3 = (x + a)(x^2 - ax + a^2)$
$x^2 + bx + c = (x + m)(x + n),$
where $m + n = b$ and $mn = c$

Line Segments and Straight Lines

Distance d from $P_1(x_1, y_1)$ to $P_2(x_2, y_2)$:
$d = \sqrt{(x_2 - x_1)^2 + (y_2 - y_1)^2}$

Midpoint $M(x, y)$ of segment P_1P_2:
$x = \frac{1}{2}(x_1 + x_2)$ and $y = \frac{1}{2}(y_1 + y_2)$

Slope m of line through P_1 and P_2:

$m = \dfrac{y_2 - y_1}{x_2 - x_1}, \quad x_1 \neq x_2$

Equations of Lines

Vertical line: $x = c$
Point-slope form: $y - y_1 = m(x - x_1)$
Slope-intercept form: $y = mx + b$
General linear form: $ax + by = c$

Quadratic Formula

The roots of the quadratic equation $ax^2 + bx + c = 0$, where $a \neq 0$, are

$$x = \frac{-b \pm \sqrt{b^2 - 4ac}}{2a}.$$

Functions

Polynomial function P:
$P(x) = a_nx^n + a_{n-1}x^{n-1} + \cdots + a_1x + a_0$

Rational function f:
$f(x) = P(x)/Q(x)$, where P and Q are polynomials

Exponential function g:
$g(x) = a^x$, where $a > 0$ and $a \neq 1$

Logarithmic function h:
$h(x) = \log_a x \Leftrightarrow a^{h(x)} = x$

Properties of Logarithms

$\log_a (xy) = \log_a x + \log_a y$

$\log_a \left(\dfrac{x}{y}\right) = \log_a x - \log_a y$

$\log_a (x^y) = y \log_a x$

$\log_a 1 = 0 \qquad \log_a a = 1$

$\log_b x = \dfrac{\log_a x}{\log_a b}$

College Algebra and Trigonometry

Basics Through Precalculus

Fall 2005 Edition

John J. Schiller
Marie A. Wurster

Temple University

Ψ *IMAGETEC PUBLISHING SYSTEMS*

Dedication

This book is dedicated to the inequalities:

$$\frac{a}{b+c} \neq \frac{a}{b} + \frac{a}{c} \quad \text{and} \quad \sqrt{a^2+b^2} \neq a+b.$$

ISBN 1-932864-35-0

By written permission from the authors, Ψ ImageTec Publishing Systems has the exclusive rights to reproduce this publication, to prepare derivative works from this work, and to publicly distribute and display this work.

Ψ ImageTec Publishing Systems

Preface

General goals, content, and organization

College Algebra and Trigonometry: Basics Through Precalculus, which first appeared in 1987, was written to deal with the diverse mathematical backgrounds of students preparing for calculus. The text differs from most precalculus texts of the late eighties and the present in two ways. It contains a very thorough review of basic algebra, and the topics in intermediate algebra are arranged according to a functional hierarchy from all linear topics, including equations, inequalities, functions, graphs, and applications to all quadratic topics broken down in the same way, followed by polynomial, rational, and general algebraic functions. The treatment of basic algebra serves as a reference for students whose hold on the subject is tenuous, and the hierarchical arrangement of intermediate algebra is intended to give students an organizational view of the subject as progressing logically from simple to more complex equations, functions, and graphs. Following algebraic functions are the core precalculus topics: exponential, logarithmic, and trigonometric functions. Trigonometric functions of angles are defined and applied to triangles before trigonometric functions of real numbers, based on the unit circle, are developed. This sequence was chosen because it moves from concrete, familiar objects to more abstract, less familiar ones. Systems of linear equations, matrix algebra, and the theory of equations are placed after trigonometry. The theory of equations requires more mathematical sophistication than previous topics, but it is portable, as are linear systems and matrices, and can be covered immediately after algebraic functions. The final chapter of the text includes a brief treatment of mathematical induction, arithmetic and geometric sequences and series, permutations and combinations, and probability.

Reform intermediate algebra and precalculus texts

Now, almost twenty years after the first edition of the text, despite an intense emphasis on standards and the use of technology in pre-college mathematics curricula nationwide, students enrolling in college programs requiring calculus continue to have a wide range of backgrounds in algebra. The need for a review of basic algebra in intermediate algebra and precalculus texts is as strong now as it was twenty years ago. However, most, if not all, of the present-day reform texts do not deal with basic algebra, nor do they develop the notion of function in any hierarchical manner. Students are exposed to a variety of functions, including transcendental functions, from the get-go. Algebraic manipulations are downplayed, while calculator-assisted computation and graphics are emphasized. There are many inviting aspects to these approaches, and students get to work through some very practical and interesting problems. But in spite of the merits of the reform texts, we do not feel that they are an appropriate preparation for the rigorous calculus required in mathematics, natural science, and engineering. In trying to make precalculus topics accessible to all students, regardless of their backgrounds, we believe the reform texts shortchange those whose major field of study is mathematics intensive. The further one goes in mathematics, the more important algebraic skills become. In our view, the reform precalculus texts link only to the reform calculus texts, neither of which develops sufficient algebraic skills for more advanced courses.

Use of the text

The text can be used for core courses in intermediate algebra (Chapters 1-4) and precalculus (Chapters 5-8). In intermediate algebra it may also be possible to cover part of Chapter 5 (exponential and logarithmic functions), and in precalculus a selection of topics from Chapters 9 (linear systems and matrix algebra), 10 (theory of equations), and 11 (sequences, series, and probability). Chapter P (basic algebra) serves as a reference for either course. We have found that an effective way of dealing with the heterogeneous backgrounds of entering students is to first go over a selection of review exercises at the *end* of chapters P and/or 1, and back into the corresponding review outlines or text as needed. After about three or four sessions of "filling in the gaps" in this manner, the class as a whole is ready to start the course proper.

Use of calculators

When the first edition of the text appeared, scientific calculators were already in common use in precalculus courses, but graphing calculators were not. In the second edition, a generous supply of graphing calculator exercises was added to the review exercises, and these have been retained in further editions. The exercises were written for the TI-81, but they can easily be adapted to any graphing calculator. For example, each screen pixel in the TI-81 has width $(x_{max} - x_{min})/95$, so that a window having $x_{max} - x_{min}$ equal to 19, 9.5, or 4.75 will give a "friendly" Δx in the trace mode equal to .2, .1, or .05, respectively. The pixel width in the TI-83 is $(x_{max} - x_{min})/94$, so that an $x_{max} - x_{min}$ equal to 18.8, 9.4, or 4.7 will then give a Δx in the trace mode equal to .2, .1, or .05, respectively. Students taking precalculus have various backgrounds in the use of graphing calculators, and we assume that the instructor will supplement or adapt, as needed, the brief instructions given in the exercises regarding the arithmetic, graphing, matrix, and programming capabilities of graphing calculators. Scientific calculators have made tables of exponentials, logarithms, and trigonometric functions obsolete, so such tables have been removed from the original text.

Supplements

For instructors, the **Instructor's Manual** contains complete step-by-step solutions to even-numbered text exercises and three different but equivalent forms of tests for each chapter with complete solutions. For students, the **Study and Solutions Manual** contains step-by-step solutions to odd-numbered text exercises and practice chapter tests, also with complete solutions.

Acknowledgements

Overall, we have tried to produce a text that is strong in fundamentals, flexible, and brings out the essential unity of the subject. This would not have been possible without the help of many people. We extend our sincere thanks to all those who provided suggestions and assistance throughout the development of the project, especially the following reviewers:

James E. Arnold, University of Wisconsin
Nancy J. Bray, San Diego Mesa College
Carl T. Carlson, Moorhead State University
Ben L. Cornelius, Oregon Institute of Technology
Nancy Harbour, Brevard Community College
Nancy G. Hyde, Broward Community College
DeWayne S. Nymann, University of Tennessee
Robert Pruitt, San Jose State University
Ken Seydel, Skyline College
Mahendra Singhal, University of Wisconsin
Shirley C. Sorensen, University of Maryland
Michael D. Taylor, University of Central Florida
Marvel D. Townsend, University of Florida
Thomas J. Woods, Central Connecticut University

In addition, the following mathematicians provided us with comments and suggestions regarding the topical coverage and organization:

Jasper Adams, Stephen F. Austin State University
Carol A. Edwards, St. Louis Community College
Allen Epstein, West Los Angeles College
E. John Hornsby, Jr., University of New Orleans
Marvin Roof, Mott Community College
John L. Whitcomb, University of North Dakota

We would also like to thank those who helped check the answers at the back of the book: John A. Dersch, Jr., Grand Rapids Junior College; Nancy Harbour, Brevard Community College; Nancy G. Hyde, Broward Community College; and Karen E. Zak, United States Naval Academy. Among our colleagues at Temple, we especially wish to acknowledge the work of Elizabeth Van Dusen, who has made significant contributions to an effective use of graphing calculators in precalculus.

Contents

Preliminaries

Core Intermediate Algebra

Core Precalculus

Additional Topics

P Preliminaries: Basic Algebra

This chapter is a review of the most basic notions in algebra: addition, subtraction, multiplication, and division of real numbers. Since many readers are already familiar with most of this material, we will emphasize the "why" of algebra rather than the "how." Fortifying our understanding of the fundamentals at this stage will help to improve our skills in the chapters that follow.

P.1 The Real Number System

"God created the integers; the rest is the work of man."
—Leopold Kronecker

The Set of Real Numbers
Order and the Real Number Line
Algebraic Operations

The Set of Real Numbers. Children are introduced to real numbers when they learn to count objects, and the counting numbers 1, 2, 3, . . . are called **positive integers.** The absence of objects to count suggests the number 0, an integer that is not positive. Parts of objects suggest fractional numbers such as 1/2, 2/3, or 3/4. These are ratios of integers, and any such number is called a **rational number.** The positive integers and 0 are special cases of rational numbers. For example, $5 = 5/1$ and $0 = 0/1$.

Geometric measurements lead to numbers other than rational ones. For example, in a square whose side measures 1 unit, the diagonal measures $\sqrt{2}$ units, and it can be shown that $\sqrt{2}$ is not a rational number. Also, a circle whose diameter is 1 unit has a circumference of π units, and π is not a rational number.[1] These are just two examples from an infinite collection of nonrational numbers called **irrational numbers.**

Both rational and irrational numbers can be expressed as **decimals.** The decimal expansion of a rational number either will terminate, as in $3/4 = .75$, or will eventually have a pattern that repeats indefinitely, as in $25/99 = .25252525 \ldots$. The decimal expansion of an irrational number neither terminates nor becomes a repeating pattern. For example, although π is approximately 3.14159, its true decimal expansion is infinitely long and has no repeating pattern.[2]

[1] *The Babylonians of 2000 B.C. set $3\frac{1}{8}$ as the value of π. In 1761 the Swiss mathematician Johann Lambert proved that π is irrational.*

[2] *In 1984 a team of mathematicians at the University of Tokyo, using a supercomputer, determined π to 16 million decimal places. At 5000 decimals per page, a printout of this expansion would require 3200 pages.*

Example 1 A standard 8-place calculator will give 1.4142136 as the decimal value of $\sqrt{2}$. This is accurate enough for most purposes, but in fact $(1.4142136)(1.4142136) = 2.0000010642496$. Hence, the rational number 1.4142136 is slightly larger than the irrational number $\sqrt{2}$. ■

Example 2 Express the repeating decimal .49999999 . . . as the ratio of two integers.

Solution: Let $x = .49999999 \ldots$. Then $10x = 4.9999999 \ldots$ and $100x = 49.9999999 \ldots$. Subtract $10x$ from $100x$ to obtain $90x = 45.0$. Hence, $x = 45/90 = 1/2$. We note that $1/2$ also has the *terminating* decimal expansion .5 (see Exercises 13–16 at the end of this section). ■

Example 3 The infinite decimal expansion 1.01001000100001 . . . has a nonrepeating pattern because the number of 0's between successive pairs of 1's is always increasing by one. Hence, the decimal represents an irrational number. ■

The **negative integers** $-1, -2, -3, \ldots$ are suggested by debts and budget deficits, and every positive rational and irrational number has its negative counterpart: for example, $3/4$ and $-3/4$, and π and $-\pi$. The total collection of rational and irrational numbers forms the **set of real numbers.** These are the numbers with which we work in algebra.

Comment Starting with the positive integers, we arrived at the entire set of real numbers in less than two pages of text. The actual development of the real numbers was a gradual process spread out over many centuries. A classic account of this development is contained in the book *Number, The Language of Science,* by Tobias Dantzig.

Order and the Real Number Line. Geometrically, the real numbers can be identified with the points on a directed line marked off with a number scale (Figure 1). Each point on the line corresponds to a decimal expansion of a real number, and conversely, a decimal expansion of a real number occupies one point on the line. Positive numbers are to the right of 0 on the line and negative numbers to the left. The number 0 is neither positive nor negative. The rational and irrational numbers are densely packed along the real number line. In fact, between any two rational numbers there is an irrational number, and between any two irrational numbers there is a rational number.

The real numbers are arranged in increasing order from left to right along the number line. Since the number 8 is to the right of 4 on the number line, we say 8 is **greater than** 4 and write $8 > 4$. Equivalently, we say 4 is **less than** 8 and write $4 < 8$. Any positive number is greater than any negative number, regardless of magnitude. Hence, $7 > -15$ and $1 > -100$. Also, $0 > -3$ and $-5 > -20$ because 0 is to the right of -3 and -5 is to the right of -20. Since any two unequal real numbers can be compared in this manner, the set of real numbers is called an **ordered set.**

The magnitude of a real number, without regard to sign, is called the **absolute value** of the number and is denoted by vertical bars $|\ \ |$. For example, $|5| = 5$ and $|-5| = 5$. The absolute value of a number is its distance from 0 on the number line.

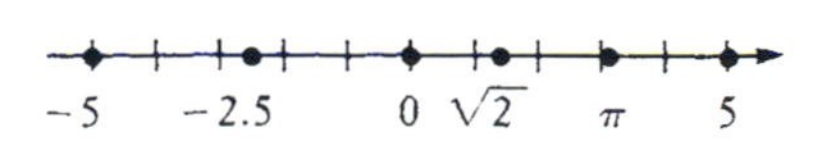

Figure 1 The Real Number Line

Example 4 Arrange the numbers 2, −11, 0, −5, −1, 6, π in increasing order.

Solution: $-11 < -5 < -1 < 0 < 2 < \pi < 6$ ■

Example 5 Arrange the numbers −3, $\sqrt{2}$, −7, 4.3, 0, $|-4|$, −2 in decreasing order.

Solution: $4.3 > |-4| > \sqrt{2} > 0 > -2 > -3 > -7$ ■

Algebraic Operations. The real numbers form a **closed system** with respect to the operations of addition, subtraction, and multiplication. That is, when any one of these operations is performed on two real numbers, the result always *remains* within the set of real numbers. Also, division of any real number by any real number *except* 0 always will result in a real number. The ordered set of real numbers, together with the four operations of addition, subtraction, multiplication, and division, make up the **real number system.**

Question 1 *Do the integers by themselves form a closed system under addition, subtraction, and multiplication?*

Answer: Yes, because the sum, difference, and product of any two integers is still an integer. ■

Question 2 *Do the odd integers form a closed system under the operation of addition?*

Answer: No, they do not form a closed system because the sum of two odd integers is not an odd integer. ■

The usual symbols + and − will be used for addition and subtraction, respectively. The product of two numbers a and b will be denoted by $a \cdot b$ or ab, and the quotient by $a \div b$, or a/b, provided $b \neq 0$. The properties of addition, subtraction, multiplication, and division within the real number system constitute our subject matter for the study of algebra. In the next section we set down nine *basic properties* of algebra. In succeeding sections and chapters we discover further properties that follow from the basic ones. Your ability to "do" algebra will be directly related to your understanding of these properties.

Exercises P.1

Fill in the blanks to make each statement true.

1. Every real number is either a ____________ number or an ____________ number.
2. A calculator can generate only numbers that are terminating decimals. Hence, all numbers generated on a calculator are ____________ numbers.
3. An irrational number has a decimal expansion that ____________.
4. The only real number that is neither positive nor negative is ____________.

5. The real numbers form a closed system with respect to the operations of ______________, ______________, and ______________.

Write true *or* false *for each statement.*

6. Every integer is a rational number.
7. The decimal .12345678910111213141 5 . . . , which continues through all the positive integers, is a rational number.
8. The decimals .4999999 . . . and .5 occupy different points on the number line.
9. Every positive number is the absolute value of two real numbers.
10. There is no largest or smallest real number.

The Set of Real Numbers

Supply the missing numbers in Exercises 11 and 12.

11. A student has just finished the seventh problem of a test, and there are three more to go. The part of the test that has been finished corresponds to the rational number ________, and the part yet to be finished corresponds to ________.
12. A pizza is cut into eight equal slices. Each slice corresponds to the rational number ________. If Terry and Pam finish the entire pizza, but Terry eats three times as many slices as Pam, then this suggests the rational number ________ for Terry's portion and ________ for Pam's.

Every terminating decimal can also be written as a repeating decimal (each represents the same point on the number line). For example, 1.28 = 1.27999999 . . . *and* .241 = .240999999 . . . . *Express each number in Exercises 13–16 as a repeating decimal.*

13. .35
14. .724
15. .667
16. 1.0

17. Express the rational number 1/3 as a decimal.
18. Express the rational number 2/7 as a decimal.

Express each decimal in Exercises 19–22 as the ratio of two integers.

19. .123
20. 1.25
21. .555555 . . .
22. .2121212121 . . .

23. Give the decimal expansion of a rational number between the rational numbers .5 and .5001.
24. Give the decimal expansion of a rational number between the irrational numbers .101001000100001 . . . and .102003000400005 . . . .

Order and the Real Number Line

25. Arrange the following numbers in increasing order.

    $\frac{22}{7}$, 3.14, π, 1.41, $\sqrt{2}$, -3, -3.5, $-\sqrt{9.1}$
26. Arrange the following numbers in decreasing order.

    $-\frac{1}{2}$, $-.501$, $-.499$, 0, .249, $\frac{1}{4}$, $\frac{1}{3}$, $|-.333|$
27. Locate the following numbers on a real number line.

    3.5, 2.75, -5, -4.25, 0, 1.5, -2, .75
28. Locate the following numbers on a real number line.

    $\frac{3}{5}$, $-\frac{1}{2}$, $-2\frac{2}{3}$, $5\frac{3}{4}$, $\frac{7}{8}$, $\frac{10}{3}$, $-\frac{7}{2}$

The middle value of an odd number of distinct values is called the **median** *of the collection. For example, the median of the numbers* 1, 3, 6, 8, *and* 15 *is* 6. *In Exercises 29–32, arrange the numbers in increasing order and find the median.*

29. weights: 110 lb, 140 lb, 135 lb, 150 lb, 100 lb
30. heights: 5′4″, 6′, 4′11″, 5′7″, 6′3″, 5′10″, 5′2″
31. ages: 17 yr, 18 yr, 22 yr, 19 yr, 20 yr, 21 yr, 16 yr
32. grades; 60, 98, 75, 45, 85, 70, 53, 67, 72, 88, 77

The median of an even number of distinct values is the **average** *of the two middle values of the collection. For example, the median of the numbers* 5, 7, 9, 14, 17, 21, 32, 38 *is* (14 + 17)/2 = 15.5. *In Exercises 33–36, arrange each list of numbers in increasing order and find the median.*

33. wages: \$325, \$475, \$250, \$625, \$125, \$750, \$350, \$500
34. salaries: \$20,000 \$15,000, \$30,000, \$10,000, \$25,000, \$35,000
35. interest rates: 8.5%, 5.5%, 5.25%, 9.75%, 10%, 10.5%
36. stock prices: $16\frac{3}{4}$, $17\frac{1}{8}$, $16\frac{7}{8}$, $17\frac{3}{4}$, $17\frac{1}{4}$, $16\frac{1}{2}$, $16\frac{1}{8}$, $17\frac{1}{2}$

Algebraic Operations

In Exercises 37–40, state the operation(s) (+, −, ·, ÷) under which the given system is closed.

37. the real numbers with 0 deleted
38. the rational numbers
39. the rational numbers that have terminating decimals
40. the real numbers between 0 and 1
41. Find a subset of the real numbers that is closed under multiplication but not under addition.
42. Find a subset of the real numbers that is closed under addition but not under multiplication.
43. When Andrew received his savings account statement for the month of June, he observed that his account was closed with respect to addition. How much money was in Andrew's account?

P.2 Basic Properties of Real Numbers

"It must be realized that the essence of algebra is in its generality."
—*Philip E. B. Jourdain*

Commutative, Associative, and Distributive Properties
Identities and Inverses
Grouping Symbols and Priority Conventions
Extended Commutative, Associative, and Distributive Properties

Commutative, Associative, and Distributive Properties. The only difference between algebra and arithmetic is that in algebra we often use letters in place of numbers. The arithmetic rules that govern the use of real numbers are the same rules that apply to letters representing real numbers. For example, we know that $7 + 5 = 5 + 7$, and if we replace 7 or 5 by other numbers, the equality still holds. In letters, we say

$$a + b = b + a, \qquad \textbf{commutative property of addition} \qquad (1)$$

where a and b stand for *any two real numbers*. This basic property illustrates the power and importance of algebra. The particular equation $7 + 5 = 5 + 7$ is a statement about the specific real numbers 7 and 5, whereas the general equation $a + b = b + a$ is a statement about the entire set of real numbers. The general equation contains infinitely more information than the particular one.

In the following basic properties, a, b, and c represent any real numbers.

$$(a + b) + c = a + (b + c) \qquad \textbf{associative property of addition} \qquad (2)$$

For example, $(8 + 4) + 16 = 8 + (4 + 16)$.

$$a \cdot b = b \cdot a \qquad \textbf{commutative property of multiplication} \qquad (3)$$

For example, $5 \cdot 12 = 12 \cdot 5$.

$$(a \cdot b) \cdot c = a \cdot (b \cdot c) \qquad \textbf{associative property of multiplication} \qquad (4)$$

For example, $(3 \cdot 4) \cdot 10 = 3 \cdot (4 \cdot 10)$.

$$a \cdot (b + c) = (a \cdot b) + (a \cdot c) \qquad \textbf{left distributive property} \qquad (5)$$

For example, $6 \cdot (5 + 7) = (6 \cdot 5) + (6 \cdot 7)$.

Question 1 *Is either subtraction or division a commutative or an associative operation?*

Answer: No. For example, $8 - 2 = 6$ but $2 - 8 = -6$, and $8/2 = 4$ but $2/8 = 1/4$. Also, $(12 - 6) - 2 = 4$ but $12 - (6 - 2) = 8$, and $(12 \div 6) \div 2 = 1$ but $12 \div (6 \div 2) = 4$. ■

Question 2 *Does the left distributive property hold if + and · are interchanged? In other words, is it true that $a + (b \cdot c) = (a + b) \cdot (a + c)$ for all values of a, b, and c?*

Answer: No. For example, $3 + (4 \cdot 5) = 23$ but $(3 + 4) \cdot (3 + 5) = 56$. ■

We can combine basic properties (3) and (5) to derive the following property.

$$(a + b) \cdot c = (a \cdot c) + (b \cdot c) \qquad \textbf{right distributive property}$$

For example, $(6 + 8) \cdot 5 = (6 \cdot 5) + (8 \cdot 5)$.

Identities and Inverses. The number 0 is called the additive identity or the neutral element for addition because for any real number a,

$$a + 0 = a = 0 + a. \qquad \textbf{additive identity property} \qquad (6)$$

Similarly, the number 1 is called the multiplicative identity or the neutral element for multiplication because for any real number a,

$$a \cdot 1 = a = 1 \cdot a. \qquad \textbf{multiplicative identity property} \qquad (7)$$

Both 0 and 1 are *unique* in that they are the only identity elements for addition and multiplication, respectively.

If -7 is added to 7, the result is the additive identity 0. For this reason -7 is called the additive inverse of 7, and conversely, 7 is the additive inverse of -7. Every real number a has a *unique* additive inverse, which is denoted by $-a$ and which satisfies

$$a + (-a) = 0 = -a + a. \qquad \textbf{additive inverse property} \qquad (8)$$

It is important to note that $-a$ is not necessarily a negative number. For example, if $a = -7$, then $-a = 7$ because $-7 + 7 = 0$. Hence, $-(-7) = 7$.

In general, a and $-a$ are the same distance from 0 on the number line but on opposite sides of 0. In other words, the effect of placing a minus sign before a number is to reverse its direction on the number line. If a is positive, then $-a$ is negative, but if a is negative, then $-a$ is positive.

Always,

$$-(-a) = a. \qquad \textbf{double inverse property}$$

That is, the additive inverse of the additive inverse of a is a itself, or a double reversal of a on the number line leads back to a.

If 5 is multiplied by its reciprocal $1/5$, the result is the multiplicative identity 1. For this reason $1/5$ is called the multiplicative inverse of 5, and conversely, 5 is the multiplicative inverse of $1/5$. The reciprocal $1/5$ is also denoted by 5^{-1}. In general, every nonzero number a has a *unique* multiplicative inverse, which is denoted by $1/a$ or a^{-1}, and which satisfies

$$a \cdot a^{-1} = 1 = a^{-1} \cdot a. \qquad \textbf{multiplicative inverse property} \qquad (9)$$

The number 0 does not have a multiplicative inverse. If it did, then 0^{-1} would have to be a real number that satisfies the equation $0 \cdot 0^{-1} = 1$. However, 0 times any real number is equal to 0, as is shown in the following example.

Example 1 Show that, for any real number a,

$$0 \cdot a = 0. \qquad \textbf{zero-multiplier property}$$

Solution: An actual proof of this apparently simple result is by no means simple. We give one here in order to illustrate the use of the basic properties. At this stage in your study of algebra you would not be expected to think up such a proof, but you should be able to follow it step by step.

By the additive identity property,

$$0 + 0 = 0.$$

Multiply both sides of the above equation by a:

$$(0 + 0) \cdot a = 0 \cdot a.$$

Apply the right distributive property to the left side:

$$(0 \cdot a) + (0 \cdot a) = 0 \cdot a.$$

Let x stand for $0 \cdot a$. The above equation says that

$$x + x = x.$$

Add $-x$ to both sides:

$$(x + x) + (-x) = x + (-x).$$

Apply the associative property for addition to the left side and the additive inverse property to the right:

$$x + [x + (-x)] = 0.$$

Apply the additive inverse property to the quantity in brackets on the left side:

$$x + 0 = 0.$$

Apply the additive identity property to the left side:

$$x = 0.$$

Hence, $0 \cdot a = 0$, which completes the proof. ∎

Another way of saying that 0 has no multiplicative inverse is to say that 1/0 is a meaningless expression or that division by 0 is not possible. Remember that the real numbers form a closed system under addition, subtraction, and multiplication, but division by 0 is not defined.

In the proof in Example 1 above, we assumed the following algebraic and logical properties of equality of real numbers.

Algebraic Properties of Equality

If $a = b$, then $a \pm c = b \pm c$.	Any number may be added or subtracted on both sides of an equation.
If $a = b$, then $a \cdot c = b \cdot c$.	An equation may be multiplied on both sides by any real number.
If $a = b$ and $c \neq 0$, then $\frac{a}{c} = \frac{b}{c}$.	An equation may be divided on both sides by a nonzero real number.

Logical Properties of Equality

$a = a$ for any real number a.	**reflexive property**
If $a = b$, then $b = a$.	**symmetric property**
If $a = b$ and $b = c$, then $a = c$.	**transitive property**

These properties also justify the familiar process of **substitution,** that is, if $a = b$, we may substitute b for a in any algebraic equation.

By the zero-multiplier property and the commutative property for multiplication we know that if either $a = 0$ or $b = 0$, then $ab = 0$. Conversely, if $ab = 0$, then either $a = 0$ or $b = 0$ (see Exercise 32). Therefore, the statement "$ab = 0$" is equivalent to the statement "$a = 0$ or $b = 0$."

We will use the double arrow $\Leftrightarrow$ *(read "if and only if") to indicate that two statements are equivalent.* Hence,

$$ab = 0 \Leftrightarrow a = 0 \text{ or } b = 0. \qquad \textbf{zero-product property}$$

For example,

$$\begin{aligned}(x-1)(x+2) = 0 &\Leftrightarrow x - 1 = 0 \quad \text{or} \quad x + 2 = 0 \\ &\Leftrightarrow \quad x = 1 \quad \text{or} \quad x = -2.\end{aligned}$$

The zero-product property extends to more than two factors. For instance,

$$\begin{aligned}x(x-3)(x+4) = 0 &\Leftrightarrow x = 0 \quad \text{or} \quad x - 3 = 0 \quad \text{or} \quad x + 4 = 0 \\ &\Leftrightarrow x = 0 \quad \text{or} \quad x = 3 \quad \text{or} \quad x = -4.\end{aligned}$$

Analogous to the equation $-(-a) = a$ for additive inverses, we have, for multiplicative inverses, the equation

$$(a^{-1})^{-1} = a \quad (a \neq 0). \qquad \textbf{double inverse property}$$

The double inverse property for multiplication follows because the expression $(a^{-1})^{-1}$ means the unique number that when multiplied by a^{-1} results in 1. Since we know that $a^{-1} \cdot a = 1$, it follows that $(a^{-1})^{-1} = a$. Hence, $(5^{-1})^{-1} = 5$, which is another way of saying that $1/(1/5) = 5$. In general, $1/(1/a) = a$ for any real number $a \neq 0$.

If $a \neq 0$, then the numbers a and $1/a$ are on the same side of 0 on the number line. That is, if a is positive, so is $1/a$, and if a is negative, so is $1/a$. However, a positive number and its reciprocal are on opposite sides of 1 on the number line (*why?*), whereas a negative number and its reciprocal are on opposite sides of -1. The reciprocal $1/a$ of a number a can be located on the number line by means of the geometric construction indicated in Figure 2 (see Exercise 50). The circle in Figure 2 has its center at 0, and its radius is 1. Line L is tangent to the circle.

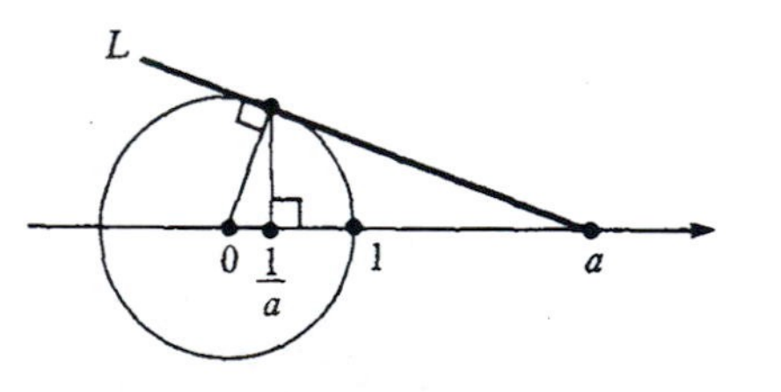

Figure 2 Construction of $1/a$

Basic properties (1) through (9) are the basis for all other algebraic properties of the real number system, including those properties related to order (see Section P.3). In Example 1, we saw how the basic properties are used to derive further ones. The following example is another illustration of this process. See if you can supply the reason for each step. Each reason is one of the nine basic properties or one of the properties already derived from them.

Example 2 Show that, for any real number a,

$$-a = (-1)a. \qquad \textbf{negative multiplier property}$$

That is, the additive inverse of a is equal to a multiplied by -1.

Solution: By definition, $-a$ is the unique number that when added to a results in 0. Hence, we would like to show that $a + (-1)a = 0$. We proceed as follows.

$$\begin{aligned} a + (-1)a &= (a \cdot 1) + (-1)a && \underline{\qquad\qquad} \text{ (basic property)} \\ &= (a \cdot 1) + a(-1) && \underline{\qquad\qquad} \text{ (basic property)} \\ &= a \cdot [1 + (-1)] && \underline{\qquad\qquad} \text{ (basic property)} \\ &= a \cdot 0 && \underline{\qquad\qquad} \text{ (basic property)} \\ &= 0 && \underline{\qquad\qquad} \text{ (derived property)} \end{aligned}$$ ■

We will not labor this process of deriving new properties in the following sections and chapters. Our main goal is to use algebra as a tool rather than to forge this tool. However, it is important to keep in mind that every algebraic property we learn can be traced to the nine basic ones.

Question 3 *Why do the basic properties involve only addition and multiplication, not subtraction or division?*

Answer: The operations of subtraction and division can be defined in terms of addition and multiplication, respectively (see Section P.3). When we say that $a - b = c$, we mean that c is the unique number that when *added* to b results in a, that is, $a - b = c$ means $c + b = a$. Similarly, $a \div b = c$ means $c \cdot b = a$. Hence, it is not necessary to state basic properties of subtraction and division. The properties of these operations can be worked out by using the established properties of addition and multiplication. ■

Grouping Symbols and Priority Conventions. When several operations are applied to numbers or letters, grouping symbols such as parentheses ( ), brackets [ ], or braces { } are used to indicate the order in which the operations are to be performed. For example, $(6 \cdot 3) + (8 \div 2) = 18 + 4 = 22$. The operations inside the parentheses are performed first. If grouping symbols appear within other grouping symbols, the operations are performed inside out; that is, the operations within the innermost grouping symbols are performed first, then the operations within the remaining innermost grouping symbols, and so on. For example,

$$\begin{aligned} 24 \div \{32 - [8 + (6 \cdot 3)]\} &= 24 \div \{32 - [8 + 18]\} \\ &= 24 \div \{32 - 26\} \\ &= 24 \div 6 \\ &= 4. \end{aligned}$$

We can reduce the number of grouping symbols required by establishing the following **order priority**[1] for the operations of addition, subtraction, multiplication, and division.

> Unless grouping symbols indicate otherwise,
>
> 1. all multiplications and divisions will be performed before any additions or subtractions; for example, $6 + 5 \cdot 3 = 6 + 15 = 21$;
> 2. multiplications and divisions will be performed in the order in which they appear from left to right in an algebraic expression; for example, $8 \div 2 \cdot 3 \div 6 = 4 \cdot 3 \div 6 = 12 \div 6 = 2$;
> 3. additions and subtractions will be performed in the order in which they appear from left to right in an algebraic expression; for example, $18 - 4 + 16 - 7 = 14 + 16 - 7 = 30 - 7 = 23$.

For visual clarity, we will sometimes use grouping symbols even though they are not absolutely necessary. For instance, we might write

$$3 \cdot 6 - 20 \div 5 + 2 \cdot 8 - 28$$

as

$$(3 \cdot 6) - (20 \div 5) + (2 \cdot 8) - 28.$$

Question 4 *On a calculator that has the above priorities, which of the following three procedures give the correct answer for* $\frac{256}{16 \cdot 8}$?

(a)	(b)	(c)
Enter 256	Enter 256	Enter 256
Enter ÷ 16	Enter ÷	Enter ÷ 16
Enter * 8	Enter (16*8)	Enter ÷ 8

Answer: Procedures (b) and (c) give the correct answer, 2. Procedure (a) gives 128, which is equal to $(256/16) \cdot 8$. ■

Extended Commutative, Associative, and Distributive Properties. The commutative and associative properties for addition and multiplication can be extended to three or more terms. That is, in a sum or product of three or more terms, the terms can be arranged or grouped in any manner whatsoever without changing the results. For example, $b + y + a + x = a + b + x + y$, and $byax = abxy$. Also, $a + b + x + y = (a + b) + (x + y) = a + (b + x) + y = (a + b + x) + y$, and similarly for multiplication. The left and right distributive properties can be extended to three or more terms. For example, $a(x + y + z) = ax + ay + az$, and $(a + b + c)x = ax + bx + cx$. The following chart illustrates how the priority conventions and extended properties simplify algebraic notation.

[1]This is the priority that is programmed into most scientific calculators. Also, special functions, such as the square root, exponential, logarithmic, and trigonometric functions, precede multiplication and division in priority.

Unsimplified Notation	Simplified Notation
$2(xy)$ or $(2x)y$	$2xy$
$2 + (x + y)$ or $(2 + x) + y$	$2 + x + y$
$[(2a) + (3b)] + [(4x) + (5y)]$	$2a + 3b + 4x + 5y$
$a[(x + y) + z]$ $= [a(x + y)] + (az)$ $= [(ax) + (ay)] + (az)$	$a(x + y + z) = ax + ay + az$
$(2x)(3y) = 2[x(3y)]$ $= 2[(3y)x]$ $= 2[3(yx)]$ $= (2 \cdot 3)(yx)$ $= 6(xy)$	$2x \cdot 3y = 6xy$

We will, of course, use the simplified notation throughout the remainder of the text except when emphasis or visual clarity suggests otherwise.

Comment As children, we learned to speak words before we were taught the alphabet or basic elements of words. Similarly, mathematicians worked with algebra long before the basic properties of real numbers were discovered. It was not until the nineteenth century that mathematicians finally took time out from *doing* algebra to ask the question, "*What is algebra?*". They went from the "*how*" of algebra to the "*why.*" The nine basic properties described in this section became known as the **field axioms** of the real number system. Other mathematical systems satisfying different basic properties also were investigated, which led to the present-day axiomatic development of abstract algebra. Hence, almost 4000 years after the Babylonians were solving equations, the guiding principles of algebra were finally determined. We might say that it took mathematicians 4000 years to put the horse before the cart!

Exercises P.2

Fill in the blanks to make each statement true.

1. The property $a + b = b + a$ is called the __________ of addition, and $(a + b) + c = a + (b + c)$ is called the __________ of addition.
2. The left distributive property is __________, and the right distributive property is __________.
3. The additive inverse of -3 is ______, and the multiplicative inverse of $1/3$ is ______.
4. The additive inverse of 0 is ______; that is, $-0 =$ ______; the multiplicative inverse of 1 is ______; that is, $1^{-1} =$ ______.

5. If $1/0$ were a real number, it would have to satisfy $0 \cdot 1/0 =$ ______. But $0 \cdot a =$ ______ for any real number a. Therefore, ______________.

Write true *or* false *for each statement.*

6. Subtraction is a commutative operation.

7. Division is not an associative operation.

8. The additive inverse of a number is always a negative quantity.

9. Every real number has an additive inverse.

10. Every real number has a multiplicative inverse.

Basic Properties

11. Name the basic property illustrated by each of the following equations.

(a) $3 \cdot 5 = 5 \cdot 3$

(b) $(4 + 6) + 3 = 4 + (6 + 3)$

(c) $2(5 + 7) = 2 \cdot 5 + 2 \cdot 7$

12. Given $a = 4$, $b = 3$, and $c = 8$, name the basic property illustrated by each of the following.

(a) $7 + 8 = 4 + 11$ (b) $12 \cdot 8 = 4 \cdot 24$

(c) $4 \cdot 11 = 12 + 32$

State the basic property or properties that justify each statement in Exercises 13–16. The values of numerical sums may be assumed.

13. $(5x + 4) + 3 = 5x + 7$

14. $(y + 6) + (x + 1) = (x + 1) + (y + 6)$

15. $2x + 3x = 5x$ **16.** $x + x = 2x$

17. Which of the following statements are equivalent to the statement "0 has no multiplicative inverse"?

(a) $0 \cdot a = 0$ for any real number a.

(b) $1/0$ has no meaning as a real number.

(c) Division by 0 is an invalid operation.

18. Which of the following statements are equivalent to the statement "a is negative"?

(a) a is to the left of 0 on the number line. (b) $-a$ is positive.

(c) a is less than 0. (d) $-a > a$

In each of the following, find all values of x that satisfy the equation.

19. $(x - 2)(x + 2) = 0$ **20.** $(x + 3)(x - 1) = 0$

21. $(|x| - 1)(|x| - 2) = 0$ **22.** $(|x| - 3)(|x| + 4) = 0$

Here is a simple application of the distributive property. If there is a 6% sales tax on an item whose price is \$15, then the total cost of the item in dollars is

$$15 + 15(.06) = 15 \cdot 1 + 15(.06) = 15(1.06) = 15.90$$

That is, instead of computing the sales tax separately and adding it to the price, just multiply the price by 1.06 to obtain the total cost. Use this method in Exercises 23–26.

23. The price of a shirt is \$25 and the sales tax is 6%. Find the total cost.

24. The price of a rare coin is \$170 and the sales tax is 7%. Find the total cost.

25. The total cost of a used car is \$2662.50, which includes a sales tax of 6.5%. Find the price of the car.

26. The total cost of a new car is \$9675, which includes a 7.5% sales tax. Find the price of the car.

27. In attempting to solve an algebraic equation, Adam arrives at the conclusion that $5 = 3$. What is wrong with Adam's "proof"?

Given: $5x + 10 = 3x + 6$.

Then $5(x + 2) = 3(x + 2)$ (left distributive property)

$[5(x + 2)](x + 2)^{-1} = [3(x + 2)](x + 2)^{-1}$ (property of equality)

$5[(x + 2)(x + 2)^{-1}] = 3[(x + 2)(x + 2)^{-1}]$ (associative property)

$5 \cdot 1 = 3 \cdot 1$ (multiplicative inverse property)

$5 = 3$. (multiplicative identity property)

Use the basic properties (and properties of equality) to prove the statements in Exercises 28–31.

28. If $x + 5 = y + 5$, then $x = y$.

29. If $a + b = a + c$, then $b = c$.

30. If $4x = 4y$, then $x = y$.

31. If $ac = bc$ $(c \neq 0)$, then $a = b$.

32. Example 1 and the commutative property for multiplication say that if either a or $b = 0$, then $ab = 0$. The converse is also true. That is,

$$\text{if } ab = 0, \text{ then either } a = 0 \text{ or } b = 0.$$

Fill in the blanks to complete the following proof. Each reason is a basic property, a property of equality, or a derived property.

Given: $ab = 0$.

If $b \neq 0$, then b has a multiplicative inverse b^{-1} and

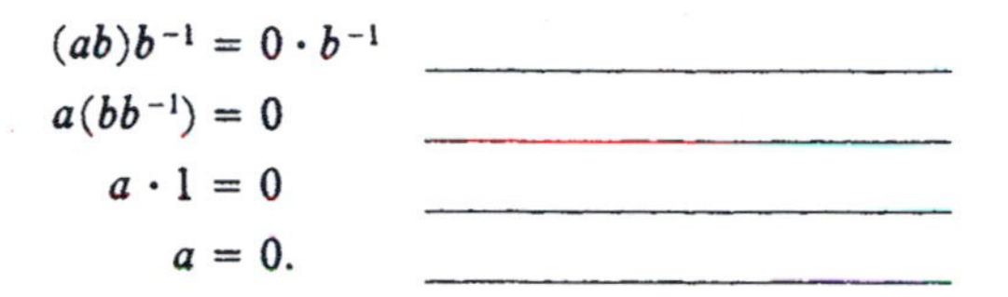

$(ab)b^{-1} = 0 \cdot b^{-1}$ ____________

$a(bb^{-1}) = 0$ ____________

$a \cdot 1 = 0$ ____________

$a = 0.$ ____________

Hence, if $b \neq 0$, then $a = 0$. Therefore, if $ab = 0$, then either $a = 0$ or $b = 0$.

33. Because $0 + 0 = 0$, it follows that $-0 = 0$. Here we show that 0 is the *only* real number with this property. Fill in the reasons for each step.

Given: $-x = x$.

Then $x + x = 0$ by the __________ (basic property).

Therefore, $2x = 0$ by Exercise _____.

Therefore, $x = 0$ by Exercise _____.

Grouping Symbols and Priority Conventions

Evaluate each expression in Exercises 34–35.

34. (a) $(48 \div 3)[(6 + 12) \div (4 - 2)]$

(b) $48 \div \{3[6 + 12 \div (4 - 2)]\}$

(c) $48 \div 3[6 + 12 \div 4 - 2]$

(d) $48 \div 3 \cdot 6 + 12 \div 4 - 2$

35. (a) $8 \cdot [(18 + 9) \div (3 \cdot 3)]$

(b) $8 \cdot [18 + 9 \div (3 \cdot 3)]$

(c) $8 \cdot [18 + 9 \div 3 \cdot 3]$

(d) $8 \cdot 18 + 9 \div 3 \cdot 3$

Use a calculator to evaluate each expression in Exercises 36–37.

36. (a) $\dfrac{152.24 + 5.018}{12.687 - 8.972}$ (b) $\dfrac{152 + 5.018}{12.687} - 8.972$

(c) $152.24 + 5.018 \div (12.687 - 8.972)$

(d) $152.24 + 5.018 \div 12.687 - 8.972$

37. (a) $(12.52 + 6.8) \cdot (5.75 \div 3.21)$

(b) $(12.52 + 6.8 \cdot 5.75) \div 3.21$

(c) $(12.52 + 6.8) \cdot 5.75 \div 3.21$

(d) $12.52 + 6.8 \cdot 5.75 \div 3.21$

Extended Properties

State the extended property or properties that justify each of the following. The values of numerical sums and products may be assumed.

38. $2y \cdot 5x = 10xy$

39. $(x + 1) + (3 + y) = x + 4 + y$

40. $4x + 5x + 7x = 16x$

41. $2x + 7 + 3x + 4 = 5x + 11$

Miscellaneous

Place all the even integers into one category called E *and all the odd integers into another category called* O. *Addition is defined in this system by:* $E + E = E$, $E + O = O$, $O + E = O$, $O + O = E$. *Multiplication is defined by:* $E \cdot E = E$, $E \cdot O = E$, $O \cdot E = E$, $O \cdot O = O$.

42. Verify that the above system satisfies basic properties (1)–(5).

43. Verify that E is the identity element for addition in the above system. What is the additive inverse of E? What is the additive inverse of O?

44. Verify that O is the identity element for multiplication in the above system. Does E have a multiplicative inverse? If so, what is it? Does O have a multiplicative inverse? If so, what is it?

In Exercises 45–49, let $a \text{ A } b$ *denote the* ***average*** *of the two numbers* a *and* b, *that is,* $a \text{ A } b = (a + b)/2$. *Also, let* $a \text{ M } b$ *denote the* ***maximum*** *of* a *and* b. *That is,* $a \text{ M } b = a$ *if* $a > b$ *or* $a = b$, *and* $a \text{ M } b = b$ *if* $b > a$ *or* $b = a$.

45. In the system above, compute each of the following.

(a) 6 A 2 (b) (6 A 2) A 12

(c) 6 A (2 A 12) (d) 6 M 2

(e) (6 M 2) M 12 (f) 6 M (2 M 12)

46. In the system above, find b in each of the following cases, if possible.

(a) $2 \text{ A } b = 2$ (b) $5 \text{ M } b = 0$

(c) $2 \text{ M } b = 2$ (d) $7 \text{ A } b = 1$

(e) $5 \text{ A } b = 0$ (f) $7 \text{ M } b = 1$

47. Answer each question for the system above.

(a) Is A a commutative operation?

(b) Is A an associative operation?

(c) Is M a commutative operation?

(d) Is M an associative operation?

48. It can be shown that the left distributive property holds when $\cdot$ is replaced by A and $+$ is replaced by M. Verify that $6 \text{ A } (2 \text{ M } 12) = (6 \text{ A } 2) \text{ M } (6 \text{ A } 12)$.

49. It can be shown that the left distributive property does *not* always hold when $\cdot$ is replaced by M and $+$ is replaced by A. Verify that $6 \text{ M } (2 \text{ A } 12) \neq (6 \text{ M } 2) \text{ A } (6 \text{ M } 12)$.

50. The location of $1/a$ as indicated by x in the figure can be demonstrated by elementary geometry as follows: right triangles $0xP$ and $0Pa$ are similar because they have acute angle $P0x$ in common. Therefore, the ratio of the length of base x to that of hypotenuse $0P$ in triangle $0xP$ is equal to the ratio of the length of base $0P$ to that of hypotenuse a in triangle $0Pa$. That is,

$$\frac{x}{\text{length of } 0P} = \frac{\text{length of } 0P}{a}.$$

Since the length of $0P = 1$, it follows that $x = 1/a$, as stated.

Now $x = 1/a$ implies that $a = 1/x$. What happens to point a as x approaches 0? If x becomes 0, what happens to the triangle $0Pa$? This shows that the algebraic property "0 has no multiplicative inverse" is related to the geometric property "parallel lines do not intersect."

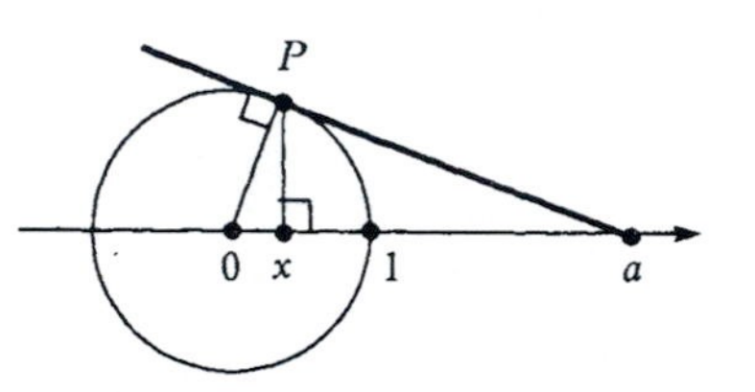

P.3 Signed Quantities, Order, and Absolute Value

"The algebraic rules of operation with negative numbers are generally admitted by everyone and acknowledged as exact, whatever idea we may have about these quantities."

—*Jean Le Rond d'Alembert*

Addition and Subtraction. Every nonzero real number has both a magnitude and a sign. However, because $+a = a$ for every real number a, the leading plus sign is rarely used. The main use of the symbol $+$ is for the operation of addition. Hence, the equation $+6 + (+5) = +11$ is normally written as $6 + 5 = 11$. The minus sign $-$ is used for negative numbers or, more generally, for additive inverses, and it is also used for the operation of subtraction. Now $-6 + (-5) = (-1)6 + (-1)5 = (-1)(6 + 5) = (-1)11 = -11$. These special cases illustrate the general rule for adding quantities with the same sign: to compute $a + b$ when a and b have the *same* *sign*, add their absolute values and precede this result by their common sign. For example,

$$-7 + (-8) = -\underbrace{(7 + 8)}_{\text{Sum of the absolute values}} = -15.$$

Sum of the absolute values

Common sign

The special cases $-5 + 9 = -5 + 5 + 4 = 0 + 4 = 4$ and $-5 + 3 = -2 + (-3) + 3 = -2 + 0 = -2$ illustrate the general rule for adding quantities with opposite signs: to compute $a + b$ when a and b have *opposite signs,* subtract the smaller absolute value from the larger, and precede this result by the sign of the term with the larger absolute value. If both terms have the same absolute value, their sum is 0. For example,

$$7 + (-8) = -\underbrace{(8 - 7)}_{\text{Difference of the absolute values}} = -1.$$

Difference of the absolute values

Sign of the term with the larger absolute value (sign of -8)

Similarly, $$-7 + 8 = +(8 - 7) = 1.$$

As discussed in Question 3 of Section P.2, subtraction is defined in terms of addition. By definition,

$$\boldsymbol{a - b = a + (-b).}$$ **definition of subtraction**

Hence, to subtract b from a, we *add* the additive inverse of b to a. For example,

$$3 - 14 = 3 + (-14) = -11,$$

and $$-4 - 12 = -4 + (-12) = -16.$$

The result $-(7 + 8) = -7 + (-8)$ can be generalized as follows.

Example 1 Show that

$$\boldsymbol{-(a + b) = -a + (-b).}$$ **inverse of a sum**

That is, show that the additive inverse of the sum of two numbers is equal to the sum of their additive inverses.

Solution:

$-(a + b) = (-1)(a + b)$	*Negative multiplier property*
$= (-1)a + (-1)b$	*Left distributive property*
$= -a + (-b)$	*Negative multiplier property*

■

Some applications of Example 1 are listed below.

1. $-(x + 6) = -x + (-6) = -x - 6$
2. $-(-3 + y) = -(-3) + (-y) = 3 - y$
3. $-(10 - y) = -[10 + (-y)] = -10 + -(-y) = -10 + y$
4. $-(-x - 2) = -[-x + (-2)] = -(-x) + -(-2) = x + 2$

The above four special cases illustrate the general rule that *when there is a minus sign before parentheses, the parentheses can be removed by changing the sign of each term within the parentheses.* This rule can be extended to more than two terms. For instance,

$$-(2x - 3y + 4) = -2x + 3y - 4$$

and $$-(-x + y - a - b) = x - y + a + b.$$

Multiplication and Division. In the statement

$$(-1)(-1) = -(-1) = 1,$$

the first equality follows from the negative multiplier property, and the second from the double inverse property $-(-a) = a$. In a similar way, we can show that for any two real numbers a and b,

$$(-a)(-b) = ab \quad \text{and} \quad (-a)b = a(-b) = -(ab).$$ **sign rules for multiplication**

In particular, these sign rules tell us that *the product of two factors having the same sign is positive, and the product of two factors having opposite signs is negative.* For example,

$$(-6)(-8) = 6 \cdot 8 = 48 \quad \text{and} \quad (-6)8 = 6(-8) = -48.$$

Also, $$(-5)(-x) = 5x \quad \text{and} \quad (-5)x = 5(-x) = -5x.$$

As previously stated, division is defined in terms of multiplication. By definition, if $b \neq 0$, then

$$\frac{a}{b} = a \cdot \frac{1}{b} = a \cdot b^{-1}.$$ **definition of division**

Hence, to divide a by b, we *multiply* a by the reciprocal or multiplicative inverse of b. From the definition of division and the sign rules for multiplication, we can show that if $b \neq 0$, then

$$\frac{-a}{-b} = \frac{a}{b} \quad \text{and} \quad \frac{-a}{b} = \frac{a}{-b} = -\frac{a}{b}.$$ **sign rules for division**

As in products, *the quotient of two terms with the same sign is positive, and the quotient of two terms with opposite signs is negative.*

For example, $$\frac{-28}{-14} = \frac{28}{14} = 2 \quad \text{and} \quad \frac{-28}{14} = \frac{28}{-14} = -2.$$

Also, $\dfrac{-1}{-x} = \dfrac{1}{x}$ and $\dfrac{-1}{x} = \dfrac{1}{-x} = -\dfrac{1}{x}$ if $x \neq 0$.

In Example 1 it was shown that the additive inverse of the sum of two numbers is equal to the sum of their additive inverses. In a similar way, we could show that the multiplicative inverse of a product of two numbers is the product of their multiplicative inverses. That is, if $a \neq 0$ and $b \neq 0$, then

$$(ab)^{-1} = a^{-1}b^{-1} \quad \text{or} \quad \frac{1}{ab} = \frac{1}{a} \cdot \frac{1}{b}. \qquad \textbf{inverse of a product}$$

We can also apply the above inverse property to quotients. For instance,

$$\begin{aligned} \left(\frac{3}{4}\right)^{-1} &= (3 \cdot 4^{-1})^{-1} && \textit{Definition of division} \\ &= 3^{-1}(4^{-1})^{-1} && \textit{Inverse of a product} \\ &= 3^{-1} \cdot 4 && \textit{Double inverse} \\ &= 4 \cdot 3^{-1} && \textit{Commutative property} \\ &= \frac{4}{3}. && \textit{Definition of division} \end{aligned}$$

In general, we can conclude that if $a \neq 0$ and $b \neq 0$, then

$$\left(\frac{a}{b}\right)^{-1} = \frac{b}{a}. \qquad \textbf{inverse of a quotient}$$

For example,

$$\left(\frac{x+2}{x-2}\right)^{-1} = \left(\frac{x-2}{x+2}\right) \quad \text{if } x \neq 2 \text{ or } -2,$$

and

$$\left(-\frac{x}{7}\right)^{-1} = \left(\frac{-x}{7}\right)^{-1} = \frac{7}{-x} = -\frac{7}{x} \quad \text{if } x \neq 0.$$

We can also apply the zero-product property from Section P.2 to quotients. The quotient a/b is equal to the product $a \cdot b^{-1}$, and therefore a/b is zero if and only if $a \cdot b^{-1}$ is zero; but neither b nor b^{-1} can equal zero, or a/b is not defined. We arrive at the following result.

$$\frac{a}{b} = 0 \Leftrightarrow a = 0 \text{ and } b \neq 0 \qquad \textbf{zero-quotient property}$$

For example,

$$\begin{aligned} \frac{2x-1}{2x+1} = 0 &\Leftrightarrow 2x - 1 = 0 \text{ and } 2x + 1 \neq 0 \\ &\Leftrightarrow x = \frac{1}{2} \text{ and } x \neq -\frac{1}{2} \\ &\Leftrightarrow x = \frac{1}{2}. \end{aligned}$$

Example 2 Show that $(x-2)/(x^2-4)$ cannot equal 0 for any value of x.

Solution: Now

$$\frac{x-2}{x^2-4} = 0 \Leftrightarrow x-2=0 \text{ and } x^2-4 \neq 0.$$

But it is impossible to have $x-2=0$ and $x^2-4 \neq 0$ (*why?*). Therefore, $(x-2)/(x^2-4)$ can never equal 0. ■

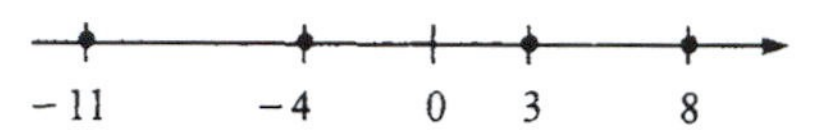

Figure 3 $8 > 3$ and $-4 > -11$

Algebraic Properties of Order. As stated in Section P.1, the real number a is greater than the real number b, denoted $a > b$ or $b < a$, if a is located to the right of b on the real number line. For example, $8 > 3$ and $-4 > -11$ (Figure 3). Note that $8 = 3 + 5$ and $-4 = -11 + 7$, where 5 and 7 are *positive* quantities. That is, adding a positive quantity increases the value of a number. From this we can deduce that our positional definition of "greater than" in Section P.1 is equivalent to the following algebraic definition.

$\boldsymbol{a > b \Leftrightarrow a = b + p}$**, where** $\boldsymbol{p}$ **is positive.** — **algebraic definition of "greater than"**

Example 3 Show that $x + 5 > x + 2$ for any value of x.

Solution: We can write

$$\underbrace{x+5}_{a} = \underbrace{x+2}_{b} + \underbrace{3}_{p},$$

and since 3 is positive, it follows that $x + 5 > x + 2$ for any value of x. ■

The above definition can be used to prove that the relationship "greater than" is preserved under addition (see Exercise 49). That is,

if $\boldsymbol{c}$ **is any real number, then** — **preservation property for addition**

$$\boldsymbol{a > b \Leftrightarrow a + c > b + c.}$$

Example 4 Show that $x - 6 > 4$ is equivalent to $x > 10$.

Solution:

$$x - 6 > 4 \Leftrightarrow x - 6 + 6 > 4 + 6 \qquad \textit{Add 6 to both sides}$$
$$\Leftrightarrow x > 10$$ ■

Question 1 *Is "greater than" preserved under subtraction? That is, is $a > b$ equivalent to $a - c > b - c$?*

Answer: Yes, since subtracting c is the same as adding $-c$. ■

If we multiply both sides of $8 > 5$ by the *positive* quantity 10, we obtain $10 \cdot 8 > 10 \cdot 5$ or $80 > 50$. On the other hand, if we multiply both sides by the *negative* quantity -10, we obtain $-10 \cdot 8 < -10 \cdot 5$ or $-80 < -50$. In general (see Exercise 50), we have the following results.

If c is any positive real number, then **preservation property for multiplication**

$$a > b \Leftrightarrow a \cdot c > b \cdot c.$$

If c is any negative real number, then **reversal property for multiplication**

$$a > b \Leftrightarrow a \cdot c < b \cdot c.$$

Example 5 Show that (a) $2x > 5$ is equivalent to $x > 5/2$ and (b) $-2x > 6$ is equivalent to $x < -3$.

Solution:

(a) $2x > 5 \Leftrightarrow 2x \cdot \frac{1}{2} > 5 \cdot \frac{1}{2}$ *Multiply both sides by the positive quantity* $1/2$

$\Leftrightarrow x > \frac{5}{2}$

(b) $-2x > 6 \Leftrightarrow -2x\left(-\frac{1}{2}\right) < 6\left(-\frac{1}{2}\right)$ *Multiply both sides by the negative quantity* $-1/2$ *and change* $>$ *to* $<$

$\Leftrightarrow x < -3$ ■

Question 2 *Does the above multiplication property hold for division by c?*

Answer: Yes, since $1/c$ has the same sign as c and division by c is the same as multiplication by $1/c$. ■

In addition to the above algebraic properties, the relationship "greater than" also has the following logical property (see Exercise 51).

If $a > b$ and $b > c$, then $a > c$. **transitive property**

For example, if $x + 2 > y$ and $y > 7$, then $x + 2 > 7$. Compare the above algebraic and logical properties of "greater than" with those of "equality" in Section P.2.

We will use the notation $a \geq b$ to mean that a is greater than or equal to b. Similarly, $b \leq a$ means that b is less than or equal to a. The above properties hold if $>$ and $<$ are replaced by $\geq$ and $\leq$, respectively.

Comment The real number system has a nonalgebraic property (one that cannot be derived from the nine basic properties) related to "greater than" which, along with its algebraic properties, makes the real number system different from any other number system. This property, called the **completion axiom,** states that if some real number is greater than or equal to every number in a given nonempty subset of real numbers, then there is a *smallest* number that is greater than or equal to all the numbers in the given subset. For example, let the subset consist of all numbers $a_1, a_2, a_3, a_4, \ldots,$ where

$$a_1 = \frac{1}{1 \cdot 2},$$
$$a_2 = \frac{1}{1 \cdot 2} + \frac{1}{3 \cdot 4},$$
$$a_3 = \frac{1}{1 \cdot 2} + \frac{1}{3 \cdot 4} + \frac{1}{5 \cdot 6}$$
$$a_4 = \frac{1}{1 \cdot 2} + \frac{1}{3 \cdot 4} + \frac{1}{5 \cdot 6} + \frac{1}{7 \cdot 8},$$
$$\vdots$$

It can be shown that 1 is greater than or equal to every number in this subset. Therefore, there is a smallest real number that is greater than or equal to every number in the subset. The completion axiom is important for calculus, but it will not be needed in our study of algebra.

Absolute Value. As stated in Section P.1, the magnitude of a real number x without regard to its sign is called the absolute value of x and is denoted by $|x|$. If x is positive or 0, then $|x| = x$. However, if x is negative, then multiplying x by -1 will produce a positive quantity of the same magnitude. Since $(-1)x = -x$, we arrive at the following algebraic definition.

$$|x| = \begin{cases} x \text{ if } x \geq 0 \\ -x \text{ if } x < 0 \end{cases} \qquad \textbf{definition of absolute value}$$

For example, $|-10| = -(-10) = 10$, and $|-2/3| = -(-2/3) = 2/3$.

The rules for multiplication and division with absolute values, which can be derived from the sign rules for multiplication and division, are as follows:

$$|ab| = |a|\,|b| \qquad \text{and} \qquad \left|\frac{a}{b}\right| = \frac{|a|}{|b|}.$$

That is, the absolute value of a product equals the product of the absolute values, and the absolute value of a quotient equals the quotient of the absolute values. For example,

$$|-2x| = |-2|\,|x| = 2|x| \qquad \text{and} \qquad \left|\frac{x}{-5}\right| = \frac{|x|}{|-5|} = \frac{|x|}{5}.$$

We note that

$$|b - a| = |(-1)(a - b)| = |-1|\,|a - b| = |a - b|.$$

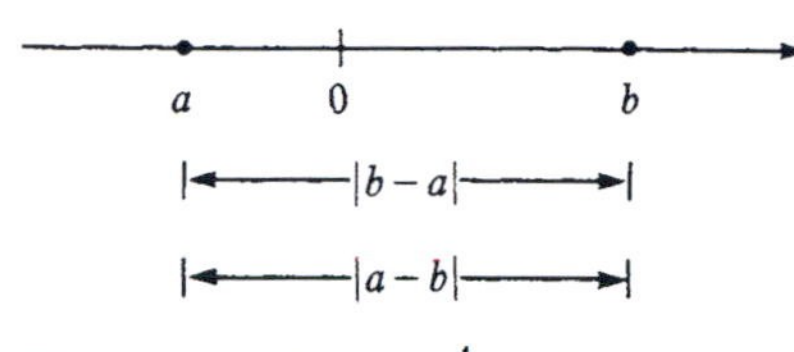

Figure 4 $|b - a| = |a - b|$

Geometrically, both $|b - a|$ and $|a - b|$ represent the **distance** between a and b on the real number line (Figure 4).

The rules for addition and subtraction with absolute values, which can be deduced from the sign rules for addition and subtraction, are as follows.

$$|a + b| \leq |a| + |b| \qquad \text{and} \qquad |a - b| \geq \big||a| - |b|\big|$$

triangle inequality

Equality of $|a + b|$ and $|a| + |b|$ occurs when either a or $b = 0$ or a and b have the same sign. Inequality occurs when they have opposite signs. The same holds true for $|a - b|$ and $\big||a| - |b|\big|$. For example, equality occurs in both cases for $a = -10$ and $b = -5$, while inequality occurs if $a = -10$ and $b = 5$.

Question 3 *What is the meaning of the quotation at the beginning of this section?*

Answer: Although today we work freely with negative numbers, their acceptance by mathematicians was a gradual process. The ancient Greeks could not conceive of negative numbers; to Diophantus (ca. 275 A.D.) the equation $4x + 20 = 4$ had no solution. The Hindus realized that negative numbers could be used to signify debts or losses, but they had misgivings. Ehaskāra (ca. 1150 A.D.) rejected -5 as a solution to an equation, stating that "people do not approve of negative numbers." Even by the seventeenth century negative numbers were not completely accepted. The French philosopher and mathematician René Descartes (1596–1650) referred to negative solutions of equations as "false" solutions. The statement by d'Alembert appeared in the French *Encyclopedia* in 1751 and shows that even at this late date mathematicians had reservations about negative numbers. ■

Exercises P.3

Fill in the blanks to make each statement true.

1. To compute $a + b$ when a and b are unequal and have opposite signs, we ____________________.
2. When there is a minus sign before parentheses, the parentheses can be removed by ____________ of each term inside the parentheses.
3. The subtraction $a - b$ is equal to the addition ____________.
4. The division a/b is equal to the multiplication ____________.
5. The product of two nonzero terms having the same sign is always ____________, and if the terms have opposite signs, their product is ____________.

Write true *or* false *for each statement.*

6. If a and b have opposite signs but the same absolute value, then $a + b = 0$.
7. Since the minus sign is used to denote negative numbers, it follows that $-a$ is always a negative number.

8. If a is to the right of b on the number line, then $a - b$ is positive.
9. If a and b have opposite signs, then $|a + b| < |a| + |b|$.
10. If $|a| > |b|$, then $a > b$.

Addition, Subtraction, Multiplication, and Division

Evaluate each of the following.

11. $-7 + (-2)$
12. $-4 + (-4)$
13. $5 - (-8)$
14. $3 + (-8)$
15. $(-3)(-5)$
16. $\frac{24}{-6}$
17. $4(-2)^{-1}$
18. $(-6)3^{-1}$
19. $(-4)(-4)(-4)$
20. $(-3)^{-1}[-(-3)]$
21. $8 - (4 - 2)$
22. $-3 - (8 - 16)$
23. $(-2)[17 + (-3)]$
24. $(-3)(8 - 12)$
25. $\frac{-2(3-7)}{3(-4) + 4(-3)}$
26. $\frac{-3(4-12)}{3(12-4)}$

Express each of the following without parentheses or brackets.

27. $-(-x - 7)$
28. $-(-x - 3)$
29. $-(6 - y)$
30. $-[-(-x) + (-4)]$
31. $-[-(-x) - (-y)]$
32. $(-5)(-y)$
33. $(-4)b$
34. $(-x)[-(-3)]$
35. $[-(-5)](-y)$
36. $-(-1)[-(-y)]$

In each of the following, find all possible values of x that satisfy the equation.

37. $\frac{x-1}{x-2} = 0$
38. $\frac{x+3}{x-4} = 0$
39. $\frac{x+5}{|x| + 5} = 0$
40. $\frac{x+1}{|x| - 1} = 0$
41. $\frac{(|x| - 2)(|x| - 3)}{x - 3} = 0$
42. $\frac{(|x| - 4)(|x| - 5)}{(x + 4)(x + 5)} = 0$

Properties of Order

In Exercises 43–48, use the algebraic properties of order to complete each statement.

43. $x + 8 > x + 3$ because $x + 8 = (x + 3) + 5$, and 5 is ________.
44. $x + 5 > 2$ is equivalent to $x >$ ________.
45. $3x > 18$ is equivalent to $x >$ ________.
46. $-3x > 18$ is equivalent to x ________ -6.
47. $-2x > -6$ is equivalent to x ________.
48. $x > x - 1$ because ________.
49. Fill in the reason for each step in the following proof that "greater than" is preserved under addition.

$a > b$, and c is any real number.	Given
$a = b + p$, where p is positive.	________
$a + c = (b + p) + c$	________
$a + c = (b + c) + p$	________
$a + c > b + c$	________

50. The first proof below shows that "greater than" is preserved under multiplication by a positive quantity. The second proof shows that a negative multiplier changes "greater than" to "less than." Fill in the reason for each step.

Given: $a > b$, and c is a *positive* real number. Then, by definition, $a = b + p$, where p is positive.

(i)

$ac = (b + p)c$	________
$ac = bc + pc$	________
pc is positive.	________
$ac > bc$	________

(ii)

$a(-c) = (b + p)(-c)$	________
$a(-c) = b(-c) + p(-c)$	________
$a(-c) + pc = b(-c)$	________
pc is positive.	________
$a(-c) < b(-c)$	________

51. The transitive property of "greater than" says that if $a > b$ and $b > c$, then $a > c$. Fill in a reason for each step in the following proof of the transitive property.

$a > b$ and $b > c$	Given
$a = b + p$, where p is positive.	________
$b = c + q$, where q is positive.	________
$a = (c + q) + p$	________
$a = c + (q + p)$,	________,
and $q + p$ is positive.	________
$a > c$	________

52. Use the algebraic definition of "greater than" to conclude that $0 > a \Leftrightarrow a$ is negative. [*Hint:* start with the equation $0 = a + (-a)$.]

Absolute Value

53. Show that $|-x| = |x|$ for any real number x.
54. Show that $a \le |a|$ for any real number a. (*Hint:* consider separately the cases $a > 0$, $a = 0$, and $a < 0$.)
55. Show that $|a^{-1} + b^{-1}| \le \dfrac{|a| + |b|}{|a|\,|b|}$.
56. Show that $|a^{-1} - b^{-1}| \ge \dfrac{||a| - |b||}{|a|\,|b|}$.
57. Show that $|a + b + c| \le |a| + |b| + |c|$. [*Hint:* write $a + b + c = (a + b) + c$.]
58. Show that $|a - b| \ge |a| - |b|$ by writing $|a| = |a - b + b|$ and using the triangle inequality.

Miscellaneous

59. *Subtraction Is Not Commutative.* Show that if $a - b = b - a$, then $a = b$. (*Hint:* let $x = a - b$ and show that $-x = b - a$.)
60. *Subtraction Is Not Associative.* Show that if $(a - b) - c = a - (b - c)$, then $c = 0$.
61. *Definition of Subtraction.* Use the fact that $a - b = c$ means $c + b = a$ to arrive at the definition $a - b = a + (-b)$. That is, starting with the equation $c + b = a$, use the basic properties to conclude that $c = a + (-b)$.
62. *Definition of Division.* Use the fact that $a/b = c$ means $c \cdot b = a$ to arrive at the definition $a/b = a \cdot b^{-1}$. That is, starting with the equation $c \cdot b = a$, use the basic properties to conclude that $c = a \cdot b^{-1}$.
63. Fill in a reason for each step in the proof of the following algebraic property of "greater than": if $a > b$ and $c > d$, then $a + c > b + d$.

$a > b$	and	$c > d$	Given
$a + c > b + c$	and	$b + c > b + d$	________
	$a + c > b + d$		________

64. Fill in a reason for each step in the proof of the following algebraic property of "greater than": if $a > b > 0$ and $c > d > 0$, then $ac > bd$. Also, show by examples that the result does not hold in general if the condition $b > 0$ is removed or if the condition $d > 0$ is removed.

$a > b > 0$	and	$c > d > 0$	Given
$ac > bc$	and	$bc > bd$	________
	$ac > bd$		________

P.4 Fractional Expressions

"After making the denominator of the divisor its numerator, the operation to be performed is as in multiplication."

—*Rule for dividing fractions in a ninth-century work by the mathematician Mahāvira*

Equality of Fractions
Addition and Subtraction
Multiplication and Division
Simplifying Complex Fractions

Equality of Fractions. We know from arithmetic that different ratios can have the same value. For example,

$$\frac{3}{4} = \frac{6}{8} = \frac{9}{12} = \cdots = \frac{3n}{4n}$$

for any integer n except 0. Similarly, if $b \ne 0$, then

$$\frac{a}{b} = \frac{ac}{bc}, \qquad \textbf{common factor property}$$

where c can be any real number except 0. This property follows from

$$\frac{ac}{bc} = ac(bc)^{-1} = acb^{-1}c^{-1} = ab^{-1}cc^{-1} = ab^{-1} = \frac{a}{b}.$$

In arithmetic, the common factor property is used to reduce rational numbers to lowest terms. For instance,

$$\frac{45}{60} = \frac{3 \cdot \cancel{15}}{4 \cdot \cancel{15}} = \frac{3}{4}$$

and

$$\frac{165}{462} = \frac{\cancel{3} \cdot 5 \cdot \cancel{11}}{2 \cdot \cancel{3} \cdot 7 \cdot \cancel{11}} = \frac{5}{14}.$$

Similarly, the property can be used to cancel common factors of the numerator and denominator of an algebraic fraction.

Example 1 (a) $\dfrac{3ax}{12xy} = \dfrac{\cancel{3}a\cancel{x}}{\cancel{3} \cdot 4\cancel{x}y} = \dfrac{a}{4y} \quad (x \neq 0, y \neq 0)$

(b) $\dfrac{ac}{ac + bc} = \dfrac{a\cancel{c}}{(a + b)\cancel{c}} = \dfrac{a}{a + b} \quad (c \neq 0, a + b \neq 0)$ ■

Example 2 (a) $\dfrac{ax + 3ay}{4bx + 12by} = \dfrac{a\cancel{(x + 3y)}}{4b\cancel{(x + 3y)}} = \dfrac{a}{4b} \quad (b \neq 0, x + 3y \neq 0)$

(b) $\dfrac{a - b}{b - a} = \dfrac{-1\cancel{(b - a)}}{1\cancel{(b - a)}} = \dfrac{-1}{1} = -1 \quad (b - a \neq 0)$ ■

Caution It is very important to note that only common *factors* of the numerator and denominator can be canceled. For example, in the expression $a/(a + b)$, a is not a factor of the denominator and therefore cannot be canceled. That is,

$$\frac{a}{a + b} \neq \frac{1}{1 + b};$$

for instance,

$$\frac{3}{3 + 7} \neq \frac{1}{1 + 7}.$$

For a to be canceled, a would have to be a factor of both terms in the denominator. That is, if $a \neq 0$, then

$$\frac{a}{a + ab} = \frac{\cancel{a}}{\cancel{a}(1 + b)} = \frac{1}{1 + b} \quad (1 + b \neq 0);$$

for instance,

$$\frac{3}{3 + 3 \cdot 7} = \frac{\cancel{3}}{\cancel{3}(1 + 7)} = \frac{1}{1 + 7}.$$

Similarly,

$$\frac{a + b}{b} \neq a + 1,$$

whereas

$$\frac{ab+b}{b} = \frac{(a+1)\cancel{b}}{\cancel{b}} = \frac{a+1}{1} = a+1 \quad (b \neq 0),$$

and

$$\frac{a+b}{a+c} \neq \frac{1+b}{1+c},$$

whereas

$$\frac{a+ab}{a+ac} = \frac{\cancel{a}(1+b)}{\cancel{a}(1+c)} = \frac{1+b}{1+c} \quad (a \neq 0,\ 1+c \neq 0).$$

In short, *if a quantity cannot be factored out, then it cannot be canceled.*

Another criterion for equality of fractions is

$$\frac{a}{b} = \frac{c}{d} \Leftrightarrow ad = bc \quad (b \neq 0,\ d \neq 0)$$

equality of fractions

which follows from

$$\frac{a}{b} = \frac{c}{d} \Leftrightarrow ab^{-1} = cd^{-1} \Leftrightarrow ab^{-1}bd = cd^{-1}bd \Leftrightarrow ad = bc.$$

For example,

$$\frac{2x+1}{7} = \frac{3}{5} \Leftrightarrow (2x+1)\cdot 5 = 7\cdot 3$$
$$\Leftrightarrow 10x + 5 = 21.$$

Addition and Subtraction. For two fractions with the same denominator, the respective procedures for addition and subtraction are

$$\frac{a}{b} + \frac{c}{b} = \frac{a+c}{b} \quad \text{and} \quad \frac{a}{b} - \frac{c}{b} = \frac{a-c}{b},$$

addition and subtraction, common denominator

where $b \neq 0$. For example,

$$\frac{3}{5} + \frac{7}{5} = \frac{10}{5}$$

and

$$\frac{x}{x-1} - \frac{2}{x-1} = \frac{x-2}{x-1} \quad (x - 1 \neq 0).$$

The rule for addition follows from

$$\frac{a}{b} + \frac{c}{b} = ab^{-1} + cb^{-1} = (a+c)b^{-1} = \frac{a+c}{b},$$

and the subtraction rule is similarly derived.

Caution There are no corresponding rules for adding or subtracting fractions with the same numerators. In general,

$$\frac{a}{b} + \frac{a}{c} \neq \frac{a}{b+c} \quad \text{and} \quad \frac{a}{b} - \frac{a}{c} \neq \frac{a}{b-c}.$$

(See Section 1.1, Exercise 101.) For example,

$$\frac{5}{3} + \frac{5}{7} \neq \frac{5}{10}$$

and

$$\frac{|x|+1}{x} - \frac{|x|+1}{2} \neq \frac{|x|+1}{x-2}.$$

Two of the most common mistakes in algebra are to equate $a/(b+c)$ with $a/b + a/c$, and $a/(b-c)$ with $a/b - a/c$. Be sure to avoid making these errors.

The general procedures for adding and subtracting fractions whose denominators are not necessarily equal are

$$\frac{a}{b} + \frac{c}{d} = \frac{ad+bc}{bd} \quad \text{and} \quad \frac{a}{b} - \frac{c}{d} = \frac{ad-bc}{bd},$$ **addition and subtraction, general rule**

where $b \neq 0$ and $d \neq 0$. The addition rule follows from

$$\frac{a}{b} + \frac{c}{d} = \frac{ad}{bd} + \frac{bc}{bd} = \frac{ad+bc}{bd},$$

and the subtraction rule is derived in a similar way.

Example 3 (a) $\frac{3}{2} + \frac{5}{x} = \frac{3x+10}{2x} \quad (x \neq 0)$

(b) $$\frac{2}{x} + \frac{4}{x+1} = \frac{2(x+1)+4x}{x(x+1)}$$

$$= \frac{6x+2}{x(x+1)} \quad (x \neq 0,\ x+1 \neq 0)$$ ■

Example 4 (a) $\frac{x}{2} - \frac{x}{3} = \frac{3x-2x}{6} = \frac{x}{6}$

(b) $$\frac{3}{x+1} - \frac{2}{x+2} = \frac{3(x+2)-2(x+1)}{(x+1)(x+2)}$$

$$= \frac{x+4}{(x+1)(x+2)} \quad (x+1 \neq 0,\ x+2 \neq 0)$$ ■

Multiplication and Division. The procedure used for multiplying fractions is

$$\frac{a}{b} \cdot \frac{c}{d} = \frac{ac}{bd},$$ **multiplication**

where $b \neq 0$ and $d \neq 0$. That is, the product of two fractions is equal to the product of their numerators divided by the product of their denominators. This follows from

$$\frac{a}{b} \cdot \frac{c}{d} = ab^{-1} \cdot cd^{-1} = ac \cdot b^{-1}d^{-1} = ac \cdot (bd)^{-1} = \frac{ac}{bd}.$$

For instance,

$$\frac{3}{2} \cdot \frac{x}{x+1} = \frac{3x}{2(x+1)}$$

and $$\frac{4}{x+1} \cdot \frac{5}{x-2} = \frac{20}{(x+1)(x-2)} \quad (x+1 \neq 0,\ x-2 \neq 0).$$

To divide one fraction by another, the procedure is

$$\frac{\frac{a}{b}}{\frac{c}{d}} = \frac{a}{b} \cdot \frac{d}{c}, \qquad \textbf{division}$$

where $b \neq 0$, $c \neq 0$, and $d \neq 0$. This is so because

$$\frac{\frac{a}{b}}{\frac{c}{d}} = \frac{a}{b} \cdot \left(\frac{c}{d}\right)^{-1} = \frac{a}{b} \cdot \frac{d}{c}.$$

Hence, to divide two fractions, *invert the fraction in the denominator and proceed as in multiplication.*

Example 5 (a) $\dfrac{\frac{1}{x}}{\frac{1}{y}} = \dfrac{1}{x} \cdot \dfrac{y}{1} = \dfrac{y}{x}$ $(x \neq 0, y \neq 0)$; for instance, $\dfrac{\frac{1}{2}}{\frac{1}{3}} = \dfrac{1}{2} \cdot \dfrac{3}{1} = \dfrac{3}{2}.$

(b) $$\frac{\frac{x+1}{x}}{\frac{x-3}{x+5}} = \frac{x+1}{x} \cdot \frac{x+5}{x-3}$$
$$= \frac{(x+1)(x+5)}{x(x-3)} \quad (x \neq 0,\ x-3 \neq 0,\ x+5 \neq 0)$$ ■

Just as any integer n can be written as a rational number $n/1$, so can any algebraic expression x be written as a fractional expression $x/1$. We use this property in divisions that involve fractional and non-fractional expressions, as is illustrated in the following example, where $b \neq 0$ and $c \neq 0$.

Example 6 (a) $\dfrac{\frac{a}{b}}{c} = \dfrac{\frac{a}{b}}{\frac{c}{1}} = \dfrac{a}{b} \cdot \dfrac{1}{c} = \dfrac{a}{bc}$; for instance, $\dfrac{\frac{2}{3}}{5} = \dfrac{2}{15}$.

(b) $\dfrac{a}{\frac{b}{c}} = \dfrac{\frac{a}{1}}{\frac{b}{c}} = \dfrac{a}{1} \cdot \dfrac{c}{b} = \dfrac{ac}{b}$; for instance, $\dfrac{2}{\frac{3}{5}} = \dfrac{10}{3}$. ■

Simplifying Complex Fractions. Ratios of algebraic combinations of fractions, such as

$$\frac{\frac{a}{b} + \frac{c}{d}}{\frac{a}{b} - \frac{c}{d}} \quad \text{and} \quad \frac{a^{-1} + b^{-1}}{a^{-1} - b^{-1}}$$

are called **complex fractions.** These can be simplified by applying the previous rules for addition, subtraction, multiplication, division, and cancellation. For example,

$$\frac{\frac{a}{b} + \frac{c}{d}}{\frac{a}{b} - \frac{c}{d}} = \frac{\frac{ad + bc}{bd}}{\frac{ad - bc}{bd}} = \frac{ad + bc}{bd} \cdot \frac{bd}{ad - bc} = \frac{ad + bc}{ad - bc}.$$

The above result can also be obtained by multiplying the numerator and denominator of the original complex fraction by bd. Also,

$$\frac{a^{-1} + b^{-1}}{a^{-1} - b^{-1}} = \frac{\frac{1}{a} + \frac{1}{b}}{\frac{1}{a} - \frac{1}{b}}$$

$$= \frac{\frac{b + a}{ab}}{\frac{b - a}{ab}} = \frac{b + a}{\cancel{ab}} \cdot \frac{\cancel{ab}}{b - a} = \frac{b + a}{b - a}.$$

Example 7 $\dfrac{a + \frac{b}{c}}{\frac{a}{c} + b} = \dfrac{\frac{a}{1} + \frac{b}{c}}{\frac{a}{c} + \frac{b}{1}}$

$$= \frac{\frac{ac + b}{c}}{\frac{a + bc}{c}} = \frac{ac + b}{\cancel{c}} \cdot \frac{\cancel{c}}{a + bc} = \frac{ac + b}{a + bc}$$ ■

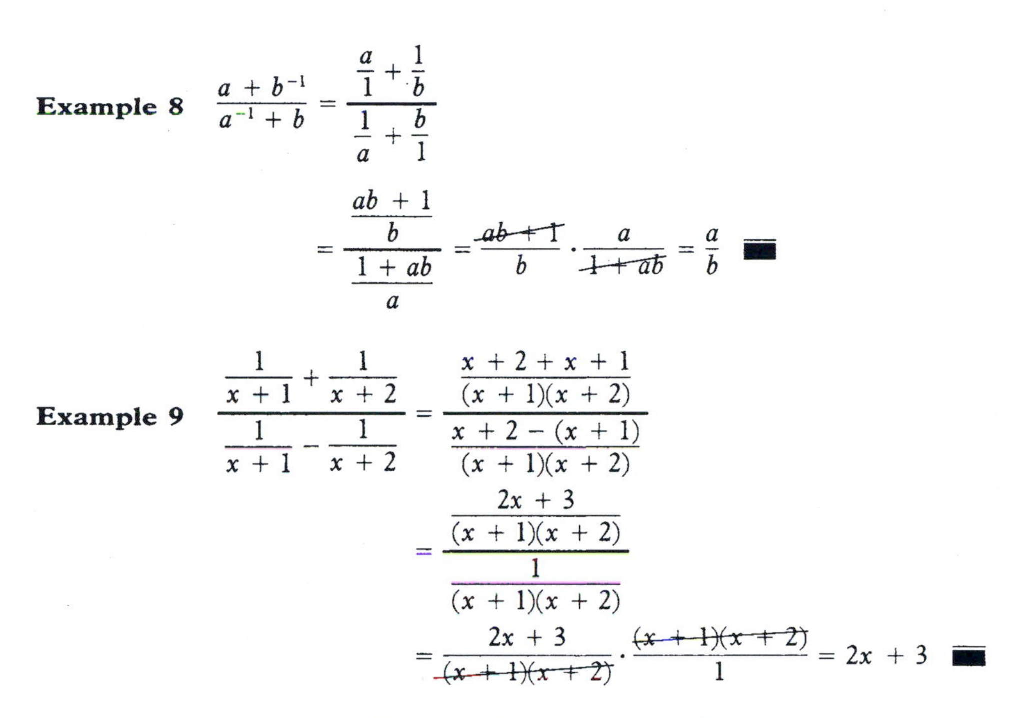

Example 8 $\dfrac{a + b^{-1}}{a^{-1} + b} = \dfrac{\dfrac{a}{1} + \dfrac{1}{b}}{\dfrac{1}{a} + \dfrac{b}{1}} = \dfrac{\dfrac{ab + 1}{b}}{\dfrac{1 + ab}{a}} = \dfrac{\cancel{ab + 1}}{b} \cdot \dfrac{a}{\cancel{1 + ab}} = \dfrac{a}{b}$ ■

Example 9 $\dfrac{\dfrac{1}{x + 1} + \dfrac{1}{x + 2}}{\dfrac{1}{x + 1} - \dfrac{1}{x + 2}} = \dfrac{\dfrac{x + 2 + x + 1}{(x + 1)(x + 2)}}{\dfrac{x + 2 - (x + 1)}{(x + 1)(x + 2)}} = \dfrac{\dfrac{2x + 3}{(x + 1)(x + 2)}}{\dfrac{1}{(x + 1)(x + 2)}} = \dfrac{2x + 3}{\cancel{(x + 1)(x + 2)}} \cdot \dfrac{\cancel{(x + 1)(x + 2)}}{1} = 2x + 3$ ■

Comment Equalities obtained by simplifying algebraic expressions are **conditional equalities.** For instance, in Example 9, the simplified expression $2x + 3$ is meaningful for all values of x, but $1/(x + 1)$ and $1/(x + 2)$ are meaningful only if $x \neq -1$ or -2, respectively. Hence, the equality is true for all x except -1 and -2. In general, *when we say that two algebraic expressions in x are equal, we mean that equality holds for all values of x for which both expressions are meaningful.* Also, whether stated explicitly or not, it is understood that the denominator of a fractional expression is restricted to nonzero values. See if you can determine the corresponding restrictions placed on a, b, and c in each of the above complex fractions.

Exercises P.4

Fill in the blanks to make each statement true.

1. If $a/b = 0$, then $a =$ ________, and $b \neq$ ________.
2. If $a/b = 1$, then $a =$ ________, and $b \neq$ ________.
3. The sum $a/b + c/b$ is equal to ____________, whereas $a/b + a/c =$ ____________.
4. The product of two fractions is equal to the product of their ____________ divided by the product of their ____________.
5. To divide one fraction by another, ____________ the fraction in the denominator and proceed as in ____________.

Write true *or* false *for each statement.*

6. $\dfrac{ac}{b + c} = \dfrac{a}{b + 1}$, provided $b + c \neq 0$ and $b + 1 \neq 0$.

7. $\frac{1}{a+b} = \frac{1}{a} + \frac{1}{b}$, provided $a \neq 0$, $b \neq 0$, and $a + b \neq 0$.

8. $\frac{1}{a^{-1} + b^{-1}} = a + b$ if $a \neq 0$ and $b \neq 0$.

9. $\frac{a}{b} \cdot \frac{c}{d} = ab^{-1}cd^{-1}$ if $b \neq 0$ and $d \neq 0$.

10. $\frac{\frac{a}{b}}{\frac{c}{d}} = ab^{-1}c^{-1}d$, provided $b \neq 0$, $c \neq 0$, and $d \neq 0$.

Equality of Fractions

In each of the following, find all possible values of x for which the equality holds.

11. $\frac{3}{2} = \frac{x}{2}$

12. $\frac{x}{15} = \frac{1}{5}$

13. $\frac{x-7}{2} = 1$

14. $\frac{1}{x+4} = 1$

15. $\frac{(x-2)(x-3)}{x-3} = 0$

16. $\frac{x-1}{x+1} = 1$

Simplify each of the following by canceling all factors common to the numerator and denominator. State any restrictions on a, b, c, x, y, or z.

17. $\frac{2ab}{6a}$

18. $\frac{-6axy}{8y}$

19. $\frac{3a + 6ab}{6ab}$

20. $\frac{x-2}{2-x}$

21. $\frac{ab - ac}{ab + ac}$

22. $\frac{2x+3}{4x+6}$

23. $\frac{2x - 3y + 4z}{4ax - 6ay + 8az}$

24. $\frac{2ax - 2x}{2ax + 2x}$

25. $\frac{ax - ay + bx - by}{x - y}$

Operations with Fractions

Perform each of the following additions and subtractions. State any restrictions on x.

26. $\frac{4}{x} + \frac{2}{x}$

27. $\frac{3}{x+1} + \frac{1}{x+1}$

28. $\frac{3}{x-3} + \frac{2}{3-x}$

29. $\frac{x}{2x-1} - \frac{x}{1-2x}$

30. $\frac{x+3}{x+2} - \frac{1}{x+2}$

31. $\frac{3x}{3x-1} - \frac{1}{3x-1}$

32. $\frac{x-2}{(2x+1)(x+3)} + \frac{x+2}{(2x+1)(x+3)}$

33. $\frac{x+5}{3x+1} - \frac{x-2}{3x+1} + \frac{2x-1}{3x+1}$

34. $\frac{x}{x-1} + \frac{3}{x-1} + \frac{2}{1-x}$

35. $\frac{3}{x(x+1)} + \frac{4x+2}{x(x+1)} - \frac{x+2}{x(x+1)}$

Perform each of the following additions and subtractions. State any restrictions on x, a, or b.

36. $\frac{x}{3} + \frac{1}{5}$

37. $\frac{x}{3} + \frac{x}{5}$

38. $\frac{3}{x} - \frac{2}{x+1}$

39. $\frac{x+1}{2} - \frac{x}{3}$

40. $\frac{1}{x+1} + \frac{1}{x-1}$

41. $1 + \frac{1}{x}$

42. $5 - \frac{2}{x+1}$

43. $\frac{4}{x+2} + \frac{3}{x-2}$

44. $\frac{1}{2x+3} - \frac{1}{3x+2}$

45. $x + \frac{2x-3}{5}$

46. $\frac{1}{x+a} + \frac{1}{x-a}$

47. $\frac{1}{x-a} - \frac{1}{x-a}$

48. $\frac{1}{ax-1} - \frac{1}{ax-2}$

49. $\frac{a}{2x+a} + \frac{b}{2x-b}$

50. $\frac{a}{ax+2} - \frac{b}{bx+2}$

Perform each of the following multiplications. Simplify answers. State any restrictions on a, b, c, or x.

51. $\frac{a}{8} \cdot \frac{2}{a}$

52. $\frac{3b}{4a} \cdot \frac{2c}{9x}$

53. $2a \cdot \frac{3}{4}$

54. $a \cdot \frac{2}{a}$

55. $\frac{2x+3}{6} \cdot \frac{3}{4x+6}$

56. $\frac{2}{x} \cdot \frac{x+3}{6}$

57. $\dfrac{4}{2x-1}\cdot\dfrac{10x-5}{16}$

58. $\dfrac{x-1}{x+2}\cdot\dfrac{x+2}{x-2}$

59. $\dfrac{2x-1}{2x-6}\cdot\dfrac{x-3}{4x-2}$

60. $\dfrac{10x}{5x+10}\cdot\dfrac{2x+1}{2x}$

Perform each of the following divisions. Simplify answers. State any restrictions on a, b, c, d, or x.

61. $\dfrac{\frac{a}{8}}{\frac{a}{2}}$

62. $\dfrac{\frac{a}{2b}}{\frac{a}{4}}$

63. $\dfrac{\frac{3a}{4b}}{\frac{2c}{9d}}$

64. $\dfrac{\frac{9a}{4b}}{\frac{3a}{10b}}$

65. $\dfrac{\frac{5a}{a}}{5}$

66. $\dfrac{\frac{10}{2x+3}}{\frac{4}{10x+5}}$

67. $\dfrac{\frac{10}{2x+3}}{5}$

68. $\dfrac{10}{\frac{2x+3}{5}}$

69. $\dfrac{\frac{4}{x+1}}{8}$

70. $\dfrac{4}{\frac{x+1}{8}}$

71. $\dfrac{\frac{x-1}{x-2}}{\frac{x+2}{x-2}}$

72. $\dfrac{\frac{3}{x-2}}{\frac{6}{x-2}}$

73. $\dfrac{\frac{2x}{x+1}}{\frac{3x}{4x+4}}$

74. $\dfrac{\frac{3x-5}{x}}{\frac{6x-10}{5x}}$

75. $\dfrac{\frac{4x+2}{3x-6}}{\frac{2x+1}{4x-8}}$

Complex Fractions

Simplify each expression. State any restrictions on a, b, c, or x.

76. $\dfrac{\frac{a}{b+c}+\frac{b}{b+c}}{\frac{c}{b+c}}$

77. $\dfrac{\frac{a-b}{a+b}}{\frac{a}{a+b}-\frac{b}{a+b}}$

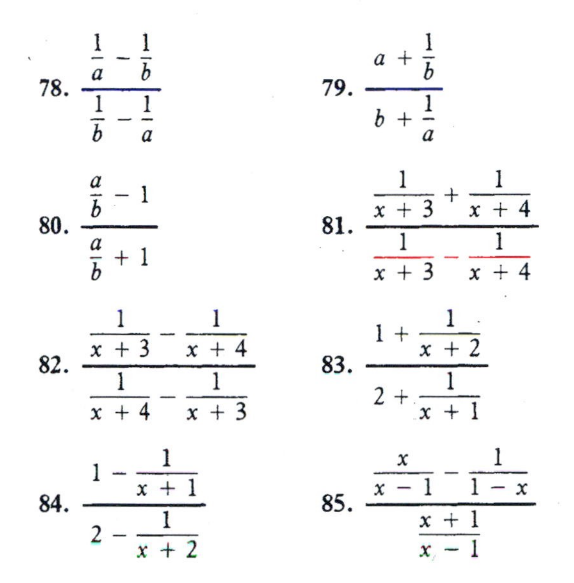

78. $\dfrac{\frac{1}{a}-\frac{1}{b}}{\frac{1}{b}-\frac{1}{a}}$

79. $\dfrac{a+\frac{1}{b}}{b+\frac{1}{a}}$

80. $\dfrac{\frac{a}{b}-1}{\frac{a}{b}+1}$

81. $\dfrac{\frac{1}{x+3}+\frac{1}{x+4}}{\frac{1}{x+3}-\frac{1}{x+4}}$

82. $\dfrac{\frac{1}{x+3}-\frac{1}{x+4}}{\frac{1}{x+4}-\frac{1}{x+3}}$

83. $\dfrac{1+\frac{1}{x+2}}{2+\frac{1}{x+1}}$

84. $\dfrac{1-\frac{1}{x+1}}{2-\frac{1}{x+2}}$

85. $\dfrac{\frac{x}{x-1}-\frac{1}{1-x}}{\frac{x+1}{x-1}}$

Express each of the following without the superscript −1 and simplify. State any restrictions on a, b, c, or x.

86. $\dfrac{a}{b^{-1}+c^{-1}}$

87. $\dfrac{a^{-1}}{b^{-1}+c^{-1}}$

88. $\dfrac{a^{-1}+b^{-1}}{c}$

89. $\dfrac{a+b^{-1}}{c^{-1}}$

90. $\dfrac{a^{-1}+b^{-1}}{a^{-1}-b^{-1}}$

91. $\dfrac{\frac{1}{a^{-1}}+\frac{1}{b^{-1}}}{\frac{1}{a^{-1}}-\frac{1}{b^{-1}}}$

92. $\dfrac{a+\frac{b^{-1}}{c}}{a-\frac{b^{-1}}{c}}$

93. $\dfrac{a+\frac{b^{-1}}{c}}{c+\frac{b^{-1}}{a}}$

94. $\dfrac{(x+1)^{-1}}{(x+2)^{-1}+(x+3)^{-1}}$

95. $\dfrac{(x+1)^{-1}+(x+2)^{-1}}{(x+3)^{-1}}$

96. $\dfrac{1+(x-1)^{-1}}{2+(x-1)^{-1}}$

97. $\dfrac{(x-1)^{-1}+(x-2)^{-1}}{(x-1)^{-1}-(x-2)^{-1}}$

Miscellaneous

98. What values of a, b, and c are excluded in the equation of Example 7 of the text? What values of a and b are excluded in Example 8?

99. Prove the subtraction rule: $\frac{a}{b} - \frac{c}{d} = \frac{ad - bc}{bd}$ $(b \neq 0,\ d \neq 0)$.

100. Prove each of the following.

(a) $\frac{a}{a+b} = \frac{1}{1+b} \Leftrightarrow a = 1 \text{ or } b = 0$
(Assume $a \neq -b$ and $b \neq -1$.)

(b) $\frac{a+b}{a+c} = \frac{1+b}{1+c} \Leftrightarrow a = 1 \text{ or } b = c$
(Assume $a \neq -c$ and $c \neq -1$.)

P.5 Powers and Roots

"Though people do a great deal of talking, the total output since the beginning of gabble to the present day, including all baby talk, love songs, and Congressional debates, totals about 10^{16}."

—*from* Mathematics and the Imagination *by Edward Kasner and James R. Newman*

Positive Exponents
Zero and Negative Exponents
Roots
Algebraic Expressions

Positive Exponents. A sum such as $2 + 2 + 2 + 2 + 2$ can be written more compactly as $5 \cdot 2$. Similarly, we write the product $2 \cdot 2 \cdot 2 \cdot 2 \cdot 2$ in more compact form as 2^5, where the **exponent** 5 indicates the number of times that the **base** 2 occurs as a factor. Thus, $3^4 = 3 \cdot 3 \cdot 3 \cdot 3 = 81$, and $(-6)^3 = (-6)(-6)(-6) = -216$. In general, if x stands for a real number and n is a positive integer, then we define the nth power of x as follows.

$$x^n = \underbrace{x \cdot x \cdot x \cdot \ldots \cdot x}_{n \text{ factors}} \qquad \textbf{nth power of } x$$

We note that $3^2 \cdot 3^5 = (3 \cdot 3)(3 \cdot 3 \cdot 3 \cdot 3 \cdot 3) = 3^7$, whereas $(3^2)^5 = (3 \cdot 3)(3 \cdot 3)(3 \cdot 3)(3 \cdot 3)(3 \cdot 3) = 3^{10}$. In general, if m and n are positive integers, then

$$x^m x^n = x^{m+n} \qquad \textbf{same-base rule (multiplication)}$$

and

$$(x^m)^n = x^{mn}. \qquad \textbf{power-of-a-power rule}$$

Hence, to multiply powers of the same base, you *add* the exponents, and to raise a power of x to another power, you *multiply* the exponents. These rules can be extended to more than two exponents. For instance,

$$x^2 \cdot x^3 \cdot x^4 = x^{2+3+4} = x^9 \quad \text{and} \quad [(x^2)^3]^4 = x^{2 \cdot 3 \cdot 4} = x^{24}.$$

Example 1

(a) $3^5 \cdot 2^4 \cdot 3^2 \cdot 2^6 = 3^5 \cdot 3^2 \cdot 2^4 \cdot 2^6 = 3^7 \cdot 2^{10}$

(b) $(-3)^{2n} = [(-3)^2]^n = 9^n$

(c) $(x^2)^5(x^3)^4 = x^{10} \cdot x^{12} = x^{22}$ ■

We note that

$$\begin{aligned}(3 \cdot 2)^4 &= (3 \cdot 2)(3 \cdot 2)(3 \cdot 2)(3 \cdot 2)\\ &= (3 \cdot 3 \cdot 3 \cdot 3)(2 \cdot 2 \cdot 2 \cdot 2)\\ &= 3^4 \cdot 2^4.\end{aligned}$$

In general, if n is any positive integer, then

$$(xy)^n = x^n y^n. \qquad \textbf{same-exponent rule (multiplication)}$$

That is, the nth power of a product is equal to the product of the nth powers of the factors. The rule extends to more than two base factors, as in

$$(xyz)^4 = x^4y^4z^4.$$

Example 2 (a) $(2y)^3 = 2^3y^3 = 8y^3$

(b) $(-x)^n = [(-1)x]^n = (-1)^n x^n = \begin{cases} x^n \text{ if } n \text{ is even} \\ -x^n \text{ if } n \text{ is odd} \end{cases}$ ■

Example 3 Express $(2^5 4^3)^2$ as a power of 2.

Solution: $(2^5 4^3)^2 = (2^5)^2(4^3)^2 = 2^{10}4^6 = 2^{10}(2^2)^6 = 2^{10}2^{12} = 2^{22}$ ■

The product of fractions is equal to the product of the numerators divided by the product of the denominators. For instance,

$$\left(\frac{3}{2}\right)^4 = \frac{3}{2}\cdot\frac{3}{2}\cdot\frac{3}{2}\cdot\frac{3}{2} = \frac{3 \cdot 3 \cdot 3 \cdot 3}{2 \cdot 2 \cdot 2 \cdot 2} = \frac{3^4}{2^4}.$$

In general, if n is any positive integer and $y \neq 0$, then

$$\left(\frac{x}{y}\right)^n = \frac{x^n}{y^n}. \qquad \textbf{same-exponent rule (division)}$$

That is, the nth power of a quotient equals the quotient of the nth powers.

Example 4 (a) $\left(\frac{x}{2}\right)^6 = \frac{x^6}{2^6} = \frac{x^6}{64}$

(b) $\left(\frac{2x+1}{3x+4}\right)^2 = \frac{(2x+1)^2}{(3x+4)^2}$, provided $3x + 4 \neq 0$ ■

Common factors of the numerator and denominator of a fraction can be canceled without changing the value of the fraction. For instance,

$$\frac{3^6}{3^2} = \frac{3 \cdot 3 \cdot 3 \cdot 3 \cdot \cancel{3} \cdot \cancel{3}}{\cancel{3} \cdot \cancel{3}} = 3^4 = 3^{6-2}.$$

More generally, if $x \neq 0$ and $m > n$, then

$$\frac{x^m}{x^n} = x^{m-n}. \qquad \textbf{same-base rule (division)}$$

Hence, to divide powers of the same base, we *subtract* the exponents. Furthermore, the restriction $m > n$ can be removed with the introduction of zero and negative exponents, as we shall soon see.

Example 5 (a) $\dfrac{(4x-5)^5}{(4x-5)^3} = (4x-5)^{5-3} = (4x-5)^2$, provided $4x - 5 \neq 0$

(b) $\dfrac{(x^2+1)^4}{x^2+1} = \dfrac{(x^2+1)^4}{(x^2+1)^1} = (x^2+1)^{4-1} = (x^2+1)^3$ ■

Zero and Negative Exponents. If we formally apply the same-base rule for quotients to the case $m = n$, we obtain

$$\text{(a)} \quad \frac{x^n}{x^n} = x^{n-n} = x^0 \quad (x \neq 0).$$

But

$$\text{(b)} \quad \frac{x^n}{x^n} = 1 \quad (x \neq 0).$$

Therefore, in order to make (a) consistent with (b), we are led to the following definition. If $x \neq 0$, then

$$\mathbf{x^0 = 1.}$$

That is, any nonzero quantity raised to the zero power is 1 by definition. The expression 0^0 is undefined.

Example 6 (a) $7^0 = 1$, $(-5)^0 = 1$, and $\left(\dfrac{3}{4}\right)^0 = 1$

(b) $(x + 12)^0 = 1$, provided $x + 12 \neq 0$ ■

If we formally apply the same-base rule to the expression x^2/x^6, we obtain

$$\text{(a)} \quad \frac{x^2}{x^6} = x^{2-6} = x^{-4} \quad (x \neq 0).$$

But

$$\text{(b)} \quad \frac{x^2}{x^6} = \frac{x \cdot x}{x \cdot x \cdot x \cdot x \cdot x \cdot x} = \frac{1}{x^4} \quad (x \neq 0).$$

In order to make (a) consistent with (b), we are led to the following definition. If n is a positive integer and $x \neq 0$, then by definition

$$\mathbf{x^{-n} = \frac{1}{x^n}.}$$

That is, x^{-n} is the reciprocal of x^n. The expression 0^{-n} is undefined.

Example 7 (a) $2^{-3} = \frac{1}{2^3} = \frac{1}{8}$

(b) $(3x + 4)^{-2} = \frac{1}{(3x + 4)^2}$, provided $3x + 4 \neq 0$ ■

Comment In Section P.2, we defined x^{-1} as the multiplicative inverse of x. This agrees with our new definition of x^{-n} when $n = 1$.

Three equivalent interpretations of x^{-n}, which can be derived from our definitions, are

$$x^{-n} = \begin{cases} 1/x^n \\ (x^n)^{-1}. \\ (x^{-1})^n \end{cases}$$

Example 8 (a) $4^{-3} = \frac{1}{4^3} = \frac{1}{64}$

(b) $4^{-3} = (4^3)^{-1} = 64^{-1} = \frac{1}{64}$

(c) $4^{-3} = (4^{-1})^3 = \left(\frac{1}{4}\right)^3 = \frac{1}{64}$ ■

It can be shown that *all of the previous exponent rules for products, quotients, and powers hold for all integers m and n, positive, negative, or zero, provided the base is not* 0 *when the exponent is* 0 *or negative*. In particular, we can rewrite the same-base rule for division as

$$\frac{x^m}{x^n} = x^{m-n} = \frac{1}{x^{n-m}} \quad (x \neq 0),$$

with no restrictions placed on m and n. The following examples illustrate how the exponent rules are used to simplify algebraic expressions.

Example 9 (Same-Base Rule)
If $x \neq 0$, then

(a) $x^5x^{-2} = x^{5+(-2)} = x^3$;

(b) $\frac{x^{-5}}{x^{-2}} = x^{-5-(-2)} = x^{-3} = \frac{1}{x^3}$ or $\frac{x^{-5}}{x^{-2}} = \frac{1}{x^{-2-(-5)}} = \frac{1}{x^3}$;

(c) $\frac{1}{x^{-2}} = \frac{x^0}{x^{-2}} = x^{0-(-2)} = x^2$ [see Example 10(c)]. ■

Example 10 (Power-of-a-Power Rule)
If $a \neq 0$, then

(a) $(a^{-3})^4 = a^{(-3)4} = a^{-12} = \frac{1}{a^{12}}$;

(b) $(a^{-3})^{-4} = a^{(-3)(-4)} = a^{12}$;

(c) $\dfrac{1}{a^{-2}} = (a^{-2})^{-1} = a^{(-2)(-1)} = a^2$. ∎

Example 11 (Same-Exponent Rule)
If $a \neq 0$, then

(a) $(2a^{-1})^3 = 2^3(a^{-1})^3 = 2^3a^{-3} = \dfrac{8}{a^3}$;

(b) $\left(\dfrac{1}{a^{-3}}\right)^{-2} = \dfrac{1^{-2}}{(a^{-3})^{-2}} = \dfrac{1}{a^6}$. ∎

Example 12 (Miscellaneous)
If $a \neq 0$, $x \neq 0$, and $y \neq 0$, then

(a)
$$\begin{aligned}(3^2a^3x)(3^{-2}a^{-1}x^{-4}) &= 3^23^{-2} \cdot a^3a^{-1} \cdot xx^{-4} \\ &= 3^{2+(-2)} \cdot a^{3+(-1)} \cdot x^{1+(-4)} \\ &= 3^0 \cdot a^2 \cdot x^{-3} \\ &= \frac{a^2}{x^3};\end{aligned}$$

(b)
$$\begin{aligned}\left(\frac{a^{-1}}{y}\right)^2 \cdot \frac{x^3y^{-3}}{a^{-4}x^{-2}} &= \frac{a^{-2}}{y^2} \cdot \frac{x^3y^{-3}}{a^{-4}x^{-2}} \\ &= \frac{a^{-2}}{a^{-4}} \cdot \frac{x^3}{x^{-2}} \cdot \frac{y^{-3}}{y^2} \\ &= a^{-2-(-4)} \cdot x^{3-(-2)} \cdot y^{-3-2} \\ &= a^2 \cdot x^5 \cdot y^{-5} \\ &= \frac{a^2x^5}{y^5}.\end{aligned}$$ ∎

Example 13 (Complex Fractions)
If $x \neq 0$, $y \neq 0$, $a \neq 0$, and $b \neq 0$, then

(a)
$$\begin{aligned}\frac{x^{-2} + y^{-3}}{y^{-2}} &= \frac{\dfrac{1}{x^2} + \dfrac{1}{y^3}}{\dfrac{1}{y^2}} = \frac{\dfrac{y^3 + x^2}{x^2y^3}}{\dfrac{1}{y^2}} \\ &= \frac{y^3 + x^3}{x^2y^3} \cdot \frac{y^2}{1} = \frac{x^3 + y^3}{x^2y};\end{aligned}$$

(b)
$$\begin{aligned}\frac{a^2 + b^{-3}}{a^{-2} + b^3} &= \frac{a^2 + \dfrac{1}{b^3}}{\dfrac{1}{a^2} + b^3} = \frac{\dfrac{a^2b^3 + 1}{b^3}}{\dfrac{1 + a^2b^3}{a^2}} \\ &= \frac{a^2b^3 + 1}{b^3} \cdot \frac{a^2}{1 + a^2b^3} = \frac{a^2}{b^3}.\end{aligned}$$ ∎

Roots. If $y^2 = x$, then y is called a **square root** of x, and if $y^3 = x$, then y is called a **cube root** of x. In general, a number y is called an ***n*th root** of x if $y^n = x$ $(n = 2, 3, 4, \ldots)$. Hence, 3 and -3 are fourth roots of 81 since $3^4 = 81$ and $(-3)^4 = 81$. Similarly, -2 is a fifth root of -32 since $(-2)^5 = -32$.

When n is odd, every real number has exactly one real nth root, which is denoted by $\sqrt[n]{x}$. For example, $\sqrt[5]{-32} = -2$. When n is even, it is "double or nothing"; that is, every *positive* real number will have *two* real nth roots, one the negative of the other, but a *negative* number has *no* even roots in the real number system. For example, 81 has the fourth roots 3 and -3, but -81 has no real fourth roots. The nth root of 0 is 0 itself for both even and odd n.

The symbol $\sqrt[n]{x}$ is also used to denote the *positive* nth root when n is even and x is positive. For example, $\sqrt[4]{81} = 3$. In general, the expression $\sqrt[n]{x}$ is called a **radical,** $\sqrt{}$ is the **radical sign,** n is the **index,** and x is the **radicand.** For square roots, the radical sign is written without an index.

Example 14 (a) $\sqrt[4]{16} = 2$, but $\sqrt[4]{-16}$ does not exist as a real number.

(b) $\sqrt[5]{32} = 2$ and $\sqrt[5]{-32} = -2$ ■

Example 15 (a) $\sqrt{3^2} = \sqrt{9} = 3$ and $\sqrt{(-3)^2} = \sqrt{9} = 3$

(b) $\sqrt[3]{4^3} = \sqrt[3]{64} = 4$ and $\sqrt[3]{(-4)^3} = \sqrt[3]{-64} = -4$ ■

Example 15 illustrates an important difference between even and odd roots. When n is even, a^n is positive whether a is positive or negative, and $\sqrt[n]{a^n}$ means the positive nth root of a^n. Therefore

$$\sqrt[n]{a^n} = |a| \text{ when } n \text{ is even.}$$

On the other hand, when n is odd, the nth root of a^n is a itself, whether a is positive or negative. That is,

$$\sqrt[n]{a^n} = a \text{ if } n \text{ is odd.}$$

We note that

$$\sqrt{25 \cdot 4} = 5 \cdot 2 = \sqrt{25}\,\sqrt{4} \qquad \text{and} \qquad \sqrt{\frac{25}{4}} = \frac{5}{2} = \frac{\sqrt{25}}{\sqrt{4}}.$$

In general,

$$\sqrt[n]{ab} = \sqrt[n]{a}\,\sqrt[n]{b} \qquad \text{and} \qquad \sqrt[n]{\frac{a}{b}} = \frac{\sqrt[n]{a}}{\sqrt[n]{b}},$$

where $a > 0$ and $b > 0$ if n is even. That is, the nth root of a product equals the product of the nth roots of the factors, and the nth root of a quotient of two terms equals the quotient of their nth roots. These rules can be used to simplify expressions that are not exact nth roots, as is illustrated in the following examples.

Example 16 (a) $\sqrt{20} = \sqrt{4 \cdot 5} = \sqrt{4}\,\sqrt{5} = 2\sqrt{5}$

(b) $\sqrt[3]{32} = \sqrt[3]{8 \cdot 4} = \sqrt[3]{8}\,\sqrt[3]{4} = 2\sqrt[3]{4}$ ■

Example 17 (a) $\sqrt[3]{a^7b^5} = \sqrt[3]{a^3a^3a \cdot b^3b^2} = \sqrt[3]{a^3}\,\sqrt[3]{a^3}\,\sqrt[3]{a} \cdot \sqrt[3]{b^3}\,\sqrt[3]{b^2}$
$= a \cdot a\sqrt[3]{a} \cdot b\sqrt[3]{b^2} = a^2b\sqrt[3]{ab^2}.$

(b) $\sqrt{\dfrac{a^3}{b^2}} = \dfrac{\sqrt{a^3}}{\sqrt{b^2}} = \dfrac{\sqrt{a^2a}}{\sqrt{b^2}} = \dfrac{|a|\sqrt{a}}{|b|} = \dfrac{a\sqrt{a}}{|b|} \qquad (b \neq 0)$

(Why is $|a|$ equal to a?) ■

Algebraic Expressions. The extraction of roots and the basic operations of addition, subtraction, multiplication, and division are called **algebraic operations.** Any expression constructed out of numbers and letters representing numbers by these operations is called an **algebraic expression.** Any process that utilizes these operations, for example, solving equations, is called an **algebraic process.**

In a given algebraic expression, some letters may represent real numbers that are unspecified but fixed in value. These letters are called **constants.** Also, any real number itself is called a constant. Letters that represent real numbers that may assume different values in a given context are called **variables.** We usually let letters toward the beginning of the alphabet, such as a, b, and c, denote constants, and letters toward the end of the alphabet, such as x, y, and z, denote variables. For example, in the algebraic expression

$$\sqrt{ax + b},$$

a and b are constants and x is a variable. Later we will extend the notions of constant and variable to apply to complex numbers and letters representing them.

Exercises P.5

1. Write each of the following equations in compact form.
   (a) $(4 \cdot 4 \cdot 4)(4 \cdot 4 \cdot 4 \cdot 4 \cdot 4) = 4 \cdot 4 \cdot 4 \cdot 4 \cdot 4 \cdot 4 \cdot 4 \cdot 4$
   (b) $(3 \cdot 3)(3 \cdot 3)(3 \cdot 3)(3 \cdot 3) = 3 \cdot 3 \cdot 3 \cdot 3 \cdot 3 \cdot 3 \cdot 3 \cdot 3$
   (c) $(5 \cdot y)(5 \cdot y)(5 \cdot y)(5 \cdot y)(5 \cdot y)$
   $= 5 \cdot 5 \cdot 5 \cdot 5 \cdot 5 \cdot y \cdot y \cdot y \cdot y \cdot y$

Fill in the blanks to make each statement true.

2. x^{-n} is the multiplicative inverse of ______, provided $x \neq 0$.

3. The number 2 is a sixth root of 64 because ______. In general, y is an nth root of x if ______.

4. $\sqrt[n]{x^n} =$ ______ if n is odd; $\sqrt[n]{x^n} =$ ______ if n is even.

5. If n is an even positive integer, then every positive real number has ______ nth root(s) and every negative real number has ______ nth root(s); if n is odd, then every positive real number has ______ nth root(s) and every negative real number has ______ nth root(s).

Write true *or* false *for each statement.*

6. To raise one power of x to another power, multiply the two exponents.
7. If x is positive, then x^{-7} is negative.
8. To divide two powers of x, divide the exponents.
9. $\sqrt[5]{ab} = \sqrt[5]{a}\sqrt[5]{b}$ for all real numbers a and b, but $\sqrt[4]{ab} = \sqrt[4]{a}\sqrt[4]{b}$ only if a and b are both ≥ 0.
10. If y is a square root of x, then y is also a fourth root of x^2.

Positive Exponents

11. Express each of the following as a power of -2.
    (a) -8 (b) 16 (c) -32
    (d) 64 (e) $(-32)(64)$
12. Express each of the following as a power of -3.
    (a) 9 (b) -27 (c) 81
    (d) -243 (e) $(-27)(-243)$
13. Express each of the following as a power of 2.
    (a) $2^3 \cdot 4$ (b) $2^2 \cdot 4^3$ (c) $2^2 \cdot 4^4 \cdot 16^2$
14. Express each of the following as a power of 3.
    (a) $27 \cdot 3^2$ (b) $3^2 \cdot 9^3 \cdot 81$ (c) $(3 \cdot 9)^2(3 \cdot 27)^3$
15. Express each of the following as a product of a power of 2 by a power of 3.
    (a) $2^3 \cdot 3^2 \cdot 6$ (b) $4 \cdot 6 \cdot 12 \cdot 16$ (c) $18^2 \cdot 24^3$

Simplify each of the following.

16. $(2^2)^4$
17. $(x^3)^5$
18. $7^2 \cdot 7^3 \cdot 7^5$
19. $(-3^7(-3)^{10}$
20. $(-5)(-5)^2(-5)^3$
21. $a \cdot a \cdot a^2 \cdot a^2 \cdot a^3 \cdot a^3$
22. $x^2 \cdot x \cdot x^3 \cdot x \cdot x^4 \cdot x$
23. $[(-2)^3]^5$
24. $[(4^2)^3]^4$
25. $[(x^2)^n]^5$
26. $2^8 \cdot 5^8$
27. $(-3)^9(-4)^9$
28. $2^6a^6b^6$
29. $(a^2b)^4(ab^2)^4$
30. $(a^2b)^5(ac)^3(b^2c^3)^4(ab)^2$

Zero and Negative Exponents

Simplify each of the following. Express answers in terms of positive exponents.

31. $3^5 \cdot 2^7 \cdot 3^{-2} \cdot 2^{-4}$
32. $(2^3a^0b^6)(2^{-1}a^{-3}b^{-4})$
33. $(x^2y^{-1})^3(x^{-1}y^4)^2$
34. $(3^{-2})^5(3^{-2})^{-3}(3^2)^0$
35. $3^8 \cdot 9^{-2}$
36. $16 \cdot 2^{-3} \cdot 4^3 \cdot 32^{-1}$
37. $\dfrac{2^0}{2^{-3}}$
38. $\dfrac{3}{3^2}$
39. $\dfrac{8^{-7}}{8^{-9}}$
40. $\dfrac{(x^{-3})^0}{x^{-1}}$
41. $\dfrac{a^{-1}}{(a^{-1})^2}$
42. $\dfrac{(b^{-2})^3}{(b^{-4})^2}$
43. $\dfrac{x^{-1}y}{x^{-2}y^{-3}}$
44. $\left(\dfrac{b^{-2}}{b^{-3}}\right)^{-4}$
45. $\left(\dfrac{a^2b^4}{ab^{-4}}\right)^2$
46. $\left(\dfrac{x^{-1}y^2}{x^3y^4}\right)^{-3}$
47. $\dfrac{(a^{-1}b^2)^{-3}}{(ab^{-2})^{-2}}$
48. $\dfrac{2^{-1}a^2b^6}{(2^{-2}a^{-1}b)^4}$

Roots

Evaluate, if possible, or simplify each of the following.

49. $\sqrt{25 \cdot 36 \cdot 49 \cdot 64}$
50. $\sqrt{16a^2b^4c^5}$
51. $\sqrt[3]{8 \cdot 27 \cdot (-a)^3b^6}$
52. $\sqrt{\dfrac{16 \cdot 36}{25 \cdot 49}}$
53. $\sqrt[3]{\dfrac{8a^6}{125b^3}}$
54. $\sqrt[4]{\dfrac{16a^4}{(-3)^4b^8}}$
55. $\sqrt[3]{(-15)^3}$
56. $\sqrt[4]{(-15)^4}$
57. $(\sqrt[3]{-15})^3$
58. $(\sqrt[4]{-15})^4$
59. $\sqrt{128}$
60. $\sqrt[3]{96}$
61. $\sqrt[3]{16a^4b^5c^6}$
62. $\sqrt{18a^3b^4}$
63. $\sqrt{a^2b^3c^4d^5}$
64. $\sqrt{\dfrac{a^3b^3}{c^3}}$
65. $\sqrt[3]{\dfrac{a^7}{b^5}}$
66. $\sqrt{\dfrac{a^3b^4}{c^2d^6}}$

Miscellaneous

Science deals with some very large and some very small numbers. In order to avoid many zeros, we use **scientific notation.** *In scientific notation, a number is written as a product of a decimal between 1 and 10 and a power of 10 (the symbol $\times$ is used for multiplication). For numbers greater than 1, positive* *exponents are used. For example,* 520 *is written as* 5.2×10^2 *and* $5{,}200{,}000 = 5.2 \times 10^6$. *For numbers between* 0 *and* 1, *negative exponents are used. For example,* $.0125 = 1.25 \times 10^{-2}$ *and* $.00000125 = 1.25 \times 10^{-6}$. *Write each of the following numbers in scientific notation.*

67. 5,280
68. 52,800
69. 176
70. .025
71. .00025
72. .76
73. 176,000
74. 4,125,271
75. 25,000,000
76. .000076
77. .0000001
78. .00000000001

Compute each of the following. Express answers in scientific notation.

79. $(1.2 \times 10^5)(1.1 \times 10^7)$

80. $(1.5 \times 10^4)(8 \times 10^6)$

81. $(2.5 \times 10^8)(2.5 \times 10^{-2})$

82. $(6.2 \times 10^{-4})(4 \times 10^5)$

83. $(2.5 \times 10^{-3})(4 \times 10^{-5})$

84. $\dfrac{3.3 \times 10^8}{6.6 \times 10^2}$

85. $\dfrac{4.5 \times 10^3}{2 \times 10^7}$

86. $\dfrac{5.1 \times 10^6}{3 \times 10^{-4}}$

87. $\dfrac{7.5 \times 10^{-5}}{2.5 \times 10^6}$

88. $\dfrac{2.13 \times 10^{-10}}{8.52 \times 10^{-5}}$

89. The speed of light is 29,977,600,000 cm/sec. Express this value in scientific notation.

90. **Avogadro's number** (the number of atoms of an element in a mass equal to its gram atomic weight) is 602,300,000,000,000,000,000,000. Write this number in scientific notation.

Each of the following numbers represents the corresponding planet's average distance from the sun. Express each value in standard decimal notation.

91. Mercury: 3.6×10^7 mi

92. Venus: 6.72×10^7 mi

93. Earth: 9.2956×10^7 mi

94. Mars: 1.416×10^8 mi

95. Jupiter: 4.834×10^8 mi

96. Saturn: 8.86×10^8 mi

97. Uranus: 1.782×10^9 mi

98. Neptune: 1.792×10^9 mi

99. Pluto: 3.664×10^9 mi

100. The electric charge of an electron is .00000000000000000016 coulombs. Express this number in scientific notation.

101. **Planck's constant** (the increase in maximum energy of a photoelectron per unit increase in the frequency of absorbed light) is .0000000000000000000000000006624. Write Planck's constant in scientific notation.

Write the given mass of each of the following in decimal notation.

102. hydrogen: 1.673×10^{-24} g

103. oxygen: 1.328×10^{-23} g

104. silver: 1.791×10^{-22} g

105. gold: 3.27×10^{-22} g

Use a calculator to do Exercises 106–7.

106. The formula for **compound interest** is

$$A = P\left(1 + \frac{R}{N}\right)^{Nt},$$

where

$$\begin{cases} P = \text{amount originally invested,} \\ A = \text{amount accumulated after } t \text{ years,} \\ R = \text{annual interest rate (annual percentage rate/100),} \\ N = \text{number of compound periods per year.} \end{cases}$$

For example, if \$250 is invested at 9.6% annual rate, compounded monthly, then the amount of the investment after ten years is

$$250\left(1 + \frac{.096}{12}\right)^{12 \cdot 10} = 250(1.008)^{120} \approx 250(2.6) = 650 \text{ dollars.}$$

Compute A in each of the following cases.

(a) $P = \$500$, $R = 16\%$, $N = 1$, $t = 10$

(b) $P = \$500$, $R = 15\%$, $N = 12$, $t = 10$

(c) $P = \$1{,}000$, $R = 12\%$, $N = 1$, $t = 5$

(d) $P = \$1{,}000$, $R = 12\%$, $N = 360$, $t = 5$

107. The **monthly payment formula** for repaying a loan is

$$P = \frac{rL(1 + r)^n}{(1 + r)^n - 1},$$

where $\begin{cases} P = \text{monthly payment}, \\ L = \text{amount of loan}, \\ r = \frac{1}{12} \cdot R,\ R = \text{annual interest rate}, \\ n = \text{the total number of payments.} \end{cases}$

Compute P in each of the following cases.

(a) auto: $L = \$8{,}000$, $R = 10.75\%$, $n = 36$

(b) auto: $L = \$8{,}000$, $R = 10.75\%$, $n = 48$

(c) house: $L = \$60{,}000$, $R = 12\%$, $n = 360$

(d) house: $L = \$60{,}000$, $R = 9\%$, $n = 360$

Chapter P Review Outline

P.1 The Real Number System

The set of real numbers is made up of the rational numbers (terminating or repeating decimals) and the irrational numbers. It corresponds to the set of points on a directed line.

The real numbers are closed under addition, subtraction, and multiplication. Division by any real number except 0 is possible.

The real numbers form an ordered set. That is, either any two real numbers are equal, or one is greater than the other.

$a > b$ (a is greater than b) if a is to the right of b on the number line.

$|a|$ (the absolute value of a) is the distance between 0 and a on the number line.

P.2 Basic Properties of Real Numbers

Basic Properties

Commutative Properties:

$a + b = b + a$

$a \cdot b = b \cdot a$

Associative Properties:

$(a + b) + c = a + (b + c)$

$(a \cdot b) \cdot c = a \cdot (b \cdot c)$

Left Distributive Property:

$a \cdot (b + c) = (a \cdot b) + (a \cdot c)$

Identities:

$a + 0 = a = 0 + a$

$a \cdot 1 = a = 1 \cdot a$

Inverses:

$a + (-a) = 0 = -a + a$

$a \cdot a^{-1} = 1 = a^{-1} \cdot a \quad (a \neq 0)$

Derived Properties

Right Distributive Property:

$(a + b) \cdot c = (a \cdot c) + (b \cdot c)$

Zero-Multiplier Property:

$0 \cdot a = 0$

Negative Multiplier Property:

$(-1)a = -a$

Double Inverse Properties:

$-(-a) = a$, and $(a^{-1})^{-1} = a \quad (a \neq 0)$

Zero-Product Property:

$ab = 0 \Leftrightarrow a = 0$ or $b = 0$

P.3 Signed Quantities, Order, and Absolute Value

Definitions

Subtraction: $a - b = a + (-b)$

Division: $\frac{a}{b} = a \cdot \frac{1}{b} = a \cdot b^{-1} \quad (b \neq 0)$

Greater Than: $a > b \Leftrightarrow a = b + p$, where p is a positive real number.

Absolute Value:

$$|x| = \begin{cases} x \text{ if } x \geq 0 \\ -x \text{ if } x < 0 \end{cases}$$

Derived Properties

Inverses of Sums, Products, and Quotients:

$-(a + b) = -a + (-b)$

$(ab)^{-1} = a^{-1}b^{-1} \quad (a \neq 0,\ b \neq 0)$

$\left(\frac{a}{b}\right)^{-1} = \frac{a^{-1}}{b^{-1}} = \frac{b}{a} \quad (a \neq 0,\ b \neq 0)$

Sign Rules:

$(-a)(-b) = ab$

$(-a)b = a(-b) = -(ab)$

$\frac{-a}{-b} = \frac{a}{b} \quad (b \neq 0)$

$\frac{-a}{b} = \frac{a}{-b} = -\frac{a}{b} \quad (b \neq 0)$

Zero-Quotient Property:

$\frac{a}{b} = 0 \Leftrightarrow a = 0 \text{ and } b \neq 0$

Properties of Greater Than:

$a > b \Leftrightarrow a + c > b + c$

If $c > 0$, then $a > b \Leftrightarrow ac > bc$.
If $c < 0$, then $a > b \Leftrightarrow ac < bc$.
If $a > b$ and $b > c$, then $a > c$.

Properties of Absolute Value:

$|ab| = |a|\,|b|$

$\left|\frac{a}{b}\right| = \frac{|a|}{|b|} \quad (b \neq 0)$

$|a + b| \leq |a| + |b|$

$|a - b| \geq \big||a| - |b|\big|$

P.4 Fractional Expressions

Equality:

$\frac{ac}{bc} = \frac{a}{b} \quad (b \neq 0,\ c \neq 0)$

$\frac{a}{b} = \frac{c}{d} \Leftrightarrow ad = bc \quad (b \neq 0,\ d \neq 0)$

Addition and Subtraction:

$\frac{a}{b} \pm \frac{c}{b} = \frac{a \pm c}{b} \quad (b \neq 0)$

$\frac{a}{b} \pm \frac{c}{d} = \frac{ad \pm bc}{bd} \quad (b \neq 0,\ d \neq 0)$

Multiplication and Division:

$\frac{a}{b} \cdot \frac{c}{d} = \frac{ac}{bd} \quad (b \neq 0,\ d \neq 0)$

$\frac{\frac{a}{b}}{\frac{c}{d}} = \frac{a}{b} \cdot \frac{d}{c} \quad (b \neq 0,\ c \neq 0,\ d \neq 0)$

P.5 Powers and Roots

Definitions (n a positive integer)

nth Power: $x^n = x \cdot x \cdot \ldots \cdot x$ (n factors)

Zero Exponent: $x^0 = 1 \quad (x \neq 0)$

Negative Exponent:

$x^{-n} = \frac{1}{x^n} \quad (x \neq 0)$

nth Root: $y = \sqrt[n]{x} \Leftrightarrow y^n = x$
(x, y positive if n is even)

Properties of nth Roots:

$\sqrt[n]{ab} = \sqrt[n]{a}\,\sqrt[n]{b}$ (a and b positive if n even)

$\sqrt[n]{\frac{a}{b}} = \frac{\sqrt[n]{a}}{\sqrt[n]{b}}$ ($b \neq 0$; a and b positive if n is even)

Derived Rules for Powers

In each of the following, m and n are any integers (positive, negative, or 0) for which both sides of the equation are defined.

Same-Base Rules:

$x^m x^n = x^{m+n}$

$\frac{x^m}{x^n} = x^{m-n}$

Same-Exponent Rules:

$(xy)^n = x^n y^n$

$\left(\frac{x}{y}\right)^n = \frac{x^n}{y^n}$

Power-of-a-Power Rule:

$(x^m)^n = x^{mn}$

Chapter P Review Exercises

1. Which of the following numbers are real numbers? Classify those that are real numbers as either rational or irrational.

$$\frac{2}{3},\ 3.14159,\ \sqrt{5},\ \sqrt{-8},\ \sqrt[3]{-8},\ \frac{0}{4},\ \frac{4}{0},\ |-3|$$

2. Express the repeating decimal .757575. . . as the ratio of two integers.
3. Express the rational number 4/7 as a decimal.
4. Show that the repeating decimal .9999. . . . is equal to 1.
5. Without actually dividing, show that $\frac{3}{8} = \frac{33}{88} = \frac{333}{888}$.
6. Name the basic property of algebra illustrated by each of the following.
   (a) $3 + 7 = 7 + 3$ (b) $(2 \cdot 3) \cdot 4 = 2 \cdot (3 \cdot 4)$
   (c) $3 \cdot (4 + 5) = 3 \cdot 4 + 3 \cdot 5$ (d) $5 + 0 = 5$
7. Use the basic properties of algebra and equality to verify each of the following.
   (a) If $7 + a = 7 + b$, then $a = b$.
   (b) If $5 \cdot a = 5 \cdot b$, then $a = b$.
8. Use the algebraic definition of "greater than" to show that
   (a) $\frac{1}{2} > \frac{1}{3}$ (b) $-\frac{1}{3} > -\frac{1}{2}$
9. Show that the inequality $2x - 18 > 5x - 3$ is equivalent to $x < -5$.
10. Answer True or False for each of the following. Explain your answer.
    (a) If $a + c = b + c$, then $a = b$.
    (b) If $a \cdot c = b \cdot c$, then $a = b$.
    (c) $a - b = a + (-b)$
    (d) $-a = (-1)a$
    (e) If $a \neq 0$, then $a^{-1} = 1/a$.
    (f) If $a < 0$, then $-a > 0$.
    (g) $\sqrt{a^2} = a$
    (h) If $a > 0$ and $b > 0$, then $\sqrt{a + b} = \sqrt{a} + \sqrt{b}$
    (i) If $a < 0$ and $b < 0$, then $|a + b| = -(a + b)$
    (j) If $a > 0$ and $b < 0$, then $|a - b| = a - b$
11. Evaluate each of the following.
    (a) $\frac{3}{4} + \frac{2}{5}$ (b) $\frac{3 + 2}{4 + 5}$ (c) $72 \div (12 \div 6)$
    (d) $72 \div 12 \div 6$ (e) $\frac{-3(5 - 9)}{3 \cdot 5 + 5(-2)}$
    (f) $[52 + 24 \div (8 - 2)] \div 8 - 2$
12. Express each of the following without grouping symbols.
    (a) $2[5(a + b) + 6c]$
    (b) $3\{-2 + 4[(a + 1) + (b + 5)]\}$
13. Show that if $a \neq b$, then $\frac{a - b}{b - a} = -1$.
14. Explain why $\frac{x + 3}{x^2 - 9} = 0$ cannot be true for any value of x.
15. Write each of the following as a single fractional expression.
    (a) $\frac{x - 7}{x - 3} + \frac{4}{x - 3}$ (b) $\frac{x - 1}{2} - \frac{x}{3}$
    (c) $\frac{1}{2} + \frac{2}{x}$
16. Write the expression $2^3 \cdot 6^2 \cdot 12^3 \cdot 18^2$ as a product of powers of 2 and 3.
17. Simplify each of the following. Express answers in terms of positive exponents. Assume $x \neq 0$, $a \neq 0$, $b \neq 0$.
    (a) $2x^2x^{-3}$ (b) $\frac{5}{x^{-2}}$ (c) $(2x)^2(x^2)^3$
    (d) $\frac{ab^{-1}}{a^{-1}b}$ (e) $\frac{a + b^{-1}}{a^{-1} + b}$
18. Evaluate (a) $\sqrt{(-2)^2}$ (b) $\sqrt[3]{(-2)^3}$ (c) $\sqrt[4]{(-2)^0}$
19. Simplify each of the following.
    (a) $\sqrt{50a^2b^3}$ (b) $\sqrt[3]{16a^5b^3}$
    (c) $\sqrt{(-x)^2}$ (d) $\sqrt[3]{(-x)^3}$
20. Simplify each of the following.
    (a) $\frac{\sqrt{50}}{\sqrt{8}}$ (b) $\sqrt{8} \cdot \sqrt{50}$ (c) $\sqrt{8} + \sqrt{50}$

Calculator Exercises

21. Approximate $\sqrt{2}$ to nine decimal places. Can you determine the digit in the 10th decimal place? Hint: Consider $\sqrt{2} - 1$.

22. Suppose x is one of the positive integers 2, 3, 4, . . . , 9, and n is any positive integer that is not a multiple of x. If x has the property that the decimal representation of n/x is always a repeating decimal, then $x =$ ________________. *Hint:* You need only consider $n = 1, 2, \ldots x - 1$. Why? If the length of the repeating pattern is one digit, then $x =$ ________. If $x = 7$, then the length of the repeating pattern is ________.

23. Compute each of the following.

(a) 24^5 (b) 32^{-1} (c) 21^{-2}
(d) $\sqrt{10}$ (e) $\sqrt[3]{25}$

24. Compute each of the following and explain your answer.

(a) $\dfrac{7.5}{(3.3)(4.5)}$ (b) $(7.5)/(3.3)/(4.5)$
(c) $(7.5)/(3.3)(4.5)$ (d) $(7.5)/(3.3/4.5)$
(e) $(7.5/4.5)/(3.3)$

25. Compute each of the following.

(a) $\dfrac{17.5}{1.4 + 2.5^{-1}}$ (b) $17.5/1.4 + 2.5^{-1}$
(c) $17.5/(1.4 + 2.5)^{-1}$

Use the information in the Miscellaneous Exercises of Section P5 to do the following exercises.

26. Enter the following numbers in standard notation. How are the numbers denoted on the calculator?

(a) the speed of light in cm/sec
(b) Avogadro's number
(c) the charge of an electron in coulombs
(d) Planck's constant

27. Enter the following numbers in scientific notation. How are the numbers denoted on the calculator?

(a) the distance between Earth and the sun in miles
(b) the distance between Mars and the sun in miles
(c) the mass of a hydrogen atom in grams
(d) the mass of an oxygen atom in grams

28. Compute each of the following.

(a) (the speed of light)(Avogadro's number)
(b) (the charge of an electron)/(Planck's constant)
(c) (the distance between the Earth and the sun)(the distance between Mars and the sun)
(d) (the mass of a hydrogen atom)/(the mass of an oxygen atom)

29. Suppose \$10,000 is invested at 5% annual interest, compounded monthly. What is the amount accumulated in

(a) 5 years (b) 10 years (c) 20 years?

30. Suppose \$10,000 is borrowed at 8% annual interest. Find the monthly payments if the loan is to be repaid in

(a) 3 years (b) 4 years (c) 5 years.

1 Algebraic Expressions

In the preceding chapter we showed how the basic properties of addition, subtraction, multiplication, and division of real numbers are used to perform operations with elementary algebraic expressions. We now proceed to more general algebraic expressions. However, no matter how complex an expression may appear, both the expression and its components represent real numbers and therefore follow the same basic rules. With practice, manipulating algebraic expressions becomes as routine as working with integers.

1.1 Operations with Polynomials

Polynomials are the integers of algebra.

Polynomials. Polynomials are algebraic expressions that arise when the operations of addition, subtraction, and multiplication are applied to numbers and letters representing numbers. For example, $x + 2$, $x^3 - 2xy + 4y^2$, and $ax^2 + bx + c$ are polynomials. Generalizing gives the following definition.

> A **polynomial of degree n in x** is an algebraic expression of the form
>
> $$a_nx^n + a_{n-1}x^{n-1} + a_{n-2}x^{n-2} + \ldots + a_1x + a_0,$$
>
> where n is a nonnegative integer and $a_n \neq 0$.

In this definition, the real numbers a_k $(k = 0, 1, 2, \ldots, n)$ are called the **coefficients** of the polynomial, and the a_kx^k $(k = 0, 1, 2, \ldots, n)$ are called the **terms** of the polynomial. Polynomials of degree 1, such as $2x + 3$, are called **linear polynomials,** and those of degree 2, such as $x^2 + 4x + 7$, are called **quadratic polynomials.** Since $x^0 = 1$, polynomials of degree 0 are simply nonzero real numbers. Polynomials can be added, subtracted, and multiplied, and the result is always a polynomial. Hence, the collection of all polynomials forms a closed system under these operations.[1] Division of polynomials is also possible, but we shall see that the result is not always a polynomial. That is, the collection of all polynomials is not a closed system under division.

[1]In order to be correct when saying that polynomials are closed under addition and subtraction, we must admit the constant 0 as a polynomial. However, we will not attach a degree to the polynomial 0. (Some authors assign the degree $-\infty$ to the polynomial 0 for reasons that do not concern us here.)

For example, the fractions

$$\frac{x+1}{x-1} \quad \text{and} \quad \frac{1}{x^2+2}$$

are *not* polynomials, nor are the radicals

$$\sqrt{x^2+4} \quad \text{and} \quad \sqrt{x+3}.$$

Addition and Subtraction. Polynomials are added and subtracted by means of the commutative and associative properties for addition and the distributive properties.

Example 1 $(3x^2 + 2x + 1) + (5x^2 - 7x + 4)$

$= 3x^2 + 5x^2 + 2x - 7x + 1 + 4$ — *Extended commutative and associative properties for addition*

$= (3 + 5)x^2 + (2 - 7)x + 5$ — *Right distributive property*

$= 8x^2 - 5x + 5$ ■

Example 2 $(4x^2 - 3x + 1) - (x^2 - 7x + 4)$

$= 4x^2 - 3x + 1 - x^2 + 7x - 4$ — *When removing parentheses following a minus sign, change the sign of each term within the parentheses*

$= 4x^2 - x^2 + (-3x + 7x) + 1 - 4$ — *Extended commutative and associative properties for addition*

$= (4 - 1)x^2 + (-3 + 7)x - 3$ — *Right distributive property*

$= 3x^2 + 4x - 3$ ■

Example 3 $(ax^2y^2 + bxy + cx + dy + e) + (2x^2y^2 + 3xy + 4x + 5y + 6)$

$= ax^2y^2 + 2x^2y^2 + bxy + 3xy + cx + 4x + dy + 5y + e + 6$

$= (a + 2)x^2y^2 + (b + 3)xy + (c + 4)x + (d + 5)y + e + 6$ ■

In the above examples, the terms of the polynomials were first rearranged and regrouped in order to bring together terms corresponding to similar powers, and then the respective similar terms were combined by means of the right distributive property. This same technique of rearranging, regrouping, and combining applies when adding or subtracting any number of polynomials. In short, to add or subtract polynomials, simply add or subtract similar terms.

Example 4 $(4x + 7) + (3x - 12) - (2x - 7) - (-3x + 4)$

$= 4x + 7 + 3x - 12 - 2x + 7 + 3x - 4$

$= 4x + 3x - 2x + 3x + 7 - 12 + 7 - 4$

$= 8x - 2$ ■

Multiplication. To multiply two polynomials, we first apply the distributive properties and then proceed as in addition.

Example 5

$$\begin{aligned} (3x + 4)(x + 6) &= 3x(x + 6) + 4(x + 6) && \textit{Right distributive property} \\ &= 3x^2 + 18x + 4x + 24 && \textit{Left distributive property} \\ &= 3x^2 + 22x + 24 && \textit{Addition} \end{aligned}$$

■

Example 6

$$\begin{aligned} &(x^2 + 2x - 1)(3x^2 - 5x + 4) \\ &\quad = x^2(3x^2 - 5x + 4) + 2x(3x^2 - 5x + 4) \\ &\qquad - 1(3x^2 - 5x + 4) && \textit{Right distributive property} \\ &\quad = 3x^4 - 5x^3 + 4x^2 + 6x^3 - 10x^2 + 8x \\ &\qquad - 3x^2 + 5x - 4 && \textit{Left distributive property} \\ &\quad = 3x^4 + x^3 - 9x^2 + 13x - 4 && \textit{Addition} \end{aligned}$$

■

In both of the above examples, the effect of applying the distributive properties is that each term of the polynomial factor on the left multiplies each term of the polynomial factor on the right. Thus, in Example 5,

$$(\underset{\substack{\uparrow\\ 1}}{3x} + \underset{\substack{\uparrow\\ 2}}{4})(\underset{\substack{\uparrow\\ a}}{x} + \underset{\substack{\uparrow\\ b}}{6}) = \underset{\substack{\uparrow\\ 1a}}{3x^2} + \underset{\substack{\uparrow\\ 1b}}{18x} + \underset{\substack{\uparrow\\ 2a}}{4x} + \underset{\substack{\uparrow\\ 2b}}{24},$$

and in Example 6,

$$(\underset{\substack{\uparrow\\ 1}}{x^2} + \underset{\substack{\uparrow\\ 2}}{2x} - \underset{\substack{\uparrow\\ 3}}{1})(\underset{\substack{\uparrow\\ a}}{3x^2} - \underset{\substack{\uparrow\\ b}}{5x} + \underset{\substack{\uparrow\\ c}}{4})$$

$$= \underset{\substack{\uparrow\\ 1a}}{3x^4} - \underset{\substack{\uparrow\\ 1b}}{5x^3} + \underset{\substack{\uparrow\\ 1c}}{4x^2} + \underset{\substack{\uparrow\\ 2a}}{6x^3} - \underset{\substack{\uparrow\\ 2b}}{10x^2} + \underset{\substack{\uparrow\\ 2c}}{8x} - \underset{\substack{\uparrow\\ 3a}}{3x^2} + \underset{\substack{\uparrow\\ 3b}}{5x} - \underset{\substack{\uparrow\\ 3c}}{4}.$$

This principle applies when multiplying any two polynomials P and Q, regardless of their respective number of terms.

> To form the **product** PQ of polynomials P and Q, first multiply each term of P by each term of Q and then proceed as in addition.

Polynomials with two terms are called **binomials,** and binomials occur frequently in mathematics. If we apply the above product rule to the binomials $a + b$ and $a - b$, we obtain the following important results.

1. $(a + b)(a + b) = a^2 + ab + ba + b^2 = a^2 + ab + ab + b^2$; therefore,

$$(a + b)^2 = a^2 + 2ab + b^2. \qquad \textbf{square of a sum}$$

2. $(a - b)(a - b) = a^2 - ab - ba + (-b)^2 = a^2 - ab - ab + b^2$; therefore,

$$(a - b)^2 = a^2 - 2ab + b^2. \qquad \textbf{square of a difference}$$

3. $(a - b)(a + b) = a^2 + ab - ba - b^2 = a^2 + ab - ab - b^2$; therefore,

$$(a - b)(a + b) = a^2 - b^2. \qquad \textbf{difference of squares}$$

Example 7 (a) $(3x + 4)^2 = 9x^2 + 24x + 16$

(b) $(3x - 4)^2 = 9x^2 - 24x + 16$

(c) $(3x - 4)(3x + 4) = 9x^2 - 16$ ■

Example 8 (a) $(2x^2 + 3y)^2 = 4x^4 + 12x^2y + 9y^2$

(b) $(2x^2 - 3y)^2 = 4x^4 - 12x^2y + 9y^2$

(c) $(2x^2 - 3y)(2x^2 + 3y) = 4x^4 - 9y^2$ ■

As illustrated in Example 8, a binomial need not be a linear polynomial in x. Any polynomial with exactly two terms is a binomial, regardless of its degree or the number of variables in each term.

Caution From the formulas for the square of a sum and the square of a difference, we see that in general,

$$(a + b)^2 \neq a^2 + b^2 \qquad \text{and} \qquad (a - b)^2 \neq a^2 - b^2.$$

Similarly,

$$\sqrt{a^2 + b^2} \neq a + b \qquad \text{and} \qquad \sqrt{a^2 - b^2} \neq a - b.$$

Treating these inequalities as equalities is a very common mistake in algebra. Be sure to avoid doing this. For example,

$$(7 + 3)^2 = 100, \quad \text{but} \quad 7^2 + 3^2 = 58,$$

and

$$(10 - 4)^2 = 36, \quad \text{but} \quad 10^2 - 4^2 = 84.$$

Also,

$$\sqrt{3^2 + 4^2} = 5 \neq 3 + 4 \qquad \text{and} \qquad \sqrt{13^2 - 5^2} = 12 \neq 13 - 5.$$

Products of three or more polynomials can be computed by taking the factors two at a time.

Example 9 $(x + 1)(x + 2)(x + 3) = (x^2 + 3x + 2)(x + 3)$
$$= x^3 + 3x^2 + 3x^2 + 9x + 2x + 6$$
$$= x^3 + 6x^2 + 11x + 6 \quad ■$$

Example 10 Verify the formula

$$(a + b)^3 = a^3 + 3a^2b + 3ab^2 + b^3. \qquad \text{cube of a sum}$$

Solution: $(a + b)^3 = (a + b)^2(a + b)$
$$= (a^2 + 2ab + b^2)(a + b)$$
$$= a^3 + a^2b + 2a^2b + 2ab^2 + b^2a + b^3$$
$$= a^3 + 3a^2b + 3ab^2 + b^3 \quad ■$$

Example 11 Use the result of Example 10 to expand the following:

(a) $(2x + 5)^3$ and (b) $(x - 2)^3$.

Solution:
(a) If we substitute $a = 2x$ and $b = 5$ in the formula in Example 10, then

$$(2x + 5)^3 = (2x)^3 + 3(2x)^2 \cdot 5 + 3(2x) \cdot 5^2 + 5^3$$
$$= 8x^3 + 60x^2 + 150x + 125.$$

(b) Here we substitute $a = x$ and $b = -2$ to obtain

$$(x - 2)^3 = [x + (-2)]^3$$
$$= x^3 + 3x^2(-2) + 3x(-2)^2 + (-2)^3$$
$$= x^3 - 6x^2 + 12x - 8. \quad ■$$

Division. We know that if one integer is divided by another, the result need not be an integer. For instance,

$$\frac{27}{4} = 6 + \frac{3}{4},$$

and we say that when 27 is divided by 4, the quotient or integer part is 6 and the remainder is 3. The above equation is equivalent to

$$27 = 6 \cdot 4 + 3,$$

and in general, if n is any integer and m is a positive integer, then there are unique integers q and r for which

$$n = qm + r \quad (0 \leq r < m).$$

The result stated in this equation is called the **division algorithm for integers** and can be seen geometrically by marking off a number line in multiples of m as shown in Figure 1. The integer n is located between qm and $(q + 1)m$, and r is the difference between n and qm.

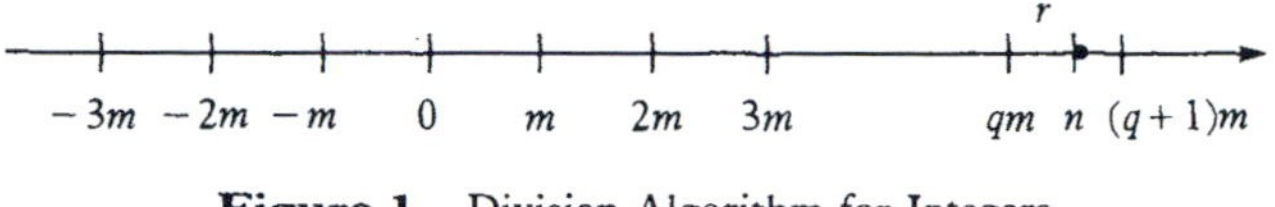

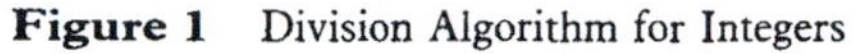

Figure 1 Division Algorithm for Integers

There is also a **division algorithm for polynomials.** It says that if P is any polynomial and T is a polynomial of degree greater than 0, then there are unique polynomials Q and R for which

$$\mathbf{P = Q \cdot T + R \quad (R = 0 \text{ or degree } R < \text{degree } T).}$$

For example, if $P = 6x^4 + 7x^3 + 16x - 11$ and $T = x + 2$, then

$$6x^4 + 7x^3 + 16x - 11 = (6x^3 - 5x^2 + 10x - 4)(x + 2) - 3,$$

which is equivalent to

$$\frac{6x^4 + 7x^3 + 16x - 11}{x + 2} = 6x^3 - 5x^2 + 10x - 4 + \frac{-3}{x + 2}.$$

Here $Q = 6x^3 - 5x^2 + 10x - 4$ and $R = -3$. We will soon see how Q and R are determined, but first some terminology. The numerator

$$6x^4 + 7x^3 + 16x - 11$$

of the ratio

$$\frac{6x^4 + 7x^3 + 16x - 11}{x + 2}$$

is called the **dividend,** and the denominator $x + 2$ is called the **divisor.** The polynomial part

$$6x^3 - 5x^2 + 10x - 4$$

of the result is called the **quotient,** and the numerator -3 of the fractional part of the result is called the **remainder.** That is,

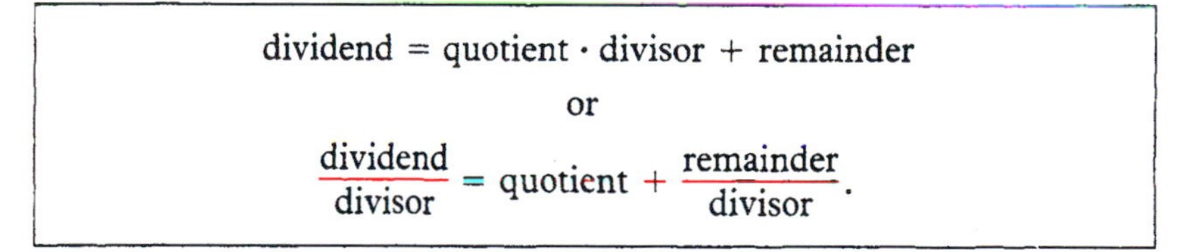

$$\text{dividend} = \text{quotient} \cdot \text{divisor} + \text{remainder}$$

or

$$\frac{\text{dividend}}{\text{divisor}} = \text{quotient} + \frac{\text{remainder}}{\text{divisor}}.$$

Hence, the ratio of dividend to divisor will itself be a polynomial precisely when the remainder is equal to 0. We now describe a step-by-step process for finding Q and R.

1. Express both the dividend and the divisor in terms of decreasing powers of x, and use 0 for the coefficient of any missing power of x in the dividend. Then position the divisor and dividend as follows:

$$x + 2\overline{)6x^4 + 7x^3 + 0x^2 + 16x - 11}.$$

2. Divide the lead term x of the divisor into the lead term $6x^4$ of the dividend, and place the result $6x^3$ above the division line:

$$\begin{array}{r|l} & 6x^3 \\ \hline x+2 & 6x^4 + 7x^3 + 0x^2 + 16x - 11. \end{array}$$

3. Multiply the divisor $x + 2$ by the result $6x^3$ of step 2, and subtract this product $6x^4 + 12x^3$ from the dividend to obtain the reduced dividend $-5x^3 + 0x^2 + 16x - 11$:

$$\begin{array}{r|l} & 6x^3 \\ \hline x+2 & 6x^4 + 7x^3 + 0x^2 + 16x - 11 \\ & 6x^4 + 12x^3 \\ \hline & \quad -5x^3 + 0x^2 + 16x - 11. \qquad \textit{Reduced dividend} \end{array}$$

4. Repeat steps 2 and 3 with the reduced dividend taking the place of the dividend until the degree of the reduced dividend becomes less than the degree of the divisor. The last reduced dividend is the remainder.

$$\begin{array}{r|rrrrrl} & 6x^3 & -\ 5x^2 & +\ 10x & -\ 4 & & \\ \hline x+2 & 6x^4 & +\ 7x^3 & +\ 0x^2 & +\ 16x & -\ 11 & \\ & 6x^4 & +\ 12x^3 & & & & \\ \hline & & -5x^3 & +\ 0x^2 & +\ 16x & -\ 11 & \textit{Reduced dividend} \\ & & -5x^3 & -\ 10x^2 & & & \\ \hline & & & 10x^2 & +\ 16x & -\ 11 & \textit{Reduced dividend} \\ & & & 10x^2 & +\ 20x & & \\ \hline & & & & -4x & -\ 11 & \textit{Reduced dividend} \\ & & & & -4x & -\ 8 & \\ \hline & & & & & -3 & \textit{Remainder} \end{array}$$

Example 12 Perform the division $\dfrac{5x^5 + 6x^4 - 2x^2 + 4}{x^2 - 3}$.

Solution:

$$\begin{array}{r|rrrrrrl} & 5x^3 & +\ 6x^2 & +\ 15x & +\ 16 & & & \\ \hline x^2-3 & 5x^5 & +\ 6x^4 & +\ 0x^3 & -\ 2x^2 & +\ 0x & +\ 4 & \\ & 5x^5 & & -\ 15x^3 & & & & \\ \hline & & 6x^4 & +\ 15x^3 & -\ 2x^2 & +\ 0x & +\ 4 & \textit{Reduced dividend} \\ & & 6x^4 & & -\ 18x^2 & & & \\ \hline & & & 15x^3 & +\ 16x^2 & +\ 0x & +\ 4 & \textit{Reduced dividend} \\ & & & 15x^3 & & -\ 45x & & \\ \hline & & & & 16x^2 & +\ 45x & +\ 4 & \textit{Reduced dividend} \\ & & & & 16x^2 & & -\ 48 & \\ \hline & & & & & 45x & +\ 52 & \textit{Remainder} \end{array}$$

Hence, the quotient is $5x^3 + 6x^2 + 15x + 16$ and the remainder is $45x + 52$.

That is, $$\frac{5x^5 + 6x^4 - 2x^2 + 4}{x^2 - 3} = 5x^3 + 6x^2 + 15x + 16 + \frac{45x + 52}{x^2 - 3}$$

or $$5x^5 + 6x^4 - 2x^2 + 4 = (5x^3 + 6x^2 + 15x + 16)(x^2 - 3) + 45x + 52.$$ ■

Note that in Example 12 the divisor $x^2 - 3$ has degree 2 and the remainder $45x + 52$ has degree 1. In general, as stated above, the remainder is either 0 or a polynomial of degree less than the degree of the divisor. Also, the degree of the quotient *plus* the degree of the divisor is equal to the degree of the dividend.

Example 13 Perform the division $\dfrac{3x^3 - 3x^2 + 5x + 7}{2x + 1}$.

Solution:

$$\begin{array}{r|l} & 1.5x^2 - 2.25x + 3.625 \\ \hline 2x + 1 & 3x^3 - 3x^2 + 5x + 7 \\ & 3x^3 + 1.5x^2 \\ \hline & -4.5x^2 + 5x + 7 \\ & -4.5x^2 - 2.25x \\ \hline & 7.25x + 7 \\ & 7.25x + 3.625 \\ \hline & 3.375 \end{array}$$

Here the quotient is $1.5x^2 - 2.25x + 3.625$, and the remainder is 3.375, which illustrates that the quotient and remainder need not have integer coefficients, even though the coefficients of the dividend and divisor are integers. ■

Polynomial Functions and Zeros of Polynomials. Every integer is a polynomial in the base 10. For example, $3258 = 3 \cdot 10^3 + 2 \cdot 10^2 + 5 \cdot 10 + 8$. That is, the integer 3258 corresponds to the polynomial $P = 3x^3 + 2x^2 + 5x + 8$ evaluated at $x = 10$. In this sense we can think of polynomials as generalized integers. For $x = -1$, the value of P is $3(-1)^3 + 2(-1)^2 + 5(-1) + 8 = 2$, and for $x = \sqrt{2}$, the value of P is $3(\sqrt{2})^3 + 2(\sqrt{2})^2 + 5\sqrt{2} + 8 = 11\sqrt{2} + 12$. **The notation $P(x)$, read "P of x," not "P times x," is used to denote the numerical dependence of P on x.** Hence, $P(10) = 3258$, $P(-1) = 2$, and $P(\sqrt{2}) = 11\sqrt{2} + 12$.

Example 14 Let $P(x) = 4x^2 - 6x - 15$. Find (a) $P(-2)$, (b) $P(0)$, (c) $P(1/2)$, and (d) $P(1 + \sqrt{3})$.

Solution:

(a) $P(-2) = 4(-2)^2 - 6(-2) - 15 = 16 + 12 - 15 = 13$

(b) $P(0) = 4 \cdot 0^2 - 6 \cdot 0 - 15 = -15$

(c) $P\left(\frac{1}{2}\right) = 4\left(\frac{1}{2}\right)^2 - 6\left(\frac{1}{2}\right) - 15 = 1 - 3 - 15 = -17$

(d) $P(1 + \sqrt{3}) = 4(1 + \sqrt{3})^2 - 6(1 + \sqrt{3}) - 15$
$= 4(1 + 2\sqrt{3} + 3) - 6 - 6\sqrt{3} - 15$
$= 4 + 8\sqrt{3} + 12 - 6 - 6\sqrt{3} - 15$
$= -5 + 2\sqrt{3}$ ■

Hence, for a given polynomial P, each value of x produces a corresponding value of $P(x)$. In this sense, we can think of a polynomial as a *rule that assigns to a real number x a corresponding real number $P(x)$*. In mathematics, such rules are called **functions,** and the concept of a function is fundamental to all branches of mathematics. We will study functions more formally in Chapter 4. In this chapter we are more concerned with polynomials as **algebraic expressions,** that is, with how polynomials are added, subtracted, multiplied, and divided.

A value of x that makes a given polynomial $P(x)$ equal to zero is called a **root of the equation $P(x) = 0$** or a **zero of $P(x)$.** For example, -3 is a root of the equation $2x + 6 = 0$ since $2(-3) + 6 = 0$. Similarly, $\sqrt{2}$ and $-\sqrt{2}$ are zeros of the polynomial $x^2 - 2$. The problem of finding roots of polynomial equations is the oldest problem in algebra. Indeed, the development of algebra from ancient to modern times is the story of mathematicians' efforts to solve polynomials for their zeros. We will trace this development throughout the rest of the text.

Example 15 Verify that $1 + \sqrt{2}$ and $1 - \sqrt{2}$ are zeros of $P(x) = x^2 - 2x - 1$.

Solution: $P(1 + \sqrt{2}) = (1 + \sqrt{2})^2 - 2(1 + \sqrt{2}) - 1$
$= 1 + 2\sqrt{2} + 2 - 2 - 2\sqrt{2} - 1 = 0$

and $P(1 - \sqrt{2}) = (1 - \sqrt{2})^2 - 2(1 - \sqrt{2}) - 1$
$= 1 - 2\sqrt{2} + 2 - 2 + 2\sqrt{2} - 1 = 0$ ■

Question *Why does $P(x) = x^2 + 1$ not have any zeros in the real number system?*

Answer: For $x^2 + 1$ to be 0, x^2 would have to equal -1. But x^2 cannot be negative for any real number x. ■

As illustrated in the question and answer above, some polynomials do not have any zeros in the real number system. This situation led mathematicians to consider the possibility of a larger number system in which all polynomials do have zeros. Efforts to develop such a system resulted in the **complex number system,** which will be introduced in Chapter 3.

Exercises 1.1

Fill in the blanks to make each statement true.

1. Polynomials are generated from real numbers and letters by the operations of ______________, ______________, and ______________.
2. If $P = 3x^2 - 5x - 2$, then $-P =$ ______________.
3. The product PQ of two polynomials P and Q is obtained by ______________.
4. dividend = divisor · ______________ + ______________
5. If $P(r) = 0$, then r is called a ______________ of the equation $P(x) = 0$ or a ______________ of $P(x)$.

Write true *or* false *for each statement.*

6. If P is a polynomial, then $P(x)$ does not mean $P \cdot x$.
7. If P and Q are polynomials, then $P - Q = P + (-Q)$.
8. $(x + 3)^2 = x^2 + 9$
9. $\sqrt{4x^2 + 25} = |2x + 5|$
10. $\dfrac{ax^2 + b}{x + b} = ax + 1$

Addition and Subtraction

Add and subtract as indicated in Exercises 11–20.

11. $(x^2 + 4xy + y) + (x^2 + xy + 3y)$
12. $(x^2 - 2y^2) + (x^2 + y^2) - (x^2 + 4y^2)$
13. $(x^2 - 2x - 3) - (4x^2 + 5x - 1) + (3x^2 + 7)$
14. $(2x^2 + 3x - 7) - (x^2 + 4) + (3x^2 + 2x)$
15. $(3x - 2) - (2x^2 + 2x + 5) + (3x^3 - 1)$
16. $(3x + 5) + (x^4 - 1) + (x^3 - 2x) + (5x^2 - 2)$
17. $(ax^2 + bx + c) + (2ax^2 + 3bx + 4c)$
18. $(ax^2 + bx + c) + (ax^2 + bx) + (ax^2 + c)$
19. $(x^3 + x^2 + x + 1) + (ax^3 + bx^2 + cx + d)$
20. $(ax + 1) + (ax^2 + 2x + 3) - (x^2 + x + a)$

Multiplication

Perform each of the following multiplications.

21. $2x(3x + 4)$
22. $a(2b + 3)$
23. $(x^2 + 1)2xy$
24. $(a^2 - 1)b^2$
25. $(x + 1)(x + 2)$
26. $(2a + 3)(4a + 5)$
27. $(x - 3)(x + 2)$
28. $(2a - 1)(3a + 5)$
29. $(xy^2 + y)(x^2y + x)$
30. $(a^2b^2 + 2ab)(2a^3 + 3b^2)$
31. $(5x + 4)^2$
32. $(5x - 4y)^2$
33. $(2x^2 + 3)^2$
34. $(3x^3 - 2y^2)^2$
35. $(2x^2 - 5)(2x^2 + 5)$
36. $(x^3 - y^3)(x^3 + y^3)$
37. $(x + 2)^3$
38. $(x - y)^3$
39. $(x + 2)^4$
40. $(x - y)^4$
41. $6x(x + 3)(x - 5)$
42. $(x - 1)(x - 2)(x - 3)$
43. $5x^2(x^2 + 1)(3x + 4)$
44. $2x(3x - 5)(4x^2 + 2x + 7)$
45. $(2x - 3)(3x - 4)(4x - 5)$
46. $x(x - 2)(x + 1)(x^2 - 3)$
47. $(x - 1)(x + 2)(x + 1)(x - 2)$
48. $3x(2x + 1)(3x - 2)(x^2 - x + 1)$
49. $x(x + 1)(x + 2)(x + 3)(x + 4)$
50. $(3x - 1)(2x^2 + 3x + 1)(x + 2)(x^2 - 2x - 1)$

Division

Perform each of the following divisions.

51. $x + 3\overline{)x^3 + 4x^2 + 2x - 3}$
52. $x - 2\overline{)3x^3 - 2x^2 + x - 1}$
53. $x - 1\overline{)x^4 + x^3 + x^2 + x + 1}$
54. $x + 5\overline{)2x^4 + 3x^2 + 2x - 2}$
55. $x - 4\overline{)x^5 - 5x^3 + 2x + 4}$
56. $x + 1\overline{)x + 2}$
57. $2x - 3\overline{)4x + 6}$
58. $2x + 3\overline{)5x^2 + 8x + 11}$
59. $2x - 1\overline{)6x^3 + 5x^2 - 8x + 2}$
60. $2x - 1\overline{)x^2 + 3x - 5}$
61. $x^2 + 2x - 1\overline{)3x^4 + 2x^3 - 10x + 5}$
62. $x^2 + 3x + 1\overline{)x^4 + 3x^3 - 5x^2 + 8x + 2}$
63. $x^2 + x + 1\overline{)x^5 + x^3 + x + 1}$
64. $x^2 - 1\overline{)4x^4 + 3x^3 + 2x^2 - x + 15}$

65. $x^2 + 1\overline{)x^8 - 4x^6 - 2x^4 - 3x^2 - 1}$

66. $x^2 + 1\overline{)4x^2 + 3x + 2}$

67. $x^2 - 2x - 1\overline{)5x^2 + 7}$

68. $x^2\overline{)x^3 - 3x^2 + 2x + 11}$

69. $x^3 + x^2 + x + 1\overline{)x^4}$

70. $x^2 - 1\overline{)x^8 - 1}$

In each of the following, fill in the blanks with the quotient and remainder.

71. $\dfrac{5x^2 + 3x + 2}{x + 1} = \underline{\qquad\qquad} + \dfrac{\underline{\quad}}{x + 1}$

72. $\dfrac{x^3 - x^2 + 1}{x + 3} = \underline{\qquad\qquad} + \dfrac{\underline{\quad}}{x + 3}$

73. $\dfrac{x^7 - 1}{x - 1} = \underline{\qquad\qquad\qquad} + \dfrac{\underline{\quad}}{x - 1}$

74. $\dfrac{x^5 + 3x^4 + 2x^3 - x^2 + 7x - 5}{x^2 + 3x + 4}$

$= \underline{\qquad\qquad} + \dfrac{\underline{\qquad}}{x^2 + 3x + 4}$

75. $\dfrac{2x^2 + 7x + 5}{x^2 + 1} = \underline{\qquad\qquad\qquad} + \dfrac{\underline{\quad}}{x^2 + 1}$

76. $4x^2 + 3x + 1 = (x + 2)(\underline{\qquad\qquad}) + \underline{\qquad\qquad}$

77. $x^3 + 2x^2 - x + 3 = (x^2 + x + 1)(\underline{\qquad\qquad})$ $+ \underline{\qquad\qquad}$

78. $x^4 - 5x^2 + 2 = (x^2 - 3)(\underline{\qquad\qquad}) + \underline{\qquad\qquad}$

79. $x^5 + 1 = (x + 1)(\underline{\qquad\qquad}) + \underline{\qquad\qquad}$

80. $x^5 + 1 = (x - 1)(\underline{\qquad\qquad}) + \underline{\qquad\qquad}$

Polynomial Functions and Zeros of Polynomials

In Exercises 81–85, find each value of P for the given polynomial.

81. $P(x) = 2x^2 - 7x + 4$
    (a) $P(1)$ (b) $P(3)$ (c) $P(0)$

82. $P(x) = x^3 + 3x^2 + 2x + 1$
    (a) $P(-2)$ (b) $P(4)$ (c) $P(10)$

83. $P(x) = x^2 - 10$
    (a) $P(5)$ (b) $P(-7)$ (c) $P(\sqrt{10})$

84. $P(x) = 3x^4 + 2x^2 + 5$
    (a) $P(\sqrt{2})$ (b) $P(-\sqrt{3})$ (c) $P(2a)$

85. $P(x) = x^2 - 4x - 1$
    (a) $P(1 + \sqrt{2})$ (b) $P(2 + \sqrt{3})$ (c) $P(2 - \sqrt{3})$

Find all possible real zeros of each of the following polynomials.

86. $P(x) = (2x - 3)(4x + 5)$

87. $P(x) = (3x + 2)(3x - 5)(x + 2)(x - 7)$

88. $P(x) = x(x^2 - 4)(x^2 - 10)$

89. $P(x) = (x^2 - 5)(x^2 + 5)$

90. $P(x) = x^2 + 10x + 25$

Miscellaneous

91. If $(ax + b)^2 = 4x^2 + 12x + 9$, then either $a = \underline{\qquad}$ and $b = \underline{\qquad}$, or $a = \underline{\qquad}$ and $b = \underline{\qquad}$.

92. Use the rule $(a + b)^2 = a^2 + 2ab + b^2$ to compute each of the following.
    (a) 16^2 (b) 17^2 (c) 21^2 (d) 26^2

93. Use the rule $(a - b)^2 = a^2 - 2ab + b^2$ to compute each of the following.
    (a) 19^2 (b) 18^2 (c) 29^2 (d) 28^2

94. Use the rule $a^2 - b^2 = (a - b)(a + b)$ to compute each of the following.
    (a) $18 \cdot 22$ (b) $27 \cdot 33$
    (c) $36 \cdot 44$ (d) $45 \cdot 55$

95. Use the formula for the difference of squares to show that

$$x^2 = y^2 \Leftrightarrow x = y \quad \text{or} \quad x = -y.$$

96. From Section P.2, Exercise 33, it follows that $x = -x \Leftrightarrow x = 0$. Now show that

$$x = x^{-1} \Leftrightarrow x = 1 \quad \text{or} \quad x = -1.$$

97. *Division Is Not Commutative.* Show that if $a/b = b/a$, then either $a = b$ or $a = -b$.

98. *Division Is Not Associative.* Show that if $(a/b)/c = a/(b/c)$, then either $a = 0$, or $c = 1$ or -1.

99. Show that + and · are not interchangeable in the left distributive property. That is, show that if $a + bc = (a + b)(a + c)$, then either $a = 0$ or $a + b + c = 1$.

100. A square with side b is placed next to a square with side a. If two rectangles, each with length a and width b, are placed as shown in the figure, a square with side $a + b$ is obtained. The geometric statement "the area of the large square is equal to the sum of the areas of its parts" becomes the algebraic equation

$$(a + b)^2 = a^2 + 2ab + b^2.$$

Draw a figure to illustrate the equation $(a - b)^2 = a^2 - 2ab + b^2$.

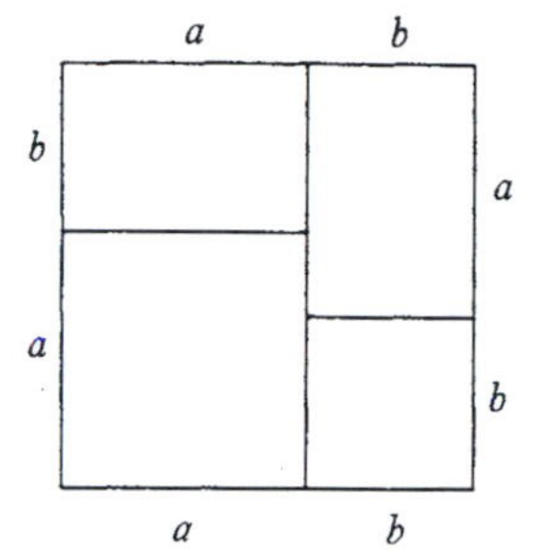

101. $a/b + a/c \neq a/(b + c)$ unless $a = 0$. The following argument shows that the assumption $a/b + a/c = a/(b + c)$ leads to a contradiction. Fill in the reasons for each step. Suppose $a \neq 0$. Then

$$\frac{a}{b} + \frac{a}{c} = \frac{a}{b + c} \Leftrightarrow \frac{1}{b} + \frac{1}{c} = \frac{1}{b + c} \qquad \underline{\hspace{3cm}}$$

$$\Leftrightarrow \frac{c + b}{bc} = \frac{1}{b + c} \qquad \underline{\hspace{3cm}}$$

$$\Leftrightarrow (b + c)^2 = bc. \qquad \underline{\hspace{3cm}}$$

Now, assuming that $b \neq 0$ and $c \neq 0$, show that $(b + c)^2 > bc$. (*Hint:* consider the two cases $bc < 0$ and $bc > 0$ separately.)

1.2 Factoring Polynomials

$$\begin{array}{c|l} a - b & = aa - ba \\ a + c & \qquad + ca - bc \\ \hline \end{array}$$

The factoring of $a^2 - ab + ac - bc$ into $(a - b)(a + c)$ was depicted this way in Artis Analyticae Praxis, *by Thomas Harriott (1631).*

Perfect Squares and the Difference of Squares
Common Factors and Grouping
Quadratics $x^2 + Bx + C$
Quadratics $Ax^2 + Bx + C$
Miscellaneous Factoring

Up to now we have multiplied factors to obtain products. Now we consider the problem in reverse: given a product, we are to find its factors. We will consider polynomials whose coefficients are integers and look first for factors $ax + b$, where a and b are also integers. Later we consider cases where a and b are arbitrary real numbers.

Perfect Squares and the Difference of Squares. From the rules for multiplying binomials, we already have three important factorizations.

$$\left.\begin{aligned} a^2x^2 + 2abx + b^2 &= (ax + b)(ax + b) \\ a^2x^2 - 2abx + b^2 &= (ax - b)(ax - b) \end{aligned}\right\} \quad \textbf{perfect squares}$$

$$a^2x^2 - b^2 = (ax - b)(ax + b) \quad \textbf{difference of squares}$$

Example 1 (a) $4x^2 + 12x + 9 = (2x + 3)(2x + 3)$

(b) $4x^2 - 12x + 9 = (2x - 3)(2x - 3)$

(c) $4x^2 - 9 = (2x - 3)(2x + 3)$ ■

Caution It is important to remember that in the square of a difference, $a^2x^2 - 2abx + b^2$, only the *middle term* $-2abx$ has the minus sign. For instance, $x^2 - 10x - 25$ is *not* the square of a difference, nor is $4x^2 - 12x - 9$. Also, we can factor the difference of squares, but there is no factorization of the sum of squares into linear factors in the real number system. For instance, the expressions $x^2 + 25$ and $4x^2 + 9$ *cannot* be factored into linear factors in the real number system.

Common Factors and Grouping. The terms of a polynomial may have a factor in common. For example, each term of $6x^2 - 2x + 8$ has 2 as a factor. Therefore, $6x^2 - 2x + 8 = 2(3x^2 - x + 4)$. After removing a common factor from the terms of a polynomial, the resulting expression may still be factorable, for example, $5x^2 + 20x + 20 = 5(x^2 + 4x + 4) = 5(x + 2)(x + 2)$.

Example 2 (a) $3x^2 - 12x + 12 = 3(x^2 - 4x + 4) = 3(x - 2)(x - 2)$

(b) $7x^2 - 7 = 7(x^2 - 1) = 7(x - 1)(x + 1)$ ∎

If a polynomial has more than three terms, it may be possible to group the terms in such a way that each grouping has a common polynomial factor. For example,

$$\begin{aligned} 2x^3 + 8x^2 + 3x + 12 &= 2x^2(x + 4) + 3(x + 4) \\ &= (2x^2 + 3)(x + 4) \end{aligned}$$

Here, after removing the common factor $2x^2$ from the first two terms and 3 from the last two, we see that the binomial $x + 4$ is a common factor for each pair in the grouping. Similarly,

$$\begin{aligned} 4x^2 - 3 + 3x - 4x &= 4x^2 - 4x + 3x - 3 \\ &= 4x(x - 1) + 3(x - 1) \\ &= (4x + 3)(x - 1). \end{aligned}$$

Here, the first and last terms have $4x$ as a common factor, and the middle two terms have 3 as a common factor. The grouped terms each have $x - 1$ as a factor.

Example 3

$$\begin{aligned} 5x^3 + x^2 - 20x - 4 &= x^2(5x + 1) - 4(5x + 1) \\ &= (x^2 - 4)(5x + 1) \\ &= (x - 2)(x + 2)(5x + 1) \end{aligned}$$ ∎

Example 4

$$\begin{aligned} x^3 + 10x^2 + 25x + 2x^2 + 20x + 50 &= x(x^2 + 10x + 25) \\ &\quad + 2(x^2 + 10x + 25) \\ &= (x + 2)(x^2 + 10x + 25) \\ &= (x + 2)(x + 5)(x + 5) \end{aligned}$$ ∎

Quadratics $x^2 + Bx + C$. We now turn our attention to quadratics other than perfect squares or the difference of squares. For example, $x^2 + 4x + 3$ is not a perfect square but factors into $(x + 3)(x + 1)$. If we start with a quadratic $x^2 + Bx + C$, where B and C are integers, we want to determine whether there are integers m and n for which $x^2 + Bx + C = (x + m)(x + n)$. Now $(x + m)(x + n) = x^2 + (m + n)x + mn$, which will equal $x^2 + Bx + C$ precisely when $m + n = B$ and $mn = C$. Hence, the quadratic $x^2 + Bx + C$ is factorable by integers if there are two integers m and n whose sum is B and whose product is C; that is,

$$x^2 + \underset{\substack{\uparrow \\ m+n}}{B}x + \underset{\substack{\uparrow \\ mn}}{C} = (x + m)(x + n). \tag{1}$$

Example 5 Factor each of the following by integers: (a) $x^2 - 5x + 4$ (b) $x^2 + x - 12$.

Solution:

(a) We want two integers whose sum is -5 and whose product is 4. The integers -4 and -1 have these properties. Hence, $x^2 - 5x + 4 = (x - 4)(x - 1)$.

(b) The integers 4 and -3 have their sum equal to 1 and their product equal to -12. Therefore, $x^2 + x - 12 = (x + 4)(x - 3)$. ■

Example 6 If possible, factor each of the following by integers: (a) $3x^2 - 6x - 45$ (b) $x^2 + 2x + 3$.

Solution:

(a) First, we can factor 3 from each term to obtain $3(x^2 - 2x - 15)$. Next, two integers whose sum is -2 and whose product is -15 are -5 and 3. Hence, $3x^2 - 6x - 45 = 3(x - 5)(x + 3)$.

(b) There are no two integers whose sum is 2 and whose product is 3. Therefore, $x^2 + 2x + 3$ is not factorable by integers. ■

Question 1 *Is it possible to have two different factorizations of $x^2 + Bx + C$ by integers?*

Answer: No. If $x + m$ and $x + n$ are factors of $x^2 + Bx + C$, then $-m$ and $-n$ are *roots* of $x^2 + Bx + C = 0$. We will see in Chapter 3 that a quadratic equation cannot have more than two roots. Therefore, except for the order in which we write the factors, the factorization is *unique*. ■

Quadratics $Ax^2 + Bx + C$. In the previous section we factored quadratics in which the coefficient of x^2 is 1. We now consider arbitrary quadratics $Ax^2 + Bx + C$ in which A, B, and C are integers. For example, $10x^2 + 9x + 2$ factors into $(5x + 2)(2x + 1)$, and $6x^2 + x - 2$ factors into

$(2x-1)(3x+2)$. In general, if Ax^2+Bx+C factors into $(px+m)(qx+n)$, where p, q, m, and n are integers, then

$$\begin{aligned} Ax^2 + Bx + C &= (px+m)(qx+n) \\ &= \underset{\underset{A}{\uparrow}}{pq}x^2 + (\underset{\underset{B}{\uparrow}}{pn+qm})x + \underset{\underset{C}{\uparrow}}{mn}. \end{aligned} \tag{2}$$

Hence, p, q, m and n must satisfy the equations

$$A = pq, \qquad B = pn + qm, \qquad C = mn.$$

The integers p, q, m, and n are often found by inspection, as in the previous case $x^2 + Bx + C$. Now, finding *four* numbers satisfying the conditions in equation (2) is much more difficult than finding *two* numbers satisfying equation (1). However, if we note in (2) that the product $(pn)(qm)$ of the terms whose sum is B is equal to the product $AC = (pq)(mn)$, then we are led to the following procedure.

> To factor $Ax^2 + Bx + C$, first find two numbers M and N whose sum is B and whose product is AC, then write
>
> $$Ax^2 + Bx + C = Ax^2 + Mx + Nx + C$$
>
> and proceed by the method of grouping.

Example 7

$$\begin{aligned} 2x^2 + 11x + 12 &= 2x^2 + 8x + 3x + 12 \\ &= 2x(x+4) + 3(x+4) \\ &= (2x+3)(x+4) \quad ■ \end{aligned}$$

$M = 8, N = 3; 8 + 3 = 11 = B$ *and* $8 \cdot 3 = 2 \cdot 12 = AC$

Example 8

$$\begin{aligned} 6x^2 + x - 2 &= 6x^2 + 4x - 3x - 2 \\ &= 2x(3x+2) - 1(3x+2) \\ &= (2x-1)(3x+2) \quad ■ \end{aligned}$$

$M = 4, N = -3$

Note that $-3x - 2 = -1(3x + 2)$

Example 9

$$\begin{aligned} 10x^2 - 29x + 10 &= 10x^2 - 25x - 4x + 10 \\ &= 5x(2x-5) - 2(2x-5) \\ &= (5x-2)(2x-5) \quad ■ \end{aligned}$$

$M = -25, N = -4$

Question 2 *How many pairs of numbers M and N can be found whose sum is B and whose product is AC?*

Answer: If $M + N = B$ and $MN = AC$, then $x^2 - Bx + AC$ factors into $(x - M)(x - N)$. Hence, M and N are roots of the quadratic equation $x^2 - Bx + AC = 0$. As noted in the answer to Question 1, a quadratic equation cannot have more than two roots. Therefore, there can be only one pair of numbers whose sum is B and whose product is AC. ■

Miscellaneous Factoring. The previous rules for factoring quadratics in x can be applied to a wide variety of polynomials. For example, $x^6 - 7x^3 + 12$ can be written as $(x^3)^2 - 7x^3 + 12$, which is a quadratic in x^3 that factors into $(x^3 - 4)(x^3 - 3)$. Similarly, $x^4 - 1$ factors into $(x^2 - 1)(x^2 + 1)$, which can be further factored into $(x - 1)(x + 1)(x^2 + 1)$. The polynomial $x^2 + 5xy + 4y^2$ in x and y factors into $(x + 4y)(x + y)$, and $x^2y^2 - 2xy - 3$ factors into $(xy - 3)(xy + 1)$. By letting $x = a$ in the Thomas Harriott example cited in the quotation at the beginning of this section, we obtain $x^2 - bx + cx - bc = x(x - b) + c(x - b) = (x + c)(x - b)$.

Example 10

$$\begin{aligned} 16x^4 - 81y^4 &= (4x^2 - 9y^2)(4x^2 + 9y^2) \\ &= (2x - 3y)(2x + 3y)(4x^2 + 9y^2) \end{aligned}$$ ■

Example 11

$$x^4 - 13x^2 + 36 = (x^2 - 9)(x^2 - 4) = (x - 3)(x + 3)(x - 2)(x + 2)$$ ■

Example 12

$$\begin{aligned} ax^2 + by^2 - bx^2 - ay^2 &= (a - b)x^2 + (b - a)y^2 \\ &= (a - b)x^2 - (a - b)y^2 \\ &= (a - b)(x^2 - y^2) \\ &= (a - b)(x - y)(x + y) \end{aligned}$$ ■

Note that $b - a = -(a - b)$

Example 13

$$\begin{aligned} 6x^2 + 7xy + y^2 &= 6x^2 + 6xy + xy + y^2 \\ &= 6x(x + y) + y(x + y) \\ &= (6x + y)(x + y) \end{aligned}$$ ■

Example 14

$$\begin{aligned} 12ax^3 - 2ax^2 - 2ax &= 2ax(6x^2 - x - 1) \\ &= 2ax(6x^2 - 3x + 2x - 1) \\ &= 2ax[3x(2x - 1) + 1(2x - 1)] \\ &= 2ax(3x + 1)(2x - 1) \end{aligned}$$ ■

As illustrated in the following example, our factoring procedures can also be applied to polynomials whose coefficients are not necessarily integers. We will consider the problem of factoring polynomials whose coefficients are arbitrary real numbers in the chapter on roots of polynomial equations.

Example 15

(a) $x^2 - \frac{4}{9} = \left(x - \frac{2}{3}\right)\left(x + \frac{2}{3}\right)$

(b) $x^2 - 5 = (x - \sqrt{5})(x + \sqrt{5})$

(c) $x^2 + .5x - .5 = (x + 1)(x - .5)$ ■

If $x^3 - a^3$ is divided by $x - a$ according to the method in Section 1.1, the quotient is $x^2 + ax + a^2$ and the remainder is zero (see Exercise 103). Therefore, we obtain the factorization

$$x^3 - a^3 = (x - a)(x^2 + ax + a^2). \qquad \textbf{difference of cubes}$$

Similarly, if we divide $x^3 + a^3$ by $x + a$, then the quotient is $x^2 - ax + a^2$ and the remainder is zero (Exercise 104). That is,

$$x^3 + a^3 = (x + a)(x^2 - ax + a^2). \qquad \textbf{sum of cubes}$$

The factorization for the sum of cubes could also be obtained by replacing a by $-a$ in the factorization of the difference of two cubes (*why?*)

Example 16 (a) $x^3 - 27 = x^3 - 3^3$
$= (x - 3)(x^2 + 3x + 9)$

(b) $x^3 + 27 = x^3 + 3^3$
$= (x + 3)(x^2 - 3x + 9)$

(c) $a^3 - 8b^3 = a^3 - (2b)^3$
$= (a - 2b)(a^2 + 2ab + 4b^2)$

(d) $a^3 + 8b^3 = a^3 + (2b)^3$
$= (a + 2b)(a^2 - 2ab + 4b^2)$ ■

Note that both $x^3 - a^3$ and $x^2 - a^2$ have $x - a$ as a factor, and $x^3 + a^3$ has $x + a$ as a factor but $x^2 + a^2$ does not. In general, if n is *any* positive integer, then $x^n - a^n$ has $x - a$ as a factor (see Exercise 105). Also, if n is an *odd* positive integer, then $x^n + a^n$ has $x + a$ as a factor (Exercise 106). However, if n is an *even* positive integer and $a \neq 0$, then $x + a$ is *not* a factor of $x^n + a^n$ (Exercise 107).

Exercises 1.2

Fill in the blanks to make each statement true.

1. If $m + n = 4$ and $mn = -5$, then $(x + m)(x + n) =$ ________.
2. $x^2 + (6 - a)x - 6a$ factors into ________.
3. $a(x - y) + b(y - x) = a(x - y)$ ____ $b(x - y)$
$= ($______$)($______$)$
4. If $m + n = B$ and $mn = C$, then $x^{2k} + Bx^k + C$
$= ($________$)($________$)$.
5. To factor $Ax^2 + Bx + C$, first find two integers M and N, where $M + N =$ ______ and $MN =$ ______, then write $Ax^2 + Bx + C = Ax^2 + Mx + Nx + C$ and proceed by the method of ________.

Write true *or* false *for each statement.*

6. The expression $x^2 + 2x + 2$ is not factorable by integers.
7. $(a - b)(x - y) + (b - a)(y - x) = 2(a - b)(x - y)$
8. If $Ax^2 + Bx + C = (px + m)(qx + n)$, then $m + n = B$ and $mn = C$.

9. If $A = 1$, then the procedure for factoring $Ax^2 + Bx + C$ will give an answer different from the one given by the procedure for factoring $x^2 + Bx + C$.

10. The quadratic $Ax^2 + (A + C)x + C$ has the same factors as $x^2 + (A + C)x + AC$.

Perfect Squares and the Difference of Squares

Some of the following are perfect squares and some are not. Write the ones that are perfect squares in factored form.

11. $x^2 + 4x + 4$
12. $x^2 + 4x - 4$
13. $x^2 + 4$
14. $x^2 - 2x + 1$
15. $x^2 - 2x - 1$
16. $x^2 + x + 1$
17. $4x^2 - 20xy + 25y^2$
18. $4x^2 + 20xy - 25y^2$
19. $x^4 + 6x^2 + 9$
20. $49x^8 - 14x^4y^2 + y^4$

Factor by integers each of the following.

21. $x^2 - 64$
22. $4x^2 - 25$
23. $x^4 - y^4$
24. $4x^4 - 625$
25. $(a + x)^2 - (a - x)^2$
26. $(a + x)^2 - (a + y)^2$

Common Factors and Grouping

Factor by integers each of the following.

27. $3x^2 - 12$
28. $125 - 20y^2$
29. $2x^2 + 8x + 8$
30. $3x^2 - 12x + 12$
31. $x^2 + x + 4x + 4$
32. $x^2 - 4x - x + 4$
33. $3x^2 + 9x - 12x - 36$
34. $4x^2 + 16x + 12x + 48$
35. $x^3 + x^2 - 9x^2 - 9x$
36. $x^3 - 9x^2 - x^2 + 9x$

Factoring Quadratics $Ax^2 + Bx + C$

Some of the following are factorable by integers and some are not. Write each factorable polynomial in factored form.

37. $x^2 - 3x + 2$
38. $x^2 + 3x - 2$
39. $x^2 - x - 2$
40. $x^2 - x + 2$
41. $x^2 - 2$
42. $2x^2 - 50$
43. $3x^2 - 3x - 6$
44. $3x^2 - 3x + 6$
45. $2x^2 - 5x + 3$
46. $2x^2 + 5x + 3$

Factor by integers each of the following.

47. $x^2 - 18x + 80$
48. $x^2 + 18x + 80$
49. $x^2 + 29x + 100$
50. $x^2 - 48x - 100$
51. $x^2 + 23x + 132$
52. $x^2 + x - 132$
53. $3x^2 + 15x + 12$
54. $4x^2 - 20x + 16$
55. $2x^2 - 16x - 18$
56. $5x^2 - 50x + 45$
57. $2x^2 + 9x + 4$
58. $3x^2 + 17x + 10$
59. $5x^2 + 21x + 4$
60. $3x^2 + 25x + 8$
61. $5x^2 + 12x + 7$
62. $6x^2 + 5x + 1$
63. $4x^2 - 12x + 5$
64. $8x^2 - 15x - 2$
65. $10x^2 - 21x - 10$
66. $12x^2 + 7x - 10$

Miscellaneous Factoring

Factor by integers each of the following.

67. $x^3 + 8$
68. $x^3 - 8$
69. $8x^3 - 27y^3$
70. $64x^3 + 125y^3$
71. $16a^6 - 54b^3$
72. $(x + a)^3 - 8a^3$
73. $x^2 - 5bx + 4b^2$
74. $x^2 + 2ax - 3a^2$
75. $x^2 - (a + b)x + ab$
76. $x^2 - (2a + 3b)x + 6ab$
77. $a^2x^2 + 5ax + 6$
78. $a^2x^2 - 9ax + 20$
79. $(a - b)x^2 - 3(a - b)x + 2(a - b)$
80. $(a + b)x^2 - 7(a + b)x + 10(a + b)$
81. $(x + a)^2 - a^2$
82. $(a^2 + 2a + 1)x^2 - 25$
83. $(x^2 + 6x + 9) - (a^2 + 2a + 1)$
84. $(a - b)x^4 + (b - a)y^4$
85. $ax^2 - ay^2 + 2x^2 - 2y^2$
86. $ax^2 - ay^2 - 2x^2 + 2y^2$
87. $(a + b)x^2 - (a + b)y^2 + (a - b)x^2 - (a - b)y^2$
88. $(a + b)x^2 - (a + b)y^2 - (a - b)x^2 + (a - b)y^2$
89. $6ax^2 + 13ax + 6a$
90. $10ax^2 + 3ax - 4a$
91. $5x^3 + 18x^2 - 8x$
92. $9x^3 - 24x^2 + 12x$
93. $3x^4 + 8x^2 + 5$
94. $4x^4 - 17x^2 + 4$
95. $2x^4 + (2a^2 + 1)x^2 + a^2$
96. $4x^4 + (12 - b^2)x^2 - 3b^2$

Factor the expressions in Exercises 97–102 by real numbers.

97. $x^2 - 2$
98. $9x^2 - \frac{1}{4}$

99. $4x^2 - 12$

100. $x^2 + \frac{6}{5}x + \frac{1}{5}$

101. $x^2 + 2x + \frac{3}{4}$

102. $x^2 - 1.5x - 1$

103. Verify that

$$\frac{x^3 - a^3}{x - a} = x^2 + ax + a^2$$

by performing the division

$$x - a\overline{)x^3 + 0x^2 + 0x - a^3}.$$

Hence, conclude that $x^3 - a^3 = (x - a)(x^2 + ax + a^2)$.

104. Verify that

$$\frac{x^3 + a^3}{x + a} = x^2 - ax + a^2$$

by performing the division

$$x + a\overline{)x^3 + 0x^2 + 0x + a^3}.$$

Hence, conclude that $x^3 + a^3 = (x + a)(x^2 - ax + a^2)$.

105. Let n be *any* positive integer. Show that

$$\frac{x^n - a^n}{x - a} = x^{n-1} + ax^{n-2} + a^2x^{n-3} + \dots + a^{n-2}x + a^{n-1}$$

by performing the division

$$x - a\overline{)x^n + 0x^{n-1} + 0x^{n-2} + \dots + 0x - a^n}.$$

Hence, $x - a$ is a factor of $x^n - a^n$.

106. Let n be an *odd* positive integer. Show that

$$\frac{x^n + a^n}{x + a} = x^{n-1} - ax^{n-2} + a^2x^{n-3} - \dots - a^{n-2}x + a^{n-1}$$

by performing the division

$$x + a\overline{)x^n + 0x^{n-1} + 0x^{n-2} + \dots + 0x + a^n}.$$

Hence, $x + a$ is a factor of $x^n + a^n$ when n is odd.

107. Let n be an *even* positive integer. Show that

$$\frac{x^n + a^n}{x + a} = x^{n-1} - ax^{n-2} + a^2x^{n-3} - \dots + a^{n-2}x - a^{n-1} + \frac{2a^n}{x + a}$$

by performing the division

$$x + a\overline{)x^n + 0x^{n-1} + 0x^{n-2} + \dots + 0x + a^n}.$$

Hence, if $a \neq 0$, then $x + a$ is *not* a factor of $x^n + a^n$ when n is even.

1.3 Operations with Rational Expressions

To add rational expressions with the least amount of work, use the least common denominator.

Rational Expressions
Least Common Multiple
Addition of Rational Expressions
Multiplication and Division of Rational Expressions

Rational Expressions. As defined in the preceding chapter, a rational number is the ratio of two integers. Similarly, the ratio of two polynomials is a **rational expression.** For instance, 3/4 and 5/6 are rational numbers, whereas

$$\frac{x + 2}{x^2 - 1} \quad \text{and} \quad \frac{1}{x^2 + 3x + 2}$$

are rational expressions. In general, any algebraic expression generated by the operations of addition, subtraction, multiplication, and division is called a rational expression. Of course, it is understood that a denominator in a rational expression cannot be equal to zero.

Example 1 Which of the following algebraic expressions are rational expressions?

(a) $\dfrac{x^2 + x + 1}{x + 1}$ (b) $\sqrt{x^2 + 3}$ (c) $(x^3 - x + 2)^2$

(d) $(x + 4)^{-1}$ (e) $\dfrac{\sqrt{x}}{x}$

Solution: (a), (c), and (d) are rational expressions. ■

Least Common Multiple. Following the rules for addition of fractions in Section P.4, we have

$$\frac{5}{6} + \frac{3}{4} = \frac{5 \cdot 4 + 6 \cdot 3}{6 \cdot 4} = \frac{38}{24} = \frac{19}{12},$$

and

$$\begin{aligned}\frac{x}{x^2 + 3x + 2} + \frac{1}{x^2 - 1} &= \frac{x(x^2 - 1) + x^2 + 3x + 2}{(x^2 + 3x + 2)(x^2 - 1)} \\ &= \frac{x^3 + x^2 + 2x + 2}{(x^2 + 3x + 2)(x^2 - 1)} \\ &= \frac{x^2(x + 1) + 2(x + 1)}{(x + 2)(x + 1)(x - 1)(x + 1)} \\ &= \frac{(x^2 + 2)\cancel{(x + 1)}}{(x + 2)(x + 1)(x - 1)\cancel{(x + 1)}} \\ &= \frac{x^2 + 2}{(x + 2)(x + 1)(x - 1)}.\end{aligned}$$

In both cases, the addition can be simplified by using the common factor property (Section P.4) to introduce a **common denominator** as follows:

$$\frac{5}{6} + \frac{3}{4} = \frac{5 \cdot 2}{6 \cdot 2} + \frac{3 \cdot 3}{4 \cdot 3} = \frac{10}{12} + \frac{9}{12} = \frac{19}{12},$$

and

$$\begin{aligned}\frac{x}{x^2 + 3x + 2} + \frac{1}{x^2 - 1} &= \frac{x}{(x + 2)(x + 1)} + \frac{1}{(x - 1)(x + 1)} && \textit{Factor each denominator} \\ &= \frac{x \cdot (x - 1)}{(x + 2)(x + 1) \cdot (x - 1)} \\ &\quad + \frac{1 \cdot (x + 2)}{(x - 1)(x + 1) \cdot (x + 2)} && \textit{Common factor property} \\ &= \frac{x(x - 1) + x + 2}{(x + 2)(x + 1)(x - 1)} && \textit{Addition with common denominator} \\ &= \frac{x^2 + 2}{(x + 2)(x + 1)(x - 1)}.\end{aligned}$$

In the first case, the common denominator 12 is the *smallest* positive integer that has for its factors the factors of 6 and 4. Hence, 12 is called the **least**

common multiple (LCM) of 6 and 4. In the second case, the common denominator $(x + 2)(x + 1)(x - 1)$ is the polynomial of *smallest degree* that has for its only factors the factors of $x^2 + 3x + 2$ and $x^2 - 1$. Hence, $(x + 2)(x + 1)(x - 1)$ is called the least common multiple (LCM) of $x^2 + 3x + 2$ and $x^2 - 1$.

Comment In the case of integers, the LCM is defined for *positive* integers only. Similarly, for polynomials of degree n, we assume that the coefficient of x^n is a *positive* integer.

Example 2 Show that the LCM of 56 and 98 is 392.

Solution: First factor each term.

$$56 = 2^3 \cdot 7 \qquad \text{and} \qquad 98 = 2 \cdot 7^2$$

Therefore, $2^3 \cdot 7^2 = 392$ is the smallest positive integer that has 2^3, 7, 2, and 7^2 as factors. ∎

Example 3 Find the LCM of $2x - 4$, $x^2 - 4x + 4$, and $x^3 - 8$.

Solution: First factor each polynomial.

$$2x - 4 = 2(x - 2)$$
$$x^2 - 4x + 4 = (x - 2)^2$$
$$x^3 - 8 = (x - 2)(x^2 + 2x + 4)$$

Therefore, $2(x - 2)^2(x^2 + 2x + 4)$ is the polynomial of smallest degree that has 2, $x - 2$, $(x - 2)^2$, $x - 2$, and $x^2 + 2x + 4$ as its only factors. ∎

As these examples illustrate, the LCM of a set of integers or polynomials is computed according to the following procedure.

1. Factor each member of the set into powers of prime factors.[1]
2. The **least common multiple (LCM)** is the product of all the different prime factors, each raised to the highest power to which it appears among all the members of the set.

Example 4 Find the LCM of 54, $24x + 12$, and $2x^2 - x - 1$.

Solution:

$$54 = 2 \cdot 3^3$$
$$24x + 12 = 12(2x + 1)$$
$$= 2^2 \cdot 3(2x + 1)$$
$$2x^2 - x - 1 = (2x + 1)(x - 1)$$

[1]A prime factor is a factor other than 1 whose only factors are itself and 1.

Therefore, the LCM $= 2^2 \cdot 3^3(2x + 1)(x - 1)$. ▬

Example 5 Find the LCM of $(x - 1)^3(2x + 3)(x + 4)$, $(x - 1)(2x + 3)^3(x + 4)^2$, and $(x - 1)^2(2x + 3)^2(x + 4)^3$.

Solution: The terms are already in factored form. Therefore,

$$\text{LCM} = (x - 1)^3(2x + 3)^3(x + 4)^3. \quad ▬$$

Addition of Rational Expressions. As observed at the beginning of this section, the easiest way to add (or subtract) rational expressions is to use as a common denominator the LCM of the given denominators. In this context, the LCM is called the **least common denominator (LCD).** Hence, to add or subtract rational expressions, we first factor each denominator and compute the LCD. Then, by multiplying each numerator and denominator by the necessary factor, we replace each rational expression by an equivalent one whose denominator is the LCD. We then follow the procedure for adding or subtracting fractional expressions having a common denominator. That is, we combine the numerators and place this result over the LCD. Of course, answers should be simplified as much as possible.

Example 6

$$\frac{x+1}{x^2+x-2} + \frac{x-2}{x^2-2x+1} = \frac{x+1}{(x+2)(x-1)} + \frac{x-2}{(x-1)^2} \qquad \textit{Factor each denominator}$$

LCD $= (x + 2)(x - 1)^2$

$$= \frac{(x+1)\cdot(x-1)}{(x+2)(x-1)\cdot(x-1)} + \frac{(x-2)\cdot(x+2)}{(x-1)^2\cdot(x+2)} \qquad \textit{Common factor property}$$

$$= \frac{x^2-1+x^2-4}{(x+2)(x-1)^2} \qquad \textit{Addition with LCD}$$

$$= \frac{2x^2-5}{(x+2)(x-1)^2} \quad ▬$$

Example 7

$$\frac{5}{2x-4} + \frac{4}{3x+6} - \frac{x}{x^2-4} = \frac{5}{2(x-2)} + \frac{4}{3(x+2)} - \frac{x}{(x-2)(x+2)}$$

LCD $= 6(x - 2)(x + 2)$

$$= \frac{5\cdot 3(x+2)}{2(x-2)\cdot 3(x+2)} + \frac{4\cdot 2(x-2)}{3(x+2)\cdot 2(x-2)} - \frac{x\cdot 6}{(x-2)(x+2)\cdot 6}$$

$$= \frac{15x+30+8x-16-6x}{6(x-2)(x+2)}$$

$$= \frac{17x+14}{6(x-2)(x+2)} \quad ▬$$

Example 8 Add $\dfrac{2x+1}{x+3}+\dfrac{1}{9-x^2}$.

Solution: Here we first multiply the numerator and denominator of $\dfrac{1}{9-x^2}$ by -1 to obtain $\dfrac{-1}{x^2-9}$ (see previous Comment). We then continue as in the preceding examples.

$$\frac{2x+1}{x+3}+\frac{1}{9-x^2}=\frac{2x+1}{x+3}+\frac{-1}{x^2-9}$$

$$\text{LCD} = (x-3)(x+3) \qquad =\frac{2x+1}{x+3}+\frac{-1}{(x-3)(x+3)}$$

$$=\frac{2x+1)\cdot(x-3)}{(x+3)\cdot(x-3)}+\frac{-1}{(x-3)(x+3)}$$

$$=\frac{2x^2-5x-3+(-1)}{(x-3)(x+3)}$$

$$=\frac{2x^2-5x-4}{(x-3)(x+3)} \quad \blacksquare$$

Multiplication and Division of Rational Expressions. Unlike addition and subtraction, multiplication of rational expressions does not require a common denominator. As discussed in Section P.4, we simply multiply the numerators and multiply the denominators. Also, to divide rational expressions, we first invert the divisor and proceed as in multiplication. Of course, we simplify the results as far as possible by canceling any common factors, *which we assume are never zero.*

Example 9 Perform the multiplication $\dfrac{x^2-2xy}{3x+3y}\cdot\dfrac{x^2-y^2}{xy-2y^2}$.

Solution:

$$\frac{x^2-2xy}{3x+3y}\cdot\frac{x^2-y^2}{xy-2y^2}=\frac{(x^2-2xy)(x^2-y^2)}{(3x+3y)(xy-2y^2)}$$

$$=\frac{x\cancel{(x-2y)}\cancel{(x+y)}(x-y)}{3\cancel{(x+y)}y\cancel{(x-2y)}}$$

$$=\frac{x(x-y)}{3y} \quad \blacksquare$$

In Example 9, we were able to simplify the result by canceling common factors *after* the multiplication was performed. As the following example illustrates, we can sometimes simplify the computations by canceling in each rational expression *before* the operations are performed.

Example 10 Perform the division $\dfrac{x^2-16}{x^2+5x+4}\div\dfrac{x^2-3x-4}{x^3+1}$.

Solution:
$$\frac{x^2-16}{x^2+5x+4} \div \frac{x^2-3x-4}{x^3+1}$$
$$= \frac{\cancel{(x+4)}(x-4)}{\cancel{(x+4)}(x+1)} \div \frac{(x-4)\cancel{(x+1)}}{\cancel{(x+1)}(x^2-x+1)}$$
$$= \frac{x-4}{x+1} \div \frac{x-4}{x^2-x+1}$$ *Simplify each expression before performing the division*
$$= \frac{x-4}{x+1} \cdot \frac{x^2-x+1}{x-4}$$ *Invert the divisor*
$$= \frac{\cancel{(x-4)}(x^2-x+1)}{(x+1)\cancel{(x-4)}}$$ *Cancel after the division*
$$= \frac{x^2-x+1}{x+1}$$ ■

Exercises 1.3

Fill in the blanks to make each statement true.

1. The LCM of two positive integers is the ______________ positive integer that has both integers as a ______________.
2. To add rational numbers, we use the LCD, which is the ______________ of the given denominators.
3. The LCD of two denominators must be a ______________ of both.
4. The rational expression $(x^2-36)(2x+12)(x^2+12x+36)$ is a multiple of x^2-36, $2x+12$, and $x^2+12x+36$, but the LCM is ______________.
5. Operations with rational expressions follow the same rules as operations with ______________.

Write true *or* false *for each statement.*

6. If m and n are integers, then $m \cdot n$ is a multiple of m and n.
7. If m and n are positive integers, then $m \cdot n$ is always their LCM.
8. In adding two rational expressions with polynomial denominators $P(x)$ and $Q(x)$, the common denominator $P(x) \cdot Q(x)$ can be used.
9. In adding two rational expressions with polynomial denominators $P(x)$ and $Q(x)$, the simplest denominator to use is the LCM of $P(x)$ and $Q(x)$.
10. When dividing rational expressions, we divide the LCM of the numerators by the LCM of the denominators.

Least Common Multiple

Find the LCM of the given terms in each of the following.

11. 14, 42, 9
12. $3^4 \cdot 2$, $3 \cdot 2^3 \cdot 5$, $3^2 \cdot 5^2$
13. x^2-9, x^2-2x-3, x^2+2x-3
14. $2x^3+16$, $6x^2+9x-6$, $4x^2-8x+16$

Find the LCD to be used when adding fractions with the given denominators.

15. $3x^2$, x^3y, xy^2
16. $3x-3y$, x^2-y^2
17. x^2-x-2, x^2-4, $2x+2$
18. x^3-8, x^2-4x+4, x^2+x-2

Addition of Rational Expressions

Combine each of the following by using the LCD, and simplify.

19. $\frac{1}{14}+\frac{5}{42}-\frac{2}{9}$
20. $\frac{3x}{20}-\frac{3x}{15}$
21. $\frac{1}{x+h}-\frac{1}{x}$
22. $\frac{1}{(x+h)^2}-\frac{1}{x^2}$

23. $\dfrac{x+1}{x^3} - \dfrac{3}{x^2}$

24. $\dfrac{4}{x^3} + \dfrac{1}{x^4}$

25. $\dfrac{x+2}{2x-2} + \dfrac{3x-2}{x^2-4x+3}$

26. $\dfrac{x}{(x^2-1)^2} + \dfrac{3}{(x+1)^2}$

27. $\dfrac{4x-3}{4x^2-1} + \dfrac{3x}{2-4x}$

28. $\dfrac{2}{2x-x^2} - \dfrac{3}{x^2-4}$

29. $\dfrac{1}{x^2+3x+2} + \dfrac{x+2}{x^2+5x+4}$

30. $\dfrac{x-2}{x^2+x-2} + \dfrac{x-3}{x^2+2x-3}$

31. $\dfrac{x+1}{2x^2-5x-3} - \dfrac{x-2}{x^2-3x}$

32. $\dfrac{x-4}{3x^2+7x+2} + \dfrac{x+4}{x^2+2x}$

Perform the indicated additions and subtractions.

33. $\dfrac{1}{x^2-4} + \dfrac{x}{4-x^2} + \dfrac{1}{x-2}$

34. $\dfrac{x}{x^2-9} - \dfrac{x}{3-x} + \dfrac{x}{x+3}$

35. $\dfrac{1}{x-2} + \dfrac{x+4}{x^2-5x+6} - \dfrac{1}{x^2-4x+4}$

36. $\dfrac{1}{x^2-1} - \dfrac{x}{x^2+2x+1} - x$

37. $\dfrac{3}{x} - \dfrac{3}{x(x-2)} + \dfrac{x}{(x-2)^2}$

38. $\dfrac{1}{x(x+1)} + \dfrac{x+1}{x^2} - \dfrac{x}{(x+1)^2}$

39. $\dfrac{x}{x+3} - \dfrac{2}{x(x+3)} + \dfrac{1}{x^2} - \dfrac{5}{(x+3)^2}$

40. $\dfrac{x+3}{x(x-1)} + \dfrac{2}{x} + \dfrac{x+1}{(x-1)^2} + \dfrac{1}{x^2(x-1)}$

41. $\dfrac{1}{x^2+3x+2} - \dfrac{1}{x+2} + \dfrac{2}{x^2+4x+3} - \dfrac{2}{x+1}$

42. $\dfrac{x}{x^2-1} - \dfrac{1}{x^2+x-2} - \dfrac{x}{x+1} + \dfrac{1}{x^2-x-2}$

43. $\dfrac{x}{x^2+(a+b)x+ab} - \dfrac{1}{x+a} + \dfrac{1}{x+b}$

44. $\dfrac{ax}{x^2-a^2} + \dfrac{1}{x-a} - \dfrac{1}{x+a}$

45. $\dfrac{x^2}{x^3-a^3} - \dfrac{x}{x^2+ax+a^2} + \dfrac{1}{x-a}$

46. $\dfrac{x^2}{x^3+a^3} - \dfrac{x}{x^2-ax+a^2} + \dfrac{1}{x+a}$

Multiplication and Division of Rational Expressions

Perform the indicated operations in each of the following, and simplify.

47. $\dfrac{3xy^2}{5ab} \cdot \dfrac{20a^3}{9x^2}$

48. $\dfrac{2x+y}{x^2-4y^2} \cdot \dfrac{x+2y}{4x^2+4xy+y^2}$

49. $\dfrac{10x^2-29x+10}{9x^2-4} \cdot \dfrac{6x^2+x-2}{4x^2-20x+25}$

50. $\dfrac{2x^2+11x+12}{2x^2-3x-20} \cdot \dfrac{4x^2+16x+15}{x^2-16}$

51. $\dfrac{3ab^2c^3}{7xy} \div \dfrac{12abc}{35y^2z}$

52. $\dfrac{6x^2+7xy+y^2}{6x+6y} \div \dfrac{(x+y)^2}{6x+y}$

53. $\dfrac{x^4-13x^2+36}{x^2-2x} \div \dfrac{x^2-x-6}{5x}$

54. $\left(\dfrac{1}{x} - x\right) \div \left(\dfrac{2}{x+1} - 1\right)$

55. $\dfrac{\dfrac{x+1}{x^2+x-2} - \dfrac{x-2}{x^2-2x+1}}{\dfrac{1}{x-1} - \dfrac{1}{x^2+x-2}}$

Miscellaneous

In each of the following, simplify and perform the indicated operations.

56. $\dfrac{1}{1+\dfrac{2}{x+1}} + \dfrac{1}{x+1-\dfrac{4}{x+1}}$

57. $\dfrac{1}{1-\dfrac{1}{x^2}} - \dfrac{1}{1+\dfrac{1}{x}}$

58. $\dfrac{x^{-1}}{1-x^{-1}} + \dfrac{1}{1+x^{-1}} + \dfrac{x^{-1}}{2}$

59. $\left(1+\cfrac{1}{1+\cfrac{1}{1+\cfrac{1}{x}}}\right)\cdot\left(3-\cfrac{2}{1-\cfrac{1}{1-\cfrac{1}{x}}}\right)$

60. $\dfrac{1+\cfrac{1}{1+\cfrac{1}{x}}}{1+\cfrac{1}{1-\cfrac{1}{x}}} \div \dfrac{1-\cfrac{1}{1+\cfrac{1}{x}}}{1-\cfrac{1}{1-\cfrac{1}{x}}}$

1.4 Rational Exponents

Definition of $x^{1/n}$
Definition of $x^{m/n}$
Power-Root Rule
Rules for Rational Exponents

With rational exponents, roots behave like powers.

The meaning of integral exponents, as in

$$3^4 = 81, \qquad 10^{-2} = \frac{1}{100}, \qquad \text{and} \qquad 2^0 = 1,$$

and the meaning of roots, such as

$$\sqrt{4} = 2, \qquad \sqrt[3]{-8} = -2, \qquad \text{and} \qquad \sqrt[4]{(-2)^4} = 2,$$

were explained in Section P.5. We now show how exponents and roots are related.

Definition of $x^{1/n}$. We know that $(\sqrt{5})^2 = 5$, and if we formally apply the exponent rule $(x^m)^n = x^{mn}$ to the expression $(5^{1/2})^2$, the result is

$$(5^{1/2})^2 = 5^{(1/2)2} = 5^1 = 5.$$

Therefore, since the squares of both $5^{1/2}$ and $\sqrt{5}$ must equal 5, consistency requires that we set $5^{1/2} = \sqrt{5}$. This example suggests the following definition.

> If n is an integer greater than or equal to 2, then by definition
> $$\mathbf{x^{1/n} = \sqrt[n]{x}.}$$

The same restrictions must be applied to $x^{1/n}$ as to $\sqrt[n]{x}$. That is, if n is *even*, x cannot be negative, and for n even and x positive, both $x^{1/n}$ and $\sqrt[n]{x}$ mean the *positive* nth root of x.

Example 1 (a) $36^{1/2} = \sqrt{36} = 6$, whereas $(-36)^{1/2}$ has no meaning in the real number system.

(b) $8^{1/3} = \sqrt[3]{8} = 2$ and $(-8)^{1/3} = \sqrt[3]{-8} = -2$

(c) $10^{1/4} = \sqrt[4]{10} = 1.778$ (to three decimal places) ■

Definition of $x^{m/n}$. Now let us figure out how to give meaning to $x^{m/n}$ for any rational number m/n. For example, the expression $9^{3/2}$ should equal $(9^{1/2})^3$ since $3/2 = (1/2)3$, and, similarly, we want $(-8)^{2/3}$ to mean $[(-8)^{1/3}]^2$. But we note a necessary restriction. For instance, since $2/3 = 4/6$, we would expect $(-8)^{2/3}$ and $(-8)^{4/6}$ to have the same meaning. But

$$(-8)^{2/3} = [(-8)^{1/3}]^2 = (-2)^2 = 4,$$

whereas $(-8)^{4/6} = [(-8)^{1/6}]^4$ can't be a real number (*why?*). To avoid this problem, we assume that the rational exponent m/n is in *lowest terms,* that is, all common factors (except 1) have been canceled from the numerator and denominator, and of course $n \neq 0$. These considerations lead us to the following definition.

> If m and n are positive integers with no common factors (except 1), then by definition
>
> $$x^{m/n} = (x^{1/n})^m,$$
>
> where x cannot be negative when n is even. If, in addition, $x \neq 0$, then
>
> $$x^{-m/n} = \frac{1}{x^{m/n}}.$$

Note that the last part of this definition agrees with the corresponding definition in Section P.5 when m/n is an integer.

Example 2 (a) $4^{5/2} = (4^{1/2})^5 = 2^5$, but $(-4)^{5/2}$ has no meaning in the real number system.

(b) $32^{-3/5} = \frac{1}{32^{3/5}} = \frac{1}{(32^{1/5})^3} = \frac{1}{2^3} = \frac{1}{8}$

and

$$(-32)^{-3/5} = \frac{1}{(-32)^{3/5}} = \frac{1}{[(-32)^{1/5}]^3} = \frac{1}{(-2)^3} = \frac{1}{(-8)} = -\frac{1}{8}$$ ■

Example 3 $x^{3/2} = (\sqrt{x})^3$ for $x \geq 0$, and $x^{2/3} = (\sqrt[3]{x})^2$ for *every* real number x. ■

Power-Root Rule. If we note that $(4^3)^{1/2} = 64^{1/2} = 8$, and also that $(4^{1/2})^3 = 2^3 = 8$, we have an example of the following general rule (see Exercise 40).

$$(x^m)^{1/n} = (x^{1/n})^m \qquad \textbf{power-root rule}$$

Hence, we can interpret $x^{m/n}$ as either $(x^{1/n})^m$ or $(x^m)^{1/n}$, provided the restrictions in the definition of $x^{m/n}$ are observed. The power-root rule in terms of radicals is

$$\sqrt[n]{x^m} = (\sqrt[n]{x})^m.$$

Example 4 By definition, $8^{4/3} = (8^{1/3})^4 = 2^4 = 16$. Also, by the power-root rule, $8^{4/3} = (8^4)^{1/3} = (4096)^{1/3} = 16$. ■

Example 5 Show that $8^{-4/3}$, $(8^{1/3})^{-4}$, and $(8^{-4})^{1/3}$ are equal to $1/16$.

Solution: $8^{-4/3} = \dfrac{1}{8^{4/3}} = \dfrac{1}{2^4} = \dfrac{1}{16}$

$$(8^{1/3})^{-4} = 2^{-4} = \frac{1}{2^4} = \frac{1}{16}$$

$$(8^{-4})^{1/3} = \sqrt[3]{\frac{1}{8^4}} = \sqrt[3]{\frac{1}{4096}} = \frac{1}{16}$$ ■

Example 6 $[(-9)^2]^{1/4} = \sqrt[4]{81} = 3$, but $[(-9)^{1/4}]^2$ is not a real number. This example does not contradict the power-root rule because $2/4$ is not yet in lowest terms. ■

Rules for Rational Exponents. All the basic rules given for integral exponents in Section P.5 continue to hold for rational exponents, provided we avoid even roots of negative numbers. These rules are as follows, where it is understood that m/n and p/q are positive or negative rational numbers in lowest terms.

For products with the same base, we *add* the exponents, and for quotients with the same base, we *subtract* the exponents:

$$(x^{m/n})(x^{p/q}) = x^{m/n+p/q} \quad \text{and} \quad \frac{x^{m/n}}{x^{p/q}} = x^{m/n-p/q}. \quad \textbf{same-base rules}$$

Example 7 (a) $8^{2/3} \cdot 8^{1/3} = 8^{2/3+1/3} = 8^1 = 8$

(b) $\dfrac{16^{-3/4}}{16^{1/2}} = 16^{-3/4-1/2} = 16^{-5/4} = \dfrac{1}{16^{5/4}} = \dfrac{1}{32}$ ■

Also, the power of a product equals the *product* of the powers, and the power of a quotient equals the *quotient* of the powers:

$$(x \cdot y)^{m/n} = (x^{m/n})(y^{m/n}) \quad \text{and} \quad \left(\frac{x}{y}\right)^{m/n} = \frac{x^{m/n}}{y^{m/n}}. \quad \textbf{same-exponent rules}$$

Example 8 (a) $(16 \cdot 9)^{3/2} = 16^{3/2} \cdot 9^{3/2} = 64 \cdot 27 = 1728$

(b) $\left(\dfrac{4}{25}\right)^{3/2} = \dfrac{4^{3/2}}{25^{3/2}} = \dfrac{8}{125}$ ■

Finally, if $x > 0$, then

$$(x^{m/n})^{p/q} = x^{(m/n)(p/q)}. \qquad \textbf{power-of-a-power rule}$$

Example 9 (a) $(1024^{5/2})^{1/5} = 1024^{(5/2)(1/5)} = 1024^{1/2} = 16$

(b) $[(-64)^{2/3}]^{3/4} = 16^{3/4} = 8$ but $(-64)^{(2/3)(3/4)} = (-64)^{1/2}$ has no meaning in the real number system. Therefore, the power-of-a-power rule may not apply when the base is negative. ■

Examples 10–14 are included to show how the definitions and rules for rational exponents can be used to simplify algebraic expressions.

Example 10

$$\begin{aligned} \sqrt{x}\cdot\sqrt[3]{x} &= x^{1/2}\cdot x^{1/3} && \textit{Definition } (x \geq 0)\\ &= x^{1/2+1/3} && \textit{Same-base rule}\\ &= x^{5/6} \\ &= (\sqrt[6]{x})^5 && \textit{Definition} \end{aligned}$$
■

Example 11

$$\begin{aligned} \sqrt{x}/\sqrt[3]{x} &= x^{1/2}/x^{1/3} && \textit{Definition } (x > 0)\\ &= x^{1/2-1/3} && \textit{Same-base rule}\\ &= x^{1/6} \\ &= \sqrt[6]{x} && \textit{Definition} \end{aligned}$$
■

Example 12

$$\begin{aligned} \sqrt{4a^2} &= (4a^2)^{1/2} && \textit{Definition}\\ &= 4^{1/2}(a^2)^{1/2} && \textit{Same-exponent rule}\\ &= 2\sqrt{a^2} && \textit{Definition}\\ &= 2|a| && \textit{Section P.5} \end{aligned}$$
■

Example 13

$$\begin{aligned} \sqrt[3]{\sqrt{2}} &= (\sqrt{2})^{1/3} && \textit{Definition}\\ &= (2^{1/2})^{1/3} && \textit{Definition}\\ &= 2^{(1/2)(1/3)} && \textit{Power-of-a-power rule}\\ &= 2^{1/6} \\ &= \sqrt[6]{2} && \textit{Definition} \end{aligned}$$
■

Example 14

$$\begin{aligned} x^{2/3} + x^{5/3} &= x^{2/3} + x^{2/3}\cdot x^{3/3} && \textit{Same-base rule}\\ &= x^{2/3}(1 + x^{3/3}) \\ &= x^{2/3}(1 + x) \end{aligned}$$
■

Exercises 1.4

Fill in the blanks to make each statement true.

1. $x^{1/n}$ means ______ with the restriction that x cannot be negative when n is ______________.
2. If n is an even positive integer and x is positive, then $x^{1/n}$ means the ______ root of x.
3. By definition, $x^{m/n}$ means ______ when m and n are positive integers with no common factor; also, $x^{-m/n}$ means ______, provided x is not ______________.
4. If m/n is a positive or negative rational number in lowest terms, then $x^{m/n}$ may be interpreted as either ______ or ______.
5. The rule $(x^{m/n})^{p/q} = x^{(m/n)(p/q)}$ is valid for all rational exponents if we restrict the base x to be ______________.

Write true *or* false *for each statement.*

6. For every real number x, $x^{2/3}$ means $(\sqrt[3]{x})^2$.
7. For every real number x, $x^{3/2}$ means $(\sqrt{|x|})^3$.
8. For every real number x, $x^{-2/3} = (1/\sqrt[3]{x})^2$.
9. For every real number x, $x^{3/5} = |x|^{3/5}$.
10. For every real number x, $x^{2/3} = |x|^{2/3}$.

Definition of $x^{m/n}$

Evaluate, if possible, each of the following.

11. (a) $64^{1/6}$ (b) $(-64)^{1/6}$ (c) $(-64)^{1/3}$
12. (a) $32^{1/5}$ (b) $(-32)^{1/5}$ (c) $(-32)^{-1/5}$
13. (a) $81^{3/2}$ (b) $81^{-3/2}$ (c) $(-81)^{3/2}$
14. (a) $125^{2/3}$ (b) $125^{-2/3}$ (c) $(-125)^{-2/3}$

Power-Root Rule

Verify that $(x^m)^{1/n} = (x^{1/n})^m$ in each of the following. Use a calculator for Exercises 18–20.

15. $x = 9$, $n = 2$, $m = 3$
16. $x = -8$, $n = 3$, $m = 2$
17. $x = -32$, $n = 5$, $m = 2$
18. $x = 100$, $n = 4$, $m = 3$
19. $x = 1250$, $n = 5$, $m = 2$
20. $x = 707$, $n = 8$, $m = 4$

Rules for Rational Exponents

Use the rules for rational exponents to simplify each of the following.

21. (a) $64^{1/2} \cdot 64^{-1/3}$ (b) $\sqrt[3]{64} \cdot \sqrt[6]{64}$ (c) $(-1)^{1/3}(-1)^{2/9}$
22. (a) $\dfrac{8^{2/3}}{8^{1/3}}$ (b) $\dfrac{16^{1/2}}{16^{1/4}}$ (c) $\dfrac{27^{-1/3}}{27^{2/3}}$
23. (a) $(8 \cdot 64)^{4/3}$ (b) $(9 \cdot 81)^{-3/2}$ (c) $(\sqrt[4]{16})^3$
24. (a) $\left(-\dfrac{125}{8}\right)^{2/3}$ (b) $\left(\dfrac{9}{225}\right)^{-1/2}$ (c) $(\sqrt[3]{16})^4$
25. $2^{1/2} \cdot 3^{1/4} \cdot 2^{-1/3} \cdot 3^{1/2} \cdot 2^{1/6}$
26. $(5^{-6/7})^{-7/2}(5^{-2/3})^3$

Assuming x and y are positive numbers, write each expression in Exercises 27–31 with only positive exponents.

27. (a) $(x^{2/3})^6$ (b) $(x^{3/2})^6$ (c) $(\sqrt{x})^{-1/2}$
28. $\dfrac{(x^{2/3})^6(x^{-3/2})^4}{(x^{-4/3})^{3/2}}$
29. $\left(\dfrac{8x^{1/3}x^{-1/2}}{x^{-1/6}}\right)^{-1/3}$
30. $(9x^{1/3}y^{-1/3})^{1/2}(y^{-2}x^6)^{-1}$
31. $\sqrt[3]{\dfrac{(7^{-2}y^{-6})^6}{5^{-2}x^4}}$

32. Explain why $[(-8)^2]^{1/6} \neq (-8)^{2 \cdot 1/6}$.
33. Find a real number x for which $(x^2)^{1/2} \neq x$.

Simplify each of the following. Use factoring where possible.

34. $x^{3/4} + x^{7/4}$ $(x > 0)$
35. $(xy)^{2/3} + y^{5/3}$
36. $(x^{3/2}y)^{1/3} + x^{1/2}$ $(x > 0)$
37. $\dfrac{\left(\dfrac{x^2}{y}\right)^3 + \left(\dfrac{x}{y}\right)^6}{\left(\dfrac{x}{y}\right)^4}$ $(x \neq 0, y \neq 0)$
38. $\dfrac{(x^{1/2}y^{1/3})^3 + (x^3y^2)^{1/2}}{x^{3/2}}$ $(x > 0, y > 0)$
39. $\dfrac{(x^{1/2}y^{1/3})^3 + (x^3y^2)^{1/2}}{x^{3/2}}$ $(x > 0, y < 0)$

40. *The Power-Root Rule.* $(x^m)^{1/n} = (x^{1/n})^m$ can be proven for $x > 0$, $m = 25$, and any positive integer n as follows.

Let $y = (x^{25})^{1/n}$. Then, by the definition of $\sqrt[n]{}$, y is the positive number that satisfies $y^n = x^{25}$. Therefore, to prove that $y = (x^{1/n})^{25}$, we must show that $[(x^{1/n})^{25}]^n = x^{25}$. But

$$\begin{aligned}
[(x^{1/n})^{25}]^n &= (x^{1/n})^{25n} && \text{Rule } (a^m)^n = a^{mn} \text{ applied to base } x^{1/n} \text{ and } \textit{integral} \text{ exponents 25 and } n\\
&= [x^{1/n})^n]^{25} && \\
&= [(\sqrt[n]{x})^n]^{25} && \text{Definition of } x^{1/n}\\
&= x^{25}. && \text{Definition of } \sqrt[n]{x}
\end{aligned}$$

Use this same method to prove that $(x^m)^{1/n} = (x^{1/n})^m$ for *any* positive integral exponents m and n, given that x is greater than 0.

41. *The Same-Exponent Rule.* $(xy)^{m/n} = x^{m/n} \cdot y^{m/n}$ can be proven for the case $m = 1$ and $n = 6$ as follows (since 6 is even, we are assuming x and y are both positive).

Let $z = (xy)^{1/6}$. Then, by the definition of $\sqrt[6]{}$, z is the positive number satisfying $z^6 = xy$. Hence, to prove that $z = x^{1/6} \cdot y^{1/6}$, we must show that $(x^{1/6} \cdot y^{1/6})^6 = xy$. But

$$\begin{aligned}
&(x^{1/6} \cdot y^{1/6})^6 && \\
&= (x^{1/6})^6 \cdot (y^{1/6})^6 && \text{Rule } (ab)^n = a^n b^n \text{ applied to bases } x^{1/6} \text{ and } y^{1/6} \text{ and } \textit{integral} \text{ exponent 6}\\
&= (\sqrt[6]{x})^6 \cdot (\sqrt[6]{x})^6 && \text{Definition of } x^{1/6} \text{ and } y^{1/6}\\
&= xy. && \text{Definition of sixth roots}
\end{aligned}$$

Use this method to prove that $(xy)^{1/n} = x^{1/n} \cdot y^{1/n}$ for *any* positive integer n, given that x and y are both positive if n is even.

1.5 Operations with Radicals

Working with radicals is the nitty-gritty of algebra.

Radicals and Rational Exponents
Rationalizing Denominators

Radicals were introduced in Section P.5. Here we make use of rational exponents in order to perform algebraic operations on radicals. For all radicals in this section, we assume that the base is positive whenever the root is even or the power-of-a-power rule is applied.

Radicals and Rational Exponents. To add, subtract, multiply, or divide algebraic expressions containing radicals, we can convert to rational exponents and apply our previous rules. Answers can then be converted back to radical form.

Example 1

$$\begin{aligned}
\sqrt{x-1}\,\sqrt[4]{(x-1)^3} &= (x-1)^{1/2}(x-1)^{3/4} && \textit{Convert to rational exponents}\\
&= (x-1)^{1/2+3/4} && \textit{Same-base rule}\\
&= (x-1)^{5/4} && \textit{Answer in exponent form}\\
&= (x-1)^{1+1/4} && \\
&= (x-1)(x-1)^{1/4} && \textit{Same-base rule}\\
&= (x-1)\sqrt[4]{x-1} && \textit{Answer in simplified radical form}
\end{aligned}$$

■

Example 2
$$\frac{\sqrt[4]{(x-1)^3}}{\sqrt{x-1}} = \frac{(x-1)^{3/4}}{(x-1)^{1/2}}$$ *Convert to rational exponents*

$$= (x-1)^{3/4-1/2}$$ *Same-base rule*

$$= (x-1)^{1/4}$$ *Answer in exponent form*

$$= \sqrt[4]{x-1}$$ *Answer in simplified radical form* ■

Example 3
$$\sqrt{\frac{a+b}{a-b}}(\sqrt{a+b})^3\sqrt{(a-b)^3} = \frac{(a+b)^{1/2}}{(a-b)^{1/2}}(a+b)^{3/2}(a-b)^{3/2}$$
$$= (a+b)^{1/2+3/2}(a-b)^{3/2-1/2}$$
$$= (a+b)^2(a-b)$$ ■

Example 4
$$x^2\sqrt{x-1} - 2x\sqrt{x-1} + 4\sqrt{x-1} = (x^2 - 2x + 4)\sqrt{x-1}$$

Here there was no need to convert to rational exponents. We simply factored out $\sqrt{x-1}$ from each term. ■

Example 5
$$2\sqrt{x-1} + 3\sqrt{(x-1)^3}$$
$$= 2(x-1)^{1/2} + 3(x-1)^{3/2}$$
$$= 2(x-1)^{1/2} + 3(x-1)(x-1)^{1/2}$$
$$= (x-1)^{1/2}[2 + 3(x-1)]$$ *Factor out $(x-1)^{1/2}$*
$$= (x-1)^{1/2}(3x-1)$$
$$= (3x-1)\sqrt{x-1}$$ ■

Rationalizing Denominators. If terms to be added have radicals in their denominators, as in

$$\frac{1}{\sqrt{x}} + \frac{1}{\sqrt{x^3}} + \frac{1}{\sqrt[4]{x^3}},$$

we can first replace each term by an equivalent one whose denominator contains only *integral powers*. That is, we can first **rationalize** each denominator as follows.

$$\frac{1}{\sqrt{x}} = \frac{1}{\sqrt{x}} \cdot \frac{\sqrt{x}}{\sqrt{x}} = \frac{\sqrt{x}}{x}$$

$$\frac{1}{\sqrt{x^3}} = \frac{1}{x^{3/2}} = \frac{1}{x^{3/2}} \cdot \frac{x^{1/2}}{x^{1/2}} = \frac{x^{1/2}}{x^2} = \frac{\sqrt{x}}{x^2}$$

$$\frac{1}{\sqrt[4]{x^3}} = \frac{1}{x^{3/4}} = \frac{1}{x^{3/4}} \cdot \frac{x^{1/4}}{x^{1/4}} = \frac{x^{1/4}}{x} = \frac{\sqrt[4]{x}}{x}$$

Therefore,

$$\frac{1}{\sqrt{x}} + \frac{1}{\sqrt{x^3}} + \frac{1}{\sqrt[4]{x^3}} = \frac{\sqrt{x}}{x} + \frac{\sqrt{x}}{x^2} + \frac{\sqrt[4]{x}}{x}. \tag{1}$$

We can now proceed with the addition by means of a common denominator. In this case, the LCD of the terms on the right side of equation (1) is x^2. Therefore,

$$\frac{1}{\sqrt{x}} + \frac{1}{\sqrt{x^3}} + \frac{1}{\sqrt[4]{x^3}} = \frac{\sqrt{x}}{x} \cdot \frac{x}{x} + \frac{\sqrt{x}}{x^2} + \frac{\sqrt[4]{x}}{x} \cdot \frac{x}{x}$$

$$= \frac{x\sqrt{x} + \sqrt{x} + x\sqrt[4]{x}}{x^2} \quad \textit{Addition with LCD}$$

$$= \frac{(x+1)\sqrt{x} + x\sqrt[4]{x}}{x^2} \quad \textit{Answer with rationalized denominator}$$

Example 6 Perform the addition $\dfrac{1}{x^2\sqrt{x-1}} + \dfrac{1}{x\sqrt{(x-1)^3}}$.

Solution: First we rationalize the denominators.

$$\frac{1}{x^2\sqrt{x-1}} = \frac{1}{x^2\sqrt{x-1}} \cdot \frac{\sqrt{x-1}}{\sqrt{x-1}} = \frac{\sqrt{x-1}}{x^2(x-1)}$$

$$\frac{1}{x\sqrt{(x-1)^3}} = \frac{1}{x(x-1)^{3/2}}$$

$$= \frac{1}{x(x-1)^{3/2}} \cdot \frac{(x-1)^{1/2}}{(x-1)^{1/2}} = \frac{\sqrt{x-1}}{x(x-1)^2}$$

Therefore,

$$\frac{1}{x^2\sqrt{x-1}} + \frac{1}{x\sqrt{(x-1)^3}} = \frac{\sqrt{x-1}}{x^2(x-1)} + \frac{\sqrt{x-1}}{x(x-1)^2}$$

$$\textit{LCD} = x^2(x-1)^2 \qquad = \frac{\sqrt{x-1}}{x^2(x-1)} \cdot \frac{x-1}{x-1} + \frac{\sqrt{x-1}}{x(x-1)^2} \cdot \frac{x}{x}$$

$$= \frac{(x-1)\sqrt{x-1} + x\sqrt{x-1}}{x^2(x-1)^2}$$

$$= \frac{(2x-1)\sqrt{x-1}}{x^2(x-1)^2}. \quad \textit{Answer with rationalized denominator}$$

■

Example 7 Perform the subtraction $\dfrac{1}{\sqrt{a} - \sqrt{b}} - \dfrac{\sqrt{a}}{a-b}$.

Solution: First we rationalize the denominator of the term on the left.

$$\frac{1}{\sqrt{a} - \sqrt{b}} = \frac{1}{\sqrt{a} - \sqrt{b}} \cdot \frac{\sqrt{a} + \sqrt{b}}{\sqrt{a} + \sqrt{b}} = \frac{\sqrt{a} + \sqrt{b}}{a - b}$$

Here we substituted $\sqrt{a} = x$, $\sqrt{b} = y$ in the rule $(x - y)(x + y) = x^2 - y^2$ to obtain $(\sqrt{a} - \sqrt{b})(\sqrt{a} + \sqrt{b}) = a - b$. The expressions $\sqrt{a} - \sqrt{b}$ and $\sqrt{a} + \sqrt{b}$ are called **conjugates** of each other. Therefore,

$$\frac{1}{\sqrt{a}-\sqrt{b}} - \frac{\sqrt{a}}{a-b} = \frac{\sqrt{a}+\sqrt{b}}{a-b} - \frac{\sqrt{a}}{a-b}$$ *Terms with rationalized denominators*

$$= \frac{\sqrt{a}+\sqrt{b}-\sqrt{a}}{a-b}$$ *Subtraction with common denominator*

$$= \frac{\sqrt{b}}{a-b}.$$ *Answer with rationalized denominator* ■

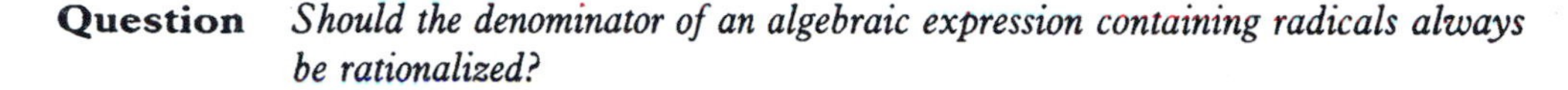

Question *Should the denominator of an algebraic expression containing radicals always be rationalized?*

Answer: No. When adding algebraic expressions with radical denominators, it is sometimes easier to obtain a common denominator by first rationalizing the denominators. However, this is not always the case, as is illustrated in Examples 8 and 9 below. Also, if rationalizing a denominator results in an expression that is considerably more complicated than the original one, then the process should be avoided. For instance, when

$$\frac{1}{\sqrt{a}+\sqrt{b}+\sqrt{c}}$$

is rationalized, the result is

$$\frac{(a-b-c)\sqrt{a}+(b-a-c)\sqrt{b}+(c-a-b)\sqrt{c}+2\sqrt{abc}}{a^2+b^2+c^2-2ab-2ac-2bc}$$

Here the unrationalized form is obviously preferable. ■

In Examples 8 and 9 below we further illustrate the various approaches to performing operations with radicals.

Example 8 Perform the addition

$$\frac{x}{\sqrt{2x+1}} + \sqrt{2x+1}$$

in two ways: (a) by first rationalizing the denominator, and (b) without rationalizing the denominator.

Solution:

(a) First rationalize the denominator, then add:

$$\frac{x}{\sqrt{2x+1}} + \sqrt{2x+1} = \frac{x\sqrt{2x+1}}{2x+1} + \frac{\sqrt{2x+1}}{1}$$ *Write each term as a fraction*

$$= \frac{x\sqrt{2x+1}+(2x+1)\sqrt{2x+1}}{2x+1}$$ *Addition of fractions*

$$= \frac{(3x+1)\sqrt{2x+1}}{2x+1}.$$ *Answer with rationalized denominator*

(b) Add without rationalizing the denominator:

$$\frac{x}{\sqrt{2x+1}} + \sqrt{2x+1} = \frac{x}{\sqrt{2x+1}} + \frac{\sqrt{2x+1}}{1} \quad \textit{Write each term as a fraction}$$

$$= \frac{x + \sqrt{2x+1}\sqrt{2x+1}}{\sqrt{2x+1}} \quad \textit{Addition of fractions}$$

$$= \frac{x + 2x + 1}{\sqrt{2x+1}}$$

$$= \frac{3x+1}{\sqrt{2x+1}}. \quad \textit{Answer in nonrationalized form}$$

Example 9 Simplify $\dfrac{\sqrt{x^2+1} - \dfrac{x}{\sqrt{x^2+1}}}{x^2+1}$ in three ways:

(a) by first rationalizing the denominator of $\dfrac{x}{\sqrt{x^2+1}}$, (b) without rationalizing, and (c) by using rational exponents.

Solution:

(a) Simplify by first rationalizing the denominator of $\dfrac{x}{\sqrt{x^2+1}}$:

$$\frac{\sqrt{x^2+1} - \dfrac{x}{\sqrt{x^2+1}}}{x^2+1} = \frac{\dfrac{\sqrt{x^2+1}}{1} - \dfrac{x\sqrt{x^2+1}}{x^2+1}}{\dfrac{x^2+1}{1}} \quad \textit{Write each term as a fraction}$$

$$= \frac{\dfrac{(x^2+1)\sqrt{x^2+1} - x\sqrt{x^2+1}}{x^2+1}}{\dfrac{x^2+1}{1}}$$

$$= \frac{(x^2+1)\sqrt{x^2+1} - x\sqrt{x^2+1}}{x^2+1} \cdot \frac{1}{x^2+1}$$

$$= \frac{(x^2-x+1)\sqrt{x^2+1}}{(x^2+1)^2}. \quad \textit{Answer in rationalized form}$$

(b) Simplify without rationalizing:

$$\frac{\sqrt{x^2+1} - \dfrac{x}{\sqrt{x^2+1}}}{x^2+1} = \frac{\dfrac{\sqrt{x^2+1}}{1} - \dfrac{x}{\sqrt{x^2+1}}}{\dfrac{x^2+1}{1}} \quad \textit{Write each term as a fraction}$$

$$= \frac{\dfrac{\sqrt{x^2+1}\sqrt{x^2+1} - x}{\sqrt{x^2+1}}}{\dfrac{x^2+1}{1}}$$

$$= \frac{\sqrt{x^2+1}\sqrt{x^2+1} - x}{\sqrt{x^2+1}} \cdot \frac{1}{x^2+1}$$

$$= \frac{x^2+1-x}{\sqrt{x^2+1}(x^2+1)}$$

$$= \frac{x^2-x+1}{(x^2+1)\sqrt{x^2+1}}.$$ *Answer in nonrationalized form*

(c) Simplify by using rational exponents:

$$\frac{\sqrt{x^2+1} - \dfrac{x}{\sqrt{x^2+1}}}{x^2+1} = \frac{\dfrac{(x^2+1)^{1/2}}{1} - \dfrac{x}{(x^2+1)^{1/2}}}{\dfrac{x^2+1}{1}}$$ *Write each term as a fraction*

$$= \frac{(x^2+1)^{1/2}(x^2+1)^{1/2} - x}{(x^2+1)^{1/2}} \cdot \frac{1}{x^2+1}$$

$$= \frac{x^2+1-x}{(x^2+1)^{1/2}(x^2+1)}$$

$$= \frac{x^2-x+1}{(x^2+1)^{3/2}}.$$ *Answer with rational exponents* ■

Exercises 1.5

All radicands in these exercises are assumed to be positive.

Fill in the blanks to make each statement true.

1. The rationalized form of $1/\sqrt{x}$ is ________.
2. The rationalized form of $1/\sqrt[3]{x}$ is ________.
3. The rationalized form of $1/x^{5/4}$ is ________.
4. The conjugate of $\sqrt{a} - \sqrt{b}$ is ________.
5. The rationalized form of $(\sqrt{x} + \sqrt{a})/(\sqrt{x} - \sqrt{a})$ is

Write true *or* false *for each statement.*

6. $(x+2)^{-1/2} = \dfrac{\sqrt{x+2}}{x+2}$
7. $\dfrac{1}{(x-1)^{2/3}} = \dfrac{\sqrt[3]{x-1}}{x-1}$
8. $x - \dfrac{1}{\sqrt{x}} = \dfrac{(x^2-1)\sqrt{x}}{x}$
9. When working with radical denominators, the first step is always to rationalize the denominator.

10. When working with radicals, it is usually convenient to switch to rational exponents.

Radicals and Rational Exponents

Express each of the following in terms of rational exponents.

11. $\sqrt{x^2+1}$
12. $\sqrt[3]{(x-2)^2}$
13. $(\sqrt[4]{x^2+3})^3$
14. $(\sqrt{(x+1)^3})^5$
15. $\sqrt{x+\sqrt{x}}$
16. $\sqrt[5]{\sqrt{x-1}}$
17. $\sqrt{(\sqrt{x}+2)^3}$
18. $\sqrt{\sqrt{\sqrt{x+1}}}$

Perform the indicated operations in each of the following. Simplify and express answers in radical form.

19. $\sqrt{x^2+1}\sqrt[3]{x^2+1}$
20. $\sqrt{(x-1)^3}\sqrt{x-1}$
21. $\sqrt{x+1}\sqrt[3]{(x+1)^2}$
22. $\sqrt[4]{(x-2)^3}(\sqrt[4]{x-2})^3$
23. $\sqrt{(x^2-1)^3}$
24. $\dfrac{\sqrt{x^2+1}}{\sqrt[3]{x^2+1}}$
25. $\dfrac{\sqrt{(x-2)^3}}{\sqrt{x-2}}$
26. $\dfrac{\sqrt{x+1}}{\sqrt[3]{(x+1)^2}}$
27. $\dfrac{\sqrt[4]{(x-2)^3}}{(\sqrt[4]{x-2})^3}$
28. $\dfrac{\sqrt{(x^2-1)^3}}{\sqrt[3]{(x^2-1)^2}}$
29. $5\sqrt[3]{x^2-1}+\sqrt[3]{x^2-1}$
30. $3\sqrt[4]{x+2}-5\sqrt[4]{x+2}+4\sqrt[4]{x+2}$
31. $2a\sqrt{x^2+x+1}+3b\sqrt{x^2+x+1}$
32. $a\sqrt{x^2+5}-2b\sqrt{x^2+5}+(3a-b)\sqrt{x^2+5}$
33. $x\sqrt[3]{x-4}+2\sqrt[3]{x-4}$
34. $(x+1)\sqrt{x+1}-(2x+3)\sqrt{x+4}+(x+4)\sqrt{x+4}$

Rationalizing Denominators

Rationalize the denominator in each of the following.

35. $\dfrac{1}{\sqrt{x}-3}$
36. $\dfrac{2}{\sqrt{2x}}$
37. $\dfrac{x^2}{\sqrt{x}}$
38. $\dfrac{x-1}{\sqrt{x}-1}$
39. $\dfrac{1}{(\sqrt[3]{x}+1)^2}$
40. $\dfrac{x+1}{\sqrt[3]{x^2}+1}$
41. $\dfrac{1}{\sqrt{x+1}+\sqrt{x-1}}$
42. $\dfrac{1}{\sqrt{a}+\sqrt{b}+\sqrt{c}}$

Perform the indicated operations in each of the following. Express answers in radical form with rationalized denominators.

43. $\dfrac{1}{\sqrt{x}}+\dfrac{x}{\sqrt{x}}$
44. $\dfrac{1}{\sqrt{a}}+\dfrac{\sqrt{a}}{a}$
45. $\dfrac{\sqrt{a}}{b}-\sqrt{\dfrac{b}{a}}$
46. $\dfrac{1}{\sqrt{x}}+\dfrac{x^2-1}{x\sqrt{x}}$
47. $\dfrac{1}{a}+\dfrac{1}{\sqrt{a}}+\dfrac{1}{a\sqrt{a}}$
48. $\dfrac{1}{\sqrt[3]{x^2}}-\dfrac{1}{\sqrt{x^3}}$
49. $\dfrac{x}{\sqrt{x}}+\dfrac{x^2}{\sqrt[4]{x}}+\dfrac{x^4}{\sqrt[6]{x}}$
50. $\dfrac{1}{\sqrt{x}-1}-\dfrac{1}{x^2-1}+\dfrac{1}{\sqrt{x}+1}$
51. $\dfrac{1}{\sqrt{a^2-b^2}}+\sqrt{\dfrac{a+b}{a-b}}+\sqrt{\dfrac{a-b}{a+b}}$
52. $\dfrac{1}{x\sqrt{x+2}}-\dfrac{1}{(x+2)\sqrt{x}}$
53. $\dfrac{1}{\sqrt{x}-1}-\dfrac{1}{\sqrt{x}+1}$
54. $\dfrac{1}{\sqrt{a}-\sqrt{b}}+\dfrac{1}{\sqrt{a}+\sqrt{b}}$

Miscellaneous

The following expressions appear in calculus. Simplify each expression and express answers in radical form.

55. $\sqrt{x^2+1}+\dfrac{x^2}{\sqrt{x^2+1}}$
56. $\sqrt{x+1}+\dfrac{x}{2\sqrt{x+1}}$
57. $\dfrac{\dfrac{x}{2\sqrt{x+5}}-\sqrt{x+5}}{x^2}$
58. $\dfrac{\dfrac{x^2}{\sqrt{x^2+5}}-\sqrt{x^2+5}}{x^2}$
59. $(x^2-1)^{1/3}+\frac{2}{3}x^2(x^2-1)^{-2/3}$
60. $(x+3)^{2/3}+\frac{2}{3}(x+2)(x+3)^{-1/3}$
61. $\dfrac{x^2(x^2+2)^{-1/2}-(x^2+2)^{1/2}}{x^2}$
62. $\dfrac{x\sqrt{x+1}(x^2+2)^{-1/2}-\sqrt{x^2+2}(x+1)^{-1/2}}{x+1}$

1.6 The Binomial Theorem

The binomial theorem is one of the jewels of algebra.

Powers of Binomials
Pascal's Triangle
Binomial Coefficients
The Binomial Theorem

Powers of Binomials. In Section 1.1, we expanded the square and the cube of a binomial, that is, $(a + b)^2$ and $(a + b)^3$. Similarly, any positive integral power of a binomial $a + b$ can be obtained by successively multiplying two factors at a time. For example,

$$\begin{aligned}(a + b)^5 &= (a + b)(a + b)(a + b)(a + b)(a + b)\\ &= (a^2 + 2ab + b^2)(a + b)(a + b)(a + b)\\ &= (a^3 + 3a^2b + 3ab^2 + b^3)(a + b)(a + b)\\ &= (a^4 + 4a^3b + 6a^2b^2 + 4ab^3 + b^4)(a + b)\\ &= a^5 + 5a^4b + 10a^3b^2 + 10a^2b^3 + 5ab^4 + b^5.\end{aligned}$$

In the process of obtaining $(a + b)^5$ we first determined $(a + b)^2$, $(a + b)^3$, and $(a + b)^4$. However, there is a formula for obtaining the expansion for any positive integral power without the need to first compute all lower powers. We can acquire some insight into this formula by examining the pattern followed by the exponents and coefficients of the terms of $(a + b)^n$ for special cases of n.

The terms of $(a + b)^5$, without regard to coefficients, are

$$a^5,\ a^4b,\ a^3b^2,\ a^2b^3,\ ab^4,\ b^5.$$

We note that the exponents of a are decreasing by 1 from 5 to 0, while those of b are increasing from 0 to 5. In each term, the sum of the exponents is 5. That is, the order of each term is 5. Also, there are six terms in all.

In general, for $(a + b)^n$, where n is any positive integer, the terms without regard to coefficients are

$$a^n,\ a^{n-1}b,\ a^{n-2}b^2,\ \ldots,\ a^{n-r}b^r,\ \ldots,\ ab^{n-1},\ b^n.$$

That is,

$$\begin{aligned}(a + b)^n = {}&\_\_\_\ a^n + \_\_\_\ a^{n-1}b + \_\_\_\ a^{n-2}b^2 + \cdots\\ &+ \_\_\_\ a^{n-r}b^r + \cdots + \_\_\_\ ab^{n-1} + \_\_\_\ b^n,\end{aligned} \tag{1}$$

where the blanks stand for the coefficients yet to be determined. Here the exponents of a decrease by 1 from n to 0, while those of b increase from 0 to n. *The* ***order*** *(sum of the exponents) of each term is n, and there are $n + 1$ terms in all.*

Example 1 What are the terms, without regard to coefficients, in the binomial expansion of $(a + b)^{20}$?

Solution: The terms are a^{20}, $a^{19}b$, $a^{18}b^2$, $a^{17}b^3$, . . . , a^3b^{17}, a^2b^{18}, ab^{19}, b^{20}. There are 21 terms in all. ■

[1]*Blaise Pascal (1623–62) was a precocious French mathematician who later became a religious thinker. He discovered many theorems in geometry and is one of the founders of probability theory. However, the triangle that bears his name was already in use by Chinese mathematicians early in the fourteenth century.*

We now investigate the coefficients of $(a + b)^n$ for various choices of n.

Pascal's Triangle. The expansions of $(a + b)^n$, for n from 0 to 5, are as follows:

$$
\begin{aligned}
(a+b)^0 &= 1\\
(a+b)^1 &= a + b\\
(a+b)^2 &= a^2 + 2ab + b^2\\
(a+b)^3 &= a^3 + 3a^2b + 3ab^2 + b^3\\
(a+b)^4 &= a^4 + 4a^3b + 6a^2b^2 + 4ab^3 + b^4\\
(a+b)^5 &= a^5 + 5a^4b + 10a^3b^2 + 10a^2b^3 + 5ab^4 + b^5.
\end{aligned}
$$

The coefficients of these expansions form the following triangular array, which makes up the first six rows of what is called **Pascal's triangle.**[1]

```
          1
         1 1
        1 2 1
       1 3 3 1
      1 4 6 4 1
    1 5 10 10 5 1
```

Pascal's triangle has the following three basic properties:

(a) Each row begins and ends with the number 1.

(b) Every interior number in a given row is the sum of the numbers immediately to the right and left in the row above.

(c) In each row, the pattern from right to left is the same as the pattern from left to right.

Example 2 Construct the next two rows of Pascal's triangle, corresponding to $(a + b)^6$ and $(a + b)^7$. Also, write out the binomial expansion in each case.

Solution: By properties (a) and (b), the next two rows are

```
  1  6  15  20  15  6  1
1  7  21  35  35  21  7  1.
```

Therefore,

$$(a + b)^6 = a^6 + 6a^5b + 15a^4b^2 + 20a^3b^3 + 15a^2b^4 + 6ab^5 + b^6,$$

and

$$
\begin{aligned}
(a + b)^7 = a^7 &+ 7a^6b + 21a^5b^2 + 35a^4b^3 + 35a^3b^4\\
&+ 21a^2b^5 + 7ab^6 + b^7. \quad \blacksquare
\end{aligned}
$$

We could verify the expansions in Example 2 by direct computation. However, it can be shown that Pascal's triangle gives the correct coefficients of $(a + b)^n$ for any nonnegative integer n. Therefore, Pascal's triangle is very useful for expanding $(a + b)^n$ when n is moderately small. But, since the construction of each new row requires the entries of the previous row, the triangle becomes impractical for large values of n.

Binomial Coefficients. Let ${}_nC_r$ denote the coefficient of $a^{n-r}b^r$ in the expansion of $(a + b)^n$, where n is a positive integer and r is a nonnegative integer less than or equal to n. In the case $n = 5$, we have

$$\begin{aligned}(a + b)^5 = \underset{\substack{\uparrow \\ {}_5C_0}}{1} \cdot a^5b^0 + \underset{\substack{\uparrow \\ {}_5C_1}}{5} \cdot a^4b^1 + \underset{\substack{\uparrow \\ {}_5C_2}}{10} \cdot a^3b^2 + \underset{\substack{\uparrow \\ {}_5C_3}}{10} \cdot a^2b^3 \\ + \underset{\substack{\uparrow \\ {}_5C_4}}{5} \cdot a^1b^4 + \underset{\substack{\uparrow \\ {}_5C_5}}{1} \cdot a^0b^5.\end{aligned}$$

The ${}_nC_r$ are called the **binomial coefficients,** and we want a general algebraic formula that expresses each ${}_nC_r$ in terms of n and r. The beforementioned three basic properties of Pascal's triangle can now be expressed in terms of the binomial coefficients as follows.

(a) ${}_nC_0 = 1 = {}_nC_n$	Each row begins and ends with 1.
(b) ${}_nC_r = {}_{n-1}C_r + {}_{n-1}C_{r-1}$	Each interior number in row n is equal to the sum of the numbers immediately to the right and left in row $n - 1$ above.
(c) ${}_nC_{n-r} = {}_nC_r$	The pattern from right to left in row n is the same as the pattern from left to right.

By means of properties (a) and (b) above, it can be proved (see the section on mathematical induction in the last chapter) that

$${}_nC_r = \frac{n!}{(n - r)!r!} \quad (0 \le r \le n), \tag{2}$$

where $n!$ (read "**n factorial**") is defined as

$$n! = n(n - 1)(n - 2) \cdot \ldots \cdot 1 \quad \text{for } n \ge 1 \text{ and } 0! = 1.$$

For example, if $n = 5$, then

$${}_5C_0 = \frac{5!}{5!0!} = \frac{5 \cdot 4 \cdot 3 \cdot 2 \cdot 1}{5 \cdot 4 \cdot 3 \cdot 2 \cdot 1 \cdot 1} = 1 \qquad \textit{Note that } 0! = 1$$

$${}_5C_1 = \frac{5!}{4!1!} = \frac{5 \cdot 4 \cdot 3 \cdot 2 \cdot 1}{4 \cdot 3 \cdot 2 \cdot 1 \cdot 1} = 5$$

$${}_5C_2 = \frac{5!}{3!2!} = \frac{5 \cdot 4 \cdot 3 \cdot 2 \cdot 1}{3 \cdot 2 \cdot 1 \cdot 2 \cdot 1} = 10$$

$${}_5C_3 = \frac{5!}{2!3!} = \frac{5 \cdot 4 \cdot 3 \cdot 2 \cdot 1}{2 \cdot 1 \cdot 3 \cdot 2 \cdot 1} = 10$$

$${}_5C_4 = \frac{5!}{1!4!} = \frac{5 \cdot 4 \cdot 3 \cdot 2 \cdot 1}{1 \cdot 4 \cdot 3 \cdot 2 \cdot 1} = 5$$

$${}_5C_5 = \frac{5!}{0!5!} = \frac{5 \cdot 4 \cdot 3 \cdot 2 \cdot 1}{1 \cdot 5 \cdot 4 \cdot 3 \cdot 2 \cdot 1} = 1.$$

These values agree with the coefficients obtained for $(a + b)^5$ at the beginning of this section.

Example 3 Find the coefficient of a^7b^3 in $(a + b)^{10}$.

Solution: The coefficient is ${}_{10}C_3$, and by equation (2),

$${}_{10}C_3 = \frac{10!}{7!3!} = \frac{10 \cdot 9 \cdot 8 \cdot 7 \cdot 6 \cdot 5 \cdot 4 \cdot 3 \cdot 2 \cdot 1}{7 \cdot 6 \cdot 5 \cdot 4 \cdot 3 \cdot 2 \cdot 1 \cdot 3 \cdot 2 \cdot 1} = 120. \quad ■$$

In Example 3, note that $10! = 10 \cdot 9 \cdot 8 \cdot 7!$ so that we can cancel $7!$ from the numerator and denominator when computing ${}_{10}C_3$. That is,

$${}_{10}C_3 = \frac{10 \cdot 9 \cdot 8 \cdot 7!}{7! \cdot 3!} = \frac{10 \cdot 9 \cdot 8}{3!}.$$

In general, $n! = n(n - 1)(n - 2) \ldots (n - r + 1) \cdot (n - r)!$, and we can cancel $(n - r)!$ from the numerator and denominator on the right side of equation (2) when computing ${}_nC_r$. That is,

$${}_nC_r = \frac{n(n-1)(n-2) \ldots (n-r+1)}{r!}. \qquad (3)$$

Equation (3) is usually more convenient than (2) when computing the binomial coefficients. (*Why?*)

The Binomial Theorem. By filling in the blanks of equation (1) with the binomial coefficients of equation (3), we arrive at the **binomial theorem.**

For any positive integer n,

$$(a+b)^n = a^n + na^{n-1}b + \frac{n(n-1)}{2}a^{n-2}b^2 + \dots + \frac{n(n-1)(n-2)\dots(n-r+1)}{r!}a^{n-r}b^r + \dots + nab^{n-1} + b^n.$$

Example 4 Use the binomial theorem to expand $(a+b)^{10}$.

Solution: We use equation (3) and property (c) of the binomial coefficients:

$${}_{10}C_1 = \frac{10}{1} = 10 = {}_{10}C_9 \qquad {}_{10}C_4 = \frac{10\cdot 9\cdot 8\cdot 7}{4\cdot 3\cdot 2\cdot 1} = 210 = {}_{10}C_6$$

$${}_{10}C_2 = \frac{10\cdot 9}{2\cdot 1} = 45 = {}_{10}C_8 \qquad {}_{10}C_5 = \frac{10\cdot 9\cdot 8\cdot 7\cdot 6}{5\cdot 4\cdot 3\cdot 2\cdot 1} = 252.$$

$${}_{10}C_3 = \frac{10\cdot 9\cdot 8}{3\cdot 2\cdot 1} = 120 = {}_{10}C_7$$

Therefore,

$$(a+b)^{10} = a^{10} + 10a^9b + 45a^8b^2 + 120a^7b^3 + 210a^6b^4 + 252a^5b^5 + 210a^4b^6 + 120a^3b^7 + 45a^2b^8 + 10ab^9 + b^{10}. \quad ■$$

Example 5 Derive an expansion for $(a-b)^5$.

Solution: Replace b by $-b$ in the binomial theorem with $n = 5$ to obtain

$$(a-b)^5 = a^5 + 5a^4(-b) + 10a^3(-b)^2 + 10a^2(-b)^3 + 5a(-b)^4 + (-b)^5$$
$$= a^5 - 5a^4b + 10a^3b^2 - 10a^2b^3 + 5ab^4 - b^5. \quad ■$$

Example 6 Expand $(2x+y)^6$.

Solution: We substitute $a = 2x$, $b = y$ and $n = 6$ in the binomial theorem to obtain

$$(2x+y)^6 = (2x)^6 + 6(2x)^5y + 15(2x)^4y^2 + 20(2x)^3y^3 + 15(2x)^2y^4 + 6(2x)y^5 + y^6$$
$$= 64x^6 + 192x^5y + 240x^4y^2 + 160x^3y^3 + 60x^2y^4 + 12xy^5 + y^6. \quad ■$$

Comment Isaac Newton (1642–1727) extended the binomial theorem to the case $(a+x)^{p/q}$, where p/q is a rational number. Although $n!$ is defined only when n is a nonnegative integer, formula (3) can be used for the coefficient of $a^{n-r}x^r$ when n is a rational number p/q. For example, the expansion of $(1+x)^{1/2}$ becomes

$$1 + \frac{1}{2}x - \frac{1\cdot 1}{2\cdot 4}x^2 + \frac{1\cdot 1\cdot 3}{2\cdot 4\cdot 6}x^3 - \frac{1\cdot 1\cdot 3\cdot 5}{2\cdot 4\cdot 6\cdot 8}x^4 + \dots.$$

When n is a positive integer, the binomial expansion has $n + 1$ terms, corresponding to $r = 0, 1, 2, \ldots, n$, respectively. However, when n is a rational number p/q other than a positive integer, the expansion contains a term corresponding to every nonnegative integer r. For example, the above expansion contains an *infinite* number of terms. Sums with an infinite number of terms were not completely understood by mathematicians in Newton's day. It was not until the nineteenth century that a precise treatment for such sums was developed. Today the subject is called *infinite series* and is studied in calculus. Note that if we substitute $x = 1$ in $(1 + x)^{1/2}$, we get $2^{1/2}$, or $\sqrt{2}$. Use a calculator to add the first 7 terms of the above expansion with $x = 1$ and see how close the sum comes to the value of $\sqrt{2}$ obtained with the square root key on the calculator.

Exercises 1.6

Whenever n, r, or ${}_nC_r$ appear in the following exercises, it is assumed that n is a positive integer and r is a nonnegative integer less than or equal to n.

Fill in the blanks to make each statement true.

1. In the expansion of $(a + b)^n$, there are ______ terms.
2. In the expansion of $(a + b)^n$, the sum of the exponents of each term is ______.
3. In the expansion of $(a + b)^n$, the coefficient of $a^{n-r}b^r$ is given by the formula ____________.
4. The coefficient of a^8b^4 in the expansion of $(a + b)^{12}$ is the same as the coefficient of ______.
5. The largest coefficient in the expansion of $(a + b)^{12}$ is ______.

Write true *or* false *for each statement.*

6. One of the terms in the expansion of $(a - b)^{10}$ is $120a^7b^3$.
7. According to Pascal's triangle, ${}_{12}C_7 = {}_{11}C_7 + {}_{11}C_6$.
8. ${}_nC_r = {}_nC_{r+1}$
9. ${}_nC_r = {}_{n+r}C_{n-r}$
10. $0! = 0$

Powers of Binomials

Expand each of the following by successively multiplying two factors at a time.

11. $(x - 1)^3$
12. $(2x + 3)^3$
13. $(x + 2)^4$
14. $(3x - 1)^4$
15. $(x - 1)^5$
16. $(2x - 3)^5$

Give all the terms, without regard to coefficients, in the expansions of each of the following.

17. $(x + 1)^5$
18. $(1 + x^2)^5$
19. $(ax + y)^4$
20. $(a^2b + xy^2)^4$

Pascal's Triangle

Use Pascal's triangle to expand each of the following in Exercises 21–26.

21. $(x + 2)^4$
22. $(x - 2)^5$
23. $(2x + 3)^4$
24. $(2x - 1)^5$
25. $(1 + x)^6$
26. $(1 - x)^7$
27. Use Pascal's triangle to find the coefficient of x^4 in the expansion of $(2x + 1)^8$.
28. Use Pascal's triangle to find the coefficient of x^5 in the expansion of $(1 - 3x)^8$.
29. First add the numbers in the nth row of Pascal's triangle for $n = 1, 2, 3, 4, 5$, then use these results to predict the sum of the coefficients for the following.
    (a) $(a + b)^6$
    (b) $(a + b)^n$, where n is any positive integer
30. First add *all* of the numbers in the first n rows of Pascal's triangle for $n = 1, 2, 3, 4, 5$, then use these results to predict the corresponding sum for the following.
    (a) $n = 6$ **(b)** any positive integer n

Binomial Coefficients

31. Evaluate each factorial.

(a) 4! (b) 5! (c) 7!

32. Evaluate each binomial coefficient.

(a) ${}_7C_0$ (b) ${}_7C_3$ (c) ${}_7C_5$

33. Simplify each expression.

(a) $\dfrac{(n+2)!}{n!}$ (b) $\dfrac{(n+1)!}{(n-1)!}$ (c) $\dfrac{[2(n+1)]!}{(2n)!}$

34. Simplify each binomial coefficient.

(a) ${}_{n+2}C_n$ (b) ${}_{n+1}C_{n-1}$ (c) ${}_{2(n+1)}C_{2n}$

Verify each statement in Exercises 35 and 36.

35. (a) ${}_5C_3 = {}_4C_3 + {}_4C_2$ (b) ${}_6C_1 = {}_5C_1 + {}_5C_0$

36. (a) ${}_{10}C_5 = {}_9C_5 + {}_9C_4$ (b) ${}_{11}C_7 = {}_{10}C_7 + {}_{10}C_6$

Binomial Theorem

Use the binomial theorem to expand each of the following.

37. $(x + y)^4$

38. $(x - y)^4$

39. $(2x - 3y)^4$

40. $(3x + 5)^4$

41. $(x + 2)^6$

42. $\left(y - \dfrac{1}{2}\right)^6$

43. $\left(2x + \dfrac{3}{4}\right)^5$

44. $\left(3x - \dfrac{1}{3}\right)^5$

Find the coefficient of x^6 in each of the following.

45. $(x + 1)^8$

46. $(2x - 1)^8$

47. $(x^2 + 1)^8$

48. $\left(x + \dfrac{1}{x}\right)^8$

49. $\left(x^2 + \dfrac{2}{x}\right)^6$

50. $(x^3 - 2x)^6$

51. Evaluate $(1.1)^5$ by means of the binomial theorem. (*Hint:* $1.1 = 1 + .1$.)

52. Use the binomial theorem to evaluate $(.9)^5$. (*Hint:* see Exercise 51.)

As in Exercises 51 and 52, use the binomial theorem to evaluate each of the following.

53. $(1.01)^4$

54. $(.99)^4$

55. $(9.8)^3$

56. $(10.2)^3$

57. Show that ${}_nC_0 + {}_nC_1 + {}_nC_2 + \ldots + {}_nC_n = 2^n$. (*Hint:* let $a = b = 1$ in the binomial theorem.)

58. Show that the sum of the coefficients in the expansion of $(a - b)^n$ is zero. (*Hint:* see Exercise 57.)

59. Expand $(x + y + z)^3$. (*Hint:* let $x + y = a$ and expand $(a + z)^3$. Then replace a by $x + y$ and complete the expansion.)

60. Use the method from Exercise 59 to expand $(1 + x + x^2)^4$.

Miscellaneous

*The formula for **compound interest** (see Exercise 106 in Section P.5) is $A = P\left(1 + \dfrac{R}{N}\right)^{Nt}$, where A is the amount accumulated in t years when P dollars are invested, interest is compounded N times per year, and R is the annual percentage interest rate divided by* 100.

In each of the following cases, use the binomial theorem to find A.

61. $P = \$1$, $N = 4$, $t = 1$, and the annual interest rate is 4%. (*Hint:* see Exercise 53.)

62. $P = \$1$, $N = 2$, $t = 2$, and the annual interest rate is 12%.

63. $P = \$100$, $N = 1$, $t = 3$, and the annual interest rate is 8%.

64. $P = \$100$, $N = 6$, $t = .5$, and the annual interest rate is 6%.

*It is shown in the last chapter of the text that ${}_nC_r$ is the number of ways of selecting r objects from a collection of n distinct objects, called the number of **combinations** of n distinct objects taken r at a time. For example, from the set* {a, b, c, d}, *2 letters can be selected in 6 ways, namely* ab, ac, ad, bc, bd, cd, *and* ${}_4C_2 = 4!/(2!2!) = 6$. *Use this interpretation of ${}_nC_r$ to do each of the following exercises.*

65. From a group of 30 students, in how many ways can a committee of 3 be selected?

66. In how many ways can a jury of 12 be selected from a panel of 25 persons?

67. If a student must answer 8 out of 10 questions on a test, how many choices are possible?

68. In how many ways can a student choose 4 required courses out of a set of 9 possible courses?

69. If 2 out of every 100 items on a production line are inspected, in how many ways can the 2 items be selected?

70. A fair coin is tossed 5 times. In how many ways can exactly 2 heads result? (*Hint:* number the tosses 1, 2, 3, 4, 5.)

Chapter 1 Review Outline

1.1 Operations with Polynomials

Definitions

A polynomial $P(x)$ of degree n has the form

$$a_n x^n + a_{n-1}x^{n-1} + \dots + a_1 x + a_0,$$

where n is a nonnegative integer and $a_n \neq 0$.

A zero of $P(x)$ is a root of the equation $P(x) = 0$.

Operations

To add or subtract polynomials, combine similar terms.

To multiply polynomials P and Q, multiply each term of P by each term of Q and proceed as in addition.

dividend = quotient · divisor + remainder

$$\frac{\text{dividend}}{\text{divisor}} = \text{quotient} + \frac{\text{remainder}}{\text{divisor}}$$

Square of a Binomial:

$(a + b)^2 = a^2 + 2ab + b^2$

$(a - b)^2 = a^2 - 2ab + b^2$

Difference of Squares:

$(a - b)(a + b) = a^2 - b^2$

Caution: $(a \pm b)^2 \neq a^2 \pm b^2$

$\sqrt{a^2 \pm b^2} \neq a \pm b$

Cube of a Sum:

$(a + b)^3 = a^3 + 3a^2b + 3ab^2 + b^3$

1.2 Factoring Polynomials

Perfect Squares:

$a^2x^2 + 2abx + b^2 = (ax + b)(ax + b)$

$a^2x^2 - 2abx + b^2 = (ax - b)(ax - b)$

Difference of Squares:

$a^2x^2 - b^2 = (ax - b)(ax + b)$

Factoring a Quadratic:

$$x^2 + Bx + C = (x + m)(x + n)$$

↑ $m + n$ ↑ mn

$$Ax^2 + Bx + C = Ax^2 + Mx + Nx + C,$$
$$M + N = B, MN = AC$$

(finish by grouping)

Sum and Difference of Cubes:

$x^3 + a^3 = (x + a)(x^2 - ax + a^2)$

$x^3 - a^3 = (x - a)(x^2 + ax + a^2)$

1.3 Operations with Rational Expressions

Definitions

A rational expression is the ratio of two polynomials.

The least common multiple (LCM) of a set of polynomials is the polynomial of smallest degree that has for its only factors the factors of each of the polynomials in the set.

The least common denominator (LCD) of a set of rational expressions is the LCM of their denominators.

Operations

To add or subtract rational expressions, first factor each denominator and determine the LCD. Second, replace each rational expression by an equivalent one with the LCD. Third, proceed as in addition and subtraction of simple fractions.

Multiplication and division of rational expressions follow the same rules as in the case of simple fractions.

1.4 Rational Exponents

Definitions for Positive Integers *n*, *m*

$x^{1/n} = \sqrt[n]{x} \quad (n = 2, 3, 4, \dots)$

(If n is even, then x must be greater than or equal to zero and $x^{1/n}$ is the nonnegative nth root.)

$x^{m/n} = (x^{1/n})^m$

(m and n in lowest terms)

$x^{-m/n} = 1/x^{m/n} \quad (x \neq 0)$

Rules for Rational Exponents

In the following, m and p can be positive or negative integers; n and q are positive integers.

Power-Root Rule:

$(x^m)^{1/n} = (x^{1/n})^m$

Same-Base Rules:

$(x^{m/n})(x^{p/q}) = x^{m/n+p/q}$

$$\frac{x^{m/n}}{x^{p/q}} = x^{m/n-p/q}$$

Same-Exponent Rules: $(x \cdot y)^{m/n} = (x^{m/n})(y^{m/n})$

$$\left(\frac{x}{y}\right)^{m/n} = \frac{x^{m/n}}{y^{m/n}}$$

Power-of-a-Power Rule: $(x^{m/n})^{p/q} = x^{(m/n)(p/q)}$

(The usual restrictions against dividing by 0 and taking even roots of negative numbers apply to all of the above.)

1.5 Operations with Radicals

To add, subtract, multiply, or divide radicals, we can convert to rational exponents, perform the operations, and then convert back to radicals.

When adding radicals, it is sometimes helpful to rationalize the denominators.

1.6 The Binomial Theorem

$$(a+b)^n = a^n + na^{n-1}b + \ldots + {}_nC_r a^{n-r}b^r + \ldots + b^n,$$

where

$${}_nC_r = \frac{n!}{(n-r)!r!} = \frac{n(n-1)\ldots(n-r+1)}{r!}$$

(n a positive integer, $r = 0, 1, \ldots, n$)

Also, Pascal's triangle can be used to determine ${}_nC_r$.

Properties of Binomial Coefficients:

${}_nC_0 = 1 = {}_nC_n$

${}_nC_r = {}_{n-1}C_r + {}_{n-1}C_{r-1}$

${}_nC_{n-r} = {}_nC_r$

Chapter 1 Review Exercises

1. Perform the operations $(3ab + 2a - 1) - (4a + 6b^2 + 8) + (5ab - 2a + 6b^2 + 10)$.
2. Multiply: $3x(x^2 - 2)(x^2 + 2)(x^4 + 4)$.
3. Divide $x^3 - 3x^2 + 5$ by $x + 3$ and express the result in two equivalent ways.
4. Find the quotient and remainder when $x^5 + 5x^3 + 10x - 3$ is divided by $x^2 + 2$.
5. Expand $(2x + 1)^2$ and $(2x - 1)^3$.
6. If $P(x) = x^3 - 2x^2 + 5$, find $P(-2)$ and $P(\sqrt{2})$.
7. Verify that $\sqrt{2} - 2$ is a zero of $x^2 + 4x + 2$.
8. Explain why the polynomial $P(x) = x^6 + 3x^4 + 4x^2 + 1$ has no zeros in the real number system.

In Exercises 9–14, factor by integers as much as possible.

9. $16x^2 - 49y^2$
10. $5x^2 - 30x + 45$
11. $6x^2 + 11x - 10$
12. $2x^3 - 16y^3$
13. $5x^2 - 30x + 2x - 12$
14. $a^4 + 2a^2 + 1 - b^4$

15. Show that $x - 2$ is a factor of $x^5 - 32$.

In Exercises 16 and 17, factor by real numbers.

16. $2x^2 - 6$
17. $x^2 + 2.1x + .2$

18. Find the LCM of $6x + 6$, $x^2 + 2x + 1$, and $8x^2 - 8$.
19. Subtract by using the LCD and then simplify:
$$\frac{x+1}{x^2+x-2} - \frac{x}{x^2+2x-3}.$$
20. Combine by using the LCD and then simplify:
$$\frac{x}{x^2-4} + \frac{3x}{x-2} - \frac{x-2}{x+2}.$$
21. Perform the indicated operations and simplify:
$$\frac{2x^2-18}{x^2+2x-15} \cdot \frac{3x+15}{4x+12} \div \frac{6x+3}{4x^2-1}.$$
22. Simplify $\left\{\frac{2}{x} + \frac{2}{1-x} + \frac{1}{(x-1)^2}\right\} \div \left\{\frac{3}{x-1} - \frac{3}{x}\right\}$.
23. Evaluate each expression as a real number if possible.
(a) $16^{3/2}$ (b) $16^{-3/2}$ (c) $(-16)^{3/2}$
(d) $(-16)^0$ (e) 0^{-16}
24. Simplify and then evaluate:
$$2^{1/2} \cdot 5^{3/4} \cdot 2^{-1/3} \cdot 10^{5/6} \div 5^{7/12}.$$
25. Evaluate $64^{-2/3}$ in three different ways.
26. Now $[(-5)^2]^{1/2} = 25^{1/2} = 5$, whereas $(-5)^{2/2} = (-5)^1 = -5$, and $[(-5)^{1/2}]^2$ is undefined. Explain why these results do not violate the rule $(x^m)^{1/n} = x^{m/n} = (x^{1/n})^m$.

27. Simplify $(x^{3/2}y^3)^{1/3} + (9xy^2)^{1/2}$.

28. Simplify $\dfrac{x^{2/3} + x^{5/3}}{1 + x}$.

29. Simplify $5\sqrt{9a^4} - 3\sqrt[4]{16a^8}$.

30. Simplify $\sqrt{\dfrac{x + y}{x - y}} \cdot \dfrac{(\sqrt{x - y})^5}{(\sqrt{x + y})^3}$.

31. Rationalize the denominator of

$$\frac{\sqrt{x + 2}}{\sqrt{x + 2} - \sqrt{2}}.$$

32. Rationalize the denominators and combine:

$$\frac{1}{\sqrt{x} - 1} - \frac{1}{\sqrt{x} + 1} + \frac{1}{x - 1}.$$

33. Construct the first 6 rows of Pascal's triangle, and use the triangle to expand $(a + b)^n$ for $n = 2, 3, 4, 5$.

34. Evaluate: (a) ${}_7C_2$ (b) ${}_5C_0$ (c) ${}_{10}C_3$.

35. Verify that ${}_8C_4 = {}_7C_4 + {}_7C_3$.

36. Expand the binomial theorem:

    (a) $(x + 2)^5$ (b) $(x - 3)^4$.

37. Use the binomial theorem to expand $(x - y + 1)^4$. (*Hint:* let $a = x - y$.)

Calculator Exercises

In each of the following, use a calculator to verify your answers to the indicated review exercises of this section.

38. Exercise 6. *Hint:* Your calculator will give a numerical answer for $P(\sqrt{2})$. Check that it is the calculator value of $1 + 2\sqrt{2}$.

39. Exercise 7

40. Exercise 23. For (b), see the hint in Exercise 38.

41. Exercise 24

42. Exercise 34

43. Exercise 35

44. Exercise 33. Find the coefficients for $(a + b)^6$ by using your calculator to compute ${}_nC_r$ instead of using Pascal's triangle.

45. Exercise 36. As in Exercise 44, obtain the coefficients on your calculator.

2 Algebra and Graphs of Linear Expressions

Chapters P and 1 were devoted to manipulating various types of algebraic expressions by means of the basic rules. In this chapter we concentrate on **linear expressions,** which are polynomials of degree 1 in one or more variables. Specifically, we consider

$$(1)\ ax + b \quad (a \neq 0)$$

and

$$(2)\ ax + by + c \quad (a \neq 0 \text{ or } b \neq 0),$$

where the coefficients a, b, and c are constant real numbers, and x and y are the variables.

From Chapters P and 1 we know how to manipulate polynomials algebraically. Our main concern here is with the *zeros* of these expressions. We will see that expression (1) has a single point on the real line for its zero, whereas the zeros of expression (2) form a straight line in the plane. Hence, from a geometric point of view, this chapter is devoted to points and lines.

2.1 Linear Equations

Solving equations is the most fundamental process in algebra.

Equations. We have been working with equations since Section P.1, but without the aid of a formal definition. We now define an **equation** as a statement that one algebraic expression is equal to another; for example,

$$5x - 1 = 3x + 6.$$

Here x is called the **unknown quantity,** and to **solve** the equation means to find all values of x for which the statement is true. We proceed as follows:

$$\begin{aligned} 5x - 1 + 1 &= 3x + 6 + 1 && \textit{Add 1 to both sides} \\ 5x &= 3x + 7 \\ 5x - 3x &= 3x + 7 - 3x && \textit{Subtract } 3x \textit{ from both sides} \\ 2x &= 7 \\ \frac{2x}{2} &= \frac{7}{2} && \textit{Divide both sides by 2} \\ x &= \frac{7}{2}. && \textit{Solution} \end{aligned}$$

Check by substituting $7/2$ for x in the original equation:

$$5\left(\frac{7}{2}\right) - 1 = \frac{35}{2} - 1 = \frac{33}{2} \quad \text{and} \quad 3\left(\frac{7}{2}\right) + 6 = \frac{21}{2} + 6 = \frac{33}{2}.$$

The method for solving the previous equation is based on the following algebraic properties of equality, which were introduced in Section P.2. Here M and N are any two algebraic expressions.

Algebraic Properties of Equality

$M = N \Leftrightarrow M \pm c = N \pm c$ **for any value of** c.	Adding or subtracting a quantity on both sides of an equation results in an equivalent equation.
If $c \neq 0$, **then** $M = N \Leftrightarrow \begin{cases} M \cdot c = N \cdot c \\ \text{and} \\ \dfrac{M}{c} = \dfrac{N}{c}. \end{cases}$	Multiplying or dividing on both sides by a *nonzero* quantity results in an equivalent equation.

For a given equation in x, the object is to utilize the above properties in order to reduce the given equation to one in which all the terms involving the unknown x are on one side and those not involving x are on the other. Since each step in this process is reversible, the reduced form of the equation and the original one are by definition **equivalent.** Equivalent equations have the same solution.

Example 1 Solve $x + 8 = 4(x - 1)$.

Solution:

$x + 8 = 4x - 4$ *Apply distributive property on right*

$x + 8 - 8 = 4x - 4 - 8$ *Subtract 8 from both sides*

$x = 4x - 12$

$x - 4x = 4x - 12 - 4x$ *Subtract 4x from both sides*

$-3x = -12$

$\dfrac{-3x}{-3} = \dfrac{-12}{-3}$ *Divide both sides by −3*

$x = 4$ *Solution*

Check: $4 + 8 = 12$ and $4(4 - 1) = 12$. ■

Comment Technically, the solution in Example 1 is 4, not $x = 4$. The statement $x = 4$ is an *equation* that is equivalent to the given equation $x + 8 = 4(x - 1)$. When we say "the solution is $x = 4$," we mean that 4 is the value of the unknown x that satisfies the equation.

Example 2 Solve $3x + \frac{1}{2} = \frac{x}{2} + \frac{1}{3}$.

Solution:

$$6\left(3x + \frac{1}{2}\right) = 6\left(\frac{x}{2} + \frac{1}{3}\right) \qquad \textit{Multiply both sides by 6 to clear the fractions}$$

$$18x + 3 = 3x + 2$$

$$18x + 3 - 3 = 3x + 2 - 3 \qquad \textit{Subtract 3 from both sides}$$

$$18x = 3x - 1$$

$$18x - 3x = 3x - 1 - 3x \qquad \textit{Subtract } 3x \textit{ from both sides}$$

$$15x = -1$$

$$\frac{15x}{15} = \frac{-1}{15} \qquad \textit{Divide both sides by 15}$$

$$x = -\frac{1}{15} \qquad \textit{Solution}$$

Check: $3\left(\frac{-1}{15}\right) + \frac{1}{2} = -\frac{1}{5} + \frac{1}{2} = \frac{3}{10}$

and $\dfrac{\frac{-1}{15}}{2} + \frac{1}{3} = -\frac{1}{30} + \frac{1}{3} = \frac{3}{10}$. ■

The Linear Form. The equations in the previous examples are called **linear equations in x** because each is equivalent to an equation of the form

$$ax + b = 0 \quad (a \neq 0).$$

For instance, Example 1 is equivalent to $-3x + 12 = 0$, and Example 2 is equivalent to $(5/2)x + 1/6 = 0$. We will see later why the word "linear" is used to describe these equations. For a given equation $ax + b = 0$, there are three possibilities:

(a) If the equation is linear ($a \neq 0$), then the solution is $-b/a$.

(b) If $a = 0$ and $b \neq 0$, then the equation has *no* solution (*why?*).

(c) If $a = 0$ and $b = 0$, then *every real number* is a solution (*why?*).

Because of the extreme nature of the "solutions" in cases (b) and (c), equations of the form $0 \cdot x + b = 0$ are not called linear equations.

Example 3 Solve $c(x - 1) = x + 1$ for x in terms of c.

Solution:

$$cx - c = x + 1 \qquad \textit{Apply distributive property on left}$$

$$cx - c + c = x + 1 + c \qquad \textit{Add } c \textit{ to both sides}$$

$$cx = x + c + 1$$

$$cx - x = x + c + 1 - x \qquad \textit{Subtract } x \textit{ from both sides}$$

$$(c - 1)x = c + 1 \qquad \textit{Factor out } x \textit{ on left side}$$

If $c \neq 1$, then the solution is $(c+1)(c-1)$, but if $c = 1$, then there is no solution. ■

We now consider several types of equations that lead to linear equations.

Equations with Fractional Expressions. If an equation contains fractional expressions in x, we can first clear the fractions by multiplying both sides by the LCM of the denominators. Then, if the resulting equation is linear, we can proceed as before.

Caution Any value of x obtained as a solution after clearing fractions must be tested in the original equation. If it makes a denominator equal to zero, the value is a false solution called an **extraneous root,** and it must be discarded.

Example 4 Solve $\dfrac{5}{x+1} - \dfrac{3}{x} = \dfrac{1}{x(x+1)}$.

Solution:

$$x(x+1)\left(\frac{5}{x+1} - \frac{3}{x}\right) = \frac{\cancel{x(x+1)}}{\cancel{x(x+1)}} \qquad \textit{Multiply both sides by } x(x+1)$$

$$\frac{5x\cancel{(x+1)}}{\cancel{x+1}} - \frac{3\cancel{x}(x+1)}{\cancel{x}} = 1$$

$$5x - 3(x+1) = 1$$

The linear equation $5x - 3(x+1) = 1$ and the original fractional equation are equivalent provided $x \neq 0$ or -1 (*why?*). We can now solve the linear equation as before.

$$5x - 3x - 3 = 1$$
$$2x = 4$$
$$x = 2$$

Since $2 \neq 0$ or -1, 2 is the solution of the original equation. ■

Example 5 Solve $\dfrac{2x}{x-1} + \dfrac{x+1}{x-2} = \dfrac{3x-1}{x-1}$.

Solution: Multiply both sides by $(x-2)(x-1)$ and then simplify.

$$(x-2)(x-1)\left(\frac{2x}{x-1} + \frac{x+1}{x-2}\right) = (x-2)(x-1)\left(\frac{3x-1}{x-1}\right)$$

$$\frac{(x-2)\cancel{(x-1)}2x}{\cancel{x-1}} + \frac{\cancel{(x-2)}(x-1)(x+1)}{\cancel{x-2}} = \frac{(x-2)\cancel{(x-1)}(3x-1)}{\cancel{x-1}}$$

$$2x^2 - 4x + x^2 - 1 = 3x^2 - 7x + 2$$
$$3x^2 - 4x - 1 = 3x^2 - 7x + 2$$
$$-4x - 1 = -7x + 2$$
$$3x = 3$$
$$x = 1$$

The solution 1 of the linear equation $-4x - 1 = -7x + 2$ is an extraneous root of the original equation. Therefore, the original equation has no solution. ■

Example 6 Solve $\dfrac{x}{x-1} = \dfrac{1}{x-1} + 1$.

Solution:

$$(x-1)\left(\frac{x}{x-1}\right) = (x-1)\left(\frac{1}{x-1} + 1\right) \qquad \textit{Multiply both sides by } x - 1$$

$$\frac{\cancel{(x-1)}x}{\cancel{x-1}} = \frac{\cancel{x-1}}{\cancel{x-1}} + x - 1$$

$$x = 1 + x - 1$$

$$x = x$$

The equation $x = x$ is true for all real numbers x. However, the original equation is meaningful only if $x \neq 1$. Therefore, the solution of the original equation consists of all real numbers except 1. ■

Equations with Radicals. The equation $\sqrt{2x+3} = 5$ can be transformed into a linear equation by squaring both sides.

$$(\sqrt{2x+3})^2 = 5^2$$

$$2x + 3 = 25$$

$$x = 11$$

Check: $\sqrt{2 \cdot 11 + 3} = \sqrt{25} = 5$.

Caution Squaring both sides of an equation is not a reversible step and may lead to false solutions (extraneous roots). For example, $\sqrt{2x+3} = -5$ also becomes $2x + 3 = 25$ when both sides are squared, but in this case 11 is not a solution. Because of the possibility of introducing extraneous roots, all values obtained by squaring must be tested for acceptability by substitution in the *original* equation.

Example 7 Solve $\sqrt{x^2+4} + x = 1$.

Solution:

$$\sqrt{x^2+4} = 1 - x \qquad \textit{Isolate the radical by subtracting } x \textit{ from both sides}$$

$$x^2 + 4 = 1 - 2x + x^2 \qquad \textit{Square both sides}$$

$$2x = -3$$

$$x = -\frac{3}{2}$$

Check: We have $\sqrt{(-3/2)^2 + 4} + (-3/2) = \sqrt{9/4 + 4} - 3/2 = \sqrt{25/4} - 3/2 = 1$, so $-3/2$ is a solution. ■

Example 8 Solve $\sqrt{x^2 - 8} + x = 2$.

Solution:

$\sqrt{x^2 - 8} = 2 - x$	*Isolate the radical by subtracting x from both sides*
$x^2 - 8 = 4 - 4x + x^2$	*Square both sides*
$4x = 12$	
$x = 3$	

Check: $\sqrt{3^2 - 8} + 3 = 4 \neq 2$; therefore, 3 is an extraneous root, and the original equation has no solution. ■

Example 9 Solve $\sqrt{x + 1} + \sqrt{x - 1} = 2$.

Solution:

$\sqrt{x + 1} = 2 - \sqrt{x - 1}$	*Isolate $\sqrt{x + 1}$ on left*
$x + 1 = 4 - 4\sqrt{x - 1} + x - 1$	*Square both sides*
$4\sqrt{x - 1} = 2$	*Isolate $4\sqrt{x - 1}$ on left*
$\sqrt{x - 1} = \frac{1}{2}$	
$x - 1 = \frac{1}{4}$	*Square again*
$x = \frac{5}{4}$	

Check: $\sqrt{5/4 + 1} + \sqrt{5/4 - 1} = \sqrt{9/4} + \sqrt{1/4} = 3/2 + 1/2 = 2$; therefore, 5/4 is a solution. ■

Two squarings were necessary in Example 9. The general procedure for solving equations with radicals is to eliminate one radical at a time by isolating it on one side of the equation and then squaring both sides.

Equations with Absolute Values. The solution to $|x| = 3$ is $x = 3$ or $x = -3$. In general, if M is an algebraic expression in x and $c \geq 0$, then the solution to $|M| = c$ is the set of all real numbers x for which either $M = c$ or $M = -c$.

Example 10 Solve $|2x + 3| = 4$.

Solution:

$2x + 3 = 4$	or	$2x + 3 = -4$	*Each equation is solved separately*
$2x = 1$		$2x = -7$	
$x = \frac{1}{2}$ (1)		$x = -\frac{7}{2}$ (2)	

Check: (1) $|2(1/2) + 3| = |4| = 4$, so $1/2$ is a solution;
(2) $|2(-7/2) + 3| = |-4| = 4$, so $-7/2$ is also a solution. ■

Similarly, if M and N are algebraic expressions in x, then the solution to $|M| = |N|$ consists of all real numbers x for which either $M = N$ or $M = -N$.

Example 11 Solve $|3x - 4| = |x - 6|$.

Solution:

$$\begin{aligned} 3x - 4 &= x - 6 \\ 2x &= -2 \\ x &= -1 \quad (1) \end{aligned} \qquad \text{or} \qquad \begin{aligned} 3x - 4 &= -(x - 6) \\ 3x - 4 &= -x + 6 \\ 4x &= 10 \\ x &= \frac{5}{2} \quad (2) \end{aligned}$$

Check: (1) $|3(-1) - 4| = |-7| = 7$ and $|-1 - 6| = |-7| = 7$;
(2) $|3 \cdot 5/2 - 4| = |7/2| = 7/2$ and $|5/2 - 6| = |-7/2| = 7/2$.

Therefore, both -1 and $5/2$ are solutions. ∎

The method of solution in each of the previous examples is based on the following property of **absolute value** that was introduced in Section P.3.

If M is an algebraic expression in x, then

$$|M| = \begin{cases} M \text{ for all } x \text{ in which } M \geq 0, \\ -M \text{ for all } x \text{ in which } M < 0. \end{cases}$$

In each example, the above property was used to replace one equation with absolute values by two separate equations not involving absolute values. For instance, the single equation $|M| = c$ becomes $M = c$ or $-M = c$; equivalently, $M = c$ or $M = -c$.

The general procedure for solving an equation with absolute values is to eliminate one absolute value $|M|$ at a time by replacing the equation with two auxiliary equations, one with M in place of $|M|$ and the other with $-M$ in place of $|M|$. If there is another absolute value $|N|$ in the original equation, then four auxiliary equations result, corresponding to M, N; $M, -N$; $-M, N$; and $-M, -N$, respectively. Three absolute value terms $|M|$, $|N|$, and $|P|$ would generate eight auxiliary equations (*why?*), and so on, until all absolute value terms have been eliminated. *The final auxiliary equations are solved separately, and all solutions must be tested in the original equation to eliminate extraneous roots.*

Example 12 Solve $|x + 1| + |2x - 5| = 4$.

Solution: This equation generates the four auxiliary equations (a), (b), (c), and (d) that are listed on the next page. We solve each of these and check our answers in the original equation.

(a) $x + 1 + 2x - 5 = 4$
$$3x - 4 = 4$$
$$3x = 8$$
$$x = \frac{8}{3}$$

Check: $|8/3 + 1| + |2(8/3) - 5| = |11/3| + |1/3| = 12/3 = 4$.
Hence, $8/3$ *is* a solution.

(b) $x + 1 - (2x - 5) = 4$
$$-x + 6 = 4$$
$$-x = -2$$
$$x = 2$$

Check: $|2 + 1| + |2 \cdot 2 - 5| = |3| + |-1| = 4$.
Hence, 2 *is* a solution.

(c) $-(x + 1) + 2x - 5 = 4$
$$x - 6 = 4$$
$$x = 10$$

Check: $|10 + 1| + |2 \cdot 10 - 5| = |11| + |15| = 26 \neq 4$.
Therefore, 10 is *not* a solution.

(d) $-(x + 1) - (2x - 5) = 4$
$$-3x + 4 = 4$$
$$-3x = 0$$
$$x = 0$$

Check: $|0 + 1| + |2 \cdot 0 - 5| = |1| + |-5| = 6 \neq 4$.
Therefore, 0 is *not* a solution.

Our checks tell us to accept $8/3$ and 2 and to reject 10 and 0. Therefore, $8/3$ and 2 are the only solutions to the original equation. ■

Exercises 2.1

Fill in the blanks to make each statement true.

1. A statement that one algebraic expression is equal to another is called an ____________.
2. If one equation can be transformed into another by a series of reversible algebraic steps, then the two equations are said to be ____________.
3. An equation $ax + b = 0$, where $a \neq 0$, is called a ____________ equation.
4. The solution to $ax + b = 0$, with $a \neq 0$, is ____________.
5. If $a = 0$, then the solution to $ax + b = 0$ is ____________ if $b = 0$, and ____________ if $b \neq 0$.

Write true *or* false *for each statement.*

6. The equation $M = N$ is equivalent to the equation $M + c = N + c$, for any real number c.
7. The equation $M = N$ is equivalent to the equation $Mc = Nc$, for any real number c.
8. The equation $M = N$ is equivalent to the equation $M - N = 0$.
9. The equation $M = N$ is equivalent to the equation $M/N = 1$.
10. The equations $M = N$ and $M^2 = N^2$ are equivalent.

Linear Equations

Solve each of the following equations.

11. $3x + 5 = 11$
12. $2x - 6 = 7$
13. $2x - 5 - (x + 5) = 0$
14. $3x + 11 - (6x - 11) = 0$
15. $\frac{2}{3}x - 7 = x + \frac{5}{2}$
16. $\frac{3}{4}x + \frac{1}{2} = 5x - \frac{1}{4}$
17. $2.4(x - 1) - 4.2 = 1.8 - 3.6(x - 2)$
18. $5.2(2x + 1) - 3.2 = -2.2(3x + 4) + 3.8$

19. $\frac{2}{5}(x-1)+\frac{5}{2}(x+2)=\frac{3}{4}(x-2)$

20. $\frac{5}{6}(2x+3)-\frac{2}{3}(2x-1)=\frac{1}{2}(x+2)$

21. $3x+4-(x+1)=2x+3$

22. $5(x-2)+2(3x-1)=4(x-3)+7x$

23. $2x+4=4(x-1)-(2x-3)$

24. $5x+2(x-1)=7(x-1)$

25. $\sqrt{3}x+2(x-4)=5x+\sqrt{3}$

26. $\sqrt{2}(3x-1)+4x=\sqrt{2}(2x-3)+1$

27. $\frac{2}{3}(x+\sqrt{5})+\frac{1}{3}(x-\sqrt{5})=1$

28. $\frac{1}{2}(x+\sqrt{2})+\frac{1}{4}(\sqrt{3}x-1)=0$

29. $x+4[x-2(x+7)]=5[1+2(x-1)]$

30. $3x-2[5-2(x-1)]=2-3(x+1)$

Solve for x in terms of the remaining constants and variables.

31. $cx+5=2x-7\quad(c\neq 2)$

32. $3x+c=cx-4\quad(c\neq 3)$

33. $c^2x+5=25x+c$

34. $c^2x-3=9x-c$

35. $ax+by+c=0\quad(a\neq 0)$

36. $\frac{x}{a}+\frac{y}{b}+\frac{1}{c}=0\quad(a\neq 0,\ b\neq 0,\ c\neq 0)$

37. $a(x+b)+b(x+a)=0\quad(a\neq -b)$

38. $a(x-b)+b(x-a)=1\quad(a\neq -b)$

39. $\frac{x}{a}+\frac{y}{b}=\frac{x+y}{a+b}\quad(a\neq 0,\ b\neq 0,\ a\neq -b)$

40. $\frac{x}{a}-\frac{y}{b}=\frac{x-y}{ab}\quad(a\neq 0,\ b\neq 0)$

Fractional Equations

Solve each of the following equations.

41. $\frac{1}{2x+3}=1$

42. $\frac{7}{3x-5}=2$

43. $\frac{3}{2x-4}-1=0$

44. $\frac{1}{x+1}-2=0$

45. $5-\frac{2}{3x+1}=0$

46. $4-\frac{6}{x-1}=0$

47. $\frac{1}{x+3}=\frac{2}{3x+1}$

48. $\frac{3}{2x-5}=\frac{1}{2x+5}$

49. $x+\frac{3}{x-1}=x-\frac{2}{x-1}$

50. $x+\frac{1}{x}=x-\frac{1}{x}$

51. $\frac{1}{x}+\frac{1}{x-2}=\frac{3}{x}-\frac{2}{x-2}$

52. $\frac{4}{4x-1}-\frac{5}{4x+1}=\frac{2}{4x-1}-\frac{1}{4x+1}$

53. $\frac{1}{x-3}+\frac{1}{x+3}=\frac{2}{x^2-9}$

54. $\frac{2}{x-5}-\frac{1}{x+5}=\frac{1}{x^2-25}$

55. $\frac{4}{x^2-4}=\frac{1}{x-2}-\frac{1}{x+2}$

56. $\frac{3x+5}{x^2+3x+2}=\frac{1}{x+2}+\frac{2}{x+1}$

57. $\frac{1}{x+3}-\frac{2}{x-2}=\frac{x+1}{x^2+x-6}$

58. $\frac{x}{x-5}-\frac{x+1}{x-4}=\frac{2x+3}{x^2-9x+20}$

59. $\frac{1}{x^2-5x+6}=\frac{x-1}{x-3}-\frac{x}{x-2}$

60. $\frac{3x+2}{x^2+6x+8}=\frac{x+1}{x+2}-\frac{x}{x+4}$

61. $\frac{4}{2x^2+7x+3}-\frac{3}{2x+1}+\frac{2}{x+3}=0$

62. $\frac{10}{5x^2+x-4}+\frac{x}{x+1}-\frac{5x}{5x-4}=0$

63. $\frac{1}{x-\frac{1}{x}}=\frac{4}{1+\frac{1}{x}}$

64. $\frac{2x}{x+1-\frac{1}{x+1}}=\frac{2}{1+\frac{2}{x}}$

65. $\frac{\frac{1}{x+1}-\frac{1}{x+4}}{\frac{x}{x^2+5x+4}}=\frac{1}{x+2}$

66. $\frac{\frac{1}{(x+2)(x+3)}}{\frac{1}{x+2}+\frac{1}{x+3}}=\frac{1}{5x+2}$

67. $\frac{1+(x+2)^{-1}}{1-(x+2)^{-1}}=\frac{x+3}{x+1}$

68. $\dfrac{(x+1)^{-1}+(x+2)^{-1}}{(x+2)^{-1}+(x+3)^{-1}}=\dfrac{x+2}{x+3}$

69. $\dfrac{\dfrac{1}{x+1-\dfrac{2}{x+2}}}{\dfrac{4}{x+3}}=1+\dfrac{4}{x}+\dfrac{4}{x^2}$

70. $\dfrac{\dfrac{-1}{(x+2)^{-1}+(x+3)^{-1}}}{\dfrac{1}{2x+5}}=x+1+\dfrac{2}{x+4}$

Solve for x in terms of the remaining constants and variables.

71. $y=\dfrac{x}{x+1}$

72. $y=\dfrac{x-1}{x}$

73. $\dfrac{1}{x}+\dfrac{1}{y}=1$

74. $\dfrac{a}{x}+\dfrac{b}{y}=1 \quad (a\neq 0,\ b\neq 0)$

75. $\dfrac{a}{x}-\dfrac{b}{x}=c \quad (a\neq b,\ c\neq 0)$

76. $\dfrac{x}{x+a}+\dfrac{y}{x+a}=c \quad (c\neq 1)$

77. $\dfrac{1}{x+a}+\dfrac{1}{x-a}=\dfrac{a}{x^2-a^2} \quad (a\neq 0)$

78. $\dfrac{a}{x+y}-\dfrac{b}{x-y}=\dfrac{1}{x^2-y^2} \quad (a\neq b)$

79. $\dfrac{x+y}{x-y}-\dfrac{x-y}{x+y}=\dfrac{1}{x-y}-\dfrac{1}{x+y}$

80. $\dfrac{x+y}{x-y}-\dfrac{x-y}{x+y}=\dfrac{2y}{x-y}+\dfrac{2y}{x+y}$

Equations with Radicals

Solve each equation.

81. $\sqrt{2x+1}=4$

82. $\sqrt{3x-2}=5$

83. $\sqrt{5x-1}+2=0$

84. $\sqrt{2x-3}+1=0$

85. $\sqrt{5-2x}-\sqrt{9-x}=0$

86. $\sqrt{4x+11}-\sqrt{1-x}=0$

87. $\sqrt{3x+2}-\sqrt{3x-1}=1$

88. $\sqrt{5x+8}-\sqrt{5x+3}=1$

89. $\sqrt{2x+6}+\sqrt{2x+1}=5$

90. $\sqrt{2x+6}-\sqrt{2x+1}=5$

91. $5+\sqrt{x^2-4}=3-x$

92. $\sqrt{x^2+2x+2}=3-x$

93. $x+\sqrt{4x^2+4}=3x-2$

94. $3+\sqrt{x^2+2x+2}=1+x$

95. $3x+\sqrt{4x^2+4}=x+2$

96. $3+\sqrt{x^2-4}=5+x$

97. $\dfrac{\sqrt{x^2+1}}{x+1}-\dfrac{1}{x+1}=1$

98. $\dfrac{\sqrt{x^2+1}}{x+1}-\dfrac{1}{x+1}=1$

99. $\sqrt{x+1}-\dfrac{x}{\sqrt{x+1}}=2$

100. $\sqrt{x+1}+2=\dfrac{x}{\sqrt{x+1}}$

101. $\dfrac{x}{\sqrt{x^2+1}}+\dfrac{1}{x\sqrt{x}}=\dfrac{\sqrt{x^2+1}}{x}$

102. $\dfrac{1}{\sqrt{x^2+x+1}}+\dfrac{1}{x}=\dfrac{2}{x\sqrt{x^2+x+1}}$

Equations with Absolute Values

Solve each equation.

103. $|x-2|=3$

104. $|x+1|=5$

105. $|3x-7|=5$

106. $|4x+3|=8$

107. $|x+2|+3=0$

108. $|2x-1|+4=0$

109. $\left|\dfrac{2}{3}x+7\right|=\dfrac{5}{6}$

110. $\left|\dfrac{4}{5}x-\dfrac{1}{5}\right|=\dfrac{3}{4}$

111. $\left|\dfrac{2x+1}{3x-4}\right|=5$

112. $\left|\dfrac{4x-3}{x+7}\right|=12$

113. $2+\dfrac{1}{|x-1|}=4$

114. $3-\dfrac{1}{|2x+1|}=1$

115. $|4x-3|=|3x+2|$

116. $|5x-7|=|7x-5|$

117. $\dfrac{1}{|x+2|}=\dfrac{2}{|3x-1|}$

118. $\dfrac{5}{|2x-3|}=\dfrac{4}{|3x+2|}$

119. $|2x+1|+|5x-7|=10$

120. $|3x-2|+|x+4|=3$

121. $|5x-3|-|2x+1|=4$

122. $|6x+7|-|3x-4|=5$

2.2 Linear Inequalities

Not all algebraic expressions are created equal.

Inequalities
Linear Inequalities and Infinite Intervals
Finite Intervals, Union, and Intersection
Inequalities with Fractional Expressions
Inequalities with Absolute Values

Inequalities. In the previous section, we applied the relationship of equality to algebraic expressions, and now we do the same for the relationship of inequality. An **inequality** is a statement that one algebraic expression is *not* equal to another. Four basic types of inequalities are listed below.

Type	*Definition*	*Example*
(1) $M > N$	M is greater than N.	$3x - 7 > x + 5$
(2) $M \geq N$	M is greater than or equal to N.	$5x \geq 3(2x + 1)$
(3) $M < N$	M is less than N.	$-3x + 7 < 2(4x + 3)$
(4) $M \leq N$	M is less than or equal to N.	$x - \frac{1}{2} \leq -\frac{x}{2} + 2$

Even though types (2) and (4) combine inequality with equality, we still call them inequalities. By contrast, types (1) and (3) are called **strict inequalities.** The method for solving an inequality is similar to that for solving an equation and is based on the following properties of "greater than" that were introduced in Section P.3.

If c is any real number, then $M > N \Leftrightarrow M + c > N + c$. **preservation property for addition**

If $c > 0$, then $M > N \Leftrightarrow M \cdot c > N \cdot c$. **preservation property for multiplication**

The preservation property for addition also holds for subtraction and that for multiplication also holds for division (*why?*). These properties also apply to inequality types (2), (3), and (4). That is, the type of an inequality is not changed by addition and subtraction, or by multiplication and division by a positive quantity.

Example 1 Solve $3x - 7 > x + 5$.

Solution:

$3x > x + 12$ *Add 7 to both sides*

$2x > 12$ *Subtract x from both sides*

$x > 6$ *Divide both sides by 2*

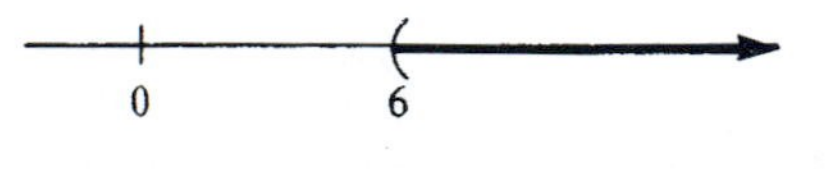

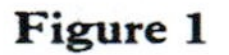
Figure 1

Hence, the solution consists of all real numbers to the right of 6 on the number line, as indicated in Figure 1. The parenthesis symbol (is used to indicate that the endpoint 6 is *not* included in the solution. ■

Example 2 Solve $x - 1/2 \leq -x/2 + 2$.

Solution:

$$2x - 1 \leq -x + 4 \qquad \textit{Multiply both sides by 2}$$
$$2x \leq -x + 5 \qquad \textit{Add 1 to both sides}$$
$$3x \leq 5 \qquad \textit{Add } x \textit{ to both sides}$$
$$x \leq \frac{5}{3} \qquad \textit{Divide both sides by 3}$$

Figure 2

The solution consists of 5/3 and all real numbers to the left of 5/3 on the number line, as indicated in Figure 2. Here the bracket symbol] is used to indicate that the endpoint 5/3 *is* included in the solution. ■

Example 3 Solve $5x + 4 \geq 5(x + 1)$.

Solution:

$$5x + 4 \geq 5x + 5 \qquad \textit{Apply distributive property on right}$$
$$5x \geq 5x + 1 \qquad \textit{Subtract 4 from both sides}$$
$$0 \geq 1 \qquad \textit{Subtract } 5x \textit{ from both sides}$$

Since 0 is not greater than or equal to 1, regardless of the value of x, we must conclude that there is *no* solution, or we can say that the solution is the empty set. ■

Example 4 Solve $x + 2(x - 1) < 3x + 5$.

Solution:

$$x + 2x - 2 < 3x + 5 \qquad \textit{Apply distributive property on left}$$
$$3x - 2 < 3x + 5$$
$$3x < 3x + 7 \qquad \textit{Add 2 to both sides}$$
$$0 < 7 \qquad \textit{Subtract } 3x \textit{ from both sides}$$

Since 0 is less than 7, regardless of x, the solution consists of the set of *all* real numbers. ■

Up to now we have multiplied and divided inequalities by positive quantities only. Now suppose $c < 0$. If $M > N$, then $M - N$ is positive and therefore the product $(M - N)c$ is negative. Hence, $Mc - Nc$ is negative, which means that $Mc < Nc$. We can reverse the argument and thereby obtain the following result.

If $c < 0$, then $M > N \Leftrightarrow M \cdot c < N \cdot c$. **reversal property for multiplication**

The reversal property also holds for division by a negative quantity (*why?*). Also, if $c < 0$, then $M \geq N$ becomes $Mc \leq Nc$, $M < N$ becomes $Mc > Nc$, and $M \leq N$ becomes $Mc \geq Nc$. Similar results apply to division. Hence, *the effect of multiplying or dividing any inequality by a negative quantity is to reverse the direction of the inequality.*

Example 5 Solve $5x - 3(2x + 1) \geq 0$.

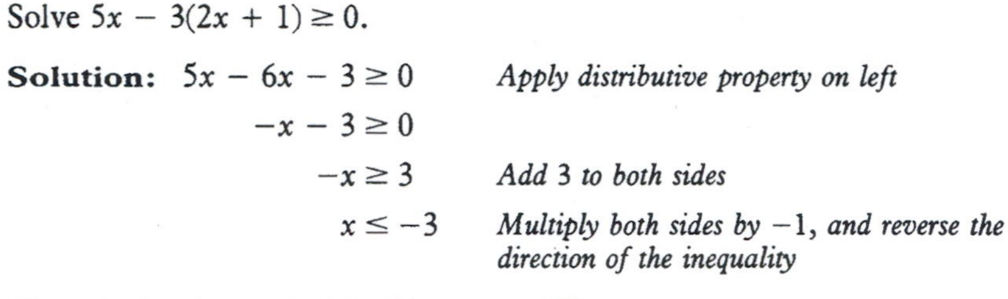

Solution:

$5x - 6x - 3 \geq 0$ *Apply distributive property on left*

$-x - 3 \geq 0$

$-x \geq 3$ *Add 3 to both sides*

$x \leq -3$ *Multiply both sides by -1, and reverse the direction of the inequality*

Figure 3

The solution is graphed in Figure 3. ■

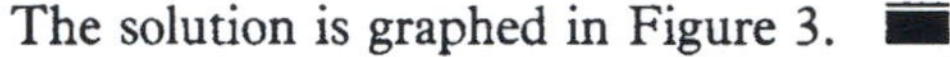

Example 6 Solve $-3x + 7 \leq 2(4x + 3)$.

Solution:

$-3x + 7 \leq 8x + 6$ *Apply distributive property on right*

$-3x \leq 8x - 1$ *Subtract 7 from both sides*

$-11x \leq -1$ *Subtract 8x from both sides*

$x \geq \frac{1}{11}$ *Divide both sides by -11, and reverse the direction of the inequality*

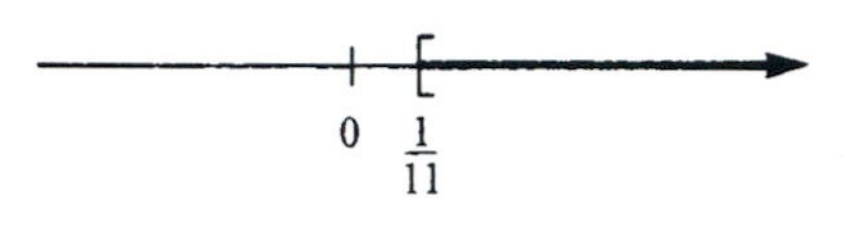

Figure 4

The solution is graphed in Figure 4. ■

Linear Inequalities and Infinite Intervals. Any inequality that can be reduced to one of the four basic inequalities in which $M = ax + b$ $(a \neq 0)$ and $N = 0$ is called a **linear inequality.** All of our previous examples except Examples 3 and 4 are linear inequalities. The solution of a linear inequality is always an **infinite interval** of points along the real line. The following chart indicates the solution for each of the four basic types of inequalities when $a > 0$. (If the coefficient of x is negative in a linear inequality, we can obtain a positive coefficient $a > 0$ by multiplying both sides by -1 and reversing the direction of the inequality. For example, $-2x + 1 > 0$ is equivalent to $2x - 1 < 0$.)

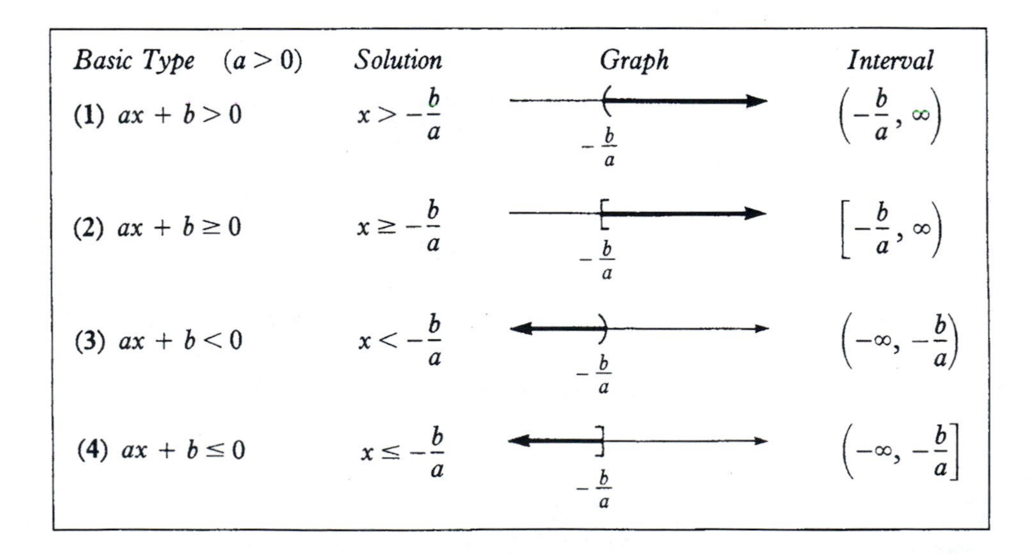

Basic Type $(a > 0)$	*Solution*	*Graph*	*Interval*
(1) $ax + b > 0$	$x > -\frac{b}{a}$	$-\frac{b}{a}$	$\left(-\frac{b}{a}, \infty\right)$
(2) $ax + b \geq 0$	$x \geq -\frac{b}{a}$	$-\frac{b}{a}$	$\left[-\frac{b}{a}, \infty\right)$
(3) $ax + b < 0$	$x < -\frac{b}{a}$	$-\frac{b}{a}$	$\left(-\infty, -\frac{b}{a}\right)$
(4) $ax + b \leq 0$	$x \leq -\frac{b}{a}$	$-\frac{b}{a}$	$\left(-\infty, -\frac{b}{a}\right]$

The symbol ∞ (read "infinity") in the chart indicates that the solution extends to all real numbers *greater than* $-b/a$. Similarly, $-\infty$ indicates that the solution extends to all real numbers *less than* $-b/a$. We note that ∞ and $-\infty$ are not real numbers (*why?*). The parentheses in the graphs and the intervals for (1) and (3) indicate that the endpoint $-b/a$ is *not* included in the solution, whereas the brackets in (2) and (4) mean that $-b/a$ *is* included in the solution.

We note that two extreme cases of "intervals" were met in Examples 3 and 4. In Example 4, the solution is the set of *all* real numbers, which can be denoted in interval notation as $(-\infty, \infty)$. Example 3 has for its solution the **empty set,** usually denoted by $\varnothing$.

Finite Intervals, Union, and Intersection. We have seen that the solution to a linear inequality is a real line interval extending infinitely in one direction. We will soon see that *two* infinite intervals can occur as the solution to a fractional inequality. A fractional inequality also can have as its solution a **finite interval.** For any two real numbers c and d, where $c < d$, there are four types of finite intervals, as illustrated in the following chart.

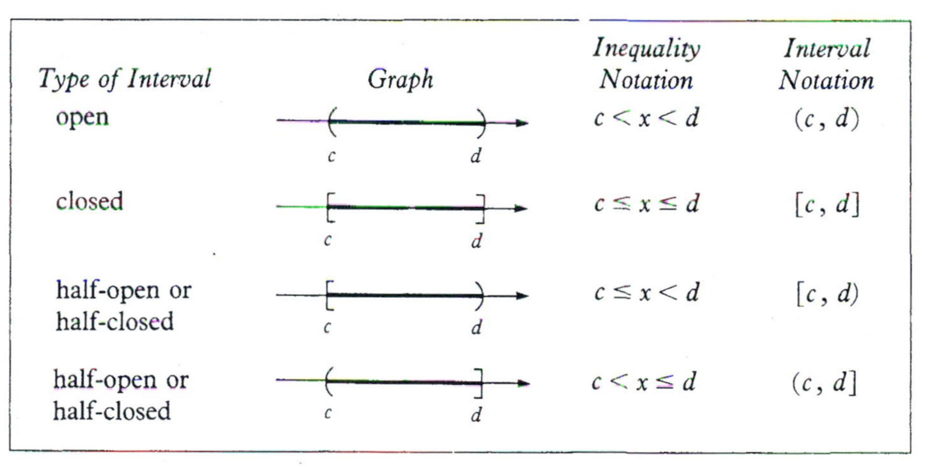

Type of Interval	*Graph*	*Inequality Notation*	*Interval Notation*
open	c d	$c < x < d$	(c, d)
closed	c d	$c \leq x \leq d$	$[c, d]$
half-open or half-closed	c d	$c \leq x < d$	$[c, d)$
half-open or half-closed	c d	$c < x \leq d$	$(c, d]$

Intervals are sets of real numbers, and as such they can be combined by the operations of *union* and *intersection,* which are defined as follows.

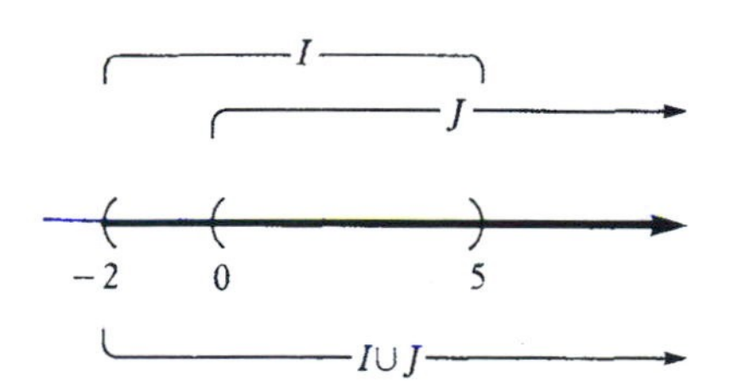

Figure 5

> The **union** of two sets I and J, which is denoted by $I \cup J$, is the set of all real numbers contained in *either* I *or* J.

For instance, $(-2, 5) \cup (0, \infty) = (-2, \infty)$, as illustrated in Figure 5.

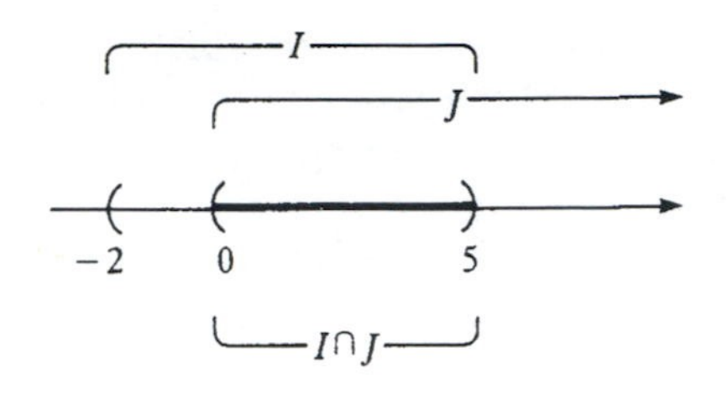

Figure 6

> The **intersection** of two sets I and J, which is denoted by $I \cap J$, is the set of all real numbers contained in *both I and J*.

For instance, $(-2, 5) \cap (0, \infty) = (0, 5)$, as illustrated in Figure 6. We will see how the operations of union and intersection can be used in the solution of fractional inequalities.

Inequalities with Fractional Expressions. A fraction is positive if its numerator and denominator are both positive or both negative. For instance,

$$\frac{7}{8} > 0 \quad \text{and} \quad \frac{-7}{-8} > 0.$$

The same principle applies to rational expressions in x. Consider the following example.

Example 7 Solve $\dfrac{2x+1}{x-2} > 0$.

Solution: We can use the operations of union and intersection to obtain the solution. The key term in the definition of union is "or," and the key term in the definition of intersection is "and." In the following, note how "or" translates to $\cup$, and how "and" translates to $\cap$.

The solution of $\dfrac{2x+1}{x-2} > 0$

= all x for which either

$$2x + 1 > 0 \quad \text{and} \quad x - 2 > 0 \quad \text{or} \quad 2x + 1 < 0 \quad \text{and} \quad x - 2 < 0$$

= all x for which either

$$x > -\frac{1}{2} \quad \text{and} \quad x > 2 \quad \text{or} \quad x < -\frac{1}{2} \quad \text{and} \quad x < 2$$

$$= \left\{\left(-\frac{1}{2}, \infty\right) \cap (2, \infty)\right\} \cup \left\{\left(-\infty, -\frac{1}{2}\right) \cap (-\infty, 2)\right\}$$

$$= (2, \infty) \cup \left(-\infty, -\frac{1}{2}\right)$$

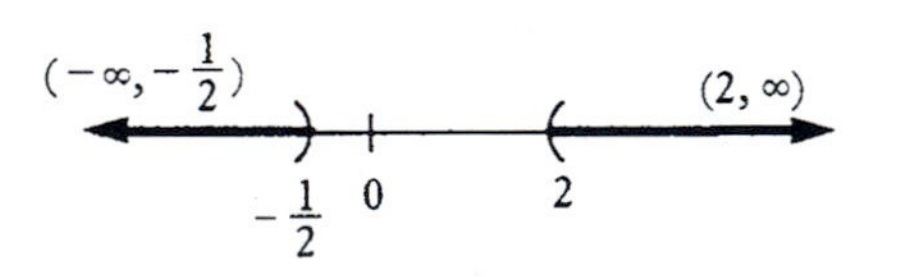

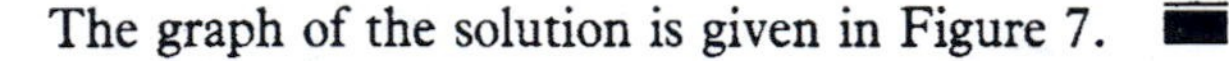

Figure 7

The graph of the solution is given in Figure 7. ■

The above **union and intersection method** can be applied also to fractional inequalities of the type $\geq$, $<$, and $\leq$. An alternative method for solving fractional inequalities, which is called the **method of testing points,** is illustrated in the following example.

Example 8 Solve the inequality $\dfrac{x+8}{x+3} > 2$.

Solution: We first transform the given inequality into one with 0 on the right side as follows:

$$\frac{x+8}{x+3} > 2 \qquad \textit{Given}$$

$$\frac{x+8}{x+3} - 2 > 0 \qquad \textit{Subtract 2 from both sides}$$

$$\frac{x+8}{x+3} - \frac{2(x+3)}{x+3} > 0 \qquad \textit{Obtain a common denominator}$$

$$\frac{x+8-2(x+3)}{x+3} > 0 \qquad \textit{Subtract fractions with a common denominator}$$

$$\frac{-x+2}{x+3} > 0 \qquad \textit{Simplify}$$

$$\frac{x-2}{x+3} < 0. \qquad \textit{Multiply both sides by } -1 \textit{ and reverse the direction of the inequality}$$

We now solve the inequality $(x-2)/(x+3) < 0$, which is equivalent to the given one. The points 2 and -3, which are the respective roots of $x - 2 = 0$ and $x + 3 = 0$, break up the real line into the intervals $(-\infty, -3)$, $(-3, 2)$, and $(2, \infty)$. In each of these, the sign of $x - 2$ and of $x + 3$ does not change (*why?*). Therefore, *the sign of* $(x-2)/(x+3)$ *does not change in each of these intervals.* As illustrated in Figure 8, we can compute the value of $(x-2)/(x+3)$ at *one* convenient test point in each interval to determine the sign of $(x-2)/(x+3)$ over the *entire* interval.

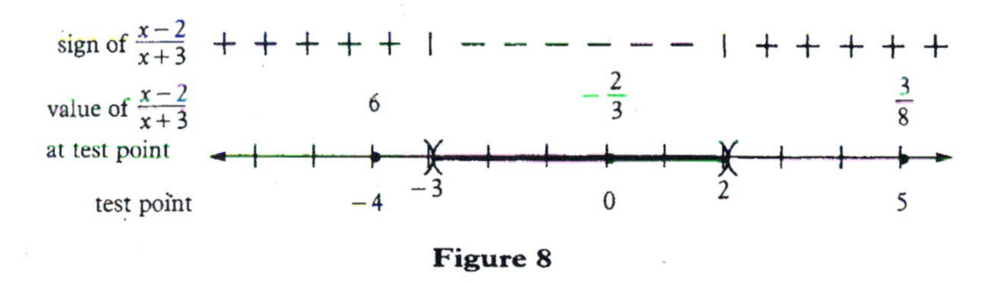

Figure 8

We choose the test points -4, 0, and 5, and conclude that $(x-2)/(x+3)$ is negative on the interval $(-3, 2)$. Hence, $(-3, 2)$ is also the solution of the given inequality $(x+8)/(x+3) > 2$. ■

Both of the previous methods apply to fractional inequalities with 0 on one side. As in Example 8, any given inequality can be put in this form. Also, if the inequality is of the type ≥ 0 or ≤ 0, then the zeros of the numerator (distinct from those of the denominator) are included in the solution. The zeros of the denominator are never included in the solution (*why?*).

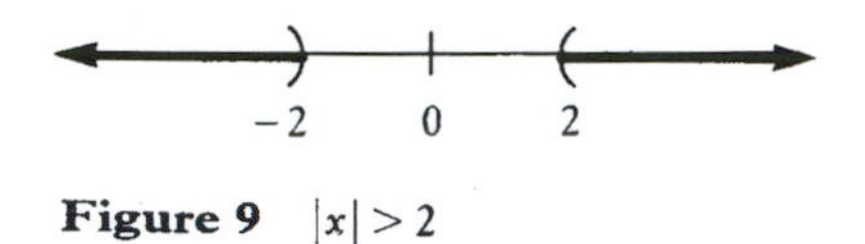

Figure 9 $|x| > 2$

Inequalities with Absolute Values. We have seen that $|x|$ is equal to the distance between 0 and x on the real number line. Therefore, the solution of the inequality $|x| > 2$ consists of all real numbers whose distance from 0 is greater than 2, namely all numbers to the right of 2 *or* to the left of -2 on the number line. That is, the solution of $|x| > 2$ is the set $(-\infty, -2) \cup (2, \infty)$, which is illustrated in Figure 9. Hence, the single inequality $|x| > 2$ is equivalent to the two *separate* inequalities $x > 2$ *or* $x < -2$. This special case leads to the following general rule, in which M is any algebraic expression in x and c is any real number.

> The solution of $|M| > c$ consists of all real numbers x that satisfy *either* $M > c$ *or* $M < -c$. In short,
>
> $$|M| > c \Leftrightarrow M > c \quad \text{or} \quad M < -c.$$

Note that if c is negative, then $|M| > c$ for *all* real numbers for which M is defined. Therefore, we are mainly concerned with nonnegative c's. Also, the general rule holds if $>$ and $<$ are replaced by $\geq$ and $\leq$, respectively.

Example 9 Solve $|5x - 1| > 6$.

Solution:

$5x - 1 > 6$	or	$5x - 1 < -6$	*Each inequality is solved separately*
$5x > 7$		$5x < -5$	
$x > \frac{7}{5}$		$x < -1$	

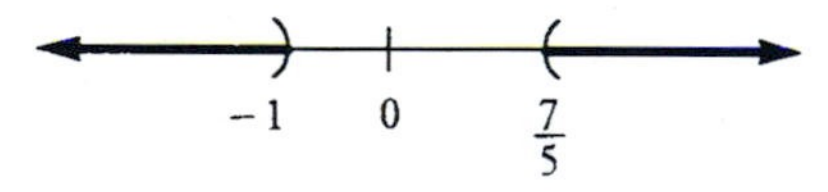

Figure 10

Thus, the solution is $(-\infty, -1) \cup (7/5, \infty)$, as shown in Figure 10. ∎

The solution of $|x| < 2$ consists of all real numbers x whose distance from 0 is less than 2, namely all x in the interval $(-2, 2)$ (Figure 11). Hence, the single inequality $|x| < 2$ is equivalent to the *simultaneous* inequalities $x < 2$ *and* $x > -2$; that is, $-2 < x < 2$.

This special case leads to the following general rule.

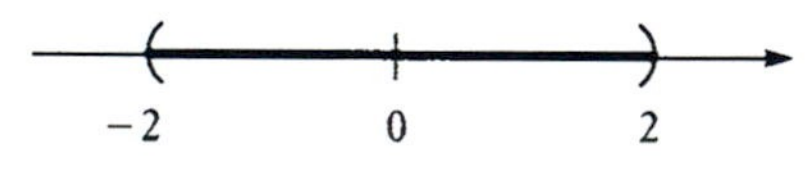

Figure 11 $|x| < 2$

> The solution of $|M| < c$ consists of all real numbers x that satisfy *both* $M < c$ *and* $M > -c$. In short,
>
> $$|M| < c \Leftrightarrow -c < M < c.$$

Note that if c is negative or zero, then the solution of $|M| < c$ is the empty set $\varnothing$. Therefore, we are mainly concerned with positive c's. Also, the general rule applies when $<$ and $>$ are replaced by $\leq$ and $\geq$, respectively.

Example 10 Solve $|2x + 7| < 9$.

Solution:

$$-9 < 2x + 7 < 9 \quad \textit{The two inequalities are solved simultaneously}$$
$$-16 < 2x < 2 \quad \textit{Subtract 7 from each expression}$$
$$-8 < x < 1 \quad \textit{Divide each expression by 2}$$

Therefore, the solution is $(-8, 1)$, as shown in Figure 12. ■

Example 11 Solve $|4x - 3| \leq 8$.

Solution:

$$-8 \leq 4x - 3 \leq 8$$
$$-5 \leq 4x \leq 11 \quad \textit{Add 3 to each expression}$$
$$\frac{-5}{4} \leq x \leq \frac{11}{4} \quad \textit{Divide each expression by 4}$$

Therefore, the solution is $[-5/4, 11/4]$, as shown in Figure 13. ■

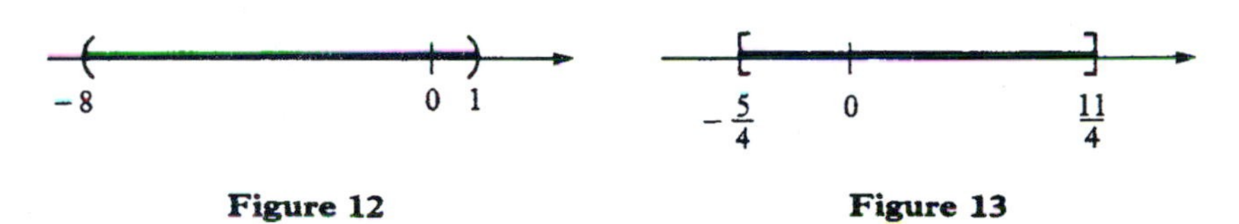

Figure 12 Figure 13

Exercises 2.2

Fill in the blanks to make each statement true.

1. A statement that one algebraic expression is not equal to another is called an ______________.
2. For algebraic expressions M and N, the four basic inequalities are M _____ N, M _____ N, M _____ N, and M _____ N.
3. If $M < N$ and $c < 0$, then $M + c$ _____ $N + c$, and Mc _____ Nc.
4. The solution of $|M| > c$ consists of all real numbers x that satisfy either ______________ or ______________.
5. The solution of $|M| < c$ consists of all real numbers x that satisfy both ______________ and ______________.

Write true *or* false *for each statement.*

6. If $M + 5 < N + 5$, then $M < N$.
7. If $-5M < -5N$, then $M < N$.
8. If $M \leq N$, then $M + 1 < N + 2$.
9. If $M < 5$, then $|M| < 5$.
10. If $|M| < 5$, then $M < 5$.

Linear Inequalities

Solve the following inequalities. Graph the solutions and express the solutions in interval notation.

11. $2x - 4 > 0$
12. $3x + 5 > 0$
13. $-2x + 7 \geq 0$
14. $-4x + 1 \geq 0$
15. $2x - 7 < 0$
16. $4x + 2 < 0$
17. $-3x - 1 \leq 0$
18. $-4x - 5 \leq 0$
19. $5x + 1 > 3x + 4$
20. $3x - 1 > x + 2$
21. $2(x + 5) \geq 3x - 4$
22. $3(x - 2) \geq 2x + 5$
23. $5x - 3 < 2(3x + 1)$
24. $2x + 2 < 5(x - 1)$
25. $3(2x - 4) \leq 6x + 5$
26. $4x + 7 \leq 2(2x + 1)$

Solve the following inequalities for x in terms of the remaining constants and variables.

27. $ax + b > 0 \quad (a < 0)$
28. $ax - by \leq 0 \quad (a > 0)$
29. $\frac{x}{a} + \frac{y}{b} < 1 \quad (a > 0, b \neq 0)$
30. $\frac{x}{a} - \frac{y}{b} \geq 1 \quad (a < 0, b \neq 0)$
31. $ax + b > cx + d \quad (a > c)$
32. $ax + b \leq cx + d \quad (a < c)$

Inequalities with Fractional Expressions

Solve the following inequalities. Express the solutions in interval notation and graph.

33. $\frac{3}{4x + 1} > 0$
34. $\frac{2}{2x - 5} < 0$
35. $\frac{x}{x - 1} \leq 0$
36. $\frac{x}{x + 2} \geq 0$
37. $\frac{x + 3}{2x - 1} < 0$
38. $\frac{x - 2}{4x + 5} > 0$
39. $\frac{1}{x - 1} \geq 2$
40. $\frac{1}{x - 1} \leq -2$
41. $\frac{4}{x + 3} < 2$
42. $\frac{3}{x - 5} > -2$
43. $\frac{x + 1}{x - 1} \geq 5$
44. $\frac{2x - 3}{x - 3} \leq 1$
45. $\frac{x + 2}{x + 3} > 1$
46. $\frac{x - 5}{x - 4} < 2$
47. $\frac{2x + 7}{x + 3} \leq -2$
48. $\frac{3x - 9}{x - 7} \leq -1$

Solve each of the following for x in terms of the remaining constants and variables. Express your answers in interval notation.

49. $\frac{1}{x - a} > 1 \quad (x > a)$
50. $\frac{1}{x - a} < -1 \quad (x < a)$
51. $\frac{x}{a} \geq \frac{y}{a} \quad (a > 0)$
52. $\frac{x}{a} \geq \frac{y}{a} \quad (a < 0)$
53. $\frac{1}{x - a} + \frac{1}{x - 1} < 0 \quad (a < x < 1)$
54. $\frac{1}{x - a} + \frac{1}{x - 1} < 0 \quad (x < a < 1)$

Inequalities with Absolute Values

Solve and graph the following inequalities. Express your answers in interval notation.

55. $|x - 1| > 2$
56. $|x - 3| > 5$
57. $|x + 2| \geq 1$
58. $|x + 4| \geq 5$
59. $|2x - 3| < 6$
60. $|3x + 4| < 3$
61. $|4x + 5| \leq 2$
62. $|2x + 3| \leq 4$
63. $|5x - 1| > 0$
64. $|2x - 7| \leq 0$
65. $|5x - 4| \geq -1$
66. $|2x - 1| < -1$

Miscellaneous

Combine the techniques in this section to solve each of the following inequalities. Express your answers in interval notation and graph.

67. $5x + 3(x - 1) > 2(x + 4) - 7$
68. $3(2x + 1) + 5 > 5(4x + 2) - 3x$
69. $-8x + 1 + 2(x - 3) \geq -4(x + 2) + 5(x - 1)$
70. $-3x + 7 + 2(4x + 1) \geq 6(x + 2) - (2x + 3)$
71. $-4x + 3[2 - (x - 4)] < 5[x - (1 - 2x)]$
72. $-5x + 3 + 2(x + 7) < 4 + 3[7 - (x - 2)]$
73. $3x + 2(3x - 7) \leq 5 + 3(3x - 1)$
74. $2(x + 7) + 3(2x - 4) \leq 4(2x + 12)$
75. $\frac{1}{x - 1} > \frac{2}{x - 1}$
76. $\frac{2}{x + 2} > \frac{1}{x + 2}$
77. $1 + \frac{1}{x} \geq 3 - \frac{2}{x}$
78. $2 - \frac{5}{x + 1} \geq 1 + \frac{4}{x + 1}$
79. $\frac{1}{x + 2} - 3 < 3 - \frac{1}{x + 2}$
80. $\frac{4}{x + 4} + 5 < \frac{-4}{x + 4} + 5$

81. $\frac{1}{2x-1} - 3 \le \frac{1}{2x-1} - 3$

82. $4 - \frac{3}{x-2} \le -4 - \frac{3}{x-2}$

83. $\frac{2}{|x+1|} > 1$

84. $\frac{4}{|x-2|} > 1$

85. $\frac{1}{|2x+3|} \ge 2$

86. $\frac{1}{|3x-4|} \ge 5$

87. $\frac{2}{|3x+1|} < 3$

88. $\frac{4}{|5x-1|} - 5 < 0$

89. $\frac{1}{|2x+7|} - 1 \le 0$

90. $\frac{5}{|3x-2|} - 6 \le 0$

2.3 Applications of Linear Equations and Inequalities

Percent
Motion
Mixtures
Work
Miscellaneous Applications

In an algebra class, one question that can always be counted on is "Will there be any word problems on the test?"

People solve linear equations every day, even though they are not always aware of it. Whenever they make change, compute the sales tax on an item, or glance at a clock to determine how much work or class time is left, they are solving linear equations. Whenever people budget their time to maintain a schedule, they are working with inequalities.

Several standard applications of linear equations and inequalities are discussed in this section. To solve a word problem, use the following procedure.

1. Read the problem very carefully, identify the quantities involved, and assign letters to the unknown quantities.
2. Express any relationships that exist among the quantities in the form of algebraic equations and/or inequalities. Draw diagrams whenever possible.
3. Solve for the unknown quantities.
4. Interpret the algebraic solution in terms of the original problem.

The above four-step transformation of a word problem to its mathematical model and back to its solution is illustrated in the following flow chart.

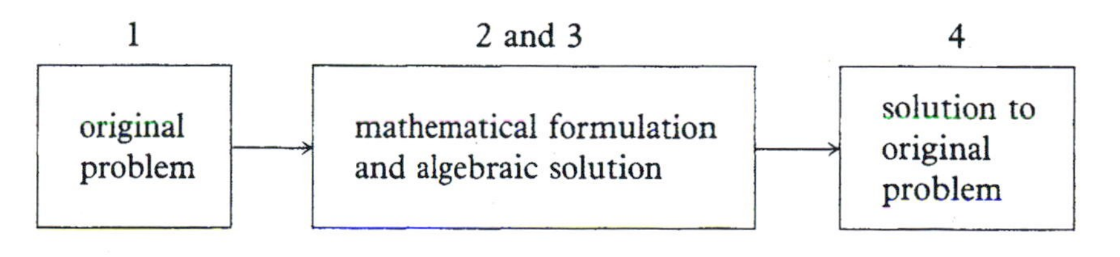

We now consider word problems involving several topics, including percent, motion, mixtures, and work.

Percent. Percent means "hundredths," and $r\%$ of a quantity Q means $(r/100)Q$. For instance, 15% of 250 means $(15/100)250 = (.15)250 = 37.5$.

Example 1 The retail price of a stereo is $225, which includes a 50% markup on the wholesale cost. (a) What is the wholesale cost of the stereo? (b) What percent of the retail price is the markup?

Solution:

(a) Let C = wholesale cost in dollars. Then

$$\begin{aligned} C + .5C &= 225 \qquad \textit{Cost + markup = retail price} \\ 1.5C &= 225 \\ C &= \frac{225}{1.5} \\ &= 150. \end{aligned}$$

Hence, the wholesale cost of the stereo is $150.

(b) The markup in dollars is $225 - 150 = 75$. As a percent of retail price, the markup is

$$\left(\frac{75}{225} \cdot 100\right)\% = \left(\frac{1}{3} \cdot 100\right)\% = 33\tfrac{1}{3}\%.$$ ■

Caution When working with percents, it is important to specify the quantity on which the percent is computed. In Example 1, the $75 markup is 50% of the wholesale cost but only 33 1/3% of the retail price.

Example 2 The sticker price of a new car is $9995, which includes a $7545 base price plus $2450 in optional equipment. The base price includes a 14% markup on the base cost, and the options price includes an 18% markup on the options cost. What is the dealer's cost of the car?

Solution: Let B = the base cost to the dealer and let E = the options cost to the dealer, both in dollars. Then

$$\begin{aligned} B + .14B &= 7545 & \text{and} \qquad E + .18E &= 2450 \\ 1.14B &= 7545 & 1.18E &= 2450 \\ B &= \frac{7545}{1.14} & E &= \frac{2450}{1.18} \\ &= 6618.42 & &= 2076.27. \end{aligned}$$

Therefore, the dealer's cost of the car is

$$\$6618.42 + \$2076.27 = \$8694.69.$$ ■

Example 3 The workers at an electronics plant want their work week reduced from 40 hours to 35 hours but with the same weekly pay. If their current hourly rate is x dollars an hour, what will the new hourly rate be? To what percent increase is the new hourly rate equivalent?

Solution: The current weekly income is $40x$ dollars. If the new hourly rate is r dollars per hour, then the new weekly pay is $35r$ dollars. For the new weekly pay to be equal to the old, we must have $35r = 40x$. Therefore,

$$r = \frac{40x}{35} = \frac{8x}{7} = x + \frac{1}{7}x.$$

Hence, the new hourly rate is an increase of one seventh over the old. As a percent, the increase is $(1/7 \cdot 100)\%$, or approximately 14.3%. ■

Example 4 A clothing retailer has a 35% markup over cost on a certain line of sport shirts. What percentage discount on the selling price will guarantee at least a 20% profit over cost?

Solution: Let C = the cost of a shirt to the retailer, and let x = the discount rate on the selling price.

The current selling price of a shirt is

$$C + .35C = 1.35C,$$

and the discount price is to be at least

$$C + .2C = 1.2C.$$

That is,

$$1.35C - (1.35C)x \geq 1.2C \qquad \textit{Selling price} - \textit{discount} \geq 20\% \textit{ over cost}$$

$$1.35C(1 - x) \geq 1.2C$$

$$1 - x \geq \frac{1.2}{1.35} \approx .89 \qquad \textit{Divide both sides by } 1.35C$$

$$1 - .89 \geq x$$

$$.11 \geq x.$$

Hence, a discount rate of 11% or less will insure a profit of at least 20%. Note that the result does not depend on the actual value of C. ■

Motion. The most basic formula for motion is

$$d = rt,$$

where

d = the distance traveled by an object along a given path,

r = the average rate of speed of the object,

and t = the time of travel.

The formula contains three quantities: d, r, and t. If any two of these are known, the formula can be solved for the remaining one.

Example 5 Paul drove 150 miles in 3 1/2 hours. His average speed for the first 90 miles was 45 miles per hour. What was his average speed for the last 60 miles?

Solution: The formula can be applied to each part of the trip as needed.

Distance (mi)	Rate (mph)	Time (hr)
150		3.5
90	45	$90/45 = 2$
60	x	t

In the table we have set

$$x = \text{rate for last 60 miles,}$$

and

$$t = \text{time for last 60 miles.}$$

Now,

$$x = \frac{60}{t}. \qquad \textit{Rate} = \frac{\textit{distance}}{\textit{time}}$$

Also,

$$\begin{aligned} 2 + t &= 3.5 \\ t &= 1.5. \end{aligned} \qquad \textit{Time for first 90 miles + time for last 60 miles = time for entire 150 miles}$$

Therefore,

$$\begin{aligned} x &= \frac{60}{1.5} \\ &= 40. \end{aligned}$$

The average speed for the last 60 miles was 40 miles per hour. ■

Example 6 At noon, cyclists Dee and Donna are 50 miles apart and pedaling toward each other. If Dee pedals at 10.5 miles per hour and Donna at 14.5 miles per hour, at what point between them and at what time do they meet?

Solution: We can apply the formula to each cyclist. If they meet in t hours, then Dee has traveled x miles and Donna $50 - x$ miles, as indicated in Figure 14 and the table.

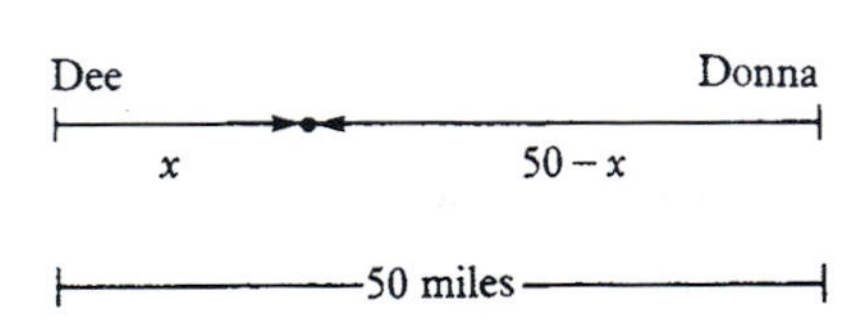

Figure 14

	Distance	Rate	Time
Dee	x	10.5	t
Donna	$50 - x$	14.5	t

Now,

$$\frac{x}{10.5} = \frac{50 - x}{14.5} \qquad \textit{Dee's time = Donna's time}$$

$$14.5x = 10.5(50 - x)$$

$$14.5x = 525 - 10.5x$$

$$25x = 525$$

$$x = 21.$$

Therefore, they meet at a point 21 miles from Dee's noontime position. The time of travel is

$$t = \frac{21}{10.5} \qquad \textit{Time} = \frac{\textit{distance traveled by Dee}}{\textit{Dee's rate of speed}}$$

$$= 2.$$

Hence, they meet at 2 P.M. ■

Example 7 An airplane traveling against a steady wind takes 2 hours to make a 500-mile trip. The return trip in the direction of the wind takes only 1 hour and 40 minutes. What is the speed of the plane in still weather, and what is the speed of the wind?

Solution: Let r = the speed of the plane in still weather, and let x = the speed of the wind. Then $r - x$ is the speed against the wind, and $r + x$ is the speed with the wind, as indicated in the table.

	Distance (mi)	Rate (mph)	Time (hr)
Against the wind	500	$r - x$	2
With the wind	500	$r + x$	$1\frac{2}{3} = \frac{5}{3}$

Now,

$$r - x = \frac{500}{2} \quad \text{and} \quad r + x = \frac{500}{\frac{5}{3}} \qquad \textit{Rate} = \frac{\textit{distance}}{\textit{time}}$$

$$r - x = 250 \qquad\qquad r + x = 300.$$

Therefore,

$$r - x + r + x = 250 + 300 \qquad \textit{Add } r - x \textit{ and } r + x \textit{ to cancel } x$$

$$2r = 550$$

$$r = 275,$$

giving $x = 25.$ *Use either* $r - x = 250$ *or* $r + x = 300$

The speed of the plane in still weather is 275 miles per hour, and the speed of the wind is 25 miles per hour. ■

Mixtures. A typical mixing problem involves combining two quantities having different concentrations to form a third quantity that has an intermediate concentration. For instance, if 10 liters of 30% acid are mixed with 15 liters of 60% acid, the result is 25 liters of acid with concentration x, where

$$\underbrace{10(.3)}_{\text{Liters of acid in 30\% solution}} + \underbrace{15(.6)}_{\text{Liters of acid in 60\% solution}} = \underbrace{25x}_{\text{Total amount of acid}}. \tag{1}$$

Therefore,

$$3 + 9 = 25x$$
$$12 = 25x$$
$$.48 = x.$$

The mixture is 48% acid.

In any mixing problem, each part to be combined contains a concentrate in some medium. In the above example, the concentrate is acid, which is measured in percent, and the medium is the solution, which is measured in liters. *We compute the amount of concentrate in each part and add these amounts to get the total amount of concentrate.* For convenience, we can list the given data for the above example in the following table.

	Part 1	*Part 2*	*Mixture*
Concentrate	30%	60%	x
Medium	10 liters	15 liters	25 liters
Amount of Concentrate	10(.3)	15(.6)	$25x$

Equation (1) is obtained from the last row of the table by adding the amount of concentrate in each part and equating the sum to the amount of concentrate in the mixture.

Example 8 Coffee worth \$2.85 a pound is mixed with coffee worth \$2.50 a pound to obtain 5 pounds of coffee worth \$2.64 a pound. How much of each type of coffee goes into the blend?

Solution: Here the concentrate is the value, which is measured in dollars per pound, and the medium is coffee, which is measured in pounds. Let $x =$ the number of pounds of \$2.85 coffee required; then $5 - x =$ the number of pounds of \$2.50 coffee needed. We have the following table of data.

	Part 1	*Part 2*	*Mixture*
Concentrate	\$2.85/lb	\$2.50/lb	\$2.64/lb
Medium	x lbs	$5 - x$ lb	5 lbs
Amount of Concentrate	$2.85x$	$2.50(5 - x)$	$2.64(5)$

From the last row of the table, we add the amount of concentrate in each part to obtain

$$\begin{aligned} 2.85x + 2.50(5 - x) &= 2.64(5) \\ 2.85x + 12.50 - 2.50x &= 13.20 \\ .35x &= .70 \\ x &= 2. \end{aligned}$$

Hence, 2 pounds of \$2.85 coffee must be mixed with 3 pounds of \$2.50 coffee to obtain the desired 5 pounds of \$2.64 coffee. ■

As illustrated in the following example, we can also apply inequalities to mixing problems.

Example 9 What is the least number of liters of 30% alcohol that must be mixed with 6 liters of 55% alcohol to make a solution that is at most 40% alcohol?

Solution: Here the concentrate is alcohol, which is measured in percent, and the medium is the solution, which is measured in liters. Let x = the number of liters of 30% alcohol added. We have the following data.

	Part 1	*Part 2*	*Mixture*
Concentrate	30%	55%	$\le 40\%$
Medium	x liters	6 liters	$x + 6$ liters
Amount of Concentrate	$.30x$	$.55(6)$	$\le .40(x + 6)$

Since the mixture is to be less than or equal to 40% alcohol, we add the amount of concentrate in each part and set the sum less than or equal to the amount of concentrate in a 40% mixture. That is,

$$\begin{aligned} .30x + .55(6) &\le .40(x + 6) \\ .3x + 3.3 &\le .4x + 2.4 \\ .9 &\le .1x \\ 9 &\le x. \end{aligned}$$

Therefore, at least 9 liters of 30% alcohol must be added. ■

Work. In a typical work problem, we are given the time required for each member of a group to do a certain job when working individually, and we are asked to determine the time required when the members work together. We will consider situations in which the following basic principle is a reasonable assumption.

> If it takes n time units to do a given job, then $1/n$ of the job is completed per time unit.

Example 10 Andy can paint a fence in 4 hours, and Ben can do the job in 6 hours. If both work together, how long will it take to paint the fence?

Solution: Let $x =$ the number of hours required to paint the fence with both working. Then, according to the above basic principle, $1/x$ of the fence is painted per hour. Andy's contribution per hour is $1/4$, and Ben's is $1/6$. Therefore,

$$\frac{1}{4} + \frac{1}{6} = \frac{1}{x} \qquad \textit{Andy's contribution per hour plus Ben's contribution per hour equals the total amount done per hour}$$

$$\frac{5}{12} = \frac{1}{x}$$

$$x = \frac{12}{5}.$$

Hence, it takes 2 2/5 hours, or 2 hours and 24 minutes, for Andy and Ben to paint the fence. ■

Example 11 By typing together, secretaries A and B can finish a report in 2 hours. It would take B, typing alone, 5 hours to do the same report. How long would it take A?

Solution: Let $x =$ the number of hours it would take A to type the report. When A and B type together, $1/2$ of the report is finished in 1 hour. B's contribution in one hour is $1/5$ of the report, and A's is $1/x$. Therefore,

$$\frac{1}{5} + \frac{1}{x} = \frac{1}{2} \qquad \textit{B's contribution per hour plus A's contribution per hour equals the total amount done per hour}$$

$$\frac{1}{x} = \frac{1}{2} - \frac{1}{5}$$

$$\frac{1}{x} = \frac{3}{10}$$

$$x = \frac{10}{3}.$$

Hence, it takes A 3 1/3 hours, or 3 hours and 20 minutes, to type the report. ■

Example 12 One machine at a stationery company can produce 1000 envelopes in t minutes. The company wants to add another machine so that the two, working together, can produce 1000 envelopes in at most $t/3$ minutes, that is, at least $1000/(t/3)$ per minute. How fast must the new machine be?

Solution: Let x = the number of minutes required for the new machine to produce 1000 envelopes. In one minute the old and the new machines, working together, produce

$$\underbrace{\frac{1000}{t}}_{\text{Old machine}} + \underbrace{\frac{1000}{x}}_{\text{New machine}}$$

envelopes, and we want this number to be at least

$$\frac{1000}{\frac{t}{3}}.$$

That is,

$$\frac{1000}{t} + \frac{1000}{x} \geq \frac{1000}{\frac{t}{3}}$$

$$\frac{1}{t} + \frac{1}{x} \geq \frac{3}{t}$$

$$x + t \geq 3x \qquad \textit{Multiply both sides by the positive quantity } tx$$

$$t \geq 2x$$

$$\frac{t}{2} \geq x.$$

Therefore, the new machine must be at least twice as fast as the old one. ■

Miscellaneous Applications. A variety of other situations give rise to linear equations and inequalities.

Example 13 A student has grades of 75, 80, and 87 on the first three math tests in a course. What grade must be obtained on the next test in order to give an average grade of at least 84 for the four tests?

Solution: Let x = the grade needed on the next test. Then

$$\frac{75 + 80 + 87 + x}{4} \geq 84 \qquad \textit{The average of n numbers equals their sum divided by n}$$

$$75 + 80 + 87 + x \geq 336$$

$$242 + x \geq 336$$

$$x \geq 94.$$

Hence, a grade of 94 or better will bring the student's average up to at least 84. ■

Example 14 Find three consecutive even integers such that twice the first plus three times the third is twenty more than four times the second.

Solution: Let x be the first of the three even integers. Then the next two are $x + 2$ and $x + 4$, and

$$\underbrace{2x + 3(x + 4)}_{\text{Twice the first plus three times the third}} = \underbrace{4(x + 2) + 20}_{\text{Twenty more than four times the second}}$$

$$5x + 12 = 4x + 8 + 20$$

$$x = 16.$$

Therefore, the numbers are 16, 18, and 20. ■

Exercises 2.3

Percent

1. A clock radio that regularly sells for $24.95 is on sale at a 20% discount. What is the sale price?
2. A color TV is selling at a discount of 25%. The sale price is $375. What was the price of the TV before the sale?
3. During Goodwear Tire Company's annual sale, anyone who buys three tires at the regular price gets a fourth one free. To what percent discount on all four tires is this sale equivalent?
4. Alex bought gloves at a 50% discount for $10, a shirt at a 20% discount for $15, and slacks at a 25% discount for $30. What overall discount did Alex receive on his purchases?
5. The regular price of a sweater is $30, which includes a 50% markup over cost. If a retailer wants to clear out the inventory of sweaters and at least break even, what is the maximum discount that can be offered?
6. Rita's Appliances had 100 kerosene heaters to sell. It sold 75 of them at a 40% profit. What discount can it offer on the remaining 25 and still make a total profit of at least 35% on the 100 heaters?
7. A company's sales decreased by 18% from 1982 to 1983, and they increased by 20% from 1983 to 1984. What was the percent change from 1982 to 1984?
8. Newspaper A charges $6.50 a line for classified advertising, and Newspaper B charges $5.00 a line. What percent greater than B's charge is A's? What percent less than A's charge is B's?
9. Toni invested cash in a bank on December 31, 1982. On December 31, 1983 she received 10% interest and reinvested it in the same bank. On December 31, 1984 she received another 10% interest, and her grand total was $285.56. How much did Toni invest originally?
10. On June 1 a customer has a charge account balance of $500. At the end of each month a 1.5% finance charge is added to the previous month's balance, and then the customer makes a payment of 10% of the total amount. If no further purchases are made, what are the customer's payments on June 30, July 31, August 31, and September 30?

Motion

11. Frank drove 100 mi. For the first 55 mi he averaged 50 mph, and for the last 45 mi he averaged 30 mph. What was his average speed for the entire trip?
12. A bus driver wants to average at least 40 mph on a 70-mi trip. If the first 35 mi take the driver 1 hr, what should be the lowest average speed for the second 35 mi?
13. From a distance 102 mi apart, two cars travel toward each other, starting at 10 A.M. One car travels at 45 mph, and the other at 40 mph. At what point between them and at what time do they meet?
14. Drivers A and B are 50 mi apart, and they plan to meet at 7 P.M. at a point between them that is 30 mi from A. Both drive at 50 mph. What time should each driver start out?

15. At 3 P.M. Jean leaves her house and drives south on Interstate 95 at 50 mph. At 3:03 P.M. her sister Tammy leaves the house and follows Jean at 55 mph. When does Tammy catch up to Jean?

16. At 1 P.M. a person starts jogging, and from the same point a second person starts jogging at 1:05 P.M. The second catches up to the first at 1:25 P.M. How much faster than the first does the second jog?

17. Runner A can run at least one and one-half times as fast as runner B. In a 100-meter race, how many meters' head start can A give B and still not finish after B?

18. Runner A can run three-fourths as fast as Runner B. In a 5-mi race, B gives A a 1.5-mi head start. Who will win the race?

19. With the aid of the current, Patrick can row a canoe 3 mi in 12 min. Against the current, he requires 18 min to row the same distance. How fast does Patrick row in still water, and how fast is the current?

20. Kristi is walking alongside a merry-go-round. When she walks in the direction of the rotation, it takes 10 sec for a given point of the merry-go-round to return to her. When she walks against the rotation, it takes only 5 sec. How much faster than Kristi is the edge of the merry-go-round moving?

Mixtures

21. If 3 liters of 30% alcohol are mixed with 2 liters of 50% alcohol, what will be the concentration of the mixture?

22. Two liters of 40% alcohol are mixed with 4 liters of another alcohol to form a 50% alcohol solution. What is the strength of the added alcohol?

23. Five liters of 55% acid are mixed with a 40% acid to form a solution that is at most 50% acid. At least how many liters of 40% acid were used?

24. At most how much water can be added to 4 liters of a solution that is 60% acid to obtain a solution that is at least 50% acid?

25. How many pounds of hamburger worth $1.89 a pound must be mixed with 2 lb of hamburger worth $1.59 a pound to make hamburger worth $1.69 a pound?

26. If 2 lb of hamburger worth $1.49 a pound are mixed with 3 lb of hamburger worth $1.79 a pound, what should be the price per pound of the mixture?

27. A 15-qt car radiator is filled with a 10% antifreeze solution. If at most 4 qt are drained off and replaced with pure antifreeze, what is the minimum strength of the new solution?

28. A 10-qt car radiator is filled with a 20% antifreeze solution. At least how many quarts must be drained off and replaced with pure antifreeze in order to bring the strength up to at least 50% antifreeze?

29. If two gasolines with octane ratings of 89 and 92 are mixed 2 parts 89 to 3 parts 92, what is the octane rating of the mixture?

30. If three gasolines with octane ratings of 89, 92, and 94, respectively, are mixed 1 part 89 to 2 parts 92 to 3 parts 94, what is the octane rating of the mixture?

Work

31. It takes Charles 1 hr to cut and trim the grass, whereas Richard can do the same job in 1/2 hr. How long will it take with both of them working?

32. One snow blower can clear the snow from a driveway in 20 min. It takes another one 30 min to do the same job. How long would it take both of them working together?

33. Machine A processes 100 forms in 4 hr. Machines A and B together can process 100 forms in 1.5 hr. How long does it take B to process 100 forms working alone?

34. Bank tellers A and B can process 25 customers in 60 minutes. A can process 25 in 150 minutes. How long does it take B?

35. If machine A works twice as fast as B, how much faster are the two of them than each one?

36. A company has three machines that manufacture computer chips. They can process a given order in 2, 3, and 4 hours, respectively. How fast can all three machines working together process the order?

37. A company has two machines that can turn out 1000 items in 4 and 12 minutes, respectively. By adding a third machine they can turn out the 1000 items in 1 minute. How fast can the new machine do the job?

38. Plants A and B produce n_1 and n_2 autos per day, respectively. In t_3 days the two plants produce as many autos as A does in t_1 days and B does in t_2 days. That is, $n_1 t_1 = n_2 t_2 = (n_1 + n_2)t_3$ (production rate · time = total production). Show that $1/t_1 + 1/t_2 = 1/t_3$.

39. In a parallel circuit, electrical current of $I_1 + I_2$ amperes splits into I_1 amperes across a resistor of R_1 ohms and I_2 amperes across a resistor of R_2 ohms. According to Ohm's law, the total resistance in the parallel circuit is R_3 ohms, where

$$I_1R_1 = I_2R_2 = (I_1 + I_2)R_3. \quad \text{(current} \cdot \text{resistance} = \text{voltage)}$$

Show that $1/R_1 + 1/R_2 = 1/R_3$.

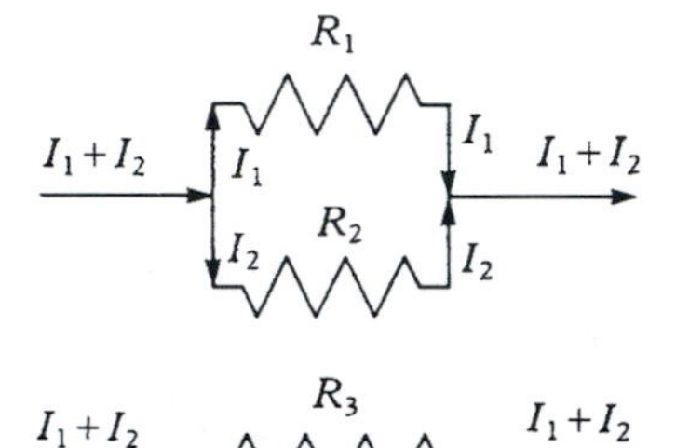

40. Compare the work equations in Exercise 38 with those in Exercise 39, and insert the corresponding electrical terms in the following chart.

Work	Parallel Circuit
Production Rate	
Time	
Total Production	

Miscellaneous

41. A math professor gives four tests each worth 1/6 of the final grade and a comprehensive exam worth 2/6 of the final grade. If Tina gets 76, 80, 82, and 88 on the four tests, what is the lowest grade she can receive on the final exam and still have a final grade of at least 85?

42. A student averaged 79 on the first two tests in a course and scored 91 on the third test. The student concluded that the average for the three tests is (79 + 91)/2 = 85. Explain why this conclusion is wrong, and find the correct average.

43. Show that for any three consecutive integers, the sum of the first and third is twice the second.

44. The square of the sum of two numbers is 20 more than the square of their difference. One of the numbers is 1. What is the other number?

45. One number is 5 more than another, and the difference of their squares is 30. What are the two numbers?

46. The sum of three numbers is 79. The second number is 2 more than 3 times the first, and the third is 3 more than 2 times the second. What are the numbers?

47. Fifty micrograms of Vitamin B-12 is 83.3% of the U.S. recommended daily allowance for adults. How many micrograms constitute 100% of the recommended allowance?

48. A 10-oz jar of instant coffee costs $2.12. What is the equivalent price per pound?

49. If the sum of two numbers is greater than or equal to 5, and the difference is greater than or equal to 7, what is the smallest possible value of the larger number?

50. If the sum of two numbers is less than or equal to 5 and the difference is greater than or equal to 7, what is the largest possible value of the smaller number?

2.4 The Coordinate Plane

Algebra and geometry come together in the coordinate plane.

Coordinates
The Distance Formula
The Midpoint Formula

Coordinates. We used the real number line to indicate solutions of linear equations and inequalities in one variable x. For equations and inequalities in two variables x and y, a two-dimensional plane is needed. To determine points in a plane, we position two perpendicular number lines to intersect at 0 on their respective number scales. We assume that one line, called the **x-axis,** is horizontal, with its positive direction to the

[1]*Rectangular coordinates are also called Cartesian coordinates in honor of René Descartes (1596–1650), one of the founders of analytic geometry.*

right, and the other, called the **y-axis,** is vertical, with its positive direction upward (Figure 15).

Now let P be a point in the plane. Through P draw a vertical line segment V and a horizontal line segment H as shown in Figure 16. The point x of intersection of V with the x-axis is called the **x-coordinate** or **abscissa** of P, and the point y of intersection of H with the y-axis is called the **y-coordinate** or **ordinate** of P. The ordered pair of numbers (x, y), with x to the left of y, are called the **rectangular coordinates**[1], or simply the **coordinates** of P. In this manner we establish a one-to-one correspondence between the points in the plane and ordered pairs of real numbers. This correspondence enables us to identify points with their coordinates, and we will use the notation P or $P(x, y)$ or simply (x, y) to denote a point in the plane.[2] In particular, (0, 0), which is the point of intersection of the x-axis and y-axis, is called the **origin** of the coordinate system.

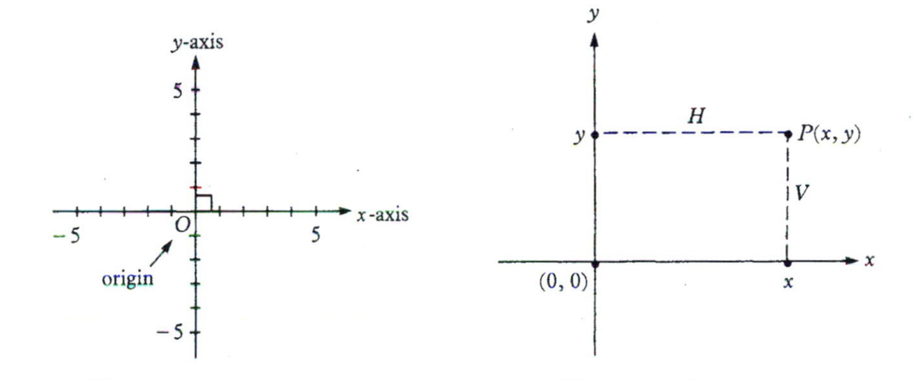

Figure 15 Coordinate Axes **Figure 16** Coordinates

Example 1 Find the coordinates of the points O, P, Q, R, S, and T in Figure 17.

Solution: $O(0, 0)$, $P(2, 5)$, $Q(5, 2)$, $R(-3, 4)$, $S(-5, -6)$, and $T(4, -4)$ ■

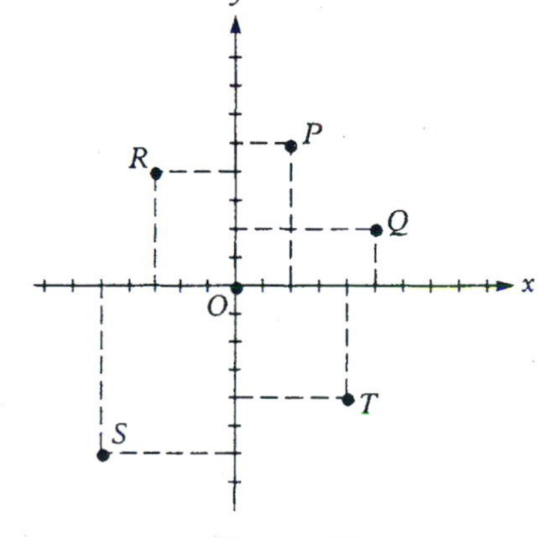

Figure 17

[2]The notation (x, y) is also used to denote an open interval. We will rely on context to distinguish coordinates from open intervals.

Example 2 Locate the points with the following coordinates: $(2, -5)$, $(-5, 2)$, $(-3, -7)$, $(4, 4)$, $(6, 0)$, $(0, 5)$.

Solution: To locate the point (x, y), start at the origin, go x directed units along the x-axis and then y directed units parallel to the y-axis. The endpoint has coordinates (x, y) (Figure 18). ■

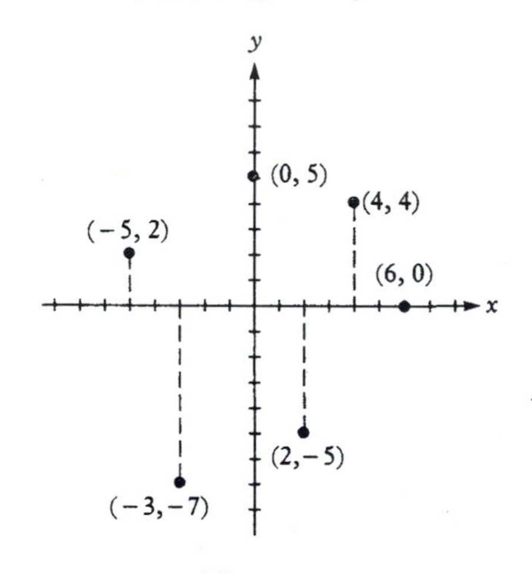

Figure 18

The coordinate axes partition the plane into four quarter planes called **quadrants** (Figure 19). For every point in quadrant I, both coordinates are positive. In quadrant II, the abscissa is negative and the ordinate positive. In III both coordinates are negative, and in IV the abscissa is positive and the ordinate negative. The coordinate axes form the boundaries of the four quadrants.

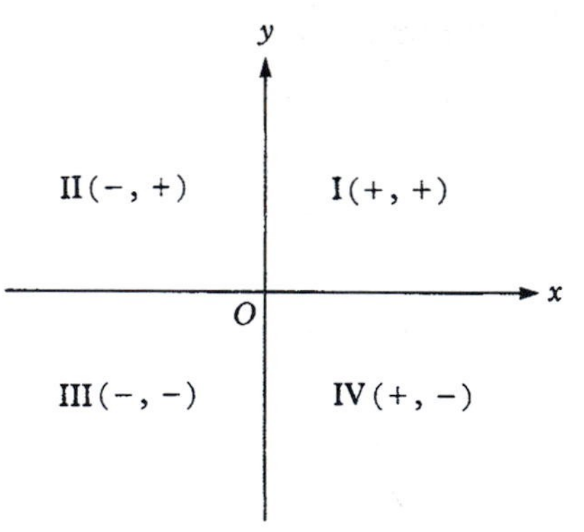

Figure 19 Quadrants

Example 3 Determine the quadrants of each of the following points: (a) $(-2, 100)$; (b) $(-10, -50)$; (c) $(a, -a)$ if $a > 0$; (d) $(b, -b)$ if $b < 0$; (e) $(-c, -c)$ if $c < 0$.

Solution:

(a) II (b) III (c) IV (d) II (e) I ■

The Distance Formula. Let x_1 and x_2 be any two points on a number line, and let d be the distance between them. If $x_2 > x_1$, there are three possibilities for their positions, as indicated in Figure 20.

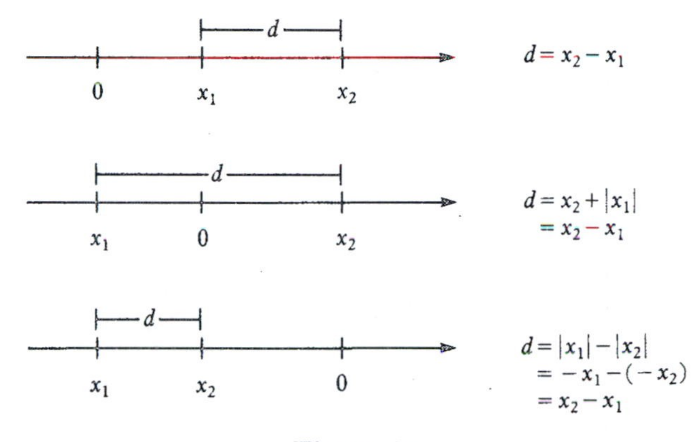

Figure 20

Hence, if $x_2 > x_1$, then $d = x_2 - x_1 = |x_2 - x_1|$ regardless of the positions of x_1 and x_2 relative to 0. Similarly, if $x_1 > x_2$, then $d = x_1 - x_2 = -(x_2 - x_1) = |x_2 - x_1|$. Hence, for any two points x_1 and x_2 on a number line,

$$d = |x_2 - x_1|. \qquad \textbf{distance formula along a number line}$$

Example 4 In each of the following cases, locate the two points on a number line and compute the distance between them.

(a) $x_1 = 2$; $x_2 = 5$ (b) $x_1 = -2$; $x_2 = 5$

(c) $x_1 = 2$; $x_2 = -5$ (d) $x_1 = -2$; $x_2 = -5$

Solution: The given points in each case are shown in Figure 21. ■

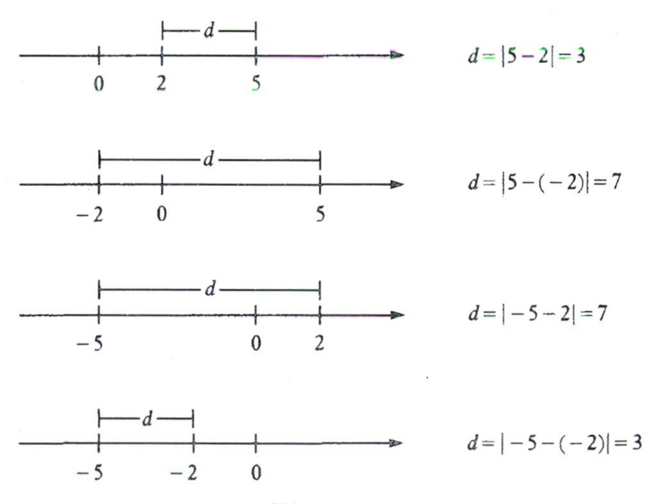

Figure 21

Now let $P_1(x_1, y_1)$ and $P_2(x_2, y_2)$ be two points in the plane where the line segment P_1P_2 from P_1 to P_2 is neither horizontal nor vertical (Figure 22). Then P_1P_2 is the hypotenuse of a right triangle whose vertices are P_1, P_2, and $P_3(x_1, y_2)$.

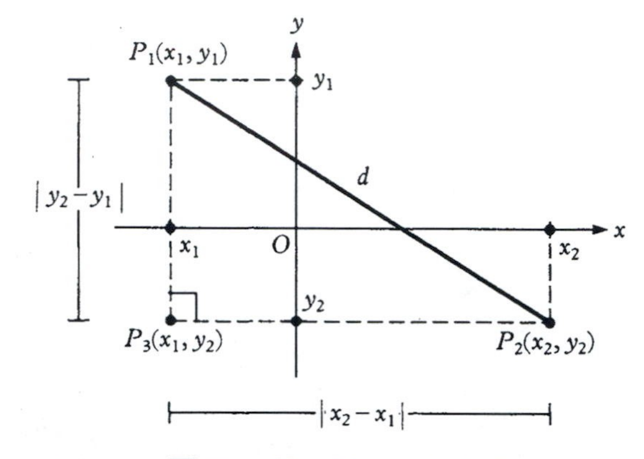

Figure 22 Distance in the Plane

By our previous formula for distance along a number line, the base P_3P_2 of the triangle has length $|x_2 - x_1|$ and the height P_3P_1 has length $|y_2 - y_1|$. Therefore, by the **Pythagorean theorem,** the distance d between P_1 and P_2 satisfies the equation

$$\begin{aligned} d^2 &= |x_2 - x_1|^2 + |y_2 - y_1|^2 \\ &= (x_2 - x_1)^2 + (y_2 - y_1)^2. \end{aligned}$$

Hence,

$$d = \sqrt{(x_2 - x_1)^2 + (y_2 - y_1)^2}. \qquad \textbf{distance formula in the plane}$$

Example 5 Compute the distance between $P_1(5, 3)$ and $P_2(7, -1)$.

Solution:
$$\begin{aligned} d &= \sqrt{(7 - 5)^2 + (-1 - 3)^2} \\ &= \sqrt{4 + 16} \\ &= \sqrt{20} \\ &= 2\sqrt{5} \end{aligned}$$ ■

If P_1P_2 is a horizontal line, then $y_1 = y_2$ and $d = |x_2 - x_1| = \sqrt{(x_2 - x_1)^2 + 0^2}$. Also, if P_1P_2 is vertical, then $x_1 = x_2$ and $d = |y_2 - y_1| = \sqrt{0^2 + (y_2 - y_1)^2}$. Therefore, the distance formula applies to *any* two points in the plane. We also use the notation $d(P_1, P_2)$ to denote the distance between P_1 and P_2, especially when more than two points are under consideration.

Question *In the distance formula, does it matter which point is labeled P_1 and which is labeled P_2?*

Answer: No. If P_1 and P_2 are interchanged, then $x_2 - x_1$ becomes $x_1 - x_2$ and $y_2 - y_1$ becomes $y_1 - y_2$, but $(x_1 - x_2)^2 = (x_2 - x_1)^2$ and $(y_1 - y_2)^2 = (y_2 - y_1)^2$. Therefore, d remains the same. ■

Example 6 Show that the points $P_1(2, 3)$, $P_2(8, 3)$, and $P_3(5, 3 + 3\sqrt{3})$ are the vertices of an equilateral triangle.

Solution:

$$d(P_1, P_2) = \sqrt{(8-2)^2 + (3-3)^2} = 6$$
$$d(P_1, P_3) = \sqrt{(5-2)^2 + (3 + 3\sqrt{3} - 3)^2} = 6$$
$$d(P_2, P_3) = \sqrt{(5-8)^2 + (3 + 3\sqrt{3} - 3)^2} = 6$$

The triangle is shown in Figure 23. Since its three sides are of equal length, it is equilateral. ■

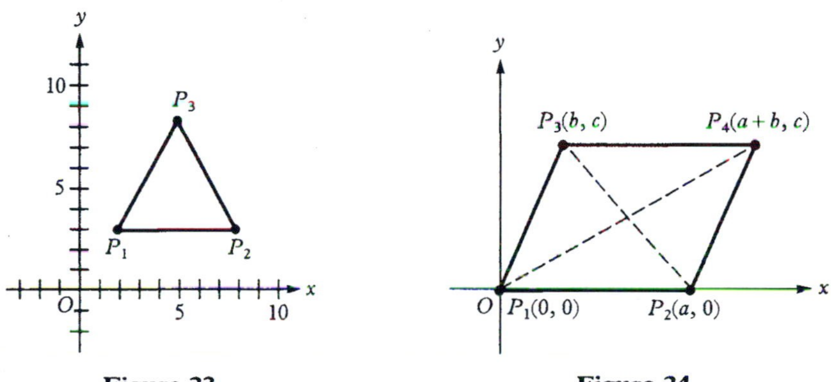

Figure 23 **Figure 24**

Example 7 Show that the sum of the squares of the sides of a parallelogram is equal to the sum of the squares of the diagonals.

Solution: Any parallelogram can be positioned as shown in Figure 24. The lengths of the sides are

$$d(P_1, P_2) = a = d(P_3, P_4)$$

and $$d(P_1, P_3) = \sqrt{b^2 + c^2} = d(P_2, P_4).$$

The lengths of the diagonals are

$$d(P_1, P_4) = \sqrt{(a+b)^2 + c^2} = \sqrt{a^2 + 2ab + b^2 + c^2}$$

and $$d(P_2, P_3) = \sqrt{(b-a)^2 + c^2} = \sqrt{b^2 - 2ab + a^2 + c^2}.$$

Therefore,

$$d^2(P_1, P_2) + d^2(P_3, P_4) + d^2(P_1, P_3) + d^2(P_2, P_4) = 2(a^2 + b^2 + c^2),$$

and

$$d^2(P_1, P_4) + d^2(P_2, P_3) = a^2 + 2ab + b^2 + c^2 + b^2 - 2ab + a^2 + c^2$$
$$= 2(a^2 + b^2 + c^2),$$

which proves the desired result. ■

The Midpoint Formula. Let x_1 and x_2 be any two points on a number line, and let x be the midpoint of the line segment from x_1 to x_2, as shown in Figure 25.

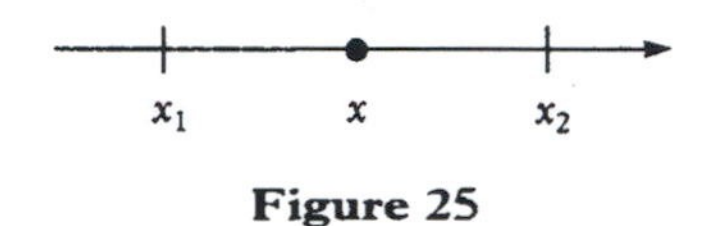

Figure 25

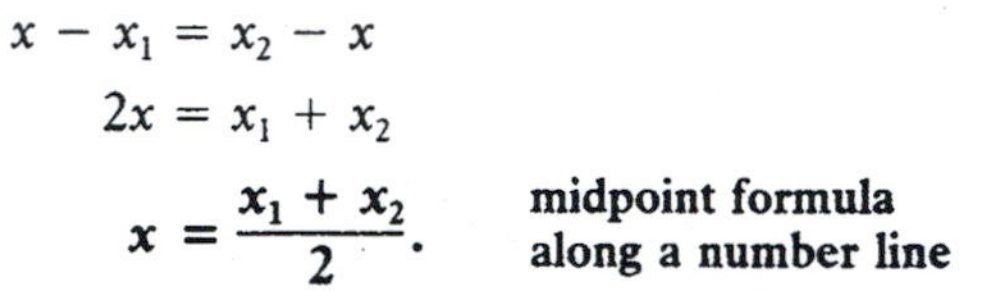

Then

$$x - x_1 = x_2 - x$$
$$2x = x_1 + x_2$$
$$x = \frac{x_1 + x_2}{2}. \quad \textbf{midpoint formula along a number line}$$

Therefore, the midpoint is located at the **average value** of x_1 and x_2. Although x_2 is to the right of x_1 in Figure 25, the location of the midpoint does not depend on which value is larger. Therefore, the formula is valid regardless of the relative positions of x_1 and x_2.

Example 8 Find the midpoint of the two points in each of the following cases.

(a) $x_1 = 3; x_2 = 7$ (b) $x_1 = -3; x_2 = 7$

(c) $x_1 = 3; x_2 = -7$ (d) $x_1 = -3; x_2 = -7$

Solution: Each case is illustrated in Figure 26. ■

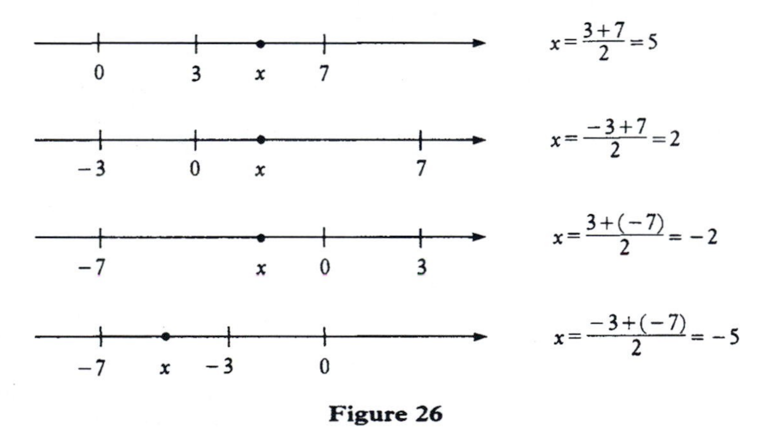

Figure 26

If we let $P_1(x_1, y_1)$ and $P_2(x_2, y_2)$ be any two points in the plane and $P(x, y)$ be the midpoint of the line segment P_1P_2 (Figure 27), then $Q(x, y_1)$ is the midpoint of the horizontal line segment from P_1 to $P_3(x_2, y_1)$, and $R(x_2, y)$ is the midpoint of the vertical line segment from P_3 to P_2 (*why?*). Therefore, by the midpoint formula for points along a number line, the coordinates of P are

$$x = \frac{x_1 + x_2}{2} \quad \text{and} \quad y = \frac{y_1 + y_2}{2}. \quad \textbf{midpoint formula in the plane}$$

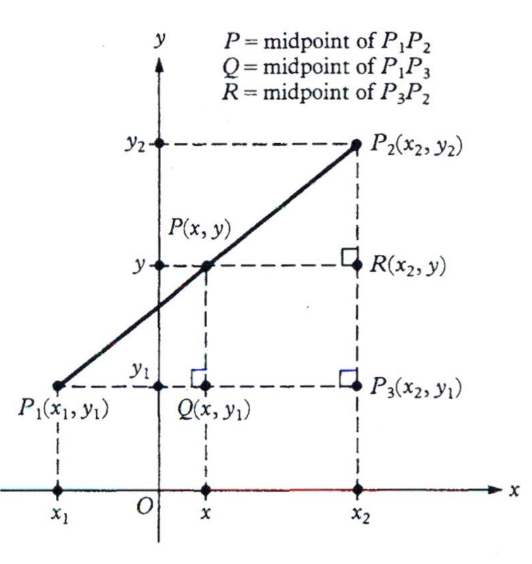

Figure 27 Midpoint of a Line Segment

Example 9 Find the distance between the point (5, 0) and the midpoint of the line segment from (1, 2) to (7, −8).

Solution: As shown in Figure 28, the midpoint P has coordinates

$$\left(\frac{1+7}{2}, \frac{2-8}{2}\right) = (4, -3).$$

The distance between (5, 0) and (4, −3) is

$$d = \sqrt{(4-5)^2 + (-3-0)^2} = \sqrt{10}.$$ ■

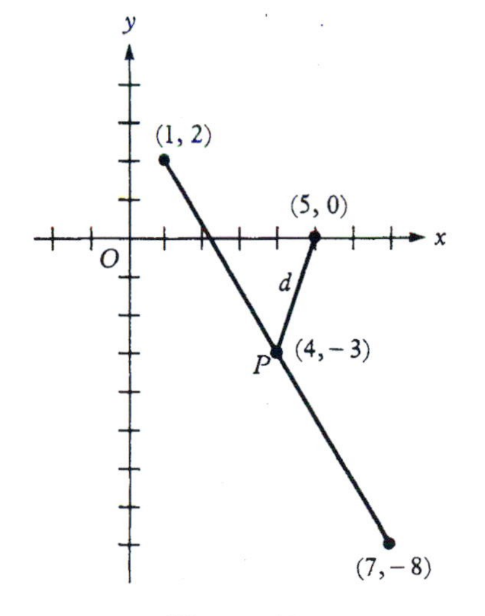

Figure 28

Example 10 Show that the diagonals of a parallelogram bisect each other.

Solution: With reference to Figure 24 in Example 7, the midpoint of diagonal P_1P_4 has coordinates

$$\left(\frac{a+b}{2}, \frac{c}{2}\right),$$

and these are also the coordinates of the midpoint of diagonal P_2P_3. Since both diagonals have the same midpoint, it follows that the diagonals bisect each other. ∎

Example 11 Show that the midpoints of the sides of any quadrilateral are the vertices of a parallelogram.

Solution: As shown in Figure 29, let P_1, P_2, P_3, P_4 be the vertices of a quadrilateral. The midpoints are M_1, M_2, M_3, M_4, where

$$M_1 = \left(\frac{x_1+x_2}{2}, \frac{y_1+y_2}{2}\right) \qquad M_3 = \left(\frac{x_3+x_4}{2}, \frac{y_3+y_4}{2}\right)$$

$$M_2 = \left(\frac{x_2+x_3}{2}, \frac{y_2+y_3}{2}\right) \qquad M_4 = \left(\frac{x_1+x_4}{2}, \frac{y_1+y_4}{2}\right).$$

We prove that $M_1M_2M_3M_4$ is a parallelogram by showing that opposite sides are equal in length. Now

$$d(M_1, M_2) = \sqrt{\left(\frac{x_3-x_1}{2}\right)^2 + \left(\frac{y_3-y_1}{2}\right)^2} = d(M_3, M_4)$$

and $$d(M_2, M_3) = \sqrt{\left(\frac{x_4-x_2}{2}\right)^2 + \left(\frac{y_4-y_2}{2}\right)^2} = d(M_1, M_4).$$

Therefore, $M_1M_2M_3M_4$ is a parallelogram. ∎

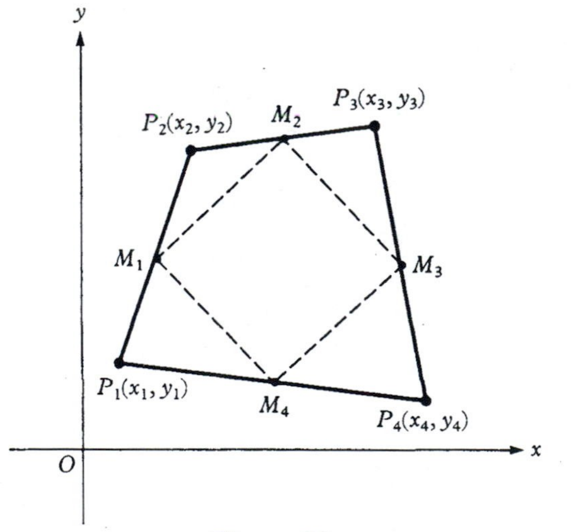

Figure 29

Comment Until the seventeenth century, algebra and geometry developed for the most part as separate subjects. Algebra was concerned mainly with solving equations, while plane geometry studied figures constructed with ruler and compass, the conic sections, and other loci. When the two subjects did meet, it was usually the case of a geometric construction to solve an algebraic equation. That is, the relationship was more geometric algebra than algebraic geometry. However, in the seventeenth century, Pierre Fermat (1601–65) and René Descartes (1596–1650) began to systematically apply the methods of algebra to geometry. Their new idea was the "coordinate plane," which enabled them to identify a curve with the locus of points that satisfy an equation in two variables x and y. This blend of algebra and geometry forged a major new branch of mathematics called **analytic geometry.** Analytic geometry, which is also called **coordinate geometry,** provides the perfect setting for the development of calculus, and it is an indispensable tool in applied mathematics and science. Much of the material in the remainder of this book is analytic geometry.

Exercises 2.4

Fill in the blanks to make each statement true.

1. The x-coordinate of a point P is also called the _______________ of P, and the y-coordinate is also called the _______________ of P.
2. The x-coordinate of every point on the y-axis is _______________.
3. If $a > 0$ and $b < 0$, then the point $(-a, b)$ is in quadrant ________, and $(a, -b)$ is in quadrant ________.
4. If (a, b) is in the second quadrant, then $(-a, -b)$ is in quadrant ________, and (b, a) is in quadrant ________.
5. The distance formula in the plane is equivalent to the _______________ theorem from geometry.

Write true *or* false *for each statement.*

6. If the ordinate of a point P is 0, then P lies on the y-axis.
7. The distance from $P_1(x_1, y_1)$ to $P_2(x_2, y_2)$ is $|x_2 - x_1| + |y_2 - y_1|$.
8. If $d(P_1, P_2) = d(P_2, P_3)$, where P_1, P_2, and P_3 are in the plane, then P_2 is the midpoint of the line segment P_1P_3,
9. If $d(P_1, P_2) + d(P_2, P_3) = d(P_1, P_3)$, where P_1, P_2, and P_3 are in the plane, then P_2 is on the line segment P_1P_3.
10. The points (a, b) and (b, a) are always in the same quadrant.

Coordinates

Accompany each of the following with an appropriate diagram.

11. On a coordinate plane, locate each of the following points.
    (a) $(5, 2)$ (b) $(-4, 3)$
    (c) $(-6, -5)$ (d) $(5, -8)$
12. If $a < 0$ and $b > 0$, determine the quadrant of each of the following points.
    (a) (a, b) (b) $(-a, b)$
    (c) $(a, -b)$ (d) $(-a, -b)$

13. Three vertices of a rectangle are $(-1, 2)$, $(4, 2)$, and $(-1, -3)$. Find the fourth vertex.

14. Three vertices of a parallelogram are $(-3, 2)$, $(1, 7)$, and $(12, 2)$. Find three choices for the fourth vertex.

15. Find two points P_1 and P_2 such that P_1, P_2, $(-4, -2)$, and $(5, 3)$ are the vertices of a rectangle whose sides are parallel to the coordinate axes.

16. Find two points P_1 and P_2 such that P_1, P_2, $(2, 0)$, and $(2, 6)$ are the vertices of a square.

17. Find a point P for which $Q = (2, 3)$, $R = (10, 3)$ and P are the vertices of an isosceles triangle (two sides equal) with base QR and height 6.

18. Find a point P for which $Q = (3, -1)$, $R = (3, 7)$, and P are the vertices of a right triangle with area 20 and height QR.

19. Find the coordinates of another point on the line through $(1, 3)$ and $(5, 6)$.

20. Find the point at which the line through $(1, 8)$ and $(2, 6)$ intersects the x-axis.

Distance

Include a coordinate diagram for each of the following.

21. Compute $d(P_1, P_2)$ for each of the following pairs of points.
    (a) $P_1(2, 1)$, $P_2(-3, 4)$ (b) $P_1(-4, -2)$, $P_2(4, 2)$
    (c) $P_1(5, 0)$, $P_2(3, 6)$ (d) $P_1(8, -5)$, $P_2(-5, 8)$

22. Find the perimeter of the triangle with vertices $P_1(3, 1)$, $P_2(-4, 2)$, and $P_3(2, -3)$.

23. Find the lengths of the diagonals of the quadrilateral with vertices $P_1(2, 0)$, $P_2(5, 0)$, $P_3(7, 4)$, and $P_4(0, 6)$.

24. Given $P_1(2, -3)$, $P_2(4, 1)$, and $P(x, 5)$, find the value of x for which $d(P, P_1) = d(P, P_2)$.

25. Given $P_1(3, 4)$, $P_2(-1, 6)$, and $P(2, y)$, find the value of y for which $d(P, P_1) = d(P, P_2)$.

26. Show that $P_1(2, 1)$, $P_2(4, -1)$, and $P_3(7, 4)$ are the vertices of an isosceles triangle (two sides equal in length).

27. Show that $P_1(0, 0)$, $P_2(1, \sqrt{3})$, and $P_3(-1, \sqrt{3})$ are the vertices of an equilateral triangle.

28. Show that $P_1(-1, 4)$, $P_2(2, 7)$, $P_3(5, 4)$, and $P_4(2, 1)$ are the vertices of a square.

29. Are $P_1(-4, 4)$, $P_2(1, 10)$, $P_3(6, 4)$, and $P_4(1, -2)$ the vertices of a square? Why or why not?

30. Show that $P_1(2, 5)$, $P_2(4, -1)$, and $P_3(-7, 2)$ are the vertices of a right triangle.

31. A diameter of a circle has endpoints $P_1(-1, 4)$ and $P_2(10, 1)$. Is the point $P(8, 7)$ on the circle? Why or why not?

32. Show that the points $P_1(5, -1)$, $P_2(6, 0)$, and $P_3(-1, 7)$ lie on a circle with center $P(2, 3)$.

33. Show that the points $P_1(-3, 4)$, $P_2(2, 1)$, and $P_3(7, -2)$ are collinear (lie on the same straight line).

34. Jason walks 2 mi east, then 4 mi north, and then $5\sqrt{2}$ mi northeast. How far is Jason from the starting point?

35. Lynn walks 1 mi east and then 2 mi northeast. She returns by walking west and then south. How long was the entire walk?

Midpoints

Work each of the following exercises with the aid of a diagram.

36. Find the midpoint of the line segment P_1P_2 in each of the following cases.
    (a) $P_1(4, 5)$, $P_2(-3, 3)$ (b) $P_1(8, 0)$, $P_2(0, 8)$
    (c) $P_1(-6, 1)$, $P_2(-2, 7)$ (d) $P_1(2, 7)$, $P_2(-2, -7)$

37. Find the midpoints of the sides of the quadrilateral with vertices $P_1(1, 5)$, $P_2(-2, 7)$, $P_3(-3, -6)$, $P_4(4, -8)$.

38. $P(2, 3)$ is the midpoint of line segment P_1P_2, where P_1 has coordinates $(-4, 5)$. Find the coordinates of P_2.

39. $P(x, 4)$ is the midpoint of the line segment from $P_1(2, 5)$ to $P_2(-7, y_2)$. Find x and y_2.

40. The points $P_1(-1, 2)$ and $P_2(3, 2)$ are on a circle with center $(1, 6)$. Find two other points on the circle.

41. $P(3, -2)$ is the center of a circle of radius 5. Find four points on the circle.

42. One diagonal of a parallelogram has endpoints $P_1(4, 5)$ and $P_2(-1, -3)$. The other diagonal is a vertical line segment of length 6. Find the other two vertices of the parallelogram.

43. $P_1(-1, 2)$ is a vertex of a parallelogram. One diagonal of the parallelogram is a horizontal segment of length 6 with one endpoint $P_2(5, 4)$. Find the other two vertices of the parallelogram (two solutions).

44. Show that the line segments joining the midpoints of opposite sides of a quadrilateral bisect each other.

45. Show that the larger of any two points x_1, x_2 on the x-axis is $(x_1 + x_2)/2 + |x_2 - x_1|/2$, and the smaller is $(x_1 + x_2)/2 - |x_2 - x_1|/2$.

Miscellaneous

46. Show that if $P(x, y)$ is the point on the line segment P_1P_2 for which

$$\frac{d(P_1, P)}{d(P_1, P_2)} = r \quad (0 \le r \le 1),$$

then

$$x = x_1 + r(x_2 - x_1)$$

and $$y = y_1 + r(y_2 - y_1). \qquad \textbf{ratio formula}$$

Show that the midpoint formula is the special case of the ratio formula corresponding to $r = 1/2$.

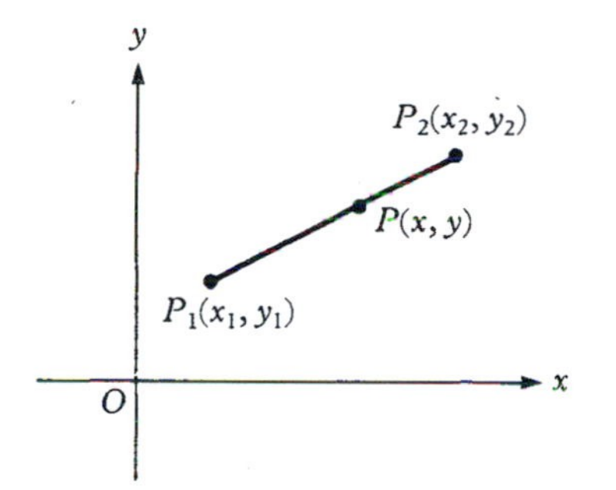

47. Use the ratio formula to show that the medians of a triangle intersect in a point whose distance from a given vertex is two thirds the length of the corresponding median. (A median of a triangle is a line segment from a vertex to the midpoint of the opposite side.)

48. Show that the sum of the squares of the medians of a triangle is equal to three fourths the sum of the squares of the sides.

49. Show that the midpoint of the hypotenuse of a right triangle is equidistant from the three vertices.

50. Show that the length of the line segment joining the midpoints of two sides of a triangle is one half the length of the remaining side.

2.5 Equations of Straight Lines

Single points have no length, but if you place enough of them side by side, you can fill out a straight line!

Slope of a Line
Point-Slope Equation
Slope-Intercept Equation
General Linear Equation

Until now we have considered equations in one unknown x whose solutions are points on the real line. We now begin a study of equations in two unknowns x and y whose solutions are sets of points in the plane. The set of all points satisfying a given equation is called the **graph** of the equation. Our study will focus on the following two problems:

1. given an equation, find its graph;
2. given a geometric figure, find an equation whose graph is the given figure.

These problems will play a central role throughout the rest of this book, a role that carries over into calculus. We start here with their application to straight lines.

Slope of a Line. Two points $P_1(x_1, y_1)$ and $P_2(x_2, y_2)$ determine a line L. Suppose L is not a vertical line, and let $P(x, y)$ be another point on L. By equating ratios of corresponding sides of the similar right triangles P_1RP and P_1QP_2 (Figure 30), we conclude that

$$\frac{y - y_1}{x - x_1} = \frac{y_2 - y_1}{x_2 - x_1}. \tag{1}$$

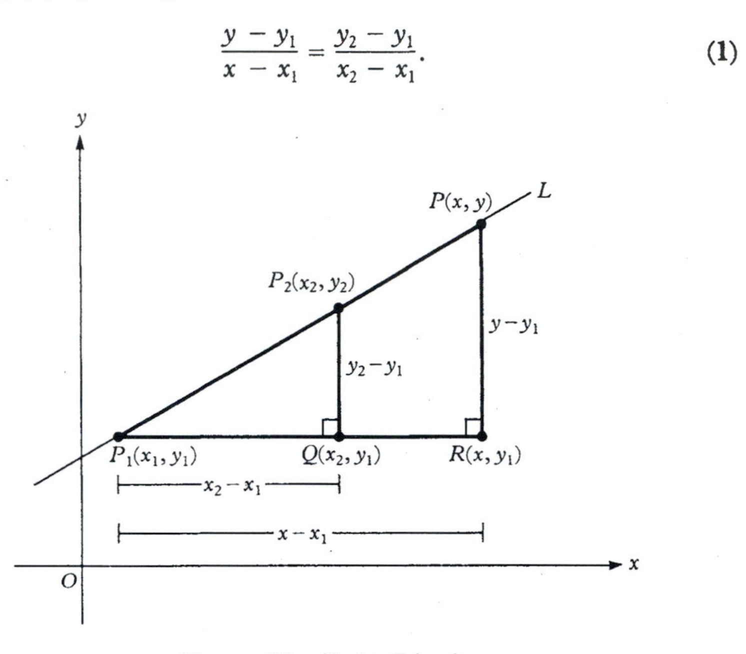

Figure 30 Similar Triangles

Conversely, if (1) is satisfied for some pair (x, y), then the point $P(x, y)$ lies on L.

Example 1 Show that the three points $P_1(-1, 2)$, $P_2(3, 5)$, and $P_3(7, 8)$ are collinear (lie on the same line).

Solution:

$$\frac{y_2 - y_1}{x_2 - x_1} = \frac{5 - 2}{3 - (-1)} = \frac{3}{4} \quad \text{and} \quad \frac{y_3 - y_1}{x_3 - x_1} = \frac{8 - 2}{7 - (-1)} = \frac{3}{4}.$$

Hence, P_3 is on the line determined by P_1 and P_2. ■

With reference to Figure 30, the quantity

$$m = \frac{y_2 - y_1}{x_2 - x_1} \tag{2}$$

is called the **slope** of L. The slope of a nonvertical line L is a geometric parameter whose value can be computed by means of any two points P_1 and P_2 on L. As Figure 31 illustrates, m is a measure of the *steepness* and *direction* of L. Lines that rise from left to right have *positive* slopes [Figures

31(a) and 31(b)], and lines that fall have *negative* slopes [Figures 31(c) and 31(d)]. Also, the steeper a line is, the larger its slope is in *magnitude*.

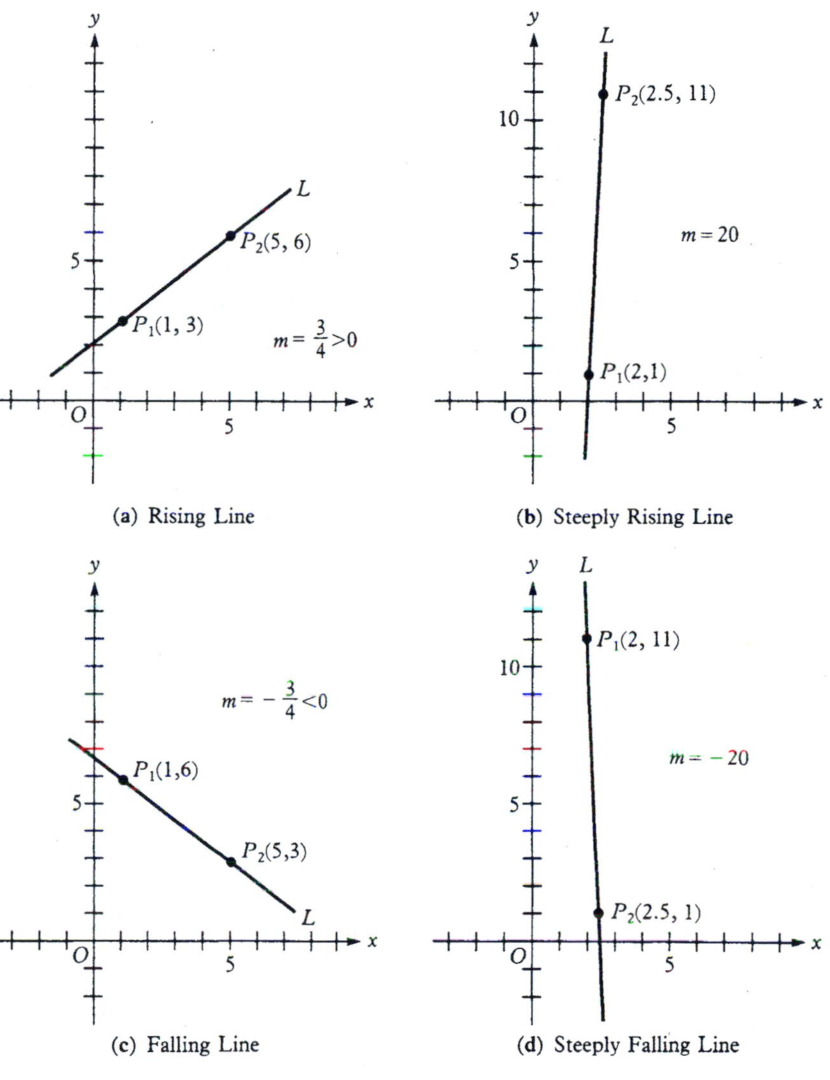

Figure 31

Question 1 *In computing the slope of L by means of two points, does it matter which of the two is labeled P_1 and which is P_2?*

Answer: No. If the points are interchanged, then $x_2 - x_1$ becomes $x_1 - x_2$, and $y_2 - y_1$ becomes $y_1 - y_2$, but

$$\frac{y_1 - y_2}{x_1 - x_2} = \frac{y_2 - y_1}{x_2 - x_1}.$$

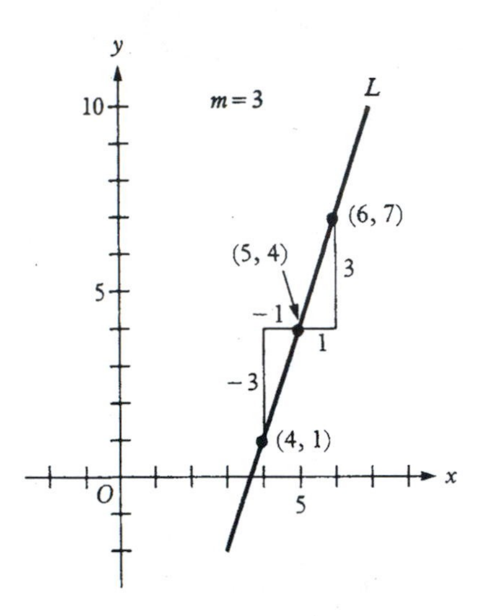

Figure 32

Therefore, the slope remains the same. Hence,

$$m = \frac{\text{change in } y}{\text{change in } x}$$

in going from *any* point on L to *any other* point on L. Note that if $x_2 - x_1 = \pm 1$, then $y_2 - y_1 = \pm m$. That is, if x changes by ± 1 unit, then y changes by $\pm m$ units. For example, if L has slope 3 and (5, 4) lies on L, then $(5 + 1, 4 + 3) = (6, 7)$ and $(5 - 1, 4 - 3) = (4, 1)$ also lie on L (Figure 32). ∎

Two extreme cases regarding steepness are *horizontal* and *vertical* lines (Figure 33). A horizontal line has slope 0 because the numerator $y_2 - y_1$ in equation (2) is 0. For a vertical line, m is *not defined* because the denominator $x_2 - x_1$ in equation (2) is 0. To put it another way, there is no real number large enough to be the slope of a vertical line.

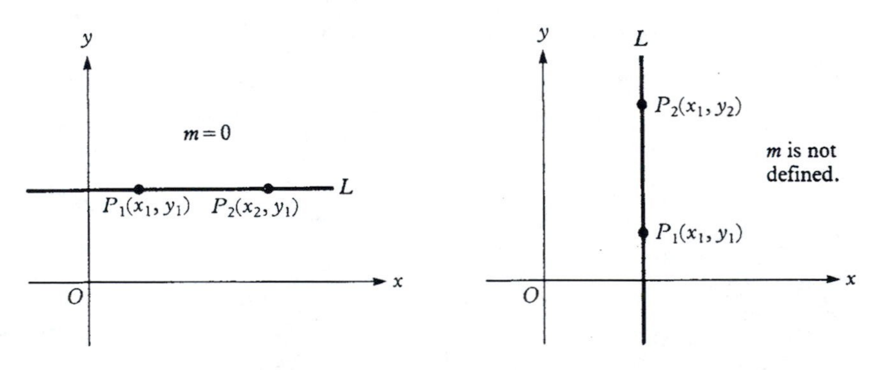

Figure 33 Horizontal Line and Vertical Line

Point-Slope Equation. Let L be a line that has slope m and passes through the point $P_1(x_1, y_1)$. By combining equations (1) and (2), we see that a point $P(x, y)$ other than P_1 lies on L if and only if

$$\frac{y - y_1}{x - x_1} = m,$$

that is,

$$y - y_1 = m(x - x_1). \tag{3}$$

Equation (3) has line L for its graph and is called a **point-slope equation** of L. We note that although equation (3) was derived under the assumption that $P(x, y)$ was a point on L other than $P_1(x_1, y_1)$, the equation is also satisfied when $x = x_1$ and $y = y_1$. That is, *every* point on L satisfies (3).

Example 2 Find a point-slope equation of the line L that has slope 2 and passes through the point (1, 3), and then graph L.

Solution: From (3), the corresponding point-slope equation of L is

$$y - 3 = 2(x - 1).$$

One point on L is, of course, (1, 3). Another is (2, 5) (*why?*). These two points can be used to graph L (Figure 34). ■

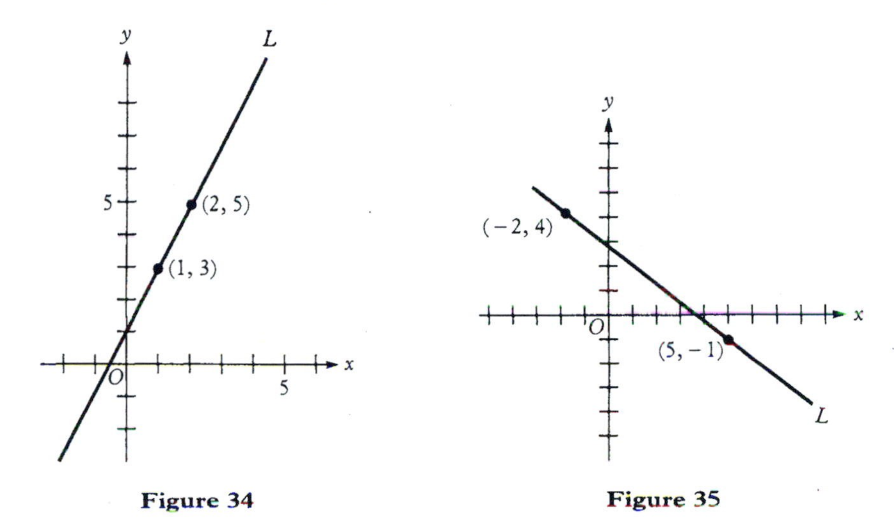

Figure 34 **Figure 35**

Example 3 Graph the line L that passes through the points (−2, 4) and (5, −1). Find a point-slope equation of L.

Solution: The graph of L through the two given points is shown in Figure 35. From (2), the slope of L is

$$m = \frac{(-1) - 4}{5 - (-2)} = -\frac{5}{7}.$$

Therefore, a point-slope equation of L can be written as either

$$y - 4 = -\frac{5}{7}(x + 2) \qquad \textit{Using the point } (-2, 4)$$

or

$$y + 1 = -\frac{5}{7}(x - 5). \qquad \textit{Using the point } (5, -1)$$

Note that both of these equations are equivalent to

$$y = -\frac{5}{7}x + \frac{18}{7}.$$ ■

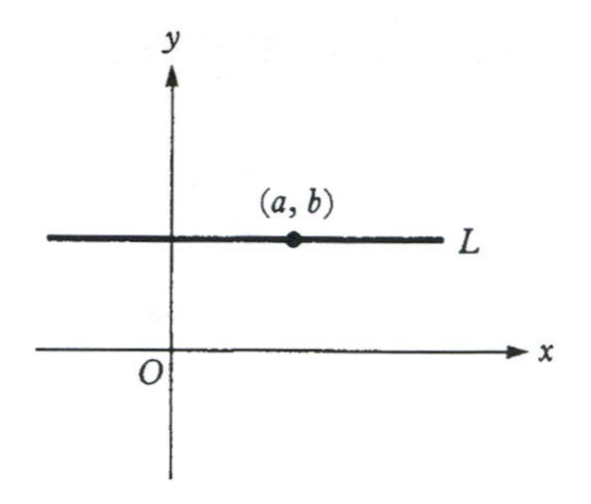

Figure 36

A horizontal line L has slope $m = 0$. If L passes through the point (a, b) as in Figure 36, then the corresponding point-slope equation of L is

$$y - b = 0(x - a),$$

or $$\mathbf{y = b.}$$ **horizontal line**

For example, the equation of the horizontal line through the point $(-2, 3)$ is $y = 3$, and the equation of the x-axis is $y = 0$. In the plane, we can interpret the equation $y = b$ as saying "y must equal the constant value b, but x can equal any value."

A vertical line L has undefined slope and therefore no point-slope equation. However, if L passes through the point (a, b) as in Figure 37, then for every point on L, the x-coordinate is equal to a. Therefore, L is the graph of the equation

$$\mathbf{x = a.}$$ **vertical line**

For example, the equation of the vertical line through $(-2, 3)$ is $x = -2$, and the equation of the y-axis is $x = 0$. In the plane, the equation $x = a$ says "x must have the constant value a, but y can have any value."

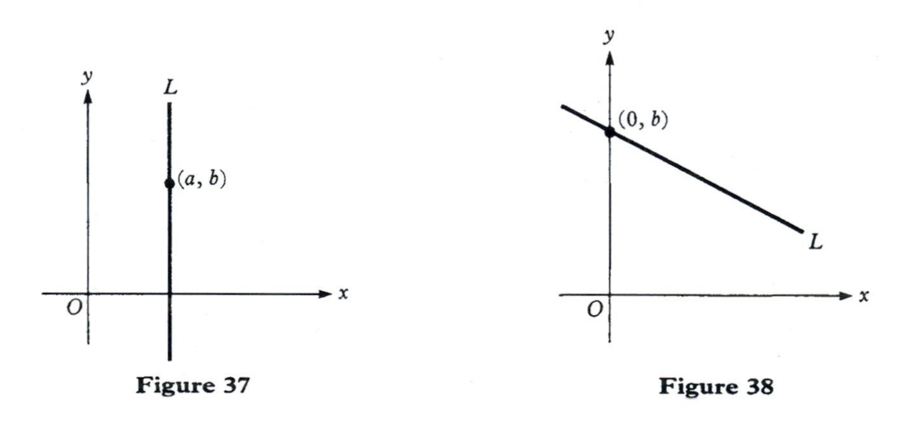

Figure 37

Figure 38

Slope-Intercept Equation. A nonvertical line L will intersect the y-axis at a point whose coordinates are $(0, b)$ for some value of b (Figure 38). The point $(0, b)$, or simply b itself, is called the ***y*-intercept** of L. By using $(0, b)$ for (x_1, y_1) in the point-slope equation (3), we obtain

$$y - b = m(x - 0),$$

or $$\mathbf{y = mx + b,} \tag{4}$$

where m is the slope of L. Equation (4) is called the **slope-intercept equation** of L.

Example 4 Let L be the line with point-slope equation $y - 5 = 3(x - 1)$. Find the y-intercept and the slope-intercept equation of L, and then graph L.

Solution: By substituting $x = 0$ in the point-slope equation, we obtain $y = 2$ as the y-intercept. Therefore, the slope-intercept equation of L is

$$y = 3x + 2.$$

The points (1, 5) and (0, 2) can be used to graph L (Figure 39).

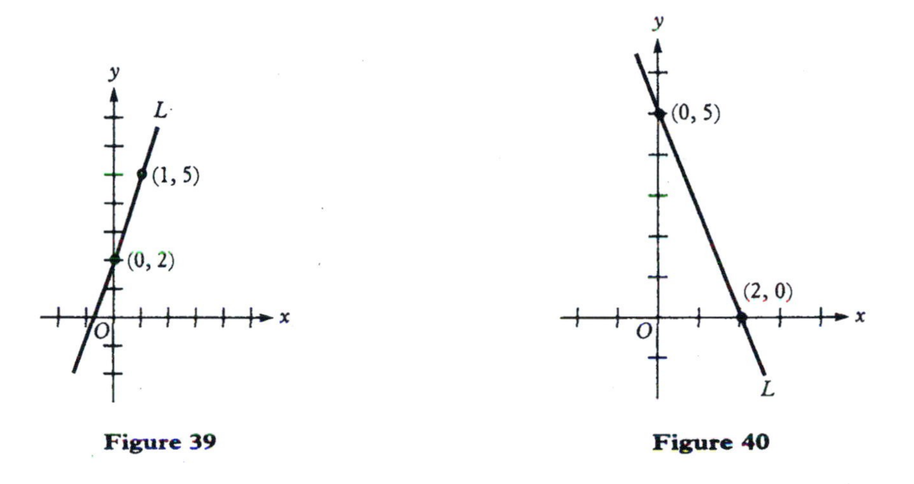

Figure 39 **Figure 40**

Alternate Solution:

$y - 5 = 3(x - 1)$	*Point-slope equation of L*
$y = 3x - 3 + 5$	*Solve for y in terms of x*
$y = 3x + 2$	*Slope-intercept equation of L* ■

We note that a nonvertical line L of slope m has exactly one y-intercept b. Therefore, the slope-intercept equation of L is *unique*, whereas a point-slope equation of L depends on which point is chosen.

Example 5 Find the slope-intercept equation of the line L that crosses the x-axis at (2, 0) and the y-axis at (0, 5), and then graph L.

Solution: The slope of L is $(5 - 0)/(0 - 2) = -5/2$, and the y-intercept is 5. Therefore, the slope-intercept equation of L is

$$y = -\frac{5}{2}x + 5.$$

The graph of L is shown in Figure 40. ■

Example 6 Show that the point-slope equations $y - 2 = 3(x + 1)$ and $y - 8 = 3(x - 1)$ each have the same graph.

Solution:

$$\begin{aligned} y - 2 &= 3(x + 1) \qquad & y - 8 &= 3(x - 1) \\ y - 2 &= 3x + 3 & y - 8 &= 3x - 3 \\ y &= 3x + 5 & y &= 3x + 5 \end{aligned}$$

Each equation has for its graph the line with slope 3 and y-intercept 5 (Figure 41). ■

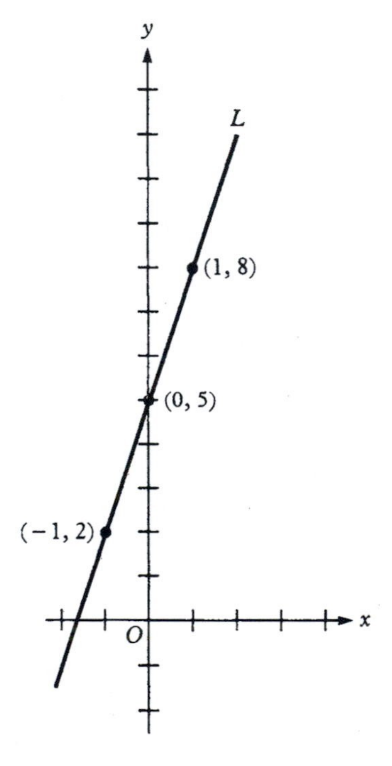

Figure 41

General Linear Equation. The equation

$$Ax + By = C, \tag{5}$$

where either $A \neq 0$ or $B \neq 0$, is called the **general linear equation.** As the name implies, the graph of (5) is a straight line. There are three possible cases.

1. If $A = 0$ and $B \neq 0$, then $y = C/B$, a horizontal line. For example, the graph of $2y = 3$ is the horizontal line $y = 3/2$.

2. If $A \neq 0$ and $B = 0$, then $x = C/A$, a vertical line. For example, the graph of $-5x = 7$ is the vertical line $x = -7/5$.
3. If $A \neq 0$ and $B \neq 0$, then $y = -(A/B)x + C/B$, a line with slope $-A/B$ and y-intercept C/B. For example, $-2x + 3y = 5$ is equivalent to $y = (2/3)x + 5/3$, whose graph is the straight line with slope $2/3$ and y-intercept $5/3$.

Hence, the general linear equation covers all possible lines, including vertical ones.

Example 7 Graph the line L: $3x - 5y = 15$.

Solution: Any two points that satisfy the equation will determine the graph. For instance, if $x = 0$, then $y = -3$. Also, if $y = 0$, then $x = 5$. Hence, L passes through the points $(0, -3)$ and $(5, 0)$ (Figure 42). ∎

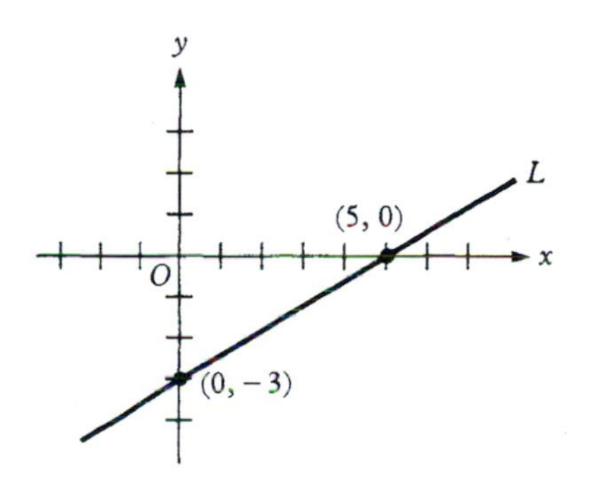

Figure 42

Example 8 Line L passes through the points $(3, -5)$ and $(-4, 6)$. Write an equation of L in (a) point-slope form, (b) slope-intercept form, and (c) general linear form.

Solution: First find the slope of L:

$$m = \frac{6 - (-5)}{-4 - 3} = -\frac{11}{7}.$$

(a) $y + 5 = -\frac{11}{7}(x - 3)$ *Use point* $(3, -5)$

$y - 6 = -\frac{11}{7}(x + 4)$ *Use point* $(-4, 6)$

(b) $y = -\frac{11}{7}x - \frac{2}{7}$ *Obtained from either equation in (a) by solving for y in terms of x*

(c) $11x + 7y = -2$ *Obtained from (b)* ∎

Question 2 *What is the graph of $Ax + By = C$ if $A = 0$ and $B = 0$?*

Answer: The graph of $0x + 0y = C$ depends on C. If $C = 0$, then *every* point (x, y) satisfies the equation, and the graph is the *whole plane*. If $C \neq 0$, then *no* point (x, y) satisfies the equation, and the graph is the *empty set*. To avoid these extreme cases, we insist that either $A \neq 0$ or $B \neq 0$ in the general linear equation. ■

Comment In Section 2.1, where we considered equations in one variable, the equation $ax + b = 0$ was called a linear equation. In this section, $Ax + By = C$ is called a linear equation. Later in the text, when we consider three variables x, y, and z, the equation $Ax + By + Cz = D$ will be called a linear equation. Of the three, only the second one actually has a line for its graph. However, all three equations have the same *form* with respect to the number of variables considered. In general, for n variables $x_1, x_2, \ldots, x_n$, a **linear equation** is of the form

$$A_1x_1 + A_2x_2 + \cdots + A_nx_n = C.$$

Exercises 2.5

Fill in the blanks to make each statement true.

1. The set of all points in the plane satisfying a given equation is called the ____________ of the equation.
2. The ____________ of a line is a geometric parameter that measures the steepness and direction of the line.
3. If a line L has slope 4, and the point (2, 1) is on L, then the points (3, ____) and (1, ____) are also on L.
4. Three forms of equations of a straight line are ____________, ____________, and ____________.
5. The horizontal line through the point (3, −4) has equation ____________, and the vertical line through the same point has equation ____________.

Write true *or* false *for each statement.*

6. Two different lines can have the same slope.
7. Two different lines can have the same y-intercept.
8. Two different lines can have the same slope and y-intercept.
9. A line with slope −10 is steeper than a line with slope 5.
10. If (2, 1) and (3, 5) are on line L, then (4, 9) is also on L.

Slope

Compute, if possible, the slope of the line through each of the pairs of points P_1 and P_2 in Exercises 11–16.

11. $P_1(4, 3)$, $P_2(3, 4)$
12. $P_1(2, -5)$, $P_2(6, -7)$
13. $P_1(8, 1)$, $P_2(0, 3)$
14. $P_1(0, 0)$, $P_2(4, 5)$
15. $P_1(3, 7)$, $P_2(3, 5)$
16. $P_1(4, -1)$, $P_2(4, 3)$

Find the value of each indicated variable in Exercises 17–20.

17. L has slope 1, and points $P_1(2, 5)$ and $P_2(4, y)$ are on L. Find y.
18. L has slope −3, and points $P_1(1, 4)$ and $P_2(x, 2)$ are on L. Find x.
19. L has slope 2, and points $P_1(0, 5)$ and $P_2(x, 0)$ are on L. Find x.

20. L has slope -1, and points $P_1(3, 0)$ and $P_2(0, y)$ are on L. Find y.

Equations

Find a point-slope equation of each line L in Exercises 21–26, and then graph L.

21. L passes through point $P_1(-2, 4)$ and has slope $m = 3$.

22. L passes through point $P_1(5, -5)$ and has slope $m = -2$.

23. L passes through the points $P_1(3, 0)$ and $P_2(0, -3)$.

24. L passes through the points $P_1(6, -3)$ and $P_2(-1, 1)$.

25. L has slope $m = 2$ and passes through the midpoint of the line segment from $(-3, 4)$ to $(3, 4)$.

26. L passes through $P(-1, 2)$ and the midpoint of the line segment from $(4, 7)$ to $(2, 3)$.

Find the slope-intercept equation of each line L in Exercises 27—32, and then graph L.

27. L passes through point $P_1(3, 5)$ and has slope $m = 2$.

28. L passes through point $P_1(-4, 3)$ and has slope $m = -2$.

29. L passes through the points $P_1(4, 4)$ and $P_2(5, 7)$.

30. L passes through the points $P_1(-2, -3)$ and $P_2(3, 2)$.

31. L passes through the midpoint of the line segment from $P_1(4, -1)$ to $P_2(4, 5)$ and the midpoint of the segment from $Q_1(2, 3)$ to $Q_2(4, 3)$.

32. L passes through $P(2, 5)$ and the midpoint of the line segment from $(6, 2)$ to $(-8, 8)$.

Graph each of the following. Also, where possible, determine the slope and the y-intercept.

33. $4x - 3y = 7$

34. $-2x + 5y = 5$

35. $4x = 3$

36. $-2y = 5$

37. $x + y = 1$

38. $x - y = 0$

Miscellaneous

The graph of the equation

$$\frac{x}{a} + \frac{y}{b} = 1 \qquad \textbf{intercept equation}$$

is a straight line L with y-intercept b and x-intercept a [i.e., L intersects the x-axis at the point (a, 0)]. Find the intercept equation of the line L determined in each of the following cases, and then graph L.

39. L has x-intercept 4 and y-intercept 3.

40. L has x-intercept 2 and y-intercept -5.

41. L passes through the points $P_1(2, 4)$ and $P_2(8, -4)$.

42. L passes through the points $P_1(-3, 5)$ and $P_2(2, 0)$.

43. L passes through point $P_1(-2, -6)$ and has slope $m = 1$.

44. L passes through point $P_1(7, 4)$ and has slope $m = -1$.

The ratio formula of Exercise 46 in Section 2.4 gives the coordinates of a point $P(x, y)$ on the line segment from P_1 to P_2. That is, the equations

$$x = x_1 + r(x_2 - x_1)$$
$$y = y_1 + r(y_2 - y_1)$$

have for their graph, as the parameter r varies from 0 to 1, the line segment from $P_1(x_1, y_1)$ to $P_2(x_2, y_2)$. If the parameter r ranges through all *real numbers, the same equations have for their graph the* entire *line L through P_1* and *P_2. They are called **parametric equations** of L. Find parametric equations for each line L determined in Exercises 45–50, and then graph L.*

45. $P_1(2, -5)$, $P_2(5, -2)$

46. $P_1(4, 3)$, $P_2(-4, -3)$

47. $P_1(6, 2)$, $P_2(6, -2)$

48. $P_1(8, 6)$, $P_2(6, 8)$

49. L passes through point $P_1(7, 2)$ and has slope $m = 3$.

50. L passes through point $P_1(-3, 5)$ and has slope $m = -3$.

51. Use the distance formula in the plane to determine the equation of the perpendicular bisector of the line segment from $P_1(x_1, y_1)$ to $P_2(-x_1, -y_1)$.

52. The graph of the equation relating temperature in Fahrenheit (F) to temperature in Celsius (C) is a straight line L. With C as abscissa and F as ordinate, find the equation of L, given that water freezes at 0° C and 32° F, and it boils at 100° C and 212° F.

53. After school, Ted walks part of the way home and jogs the rest, both along the same straight line. One half hour after starting to jog, Ted is 3 mi from school, and after three fourths of an hour, he is 4 1/4 mi from school.

(a) How fast does Ted jog?

(b) How far from school was Ted when he started to jog?

(c) Let x = time spent jogging and y = distance from school. Express y in terms of x and graph the equation.

54. A gas station charges a certain rate per gallon for the first 10 gal and a reduced rate for every gallon over 10. Amy pays \$12.00 for 11 gal and \$16.00 for 15 gal.

(a) What is the rate per gallon for the first 10 gal?

(b) What is the rate per gallon after the first 10?

(c) Let y = total amount paid for gas and let x = the number of gallons over 10. Express y in terms of x, and graph the equation.

55. If P_0 dollars are invested at a monthly simple interest rate of r, the accumulated amount P after t months is

$$P = P_0 + P_0rt.$$

Lou invests P_0 dollars on December 31. On March 31 Lou has accumulated \$155.53, and on October 21 the amount is \$166.10.

(a) How much did Lou invest?

(b) What was the monthly interest rate?

(c) Graph the equation relating P to t.

2.6 Systems of Linear Equations

Two points determine a line, and in turn, two lines determine a point, possibly at infinity.

Parallel Lines
Perpendicular Lines
Intersecting Lines
Determinants
Applications

Two distinct lines in the plane either intersect in a single point or are parallel. Intersecting lines may or may not be perpendicular to each other. In this section we investigate these and other geometric properties of lines by means of their algebraic equations.

Parallel Lines. Distinct lines L_1 and L_2 are parallel if and only if they are both vertical or their slopes m_1 and m_2 are equal:

$$m_1 = m_2. \qquad \textbf{parallel lines (nonvertical)} \qquad (1)$$

Example 1 Which of the following pairs of equations represent parallel lines?

(a) $2x - 3y = 4$
$-4x + 6y = 5$

(b) $y = 3x - 1$
$y = -3x + 1$

(c) $x = -3$
$x = 3$

(d) $y - 1 = 2(x + 1)$
$y - 5 = 2(x - 1)$

Solution:

(a) The equations are equivalent to $y = (2/3)x - 4/3$ and $y = (2/3)x + 5/6$, respectively. Both of the corresponding lines have slope 2/3, so they are parallel.

(b) The lines have unequal slopes, so they are not parallel.

(c) The corresponding graphs are parallel vertical lines.
(d) Both equations are equivalent to $y = 2x + 3$, and therefore the lines coincide. Coincident lines are not considered to be parallel; by definition, a line is not parallel to itself. ■

Example 2 Find an equation of the line L through the point $(-2, 1)$ and parallel to the line $2x - 5y = 7$.

Solution: The line $2x - 5y = 7$ has slope $2/5$ (*why?*). Therefore, a point-slope equation of L is

$$y - 1 = \frac{2}{5}(x + 2) \qquad \text{or, equivalently,} \qquad 2x - 5y = -9.$$ ■

If L_1 and L_2 have general linear equations

$$L_1\colon A_1x + B_1y = C_1$$
$$L_2\colon A_2x + B_2y = C_2$$

then $m_1 = -A_1/B_1$ and $m_2 = -A_2/B_2$, assuming $B_1 \neq 0$ and $B_2 \neq 0$. Hence, the equation $m_1 = m_2$ is equivalent to $-A_1/B_1 = -A_2/B_2$, or

$$\mathbf{A_1B_2 - A_2B_1 = 0.} \qquad \textbf{parallel lines} \tag{2}$$

Furthermore, if L_1 and L_2 are vertical lines, and therefore parallel, then $B_1 = 0$ and $B_2 = 0$ and equation (2) still holds. Therefore, equation (2) is the algebraic condition for any two lines, vertical or not, to be parallel.

Caution As stated in Example 1 (d), the designation "parallel" does not apply to lines that coincide. Parallel lines are *distinct* lines in the plane that *do not intersect*. Coincident lines fail on both counts. To determine whether nonvertical lines coincide, write their respective equations in slope-intercept form by solving for y in terms of x. If these equations are identical, the lines coincide. Vertical lines are coincident if their equations $x = c$ are identical.

Example 3 Determine which of the following pairs of equations represent parallel lines and which represent coincident lines.

(a) $2x - 3y = 1$
$-8x + 12y = -4$

(b) $5x - 2y = 3$
$-10x + 4y = 6$

(c) $3x = 2$
$6x = 4$

(d) $2y = 5$
$10y = 15$

Solution:
(a) Each equation is equivalent to $y = (2/3)x - 1/3$. Therefore, the lines are coincident.
(b) The first equation is equivalent to $y = (5/2)x - 3/2$ and the second to $y = (5/2)x + 3/2$. The lines are parallel.

(c) Each equation is equivalent to $x = 2/3$. The graphs are coincident vertical lines.
(d) The first equation is equivalent to $y = 5/2$ and the second to $y = 3/2$. The graphs are parallel horizontal lines. ■

Perpendicular Lines. Now suppose that lines L_1 and L_2 are perpendicular and that $P_0(x_0, y_0)$ is their point of intersection. If one of the lines is vertical, then the other is horizontal. If neither line is vertical, then one of the lines will have a positive slope and the other will have a negative slope (*why?*). In this case let L_1 have positive slope m_1 and let L_2 have negative slope m_2. As indicated in Figure 43, if x changes from x_0 to $x_0 + 1$, then along L_1, P_0 changes to $P_1(x_0 + 1, y_0 + m_1)$, and along L_2, P_0 changes to $P_2(x_0 + 1, y_0 + m_2)$. (Along a straight line, when x increases by 1 unit, y changes by m units.)

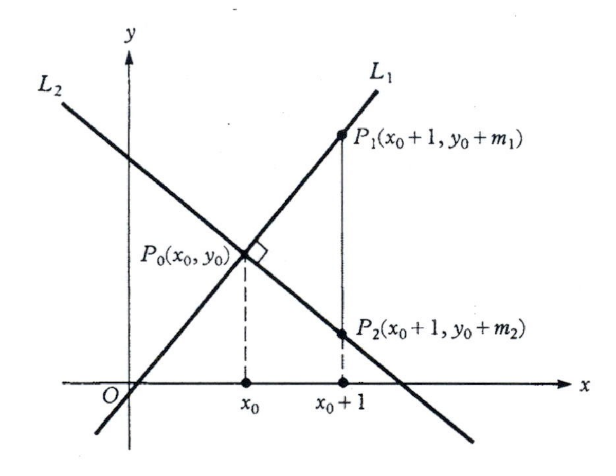

Figure 43 Perpendicular Lines

By the Pythagorean theorem applied to right triangle $P_0P_1P_2$, we obtain

$$d(P_1, P_2)^2 = d(P_0, P_1)^2 + d(P_0, P_2)^2,$$

which becomes

$$[y_0 + m_2 - (y_0 + m_1)]^2 = (x_0 + 1 - x_0)^2 + (y_0 + m_1 - y_0)^2 + (x_0 + 1 - x_0)^2 + (y_0 + m_2 - y_0)^2$$

or

$$(m_2 - m_1)^2 = 1 + m_1^2 + 1 + m_2^2$$

and simplifies to

$$m_1 m_2 = -1. \qquad \text{perpendicular lines (neither one vertical)} \qquad (3)$$

If the equation of L_1 is $A_1x + B_1y = C_1$ and that of L_2 is $A_2x + B_2y = C_2$, then $m_1 = -A_1/B_1$ and $m_2 = -A_2/B_2$, and equation (3) is equivalent to

$$A_1A_2 + B_1B_2 = 0. \qquad \text{perpendicular lines} \qquad (4)$$

Conversely, if (3) or (4) is satisfied, then L_1 is perpendicular to L_2. Furthermore, equation (4) also includes the case where one of the lines is vertical (B_1 or $B_2 = 0$) and the other is horizontal (A_2 or $A_1 = 0$).

Example 4 Which of the following pairs of equations represent perpendicular lines?

(a) $y = (1/3)x - 1$
$y = -3x + 2$

(b) $y - 2 = 4(x - 1)$
$y + 1 = -4(x + 2)$

(c) $y = 2$
$x = 5$

(d) $4x - 3y = -1$
$3x + 4y = 7$

Solution:

(a) Slope $m_1 = 1/3$ and $m_2 = -3$; the lines are perpendicular.
(b) Slope $m_1 = 4$ and $m_2 = -4$; the lines are not perpendicular.
(c) One line is horizontal and one is vertical; therefore, they are perpendicular.
(d) Slope $m_1 = 4/3$ and $m_2 = -3/4$; the lines are perpendicular. Alternatively, using (4), $A_1A_2 + B_1B_2 = 4 \cdot 3 + (-3)4 = 0$. ■

Example 5 Find an equation of the perpendicular bisector of the line segment from $(-2, 3)$ to $(6, 7)$.

Solution: The midpoint of the line segment is

$$\left(\frac{-2+6}{2}, \frac{3+7}{2}\right) = (2, 5),$$

and the slope of the line segment is

$$\frac{7-3}{6+2} = \frac{1}{2}.$$

Therefore, the perpendicular bisector passes through the point $(2, 5)$ and has slope -2 (Figure 44). Its corresponding point-slope equation is $y - 5 = -2(x - 2)$. ■

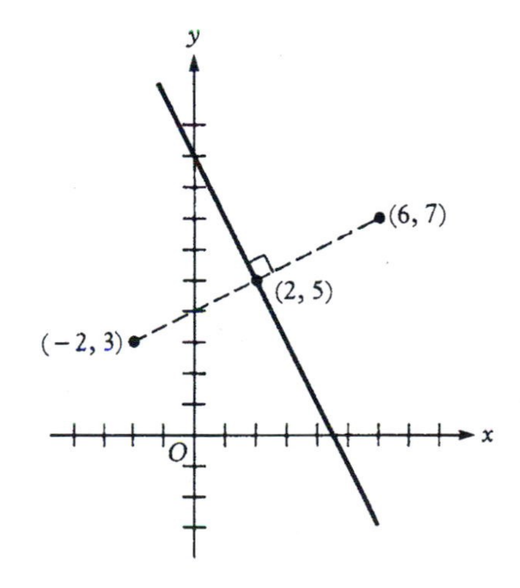

Figure 44

Example 6 Show that the (shortest) distance from the origin to the line L having equation $Ax + By = C$ is

$$\frac{|C|}{\sqrt{A^2 + B^2}}.$$

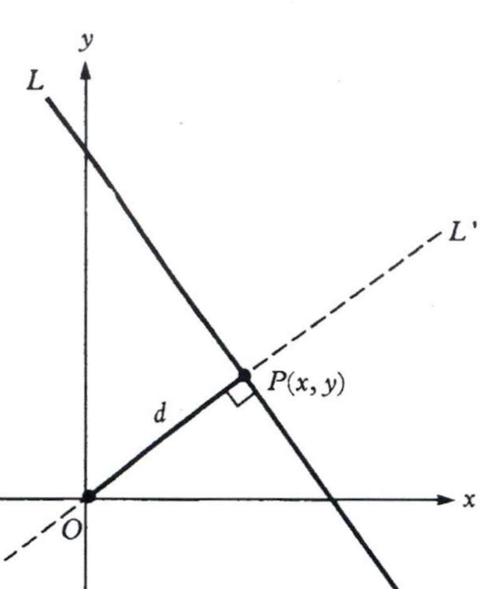

Figure 45

Solution: Let L' be the line through the origin and perpendicular to L. The shortest distance from the origin to L is the length of segment OP, where P is the point of intersection of L and L' (Figure 45). An equation of L' is $Bx - Ay = 0$. (*why?*), and the coordinates of P can be obtained as follows:

$$A^2x + ABy = AC \qquad \textit{Multiply equation of } L \textit{ by } A$$
$$B^2x - ABy = 0 \qquad \textit{Multiply equation of } L' \textit{ by } B$$
$$(A^2 + B^2)x = AC \qquad \textit{Add the two equations}$$
$$x = \frac{AC}{A^2 + B^2}.$$

Similarly, by multiplying L by B and L' by A and subtracting, we obtain

$$y = \frac{BC}{A^2 + B^2}.$$

Therefore, by the distance formula, the length of OP is

$$d = \sqrt{\left(\frac{AC}{A^2 + B^2}\right)^2 + \left(\frac{BC}{A^2 + B^2}\right)^2}$$
$$= \sqrt{\frac{A^2C^2 + B^2C^2}{(A^2 + B^2)^2}}$$
$$= \sqrt{\frac{\cancel{(A^2 + B^2)}C^2}{(A^2 + B^2)^{\cancel{2}}}} = \frac{|C|}{\sqrt{A^2 + B^2}}. \qquad \blacksquare$$

Intersecting Lines. In Example 6 we determined the point of intersection of the perpendicular lines $Ax + By = C$ and $Bx - Ay = 0$. A similar procedure can be used to determine the point $P(x, y)$ of intersection of any two intersecting lines. Let L_1 and L_2 have the equations

$$L_1\colon A_1x + B_1y = C_1 \tag{5a}$$
$$L_2\colon A_2x + B_2y = C_2. \tag{5b}$$

We obtain the x-coordinate of P as follows:

$$A_1B_2x + B_1B_2y = C_1B_2 \qquad \textit{Multiply } L_1 \textit{ by } B_2$$
$$A_2B_1x + B_2B_1y = C_2B_1 \qquad \textit{Multiply } L_2 \textit{ by } B_1$$
$$(A_1B_2 - A_2B_1)x = C_1B_2 - C_2B_1 \qquad \textit{Subtract}$$
$$x = \frac{C_1B_2 - C_2B_1}{A_1B_2 - A_2B_1}. \tag{6a}$$

Similarly, by multiplying L_1 by A_2 and L_2 by A_1 and subtracting, we have

$$y = \frac{A_1C_2 - A_2C_1}{A_1B_2 - A_2B_1}. \tag{6b}$$

These formulas for x and y need not be memorized. In any given problem, the point of intersection can be obtained by repeating the above procedure. However, the formulas show that there is a unique pair (x, y) of real numbers that satisfies both equations (5a) and (5b) if and only if $A_1B_2 - A_2B_1 \neq 0$. This algebraic statement is equivalent to the evident geometric property that two distinct lines in the plane will intersect in a unique point if and only if the lines are not parallel [see equation (2)].

Example 7 Find the point of intersection of the line L_1: $3x - 4y = 7$ and L_2: $2x + 5y = -3$.

Solution:

$$\begin{array}{rl} 15x - 20y = 35 & \textit{Multiply } L_1 \textit{ by } 5 \\ 8x + 20y = -12 & \textit{Multiply } L_2 \textit{ by } 4 \\ \hline 23x = 23 & \textit{Add to eliminate } y \\ x = 1 & \end{array}$$

We can obtain y by a similar procedure or by substituting $x = 1$ into either L_1 or L_2. For instance, from L_1 we obtain

$$3 \cdot 1 - 4y = 7$$
$$y = -1$$

Therefore, the point of intersection of L_1 and L_2 is $(1, -1)$, as shown in Figure 46. ■

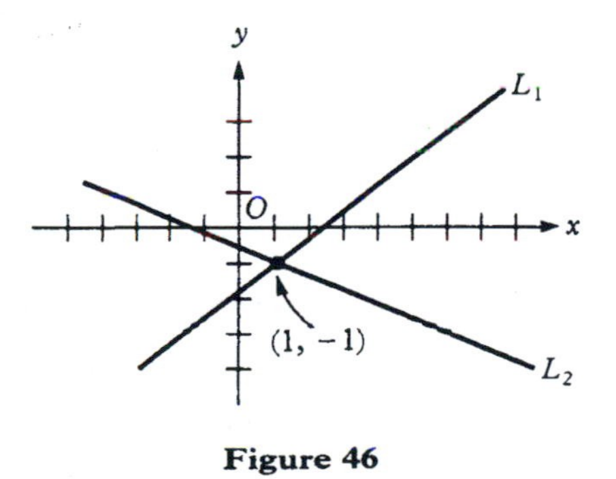

Figure 46

An alternate method for solving a **system of two linear equations in two unknowns,** such as equations (5a) and (5b), is illustrated in the following example.

Example 8 Solve the system of equations

$$-2x + 3y = 18$$
$$3x - 2y = -17.$$

Solution: From the first equation, express y in terms of x:

$$y = 6 + \frac{2}{3}x.$$

Substitute this expression for y into the second equation and solve for x.

$$3x - 2\left(6 + \frac{2}{3}x\right) = -17$$

$$3x - 12 - \frac{4}{3}x = -17$$

$$\frac{5}{3}x = -5$$

$$x = -3$$

Now obtain the value of y by substituting $x = -3$ in the above expression for y.

$$y = 6 + \frac{2}{3}(-3)$$

$$y = 4$$

Hence, the solution is $x = -3$ and $y = 4$. ■

The procedure for solving the equations in Example 7 is called the method of **elimination,** and the procedure in Example 8 is called the method of **substitution.** Equations (6a) and (6b), which were derived by the method of elimination, provide formulas for obtaining the solution directly. These formulas are recast in the next section in a way that makes them easier to remember.

Determinants. The expression

$$A_1B_2 - A_2B_1 \qquad (7)$$

has emerged as an important quantity in our investigation of the relationship between the lines

$$L_1\colon A_1x + B_1y = C_1$$

and

$$L_2\colon A_2x + B_2y = C_2.$$

For any four numbers A_1, B_1, A_2, and B_2, we denote quantity (7) by

$$\begin{vmatrix} A_1 & B_1 \\ A_2 & B_2 \end{vmatrix}, \tag{7'}$$

which is called a **determinant of order two.** For example,

$$\begin{vmatrix} 3 & 5 \\ 2 & 6 \end{vmatrix} = 3 \cdot 6 - 2 \cdot 5 = 8.$$

From equations (6a) and (6b), we see that the coordinates of the **point $P(x, y)$ of intersection of L_1 and L_2** can be expressed in determinants as follows:

$$x = \frac{\begin{vmatrix} C_1 & B_1 \\ C_2 & B_2 \end{vmatrix}}{\begin{vmatrix} A_1 & B_1 \\ A_2 & B_2 \end{vmatrix}} \quad \text{and} \quad y = \frac{\begin{vmatrix} A_1 & C_1 \\ A_2 & C_2 \end{vmatrix}}{\begin{vmatrix} A_1 & B_1 \\ A_2 & B_2 \end{vmatrix}}. \tag{8}$$

Of course, for x and y to be defined, the denominators in equations (8) cannot equal 0, which is another way of saying that distinct lines L_1 and L_2 will intersect if and only if they are not parallel [see equation (2)].

Example 9 Use determinants to find the point of intersection of the lines

$$-4x + 5y = 31$$
$$3x + 7y = 9.$$

Solution: By substituting in (8), we obtain

$$x = \frac{\begin{vmatrix} 31 & 5 \\ 9 & 7 \end{vmatrix}}{\begin{vmatrix} -4 & 5 \\ 3 & 7 \end{vmatrix}} = \frac{31 \cdot 7 - 9 \cdot 5}{(-4)7 - 3 \cdot 5} = \frac{172}{-43} = -4$$

and

$$y = \frac{\begin{vmatrix} -4 & 31 \\ 3 & 9 \end{vmatrix}}{\begin{vmatrix} -4 & 5 \\ 3 & 7 \end{vmatrix}} = \frac{(-4)9 - 3 \cdot 31}{(-4)7 - 3 \cdot 5} = \frac{-129}{-43} = 3.$$

The point of intersection is $(-4, 3)$. ■

Comment Later in the text, we will introduce determinants of order 3 and show how they can be used to find the intersection of three *planes* in space. In addition, we will discuss general properties of determinants at that time.

Applications. The following example illustrates how two equations in two unknowns can be used to solve practical problems.

Example 10 In order to start a small business, a young couple borrowed \$25,000, some from parents at 5% simple interest and the rest from a bank at 11% interest. If their interest payment is \$2450, how much did they borrow from each lender?

Solution: Let x = the number of dollars borrowed from parents, and y = the number of dollars borrowed from the bank. Then

$$\begin{aligned} x + \quad y &= 25{,}000 \\ .05x + .11y &= 2450. \end{aligned}$$

Equivalently,

$$\begin{aligned} 5x + \ 5y &= 125{,}000 \\ 5x + 11y &= 245{,}000. \end{aligned}$$

By subtracting the first equation from the second, we get

$$6y = 120{,}000 \quad \text{or} \quad y = 20{,}000,$$

and therefore

$$x = 5000.$$

Hence, the couple borrowed \$5000 from parents and \$20,000 from the bank. ■

The following example combines the notions of distance, slope, equations of a line, perpendicular lines, point of intersection, and determinants.

Example 11 Show that the (shortest) distance from a point $P_1(x_1, y_1)$ to a line L with equation $Ax + By = C$ is

$$d = \frac{|Ax_1 + By_1 - C|}{\sqrt{A^2 + B^2}}.$$

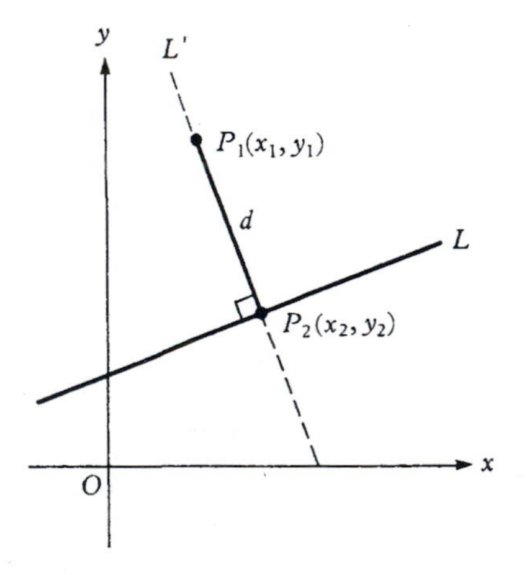

Figure 47

Solution: Let L' be the line through P_1 and perpendicular to L. If $P_2(x_2, y_2)$ is the point of intersection of L' and L, then $d(P_1, P_2)$ is the shortest distance from P_1 to L, as indicated in Figure 47. If L is not vertical, then its slope is $m = -A/B$ and therefore the slope of L' is $m' = B/A$. A point-slope equation of L' is

$$L'\colon y - y_1 = \frac{B}{A}(x - x_1),$$

which is equivalent to the equation

$$L'\colon Bx - Ay = Bx_1 - Ay_1.$$

We find the point $P_2(x_2, y_2)$ of intersection of L' and L by the method of determinants [equations (8)]:

$$x_2 = \frac{\begin{vmatrix} C & B \\ Bx_1 - Ay_1 & -A \end{vmatrix}}{\begin{vmatrix} A & B \\ B & -A \end{vmatrix}} = \frac{-AC - B(Bx_1 - Ay_1)}{-A^2 - B^2} = \frac{AC + B^2x_1 - ABy_1}{A^2 + B^2}$$

$$y_2 = \frac{\begin{vmatrix} A & C \\ B & Bx_1 - Ay_1 \end{vmatrix}}{\begin{vmatrix} A & B \\ B & -A \end{vmatrix}} = \frac{A(Bx_1 - Ay_1) - BC}{-A^2 - B^2} = \frac{BC + A^2y_1 - ABx_1}{A^2 + B^2}.$$

By substituting these values for x_2 and y_2 in the distance formula, we get

$$d(P_1, P_2) = \sqrt{\left(\frac{AC + B^2x_1 - ABy_1}{A^2 + B^2} - x_1\right)^2 + \left(\frac{BC + A^2y_1 - ABx_1}{A^2 + B^2} - y_1\right)^2}.$$

We leave it for the reader to verify that the above expression simplifies to

$$d(P_1, P_2) = \sqrt{\frac{(A^2 + B^2)(Ax_1 + By_1 - C)^2}{(A^2 + B^2)^2}} = \frac{|Ax_1 + By_1 - C|}{\sqrt{A^2 + B^2}}.$$

In Exercise 58 you will be asked to verify the same result for the case in which L is a vertical line. ■

Exercises 2.6

Fill in the blanks to make each statement true.

1. The lines $y = m_1x + b_1$ and $y = m_2x + b_2$ ($b_1 \neq b_2$) are parallel if and only if ______________.
2. If L_1: $A_1x + B_1y = C_1$ is parallel to L_2: $A_2x + B_2y = C_2$, then ______________.
3. The lines $y = m_1x + b_1$ and $y = m_2x + b_2$ are perpendicular if and only if ______________.
4. L_1: $A_1x + B_1y = C_1$ is perpendicular to L_2: $A_2x + B_2y = C_2$ if and only if ______________.
5. Lines $A_1x + B_1y = C_1$ and $A_2x + B_2y = C_2$ intersect in a unique point if and only if ______________.

Write true *or* false *for each statement. Assume L_1, L_2, and L_3 are distinct lines in the plane.*

6. If L_1 is parallel to L_2 and L_2 is parallel to L_3, then L_1 is parallel to L_3.
7. If L_1 is perpendicular to L_2 and L_2 is perpendicular to L_3, then L_1 is perpendicular to L_3.
8. If L_1 is parallel to L_2 and L_2 is perpendicular to L_3, then L_1 is perpendicular to L_3.
9. If L_1: $A_1x + B_1y = C_1$ and L_2: $A_2x + B_2y = C_2$ satisfy $A_1B_2 - A_2B_1 = 0$, then L_1 is parallel to L_2.
10. If L_1: $A_1x + B_1y = C_1$ and L_2: $A_2x + B_2y = C_2$ satisfy $A_1B_2 - A_2B_1 = 0$ and $A_1 = 0$, then $A_2 = 0$.

Parallel and Perpendicular Lines

Determine which pairs of lines in Exercises 11–22 are parallel, which are perpendicular, and which are coincident.

11. $y = 2x + 3$
    $y = 2x - 5$
12. $y = 3x - 4$
    $y = 2x + 3$
13. $y = x - 2$
    $y = -x + 2$
14. $y = x - 4$
    $y = x + 4$
15. $y - 2 = -\frac{3}{2}(x - 1)$
    $y - 1 = \frac{2}{3}(x + 4)$
16. $y - 5 = -3(x + 2)$
    $y - 5 = 3(x + 2)$

17. $y - 3 = 5(x - 1)$
$y + 7 = 5(x + 1)$

18. $y + 4 = 2(x + 1)$
$y - 4 = 2(x + 1)$

19. $4x - 2y = 3$
$-2x + y = 1$

20. $2x - 3y = 1$
$-4x + 6y = -2$

21. $3x = 4$
$5y = 2$

22. $5x - 4y = 2$
$4x + 5y = 1$

23. Find a general linear equation of the line through the point (3, 5) and parallel to the line $4x - 7y = 8$.

24. Find a general linear equation of the line through the point (−4, 6) and perpendicular to the line $5x + 2y = 7$.

25. Show that a general linear equation of the line through the point (x_0, y_0) and parallel to the line L: $Ax + By = C$ is $Ax + By = Ax_0 + By_0$. Assume (x_0, y_0) is not on L.

26. Show that a general linear equation of the line through the point (x_0, y_0) and perpendicular to the line $Ax + By = C$ is $Bx - Ay = Bx_0 - Ay_0$.

27. Show that the line L_1 through the points (0, 0) and (3, 4) is perpendicular to the line L_2: $3x + 4y = 1$.

28. Show that the line through the points (0, 0) and (5, 2) is parallel to the line $-2x + 5y = 1$.

29. Show that the line through the points (0, 0) and (A, B) is perpendicular to the line $Ax + By = C$.

30. Show that the line through the points (0, 0) and $(B, -A)$ is parallel to the line $Ax + By = C$. Assume $C \neq 0$.

31. Given that L_1: $A_1x + B_1y = C_1$ and L_2: $A_2x + B_2y = C_2$ are parallel, show that the line through the points (A_1, B_1) and (A_2, B_2) is perpendicular to both L_1 and L_2. Assume $A_1 \neq A_2$.

32. Given that L_1: $A_1x + B_1y = C_1$ and L_2: $A_2x + B_2y = C_2$ are parallel, show that the line through the points (B_1, A_2) and (B_2, A_1) is parallel to both L_1 and L_2. Assume neither point is on L_1 or L_2.

Intersecting Lines

Use the method of elimination to solve each of the following systems of equations.

33. $2x - y = 4$
$3x + 5y = 8$

34. $4x - 5y = 0$
$x + 2y = 6$

35. $-2x + 7y = 4$
$3x - 4y = 7$

36. $7x - 5y = -1$
$3x + 2y = 12$

37. $3x - 7 = 2y + 4$
$x + 4 = y - 3$

38. $2x - 6y - 5 = 0$
$4x - 3y + 6 = 0$

Use the method of substitution to solve each of the following systems of equations.

39. $2x + y = 5$
$-4x + 3y = 10$

40. $x - 2y = 4$
$6x + 5y = 2$

41. $4x + 5y = 7$
$3x + 4y = -2$

42. $6x - 4y = 0$
$3x - 5y = 7$

43. $7x - 3y = x + y + 2$
$2x + 3y = 3x + 2y + 4$

44. $5x - y = x - 5y + 1$
$3x + 2y = 2x + 3y - 2$

Determinants

Use the method of determinants to solve each of the following systems of equations.

45. $4x + 5y = 3$
$2x + 4y = 7$

46. $3x + y = 9$
$2x + 8y = 7$

47. $3x - 2y = 4$
$7x + 6y = -3$

48. $-2x + 6y = 1$
$2x - 4y = -3$

49. $2x = 7$
$4x - 3y = 1$

50. $5x - 7y = 4$
$2y = 3$

Applications

51. If a chemist has one solution of 20% acid and another of 50% acid, how many milliliters of each are needed to obtain 18 ml of 30% acid?

52. The cost of 40 lb of sugar and 16 lb of flour is $26, as is the cost of 30 lb of sugar and 25 lb of flour. Find the cost per pound of each item.

53. There is a fixed charge for a telegram of up to 15 words and an additional charge for each word after the fifteenth. A telegram of 20 words costs $32.90, and one of 25 words costs $37.90. What are the fixed charge and the rate for each word after the fifteenth?

54. Flying with a tail wind, a plane covers 1800 mi in 3 hr, but returns against the wind in 3 hr and 36 min. Find the speed of the plane and the speed of the wind.

55. A company makes two styles of blouses. The first style requires 12 min for cutting and 15 min for sewing, while the second needs 18 min for cutting and 10 min for sewing. How many of each type are produced if a total of 850 hr are spent on cutting and 750 hr on sewing?

Miscellaneous

56. Find the distance between the origin and the line $5x + 4y = -3$.

57. Find the distance between the point $(-5, 2)$ and the line $3x - 4y = 7$.

58. Show that the result in Example 11 holds when L is a vertical line, that is, when $B = 0$ in the equation of L.

In Exercises 59–61, L_1 and L_2 are lines, not necessarily distinct, with the general equations

$$L_1\colon A_1x + B_1y = C_1$$
$$L_2\colon A_2x + B_2y = C_2.$$

59. Show that if $A_1B_2 - A_2B_1 = 0$, then L_1 and L_2 are either parallel or coincident.
[*Hint:* show that the only possibilities are
(a) $B_1 = 0$ and $B_2 = 0$ or
(b) $B_1 \neq 0$ and $B_2 \neq 0$.]

60. Show that if $A_1C_2 - A_2C_1 = 0$, then either L_1 and L_2 are parallel horizontal lines, or they intersect at some point on the x-axis, or they are coincident.
[*Hint:* consider the following four cases separately:
(a) $A_1 = 0, A_2 = 0$; (b) $A_1 = 0, A_2 \neq 0$;
(c) $A_1 \neq 0, A_2 = 0$; (d) $A_1 \neq 0, A_2 \neq 0$.]

61. Show that if $B_1C_2 - B_2C_1 = 0$, then either L_1 and L_2 are parallel vertical lines, or they intersect at some point on the y-axis, or they are coincident.
[*Hint:* consider the following four cases separately:
(a) $B_1 = 0, B_2 = 0$; (b) $B_1 = 0, B_2 \neq 0$;
(c) $B_1 \neq 0, B_2 = 0$; (d) $B_1 \neq 0, B_2 \neq 0$.]

62. Given $L_1\colon A_1x + B_1y = C_1$ and $L_2\colon A_2x + B_2y = C_2$, use the results of Exercises 59, 60, and 61 to classify each of the following cases as either intersecting lines, parallel vertical lines, parallel horizontal lines, parallel lines that are neither vertical nor horizontal, or coincident lines.

(a) $\begin{vmatrix} A_1 & B_1 \\ A_2 & B_2 \end{vmatrix} \neq 0$

(b) $\begin{vmatrix} A_1 & B_1 \\ A_2 & B_2 \end{vmatrix} = 0, \quad \begin{vmatrix} C_1 & B_1 \\ C_2 & B_2 \end{vmatrix} \neq 0, \quad \begin{vmatrix} A_1 & C_1 \\ A_2 & C_2 \end{vmatrix} \neq 0$

(c) $\begin{vmatrix} A_1 & B_1 \\ A_2 & B_2 \end{vmatrix} = 0, \quad \begin{vmatrix} C_1 & B_1 \\ C_2 & B_2 \end{vmatrix} \neq 0, \quad \begin{vmatrix} A_1 & C_1 \\ A_2 & C_2 \end{vmatrix} = 0$

(d) $\begin{vmatrix} A_1 & B_1 \\ A_2 & B_2 \end{vmatrix} = 0, \quad \begin{vmatrix} C_1 & B_1 \\ C_2 & B_2 \end{vmatrix} = 0, \quad \begin{vmatrix} A_1 & C_1 \\ A_2 & C_2 \end{vmatrix} \neq 0$

(e) $\begin{vmatrix} A_1 & B_1 \\ A_2 & B_2 \end{vmatrix} = 0, \quad \begin{vmatrix} C_1 & B_1 \\ C_2 & B_2 \end{vmatrix} = 0, \quad \begin{vmatrix} A_1 & C_1 \\ A_2 & C_2 \end{vmatrix} = 0$

2.7 Linear Inequalities in the Plane with Applications

A thin line separates the "greater thans" from the "less thans."

Linear Inequalities and Half-Planes. We have seen that the solution of a linear inequality in one variable x is an infinite interval along the real line. We now show that the solution of a linear inequality in two variables x and y is a *half-plane*. To see why this is so, let L be a nonhorizontal line. Then

L partitions the plane into a right **half-plane** $\mathcal{R}$ and a left **half-plane** $\mathcal{L}$. $\mathcal{R}$ consists of all points $(x_0 + h, y_0)$, where (x_0, y_0) is on L and $h > 0$; all points $(x_0 - h, y_0)$ make up $\mathcal{L}$ (Figure 48). The line L itself is the boundary of both $\mathcal{R}$ and $\mathcal{L}$.

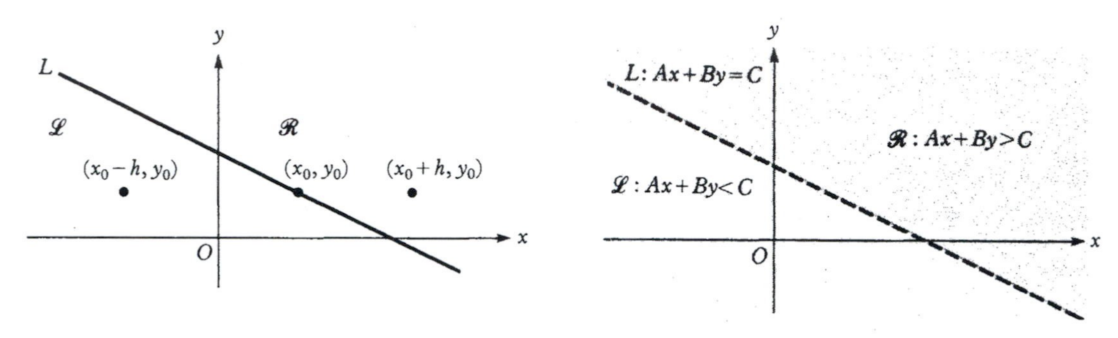

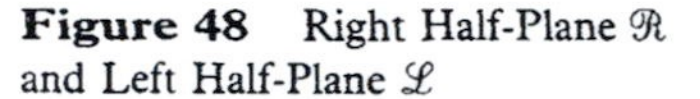
Figure 48 Right Half-Plane $\mathcal{R}$ and Left Half-Plane $\mathcal{L}$

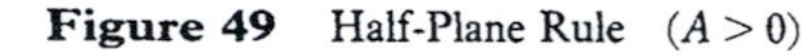
Figure 49 Half-Plane Rule $(A > 0)$

The equation of line L can be written in the form

$$Ax + By = C,$$

where $A > 0$ (*why may we assume that* $A > 0$?). Suppose that (x_0, y_0) lies on L and that $h > 0$. Then

$$A(x_0 + h) + By_0 = Ax_0 + By_0 + Ah = C + Ah > C$$

and $$A(x_0 - h) + By_0 = Ax_0 + By_0 - Ah = C - Ah < C.$$

Therefore, as indicated in Figure 49, we obtain the following **half-plane rule.**

> Given line $L\colon Ax + By = C$, where $A > 0$, then
>
> $Ax + By > C$ for all points (x, y) in the right half-plane $\mathcal{R}$,
>
> and
>
> $Ax + By < C$ for all points (x, y) in the left half-plane $\mathcal{L}$.

The solution to $Ax + By \geq C$ consists of all points in $\mathcal{R}$ and on L, which we denote by $\mathcal{R} \cup L$. Similarly, the solution to $Ax + By \leq C$ is $\mathcal{L} \cup L$.

Example 1 Graph the solution to each of the following inequalities.

(a) $2x + 3y > 6$ (b) $2x + 3y < 6$

(c) $2x + 3y \geq 6$ (d) $2x + 3y \leq 6$

Solution: In Figure 50, a solid line for L indicates that L is included in the solution; a broken line indicates that L is not in the solution. ∎

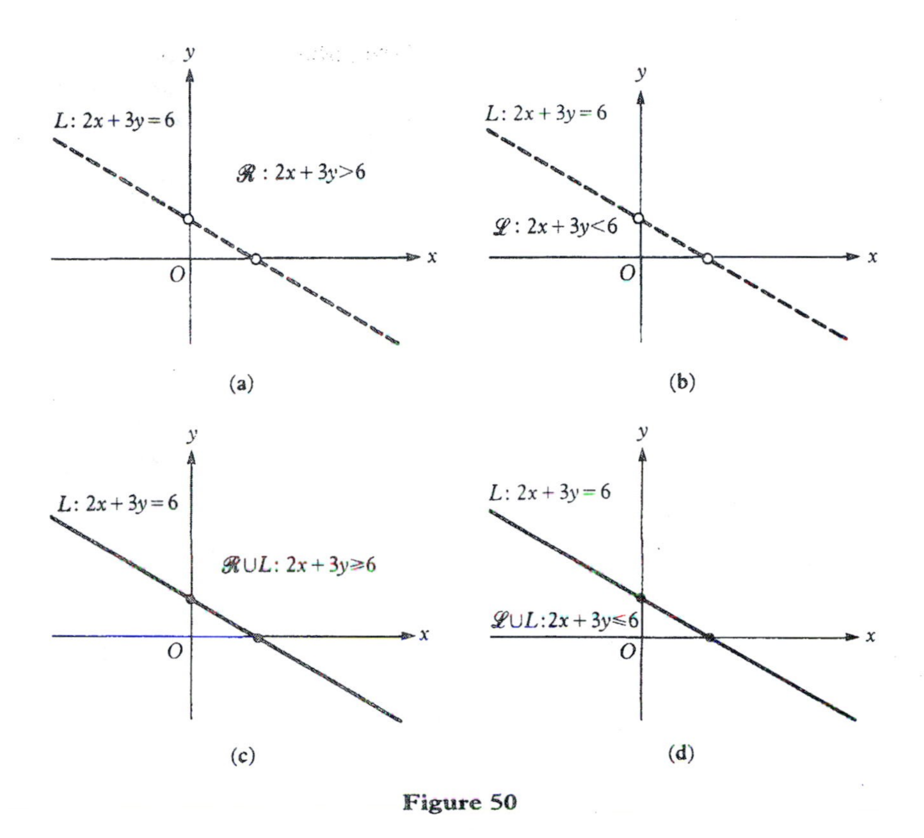

Figure 50

Example 2 Graph the solution to each of the following inequalities.
(a) $-3x + 4y > 12$ (b) $-3x + 4y \leq 12$

Solution: The half-plane rule requires that the coefficient of x be positive. We therefore multiply both sides of each inequality by -1, which changes all the signs and reverses the direction of each inequality. That is, we replace (a) and (b) by the equivalent inequalities (a′) $3x - 4y < -12$ and (b′) $3x - 4y \geq -12$. The graphs of (a) and (b) are equal to the graphs of (a′) and (b′), respectively. See Figure 51. ■

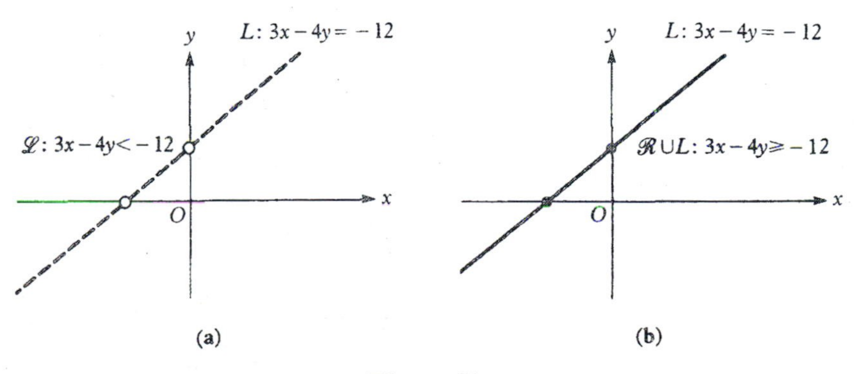

Figure 51

A horizontal line L given by an equation $y = c$ does not partition the plane into a right half and a left half, but into an **upper half-plane** and a **lower half-plane.** All points in the upper half-plane satisfy $y > c$, and those in the lower half-plane satisfy $y < c$ (Figure 52). Finally, the solution to $y \geq c$ consists of all points on or above the line $y = c$, and $y \leq c$ consists of all points on or below the line.

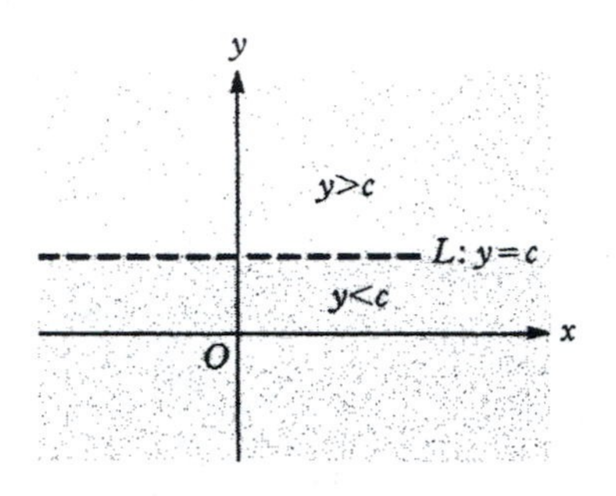

Figure 52

Example 3 Graph the solution to each of the following inequalities.
(a) $y \geq -2$ (b) $y < 4$

Solution: The half-plane solutions are indicated in Figure 53. ■

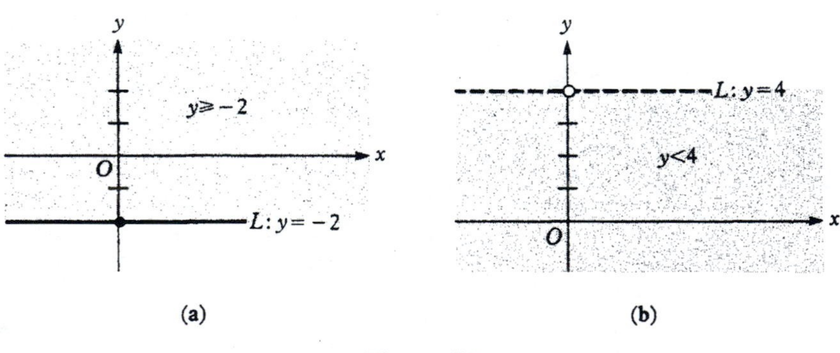

Figure 53

Systems of Linear Inequalities. We have seen that inequalities involving linear expressions in x and y have for their solutions half-planes with or without their boundaries. For a system of two or more such inequalities, the solution is the region of intersection of the individual solutions.

Example 4 Graph the solution of the system

$$\begin{aligned} 2x - y &> 4 \\ x + y &< 6. \end{aligned}$$

Solution: In Figure 54, the area shaded in color denotes the solution of $2x - y > 4$, and the gray shaded area indicates the solution of $x + y < 6$.

The region of intersection, the double-shaded portion of the figure, is the solution of the system. ■

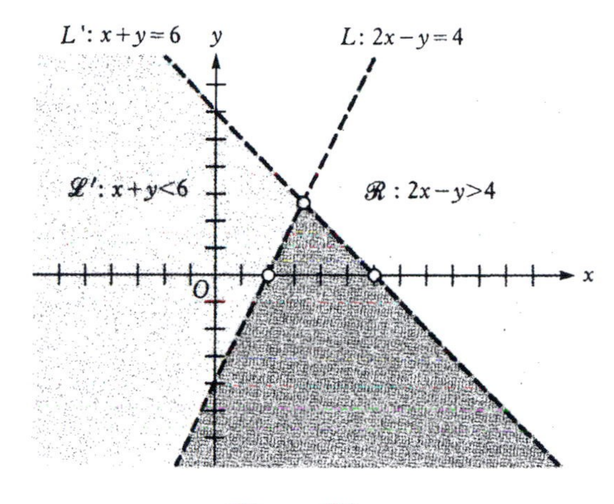

Figure 54

Example 5 Graph the solution to the system

$$\begin{aligned} 2x + 3y &\le 12 \\ 3x + y &\le 9 \\ x &\ge 0 \\ y &\ge 0. \end{aligned}$$

Solution: The last two inequalities tell us that the solution is confined to the first quadrant and its boundary. The shaded region in Figure 55, including its boundary, indicates the solution of the first two inequalities in the first quadrant. ■

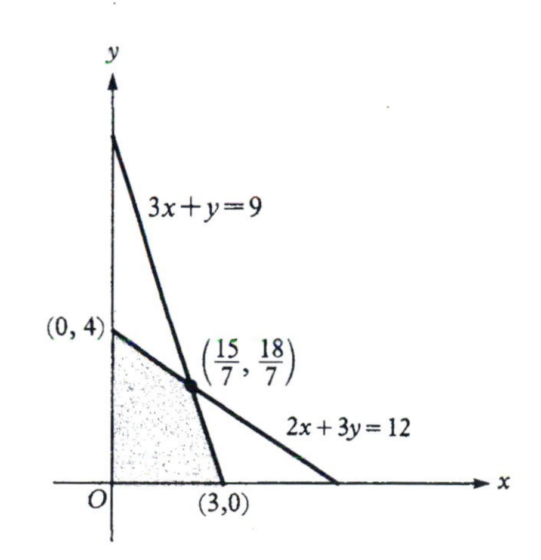

Figure 55

Applications. The following two examples illustrate some applications of systems of linear inequalities in the plane.

Example 6 A clock company manufactures two types of clocks, A and B. Each type A clock costs the company $25 in parts and requires 5 hours of labor. Each type B clock costs $40 in parts but requires only 4 hours of labor. If $2000 is available for parts, what combinations of A and B clocks can be produced with 320 hours of labor?

Solution: Let x = the number of type A clocks, and y = the number of type B clocks.

Then

$$\begin{aligned} 25x + 40y &\leq 2000 && \textit{Parts} \\ 5x + 4y &\leq 320 && \textit{Labor} \\ x &\geq 0 \\ y &\geq 0. \end{aligned}$$

Any pair of integers (x, y) in the shaded region of Figure 56 represents a combination of the number of A and B clocks that can be manufactured within the given constraints. 

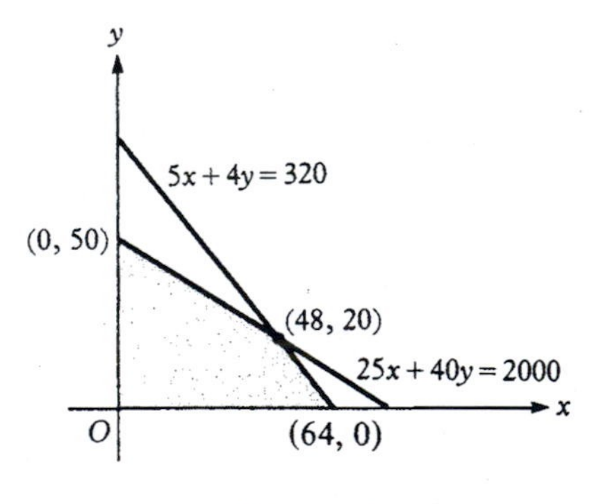

Figure 56

Example 7 A construction company buys lumber in 8-foot and 14-foot board lengths. If at least 75 5-foot lengths and 125 3-foot lengths are needed to complete a project, what combinations of 8-foot and 14-foot boards will be sufficient? Assume that from each 8-foot board, one 5-foot and one 3-foot length will be cut. Also, from each 14-foot board, one 5-foot and three 3-foot lengths will be cut.

Solution: Let x = the number of 8-foot boards, and y = the number of 14-foot boards.

Then

$$
\begin{aligned}
x + \ \ y &\geq 75 && \textit{5-foot lengths} \\
x + 3y &\geq 125 && \textit{3-foot lengths} \\
x &\geq 0 \\
y &\geq 0.
\end{aligned}
$$

Each pair of integers (x, y) in the shaded region of Figure 57 represents a combination of 8-foot and 14-foot boards that will be sufficient for the job. ■

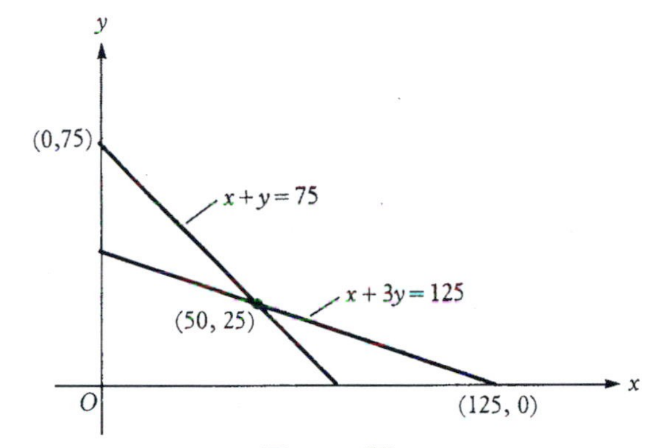

Figure 57

Linear Programming. Linear programming is a method for making the optimum choice in a situation where several choices are possible. We can apply linear programming to the previous two examples.

Suppose, in Example 6, that the clock company makes \$10 profit on each type A clock and \$12 profit on each type B clock. Then the company would want to know which combinations of type A and B clocks would provide the *most profit*. One way to proceed would be to compute the profit for each production point in the shaded region. A more efficient way is to begin by setting up an equation for the profit P for x type A clocks and y type B clocks:

$$P = 10x + 12y. \tag{1}$$

Now for each value of P, the graph of equation (1) is a straight line L_P with slope $m = -5/6$; different values of P correspond to parallel lines. Furthermore, by Example 6 of Section 2.6, the distance from the origin to L_P is $P/\sqrt{244}$. Hence, *the maximum profit will occur on the line with slope* $-5/6$ *that is farthest from the origin but still touches the shaded region of Figure 56*. As indicated in Figure 58, the line of maximum profit is L_{720}, which passes through the vertex (48, 20). Hence, the maximum profit of \$720 is achieved by producing 48 type A clocks and 20 type B clocks.

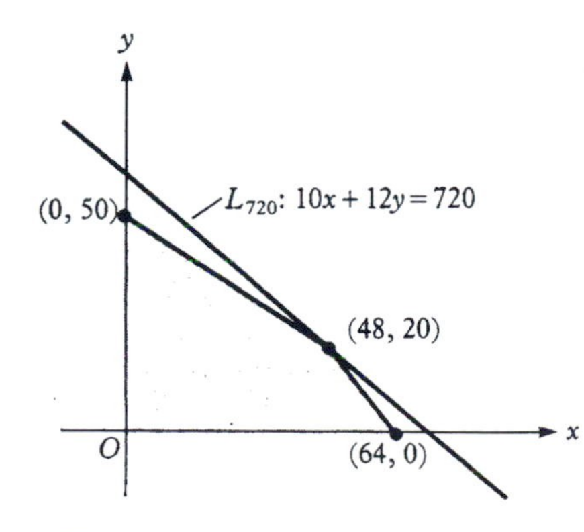

Figure 58

If the respective profit for type A or type B clocks changes, then the slope of L_P may change. However, because of the shape of the shaded region, *the maximum P will always occur among the vertices* (0, 50), (48, 20), *or* (64, 0). *Hence, only these three points need be tested.*

Example 6 (continued) Let P_A and P_B stand for the profit that the company makes on each A and B clock, respectively. Find the maximum profit in each of the following.

(a) $P_A = 10, \quad P_B = 18$ (b) $P_A = 10, \quad P_B = 16$

(c) $P_A = 14, \quad P_B = 12$ (d) $P_A = 10, \quad P_B = 8$

(e) $P_A = 14, \quad P_B = 10$

Solution: The profit P at each vertex (x, y) for each case is listed in the following table. The maximum in each case is indicated by a star. The line of maximum profit in each case is shown in Figure 59 below the table.

Vertex	(a) $P = 10x + 18y$	(b) $P = 10x + 16y$	(c) $P = 14x + 12y$	(d) $P = 10x + 8y$	(e) $P = 14x + 10y$
(0, 50)	900★	800★	600	400	500
(48, 20)	840	800★	912★	640★	872
(64, 0)	640	640	896	640★	896★

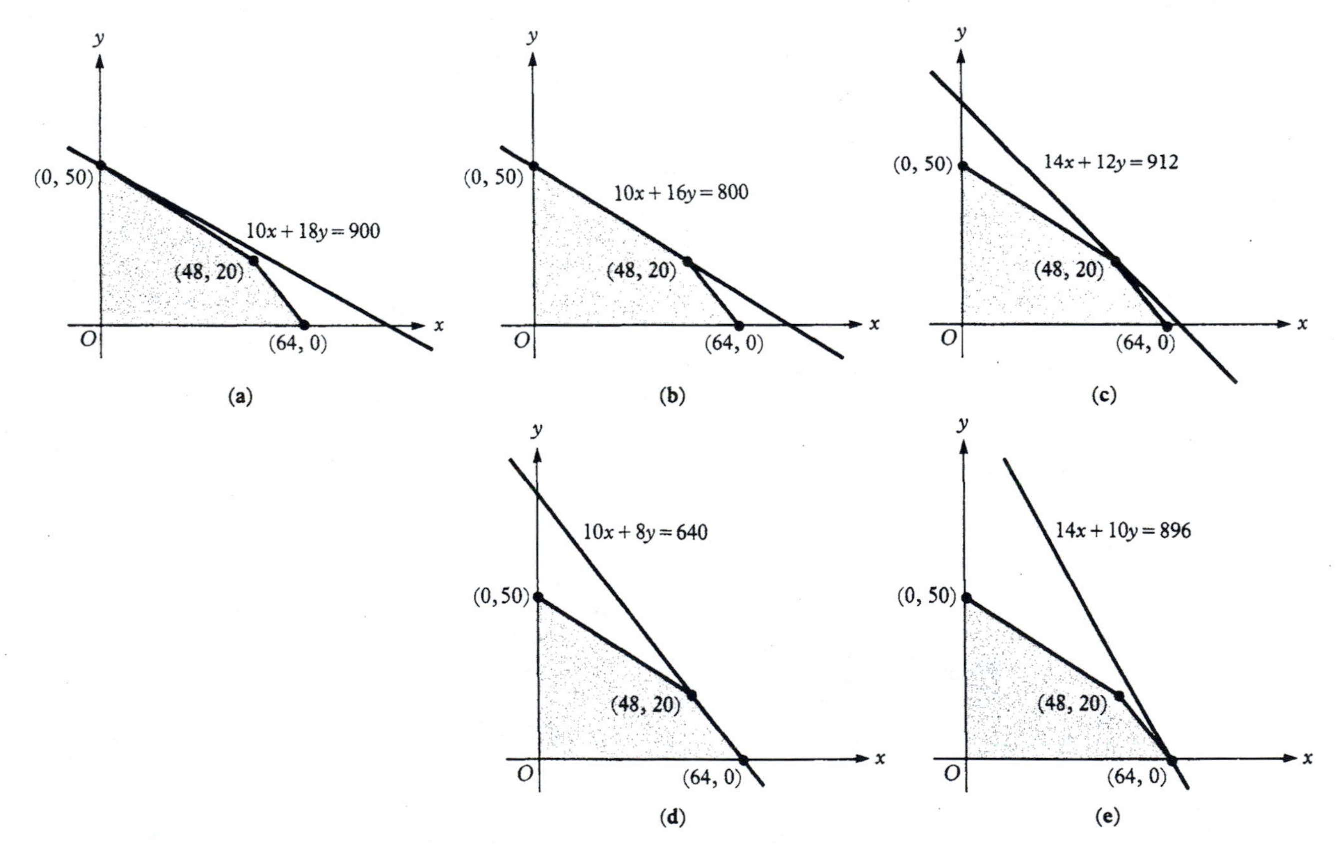

Figure 59

We note that in case (b) the maximum profit of \$800 occurs at any production point on the line $10x + 16y = 800$ between (0, 50) and (48, 20), namely at the points (0, 50), (8, 45), (16, 40), (24, 35), (32, 30), (40, 25), and (48, 20). Similarly, in case (d) the maximum profit of \$640 occurs at any production point on the line $10x + 8y = 640$ between (48, 20) and (64, 0), namely at (48, 20), (52, 15), (56, 10), (60, 5), and (64, 0). In all the other cases, the maximum profit occurs at a unique point. ■

In Example 7, the objective is not to maximize profits but to *minimize costs*. For instance, if the company pays 50 cents per foot for each board, then the cost C (in dollars) for x 8-foot and y 14-foot boards is

$$C = 4x + 7y. \tag{2}$$

For each value of C, equation (2) is a straight line L_C with slope $m = -4/7$. *The minimum cost will correspond to the line of slope $-4/7$ that is closest to the origin but still touches the shaded region of Figure 57 (why?).* These properties are satisfied by the line

$$L_{375}\colon 4x + 7y = 375$$

which passes through the vertex (50, 25), as shown in Figure 60. Hence, the minimum cost is \$375, which is the price of 50 8-foot and 25 14-foot boards.

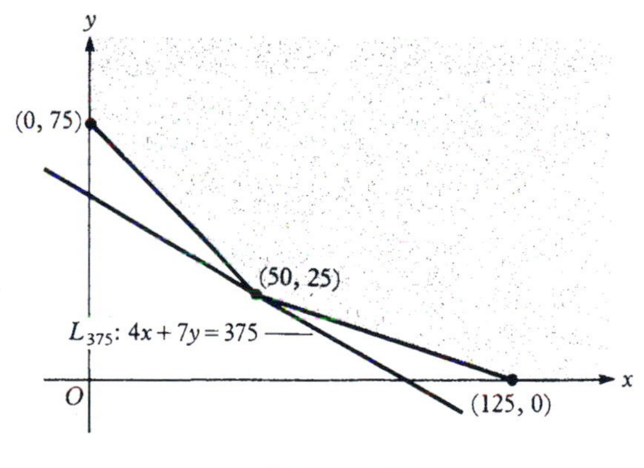

Figure 60

If either of the respective prices of the 8-foot and 14-foot boards changes, then the slope of the cost line L_C may change. However, *the minimum cost will always occur among the three vertices* (0, 75), (50, 25), *and* (125, 0) *(why?).*

Example 7 (continued) In order to reduce inventory, the lumber supplier occasionally lowers the prices on either the 8-foot or the 14-foot boards. In each of the following cases, determine the minimum cost and graph the line L_C of minimum cost.

(a) 8-foot board: \$4
14-foot board: \$4

(b) 8-foot board: \$2
14-foot board: \$6

(c) 8-foot board: \$2
14-foot board: \$7

Solution: The cost at each vertex for each case is listed in the following table. The minimum in each case is indicated by a star. The line of minimum cost in each case is shown in Figure 61 below the table.

Vertex	(a) $C = 4x + 4y$	(b) $C = 2x + 6y$	(c) $C = 2x + 7y$
(0, 75)	300*	450	525
(50, 25)	300*	250*	275
(125, 0)	500	250*	250*

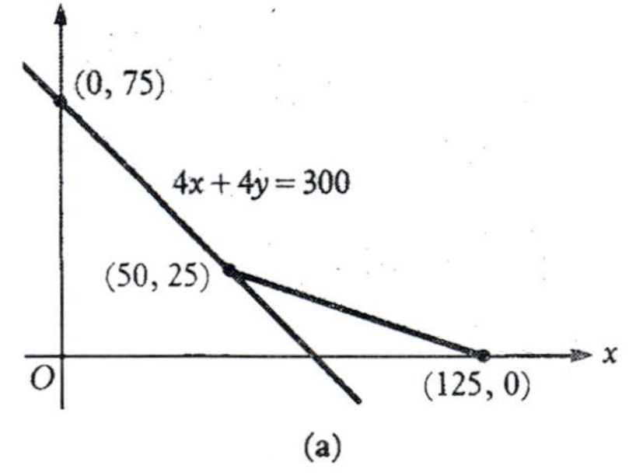

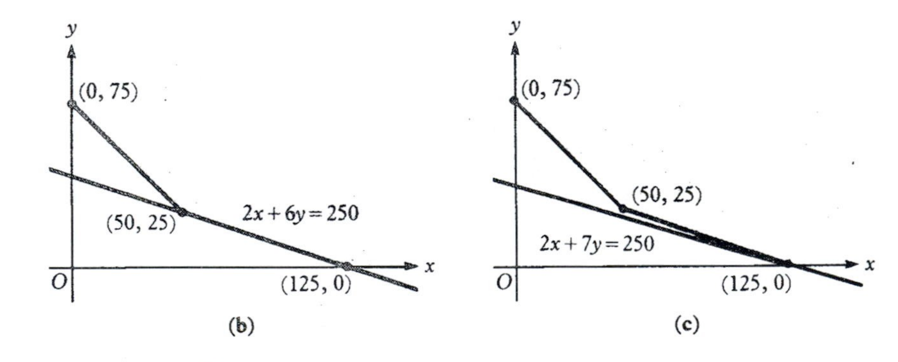

Figure 61

In case (a) the minimum cost of \$300 occurs at any purchase point on the line $4x + 4y = 300$ from (0, 75) to (50, 25), namely at the points (0, 75), (5, 70), (10, 65), (15, 60), . . . , (45, 30), (50, 25). Similarly, in case (b) the minimum cost of \$250 occurs at any purchase point on the line $2x + 6y = 250$ from (50, 25) to (125, 0), namely at (50, 25), (53, 24), (56, 23), (59, 22), . . . , (122, 1), (125, 0). In case (c) the minimum cost of \$250 occurs only at (125, 0). ■

Comment The basic idea in linear programming is that only the corners (vertices) of the region need be tested. The number of corners depends on the number of inequalities, and the dimension of the region is equal to the number of variables involved. We have considered problems with two variables, hence, two dimensions. Large corporations, such as airlines and telephone companies, routinely consider problems with thousands of variables. In these cases the corner points are tested by techniques carried out on high-speed computers.

Exercises 2.7

Linear Inequalities and Half-Planes

In the plane, graph the solution to each of the following inequalities.

1. $x + y > 1$
2. $x - y < 1$
3. $-x + y \geq 1$
4. $-x - y \leq 1$
5. $x > 2$
6. $y \leq 5$
7. $2x + 3y \geq 6$
8. $4x - 5y \leq 20$
9. $3x + 4y < 12$
10. $2x + 5y > 10$
11. $-x + 4y > 3$
12. $-3x - 5y < 4$

Systems of Linear Inequalities

In the plane, graph the solution to each of the following systems of inequalities.

13. $x + y > 2$, $x - 2y < 2$
14. $x - y \leq 2$, $2x - y \geq 4$
15. $-x + y < 3$, $2x - y < 5$
16. $x + y \geq 1$, $x + 2y \geq 2$
17. $2x - y > 0$, $x - 2y < 0$
18. $2x - 3y < 0$, $-3x + 4y > 0$
19. $x \geq 3$, $y \leq 5$, $y \geq 0$, $x + y \leq 8$
20. $x < 2$, $y < 4$, $x > 0$, $y > 0$
21. $2x + 3y \leq 6$, $4x + 5y \leq 20$, $x \geq 0$, $y \geq 0$
22. $3x + 4y > 12$, $x + y < 5$, $x > 0$, $y > 0$
23. $3x + 5y \leq 30$, $x + y \geq 4$, $x \geq 2$, $y \geq 2$
24. $4x + 7y \geq 28$, $x + y \leq 10$, $x \leq 7$, $y \geq 3$

In each of the following cases, find a system of inequalities whose solution is the given graph.

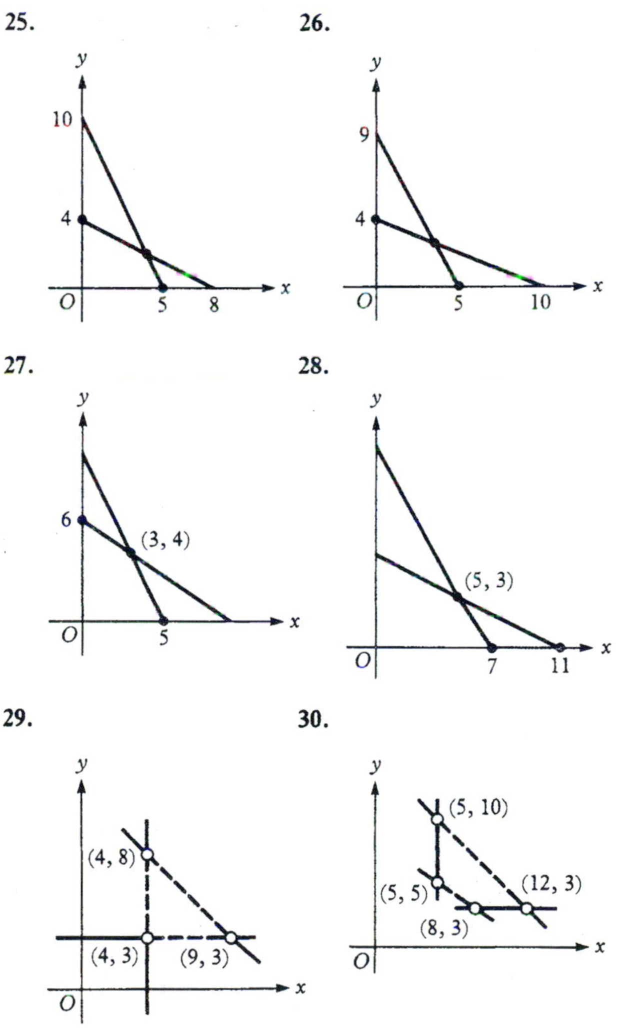

Applications

Graph the solution for each problem.

31. A company packages and sells 16-oz containers of mixed nuts. Brand A contains 12 oz of peanuts and 4 oz of cashews. Brand B contains 8 oz of peanuts and 8 oz of cashews. If 120 lb of peanuts and 96 lb of cashews are available, how many containers of each brand can be packaged?

32. A company packages and sells 32-oz cartons of mixed fruit juice. Type 1 contains 24 oz of pineapple juice and 8 oz of grapefruit juice. Type 2 contains 12 oz of pineapple juice and 20 oz of grapefruit juice. If 150 gal of pineapple juice and 120 gal of grapefruit juice are available, how many cartons of each type can be packaged?

33. A gardener has a 2304-sq-in flower garden in which she wants to plant begonias and marigolds. Each begonia requires 64 sq in of space, and each marigold requires 144 sq in. If the gardener wants to put in at least 26 plants, what combinations of begonias and marigolds are possible?

34. A 60-sq-ft vegetable garden is to be planted in tomatoes and peppers. Each tomato plant requires 4 sq ft of space, and each pepper plant requires 1 sq ft. If there must be at least 24 plants, what combinations of tomatoes and peppers are possible?

35. A stadium has 10,000 seats for $5 each and 4,000 seats for $8. To make a profit on a given event, the stadium must take in at least $50,000. What seating combinations will result in a profit?

36. A theater has 50 seats at $20 each and 150 seats at $12 each. To make a profit on a given performance, the theater must take in at least $1,500. What seating combinations will result in a profit?

37. A small toy company makes two kinds of stuffed animals. A stuffed puppy costs $2 in materials and requires 4 hr of labor. A teddy bear costs $2.50 and takes 3 hr. The company has $40 available for materials and has orders for 5 animals of each type. How many stuffed animals can the company make in 64 hr and still meet its orders?

38. A retired carpenter makes toy airplanes and trains out of wood. An airplane costs $3 for materials and requires 2 hr to cut, shape, and finish. A train costs $8 and requires 4 hr of work. The carpenter has orders for 4 airplanes and 3 trains. With a budget of $72, how many of each type can be made in 40 hr and still meet the orders?

39. Marie plans to have a spring garden with at least 60 daffodils and 48 tulips but no more than 128 bulbs in all. A type A package contains 5 daffodil and 3 tulip bulbs. A type B package contains 2 daffodil and 2 tulip bulbs. What combinations of type A and B packages will be sufficient?

40. Sarah's fall garden is to have at least 15 yellow chrysanthemums and 12 white ones but no more than 36 plants in all. A type A variety pack contains 3 yellow and 2 white plants, and a type B pack contains 3 yellow and 3 white. What combinations of type A and B variety packs will be sufficient?

Linear Programming

41. In Exercise 31, let P_A equal the profit in selling a 16-oz container of brand A, and let P_B be the profit for brand B. Find the maximum profit in each of the following cases.

(a) $P_A = \$.75, \quad P_B = \$1.00$

(b) $P_A = \$1.00, \quad P_B = \$.25$

(c) $P_A = \$1.00, \quad P_B = \$.75$

42. In Exercise 32, let P_1 equal the profit in selling a 32-oz carton of type 1 juice, and let P_2 be the profit for type 2. Find the maximum profit in each of the following cases.

(a) $P_1 = \$.50, \quad P_2 = \$.25$

(b) $P_1 = \$.50, \quad P_2 = \$.10$

(c) $P_1 = \$.50, \quad P_2 = \$.50$

43. In Exercise 33, let C_B equal the cost of each begonia, and let C_M be the cost of each marigold. Find the minimum cost to plant the garden in each of the following cases.

(a) $C_B = \$3, \quad C_M = \$2$

(b) $C_B = \$2, \quad C_M = \$3$

(c) $C_B = \$2, \quad C_M = \$2$

44. In Exercise 34, let C_T equal the cost of each tomato plant, and let C_P be the cost of each pepper plant. Find the minimum cost to plant the garden in each of the following cases.

(a) $C_T = \$1.00, \quad C_P = \$.50$

(b) $C_T = \$.25, \quad C_P = \$.50$

(c) $C_T = \$.75, \quad C_P = \$.75$

45. In Exercise 35, what is the least number of tickets that must be sold in order to make a profit?

46. In Exercise 36, what is the least number of tickets that must be sold in order to make a profit?

47. In Exercise 37, let P_P equal the profit made on each stuffed puppy, and let P_T be the profit on each teddy bear. Find the maximum profit in each of the following cases.

 (a) $P_P = \$12, \quad P_T = \$16$

 (b) $P_P = \$15, \quad P_T = \$10$

 (c) $P_P = \$15, \quad P_T = \$12$

48. In Exercise 38, let P_A equal the profit made on each toy airplane, and let P_T be the profit on each train. Find the maximum profit in each of the following cases.

 (a) $P_A = \$10, \quad P_T = \$30$

 (b) $P_A = \$15, \quad P_T = \$28$

 (c) $P_A = \$12, \quad P_T = \$30$

49. In Exercise 39, let C_A be the cost of each type A package of bulbs, and let C_B equal the cost of a type B package. Find the minimum cost in each of the following cases.

 (a) $C_A = \$8, \quad C_B = \$3$

 (b) $C_A = \$6, \quad C_B = \$5$

 (c) $C_A = \$7, \quad C_B = \$4$

50. In Exercise 40, let C_A be the cost of a type A variety pack, and let C_B equal the cost of a type B pack. Find the minimum cost in each of the following cases.

 (a) $C_A = \$12, \quad C_B = \$10$

 (b) $C_A = \$9, \quad C_B = \$15$

 (c) $C_A = \$10, \quad C_B = \$12$

Chapter 2 Review Outline

2.1 Linear Equations

Definitions

An equation in x is a statement that two algebraic expressions in x are equal. A solution is a value of x for which the statement is true.

Two equations are equivalent if one can be transformed into the other by the algebraic properties of equality. (Equivalent equations have the same solution.)

A linear equation in x is one equivalent to $ax + b = 0$, where $a \neq 0$.

Linear Equations

The following properties of equality are used to reduce a given linear equation to an equivalent one in which all terms involving x are on one side.

(1) $M = N \Leftrightarrow M \pm c = N \pm c$

(2) If $c \neq 0$, then $M = N \Leftrightarrow Mc = Nc$ and $M/c = N/c$.

Equations with Fractions

First clear the fractions and then proceed as above. *Check answers for extraneous roots.*

Equations with Radicals

Eliminate one radical at a time by isolating it on one side and squaring both sides. *Check answers for extraneous roots.*

Equations with Absolute Values

First eliminate one absolute value $|M|$ by replacing the equation with two auxiliary equations, one with M in place of $|M|$ and the other with $-M$ in place of $|M|$. If $|N|$ is in the auxiliary equations, repeat the process for N in each auxiliary equation. Continue until all absolute values have been eliminated. Solve each of the final auxiliary equations for x, and *check answers for extraneous roots in the original equation.*

2.2 Linear Inequalities

Inequalities in which M and N are linear expressions in x can be solved by applying the following properties:

(1) $M > N \Leftrightarrow M \pm c > N \pm c$

(2) If $c > 0$, then $M > N \Leftrightarrow Mc > Nc$ and $M/c > N/c$

(3) If $c < 0$, then $M > N \Leftrightarrow Mc < Nc$ and $M/c < N/c$.

Any inequality with fractional expressions can be transformed to one of the form $M/N > 0, \geq 0, < 0$, or ≤ 0, which can be solved by the union and intersection method or the method of testing points.

The properties for solving inequalities with absolute values are as follows:

(1) $|M| > c \Leftrightarrow M > c$ or $M < -c$

(2) $|M| < c \Leftrightarrow -c < M < c$.

2.3 Applications of Linear Equations and Inequalities

In a word problem, first translate the wording into equations and/or inequalities; then solve algebraically and interpret the results in terms of the original problem.

Percent: $r\%$ of $Q = \frac{r}{100} \cdot Q$

Motion: distance = rate · time

Mixtures: In any mixing problem, each quantity to be combined, contains a certain concentrate in some mixture. Compute the amount of concentrate in each quantity to be combined, and add these amounts to get the total amount of concentrate.

Work: If it takes n time units to do a given job, then we assume that $1/n$ of the job is completed in each time unit.

2.4 The Coordinate Plane

Every point P in the plane corresponds to an ordered pair (x, y) of real numbers.

Distance Formula:

$$d(P_1, P_2) = \sqrt{(x_2 - x_1)^2 + (y_2 - y_1)^2}$$

$$\textit{midpoint } (P_1, P_2) = \left(\frac{x_1 + x_2}{2}, \frac{y_1 + y_2}{2}\right)$$

2.5 Equations of Straight Lines

The slope m of a nonvertical line through the points $P_1(x_1, y_1)$ and $P_2(x_2, y_2)$ is defined as

$$m = \frac{y_2 - y_1}{x_2 - x_1} \quad (x_1 \neq x_2).$$

A rising (falling) line has positive (negative) slope. A horizontal line has slope 0; the slope is not defined for a vertical line. The steeper the line, the larger the magnitude of its slope. When x changes by ± 1 unit along a line, y changes by $\pm m$ units.

Equations of a Line

Point-Slope: $y - y_1 = m(x - x_1)$

Slope-Intercept: $y = mx + b$

Vertical Line: $x = c$

Horizontal Line: $y = c$

General Linear: $Ax + By = C \quad (A \text{ or } B \neq 0)$

2.6 Systems of Linear Equations

Given two distinct lines

$$L_1\colon A_1x + B_1y = C_1$$
$$L_2\colon A_2x + B_2y = C_2$$

then

(1) $L_1 \parallel L_2 \Leftrightarrow m_1 = m_2$ (nonvertical)
$\Leftrightarrow A_1B_2 - A_2B_1 = 0$

(2) $L_1 \perp L_2 \Leftrightarrow m_1m_2 = -1$ (neither line vertical)
$\Leftrightarrow A_1A_2 + B_1B_2 = 0.$

(3) If L_1 and L_2 are not parallel, their point of intersection may be found by the elimination or substitution methods, or by the formulas

$$x = \frac{\begin{vmatrix} C_1 & B_1 \\ C_2 & B_2 \end{vmatrix}}{\begin{vmatrix} A_1 & B_1 \\ A_2 & B_2 \end{vmatrix}}, \quad y = \frac{\begin{vmatrix} A_1 & C_1 \\ A_2 & C_2 \end{vmatrix}}{\begin{vmatrix} A_1 & B_1 \\ A_2 & B_2 \end{vmatrix}},$$

where $\begin{vmatrix} A & B \\ C & D \end{vmatrix} = AD - BC$

for any real numbers A, B, C, and D.

The distance from a point $P(x_1, y_1)$ to a line $L\colon Ax + By = C$ is

$$d = \frac{|Ax_1 + By_1 - C|}{\sqrt{A^2 + B^2}}.$$

2.7 Linear Inequalities in the Plane with Applications

A line L with equation $Ax + By = C \quad (A > 0)$ divides the plane into a right half-plane $\mathcal{R}$ and a left half-plane $\mathcal{L}$, and

$$Ax + By > C \text{ for all } (x, y) \text{ in } \mathcal{R},$$
$$Ax + By < C \text{ for all } (x, y) \text{ in } \mathcal{L}.$$

The solution of a system of linear inequalities is the intersection of their respective half-plane solutions. In linear programming, a point of maximum profit (minimum cost) occurs at a vertex of the region of solution.

Chapter 2 Review Exercises

Solve each equation or inequality in Exercises 1–14.

1. $3x - 2 = 7x + 12$
2. $4.5x - 3 = 1.5(x - 2) + 7.25$
3. $3x - 2 < 7x + 12$
4. $5x + 4 \geq 4x + 5$
5. $\dfrac{3}{2x - 4} - \dfrac{1}{2} = \dfrac{1}{2 - x}$
6. $\dfrac{3x}{x + 1} = 1 - \dfrac{3}{x + 1}$
7. $\sqrt{x^2 + 25} = x + 5$
8. $\sqrt{x + 7} = 1 + \sqrt{x}$
9. $\sqrt{x + 1} + \sqrt{x - 1} + 2 = 0$
10. $|2x + 4| < 8$
11. $|3x - 5| \geq 4$
12. (a) $|4x - 7| > -2$ (b) $|4x - 7| \leq -2$
13. $|x - 2| = |2x - 7|$
14. $|3x - 2| + |x + 5| = 11$

15. Solve the equation $a^2x - 4 = a + 16x$ for x in terms of a.
16. Solve $2/(x + a) + 3/x = 5/(x^2 + ax)$ for x in terms of a.
17. Solve the inequality $(x - 2)/(2x + 1) > 1$.
18. Solve $(x - 2)/(2x + 1) \leq 1$, and express the solution in interval notation.
19. Solve $1/(x + a) + 1/(x - a) > 0$ for x, given that $a > 0$ and $-a < x < a$.
20. Let $x_1 = 6$ and $x_2 = -7$ be two points on the real line. Find their midpoint and the distance between them.
21. For the points $(6, -2)$ and $(-7, 3)$ in the plane, find their midpoint and the distance between them.
22. Show that $P_1(8, -2)$, $P_2(-1, -1)$, and $P_3(3, -6)$ are vertices of a right triangle by using the following procedures.
    (a) distance formula (b) slopes
23. Find a value of x for which the point $P(x, 7)$ is equidistant from $P_1(-2, 4)$ and $P_2(3, 6)$.
24. If $P(3, 4)$ is the midpoint of the line segment from $P_1(-1, 5)$ to $P_2(x, y)$, find x and y.
25. Show that all points $P(x, y)$ equidistant from the two points $P_1(5, 2)$ and $P_2(2, 6)$ lie on a line L. What is an equation of L?
26. Find the equation of the horizontal line through the point $P(6, 4)$ and also the equation of the vertical line through P.
27. Find the slope-intercept equation of the line through $(3, -2)$ and $(-7, 4)$.
28. Find the slope-intercept equation for the line $3x - 2y = 4$.
29. Find a general linear equation of the perpendicular bisector of the line segment from $P_1(5, 2)$ to $P_2(2, 6)$. Compare with Exercise 25.
30. Find an equation of the line that passes through $(-1, 5)$ and is parallel to $2x - 3y = 7$.
31. By the method of elimination, solve the system
$$\begin{cases} 3x - 2y = 18 \\ 4x + 3y = 7 \end{cases}.$$

32. Solve the system in Exercise 31 by the method of substitution.

33. Solve the system in Exercise 31 by the method of determinants.

34. Use any method to solve the system

$$\begin{cases} 5x - 6y = 8 \\ y = -x + 6. \end{cases}$$

***35.** In the plane, graph the solution of the system $x \geq 3$, $y \geq 2$, $x + y \leq 10$.

***36.** In the plane, graph the solution of the system

$$\begin{cases} y - 2x \leq 4 \\ x + y < 3. \end{cases}$$

37. For a rock concert, 20,000 tickets were sold for $560,000. If tickets cost either $25 or $40, how many were sold at each price?

38. The price of a car is $10,269.50 after a 5% discount on the sticker price. If the sticker price includes the dealer's cost plus a 15% markup on that cost, what is the dealer's cost?

39. Steve drove his car 165 mi. He drove the first 1.5 hr at one speed and the next 2.5 hr at 10 mph faster. Find his speeds for each part of the trip.

***40.** A college bookstore must order two books A and B, with more of A than B and a total of 60 or less. The order can include at most 40 of A. If A nets a $5 profit and B a $2 profit, how many of each should be ordered to obtain the maximum profit?

Graphing Calculator Exercises

In each of the following, verify your answers to the indicated review exercises of this section by using a graphing calculator for the given functions and windows $[x_{min}, x_{max}]$, $[y_{min}, y_{max}]$.

41. Exercise 3: $Y_1 = 7x + 12 - (3x - 2)$, $[-7, 2.5]$, $[-15, 15]$

42. Exercise 5: $Y_1 = 3/(2x - 4) - 1/2 - 1/(2 - x)$, $[0, 19]$, $[-1, 1]$

43. Exercise 7: $Y_1 = \sqrt{x^2 + 25} - (x + 5)$, $[-4, 5.5]$, $[-5, 5]$

44. Exercise 8: $Y_1 = \sqrt{x + 7} - (1 + \sqrt{x})$, $[0, 19]$, $[-1, 1]$

45. Exercise 13: $Y_1 = |x - 2| - |2x - 7|$, $[0, 9.5]$, $[-5, 5]$

46. Exercise 31: $Y_1 = 3x/2 - 9$, $Y_2 = -4x/3 + 7/3$, $[0, 9.5]$, $[-5, 5]$

47. Exercise 34: $Y_1 = 5x/6 - 8/6$, $Y_2 = -x + 6$, $[0, 9.5]$, $[-5, 5]$

*Asterisks indicate topics that may be omitted, depending on individual course needs.

3 Algebra and Graphs of Quadratic Expressions

Chapter 2 was devoted to the algebra and geometry of linear expressions. Now we consider **quadratic expressions,** which are polynomials of degree 2 in one or more variables. We will restrict our attention to the cases

(1) $ax^2 + bx + c \quad (a \neq 0)$

and (2) $ax^2 + by^2 + cx + dy + e \quad (a \neq 0 \text{ or } b \neq 0)$,

where the coefficients a, b, c, d, and e are constant real numbers, and x and y are the variables. In seeking the zeros of (1), we are led to the quadratic formula and to the complex number system. The zeros of (2) can be any of a variety of curves in the plane called **conic sections.**

3.1 Quadratic Equations

People have been solving quadratic equations since the time of ancient Babylonia.

The Quadratic Form
Solutions by Factoring
The Quadratic Formula
The Discriminant
Factoring by Roots

The Quadratic Form. The standard form of a **quadratic equation** is

$$\boldsymbol{ax^2 + bx + c = 0,} \tag{1}$$

where the coefficients a, b, and c are real numbers and a is not zero. Some examples of quadratic equations and their solutions are shown below.

	Equation	*Solution (roots)*
(a)	$x^2 + 2x - 3 = 0$	$-3, 1$
(b)	$2x^2 - 5x + 2 = 0$	$1/2, 2$
(c)	$x^2 - 10x + 25 = 0$	5

We note that a quadratic equation can have *two* roots, while a linear equation always has exactly one. We now consider the problem of finding the roots of a quadratic equation.

Solutions by Factoring. When the left side of the quadratic equation (1) is factorable by integers, we can obtain its roots by setting each factor equal to zero. This method is based on the zero-product property, which states that a product of real numbers is zero if and only if at least one of its factors is zero.

Example 1 Find the roots of equations (a), (b), and (c) by factoring.

Solution:

(a) $x^2 + 2x - 3 = 0$

$$(x + 3)(x - 1) = 0$$

$$x + 3 = 0 \quad \text{or} \quad x - 1 = 0$$

$$x = -3 \qquad\qquad x = 1$$

Therefore, the roots are -3 and 1.

(b) $2x^2 - 5x + 2 = 0$

$$(2x - 1)(x - 2) = 0$$

$$2x - 1 = 0 \quad \text{or} \quad x - 2 = 0$$

$$x = \frac{1}{2} \qquad\qquad x = 2$$

The roots are 1/2 and 2.

(c) $x^2 - 10x + 25 = 0$

$$(x - 5)(x - 5) = 0$$

$$x = 5$$

The only root is 5. ■

When a quadratic equation has two distinct roots, as in equations (a) and (b) above, each root is called a **simple root** or a **root of order 1.** In the case where a quadratic equation has only one distinct root (i.e., two equal roots) as in (c) above, the root is called a **double root** or a **root of order 2.**

If we are given one or two real numbers, we can construct a quadratic equation having these numbers for roots as illustrated in the following example.

Example 2 In each of the following cases, find a quadratic equation having the given roots.

(a) 2, 3 (b) 3/2, -5 (c) 6 only (d) $2 + \sqrt{3}$, $2 - \sqrt{3}$

Solution: For each simple root a_k, we construct the linear factor $x - a_k$, and for a double root a_k, we construct *two* linear factors $x - a_k$. We then multiply the factors and set the product equal to zero.

(a) $(x - 2)(x - 3) = 0$

$$x^2 - 5x + 6 = 0$$

(b) $\left(x - \frac{3}{2}\right)(x + 5) = 0$

$$x^2 + \frac{7}{2}x - \frac{15}{2} = 0$$

or $2x^2 + 7x - 15 = 0$ *Multiply both sides by 2*

(c) $(x - 6)(x - 6) = 0$

$$x^2 - 12x + 36 = 0$$

(d) $[x - (2 + \sqrt{3})][x - (2 - \sqrt{3})] = 0$

$$[(x - 2) - \sqrt{3}][(x - 2) + \sqrt{3}] = 0$$
$$(x - 2)^2 - 3 = 0$$
$$x^2 - 4x + 1 = 0 \quad \blacksquare$$

The Quadratic Formula. The method of factoring to find roots works well for quadratics that are easily factored. A more general method can be used to solve *any* quadratic equation. First, from the equation

$$x^2 + bx + \frac{b^2}{4} = \left(x + \frac{b}{2}\right)^2$$

we can deduce the following result, which is used in the process called **completing the square.**

If $\frac{b^2}{4}$ is added to $x^2 + bx$, then a perfect square $\left(x + \frac{b}{2}\right)^2$ is obtained.

Note that the coefficient of x^2 must be 1 in the expression $x^2 + bx$ that is to be completed, and the quantity added is $b^2/4 = (b/2)^2$, the square of one-half the coefficient of x. We now apply completing the square to the quadratic equation $2x^2 - 5x + 1 = 0$ in which the left side cannot be factored by integers.

$2x^2 - 5x + 1 = 0$	*Given*
$x^2 - \frac{5}{2}x + \frac{1}{2} = 0$	*Divide both sides by 2*
$x^2 - \frac{5}{2}x = -\frac{1}{2}$	*Subtract $\frac{1}{2}$ from both sides*
$x^2 - \frac{5}{2}x + \frac{25}{16} = -\frac{1}{2} + \frac{25}{16}$	*Complete the square on left by adding $\left(-\frac{5}{4}\right)^2$ to both sides*
$\left(x - \frac{5}{4}\right)^2 = \frac{17}{16}$	*Add on right side with LCD*
$x - \frac{5}{4} = \pm\frac{\sqrt{17}}{4}$	*Take square roots on both sides*
$x = \frac{5 \pm \sqrt{17}}{4}$	*Add $\frac{5}{4}$ to both sides*

We could use the method of completing the square to solve any quadratic equation, but it is more efficient to use it to derive the quadratic formula, which can then be used to solve any quadratic equation without the need to repeat the steps in the above example. To obtain the quadratic formula, we apply the method of completing the square to the general quadratic equation $ax^2 + bx + c = 0$ as follows:

$ax^2 + bx + c = 0$	*Given*
$x^2 + \frac{b}{a}x + \frac{c}{a} = 0$	*Divide both sides by a*
$x^2 + \frac{b}{a}x = -\frac{c}{a}$	*Subtract $\frac{c}{a}$ from both sides*
$x^2 + \frac{b}{a}x + \frac{b^2}{4a^2} = -\frac{c}{a} + \frac{b^2}{4a^2}$	*Complete the square on left by adding $\left(\frac{b}{2a}\right)^2$ to both sides*
$\left(x + \frac{b}{2a}\right)^2 = \frac{b^2 - 4ac}{4a^2}$	*Add on right side with LCD = $4a^2$*
$x + \frac{b}{2a} = \pm\frac{\sqrt{b^2 - 4ac}}{2a}$	*Take square roots on both sides*
$x = \frac{-b \pm \sqrt{b^2 - 4ac}}{2a}$	*Add $-\frac{b}{2a}$ to both sides*

Hence, we obtain the **quadratic formula:**

> The roots of the quadratic equation $ax^2 + bx + c = 0 \quad (a \neq 0)$, are
>
> $$\boldsymbol{a_1 = \frac{-b + \sqrt{b^2 - 4ac}}{2a}} \quad \textbf{and} \quad \boldsymbol{a_2 = \frac{-b - \sqrt{b^2 - 4ac}}{2a}}.$$

Example 3 Find the roots of the equations in Example 1 by using the quadratic formula.

Solution:

(a) $x^2 + 2x - 3 = 0 \quad (a = 1, b = 2, c = -3)$

$$x = \frac{-2 \pm \sqrt{4 - (-12)}}{2} = \frac{-2 \pm \sqrt{16}}{2} = \frac{-2 \pm 4}{2} = -1 \pm 2$$

The roots are $a_1 = -1 + 2 = 1$ and $a_2 = -1 - 2 = -3$.

(b) $2x^2 - 5x + 2 = 0$ $(a = 2, b = -5, c = 2)$

$$x = \frac{-(-5) \pm \sqrt{(-5)^2 - 16}}{4}$$

$$= \frac{5 \pm \sqrt{9}}{4} = \frac{5 \pm 3}{4}$$

The roots are $a_1 = (5 + 3)/4 = 2$ and $a_2 = (5 - 3)/4 = 1/2$.

(c) $x^2 - 10x + 25 = 0$ $(a = 1, b = -10, c = 25)$

$$x = \frac{-(-10) \pm \sqrt{(-10)^2 - 100}}{2}$$

$$= \frac{10 \pm \sqrt{0}}{2} = 5$$

Here the only root is the double root $a_1 = 5$. ■

Example 4 Solve each equation by using the quadratic formula.

(a) $\frac{1}{2}x^2 + 2x + 1 = 0$ (b) $4x^2 + 8x + 8 = 0$

Solution:

(a) We first multiply both sides by 2 in order to clear fractions and thereby obtain the equivalent equation $x^2 + 4x + 2 = 0$. We now use the quadratic formula with $a = 1$, $b = 4$, and $c = 2$ to get

$$x = \frac{-4 \pm \sqrt{16 - 8}}{2}$$

$$= \frac{-4 \pm \sqrt{8}}{2}$$

$$= \frac{-4 \pm 2\sqrt{2}}{2}$$

$$= -2 \pm \sqrt{2}.$$

The roots are $a_1 = -2 + \sqrt{2}$ and $a_2 = -2 - \sqrt{2}$.

(b) We can first divide both sides by 4 to obtain the equivalent equation $x^2 + 2x + 2 = 0$. We now use the quadratic formula with $a = 1$, $b = 2$, and $c = 2$ to get

$$x = \frac{-2 \pm \sqrt{4 - 8}}{2}$$

$$= \frac{-2 \pm \sqrt{-4}}{2}.$$

However, $\sqrt{-4}$ has no meaning in the real number system. Therefore, the quadratic equation $x^2 + 2x + 2 = 0$ has *no real roots;* that is, it has *no solution in the real number system.* ■

Example 5 Put each of the following quadratic equations in standard form and solve by any method.

(a) $2x^2 - x = 5x + 3$

(b) $3x^2 + 7x - 8 = x^2 + 2x + 4$

(c) $4x^2 + 6x + 18 = 3x^2 - 6x - 18$

(d) $x^2 + 5x + 4 = x - 2$

Solution:

(a) Subtract $5x + 3$ from both sides to obtain $2x^2 - 6x - 3 = 0$. By the quadratic formula,

$$\begin{aligned} x &= \frac{6 \pm \sqrt{36 + 24}}{4} \\ &= \frac{6 \pm \sqrt{60}}{4} \\ &= \frac{6 \pm 2\sqrt{15}}{4} \\ &= \frac{2(3 \pm \sqrt{15})}{4} \\ &= \frac{3 \pm \sqrt{15}}{2}. \end{aligned}$$

Therefore, the roots are $a_1 = \dfrac{3 + \sqrt{15}}{2}$ and $a_2 = \dfrac{3 - \sqrt{15}}{2}$.

(b) Subtract $x^2 + 2x + 4$ from both sides to obtain $2x^2 + 5x - 12 = 0$, which can be factored by integers into $(2x - 3)(x + 4) = 0$. Therefore, the roots are $a_1 = 3/2$ and $a_2 = -4$.

(c) Subtract $3x^2 - 6x - 18$ from both sides to obtain $x^2 + 12x + 36 = 0$, which can be written as $(x + 6)^2 = 0$. The only root is the double root $a_1 = -6$.

(d) Subtract $x - 2$ from both sides to obtain $x^2 + 4x + 6 = 0$. By the quadratic formula,

$$\begin{aligned} x &= \frac{-4 \pm \sqrt{16 - 24}}{2} \\ &= \frac{-4 \pm \sqrt{-8}}{2}. \end{aligned}$$

Since $\sqrt{-8}$ has no meaning as a real number, there are no roots in the real number system. ■

The Discriminant. As the previous examples show, a quadratic equation can have two distinct real roots, two equal roots, or no real root at all. To discover which of these possibilities is the case, we look at the sign of the quantity

$$b^2 - 4ac,$$

which is called the **discriminant** of $ax^2 + bx + c$. By the quadratic formula,

> the quadratic equation $ax^2 + bx + c = 0$ will have
>
> (a) two distinct real roots $\Leftrightarrow b^2 - 4ac > 0$,
>
> (b) two equal real roots $\Leftrightarrow b^2 - 4ac = 0$,
>
> (c) no real roots $\Leftrightarrow b^2 - 4ac < 0$.

Example 6 Without actually solving for the roots, determine the exact number of distinct real roots in each of the following cases.

(a) $x^2 + 50x + 750 = 0$ (b) $3x^2 + 70x - 125 = 0$

(c) $4x^2 - 50x + 156.25 = 0$

Solution:

(a) $a = 1,\ b = 50,\ c = 750$

$$\begin{aligned} b^2 - 4ac &= 2500 - 3000 \\ &= -500 \end{aligned}$$

Since -500 is less than zero, there are *no* real roots.

(b) $a = 3,\ b = 70,\ c = -125$

$$\begin{aligned} b^2 - 4ac &= 4900 + 1500 \\ &= 6400 \end{aligned}$$

There are *two* distinct real roots.

(c) $a = 4,\ b = -50,\ c = 156.25$

$$\begin{aligned} b^2 - 4ac &= 2500 - 2500 \\ &= 0 \end{aligned}$$

The equation has *one* distinct real root (two equal real roots). ■

Factoring by Roots. We have seen that if $ax^2 + bx + c$ factors into $a(x - a_1)(x - a_2)$, then a_1 and a_2 are the roots of $ax^2 + bx + c = 0$. We now use the quadratic formula to show that the converse is also true.

Let a_1 and a_2 be the roots of $ax^2 + bx + c = 0$ obtained by the quadratic formula. Now

$$\begin{aligned} a(x - a_1)(x - a_2) &= a(x^2 - a_1x - a_2x + a_1a_2) \\ &= a[x^2 - (a_1 + a_2)x + a_1a_2]. \qquad (2) \end{aligned}$$

By expressing a_1 and a_2 in terms of a, b, c as given in the quadratic formula, we obtain the following formulas for the sum and product of the roots of $ax^2 + bx + c = 0$ (see Exercise 86):

$$a_1 + a_2 = -\frac{b}{a} \quad \text{and} \quad a_1 a_2 = \frac{c}{a}. \tag{3}$$

We now substitute from (3) into (2) to obtain

$$\begin{aligned} a(x - a_1)(x - a_2) &= a[x^2 - (a_1 + a_2)x + a_1 a_2] \\ &= a\left[x^2 + \frac{b}{a}x + \frac{c}{a}\right] \\ &= ax^2 + bx + c. \end{aligned}$$

Hence, we have the following result for factoring a quadratic by its zeros.

> If a_1 and a_2 are the roots of $ax^2 + bx + c = 0$, then
> $$ax^2 + bx + c = a(x - a_1)(x - a_2). \tag{4}$$

We note, however, that if $b^2 - 4ac < 0$, then a_1 and a_2 are not real numbers and therefore $ax^2 + bx + c$ cannot be factored by real numbers. In this case we say that $ax^2 + bx + c$ is **irreducible** in the real number system.

Example 7 Use result (4) to factor each of the following quadratics by their real zeros.

(a) $x^2 - 6x + 7$ (b) $4x^2 - 20x + 25$

(c) $6x^2 - 5x + 1$ (d) $x^2 + 4x + 7$

Solution:

(a) The roots of $x^2 - 6x + 7 = 0$ are $a_1 = 3 + \sqrt{2}$ and $a_2 = 3 - \sqrt{2}$. Therefore, $x^2 - 6x + 7 = [x - (3 + \sqrt{2})][x - (3 - \sqrt{2})]$.

(b) The equation $4x^2 - 20x + 25 = 0$ has $a_1 = 5/2$ as a double root. Therefore $4x^2 - 20x + 25 = 4(x - 5/2)^2$.

(c) The roots of $6x^2 - 5x + 1 = 0$ are $a_1 = 1/2$ and $a_2 = 1/3$. Therefore, $6x^2 - 5x + 1 = 6(x - 1/2)(x - 1/3)$.

(d) The discriminant of $x^2 + 4x + 7$ is the negative quantity -12. Therefore, $x^2 + 4x + 7$ is irreducible in the real number system. ■

Question *Can two different quadratic equations have the same roots?*

Answer: Yes, but only if one of the quadratic expressions is a constant multiple of the other. Suppose $ax^2 + bx + c = 0$ and $Ax^2 + Bx + C = 0$ have the same roots, a_1 and a_2. Then, by result (4),

$$ax^2 + bx + c = a(x - a_1)(x - a_2)$$

and

$$Ax^2 + Bx + C = A(x - a_1)(x - a_2).$$

Therefore,

$$Ax^2 + Bx + C = \frac{A}{a}[a(x - a_1)(a - a_2)] = \frac{A}{a}(ax^2 + bx + c). \blacksquare$$

Comment In Chapter 2 we saw that every linear equation has exactly one root in the real number system. However, to a Greek mathematician in 250 B.C., the linear equation $2x + 1 = x$ would have *no* solution, since only *positive* numbers had a geometric interpretation and therefore meaning in ancient Greek mathematics. If we want every linear equation to have a solution, then we must extend the set of positive real numbers to the larger system of *all* real numbers, positive, negative, and zero. In a similar way, if we want every quadratic equation to have a solution, then we must extend the real number system to a larger system in which it is permissible to take a square root of a negative quantity. Mathematicians began to do this in the seventeenth century, but it was not until the latter part of the nineteenth century that this larger system, called the **complex number system,** was completely understood and fully accepted by all mathematicians.

Exercises 3.1

Fill in the blanks to make each statement true.

1. The standard form of a quadratic equation is ____________, where the coefficients a, b, and c are __________ and _______ $\neq 0$.
2. A quadratic equation can have as many as _______ roots, whereas a linear equation always has exactly _______ root(s).
3. Three methods for finding the roots of a quadratic equation are __________, __________, and __________.
4. By the quadratic formula, the roots of $ax^2 + bx + c = 0$ are ____________.
5. The quadratic equation $ax^2 + bx + c = 0$ will have two distinct real roots if __________, two equal real roots if __________, and no real roots if __________.

Write true *or* false *for each statement.*

6. If the quadratic expression $ax^2 + bx + c$ is factorable by integers, then the roots of $ax^2 + bx + c = 0$ are rational numbers.
7. All the roots of a given quadratic equation can be obtained by the quadratic formula.
8. If a_1 and a_2 are the roots of $ax^2 + bx + c = 0$, then $ax^2 + bx + c = a(x - a_1)(x - a_2)$.
9. If $ax^2 + bx + c = a(x - a_1)(x - a_2)$, then the roots of $ax^2 + bx + c = 0$ are a_1 and a_2.
10. If $a > 0$ and $c < 0$, then the quadratic equation $ax^2 + bx + c = 0$ has two distinct real roots.

Factoring

Solve each quadratic equation by the method of factoring.

11. $x^2 - 7x + 12 = 0$
12. $x^2 - 7x + 10 = 0$
13. $x^2 - 6x - 16 = 0$
14. $x^2 - 2x - 35 = 0$
15. $x^2 - 8x + 16 = 0$
16. $x^2 - 25 = 0$
17. $4x^2 - 9 = 0$
18. $4x^2 + 12x + 9 = 0$
19. $3x^2 + 5x + 2 = 0$
20. $6x^2 - x - 1 = 0$
21. $x^2 + (a + b)x + ab = 0$
22. $ax^2 + bx = 0$

Find a quadratic equation with integral coefficients that has the given roots in each of the following cases.

23. $3, -3$
24. $\frac{2}{5}, -\frac{2}{5}$
25. $\frac{3}{2}, \frac{4}{3}$
26. $5, 2$
27. $\frac{7}{2}$ only
28. -8 only
29. $3 + \sqrt{2}, 3 - \sqrt{2}$
30. $\frac{1 + \sqrt{5}}{2}, \frac{1 - \sqrt{5}}{2}$

The Quadratic Formula

Use the quadratic formula to solve each of the following equations.

31. $x^2 - x - 1 = 0$
32. $x^2 + x - 1 = 0$
33. $x^2 + 4x + 2 = 0$
34. $x^2 + 4x - 2 = 0$
35. $2x^2 - 6x + 3 = 0$
36. $2x^2 + 6x - 3 = 0$
37. $3x^2 - 10x + 4 = 0$
38. $3x^2 + 10x - 4 = 0$
39. $2x^2 - 5x - 2 = 0$
40. $2x^2 - 5x + 2 = 0$
41. $x^2 - (a + b)x + ab = 0$
42. $x^2 - (a + 1)x + a = 0$

Use factoring or the quadratic formula to find all real roots of each of the following.

43. $x^2 - 3x + 2 = 3x + 2$
44. $x^2 - 4 = x - 2$
45. $3x^2 - 2x + 4 = x^2 - 5x + 3$
46. $5x^2 + 6x - 4 = x^2 - 2x + 1$
47. $2x^2 - 4x + 5 = -2x + 6$
48. $x^2 + 4x + 2 = 2x^2 + 1$
49. $4x^2 + 3x + 3 = -2x^2 + 4x + 4$
50. $3x^2 + 7x + 8 = -3x^2 - 6x + 2$

The Discriminant

Without solving the equation, find the number of distinct real roots in each of the following.

51. $x^2 - 100x + 275 = 0$
52. $x^2 - 100x - 275 = 0$
53. $5x^2 + 120x + 720 = 0$
54. $5x^2 - 120x + 720 = 0$
55. $x^2 + 25x + 175 = 0$
56. $x^2 - 25x - 175 = 0$
57. $321x^2 + 225x - 720 = 0$
58. $321x^2 + 225x + 720 = 0$
59. $36x^2 - 204x + 289 = 0$
60. $36x^2 + 204x - 289 = 0$

Factoring by Roots

If possible, factor each of the following quadratics $ax^2 + bx + c$ as $a(x - a_1)(x - a_2)$, where a_1 and a_2 are the real roots of $ax^2 + bx + c = 0$.

61. $2x^2 - 13x + 20$
62. $6x^2 + 13x + 6$
63. $x^2 - 10x + 22$
64. $x^2 - 6x + 4$
65. $4x^2 - 12x + 9$
66. $9x^2 + 30x + 25$
67. $2x^2 + 5x + 5$
68. $3x^2 + 4x + 2$
69. $4x^2 - 4x - 1$
70. $9x^2 - 6x - 1$

Miscellaneous

Solve each of the following quadratic equations by completing the square. (Go through each step used in the derivation of the quadratic formula.)

71. $x^2 - 5x + 5 = 0$
72. $x^2 + 6x + 4 = 0$
73. $2x^2 + 4x - 3 = 0$
74. $3x^2 + 6x + 1 = 0$
75. $4x^2 + 4x - 1 = 0$
76. $2x^2 + 4x - 3 = 0$
77. $3x^2 - 2x - 1 = 0$
78. $4x^2 - 10x + 5 = 0$

The process of completing the square can be used to write any quadratic expression as a constant times a perfect square plus a constant. For example,

$$2x^2 + 5x + 7 = 2\left(x^2 + \frac{5}{2}x + \_\_\right) + (7 - \_\_)$$

$$= 2\left(x^2 + \frac{5}{2}x + \frac{25}{16}\right) + \left(7 - \underset{\uparrow}{2} \cdot \frac{25}{16}\right)$$

(*Why the factor 2?*)

$$= 2\left(x + \frac{5}{4}\right)^2 + \frac{31}{8}.$$

Apply this procedure to each of the quadratic expressions in Exercises 79–84.

79. $2x^2 + 12x + 3$
80. $3x^2 - 3x + 1$
81. $4x^2 + 2x - 5$
82. $x^2 - 18x - 17$
83. $x^2 + x + 1$
84. $2x^2 + 4x$

85. *Perfect Square.* Show that $ax^2 + bx + c \quad (a > 0)$ is a perfect square $\Leftrightarrow b^2 - 4ac = 0$. (*Hint:* complete the square for $ax^2 + bx + c$.)

86. *Sum and Product of Roots.* Given that a_1 and a_2 are the roots of $ax^2 + bx + c = 0$, use the quadratic formula to show that

$$a_1 + a_2 = -\frac{b}{a} \quad \text{and} \quad a_1 a_2 = \frac{c}{a}.$$

(In the case in which $b^2 - 4ac = 0$, take $a_1 = a_2$.)

87. Use the result from Exercise 86 to find the sum and product of the roots in Exercises 51, 53, 57, and 59.

88. Use the result from Exercise 86 to find the sum and product of the roots in Exercises 52, 54, 56, and 60.

89. *The Case of the Missing Root.* Consider the linear and quadratic equations

$$\text{(i) } bx + c = 0 \quad \text{and} \quad \text{(ii) } ax^2 + bx + c = 0$$

where $a > 0$, $b > 0$, and $c < 0$. Now (ii) has two distinct roots (*why?*), and (i) has the single root $-c/b$. This suggests that if the value of the coefficient a approaches 0 (written: $a \to 0$), then one of the roots of (ii) approaches the value $-c/b$. Verify that such is the case, and also determine the fate of the other root of (ii) as $a \to 0$. (*Hint:* the two roots of (ii) are

$$a_1 = \frac{-b + \sqrt{b^2 - 4ac}}{2a} \quad \text{and} \quad a_2 = \frac{-b - \sqrt{b^2 - 4ac}}{2a}.)$$

Now complete the following steps:

1. $\dfrac{a_1 a_2}{a_1 + a_2} =$ ______________ by Exercise 86.
2. Also, $\dfrac{a_1 a_2}{a_1 + a_2} =$ ______________ by dividing numerator and denominator by a_2.
3. As $a \to 0$, the numerator of a_2 in the quadratic formula approaches $-2b$ and the denominator approaches 0. Hence, $a_2 \to -$________ (think big), and by steps 1 and 2, $a_1 \to$ ________.

3.2 Complex Numbers

Mathematicians will do anything to solve a problem, even if it requires taking square roots of negative numbers!

The Complex Number System
Complex Roots of a Quadratic Equation
Geometric Representation of Complex Numbers

The Complex Number System. Square roots of negative numbers arise when we formally apply the quadratic formula to certain quadratic equations, such as $x^2 + 2x + 2 = 0$. In order that such quadratic equations have roots, we extend the real number system to a larger one in which square roots of negative numbers exist. The larger system is called the **complex number system,** and a number in this system is defined as follows.

> A **complex number** is a number
>
> $$z = a + bi,$$
>
> where a and b are real numbers and $i = \sqrt{-1}$; a is called the **real part** of z, and b is called the **imaginary part** of z.

For example, $4 + 3i$ and $2 - \sqrt{5}i$ are complex numbers. Also, the real number 4 is a complex number with zero imaginary part, and the number $3i$ is a complex number with zero real part. Complex numbers with zero real part, such as $3i$, are called **pure imaginary numbers.**

Two complex numbers are said to be equal if and only if their real parts are equal and their imaginary parts are equal. That is,

$$\mathbf{a + bi = c + di \Leftrightarrow a = b \text{ and } c = d.}$$

equality of complex numbers

The operations of addition, subtraction, multiplication, and division of complex numbers are defined in such a way that the basic properties (1) through (9) of Chapter P continue to hold. Hence, by definition,

$$\mathbf{(a + bi) + (c + di) = (a + c) + (b + d)i}$$

addition of complex numbers

and

$$\mathbf{(a + bi) - (c + di) = (a - c) + (b - d)i.}$$

subtraction of complex numbers

For example, $(4 + 3i) + (2 + 5i) = 6 + 8i$ and $(4 + 3i) - (2 + 5i) = 2 - 2i$.

To arrive at the definition for multiplication, we formally apply the usual rules for multiplying two binomials.

$$\begin{aligned}(a + bi)(c + di) &= ac + adi + bci + bdi^2 \\ &= (ac - bd) + (ad + bc)i \qquad i^2 = -1\end{aligned}$$

Therefore, we define

$$\mathbf{(a + bi)(c + di) = (ac - bd) + (ad + bc)i.}$$

multiplication of complex numbers

For example,

$$\begin{aligned}(4 + 3i)(2 + 5i) &= (8 - 15) + (20 + 6)i \\ &= -7 + 26i.\end{aligned}$$

To obtain a rule for dividing complex numbers, we recall that for real numbers the division $a/b = c$ means $bc = a$. Similarly, for complex numbers,

$$\frac{a + bi}{c + di} = x + yi \qquad \text{means} \qquad (c + di)(x + yi) = a + bi.$$

Therefore, by the above rule for multiplication,

$$(cx - dy) + (cy + dx)i = a + bi,$$

and by the rule for equality of complex numbers,

$$cx - dy = a$$

and

$$dx + cy = b,$$

which are two equations in two unknowns x and y. By the method of determinants from Section 2.6,

$$x = \frac{\begin{vmatrix} a & -d \\ b & c \end{vmatrix}}{\begin{vmatrix} c & -d \\ d & c \end{vmatrix}} = \frac{ac + bd}{c^2 + d^2} \quad \text{and} \quad y = \frac{\begin{vmatrix} c & a \\ d & b \end{vmatrix}}{\begin{vmatrix} c & -d \\ d & c \end{vmatrix}} = \frac{bc - ad}{c^2 + d^2}.$$

Hence,

$$\frac{a + bi}{c + di} = \frac{ac + bd}{c^2 + d^2} + \frac{bc - ad}{c^2 + d^2}i. \qquad \textbf{division of complex numbers}$$

For example,

$$\frac{4 + 3i}{5 + 2i} = \frac{20 + 6}{25 + 4} + \frac{15 - 8}{25 + 4}i = \frac{26}{29} + \frac{7}{29}i.$$

In the complex number system, zero is $0 + 0i$, and division by zero is prohibited just as in the real number system.

The rules for multiplication and division of complex numbers need not be memorized. To multiply two complex numbers, we can just apply the usual rules for multiplying two binomials, keeping in mind that $i^2 = -1$. To divide two complex numbers, the following technique can be used.

$$\frac{a + bi}{c + di} = \frac{a + bi}{c + di} \cdot \frac{c - di}{c - di} \qquad \textit{Multiply numerator and denominator by } c - di$$

$$= \frac{ac - adi + bci - bdi^2}{c^2 - \cancel{cdi} + \cancel{dci} - d^2i^2}$$

$$= \frac{(ac + bd) + (bc - ad)i}{c^2 + d^2} \qquad \textit{The denominator is a real number}$$

$$= \frac{ac + bd}{c^2 + d^2} + \frac{bc - ad}{c^2 + d^2}i.$$

For example,

$$\frac{4 + 3i}{5 + 2i} = \frac{4 + 3i}{5 + 2i} \cdot \frac{5 - 2i}{5 - 2i} = \frac{(20 + 6) + (-8 + 15)i}{25 + 4} = \frac{26}{29} + \frac{7}{29}i.$$

The number $c - di$ is called the **complex conjugate** of $c + di$, and

$$(c + di)(c - di) = c^2 + d^2.$$

Hence, to divide two complex numbers, multiply numerator and denominator by the complex conjugate of the denominator.

Example 1 Compute each of the following.

(a) $(2 - 3i)(1 + i)$ (b) $(4 + i)i$ (c) $\dfrac{5 + i}{1 + i}$

(d) $\dfrac{1}{i}$ (e) $(2i)^2$

Solution:

(a) $(2 - 3i)(1 + i) = 2 + 2i - 3i - 3i^2$ $\quad i^2 = -1$

$= 5 - i$

(b) $(4 + i)i = 4i + i^2$

$= -1 + 4i$

(c) $\dfrac{5 + i}{1 + i} = \dfrac{(5 + i)(1 - i)}{(1 + i)(1 - i)}$ *Multiply numerator and denominator by the conjugate of the denominator*

$= \dfrac{5 - 5i + i - i^2}{1 + 1}$

$= \dfrac{6 - 4i}{2}$

$= 3 - 2i$

(d) $\dfrac{1}{i} = \dfrac{1(-i)}{i(-i)}$ *The conjugate of i is $-i$*

$= \dfrac{-i}{-i^2}$

$= \dfrac{-i}{-(-1)}$

$= -i$ *The reciprocal of i is $-i$*

(e) $(2i)^2 = (2i)(2i) = 4i^2 = 4(-1) = -4$ ■

From part (e) of Example 1, we see that $(2i)^2 = -4$. Similarly, $(-2i)^2 = -4$, so that both $2i$ and $-2i$ are square roots of -4. We use the notation $\sqrt{-4}$ to denote $2i$. In general, if p is any positive number, then by definition,

$$\sqrt{-p} = \sqrt{p}\,\sqrt{-1} = \sqrt{p}\,i \quad (p > 0),$$

where $\sqrt{p}$ denotes the positive square root of p.

Complex Roots of a Quadratic Equation. If $ax^2 + bx + c = 0$ cannot be solved by factoring with real numbers, then we use the quadratic formula to find its roots in the complex number system.

Example 2 Find the roots of each of the following quadratic equations.

(a) $x^2 + 2x + 2 = 0$ (b) $x^2 + 4x + 6 = 0$

Solution:

(a) By the quadratic formula, with $a = 1$, $b = 2$, and $c = 2$, we have

$$\begin{aligned} x &= \frac{-2 \pm \sqrt{4-8}}{2} \\ &= \frac{-2 \pm \sqrt{-4}}{2} \\ &= \frac{-2 \pm 2i}{2} \qquad \sqrt{-4} = \sqrt{4}\,\sqrt{-1} = 2i \\ &= -1 \pm i. \end{aligned}$$

(b) Here, $a = 1$, $b = 4$, and $c = 6$. Therefore,

$$\begin{aligned} x &= \frac{-4 \pm \sqrt{16-24}}{2} \\ &= \frac{-4 \pm \sqrt{-8}}{2} \\ &= \frac{-4 \pm \sqrt{8}\,i}{2} = \frac{-4 \pm 2\sqrt{2}\,i}{2} = -2 \pm \sqrt{2}\,i. \end{aligned}$$ ■

In Example 2, the roots $-1 + i$ and $-1 - i$ of part (a) are complex conjugates. Similarly, in (b) the roots $-2 + \sqrt{2}\,i$ and $-2 - \sqrt{2}\,i$ are complex conjugates. In general, if $a + bi$ is a complex root of a quadratic equation with real coefficients, then so is its complex conjugate $a - bi$ (*why?*). We can therefore reformulate our previous result (Section 3.1), which related the roots of the equation $ax^2 + bx + c = 0$ to its discriminant as follows.[1]

[1] *The relationship between the roots and the discriminant of a quadratic equation was discovered by Sir Isaac Newton (1642–1727).*

> The quadratic equation $ax^2 + bx + c = 0$, where a, b, and c are real numbers, has
>
> (a) two distinct real roots $\Leftrightarrow b^2 - 4ac > 0$,
>
> (b) two equal real roots $\Leftrightarrow b^2 - 4ac = 0$,
>
> (c) two complex conjugate roots $\Leftrightarrow b^2 - 4ac < 0$.

The result $ax^2 + bx + c = a(x - a_1)(x - a_2)$, where a_1 and a_2 are the roots of $ax^2 + bx + c = 0$, applies when a_1 and a_2 are complex numbers, as is illustrated in the next example.

Example 3 Factor each of the following quadratics by their complex zeros.

(a) $x^2 + 4$ (b) $2x^2 + 2x + 1$

Solution:

(a) The quadratic equation $x^2 + 4 = 0$ has roots $a_1 = 2i$ and $a_2 = -2i$. Therefore,

$$\begin{aligned} x^2 + 4 &= (x - 2i)[x - (-2i)] \\ &= (x - 2i)(x + 2i). \end{aligned}$$

(b) The roots of $2x^2 + 2x + 1 = 0$ are $a_1 = -\frac{1}{2} + \frac{i}{2}$ and $a_2 = -\frac{1}{2} - \frac{i}{2}$. Therefore,

$$\begin{aligned} 2x^2 + 2x + 1 &= 2\left[x - \left(-\frac{1}{2} + \frac{i}{2}\right)\right]\left[x - \left(-\frac{1}{2} - \frac{i}{2}\right)\right] \\ &= 2\left(x + \frac{1}{2} - \frac{i}{2}\right)\left(x + \frac{1}{2} + \frac{i}{2}\right). \end{aligned}$$ ■

Question *Does the quadratic formula apply to quadratic equations whose coefficients are complex numbers?*

Answer: Yes. The algebraic properties of real numbers needed to derive the quadratic formula in the previous section also hold for complex numbers. However, we will not encounter quadratic equations with complex coefficients in this course. ■

Geometric Representation of Complex Numbers. The complex number $z = a + bi$ can be represented in the plane by the directed line segment from the origin to the point with coordinates (a, b). The real numbers a and b are called the **rectangular coordinates of z** (Figure 1).

Addition of complex numbers follows the **parallelogram rule.** As shown in Figure 2, the sum of two complex numbers z_1 and z_2 is the diagonal (from the origin) of the parallelogram formed by z_1 and z_2 (see Exercise 55).

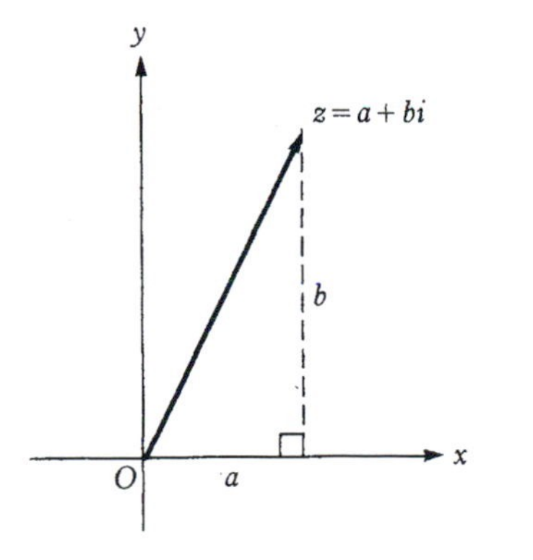

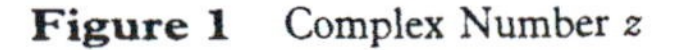

Figure 1 Complex Number z

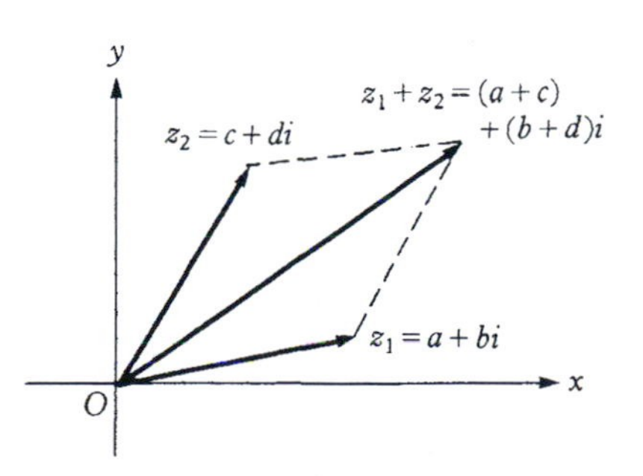

Figure 2 Addition of Complex Numbers

Example 4 In each of the following cases, compute $z_1 + z_2$ and illustrate the result with a corresponding parallelogram.

(a) $z_1 = 3 + 2i;\ z_2 = 1 + 4i$ (b) $z_1 = 3 + 2i;\ z_2 = 3 - 2i$

Solution: Each result is indicated in Figure 3. ■

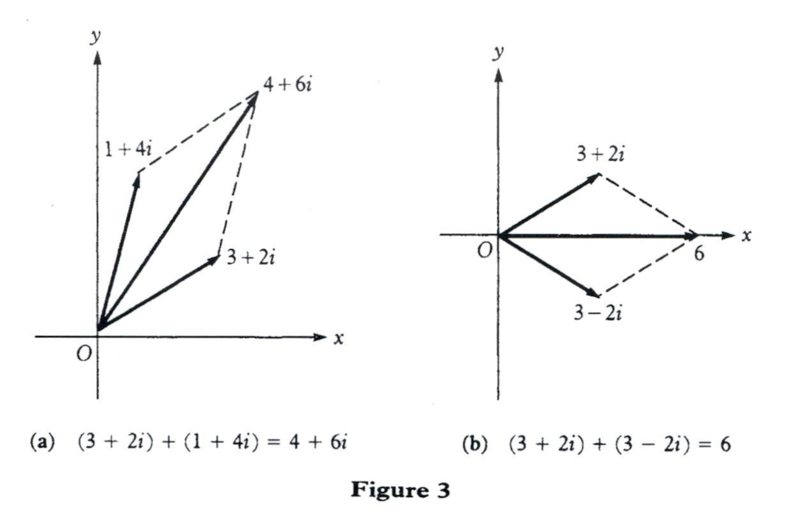

(a) $(3 + 2i) + (1 + 4i) = 4 + 6i$ (b) $(3 + 2i) + (3 - 2i) = 6$

Figure 3

Although addition of complex numbers is easily represented geometrically in the rectangular coordinate system of the plane, a convenient representation of multiplication requires the use of trigonometry (see also Exercises 56 and 57).

Comment As mentioned earlier, complex numbers were developed by mathematicians in order to solve equations such as $x^2 + 2 = 1$, just as negative numbers were introduced to solve equations such as $x + 2 = 1$. The acceptance of both types of numbers was a very gradual process. However, our geometric interpretation of real numbers as points on a directed line and complex numbers as points in the plane, along with a geometric interpretation of addition and multiplication, serve to dispel any mystery about either type of number (at least among mathematicians).

Exercises 3.2

Fill in the blanks to make each statement true.

1. A complex number is a number of the form $a + bi$, where a and b are ____________ and $i =$ ______.
2. The complex conjugate of $a + bi$ is ____________.
3. A real number has imaginary part ______.
4. A complex number with zero real part is called a ____________ number.
5. Geometrically, the sum of two complex numbers z_1 and z_2 is a ____________ of the parallelogram with sides z_1 and z_2.

Write true *or* false *for each statement.*

6. A real number is not a complex number.
7. A complex number is a point in the plane; points along the x-axis correspond to real numbers, and points along the y-axis correspond to pure imaginary numbers
8. If $3 + 2i$ is a root of a quadratic equation with real coefficients, then $3 - 2i$ is also a root of the equation.
9. Two complex numbers are equal if and only if their rectangular coordinates are equal.
10. To divide two complex numbers, multiply the numerator and denominator by the conjugate of the numerator.

The Complex Number System

Perform the indicated operations in Exercises 11–26.

11. $(2 + 3i) + (5 - 7i)$
12. $(3 - 5i) - (-3 + 4i)$
13. $(3 + 2i) + 4i$
14. $6 + (4 - 2i)$
15. $(4 + i)(3 - 2i)$
16. $(-3 + 5i)(2 + 7i)$
17. $(4 + 3i)(4 - 3i)$
18. $(-2 + 5i)(-2 - 5i)$
19. $3i(1 - i)$
20. $(4 + 2i)5i$
21. $\dfrac{3 + 2i}{1 + i}$
22. $\dfrac{2 - i}{2 + i}$
23. $\dfrac{i}{5 - 2i}$
24. $\dfrac{6i}{2 - 3i}$
25. $\dfrac{7 + 4i}{i}$
26. $\dfrac{-5 + 6i}{-3i}$

27. Show that $i^4 = 1$ and use this result to compute each of the following.
 (a) i^5 (b) i^{10}
 (c) i^{15} (d) i^{20}
28. Any positive integer n can be written as $4q + r$, where q and r are nonnegative integers and $r \leq 3$. Show that $i^n = i^r$.

Complex Roots of a Quadratic Equation

Solve each of the following quadratic equations.

29. $x^2 + 9 = 0$
30. $4x^2 + 9 = 0$
31. $x^2 - 4x + 5 = 0$
32. $x^2 - 2x + 5 = 0$
33. $x^2 + 8x + 25 = 0$
34. $x^2 + 10x + 26 = 0$
35. $2x^2 - 2x + 1 = 0$
36. $2x^2 + 2x + 1 = 0$
37. $2x^2 - 4x + 3 = 0$
38. $2x^2 + 4x + 3 = 0$

Factor each of the following quadratics $ax^2 + bx + c$ as $a(x - a_1)(x - a_2)$, where a_1 and a_2 are the complex roots of $ax^2 + bx + c = 0$.

39. $x^2 + 9$
40. $x^2 - 4x + 5$
41. $x^2 + 8x + 25$
42. $x^2 + x + 1$
43. $x^2 + 10x + 26$
44. $2x^2 + 2x + 1$
45. $2x^2 + 4x + 3$
46. $3x^2 + 10x + 13$

Geometric Representation of Complex Numbers

In each of the following cases, compute $z_1 + z_2$ and illustrate the result by means of the parallelogram rule.

47. $z_1 = 3 + 2i$; $z_2 = 2 + 3i$
48. $z_1 = 4 - 2i$; $z_2 = 2 + 5i$
49. $z_1 = -3 + 4i$; $z_2 = 1 + 2i$
50. $z_1 = 5 - 3i$; $z_2 = -3 - i$
51. $z_1 = 4 + 2i$; $z_2 = -4 + 2i$
52. $z_1 = 2 - 3i$; $z_2 = 2 + 3i$
53. $z_1 = -3$; $z_2 = 5 + 2i$
54. $z_1 = 2i$; $z_2 = -4 + 2i$

55. *Parallelogram Rule for Addition.* Prove the parallelogram rule for addition of complex numbers as follows:
 (a) Use congruent triangles to show that the coordinates of Q are $(a + c, b)$.
 (b) Use part (a) to show that the coordinates of P_3 are $(a + c, b + d)$.

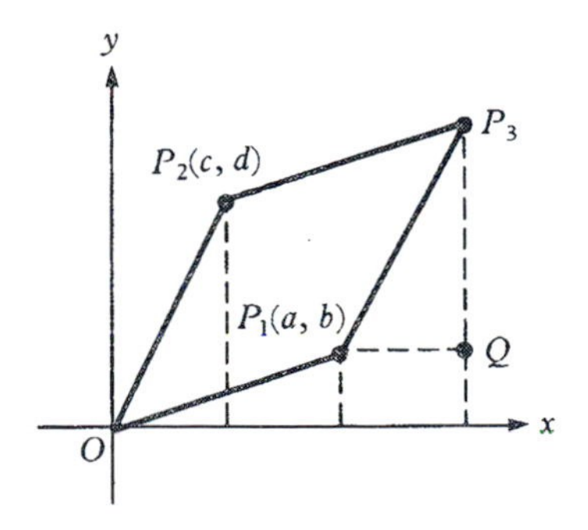

56. *Multiplication by i.* If the complex number $c + di$ is multiplied by i, we get

$$i(c + di) = -d + ci.$$

As indicated in the figure, the directed line segment $-d + ci$ is perpendicular to $c + di$. Also, both line segments have the same length, $\sqrt{c^2 + d^2}$, and when $c + di$ is in quadrant I, II, III, or IV, $-d + ci$ is in II, III, IV, or I, respectively. Therefore, *the effect of multiplying by i is to rotate $c + di$ through* 90°. Draw a figure to illustrate each of the following products.

(a) $i(3 + 4i)$ (b) $i(-3 + 4i)$
(c) $i \cdot i$ (d) $i \cdot 5$

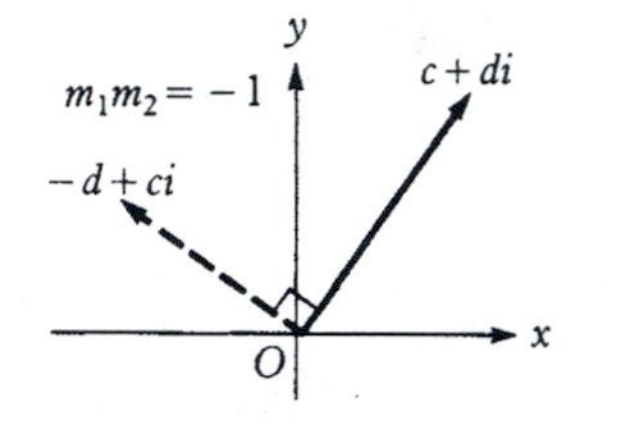

57. *Construction of $(a + bi)(c + di)$.* As indicated in the figure, we can construct the product of two complex numbers as the sum of two perpendicular directed line segments. Now

$$(a + bi)(c + di) = a(c + di) + bi(c + di)$$
$$= a(c + di) + b(-d + ci).$$

As in Exercise 56, $-d + ci$ is obtained by rotating $c + di$ through 90°. The effect of the real factor a is to multiply the length of the line segment $c + di$ by $|a|$, and if a is negative, to reverse its direction. The effect of the real factor b is similar. Draw a figure to illustrate each of the following products.

(a) $(1 + i)(3 + 4i)$ (b) $(1 - i)(3 + 4i)$
(c) $(-1 + i)(3 + 4i)$

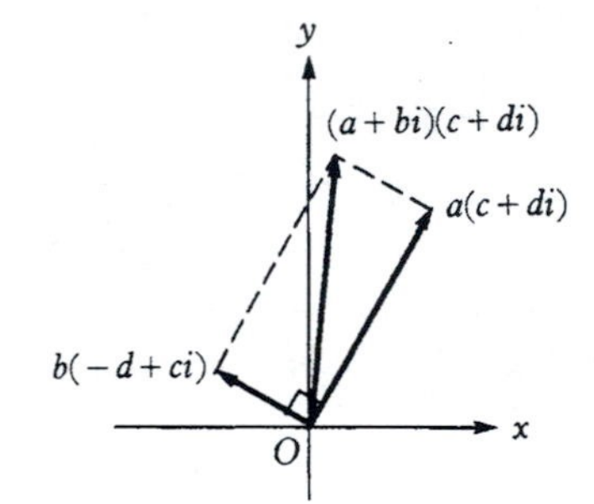

3.3 Equations Reducible to Quadratic Equations

The equations in this section are just quadratic equations in disguise.

Quadratics Obtained by Substitution
Equations with Fractional Expressions
Equations with Radicals

Quadratics Obtained by Substitution. It is sometimes possible to use substitution to transform a nonquadratic equation into a quadratic equation. Some examples are listed below.

	Original Equation	*Substitution*	*Quadratic Equation*
(a)	$x^4 - 6x^2 + 5 = 0$	$w = x^2$	$w^2 - 6w + 5 = 0$
(b)	$(x - 1)^6 - 7(x - 1)^3 - 8 = 0$	$w = (x - 1)^3$	$w^2 - 7w - 8 = 0$
(c)	$2x + 3\sqrt{x} - 2 = 0$	$w = \sqrt{x}$	$2w^2 + 3w - 2 = 0$

The resulting quadratic equation in each case can be solved for w either by factoring or by the quadratic formula. Then, by reversing the substitution to express x in terms of w, we obtain the roots of the original equation.

Caution Some substitutions place a restriction on the w-roots. For instance, in equations (a) and (c) above, only *nonnegative* values of w are acceptable in the real number system. All values of x obtained by a substitution process should be checked for acceptability in the original equation.

Example 1 Find all real solutions of the above equations (a), (b), and (c) by means of the substitutions given there.

Solution:

(a) The roots of $w^2 - 6w + 5 = 0$ are $w_1 = 1$ and $w_2 = 5$. The substitution $w = x^2$ implies that $x = \pm\sqrt{w}$. Therefore, the real roots of $x^4 - 6x^2 + 5 = 0$ are $a_1 = 1$, $a_2 = -1$, $a_3 = \sqrt{5}$, and $a_4 = -\sqrt{5}$.

Check each root in $x^4 - 6x^2 + 5 = 0$:

a_1: $1^4 - 6 \cdot 1^2 + 5 = 1 - 6 + 5 = 0$

a_2; $(-1)^4 - 6(-1)^2 + 5 = 1 - 6 + 5 = 0$

a_3: $(\sqrt{5})^4 - 6(\sqrt{5})^2 + 5 = 25 - 30 + 5 = 0$

a_4: $(-\sqrt{5})^4 - 6(-\sqrt{5})^2 + 5 = 25 - 30 + 5 = 0.$

(b) The roots of $w^2 - 7w - 8 = 0$ are $w_1 = -1$ and $w_2 = 8$. Now $w = (x - 1)^3$ implies that $x - 1 = \sqrt[3]{w}$ or $x = 1 + \sqrt[3]{w}$. Hence, the real roots of $(x - 1)^6 - 7(x - 1)^3 - 8 = 0$ are $a_1 = 1 + \sqrt[3]{-1} = 0$ and $a_2 = 1 + \sqrt[3]{8} = 3$.

Check each root in $(x - 1)^6 - 7(x - 1)^3 - 8 = 0$:

a_1: $(0 - 1)^6 - 7(0 - 1)^3 - 8 = 1 + 7 - 8 = 0$

a_2: $(3 - 1)^6 - 7(3 - 1)^3 - 8 = 64 - 56 - 8 = 0.$

(c) The roots of $2w^2 + 3w - 2 = 0$ are $w_1 = -2$ and $w_2 = 1/2$. The root $w_1 = -2$ is unacceptable since the substitution $w = \sqrt{x}$ implies that w is nonnegative. From the root $w_2 = 1/2$, we obtain $x = w_2{}^2 = 1/4$ as the only real root of the original equation.

Check this root in $2x + 3\sqrt{x} - 2 = 0$:

$2(1/4) + 3\sqrt{1/4} - 2 = 1/2 + 3/2 - 2 = 0.$ ■

Equations with Fractional Expressions. If an equation contains fractional expressions in x, we proceed as in Section 2.1. That is, we first clear the fractions by multiplying both sides of the equation by the factors of the denominators. If the resulting equation is quadratic, we can then solve it by factoring or by the quadratic formula.

Caution As stated in Section 2.1, any value of x obtained as a solution after clearing fractions must be tested in the original equation. If the value makes a denominator equal to zero, it is an extraneous root and cannot be accepted.

Example 2 Solve $\frac{x}{x-1} + \frac{6}{x} = \frac{7}{x-1}$.

Solution: $\frac{x \cdot x(x-1)}{x-1} + \frac{6 \cdot x(x-1)}{x} = \frac{7 \cdot x(x-1)}{x-1}$ *Multiply both sides by* $x(x-1)$

$$x^2 + 6(x-1) = 7x$$
$$x^2 - x - 6 = 0$$
$$(x+2)(x-3) = 0$$
$$x = -2 \quad \text{or} \quad x = 3$$

Since neither of these values of x makes any denominator of the original equation equal to zero, the solutions of the original equation are -2 and 3. ■

Example 3 Solve $\frac{x^2}{x-5} + 3 = \frac{5x}{x-5}$.

Solution: $\frac{x^2 \cdot (x-5)}{x-5} + 3 \cdot (x-5) = \frac{5x \cdot (x-5)}{x-5}$ *Multiply both sides by* $x - 5$

$$x^2 + 3x - 15 = 5x$$
$$x^2 - 2x - 15 = 0$$
$$(x+3)(x-5) = 0$$
$$x = -3 \quad \text{or} \quad x = 5$$

The value 5 is an extraneous root of the original quadratic equation (*why?*), but -3 is an acceptable root. Therefore, the solution to the original equation is -3. ■

Equations with Radicals. As stated in Section 2.1, the general procedure for solving equations with radicals is to eliminate one radical at a time by squaring both sides. Again, because squaring may result in extraneous roots, all values obtained by this method must be checked for acceptability by substitution in the original equation.

Example 4 Solve $2x - 1 = \sqrt{x+1}$.

Solution: $4x^2 - 4x + 1 = x + 1$ *Square both sides*

$$4x^2 - 5x = 0$$
$$x(4x - 5) = 0$$
$$x = 0 \quad \text{or} \quad x = \frac{5}{4}$$

By substituting these values into the original equation, we see that 0 is an extraneous root (*why?*), but 5/4 is acceptable. Therefore, the solution to the original equation is 5/4. ■

Example 5 Solve $2\sqrt{3x+1} = \dfrac{2}{\sqrt{3x+1}} - 3.$

Solution: First clear the fraction by multiplying both sides by $\sqrt{3x+1}$. Then eliminate the radical by squaring.

$$2\sqrt{3x+1}\cdot\sqrt{3x+1} = \frac{2\sqrt{3x+1}}{\sqrt{3x+1}} - 3\sqrt{3x+1} \qquad \textit{Clear fraction}$$

$$2(3x+1) = 2 - 3\sqrt{3x+1}$$

$$2x = -\sqrt{3x+1}$$

$$4x^2 = 3x + 1 \qquad \textit{Square both sides}$$

$$4x^2 - 3x - 1 = 0$$

$$x = 1 \quad \text{or} \quad x = -\frac{1}{4}$$

By substituting these values into the original equation, we see that 1 is an extraneous root (*why?*), but $-1/4$ is acceptable. Hence, the solution to the original equation is $-1/4$. ■

Exercises 3.3

Fill in the blanks to make each statement true.

1. A nonquadratic equation can sometimes be transformed into a quadratic equation by means of a ______________.
2. An equation containing fractions cannot have as a solution any value that makes one of its denominators equal to ________.
3. A value obtained after squaring both sides of an equation must be tested for acceptability by substitution into ______________.
4. If an equation is solved for w after making the substitution $w = x^{2/3}$, then only ______________ values of w are acceptable in the real number system.
5. The equation $(x+1)^3 + (x+1)^{3/2} - 2 = 0$ becomes a quadratic equation in w if we let $w =$ ______________.

Write true *or* false *for each statement.*

6. If an equation in x is transformed into a quadratic equation in w by means of a substitution, then the original equation cannot have more than two solutions for x.
7. If an equation in x becomes a quadratic equation in w by means of the substitution $w = (x-1)^2$, then any real solution x of the original equation must satisfy $x \geq 1$.
8. An equation obtained by clearing fractions in a given equation is always equivalent to the original equation.
9. If an equation in x becomes a quadratic equation in x after clearing fractions, both solutions of the quadratic equation may be solutions of the original equation.
10. An equation with radicals always has an extraneous root.

Substitution

Find all real solutions of the following equations by the substitution process.

11. $x^4 - 3x^2 + 2 = 0$
12. $x^6 - 6x^3 + 5 = 0$
13. $4x - 5\sqrt{x} - 9 = 0$
14. $3(x-1) + 2\sqrt{x-1} - 8 = 0$
15. $\left(x + \frac{1}{x}\right)^2 - 2\left(x + \frac{1}{x}\right) - 3 = 0$

16. $\left(x - \frac{1}{x}\right)^2 - 13\left(x - \frac{1}{x}\right) + 36 = 0$

17. $(x^2 - 1)^2 - 8(x^2 - 1) + 16 = 0$

18. $(x^2 + 1)^2 - 12(x^2 + 1) + 36 = 0$

19. $(2x + 3)^{-2} + 9(2x + 3)^{-1} + 20 = 0$

20. $(3x + 5)^{-4} + 4(3x + 5)^{-2} + 3 = 0$

Fractional Expressions

Find all solutions of the following equations.

21. $x + \frac{6}{x - 5} = 0$

22. $x + \frac{2}{x + 3} = 0$

23. $\frac{x}{x + 1} + \frac{9}{x(x + 1)} = \frac{5}{x}$

24. $\frac{1}{x - 6} + \frac{1}{x - 5} + \frac{12}{x(x - 5)(x - 6)} = 0$

25. $\frac{1}{x - 2} + \frac{3}{x(x - 5)} = \frac{9}{x(x - 2)(x - 5)}$

26. $\frac{x}{x - 3} + \frac{3}{(x - 3)(x - 4)} = \frac{3}{x - 4}$

27. $\frac{1}{x + 2} + \frac{x + 1}{3x - 2} = 1$

28. $\frac{x + 4}{x + 5} + \frac{x + 3}{2x + 7} = 1$

29. $\frac{2x}{x - 7} - \frac{1}{x - 5} = \frac{1}{(x - 7)(x - 5)}$

30. $\frac{x^2}{(x + 2)(x + 4)} + \frac{2x}{x + 4} = \frac{2}{x + 2}$

Equations with Radicals

Find all real solutions of the following equations.

31. $x - 3 = \sqrt{3x - 11}$

32. $2(x - 1) = \sqrt{2x^2 - x - 1}$

33. $x + 1 = \sqrt{\frac{x + 5}{3}}$

34. $x - 2 = \sqrt{16 - 3x}$

35. $\sqrt{2x + 1} - \sqrt{x + 4} = 1$

36. $\sqrt{3x + 4} - \sqrt{x} = 2$

37. $\sqrt{x - 1} + \sqrt{x - 4} = \sqrt{x + 4}$

38. $\sqrt{2x - 6} + \sqrt{2x - 1} = \sqrt{5x}$

39. $\sqrt{\frac{x + 2}{3x + 7}} - \frac{1}{\sqrt{2x + 6}} = 0$

40. $\frac{1}{\sqrt{x + 6}} - \frac{\sqrt{x + 3}}{\sqrt{2 - x}} = 0$

3.4 Quadratic Inequalities

The sign of $ax^2 + bx + c$ depends on the sign of a and the roots of $ax^2 + bx + c = 0$.

Quadratics with Distinct Real Roots
Quadratics with Equal Real Roots
Quadratics with Complex Roots

In Section 2.2 we solved the four basic inequalities

$$M > 0, \quad M \geq 0, \quad M < 0, \quad \text{and} \quad M \leq 0,$$

where M was a linear expression $ax + b$. We now consider the same inequalities where M is a quadratic expression $ax^2 + bx + c$. We consider separately the cases in which $ax^2 + bx + c = 0$ has distinct real roots, equal real roots, and complex roots. These cases correspond to $b^2 - 4ac > 0$, $b^2 - 4ac = 0$, and $b^2 - 4ac < 0$, respectively.

Quadratics with Distinct Real Roots. In the case of distinct real roots, we can factor $ax^2 + bx + c$ and proceed as we did for fractional inequalities in Section 2.2. That is, we can apply either the **union and intersection method** or the **method of testing points.** These are illustrated in the following examples.

Example 1 Solve $-x^2 + 3x + 10 \geq 0$ by the union and intersection method.

Solution: By first multiplying both sides by -1, we obtain the equivalent inequality

$$x^2 - 3x - 10 \leq 0.$$

The quadratic $x^2 - 3x - 10$ factors into $(x + 2)(x - 5)$. We are dealing with the case of two distinct real roots -2 and 5. Since a product of two factors is negative if and only if one of the factors is positive and the other is negative,

$$\begin{aligned}
&\text{the solution of } (x+2)(x-5) < 0\\
&\quad = \text{all } x \text{ for which either}\\
&\qquad x+2>0 \text{ and } x-5<0 \quad \text{or} \quad x+2<0 \text{ and } x-5>0\\
&\quad = \text{all } x \text{ for which either}\\
&\qquad x>-2 \text{ and } x<5 \quad \text{or} \quad x<-2 \text{ and } x>5\\
&\quad = \{(-2, \infty) \cap (-\infty, 5)\} \cup \{(-\infty, -2) \cap (5, \infty)\}\\
&\quad = (-2, 5) \cup \varnothing\\
&\quad = (-2, 5).
\end{aligned}$$

Hence, the expression $(x + 2)(x - 5)$ is negative for x in the open interval $(-2, 5)$. Therefore, $(x + 2)(x - 5) \leq 0$ for x in closed interval $[-2, 5]$, which is also the solution of the given inequality $-x^2 + 3x + 10 \geq 0$ (Figure 4). ■

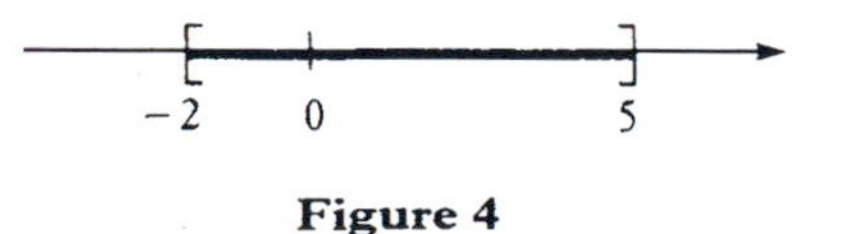

Figure 4

Example 2 Solve $6x^2 - 7x - 20 \geq 0$ by the method of testing points.

Solution: The quadratic $6x^2 - 7x - 20$ factors into

$$(3x + 4)(2x - 5) = 6\left(x + \frac{4}{3}\right)\left(x - \frac{5}{2}\right).$$

Here the equation $6x^2 - 7x - 20 = 0$ has distinct real roots $-4/3$ and $5/2$.

As shown in Figure 5, $-4/3$ and $5/2$ break up the number line into the intervals $(-\infty, -4/3)$, $(-4/3, 5/2)$, and $(5/2, \infty)$. In each of these, the sign of $x + 4/3$ and of $x - 5/2$ does not change (*why?*). Hence, *the sign of* $6(x + 4/3)(x - 5/2)$ *does not change in each of these intervals.*

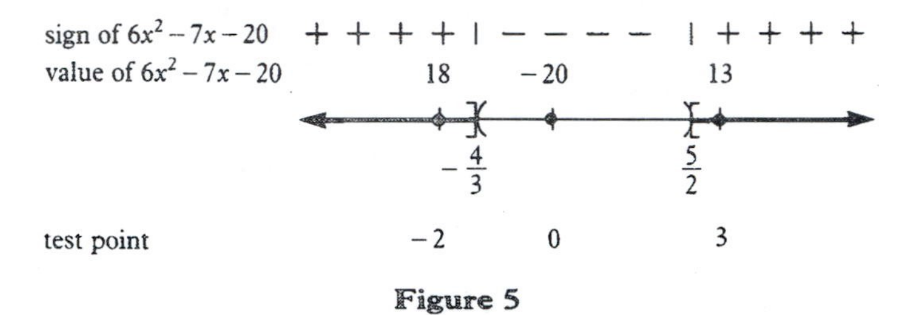

Figure 5

We choose three convenient test points, -2, 0, and 3 (one in each interval), and evaluate $6x^2 - 7x - 20$ at each of the test points. As indicated in Figure 5, we find that $6x^2 - 7x - 20$ is positive on $(-\infty, -4/3) \cup (5/2, \infty)$. Finally, since $6x^2 - 7x - 20 = 0$ for $x = -4/3$ and $x = 5/2$, we conclude that the solution of $6x^2 - 7x - 20 \geq 0$ is $(-\infty, -4/3] \cup [5/2, \infty)$. ■

Quadratics with Equal Real Roots. The quadratic $4x^2 - 12x + 9$ factors into

$$(2x - 3)^2 = 4\left(x - \frac{3}{2}\right)^2.$$

Here the equation $4x^2 - 12x + 9 = 0$ has a double root $3/2$. The factored form is zero when $x = 3/2$ and is positive for all other values of x. Therefore, $4x^2 - 12x + 9 \geq 0$ for all x.

In general, if a_1 is a double root of $ax^2 + bx + c = 0$, then

$$ax^2 + bx + c = a(x - a_1)^2.$$

Therefore, $ax^2 + bx + c = 0$ for $x = a_1$ and has the same sign as a for all other values of x.

Example 3 Solve $-4x^2 + 4x - 1 \geq 0$.

Solution: We first multiply both sides by -1 to obtain the equivalent inequality

$$4x^2 - 4x + 1 \leq 0.$$

Now the equation $4x^2 - 4x + 1 = 0$ has a double root $1/2$, and the factored form

$$(2x - 1)^2 = 4\left(x - \frac{1}{2}\right)^2$$

is positive for all other values of x. Hence, $4x^2 - 4x + 1 \leq 0$ is satisfied for $x = 1/2$ only. Therefore, the solution of the given inequality consists of the single point $1/2$ on the number line (Figure 6). ■

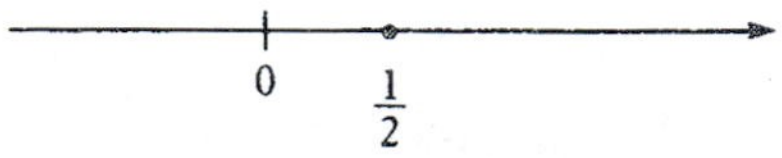

0 $\frac{1}{2}$

Figure 6

Quadratics with Complex Roots. The equation $x^2 - 2x + 2 = 0$ has complex roots $1 + i$ and $1 - i$. (Note that the discriminant $b^2 - 4ac = -4$ is negative.) Therefore,

$$\begin{aligned} x^2 - 2x + 2 &= [x - (1 + i)][x - (1 - i)] \\ &= [(x - 1) - i][(x - 1) + i] \\ &= (x - 1)^2 + 1. \end{aligned}$$

This same result could be obtained by completing the square in x. Now $(x - 1)^2 + 1 > 0$ for all real x, and therefore $x^2 - 2x + 2 > 0$ for all real numbers x.

In general, if the equation $ax^2 + bx + c = 0$ has complex roots $a_1 + b_1 i$, $a_1 - b_1 i$, then

$$ax^2 + bx + c = a[(x - a_1)^2 + b_1^2].$$

Since $(x - a_1)^2 + b_1^2 > 0$ for all real x, it follows that $ax^2 + bx + c$ has the same sign as a for all real numbers x. That is, *a quadratic with complex roots is either positive for all real numbers x or negative for all real numbers x.*

Example 4 Solve $x^2 + 2x + 3 > 0$.

Solution: The equation $x^2 + 2x + 3 = 0$ has complex roots since the discriminant $b^2 - 4ac = 4 - 12 = -8 < 0$. Therefore, $x^2 + 2x + 3$ is either positive for all real values of x or is negative for all real x. Now when $x = 0$, we have $0^2 + 2 \cdot 0 + 3 = 3 > 0$. Thus, $x^2 + 2x + 3 > 0$ for all real numbers x, and the solution is the entire real line $(-\infty, \infty)$. ■

Example 5 Solve $-2x^2 + 5x - 5 \geq 0$.

Solution: The equation $-2x^2 + 5x - 5 = 0$ has complex roots, since $b^2 - 4ac = 25 - 40 = -15 < 0$. Therefore, $-2x^2 + 5x - 5$ has the same sign for all x. For $x = 0$, we get $-2 \cdot 0^2 + 5 \cdot 0 - 5 = -5 < 0$. Thus, $-2x^2 + 5x - 5 < 0$ for all real numbers x, and the solution of the given inequality is the empty set. ■

We can summarize our results for quadratic inequalities as follows:

1. If $ax^2 + bx + c = 0$ has distinct real roots a_1 and a_2 $(a_1 < a_2)$, then $ax^2 + bx + c = a(x - a_1)(x - a_2)$, which does not change sign within each of the intervals $(-\infty, a_1)$, (a_1, a_2), and (a_2, ∞).
2. If $ax^2 + bx + c = 0$ has a double root a_1, then $ax^2 + bx + c = a(x - a_1)^2$, which has the same sign as a for all values of x other than a_1.
3. If $ax^2 + bx + c = 0$ has complex roots, then $ax^2 + bx + c$ is irreducible in the real number system and has the same sign as a for all real values of x.

Exercises 3.4

Fill in the blanks to make each statement true.

1. The solution of a quadratic inequality $ax^2 + bx + c > 0$ is completely determined by the sign of __________ and the __________ of $ax^2 + bx + c = 0$.
2. If $a > 0$ and $ax^2 + bx + c = 0$ has roots a_1 and a_2 $(a_1 < a_2)$, then $ax^2 + bx + c = a(x - a_1)(x - a_2)$ is __________ 0 for x in the interval (a_1, a_2).
3. If $b^2 - 4ac < 0$, then the solution to $ax^2 + bx + c > 0$ is either __________ or __________.
4. The quadratic inequality $a(x - a_1)^2 > 0$ has the empty set as its solution if __________.
5. If $ax^2 + bx + c = 0$ has roots a_1 and a_2 $(a_1 < a_2)$, and $ax^2 + bx + c > 0$ for $x = (a_1 + a_2)/2$, then the solution to $ax^2 + bx + c \leq 0$ is __________.

Write true *or* false *for each statement.*

6. If $b^2 - 4ac = 0$, the solution of $ax^2 + bx + c > 0$ is a single point on the real line.
7. If $b^2 - 4ac < 0$, then the inequality $ax^2 + bx + c > 0$ has no solution in the real number system.
8. If $b^2 - 4ac > 0$, and a_1 and a_2 $(a_1 < a_2)$ are the roots of $ax^2 + bx + c = 0$, then the solution of $ax^2 + bx + c > 0$ is either $(-\infty, a_1) \cup (a_2, \infty)$ or (a_1, a_2).
9. If we know the roots of $ax^2 + bx + c = 0$, then we know the solution of $ax^2 + bx + c < 0$.
10. If we know the sign of a, then we know the solution of $ax^2 + bx + c \geq 0$.

Quadratics with Distinct Real Roots

Solve each of the following inequalities, and graph the solution on the real number line.

11. $x^2 - 4x + 3 \geq 0$
12. $x^2 + 7x + 10 > 0$
13. $x^2 + 2x - 8 < 0$
14. $x^2 - 3x - 4 \leq 0$
15. $-2x^2 + 7x - 3 \leq 0$
16. $-3x^2 + 17x - 10 < 0$
17. $3x^2 - 10x - 8 \leq 0$
18. $4x^2 - 5x - 6 < 0$
19. $-6x^2 + 13x + 8 \leq 0$
20. $-6x^2 + 7x + 20 < 0$

Quadratics with Equal Real Roots

Solve each of the following inequalities.

21. $x^2 - 4x + 4 \geq 0$
22. $x^2 + 10x + 25 > 0$
23. $x^2 + 6x + 9 < 0$
24. $x^2 - 2x + 1 \leq 0$
25. $-4x^2 + 12x - 9 > 0$
26. $-9x^2 + 12x - 4 \geq 0$
27. $32x^2 - 16x + 2 \leq 0$
28. $50x^2 + 40x + 8 < 0$
29. $-48x^2 + 72x - 27 < 0$
30. $-20x^2 + 20x - 5 \geq 0$

Quadratics with Complex Roots

Solve each of the following inequalities in the real number system.

31. $x^2 + x + 1 \geq 0$
32. $x^2 + 2x + 4 < 0$
33. $-x^2 + 5x - 7 \leq 0$
34. $-2x^2 + 6x - 5 > 0$
35. $3x^2 - 4x + 2 \leq 0$
36. $5x^2 - 10x + 7 > 0$

Miscellaneous

Solve each of the inequalities in Exercises 37–48.

37. $12x^2 - 27 \geq 0$
38. $75 - 12x^2 \leq 0$
39. $(2x + 1)(x - 3) < 0$
40. $(3x - 2)(x + 1) > 0$
41. $x^2 + 3x - 4 < 4 - 3x - 4x^2$
42. $3x^2 + 12x - 3 > 2 - 6x^2$
43. $(x - 5)(x - 1) > x(5 - x)$
44. $(x - 3)(x + 1) \leq x(3 - x)$
45. $\dfrac{x^2 + 10}{x^2 + 4} > 2$
46. $\dfrac{x + 1}{x^2 + 1} > 1$
47. $\dfrac{1}{x^2 + 2} + 1 \leq \dfrac{4x}{x^2 + 2}$
48. $\dfrac{x}{x^2 + 1} \geq \dfrac{x + 2}{x^2 + x + 1}$
49. Find all points on the x-axis whose distance from the point $P(8, 2)$ is at least twice its distance from $Q(2, 1)$.
50. Find all points on the x-axis whose distance from $P(3, 4)$ is at least 5.
51. A ball is tossed upward. After t sec its height is y ft, where $y = -16t^2 + 32t$. For what values of t is the ball at least 7 ft above the ground?
52. For what values of t is the ball in Exercise 51 less than 12 ft above the ground?

3.5 Applications of Quadratic Equations and Inequalities

Why is it that when a quadratic equation is used to solve a problem, usually only one of the roots makes sense?

Percent
Motion Based on Average Speed
Motion Due to Gravity
Work
Miscellaneous Applications

In this section we investigate word problems whose mathematical interpretation results in a quadratic equation or inequality. In addition to the topics of percent, average speed, and work, which were introduced in Section 2.3, we also discuss motion due to gravity and some miscellaneous geometric and numeric problems. Although the expressions here are quadratic, they are arrived at in the same manner as in the linear case of Section 2.3, and you should follow the guidelines and flow chart given there.

Percent. When a percentage increase or decrease is applied to a quantity, a linear equation results. If a second increase or decrease is applied, a quadratic equation may result.

After graduating in June with an engineering degree, James gets a job starting at \$24,000 per year. In December he receives an $r\%$ increase, and six months later he gets a $2r\%$ increase, making his new salary \$27,720. What is r?

Solution: Let x be the decimal equivalent of $r\%$: $x = r/100$. We have the following:

$$\begin{aligned}\text{salary in December} &= \text{starting salary} + (\text{starting salary})x \\ &= 24{,}000 + 24{,}000 \cdot x \\ &= 24{,}000(1 + x);\end{aligned}$$

$$\begin{aligned}\text{salary six months later} &= \text{December salary} + (\text{December salary})2x \\ &= 24{,}000(1 + x) + 24{,}000(1 + x)2x \\ &= 24{,}000(1 + x)(1 + 2x) \\ &= 24{,}000(1 + 3x + 2x^2).\end{aligned}$$

Therefore,
$$\begin{aligned}24{,}000(1 + 3x + 2x^2) &= 27{,}720 \\ 1 + 3x + 2x^2 &= 1.155 \qquad 27{,}720/24{,}000 = 1.155 \\ 2x^2 + 3x - .155 &= 0.\end{aligned}$$

The roots are $x = .05$ and $x = -1.55$, but only the positive value is meaningful here. Hence, $x = .05$ and $r = 5$, so that the first increase is 5% and the second is 10%. ∎

Example 2 A clothing store wants to sell 100 suits originally priced at \$225 each. Seventy-five of them are sold at an $r\%$ discount, and the remaining 25 are marked down by another $r\%$. All 100 suits are sold for a total of \$14,568.75. What is r?

Solution: As in Example 1, let $x = r/100$, the decimal equivalent of $r\%$. Then

$$\begin{aligned}\text{selling price of first 75 suits} &= \text{original price} - (\text{original price})x \\ &= 225 - 225 \cdot x \\ &= 225(1 - x);\end{aligned}$$

$$\begin{aligned}\text{selling price of last 25 suits} &= \text{selling price of first 75 suits} \\ &\quad - (\text{selling price of first 75 suits})x \\ &= 225(1 - x) - 225(1 - x)x \\ &= 225(1 - x)(1 - x) \\ &= 225(1 - x)^2.\end{aligned}$$

Therefore,

$$\underbrace{75 \cdot 225(1 - x)}_{\textit{Income from first 75 suits}} + \underbrace{25 \cdot 225(1 - x)^2}_{\textit{Income from second 25 suits}} = \underbrace{14{,}568.75}_{\textit{Total income}}. \quad (1)$$

Equation (1) is a quadratic in $w = 1 - x$. That is,

$$25 \cdot 225w^2 + 75 \cdot 225w - 14{,}568.75 = 0$$

or

$$w^2 + 3w - 2.59 = 0.$$

By the quadratic formula, $w = .7$ and $w = -3.7$. Hence, $x = .3$ and $x = 4.7$. Only the root $x = .3$ has a meaningful interpretation (*why?*). Therefore, $r = 30\%$. ■

Motion Based on Average Speed. Quadratic equations often arise in basic motion problems in which an event taking t time units consists of two parts of t_1 and t_2 time units, respectively. The event is described by the equation

$$t_1 + t_2 = t,$$

or

$$\frac{d_1}{r_1} + \frac{d_2}{r_2} = t,$$

where d_1 and d_2 are the distances and r_1 and r_2 are the average rates for the first and second parts, respectively.

Example 3 A bus travels 60 miles at a certain average rate of speed and another 75 miles at an average rate 10 miles per hour faster. The entire trip takes 3 hours. What are the two rates of speed?

Solution: Let x = the rate of speed in miles per hour for the first 60 miles. Then $x + 10$ = the rate of speed for the last 75 miles. Also, $t_1 = 60/x$ is the time (in hours) for the first 60 miles, and $t_2 = 75/(x + 10)$ is the time for the last 75 miles. We have

$$t_1 + t_2 = 3,$$

or

$$\frac{60}{x} + \frac{75}{x + 10} = 3. \qquad (2)$$

After clearing fractions in equation (2), we obtain the quadratic equation

$$x^2 - 35x - 200 = 0$$

whose roots are $x = 40$ and $x = -5$. Only the positive root $x = 40$ is meaningful. Therefore, the average rate of speed for the first 60 miles is 40 miles per hour, and that for the last 75 miles is 50 miles per hour. ■

Example 4 Herb and Ray row at the same speed. Herb rows 3 miles with the aid of a 1-mile-per-hour current and then back 3 miles against the same current. Ray rows 6 miles in still water. If Herb's trip takes 4 hours, how long does Ray's take?

Solution: Let x = the rate (in miles per hour) at which both Herb and Ray row,
t_1 = the time for Herb's trip downstream, and
t_2 = the time for Herb's trip upstream. Then

$$t_1 + t_2 = 4,$$

or

$$\frac{3}{x + 1} + \frac{3}{x - 1} = 4.$$

Herb's speed downstream is $x + 1$, and upstream it is $x - 1$

When fractions are cleared in the above equation, the result is the quadratic equation

$$2x^2 - 3x - 2 = 0$$

whose roots are $x = 2$ and $x = -1/2$. Only the positive root $x = 2$ is meaningful. Therefore, both Herb and Ray row at 2 miles per hour in still water. Hence, the time for Ray's trip is $6/2 = 3$ hours. ■

Motion Due to Gravity. If an object is projected with a vertical velocity of v_0 feet per second from a point y_0 feet above the ground (Figure 7), then until it hits the ground its **height** y is given by

$$y = -16t^2 + v_0 t + y_0. \qquad (3)$$

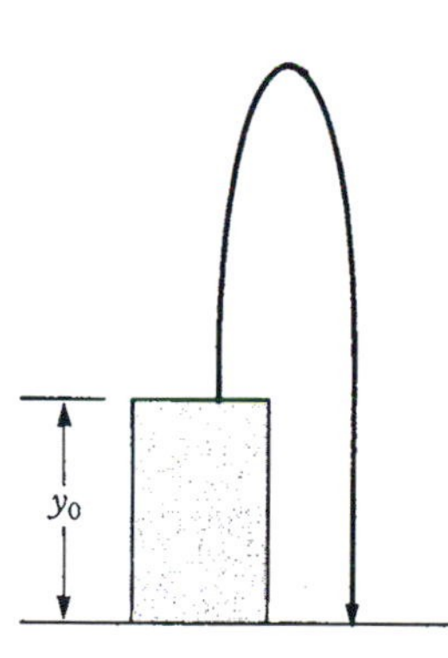

Figure 7

The height y is measured in feet, and the time t is measured in seconds. Formula (3) can be proven using calculus, and is based on the assumption that gravity is the only force acting on the object and that the acceleration due to gravity is a constant -32 ft/sec^2. Also, the **vertical velocity** v of the object at time t is given by the formula

$$v = -32t + v_0. \qquad (4)$$

The velocity is measured in feet per second; it is positive when the object is rising and negative when the object is falling.

Example 5 An object is projected upward from ground level with an initial velocity of 128 feet per second. For what period of time is the object at least 192 feet above the ground?

Solution: We are given $y_0 = 0$ and $v_0 = 128$. Therefore, formula (3) becomes

$$y = -16t^2 + 128t.$$

Set $y \geq 192$ and solve for t:

$$-16t^2 + 128t \geq 192$$
$$-t^2 + 8t - 12 \geq 0$$
$$t^2 - 8t + 12 \leq 0$$
$$(t-2)(t-6) \leq 0.$$

By the method of Section 3.4, the solution of the inequality $(t-2)(t-6) \leq 0$ is $2 \leq t \leq 6$. Therefore, the object is at least 192 feet above the ground from 2 seconds to 6 seconds after it leaves the ground. ■

Example 6 Jacqui drops a coin into a deep wishing well. She hears the splash 2.86 seconds later. How far down is the water level of the well? (Sound travels at 1100 feet per second.)

Solution: As indicated in Figure 8, let d = the depth of the water level in feet, t_1 = the time in seconds for the coin to hit the water, and t_2 = the

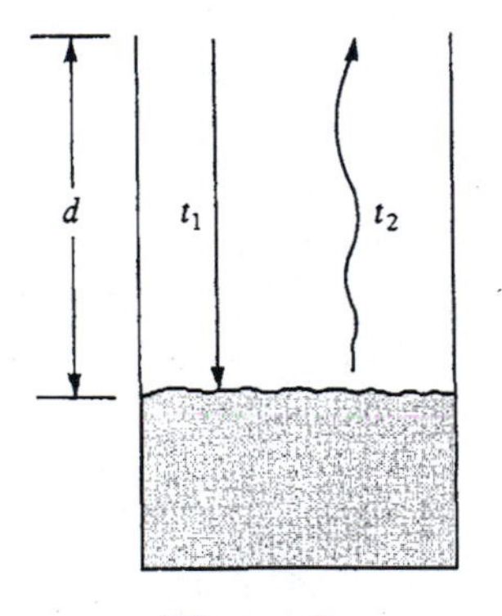

Figure 8

time for the sound to return to the top of the well. The motion downward is due to gravity with zero initial velocity (since the coin was just dropped) and an initial height of d feet. Therefore, from the formula (3),

$$-16t_1^2 + d = 0$$

or

$$t_1 = \frac{\sqrt{d}}{4}.$$

The motion upward is sound traveling at 1100 ft/sec. Therefore,

$$d = 1100t_2,$$

or

$$t_2 = \frac{d}{1100}.$$

We are given that

$$t_1 + t_2 = 2.86.$$

Therefore,

$$\frac{\sqrt{d}}{4} + \frac{d}{1100} = 2.86. \tag{5}$$

After clearing the fractions and substituting $w = \sqrt{d}$, equation (5) becomes

$$w^2 + 275w - 3146 = 0,$$

whose roots are $w = 11$ and $w = -286$. (Only $w = 11$ is acceptable.) Since $d = w^2$, the water level of the well is $11^2 = 121$ feet down. ■

Work. Here, as in the linear case, we consider work problems in which the following principle applies.

> If it takes n time units to do a given job, then $1/n$ of the job is completed per time unit.

Example 7 It takes secretary A 3 hours more than secretary B to prepare a certain report. Working together, the two can complete the job in 2 hours. How long does it take each secretary to finish the report when working alone?

Solution: Let x = the number of hours required for B to prepare the report. Then $x + 3$ = the number of hours required for A. When the two work together, 1/2 of the job is completed per hour. B's contribution in one hour is $1/x$ and A's is $1/(x + 3)$. Therefore,

$$\frac{1}{x} + \frac{1}{x+3} = \frac{1}{2}.$$

We clear fractions by multiplying both sides by $2x(x + 3)$ to obtain

$$2(x + 3) + 2x = x(x + 3)$$

or

$$x^2 - x - 6 = 0,$$

whose roots are $x = 3$ and $x = -2$. Only the positive root is meaningful here. Therefore, it takes B 3 hours to prepare the report when working alone, and it takes A 6 hours for the same report. ■

Example 8 Machine A can do a job in 3 1/3 hours. Machines A and B do the same job in 3 hours less time than it takes B working alone. How long does it take B to do the job?

Solution: Let x = number of hours for B to do the job. Then $x - 3$ = number of hours for A and B working together. Each hour A does 3/10 of the job, B contributes $1/x$, and A and B together contribute $1/(x - 3)$. Therefore,

$$\frac{3}{10} + \frac{1}{x} = \frac{1}{x - 3},$$

which, after clearing fractions, becomes

$$\begin{aligned} 3x(x - 3) + 10(x - 3) &= 10x \\ 3x^2 - 9x - 30 &= 0 \\ 3(x^2 - 3x - 10) &= 0 \\ 3(x - 5)(x + 2) &= 0 \\ x = 5 \quad \text{or} \quad x &= -2. \end{aligned}$$

Therefore, it takes B 5 hours to do the job. ■

Miscellaneous Applications. Here we treat some geometric and numeric problems that can be solved by means of quadratic equations.

Example 9 A string 12 inches long is cut into two pieces. Each piece is formed into a circle, and the sum of the areas of the two circles is $61/(2\pi)$ square inches. What is the length of each piece?

Figure 9

Solution: Let x = the length of one of the pieces. Then $12 - x$ = the length of the other piece (Figure 9). We can express the area of each circle (Figure 10) in terms of x as follows:

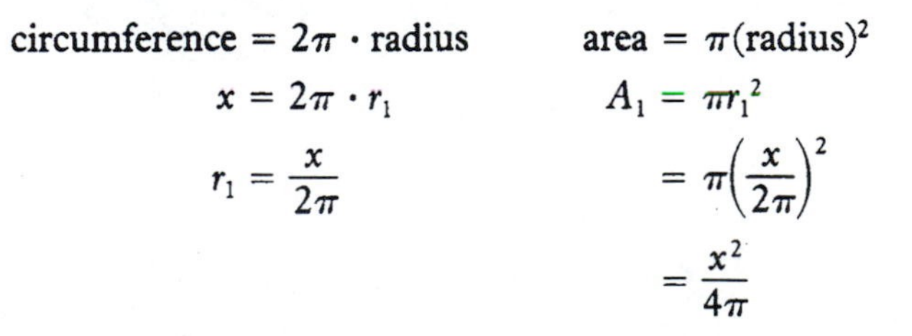

$$\begin{aligned} \text{circumference} &= 2\pi \cdot \text{radius} \\ x &= 2\pi \cdot r_1 \\ r_1 &= \frac{x}{2\pi} \end{aligned} \qquad \begin{aligned} \text{area} &= \pi(\text{radius})^2 \\ A_1 &= \pi r_1^2 \\ &= \pi\left(\frac{x}{2\pi}\right)^2 \\ &= \frac{x^2}{4\pi} \end{aligned}$$

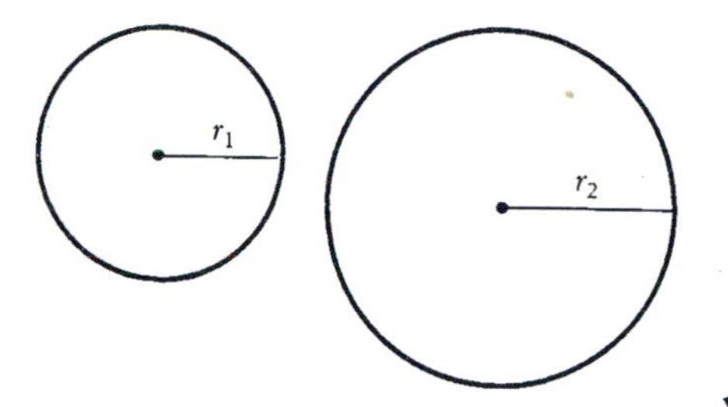

Figure 10

$$\text{circumference} = 2\pi \cdot \text{radius} \qquad \text{area} = \pi(\text{radius})^2$$

$$12 - x = 2\pi \cdot r_2 \qquad A_2 = \pi r_2^2$$

$$r_2 = \frac{12 - x}{2\pi} \qquad = \pi\left(\frac{12 - x}{2\pi}\right)^2$$

$$= \frac{144 - 24x + x^2}{4\pi}$$

We are given that

$$A_1 + A_2 = \frac{61}{2\pi}.$$

Therefore,
$$\frac{x^2}{4\pi} + \frac{144 - 24x + x^2}{4\pi} = \frac{61}{2\pi},$$

which simplifies to

$$x^2 - 12x + 11 = 0.$$

The roots are $x = 11$ and $x = 1$. In the first case $12 - x = 1$, and in the second case $12 - x = 11$. Therefore, one piece of the cut string is 11 inches, and the other piece is 1 inch. ■

Example 10 It will be shown in the final chapter that for any positive integer n,

$$1^2 + 2^2 + 3^2 + \cdots + n^2 = \frac{n(n + 1)(2n + 1)}{6}.$$

What is n if the sum of the squares from 1 to n is $20n$?

Solution:
$$\frac{n(n + 1)(2n + 1)}{6} = 20n \qquad \textit{Divide both sides by } n$$

$$2n^2 + 3n + 1 = 120 \qquad \textit{Multiply both sides by } 6$$

$$2n^2 + 3n - 119 = 0$$

By the quadratic formula,

$$n = \frac{-3 \pm \sqrt{9 + 952}}{4}$$

$$= \frac{-3 \pm \sqrt{961}}{4}$$

$$= \frac{-3 \pm 31}{4},$$

$$n = 7 \quad \text{or} \quad n = \frac{-34}{4}.$$

Since $-34/4$ is not a positive integer, the only answer is $n = 7$.

Check: $1^2 + 2^2 + 3^2 + 4^2 + 5^2 + 6^2 + 7^2 = 1 + 4 + 9 + 16 + 25 + 36 + 49 = 140 = 20 \cdot 7$. ■

Exercises 3.5

Most of the quadratic equations generated in the following exercises can be solved by factoring. However, in some cases, especially in the exercises involving percent, it will be helpful to use the quadratic formula with the aid of a calculator.

Percent

1. A union contract calls for an $r\%$ salary increase the first year and another $r\%$ increase the second year. As a result, a worker's salary of \$24,000 will increase to \$26,460 after two years. What is r?
2. An IRA earned $r\%$ interest the first year and $r\%$ the second. If an investment of \$2000 grew to \$2420 at the end of the second year, what is r?
3. In a closeout sale, ARA Audio cuts prices by $r\%$ the first week and another $r/2\%$ the second. If a stereo that sold for \$500 before the sale sells for \$240 in the second week of the sale, what is r?
4. A stock selling for \$50 goes up $r\%$ one day and $2r\%$ the next, resulting in a price of \$51.51. What is r?
5. A \$1000 investment grew to \$1875 in two years. The percent appreciation for the second year was double that of the first year. What was the percent appreciation for each year?
6. Stock in the MICRO Computer Company appreciated 75% in one year. If the percent appreciation over the second half of the year was triple that of the first half, what was the appreciation in each 6-month period?
7. A sum of \$1000 was invested in a mutual fund. After 6 months \$500 was withdrawn, and at the end of the year \$609.90 remained. If the money earned $r\%$ the first 6 months and another $r\%$ the second, what is r?
8. Trish deposited \$1200 in a savings account. Eight months later she deposited another \$600, and at the end of the year she had \$1993.86. If her investment earned r percent the first 8 months and $r/2$ percent the last 4 months, what is r?

Motion Based on Average Speed

9. Jean rides a bicycle 5 mi to school each morning. Her roommate Margo rides home from school at a rate 5 mph faster than Jean. The total time for both trips is 50 min. How fast does each woman travel?
10. When driving through a 10-mi construction area, a car's speed was 15 mph slower than for the remaining 60 mi of the trip. If the entire trip took 1 hr and 54 min, what was the car's speed through the construction area?
11. An airplane flies 990 mi against a 30-mph wind and 990 mi back in the direction of the same wind. The round trip takes 6 2/3 hr. What is the speed of the plane in still weather?
12. Cindy and Lynne row at the same speed in still water. Cindy rows 2 mi against a steady current and back 2 mi in the direction of the current. Lynne makes the same trip in still water. Whose trip takes more time?
13. A jogger runs 5 mi in one hour. If he runs x mph faster than his average for the first 3 mi and x mph slower for the last 2 mi, what is x?
14. For what values of x in Exercise 13 will the jogger run the five miles in less than one hour?

Motion Due to Gravity

15. From ground level an object is fired upward with an initial velocity of 144 ft/sec.
    (a) When does the object reach its highest point? (*Hint:* when the highest point is reached, the vertical velocity of the object is zero.)
    (b) What is the highest point reached by the object?
    (c) When does the object return to ground level?
16. An object is tossed upward from a point 16 ft above the ground with an initial velocity of 96 ft/sec. For what period of time is the object at least 144 ft above the ground?
17. If a baseball is hit from a point 4 ft above the ground with an initial vertical velocity of 64 ft/sec, then its height y above the ground at time t is

$$y = -16t^2 + 64t + 4.$$

If the ball's horizontal velocity is a constant 90 ft/sec

and there is a 10-ft-high fence 330 ft away, will the ball go over the fence?

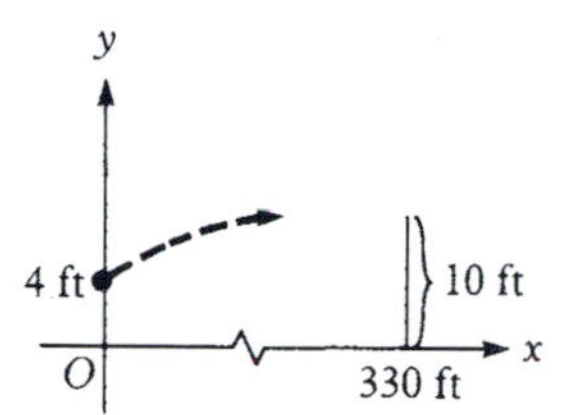

18. A football is place-kicked from ground level at the 50-yard line with an initial vertical velocity of 80 ft/sec. If at time t the horizontal distance traveled before hitting the ground is

$$x = 15t - t^2,$$

will the ball clear 10-ft-high goal posts on the goal line?

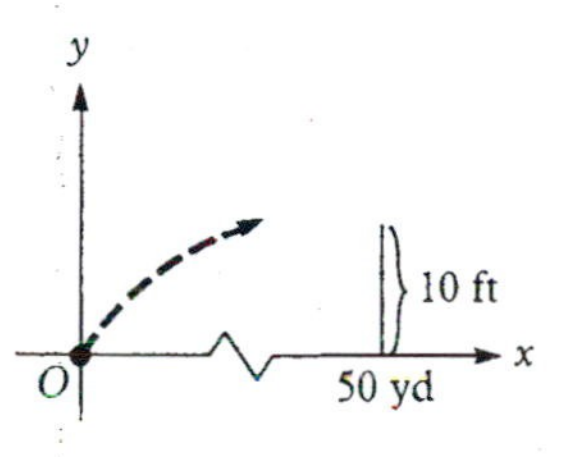

19. In a test of strength at a carnival, a hammer drives a metal weight up a rod to ring a bell at the top. If the bell is 24 ft above the starting position of the weight, what is the minimum initial vertical velocity needed to ring the bell? (*Hint:* when the weight reaches its highest point, its vertical velocity is zero.)

20. Formula (3) in this section is valid under the conditions stated in the text. Is it possible, under these conditions, for an object to be launched with an initial velocity large enough so that the object never returns to earth?

Work

21. It takes one mail carrier 1 hour more than another to deliver mail along a certain route. Working together, the two can deliver the mail in 1 hr and 12 min. How long does it take each carrier to deliver the mail?

22. Don and Rich together can set up all the banquet tables in a VFW hall in 20 min less time than Don can and 45 min less than Rich can, each working alone. How long does it take the two of them to set up the tables?

23. Machine A takes 4 min longer and machine B takes 9 min longer than it takes both A and B together to do a certain job. How long does it take each to do the job?

24. Tom can deliver papers in 15 min less time than his sister, and his father can deliver them twice as fast as Tom. With all three working, the morning delivery takes 12 min. How long does it take each person?

25. Printer A can do a job in the same time as printers B and C working together. If C takes 4 min longer than B, and it takes all three 2 min and 24 sec, how long does it take each printer to do the job working alone?

26. Machine A does a job in the same time as B and C working together, and B does the job in the same time as C and D working together. If D takes 1 hr longer than C and all four can do the job in 20 min, what is the time required for each machine?

Miscellaneous

27. One positive number is 5 more than another, and their product is 14. Find the numbers.

28. The difference between a number and its reciprocal is 1/2. Find all such pairs of numbers and check your answer.

29. A gardener wants to use 350 yd of fencing to enclose a rectangular garden with an area of 7500 sq yd. What should be the dimensions of the garden?

30. In Exercise 29, the gardener now wants to enclose his garden and also divide the total area in half with a piece of fence in the middle. Show that this is impossible with only 350 yd of fencing.

31. A page in a book is 7 in wide and 9 in long. It must contain 35 sq in of printed material and have a margin of the same width on top, bottom, and both sides. How wide should the margin be?

32. If the page of the book in Exercise 31 must have a margin at top and bottom that is twice as wide as the margin at the sides, how wide should the margin be?

3.6 Circles and Parabolas

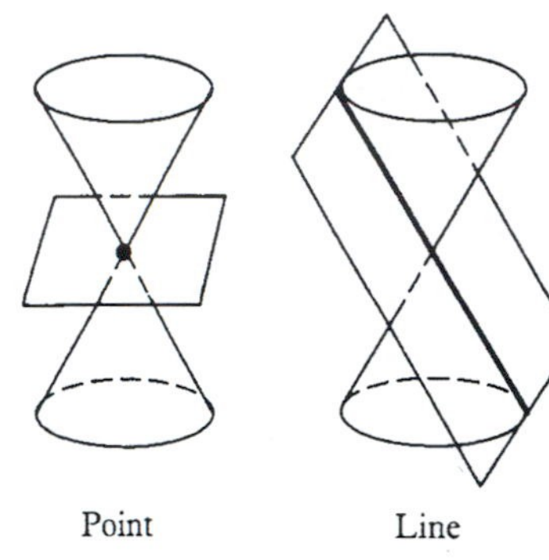
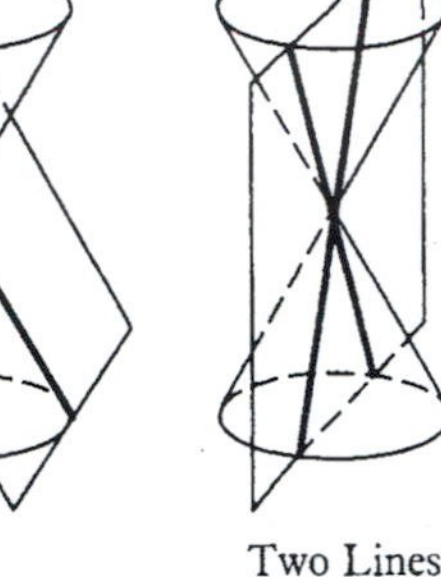
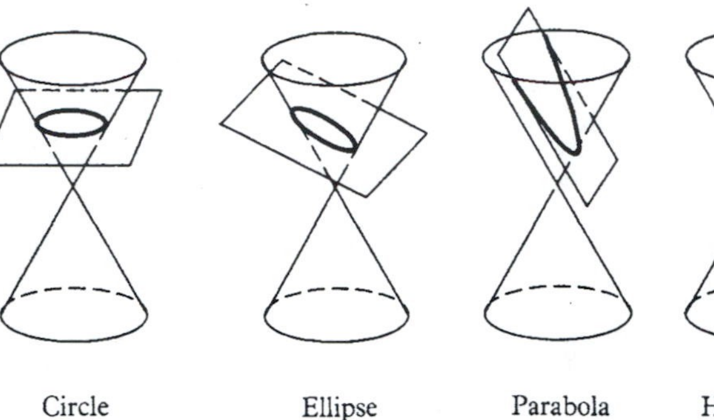

Second-Degree Equations
Circles
Parabolas
Applications

Second-Degree Equations. In Section 2.5 we studied the first-degree (linear) equation

$$Ax + By = C \quad (A \text{ or } B \neq 0), \tag{1}$$

and now we investigate the **second-degree equation**[1]

$$\mathbf{Ax^2 + By^2 + Cx + Dy = E} \quad (\mathbf{A} \text{ or } \mathbf{B \neq 0}). \tag{2}$$

The graph of equation (1) is always a straight line, but equation (2) can have for its graph a point, a line, two lines, a circle, an ellipse, a parabola, a hyperbola, or the empty set. All of these curves, except for parallel lines and the empty set, can be obtained as the intersection of a cone and a plane (see diagram above). Hence, they are called **conic sections.** However, the term *conic section* is usually restricted to the circle, parabola, ellipse, and hyperbola. We consider the circle and parabola in this section, and the ellipse and hyperbola in Section 3.8.

Circles. The set of all points $P(x, y)$ in the plane at a fixed distance $r > 0$ from a fixed point $P_0(x_0, y_0)$ is a **circle** with **center** (x_0, y_0) and **radius** r, as shown in Figure 11. By the distance formula in the plane (Section 2.4), a point $P(x, y)$ will lie on the circle of Figure 11 if and only if x and y satisfy

$$\mathbf{(x - x_0)^2 + (y - y_0)^2 = r^2}, \tag{3}$$

which is the **standard equation of a circle.** If the center is at the origin (0, 0), then equation (3) simplifies to

$$x^2 + y^2 = r^2. \tag{4}$$

Figure 11 Circle

[1]The most general second-degree equation is $Ax^2 + By^2 + Cxy + Dx + Ey = F$. However, the Cxy term can be eliminated by a rotation of axes. The equations for the rotation require trigonometry.

Example 1 In each of the following cases, find the standard equation of the circle with the given center and radius. Also, graph the circle.

(a) (0, 0), 3 (b) (3, 1), 2 (c) (−4, 0), 1

Solution:

(a) With the center at the origin and $r = 3$ [Figure 12(a)], we get

$$x^2 + y^2 = 9.$$

(b) With center (3, 1) and radius 2 [Figure 12(b)], the equation is

$$(x - 3)^2 + (y - 1)^2 = 4.$$

(c) Here [Figure 12(c)] we get

$$[x - (-4)]^2 + y^2 = 1^2 \text{ or } (x + 4)^2 + y^2 = 1.$$ ■

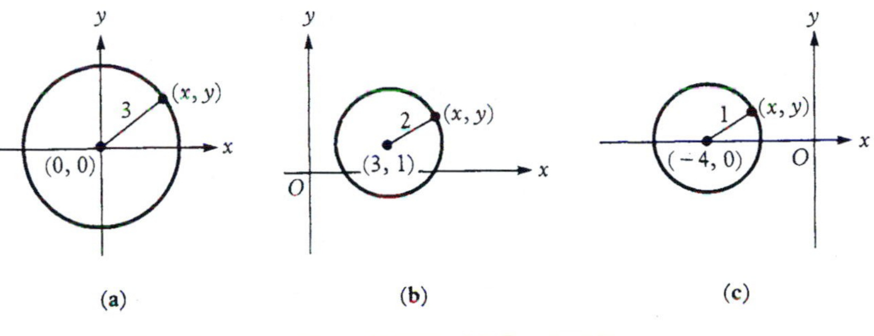

Figure 12

Example 2 Find the standard equation of the circle that has center P_0 (−2, 3) and passes through the point $P_1(1, 2)$. Also, graph the circle.

Solution: The radius of the circle is

$$d(P_0, P_1) = \sqrt{(1 + 2)^2 + (2 - 3)^2} = \sqrt{10}.$$

Therefore, the equation is

$$(x + 2)^2 + (y - 3)^2 = 10,$$

whose graph is given in Figure 13. ■

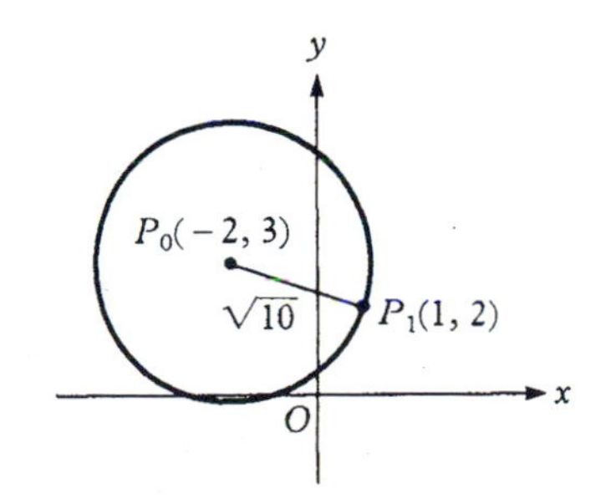

Figure 13

Example 3 Use the result of Example 6 in Section 2.6 to find the equation of the circle with center at the origin and tangent to the line $3x + 4y = 10$.

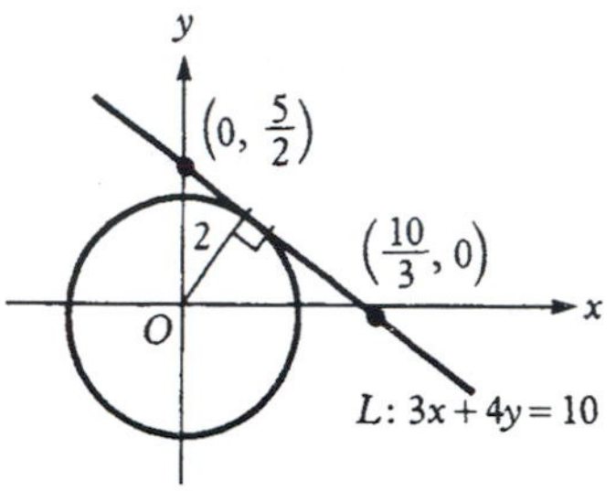

Figure 14

Solution: By Example 6 in Section 2.6, the shortest distance from the origin (0, 0) to the line $3x + 4y = 10$ is $10/\sqrt{3^2 + 4^2} = 2$. This distance is achieved by the point of intersection of a *radial line* through (0, 0) and the line $3x + 4y = 10$ (Figure 14). Hence, the radius is 2 and equation (4) of the circle becomes

$$x^2 + y^2 = 4. \quad \blacksquare$$

If we start with any second-degree equation (2) in which $A = B \neq 0$, we can proceed to an equation $(x - x_0)^2 + (y - y_0)^2 = c$ by completing the square in x and in y as will be shown in Example 4. Example 4 illustrates that there are three possible graphs: a circle ($c > 0$), a single point ($c = 0$), and the empty set ($c < 0$).

Example 4 In each of the following, complete the square in x and in y to determine the graph.

(a) $x^2 + y^2 - 6x - 2y = -6$ (b) $x^2 + y^2 + 2x - 2y = -2$

(c) $x^2 + y^2 - 2x - 4y = -6$

Solution:

(a)

$$x^2 + y^2 - 6x - 2y = -6 \qquad \textit{Given}$$

$$x^2 - 6x + y^2 - 2y = -6 \qquad \textit{Combine } x \textit{ terms and } y \textit{ terms}$$

$$x^2 - 6x + 9 + y^2 - 2y + 1 = -6 + 9 + 1 \qquad \textit{Complete the square in } x \textit{ and in } y$$

$$(x - 3)^2 + (y - 1)^2 = 4$$

The graph is a circle with center (3, 1) and radius 2.

(b)

$$x^2 + y^2 + 2x - 2y = -2 \qquad \textit{Given}$$

$$x^2 + 2x + y^2 - 2y = -2 \qquad \textit{Combine } x \textit{ terms and } y \textit{ terms}$$

$$x^2 + 2x + 1 + y^2 - 2y + 1 = -2 + 1 + 1 \qquad \textit{Complete the square in } x \textit{ and in } y$$

$$(x + 1)^2 + (y - 1)^2 = 0$$

Here the graph is the single point $(-1, 1)$ (*why?*).

(c)

$$x^2 + y^2 - 2x - 4y = -6 \qquad \textit{Given}$$

$$x^2 - 2x + y^2 - 4y = -6 \qquad \textit{Combine } x \textit{ terms and } y \textit{ terms}$$

$$x^2 - 2x + 1 + y^2 - 4y + 4 = -6 + 1 + 4 \qquad \textit{Complete the square in } x \textit{ and in } y$$

$$(x - 1)^2 + (y - 2)^2 = -1$$

The graph is the empty set (*why?*). $\blacksquare$

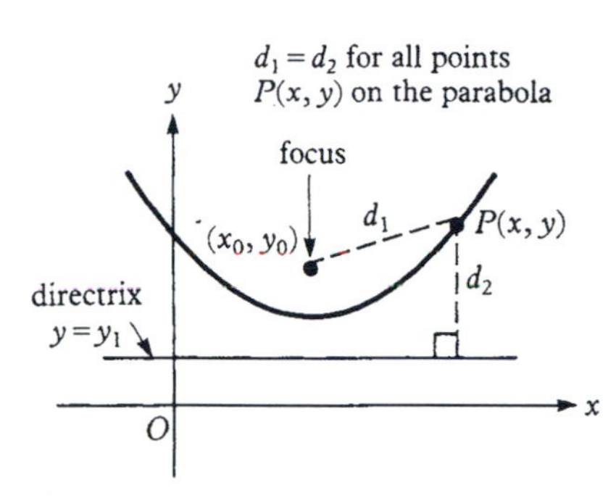

Figure 15 Parabola

Parabolas. The set of all points $P(x, y)$ equidistant from a fixed point $P_0(x_0, y_0)$ and a fixed line L not containing P_0 is called a **parabola** with **focus** (x_0, y_0) and **directrix** L (Figure 15). With reference to Figure 15, the condition $d_1 = d_2$ becomes the equation

$$\sqrt{(x - x_0)^2 + (y - y_0)^2} = |y - y_1|,$$

which simplifies (see Exercise 57) to

$$y = ax^2 + bx + c, \tag{5}$$

where $a = \dfrac{-1}{2(y_1 - y_0)} \neq 0$, $b = \dfrac{x_0}{y_1 - y_0}$, and $c = \dfrac{y_1^2 - y_0^2 - x_0^2}{2(y_1 - y_0)}$. **(6)**

Equation (5) is called the **standard equation of a parabola.**

Example 5 In each of the following cases, find the standard equation of the parabola with the given focus and directrix.

(a) (3, 0), $L: y = -1$ (b) (3, 0), $L: y = 1$

Solution:

(a) From the equation $d_1 = d_2$ [Figure 16(a)] we have

$$\sqrt{(x - 3)^2 + y^2} = |y + 1|.$$

After squaring both sides and simplifying, we obtain

$$y = \frac{1}{2}x^2 - 3x + 4$$

as the equation of the parabola.

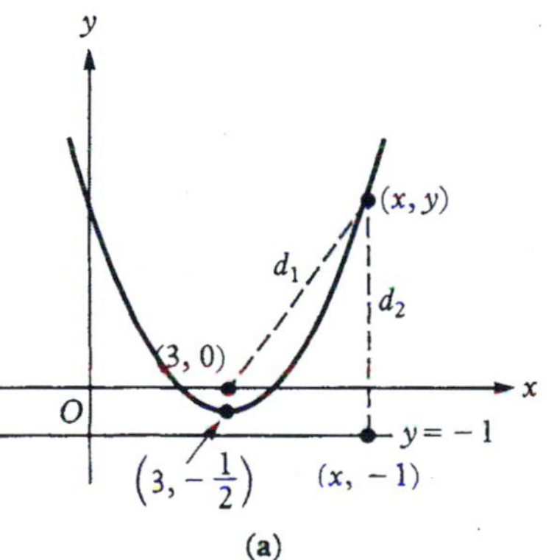
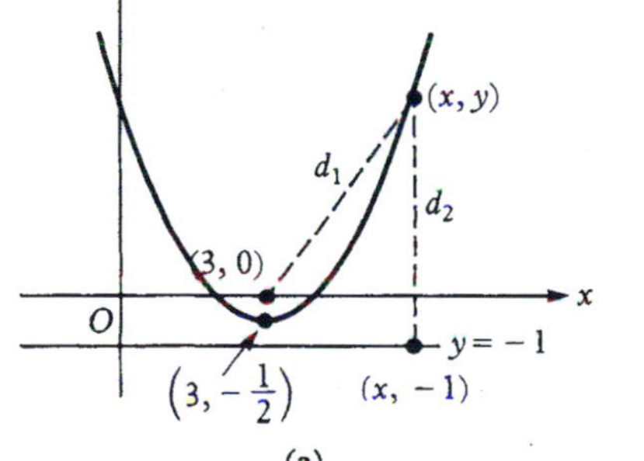

(b) Here, from $d_1 = d_2$ [Figure 16(b)] we have

$$\sqrt{(x - 3)^2 + y^2} = |y - 1|.$$

If we again square both sides, we get

$$y = -\frac{1}{2}x^2 + 3x - 4$$

as the equation of the parabola. ■

If $a > 0$ in equation (5), then the parabola opens upward as in Example 5(a); if $a < 0$, it opens downward as in Example 5(b). Either the low point $(3, -1/2)$ in Example 5(a) or the high point $(3, 1/2)$ in Example 5(b) is called the **vertex** of the parabola. In general, the vertex occurs at

$$\left(x_0, \frac{y_0 + y_1}{2}\right),$$

which is the midpoint of the vertical line segment from the focus to the directrix [see Exercise 60(a)].

Figure 16

The part of the parabola to the right of the vertical line $x = x_0$ is a reflection of the part to the left. This symmetry property and the location of the vertex are immediate consequences of the condition $d_1 = d_2$ [see Exercise 60(b)]. From the first two of equations (6), we get $x_0 = -b/2a$. Therefore, the **axis of symmetry** of the parabola (5) is the vertical line

$$x = -\frac{b}{2a}. \tag{7}$$

The vertex lies on the axis of symmetry. Hence, we can obtain the y-coordinate of the vertex by substituting $x = -b/2a$ in the equation $y = ax^2 + bx + c$.

Example 6 Find the axis of symmetry and the vertex for the parabola $y = x^2 - 4x + 3$. Also, graph the parabola.

Solution: Since $a = 1$ and $b = -4$, the axis of symmetry is the vertical line $x = -b/2a = 2$. Also, when $x = 2, y = 2^2 - 4 \cdot 2 + 3 = -1$. Therefore, the vertex is the point $(2, -1)$. To obtain other points on the graph, we select a few values of x on each side of the vertex and compute the corresponding values of y from the equation of the parabola. We then draw a smooth curve through the determined points (Figure 17). ■

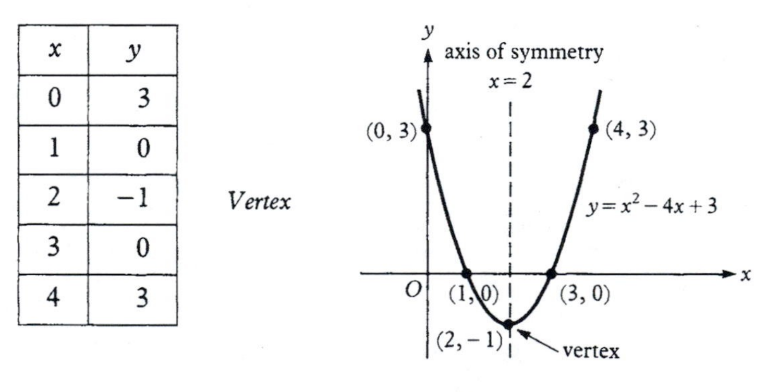

x	y
0	3
1	0
2	−1
3	0
4	3

Figure 17

We point out that *although the focus and directrix appear in the definition of a parabola, the two most important items for graphing a parabola are the vertex and the axis of symmetry*. Also, it is not necessary to memorize $x = -b/2a$ for the axis of symmetry of the parabola $y = ax^2 + bx + c$. For instance, in Example 6, the parabola crosses the x-axis at 1 and 3, which are the roots of the quadratic equation $x^2 - 4x + 3 = 0$, and the axis of symmetry $x = 2$ is the vertical line through the *midpoint* of the line segment joining the two roots. For a general parabola $y = ax^2 + bx + c$, if a_1 and a_2 are the roots of $ax^2 + bx + c = 0$, then the **axis of symmetry** is the vertical line

$$x = \frac{a_1 + a_2}{2}, \tag{7'}$$

which reduces to $x = a_1$ if $a_1 = a_2$. [See Exercise 59, which shows that the axis of symmetry is given by (7′), even if a_1 and a_2 are complex numbers.]

Example 7 Graph each of the following parabolas, showing the axis of symmetry, vertex, and, if possible, the points where the graph intersects the x-axis.

(a) $y = x^2 + 2x - 3$ (b) $y = -4x^2 + 20x - 25$

(c) $y = \frac{1}{2}x^2 - 3x + 5$

Solution:

(a) The roots of $x^2 + 2x - 3 = 0$ are $a_1 = -3$ and $a_2 = 1$. Therefore, the axis of symmetry is the vertical line

$$x = \frac{-3 + 1}{2} = -1.$$

When $x = -1$, $y = (-1)^2 + 2(-1) - 3 = -4$. Hence, the vertex is $(-1, -4)$. The points where the graph crosses the x-axis are $(-3, 0)$ and $(1, 0)$ [Figure 18(a)].

(b) The equation $-4x^2 + 20x - 25 = 0$ has $5/2$ as a double root. Setting $a_1 = 5/2 = a_2$, and proceeding as in part (a), we see that the axis of symmetry is $x = 5/2$ and the vertex is $(5/2, 0)$. Some other points on the graph are $(1, -9)$, $(2, -1)$, $(3, -1)$, and $(4, -9)$ [Figure 18(b)].

(c) The equation $(1/2)x^2 - 3x + 5 = 0$ has the complex roots $a_1 = 3 + i$ and $a_2 = 3 - i$. Hence, the axis of symmetry is

$$x = \frac{a_1 + a_2}{2} = 3.$$

Also, if $x = 3$, then $y = 1/2$, so the vertex is $(3, 1/2)$. The graph does not cross the x-axis (*why?*), but some other points on the graph are $(0, 5)$, $(2, 1)$, $(4, 1)$, and $(6, 5)$ [Figure 18(c)]. ■

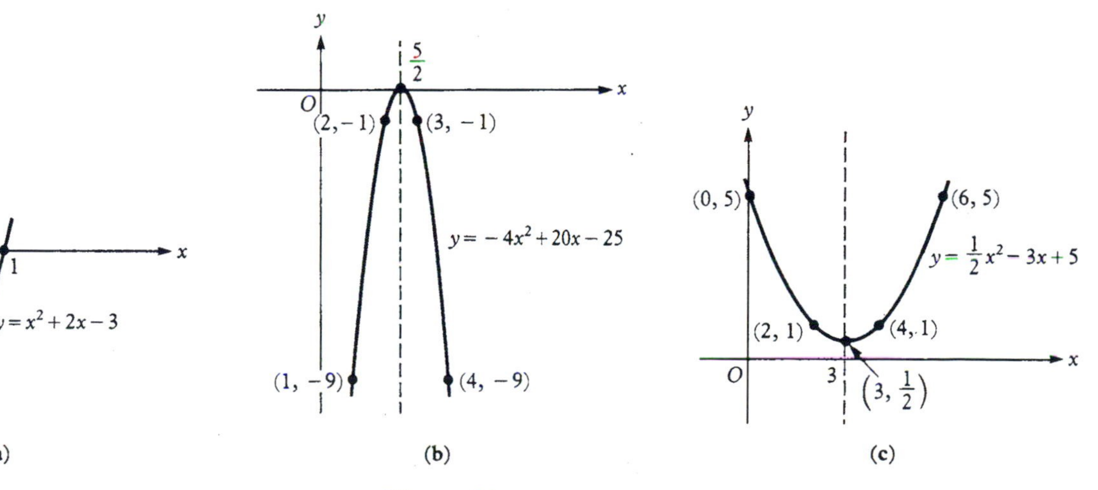

Figure 18

A third way to determine the vertex and axis of symmetry of a parabola is by **completing the square.** For instance, in Example 7(c), the axis $x = 3$ may be obtained by completing the square in x as follows.

$$y = \frac{1}{2}x^2 - 3x + 5 \qquad \textit{Given}$$

$$2y = x^2 - 6x + 10 \qquad \textit{Multiply both sides by 2}$$

$$2y = x^2 - 6x + 9 + 1 \qquad \textit{Complete the square in } x$$

$$2y = (x - 3)^2 + 1$$

$$y = \frac{1}{2}(x - 3)^2 + \frac{1}{2} \qquad \textit{Divide both sides by 2}$$

Since $(1/2)(x - 3)^2 \geq 0$ for all values of x, it follows that $y \geq 1/2$ for all values of x. That is, the smallest value of y is $1/2$, and it occurs when $x = 3$. Hence, the lowest point or vertex of the graph is $(3, 1/2)$ and the axis of symmetry is $x = 3$.

If we interchange x and y in equation (5), we obtain

$$x = ay^2 + by + c,$$

whose graph is also a **parabola** (Figure 19) but with the roles of x and y interchanged. That is, the axis of symmetry is the *horizontal line* $y = -b/(2a) = (a_1 + a_2)/2$, where a_1 and a_2 are the roots of $ay^2 + by + c = 0$. Also, the vertex is the *leftmost point* [Figure 19(a)] or the *rightmost point* [Figure 19(b)] of the graph. The graph opens to the right if $a > 0$ [Figure 19(a)] and to the left if $a < 0$ [Figure 19(b)].

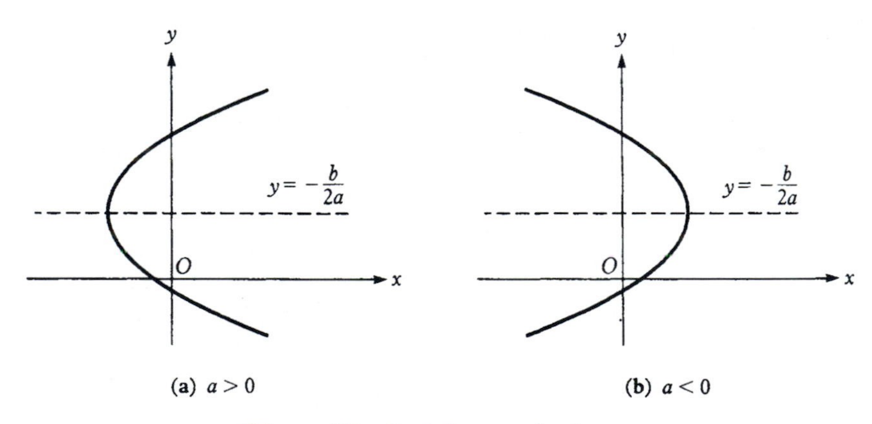

Figure 19 Parabola $x = ay^2 + by + c$

Example 8 Find the vertex and axis of symmetry, and sketch the graph of the parabola $x = y^2 + 2y - 3$. [Compare with Example 7(a).]

Solution: The equation $y^2 + 2y - 3 = 0$ has roots $a_1 = -3$ and $a_2 = 1$. Therefore, the axis of symmetry is the horizontal line $y = (-3 + 1)/2 = -1$. Also, when $y = -1$, $x = (-1)^2 + 2(-1) - 3 = -4$. Hence, the vertex is $(-4, -1)$. We can determine several other points on the curve by assigning values to y and computing the corresponding values of x. For instance, when $y = -3, -2, 0, 1$, and 2, we get $x = 0, -3, -3, 0$, and 5, respectively. Hence, the points $(0, -3)$, $(-3, -2)$, $(-3, 0)$, $(0, 1)$, and $(5, 2)$ are also on the graph (Figure 20). ■

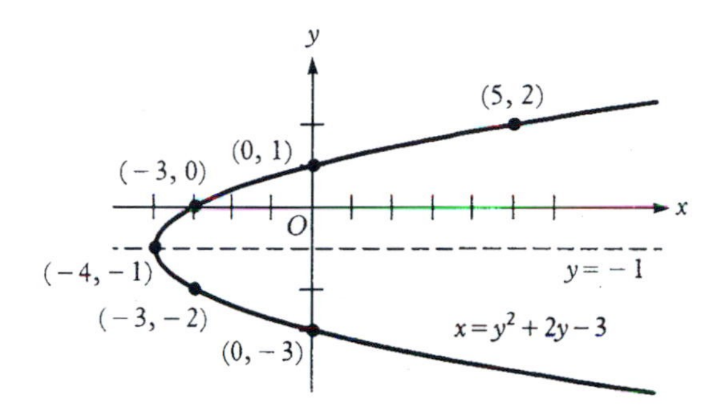

Figure 20

In general, the second-degree equation (2) represents a parabola when either $B = 0$, $A \neq 0$, and $D \neq 0$ (vertical axis of symmetry) or $A = 0$, $B \neq 0$, and $C \neq 0$ (horizontal axis of symmetry).

Question *If either $A = 0$ or $B = 0$ (but not both) in the second-degree equation (2), what graphs other than a parabola are possible?*

Answer: (We consider the case $B = 0, A \neq 0$, and $D = 0$. The other cases are similar.) Suppose $B = 0, A \neq 0$, and $D = 0$. Then equation (2) becomes $Ax^2 + Cx = E$ or $Ax^2 + Cx - E = 0$, which is a quadratic equation in x. Depending on the sign of the discriminant $C^2 + 4AE$, the quadratic equation could have (a) two distinct real roots a_1 and a_2, (b) one distinct real root a_1, or (c) two complex conjugate roots. In case (a), the graph consists of two vertical lines $x = a_1$ and $x = a_2$; in case (b), the graph is one vertical line $x = a_1$; and in case (c), the graph is the empty set. ■

Comment In graphing a parabola $y = ax^2 + bx + c$, as in Examples 6 and 7, for each value substituted for x we obtain exactly one corresponding value of y. Therefore, as discussed in Section 1.1, y is called a **function of x** (see Chapter 4 for a further discussion of functions). In particular, $y = ax^2 + bx + c \quad (a \neq 0)$ defines a **quadratic function of x** or a polynomial function of degree 2.

Applications. Parabolas appear in nature, science, and engineering. When a baseball or tennis ball is hit, a basketball is shot, or a football is kicked, the resulting path is a parabola. Optical, sonic, and electronic reflectors use parabolas because rays parallel to the axis of symmetry are reflected through the focus. Automobile headlights and some telescopes also use this principle. In the following examples, the vertex of a parabola is used to find the maximum or minimum value of a quadratic function.

Example 9 A paper airplane, in one of its dips, flies in a parabolic path whose equation is $y = x^2 - 4x + 5$, where x and y are measured in feet. What is the lowest point of the dip?

Solution: Since the coefficient $a = 1$ of x^2 is positive, the parabola opens upward, and the lowest point is at the vertex. The vertex is located at $x = -b/2a = -(-4/2) = 2$, and when $x = 2$ feet, $y = 2^2 - 4 \cdot 2 + 5 = 1$ foot (Figure 21). Hence, the lowest point of the dip is 1 foot above the ground. ■

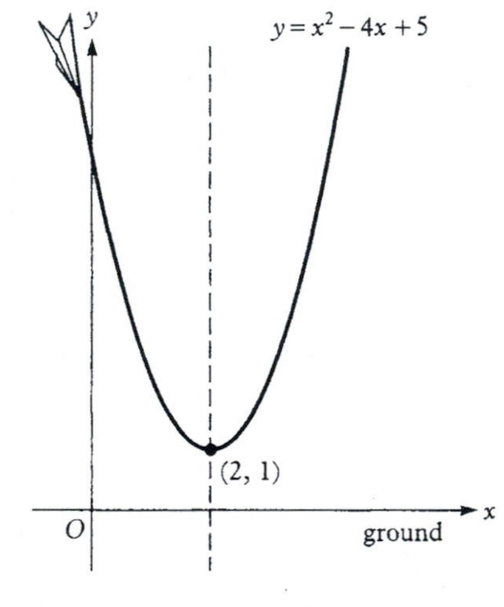

Figure 21

Example 10 A wholesaler sells dresses to a retail store for \$150 each. If more than 100 are ordered, the price per dress is reduced by \$1 for each dress over 100 ordered. What size order will give the largest amount of money to the wholesaler?

Solution: Let x be the number over 100 of dresses ordered. Then $100 + x$ dresses are ordered and the price per dress is $150 - x$ dollars. If y is the amount of money paid to the wholesaler, then

$$\begin{aligned} y &= (\text{number of dresses}) \cdot (\text{price per dress}) \\ &= (100 + x) \cdot (150 - x) \\ &= -x^2 + 50x + 15{,}000. \end{aligned}$$

Since $a = -1$ is negative, the graph is a parabola opening downward, and its highest point is its vertex. The roots of $-x^2 + 50x + 15{,}000 = 0$, as seen from the factored form, are $a_1 = -100$ and $a_2 = 150$. Hence, the axis of symmetry is $x = (a_1 + a_2)/2 = 25$, and when $x = 25$, $y = 15{,}625$. Therefore, an order for 125 dresses will give the wholesaler \$15,625, the largest amount possible. ■

Exercises 3.6

Fill in the blanks to make each statement true.

1. If the graph of $Ax^2 + By^2 + Cx + Dy = E$ is a circle, then ________ = ________.
2. The center and radius of a circle $x^2 + y^2 + Cx + Dy = E$ are usually found by the process of ____________.
3. If the graph of $Ax^2 + By^2 + Cx + Dy = E$ is a parabola, then either ________ = 0 or ________ = 0.
4. If $y = ax^2 + bx + c$ is a parabola and $ax^2 + bx + c = 0$ has roots a_1 and a_2, then the line $x = (a_1 + a_2)/2$ is the ______________ for the parabola.
5. The graph of $y = ax^2 + bx + c$ is a parabola that opens upward if ______________ and downward if ______________.

Write true *or* false *for each statement.*

6. The graph of $Ax^2 + By^2 + Cx + Dy = E$ can be a single point.
7. If $A = B \neq 0$, then the graph of $Ax^2 + By^2 + Cx + Dy = E$ is a circle.
8. If $b^2 - 4ac < 0$, then the parabola $y = ax^2 + bx + c$ does not cross the x-axis.
9. The parabola $y = ax^2 + bx + c$ has a horizontal directrix.
10. The parabola $x = ay^2 + by + c$ has a horizontal axis of symmetry.

Circles

Find the standard equation of the circle in each of the following cases.

11. center: (2, 1); radius: 3
12. center: (−1, 3); radius: 5
13. center: (2, −2); passes through the point (0, 0)
14. center: (−1, −1); passes through the point (1, 1)
15. center: (−1, 0); tangent to the y-axis
16. center: (0, 2); tangent to the x-axis
17. center: (0, 0); tangent to the line $x + y = 1$
18. center: (2, 1); tangent to the line $y = 4$
19. the circle having the line segment from (−2, −1) to (6, 3) as a diameter
20. the circle having the diagonals of the square with vertices (0, 0), (3, 0), (3, 3), and (0, 3) as diameters

By completing the square in x and y, determine the nature of the graph (circle, point, or empty set) of each of the following second-degree equations. Sketch the graph, if possible.

21. $x^2 + y^2 - 4x + 4y = 1$
22. $x^2 + y^2 + 6x - 2y = 15$
23. $4x^2 + 4y^2 + 12x - 4y = -1$
24. $2x^2 + 2y^2 - 8x + 8y = -7$
25. $x^2 + y^2 + 6x - 2y = -10$
26. $x^2 + y^2 - x - y = -2$
27. $x^2 + y^2 + x - 4y = -5$
28. $2x^2 + 2y^2 - 4x - 2y = -3$
29. $3x^2 + 3y^2 + 3x + 6y = -4$
30. $3x^2 + 3y^2 + 3x + 6y = 21/4$

Parabolas

In Exercises 31–34, find the equation of the parabola with the given focus and directrix by setting $d_1 = d_2$.

31. focus: (0, 1); directrix: $y = 2$
32. focus: (0, 2); directrix: $y = 0$
33. focus: (2, 1); directrix: $y = 3$
34. focus: (−1, 0); directrix: $y = 1$

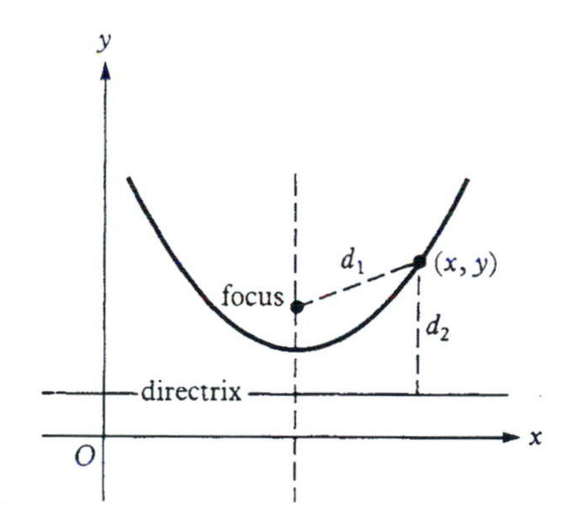

Find the vertex and axis of symmetry for each of the parabolas in Exercises 35–44. Graph each parabola.

35. $y = x^2$
36. $y = -\frac{1}{2}x^2$
37. $y = x^2 + 2$
38. $y = x^2 - 2$
39. $y = x^2 + 6x + 10$
40. $y = x^2 - 6x + 5$

41. $y = x^2 - 4x$

42. $y = -\frac{1}{2}x^2 + x - 1$

43. $2y = x^2 + 4x + 2$

44. $12y = x^2 - x - 1$

Find the vertex and axis of symmetry for each of the parabolas in Exercises 45–50. Graph each parabola.

45. $x = \frac{1}{2}y^2$

46. $x = \frac{1}{2}y^2 + 1$

47. $x = y^2 - 6y$

48. $x = y^2 + y$

49. $2x = y^2 - 3y - 4$

50. $2x = y^2 + 4y + 2$

Applications

51. A rectangular parking lot with a straight road as one side is to be fenced on the other three sides by 1000 ft of aluminum fencing (see the figure). If the area of the lot is to be maximized, what should be its length and width? (*Hint:* write the area of the lot as $ax^2 + bx + c$.)

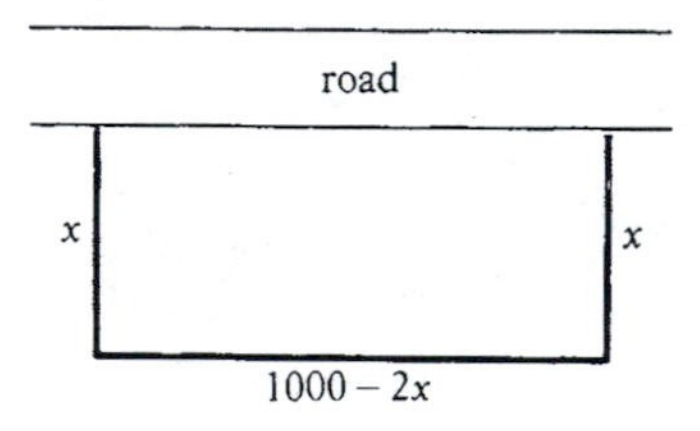

52. Find the dimensions and the area of the parking lot in Exercise 51 if the fourth side along the road is also to be fenced except for an entrance gate 10 ft wide.

53. A farmer has a crop of 600 bushels that he can sell now at \$1 per bushel. He estimates that for the next few weeks, his crop will increase by 100 bushels each week but the price per bushel will decrease 10 cents each week. How many weeks should he wait in order to make the most money? Assume that the entire crop must be harvested and sold at one time.

54. A club charters a bus containing 50 seats for an outing. The bus company charges \$30 for each passenger if the bus is filled, but increases the price by \$1 per passenger for each empty seat. How many empty seats will produce the largest income for the bus company?

55. A toy rocket launched from origin O has a parabolic path as indicated in the figure. At time t sec from launch to impact, the coordinates (x, y) of the rocket satisfy the equations

$$x = 64t \quad \text{and} \quad y = -16t^2 + 128t.$$

(a) At what time t_1 will the rocket land?

(b) How far will the rocket travel horizontally by the time of impact?

(c) What is the maximum height of the rocket?

(d) Find the equation of the parabolic path by expressing y in terms of x.

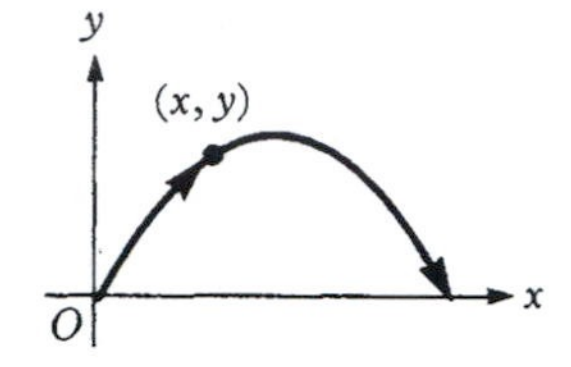

Miscellaneous

56. Each of the following is the graph of a parabola $y = ax^2 + bx + c$. Determine whether the discriminant is positive, negative, or zero. Also determine the sign of the coefficient a.

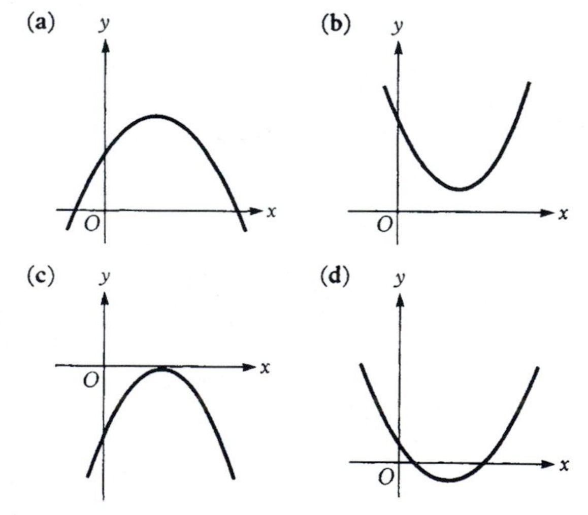

57. By squaring both sides and simplifying, show that the equation $\sqrt{(x - x_0)^2 + (y - y_0)^2} = |y - y_1|$ is equivalent to $y = ax^2 + bx + c$, where

$$a = -\frac{1}{2(y_1 - y_0)}, \qquad b = \frac{x_0}{y_1 - y_0},$$

and

$$c = \frac{y_1^2 - y_0^2 - x_0^2}{2(y_1 - y_0)}.$$

58. Use the results in Exercise 57 to show that

$$x_0 = -\frac{b}{2a}, \qquad y_0 = \frac{4ac - b^2 + 1}{4a},$$

and

$$y_1 = \frac{4ac - b^2 - 1}{4a}.$$

59. A given parabola has the equation $y = ax^2 + bx + c$. Complete the square in x to show that the axis of symmetry is the vertical line $x = -b/(2a)$. Also, use the quadratic formula to show that if the roots of $ax^2 + bx + c = 0$ are a_1 and a_2 (real or complex), then

$$\frac{a_1 + a_2}{2} = -\frac{b}{2a},$$

and therefore, the line $x = (a_1 + a_2)/2$ is the axis of symmetry.

60. A given parabola has focus (x_0, y_0) and directrix $L: y = y_1$ (see the figure in Exercises 31–34).
   (a) Use the property $d_1 = d_2$ to show that $(x_0, [y_0 + y_1]/2)$ is the point on the parabola that is closest to the directrix.
   (b) Use the property $d_1 = d_2$ to show that if $(x_0 + h, y)$ is on the parabola, then so is $(x_0 - h, y)$ for any real number h.

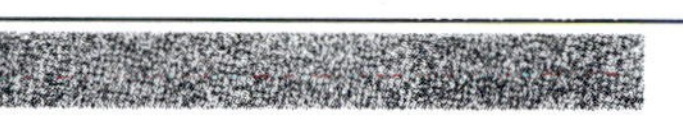

3.7 Translation of Axes and Intersections

To simplify a second-degree equation, complete the square in x and/or y, and then make the corresponding substitutions.

Translation of Axes
Points of Intersection
Regions of Intersection

Translation of Axes. The process of completing the square used in Example 4 of Section 3.6 to put the equation of a circle in standard form has a geometric application that is useful for many graphs. If we start with the equation of a circle

$$(x - x_0)^2 + (y - y_0)^2 = r^2, \tag{1}$$

and make the **substitution**

$$x' = x - x_0 \quad \text{and} \quad y' = y - y_0, \tag{2}$$

we then obtain the equation

$$x'^2 + y'^2 = r^2. \tag{3}$$

Now equation (3) is the equation of a circle with center at the origin, provided we consider new $x'y'$-axes with the origin located at (x_0, y_0) (Figure 22).

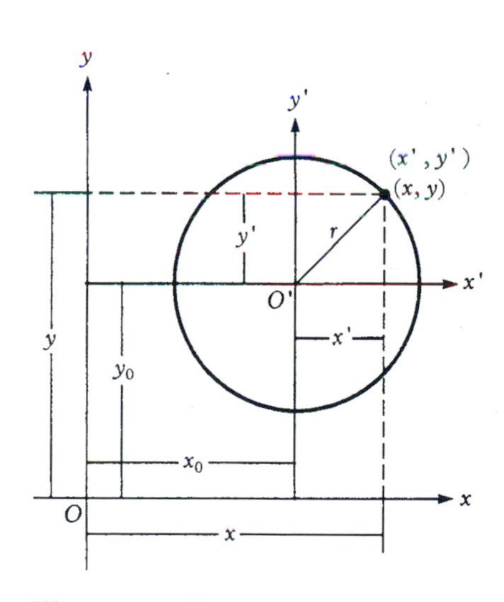

Figure 22 Translation of Axes

As indicated in Figure 22, equations (1) and (3) represent the same circle, but the coordinates have changed from xy-coordinates to $x'y'$-coordinates. Geometrically, we can imagine the $x'y'$-axes as being obtained by moving the xy-axes parallel to themselves so that the xy-origin $(0, 0)$ moves to the point (x_0, y_0). In this sense, the $x'y'$-axes are called a **translation** of the xy-axes. Every point in the plane has both xy-coordinates and $x'y'$-coordinates, which are related by the **substitution equations** (2). Similarly, every curve that has an equation in the variables x and y also has an equation in the variables x' and y', and conversely.

Example 1 The $x'y'$-axes are a translation of the xy-axes in which the point $(x_0, y_0) = (6, 3)$ becomes the $x'y'$-origin O'. Find the $x'y'$-coordinates of the $P:(x, y) = (8, 7)$, and find the xy-coordinates of $Q:(x', y') = (-5, 2)$.

Solution: With reference to Figure 23, the substitution equations (2) become

$$x' = x - 6 \quad \text{and} \quad y' = y - 3.$$

Hence, the $x'y'$-coordinates of P are

$$x' = 8 - 6 = 2 \quad \text{and} \quad y' = 7 - 3 = 4.$$

The xy-coordinates of Q must satisfy

$$-5 = x - 6 \quad \text{and} \quad 2 = y - 3.$$

Hence, $x = 1, y = 5$. ■

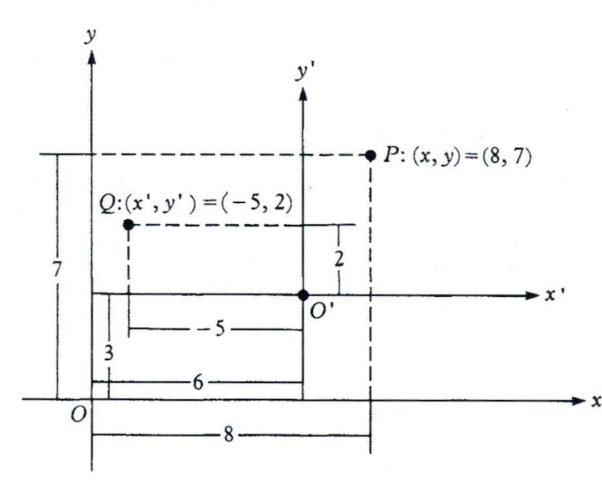

Figure 23

Example 2 A given line L has equation $y = 2x + 1$. Translate the xy-axes to $x'y'$-axes in such a way that the point $(1, 3)$ on L becomes the $x'y'$-origin. Find the equation of L in the variables x' and y', and graph L in both coordinate systems.

Solution: The substitution equations (2) are

$$x' = x - 1 \quad \text{and} \quad y' = y - 3$$

or

$$x = x' + 1 \quad \text{and} \quad y = y' + 3.$$

Hence, by substituting for x and y in $y = 2x + 1$, the equation of L in the $x'y'$-coordinates is

$$y' + 3 = 2(x' + 1) + 1$$

or

$$y' = 2x'.$$

The graph of L in both coordinate systems is shown in Figure 24. ■

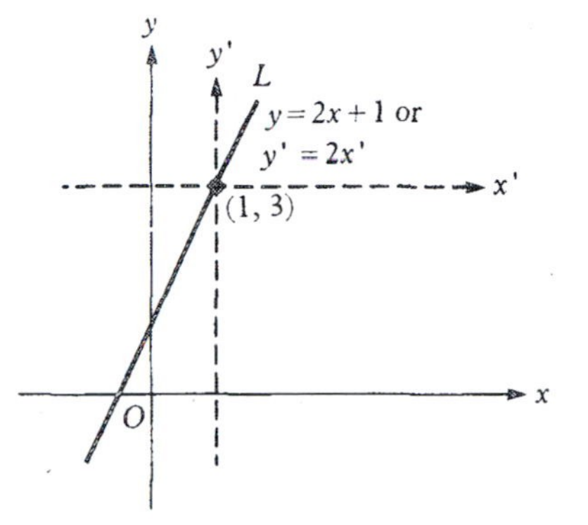

Figure 24

Example 3 A given parabola has equation $y = 2x^2 + 4x + 5$. Transform the xy-axes to $x'y'$-axes in such a way that the vertex of the parabola becomes the $x'y'$-origin. Find the equation of the parabola in terms of x' and y', and graph the parabola in both coordinate systems.

Solution: The axis of symmetry of a parabola $y = ax^2 + bx + c$ is the vertical line $x = -b/2a$. Here the axis is $x = -4/4 = -1$, and when $x = -1$, we have $y = 3$. Hence, the vertex is $(-1, 3)$, and the substitution equations (2) become

$$x' = x + 1 \quad \text{and} \quad y' = y - 3$$

or

$$x = x' - 1 \quad \text{and} \quad y = y' + 3.$$

The equation of the parabola in the $x'y'$-coordinates is

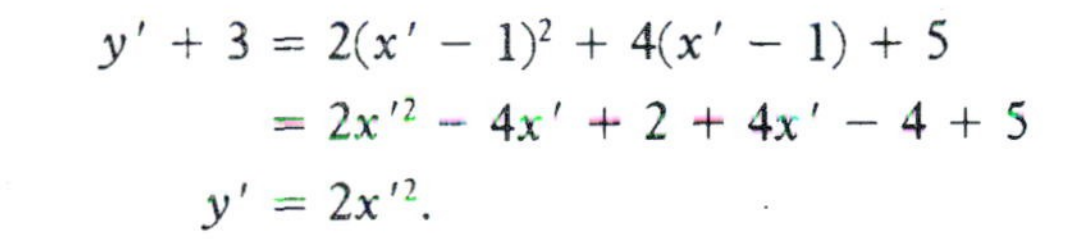

$$\begin{aligned} y' + 3 &= 2(x' - 1)^2 + 4(x' - 1) + 5 \\ &= 2x'^2 - 4x' + 2 + 4x' - 4 + 5 \end{aligned}$$

or

$$y' = 2x'^2.$$

The graph of the parabola in both coordinate systems is given in Figure 25. Note that the simplified equation $y' = 2x'^2$ could also be obtained by completing the square in x in the given equation $y = 2x^2 + 4x + 5$ to get $y - 3 = 2(x + 1)^2$ and then substituting y' for $y - 3$ and x' for $x + 1$. ■

Figure 25

In Examples 2 and 3, a translation of axes was performed in order to obtain a simpler equation of the given curve. The curve remained fixed, but the axes and variables changed. We can also obtain a simpler equation by keeping the axes fixed and moving the curve to a more convenient position. For example, if we start with the circle

$$(x - x_0)^2 + (y - y_0)^2 = r^2 \tag{1'}$$

and make the **replacements**

$$x - x_0 \text{ by } x \quad \text{and} \quad y - y_0 \text{ by } y, \tag{2'}$$

we obtain the equation

$$x^2 + y^2 = r^2. \tag{3'}$$

Equation (3′) is the equation of a circle with center (0, 0) and radius r. That is, circle (3′) is **congruent** to circle (1′). Here the axes and variables remain the same, but the circle is replaced by a congruent one with a simpler equation (Figure 26.)

We call circle (3′) a **translation** of circle (1′). These two procedures, translation of axes by the substitution equations (2) and translation of curves by the replacements (2′), are equivalent ways to achieve more convenient equations. For a second-degree equation, values of x_0 and y_0 can be obtained by completing the square in x, y, or both, as noted in Example 3.

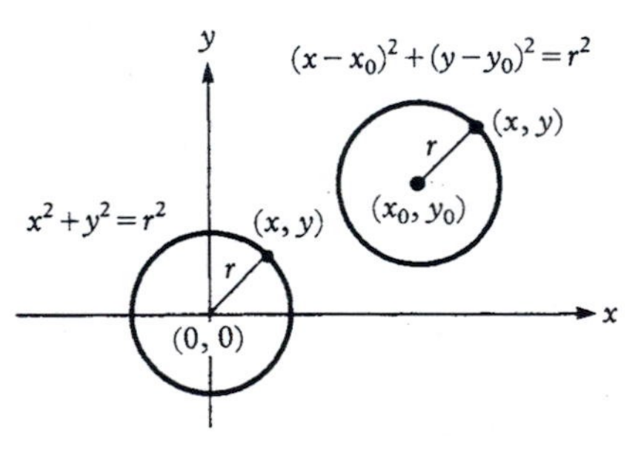

Figure 26 Translation of a Curve

Points of Intersection. In the previous chapter we discussed the intersection of two straight lines represented by linear equations. We now consider intersections involving lines, circles, and parabolas.

Example 4 Find all points of intersection of the line $x + y = 3$ and the circle $(x - 1)^2 + y^2 = 4$.

Solution: The intersections occur at those points $P(x, y)$ whose coordinates satisfy *both* equation $x + y = 3$ and equation $(x - 1)^2 + y^2 = 4$. We solve the first of these for y in terms of x, obtaining $y = 3 - x$, and then substitute this expression for y in the second equation. We get

$$(x - 1)^2 + (3 - x)^2 = 4$$
$$x^2 - 2x + 1 + 9 - 6x + x^2 = 4$$
$$2x^2 - 8x + 6 = 0$$
$$x^2 - 4x + 3 = 0$$
$$x = 1 \quad \text{or} \quad x = 3.$$

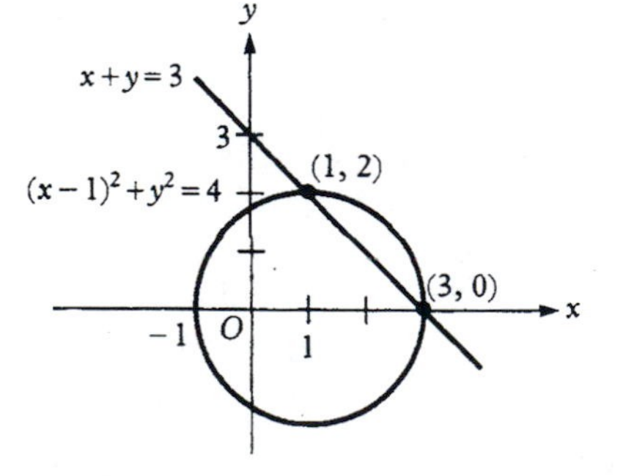

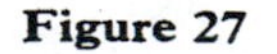
Figure 27

When $x = 1$, we have $y = 3 - x = 2$, and when $x = 3$, we get $y = 0$. Hence, the points of intersection are (1, 2) and (3, 0) (Figure 27). ■

Example 5 Find all points of intersection of the parabolas $y = x^2 - 4$ and $y = -x^2 + 3x - 2$.

Solution: Here we equate the y values of the curves, obtaining

$$x^2 - 4 = -x^2 + 3x - 2$$
$$2x^2 - 3x - 2 = 0$$
$$(2x + 1)(x - 2) = 0$$
$$x = -1/2 \qquad \text{or} \qquad x = 2.$$

When $x = -1/2$, we obtain $y = x^2 - 4 = -15/4$, and when $x = 2$, we get $y = 0$. Hence, the curves intersect in the points $(-1/2, -15/4)$ and $(2, 0)$, as shown in Figure 28. ■

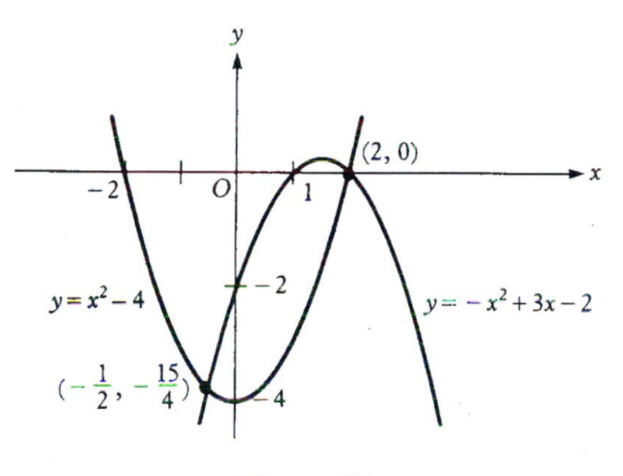

Figure 28

Comment Geometrically, it is evident that a straight line can intersect a circle or parabola in at most two points. Also, two parabolas that are both of the form $y = ax^2 + bx + c$ or both of the form $x = ay^2 + by + c$ can intersect in at most two points. Hence, it is not surprising that a quadratic equation results when the points of intersection are determined algebraically, as in Examples 4 and 5. We can also see geometrically that two circles with different centers can intersect in at most two points. In this case, the algebraic solution involves working with radicals, but eventually a quadratic equation results (see Exercises 49 and 50). On the other hand, a circle and a parabola can intersect in up to four points, as can a parabola of the type $y = ax^2 + bx + c$ with one of the type $x = ay^2 + by + c$. In these cases, an algebraic process to determine the points of intersection can result in a fourth-degree equation. For example, the system

$$x^2 + y^2 = 5$$
$$y = 3x^2 + 3x - 6$$

results in the fourth-degree equation

$$9x^4 + 18x^3 - 26x^2 - 36x + 31 = 0.$$

At present, we can solve only very special cases of fourth-degree equations, for example, those that are actually quadratics in x^2. Hence, we cannot yet determine the points of intersection of arbitrary circles and parabolas. However, we will discuss methods to approximate roots of polynomials of degree 4 and higher in a later chapter.

Regions of Intersection. In Example 4, we saw that the line L with equation $x + y = 3$ intersects the circle C with equation $(x - 1)^2 + y^2 = 4$ in the two points $(1, 2)$ and $(3, 0)$. The points *above* L satisfy $y > 3 - x$ and those *below* L satisfy $y < 3 - x$ (*why?*). Similarly, the points *outside* C satisfy $(x - 1)^2 + y^2 > 4$ and those *inside* C satisfy $(x - 1)^2 + y^2 < 4$. Hence, the system of inequalities

$$y > 3 - x$$
$$(x - 1)^2 + y^2 < 4$$

is satisfied by all points in the region above L and inside C [Figure 29(a)], and the system

$$y < 3 - x$$
$$(x - 1)^2 + y^2 < 4$$

is satisfied by all points in the region below L and inside C [Figure 29(b)].

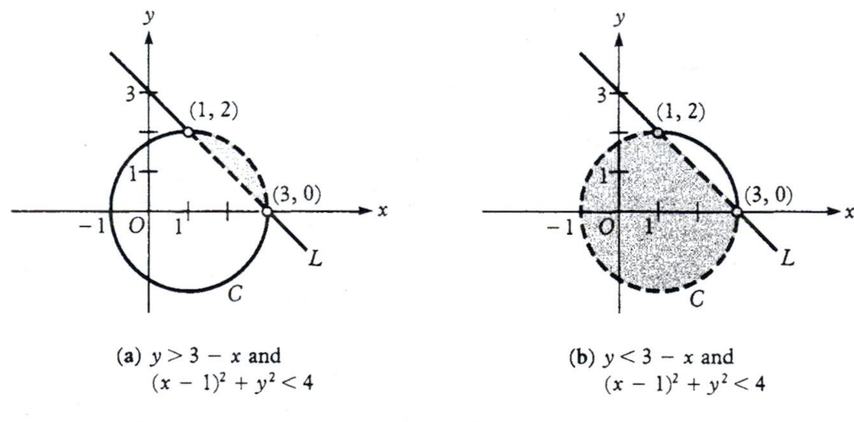

(a) $y > 3 - x$ and $(x - 1)^2 + y^2 < 4$

(b) $y < 3 - x$ and $(x - 1)^2 + y^2 < 4$

Figure 29

Example 6 Graph the region satisfied by the system of inequalities

$$y \geq x^2 - 4$$
$$y \leq -x^2 + 3x - 2.$$

Solution: As illustrated in Figure 28 in Example 5, parabolas P_1 with equation $y = x^2 - 4$ and P_2 with equation $y = -x^2 + 3x - 2$ intersect in the two points $(2, 0)$ and $(-1/2, -15/4)$. The points satisfying $y \geq x^2 - 4$

lie on or above P_1, and those satisfying $y \leq -x^2 + 3x - 2$ lie on or below P_2. Hence, the points satisfying *both* of the given inequalities lie in the shaded region of Figure 30. ■

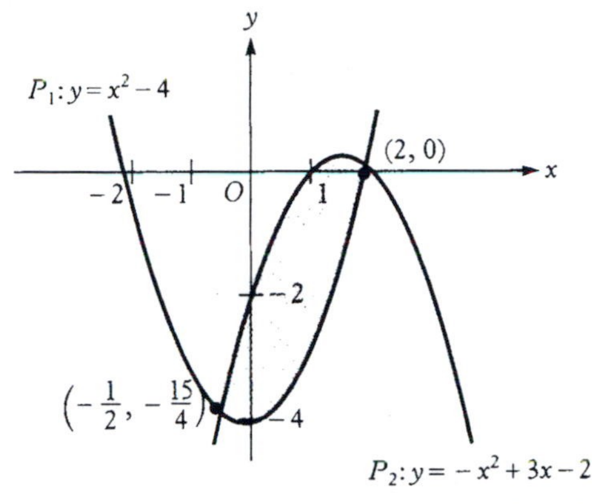

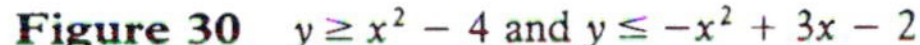

Figure 30 $y \geq x^2 - 4$ and $y \leq -x^2 + 3x - 2$

Exercises 3.7

Fill in the blanks to make each statement true.

1. If the xy-axes are translated to $x'y'$-axes so that the $x'y'$-origin is at the point (x_0, y_0), then the xy-coordinates of a point P are related to the $x'y'$-coordinates of P by the equations ________________.
2. If (x_0, y_0) is the origin of the $x'y'$-coordinate system, then the equation of the circle $(x - x_0)^2 + (y - y_0)^2 = r^2$ in $x'y'$-coordinates is ________________.
3. If the circle $(x - x_0)^2 + (y - y_0)^2 = r^2$ is translated so that its center moves to the point (x_0, y_0), then the equation of the translated circle is ________________.
4. A second-degree equation sometimes can be simplified by the process of ________________.
5. A parabola can intersect a circle in at most ________ points.

Write true *or* false *for each statement.*

6. If the xy-axes are translated to $x'y'$-axes so that the linear equation $y = mx + b$ becomes $y' = m'x' + b'$, then $m' = m$.
7. If the xy-axes are translated to $x'y'$-axes so that the equation $(x - x_0)^2 + (y - y_0)^2 = r^2$ becomes $(x')^2 + (y')^2 = (r')^2$, then $r' = r$.
8. It is possible to translate axes so that the parabola $y = ax^2 + bx + c$ has equation $y' = a(x')^2$ in the new coordinates.
9. It is possible to translate axes so that the line $y = x$ has the equation $y' = -x'$.
10. It is possible to translate axes so that the circle $(x - 1)^2 + (y - 2)^2 = 25$ has the equation $(x' - 2)^2 + (y' - 1)^2 = 25$.

Translation of Axes

11. The $x'y'$-axes are a translation of the xy-axes in which the xy-point $(2, 5)$ becomes the $x'y'$-origin. Find the $x'y'$-coordinates of the following points.
    - (a) $P: (x, y) = (3, -2)$
    - (b) $Q: (x, y) = (0, 0)$
    - (c) $R: (x, y) = (2, 5)$
12. For the translation of Exercise 11, find the xy-coordinates of each of the following points.
    - (a) $P: (x', y') = (4, 3)$
    - (b) $Q: (x'y') = (0, 0)$
    - (c) $R: (x', y') = (2, 5)$

Find the substitution equations in each of the following cases.

13. The xy-origin is the $x'y'$-point $(2, -4)$.
14. The xy-point $(5, 1)$ is the $x'y'$-point $(1, 5)$.
15. The $x'y'$-point $(3, 0)$ is the xy-point $(2, 2)$.
16. The $x'y'$-point $(-1, 2)$ is the xy-point $(0, 0)$.
17. The $x'y'$-origin is the center of the circle $x^2 + y^2 + 2x - 4y = 5$.
18. The $x'y'$-origin is the vertex of the parabola $y = 2x^2 + 4x - 3$.
19. The xy-origin is the y'-intercept of the line $2x' + 3y' = 6$.
20. The xy-origin is the x'-intercept of the line $4x' - 5y' = 12$.

If the $x'y'$-axes are a translation of the xy-axes with the point (x_0, y_0) as the $x'y'$-origin, find the $x'y'$-equation of each of the following curves. Sketch a graph showing both coordinate systems.

21. $2x + 3y = 7$; $(x_0, y_0) = (4, -3)$
22. $y - 2 = 3(x + 5)$; $(x_0, y_0) = (-5, 2)$
23. $x^2 + y^2 = 5$; $(x_0, y_0) = (2, -4)$
24. $x^2 + y^2 + 2x = 4$; $(x_0, y_0) = (-1, 0)$
25. $y = x^2 - 10x + 5$; $(x_0, y_0) = (5, -15)$
26. $y = 3x^2 + 4x - 2$; $(x_0, y_0) = (-2, 1)$

Translate each of the following curves as indicated. Sketch both curves in xy-coordinates.

27. Translate the circle $(x - 3)^2 + (y + 2)^2 = 25$ so that its center is at the origin.
28. Translate the circle $x^2 + 8x + y^2 - 8y = 4$ so that its center is at the point $(1, 2)$.
29. Translate the parabola $y = x^2 + 6x + 5$ so that its vertex is at the point $(0, 1)$.
30. Translate the parabola $x = y^2 - 2y + 2$ so that its vertex is at the origin.

Simplify the equation of each of the following curves by completing the square in x and/or y and making the corresponding $x'y'$-substitutions.

31. $y = x^2 + 3x + 1$
32. $y = 4x^2 - 2x + 8$
33. $x = 2y^2 + 2y - 4$
34. $x^2 + y^2 + x + 5y = 0$
35. $2x^2 + 2y^2 + 6x + 8y = 5$
36. $3x^2 + 3y^2 + 2x + y = 10$

In Exercises 37–40, first substitute $x' + x_0$ for x and $y' + y_0$ for y and then choose values of x_0 and y_0 that simplify the equation. Give the simplified equation.

37. $y = 2x^2 + 3x - 5$
38. $x = y^2 - y + 4$
39. $4x^2 + 4y^2 - 2x + 4y = 5$
40. $x^2 + y^2 - 3x - 6y = 0$

Points of Intersection

Find all points of intersection of each of the following pairs of curves. Sketch a graph for each.

41. $y = 4 - x^2$, $y = 1 - 2x$
42. $x + 2y - 3 = 0$, $x - y^2 = 0$
43. $5x - 3y = 0$, $x^2 + y^2 = 34$
44. $x^2 + y^2 = 25$, $y = x + 1$
45. $y = x^2 - 1$, $y = -x^2 + 3$
46. $y = x^2 + 3x + 2$, $y = -x^2 - x + 2$
47. $x = 4 - (y - 2)^2$, $x = y^2 - 4y + 6$
48. $x = y^2 - 4$, $2x = y^2 - 4$
49. $x^2 + y^2 = 4$, $(x - 1)^2 + (y + 1)^2 = 6$
50. $(x - 2)^2 + y^2 = 5$, $x^2 + (y - 1)^2 = 10$

Regions of Intersection

Use the results of Exercises 41–46 to graph the regions determined by each of the following systems of inequalities.

51. $y < 4 - x^2$, $y > 1 - 2x$
52. $x < 3 - 2y$, $x > y^2$
53. $3y \le 5x$, $x^2 + y^2 \le 34$
54. $x^2 + y^2 \le 25$, $y \le x + 1$
55. $y \ge x^2 - 1$, $y \le -x^2 + 3$
56. $y > x^2 + 3x + 2$, $y < -x^2 - x + 2$

3.8 Ellipses and Hyperbolas

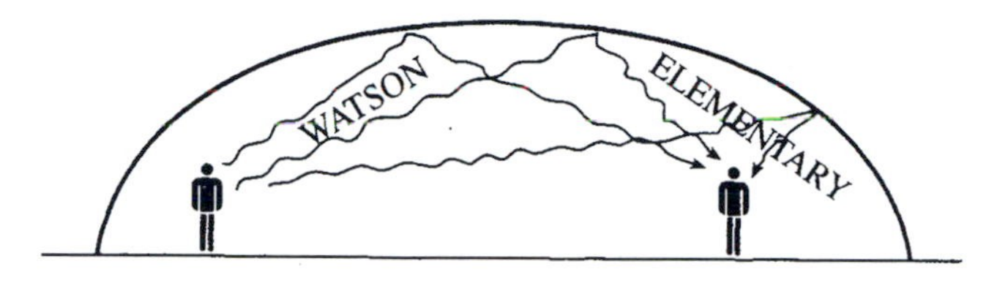

Words whispered at one focus of an elliptic "whispering chamber" can be heard loud and clear at the other focus.

Ellipses
Hyperbolas
Asymptotes

The ellipse and the hyperbola, like the parabola, appear in the physical universe, science, and engineering. Planets and satellites travel in elliptical orbits. Elliptic and hyperbolic gears are used in various kinds of machinery. Hyperbolas are used in the design of telescopes because light from one focus reflects back along the line from the other focus. Electronic instruments on a ship locate its position at sea as the intersection of two hyperbolas.

Ellipses. Given two fixed points P_1 and P_2 in the plane and a positive number c, the set of all points P for which

$$d(P_1, P) + d(P_2, P) = c \tag{1}$$

is called an **ellipse.** Each of the points P_1 and P_2 is called a **focus,** and each of the distances $d_1 = d(P_1, P)$ and $d_2 = d(P_2, P)$ is called a **focal radius.** Figure 31 is an ellipse in which $P_1 = (c_1, 0)$ and $P_2 = (-c_1, 0)$. We note that the foci P_1 and P_2 and the positive quantity c are *fixed,* whereas the focal radii d_1 and d_2 can vary in value, but their *sum* is constant.

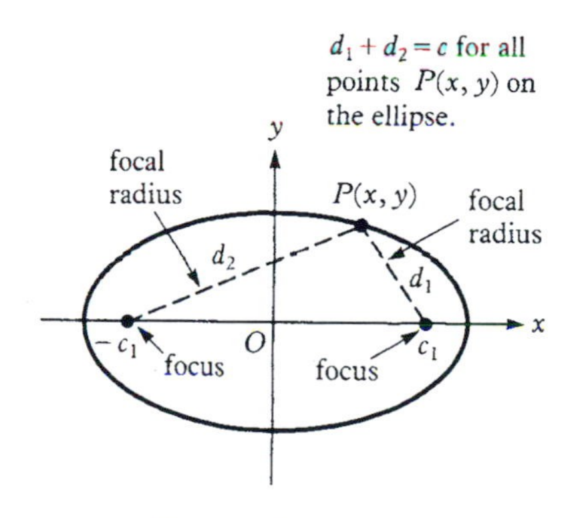

Figure 31 Ellipse

With reference to Figure 31, equation (1) becomes

$$\sqrt{(x - c_1)^2 + y^2} + \sqrt{(x + c_1)^2 + y^2} = c,$$

which simplifies (see Exercise 29) to

$$\frac{x^2}{a^2} + \frac{y^2}{b^2} = 1, \tag{2}$$

where $a = c/2$ and $b = \sqrt{a^2 - c_1^2}$ (Figure 32). Equation (2) is called the **standard equation of an ellipse.**

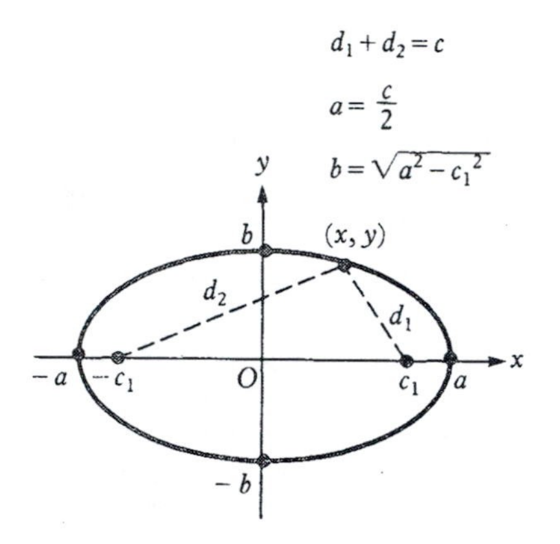

Figure 32 Ellipse $x^2/a^2 + y^2/b^2 = 1$

Example 1 Find the equation of the ellipse with foci $(2\sqrt{3}, 0)$ and $(-2\sqrt{3}, 0)$ and the sum of focal radii equal to 8. Also, graph the ellipse.

Solution: Here $c_1 = 2\sqrt{3}$ and $c = 8$. We could substitute $a = c/2 = 4$ and $b = \sqrt{a^2 - c_1^2} = 2$ into equation (2). However, in order to illustrate the way in which equation (2) follows from definition (1), we substitute the given data into (1) and simplify. That is, the condition $d_1 + d_2 = c$ becomes

$$\sqrt{(x - 2\sqrt{3})^2 + y^2} + \sqrt{(x + 2\sqrt{3})^2 + y^2} = 8,$$

which, in two squarings, simplifies to

$$x^2 + 4y^2 = 16.$$

Then, by dividing both sides by 16, we obtain

$$\frac{x^2}{16} + \frac{y^2}{4} = 1$$

for the equation of the ellipse. (Check this result with the equations in Figure 32.) For graphing the ellipse, we can first plot the points where the graph crosses the coordinate axes, as shown in the table, and then draw a smooth curve through these points (Figure 33).

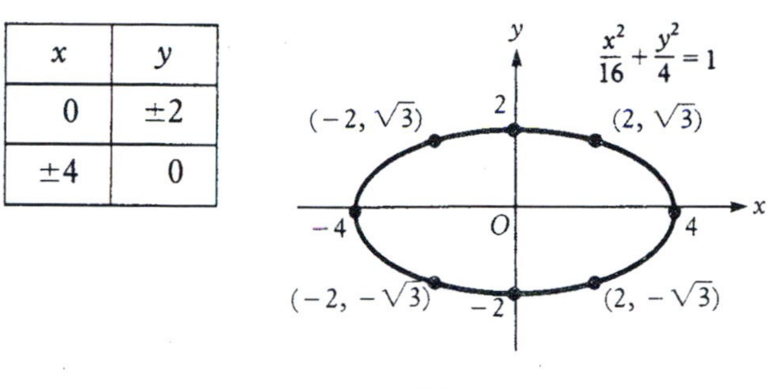

x	y
0	± 2
± 4	0

Figure 33

For more accuracy in the graph, several other points can be plotted. For instance, if $x = \pm 2$, then from the equation $x^2/16 + y^2/4 = 1$, we obtain $y = \pm\sqrt{3}$. Hence, the points $(2, \sqrt{3})$, $(2, -\sqrt{3})$, $(-2, \sqrt{3})$, and $(-2, -\sqrt{3})$ are also on the graph. ■

For an ellipse [equation (2)] with its foci $(c_1, 0)$ and $(-c_1, 0)$ on the x-axis, we note that $a > b$ (*why?*). The line segment through the two foci from $-a$ to a along the x-axis is called the **major axis** of the ellipse, and the segment from $-b$ to b along the y-axis is called the **minor axis.**

If the foci are positioned at $(0, c_1)$ and $(0, -c_1)$ on the y-axis, then the equation of the ellipse remains the same as (2), but the roles of x and y are interchanged, as are those of a and b. That is, $b > a$, the major axis is along the y-axis from $-b$ to b, and the minor axis is from $-a$ to a along the x-axis (Figure 34).

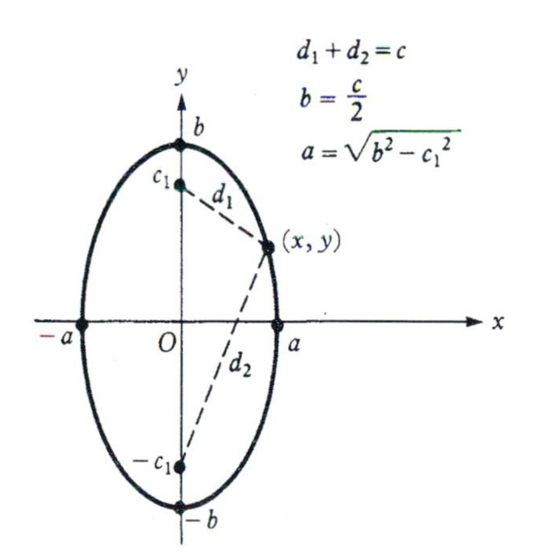

Figure 34 Ellipse $x^2/a^2 + y^2/b^2 = 1 \quad (b > a)$

Example 2 Show that each of the following equations represents an ellipse. Graph each ellipse and identify its major and minor axis.

(a) $4x^2 + 5y^2 = 20$ (b) $25x^2 + 9y^2 = 16$

Solution:

(a) If both sides of the equation are divided by 20, we obtain

$$\frac{x^2}{5} + \frac{y^2}{4} = 1,$$

which is an ellipse with $a = \sqrt{5}$ and $b = 2$. Since $a > b$, the major axis is along the x-axis from $-\sqrt{5}$ to $\sqrt{5}$ [Figure 35(a)].

(b) By dividing both sides of the given equation by 16, we get

$$\frac{25x^2}{16} + \frac{9y^2}{16} = 1 \qquad \text{or} \qquad \frac{x^2}{\left(\frac{4}{5}\right)^2} + \frac{y^2}{\left(\frac{4}{3}\right)^2} = 1,$$

which is an ellipse with $a = 4/5$ and $b = 4/3$. Since $b > a$, the major axis is along the y-axis from $-4/3$ to $4/3$ [Figure 35(b)]. ■

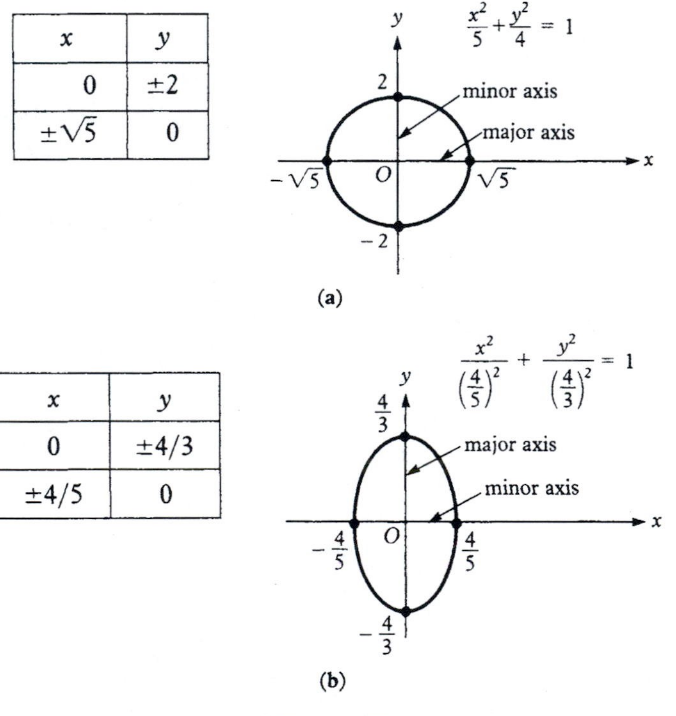

x	y
0	± 2
$\pm\sqrt{5}$	0

x	y
0	$\pm 4/3$
$\pm 4/5$	0

Figure 35

When an ellipse is given by equation (2), the origin (0, 0) is called the **center** of the ellipse. We can translate the center to a point (x_0, y_0) by replacing x by $x - x_0$ and y by $y - y_0$ in (2). We then get

$$\frac{(x - x_0)^2}{a^2} + \frac{(y - y_0)^2}{b^2} = 1 \tag{3}$$

as the equation of an **ellipse with center** (x_0, y_0) (Figure 36).

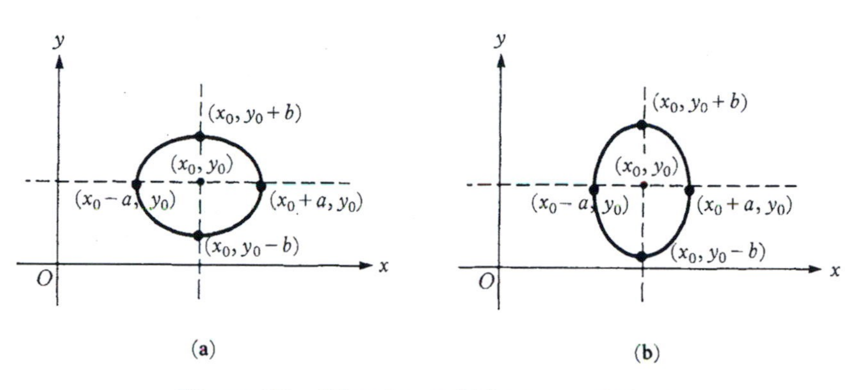

Figure 36 Ellipse $(x - x_0)^2/a^2 + (y - y_0)^2/b^2 = 1$

To determine whether a second-degree equation

$$Ax^2 + By^2 + Cx + Dy = E \tag{4}$$

represents an ellipse, we complete the square in x and in y, as illustrated in the following example.

Example 3 Show that the equation $4x^2 + 9y^2 - 16x - 54y = -61$ represents an ellipse and sketch its graph.

Solution: As indicated above, we complete the square in x and in y.

$$4x^2 - 16x + 9y^2 - 54y = -61 \qquad \textit{Combine x terms and y terms}$$

$$4(x^2 - 4x \qquad) + 9(y^2 - 6y \qquad) = -61 \qquad \textit{Factor leading coefficient from x terms and y terms, and leave space for completing the square}$$

$$4(x^2 - 4x + 4) + 9(y^2 - 6y + 9) = -61 + 4 \cdot 4 + 9 \cdot 9 \qquad \textit{Complete square in x and in y inside parentheses (Why } 4 \cdot 4 \textit{ and } 9 \cdot 9 \textit{ on right?)}$$

$$4(x - 2)^2 + 9(y - 3)^2 = 36 \qquad \textit{Simplify}$$

$$\frac{(x - 2)^2}{9} + \frac{(y - 3)^2}{4} = 1 \qquad \textit{Divide both sides by 36}$$

The last equation is of type (3) with $x_0 = 2, y_0 = 3, a = 3$, and $b = 2$. Hence, the graph (Figure 37) is an ellipse with center (2, 3) and with a horizontal major axis (*why horizontal?*). ■

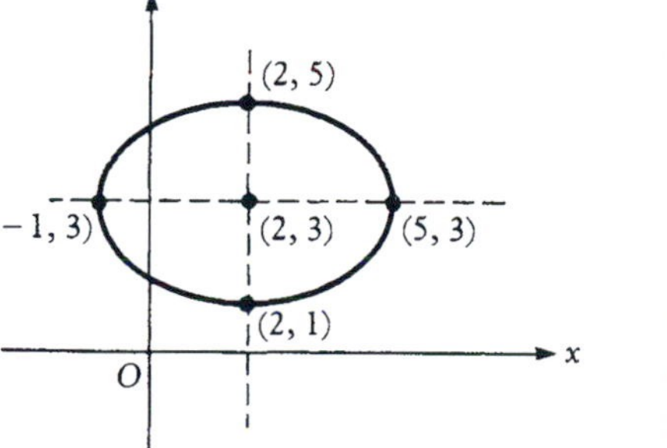

Figure 37

We saw in Section 3.6 that if a second-degree equation (4) represents a circle, then $A = B \neq 0$. Example 3 illustrates that if a second-degree equation represents an ellipse, then $A \neq B$ but A and B are either both positive or both negative ($AB > 0$). We also saw in Section 3.6 that the condition $A = B \neq 0$ does not *guarantee* a circle. Similarly, the conditions $A \neq B$ and $AB > 0$ do not guarantee an ellipse (see Exercises 23–28).

Hyperbolas. Given two points P_1 and P_2 in the plane and a positive number c, then the set of all points P for which

$$d(P_2, P) - d(P_1, P) = \pm c \tag{5}$$

is called a **hyperbola.** A hyperbola consists of two branches, one for $+c$ and the other for $-c$. As in the case of an ellipse, points P_1 and P_2 are **foci,** and distances $d_1 = d(P_1, P)$ and $d_2 = d(P_2, P)$ are **focal radii.** Figure 38 is a hyperbola in which $P_1 = (c_1, 0)$ and $P_2 = (-c_1, 0)$. With reference to Figure 38, equation (5) becomes

$$\sqrt{(x + c_1)^2 + y^2} - \sqrt{(x - c_1)^2 + y^2} = \pm c,$$

which simplifies (see Exercise 41) to

$$\frac{x^2}{a^2} - \frac{y^2}{b^2} = 1, \tag{6}$$

where $a = c/2$ and $b = \sqrt{c_1^2 - a^2}$ (Figure 39). Equation (6) is called the **standard equation of a hyperbola.** Compare these equations with the corresponding ones for an ellipse in Figure 32.

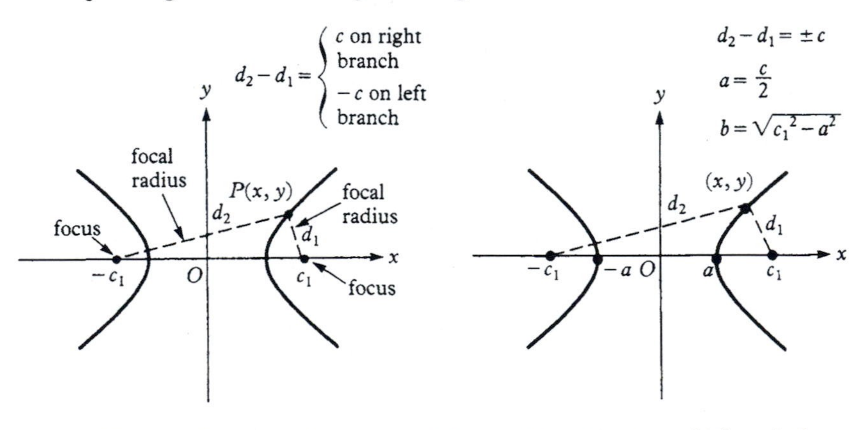

Figure 38 Hyperbola

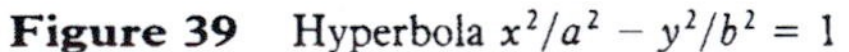
Figure 39 Hyperbola $x^2/a^2 - y^2/b^2 = 1$

Example 4 Find the equation of the hyperbola with foci (5, 0) and (−5, 0) and difference of focal radii equal to ±8. Also, graph the hyperbola.

Solution: From Figure 38, the condition $d_2 - d_1 = \pm 8$ becomes

$$\sqrt{(x + 5)^2 + y^2} - \sqrt{(x - 5)^2 + y^2} = \pm 8.$$

After two squarings to eliminate the radicals, the above equation simplifies to

$$9x^2 - 16y^2 = 144.$$

Then, by dividing both sides by 144, we obtain

$$\frac{x^2}{16} - \frac{y^2}{9} = 1$$

for the equation of the hyperbola. Here $a = 4$ and $b = 3$. (Check these values with the equations in Figure 39.) To graph the hyperbola, we plot the points of intersection with the x-axis and several other convenient points, as indicated in the table, and draw smooth branches through these points (Figure 40). ■

x	y
± 4	0
± 8	$\pm 3\sqrt{3}$
$\pm 4\sqrt{10}$	± 9

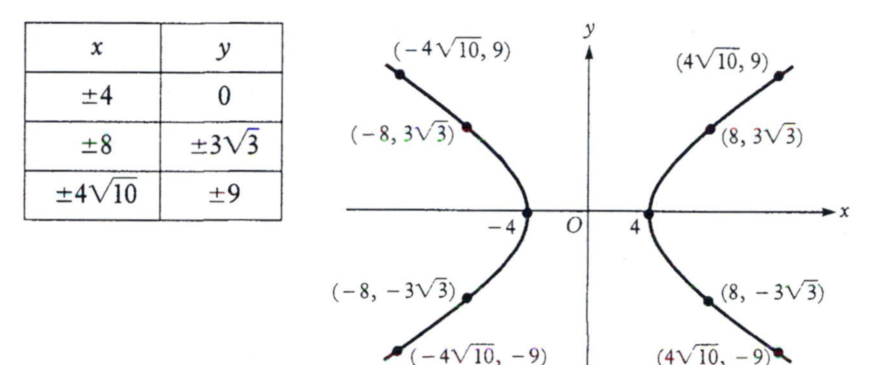

Figure 40 Hyperbola $x^2/16 - y^2/9 = 1$

If the foci of a hyperbola are positioned at $(0, c_1)$ and $(0, -c_1)$ on the y-axis, then the roles of x and y are interchanged, as are those of a and b. The equation becomes

$$\frac{y^2}{b^2} - \frac{x^2}{a^2} = 1, \tag{7}$$

where $b = c/2$ and $a = \sqrt{c_1^2 - b^2}$ (Figure 41).

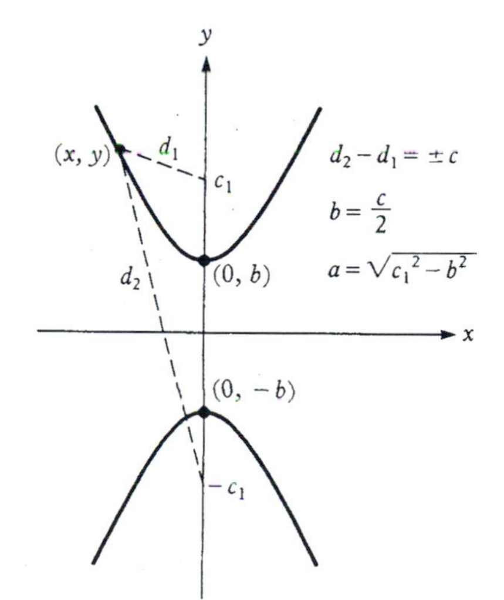

Figure 41 Hyperbola $y^2/b^2 - x^2/a^2 = 1$

When a hyperbola is given by equation (6) or (7), the origin (0, 0) is called the **center** of the hyperbola. We can translate the center to a point (x_0, y_0) by replacing x by $x - x_0$ and y by $y - y_0$ in (6) or (7). We then obtain

$$\frac{(x - x_0)^2}{a^2} - \frac{(y - y_0)^2}{b^2} = 1 \tag{8}$$

or

$$\frac{(y - y_0)^2}{b^2} - \frac{(x - x_0)^2}{a^2} = 1 \tag{9}$$

as the equation of a **hyperbola with center** (x_0, y_0) (Figure 42).

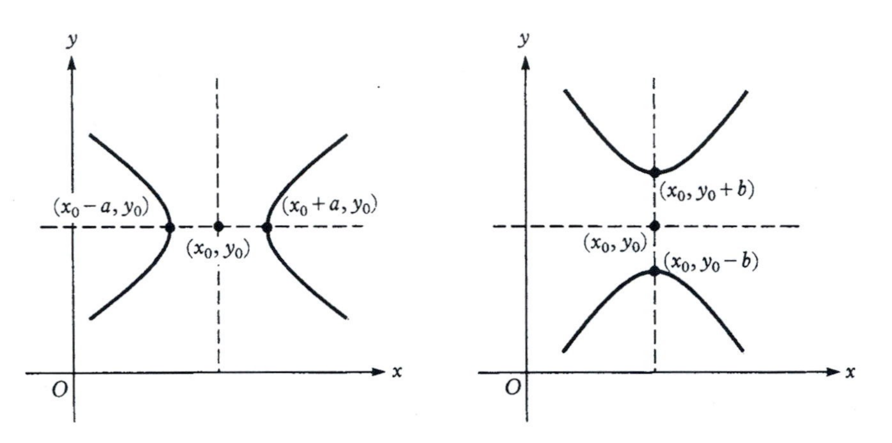

Figure 42 Hyperbola with Center (x_0, y_0)

As with the circle and ellipse, we can determine whether a second-degree equation represents a hyperbola by completing the square in x and in y, as illustrated in the following example.

Example 5 Show that the equation $3x^2 - 4y^2 - 6x - 16y = 1$ represents a hyperbola and sketch its graph.

Solution: We proceed as in Example 3.

$3x^2 - 6x - 4y^2 - 16y = 1$	*Combine x terms and y terms*
$3(x^2 - 2x \quad) - 4(y^2 + 4y \quad) = 1$	*Factor out leading coefficients*
$3(x^2 - 2x + 1) - 4(y^2 + 4y + 4)$ $= 1 + 3 \cdot 1 - 4 \cdot 4$	*Complete squares inside parentheses and balance equation*
$3(x - 1)^2 - 4(y + 2)^2 = -12$	*Simplify*
$\frac{3(x - 1)^2}{-12} - \frac{4(y + 2)^2}{-12} = 1$	*Divide both sides by −12*
$\frac{(y + 2)^2}{3} - \frac{(x - 1)^2}{4} = 1$	*Simplify*

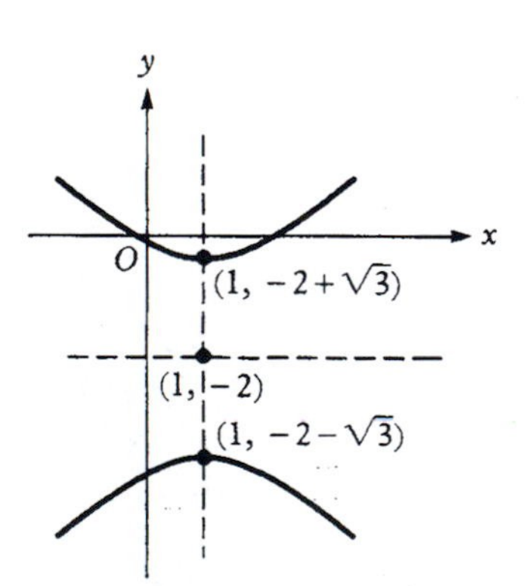

Figure 43
Hyperbola $3x^2 - 4y^2 - 6x - 16y = 1$

The last equation is of type (9) with $y_0 = -2$, $x_0 = 1$, $b = \sqrt{3}$, and $a = 2$. Hence, the graph is a hyperbola with center at $(1, -2)$ (Figure 43). ■

As illustrated in Example 5, if a second-degree equation (4) represents a hyperbola, then A and B have opposite signs ($AB < 0$). However, a second-degree equation could satisfy $AB < 0$ and have two lines for its graph (see Exercises 35–40).

Asymptotes. As shown in Figure 44, the hyperbola

$$\frac{x^2}{a^2} - \frac{y^2}{b^2} = 1 \tag{10}$$

crosses the x-axis at $x = \pm a$. To obtain a geometric interpretation of the positive constant b, we first factor the left side of equation (10):

$$\left(\frac{x}{a} - \frac{y}{b}\right)\left(\frac{x}{a} + \frac{y}{b}\right) = 1.$$

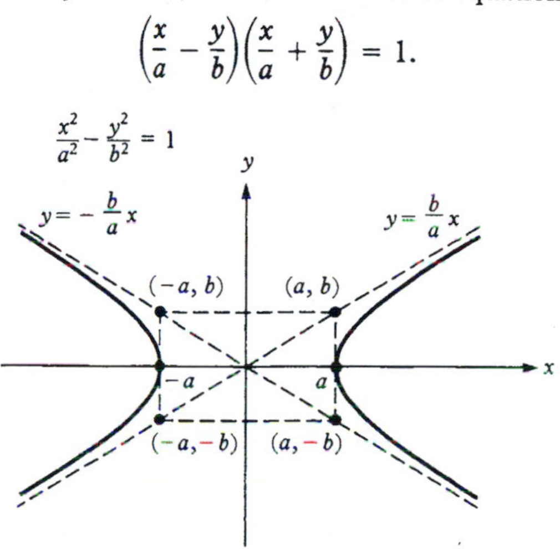

Figure 44 Asymptotes and Fundamental Rectangle

If x and y have the same sign (quadrants I and III) but get very large in magnitude, then the factor $x/a + y/b$ also gets very large in magnitude, and therefore, the factor $x/a - y/b$ must get very small in magnitude (since the product is 1 for all points on the hyperbola). That is,

$$\frac{x}{a} - \frac{y}{b} \to 0 \quad \text{or} \quad y \to \frac{b}{a}x.$$

Hence, the hyperbola approaches the line

$$y = \frac{b}{a}x \tag{11}$$

in quadrants I and III, as indicated in Figure 44. By similar reasoning, the hyperbola approaches the line

$$y = -\frac{b}{a}x \tag{12}$$

in quadrants II and IV. Lines (11) and (12) are called the **asymptotes** of hyperbola (10). The rectangle in Figure 44 with vertices (a, b), $(-a, b)$,

$(-a, -b)$, $(a, -b)$, is called the **fundamental rectangle** for the hyperbola (10). By drawing the asymptotes through the diagonals of the fundamental rectangle, we can get an accurate graph of the hyperbola.

Example 6 Use the fundamental rectangle and asymptotes to graph each of the following hyperbolas.

(a) $4x^2 - 9y^2 = 36$ (b) $4y^2 - 25x^2 = 100$

Solution:

(a) By dividing both sides of the given equation by 36, we obtain

$$\frac{x^2}{9} - \frac{y^2}{4} = 1.$$

Here $a = 3$ and $b = 2$, and the asymptotes are the lines $y = (2/3)x$ and $y = -(2/3)x$. The fundamental rectangle, asymptotes, and graph of the hyperbola are shown in Figure 45(a).

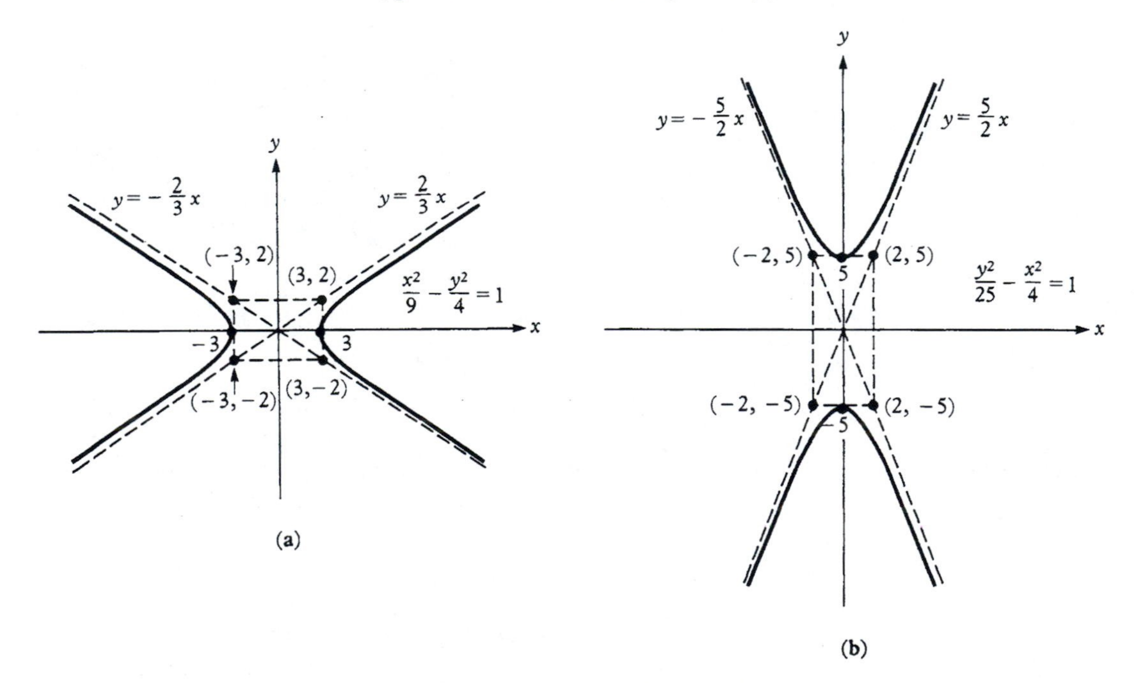

Figure 45

(b) We divide both sides of the given equation by 100 to obtain

$$\frac{y^2}{25} - \frac{x^2}{4} = 1.$$

Here $a = 2$ and $b = 5$, and the asymptotes are $y = (5/2)x$ and $y = -(5/2)x$. These asymptotes, along with the fundamental rectangle and the graph of the hyperbola, are illustrated in Figure 45(b). ■

Comment The conic sections were discovered around 350 B.C. by the Greek geometer Menæchmus while working on a solution to the problem of "duplicating the cube," that is, constructing a cube whose volume is twice the volume of a given cube. About 120 years later, Apollonius made an exhaustive study of the ellipse, parabola, and hyperbola in the *Conics*, a collection of eight books. Apollonius also devised a widely accepted system of *circular* planetary motion, apparently not realizing that the planets actually travel in elliptical orbits. The elliptical orbits of the planets were discovered empirically in 1609 by Johannes Kepler. About 70 years later, Sir Isaac Newton formulated his laws of gravitation and motion. From these laws it follows that planetary motion in the universe can be along any one of the three conics. Hence, the parabola, ellipse, and hyperbola, discovered in an investigation of a classical problem in geometry, have universal application.

Exercises 3.8

Fill in the blanks to make each statement true.

1. Let P_1 and P_2 be two fixed points in the plane and let c be a positive number. The set of all points P in the plane for which ______________ is called an ellipse, and the set of all P for which ______________ is called a hyperbola.
2. Each of the fixed points P_1 and P_2 in the definition of an ellipse and hyperbola is called a ______________.
3. The graph of $(x - 3)^2/a^2 - (y + 5)^2/b^2 = 1$ is a(n) ______________ with center __________.
4. The length of the major axis of the ellipse $x^2/4 + y^2/10 = 1$ is ________ and that of the minor axis is ________.
5. If the graph of a second-degree equation $Ax^2 + By^2 + Cx + Dy = E$ is an ellipse or hyperbola, then $AB \neq$ ________. For an ellipse, AB ________ 0, and for a hyperbola, AB ________ 0.

Write true *or* false *for each statement.*

6. A focal radius of an ellipse has constant length.
7. The major axis of an ellipse is always along the x-axis.
8. The lines $y = \pm(b/a)x$ are asymptotes for the hyperbola $x^2/a^2 - y^2/b^2 = 1$.
9. The lines $y = \pm(a/b)x$ are asymptotes for the hyperbola $y^2/b^2 - x^2/a^2 = 1$.
10. If $A \neq B$, then the graph of the second-degree equation $Ax^2 + By^2 + Cx + Dy = E$ is either an ellipse or a hyperbola.

Ellipses

In Exercises 11–14, find the equation of the ellipse with the given foci and focal radii.

11. foci (1, 0) and (−1, 0); $d_1 + d_2 = 4$
12. foci (2, 0) and (−2, 0); $d_1 + d_2 = 5$
13. foci (0, 2) and (0, −2); $d_1 + d_2 = 6$
14. foci (0, 1) and (0, −1); $d_1 + d_2 = 5$

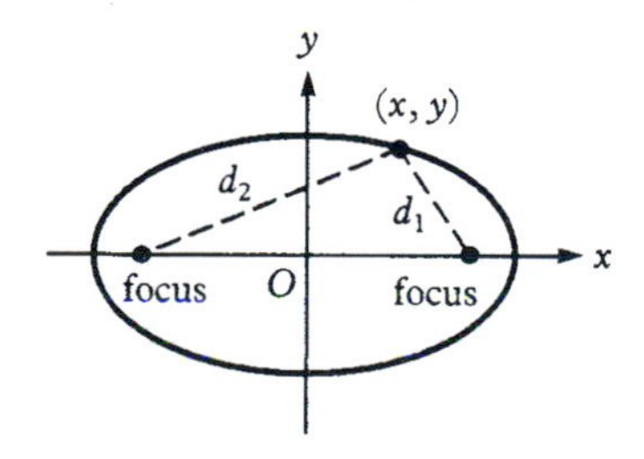

Graph each of the following ellipses.

15. $\dfrac{x^2}{16} + \dfrac{y^2}{25} = 1$

16. $\dfrac{x^2}{25} + \dfrac{y^2}{36} = 1$

17. $4x^2 + 9y^2 = 36$

18. $9x^2 + 16y^2 = 144$

19. $\dfrac{(x-2)^2}{16} + \dfrac{(y-5)^2}{25} = 1$

20. $\dfrac{(x+3)^2}{25} + \dfrac{(y+2)^2}{36} = 1$

21. $25(x-1)^2 + 4(y-2)^2 = 100$

22. $2(x+3)^2 + 8(y-3)^2 = 32$

By completing the square in x and y, determine the graph (ellipse, point, empty set) of each of the second-degree equations in Exercises 23–28.

23. $x^2 + 4y^2 + 4x - 16y = 80$

24. $2x^2 + y^2 + 4x + 10y = -30$

25. $3x^2 + 2y^2 + 6x + 8y = -11$

26. $2x^2 + 5y^2 - 4x - 10y = -7$

27. $4x^2 + 9y^2 + 16x - 18y = 11$

28. $2x^2 + 3y^2 + 2x = -1$

29. Eliminate the radicals in two squarings to show that the equation

$$\sqrt{(x-c_1)^2 + y^2} + \sqrt{(x+c_1)^2 + y^2} = c \quad (c > 2c_1 > 0)$$

simplifies to

$$\frac{x^2}{a^2} + \frac{y^2}{b^2} = 1,$$

where $a = \dfrac{c}{2}$ and $b = \sqrt{a^2 - c_1^2}$.

30. The moon travels in an elliptical orbit about the earth as one of the foci. From the data in the figure, find the equation of the ellipse and the location of the other focus.

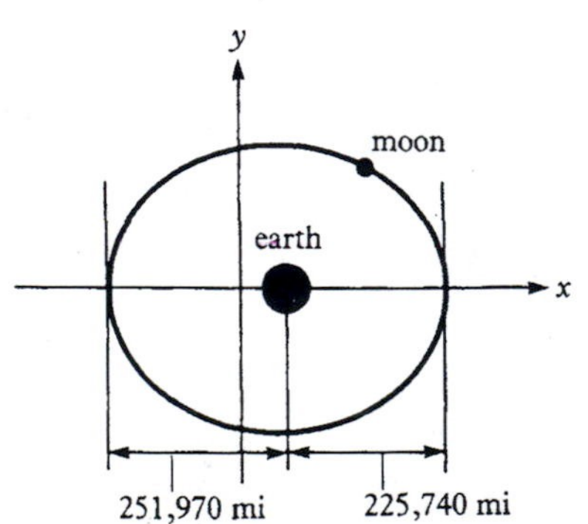

Hyperbolas

In Exercises 31–34, find the equation of the hyperbola with the given foci and focal radii.

31. foci (1, 0) and (−1, 0); $d_2 - d_1 = \pm 1$

32. foci (2, 0) and (−2, 0); $d_2 - d_1 = \pm 4$

33. foci (0, 2) and (0, −2); $d_2 - d_1 = \pm 2$

34. foci (0, 3) and (0, −3); $d_2 - d_1 = \pm 3$

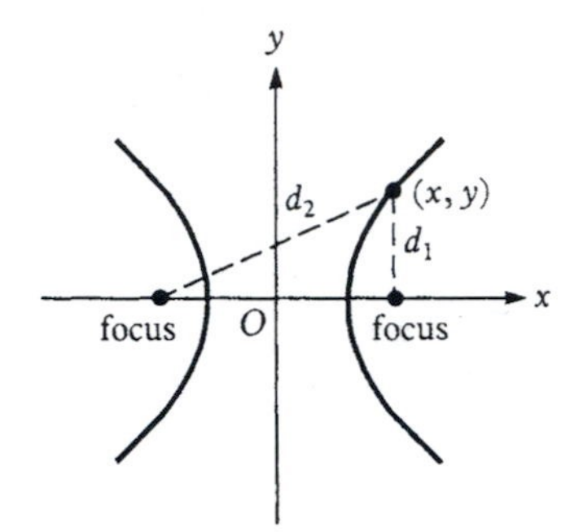

By completing the square in x and y, determine the graph (hyperbola or two lines) of each of the second-degree equations in Exercises 35–40.

35. $4x^2 - y^2 + 2x - y = 0$

36. $9x^2 - 16y^2 - 9x = 0$

37. $2x^2 - 5y^2 - 4x - 10y = -2$

38. $y^2 - x^2 + 3y - 3x = 1$

39. $8y^2 - x^2 - 8y = -1$

40. $y^2 - 8x^2 - 8x = 2$

41. Eliminate the radicals in two squarings to show that the equation

$$\sqrt{(x+c_1)^2 + y^2} - \sqrt{(x-c_1)^2 + y^2} = \pm c \quad (2c_1 > c > 0)$$

simplifies to

$$\frac{x^2}{a^2} - \frac{y^2}{b^2} = 1,$$

where $a = \dfrac{c}{2}$ and $b = \sqrt{c_1^2 - a^2}$.

Asymptotes

Graph each of the following hyperbolas, showing the fundamental rectangle and asymptotes.

42. $\dfrac{x^2}{16} - \dfrac{y^2}{25} = 1$

43. $\dfrac{x^2}{25} - \dfrac{y^2}{36} = 1$

44. $16y^2 - 9x^2 = 144$ **45.** $y^2 - x^2 = 1$

46. $x^2 - 4y^2 = 16$ **47.** $9x^2 - y^2 = 81$

48. $\dfrac{(x-2)^2}{9} - \dfrac{(y-5)^2}{16} = 1$

49. $\dfrac{(x-1)^2}{4} - \dfrac{(y-2)^2}{9} = 1$

Miscellaneous

50. *Eccentricity of an Ellipse.* If $a > b > 0$, the ellipse

$$\frac{x^2}{a^2} + \frac{y^2}{b^2} = 1$$

has its foci at $(\pm\sqrt{a^2 - b^2}, 0)$. The ratio

$$e = \frac{\sqrt{a^2 - b^2}}{a} \quad (0 < e < 1)$$

is called the **eccentricity** of the ellipse, and the vertical line

$$L: x = \frac{a}{e}$$

is called a **directrix.** In terms of the eccentricity, the foci are at $(\pm ae, 0)$. With reference to the figure, show that a point $P(x, y)$ is on the ellipse if and only if

$$\frac{d_1}{d} = e.$$

(The line $x = -a/e$ is also called a directrix, and the above equation holds when $(ae, 0)$ and $x = a/e$ are replaced by $(-ae, 0)$ and $x = -a/e$, respectively.)

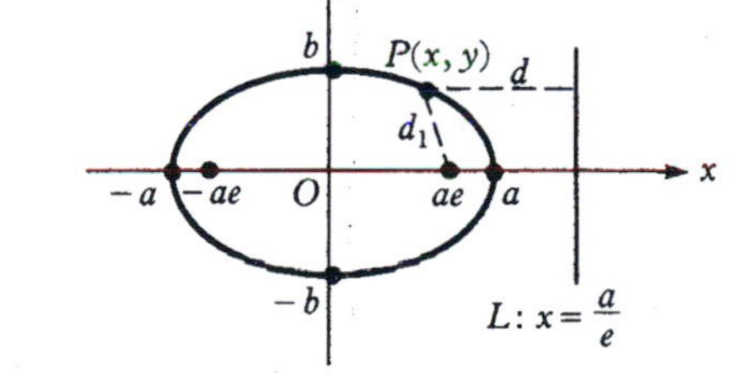

51. Discuss the change in shape of the ellipse $x^2/a^2 + y^2/b^2 = 1$ in each of the following cases.

(a) a is held fixed but the focus $(ae, 0)$ moves toward $(0, 0)$.

(b) a is held fixed but the focus $(ae, 0)$ moves toward $(a, 0)$.

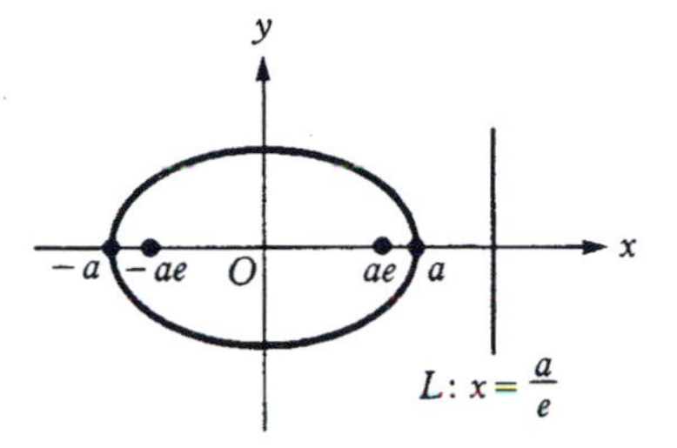

52. *Eccentricity of a Hyperbola.* The hyperbola

$$\frac{x^2}{a^2} - \frac{y^2}{b^2} = 1$$

has its foci at $(\pm\sqrt{a^2 + b^2}, 0)$. The ratio

$$e = \frac{\sqrt{a^2 + b^2}}{a} \quad (e > 1)$$

is called the **eccentricity** of the hyperbola, and the vertical line

$$L: x = \frac{a}{e}$$

is called a **directrix.** In terms of the eccentricity, the foci are at $(\pm ae, 0)$. With reference to the figure, show that a point $P(x, y)$ is on the hyperbola if and only if

$$\frac{d_1}{d} = e.$$

[As in the case of an ellipse, the line $x = -a/e$ is also called a directrix, and the above equation holds when $(ae, 0)$ and $x = a/e$ are replaced by $(-ae, 0)$ and $x = -a/e$, respectively.]

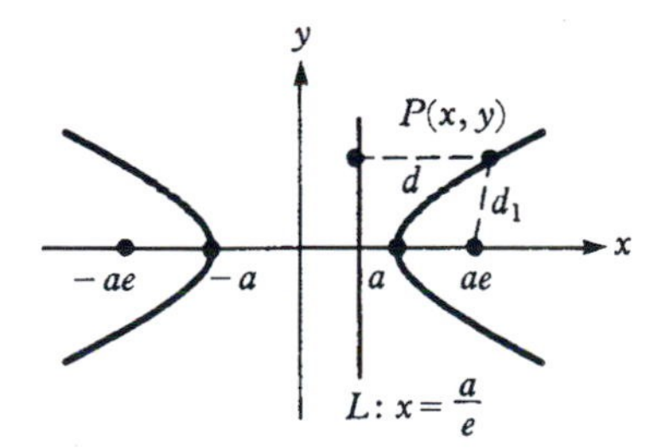

53. Discuss the change in shape of the hyperbola in each of the following cases. (*Hint:* draw the fundamental rectangle.)
    (a) a is held fixed but ae approaches ∞.
    (b) a is held fixed but ae approaches a.

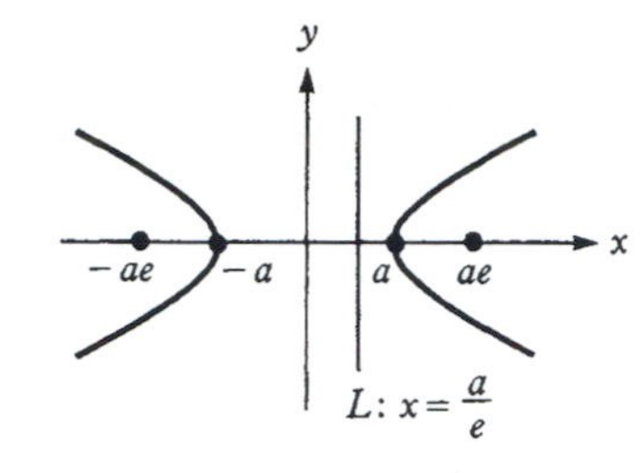

Chapter 3 Review Outline

3.1 Quadratic Equations

Quadratic Formula:

The roots of $ax^2 + bx + c = 0 \quad (a \neq 0)$ *are*

$$\frac{-b \pm \sqrt{b^2 - 4ac}}{2a}.$$

Completing the Square:

If $b^2/4$ is added to $x^2 + bx$, the result is a perfect square $(x + b/2)^2$.

Factoring by Roots:

The roots of $ax^2 + bx + c = 0$ are a_1 and $a_2 \Leftrightarrow ax^2 + bx + c = a(x - a_1)(x - a_2)$.

If the discriminant $b^2 - 4ac < 0$, then $ax^2 + bx + c$ is irreducible (cannot be factored in the real number system).

3.2 Complex Numbers

A complex number is a number of the form $a + bi$, where a and b are real numbers and $i = \sqrt{-1}$.

$a + bi = c + di \Leftrightarrow a = c$ and $b = d$

Operations:

$(a + bi) \pm (c + di) = (a \pm c) + (b \pm d)i$

$(a + bi)(c + di) = (ac - bd) + (ad + bc)i$

To divide complex numbers, proceed as follows:

$$\frac{a + bi}{c + di} = \frac{a + bi}{c + di} \cdot \frac{c - di}{c - di} = \frac{ac + bd}{c^2 + d^2} + \frac{bc - ad}{c^2 + d^2}i.$$

($c - di$ is the complex conjugate of $c + di$)

Classification of Roots:

The quadratic equation $ax^2 + bx + c = 0$ has

two distinct real roots $\Leftrightarrow b^2 - 4ac > 0$,
two equal real roots $\Leftrightarrow b^2 - 4ac = 0$,
two complex conjugate roots $\Leftrightarrow b^2 - 4ac < 0$.

3.3 Equations Reducible to Quadratic Equations

Some equations in x can be transformed into quadratic equations in w by substituting w for an expression in x.

Equations with fractional expressions or radicals can sometimes be transformed into quadratic equations by clearing fractions or squaring.

All solutions obtained by subsitution, clearing fractions, or squaring must be tested for acceptability by substitution in the original equation.

3.4 Quadratic Inequalities

If $ax^2 + bx + c = 0$ has distinct real roots a_1 and a_2 $(a_1 < a_2)$, then $ax^2 + bx + c$ does not change sign in $(-\infty, a_1)$, (a_1, a_2), and (a_2, ∞).

If $ax^2 + bx + c = 0$ has a double root a_1, then $ax^2 + bx + c$ has the same sign as a for all other values of x.

If $ax^2 + bx + c = 0$ has complex roots, then $ax^2 + bx + c$ has the same sign as a for all values of x.

3.5 Applications of Quadratic Equations and Inequalities

The four-step approach to solving word problems and the basic principles for percent, average speed, and work problems are the same here as in Section 2.3.

The height y of an object moving under the influence of gravity only is

$$y = -16t^2 + v_0 t + y_0,$$

where v_0 is the object's vertical velocity at time $t = 0$, and y_0 is its height at time 0 (distance is measured in feet and time in seconds). The vertical velocity of the object is

$$v = -32t + v_0.$$

3.6 Circles and Parabolas

The graph of the second-degree equation $Ax^2 + By^2 + Cx + Dy = E$ (A or $B \neq 0$) is a conic section or the empty set. Circles and parabolas are two types of conic sections.

Circle: $(x - x_0)^2 + (y - y_0)^2 = r^2$, center (x_0, y_0), radius r

Parabola: $y = ax^2 + bx + c$, axis of symmetry $x = -\dfrac{b}{2a} = \dfrac{a_1 + a_2}{2}$, where a_1 and a_2 are the roots of $ax^2 + bx + c = 0$ (With x and y interchanged, the equations are similar.)

A second-degree equation with $A = B$ is either a circle or the empty set. To determine which is the case, complete the square in x and y.

A second-degree equation with either $B = 0$, $A \neq 0$, and $D \neq 0$ or $A = 0$, $B \neq 0$, and $C \neq 0$ is a parabola.

3.7 Translation of Axes and Intersections

If the $x'y'$-axes are a translation of the xy-axes in which (x_0, y_0) becomes the $x'y'$-origin, then a point in the plane has both xy-coordinates and $x'y'$-coordinates that are related by the substitution equations

$$x' = x - x_0 \quad \text{and} \quad y' = y - y_0.$$

To simplify a second-degree equation, complete the square in x and/or y and make the corresponding substitutions.

A line can intersect a parabola or a circle in at most two points. Substituting from the linear equation into the second-degree equation leads to a quadratic equation.

Two parabolas of the type $y = ax^2 + bx + c$ or of the type $x = ay^2 + by + c$ can intersect in at most two points. Other pairs of parabolas, or a circle and a parabola, can intersect in up to four points; algebraically, a fourth-degree equation may result.

3.8 Ellipses and Hyperbolas

Two other conic sections are ellipses and hyperbolas.

Ellipse: $\dfrac{(x - x_0)^2}{a^2} + \dfrac{(y - y_0)^2}{b^2} = 1$, center (x_0, y_0)

Hyperbola: $\dfrac{(x - x_0)^2}{a^2} - \dfrac{(y - y_0)^2}{b^2} = 1$

or $\dfrac{(y - y_0)^2}{b^2} - \dfrac{(x - x_0)^2}{a^2} = 1$,

center (x_0, y_0)

For a hyperbola with center $(0, 0)$, the points (a, b), $(-a, b)$, $(-a, -b)$, $(a, -b)$ form the vertices of the fundamental rectangle of the hyperbola; the lines $y = (b/a)x$ and $y = -(b/a)x$ are the asymptotes of the hyperbola.

Chapter 3 Review Exercises

Solve each quadratic equation in Exercises 1–6.

1. $x^2 - 5x - 50 = 0$
2. $3x^2 - 10x - 8 = 0$
3. $3x^2 + 18x + 27 = 0$
4. $3x^2 + 2x + 2 = 0$
5. $3x^2 + 2x - 2 = 0$
6. $x^2 - \dfrac{5}{6}x + \dfrac{1}{3} = 0$
7. Factor the quadratic $8x^2 + 2x - 3$ in the form $a(x - a_1)(x - a_2)$.
8. Show that the quadratic $x^2 - 6x + 11$ is irreducible (has no factors) in the real number system.
9. Find a quadratic equation with integral coefficients whose roots are $2/3$ and $-3/4$.
10. Find all real roots of $x^4 - 6x^2 + 9 = 0$.
11. Solve $2\sqrt{x + 5} - \sqrt{2x + 1} = 3$.
12. Solve $x/(x + 1) + 2/x = 26/(5x)$.
13. Find the sum of $3 + 2i$ and $5 - 3i$.
14. Represent the numbers $3 + 2i$, $5 - 3i$, and their sum by directed line segments in the plane.
15. Compute $(4 + 3i)(2 - i)$ and $(4 + 3i)/(2 - i)$.

Solve each of the following inequalities.

16. $x^2 - 4x + 3 \geq 0$
17. $3x^2 + 10x - 8 > 0$
18. $4x^2 - 20x + 25 \leq 0$
19. $x^2 + 5x + 7 < 0$

20. Sketch the graph of $y = 3x^2 + 10x - 8$.
21. Sketch the graph of $x = y^2 + y - 12$.
22. Find the standard equation of the circle that has the line segment from $(3, -4)$ to $(5, 8)$ as its diameter.

Sketch the graph of each of the following. For hyperbolas, show the fundamental rectangle and the asymptotes.

23. $4x^2 + 4y^2 = 9$
24. $4(x - 2)^2 + 4(y + 3)^2 = 9$
25. $4(x - 2)^2 + 9(y + 3)^2 = 0$

*26. $4x^2 + 9y^2 = 9$

*27. $4x^2 - 9y^2 = 9$

*28. $4y^2 - 9x^2 = 9$

Find the center and identify the type of each of the following curves.

29. $2x^2 + 2y^2 - 4x - 24y = -42$

*30. $x^2 - y^2 + 10x + 8y = 0$

In each of the following, xy-axes are translated to x'y'-axes with the point (x_0, y_0) becoming the x'y'-origin. Find the x'y'-equation of the given curve.

*31. (x_0, y_0) is the center of the circle $x^2 + y^2 + 2x - 4y = 0$.

*32. (x_0, y_0) is the center of the hyperbola $4x^2 - y^2 - 20x = 10$.

*33. (x_0, y_0) is the vertex of the parabola $y = 2x^2 - 5x - 12$.

Find all points of intersection of each pair of curves in Exercises 34–35.

*34. $x^2 + y^2 = 25$
$x - 3y = 5$

*35. $y = x^2 - 2x - 3$
$2y = x - 3$

*36. Graph the region in the plane determined by the system of inequalities.

$$y \geq x^2 - 9$$
$$y \leq -x^2 + 3x.$$

37. A car is priced at \$10,000 on January 1. On May 1, the price is increased by r%, and on September 1 there is another r% increase, bringing the price up to \$10,609. Find r.

38. Dick jogs 1/2 mph faster than Frank and covers 10 mi in 5 min less. Find the jogging speed of each man.

39. It takes Janet and Dan 6 min to wash a car when working together. Working alone, it takes Janet 5 min more than Dan. How long does it take each when working alone?

40. Two thousand dollars are deposited in a bank account on January 1 and \$1,000 more on February 1. If the monthly interest rate is r% and the deposits plus interest amount to \$3,025.05 on March 1, what is r?

Graphing Calculator Exercises

In each of the following, verify your answers to the indicated review exercises of this section by using a graphing calculator for the given functions and windows $[x_{min}, x_{max}]$, $[y_{min}, y_{max}]$.

41. Exercise 2: $Y_1 = 3x^2 - 10x - 8$, $[-3.5,6]$, $[-20,5]$
42. Exercise 3: $Y_1 = 3x^2 + 18x + 27$, $[-5.5,4]$, $[-1,30]$
43. Exercise 8: $Y_1 = x^2 - 6x + 11$, $[-3.5,6]$, $[-1,15]$
44. Exercise 12: $Y_1 = x/(x + 1) + 2/x - 26/(5x)$, $[-2.5,7]$, $[-5,5]$
45. Exercise 16: $Y_1 = x^2 - 4x + 3$, $[-2,7.5]$, $[-2,10]$
46. Exercise 19: $Y_1 = x^2 + 5x + 7$, $[-7,2.5]$, $[-2,5]$
47. Exercise 20: $Y_1 = 3x^2 + 10x - 8$, $[-5.5,4]$, $[-20,20]$

*48. Exercise 26: $Y_1 = \sqrt{1 - 4x^2/9}$, $Y_2 = -Y_1$, $[-2.25,2.5]$, $[-2,2]$

*49. Exercise 28: $Y_1 = 1.5\sqrt{1 + x^2}$, $Y_2 = -Y_1$, $[-4.5,5]$, $[-5,5]$

*50 Exercise 35: $Y_1 = x^2 - 2x - 3$, $Y_2 = (x - 3)/2$, $[-4,5.5]$, $[-5,5]$

*Asterisks indicate topics that may be omitted, depending on individual course needs.

4 Functions and their Graphs

In Chapter 2 we saw that the graph of the equation

$$ax + by + c = 0 \quad (a \text{ or } b \neq 0)$$

is a straight line, and in Chapter 3 we saw that the graph of the equation

$$ax^2 + by^2 + cx + dy + e = 0 \quad (a \text{ or } b \neq 0)$$

is a conic section. In this chapter we first refine the idea of "equation" to that of "*function,*" which is one of the most fundamental concepts in mathematics. We then proceed to investigate the graphs of algebraic functions, including polynomial and rational functions.

4.1 Functions

The Concept of a Function
Graphs of Functions
Algebra of Functions
Composition of Functions

Functions are the machines of algebra.

The Concept of a Function. Illustrations of the concept of a function occur in many aspects of our daily lives. For example,

(a) to each person in your math class, there corresponds a name on the professor's class list;
(b) to each item in the supermarket, there corresponds a price;
(c) to each of the fifty states in the U.S., there corresponds the name of its capital city;
(d) to each real number, there corresponds its square;
(e) to each nonvertical straight line in the coordinate plane, there corresponds its slope.

Notice that every example involves a rule for pairing each member of one set with exactly one member of a second set. These examples involve sets of people, names, prices, and lines as well as numbers, but in algebra we shall be most interested in rules of correspondence between sets of real numbers. Let us formalize the basic idea.

> A **function** from a set X to a set Y is a correspondence that assigns to each element of X exactly one element of Y. The set X is called the **domain** of the function, and the set of correspondents in Y is called the **range** of the function.

Example 1 Determine the domain and range of each of the functions (a) through (e).

Solution:

	Domain	Range
(a)	The set of students in your math class.	The set of names on the class list.
(b)	All items for sale in the supermarket.	All of their prices.
(c)	The fifty states in the U.S.	Their fifty capitals.
(d)	The set of real numbers.	The set of nonnegative real numbers.
(e)	All nonvertical lines in the plane.	All real numbers. ■

It is customary to denote functions by letters such as ***f***, ***g***, and ***h***. If x is in the domain of a function f, then the corresponding element in the range of f is denoted by $f(x)$, read "f of x." We call x the **independent variable** and $y = f(x)$ the **dependent variable.**

Example 2 Let $f(x) = x^3 - 1$ define a function whose domain is all real numbers. Find each of the following values of f.
(a) $f(0)$ (b) $f(2)$ (c) $f(-2)$ (d) $f(\sqrt[3]{5})$

Solution:
(a) $f(0) = 0^3 - 1 = -1$
(b) $f(2) = 2^3 - 1 = 7$
(c) $f(-2) = (-2)^3 - 1 = -9$
(d) $f(\sqrt[3]{5}) = (\sqrt[3]{5})^3 - 1 = 4$ ■

We can think of a function as a machine that accepts an input x from its domain and produces an output $f(x)$ in its range (Figure 1). For instance, if 2 is fed into the "machine" $f(x) = x^3 - 1$, then the output is 7.

Figure 1 A Function as a Machine

Many calculators have several functions, such as the "squaring function" and the "square root function," built into them. If you push the 4 key and the x^2 key, the display (output) is 16; if you push 4 and the $\sqrt{x}$ key, the display reads 2.

Unless otherwise specified, the domain of a function f defined by an algebraic expression, such as $f(x) = \sqrt{x - 3}$, is the set of all real numbers for which $f(x)$ is also a real number. We will call this set the **implied domain.** The implied or specified domain of f will be denoted by $\boldsymbol{D_f}$ and the range by $\boldsymbol{R_f}$.

Example 3 Find the implied domain of each of the following functions.

(a) $f(x) = \sqrt{x - 3}$ (b) $g(x) = 2x + 3$ (c) $h(x) = 1/(3x - 4)$

Solution:

(a) $\sqrt{x - 3}$ is a real number if and only if $x - 3 \geq 0$. Therefore, $D_f = [3, \infty)$.

(b) $2x + 3$ is a real number for all real numbers x. Therefore $D_g = (-\infty, \infty)$.

(c) $1/(3x - 4)$ is a real number if and only if $3x - 4 \neq 0$, that is $x \neq 4/3$. Therefore, $D_h = (-\infty, 4/3) \cup (4/3, \infty)$. ■

The correspondence between domain and range that defines a function can be denoted by various choices of letters. For example, $f(x) = x^2$ could also be denoted by $g(x) = x^2$ or by $f(t) = t^2$. All three expressions define the same function. On the other hand, we can modify $f(x)$ by replacing x with another algebraic expression, as illustrated in the following example.

Example 4 Let $f(x) = x^2 + 5x - 3$. Find each of the following.

(a) $f(2x)$ (b) $f(x^2)$ (c) $f(x + 2)$

Solution:

(a) To determine $f(2x)$, we replace x by $2x$ in the expression for $f(x)$. That is, $f(2x) = (2x)^2 + 5(2x) - 3 = 4x^2 + 10x - 3$.

(b) Here we replace x by x^2 to obtain $f(x^2) = (x^2)^2 + 5x^2 - 3 = x^4 + 5x^2 - 3$.

(c) Proceeding as before, we get $f(x + 2) = (x + 2)^2 + 5(x + 2) - 3 = x^2 + 4x + 4 + 5x + 10 - 3 = x^2 + 9x + 11$. ■

Graphs of Functions. Each function f generates a set of ordered pairs $(x, f(x))$. For example, if $f(x) = x^2$, then

$$f(0) = 0, \quad f(1) = 1, \quad f(2) = 4, \quad f(-1) = 1,$$

and the corresponding ordered pairs are

$$(0, 0), \quad (1, 1), \quad (2, 4), \quad (-1, 1),$$

which may be interpreted as points in the coordinate plane. The set of all pairs (x, y) with x in the domain of f and $y = f(x)$ constitutes the **graph** of the function f (Figure 2). Notice that if all points (x, y) on the graph of f are projected vertically onto the x-axis, the resulting set of x's is the domain of f, and, similarly, if all (x, y) on the graph of f are projected horizontally onto the y-axis, the set of all y's is the range of f. For the example $f(x) = x^2$, these projections are the entire x-axis and the non-negative y-axis, respectively.

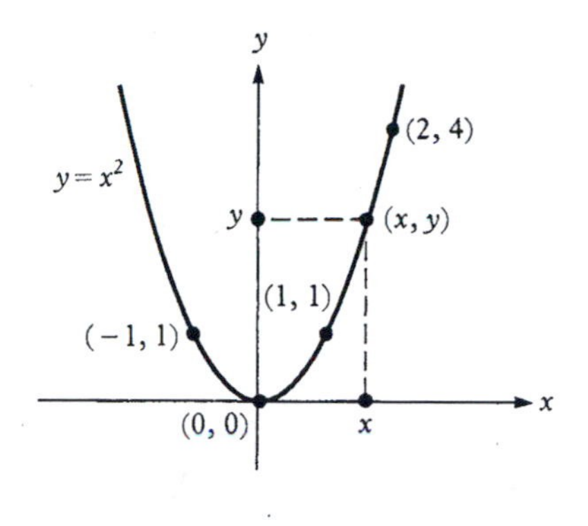

Figure 2 Graph of $f(x) = x^2$

Question 1 *Why does the equation $y^2 = x + 4$ not define a function $y = f(x)$?*

Answer: Since $y = \pm\sqrt{x+4}$, it follows that for each $x > -4$, there are *two* corresponding values of y. Therefore, we should not write $f(x) = \pm\sqrt{x+4}$ because the definition of a function f requires exactly *one* number $f(x)$ for each x in the domain. ■

Let us now compare the graph of a function [Figure 3(a)] with one that is not of a function [Figure 3(b)]. As illustrated in Figure 3, we can determine whether a graph is that of a function by applying the following **vertical line test.**

If every vertical line intersects the graph in at most one point, the graph is that of a function. On the other hand, if some vertical line intersects the graph in more than one point, the graph is not that of a function.

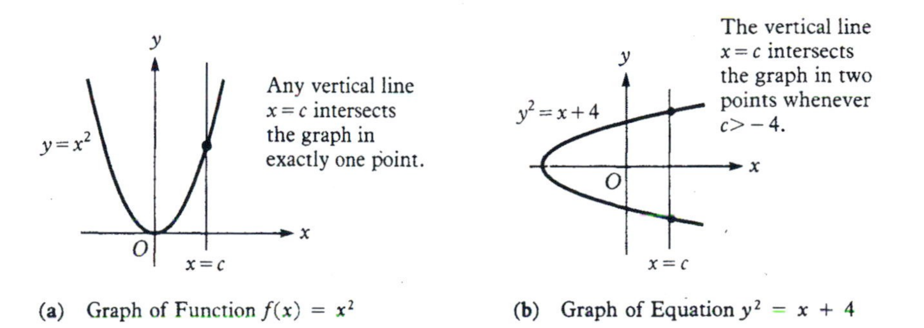

(a) Graph of Function $f(x) = x^2$

(b) Graph of Equation $y^2 = x + 4$

Figure 3

Example 5 Use the vertical line test to determine which of the graphs in Figure 4 are graphs of functions.

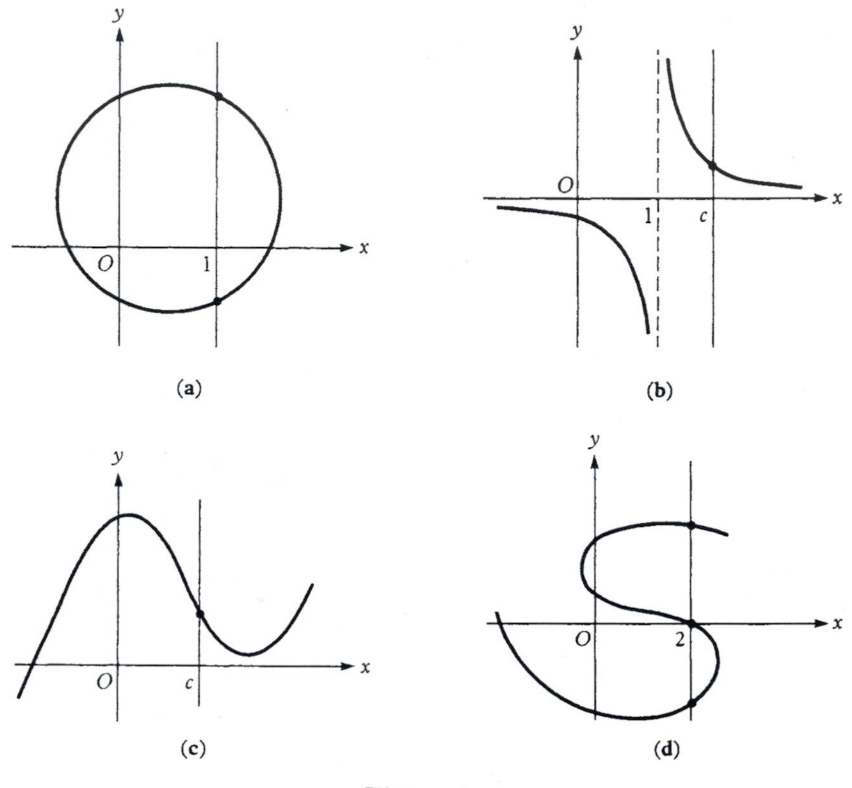

Figure 4

Solution:

(a) This is not the graph of a function, because the line $x = 1$ intersects the graph in two points.

(b) This is the graph of a function. Every vertical line $x = c \quad (c \neq 1)$ intersects the graph in exactly one point, and the vertical line $x = 1$ does not intersect the graph at all.

(c) This is the graph of a function. Every vertical line $x = c$ intersects the graph in exactly one point.

(d) This is not the graph of a function. The vertical line $x = 2$ intersects the graph in three points. ■

Algebra of Functions. Functions may be added, subtracted, multiplied, and divided in much the same way as numbers. Starting with two functions f and g, these operations produce new functions $f + g$, $f - g$, fg, and f/g that are defined as follows.

$(f+g)(x) = f(x) + g(x)$	**sum function**
$(f-g)(x) = f(x) - g(x)$	**difference function**
$(fg)(x) = f(x)g(x)$	**product function**
$(f/g)(x) = f(x)/g(x)$	**quotient function**

To be in the domain of $f+g$, a real number must be in both the domain of f and the domain of g (*why?*). The same principle holds for $f-g$, fg, and f/g. For f/g, we must also have $g(x) \neq 0$. Therefore,

$$D_{f+g} = D_f \cap D_g = D_{f-g} = D_{fg}$$

and

$$D_{f/g} = \text{all } x \text{ in } D_f \cap D_g \text{ for which } g(x) \neq 0.$$

Example 6 In each of the following cases, determine $(f+g)(x)$, $(f-g)(x)$, $(fg)(x)$, $(f/g)(x)$, and their respective domains.

(a) $f(x) = x^2 + 1, \quad g(x) = x - 2$

(b) $f(x) = \sqrt{x-3}, \quad g(x) = \sqrt{5-x}$

Solution:

(a) $(f+g)(x) = f(x) + g(x) = x^2 + 1 + x - 2 = x^2 + x - 1$

$(f-g)(x) = f(x) - g(x) = x^2 + 1 - (x-2) = x^2 - x + 3$

$(fg)(x) = f(x)g(x) = (x^2+1)(x-2) = x^3 - 2x^2 + x - 2$

$$(f/g)(x) = f(x)/g(x) = \frac{x^2+1}{x-2}$$

Now $D_f = D_g = (-\infty, \infty)$. Therefore, $D_{f+g} = (-\infty, \infty) = D_{f-g} = D_{fg}$. Also, $D_{f/g}$ = all x except 2, that is, $(-\infty, 2) \cup (2, \infty)$.

(b) $(f+g)(x) = \sqrt{x-3} + \sqrt{5-x}$

$(f-g)(x) = \sqrt{x-3} - \sqrt{5-x}$

$(fg)(x) = \sqrt{x-3}\,\sqrt{5-x}$

$$(f/g)(x) = \frac{\sqrt{x-3}}{\sqrt{5-x}}$$

Now D_f = all $x \geq 3 = [3, \infty)$, and D_g = all $x \leq 5 = (-\infty, 5]$. Hence, $D_{f+g} = D_f \cap D_g = [3,\infty) \cap (-\infty, 5] = [3, 5] = D_{f-g} = D_{fg}$. For f/g, we cannot have $x = 5$ (*why?*). Therefore, $D_{f/g} = [3, 5)$. ■

The algebraic operations can be applied to more than two functions. For example, if $f(x) = x^2$, $g(x) = 2x + 3$, $h(x) = x^2 - 1$, and $F(x) = x^2 + 1$, then

$$\frac{f(x)g(x) - h(x)}{F(x)} = \frac{x^2(2x+3) - (x^2-1)}{x^2+1} = \frac{2x^3 + 2x^2 + 1}{x^2+1}.$$

Caution When combining functions, it is important to keep in mind the restrictions placed on the domain of the combination. For example, if $f(x) = 2x^2$ and $g(x) = 1/x$, then $(fg)(x) = 2x$ for $x \neq 0$ but is *undefined* for $x = 0$. It is incorrect to simply say $(fg)(x) = 2x$. The graph of fg is the line $y = 2x$ with the origin (0, 0) removed. Similarly, if $f(x) = 1 + 1/(x - 3)$ and $g(x) = 1/(x - 3)$, then $(f - g)(x) = 1$ for $x \neq 3$ but is undefined for $x = 3$. Here the graph of $f - g$ is the horizontal line $y = 1$ with the point (3, 1) deleted.

Composition of Functions. Another way to combine functions involves following one by another. That is, *we let the output of one function serve as the input of another function.* For example, if $f(x) = x + 3$ and $g(x) = \sqrt{x}$, then f followed by g results in the function $\sqrt{x + 3}$ (Figure 5).

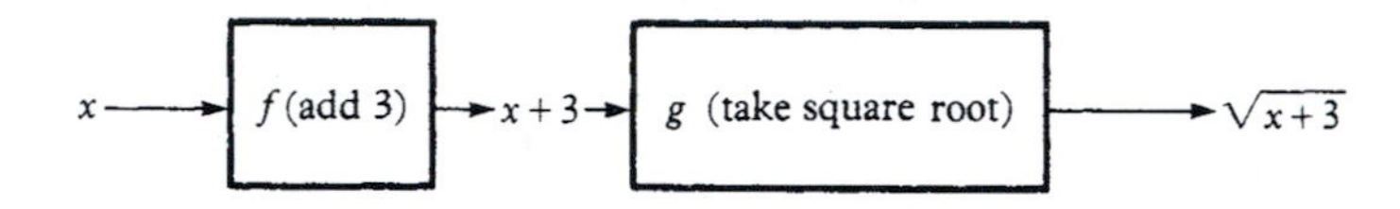

Figure 5 The Function $f(x) = x + 3$ followed by $g(x) = \sqrt{x}$

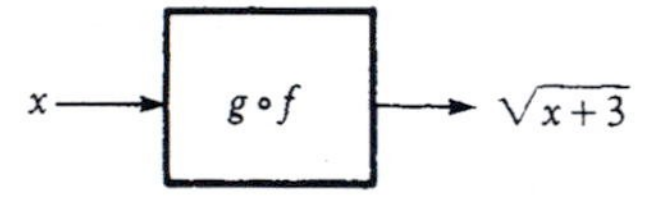

Figure 6 Composition $g \circ f$ for $f(x) = x + 3$ and $g(x) = \sqrt{x}$

In symbols, first $f(x) = x + 3$, then $g(f(x)) = g(x + 3) = \sqrt{x + 3}$. This combination of f and g is denoted by $g \circ f$ and called the *composition* of g with f (Figure 6). In general, if f and g are any two functions for which the range (output) of f is contained in the domain (input) of g, then the **composition of g with f** is defined by

$$(g \circ f)(x) = g(f(x)).$$

Example 7 Find $(g \circ f)(x)$ in each of the following cases.

(a) $f(x) = x^2, \quad g(x) = 3x + 5$

(b) $f(x) = 3x + 5, \quad g(x) = 1/x$

(c) $f(x) = -(x^2 + 1), \quad g(x) = \sqrt{x}$

Solution:

(a) $(g \circ f)(x) = g(f(x)) = g(x^2) = 3x^2 + 5$

(b) $(g \circ f)(x) = g(f(x)) = g(3x + 5) = \dfrac{1}{3x + 5}$

(c) The range of f is $R_f = (-\infty, -1]$, and the domain of g is $D_g = [0, \infty)$. Since R_f is not contained in D_g, the composition $g \circ f$ is not defined. [Note that if we formally apply the definition of composition, we obtain $(g \circ f)(x) = \sqrt{-(x^2 + 1)}$, which is meaningless in the real number system.] ■

Example 7(c) calls to our attention the importance of the domain in the definition of $g \circ f$. First x must be in D_f, then $f(x)$ must be in D_g. That is,

$$D_{g \circ f} \text{ consists of all } x \text{ in } D_f \text{ for which } f(x) \text{ is in } D_g.$$

In Example 7(c) $g \circ f$ is not defined because square roots of negative numbers do not exist in the real number system. Even when $g \circ f$ is defined, we must keep in mind any restrictions placed on its domain. For example, if $f(x) = \sqrt{x - 1}$ and $g(x) = x^2 + 3$, then $(g \circ f)(x) = x + 2$, but the domain of $g \circ f$ is only $[1, \infty)$ (*why?*). The graph of $g \circ f$ consists of that part of the line $y = x + 2$ above the interval $[1, \infty)$.

Example 8 Given $f(x) = 2x + 3$ and $g(x) = \sqrt{x}$, find (a) $(g \circ f)(x)$, (b) $(f \circ g)(x)$, and their respective domains.

Solution:

(a) $(g \circ f)(x) = g(2x + 3) = \sqrt{2x + 3}$
Now $D_f = (-\infty, \infty)$ and $D_g = [0, \infty)$. Therefore, $D_{g \circ f}$ consists of all real numbers x for which $f(x) = 2x + 3 \geq 0$, that is, $x \geq -3/2$ or $[-3/2, \infty)$.

(b) $(f \circ g)(x) = f(\sqrt{x}) = 2\sqrt{x} + 3$
Since D_g consists of all nonnegative real numbers and D_f contains all real numbers, it follows that $D_{f \circ g}$ consists of all nonnegative real numbers, that is, $[0, \infty)$. ■

Caution Example 8 shows that $g \circ f$ need not equal $f \circ g$; in other words, composition of functions is not a commutative operation.

Question 2 *Can* $g \circ f = f \circ g$ *for* some *functions* f *and* g?

Answer: Yes. For example, if $f(x) = 2x + 1$ and $g(x) = (x - 1)/2$, then $(g \circ f)(x) = x = (f \circ g)(x)$. ■

Exercises 4.1

Fill in the blanks to make each statement true.

1. A function from set X to set Y assigns to each member of X ______________ of set Y.
2. If f is a function from X to Y, then X is the ______________ of f.
3. If $f(x) = x^3$, then $f(2) =$ __________ and $f(-2) =$ __________ are numbers in the __________ of f.
4. The domain of fg is the set of numbers that are ______________.
5. To be in the domain of $g \circ f$, x must be in the domain of __________, and __________ must be in the domain of g.

Write true *or* false *for each statement.*

6. The domain and range of a function must be sets of real numbers.
7. If $f(x) = x^2 + 4$, then the domain of f is $(-\infty, \infty)$, and the range of f is $[0, \infty)$.
8. The point $(1, 5)$ belongs to the graph of $f(x) = x^2 + 4$.
9. Every vertical line $x = c$ intersects the graph of a function in exactly one point.
10. If x is in the domain of both f and g, then x is in the domain of f/g.

Functions

11. Given $f(x) = 2x^2 + 1$, find each value of f.
    (a) $f(0)$ (b) $f(2)$ (c) $f(-2)$
    (d) $f(\sqrt{2})$ (e) $f(-\sqrt{2})$ (f) $f(\pi)$
12. Given $g(x) = (3x + 1)/(5x - 2)$, find each value of g.
    (a) $g(6)$ (b) $g(2)$ (c) $g(-1)$
    (d) $g(1/3)$ (e) $g(1.5)$ (f) $g(-5/2)$
13. If $h(x) = \begin{cases} 1 & \text{for each rational number } x \\ -1 & \text{for each irrational number } x, \end{cases}$
    find each value of h.
    (a) $h(3)$ (b) $h(-3)$ (c) $h(2/3)$
    (d) $h(\sqrt{2}/3)$ (e) $h(\pi)$ (f) $h(3.14)$
14. If $F(x) = |x|/x$, find each value of F.
    (a) $F(2)$ (b) $F(10)$ (c) $F(-2)$
    (d) $F(-10)$ (e) $F(x)$ for $x > 0$ (f) $F(x)$ for $x < 0$

Find the implied domain of each of the following functions.

15. $f(x) = 2x - 5$
16. $g(x) = \dfrac{1}{2x - 5}$
17. $h(x) = \sqrt{2x - 5}$
18. $F(x) = \sqrt[3]{2x - 5}$
19. $G(x) = \dfrac{1}{\sqrt[3]{2x - 5}}$
20. $H(x) = \sqrt{x^2 + 1}$
21. $\alpha(x) = \sqrt{x^2 - 1}$
22. $\beta(x) = |x|$
23. $\gamma(x) = \dfrac{|x|}{x}$
24. $f_1(x) = \dfrac{1}{(x - 1)(x - 2)}$
25. $f_2(x) = \sqrt{(x - 1)(x - 2)}$

In Exercises 26–31, find all x in D_f satisfying $f(x) = 4$.

26. $f(x) = x - 4$
27. $f(x) = x^2$
28. $f(x) = x^2 - 5x$
29. $f(x) = x^2$; $D_f = [0, \infty)$
30. $f(x) = x^2$; $D_f = [-3, 3]$
31. $f(x) = x^2$; $D_f = [-1, 1]$

32. If $f(x) = x^2 + 2x + 3$, find each of the following.
    (a) $f(t)$ (b) $f(\sqrt{x})$ (c) $f(x^2)$
    (d) $f(x + 1)$ (e) $\dfrac{f(x) - f(1)}{x - 1}$ (f) $\dfrac{f(x + h) - f(x)}{h}$
33. If $g(x) = (x - 1)/(x + 1)$, find each of the following.
    (a) $g(2x)$ (b) $g(x^2)$ (c) $g(x + 1)$
    (d) $g(1/x)$ (e) $g(x + h)$ (f) $g(x - h)$

Decide whether each of the following equations can be put in the form $y = f(x)$, where f is a function. If it can be done, find $f(x)$.

34. $y - 3 = 4(x + 1)$
35. $y^2 - 3x - 1 = 0$
36. $x^2 + y^2 = 4$
37. $y^3 = x$
38. $|y| = x$
39. $xy - y = x^2$

Graphs of Functions

Sketch the graph of each of the following equations and determine whether it is the graph of a function $f(x)$.

40. $2x + 3y = 6$
41. $y - x^2 = 4$
42. $x^2 + y^2 = 9$
43. $x - y^2 = 4$
44. $x^2 - y^2 = 4$

Decide whether each of the following is the graph of a function $f(x)$.

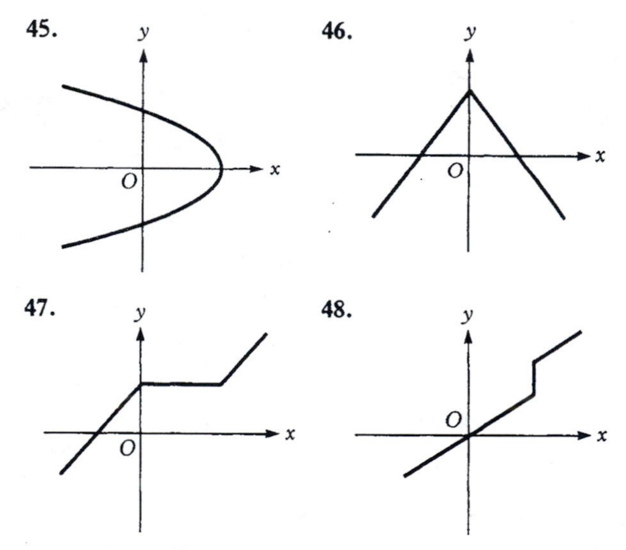

Algebra of Functions

In Exercises 49–56, find $f + g$, $f - g$, fg, and f/g, and determine their respective domains.

49. $f(x) = 2x$; $g(x) = 3x - 7$
50. $f(x) = 2x$; $g(x) = x^2 + 1$
51. $f(x) = x^2$; $g(x) = \sqrt{x}$
52. $f(x) = \dfrac{1}{x}$; $g(x) = \dfrac{1}{x - 2}$

53. $f(x) = x^3;\ g(x) = \dfrac{1}{x^3}$

54. $f(x) = \sqrt{x+1};\ g(x) = x^2$

55. $f(x) = \dfrac{x}{x^2+1};\ g(x) = x^2 + 1$

56. $f(x) = x^2;\ g(x) = |x|$

Composition of Functions

For each of the following, find $g \circ f$ and its domain.

57. $f(x) = 2x;\ g(x) = x^2$

58. $f(x) = x^2 + 1;\ g(x) = 2x$

59. $f(x) = x + 2;\ g(x) = x - 2$

60. $f(x) = \sqrt{x};\ g(x) = x^2$

61. $f(x) = x^2;\ g(x) = \sqrt{x}$

62. $f(x) = \dfrac{1}{x};\ g(x) = x$

63. $f(x) = x + 1;\ g(x) = \sqrt{x}$

64. $f(x) = x^2 + 1;\ g(x) = \sqrt{x}$

For each of the following, find $g \circ f$, $f \circ g$, and their respective domains.

65. $f(x) = x^2 - 4;\ g(x) = 1/x$

66. $f(x) = 3x - 2;\ g(x) = (x + 2)/3$

67. Let $f(x) = ax + b$ and $g(x) = cx + d$. Show that $(g \circ f)(x) = (f \circ g)(x) \Leftrightarrow ad + b = bc + d$.

Miscellaneous

In Exercises 68–77, write a formula for the indicated function.

68. $A(x)$ = the area of a square of side x

69. $A(r)$ = the area of a circle of radius r

70. $A(w)$ = the area of a rectangle whose length equals twice its width w

71. $V(s)$ = the volume of a cube of side s

72. $V(h)$ = the volume of a box whose width is half its height h and whose base is square

73. $C(x)$ = the cost of manufacturing x automobiles per day, if there is a fixed cost each day of \$5000 and the labor cost for each car is \$1500

74. $C(x)$ = the cost of renting a car per day, if the cost is \$20 per day plus 15 cents per mile for x miles

75. $S(x)$ = the surface area of a closed box whose base is a square of side x and whose height is twice its length

76. $A(x)$ = the area of a rectangular field to be fenced in with 3000 ft of fencing, if no fence is required on one side that is along a river and the other sides perpendicular to the river are each x ft long

77. $L(x)$ = the length of a fence around a rectangular field of length x ft and area 5000 sq ft

78. Given that $D(t) = 16t^2$ is the distance in feet that an object falls in t seconds, find each of the following.
   (a) distance the object falls in three seconds
   (b) distance the object falls in the third second
   (c) time it takes the object to reach the ground if it starts from 256 ft above the ground

79. For the distance function $D(t) = 16t^2$ given in Exercise 78, compute each of the following.
   (a) $\dfrac{D(4) - D(1)}{3}$ (b) $\dfrac{D(1 + \Delta t) - D(1)}{\Delta t}$ $(\Delta t \neq 0)$

80. Give a physical interpretation of each of the answers in Exercise 79.

4.2 Polynomial Functions

Definition of a Polynomial Function
Intercepts and Symmetry
Three-Step Procedure for Graphing Polynomial Functions

The most basic numbers are integers, and the most basic functions are polynomials.

Definition of a Polynomial Function. The simplest polynomial functions have already been discussed in Chapters 2 and 3, namely

$$f(x) = ax + b \qquad \textit{Linear functions}$$

and

$$f(x) = ax^2 + bx + c \quad (a \neq 0). \qquad \textit{Quadratic functions}$$

Their graphs are straight lines and parabolas, respectively. Here we discuss the general polynomial function and how to sketch its graph.

> The function f defined by
>
> $$f(x) = a_n x^n + a_{n-1} x^{n-1} + \ldots + a_1 x + a_0,$$
>
> where n is a nonnegative integer and $a_n \neq 0$, is called a **polynomial function of degree n.**

The domain of any polynomial function is the set of all real numbers. As indicated above, polynomial functions of degree 1 are linear and those of degree 2 are quadratic. If $n = 0$, then $f(x) = a_0$ and f is a nonzero **constant function.** We include the constant function $f(x) = 0$ with the polynomials, but we do not assign a degree to it. In general, any algebraic expression in x generated by addition, subtraction, and multiplication determines a polynomial function.

Example 1 Which of the following define polynomial functions?

(a) $f(x) = 3x^4 - 2x^3 + x - 1$ (b) $g(x) = 1 - 5x$

(c) $h(x) = \sqrt{x} + 10$ (d) $F(x) = \dfrac{3}{x} + x^2$

(e) $G(x) = \dfrac{2x + 5}{x^3 + 2x^2 + 1}$ (f) $H(x) = 6$

Solution: The functions in (a), (b), and (f) are polynomial functions of degrees 4, 1, and 0, respectively. (Note that $H(x) = 6x^0$.) The functions $h(x) = x^{1/2} + 10$ and $F(x) = 3x^{-1} + x^2$ in (c) and (d) are not polynomials since they involve fractional and negative exponents, respectively. Also, $G(x)$ in part (e) is not a polynomial but rather the quotient of two polynomials. ■

Intercepts and Symmetry. The most elementary way to draw the graph of a function f is to find a set of coordinates (x, y) satisfying the equation $y = f(x)$, plot the corresponding points, and then, if possible, draw a smooth curve through those points. It can be shown by methods of calculus that the graph of any polynomial function is *continuous* (has no breaks in it) and *smooth* (has no sharp turns).

Example 2 Plot the graph of $f(x) = 4 - x^2$.

Solution: First a table of coordinates is obtained by computing $y = 4 - x^2$ for $x = 0, \pm 1, \pm 2, \pm 3$; then a smooth curve is drawn through the corresponding seven points. Since f is a quadratic function, its graph is a parabola, as shown in Figure 7. ■

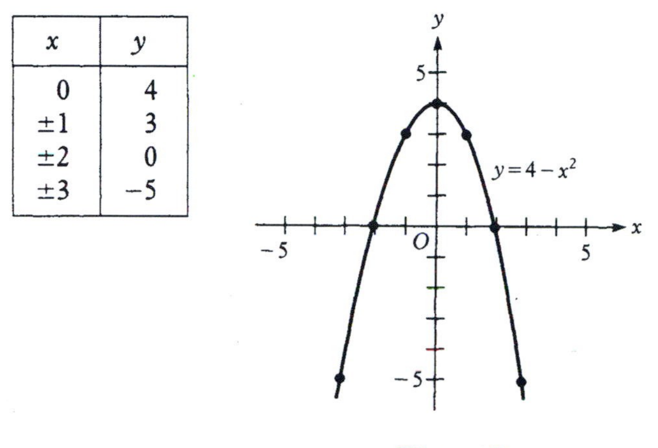

x	y
0	4
± 1	3
± 2	0
± 3	-5

Figure 7

Example 2 illustrates that it is desirable to include intercepts on the graph of a function. These are defined as follows.

> If the graph of $f(x)$ crosses the x-axis at $(a, 0)$, then the point $(a, 0)$ or the number a is called an **x-intercept.** If the graph crosses the y-axis at $(0, b)$, then the point $(0, b)$ or the number b is called a **y-intercept.**

We note that the graph of a function can have any number of x-intercepts, but at most one y-intercept. Example 2 also illustrates the simplification possible when a graph is symmetric with respect to the y-axis. Notice the pairs of points $(1, 3)$ and $(-1, 3)$, $(3, -5)$, and $(-3, -5)$ on the graph. In general, we have the following definition.

> If for every point (x, y) on a graph, the point $(-x, y)$ is also on the graph, then the graph is **symmetric with respect to the y-axis.**

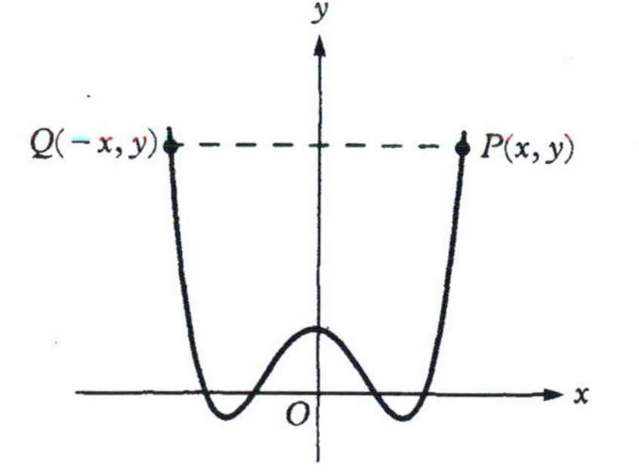

Figure 8 Symmetry with Respect to the y-axis

Figure 8 illustrates that when the graph of a function is symmetric with respect to the y-axis, then each point $P(x, y)$ on the graph can be paired with $Q(-x, y)$ also on the graph so that the y-axis is the perpendicular bisector of PQ.

Example 3 Sketch the graph of $F(x) = x^3 - 4x$.

Solution: First find intercepts. If $x = 0$, then $y = F(0) = 0$. Therefore, $(0, 0)$ is the y-intercept. If $y = 0$, then

$$\begin{aligned} x^3 - 4x &= 0 \\ x(x^2 - 4) &= 0 \\ x(x - 2)(x + 2) &= 0 \\ x &= 0,\ 2,\ -2. \end{aligned}$$

Therefore, $(0, 0)$, $(2, 0)$, and $(-2, 0)$ are the x-intercepts.

Now construct a table of coordinate values including the intercepts and several other points on the graph. Then plot the points from the table and draw a smooth curve through them (Figure 9). ■

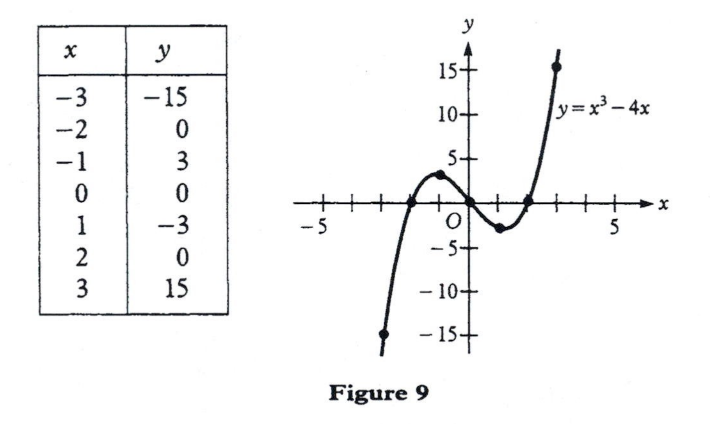

x	y
−3	−15
−2	0
−1	3
0	0
1	−3
2	0
3	15

Figure 9

The graph in Figure 9 illustrates symmetry with respect to the origin. Note that the pairs $(1, -3)$, $(-1, 3)$ and $(3, 15)$, $(-3, -15)$ are on the graph. In general, we have the following definition.

> If for every point (x, y) on a graph, the point $(-x, -y)$ is also on the graph, then the graph is **symmetric with respect to the origin.**

Figure 10 illustrates that when the graph of a function is symmetric with respect to the origin, then each point $P(x, y)$ on the graph can be paired with $Q(-x, -y)$ also on the graph so that the origin is the midpoint of the line segment PQ.

Some graphs exhibit **symmetry with respect to the x-axis.** That is, for each point $P(x, y)$ on the graph, the point $Q(x, -y)$ is also on the graph.

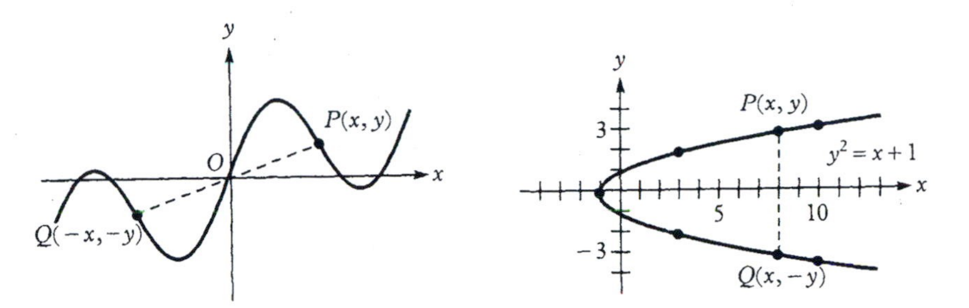

Figure 10 Symmetry with Respect to the Origin

Figure 11 Symmetry with Respect to the x-axis

For example, the graph of $y^2 = x + 1$ (Figure 11) is symmetric with respect to the x-axis. But *the graph of a function $y = f(x)$ cannot be symmetric with respect to the x-axis unless the graph lies entirely on the x-axis (why?).*

The previous examples illustrate how symmetry and intercepts can be utilized in graphing functions. We incorporate these two notions in the following systematic procedure for graphing any polynomial function.

Three-Step Procedure for Graphing Polynomial Functions

1. *Test for symmetry.*
   (a) The graph of $y = f(x)$ is symmetric with respect to the y-axis if and only if replacing x by $-x$ does not change the value of y. That is, $f(-x) = f(x)$. Such functions are called **even functions.** A polynomial function f is even and therefore its graph is symmetric with respect to the y-axis precisely when f contains only *even* powers of x.

   (b) The graph of $y = f(x)$ is symmetric with respect to the origin if and only if replacing x by $-x$ changes y to $-y$; that is, $f(-x) = -f(x)$. Such functions are called **odd functions.** A polynomial function f is odd and therefore its graph is symmetric with respect to the origin precisely when f contains only *odd* powers of x.

2. *Solve for intercepts.*
   (a) To find the x-intercepts, set $y = 0$ and solve for x, if possible.
   (b) To find the y-intercept, set $x = 0$ and solve for y.

In addition to steps 1 and 2, the following step is useful in graphing polynomial functions.

3. *Determine the sign of $f(x)$.*
   We saw in Chapter 3 that a quadratic function (parabola) can change sign only where its graph crosses the x-axis. More generally, if f is any polynomial function and hence has a continuous graph, then $f(x)$ can change sign only where $f(x) = 0$, that is, at the x-intercepts. These intercepts partition the x-axis into closed intervals, and inside each

interval the sign of $f(x)$ does not change. (We use closed intervals $[a, b]$ to distinguish them from points (a, b). Of course, $f(x) = 0$ at the endpoints a and b.)

As an illustration of step 3, we look back at Example 3, where $y = F(x) = x^3 - 4x$. Here the x-intercepts are -2, 0, and 2, which partition the x-axis into the four intervals $(-\infty, -2]$, $[-2, 0]$, $[0, 2]$, and $[2, \infty)$. In each of these, the sign of y can be determined by selecting a point x in the interval and computing $y = F(x)$. The sign of $F(x)$ behaves as shown in the following table, which agrees with the graph (Figure 9) for Example 3.

Interval	$(-\infty, -2]$	$[-2, 0]$	$[0, 2]$	$[2, \infty)$
Selected x	-3	-1	1	3
Corresponding y	-15	3	-3	15
Sign of $F(x) = x^3 - 4x$ in interval	$-$	$+$	$-$	$+$

Question 1 *How can the fact that the graph in Example 3 (Figure 9) is symmetric with respect to the origin be used to shorten the above procedure for determining the sign of $F(x)$?*

Answer: Because its graph is symmetric with respect to the origin, the sign of $F(x)$ inside an interval $[a, b]$ is the opposite of the sign at corresponding points in $[-b, -a]$. Hence, we need only determine the sign inside $[0, 2]$ and $[2, \infty)$ to conclude what the sign is inside $[-2, 0]$ and $(-\infty, -2]$. ■

Question 2 *Step 3 above says that if f changes sign at x, then $f(x) = 0$. Is the converse true; that is, if $f(x) = 0$, must f change signs at x?*

Answer: No. See Example 4 for an explanation. ■

Example 4 Sketch the graph of $f(x) = x^3 - 3x + 2 = (x - 1)^2(x + 2)$.

Solution:

1. *Test for symmetry.*
   $f(-x) = (-x)^3 - 3(-x) + 2 = -x^3 + 3x + 2$
   Since $f(-x) \neq f(x)$, f is not even and the graph is *not* symmetric with respect to the y-axis.
   Since $f(-x) \neq -f(x)$, f is not odd and the graph is *not* symmetric with respect to the origin.

2. *Solve for intercepts.*
   If $y = 0$, then $x^3 - 3x + 2 = (x - 1)^2(x + 2) = 0$. Therefore, $x = 1$ and $x = -2$ are the x-intercepts.
   If $x = 0$, then $y = 2$.
3. *Determine the sign of $f(x)$.*
   The x-intercepts -2 and 1 determine three intervals on the x-axis, in each of which the sign of y is indicated in the following chart.

Interval	$(-\infty, -2]$	$[-2, 1]$	$[1, \infty)$
Selected x	-3	-1	2
Corresponding y	-16	4	4
Sign of $f(x)$	$-$	$+$	$+$

We plot the intercepts $(1, 0)$, $(-2, 0)$, and $(0, 2)$, as well as the points $(-1, 4)$ and $(2, 4)$ from the chart, and connect them with a smooth curve (Figure 12). ■

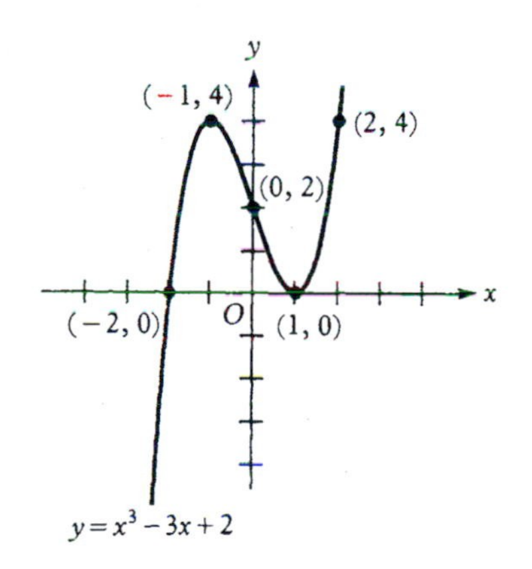

Figure 12

Example 5 Sketch the graph of $f(x) = x^6 - 4x^4 - x^2 + 4$.

Solution:

1. *Test for symmetry.*
   $f(-x) = (-x)^6 - 4(-x)^4 - (-x)^2 + 4$
   $= x^6 - 4x^4 - x^2 + 4 = f(x)$
   Therefore, the graph is symmetric with respect to the y-axis. Hence, it is sufficient to consider only $x \geq 0$ (*why?*).

2. *Solve for intercepts.*
   Setting $y = 0$, we obtain

$$\begin{aligned} x^6 - 4x^4 - x^2 + 4 &= x^4(x^2 - 4) - (x^2 - 4) \\ &= (x^4 - 1)(x^2 - 4) \\ &= (x^2 + 1)(x + 1)(x - 1)(x + 2)(x - 2) = 0. \end{aligned}$$

   Therefore, $x = -1, 1, -2$, and 2 are the x-intercepts.
   Setting $x = 0$, we obtain $y = 4$ as the y-intercept.

3. *Determine the sign of $f(x)$.*
   We consider only the intervals determined by the x-intercepts for $x \geq 0$.

Interval	[0, 1]	[1, 2]	[2, ∞)
Selected x	.5	1.5	3
Corresponding y	≈ 3.5	≈ −7.1	400
Sign of $f(x)$	+	−	+

We use the intercepts (1, 0), (2, 0), and (0, 4) and the points (.5, 3.5) and (1.5, −7.1) from the chart in order to plot the graph for $x \geq 0$, and then use the symmetry with respect to the y-axis to complete the graph for $x < 0$ (Figure 13). ■

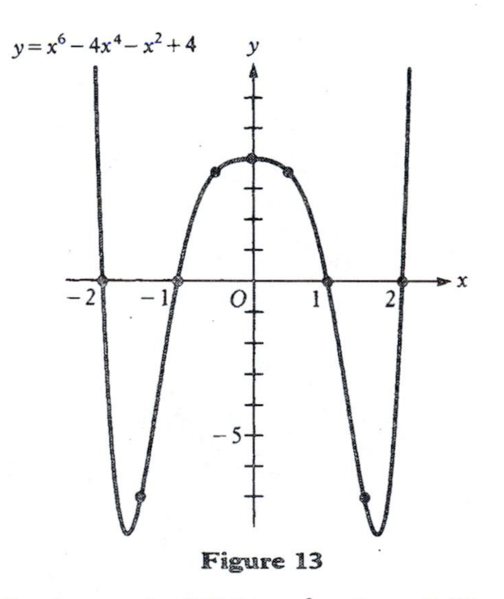

Figure 13

Comment On the graph of $f(x) = x^3 - 3x + 2$ (Example 4), the local "maximum" point (−1, 4) and the local "minimum" point (1, 0) were included because they happen to occur at integer values of x. Also, in the special case of quadratic functions (Example 2), the vertex of a parabola can be found by the methods of Section 3.6. But in general, such important points on graphs can be found only by calculus. That is the case in Examples 3 and 5. Also, when finding the x-intercepts of a polynomial function $y = f(x)$, we must solve the equation $f(x) = 0$ for its roots. We can certainly do this for polynomials of degree 1 or 2. Methods to determine or approximate the roots for polynomials of degree greater than 2 will be discussed in the chapter on roots of polynomial equations.

Exercises 4.2

Fill in the blanks to make each statement true.

1. A polynomial function of degree 4 is defined by $f(x) = a_4x^4 + a_3x^3 + a_2x^2 + a_1x + a_0$, provided __________.
2. The domain of any polynomial is __________.
3. If $f(-x) = f(x)$ for all x in the domain of f, then f is called an ________ function and its graph is __________.
4. For $f(x) = x(x + 2)(x - 3)(x^2 + 4)$, the x-intercepts are __________.
5. For $g(x) = x^3 - 3x^2 + 7x + 4$, the y-intercept is __________.

Write true *or* false *for each statement.*

6. The graph of every polynomial function has exactly one y-intercept.
7. The graph of every polynomial function has at least one x-intercept.
8. The graph of a quadratic polynomial is always symmetric with respect to a vertical line.
9. A graph that is symmetric with respect to the x-axis cannot be the graph of a function $y = f(x)$ unless $f(x) = 0$ for all x in D_f.
10. If $f(1) = 0$ and $f(3) = 0$, then $f(x)$ cannot change sign for $1 < x < 3$.

Polynomial Functions

11. Which of the following are polynomial functions?
    (a) $f(x) = x^{10}$
    (b) $g(x) = (x + 1)^{10}$
    (c) $h(x) = x^2 + 2x + 3\sqrt{x} + 4$
    (d) $F(x) = 3x^2 - x^{-2}$
    (e) $G(x) = x^{2.1} + 5$
    (f) $H(x) = \dfrac{1}{x^{-3} + 3}$
12. Which of the following are *not* polynomial functions?
    (a) $f_1(x) = (x + 3)^{1/3}$
    (b) $f_2(x) = (x + 4)^3$
    (c) $f_3(x) = (x + 4)^{-3}$
    (d) $f_4(x) = x^2(x + 1)^3$
    (e) $f_5(x) = x^{-2}(x + 1)^3$
    (f) $f_6(x) = 16$

Symmetry

13. Which of the following have graphs symmetric with respect to the y-axis?
    (a) $f(x) = x^2 - 1$
    (b) $g(x) = x^3 - 1$
    (c) $h(x) = x^6 - 1$
    (d) $F(x) = 4$
    (e) $G(x) = x^4 - 2x^2 + 1$
    (f) $H(x) = x^5 - 2x^3 + x$
14. Which of the functions in Exercise 13 are even? Which are odd?
15. Which of the following are odd functions?
    (a) $F_1(x) = 2x^3 + x$
    (b) $F_2(x) = 2x^3 + x + 1$
    (c) $F_3(x) = \sqrt[3]{x}$
    (d) $F_4(x) = (x^3 + x)^3$
    (e) $F_5(x) = (x^3 + x)^2$
    (f) $F_6(x) = 6$

16. Which of the functions in Exercise 15 has a graph that is symmetric with respect to the origin?

Intercepts

17. Find the x- and y-intercepts for the graphs of the following functions.
    (a) $f(x) = x^2 + 2x + 1$
    (b) $g(x) = (x - 1)^2$
    (c) $h(x) = 7$
    (d) $F(x) = (x - 1)^{-1}$
    (e) $G(x) = x^2 - 4$
    (f) $H(x) = x^2 + 4$
18. Find x- and y-intercepts for the graphs of the following functions.
    (a) $f_1(x) = x(x - 3)(x^2 - 3)(x^3 - 1)$
    (b) $f_2(x) = (x + 3)(x^2 + 3)(x^3 + 3)$
    (c) $f_3(x) = x^3 - 6x^2 + 11x - 6$
    (d) $f_4(x) = x^5 + x^4 - 16x - 16$

Graphs of Polynomial Functions

Using the three-step procedure, sketch the graphs of the following polynomial functions.

19. $f(x) = 9 - x^2$
20. $g(x) = x^2 - 4x + 3$
21. $h(x) = x^3 - 3x^2$
22. $F(x) = x^3 - 3x$
23. $G(x) = (x^2 - 1)^2$
24. $H(x) = x^4 + 2x^3$
25. $p(x) = x^4 - 2x^2$
26. $q(x) = 3x^4 - 4x^3$
27. $r(x) = x^4 - 4$
28. $s(x) = x^5 + x^4 - 16x - 16$ (*Hint:* factor by grouping.)
29. $t(x) = x^3 - 6x^2 + 11x - 6$ (*Hint:* $x - 1$ is a factor.)
30. $u(x) = x^3 - 3x^2 - 9x + 11$ (*Hint:* $x - 1$ is a factor.)

4.3 Rational Functions

Rational functions are related to polynomials just as rational numbers are related to integers.

Definition of a Rational Function. A **rational function** is the ratio of two polynomial functions. That is, f is a rational function if defined by

$$f(x) = \frac{P(x)}{Q(x)},$$

where $P(x)$ and $Q(x)$ are polynomials. Any algebraic expression in x generated by addition, subtraction, multiplication, and division determines a rational function. The class of rational functions includes the polynomial functions (*why?*).

Example 1 Which of the following are rational functions?

(a) $f(x) = \dfrac{x^3 + 6}{1 - 2x}$ (b) $g(x) = \dfrac{\sqrt{2}}{x^2 + 1}$

(c) $h(x) = \dfrac{x^{1/2} + x^{-1/2}}{x^2 - 1}$ (d) $F(x) = \dfrac{x^{-2} - x^{-3}}{x^2 + x^3}$

Solution: The functions in (a), (b), and (d) are rational functions. Note that

$$F(x) = \frac{\frac{1}{x^2} - \frac{1}{x^3}}{x^2 + x^3} = \frac{\frac{x-1}{x^3}}{x^2 + x^3} = \frac{x - 1}{x^5 + x^6}.$$

The function in (c) is not a rational function because the numerator, $x^{1/2} + x^{-1/2}$, has terms with fractional exponents. ■

While the domain of a polynomial function is $(-\infty, \infty)$, a rational function is undefined wherever its denominator is zero.

Example 2 Find the domain of each rational function in Example 1.

Solution:

(a) $f(x) = (x^3 + 6)/(1 - 2x)$ is undefined at $x = 1/2$. $D_f = (-\infty, 1/2) \cup (1/2, \infty)$.

(b) $g(x) = \sqrt{2}/(x^2 + 1)$ is defined for every real number since the equation $x^2 + 1 = 0$ has no real solution. Hence, $D_g = (-\infty, \infty)$.

(c) $h(x) = (x^{1/2} + x^{-1/2})/(x^2 - 1)$ is not a rational function.

(d) $F(x) = (x - 1)/[x^5(x + 1)]$ is undefined at $x = 0$ and $x = -1$. Therefore, $D_F = (-\infty, -1) \cup (-1, 0) \cup (0, \infty)$. ■

Vertical Asymptotes. The techniques used in graphing polynomial functions [symmetry, intercepts, and sign of $f(x)$] are also useful for rational functions, but in addition, it is necessary to pay special attention to the values of x for which the denominator is zero. In the following definition, we assume that $P(x)$ and $Q(x)$ have no zeros in common.

> If $Q(a) = 0$, then the vertical line $x = a$ is called a **vertical asymptote** for the rational function $f(x) = P(x)/Q(x)$.

In the following example, we examine the behavior of the graph of a rational function near a vertical asymptote.

Example 3 For $f(x) = 2/(x - 1)$, find any vertical asymptote and sketch the graph of f near that line.

Solution: The vertical line $x = 1$ is the only vertical asymptote for $f(x)$. The table shows values of x approaching 1 and the corresponding values of y.

x	$f(x)$	x	$f(x)$
0	-2	2	2
.9	-20	1.1	20
.99	-200	1.01	200
.999	-2000	1.001	2000
.9999	$-20{,}000$	1.0001	20,000
$\downarrow$	$\downarrow$	$\downarrow$	$\downarrow$
1^-	$-\infty$	1^+	∞

As x approaches 1 from the left (written: $x \to 1^-$), $f(x)$ is negative but $|f(x)|$ gets larger and larger $[f(x) \to -\infty]$. As x approaches 1 from the right $(x \to 1^+)$, $f(x)$ is positive and gets larger and larger $[f(x) \to \infty]$. Hence, the graph of f falls as it approaches the line $x = 1$ from the left and rises as it approaches $x = 1$ from the right (Figure 14). ■

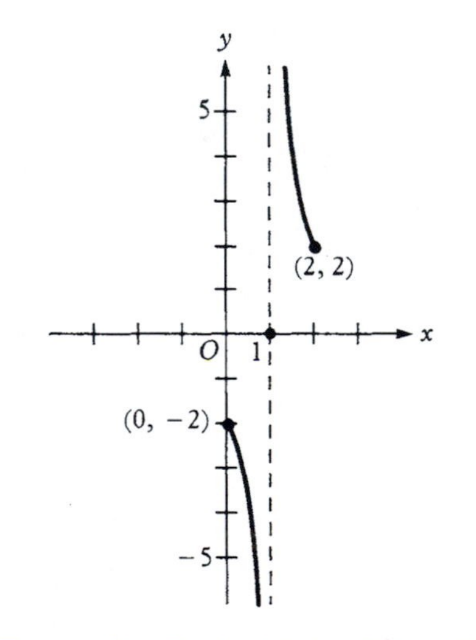

Figure 14 Graph of $f(x) = 2/(x - 1)$ near Vertical Asymptote $x = 1$

Example 3 illustrates the following **characteristic property of a vertical asymptote.**

> The line $x = a$ is a vertical asymptote for f if and only if $|f(x)|$ increases without bound as x approaches the finite value a.

Horizontal Asymptotes. We now consider the possibility that as $|x|$ increases without bound, $y = f(x)$ may approach a finite value b.

Example 3 (continued) Complete the graph of $f(x) = 2/(x - 1)$.

Solution: We consider values of x that approach ∞ and $-\infty$ and the corresponding values of y, as shown in the table.

x	$y = f(x)$	x	$y = f(x)$
3	1	-1	-1
11	.2	-9	$-.2$
101	.02	-99	$-.02$
1001	.002	-999	$-.002$
10,001	.0002	$-9{,}999$	$-.0002$
$\downarrow$	$\downarrow$	$\downarrow$	$\downarrow$
∞	0^+	$-\infty$	0^-

As x approaches ∞, $f(x)$ is positive and approaches 0 ($y \to 0^+$). As x approaches $-\infty$, y is negative and approaches 0 ($y \to 0^-$). Hence, the graph of f approaches the x-axis ($y = 0$) as x increases in magnitude (Figure 15). The line $y = 0$ is called a horizontal asymptote. ■

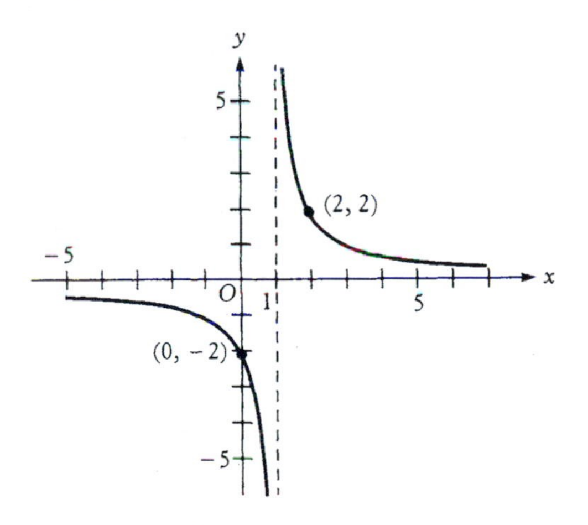

Figure 15 Graph of $f(x) = 2/(x - 1)$ Showing Vertical Asymptote $x = 1$ and Horizontal Asymptote $y = 0$

Example 3 illustrates the following definition and property of a horizontal asymptote.

> The horizontal line $y = b$ is a **horizontal asymptote** for f if and only if $f(x)$ approaches the finite value b as x approaches ∞ or as x approaches $-\infty$.

As in Example 3, we will see that $y = 0$ is a horizontal asymptote for a rational function whenever the degree of the numerator is *less than* the degree of the denominator. We now consider an example in which the degree of the numerator is *equal* to the degree of the denominator.

Example 4 Find any horizontal asymptote and sketch the graph for $f(x) = 2x/(x - 1)$.

Solution: Although we could construct a table similar to the one in Example 3, it is more efficient to write $f(x)$ in a different form by dividing the numerator and denominator by x, the highest power occurring in $f(x)$.

$$f(x) = \frac{\frac{2x}{x}}{\frac{x-1}{x}} = \frac{2}{1 - \frac{1}{x}} \qquad (x \neq 0)$$

Now as $|x| \to \infty$, we have $1/x \to 0$ so that $f(x) \to 2/1 = 2$. Hence the line $y = 2$ is a horizontal asymptote. By observing that $x = 1$ is a vertical asymptote and plotting a few points as shown in the table, we can easily draw the graph (Figure 16). ■

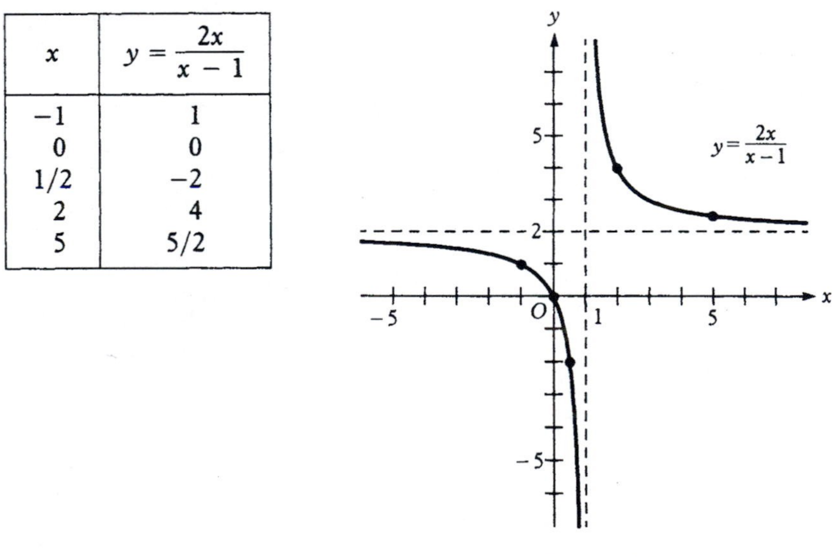

x	$y = \frac{2x}{x-1}$
-1	1
0	0
1/2	-2
2	4
5	5/2

Figure 16

In our next example, the degree of the numerator is *greater than* that of the denominator.

Example 5 Sketch the graph of $f(x) = (x^2 + 1)/x$.

Solution: $f(x) = \frac{x^2 + 1}{x} = x + \frac{1}{x} \quad (x \neq 0)$

Since $1/x \to 0$ as $|x| \to \infty$, we conclude that $f(x)$ is approximately equal to x for large values of x. The line $y = x$ is called an **oblique asymptote** for f. Hence, $f(x)$ does not approach any constant b as x increases in magnitude, and therefore $f(x)$ has *no* horizontal asymptote. By noting that the line $x = 0$ (the y-axis) is a vertical asymptote for $y = (x^2 + 1)/x$ and constructing the following table of values of x with corresponding values of y, we obtain the graph of f (Figure 17).

x	y	x	y
1/4	17/4	−1/4	−17/4
1/2	5/2	−1/2	−5/2
1	2	−1	−2
2	5/2	−2	−5/2
3	10/3	−3	−10/3
4	17/4	−4	−17/4

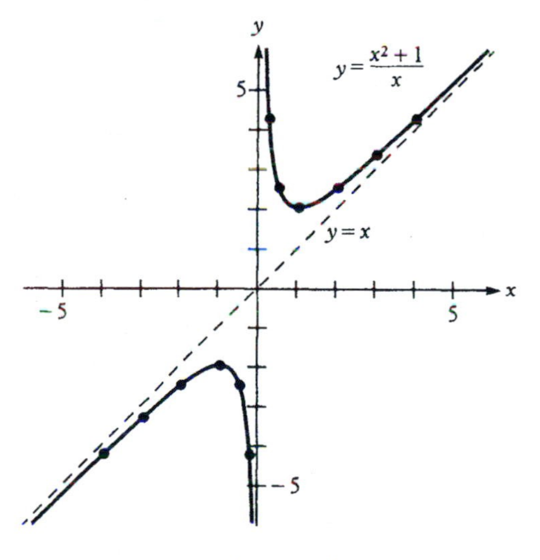

Figure 17

Note that only half of the table is really needed, since the graph is symmetric with respect to the origin. That is,

$$f(-x) = \frac{(-x)^2 + 1}{-x} = -\frac{x^2 + 1}{x} = -f(x). \quad ■$$

Examples 3, 4, and 5 illustrate the following result concerning **horizontal asymptotes.**

Let $f(x) = \dfrac{a_m x^m + a_{m-1}x^{m-1} + \ldots + a_1 x + a_0}{b_n x^n + b_{n-1}x^{n-1} + \ldots + b_1 x + b_0}$

be a rational function, where $a_m \neq 0$ and $b_n \neq 0$. Then

1. if $m < n$, the line $y = 0$ (x-axis) is a horizontal asymptote for f;
2. if $m = n$, the line $y = a_m/b_n$ is a horizontal asymptote for f;
3. if $m > n$, f has no horizontal asymptote.

Note that a rational function of x can have at most one horizontal asymptote. The above result for horizontal asymptotes need not be memorized. For any particular rational function $f(x) = P(x)/Q(x)$, we can determine the existence of a horizontal asymptote by *first dividing $P(x)$ and $Q(x)$ by the highest power of x that is present in $f(x)$, and then examining the resulting expression as $|x| \to \infty$*. To see how this works, we will try a few examples.

1. $\dfrac{3x^2 + 2x - 4}{5x^3 + 1} = \dfrac{\dfrac{3}{x} + \dfrac{2}{x^2} - \dfrac{4}{x^3}}{5 + \dfrac{1}{x^3}}$ *Divide numerator and denominator by x^3*

   As $|x| \to \infty$, the value of the resulting expression above approaches

   $$\frac{0 + 0 - 0}{5 + 0} = 0,$$

   so $y = 0$ is a horizontal asymptote.

2. $\dfrac{3x^2 + 2x - 4}{5x^2 + 1} = \dfrac{3 + \dfrac{2}{x} - \dfrac{4}{x^2}}{5 + \dfrac{1}{x^2}}$ *Divide numerator and denominator by x^2*

   As $|x| \to \infty$, the value of the resulting expression above approaches

   $$\frac{3 + 0 - 0}{5 + 0} = \frac{3}{5},$$

   so $y = 3/5$ is a horizontal asymptote.

3. $\dfrac{3x^2 + 2x - 4}{5x + 1} = \dfrac{3 + \dfrac{2}{x} - \dfrac{4}{x^2}}{\dfrac{5}{x} + \dfrac{1}{x^2}}$ *Divide numerator and denominator by x^2*

   As $|x| \to \infty$, the value of the resulting expression above approaches

   $$\frac{3 + 0 - 0}{0 + 0}, \qquad \textit{Not defined}$$

   so there is no horizontal asymptote.

By including asymptotes along with the items previously considered for polynomials, we arrive at the following step-by-step method for graphing a rational function $f(x) = P(x)/Q(x)$. We assume that $P(x)$ and $Q(x)$ have no zeros in common.

Four-Step Procedure for Graphing Rational Functions

1. *Test for symmetry.*
   Test for possible symmetry with respect to the y-axis or the origin by computing $f(-x)$. If either type of symmetry applies, it is sufficient to consider only $x \geq 0$.
2. *Solve for x-intercepts and vertical asymptotes.*[1]
   Find the zeros of the numerator $P(x)$ to obtain the x-intercepts (if any). Find the zeros of the denominator $Q(x)$ to obtain the vertical asymptotes (if any).
3. *Determine the sign of $f(x)$.*
   The values of x determining x-intercepts and vertical asymptotes partition the x-axis into intervals, and inside each of these intervals the sign of $f(x)$ does not change. Determine the sign of $f(x)$ for a selected value of x in each interval.
4. *Solve for y-intercept and horizontal asymptote.*
   (a) Set $x = 0$ to find the y-intercept. (If 0 is not in D_f, then the line $x = 0$ is a vertical asymptote.)
   (b) Determine the horizontal asymptote, if any, by comparing the degree of $P(x)$ to that of $Q(x)$ or by considering the behavior of y as $|x| \to \infty$.

   To draw the graph, we first draw broken lines for any asymptotes, horizontal or vertical. Next we plot all intercepts and the points determined in each interval by step 3. Finally, we join the points by a smooth curve broken only at the vertical asymptotes and approaching any horizontal asymptote as $|x| \to \infty$.

Example 6 Sketch the graph of $f(x) = x^2/(1 - x^2)$ by using the four-step procedure for graphing rational functions.

Solution:

1. *Test for symmetry.*

$$f(-x) = \frac{(-x)^2}{1-(-x)^2} = \frac{x^2}{1-x^2} = f(x)$$

   Therefore, f is an even function and the graph is symmetric with respect to the y-axis.

[1]Methods for locating the zeros of polynomials that are of degree greater than 2 and are not easily factored are discussed in a later chapter.

2. *Solve for x-intercepts and vertical asymptotes.*
   If the numerator $x^2 = 0$, then $x = 0$. Hence, $(0, 0)$ is the only x-intercept.
   If the denominator $1 - x^2 = 0$, then $x = 1$ or $x = -1$. Therefore, each of the lines $x = 1$ and $x = -1$ is a vertical asymptote.
3. *Determine the sign of $f(x)$.*
   Because of the symmetry with respect to the y-axis, it is necessary to consider only the intervals $[0, 1]$ and $[1, \infty)$ on the x-axis. (See the table.) Symmetry with respect to the y-axis immediately yields the point $(-1/2, 1/3)$ in $[-1, 0]$ and the point $(-2, -4/3)$ in $(-\infty, -1]$.
4. *Solve for y-intercept and horizontal asymptote.*
   0 is in D_f and $(0, 0)$ is the y-intercept (as well as the only x-intercept). Since the numerator x^2 and the denominator $1 - x^2$ are both of degree 2, the horizontal asymptote is $y = 1/(-1) = -1$. Alternatively, we note that

$$f(x) = \frac{x^2}{1 - x^2} = \frac{1}{\dfrac{1}{x^2} - 1} \to \frac{1}{0 - 1} = -1 \text{ as } |x| \to \infty.$$

To draw the graph (Figure 18), first draw broken lines for the vertical asymptotes $x = 1$ and $x = -1$ and for the horizontal asymptote $y = -1$.

Interval	[0, 1]	[1, ∞]
Selected x	1/2	2
Corresponding y	1/3	−4/3
Sign of $f(x)$	+	−

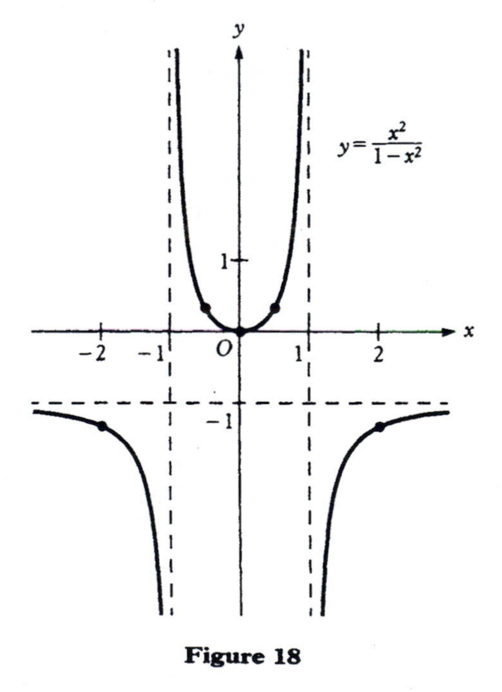

Figure 18

Next, plot the points (0, 0), (1/2, 1/3), (2, −4/3), (−1/2, 1/3), and (−2, −4/3). Finally, keeping in mind symmetry with respect to the y-axis and the sign of $f(x)$, join the points by a smooth curve broken at $x = 1$ and $x = -1$ and approaching $y = -1$ as $|x| \to \infty$. ■

Comment As illustrated in Example 5, it is sometimes necessary to compute more points on the graph than those determined by "intercepts" and "sign of $f(x)$." Also, it is sometimes helpful to write the given expression for $f(x)$ in a different form. For instance, the function in Example 5 was written as

$$f(x) = x + \frac{1}{x}.$$

In this form we saw that for x large in magnitude, the term $1/x$ is relatively insignificant and the function behaves like

$$g(x) = x.$$

On the other hand, for x small in magnitude, x becomes the insignificant term and the function behaves like

$$h(x) = \frac{1}{x}.$$

Similarly, we can decompose the function in Example 6 into a sum of simpler functions as follows:

$$f(x) = \frac{x^2 - 1 + 1}{1 - x^2} \qquad \textit{Subtract and add } -1 \textit{ in the numerator}$$

$$= \frac{x^2 - 1}{1 - x^2} + \frac{1}{1 - x^2}$$

$$= -1 + \frac{1}{1 - x^2}$$

$$= \underbrace{-1}_{\substack{\text{Most important term for } x \\ \text{large in magnitude}}} + \underbrace{\frac{1}{2}\cdot\frac{1}{1 - x}}_{\substack{\text{Most important term} \\ \text{for } x \text{ close to } 1}} + \underbrace{\frac{1}{2}\cdot\frac{1}{1 + x}}_{\substack{\text{Most important term} \\ \text{for } x \text{ close to } -1}} . \qquad \textit{See the section on partial fractions}$$

In each example, the decomposition of $f(x)$ enables us to focus on a simple term that is the most significant contributor to the behavior of $f(x)$, depending on the location of x. From this point of view, go back and examine the graph in Example 6 and see if it becomes easier to understand.

Exercises 4.3

In Exercises 1–10, assume $P(x)$ and $Q(x)$ have no common zeros.

Fill in the blanks to make each statement true.

1. A rational function is the ratio of ____________.
2. The line $x = a$ is a vertical asymptote for the graph of $f(x) = P(x)/Q(x)$ if ____________.
3. The graph of $f(x) = (3x^2 + 2x - 1)/(2x^2 + 3x + 1)$ has the horizontal asymptote $y =$ ________.
4. The x-axis is a horizontal asymptote for the graph of a rational function $f(x) = P(x)/Q(x)$ if

   ____________.
5. The graph of the rational function $f(x) = P(x)/Q(x)$ has no horizontal asymptote if ____________.

Write true *or* false *for each statement.*

6. A polynomial function is a rational function.
7. The domain of a rational function may be $(-\infty, \infty)$.
8. The graph of every rational function has at least one vertical asymptote.
9. The sign of a rational function $f(x) = P(x)/Q(x)$ always changes where $Q(x) = 0$.
10. For a rational function $f(x) = P(x)/Q(x)$, the roots of $P(x) = 0$ are x-intercepts.

Rational Functions

11. Which of the following are rational functions?
    (a) $f(x) = x^2 + 3x - 1$
    (b) $g(x) = \dfrac{x^2 + 3x - 1}{x^3 + 4}$
    (c) $h(x) = x^{-2} + 3x^{-1} + x^2$
    (d) $F(x) = (x^2 + 4)^{-3}$
    (e) $G(x) = (x^2 + 4)^{1/3}$
    (f) $H(x) = \sqrt[3]{2}$
12. Which of the following are not rational functions and why?
    (a) $F_1(x) = x + \sqrt{2}$
    (b) $F_2(x) = \sqrt{x} + 2$
    (c) $F_3(x) = \dfrac{x + \sqrt{2}}{x - \sqrt{2}}$
    (d) $F_4(x) = \dfrac{x^{-2}}{x^{-3} + 1}$
    (e) $F_5(x) = \dfrac{x^{-1/2}}{x^{-3/2} + 1}$
    (f) $F_6(x) = \dfrac{4}{x + 1}$

Vertical Asymptotes

For each of the rational functions in Exercises 13–20, find the equations of the vertical asymptotes, if any.

13. $f(x) = \dfrac{x^2}{x + 1}$
14. $g(x) = \dfrac{x^2}{x^2 + 1}$
15. $h(x) = \dfrac{x^3}{x^2 - 1}$
16. $p(x) = \dfrac{x^2}{x^2 - 4x + 3}$
17. $q(x) = \dfrac{x^2}{x^2 + 4x + 3}$
18. $r(x) = \dfrac{x^2}{x^2 + 3x + 4}$
19. $s(x) = \dfrac{x^2}{x^3 - 1}$
20. $t(x) = \dfrac{x^3}{x^3 - 6x^2 + x - 6}$

Horizontal Asymptotes

For each of the rational functions in Exercises 21–28, find the equation of the horizontal asymptote, if any.

21. $f(x) = \dfrac{2x}{x + 1}$
22. $g(x) = \dfrac{2x}{x^2 + 1}$
23. $r(x) = \dfrac{3x^2 - 4x + 5}{1 - 2x^2}$
24. $h(x) = \dfrac{2x^2}{x + 1}$
25. $s(x) = 3x^2 - 4x + 5$
26. $t(x) = \dfrac{3x^2 - 4x + 5}{1 - 2x^3}$

27. $u(x) = \dfrac{3x^2 - 4x + 5}{1 - 2x}$

28. $v(x) = x^2 - \dfrac{1}{x + 1}$

Graphs of Rational Functions

Use the four-step procedure discussed in this section to sketch a graph for each of the following rational functions.

29. $f(x) = \dfrac{2}{x + 1}$

30. $g(x) = \dfrac{2x}{x + 1}$

31. $h(x) = \dfrac{1 - x}{1 + x}$

32. $F(x) = \dfrac{1}{x^2}$

33. $G(x) = \dfrac{1}{(x + 1)^2}$

34. $H(x) = \dfrac{1}{(x - 1)^2}$

35. $p(x) = \dfrac{x^2}{x^2 - 1}$

36. $q(x) = x - \dfrac{1}{x}$

37. $r(x) = \dfrac{2x}{x - 3}$

38. $s(x) = \dfrac{x}{x^2 + 1}$

39. $t(x) = \dfrac{x}{x^2 - 1}$

40. $u(x) = \dfrac{4 - x}{x - 2}$

4.4 Algebraic Functions

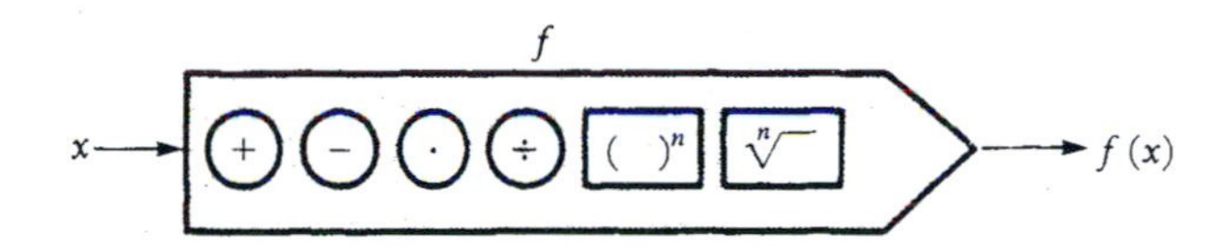

Definition of an Algebraic Function
Graphing Algebraic Functions
Piecewise Algebraic Functions

Definition of an Algebraic Function. A function obtained by the algebraic operations of addition, subtraction, multiplication, division, and the taking of roots is called an **algebraic function.**[1] The rational functions are included among the algebraic ones, but some algebraic functions, such as $f(x) = \sqrt{x}$ and $g(x) = \sqrt[3]{(x - 1)/(x + 2)}$, are not rational.

Example 1 Which of the following are algebraic functions?

(a) $f(x) = (x - 3)^2$

(b) $g(x) = (x - 3)^x$

(c) $h(x) = x^2 + \dfrac{2}{x} + \sqrt{\dfrac{x + 2}{x - 5}}$

(d) $F(x) = \sqrt[3]{x^{1/2} + x^{2/3} + 1}$

(e) $G(x) = x^{3.14} + x^{1.414}$

(f) $H(x) = x^{\pi} + x^{\sqrt{2}}$

Solution: First note that if m/n is a rational number, then $x^{m/n}$ is an algebraic expression because $x^{m/n} = \sqrt[n]{x^m}$. On the other hand, if r is an irrational number, then x^r is *not* an algebraic expression. Therefore, the functions in (a), (c), (d), and (e) are algebraic functions, while those in (b) and (f) are not. ■

[1]More generally, if y satisfies an equation

$$P_n(x)y^n + P_{n-1}(x)y^{n-1} + \cdots + P_1(x)y + P_0(x) = 0,$$

where the coefficients $P_0(x), P_1(x), \ldots, P_{n-1}(x), P_n(x)$ are polynomials in x, then y is called an algebraic function of x.

Two limitations are placed on the domain of an algebraic function. First, if the function contains a term with a denominator, then the function is undefined wherever the denominator is zero. Second, if the function contains a term that is an even root, then the function is undefined wherever the radicand is negative.

Example 2 Find the domain of each of the following algebraic functions.

(a) $f(x) = \sqrt{x - 2}$ (b) $g(x) = \sqrt[3]{x} + \dfrac{1}{\sqrt{x + 4}}$

(c) $F(x) = x^{2/3} + x^{3/2}$ (d) $G(x) = x^{1.2} - x^{2.1}$

Solution:

(a) $\sqrt{x - 2}$ is defined for $x - 2 \geq 0$ or $x \geq 2$. Therefore, $D_f = [2, \infty)$.

(b) $\sqrt[3]{x}$ is defined for all real numbers x. However, $1/\sqrt{x + 4}$ is defined and its denominator is not zero only for $x + 4 > 0$ or $x > -4$. Therefore, $D_g = (-4, \infty)$.

(c) $x^{2/3}$ is defined for all real numbers x, and $x^{3/2}$ is defined for $x \geq 0$. Therefore, $D_F = [0, \infty)$.

(d) As a rational number, $1.2 = 12/10 = 6/5$. Hence $x^{1.2} = \sqrt[5]{x^6}$ is defined for all real numbers x. However, $2.1 = 21/10$, which means that $x^{2.1} = \sqrt[10]{x^{21}}$ is defined only for $x \geq 0$. Therefore, $D_G = [0, \infty)$. ■

Graphing Algebraic Functions. We can apply the notions of symmetry, intercepts, asymptotes, and sign of $f(x)$ to the graphing of algebraic functions. However, special attention must be paid to the limitations placed on the *domain* of an algebraic function. Also, for algebraic functions, we define both vertical and horizontal asymptotes in terms of their characteristic properties.

> The vertical line $x = a$ is a **vertical asymptote** for an algebraic function $f(x)$ if $|f(x)| \to \infty$ as $x \to a^+$ or a^-. The horizontal line $y = b$ is a **horizontal asymptote** for $f(x)$ if $f(x) \to b$ as $x \to \infty$ or $-\infty$.

These definitions are consistent with those given for rational functions, and our methods for finding asymptotes for rational functions can be adapted to general algebraic functions. We now consider the graphs of some algebraic functions, starting with a very simple case and working up to more complicated ones.

Example 3 Sketch the graph of $f(x) = \sqrt{x}$.

Solution: $\sqrt{x}$ is defined only for $x \geq 0$; also, $\sqrt{x} \geq 0$ so the graph is contained in the first quadrant. The only intercept is (0, 0), and there are no asymptotes since $f(x) \to \infty$ if and only if $x \to \infty$. The graph [Figure 19(a)] is the top half of the parabola $y^2 = x$ [Figure 19(b)]. ■

x	$f(x)$
0	0
1	1
4	2
9	3
16	4

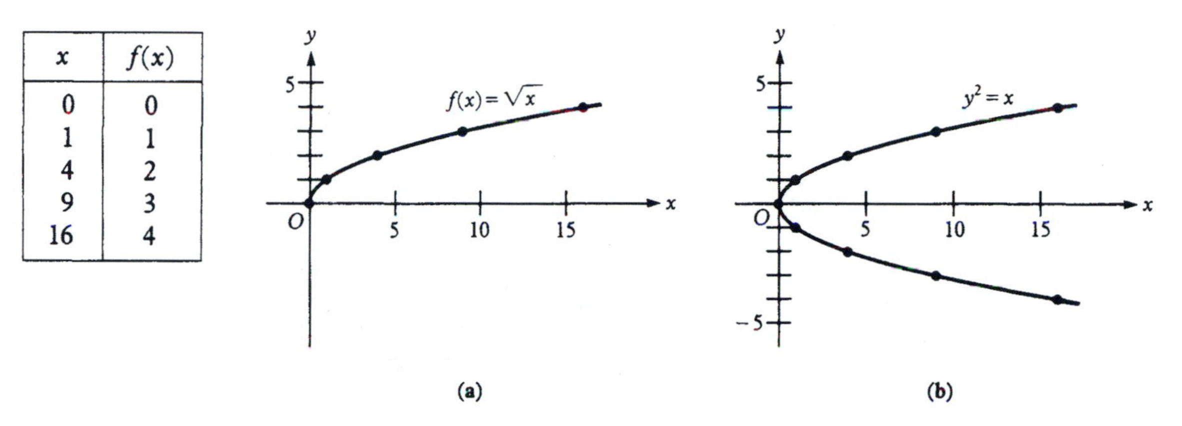

Figure 19

Example 4 Sketch the graph of $g(x) = \sqrt[3]{x}$.

Solution:
Domain: $(-\infty, \infty)$
Symmetry: $g(-x) = \sqrt[3]{-x} = -\sqrt[3]{x} = -g(x)$, so the graph is symmetric with respect to the origin.
Intercepts: (0, 0)
Asymptotes: none, since $|g(x)| \to \infty \Leftrightarrow |x| \to \infty$
Sign of $g(x)$: $g(x) > 0$ for $x > 0$, and $g(x) < 0$ for $x < 0$.
Since our analysis has given only one point, (0, 0), we must first compute coordinates for a few more points, as shown in the table. Then, keeping in mind symmetry with respect to the origin, we plot the graph as shown in Figure 20. ■

x	y	x	y
0	0		
1	1	−1	−1
8	2	−8	−2
27	3	−27	−3

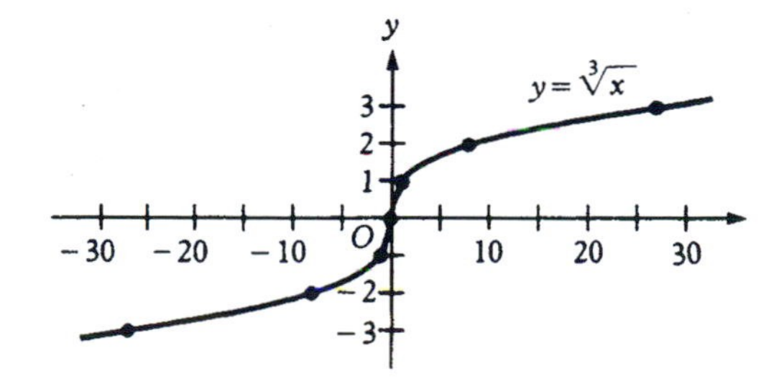

Figure 20

Example 5 Sketch the graph of $h(x) = \sqrt{(x-1)/(x+2)}$.

Solution:
Domain: $\sqrt{(x-1)/(x+2)}$ is defined only for $(x-1)/(x+2) \geq 0$. The solution to this fractional inequality, by the methods of Section 2.2, is $D_h = (-\infty, -2) \cup [1, \infty)$.
Symmetry: none, since $h(-x) \neq h(x)$ or $-h(x)$
Intercepts: The x-intercept is $(1, 0)$. There is no y-intercept, since 0 is not in D_h.
Asymptotes: $h(x) \to \infty$ as $x \to -2^-$, so the line $x = -2$ is a vertical asymptote. As $x \to \infty$ or $-\infty$,

$$h(x) = \sqrt{\frac{1 - 1/x}{1 + 2/x}} \to \sqrt{\frac{1-0}{1+0}} = 1.$$

Therefore, the line $y = 1$ is a horizontal asymptote.
Sign of $h(x)$: By definition, $h(x)$ is nonnegative for all x in its domain.
We can now graph the function h (Figure 21). It is worth noting that to graph h only three points were actually computed. By analyzing the properties of h, we were able to obtain its graph with a minimum number of computed points. ■

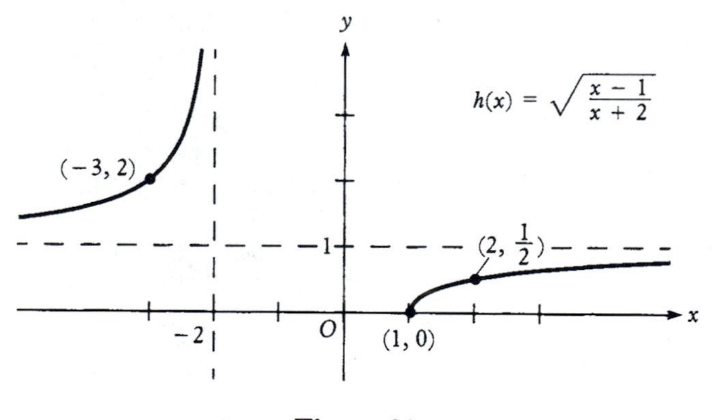

Figure 21

Comment In general, by analyzing domain, symmetry, intercepts, asymptotes, and the sign of $f(x)$, we have been able to graph some complicated functions without the need to compute a great number of points. However, for even more detailed information concerning the graph of a function, the power of calculus is needed. Calculus can give precise information on the following important properties of graphs. It can determine

(i) where the graph is rising and where it is falling: .

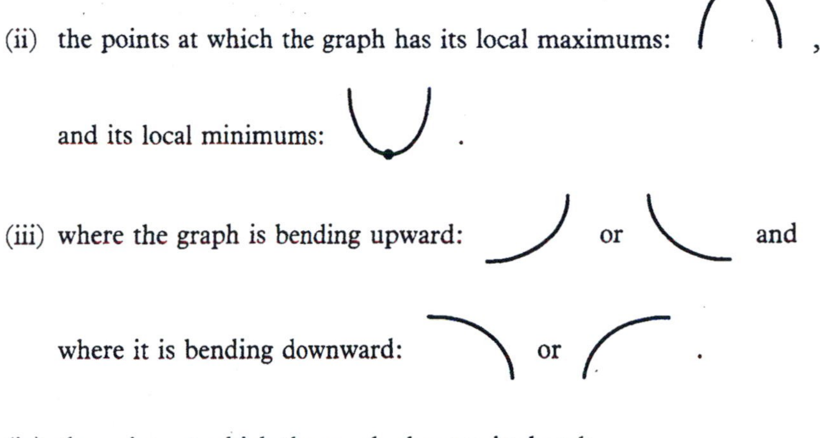

(ii) the points at which the graph has its local maximums: ,

and its local minimums: .

(iii) where the graph is bending upward: or and

where it is bending downward: or .

(iv) the points at which the graph changes its bend:

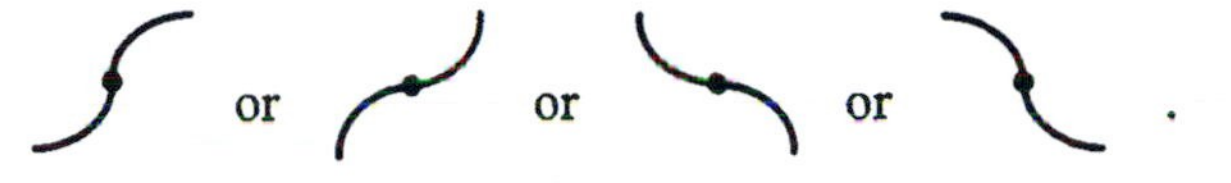

or or or .

If we were given a cluster of points as in Figure 22(a), we would normally assume that the corresponding graph behaves as in Figure 22(b). However, a closer analysis, using items (i)–(iv) above, might show that the actual behavior is as in Figure 22(c).

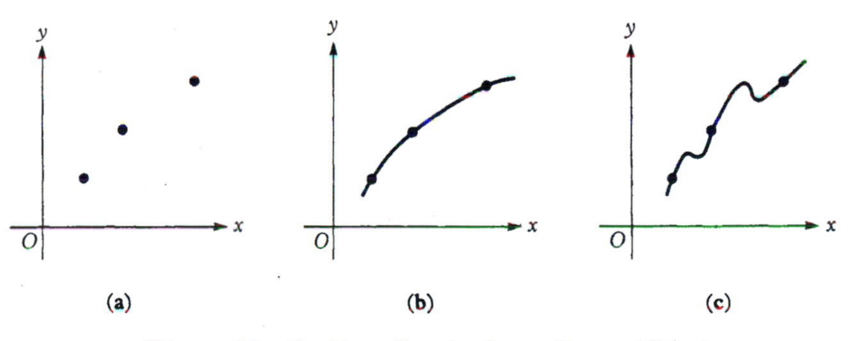

Figure 22 Graphing a Function from a Cluster of Points

Piecewise Algebraic Functions. The absolute value function $f(x) = |x|$ is an algebraic function since it can be written as $f(x) = \sqrt{x^2}$. More often, we write

$$f(x) = |x| = \begin{cases} x \text{ for } x \geq 0 \\ -x \text{ for } x < 0, \end{cases}$$

from which we can see that the graph of f is as in Figure 23. The graph of the absolute value function consists of two pieces, $y = x$ for $x \geq 0$ and $y = -x$ for $x < 0$. In general, if the graph of a function consists of several pieces, each one defined by a different algebraic expression, then the function is called a **piecewise algebraic function.** An important example is the **greatest integer function** $G(x) = [x]$, which is defined as the greatest integer less than or equal to x. That is,

$$G(x) = [x] = n \text{ for } n \leq x < n + 1,$$

where n is any integer. For instance, $[3] = 3$, $[3.95] = 3$, and $[-3.15] = -4$. The domain of G consists of the entire x-axis broken down into left-closed, right-open intervals $[n, n + 1)$, and the value of $G(x)$ on $[n, n + 1)$ is n, as shown in Figure 24, where points marked by closed dots are included, but those indicated by open dots are not.

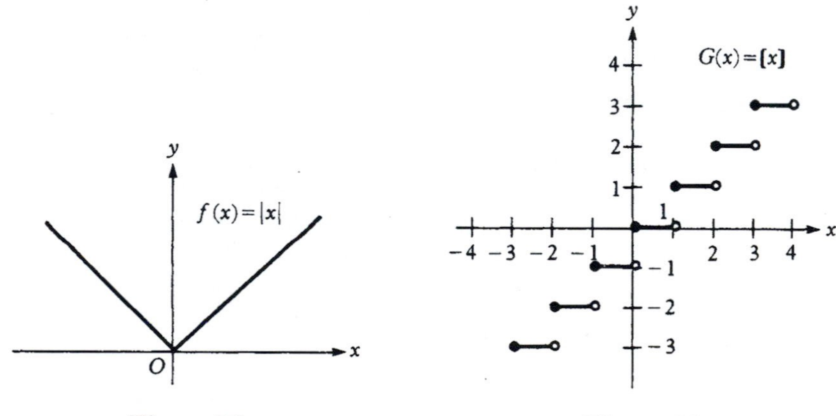

Figure 23 **Figure 24**

The graph of $G(x) = [x]$ shows why it is called a **step function.** Such functions occur often in applications, and one is described in Example 6 below.

Example 6 $P(x)$, the U.S. postage required in recent years for priority mailing of a letter weighing x ounces, has been defined as follows: the postage is 22 cents for a letter weighing one ounce or less plus 17 cents for each additional ounce or fraction thereof. Sketch the graph of the function P.

Solution: The graph of P is a step function with domain $x > 0$, as shown in Figure 25. ■

Example 7 Sketch the graph of H and give its domain and range if

$$H(x) = \begin{cases} x \text{ for } x \leq -1 \\ 1 \text{ for } -1 < x < 1 \\ x^2 \text{ for } x \geq 1. \end{cases}$$

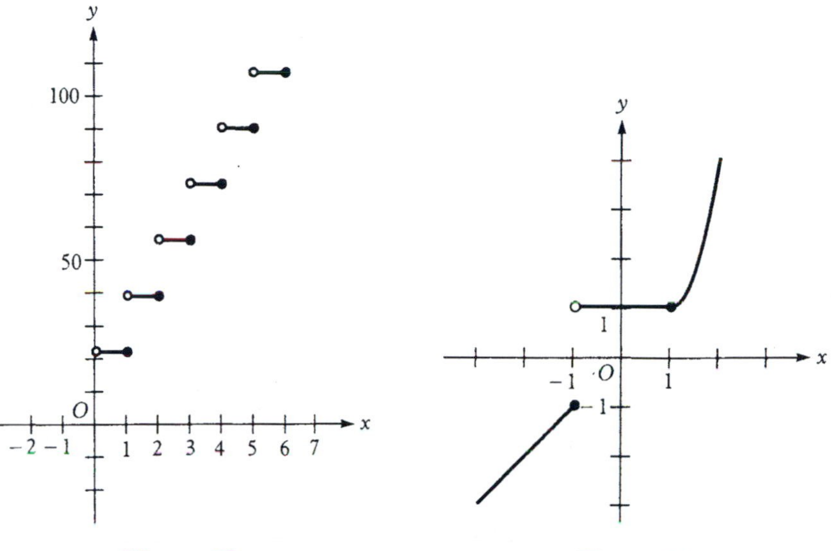

Figure 25

Figure 26

Solution: The definition of H shows that $D_H = (-\infty, \infty)$, and the graph of H (Figure 26) shows that $R_H = (-\infty, -1] \cup [1, \infty)$. ■

Comment The word *function* was first used in 1673 by Gottfried Wilhelm Leibniz to mean any quantity associated with the points on a curve. Later, function came to mean a relationship given by a single equation such as an algebraic function. The notation $f(x)$ was introduced by Leonhard Euler in 1734. The definition of a function as a correspondence that is not necessarily given by a single equation (for example, a piecewise algebraic function) is due to Peter Gustav Dirichlet (1805–1859). Today the function concept appears in every branch of mathematics.

Exercises 4.4

Fill in the blanks to make each statement true.

1. Rational functions are generated by the operations of ________, ________, ________, and ________. An algebraic function may also involve ________.
2. If an algebraic function contains a term that is an even root, then we must exclude from its domain all real numbers for which ____________.
3. For $f(x) = \sqrt[n]{x}$ ($n = 3, 5, 7, \ldots$), the domain is ________.
4. If $F(x) = |x - 1|$, then $F(x) = x - 1$ for x _______ and $F(x) = 1 - x$ for x _______.
5. Included among the ________ functions are the rational functions, and included among the rational functions are the ________ functions.

Write true *or* false *for each statement.*

6. If $f(x)$ involves a square root, then its domain cannot be $(-\infty, \infty)$.
7. If $(-\infty, -1) \cup (1, \infty)$ is the implied domain of an algebraic function $f(x)$, then $f(x)$ must have a term that is an even root.
8. Every polynomial function is algebraic.

9. Some algebraic functions are not rational.
10. If f is a piecewise algebraic function, then its graph must be disconnected for some values of x.

Algebraic Functions

11. Which of the following algebraic functions are not rational?
    (a) $f(x) = \sqrt{3}$ (b) $g(x) = \sqrt{x}$
    (c) $h(x) = (x + 3)^{1/3}$ (d) $F(x) = \dfrac{x^2 - 2x + 5}{x + 3}$
    (e) $G(x) = \dfrac{x^{1/2} - 2x + 5}{x + 3}$
    (f) $H(x) = \sqrt{\dfrac{x^2 - 2x + 5}{x + 3}}$
12. Which of the following are *not* algebraic functions?
    (a) $f(x) = x^2 + 2^x$ (b) $g(x) = x^{\sqrt{3}} + \sqrt{3}$
    (c) $h(x) = x^{3/2} - x^{3.2}$ (d) $F(x) = \sqrt[3]{x - 4}$
    (e) $G(x) = \dfrac{(x - 4)^{1/3}}{x^2}$
    (f) $H(x) = \dfrac{x^3 + 3^x}{\sqrt{x}}$
13. Find the domain of each of the following algebraic functions.
    (a) $f(x) = \dfrac{x^3 - 3x^2 + 7x - 5}{x^2 - 2x + 1}$
    (b) $g(x) = \dfrac{x + 2}{x - 1}$
    (c) $h(x) = \sqrt{\dfrac{x + 2}{x - 1}}$
    (d) $F(x) = \sqrt[3]{\dfrac{x + 2}{x - 1}}$
    (e) $G(x) = \sqrt{\dfrac{x^2 - 1}{x^2 + 1}}$
    (f) $H(x) = \sqrt{\dfrac{x^2 - 1}{x^2 - 4}}$

Sketch the graph of each of the following algebraic functions.

14. $f(x) = \sqrt{x - 1}$
15. $g(x) = \sqrt{x + 2}$
16. $h(x) = \sqrt{x} + 2$
17. $F(x) = \sqrt[3]{x + 1}$
18. $G(x) = \sqrt[3]{x - 1}$
19. $H(x) = \sqrt[3]{x} - 1$
20. $f(x) = \sqrt{\dfrac{x - 2}{x + 1}}$
21. $g(x) = \sqrt{\dfrac{x + 2}{x - 1}}$
22. $h(x) = \sqrt{\dfrac{x - 1}{x - 2}}$

Piecewise Algebraic Functions

Sketch the graphs of the following piecewise algebraic functions.

23. $f(x) = |x| - 2$
24. $g(x) = |x| + 2$
25. $F(x) = |x - 2|$
26. $G(x) = |x + 2|$
27. $u(x) = [x] + 2$
28. $v(x) = [x + 2]$
29. $w(x) = [x] - x$
30. $f(x) = \begin{cases} 1 \text{ for } x \le 0 \\ 2 \text{ for } x > 0 \end{cases}$
31. $g(x) = \begin{cases} x \text{ for } x \le 1 \\ 1 \text{ for } x > 1 \end{cases}$
32. $h(x) = \begin{cases} -1 \text{ for } x < 0 \\ 0 \text{ for } x = 0 \\ 1 \text{ for } x > 0 \end{cases}$
33. $F(x) = \begin{cases} x \text{ for } x \le 0 \\ x^2 \text{ for } x > 0 \end{cases}$
34. $G(x) = \begin{cases} x \text{ for } x \le 1 \\ x^2 \text{ for } x > 1 \end{cases}$
35. $H(x) = \begin{cases} x \text{ for } x \le 2 \\ x^2 \text{ for } x > 2 \end{cases}$
36. According to the I.R.S. in 1985, the income tax $T(x)$ for single taxpayers on x dollars of taxable income, with $\$9{,}500 \le x < \$10{,}000$, was given by the following table. Sketch a graph of the function T.

Interval containing x	$T(x)$
[9500, 9550)	999
[9550, 9600)	1007
[9600, 9650)	1015
[9650, 9700)	1023
[9700, 9750)	1031
[9750, 9800)	1039
[9800, 9850)	1047
[9850, 9900)	1055
[9900, 9950)	1063
[9950, 10000)	1071

4.5 One-to-One and Inverse Functions

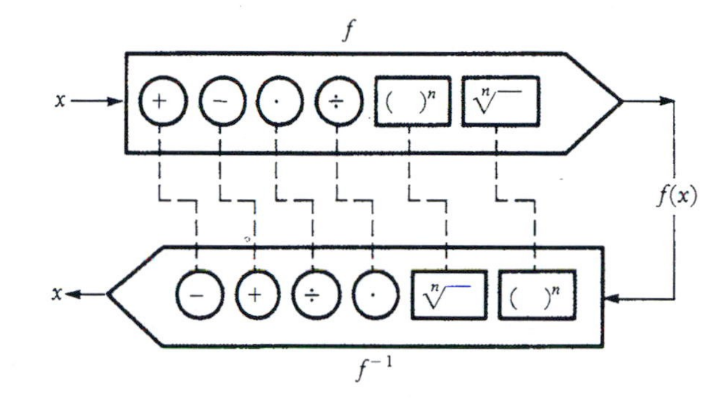

One-to-One Functions
Graphs of One-to-One Functions
Inverse Functions
Graphs of Inverse Functions

In Section P.2 we learned that every real number a has an additive inverse $-a$, and every $a \neq 0$ has a multiplicative inverse $a^{-1} = 1/a$. We now investigate the idea of an inverse for a function with respect to the operation of composition.

One-to-One Functions. If we consider the functions defined by

$$y = f(x) = 3x - 2 \tag{1}$$

and

$$y = F(x) = x^2 - 4, \tag{2}$$

and solve equations (1) and (2) for x, we obtain

$$x = \frac{y + 2}{3} \tag{3}$$

and

$$x = \pm\sqrt{y + 4}. \tag{4}$$

In equation (3), each real number y determines exactly one real number x, so that we may write

$$x = g(y) = \frac{y + 2}{3},$$

that is, x is a function of y. But in equation (4), each $y > -4$ determines *two* real numbers x, and therefore x is *not* a function of y.

By the definition of a function $y = f(x)$, each x in the domain of f must correspond to exactly one y in the range of f. If in addition each y in the range of f corresponds to exactly one x in the domain of f, then f is called a **one-to-one function.** This means that f is one-to-one if the equation $\boldsymbol{f(x_1) = f(x_2)}$ **implies** $\boldsymbol{x_1 = x_2}$. On the other hand, f is not one-to-one if the equation $f(x_1) = f(x_2)$ can be satisfied for different values x_1 and x_2 of x.

Example 1 Which of the following are one-to-one functions?

(a) $f(x) = x^3$ (b) $g(x) = x^4 - 2x^2$

(c) $F(x) = \sqrt{x - 4}, \quad x \geq 4$ (d) $G(x) = x^2, \quad x \leq 0$

Solution:

(a) Solving $y = x^3$ for x, we obtain $x = \sqrt[3]{y}$. Since every real number y has exactly one real cube root, f is one-to-one. To put it another way, if $x_1^3 = x_2^3$, then $x_1 = x_2$; that is, $f(x_1) = f(x_2)$ implies $x_1 = x_2$.

(b) Since $g(1) = g(-1) = -1$, g is not a one-to-one function.

(c) From $y = \sqrt{x - 4}$ $(x \geq 4)$, we find that $x = y^2 + 4$. Thus, each y determines a unique x, so that F is one-to-one. Alternatively, if $\sqrt{x_1 - 4} = \sqrt{x_2 - 4}$, then $x_1 - 4 = x_2 - 4$ and $x_1 = x_2$. Therefore, $F(x_1) = F(x_2)$ implies $x_1 = x_2$.

(d) If $y = G(x) = x^2$ for $x \leq 0$, then $x = -\sqrt{y}$ for $y \geq 0$. G is one-to-one because each nonnegative y has exactly one nonpositive square root. Alternately, if x_1 and x_2 are *nonpositive* and $x_1^2 = x_2^2$, then $x_1 = x_2$. Hence, $G(x_1) = G(x_2)$ implies $x_1 = x_2$. ■

Graphs of One-to-One Functions. In Section 4.1 we used the vertical line test to distinguish between the graph of a function f and the graph of an equation that could not be put in the form $y = f(x)$. Similarly, we may use the **horizontal line test** to determine whether a function f is one-to-one.

> **f is a one-to-one function if and only if any horizontal line $y = c$ intersects the graph of f in at most one point.**

Example 2 Use the horizontal line test to determine which of the functions in Example 1 are one-to-one.

Solution: We first sketch the graphs (Figures 27–30) and then apply the horizontal line test to each.

(a) In Figure 27, any horizontal line $y = c$ intersects the graph of f in exactly one point. Therefore, f is one-to-one.

(b) In Figure 28, the line $y = c$ $(-1 < c < 0)$ intersects the graph of g in four points. Therefore, g is not one-to-one.

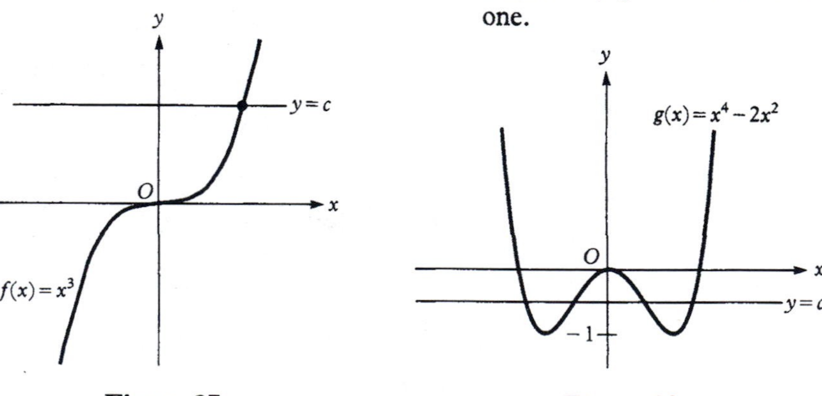

Figure 27

Figure 28

(c) In Figure 29, any horizontal line $y = c$ intersects the graph of F in at most one point. Therefore, F is one-to-one.

(d) In Figure 30, any horizontal line $y = c$ intersects the graph of G in at most one point. Therefore, G is one-to-one. ■

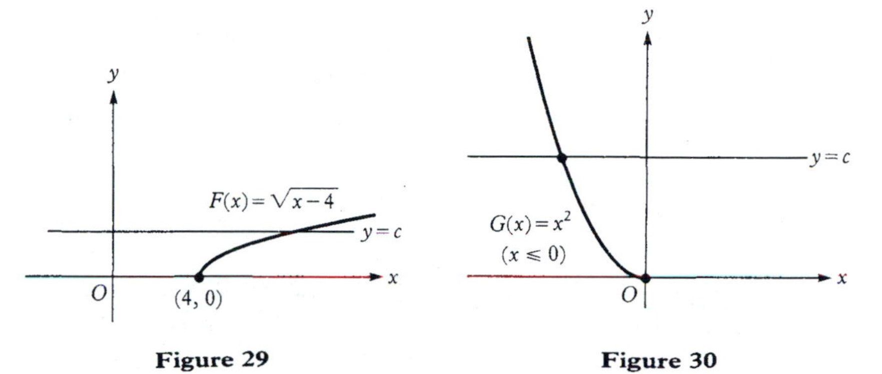

Figure 29 **Figure 30**

Note that in Figures 27 and 29 the graphs rise from left to right. Hence, the corresponding functions can be described as *increasing* functions. Similarly, the function in Figure 30 is a *decreasing* function. The algebraic definition of such functions is as follows.

> Let x_1 and x_2 be two points in D_f. Then
>
> f is an **increasing function** if $f(x_1) > f(x_2)$ whenever $x_1 > x_2$,
>
> and f is a **decreasing function** if $f(x_1) < f(x_2)$ whenever $x_1 > x_2$.

Example 3 Let $f(x) = 2x + 3$ and $g(x) = 1 - x$, where $D_f = (-\infty, \infty) = D_g$. Use the algebraic definition above to show that
(a) $f(x)$ is an increasing function and (b) $g(x)$ is a decreasing function.

Solution:

(a) Suppose x_1 and x_2 are any two real numbers satisfying $x_1 > x_2$. Then

$x_1 > x_2$	*Given*
$2x_1 > 2x_2$	*Multiplication by a positive number 2 preserves the direction of an inequality*
$2x_1 + 3 > 2x_2 + 3$	*Addition preserves the direction of an inequality*
$f(x_1) > f(x_2)$.	

Therefore, f is an increasing function.

(b) Again suppose that $x_1 > x_2$. We have

$$\begin{aligned} x_1 &> x_2 && \textit{Given} \\ -x_1 &< -x_2 && \textit{Multiplication by a negative number } -1 \textit{ reverses the direction of an inequality} \\ 1 - x_1 &< 1 - x_2 && \textit{Addition preserves the direction of an inequality} \\ g(x_1) &< g(x_2). \end{aligned}$$

Therefore, g is a decreasing function. ■

If f is an increasing or a decreasing function, then f is one-to-one (*why?*). For a continuous function (one whose graph has no breaks) with an interval domain, the converse is also true. That is, if a continuous function f is one-to-one and D_f is an interval, then f is either an increasing or a decreasing function. Note that the function g in Figure 28 is neither increasing nor decreasing, and g is also not one-to-one.

Inverse Functions. It was observed above that the function f defined in equation (1) by

$$y = f(x) = 3x - 2$$

is one-to-one, and therefore x is a function of y; that is,

$$x = g(y) = \frac{y + 2}{3}.$$

Using the definition of composition of functions given in Section 4.1, we have

$$(g \circ f)(x) = g(f(x)) = g(3x - 2) = \frac{(3x - 2) + 2}{3} = x.$$

This result is shown schematically in Figure 31.

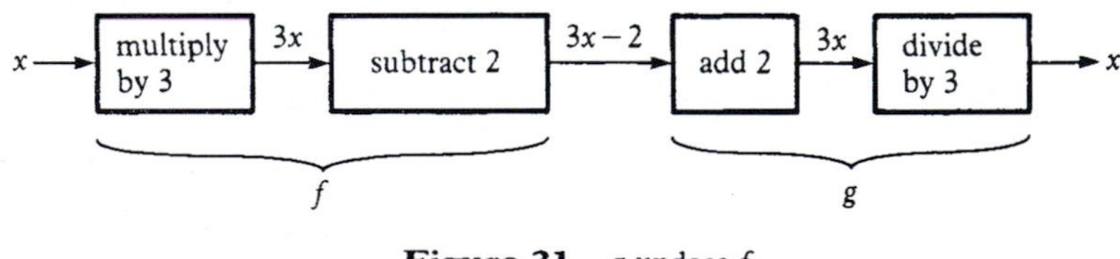

Figure 31 g undoes f

Also,

$$(f \circ g)(x) = f(g(x)) = f\left(\frac{x + 2}{3}\right) = 3\left(\frac{x + 2}{3}\right) - 2 = x,$$

which has the corresponding diagram (Figure 32).

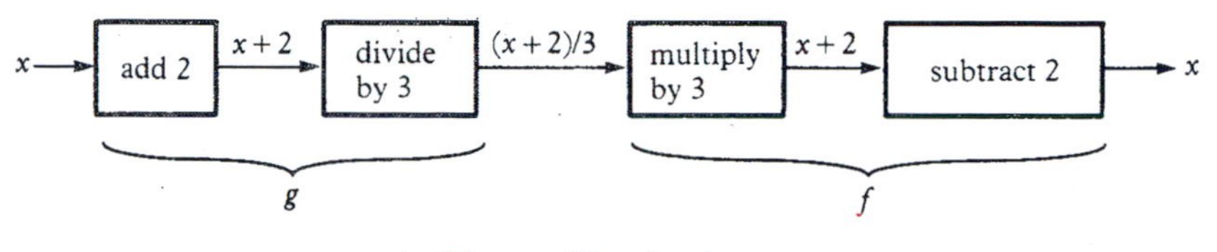

Figure 32 f undoes g

This example illustrates that when f is a one-to-one function, there is another function g for which

$$(g \circ f)(x) = x \qquad \text{and} \qquad (f \circ g)(x) = x. \tag{5}$$

The function g that satisfies both equations in (5) is called the **inverse of** f and is denoted by f^{-1}. Thus, in general,

$$(f^{-1} \circ f)(x) = x \qquad \text{and} \qquad (f \circ f^{-1})(x) = x, \tag{6}$$

as shown in Figure 33. Also, $D_{f^{-1}} = R_f$ and $R_{f^{-1}} = D_f$.

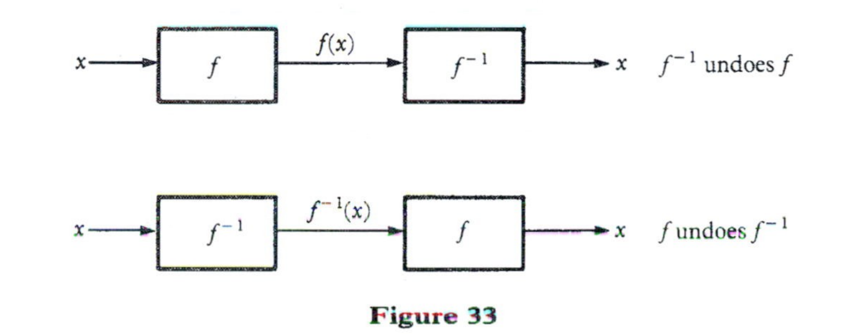

Figure 33

Using the example above, we can compute $f^{-1}(x)$ from $f(x)$ in the following three steps.

1. $y = 3x - 2 = f(x)$ *Given function*
2. $x = \dfrac{y + 2}{3} = f^{-1}(y)$ *Solve equation in step 1 for x in terms of y and set equal to $f^{-1}(y)$*
3. $y = \dfrac{x + 2}{3} = f^{-1}(x)$ *Interchange x and y in step 2*

We interchange x and y to solve for f^{-1} because we prefer x to denote the variable in the domain and y the variable in the range of f^{-1}.

Example 4 Find the inverse function for each of the one-to-one functions in Example 1, including the domain and range of each inverse function. Also verify that equations (6) are satisfied in each case.

Solution:

(a) 1. $y = x^3 = f(x)$, with $D_f = (-\infty, \infty) = R_f$

2. $x = \sqrt[3]{y} = f^{-1}(y)$

3. $y = \sqrt[3]{x} = f^{-1}(x)$, with $D_{f^{-1}} = R_f = (-\infty, \infty)$ and $R_{f^{-1}} = D_f = (-\infty, \infty)$

Here, equations (6) become

$$(f^{-1} \circ f)(x) = f^{-1}(f(x)) = f^{-1}(x^3) = \sqrt[3]{x^3} = x$$

and $$(f \circ f^{-1})(x) = f(f^{-1}(x)) = f(\sqrt[3]{x}) = (\sqrt[3]{x})^3 = x.$$

(b) $g(x) = x^4 - 2x^2$ is not one-to-one and therefore has no inverse.

(c) 1. $y = \sqrt{x - 4} = F(x)$, with $D_F = [4, \infty)$ and $R_F = [0, \infty)$

2. $x = y^2 + 4 = F^{-1}(y)$

3. $y = x^2 + 4 = F^{-1}(x)$, with $D_{F^{-1}} = R_F = [0, \infty)$, and $R_{F^{-1}} = D_F = [4, \infty)$

Equations (6) become

$$\begin{aligned}(F^{-1} \circ F)(x) &= F^{-1}(F(x)) \\ &= F^{-1}(\sqrt{x - 4}) = (\sqrt{x - 4})^2 + 4 \\ &= x - 4 + 4 = x\end{aligned}$$

and $$\begin{aligned}(F \circ F^{-1})(x) &= F(F^{-1}(x)) = F(x^2 + 4) \\ &= \sqrt{(x^2 + 4) - 4} = \sqrt{x^2} = |x| = x, \text{ since } x \geq 0.\end{aligned}$$

(d) 1. $y = G(x) = x^2$, with $D_G = (-\infty, 0]$ and $R_G = [0, \infty)$

2. $x = -\sqrt{y} = G^{-1}(y)$ (*Why the negative square root?*)

3. $y = -\sqrt{x} = G^{-1}(x)$, with $D_{G^{-1}} = R_G = [0, \infty)$ and $R_{G^{-1}} = D_G = (-\infty, 0]$

Equations (6) become

$$\begin{aligned}(G^{-1} \circ G)(x) &= G^{-1}(G(x)) \\ &= G^{-1}(x^2) = -\sqrt{x^2} = -|x| = x, \text{ since } x \leq 0\end{aligned}$$

and $$(G \circ G^{-1})(x) = G(G^{-1}(x)) = G(-\sqrt{x}) = (-\sqrt{x})^2 = x.$$ ■

Question *An inverse element for a given operation is usually defined in terms of the identity element for that operation (see basic properties (6) through (9) in Section P.2). By definition, f^{-1} is the inverse of f with respect to the operation of composition. What is the identity element for composition?*

Answer: The identity element for composition is the function $I(x) = x$ because $f \circ I = f = I \circ f$ for any function f (see Exercise 29). ■

Graphs of Inverse Functions. We now draw the graphs of some one-to-one functions and their inverses on the same coordinate axes. From Example 4(a), $f(x) = x^3$ and $f^{-1}(x) = \sqrt[3]{x}$ (Figure 34).

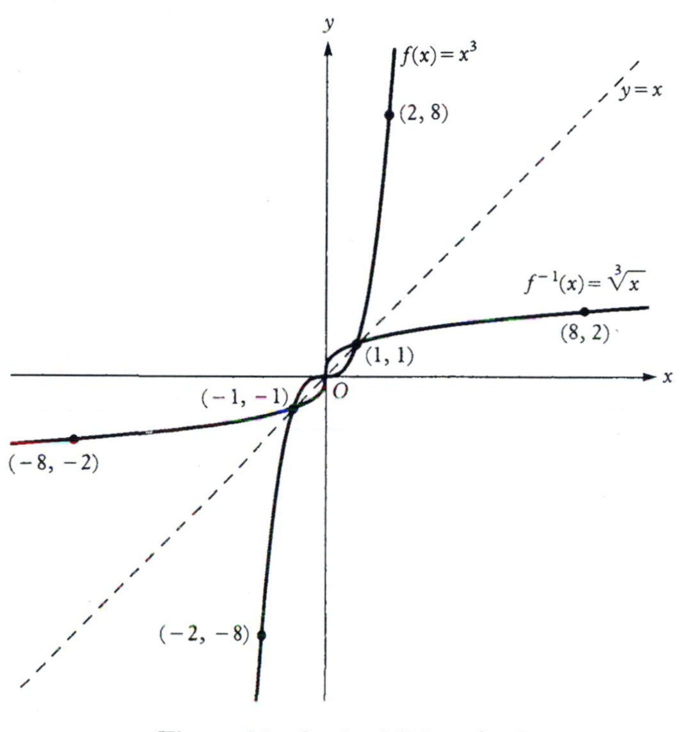

Figure 34 Graphs of $f(x) = x^3$ and $f^{-1}(x) = \sqrt[3]{x}$

The graph of f contains the points (1, 1), (2, 8), (−1, −1), and (−2, −8), while the graph of f^{-1} contains the corresponding points (1, 1), (8, 2), (−1, −1), and (−8, −2). In general, if the point (a, b) is on the graph of f, then the point (b, a) is on the graph of f^{-1}. This follows from the fact that f^{-1} is obtained from f by interchanging x and y, and it means that

> **the graphs of f and f^{-1} are symmetric with respect to the line $y = x$.**

If we were to fold the page along the line $y = x$ in Figure 34, the graphs of f and f^{-1} would coincide. The line segment joining point (a, b) on the graph of f to point (b, a) on the graph of f^{-1} has the line $y = x$ as its perpendicular bisector. This symmetry property is true not only for the function in Example 4(a) but for *any* one-to-one function and its inverse.

Example 5 Sketch the graphs of $F(x) = \sqrt{x-4}$ and its inverse $F^{-1}(x) = x^2 + 4$ $(x \geq 0)$ on the same axes [see Example 4(c)].

Solution: The graphs of F and F^{-1} are symmetric with respect to the line $y = x$, as shown in Figure 35. ■

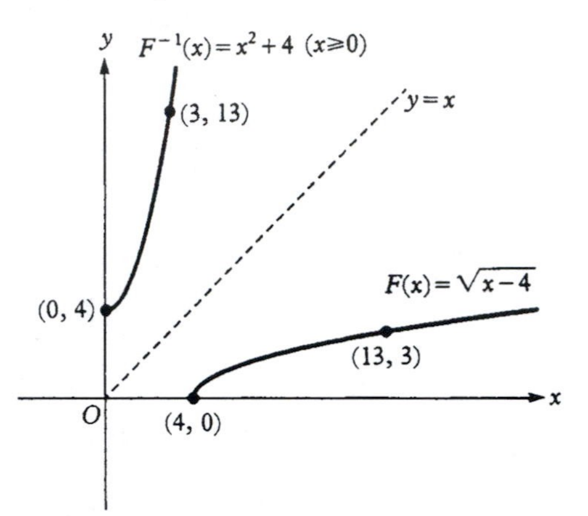

Figure 35 Graphs of $F(x) = \sqrt{x-4}$ and $F^{-1}(x) = x^2 + 4$ $(x \geq 0)$

Example 6 Sketch the graphs of $G(x) = x^2$ $(x \leq 0)$ and its inverse $G^{-1}(x) = -\sqrt{x}$ on the same axes [see Example 4(d)].

Solution: First the graph of $G(x) = x^2$ $(x \leq 0)$ is sketched; then the graph of G^{-1} is obtained by reflection about the line $y = x$ (Figure 36). ■

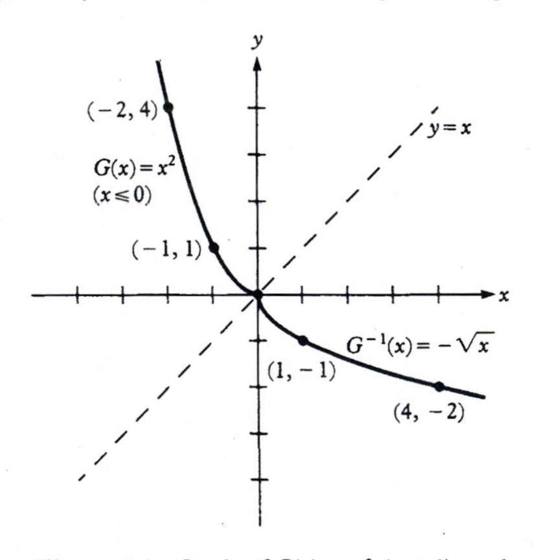

Figure 36 Graphs of $G(x) = x^2$ $(x \leq 0)$ and $G^{-1}(x) = -\sqrt{x}$

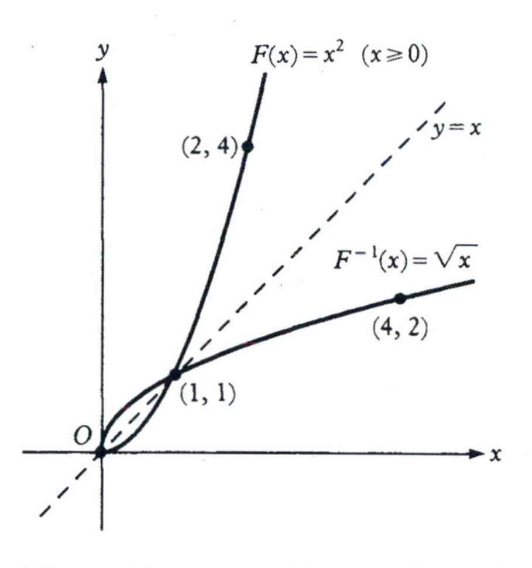

Figure 37 Graphs of $F(x) = x^2$ $(x \geq 0)$ and $F^{-1}(x) = \sqrt{x}$

Some functions that are not one-to-one and therefore have no inverse can be made one-to-one by restricting their domains. This was done in Example 1(d) where the graph of $G(x) = x^2 \quad (x \leq 0)$ (Figures 30 and 36) is the left branch of the parabola $y = x^2$, and $G^{-1}(x) = -\sqrt{x}$. Similarly, the graph of $F(x) = x^2 \quad (x \geq 0)$ (Figure 37), is the right branch of the same parabola $y = x^2$, and $F^{-1}(x) = \sqrt{x}$.

Comment In Example 4 we were able to determine the inverses of several basic one-to-one functions. However, there are many one-to-one functions whose inverses cannot be computed algebraically. For example, the function

$$f(x) = 1 + x + \frac{x^3}{27} + \frac{x^5}{256}$$

is 1 plus the sum of the increasing functions x, $x^3/27$, and $x^5/256$ and is therefore itself an increasing function. As an increasing function, f has an inverse, but to determine f^{-1} it is necessary to solve the fifth-degree equation

$$y = 1 + x + \frac{x^3}{27} + \frac{x^5}{256}$$

for x in terms of y. We cannnot solve this equation, but we can obtain a graph of f^{-1} by first graphing f and then reflecting about the line $y = x$ (Figure 38).

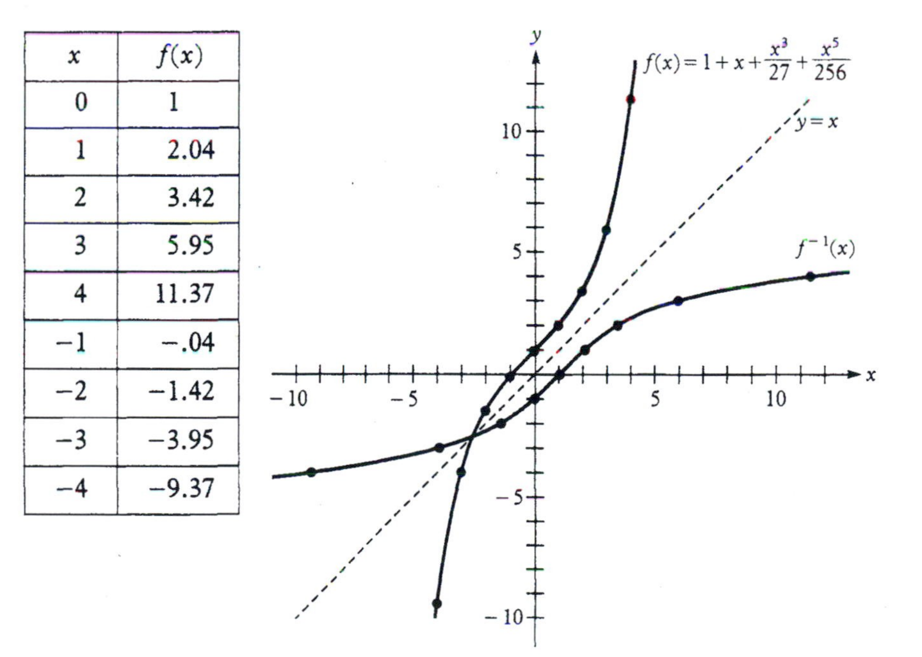

x	$f(x)$
0	1
1	2.04
2	3.42
3	5.95
4	11.37
−1	−.04
−2	−1.42
−3	−3.95
−4	−9.37

Figure 38 Graphs of $f(x) = 1 + x + x^3/27 + x^5/256$ and $f^{-1}(x)$

Exercises 4.5

Fill in the blanks to make each statement true.

1. For any function $y = f(x)$, each value of ________ in D_f corresponds to exactly one value of ________ in R_f.
2. If $y = f(x)$ has the property that each value of ________ in R_f corresponds to exactly one value of ________ in D_f, then f is called one-to-one.
3. f is one-to-one if the line $y = c$ intersects the graph of f in ____________.
4. If the function g is the inverse of the function f, then ________ $= x$ for all x in D_f, and ________ $= x$ for all x in D_g.
5. The domain of f^{-1} equals ____________ of f, and the range of f^{-1} equals ____________ of f.

Write true *or* false *for each statement.*

6. If the equation $x_1 = x_2$ always implies that $f(x_1) = f(x_2)$, then f is a one-to-one function.
7. If the equation $f(x_1) = f(x_2)$ always implies that $x_1 = x_2$, then f is a one-to-one function.
8. If f is not a one-to-one function, then f has no inverse.
9. If f is a one-to-one function, then every horizontal line $y = c$ must intersect the graph of f in one point.
10. The graphs of f and f^{-1} are symmetric with respect to the line $y = x$.

One-to-One Functions

11. Which of the following functions are one-to-one?
    (a) $f(x) = 2x + 4$
    (b) $g(x) = \dfrac{1}{2x + 4}$
    (c) $h(x) = x^2 + 4$
    (d) $F(x) = \sqrt{x^2 + 4}$
    (e) $G(x) = x^3 + 4$
    (f) $H(x) = \sqrt[3]{x - 4}$
12. Which of the following are *not* one-to-one functions and why?
    (a) $p(x) = x^2$
    (b) $q(x) = x^2 \quad (x \geq 1)$
    (c) $r(x) = x^2 \quad (x \leq 1)$
    (d) $s(x) = x^4$
    (e) $t(x) = x^4 \quad (x \geq 0)$
    (f) $u(x) = x^5$
13. Use the horizontal line test to determine which of the following are graphs of one-to-one functions.

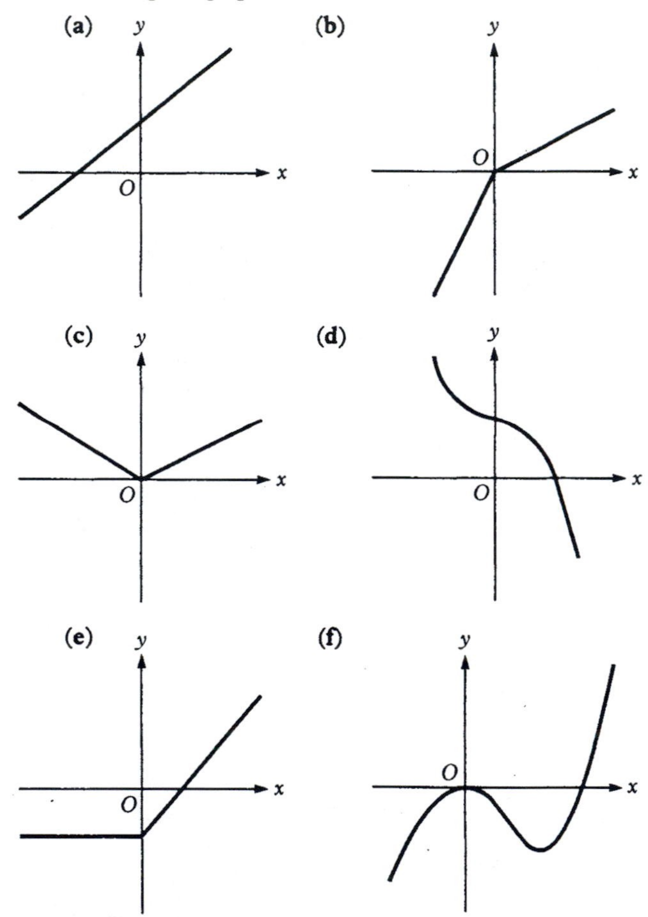

Inverse Functions

Determine $f^{-1}(x)$, including its domain and range, for each of the following.

14. $f(x) = 2x + 4$
15. $f(x) = x^3 - 1$

16. $f(x) = \dfrac{x^5}{5}$

17. $f(x) = \sqrt{x + 2}$

18. $f(x) = \sqrt[3]{x + 2}$

19. $f(x) = \sqrt[3]{x} + 2$

20. For each of the functions in Exercises 14, 16, and 18, verify that $(f^{-1} \circ f)(x) = x$ and $(f \circ f^{-1})(x) = x$.

21. For each of the functions in Exercises 15, 17, and 19, sketch the graphs of f and f^{-1} on the same axes.

Determine $g^{-1}(x)$, including its domain and range, for each of the following.

22. $g(x) = x^2 \quad (x \geq 1)$

23. $g(x) = x^2 \quad (x \leq -1)$

24. $g(x) = x^2 - 1 \quad (x \geq 1)$

25. $g(x) = x^2 - 1 \quad (x \leq -1)$

26. For each of the functions in Exercises 22 and 24, sketch the graphs of g and g^{-1} on the same axes.

27. For each of the functions in Exercises 23 and 25, verify the equations $(g^{-1} \circ g)(x) = x$ and $(g \circ g^{-1})(x) = x$.

28. Let $F(x) = \dfrac{x + 1}{x} \quad (x \neq 0)$.
    (a) Show that F is a one-to-one function.
    (b) Solve for $F^{-1}(x)$.
    (c) Verify that $(F^{-1} \circ F)(x) = x$ and $(F \circ F^{-1})(x) = x$.
    (d) Sketch the graphs of F and F^{-1} on the same axes.

29. Show that if $I(x) = x$, then $(f \circ I)(x) = f(x)$ and $(I \circ f)(x) = f(x)$ for all x in D_f. Hence, conclude that $f \circ I = f$ and $I \circ f = f$.

4.6 Variation

Formulas in geometry and science give us a precise expression of the variation among related quantities.

Direct and Inverse Variation
Joint and Combined Variation

In mathematics, the term **variation** refers to a relationship of one quantity to another. For example, the equation

$$A = \pi r^2$$

says that the area of a circle varies directly as the square of its radius. Equivalently, we say that the area is *directly proportional* to the square of the radius. Here, π is called the constant of proportionality, and the word *directly* means that when r increases in magnitude, so does A. On the other hand, Newton's law of gravitation

$$F = k\left(\frac{m_1 m_2}{r^2}\right)$$

says that the force of attraction between two bodies is directly proportional to the product of their masses and is *inversely* proportional to the square of the distance between them. Here, F decreases as r increases. In either case, the variation between the quantities involved is defined by a certain kind of algebraic function.

Direct and Inverse Variation. As illustrated above, the statement that a quantity y **varies directly** as a quantity x, or y is **directly proportional** to x, means that for some nonzero number k,

$$y = kx,$$

and k is called the **constant of variation,** or the **proportionality constant.**

Example 1 Hooke's law states that the force needed to stretch a spring d units is directly proportional to d. Translate Hooke's law into an equation. If a 6-pound weight stretches a spring 2 inches, find the constant of proportionality (spring constant).

Solution: Let F denote the force in pounds needed to stretch the spring d inches. Hooke's law says

$$F = kd,$$

where k is the proportionality constant. By substituting $F = 6$ pounds and $d = 2$ inches, we obtain

$$6 \text{ pounds} = k \cdot 2 \text{ inches}$$

or

$$k = 3 \text{ pounds per inch.}$$ ■

Example 2 If an object is dropped from above the ground, then the distance it falls before hitting the ground varies directly as the square of its falling time. A ball dropped from a hot-air balloon falls 1600 feet in 10 seconds. Find the constant of variation and the distance the ball falls in 15 seconds.

Solution: Let d denote the distance fallen and t the falling time. We have

$$d = kt^2,$$

where k is the constant of variation. By substituting $d = 1600$ feet and $t = 10$ seconds, we obtain

$$1600 \text{ feet} = k \cdot 100 \text{ seconds}^2,$$

or

$$k = 16 \text{ feet per second}^2.$$

Therefore, the distance fallen in 15 seconds is

$$\begin{aligned} d &= 16 \cdot 15^2 \\ &= 3600 \text{ feet.} \end{aligned}$$ ■

As in the case of Newton's inverse square law, we say that a quantity y **varies inversely** as a quantity x, or y is **inversely proportional** to x, if

$$y = \frac{k}{x},$$

and again k is the constant of variation, or the proportionality constant.

Example 3 At a constant temperature, the volume V of a gas varies inversely as the pressure P. If the volume is 900 cubic inches when the pressure is 10 pounds per square inch, what pressure will reduce the volume to 500 cubic inches?

Solution: We are told that

$$V = \frac{k}{P},$$

where k is a constant. Also,

$$900 \text{ cubic inches} = \frac{k}{10 \text{ pounds per square inch}}$$

implies that

$$\begin{aligned} k &= 900 \text{ cubic inches} \cdot 10 \text{ pounds per square inch} \\ &= 9000 \text{ inch-pounds.} \end{aligned}$$

Therefore,

$$V = \frac{9000}{P},$$

and when $V = 500$ cubic inches, we have

$$\begin{aligned} P &= \frac{9000 \text{ inch-pounds}}{500 \text{ cubic inches}} \\ &= 18 \text{ pounds per square inch.} \end{aligned}$$

As implied by the inverse variation, the pressure must be increased in order to decrease the volume. ■

Joint and Combined Variation. If one quantity y is directly proportional to the *product* of two or more other quantities, we say that y **varies jointly** as each of the other quantities. Examples of **joint variation** are:

$T = kPV$ (the temperature T of a gas varies jointly as the pressure P and the volume V);

$V = RI$ [the voltage drop V across a resistor varies jointly ($k = 1$) as the resistance R and the current I];

$v = \pi r^2 h$ [the volume v of a cylinder varies jointly ($k = \pi$) as the radius r of the base squared and the height h].

If a relationship among quantities involves both direct and inverse variation, we say the relationship is one of **combined variation.** Some examples of combined variation are:

$C = k\frac{A}{d}$ (the capacitance C of a parallel plate capacitor varies directly as the area A of the plates and inversely as the distance d between the plates);

$E_p = k\frac{q_1 q_2}{r}$ (the potential energy E_p of a system of two point charges varies jointly as the values q_1 and q_2 of the charges and inversely as the distance r between them);

$F = k\frac{m_1 m_2}{r^2}$ (the force of attraction between two bodies varies jointly as their masses and inversely as the square of the distance between them).

Example 4 The weight of an object on the surface of the earth is defined as the force of attraction between the object and the earth. We assume that the mass of the earth is concentrated at its center. According to Newton's law of gravitation,

$$w = k\frac{Mm}{R^2}, \quad \text{where} \quad \begin{cases} w = \text{weight of the object} \\ m = \text{mass of the object} \\ M = \text{mass of the earth} \\ R = \text{radius of the earth} \\ k = \text{proportionality constant.} \end{cases}$$

Now k and M are constants, and for a given object, m is also constant. However, the earth is not a perfect sphere, so R and therefore w vary according to the object's position on the earth. If a person weighs 110 pounds at the north pole ($R \approx 3950$ feet), what does the person weigh at the equator ($R \approx 3963$ feet)?

Solution: Since the product kMm is the same at the north pole and the equator, we can replace it with a single constant c:

$$w = \frac{c}{R^2}.$$

By using the person's weight at the north pole, we have

$$110 = \frac{c}{3950^2}$$
$$c = 1{,}716{,}275{,}000.$$

Therefore, at the equator,

$$w = \frac{1{,}716{,}275{,}000}{3963^2}$$
$$= 109.28 \text{ pounds}$$
$$\approx 109 \text{ pounds, 4 ounces.}$$

Hence, the person loses approximately 12 ounces by traveling to the equator. ■

Example 5 While at her bank, Heather notices that, on the average, she has to wait in line for 5 minutes when there are 4 people ahead of her and 2 tellers working. If she arrives at the bank with 12 people ahead of her and 3 tellers working, how long can she expect to wait?

Solution: Let

W = Heather's waiting time,
p = the number of people ahead of her, and
t = the number of tellers working.

We assume that for some constant k,

$$W = k\left(\frac{p}{t}\right).$$

(*Why is our assumption reasonable?*) We are given $W = 5$ when $p = 4$ and $t = 2$. Therefore,

$$5 = k\left(\frac{4}{2}\right),$$

which means that $k = 2.5$. Therefore, when $p = 12$ and $t = 3$, we get

$$W = 2.5\left(\frac{12}{3}\right) = 10.$$

Hence, Heather can expect to wait about 10 minutes. ■

Exercises 4.6

In the following exercises, whenever the constant k appears, it is assumed that $k \neq 0$.

Fill in the blanks to make each statement true.

1. If y varies directly as x, then $y/x =$ ________.
2. If $y = k/(x^2)$, then y varies ________ as ________.
3. When y varies ________ as x, the magnitude of y increases as that of x increases, but when y varies ________ as x, the magnitude of y decreases as that of x increases.
4. The formula $A = (1/2)bh$ for the area of a triangle may be read as "the area varies ________," where 1/2 is the ________.
5. The statement that v varies jointly as the square of x and the cube of y and inversely as z translates into the formula ________.

Write true *or* false *for each statement.*

6. Direct variation of y with x is represented by a linear equation.
7. Every linear equation in two variables x and y represents direct variation between x and y.
8. The equation $xy = k$ represents inverse variation between y and x.
9. If y varies directly as x, then x varies directly as y.
10. If y varies inversely as x, then x varies directly as y.

Direct and Inverse Variation

Translate each of the following statements into an equation, using k as the constant of variation.

11. d varies directly as t.
12. P varies inversely as V.
13. I varies directly as R.
14. F varies inversely as d^2.
15. T varies directly as the square root of L.

Translate each of the formulas in Exercises 16–20 into a statement of variation.

16. $I = kP$, where I is simple interest and P is principal invested
17. $F = kv^6$, where F is the force exerted by flowing water and v is its velocity.
18. $f = k\sqrt{T}$, where f is frequency and T is tension in a stretched string
19. $t = k/r$, where t is the time to travel between two cities and r is the average rate of speed
20. $R = k/D^2$, where R is the resistance of a conducting wire and D is its diameter
21. If y varies directly as x^2 and $y = 12$ when $x = 2$, find y when $x = 6$.
22. If y is inversely proportional to x^2 and $y = 12$ when $x = 2$, find y when $x = 6$.
23. For a given principal, simple interest varies directly as the rate of interest. If the interest earned is \$80 at 5%, what is the interest at 7%?
24. For a fixed amount of simple interest, the present value of an investment varies inversely as the interest rate. Find the present value corresponding to 8% interest if the present value is \$1000 at 6%.
25. The distance required for an automobile to stop varies as the square of its speed when the brakes are applied. If a car going 55 mph brakes to a stop in 151.25 ft, how fast is the car traveling if 175 ft are required to stop?
26. The cost of manufacturing a radio varies inversely as the number of radios produced. If the cost is \$35 per radio to produce 1000 radios, how many must be produced to lower the cost to \$25?
27. Show that if y varies directly with x, then $y_1/x_1 = y_2/x_2$ for any corresponding nonzero values x_1, y_1, and x_2, y_2.
28. Use Exercise 27 to solve Exercise 23.
29. Show that if y varies inversely as x, then $x_1y_1 = x_2y_2$ for any corresponding pairs x_1, y_1 and x_2, y_2.
30. Use Exercise 29 to solve Exercise 24.

Joint and Combined Variation

Translate each of the statements in Exercises 31–35 into an equation.

31. The quantity z varies jointly as x and the cube of y.
32. A force f varies directly with mass m and inversely with the square of the distance d.
33. Electrical current I varies directly with voltage V and inversely with resistance R.
34. The frequency f of vibration of a string varies directly with the square root of the tension t of the string and inversely with its length l.
35. Profit P varies directly with number n of sales and inversely with the square root of the cost c.

Translate each of the equations in Exercises 36–40 into a statement about variation.

36. $V = \pi r^2 h$
37. $h = \dfrac{V}{\pi r^2}$
38. $H = k\left(\dfrac{AT}{d}\right)$
39. $W = k\left(\dfrac{wd^2}{l}\right)$
40. $IQ = k\left(\dfrac{M}{C}\right)$
41. The volume V of a gas varies directly as the temperature T and inversely as the pressure P. If V is 30 cubic inches when T is 70° and $P = 4$ lb/in^2, find V for $T = 65°$ and $P = 3$ lb/in^2.

42. The weight that can be supported by a wooden beam varies jointly as its width w and the square of its depth d and inversely as its length l. A beam 4 ft long, 1 ft wide, and 2 in deep can support a weight of 90 lb. How much can a beam of the same wood support if it is 5 ft long, 2 ft wide, and 3 in deep?

43. According to Ohm's law, the resistance to the flow of current in an electric circuit varies directly with the voltage V and inversely with the current I. If $R = 11$ ohms when $V = 220$ volts and $I = 20$ amperes, find R when $V = 110$ volts and $I = 30$ amperes.

44. By using Ohm's law as defined in Exercise 43, show that

$$\frac{R_2}{R_1} = \frac{V_2 I_1}{V_1 I_2}$$

for any related triples R_1, V_1, I_1, and R_2, V_2, I_2. Also, what is the effect on the resistance when the voltage is doubled and the current is cut in half?

Chapter 4 Review Outline

4.1 Functions

A function f with domain D_f and range R_f is a rule that assigns to each element x in D_f a unique element $y = f(x)$ in R_f; x is called the independent variable, and y is called the dependent variable.

Operations with Functions:

$(f \pm g)(x) = f(x) \pm g(x)$

$(fg)(x) = f(x)g(x)$

$\left(\frac{f}{g}\right)(x) = \frac{f(x)}{g(x)} \quad [g(x) \neq 0]$

$(g \circ f)(x) = g(f(x))$

Graph of a Function:
The graph of a function f is the set of all points (x, y) with x in D_f and $y = f(x)$.

A graph is that of a function if every vertical line intersects the graph in at most one point.

4.2 Polynomial Functions

A polynomial function f of degree n is defined by

$$f(x) = a_n x^n + a_{n-1}x^{n-1} + \cdots + a_1 x + a_0,$$

where n is a nonnegative integer and $a_n \neq 0$.

Symmetry:
The graph of a function f is symmetric with respect to the y-axis $\Leftrightarrow f(-x) = f(x)$ for all x in D_f (f is an even function).

A polynomial P is an even function $\Leftrightarrow P(x)$ contains only even powers of x.

The graph of a function f is symmetric with respect to the origin $\Leftrightarrow f(-x) = -f(x)$ for all x in D_f (f is an odd function).

A polynomial P is an odd function $\Leftrightarrow P(x)$ contains only odd powers of x.

Graphing Polynomial Functions:
The three-step procedure for graphing a polynomial P includes (1) symmetry, (2) intercepts, and (3) the sign of $P(x)$.

4.3 Rational Functions

A rational function is the ratio of two polynomial functions.

Asymptotes:
The vertical line $x = a$ is a vertical asymptote for $f(x) = P(x)/Q(x)$ if $Q(a) = 0$ and $P(a) \neq 0$, which means that $|f(x)| \to \infty$ as $x \to a$.

The horizontal line $y = b$ is a horizontal asymptote for f if $f(x) \to b$ as $x \to \infty$ or as $x \to -\infty$.

Test for Horizontal Asymptotes:

Let $f(x) = \dfrac{a_m x^m + a_{m-1}x^{m-1} + \cdots + a_0}{b_n x^n + b_{n-1}x^{n-1} + \cdots + b_0} \quad (a_m \neq 0,\ b_n \neq 0)$

The line $y = 0$ is a horizontal asymptote for $f \Leftrightarrow m < n$.

The line $y = a_m/b_n$ is a horizontal asymptote for $f \Leftrightarrow m = n$.

f has no horizontal asymptote $\Leftrightarrow m > n$.

Graphing Rational Functions:
The four-step procedure for graphing a rational function f includes (1) symmetry, (2) x-intercepts and vertical asymptotes, (3) the sign of $f(x)$, and (4) y-intercept and horizontal asymptote.

4.4 Algebraic Functions

An algebraic function is one generated by addition, subtraction, multiplication, division, and extraction of roots. (Polynomials and rational functions are algebraic functions.)

Graphing Algebraic Functions:
To graph an algebraic function f, consider symmetry, intercepts, the sign of $f(x)$, asymptotes, and restrictions on the domain of f.

4.5 One-to-One and Inverse Functions

Definitions

f is one-to-one $\Leftrightarrow$ $f(x_1) = f(x_2)$ always implies $x_1 = x_2$.
f is an increasing function $\Leftrightarrow$ $x_1 > x_2$ implies $f(x_1) > f(x_2)$ whenever x_1 and x_2 are in D_f.
f is a decreasing function $\Leftrightarrow$ $x_1 > x_2$ implies $f(x_1) < f(x_2)$ whenever x_1 and x_2 are in D_f.
g is the inverse of f (denoted f^{-1}) $\Leftrightarrow$

$(g \circ f)(x) = x$ for all x in D_f and
$(f \circ g)(x) = x$ for all x in D_g.

Horizontal Line Test:
f is one-to-one $\Leftrightarrow$ every horizontal line intersects the graph of f in at most one point.

Relationships:
f has an inverse $\Leftrightarrow$ f is one-to-one.
If f is an increasing or a decreasing function, then f is one-to-one.
The graphs of f and f^{-1} are symmetric with respect to the line $y = x$.

4.6 Variation

Definitions ($k \neq 0$)

Direct Variation: $y = kx$

Inverse Variation: $y = \dfrac{k}{x}$

Joint Variation: $y = kx_1x_2$

Combined Variation: $y = \dfrac{kx_1x_2}{x_3}$

Chapter 4 Review Exercises

1. Given $f(x) = 2x^2 - 2$, find each of the following.
   (a) $f(4)$ (b) $f(-5)$ (c) $f(\sqrt{3})$
2. Given $g(x) = (3x + 4)/(x^2 + 5)$, find each of the following.
   (a) $g(2x)$ (b) $g(\sqrt{x})$ (c) $g(x - 1)$
3. Find the domain and range of each of the following functions.
   (a) $f(x) = 3x - 4$ (b) $g(x) = x^2 + 1$
   (c) $h(x) = |x|$
4. Find the domain for each of the following functions.
   (a) $f(x) = \dfrac{5}{3x - 4}$ (b) $g(x) = \dfrac{5}{\sqrt{3x - 4}}$
   (c) $h(x) = \dfrac{1}{x^2 - 4}$
5. Given $f(x) = x^2 + 2$ and $g(x) = \sqrt{x - 2}$, find each of the following.
   (a) $(f + g)(x)$ (b) $(fg)(x)$ (c) $(f/g)(x)$
6. For f and g given in Exercise 5, find each of the following.
   (a) $(f + g)(5)$ (b) $(fg)(2x)$ (c) $(f/g)(x^2)$
7. For f and g given in Exercise 5, find the domain of $f + g$, fg, and f/g, respectively.
8. Given $f(x) = 1 + x$ and $g(x) = \sqrt{x}$, find each of the following.
   (a) $(g \circ f)(x)$ (b) $(f \circ g)(x)$

9. For f and g given in Exercise 8, find each of the following.
(a) $(g \circ f)(5)$ (b) $(f \circ g)(2x)$

10. For f and g given in Exercise 8, find the respective domains of $g \circ f$ and $f \circ g$.

11. Which of the following are polynomial functions?
(a) $f(x) = 3x^4 - 1.2x^2 - .8x + \sqrt{2}$
(b) $g(x) = 2$
(c) $h(x) = \dfrac{x-2}{x^2+9}$

12. Which of the following are rational functions?
(a) $F(x) = \sqrt[3]{\dfrac{x-2}{x^2+9}}$
(b) $G(x) = x^{-2} + 2x + 2$
(c) $H(x) = \dfrac{x^2+2x}{x^3-1}$

13. Which of the following are even functions?
(a) $f(x) = x^2 - 3x^4$ (b) $g(x) = x^4 + 1$
(c) $h(x) = |x|$

14. Which of the following are odd functions?
(a) $F(x) = x^3 - 2x - 2$ (b) $G(x) = \sqrt[3]{x}$
(c) $H(x) = \dfrac{x^3+5x}{x^2+1}$

15. Which of the functions in Exercise 13 have graphs that are symmetric with respect to the y-axis?

16. Which of the functions in Exercise 14 have graphs that are symmetric with respect to the origin?

17. Find all vertical asymptotes for each of the following rational functions.
(a) $f(x) = \dfrac{x+4}{2x-5}$ (b) $g(x) = \dfrac{x}{x^2-4}$
(c) $h(x) = \dfrac{x^3+6}{x^2-4x+3}$

18. Find the horizontal asymptote, if any, for each of the rational functions in Exercise 17.

19. Find all vertical and horizontal asymptotes for the algebraic function $F(x) = (\sqrt{x^2+1})/(x-2)$.

Sketch the graph of each of the functions in Exercises 20–22.

20. $f(x) = x^3 - 4x^2$ 21. $g(x) = \dfrac{3-x}{x+2}$

22. $h(x) = \sqrt{\dfrac{3-x}{x+2}}$

23. Use the algebraic definition of a one-to-one function to show that $f(x) = 3x - 5$ is one-to-one.

24. Use the horizontal line test to show that the function $g(x) = x^3 + 1$ is one-to-one.

25. Use the horizontal line test to show that the function in Exercise 20 is not one-to-one.

26. Sketch a graph of $f(x) = 3x - 5$ and its inverse on the same axes.

27. Sketch a graph of $g(x) = x^3 + 1$ and its inverse on the same axes.

28. Which, if any, of the following functions are increasing and which are decreasing?

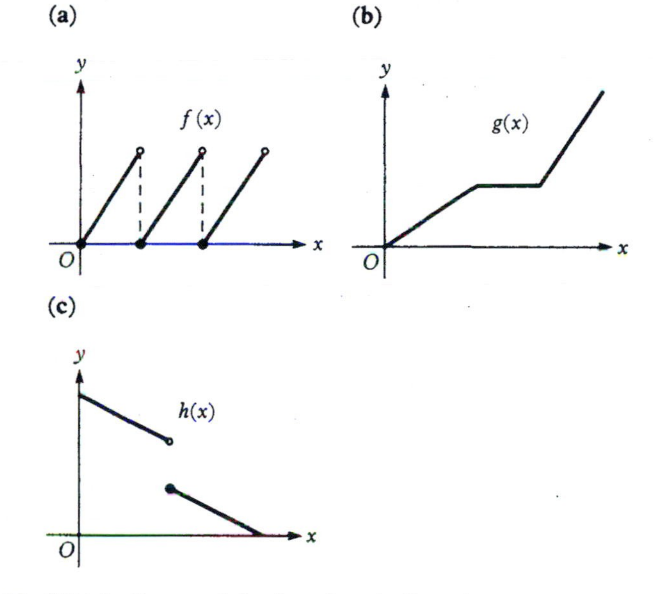

29. Which, if any, of the functions in Exercise 28 are one-to-one?

30. Let $f(x) = 3x + 5$. Use the algebraic definition to show that f is an increasing function.

31. Let $g(x) = -3x + 5$. Use the algebraic definition to show that g is a decreasing function.

Sketch the graph of each of the functions in Exercises 32–34.

32. $f(x) = |x - 1|$ 33. $g(x) = [x + 1]$

34. $h(x) = \begin{cases} x - 1 & \text{if } x \leq 1 \\ x^2 & \text{if } x > 1 \end{cases}$

35. What are the domain and range of the function g in Exercise 33?

36. What are the domain and range of the function h in Exercise 34?

*37. Express as an equation the statement that y varies jointly as u and v and inversely as w.

*38. Write the equation $L = k(wd^2)/r$ in the terminology of variation.

*39. The measure of intelligence known as the I.Q. varies directly with mental age M and inversely with chronological age C. If a 10-year-old with a mental age of 12 has an I.Q. of 120, what is the I.Q. of a 15-year-old with a mental age of 12?

*40. The time t that it takes the earth to complete one orbit about the sun is directly proportional to the square root of d^3, where d is its maximum distance from the sun. Using $d = 93$ million miles, and $t = 1$ year, compute the proportionality constant.

Graphing Calculator Exercises

In each of the following, verify your answers to the indicated review exercises of this section by using a graphing calculator for the given functions and windows $[x_{min}, x_{max}]$, $[y_{min}, y_{max}]$.

41. Exercise 13:
    (a) $Y_1 = x^2 - 3x^4$, [−.95,.95], [−.5,.5]
    (b) $Y_1 = x^4 + 1$, [−4.5,5], [−5,5]
    (c) $Y_1 = \text{abs}(x)$, [−4.5,5], [−5,5]

42. Exercise 14:
    (a) $Y_1 = x^3 - 2x - 2$, [−4.5,5], [−10,10]
    (b) $Y_1 = \sqrt[3]{x}$, [−4.5,5], [−3,3]
    (c) $Y_1 = (x^3 + 5x)/(x^2 + 1)$, [−4.5,5], [−5,5]

43. Exercises 17 and 18:
    (a) $Y_1 = (x + 4)/(2x - 5)$, [−2.5,7], [−15,15]
    (b) $Y_1 = x/(x^2 - 4)$, [−4.5,5], [−5,5]
    (c) $Y_1 = (x^3 + 6)/(x^2 - 4x + 3)$, [−2.5,7], [−30,30]

44. Exercise 19: $Y_1 = \sqrt{x^2 + 1}/(x - 2)$, [−2.5,7], [−15,15]

45. Exercise 20: $Y_1 = x^3 - 4x^2$, [−4.5,5], [−10,10]

46. Exercise 21: $Y_1 = (3 - x)/(x + 2)$, [−5.5,4], [−10,10]

47. Exercise 22: $Y_1 = \sqrt{((3 - x)/(x + 2))}$, [−4.5,5], [0,5]

48. Exercise 26: $Y_1 = 3x - 5$, $Y_2 = (x + 5)/3$, [−9.5,9.5], [−7,7]

49. Exercise 27: $Y_1 = x^3 + 1$, $Y_2 = \sqrt[3]{x - 1}$, [−4.5,5], [−5,5]

50. Exercise 32: $Y_1 = \text{abs}(x - 1)$, [−2.5,7], [−1,6]

51. Exercise 33: $Y_1 = \text{Int}(x + 1)$, [−4.5,5], [−5,5] (Dot Mode)

52. Exercise 34: $Y_1 = (x - 1)(x \leq 1) + x^2(x > 1)$, [−2.25,2.5], [−3,6] (Dot Mode)

*Asterisks indicate topics that may be omitted, depending on individual course needs.

5 Exponential and Logarithmic Functions

In this chapter we study two new functions that are quite different from the algebraic functions discussed previously. Using our knowledge of rational exponents, we define first the *exponential function* and then its inverse, the *logarithmic function*. The exponential and logarithmic functions are used extensively in calculus, as well as in many areas of applied mathematics, including growth of populations, decay of radioactive materials, and compound interest.

5.1 The Exponential Function and Its Graph

Motion in the universe is along conic sections, but growth and decay in nature are exponential.

The Exponential Function
Further Properties of the Exponential
The Natural Base *e*

In the last chapter we graphed polynomial functions, such as $f(x) = x^3 - 3x^2 + 2x + 1$, and some other algebraic functions, such as $g(x) = \sqrt{x}$ and $h(x) = x^{2/3}$. All of these functions involve terms of the form x^a, where the base x is a variable and the exponent a is a constant. Now we reverse that situation to consider a^x with constant base a and variable exponent x. The function $f(x) = a^x$ is called an **exponential function.**

The Exponential Function. Since we learned the meaning of rational exponents in Section 1.4, most of us, if asked to draw the graph of the equation $y = 2^x$, would probably construct a table and plot the corresponding points on the graph as shown in Figure 1.

x	$y = 2^x$
−3	1/8
−2	1/4
−1	1/2
0	1
1	2
2	4
3	8

Figure 1 $y = 2^x$ for Integer Values of x

By using some nonintegral values of x, we might add additional points such as $(-3/2, .35)$ and $(1/3, 1.26)$, because we know that

$$\text{if } x = -3/2, \text{ then } 2^x = 2^{-3/2} = \sqrt{2^{-3}} = \sqrt{1/8} \approx .35,$$

and

$$\text{if } x = 1/3, \text{ then } 2^x = 2^{1/3} = \sqrt[3]{2} \approx 1.26.$$

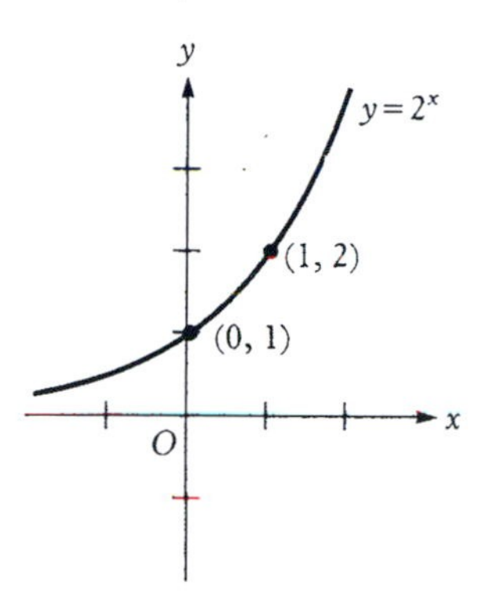

Figure 2

It is then quite natural to join the plotted points by a smooth curve, obtaining the graph in Figure 2.

In drawing such a curve, we are assigning values to 2^x for every real number, although we have not yet given meaning to 2^x when x is an irrational number. We are forced to ask: *What is meant by expressions such as*

$$2^{\sqrt{2}}, \quad 2^{\sqrt[3]{5}} \quad \text{or} \quad 2^{\pi}?$$

It is easy to guess that $2^{\sqrt{2}} \approx 2^{1.4} \approx 2.6390$ and that successively better approximations are[1]

$$2^{\sqrt{2}} \approx 2^{1.41} \approx 2.657$$
$$2^{\sqrt{2}} \approx 2^{1.414} \approx 2.665$$
$$2^{\sqrt{2}} \approx 2^{1.4142} \approx 2.665,$$

so that the point with $x = \sqrt{2}$ and $y \approx 2.67$ is included in Figure 2.

The above reasoning can lead to a mathematically precise definition of $f(x) = 2^x$ for *all real* x, but it would require concepts from calculus. Therefore, we shall *assume that* f is defined by the process described above. That is, we assume that the domain of 2^x can be extended from the rational numbers to the set of all real numbers in such a way that the graph of $f(x) = 2^x$ is continuous and smooth.[2]

More generally,

> the **exponential function with base a** is defined by
>
> $$f(x) = a^x \quad (a > 0, a \neq 1)$$
>
> with the following properties:
>
> 1. f is a function with $D_f = (-\infty, \infty)$. That is, for each real number x, a^x is a unique real number.
> 2. f is a continuous and smooth extension of $g(x) = a^x$, where D_g is the set of rational numbers.

The restriction $a > 0$ was already explained in Section 1.4. It is needed to avoid nonreal numbers such as $(-4)^{1/2}$, and to avoid exceptions to the power-of-a-power rule $(a^x)^y = a^{xy}$. The restriction $a \neq 1$ is made since $1^x = 1$ for all real x implies that a^x is a constant function for $a = 1$, and we do not consider constant functions to be exponential.

The graph of $y = 2^x$ (Figure 2) illustrates that $f(x) = a^x$ is an *increasing* function for base $a > 1$. That is, if $a > 1$, then $a^{x_1} > a^{x_2}$ whenever $x_1 > x_2$.

[1] The rational powers of 2 were computed with a scientific calculator and rounded off to four significant digits. (See the definition of significant digits in Appendix A.)

[2] By *continuous* we mean there are no breaks in the curve, and by *smooth* we mean that the curve also has no sharp turns. More precise definitions are given in calculus.

The following example illustrates that $f(x) = a^x$ is *decreasing* when $0 < a < 1$. If $0 < a < 1$, then $a^{x_1} < a^{x_2}$ whenever $x_1 > x_2$.

Example 1 Sketch the graph of $y = (1/2)^x$.

Solution: We first make a table of coordinates by substituting some rational values for x and computing the corresponding values of y. We then plot the corresponding points and join them by a smooth curve, as shown in Figure 3. ■

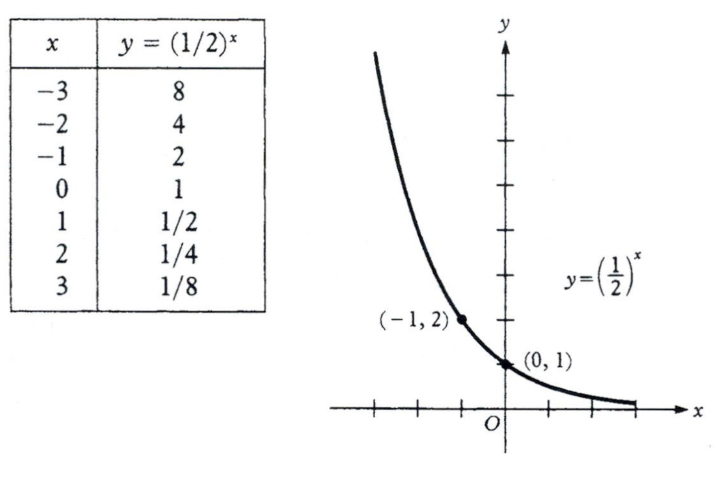

x	$y = (1/2)^x$
-3	8
-2	4
-1	2
0	1
1	1/2
2	1/4
3	1/8

Figure 3

Figures 2 and 3 illustrate the following important **properties of the graph of an exponential function $f(x) = a^x$.**

1. $f(x) > 0$ for all real numbers x.
2. $f(0) = 1$
3. If $a > 1$, then $f(x) \to \infty$ as $x \to \infty$, and $f(x) \to 0$ as $x \to -\infty$.

3′. If $0 < a < 1$, then $f(x) \to 0$ as $x \to \infty$, and $f(x) \to \infty$ as $x \to -\infty$.

Property 1 says that the graph of any exponential function is always above the x-axis, and property 2 says that the curve always goes through the point (0, 1). Property 3 (3′) says that for any exponential function, the range R_f is $(0, \infty)$, and the x-axis is a horizontal asymptote.

We could restrict our attention to bases greater than 1 by introducing a negative sign in the exponent. That is, by definition,

$$a^{-x} = \left(\frac{1}{a}\right)^x \quad (a > 0).$$

Hence, in Example 1, we may write $y = (1/2)^x = 2^{-x}$. In general, if $f(x) = b^x$ with $0 < b < 1$, then we can write $f(x) = a^{-x}$ with $a = 1/b > 1$.

Example 2 Sketch $f(x) = 2^x$ and $g(x) = 2^{-x}$ on the same axes.

Solution: By combining Figures 2 and 3 we obtain the graph shown in Figure 4. ■

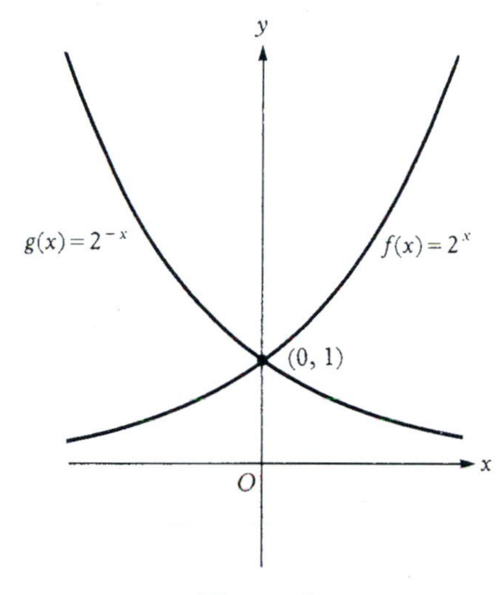

Figure 4

Figure 4 shows that the graphs of f and g, taken together, are symmetric with respect to the y-axis. More generally, to show that the graphs of $f(x) = a^x$ and $g(x) = a^{-x}$, taken together, are symmetric with respect to the y-axis, it is sufficient to observe that

$$f(-x) = a^{-x} = g(x).$$

Example 3 Sketch the graphs of $y = 3^x$ and $y = 2^x$ on the same axes.

Solution: For $y = 3^x$ we first construct a table of coordinates, and for $y = 2^x$ we use Figure 2. The graphs are shown in Figure 5. ■

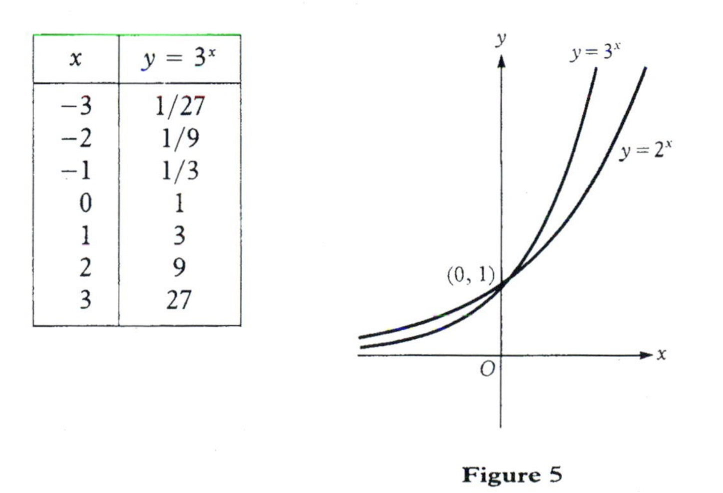

x	$y = 3^x$
−3	1/27
−2	1/9
−1	1/3
0	1
1	3
2	9
3	27

Figure 5

Figure 5 shows that the graph of 3^x starts lower but rises more rapidly than that of 2^x. The graph of 10^x would start even lower but rise more rapidly than both 2^x and 3^x. All functions $f(x) = a^x$, $a > 1$, are said to be **exponentially increasing,** and $b > a > 1$ implies that b^x increases more rapidly than a^x. Similarly, any function $g(x) = a^{-x}$, $a > 1$, is said to be **exponentially decreasing,** and $b > a > 1$ implies that b^{-x} decreases more rapidly than a^{-x}. As stated earlier, for both $f(x) = a^x$ and $g(x) = a^{-x}$, the domain is the set of all real numbers, the range is the set of all positive real numbers, and the x-axis is a horizontal asymptote.

Question *Which function increases more rapidly, 2^x or x^2?*

Answer: The 2^x function increases more rapidly. For instance, the equation $2^{x+1} = 2 \cdot 2^x$ says that 2^x doubles in value whenever x increases by 1. On the other hand, $(\sqrt{2}x)^2 = 2 \cdot x^2$ says that x must increase by a *factor* of $\sqrt{2}$ in order for x^2 to double in value. ■

Further Properties of the Exponential. Since we have seen that $f(x) = a^x$ is either increasing $(a > 1)$ or decreasing $(0 < a < 1)$, it follows from Section 4.5 that a^x is a *one-to-one function*. That is,

$$a^x = a^y \Leftrightarrow x = y.$$

Example 4 Solve for x in each of the following exponential equations.

(a) $2^x = 64$ (b) $2^{-x} = 4^x$ (c) $8^{x-1} = 16^{x-5/6}$

Solution:

(a) $2^x = 64 \Leftrightarrow 2^x = 2^6 \Leftrightarrow x = 6$ *2^x is one-to-one*

(b) $2^{-x} = 4^x \Leftrightarrow 2^{-x} = 2^{2x} \Leftrightarrow -x = 2x \Leftrightarrow 0 = 3x \Leftrightarrow x = 0$

(c) $8^{x-1} = 16^{x-5/6} \Leftrightarrow (2^3)^{x-1} = (2^4)^{x-5/6} \Leftrightarrow 2^{3x-3} = 2^{4x-10/3}$

$$\Leftrightarrow 3x - 3 = 4x - \frac{10}{3} \Leftrightarrow x = 1/3$$

Let us check this last answer by substituting $x = 1/3$ in the given equation:

$$8^{(1/3)-1} = 16^{(1/3)-(5/6)}$$
$$8^{-2/3} = 16^{-1/2}$$
$$\frac{1}{(\sqrt[3]{8})^2} = \frac{1}{\sqrt{16}}$$
$$\frac{1}{2^2} = \frac{1}{4}.$$ ■

In each part of Example 4, we were able to express both sides of the equation to the *same* base and then apply the one-to-one property. This is always possible, but in general we must use logarithms when several bases are involved, as we will see later in this chapter.

Since the function a^x is a continuous extension from the domain of rational numbers to the domain of all real numbers, the five **basic rules for exponents** (Section 1.4) hold for all positive bases and all real exponents. These rules are listed below.

$$a^x a^y = a^{x+y} \quad \text{and} \quad \frac{a^x}{a^y} = a^{x-y} \qquad \textbf{same-base rules}$$

$$(ab)^x = a^x b^x \quad \text{and} \quad \left(\frac{a}{b}\right)^x = \frac{a^x}{b^x} \qquad \textbf{same-exponent rules}$$

$$(a^x)^y = a^{xy} \qquad \textbf{power-of-a-power rule}$$

Example 5 Use the rules for exponents to simplify each of the following.

(a) $a^{\sqrt{2}}a^{3\sqrt{2}}$ (b) $\dfrac{a^{\pi}}{a^{\pi-2}}$ (c) $(a^{\sqrt{2}})^{\sqrt{2}}$ (d) $10^{\sqrt{3}}100^{\sqrt{3}}$

Solution:

(a) $a^{\sqrt{2}}a^{3\sqrt{2}} = a^{\sqrt{2}+3\sqrt{2}} = a^{4\sqrt{2}}$ *Same-base rule*

(b) $\dfrac{a^{\pi}}{a^{\pi-2}} = a^{\pi-(\pi-2)} = a^2$ *Same-base rule*

(c) $(a^{\sqrt{2}})^{\sqrt{2}} = a^{\sqrt{2}\cdot\sqrt{2}} = a^2$ *Power-of-a-power rule*

(d) $10^{\sqrt{3}}100^{\sqrt{3}} = 10^{\sqrt{3}}(10^2)^{\sqrt{3}}$

$= 10^{\sqrt{3}}10^{2\sqrt{3}} = 10^{\sqrt{3}+2\sqrt{3}} = 10^{3\sqrt{3}}$ ■

The Natural Base *e*. In Figure 5, the graphs of $y = 2^x$ and $y = 3^x$ are shown, but both in theory and in applications, the most useful base is a number between 2 and 3 denoted by the letter e [so named after the Swiss mathematician Euler (1707–1783)]. This number is irrational and hence is represented by a nonending, nonrepeating decimal, but e can be approximated as

$$e \approx 2.71828. \qquad \textbf{natural base}$$

The number e arises most naturally in calculus, but, as we will see in Section 5.4, e also appears in the context of compound interest.

The natural base e can be defined as the value approached by the expression

$$\left(1 + \frac{1}{n}\right)^n \tag{1}$$

as n goes through the sequence of integers 1, 2, 3, ... That is, for large values of n,

$$\left(1 + \frac{1}{n}\right)^n \approx e.$$

Most scientific calculators have an e^x (or exp x) key by which this function can be computed for any value of x.

Example 6 Use a calculator to compute (1) for the following values of n.

(a) $n = 10$ (b) $n = 100$ (c) $n = 10^6$

Solution: The following computations were made on a scientific calculator and rounded off to four significant digits.

(a) $\left(1 + \frac{1}{10}\right)^{10} = (1.1)^{10} \approx 2.594$

(b) $\left(1 + \frac{1}{100}\right)^{100} = (1.01)^{100} \approx 2.705$

(c) $\left(1 + \frac{1}{10^6}\right)^{10^6} = (1.000001)^{1{,}000{,}000} \approx 2.718$ ■

Exercises 5.1

Fill in the blanks to make each statement true.

1. The exponential function a^x is defined for __________ values of $a \neq 1$.
2. The exponential function a^x has domain __________ and range __________.
3. The exponential function is increasing if the base a is __________ and decreasing if a is __________.
4. If $f(x) = a^x$, then $f(x) = b^{-x}$, where $b =$ __________.
5. Because the exponential is a __________ function, $a^x = a^y \Leftrightarrow x = y$.

Write true *or* false *for each statement.*

In Exercises 6–10, it is assumed that the base a is positive and $a \neq 1$.

6. The exponential function a^x is defined for every real number x.
7. The x-axis is an asymptote for the graph of $y = a^x$.
8. $a^x \to 0$ as $x \to \infty$
9. The graphs of $y = a^x$ and $y = a^{-x}$, taken together, are symmetric with respect to the y-axis.
10. The natural base e, approximated to one decimal place, is 2.7.

Graphs of Exponential Functions

Sketch the graph of each of the following exponential functions.

11. $f(x) = 4^x$
12. $g(x) = 10^x$
13. $h(x) = \left(\frac{3}{2}\right)^x$
14. $\epsilon(x) = e^x$
15. $F(x) = \left(\frac{1}{4}\right)^x$
16. $G(x) = \left(\frac{1}{10}\right)^x$
17. $H(x) = \left(\frac{2}{3}\right)^x$
18. $E(x) = \left(\frac{1}{e}\right)^x$
19. $F(x) = 4^{-x}$
20. $G(x) = 10^{-x}$
21. $H(x) = \left(\frac{3}{2}\right)^{-x}$
22. $E(x) = e^{-x}$
23. How do the graphs in Exercises 15–18 compare with those in Exercises 11–14, respectively?
24. How do the graphs in Exercises 19–22 compare with those in Exercises 15–18, respectively?

Sketch the graph of each function in Exercises 25–32.

25. $y = 2^x - 1$
26. $y = 2^{x-1}$
27. $y = 2^{2x}$
28. $y = 2^{(x^2)}$
29. $y = 2^{-x^2}$
30. $y = e^{-x^2}$
31. $y = 1.5e^{2t}$
32. $y = 1.5e^{-2t}$
33. Sketch the graph of $y = (e^x + e^{-x})/2$. This curve is called a **catenary** and is the shape of a cable hanging suspended between two supports.

Properties of Exponentials

Solve each of the following equations for x.

34. $2^x = 4^{-x}$
35. $2^{2x} = 4^x$

36. $3^{x+1} = 81^{x+(5/2)}$

37. $3^x 9^{x+1} = 27^{1-x}$

Solve each of the following equations for a^x. (Hint: let $a^x = y$ and find a quadratic equation in y that is equivalent to the given equation.)

38. $a^x + a^{-x} = 3$

39. $a^x - 4a^{-x} = 3$

40. $\dfrac{a^x - a^{-x}}{a^x + a^{-x}} = \dfrac{1}{3}$

41. $\dfrac{a^x + a^{-x}}{a^x - a^{-x}} = -3$

Use the rules for exponents to simplify each of the following.

42. $(4^{\sqrt{2}})^{\sqrt{2}}$

43. $4^{\sqrt{2}} + 2^{\sqrt{2}}$

44. $\dfrac{3^{2x}}{9^x}$

45. $\dfrac{4^{-2x}}{32^{-x}}$

46. $\left(\dfrac{a^{3/\sqrt{2}}}{b^{\sqrt{2}/2}}\right)^{\sqrt{2}}$

47. $\dfrac{[(a^2)^x]^{1/2}}{(a^{-x/2})^{-4}}$

48. $\left(\dfrac{a^x + a^{-x}}{2}\right)^2 - \left(\dfrac{a^x - a^{-x}}{2}\right)^2$

49. $\dfrac{a^x + 1}{a^x + a^{-x} + 2}$

The Natural Base *e*

Exercises 50–60 require the use of a calculator.

50. Approximate e^2 by $(1 + 1/n)^{2n}$ for $n = 10^6$. Compare this value with $(1 + 2/n)^n$ for $n = 10^6$ and with $e^2 \approx 7.3890561$.

51. Approximate $1/e$ by $(1 + 1/n)^{-n}$ for $n = 10^6$. Compare this value with $(1 - 1/n)^n$ for $n = 10^6$ and with $1/e \approx .36787944$.

It is shown in calculus that for small $|x|$

$$e^x \approx 1 + x + \frac{x^2}{2} + \frac{x^3}{6} + \frac{x^4}{24} + \frac{x^5}{120}.$$

Use this result to approximate the following numbers and compare with the value obtained on a scientific calculator.

52. e

53. $e^{0.1}$

54. $e^{0.01}$

Replacing x by $-x$ in the above approximation for e^x gives

$$e^{-x} \approx 1 - x + \frac{x^2}{2} - \frac{x^3}{6} + \frac{x^4}{24} - \frac{x^5}{120}.$$

Use this result to approximate the following numbers and Compare with the value obtained on a scientific calculator.

55. e^{-1}

56. $e^{-0.1}$

57. $e^{-0.01}$

58. Approximate $3^{\sqrt{2}}$ by finding the successive values of $3^{1.4}$, $3^{1.41}$, $3^{1.414}$, and $3^{1.4142}$.

59. Approximate $3^{\sqrt[3]{2}}$ by finding the successive values of $3^{1.2}$, $3^{1.25}$, $3^{1.259}$, and $3^{1.2599}$.

60. It is shown in calculus that for small $|x|$

$$3^x \approx 1 + 1.0986x + \frac{(1.0986x)^2}{2} + \frac{(1.0986x)^3}{6} + \frac{(1.0986x)^4}{24}.$$

Use this formula to approximate $3^{\sqrt{2}}$ and compare with Exercise 58.

5.2 The Logarithmic Function and Its Graph

Before calculators there were slide rules, and before slide rules there were tables of logarithms.

Definition of Logarithm
Graph of the Logarithmic Function
Change of Base

Definition of Logarithm. In the exponential function $y = a^x$, the independent variable is the exponent x, and the dependent variable is the value $y = a^x$. In the logarithmic function, the roles of independent variable and dependent variable are reversed. The independent variable is now the value $x = a^y$, and the dependent variable is now the exponent y. That is, if $a^y = x$, then y is called the **logarithm of x to the base a**, and we write

$$\log_a x = y.$$

For example, $\log_5 25 = 2$ because $5^2 = 25$,

and $\log_2 \left(\frac{1}{8}\right) = -3$ because $2^{-3} = \frac{1}{8}$.

Since the logarithm of a number to a given base is the **exponent** to which the base must be raised to yield the number, we can write

$$\boldsymbol{a^{(\log_a x)} = x.} \tag{1}$$

Similarly, since x is the exponent to which the base a must be raised to give the number a^x,

$$\boldsymbol{\log_a (a^x) = x.} \tag{2}$$

Equations (1) and (2) tell us that a^x and $\log_a x$ are **inverse functions.** That is, if $f(x) = a^x$, then equation (1) says that

$$f(\log_a x) = x, \tag{1'}$$

and equation (2) says that

$$\log_a (f(x)) = x. \tag{2'}$$

Therefore, $\log_a x = f^{-1}(x)$.

Schematically, equation (1) can be represented as shown in Figure 6.

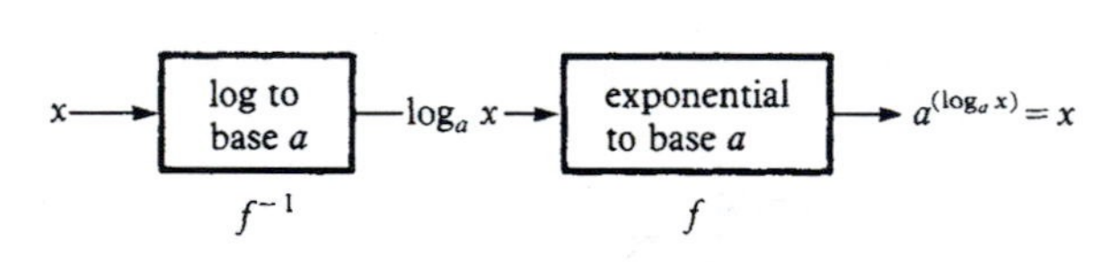

Figure 6

Equation (2) can be represented as shown in Figure 7.

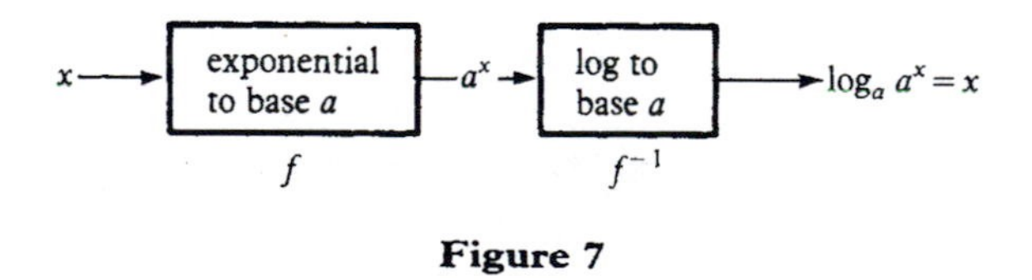

Figure 7

Figures 6 and 7 again show that a^x and $\log_a x$ are inverse functions.

The base a is always assumed to be positive and $a \neq 1$. If $a = e$, the natural base, then $\log_a x$ is called the **natural logarithm** of x and is denoted by $\ln x$. That is, we define $\ln x = \log_e x$, so that

$$\boldsymbol{y = \ln x \Leftrightarrow x = e^y},$$

which is equivalent to saying

$$\boldsymbol{f(x) = e^x \Leftrightarrow f^{-1}(x) = \ln x.}$$

The notation **log x** means $\log_{10} x$ and is called a **common logarithm.**

As illustrated below, by using the inverse property, we can convert any statement about exponents into an equivalent one about logarithms, and conversely. The most important fact to remember is that *a logarithm is an exponent.*

Exponential Statement	Equivalent Logarithmic Statement
$2^3 = 8$	$3 = \log_2 8$
$100^{-3/2} = .001$	$-3/2 = \log_{100} (.001)$
$a^0 = 1 \quad (a \neq 0)$	$0 = \log_a 1 \quad (a \neq 0)$
$e^1 = e$	$1 = \ln e$

The technique of equating logarithmic statements and exponential ones is so important that we now give some additional examples.

Example 1 Write the equivalent logarithmic statement in each of the following cases.

(a) $3^2 = 9$ (b) $a^1 = a$ (c) $a^{2/3} = \sqrt[3]{a^2}$ (d) $e^{-1} \approx .3679$

Solution:

(a) $3^2 = 9 \Leftrightarrow 2 = \log_3 9$ (b) $a^1 = a \Leftrightarrow 1 = \log_a a$

(c) $a^{2/3} = \sqrt[3]{a^2} \Leftrightarrow \dfrac{2}{3} = \log_a (\sqrt[3]{a^2})$ (d) $e^{-1} \approx .3679 \Leftrightarrow -1 \approx \ln (.3679)$ ■

Example 2 Convert each of the following into an equivalent statement in exponential form.

(a) $\log_5 125 = 3$ (b) $\log_{10} (.0001) = -4$

(c) $\ln 10 \approx 2.3026$ (d) $\log_a (a^2) = 2$

Solution:

(a) $\log_5 125 = 3$ means $5^3 = 125$.

(b) $\log_{10} (.0001) = -4$ means $10^{-4} = .0001$.

(c) $\ln 10 \approx 2.3026$ means $e^{2.3026} \approx 10$.

(d) $\log_a (a^2) = 2$ states that the exponent of a^2 is 2, and is equivalent to saying $a^2 = a^2$. ■

Example 3 Solve for x in each of the following.

(a) $x = \log_2 32$ (b) $-2 = \log_3 x$ (c) $1/2 = \log_x 100$

Solution:

(a) $x = \log_2 32 \Leftrightarrow 2^x = 32 \Leftrightarrow 2^x = 2^5 \Leftrightarrow x = 5$ (*why?*)

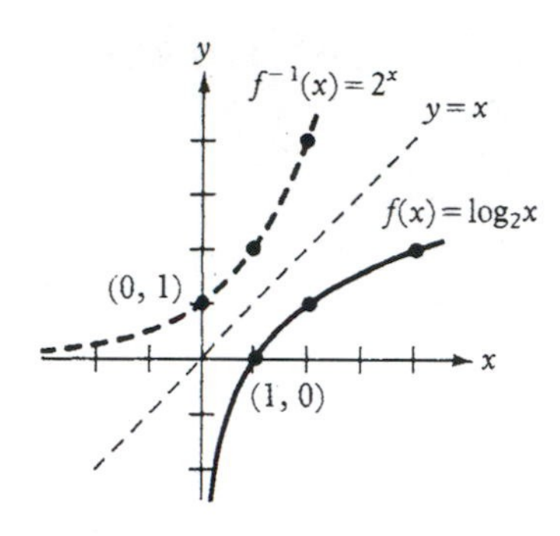

Figure 8

(b) $-2 = \log_3 x \Leftrightarrow 3^{-2} = x$ or $x = 1/9$

(c) $1/2 = \log_x 100 \Leftrightarrow x^{1/2} = 100 \Leftrightarrow x = 10{,}000$ ■

Graph of the Logarithmic Function. The graph of the function $f(x) = \log_a x$ can be obtained from the graph of its inverse $f^{-1}(x) = a^x$ by symmetry with respect to the line $y = x$. For example, the graph of $f(x) = \log_2 x$ (Figure 8) can be obtained from the graph of $f^{-1}(x) = 2^x$ given in Figure 2 in Section 5.1.

Figure 8 is typical of the graph of $f(x) = \log_a x$ for $a > 1$. Note that

$$D_f = (0, \infty) = R_{f^{-1}} \quad \text{and} \quad R_f = (-\infty, \infty) = D_{f^{-1}}.$$

In the following example, we consider $\log_a x$ with $0 < a < 1$. We do a direct point plot without reference to the graph of the inverse function.

Example 4 Sketch the graph of $f(x) = \log_{1/2} x$ by plotting points.

Solution: Since $y = \log_{1/2} x$ is equivalent to $x = (1/2)^y$, we can use this equation to make a coordinate table for the graph of f (Figure 9). ■

x	$y = \log_{1/2} x$
8	−3
4	−2
2	−1
1	0
1/2	1
1/4	2
1/8	3

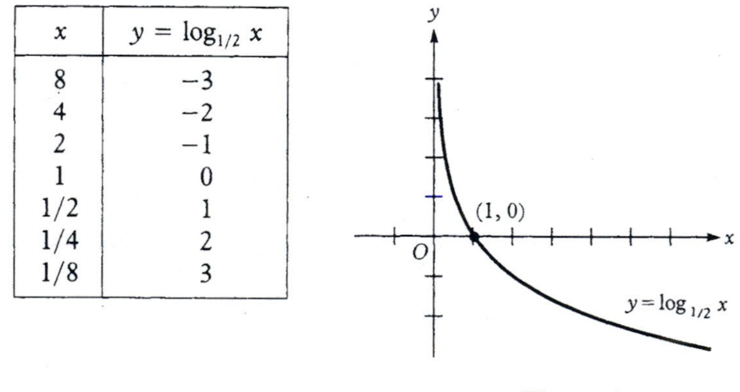

Figure 9

In the base a, we can easily compute *exact* values of logarithms of powers of a, as we did in the table in Example 4, where $a = 1/2$. We can then use the graph to *approximate* the value of $\log_a x$ when x is not an integral power of a. However, for a more precise value of $\log_a x$, say to four significant digits, we need a calculator.

Figure 9 is typical of the graph of $\log_a x$ for $0 < a < 1$. Figures 8 and 9 illustrate the following important **properties of the logarithmic function.**

1. $\log_a x$ is defined only for $x > 0$.
2. $\log_a 1 = 0$
3. If $a > 1$, then $\log_a x \to \infty$ as $x \to \infty$
   and $\log_a x \to -\infty$ as $x \to 0^+$.

3′. If $0 < a < 1$, then $\log_a x \to -\infty$ as $x \to \infty$
and $\log_a x \to \infty$ as $x \to 0^+$.

Property 1 says that the domain of the logarithmic function is $(0, \infty)$, and 2 says that for any base, the graph contains the point $(1, 0)$. Property 3 (3′) says that the range of the logarithmic function is $(-\infty, \infty)$ and the y-axis is a vertical asymptote.

Change of Base. As stated in Section 5.1, the natural base $e \approx 2.71828$ is the most useful base in mathematics. Also, the natural logarithm of a number x can be converted to the logarithm of x to a different positive base $b \neq 1$ by means of the formula

$$\log_b x = \frac{\ln x}{\ln b}. \tag{3}$$

For example, by using a calculator, we find

$$\log_2 17 = \frac{\ln 17}{\ln 2} = \frac{2.8332}{0.6931} \approx 4.0877.$$

Formula (3) is a special case of the more general result

$$\boldsymbol{\log_b x = \frac{\log_a x}{\log_a b},} \qquad \textbf{change of base rule} \tag{4}$$

which enables us to convert logarithms from *any* positive base $a \neq 1$ to *any other* positive base $b \neq 1$ [see the exercises for a derivation of (4)]. For example,

$$\log_2 17 = \frac{\log_{10} 17}{\log_{10} 2} = \frac{1.2304}{0.3010} \approx 4.0877.$$

Exercises 5.2

Whenever the expression $\log_a x$ *appears, it is assumed that* $a > 0$ *and* $a \neq 1$.

Fill in the blanks to make each statement true.

1. The equation $y = \log_a x$ is equivalent to $x =$ ________.
2. The function $f(x) = \log_a x$ has the inverse $f^{-1}(x) =$ ________.
3. The domain of $f(x) = \log_a x$ is ________, and its range is ________.
4. The function $f(x) = \log_a x$ is increasing if a is ________ and is decreasing if a is ________.
5. The notation $\log x$ is used for a logarithm with base ________, and $\ln x$ is used for base ________.

Write true *or* false *for each statement.*

6. $\log_a a^x = x$ for positive x only.
7. $a^{\log_a x} = x$ for positive x only.
8. The x-axis is an asymptote for the graph of $y = \log_a x$.
9. The y-axis is an asymptote for the graph of $y = \log_a x$.
10. The graphs of $y = \log x$ and $y = 10^x$, taken together, are symmetric with respect to the line $y = x$.

Definition of Logarithm

Write a logarithmic statement equivalent to each of the following.

11. $2^4 = 16$
12. $2^{-5} = \frac{1}{32}$
13. $10^{2/3} = \sqrt[3]{100}$
14. $10^{-3} = .001$
15. $16^{3/2} = 64$
16. $a^b = c$

Write each of the following in exponential form.

17. $\log_{10} 100 = 2$
18. $\log_3 81 = 4$
19. $\log_3 \frac{1}{27} = -3$
20. $\log_{10} .00001 = -5$
21. $\log_a 1 = 0$
22. $\log_a x = 2$

Solve each of the following equations for x.

23. $\log_2 16 = x$
24. $\log_8 x = -2$
25. $\log_x 32 = 5$
26. $\log_3 (x + 2) = 4$
27. $\log_3 [(x + 2)^2] = -4$
28. $\log_3 [(x - 2)^2] = -4$

Compute each of the following without using a calculator.

29. $10^{(\log_{10} 100)}$
30. $2^{\log_2 2}$
31. $\log_5 5^{-1000}$
32. $2 \log_a a^2$
33. $\log_{10} 100{,}000$
34. $\log_{10} 0.1$
35. $\log_{10} 1$
36. $\log_{10} (\sqrt[3]{10})$
37. $\log_{1/2} 16$
38. $\log_3 \left(\frac{1}{243}\right)$

Graphs of Logarithmic Functions

Sketch each of the following graphs by first converting $y = \log_a x$ to $x = a^y$ and then plotting points. In some cases, a calculator will be helpful.

39. $y = \log_2 x$
40. $y = \log_3 x$
41. $y = \log_{1/3} x$
42. $y = \ln x$

Obtain the graph of each of the following functions by first graphing the inverse function. A calculator may be helpful.

43. $f(x) = \log_{1.5} x$
44. $g(x) = \log_{3/4} x$
45. $F(x) = \log_2 (x + 1)$
46. $G(x) = \log_{1/2} (x - 2)$

Change of Base

47. Given two bases a and b, show that for any $x > 0$,

$$\log_b x = \frac{\log_a x}{\log_a b}.$$

(*Hint:* let $u = \log_a x$ and $v = \log_b x$. Then $a^u = b^v$. Now take log to base a of both sides of this exponential equation.)

Use the result in Exercise 47 to write each of the following as a ratio of common logarithms; then use a calculator to perform the division.

48. $\log_5 28$
49. $\log_2 100$
50. $\log_{.5} .15$
51. $\log_6 .25$
52. $\log_{25} 1055$
53. $\log_{17} 4981$

54. Use the result in Exercise 47 to show that $\log_b a = 1/(\log_a b)$.
55. Use the result in Exercise 47 to show that $\ln x = \log x / \log e$.
56. Use the result in Exercise 47 to show that $\log x = \ln x / \ln 10$.

Miscellaneous

57. Use Exercise 47 to show that if $b > a > 1$, then $\log_b x < \log_a x$.
58. Use the property that the exponential function $f(x) = a^x$ is increasing for $a > 1$ to show that $\log_a x$ is increasing for $a > 1$.
59. Use the property that the exponential function $f(x) = a^x$ is decreasing for $0 < a < 1$ to show that $\log_a x$ is decreasing for $0 < a < 1$.

*The Equation $a^x = x$

The equation $a^x = x$ will have a solution for x if and only if the graph of $y = a^x$ intersects the graph of $y = x$.

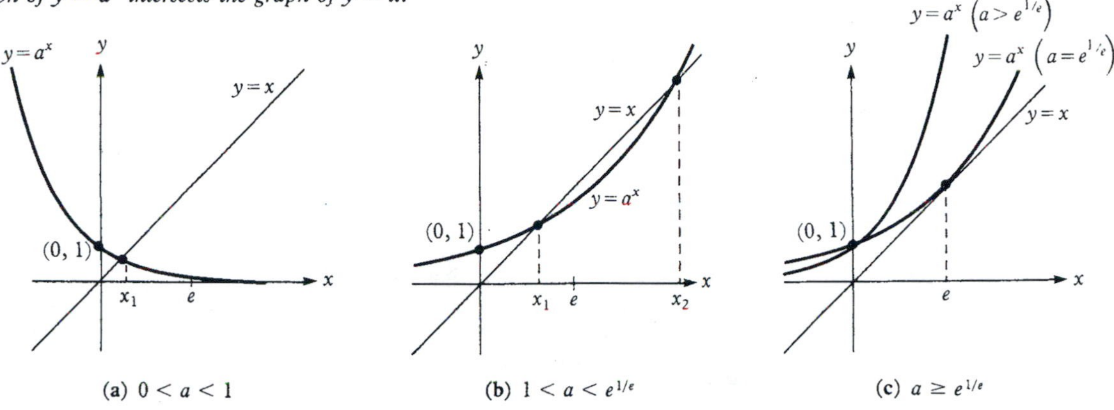

(a) $0 < a < 1$ (b) $1 < a < e^{1/e}$ (c) $a \geq e^{1/e}$

We can see from part (a) of the figure that for $0 < a < 1$, there is exactly one solution $x_1 < 1$. It can be shown, using calculus, that for $1 < a < e^{1/e}$, there are two solutions x_1 and x_2, where $x_1 < e < x_2$. For $a = e^{1/e}$, there is one solution, namely $x = e$, and for $a > e^{1/e}$, there are no solutions.

*Exercises 60–63 are optional.

Use a calculator to search for approximate solutions (to three significant digits) to each of the following equations.

60. $(.5)^x = x$ (one solution)

61. $(.75)^x = x$ (one solution)

62. $(1.1)^x = x$ (two solutions)

63. $(1.3)^x = x$ (two solutions)

5.3 Properties of the Logarithm

Logarithms can simplify computations by changing powers into products, products into sums, and quotients into differences.

Further Properties of Logarithms
Use of Logarithmic Tables

We saw several properties of the graph of the logarithmic function in Section 5.2. Here we consider some further functional properties of the logarithm.

Further Properties of Logarithms. Since logarithms are exponents, the basic rules for exponents can be translated into the following **properties of logarithms.** (We derive the first one in Example 3 and the other two in the exercises.)

$$\log_a (xy) = \log_a x + \log_a y \qquad \textbf{logarithm-of-a-product property}$$

$$\log_a \left(\frac{x}{y}\right) = \log_a x - \log_a y \qquad \textbf{logarithm-of-a-quotient property}$$

$$\log_a (x^y) = y \log_a x \qquad \textbf{logarithm-of-a-power property}$$

We will apply these logarithmic properties to a variety of examples.

Since there is no algebraic process to compute exact values of logarithmic or exponential functions, we use approximate decimal values obtained from a scientific calculator. *We shall consider two numbers to be equal if they are identical to four significant digits.* (See the definition of significant digits in the Appendix.) Before calculators became readily available, tables and slide rules with logarithmic scales were used for computing logarithms. Both the construction of logarithmic and exponential tables and the internal programming of a scientific calculator involve methods covered in calculus.

Example 1 Using $\log_{10} 2 = .3010$ and $\log_{10} 3 = .4771$, compute the following.

(a) $\log_{10} 6$ (b) $\log_{10} 8$ (c) $\log_{10} (.5)$

Solution:

(a)
$$\begin{aligned}\log_{10} 6 &= \log_{10} (2 \cdot 3)\\ &= \log_{10} 2 + \log_{10} 3 && \textit{Logarithm-of-a-product property}\\ &= .3010 + .4771\\ &= .7781\end{aligned}$$

(b)
$$\begin{aligned}\log_{10} 8 &= \log_{10} (2^3)\\ &= 3 \log_{10} 2 && \textit{Logarithm-of-a-power property}\\ &= 3(.3010)\\ &= .9030\end{aligned}$$

(c)
$$\begin{aligned}\log_{10} (.5) &= \log_{10} \left(\frac{1}{2}\right)\\ &= \log_{10} 1 - \log_{10} 2 && \textit{Logarithm-of-a-quotient property}\\ &= 0 - .3010\\ &= -.3010 \quad \blacksquare\end{aligned}$$

Example 2 (a) Express $\log_a (xy/\sqrt{z})^3$ in terms of $\log_a x$, $\log_a y$, and $\log_a z$.

(b) Express $5 \log_a x - 1/3(\log_a y + 2 \log_a z)$ as a single logarithm.

Solution:

(a)
$$\begin{aligned}\log_a \left(\frac{xy}{\sqrt{z}}\right)^3 &= \log_a \left(\frac{x^3y^3}{z^{3/2}}\right) && (\sqrt{z})^3 = (z^{1/2})^3 = z^{3/2}\\ &= \log_a (x^3y^3) - \log_a (z^{3/2}) && \textit{Logarithm-of-a-quotient property}\\ &= \log_a x^3 + \log_a y^3 - \log_a (z^{3/2}) && \textit{Logarithm-of-a-product property}\\ &= 3 \log_a x + 3 \log_a y - \frac{3}{2} \log_a z && \textit{Logarithm-of-a-power property}\end{aligned}$$

(b) Give a reason for each of the following steps.

$$5 \log_a x - \frac{1}{3}(\log_a y + 2 \log_a z) = \log_a (x^5) - \frac{1}{3}[\log_a y + \log_a (z^2)]$$

____

$$= \log_a (x^5) - \frac{1}{3} \log_a (yz^2)$$

____

$$= \log_a (x^5) - \log_a [(yz^2)^{1/3}]$$

____

$$= \log_a \left[\frac{x^5}{(yz^2)^{1/3}}\right]$$

______________ ∎

Example 3 Show that the property

$$\log_a (xy) = \log_a x + \log_a y$$

for the logarithm of a product follows from the corresponding rule for exponents. (Similar derivations for the logarithm of a power and the logarithm of a quotient are included in the exercises.)

Solution: Let $p = \log_a x$ and $q = \log_a y$.

Then	$x = a^p$ and $y = a^q$	*Definition of logarithm*
	$xy = a^p \cdot a^q = a^{p+q}$	*Same-base rule for exponents*
	$\log_a (xy) = p + q$	*Definition of logarithm*
	$\log_a (xy) = \log_a x + \log_a y.$	*Substitution for p and q* ∎

Another important property of the logarithmic function, which follows because it is the inverse of the exponential, is the fact that $f(x) = \log_a x$ is a *one-to-one function*. That is, for x and $y > 0$,

$$\log_a x = \log_a y \Leftrightarrow x = y.$$

For instance,

$\log_a (5^x) = \log_a (.04) \Leftrightarrow 5^x = .04$	*The logarithmic function is one-to-one*
$\Leftrightarrow 5^x = 5^{-2}$	*(Why?)*
$\Leftrightarrow x = -2.$	*The exponential function is one-to-one*

The following example uses several of the properties of logarithms.

Example 4 Solve for x in the equation

$$\log_a x = 2 \log_a 3 - \frac{2}{3} \log_a 8 - \log_a 5.$$

Solution:

$\log_a x = \log_a (3^2) - \log_a (8^{2/3}) - \log_a 5$	*Logarithm-of-a-power property*
$= \log_a 9 - (\log_a 4 + \log_a 5)$	
$= \log_a 9 - \log_a 20$	*Logarithm-of-a-product property*
$= \log_a \left(\frac{9}{20}\right)$	*Logarithm-of-a quotient property*
Therefore, $x = \frac{9}{20}$.	*One-to-one property for logarithms* ■

Use of Scientific Notation As mentioned above, approximate decimal values of logarithms can be obtained by pushing a button on a scientific calculator. However, for many years, beginning with John Napier (1550–1617) who invented logarithms for computational purposes, it was necessary to express a number in scientific notation (see Exercises P.5) and to use the rules and tables to find values of logarithms. Since scientific notation uses the base 10, it became customary to omit the subscript and to write $\log x$ for $\log_{10} x$. Most scientific calculators have a button marked "log" for base 10 logarithms (**common logarithms**) and a button labeled "ln" for base e (**natural**) logarithms.

Today the calculator has made logarithms much less important as a tool for computation, but the logarithmic function and its properties are widely used in both theoretical and applied mathematics. We include some further examples to illustrate the use of scientific notation. A calculator should be used to check the results.

Example 5 Given $\log 1.2 = 0.0792$, compute:

(a) $\log 1200$ (b) $\log .00012$ (c) $\log (-12)$

Solution: We begin by writing each number in scientific notation, and we then apply the logarithmic properties for products and posers.

(a)
$$\begin{aligned} \log 1200 &= \log (1.2 \times 10^3) \\ &= \log (1.2) + \log (10^3) \\ &= \log (1.2) + 3 \log 10 \\ &= 0.0792 + 3 \cdot 1 \\ &= 3.0792 \end{aligned}$$

(b) $\log .00012 = \log (1.2 \times 10^{-4})$
$= \log (1.2) + \log (10^{-4})$
$= \log (1.2) - 4 \log 10$
$= 0.0792 - (4 \cdot 1)$
$= -3.9208$

(c) $\log (-12)$ does not exist in the real number system since $\log x$ is defined for positive numbers only. If a calculator is used to try to compute $\log (-12)$, an error signal will appear. ■

Example 6 Given ln 1.2 =.1823 and ln 10 = 2.3026, find ln 1200.

Solution: $\ln 1200 = \ln (1.2 \times 10^3)$
$= \ln (1.2) + 3 \ln 10$
$= .1823 + 3(2.3026)$
$= 7.0901$ ■

Example 7 Given log 3.46 = .5391, solve the following equations for x.
(a) $\log x = 1.0782$ (b) $\log x = 3.5391$ (c) $\log x = -4.4609$

Solution:

(a) 1.0782 = 2(0.5391). From the rule $2 \log x = \log x^2$, we conclude that $x = (3.46)^2$. *Calculator check:* $(3.46)^2 = 11.9716$ and $10^{1.0782} = 11.9729$. The difference between 11.9716 and 11.9729 is due to round-off.

(b) We write

$$\log x = .5391 + 3$$
$$= \log 3.46 + \log (10^3)$$
$$= \log (3.46 \times 10^3).$$

Therefore, $x = 3.46 \times 10^3 = 3460$.
calculator check: $10^{3.5391} = 3460.1904$

(c) We first write

$$-4.4609 = .5391 - 5.$$ *Add and subtract 5 to −4.4609 (why 5?)*

Then $\log x = .5391 - 5$
$= \log 3.46 - \log (10^5)$
$= \log \left(\frac{3.46}{10^5}\right).$

Therefore, $x = 3.46/10^5 = .0000346$.
calculator check: $10^{-4.4609} = .0000346$ ■

Comment Logarithms were the invention of John Napier, a Scottish laird who worked for 20 years on the idea before publishing his work in 1614. His purpose was to simplify trigonometric computations required in astronomy. Napier's idea was to establish a one-to-one correspondence between the numbers of two sequences so that multiplication in one sequence corresponds to addition in the other. He established the correspondence by means of the dynamic model illustrated in Figure 10.

In the model, points P and Q start at 0 on their respective scales. Point Q moves with constant speed 10^7. P starts with speed 10^7, and thereafter its speed is x, where $x = d(P, 10^7)$. Point P corresponds to point Q.

It can be shown by calculus that the correspondence between P and Q in Figure 10 satisfies

$$\frac{x}{10^7} = e^{-y/10^7} \qquad \text{or} \qquad \underbrace{y = -10^7 \ln\left(\frac{x}{10^7}\right)}_{y \,=\, Nap.log\, x}.$$

Napier defined his logarithms in terms of the dynamic model, not with respect to a base as we do today. Also, Napier's logarithms satisfy

$$\text{Nap.log}\,\frac{x_1 x_2}{10^7} = \text{Nap.log}\,x_1 + \text{Nap.log}\,x_2,$$

which differs from our present-day relationship by the scaling factor 10^{-7}. Napier chose the scaling factor 10^{-7} because he was concerned primarily with computations in trigonometry, and trigonometric tables at the time were computed in terms of a circle of radius 10^7. Shortly after their introduction, Napier's logarithms were modified to become our common logarithms. Logarithms have been the primary tool in scientific computations from their introduction early in the seventeenth century up to the advent of the electronic computer.

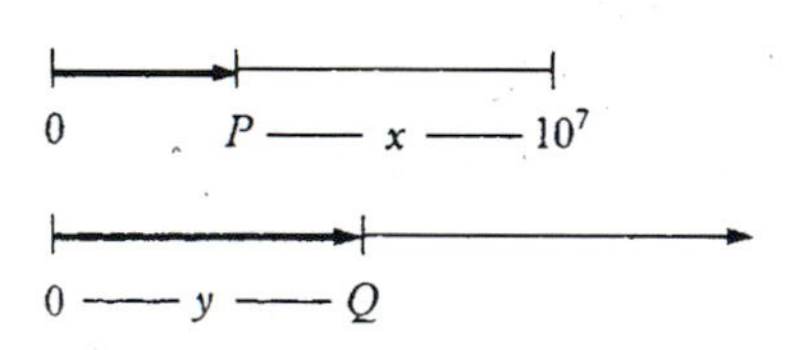

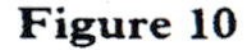
Figure 10

Exercises 5.3

Whenever $\log_a u$ *appears, it is assumed that* $a > 0$, $a \neq 1$, *and* $u > 0$.

Fill in the blanks to make each statement true.

1. If $\log_a (xy) = 2 \log_a x$, then $y =$ ________.
2. If $\log_a (y/x) = 2 \log_a x$, then $y =$ ________.
3. If $\log_a y = x \log_a 2$, then $y =$ ________.
4. If $\log_a (x^3) = \log_a (y^2)$, then $y =$ ________.
5. If $\log_a (x^2) = (\log_a x)^2$, then $x =$ ________ or ________.

Write true *or* false *for each statement.*

6. $\log_a (x + y) = \log_a x + \log_a y$
7. $\log_a (x + y) = \log_a x \cdot \log_a y$
8. $\log_a (xyz) = \log_a x + \log_a y + \log_a z$
9. If $\log_a x = x \log_a y$, then $y = x^{1/x}$.
10. If $\log_a \left(\dfrac{x}{y}\right) = \log_a x + \log_a y$, then $y = 1$.

Further Properties of Logarithms

Use $\log 5 = 0.6990$ *and* $\log 7 = 0.8451$ *to compute each of the following.*

11. $\log 35$
12. $\log 3.5$
13. $\log (.00035)$
14. $\log \left(\dfrac{7^2}{5^3}\right)$
15. $\log (.2)$
16. $\log (\sqrt[3]{7})$

Express each of the following as a single logarithm.

17. $3 \log_a x + \dfrac{1}{2} \log_a (x - 1) - 5 \log_a (x^2)$
18. $\dfrac{1}{3} \log_a \left(\dfrac{x}{y}\right) - \log_a \left(\dfrac{y}{x}\right)$
19. $\log (x^2 - 4) - 2 \log (x - 2)$
20. $\log (x^3) - \log (x^2) - \log x$

Express each of the following in terms of $\log b$, $\log c$, *and* $\log d$.

21. $\log \left(\dfrac{b^2 c}{d^4}\right)$
22. $\log (\sqrt[3]{b} \cdot \sqrt{c} \cdot d^4)$
23. $\log (\sqrt[3]{b \cdot c^{1/2} \cdot d^{-1}})$
24. $\log \left(\dfrac{b}{\frac{c}{d}}\right) - \log \left(\dfrac{\frac{b}{c}}{d}\right)$

Solve for x.

25. $\log_a x = \log_a 7$
26. $2 \log_a x = \log_a 25$
27. $\log_a x = 3 \log_a 2 - \dfrac{1}{3} \log_a 8 + 2 \log_a 5$
28. $(\log_2 x)^2 - 4 \log_2 x + 4 = 0$
29. $\log_x 20 - \log_x 5 = 2$
30. $\log_x 2 + \log_x 3 = 5$
31. $\log_x (x^2) = x$
32. $\log_x (x^4) = 2x + 1$

33. Fill in the reason for each step in the following derivation of the property $\log_a (x^y) = y \log_a x$.

Let $p = \log_a x$. Then

$x = a^p$ ________________
$x^y = (a^p)^y$ ________________
$x^y = a^{py}$ ________________
$\log_a (x^y) = py$ ________________
$\log_a (x^y) = (\log_a x)y$ ________________
$\log_a (x^y) = y \log_a x$. ________________

34. Derive the property $\log_a (x/y) = \log_a x - \log_a y$ from the corresponding exponent rule $a^p/a^q = a^{p-q}$ (see Example 3 and the previous exercise).

Use $\log 2.40 = .3802$ *and* $\log 1.47 = .1673$ *to compute each of the following. Check your results on a calculator.*

35. $\log 240$
36. $\log .024$
37. $\log 1{,}470{,}000$
38. $\log (1.47 \times 10^{-20})$

Use $\ln 2.40 = .8755$, $\ln 1.50 = .4055$, *and* $\ln 10 = 3.3026$ *to compute each of the following. Check your results on a calculator.*

39. $\ln 240$
40. $\ln .024$
41. $\ln 1{,}500{,}000$
42. $\ln (1.5 \times 10^{-20})$

Given $\log 6.12 = .7868$ to solve each of the following for x. *Check your results on a calculator.*

43. $\log x = 3.7868$ **44.** $\log x = 1.7868$

45. $\log x = -0.2132$ **46.** $\log x = -2.2132$

Miscellaneous

47. Let $y = ca^{kx}$, where c and k are constants and $c \neq 0$. Show that if $Y = \log_a y$, then $Y = mx + b$, where m and b are constants. That is, "the logarithm of an exponential curve is a straight line."

48. With reference to Exercise 47, graph the straight line associated with $y = (1/4)10^{x/2}$.

*The Equation $\log_a x = x$

The equation $\log_a x = x$ *is equivalent to the exponential equation* $a^x = x$, *which was discussed in the exercises for Section 5.2. With reference to the figures and using a calculator, search for approximate solutions (to three significant digits) to each of the following equations.*

(a) $0 < a < 1$

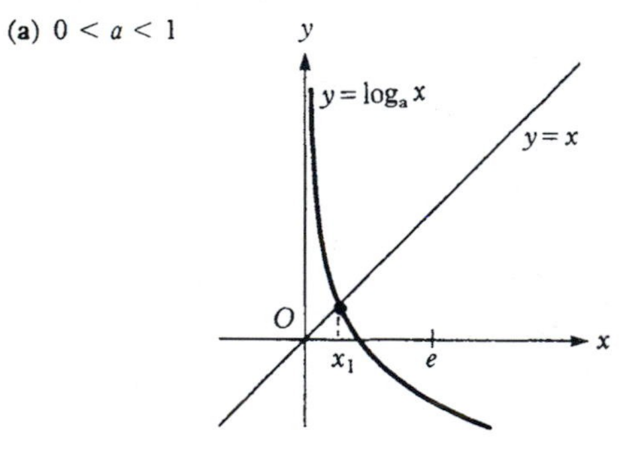

(b) $1 < a < e^{1/e}$

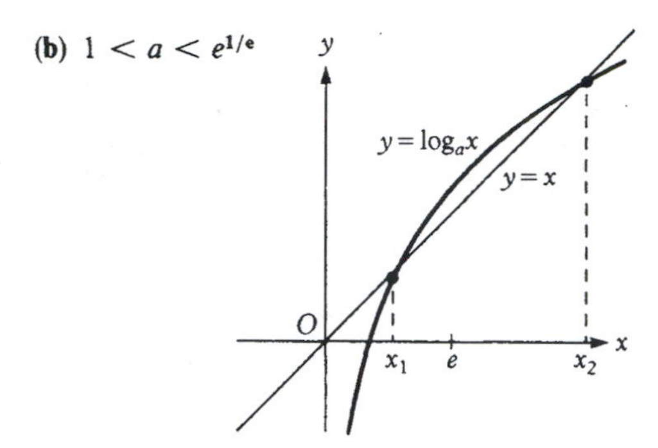

(c) $a \geq e^{1/e}$

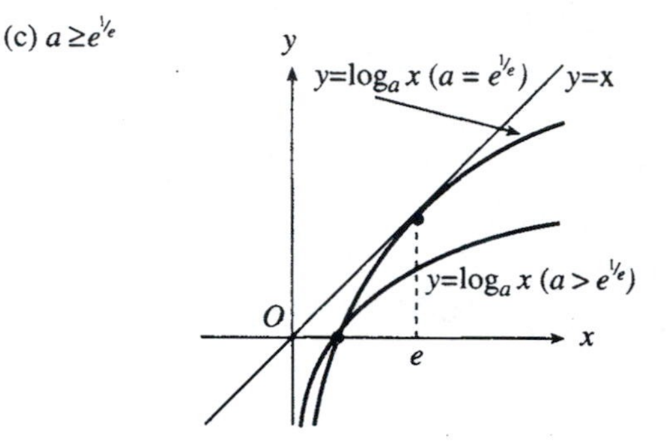

49. $\log_{.2} x = x$ (one solution)

50. $\log_{.8} x = x$ (one solution)

51. $\log_{1.2} x = x$ (two solutions)

52. $\log_{1.35} x = x$ (two solutions)

*Exercises 49–52 are optional.

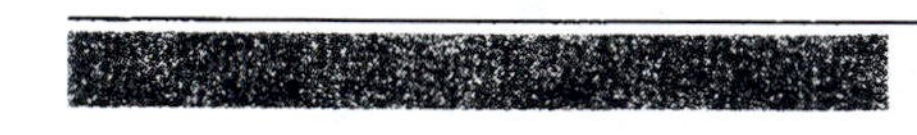

5.4 Applications and Exponential and Logarithmic Equations

The exponential is nature's curve. The logarithm is man-made.

Compound Interest
Growth and Decay
Intensity of Sound
Richter Scale

In earlier chapters we were able to solve problems concerning percent, motion, and work by using algebraic equations, namely linear and quadratic ones. However, many quantities in science and nature increase or decrease exponentially, and problems involving them require the solution of exponential equations. Also, because exponential functions can increase so rapidly, it is often more practical to convert exponential relationships to a logarithmic scale, such as the Richter scale for earthquakes.

In this section we will solve equations involving exponential and logarithmic expressions. Our tools for doing so include the usual algebraic operations of addition, subtraction, multiplication, and division on both sides of an equation, as well as the *functional* operations of taking logarithms or exponentiating on both sides of an equation. Because the logarithmic and exponential functions are one-to-one, any equation

$$f(x) = g(x)$$

is equivalent to

$$\log_a f(x) = \log_a g(x)$$

(assuming $f(x) > 0$ and $g(x) > 0$) and also to

$$a^{f(x)} = a^{g(x)}.$$

In the following examples, all logarithms and exponentials are evaluated by means of a scientific calculator.

Compound Interest. If \$1.00 is invested at 7% compounded annually, then at the end of one year the amount of dollars accumulated is

$$\begin{aligned} A &= 1.00 + .07(1.00) \\ &= 1.07. \end{aligned}$$

At the end of two years,

$$\begin{aligned} A &= 1.07 + .07(1.07) \\ &= 1.07(1 + .07) \\ &= (1.07)^2, \end{aligned}$$

and at the end of t years, the amount becomes

$$A = (1.07)^t.$$

If interest is compounded four times a year, then the interest rate per compound period is .07/4 and the number of compound periods in t years is $4t$. Therefore, the amount in t years is

$$A = \left(1 + \frac{.07}{4}\right)^{4t}.$$

More generally, for a principal of P dollars invested at an annual interest rate of R (% rate ÷ 100) with interest compounded N times a year, the amount at the end of t years is

$$\mathbf{A = P\left(1 + \frac{R}{N}\right)^{Nt}.} \qquad \textbf{compound interest} \qquad \textbf{(1)}$$

To see how this compound interest formula is related to the natural base e, we proceed as follows. First, we write

$$A = P\left(1 + \frac{R}{N}\right)^{Nt}$$
$$= P\left[\left(1 + \frac{R}{N}\right)^{N/R}\right]^{Rt}.$$

Next, we substitute $n = N/R$ to obtain

$$A = P\left[\left(1 + \frac{1}{n}\right)^{n}\right]^{Rt}.$$

In Section 5.1, we stated that the number e is approximated for large values of n by $(1 + 1/n)^n$, and the following table shows that these numbers actually increase and get close together as n increases, so that we may reasonably guess that they approach a finite value.

n	1	2	10	1000	10,000	100,000
$(1 + 1/n)^n$	2	2.25	2.59374	2.71692	2.71815	2.71827

In fact, the number e may be defined by

$$\left(1 + \frac{1}{n}\right)^{n} \to e \quad \text{as} \quad n \to \infty. \qquad \textbf{natural base}$$

Therefore, we conclude that as N, the number of compound periods per year, increases, so does $n = N/R$, and

$$A = P\left[\left(1 + \frac{1}{n}\right)^{n}\right]^{Rt} \to Pe^{Rt}.$$

For this reason, if P dollars are invested at an annual interest rate of R, and the amount A accumulated in t years is computed by the formula

$$A = Pe^{Rt}, \tag{2}$$

then the interest is said to be **compounded continuously.**

Example 1 Jennifer is six years old, and her parents want to have \$30,000 available when she starts college in 12 years. How much must they deposit now at 8% annual interest compounded quarterly in order to achieve this goal?

Solution: We substitute the numbers $A = 30{,}000$, $R = .08$, $N = 4$, and $t = 12$ into (1) and then solve for P.

$$30{,}000 = P\left(1 + \frac{.08}{4}\right)^{4(12)}$$
$$= P(1.02)^{48}$$
$$= P(2.5870704)$$

Therefore,

$$P = \frac{30{,}000}{2.5870704}$$
$$\approx 11596.13$$

Thus, a deposit of \$11,596.13 will grow to \$30,000 in 12 years. ■

Example 2 How long will it take to double the principal if money is invested at 9% (a) compounded annually and (b) compounded continuously?

Solution:

(a) We have $R = .09$ and $N = 1$, and if P dollars are invested, we want to find the value of t for which $A = 2P$ in t years. Therefore, by substitution into (1),

$$2P = P(1.09)^t$$
$$2 = (1.09)^t \qquad \textit{Divide both sides by } P$$
(key step) $$\log 2 = \log (1.09)^t \qquad \textit{Take the } \log_{10} \textit{ of both sides}$$
$$\log 2 = t \log (1.09) \qquad \textit{Logarithm-of-a-power property}$$
$$t = \frac{\log 2}{\log (1.09)} \qquad \textit{Solve for } t$$
$$= \frac{.3010300}{.0374265}$$
$$\approx 8.04.$$

Hence, it takes slightly more than 8 years for money to double at 9% compounded annually. Note that the answer does not depend on the value of P (*why?*).

(b) We substitute $R = .09$ and $A = 2P$ into formula (2), and then solve for t.

$$2P = Pe^{.09t}$$
$$2 = e^{.09t} \qquad \textit{Divide both sides by } P$$
(key step) $$\ln 2 = \ln (e^{.09t}) \qquad \textit{Take natural logs of both sides}$$
$$\ln 2 = .09t \qquad \textit{(Why?)}$$
$$t = \frac{\ln 2}{.09} \qquad \textit{Solve for } t$$
$$= \frac{.69314718}{.09}$$
$$\approx 7.70 \text{ years}$$ ■

Comment In solving Example 2(a), we used logarithms to the base 10, but we could have used base e, in which case we would have obtained

$$t = \frac{\ln 2}{\ln (1.09)} = \frac{.69314718}{.0861777} \approx 8.04$$

as before.

Example 3 Max inherits \$50,000 and decides to use it to buy a summer home that he estimates will cost \$75,000 five years from now. If he invests the money now, what rate of interest compounded daily will give him the necessary \$75,000?

Solution: We substitute $P = 50{,}000$, $A = 75{,}000$, $N = 365$, and $t = 5$ into (1), and then solve for R.

$$75{,}000 = 50{,}000\left(1 + \frac{R}{365}\right)^{365(5)}$$

$$\frac{75{,}000}{50{,}000} = \left(1 + \frac{R}{365}\right)^{1825}$$

$$1.5 = \left(1 + \frac{R}{365}\right)^{1825}$$

(key step) $$\log (1.5) = 1825 \log \left(1 + \frac{R}{365}\right) \qquad \textit{(Why?)}$$

$$\frac{\log (1.5)}{1825} = \log \left(1 + \frac{R}{365}\right)$$

$$.00009649 = \log \left(1 + \frac{R}{365}\right)$$

(key step) $$10^{.00009649} = 1 + \frac{R}{365} \qquad \textit{Use } 10^{\log x} = x$$

$$1.0002222 = 1 + \frac{R}{365}$$

$$R \approx .0811$$

Therefore, the required interest rate is 8.11%. ■

Growth and Decay. Some quantities, such as the number of bacteria in a culture or individuals in a population, increase exponentially, while others, such as quantities of radioactive elements, decrease exponentially.

Example 4 If there are 10,000 bacteria in a culture at some initial time ($t = 0$) and the number doubles every hour, then there are

$$10{,}000 \cdot 2, \qquad 10{,}000 \cdot 4, \qquad \text{and} \qquad 10{,}000 \cdot 8$$

bacteria at the end of 1, 2, and 3 hours, respectively, and at the end of t hours, the number $N(t)$ of bacteria is

$$N(t) = 10{,}000 \cdot 2^t.$$

Find the amount of time it takes for the number of bacteria to increase from 320,000 to 1,280,000.

Solution: Let t_1 be the number of hours required for the number to reach 320,000 and t_2 the number of hours to reach 1,280,000. Then

$$320{,}000 = 10{,}000 \cdot 2^{t_1} \qquad \text{and} \qquad 1{,}280{,}000 = 10{,}000 \cdot 2^{t_2}.$$

Hence, $\quad 32 = 2^{t_1} \qquad \text{and} \qquad 128 = 2^{t_2},$

which means $t_1 = 5$ and $t_2 = 7$. The time required is 2 hours. ■

The base 2 is naturally involved in any situation like Example 4 in which a quantity doubles in a given unit of time. Similarly, if a quantity triples in a given unit of time, then we could write $N(t) = N_0 3^t$, where N_0 denotes the value of N at time zero; that is $N_0 = N(0)$. However, any positive base a can be written as

$$a = e^k, \qquad k = \ln a$$

and therefore,

$$a^t = e^{kt}. \qquad \textit{Power-of-a-power rule}$$

Hence, any quantity that always changes by a fixed multiple over the same increment of time satisfies the formula

$$\mathbf{N(t) = N_0 e^{kt},} \qquad \textbf{exponential growth or decay} \qquad \textbf{(3)}$$

where $N(t)$ denotes the amount of the quantity at time t, N_0 is the amount at time zero, and k is a constant. Such quantities are said to grow ($k > 0$) or decay ($k < 0$) exponentially.

Radioactive elements disintegrate according to formula (3), and the time it takes for one half of the initial amount to disintegrate is called the **half-life** of the element.[1]

Example 5 The half-life of radium is 1600 years. If 100 grams of this element are present in a laboratory now and remain untouched for 25 years, how many grams will there be then?

[1]The term half-life means the life of one half of the element, not one half the life of the element. Theoretically, the lifetime of the element is infinite (*why?*).

Solution: First, we use the given information about the half-life to evaluate the constant k in formula (3). That is, $N_0 = 100$ and

$$N(1600) = \frac{1}{2}(100) = 50.$$

Therefore, by formula (3),

$$50 = 100e^{k(1600)}$$

$$\frac{1}{2} = e^{1600k}$$

$$\text{(key step)} \qquad \ln\left(\frac{1}{2}\right) = 1600k \qquad \textit{(Why?)}$$

$$\ln 1 - \ln 2 = 1600k \qquad \textit{Logarithm-of-a-quotient property}$$

$$-\ln 2 = 1600k \qquad \ln 1 = 0$$

$$k = \frac{-\ln 2}{1600} \qquad \textit{Solve for } k$$

$$k = -.00043322.$$

Hence, in this particular case (3) becomes

$$N(t) = 100e^{-.00043322t},$$

and the amount of radium left after 25 years is

$$\begin{aligned} N(25) &= 100e^{-.00043322(25)} \\ &= 100e^{-.0108305} \\ &= 100(.989228) \\ &\approx 98.9 \text{ grams.} \end{aligned}$$

Note that only 1.1 grams disappear in 25 years. Other radioactive elements disintegrate even more slowly than radium. This is why we have a problem with radioactive wastes. ■

When a population increases exponentially according to formula (3), it is said to satisfy the **Malthusian model.**[2] The model is reasonably accurate for organisms that reproduce in a culture under controlled laboratory conditions, and also for human populations over a short time period. But if we take into account such factors as limited food supply and limited space, then a more realistic model for population growth is the **logistic law,** which is given by the formula

$$N(t) = \frac{cN_0}{N_0 + (c - N_0)e^{-kt}}, \qquad \textbf{logistic law} \qquad (4)$$

where c and k are positive constants. The graph of (4) is given in Figure 11 and shows that the growth starts out exponentially, then tapers off after N reaches $c/2$ and approaches c as a limiting value.

[2] *Thomas Malthus (1766–1834) was an English economist.*

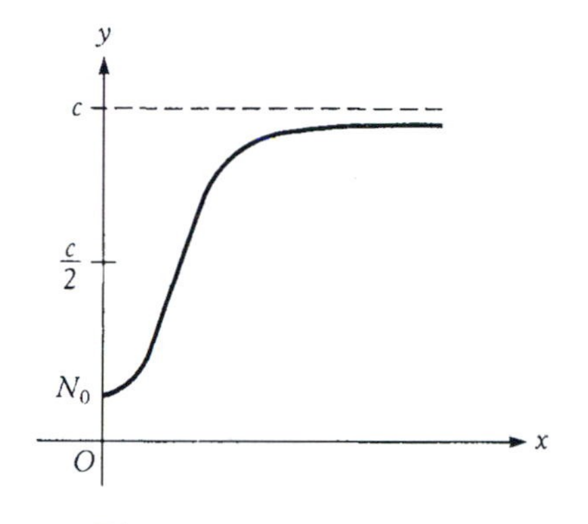

Figure 11 Logistic Growth

We can see directly from (4) that $N(t) \approx c$ for large values of t since $e^{-kt} \approx 0$ for large t. On the other hand, to investigate the graph of (4) for small values of t, we first multiply numerator and denominator by e^{kt} and rearrange the terms in the denominator to obtain

$$N(t) = \frac{cN_0e^{kt}}{N_0(e^{kt} - 1) + c}. \tag{4'}$$

Now, since $e^{kt} \approx 1$ for small values of t, equation (4′) shows that, initially, $N(t) \approx N_0e^{kt}$.

Example 6 Assume that the population growth of the United States satisfies the logistic law (4), where N is in millions and $k = .03$. If N was 150.7 in 1950 and 226 in 1980, find:

(a) the constant c,

(b) the population predicted by the formula for the year 2000.

Solution:

(a) Let $t = 0$ correspond to the year 1950. Then 1980 corresponds to $t = 30$, and we substitute $N_0 = 150.7$, $t = 30$, $N(30) = 226$, and $k = .03$ in (4) to obtain

$$\begin{aligned}
226 &= \frac{c(150.7)}{150.7 + (c - 150.7)e^{-.03(30)}} \\
226[150.7 + (c - 150.7)e^{-.9}] &= 150.7c \\
34058.2 + 226ce^{-.9} - 34058.2e^{-.9} &= 150.7c \\
c(226e^{-.9} - 150.7) &= 34058.2(e^{-.9} - 1) \\
c &= \frac{34058.2(e^{-.9} - 1)}{226e^{-.9} - 150.7} \\
c &\approx 344.
\end{aligned}$$

(b) The year 2000 corresponds to $t = 50$. Hence, using the value $c = 344$ obtained from part (a), we substitute $t = 50$ in (4) to obtain

$$N(50) = \frac{344(150.7)}{150.7 + (344 - 150.7)e^{-.03(50)}} \approx 267.$$

Therefore, the logistic law predicts a population of approximately 267,000,000 by the year 2000. ■

[3] *Named in honor of Alexander Graham Bell (1847–1922), the inventor of the telephone.*

Intensity of Sound. An object that is vibrating in some medium, such as air or water, produces sound waves. The intensity of the sound depends on the amount of energy transmitted by the waves, and the unit for measuring sound intensity is the **decibel.**[3] A sound wave that transmits to the ear energy measuring 10^{-16} watts per square centimeter is assigned a value of zero decibels. In general, if the energy transmitted by sound waves is I watts per square centimeter, then the decibel level D of the sound is defined as

$$\mathbf{D = 10 \log (10^{16} I).} \qquad \textbf{decibel level} \qquad (5)$$

Note that 10^{-16} watts per square inch corresponds to $D = 0$ (*why?*).

Example 7 Find the decibel levels for the following:

(a) a human voice for which $I = 10^{-10}$ watts per square centimeter,

(b) a jet plane for which $I = 2 \times 10^{-5}$ watts per square centimeter.

Solution:

(a)
$$\begin{aligned} D &= 10 \log (10^{16}\, 10^{-10}) \\ &= 10 \log (10^6) \\ &= 10 \cdot 6 \\ &= 60 \text{ decibels} \end{aligned}$$

(b)
$$\begin{aligned} D &= 10 \log [10^{16}(2 \times 10^{-5})] \\ &= 10 \log (2 \times 10^{11}) \\ &= 10[\log 2 + \log (10^{11})] \\ &= 10[.30103 + 11] \\ &\approx 113 \text{ decibels} \end{aligned}$$

The 113 decibels of the jet is close to the human tolerance level, which is about 120 decibels. ■

Example 8 A machinist operates two machines that produce noise levels of 40 and 50 decibels, respectively. If it is unsafe for the operator to be continuously subjected to more than 85 decibels, are earplugs needed?

Solution: When both machines are running, the energy transmitted to the ear is equal to $I_1 + I_2$, where I_1 is the energy transmitted by the machine producing 40 decibels, and I_2 is that of the machine producing 50 decibels. Now

$$\begin{aligned} 40 &= 10 \log (10^{16} I_1) \\ 4 &= \log (10^{16} I_1) \\ 10^4 &= 10^{16} I_1 \\ I_1 &= 10^{-12} \end{aligned} \qquad \text{and} \qquad \begin{aligned} 50 &= 10 \log (10^{16} I_2) \\ 5 &= \log (10^{16} I_2) \\ 10^5 &= 10^{16} I_2 \\ I_2 &= 10^{-11}. \end{aligned}$$

Therefore,

$$\begin{aligned} I_1 + I_2 &= 10^{-12} + 10^{-11} \\ &= 10^{-12}(1 + 10) \\ &= 11 \cdot 10^{-12}, \end{aligned}$$

and the combined decibel level is

$$\begin{aligned} D &= 10 \log (10^{16} \cdot 11 \cdot 10^{-12}) \\ &= 10 \log (10^4 \cdot 11) \\ &= 10(4 + \log 11) \\ &= 10(5.0414) \\ &\approx 50.4 \text{ decibels.} \end{aligned}$$

The combined level of 50.4 decibels is well below 85, so no earplugs are needed. ■

Richter Scale. The Richter scale, named for American seismologist Charles Richter (1900–1985), is mentioned in news reports whenever a serious earthquake occurs anywhere in the world. The scale measures the magnitude R of an earthquake according to the formula

$$\mathbf{R = \frac{2}{3} \log (10^{-4.7} E)}, \qquad \textbf{Richter scale} \qquad (6)$$

where E is the energy, measured in joules, released by the earthquake.

Example 9 If the energy released by an earthquake is 10^{13} joules, what is its magnitude on the Richter scale?

Solution: We substitute $E = 10^{13}$ in (6), obtaining

$$\begin{aligned} R &= \frac{2}{3} \log (10^{-4.7}\, 10^{13}) \\ &= \frac{2}{3} \log (10^{8.3}) \\ &= \frac{2}{3}(8.3) \\ &\approx 5.5. \end{aligned}$$ ■

Example 10 The magnitude of the San Francisco earthquake in 1906 has been estimated at 8.2 on the Richter scale. What was the amount of energy released?

Solution: By substituting $R = 8.2$ in equation (6), we obtain

$$8.2 = \frac{2}{3} \log (10^{-4.7} E)$$
$$12.3 = \log (10^{-4.7} E)$$
$$10^{12.3} = 10^{-4.7} E$$
$$E = 10^{17}.$$

Hence, the energy released was 10^{17} joules. ■

Exercises 5.4

Compound Interest

1. Find the amount accumulated at the end of 5 years if \$100 is invested at an annual rate of 7% and compounded
   (a) monthly (b) daily (c) continuously.
2. How much more will be earned in 10 years on \$1000 at 8% when interest is compounded continuously than when it is compounded daily?
3. Which amounts to more in 10 years: \$1000 invested at 8% compounded semiannually or \$1000 invested at 7.95% compounded daily?
4. How much more interest will be earned in 1 year by transferring \$1000 from a passbook savings account paying 5.5% compounded monthly to a money market account paying 7.3% compounded daily?
5. A bank advertises that its interest rate of 6.91% compounded daily is equivalent to an effective annual rate of 7.154%. Is this correct? (*Hint:* compare the amount earned on \$1 for 1 year at 7.154% compounded annually with the amount earned at the given compound interest rate.)
6. What is the effective annual rate equivalent to 8.05% compounded daily? (See Exercise 5.)
7. If an interest rate of 18% compounded monthly on the unpaid balance is charged on a credit card, how much is owed on a \$300 balance paid in one payment at the end of four months?
8. It is said that native Americans sold Manhattan Island for \$24 in 1626. If that \$24 had been put in a bank paying 7% interest compounded continuously, what would it amount to in 1988?

Growth and Decay

9. If the number of bacteria in a culture grows exponentially and doubles every hour, starting with only 10 bacteria, find the number at the end of 12 hr.
10. Solve Exercise 9 if the number of bacteria doubles every half hour.
11. The number of bacteria in a colony increases exponentially according to formula (3), and at a certain time there are 10,000 bacteria present. If the number 2 hr after that time is 80,000, what is the number 4 hr after that time?
12. Suppose the population growth of a small country follows the Malthusian law (3) and doubles every 50 yr. In how many years will the population triple?
13. Assume that the United States population, which was 226 million in 1980, grows according to the Malthusian law at 1.2% a year. Find the population predicted for the year 2000 and compare the result with that in Example 6. [*Hint:* use formula (3) with $k = 1.2\% = .012$.]
14. Show that if the half-life of a radioactive substance is H, then
    (a) the constant k in formula (3) is given by $k = -(\ln 2)/H$, and

(b) the amount of the radioactive substance at time t is $N(t) = N_0 2^{-t/H}$.

15. The half-life of radioactive polonium is 140 days. In how many days will there be 80% of the original amount left?

16. In each living thing the ratio of the amount of radioactive carbon-14 to the amount of ordinary carbon-12 remains approximately constant. However, when a plant or animal dies, the amount of carbon-14 decreases exponentially, and its half-life is about 5700 yr.
   (a) Find k in formula (3) for carbon-14.
   (b) Determine the age of a piece of charcoal if 20% of the original amount of carbon-14 remains.

17. In 1950 Willard Libby discovered the method of carbon-14 dating described in Exercise 16, and in 1960 he received a Nobel Prize. In a famous example, paintings discovered in the Lascaux Caves in France were found to contain, in 1950, 15% of the usual amount of carbon-14 in such paint. What was the approximate age of the paintings at that time?

18. The amount of a drug in a patient's bloodstream decreases exponentially according to formula (3), where t is the number of hours after the drug is administered. If $k = -.2$ and $N_0 = .5$ ml, find the amount of the drug left after 2 hr.

19. Verify that in the logistic law (4), $N(0) = N_0$.

20. Show algebraically that equation (4) is equivalent to (4′).

21. Assume that the population of deer in a certain forest follows the logistic law (4), with $k = .2$ and $N_0 = 100$, and that after 3 yr there are 170 deer.
   (a) Show that the number of deer can never exceed 1144. (*Hint:* compute c.)
   (b) Find the number of deer at the end of 6 yr.

22. For the logistic law (4), show that each of the following is true.
   (a) $t = \frac{1}{k} \ln \left[\frac{N(c - N_0)}{N_0(c - N)}\right]$
   (b) $N = \frac{c}{2}$ for $t = \frac{1}{k} \ln \left(\frac{c - N_0}{N_0}\right)$

Intensity of Sound

23. Find the number of decibels for the following sounds.
   (a) a whisper with $I = 10^{-14}$
   (b) rock music with $I = 10^{-6}$

24. If a sound of 130 decibels produces pain in the ear, what is the corresponding intensity I?

25. If two machines operate at noise levels of 50 and 60 decibels, respectively, what is the total number of decibels produced when both machines are operating?

26. Solve equation (5) for I in terms of D.

27. If 10^k watts produce D decibels, how many watts produce $2D$ decibels?

28. If one machine produces D decibels, how many similar machines are needed to produce $2D$ decibels?

Richter Scale

29. Find the amount of energy (in joules) released for each of the following Richter scale readings.
   (a) 0 (b) 5

30. A Himalayan earthquake in 1950 registered 8.7 on the Richter scale. What was its released energy?

31. If 10^k joules of released energy produces a Richter scale reading of R, how many joules will produce $2R$?

32. Solve equation (6) for E in terms of R.

Miscellaneous

If a cake is removed from an oven and allowed to cool, then the rate at which the cake cools is proportional to the difference between the temperature T of the cake and the temperature M of the area in which it is placed. In general, if an object of temperature T_0 is placed in a cooler medium of constant temperature M, then the temperature of the object at time t is given by the formula

$$T(t) = (T_0 - M)e^{-kt} + M, \qquad \textbf{(cooling formula)}$$

where k is a positive constant.

33. Verify that $T(0)$, the temperature at time 0, is equal to T_0.

34. Suppose that the temperature of a body is 100° F and the surrounding air has temperature 50° F. If the body cools from 100° to 80° in 15 min, when will it be 60° F?

35. A metal object is heated to 250° F and then placed in a room with temperature 70° F. If the object cools to 150° F in 10 min, what will be its temperature in 10 more minutes?

36. A cup of coffee is placed on a table, and the temperature of the surrounding air is 30° C. If the coffee cools to 35° in 6 min, what was the original temperature of the coffee?

Chemists define the pH of a liquid by the equation

$$\text{pH} = -\log\,[\text{H}^+], \qquad \textbf{(pH formula)}$$

where $[\text{H}^+]$ *is the liquid's concentration of hydrogen ions measured in moles per liter. For pure water,* pH = 7. *An* ***acid*** *has* $\text{pH} < 7$, *and a* ***base*** *has* $\text{pH} > 7$.

37. Find the pH for each of the following.

(a) milk with $[\text{H}^+] = 4 \times 10^{-7}$ moles per liter

(b) vinegar with $[\text{H}^+] = 6.3 \times 10^{-3}$ moles per liter

38. Find $[\text{H}^+]$ for each of the following.

(a) soda with pH = 2.6

(b) beer with pH = 4.8

In the process of learning some skills, progress is rapid at the beginning and then slows down. A learning curve that describes this situation may have an equation of the form

$$y = y_0 + c(1 - e^{-kx}), \qquad \textbf{(learning curve)}$$

where c and k are positive constants.

39. Show that $y = y_0$ when $x = 0$, and that $y \to y_0 + c$ as $x \to \infty$. Also sketch the graph of the learning curve for the case $y_0 = 2$, $c = 5$, $k = .04$, and $x \geq 0$.

40. Let y in the learning curve denote the number of words per minute of a typist training for x days, where $y_0 = 10$, $c = 40$, and $k = .03$.

(a) How many words per minute can the person type after training 50 days?

(b) What is the maximum number of words per minute the typist will ever do?

Chapter 5 Review Outline

5.1 The Exponential Function and Its Graph

Definitions

The exponential function $f(x) = a^x$, where the base a is a fixed positive number ($a \neq 1$) and the exponent x is any real number, is a continuous and smooth extension of $a^{m/n}$, which was defined for rational exponents in Chapter 1.

The natural base e is the value approached by the expression

$$\left(1 + \frac{1}{n}\right)^n$$

as n goes through the sequence 1, 2, 3, 4, ...

$e \approx 2.71828$

Properties of a^x

$f(x) = a^x$ is a one-to-one function that increases for $a > 1$ and decreases for $0 < a < 1$, with $D_f = (-\infty, \infty)$ and $R_f = (0, \infty)$.

$$a^x a^y = a^{x+y} \qquad \frac{a^x}{a^y} = a^{x-y}$$

$$(ab)^x = a^x b^x \qquad \left(\frac{a}{b}\right)^x = \frac{a^x}{b^x}$$

$$(a^x)^y = a^{xy}$$

5.2 The Logarithmic Function and Its Graph

Definition

$f(x) = \log_a x$ is the inverse of $g(x) = a^x$; that is,

$$\log_a (a^x) = x \text{ for all real numbers } x,$$

and

$$a^{(\log_a x)} = x \text{ for all } x > 0.$$

Equivalently,

$$y = \log_a x \Leftrightarrow a^y = x.$$

Properties of $\log_a x$

$f(x) = \log_a x$ is a one-to-one function that increases for $a > 1$ and decreases for $0 < a < 1$, with $D_f = (0, \infty)$ and $R_f = (-\infty, \infty)$.

$$\log_b x = \frac{\log_a x}{\log_a b}$$

5.3 Properties of the Logarithm

$$\log_a (xy) = \log_a x + \log_a y$$

$$\log_a (x/y) = \log_a x - \log_a y$$

$$\log_a (x^y) = y \log_a x$$

5.4 Applications and Exponential and Logarithmic Equations

Since the logarithmic and exponential functions are one-to-one, the equation

$$f(x) = g(x)$$

is equivalent to each of the equations

$$\log_a f(x) = \log_a g(x) \qquad [f(x) > 0,\ g(x) > 0]$$

and $$a^{f(x)} = a^{g(x)}.$$

Applications include compound interest:

$$A = P\left(1 + \frac{R}{N}\right)^{Nt},$$

continuous compound interest:

$$A = Pe^{Rt},$$

exponential growth and decay:

$$N(t) = N_0 e^{kt}$$

($k > 0$) for growth, < 0 for decay),

intensity of sound:

$$D = 10 \log (10^{16} I),$$

and the Richter scale:

$$R = \frac{2}{3} \log (10^{-4.7} E).$$

Chapter 5 Review Exercises

1. Sketch the graph of $y = (3/2)^x$ as in Section 5.1.
2. Sketch the graph of $y = (2/3)^x$ as in Section 5.1.
3. Solve $2^{3x} = 16^{x-1}$ for the exact value of x.
4. Solve $\dfrac{25^x}{5^{3x+4}} = \dfrac{5^{2x+1}}{25^{x+3}}$ for the exact value of x.
5. Simplify $\dfrac{a^x - 3 - 4a^{-x}}{1 + a^{-x}}$. (*Hint:* multiply numerator and denominator by a^x.)
6. Use a calculator to compute $\left(1 + \dfrac{1}{n}\right)^n$ for $n = 10, 10^2, 10^3$, and 10^4.
7. Sketch the graphs of $y = x$, $y = (1.1)^x$, and $y = (1.5)^x$ on the same axes as in Section 5.1.
8. Write a logarithmic statement equivalent to $8^{-2/3} = 1/4$.
9. Write $\log_2 (2\sqrt[3]{4}) = 5/3$ in equivalent exponential form.
10. Find the exact value of $5^{\log_5 11}$ and of $\log_3 (3^{\sqrt{2}})$.
11. Solve for x: $\dfrac{1}{2} \log_a x = 2 \log_a 10 - 10 \log_a 2 + 3 \log_a 5$.
12. Solve for x: $\log_x 20 = 4 + \log_x 5$.
13. Write $\log [x^2\sqrt{y}/z^3]^{1/4}$ as an algebraic sum in $\log x$, $\log y$, and $\log z$.
14. Express $2 \log_a x - 3 \log_a y + \log_a (x + y)$ as a single logarithm.
15. Show that $\log_3 x = 2 \log_9 x$ for any $x > 0$.
16. Write $\log_6 28$ as a ratio of common logarithms and perform the division with a calculator.
17. Express $3 \log_{10} x + 4 \log_5 x$ in terms of $\ln x$.
18. On the same axes, sketch the graph of $f(x) = (2.7)^x$ and the graph of $f^{-1}(x) = \log_{2.7} x$.
19. Sketch the graphs of $y = x$, $y = \log_{1.1} x$, and $y = \log_{1.5} x$ on the same axes.
20. Write the equation $\ln x = -2$ in exponential form.
21. Write the equation $x = e^3 \cdot e^{-5}$ in logarithmic form.
22. Find the exact value of x if $\log_2 x = -3$.
23. Find the exact value of x if $4^x \cdot 3^{2x} = 5 \cdot 2^{3x}$.
24. Find an approximate value of x if $\log x = 2.7404$.
25. Find an approximate value of x if $\log x = -1.2596$.
26. Find an approximate value of x if $\ln x = 2.1972$.
27. Solve $2^{3x} = 10^{x-1}$ for an approximate value of x.
28. Using $\log 5 = .69897$ and $\log 7 = .84510$, compute $\log (5^4/\sqrt{7})$.

29. Using the values given in Exercise 28, compute log 1.4.

30. Use the change-of-base rule to show that $\log_{a^n}(x) = (1/n)\log_a x$.

31. Solve $(1.02)^{5t} = 10{,}000$ for t.

32. Solve $P(1.05)^{50} = 10{,}000$ for P.

33. Solve $P(1 + r)^7 = 2P$ for r.

34. Find the amount accumulated at the end of 15 years if \$1,000 is invested at an annual rate of 8.35% compounded monthly.

35. Find the amount in Exercise 34 if interest is compounded continuously.

36. To accumulate \$10,000 in 5 years at 9% compounded daily, what principal must be invested now?

37. If money is invested at 8% compounded quarterly, how long will it take to triple the principal?

38. If a radioactive substance has a half-life of 2,000 yr, how long will it take for one fourth of it to disintegrate?

39. Using the formula for decibels, $D = 10 \log (10^{16} I)$, show that if $I_2 = 10I_1$, then $D_2 = D_1 + 10$.

40. If an earthquake on the Richter scale $R = (2/3)[\log (10^{-4.7} E)]$ measures $R = 6.2$, what is the amount of energy released?

Graphing Calculator Exercises

In each of the following, verify your answers to the indicated review exercises of this section by using a graphing calculator for the given functions and windows $[x_{min}, x_{max}]$, $[y_{min}, y_{max}]$.

41. Exercise 1: $Y_1 = (3/2)^x$, $[-4.5,5]$, $[-1,10]$

42. Exercise 2: $Y_1 = (2/3)^x$, $[-4.5,5]$, $[-1,10]$

43. Exercise 7: $Y_1 = x$, $Y_2 = (1.1)^x$, $Y_3 = (1.5)^x$, $[-6,13]$, $[-1,10]$

44. Exercise 18: $Y_1 = (2.7)^x$, $Y_2 = \ln x/\ln 2.7$, $[-2,2.75]$, $[-3,4]$

45. Exercise 19: $Y_1 = x$, $Y_2 = \ln x/\ln 1.1$; $Y_3 = \ln x/\ln 1.5$, $[0,9.5]$, $[-5,15]$

46. Exercise 24: $Y_1 = \log x$, $Y_2 = 2.7404$, $[0,950]$, $[0,4]$

47. Exercise 25: $Y_1 = \log x$, $Y_2 = -1.2596$, $[0,.095]$, $[-2,0]$

48. Exercise 26: $Y_1 = \ln x$, $Y_2 = 2.1972$, $[0,19]$, $[-2,3]$

6 Trigonometric Functions of Angles

Trigonometry means triangle measurement. The subject has been traced to the Greek astronomer Hipparchus, who was born in about 160 B.C. The ancient Greeks used trigonometry for the indirect measurement of distances on earth and in the heavens. In the sixteenth and seventeenth centuries trigonometry was applied to the study of periodic phenomena. Kepler studied planetary orbits, Galileo observed the vibrations of a pendulum, Huygens studied light waves, and Mersenne investigated the vibrations of a violin string. Today triangle trigonometry is still used in surveying, navigation, and astronomy, while the periodic trigonometric functions have become important tools in areas such as acoustics, optics, electronics, seismology, and medicine, and in the study of business cycles.

We will follow the historical development of the subject, beginning with trigonometric functions of angles in this chapter and then taking up the more abstract idea of periodic trigonometric functions of real numbers in the next chapter. Trigonometry gives us a good example of a branch of mathematics that acquired more applications at the same time that it became more abstract.

6.1 Angles and Their Measurement

Why was the number 360 chosen to designate one complete rotation?

Angles
Degree and Radian Measure
Standard Position

Angles. Since we want to define trigonometric functions whose domains are angles, we begin with the definition of angles and their measurement. An **angle** is formed by two rays having a common endpoint, but it is more useful to think of an angle as a *rotation*. For example, the angle θ in Figure 1 is obtained by rotating the **initial side** OP about the **vertex** O until it reaches the position of its **terminal side** OQ. If the rotation is counterclockwise, as indicated for θ in Figure 1, the angle is *positive*, while a clockwise rotation, as indicated for α, is *negative*. Angles such as θ and α that have the same initial and terminal sides are called **coterminal.**

Degree and Radian Measure. In elementary geometry, angles are usually measured in **degrees.** One complete counterclockwise rotation, until the terminal side coincides with the initial side, produces an angle of 360 degrees, denoted by 360°. An angle of 1° (one degree) is 1/360 of such a rotation. A degree is divided into 60 equal parts called **minutes,** denoted by ′, and each minute equals 60 **seconds,** denoted by ″. That is,

$$1' = \left(\frac{1}{60}\right)^\circ \qquad 1 \text{ minute} = \frac{1}{60} \text{ degree}$$

and $$1'' = \left(\frac{1}{60}\right)' = \left(\frac{1}{3600}\right)^\circ. \qquad 1 \text{ second} = \frac{1}{60} \text{ minute} = \frac{1}{3600} \text{ degree}$$

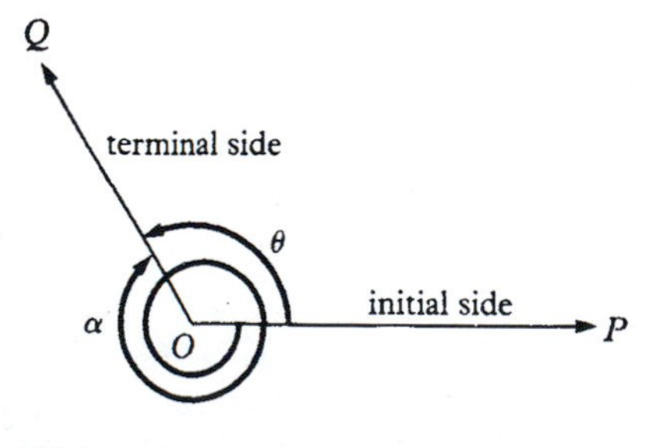

Figure 1 Angle about Vertex O

For example,

$$
\begin{aligned}
17°30'12'' &= 17 \text{ degrees, 30 minutes, and 12 seconds} \\
&= \left(17 + \frac{30}{60} + \frac{12}{3600}\right)^\circ \\
&\approx 17.5033°.
\end{aligned}
$$

As in the last line above, many scientific calculators express degrees in decimal form, rather than in degrees, minutes, and seconds. The above example shows that to convert from minutes to a decimal part of a degree, we *divide* by 60. For example,

$$24' = \left(\frac{24}{60}\right)^\circ = .4°$$

and

$$47' = \left(\frac{47}{60}\right)^\circ \approx .7833°.$$

To convert from a decimal part of a degree to minutes, we *multiply* by 60. For example,

$$.75° = (.75 \cdot 60)' = 45'$$

and

$$.82° = (.82 \cdot 60)' \approx 49'.$$

For the applications of trigonometry in this chapter, degree measure would suffice, but for the study of periodic phenomena in the next chapter, it is essential to use radian measure. To define radian measure, we now think of an angle θ with its initial and terminal sides along radii of a circle subtending an arc PQ on the circle's circumference (Figure 2). Such an angle is called a **central angle.** If the radius of the circle is r and the arc subtended is s, then the **radian measure** of θ is defined as

$$\boldsymbol{\theta = \frac{s}{r}}. \qquad \textbf{(1)}$$

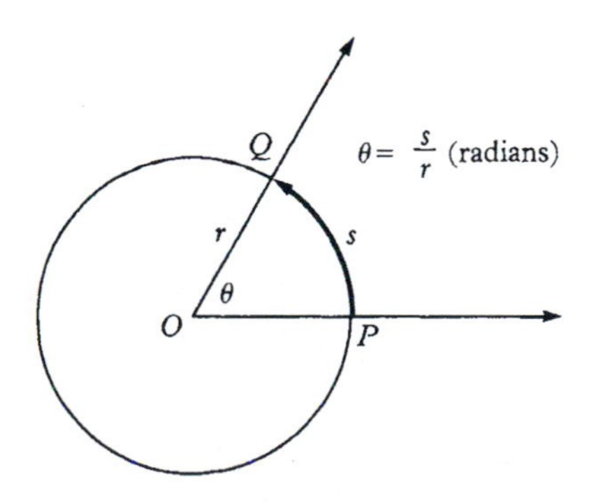

Figure 2 Radian Measure of a Central Angle

The radius and arc length must be measured in the same units, and then the radian measure of θ is a *dimensionless* real number. For example, if $r = 4$ centimeters and $s = 6$ centimeters, then

$$\theta = \frac{6 \text{ centimeters}}{4 \text{ centimeters}} = 1.5.$$

It is also important to observe that the radian measure of an angle is actually independent of the size of the radius. In Figure 3, we have two concentric circles of radii r_1 and r_2, respectively, and the angle θ subtends corresponding arcs P_1Q_1 and P_2Q_2 of respective lengths s_1 and s_2. The circular sector OP_1Q_1 is *similar* to the circular sector OP_2Q_2, and therefore each circle determines the same radian measure of θ.

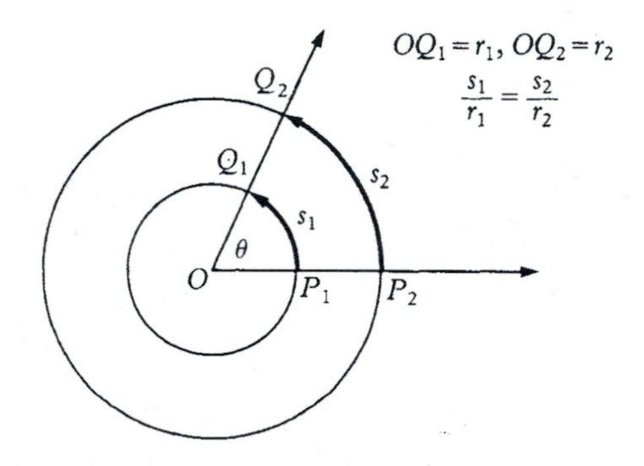

Figure 3 Concentric Circles, Same Central Angle

We sometimes need to convert from degree measure of an angle to radian measure, and conversely. Now, one complete counterclockwise rotation of a radius of a circle sweeps out a central angle of 360° and at the same time subtends an arc equal to the circumference $2\pi r$. Hence, by equation (1), the radian measure of one complete counterclockwise rotation is equal to

$$\frac{2\pi r}{r} = 2\pi.$$

We conclude that $360° = 2\pi$ radians, or, equivalently,

$$\mathbf{180° = \pi \text{ radians.}}$$

Therefore,

$$\mathbf{1° = \left(\frac{\pi}{180}\right) \text{radians,}} \qquad \textbf{degrees to radians} \tag{2}$$

and

$$\mathbf{1 \text{ radian} = \left(\frac{180}{\pi}\right)^{\circ}.} \qquad \textbf{radians to degrees} \tag{3}$$

It is customary to omit the word "radians" and write, for example, $180° = \pi$.

Example 1 Convert the following angles from degrees to radians.

(a) $90°$ (b) $12.25°$ (c) $45°36'$

Solution: We use equation (2) above.

(a) $90° = 90\left(\frac{\pi}{180}\right) = \frac{\pi}{2}$ (radians)

(b) $12.25° = 12.25\left(\frac{\pi}{180}\right) \approx .2138$ (radians)

(c) Here we first convert degrees and minutes to degrees in decimals. That is,

$$45°36' = \left(45 + \frac{36}{60}\right)^{\circ} = 45.6°.$$

Then $$45.6° = 45.6\left(\frac{\pi}{180}\right) \approx .7959 \text{ (radians)}.$$ ■

Example 2 Convert the following angles from radians to degrees.

(a) $\frac{\pi}{3}$ (b) 2 (c) -1.5π

Solution: Now we use equation (3) above.

(a) $\frac{\pi}{3} = \frac{\pi}{3}\left(\frac{180}{\pi}\right)^{\circ} = 60°$

(b) $2 = 2\left(\frac{180}{\pi}\right)^{\circ} = \left(\frac{360}{\pi}\right)^{\circ} \approx 114.6°$

(c) $-1.5\pi = -1.5\pi\left(\frac{180}{\pi}\right)^{\circ} = -270°$ ■

Hence, to convert from degrees to radians, we multiply by $\pi/180$, and to convert from radians to degrees, we multiply by $180/\pi$. To remember which factor to use, note that the degree measurement of an angle is always larger in magnitude than the radian measure. Therefore, multiply by the *larger* factor, $180/\pi$, in going from radians to degrees, and by the *smaller* one, $\pi/180$, when going the other way.

Example 3 A bicycle wheel with a radius of 15 inches makes 5000 rotations.

(a) Find the total angle in radians through which a spoke of the wheel rotates.

(b) Find the distance traveled if the wheel moves along the ground without slipping.

Solution:

(a) Each rotation corresponds to an angle of 2π radians, and therefore 5000 rotations equal $5000 \cdot 2\pi = 10{,}000\pi$ radians.

(b) The distance traveled equals the total arc length subtended, which by equation (1) is

$$\begin{aligned} s &= r\theta \\ &= 15(10{,}000\pi) \text{ inches} \\ &= 150{,}000\pi \text{ inches} \\ &\approx 7.44 \text{ miles.} \end{aligned}$$

Note that θ must be in radians if equation (1) is used to compute arc length. ■

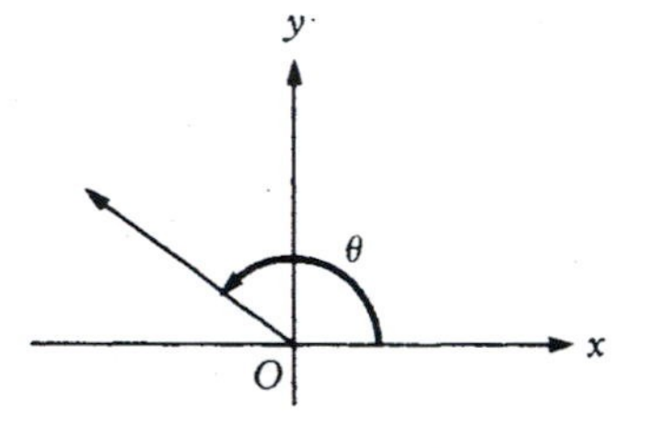

Figure 4 Angle in Standard Position

Standard Position. In order to define the trigonometric functions in the next section, we will find it helpful to utilize coordinates in connection with angles. We define an angle to be in **standard position** in a coordinate system if its vertex is at the origin and its initial side is along the positive x-axis (Figure 4). An angle in standard position whose terminal side lies on one of the axes, such as 0° or 270°, is called a **quadrantal angle. A non-quadrantal angle** is said to be in the quadrant in which its terminal side lies. For example, $\pi/6$ is in quadrant I [Figure 5(a)], and $-3\pi/4$ is in quadrant III [Figure 5(b)].

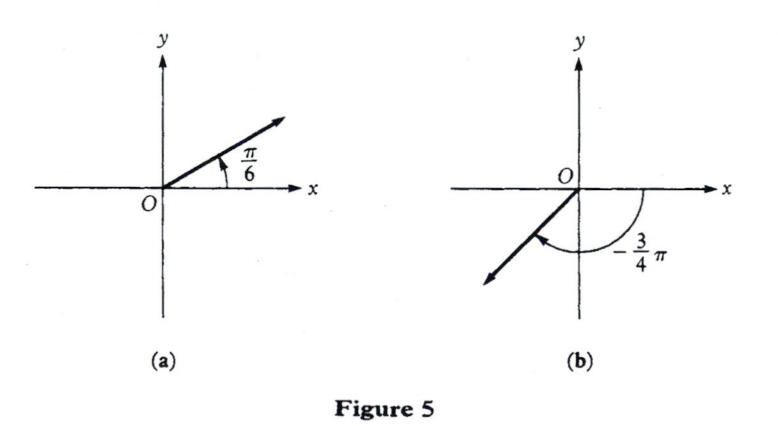

Figure 5

Exercises 6.1

Fill in the blanks to make each statement true.

1. An angle may be thought of as a ________.
2. The number of degrees for one half a counterclockwise rotation is ______.
3. The number of radians for one complete counterclockwise rotation is ______.
4. An angle is in standard position if its vertex is at the origin and ____________.
5. An angle in standard position is a third-quadrant angle if ____________.

Write true *or* false *for each statement.*

6. The number of radians measuring an angle can be any real number.
7. In order for two angles to be coterminal, one of them must be positive and greater than 360°.
8. If $\theta = x$ radians, then $\theta = 180x/\pi$ degrees.
9. The formula $s = r\theta$ applies only if θ is measured in radians.
10. An angle of 2 radians is larger than an angle of 2°.

Angles

11. Find the degree measure of the following.
    (a) 1/4 rotation (b) 3 1/2 rotations
12. Find the radian measure of each of the following.
    (a) 3/4 rotation (b) 5 2/3 rotations
13. Which of the following angles in standard position are coterminal?
    45° 135° 405° −225° −315°
14. Which of the following angles in standard position are coterminal?
    0 π 2π
    3π $-\pi$ -2π
15. Sketch each of the following angles in standard position and find its quadrant.
    $\pi/6$ $5\pi/6$ $-7\pi/6$ 4
16. Find the quadrant for each of the following angles, assuming them to be in standard position. Make a sketch.
    78° −78° 195° −195°

Degree and Radian Measure

17. Find the radian measure in terms of π of each of the following angles.
    45° 270° −315°
18. Find the degree measure of each of the following angles.
    $\pi/6$ $-3\pi/4$ 10
19. Express 47°15′45″ in terms of degrees only.
20. Express 110°30′ in radian measure.
21. Find the number of radians in a central angle θ if it subtends an arc of length s on a circle of radius r, given the following.
    (a) $s = 2$ in and $r = 2$ in
    (b) $s = 1$ cm and $r = 1$ m
22. Find the length of the arc subtended by a central angle θ on a circle of radius r, given the following.
    (a) $\theta = \pi/4$ and $r = 6$ in
    (b) $\theta = 30°$ and $r = 6$ in

Miscellaneous

23. If a bicycle wheel has a radius of 14 in, how many rotations does it make in going 1 mi (63,360 in)?
24. If the moon makes one circular orbit about the earth in 30 days, through what angle will a telescope tracking it turn in one week?
25. Assume that the earth is a sphere of radius 4000 mi. Two points P and Q on the earth have the same *longitude* if they lie on the same semicircle through the north and south poles with center C at the center of the earth. If a point P lies north (south) of the equator, its *latitude* is $L°$ North (South) where $L°$ is the angle PCE and E is the point on the equator with the same longitude as P. Find the distance along the earth between P and Q of the same longitude if their latitudes are as follows.
    (a) P is 40° N, and Q is on the equator.
    (b) P is 40° N, and Q is 25° S.

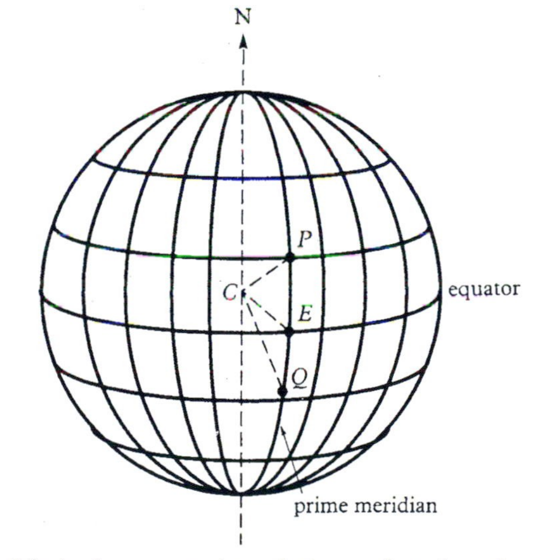

26. A semicircle that passes through the north and south poles and the original site of the Royal Observatory in Greenwich, England is called the prime meridian. If two points P and Q are on the equator and P has longitude (See exercise 25) 50° west of the prime meridian while Q has longitude 30° east, find the distance on the surface of earth between P and Q.

If a wheel is rotating, its **angular velocity**, *denoted by* ω, *is the rate at which the angle* θ *generated by a spoke on the wheel is changing per unit of time. For example, a wheel that makes a constant 2 revolutions per minute has an angular velocity* ω *equal to* 4π *radians per minute. In general, if* ω *has a constant value, then a spoke on the wheel rotates through an angle* $\theta = \omega t$ *in* t *time units.*

27. What is the angular velocity of a wheel that makes 100 revolutions per minute?
28. A wheel has radius 5 in and rotates at 10 revolutions per minute. Find its angular velocity and the distance a point on the edge of the wheel travels in 1 min.
29. If a disk rotates at 45 revolutions per minute, through what angle does a radius rotate in 10 min? How far does a point on the outside of the disk move in 10 min if its radius is 3 in?
30. The minute hand of a clock makes one revolution each hour. Through what angle does it turn in 12 hr? How far does the tip of the minute hand travel in 12 hr if its length is 3 in?
31. Find a formula for the area A of a sector of a circle of radius r subtended by an angle θ, measured in radians. (*Hint:* the area A of the sector and the area of the circle have the same ratio as their central angles.)

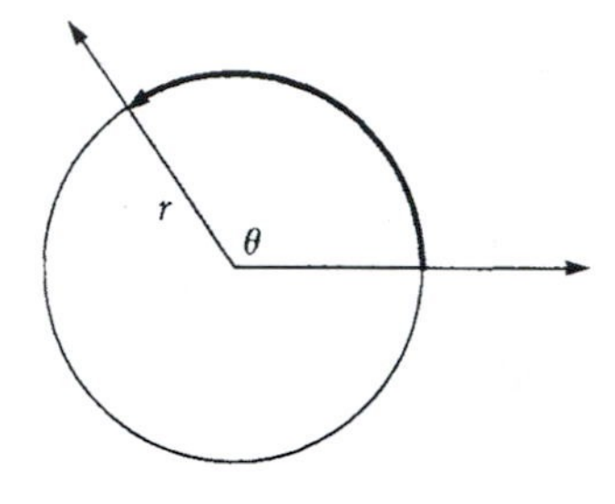

6.2 The Trigonometric Functions

Hipparchus needed a tool to measure the radius of the earth, so he invented trigonometry.

Definitions of the Trigonometric Functions
Special Angles
The Use of Calculators

Definitions of the Trigonometric Functions. With reference to Figure 6, let θ be an angle in standard position, and let $P(x, y)$ be any point except the origin on the terminal side of θ. The length of OP is denoted by r, and we use the three numbers x, y, and r to form six ratios that define the **trigonometric functions** of θ. Their names and corresponding abbreviations are *sine* (sin), *cosine* (cos), *tangent* (tan), *cotangent* (cot), *secant* (sec), and *cosecant* (csc), and they are defined as shown below.

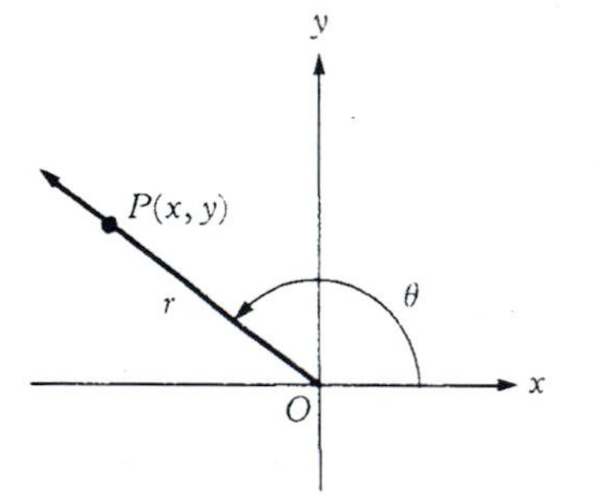

Figure 6

$$\sin\theta = \frac{y}{r} \qquad \cos\theta = \frac{x}{r} \qquad \tan\theta = \frac{y}{x}$$

$$\csc\theta = \frac{r}{y} \qquad \sec\theta = \frac{r}{x} \qquad \cot\theta = \frac{x}{y} \qquad (1)$$

These ratios define functions of θ only, because they are independent of the *particular point* chosen on the terminal side of the angle. To see this, let θ

be a nonquadrantal angle in standard position, as shown in Figure 7. If $P(x, y)$ is a point other than O on the terminal side of θ and Q is the projection of P on the x-axis, then OQP is a right triangle whose sides determine the six ratios given above. If $P'(x', y')$ is another point other than O on the terminal side of θ, then the corresponding right triangle $OQ'P'$ is *similar* to OQP and therefore determines the same values for the six ratios. If θ is a quadrantal angle, there is no associated triangle, but the trigonometric functions are still independent of $P(x, y)$, as shown in Example 2 below.

The sine and cosine functions are called **cofunctions,** as are the tangent and cotangent and also the secant and cosecant. If $x \neq 0$ and $y \neq 0$, that is, if θ is a nonquadrantal angle, then all six trigonometric functions of θ are defined.

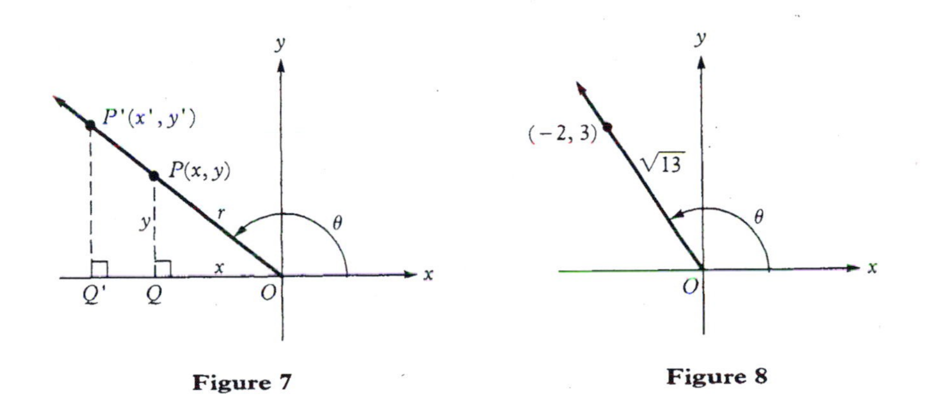

Figure 7 **Figure 8**

Example 1 Find the six trigonometric functions of an angle θ if it is in standard position and the point $(-2, 3)$ is on its terminal side (Figure 8).

Solution: We are given $x = -2$ and $y = 3$, and by the distance formula,

$$r = \sqrt{(-2)^2 + 3^2} = \sqrt{13}.$$

Therefore, by equations (1),

$$\sin \theta = \frac{3}{\sqrt{13}} \qquad \cos \theta = \frac{-2}{\sqrt{13}} \qquad \tan \theta = \frac{3}{-2}$$

$$\csc \theta = \frac{\sqrt{13}}{3} \qquad \sec \theta = \frac{\sqrt{13}}{-2} \qquad \cot \theta = \frac{-2}{3}.$$

For quadrantal angles in which $x = 0$, the secant and tangent are not defined (*why?*). Similarly, the cosecant and cotangent are not defined when $y = 0$.

Example 2 Find the trigonometric functions of each of the following angles.

(a) $\theta = \pi/2$ (b) $\theta = \pi$

Solution:

(a) Any point $P(x, y)$ on the positive y-axis is on the terminal side of $\theta = \pi/2$. Therefore, as indicated in Figure 9(a), $x = 0$ and $y = r$. Hence, from equations (1),

$$\sin\left(\frac{\pi}{2}\right) = \frac{r}{r} = 1 \qquad \cos\left(\frac{\pi}{2}\right) = \frac{0}{r} = 0 \qquad \tan\left(\frac{\pi}{2}\right) = \frac{r}{0} \text{ (undefined)}$$

$$\csc\left(\frac{\pi}{2}\right) = \frac{r}{r} = 1 \qquad \sec\left(\frac{\pi}{2}\right) = \frac{r}{0} \text{ (undefined)} \qquad \cot\left(\frac{\pi}{2}\right) = \frac{0}{r} = 0.$$

(b) Any point $P(x, y)$ on the negative x-axis is on the terminal side of $\theta = \pi$. Here, as indicated in Figure 9(b), $y = 0$ and $x = -r$. Therefore, by equations (1),

$$\sin \pi = \frac{0}{r} = 0 \qquad \cos \pi = \frac{-r}{r} = -1 \qquad \tan \pi = \frac{0}{-r} = 0$$

$$\csc \pi = \frac{r}{0} \text{ (undefined)} \qquad \sec \pi = \frac{r}{-r} = -1 \qquad \cot \pi = \frac{-r}{0} \text{ (undefined)}$$ ■

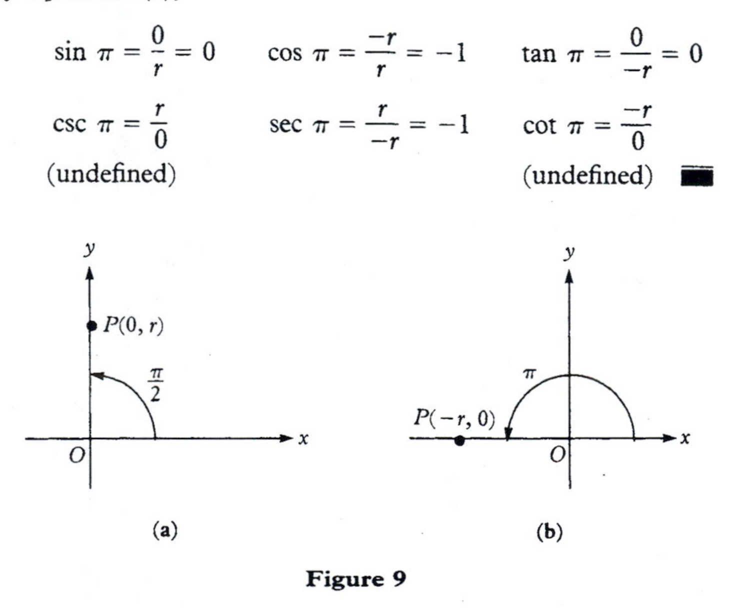

Figure 9

Example 2 illustrates that for the quadrantal angles $\pi/2$ and π, the trigonometric functions are also independent of the particular point chosen on the terminal side, even though there are no associated right triangles for these angles. Similar results can be obtained for other quadrantal angles and are listed in the following table.

Terminal Side	Positive x-axis	Positive y-axis	Negative x-axis	Negative y-axis
θ	0	$\pi/2$	π	$3\pi/2$
$\sin\theta$	0	1	0	-1
$\cos\theta$	1	0	-1	0
$\tan\theta$	0	undefined	0	undefined
$\cot\theta$	undefined	0	undefined	0
$\sec\theta$	1	undefined	-1	undefined
$\csc\theta$	undefined	1	undefined	-1

This table need not be memorized. The entries can be obtained directly from equations (1).

From equations (1) and the table, we obtain the following results concerning the **domains** of the trigonometric functions.

1. The domain of the sine and cosine functions consists of *all* angles.
2. The domain of the tangent and secant functions is all angles except $\pi/2 + n\pi$, where n is any integer. The excluded angles are all those in standard position whose terminal sides lie on the y-axis.
3. The domain of the cotangent and cosecant functions includes all angles except $n\pi$ for any integer n. The excluded angles are all those in standard position with terminal sides on the x-axis.

Note that the six trigonometric functions are not independent of each other. For example,

$$\tan\theta = \frac{\sin\theta}{\cos\theta}, \quad \cot\theta = \frac{1}{\tan\theta}, \quad \sec\theta = \frac{1}{\cos\theta}, \quad \csc\theta = \frac{1}{\sin\theta}. \tag{2}$$

Each equation in (2) holds for all values of θ for which both sides are defined. Such equations are called **identities;** we will investigate other trigonometric identities in the next chapter.

Special Angles. The angles 45°, 60°, and 30° are called **special angles** because, as indicated in Figure 10, they are associated with special triangles, namely *isosceles* (two equal sides) and *equilateral* (three equal sides) triangles. In the next two examples, we compute the trigonometric functions for these special angles.

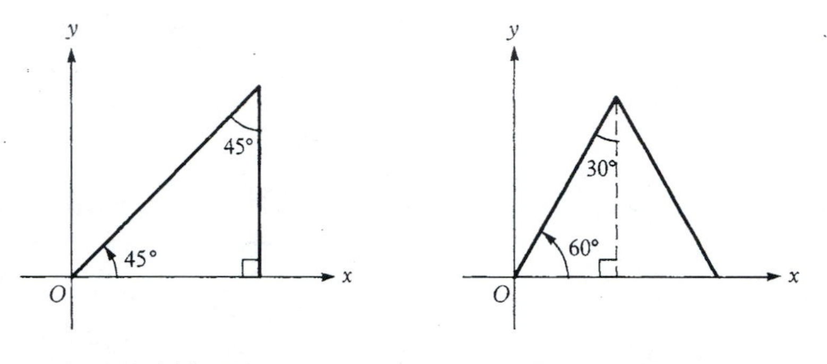

Figure 10

Example 3 Find the six trigonometric functions of $\theta = 45°$.

Solution: The ray OP in Figure 11 lies on the line $y = x$ (*why?*). Hence, we can choose the point $P(1, 1)$ on the terminal side. Then, by dropping a perpendicular from P to Q on the x-axis and applying the Pythagorean theorem to right triangle OQP, we have

$$r^2 = 1^2 + 1^2$$
$$r^2 = 2$$
$$r = \sqrt{2}.$$

Therefore, by equations (1),

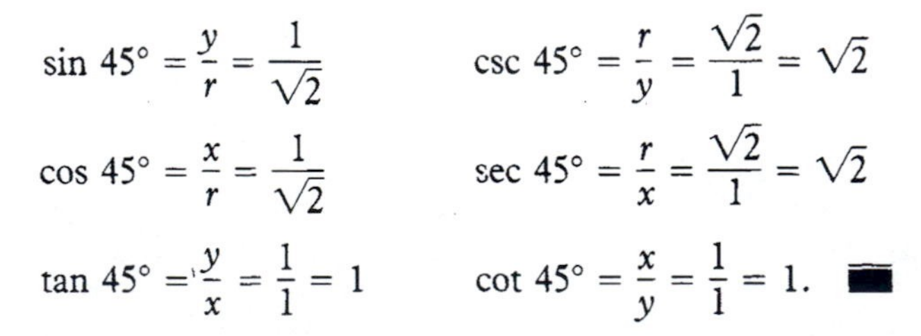

$$\sin 45° = \frac{y}{r} = \frac{1}{\sqrt{2}} \qquad \csc 45° = \frac{r}{y} = \frac{\sqrt{2}}{1} = \sqrt{2}$$

$$\cos 45° = \frac{x}{r} = \frac{1}{\sqrt{2}} \qquad \sec 45° = \frac{r}{x} = \frac{\sqrt{2}}{1} = \sqrt{2}$$

$$\tan 45° = \frac{y}{x} = \frac{1}{1} = 1 \qquad \cot 45° = \frac{x}{y} = \frac{1}{1} = 1. \quad \blacksquare$$

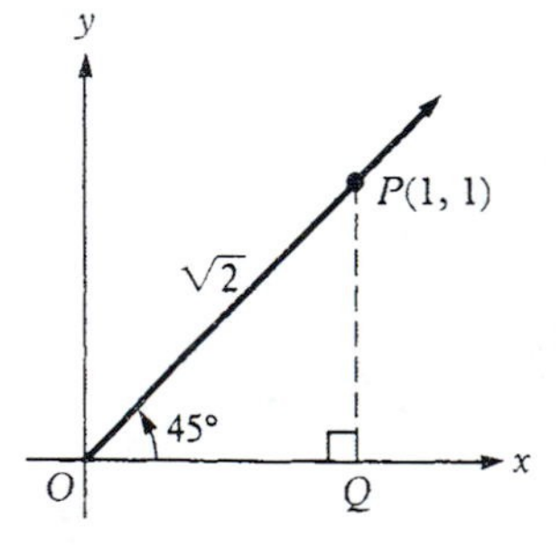

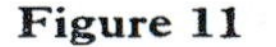
Figure 11

Example 3 illustrates an important point. That is, if θ is an **acute angle** ($0° < \theta < 90°$), then x and y are positive and hence all trigonometric functions of θ are positive. Also, in the right triangle associated with θ

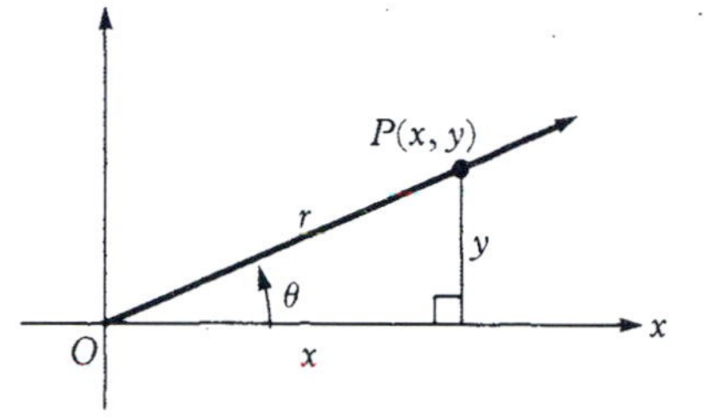

Figure 12 Acute Angle θ

(Figure 12), x is the side *adjacent* to θ, and y is its *opposite* side. Therefore, for an acute angle θ, definitions (1) read as follows.

$$\sin\theta = \frac{\text{opposite side}}{\text{hypotenuse}} \qquad \csc\theta = \frac{\text{hypotenuse}}{\text{opposite side}}$$
$$\cos\theta = \frac{\text{adjacent side}}{\text{hypotenuse}} \qquad \sec\theta = \frac{\text{hypotenuse}}{\text{adjacent side}}$$
$$\tan\theta = \frac{\text{opposite side}}{\text{adjacent side}} \qquad \cot\theta = \frac{\text{adjacent side}}{\text{opposite side}} \tag{3}$$

Equations (3) may be used to compute the trigonometric functions for either of the acute angles in any right triangle, without reference to standard position.

Example 4 Find $\sin\theta$, $\cos\theta$, and $\tan\theta$ for $\theta = 30°$ and $\theta = 60°$.

Solution: From geometry, we know that each angle of an equilateral triangle is 60°. In Figure 13, ABC is an equilateral triangle, and the altitude CD bisects both the base AB and the opposite angle at C. Thus, angle $ACD = 30°$, and if we let each side have length 2, then $AD = 1$. Also, the Pythagorean theorem applied to right triangle ACD yields $DC = \sqrt{3}$. Finally, by equations (3), we get the following:

$$\sin 60° = \frac{\text{opposite side}}{\text{hypotenuse}} = \frac{\sqrt{3}}{2},$$
$$\cos 60° = \frac{\text{adjacent side}}{\text{hypotenuse}} = \frac{1}{2},$$

and

$$\tan 60° = \frac{\text{opposite side}}{\text{adjacent side}} = \frac{\sqrt{3}}{1} = \sqrt{3}.$$

Also,

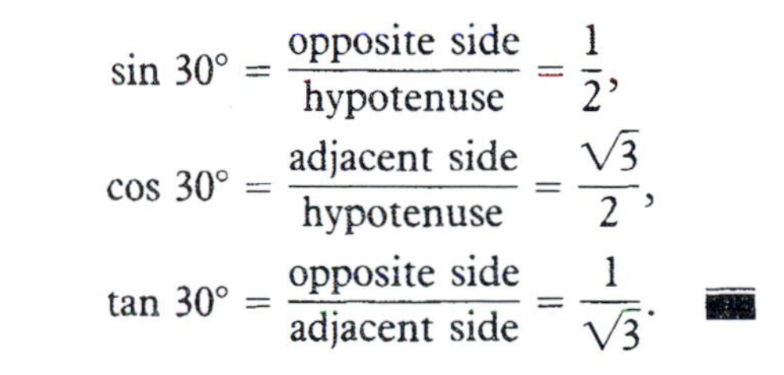

$$\sin 30° = \frac{\text{opposite side}}{\text{hypotenuse}} = \frac{1}{2},$$
$$\cos 30° = \frac{\text{adjacent side}}{\text{hypotenuse}} = \frac{\sqrt{3}}{2},$$

and

$$\tan 30° = \frac{\text{opposite side}}{\text{adjacent side}} = \frac{1}{\sqrt{3}}. \quad ■$$

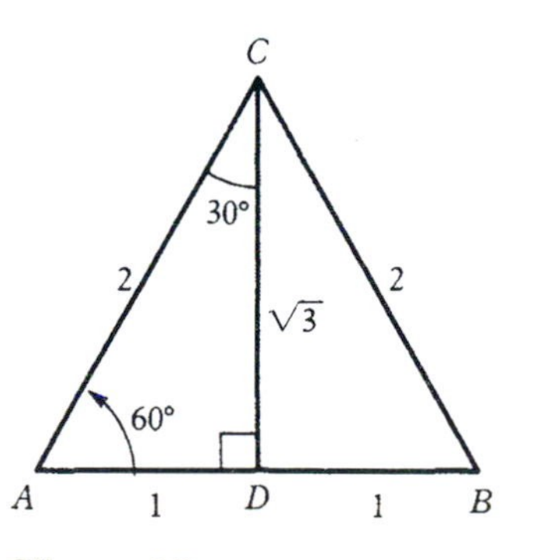

Figure 13

The special angles 45°, 60°, and 30° are used often in examples and applications. Try either to memorize the values of the sine and cosine for these angles or to remember the construction of the triangles used for computing the values. The remaining four trigonometric functions can be obtained from the sine and cosine, as previously observed. For convenient reference, values of the trigonometric functions for the special angles are listed in the following table.

θ	$\sin\theta$	$\cos\theta$	$\tan\theta$	$\cot\theta$	$\sec\theta$	$\csc\theta$
30°	$1/2$	$\sqrt{3}/2$	$1/\sqrt{3}$	$\sqrt{3}$	$2/\sqrt{3}$	2
45°	$1/\sqrt{2}$	$1/\sqrt{2}$	1	1	$\sqrt{2}$	$\sqrt{2}$
60°	$\sqrt{3}/2$	$1/2$	$\sqrt{3}$	$1/\sqrt{3}$	2	$2/\sqrt{3}$

We note that $\sin 30° = \cos 60°$, and $\sin 60° = \cos 30°$. This is no accident. If θ is one acute angle of a right triangle, then the other acute angle is $90° - \theta$. Furthermore, the opposite side for θ is the adjacent side for $90° - \theta$, and the adjacent side for θ is the opposite side for $90° - \theta$. Hence, if we examine equations (3), we see that the value of any trigonometric function at θ is equal to the value of its corresponding cofunction at $90° - \theta$. In fact, the angles θ and $90° - \theta$ are called **complementary angles,** and the "co" in cofunction is short for "complementary." We will see later that this relationship between a trigonometric function and its cofunction holds even if θ is not an acute angle.

Example 5 Find the trigonometric functions of $\theta = 120°$.

Solution: Since 120° is not an acute angle, we must put it in standard position and use the general definitions (1). From Figure 14, angle $QOP = 60°$, and if we let $r = 2$, then we can use the triangle from Example 4 to conclude that $x = -1$ and $y = \sqrt{3}$. Hence, by equations (1),

$$\sin 120° = \frac{\sqrt{3}}{2} \qquad \csc 120° = \frac{2}{\sqrt{3}}$$

$$\cos 120° = \frac{-1}{2} \qquad \sec 120° = \frac{2}{-1} = -2$$

$$\tan 120° = \frac{\sqrt{3}}{-1} = -\sqrt{3} \qquad \cot 120° = \frac{-1}{\sqrt{3}}. \quad \blacksquare$$

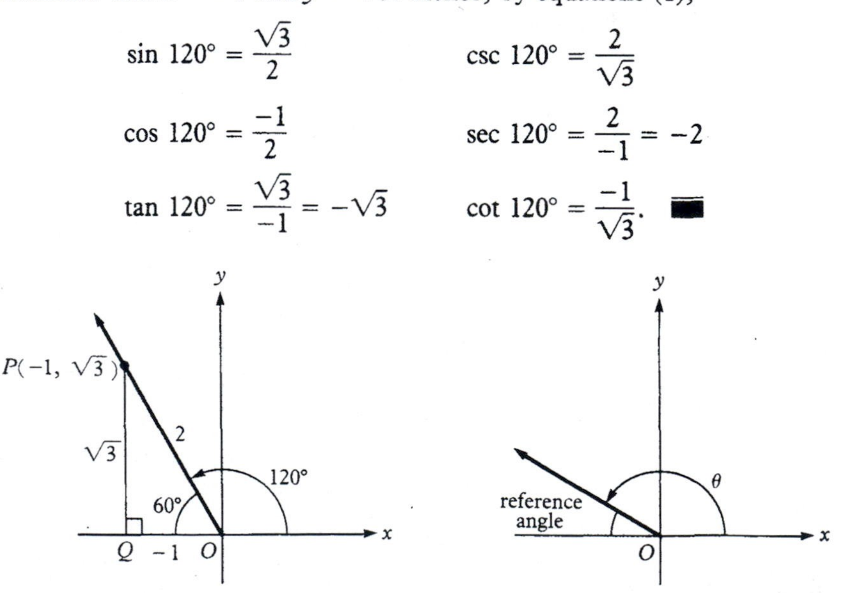

Figure 14

Figure 15

Note that the terminal side of 120° in standard position makes an angle of 60° with the x-axis and that each trigonometric function of 120° has the same absolute value as the corresponding function of 60° (see Example 4). In general, if θ is a nonquadrantal angle in standard position, then the *positive acute angle* that the terminal side of θ makes with the x-axis is called the **reference angle** for θ (Figure 15). Also, the magnitude of a given trigonometric function at θ is equal to the value of the given trigonometric function at the reference angle for θ (*why?*). Of course, if θ is an acute angle, then θ is its own reference angle.

Example 6 Find the reference angle for each of the following angles, and express the sine and cosine of each angle in terms of the reference angle.

(a) 336° (b) −245° (c) $5\pi/4$

Solution:

(a) As indicated in Figure 16(a), the reference angle for 336° is 24°. Also, 336° is in the fourth quadrant, where the sine is negative and the cosine is positive. Hence,

$$\sin 336° = -\sin 24° \quad \text{and} \quad \cos 336° = \cos 24°.$$

(b) The reference angle for −245° is 65° [Figure 16(b)]. The angle −245° is in the second quadrant, where the sine is positive and the cosine is negative. Hence,

$$\sin(-245°) = \sin 65° \quad \text{and} \quad \cos(-245°) = -\cos 65°.$$

(c) For $5\pi/4$, the reference angle is $\pi/4$ [Figure 16(c)]. Also, the sine and cosine of $5\pi/4$ are both negative. Hence,

$$\sin\frac{5\pi}{4} = -\sin\frac{\pi}{4} \quad \text{and} \quad \cos\frac{5\pi}{4} = -\cos\frac{\pi}{4}. \quad ■$$

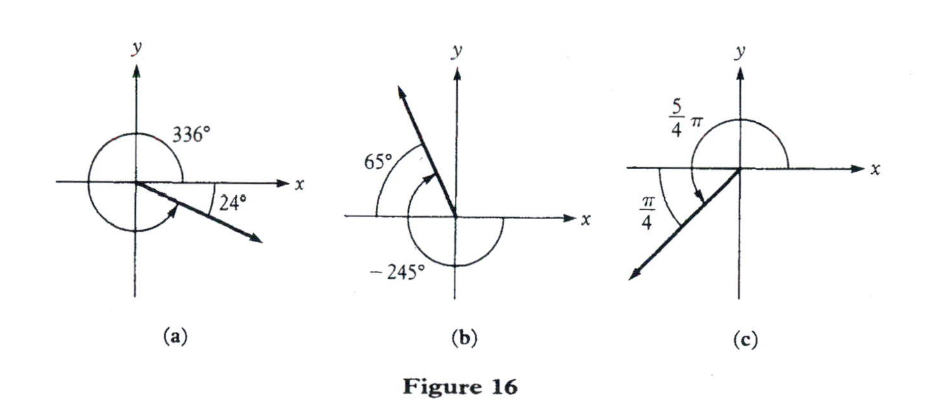

Figure 16

With regard to the *algebraic sign* of a trigonometric function, we see from equations (1) that $\sin\theta$ has the same algebraic sign as y, and $\cos\theta$ has

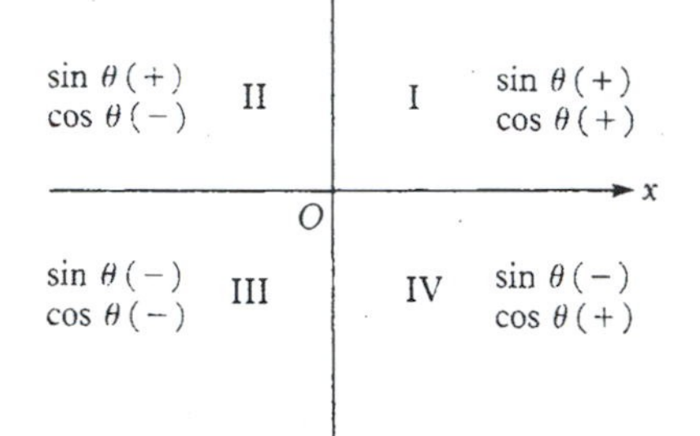

Figure 17

the same sign as x. Hence, as observed in Example 6, the respective signs of $\sin\theta$ and $\cos\theta$ depend only on the *quadrant* in which θ lies when in standard position, as indicated in Figure 17. Also, the respective signs of the remaining four trigonometric functions can be obtained from those of $\sin\theta$ and $\cos\theta$ by means of equations (2).

The Use of Calculators and Tables. We are able to compute exact values of the trigonometric functions for quadrantal angles, the special angles 30°, 45°, 60°, and for certain angles related to special angles, such as 120°. However, for most angles there is no algebraic or geometric method for finding the exact values of the trigonometric functions. For these angles, we must rely on approximation procedures, which can involve a fair amount of computation. Fortunately these procedures are built into scientific calculators, so we merely press a few keys to get values to a high degree of accuracy. Before calculators, trigonometric tables, which were tediously constructed, were used. We illustrate the use of calculators in the next example, and also briefly discuss tables.

Example 7 Use a calculator to find $\sin\theta$, $\cos\theta$, and $\tan\theta$ for each angle.

(a) $\theta = 26°30'$ (b) $\theta = 206°30'$

Solution: Most scientific calculators have available three units of measurement for angles, namely degrees, radians, and (one we shall not use) grads. Therefore, you must first be sure that your calculator is in the degree mode. Also, most calculators use decimal parts of a degree; so we convert $26°30'$ to $26.5°$ and $206°30'$ to $206.5°$. We then perform the following steps on the calculator.

	Press	Enter	Display (to four places)
(a)	sin	26.5	.4462
	cos	26.5	.8949
	tan	26.5	.4986
(b)	sin	206.5	−.4462
	cos	206.5	−.8949
	tan	206.5	.4986

■

Question 1 *What is a grad?*

Answer: A grad is a unit of angle measurement that is sometimes used in navigation. Four hundred grads equal $360°$, so that 1 grad $= .9°$. ■

To see how tables are used to find approximate values of the trigonometric functions, we first examine an excerpt from a typical table.

θ (degrees)	θ (radians)	$\sin\theta$	$\cos\theta$	$\tan\theta$	$\cot\theta$	$\sec\theta$	$\csc\theta$		
26°00′	.4538	.4384	.8988	.4877	2.050	1.113	2.281	1.1170	**64°00′**
10	.4567	.4410	.8975	.4913	2.035	1.114	2.268	1.1141	50
20	.4596	.4436	.8962	.4950	2.020	1.116	2.254	1.1112	40
30	.4625	.4462	.8949	.4986	2.006	1.117	2.241	1.1083	30
40	.4654	.4488	.8936	.5022	1.991	1.119	2.228	1.1054	20
50	.4683	.4514	.8923	.5059	1.977	1.121	2.215	1.1025	10
27°00′	.4712	.4540	.8910	.5095	1.963	1.122	2.203	1.0996	**63°00′**
		$\cos\theta$	$\sin\theta$	$\cot\theta$	$\tan\theta$	$\csc\theta$	$\sec\theta$	θ (radians)	θ (degrees)

The left column in the excerpt contains angles in degrees ranging from 26° to 27° in increments of 10′. As we go down the column, the angles increase. Next to each degree measure is the corresponding radian measure of the angle followed from left to right by the values of sine, cosine, tangent, cotangent, secant, and cosecant, respectively, as indicated above the columns. For example, we read

$$\sin 26°20' = .4436,$$
$$\cos 26°40' = .8936,$$
$$\tan .4683 = .5059.$$

In order to cut down on the number of pages required, trigonometric tables use the property that the value of a trigonometric function at a given angle is equal to the value of its cofunction at the complement of the given angle. The complements of the angles in the two leftmost columns of the excerpt are contained in the two rightmost columns. Here we read from the bottom upward for the angle and then from right to left for the values of secant, cosecant, tangent, cotangent, sine, and cosine, respectively, as indicated at the *bottom* of the columns. For example, we read

$$\cot 63°10' = .5059,$$
$$\sin 63°30' = .8949,$$
$$\cos 1.1170 = .4384.$$

To cut down further on the number of pages, trigonometric tables list values of the trigonometric functions for angles from 0° to 90° only. For a nonquadrantal angle outside this range, the corresponding reference angle must be used, as is illustrated in the following example.

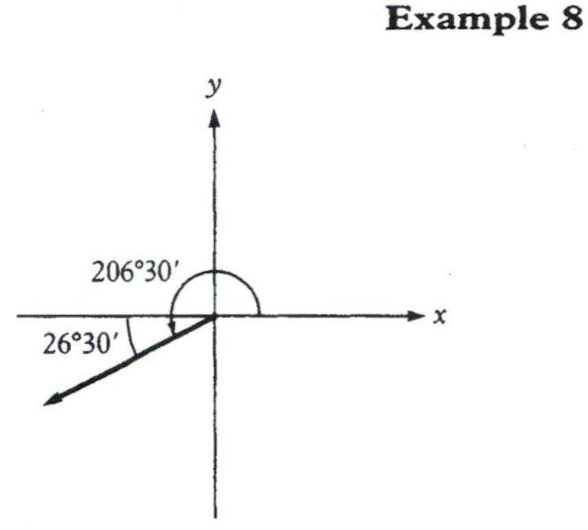

Figure 18

Example 8 Use the table excerpt to find the sine, cosine, and tangent of 206°30′.

Solution: As indicated in Figure 18, 206°30′ is in the third quadrant, and its reference angle is 26°30′. In the third quadrant, where x and y are both negative, the sine and cosine are negative and the tangent is positive. Hence,

$$\sin 206°30' = -\sin 26°30' = -.4462$$
$$\cos 206°30' = -\cos 26°30' = -.8949$$
$$\tan 206°30' = \tan 26°30' = .4986. \blacksquare$$

To summarize the above discussion on the use of trigonometric tables, we list the following important points.

1. Trigonometric tables contain values of the trigonometric functions for angles from 0° to 90° only.
2. For angles from 0° to 45°, locate the angle from the *left* column and the trigonometric function from the *top* of the column. For 45°10′ to 90°, locate the angle from the *right* column and the trigonometric function from the *bottom* of the column.
3. For a nonquadrantal angle outside the range 0° to 90°, use the corresponding reference angle and the tables to find the *magnitude* of the trigonometric function, and use the quadrant of the angle to find the *sign*.
4. For a quadrantal angle other than 0° or 90°, use definitions (1) or the first table given in this section.

Practice at working with trigonometric functions will make it easy to recall their values for angles whose reference angles are special and for quadrantal angles, but for other angles, we have to rely on calculators or tables.

Question 2 *How could Hipparchus use trigonometry to measure the radius of the earth?*

Answer: If Hipparchus stood at point P on the top of Mount Olympus, whose height is 9,550 feet above sea level, and sighted on the horizon, his angle of sight with the vertical would be 88°16′. From Figure 19, where R is the radius of the earth in feet, we have the equation

$$\frac{R}{R + 9550} = \sin 88°16' = .99954,$$

whose solution is

$$R \approx 20{,}751{,}320 \text{ feet}$$
$$\approx 3{,}930 \text{ miles.}$$

We will discuss a procedure for measuring the height of Mount Olympus in Section 6.4. $\blacksquare$

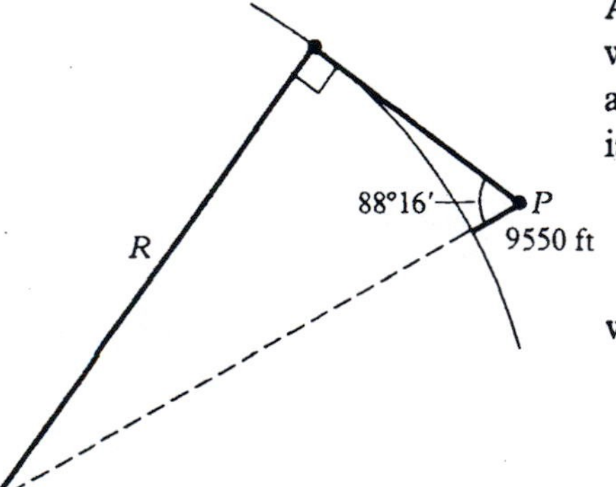

Figure 19

Exercises 6.2

Fill in the blanks to make each statement true.

1. To define the trigonometric functions of an arbitrary angle θ, we put θ in ______________ in the coordinate system.
2. If $P(x, y)$ is a point on the terminal side of θ and r is the positive distance from the origin to P, then the six trigonometric functions of θ are the six possible __________ involving any two of x, y, and r.
3. The definition $\tan \theta = \dfrac{\text{opposite side}}{\text{adjacent side}}$ is correct if θ is an _______ angle in a right triangle.
4. The angle 45° is special because it is one of the angles in an ______________ triangle, and 60° is special because it is one of the angles in an ______________ triangle.
5. To determine $\sin \theta$ for an angle θ not listed in a trigonometric table, we use the ______________ for θ to determine the magnitude of $\sin \theta$ and the ______________ of θ to determine the sign of $\sin \theta$.

Write true *or* false *for each statement.*

6. $\sin \theta$ is defined for every angle θ.
7. $\tan \theta$ is undefined for $\theta = n\pi/2$, where n is any integer.
8. The cofunction of a given trigonometric function is the reciprocal of the given function.
9. The value of one trigonometric function is not related to the value of any other trigonometric function.
10. When defining the trigonometric functions of an angle θ, we may measure θ in either degrees or radians.

Definitions of the Trigonometric Functions

In Exercises 11–14, place θ in standard position with $P(x, y)$ on its terminal side, and let r = the length OP. Then compute the six trigonometric functions of θ.

11. $x = 3$ and $y = 4$
12. $x = 5$ and $y = -12$
13. $x = -4$, $r = 5$, and θ is in quadrant II
14. $x = -4$, $r = 5$, and θ is in quadrant III

Find the exact values of the other five trigonometric functions of θ, given the following information.

15. θ is in quadrant I and $\sin \theta = 3/5$.
16. θ is in quadrant IV and $\cos \theta = 4/5$.
17. θ is in quadrant III and $\tan \theta = 5/12$.
18. θ is in quadrant II and $\cot \theta = -12/5$.

Use definitions (1) to find the values of the trigonometric functions for each of the following quadrantal angles.

19. $\theta = 0°$
20. $\theta = 90°$
21. $\theta = 270°$
22. $\theta = 540°$
23. $\theta = -630°$
24. $\theta = \dfrac{-\pi}{2}$
25. $\theta = -\pi$
26. $\theta = \dfrac{5\pi}{2}$
27. $\theta = -5\pi$
28. $\theta = \dfrac{11\pi}{2}$

Special Angles

Draw a figure to show that each of the following angles has 30°, 45°, or 60° as its reference angle. Also, evaluate the trigonometric functions at each angle.

29. 225°
30. 150°
31. −60°
32. 405°
33. 420°
34. $\dfrac{4\pi}{3}$
35. $\dfrac{-5\pi}{6}$
36. $\dfrac{9\pi}{4}$
37. $\dfrac{5\pi}{6}$
38. $\dfrac{-10\pi}{3}$

39. For what positive integers n does $n\pi/4$ have $\pi/4$ as its reference angle?
40. Answer Exercise 39 for *negative* integers n.
41. For what positive integers n does $n\pi/3$ have $\pi/3$ as its reference angle?
42. Answer Exercise 41 for *negative* integers n.
43. For what positive integers n does $n\pi/6$ have $\pi/6$ as its reference angle?
44. Answer Exercise 43 for *negative* integers n.

Use of a Calculator

Use a calculator to find each of the following.

45. $\sin 27^\circ$
46. $\cos 63^\circ$
47. $\tan 237^\circ$
48. $\sin 207^\circ$
49. $\cos(-207^\circ)$
50. $\tan 223^\circ 20'$
51. $\cot 226^\circ 40'$
52. $\cos \frac{\pi}{5}$
53. $\sec \frac{11\pi}{5}$
54. $\csc\left(\frac{-11\pi}{5}\right)$
55. $\sin .6894$
56. $\sin 1.5708$
57. $\cos .9308$
58. $\tan .8029$
59. $\sec 1.4137$

Miscellaneous

60. Let θ and ϕ be the two acute angles in a right triangle. Use equations (3) to show that each of the following statements is true.

    (a) $\sin\theta = \cos\phi$ (b) $\tan\theta = \cot\phi$

    (c) $\sec\theta = \csc\phi$

61. Let θ be any acute angle in radians. Show that each statement is true.

    (a) $\sin\theta = \cos\left(\frac{\pi}{2} - \theta\right)$

    (b) $\tan\theta = \cot\left(\frac{\pi}{2} - \theta\right)$

    (c) $\sec\theta = \csc\left(\frac{\pi}{2} - \theta\right)$

6.3 Solving Right Triangles

The Athenian Greeks of 400 B.C. used triangles to develop geometry; the Alexandrian Greeks of 200 B.C. used them to measure the universe.

Two Sides Given
One Side and One Angle Given
Applications

Historically, trigonometry began with triangle measurement, an aspect of the subject that is still useful. **Solving a triangle** means finding all its sides and angles. This is especially easy for a *right triangle*. Here the sides are related by the Pythagorean theorem, and since one angle measures 90°, the other two angles must be acute and complementary (Figure 20). Note that capital letters A, B, and C are used to denote the angles as well as the vertices in Figure 20, where C is the right angle, and the sides opposite A, B, and C are denoted by the lower-case letters a, b, and c, respectively. This will be our standard notation when solving right triangles.

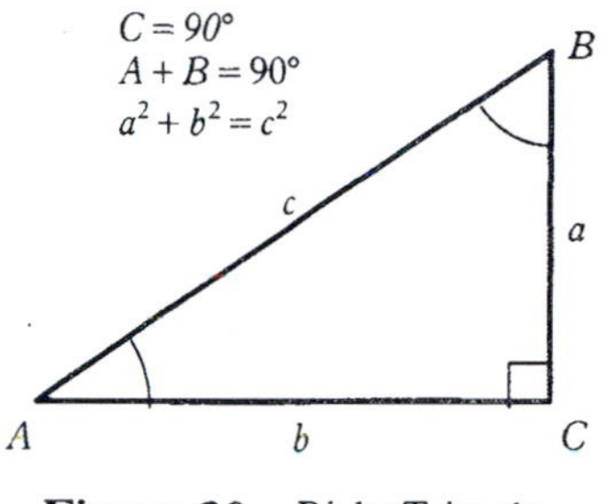

Figure 20 Right Triangle

There are two basic cases in which a right triangle can be solved:

1. two sides given,
2. one side and one acute angle given.

Variations of these basic cases are also possible. Hipparchus' method for measuring the radius of the earth (cited at the end of Section 6.2) is a variation of case 2. There, one acute angle $B = 88°16'$ was given and the sides b and c were related by the equation $c = b + 9550$.

We next consider examples of the two basic cases and some applications. In solving triangle problems in this section and the next two, we will use the rules for significant digits given in the Appendix. We will also adopt the following conventions concerning the accuracy of sides and angles of triangles.

Accuracy of Sides	Accuracy of Angles
Two significant digits	Nearest degree
Three significant digits	Nearest tenth of a degree (or nearest ten minutes)
Four significant digits	Nearest hundredth of a degree (or nearest minute)

For example, if an angle is given to the nearest ten minutes and a side to four significant digits, the other sides are computed to three significant digits and the angles to the nearest ten minutes (or tenth of a degree).

Two Sides Given. Here one acute angle can be found by using a trigonometric function involving the two given sides. The second acute angle is the complement of the first, and the remaining side can be found either by trigonometry or the Pythagorean theorem.

Example 1 Solve the right triangle in which $a = 10$ feet and $b = 5.0$ feet.

Solution: Draw right triangle ABC and assign the given values to sides a and b (Figure 21).

1. To find angle A, observe that

$$\tan A = \frac{\text{opp}}{\text{adj}} = \frac{a}{b} = \frac{10}{5} = 2.$$

From the table excerpt above, the angle A whose tangent is closest to 2 is 63°30′ or 63.5°, which we round to 64°. With calculators the method varies, but with some, angle A can be found by entering 2 and then pressing the "inv" key followed by the "tan" key. With the calculator in the degree mode, the answer is about 63.4°, which rounds to $A = 63°$.

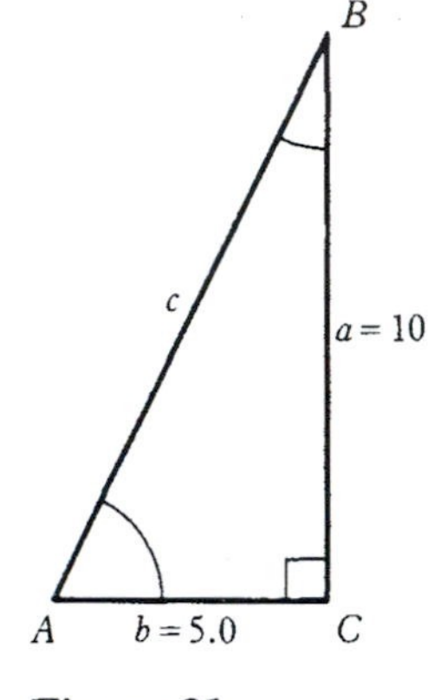

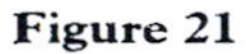
Figure 21

2. To find angle B, we use the property that the sum of the two acute angles in a right triangle is 90°. Hence, using 63° for A,

$$B \approx 90° - 63° = 27°.$$

3. To find side c, we have a choice. First, by the Pythagorean theorem,

$$c^2 = a^2 + b^2 = 100 + 25$$
$$c = \sqrt{125} \approx 11 \text{ feet.}$$

A second way to find c is by means of a trigonometric function. For example, since

$$\sin A = \frac{\text{opp}}{\text{hyp}} = \frac{10}{c},$$

we have

$$c = \frac{10}{\sin 63°} \approx \frac{10}{.8910} \approx 11 \text{ feet.}$$

We could also obtain c from the equation

$$\cos A = \frac{\text{adj}}{\text{hyp}} = \frac{5}{c},$$

or from similar equations in terms of angle B. ■

Comment As illustrated in Example 1, when solving right triangles, we are working with *approximate values*, and different methods (by tables or by calculators) can lead to slightly different answers. Also, different approaches (the Pythagorean theorem versus a trigonometric function) can yield slightly different results. This is just the nature of things when working with approximate numbers. One way to deal with the situation is to take as the answer the average value obtained by the different methods and approaches.

One Side and One Angle Given. The two acute angles in any right triangle are complementary. Therefore, if one of them is given, the second one can be obtained by subtraction from 90°. If one side is given, one of the remaining two sides can be obtained by using a trigonometric function of the given angle. The third side can then be obtained either by trigonometry or by the Pythagorean theorem.

Example 2 Solve right triangle ABC if $c = 20.4$ inches and $B = 14°10'$.

Solution: First draw the figure, label the parts, and insert the given values (Figure 22).

1. Now, angle $A = 90° - 14°10'$
$= 75°50'$.

2. For finding side b, we use the equation

$$\sin B = \frac{b}{c}.$$

We get

$$b = c \sin B = 20.4 \sin 14°10'$$
$$= 20.4(.2447)$$
$$\approx 4.99 \text{ inches.}$$

3. For finding side a, we use the equation

$$\cos B = \frac{a}{c}.$$

Hence,
$$a = c \cos B = 20.4 \cos 14°10'$$
$$= 20.4(.9696)$$
$$\approx 19.8 \text{ inches.}$$

We could also obtain side a from the Pythagorean theorem. That is,

$$a = \sqrt{c^2 - b^2}$$
$$= \sqrt{(20.4)^2 - (4.99)^2}$$
$$\approx \sqrt{391.26}$$
$$\approx 19.8 \text{ inches}$$

as before. ■

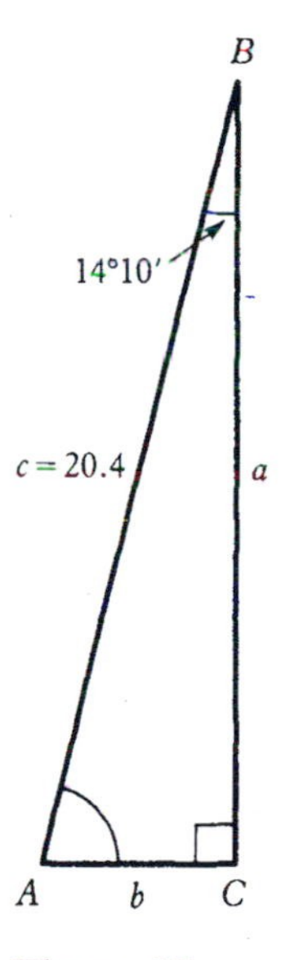

Figure 22

Applications. Our methods for solving right triangles can be applied to a variety of directional problems in areas such as astronomy, surveying, and navigation. For problems involving vertical angles, we define the

angle of elevation when looking *up* from a point P to a point Q and the **angle of depression** when looking *down* from Q to P as indicated in Figure 23.

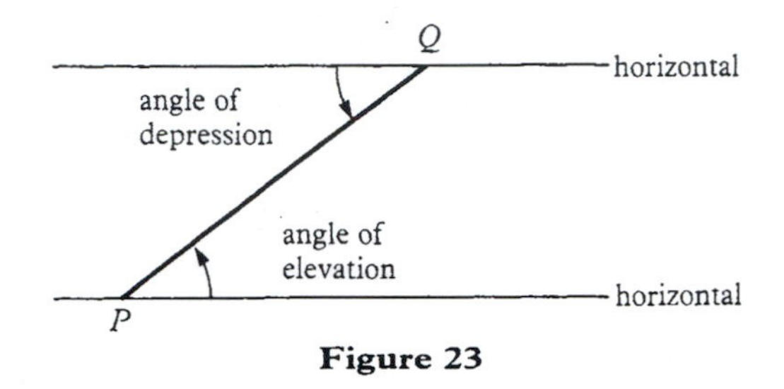

Figure 23

In navigation, two methods of indicating direction are used. First, the **bearing** of one point from another is an angle between 0° and 360° measured *clockwise* from an axis pointing north as in Figure 24.

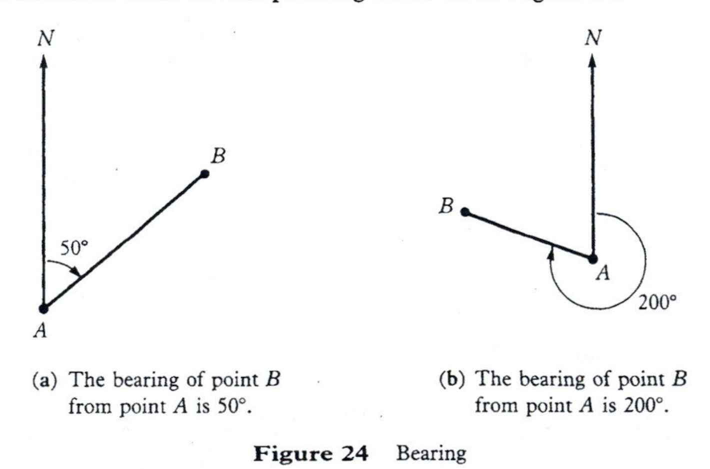

(a) The bearing of point B from point A is 50°.

(b) The bearing of point B from point A is 200°.

Figure 24 Bearing

As illustrated in Figure 25, the second method first cites the general direction of a line of travel as north or south and then specifies the acute angle that the general direction makes with a north-south axis.

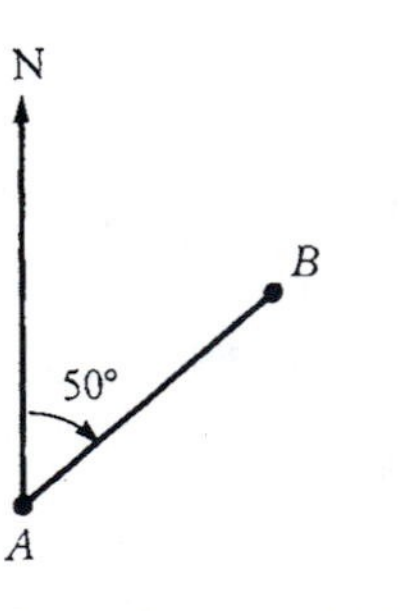

(a) Line AB has direction N 50° E (50° east of north).

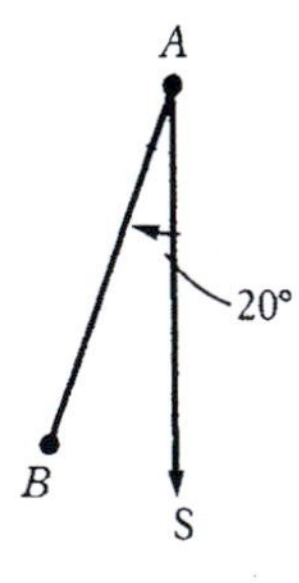

(b) Line AB has direction S 20° W (20° west of south).

Figure 25

Example 3 At the instant a plane is flying directly over a point 26,410 feet from an airport, its altitude is 5002 feet.

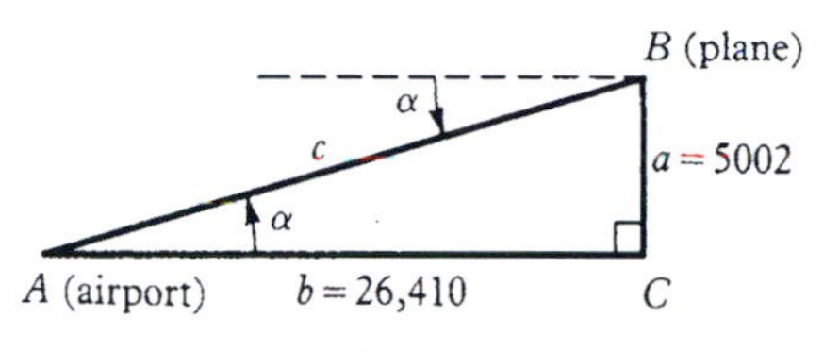

Figure 26

(a) What should be its angle of descent from the horizontal, assuming the plane descends to the airport in a straight line?

(b) What distance does the plane travel until it touches down?

Solution: After placing the plane at point B, we label Figure 26 as indicated. We are required to compute angle α and side c.

(a) From the equation

$$\tan \alpha = \frac{a}{b} = \frac{5002}{26{,}410} \approx .1894$$

we obtain, using a calculator,

$$\alpha \approx 10.72°.$$

(b) From the equation

$$\sin \alpha = \frac{a}{c} = \frac{5002}{c}$$

we get

$$c = \frac{5002}{\sin 10.72°} \approx 26{,}890 \text{ feet.}$$

Or, from the Pythagorean theorem,

$$c = \sqrt{a^2 + b^2} = \sqrt{5002^2 + 26{,}410^2}$$
$$= \sqrt{722{,}508{,}104} \approx 26{,}880 \text{ feet.} \quad ■$$

Example 4 Coast Guard station A is located 75.0 miles south of station B. A ship is on a bearing of 57.1° from A and 147.1° from B. How far is the ship from station A?

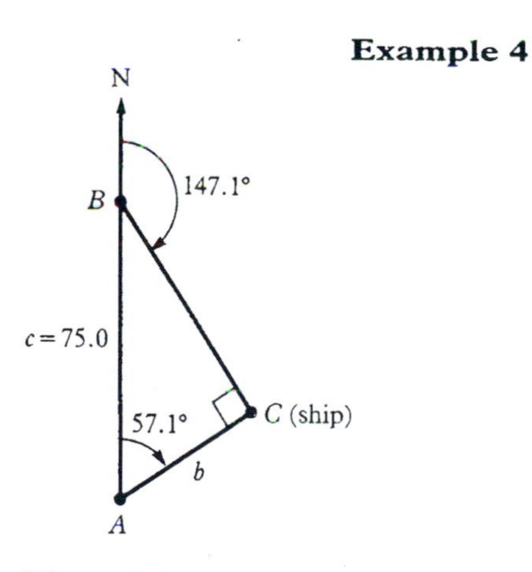

Figure 27

Solution: We form a triangle ABC with the ship at point C. As indicated in Figure 27, angle $A = 57.1°$ and $B = 180° - 147.1° = 32.9°$. Since $A + B = 90°$, the angle C must be 90°, and ABC is a right triangle with hypotenuse $c = 75.0$. The distance b from A to C satisfies the equation

$$\cos 57.1° = \frac{b}{75.0}$$

and therefore,

$$b = 75.0 \cos 57.1° \approx 40.7.$$

Hence, the ship is 40.7 miles from station A. ■

Example 5 A television antenna is mounted on the edge of the roof at one side of a house. At a point on the ground 30 feet from the side of the house, the angle of elevation to the bottom of the antenna is 50°. From the same point, the angle of elevation to the top of the antenna is 55°. What is the length of the antenna?

Solution: In Figure 28, BD represents the antenna, and CB is the side of the house. Now,

$$BD = CD - CB.$$

Also,

$$CD = 30 \tan 55° \approx 42.8$$

and

$$CB = 30 \tan 50° \approx 35.8.$$

Therefore,

$$BD \approx 42.8 - 35.8 = 7.0 \text{ feet},$$

which is the length of the antenna. ■

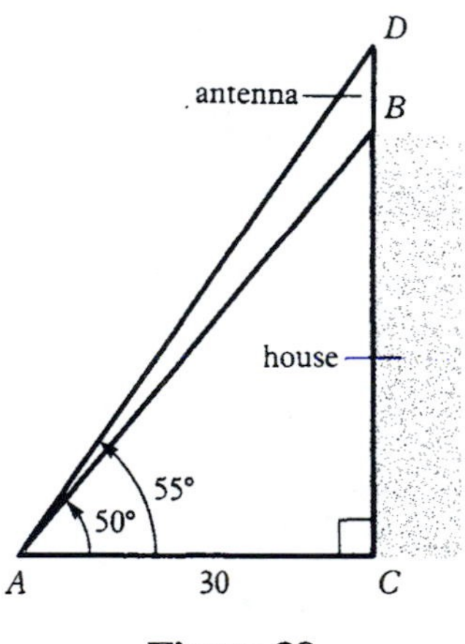

Figure 28

Exercises 6.3

Fill in the blanks to make each statement true.

1. In any triangle, the sum of the angles is ________ degrees, and therefore the sum of the acute angles in a right triangle is ________ degrees.
2. A right triangle can be solved if either ____________ or ____________ are known.
3. If the two sides a and b of right triangle ABC are known, we can find angle A by means of the ________ function.
4. If the sides a and c of right triangle ABC are known, we can find side b by the ________ theorem.
5. An angle of elevation or depression is the angle that the line of sight makes with a ________ line.

Write true *or* false *for each statement.*

6. The acute angles of a right triangle are always complementary.

7. If two angles of a triangle are complementary, then the triangle is a right triangle.
8. If the two acute angles of a right triangle are given, then the triangle can be solved.
9. If the hypotenuse and one other side of a right triangle are given, then the triangle can be solved.
10. If one acute angle and the sum of two sides of a right triangle are given, then the triangle can be solved.

Solving Right Triangles

Solve each right triangle ABC based on the information given in Exercises 11–18.

11. $a = 20$; $A = 32°$
12. $a = 20$; $B = 52°20'$
13. $b = 20$; $B = 15°$
14. $c = 20$; $B = 15°$
15. $c = 20.0$; $A = 72°30'$
16. $a = 7.5$; $c = 9.0$
17. $a = 7.5$; $b = 9.0$
18. $b = 15$; $c = 25$

Applications

19. A person walks 100 ft in a straight line from the base of an electrical relay tower and measures the angle of elevation of the top of the tower to be 77°30′. Find the height of the tower, taking into account that the person's eye level is exactly 5 ft from the ground.
20. A car travels on a road that makes an angle of 5° with the horizontal. When it travels 5 mi along the road, how far does the car travel horizontally?
21. A plane takes off with its angle of flight at 15.0° with the ground. How far must the plane travel along its flight path to reach an altitude of 2.00×10^4 ft?
22. A boy climbs a tree to a height of 8 ft. He spots his sister in the distance and measures the angle of depression to the top of her head as 50′. If the boy's voice will carry 600 ft, will his sister hear his call?
23. A surveyor wishes to determine the distance from point C on one side of a river to point A directly across it. From C, he walks 50.5 ft along the straight bank in a direction perpendicular to CA until he reaches point B. From B, he measures angle CBA as 87°30′. Find the distance from C to A.

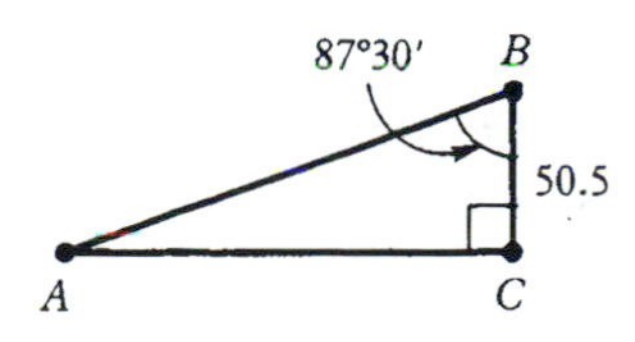

24. A 30-ft ladder is to be used to reach the roof of a house that is 24 ft high. What angle should the ladder make with the ground?
25. When the angle of elevation of the sun, measured from the top of a flagpole, is 32°40′, the flagpole casts a 110-ft horizontal shadow. Find the height of the flagpole.
26. From the roof of a building, an observer can see the top of a skyscraper with an angle of elevation of 60° and the base of the skyscraper with an angle of depression of 53°. If the two buildings are 250 ft apart, how high is the skyscraper?
27. It is desired to locate the source of a certain radio signal. Its direction from point A is measured as N 44°40′E, and from point B as N 45°20′W. If B is 10.1 mi east of A, how far is the source of the signal from A?
28. A plane has a bearing of 310°12′ from radar station A and a bearing of 40°12′ from station B. If B is 50.0 mi west of A, find the horizontal distance between the plane and station B.

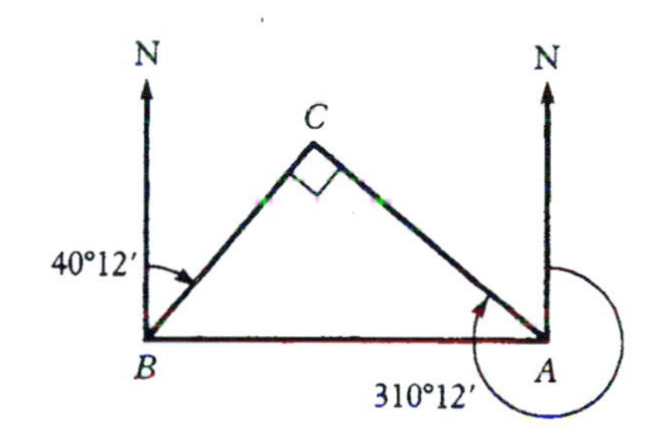

29. Two ships A and B leave port C at the same time. Ship A sails due north at 25 knots (nautical miles per hour), and ship B sails due east at 30 knots. What is the bearing of ship B from ship A one hour later?
30. A military jet plane flying at 3.00×10^4 ft photographs missile launchers A and B at the same time with angles of depression equal to 72°0′ and 70°20′, respectively. Find the distance between the missile launchers, assuming the jet is between A and B and all three are in the same vertical plane.

6.4 The Law of Sines

Derivation of the Law of Sines
One Side and Two Angles Given
Two Sides and an Opposite Angle Given (Ambiguous Case)
Applications

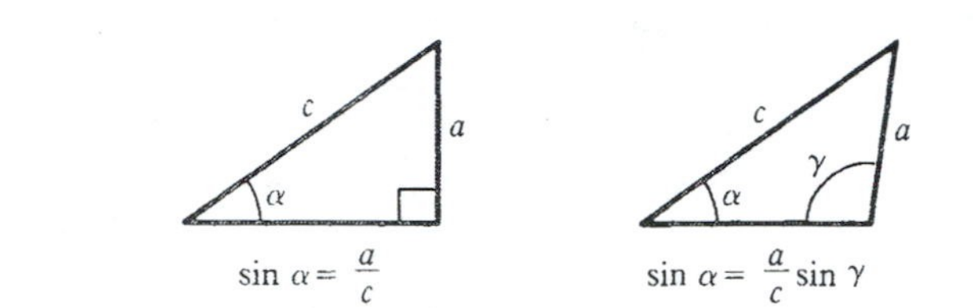

In the previous section we solved right triangles, and here we consider arbitrary triangles. A triangle that contains no right angle is called an **oblique triangle.** An oblique triangle contains either three acute angles, as in Figure 29(a), or two acute angles and one **obtuse angle** (between 90° and 180°) as in Figure 29(b).

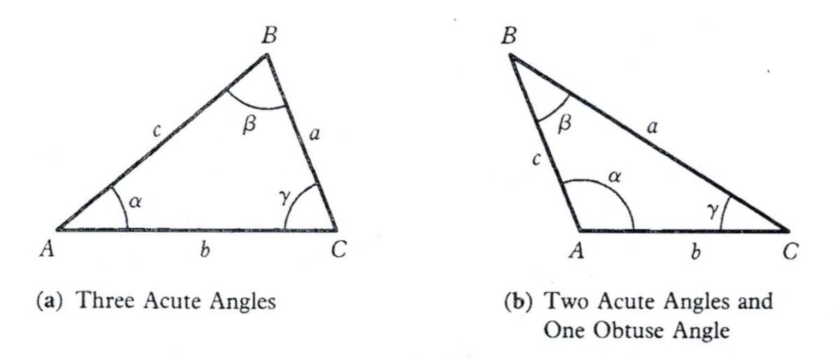

(a) Three Acute Angles

(b) Two Acute Angles and One Obtuse Angle

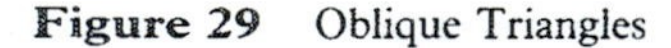

Figure 29 Oblique Triangles

For solving right triangles, there are two basic cases. For oblique triangles, we have the following *three* basic cases:

1. one side and two angles given,
2. two sides and one angle given,
3. three sides given.

Three parts, including one side, are given in each of the three basic cases. As in right triangles, some variations of the basic cases are possible. For example, a variation of case 1 would be a triangle in which two angles and an algebraic relationship between two sides were given. Note that when two angles are known, as in case 1, the third can be determined immediately since the sum of the three angles is 180°.

For solving right triangles, our basic tools are the trigonometric functions and the Pythagorean theorem. In this section, we will derive the *law of sines* and in the next section the *law of cosines* for solving oblique triangles, but we first make a slight change from our standard notation for right triangles.

The notation in Figure 29 will be our standard for oblique triangles. Note that each triangle ABC has vertices A, B, and C with corresponding angles α, β, and γ and opposite sides a, b, and c, respectively. (In right triangles, the capital letters A, B, C denote both vertices and angles, and C is always 90°.) Since all three angles can vary in an oblique triangle, we will find it helpful to use the Greek letters to distinguish angles from vertices.

Derivation of the Law of Sines. We place our standard oblique triangles in a coordinate plane with angle α in standard position (Figure 30). From vertex B, we drop a perpendicular to the x-axis at point D. Then triangles ABD and BCD are right triangles in both cases (a) and (b).

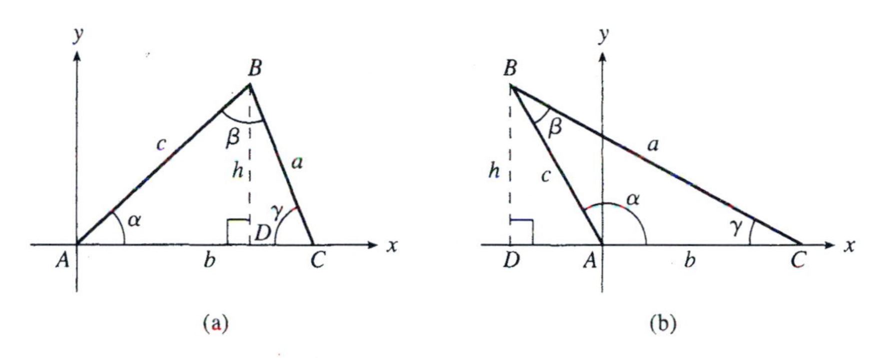

Figure 30

Now in both cases, we have

$$\sin \alpha = \frac{h}{c} \tag{1}$$

and

$$\sin \gamma = \frac{h}{a}. \tag{2}$$

Equation (1) says that $h = c \sin \alpha$, and equation (2) says $h = a \sin \gamma$. Hence,

$$a \sin \gamma = c \sin \alpha,$$

which is equivalent to

$$\frac{a}{\sin \alpha} = \frac{c}{\sin \gamma}. \tag{3}$$

Similarly, after repositioning the triangles in Figure 30 so that α remains in standard position but side AB is along the x-axis, we obtain the equation

$$\frac{a}{\sin \alpha} = \frac{b}{\sin \beta}. \tag{4}$$

By combining equations (3) and (4), we arrive at the **law of sines:**

$$\boldsymbol{\frac{a}{\sin \alpha} = \frac{b}{\sin \beta} = \frac{c}{\sin \gamma}}. \tag{5}$$

We note that the law of sines is actually *three* equations (*which is the third one?*), and any two of them imply the remaining one. Also, the law of sines holds for *all* triangles, including right triangles. For example, from equation (3) we have

$$\sin \alpha = \frac{a}{c} \sin \gamma. \tag{6}$$

Now, for a right triangle in standard position, $\gamma = 90°$ so $\sin \gamma = 1$, c is the hypotenuse, and equation (6) reduces to the definition of $\sin \alpha$. However, the importance of the law of sines lies in its application to oblique triangles.

One Side and Two Angles Given. As illustrated in the following example, we can use the law of sines to solve a triangle in which one side and two angles are given.

Example 1 Solve triangle ABC, given $a = 42.5$, $\beta = 60°10'$, and $\gamma = 43°30'$.

Solution: For any triangle,

$$\alpha + \beta + \gamma = 180°$$

Therefore,

$$\alpha = 180° - (60°10' + 43°30') = 76°20'.$$

We draw and label Figure 31 as indicated. Then, by the law of sines, we have

$$\frac{b}{\sin \beta} = \frac{a}{\sin \alpha},$$

$$b = \frac{42.5 \sin 60°10'}{\sin 76°20'}$$

$$\approx 37.9.$$

Similarly,

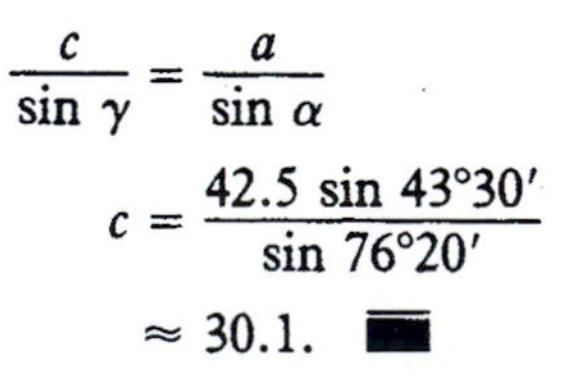

$$\frac{c}{\sin \gamma} = \frac{a}{\sin \alpha}$$

$$c = \frac{42.5 \sin 43°30'}{\sin 76°20'}$$

$$\approx 30.1. \blacksquare$$

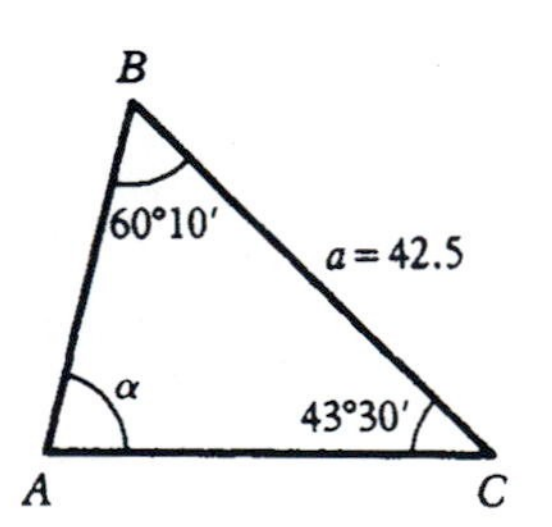

Figure 31

Two Sides and an Opposite Angle Given (Ambiguous Case). Example 1 illustrates the general result that if one side and two angles whose sum is less than 180° are given, then a *unique* triangle is determined. However, if two sides and an angle opposite one of them are given, then zero, one, or two triangles may be determined, as shown in Figure 32, where a, c, and α are given. Therefore, the case of two sides and an opposite angle is called the **ambiguous case.**

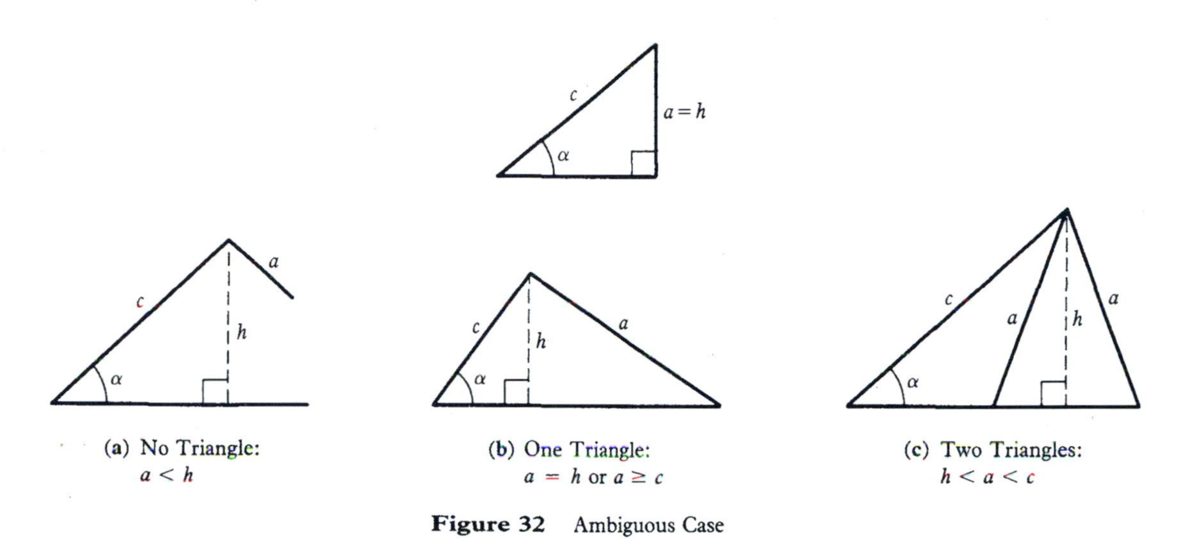

(a) No Triangle: $a < h$

(b) One Triangle: $a = h$ or $a \geq c$

(c) Two Triangles: $h < a < c$

Figure 32 Ambiguous Case

Example 2 Solve triangle ABC, given $a = 2$, $c = 5$, and $\alpha = 30°$.

Solution: From the given data, we can draw the triangle only partially (Figure 33). This is typical of the ambiguous case. Now, from the law of sines,

$$\frac{a}{\sin \alpha} = \frac{c}{\sin \gamma}$$

$$\frac{2}{\sin 30°} = \frac{5}{\sin \gamma}$$

$$\sin \gamma = \frac{5}{2} \sin 30°$$

$$= \frac{5}{4}.$$

If we go back to the definition of the sine function as the ratio y/r, where $|y| \leq r$, we see that the sine can never be greater than 1. Hence, *no solution* is consistent with the given data. ■

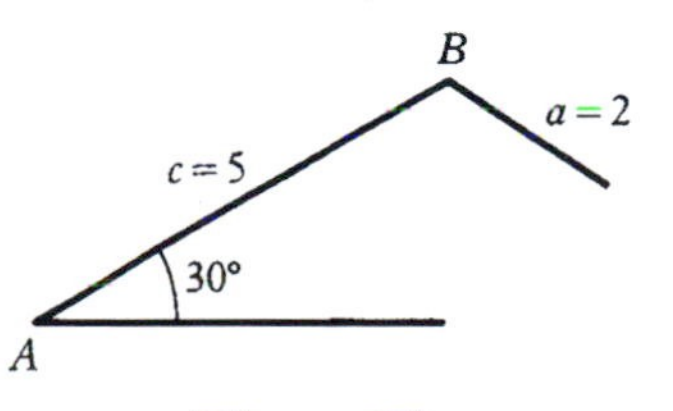

Figure 33

In the following example, we get one solution in part (a) and two in part (b).

Example 3 Solve the triangle ABC, given the following.

(a) $a = 25.2$, $c = 15.6$, and $\alpha = 44.5°$

(b) $a = 12.1$, c and α as in case (a)

Solution: (We will work with degrees in decimals, using a calculator.)
(a) By the law of sines,

$$\frac{a}{\sin \alpha} = \frac{c}{\sin \gamma}$$

$$\sin \gamma = \frac{c \sin \alpha}{a}$$

$$= \frac{15.6 \sin 44.5°}{25.2}$$

$$\approx .4339.$$

There are two angles in the 0° to 180° range satisfying $\sin \gamma = .4339$, namely $\gamma_1 \approx 25.7°$ and $\gamma_2 = 180° - \gamma_1 \approx 154.3°$. For $\gamma_1 \approx 25.7°$, the third angle β_1 must satisfy the equation

$$44.5° + \beta_1 + \gamma_1 = 180°,$$

which implies

$$\beta_1 \approx 109.8°.$$

But for $\gamma_2 \approx 154.3°$, the third angle β_2 would have to be *negative* to satisfy the equation $\alpha + \beta_2 + \gamma_2 = 180°$. Therefore, we reject γ_2 and conclude that the angles are

$$\alpha = 44.5°, \beta \approx 109.8°, \text{ and } \gamma \approx 25.7°.$$

We can now draw the triangle [Figure 34(a)], and to find the remaining side b, we again apply the law of sines.

$$b = \frac{a \sin \beta}{\sin \alpha}$$

$$\approx \frac{25.2 \sin 109.8°}{\sin 44.5°}$$

$$\approx 33.8$$

(b) With $a = 12.1$, the law of sines gives

$$\sin \gamma = \frac{c \sin \alpha}{a}$$

$$= \frac{15.6 \sin 44.5°}{12.1}$$

$$\approx .9037,$$

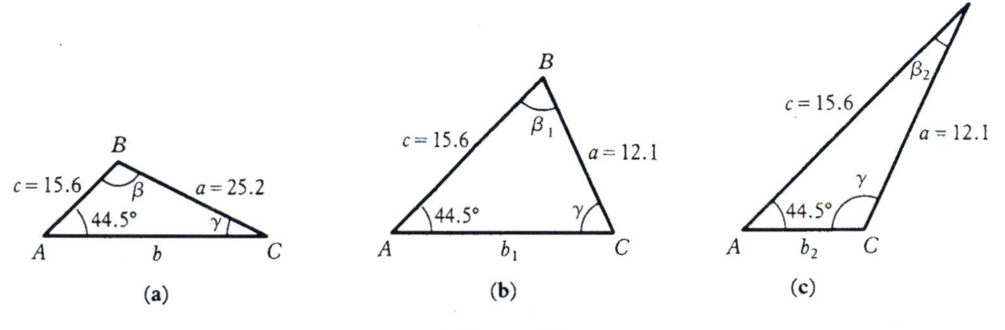

Figure 34

whose solutions in the 0° to 180° range are

$$\gamma_1 \approx 64.6^\circ \qquad \text{and} \qquad \gamma_2 \approx 115.4^\circ.$$

From the equation $\alpha + \beta + \gamma = 180^\circ$, we get

$$\beta_1 \approx 70.9^\circ \qquad \text{and} \qquad \beta_2 \approx 20.1^\circ.$$

We conclude that two triangles satisfy the given conditions.

For $\gamma_1 \approx 64.6^\circ$ and $\beta_1 \approx 70.9^\circ$ [Figure 34(b)] we have

$$\begin{aligned} b_1 &= \frac{a \sin \beta_1}{\sin \alpha} \\ &\approx \frac{12.1 \sin 70.9^\circ}{\sin 44.5^\circ} \\ &\approx 16.3. \end{aligned}$$

For $\gamma_2 \approx 115.4^\circ$ and $\beta_2 \approx 20.1^\circ$ [Figure 34(c)], we get

$$\begin{aligned} b_2 &= \frac{a \sin \beta_2}{\sin \alpha} \\ &\approx \frac{12.1 \sin 20.1^\circ}{\sin 44.5^\circ} \\ &\approx 5.9. \end{aligned}$$ ■

To summarize the results of the last three examples, we have seen that the law of sines can be used to solve a triangle if one side and two angles (whose sum is less than 180°) are given, or if two sides and an opposite angle are given. In the former case, the solution is *unique*, and in the latter case, there may be *zero*, *one*, or *two* solutions.

Applications. The area of any triangle is equal to one half the base times the height of the triangle. Suppose, as in Figure 30, we are given two sides b and c of a triangle and their included angle α. If we choose b as the base of the triangle, then the height h will be equal to $c \sin \alpha$, whether α is acute [Figure 30(a)] or obtuse [Figure 30(b)]. We therefore have the following formula.

> Given two sides b and c of a triangle and their included angle α, the area of the triangle is
> $$A = \frac{1}{2}bc \sin \alpha.$$

Example 4 Find the area of the triangle ABC, given $b = 25.5$ centimeters, $c = 18.3$ centimeters, and $\alpha = 40°30'$.

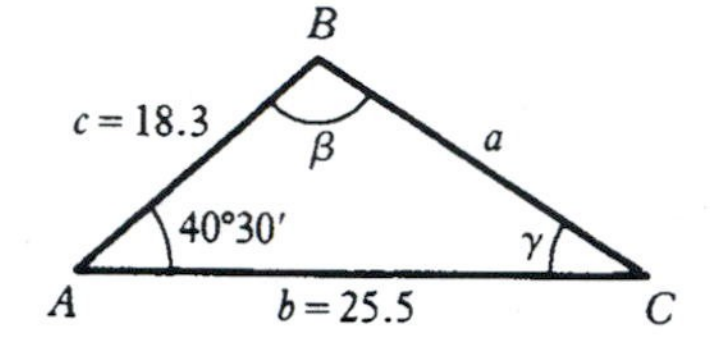

Figure 35

Solution: The given triangle ABC is illustrated in Figure 35. By the above formula, we have

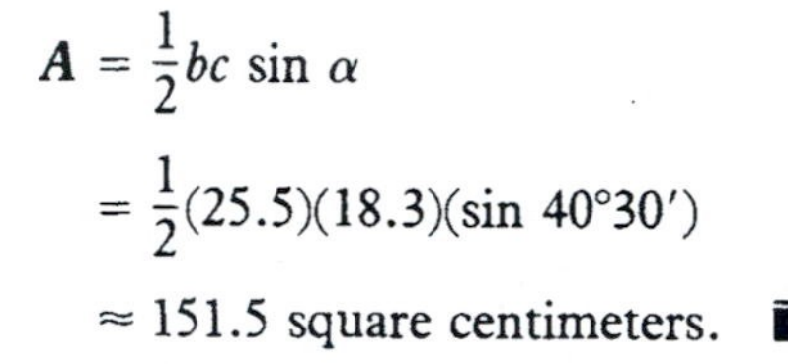

$$\begin{aligned} A &= \frac{1}{2}bc \sin \alpha \\ &= \frac{1}{2}(25.5)(18.3)(\sin 40°30') \\ &\approx 151.5 \text{ square centimeters.} \end{aligned}$$ ■

Although the formula for the area of a triangle is stated in terms of sides b and c and included angle α, the formula holds for *any two sides and their included angle.* That is,

$$\frac{1}{2}bc \sin \alpha = \frac{1}{2}ac \sin \beta = \frac{1}{2}ab \sin \gamma.$$

The above equations can be deduced geometrically, or we can divide through by $\frac{1}{2}abc$ and apply the law of sines.

Recall that in Section 6.2 we saw how Hipparchus could use trigonometry to measure the radius of the earth, provided he knew the height of Mount Olympus. We now apply the law of sines to find the height.

Example 5 Show how Hipparchus could have used trigonometry to measure the height of Mount Olympus.

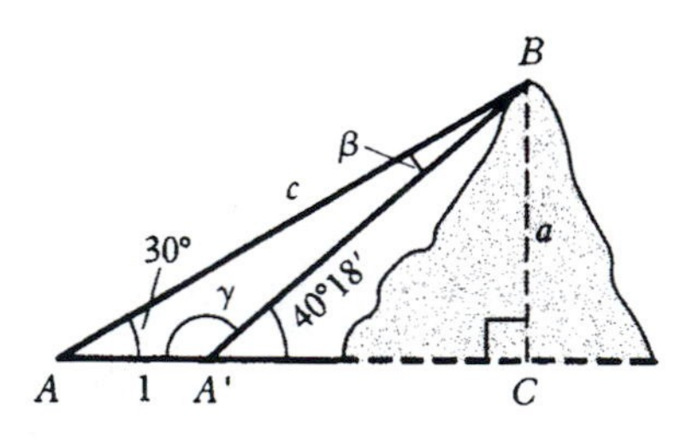

Figure 36

Solution: As indicated in Figure 36, Hipparchus could have first positioned himself at a point A (sea level) for which the angle of elevation to the top of Mount Olympus was $30°0'$. Suppose he then walked exactly one mile toward the mountain to point A' (also at sea level) and measured the angle of elevation to be $40°18'$. Then the angles of triangle ABA' are

$$\begin{aligned} \alpha &= 30°0', \\ \gamma &= 180° - 40°18' \\ &= 139°42', \end{aligned}$$

and

$$\begin{aligned} \beta &= 180° - (\alpha + \gamma) \\ &= 10°18'. \end{aligned}$$

By the law of sines,

$$\frac{c}{\sin 139°42'} = \frac{1}{\sin 10°18'},$$

$$c = \frac{\sin 139°42'}{\sin 10°18'} \approx 3.617.$$

Now triangle ABC is a right triangle, and the length of side a is the height of the mountain. We have

$$\frac{a}{c} = \sin 30°,$$

$$a = c \sin 30°$$

$$\approx 3.617(.5)$$

$$\approx 1.808.$$

Therefore, the height of Mount Olympus is approximately 1.808 miles, which is approximately 9,550 feet. ■

Exercises 6.4

Fill in the blanks to make each statement true.

1. An oblique triangle can be solved if ________ parts are given, including at least one ____________.
2. If one side and two angles of a triangle are given, then ____________ can be used to solve the triangle.
3. The ambiguous case is the case in which the given parts are ________________.
4. In the ambiguous case, there may be ________, ________, or ________ solutions.
5. If a triangle has sides b and c whose included angle is α, then the area of the triangle is equal to ____________.

Write true *or* false *for each statement.*

6. If a triangle has an obtuse angle, then the triangle is oblique.
7. If a triangle is oblique, then the triangle has an obtuse angle.
8. If we are given any three parts of an oblique triangle, then we can solve the triangle.
9. If two angles and one side of a triangle are known, then the triangle is uniquely determined.
10. If two sides and an opposite angle of a triangle are known, then the triangle is uniquely determined.

Solve each triangle ABC in Exercises 11–18, if possible.

One Side and Two Angles Given

11. $c = 48,\ \alpha = 37°,\ \beta = 65°$
12. $a = 50,\ \beta = 15°,\ \gamma = 102°$
13. $b = 15,\ \alpha = 25°,\ \gamma = 108°$
14. $b = 19,\ \alpha = 85°,\ \beta = 14°$

Two Sides and an Opposite Angle Given

15. $b = 5.0,\ c = 3.0,\ \beta = 40°$
16. $a = 5.0,\ c = 9.0,\ \alpha = 40°$
17. $a = 20,\ c = 10,\ \gamma = 100°$
18. $b = 10,\ c = 20,\ \gamma = 100°$

Applications

19. In town A, a straight road to town B intersects a straight road to town C at a 45° angle. The distance from A to B is 25 mi, and the distance from B to town D is 20 mi. Could town D possibly be located on the road from A to C?

20. Suppose town D in Exercise 19 is located 15 mi from B. Could D possibly be located on the road from A to C?

21. A surveyor is at point P on one bank of a river directly across from tree T on the opposite bank. He walks 200 m to point Q and then measures angle TPQ as 101.0° and angle PQT as 32.5°. How wide is the river between P and T?

22. Find the area of triangle ABC if $b = 14.4$, $c = 25.5$, and $\alpha = 25°$.

23. In triangle ABC, side b is twice as long as side a and their included angle γ is 30°. If the area of the triangle is 50 sq in, find sides a and b.

24. Two radar stations A and B are 10.25 mi apart on an east-west axis. From A a plane is spotted on a bearing of 52.75° and, at the same time, from B on a bearing of 327.1°. How far is the plane from A?

25. Two spotters, who are 200 yd apart on the ground, sight an object between them at an angle of elevation of 49° and 59°, respectively. If the object and both spotters are in a vertical plane, how high is the object?

6.5 The Law of Cosines

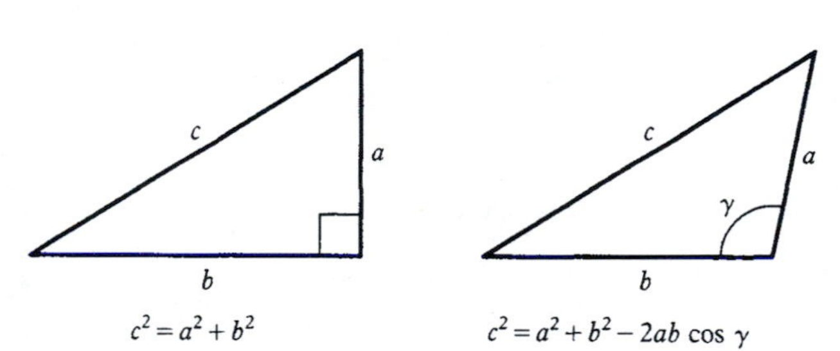

In the previous section we used the law of sines to solve oblique triangles in which either one side and two angles were given or two sides and an opposite angle were given. Here we derive the law of cosines and apply it to oblique triangles in which we are given either two sides and their *included* angle or three sides.

Derivation of the Law of Cosines. As indicated in Figure 37, let ABC be a triangle with angle α in standard position and vertex $B(x, y)$ on its terminal side. By the definition of $\cos \alpha$ and $\sin \alpha$, we have

$$\cos \alpha = \frac{x}{c} \quad \text{and} \quad \sin \alpha = \frac{y}{c}.$$

Hence, the coordinates of B are $x = c \cos \alpha$ and $y = c \sin \alpha$. Therefore, by the distance formula,

$$\begin{aligned} a = d(B, C) &= \sqrt{(c \cos \alpha - b)^2 + (c \sin \alpha - 0)^2}, \\ a^2 &= c^2(\cos \alpha)^2 - 2bc \cos \alpha + b^2 + c^2(\sin \alpha)^2 \\ &= b^2 + c^2[(\cos \alpha)^2 + (\sin \alpha)^2] - 2bc \cos \alpha. \end{aligned}$$

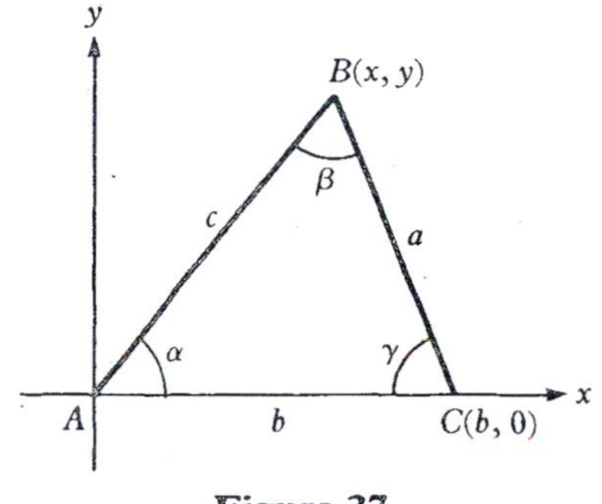

Figure 37

Now

$$(\cos\alpha)^2 + (\sin\alpha)^2 = \frac{x^2}{c^2} + \frac{y^2}{c^2} = \frac{x^2 + y^2}{c^2} = \frac{c^2}{c^2} = 1.$$

Therefore,
$$a^2 = b^2 + c^2 - 2bc\cos\alpha. \tag{1}$$

We can reposition triangle ABC so that angle β is in standard position and BA is along the positive x-axis. Then, by repeating the above argument, we will obtain the equation

$$b^2 = a^2 + c^2 - 2ac\cos\beta. \tag{2}$$

Finally, with angle γ in standard position and CB along the positive x-axis, we get

$$c^2 = a^2 + b^2 - 2ab\cos\gamma. \tag{3}$$

Equations (1), (2), and (3), either individually or collectively, are called the **law of cosines.** In words, the law of cosines says that the square of one side of a triangle equals the sum of the squares of the other two sides minus twice their product times the cosine of their included angle.

Comment If a triangle has an obtuse angle θ ($90° < \theta < 180°$), then $\cos\theta < 0$, and it follows from the law of cosines that the side opposite θ is the *largest* side of the triangle.

As with the law of sines, the law of cosines holds for *all* triangles, including right triangles. For a right triangle in standard position, $\gamma = 90°$ so that $\cos\gamma = 0$ and equation (3) reduces to the Pythagorean theorem. Hence, we may think of the law of cosines as a generalization of the Pythagorean theorem to oblique triangles. We now illustrate the role of the law of cosines in the solution of oblique triangles.

Two Sides and an Included Angle Given. In the last section, we saw that if two sides and an opposite angle were given, then either zero, one, or two triangles could result. The following example illustrates that when two sides and their *included* angle are given, a *unique* triangle is determined.

Example 1 Solve the triangle ABC, given $b = 25$, $c = 32$, and $\alpha = 25°$.

Solution: From the given data, we can draw Figure 38 as shown. Then a can be obtained from the law of cosines.

$$
\begin{aligned}
a^2 &= b^2 + c^2 - 2bc \cos \alpha \\
&= 25^2 + 32^2 - (2)(25)(32)(\cos 25°) \\
&\approx 199 \\
a &\approx \sqrt{199} \approx 14
\end{aligned}
$$

We can now find β by using either the law of cosines or the law of sines. Using the latter, we get

$$\sin \beta = \frac{b \sin \alpha}{a} \approx \frac{25 \sin 25°}{14} \approx .7547.$$

Since b is not the largest side of the triangle, it follows that β is not an obtuse angle (see previous Comment). The only acute angle whose sine is .7547 is 49° (nearest degree). Therefore,

$$\beta \approx 49°.$$

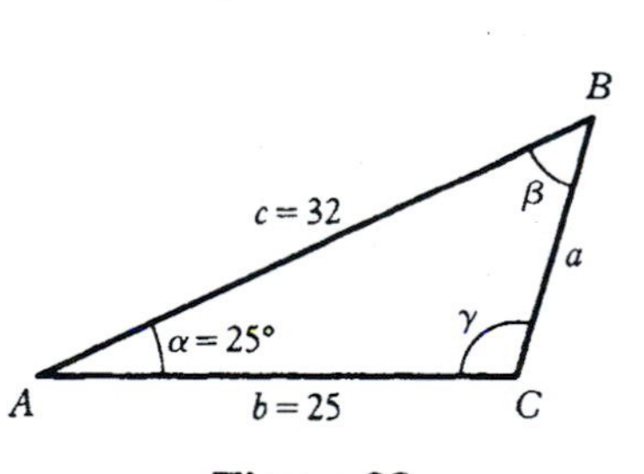

Figure 38

Finally,

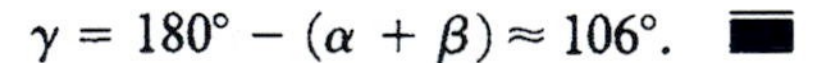

$$\gamma = 180° - (\alpha + \beta) \approx 106°. \blacksquare$$

Note that after a was obtained in Example 1, we could have used either the law of cosines or the law of sines to find β. We used the law of sines because it involves less computation.

Three Sides Given. The last case to consider in solving oblique triangles is the case in which three sides are given. Three positive numbers determine the sides of a triangle, provided the sum of any two of the numbers is greater than the remaining one. Furthermore, the triangle is *uniquely* determined.

Example 2 Solve the triangle ABC, given $a = 7.0$, $b = 11$, and $c = 6.0$.

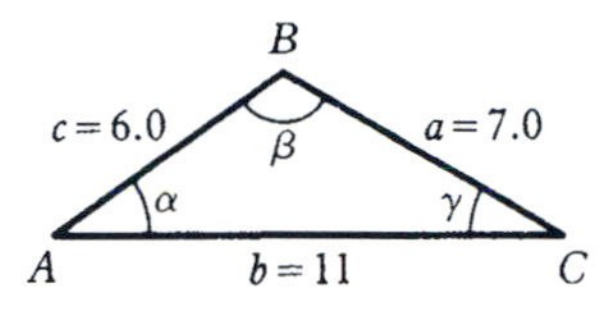

Figure 39

Solution: We first use the given data to draw Figure 39. Then, by the law of cosines,

$$a^2 = b^2 + c^2 - 2bc \cos \alpha .$$

Therefore,

$$\begin{aligned}\cos \alpha &= \frac{b^2 + c^2 - a^2}{2bc} \\ &= \frac{11^2 + 6^2 - 7^2}{2 \cdot 11 \cdot 6} \\ &\approx .8182 \\ \alpha &\approx 35^\circ .\end{aligned}$$

We can now use either the law of cosines or the law of sines to find β. We choose the law of sines because it involves less computation.

$$\sin \beta = \frac{b \sin \alpha}{a} \approx \frac{11 \sin 35^\circ}{7} \approx .9013$$

Also, since $11^2 > 7^2 + 6^2$, that is, $b^2 > a^2 + c^2$, it follows from the law of cosines that $\cos \beta < 0$ and therefore β is an obtuse angle. The table excerpt or a calculator will give the *reference angle* for β, which is approximately 64°. Hence,

$$\beta \approx 180^\circ - 64^\circ = 116^\circ .$$

Finally,

$$\gamma = 180^\circ - (\alpha + \beta) \approx 29^\circ . \blacksquare$$

Question *Suppose we are given three numbers in which the sum of two of them is not greater than the remaining one, for example, $a = 1$, $b = 2$, and $c = 5$. How can we tell from the law of cosines that a triangle is not determined?*

Answer: If we apply the law of cosines (3) to $a = 1$, $b = 2$, and $c = 5$, we get $25 = 1 + 4 - 4 \cos \gamma$, which implies that $\cos \gamma = -5$. But the cosine of any angle cannot exceed 1 in magnitude. Hence, no triangle is possible. $\blacksquare$

Applications. As with the law of sines, we can apply the law of cosines to problems of measurement.

Example 3 Before building a pipeline from point A to point B through a swamp, engineers wish to compute the length of pipe needed. They measure the distance from A to a point C outside the swamp as 1000 yards, the distance from B to C as 650 yards, and the angle ACB as $100^\circ 30'$. Find the distance from A to B.

Solution: The given information is shown in Figure 40. Since two sides and their included angle are known, we use the law of cosines to compute c, the distance from A to B.

$$\begin{aligned} c^2 &= a^2 + b^2 - 2ab \cos 100°30' \\ &\approx 650^2 + 1000^2 - 2 \cdot 650 \cdot 1000(-.1822) \\ &\approx 1{,}659{,}360 \\ c &\approx 1{,}290 \end{aligned}$$

Hence, the pipeline from A to B is approximately 1,290 yards long. ∎

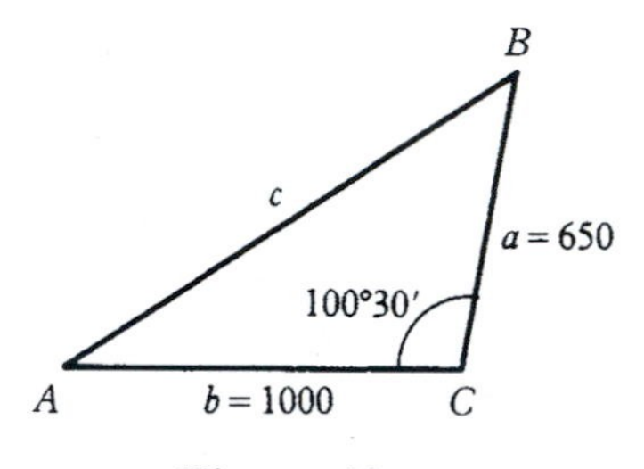

Figure 40

If three sides of a triangle are given, we can use the law of cosines to solve the triangle. We can also compute the area of the triangle by using **Heron's formula,**[1] shown below.

Given three sides a, b, and c of a triangle, the area of the triangle is

$$\mathbf{A} = \sqrt{s(s-a)(s-b)(s-c)},$$

where $s = \frac{1}{2}(a + b + c)$.

[1] *Heron (ca. 75 A.D.) was a Greek engineer, surveyor, and mathematician. He derived the area formula geometrically, but it is believed that the result was originally discovered by Archimedes.*

Example 4 Find the area of the triangle whose sides are $a = 10.2$, $b = 20.5$, and $c = 14.3$.

Solution: We first compute

$$\begin{aligned} s &= \frac{1}{2}(a + b + c) \\ &= \frac{1}{2}(10.2 + 20.5 + 14.3) \\ &= 22.5. \end{aligned}$$

Then, by Heron's formula, the area $\mathbf{A}$ is

$$\begin{aligned} \mathbf{A} &= \sqrt{22.5(22.5 - 10.2)(22.5 - 20.5)(22.5 - 14.3)} \\ &= \sqrt{4538.7} \\ &\approx 67.4. \end{aligned}$$ ∎

Both the law of cosines and the law of sines are used to derive Heron's formula. Also needed are half-angle formulas, which are derived in the next chapter. We defer the proof of Heron's formula until these results are obtained.

Exercises 6.5

Fill in the blanks to make each statement true.

1. The law of cosines is a generalization of the ____________ theorem.
2. We use the law of cosines to solve a triangle if the given parts are ____________ or ____________.
3. Three positive numbers can be the lengths of the sides of a triangle, provided ____________.
4. An oblique triangle can have at most ______ obtuse angle(s).
5. A side opposite an obtuse angle in a triangle is ______ than the remaining sides.

Write true *or* false *for each statement.*

6. The law of cosines does not hold for right triangles.
7. If three sides of a triangle are given and the law of cosines is used to find one angle, then the law of sines can be used to find a second angle.
8. If the sides a, b, and c of a triangle satisfy $c^2 > a^2 + b^2$, then the angle opposite c is obtuse.
9. The sides a, b, and c of a triangle can never satisfy $c > a + b$.
10. If two sides and their included angle are given, then a unique triangle is determined.

Solve each triangle ABC in Exercises 11–18, if possible.

Two Sides and an Included Angle Given

11. $a = 10,\ b = 5,\ \gamma = 60°$
12. $b = 12,\ c = 5.0,\ \alpha = 160°$
13. $a = 10,\ c = 30,\ \beta = 22°$
14. $b = 12,\ c = 10,\ \alpha = 40°$

Three Sides Given

15. $a = 10,\ b = 15,\ c = 20$
16. $a = 12,\ b = 15,\ c = 25$
17. $a = 15,\ b = 20,\ c = 40$
18. $a = 5,\ a + b = 12,\ a + b + c = 22$

Applications

19. Two roads intersect at an angle of 72°. A bus that travels at an average speed of 45 mph on one road and a car that travels at an average speed of 55 mph on the other road leave the intersection at the same time. How far apart are they in two hours? (There are two possible answers.)
20. A plane travels due north for 202 mi from A to B, then 304 mi to C on a bearing of 53°. How far is C from A?
21. A plane travels 500 mi from A to B on a bearing of 47°, then 400 mi from B to C on a bearing of 140°. How far is C from A, and what is the bearing of C from A?
22. Town A is 100 mi due south of town B. Town C is 150 mi from A in the direction N 39°E. What is the direction from B to C, and what is the distance between B and C?
23. Two ships leave port at the same time. One ship sails due east at 30 knots (nautical miles per hour), while the other sails on a bearing of 58° at 25 knots. After 20 hr, how far apart are the ships and what is the bearing of the second ship from the one sailing due east?
24. A surveyor measures the three sides of a triangular lot as 500 m, 450 m, and 760 m. Find the area of the lot.
25. The area of triangle ABC is $10\sqrt{2}$. If the sides a, b, and c satisfy $(a + b + c)/2 = b + 5$ and $(a + b + c)/2 = c + 4$, find a, b, and c.

Chapter 6 Review Outline

6.1 Angles and Their Measurement

An angle is generated by rotating one ray (initial side) about a vertex to a second ray (terminal side).

An angle is positive if the rotation is counterclockwise and negative if the rotation is clockwise.

Degree Measure:
One complete counterclockwise rotation defines an angle of 360°.

Radian Measure:
If a central angle θ in a circle of radius r subtends a directed arc s, then the radian measure of θ is s/r.

Standard Position:
An angle is in standard position in a coordinate system if its vertex is at the origin and its initial side is on the positive x-axis.

An angle in standard position is a quadrantal angle if its terminal side is on one of the axes; otherwise, it is in the quadrant of its terminal side.

Degrees and Radians:
π radians $= 180°$
x radians $= (180/\pi)x$ degrees
x degrees $= (\pi/180)x$ radians

6.2 The Trigonometric Functions

Definitions

If θ is in standard position and $P(x, y)$ is any point on the terminal side of θ at a distance $r > 0$ from the origin, then

$$\sin\theta = y/r, \quad \cos\theta = x/r, \quad \tan\theta = y/x,$$
$$\csc\theta = r/y, \quad \sec\theta = r/x, \quad \cot\theta = x/y.$$

Acute Angles:
An acute angle θ is one that satisfies $0° < \theta < 90°$; if θ is in a right triangle, then

$$\sin\theta = \frac{\text{opposite side}}{\text{hypotenuse}},$$
$$\cos\theta = \frac{\text{adjacent side}}{\text{hypotenuse}},$$
and
$$\tan\theta = \frac{\text{opposite side}}{\text{adjacent side}}.$$

Reference Angle:
If θ is a nonquadrantal angle in standard position, its reference angle is the (positive) acute angle that the terminal side of θ makes with the x-axis.

Special Angles:
The angle 45° is special because it is associated with isosceles right triangles; 30° and 60° are special because they are associated with equilateral triangles.

Exact values of the trigonometric functions can be found for special angles and for quadrantal angles directly from the definitions.

Calculators and Tables:
Calculators and tables give approximate values of the trigonometric functions of an angle θ. When using a table, find the magnitude of a function of θ from its reference angle and the sign from the quadrant of θ.

6.3 Solving Right Triangles

If ABC is a right triangle with $C = 90°$, then

$$A + B = 90° \quad \text{and} \quad a^2 + b^2 = c^2,$$

where a, b, and c are the sides opposite A, B, and C, respectively.

The above properties and the definitions of the trigonometric functions are used to solve right triangles (to find all sides and angles) if

(a) two sides are given, or
(b) one side and one acute angle are given.

6.4 The Law of Sines

Oblique Triangle:
An oblique triangle is one containing no right angle; in standard notation, angles at vertices A, B, and C are labeled α, β, and γ, with opposite sides a, b, and c, respectively.

For any triangle ABC, $\alpha + \beta + \gamma = 180°$.

Law of Sines:

$$\frac{a}{\sin\alpha} = \frac{b}{\sin\beta} = \frac{c}{\sin\gamma}$$

The law of sines can be used to solve an oblique triangle if

(a) one side and two angles are given, or
(b) two sides and an opposite angle are given (ambiguous case).

The solution in (a) is unique; in the ambiguous case (b), one, two, or no triangles may be determined.

Area of a Triangle:
The area A of any triangle satisfies $A = \frac{1}{2}bc \sin \alpha = \frac{1}{2}ac \sin \beta = \frac{1}{2}ab \sin \gamma$.

6.5 The Law of Cosines

Law of Cosines:

$$a^2 = b^2 + c^2 - 2bc \cos \alpha$$
$$b^2 = a^2 + c^2 - 2ac \cos \beta$$
$$c^2 = a^2 + b^2 - 2ab \cos \gamma$$

The law of cosines is a generalization of the Pythagorean theorem to oblique triangles. It can be used to solve an oblique triangle if

(a) two sides and their included angle are given, or
(b) three sides are given.

In case (a), a unique triangle is determined; in case (b), a unique triangle is determined provided the sum of any two sides is greater than the remaining one.

Obtuse Angle:
An obtuse angle θ is one that satisfies $90° < \theta < 180°$. If a triangle has an obtuse angle θ, then the side opposite θ is the largest side of the triangle.

Heron's Formula:
The area A of any triangle with sides a, b, and c satisfies

$$A = \sqrt{s(s - a)(s - b)(s - c)},$$

where $s = \frac{1}{2}(a + b + c)$.

Chapter 6 Review Exercises

In some of the following exercises, you may find it helpful to draw angles or triangles to illustrate the given information.

1. What is the radian measure (in terms of π) of $22°30'$?
2. Find the degree measure of $\pi/9$ radians.
3. Express $35°12'$ in decimal form.
4. Express $27.35°$ to the nearest minute.
5. Determine the quadrant of each of the following angles.
   (a) $7\pi/4$ (b) $-11\pi/6$
   (c) $576°$ (d) $-433°$
6. Find one positive and one negative angle coterminal with $290°$, where all three angles are in standard position.
7. If two spokes of a wheel of radius 2 ft determine an angle of $18°$ at its center, what length of arc is determined along the rim?
8. What is the angular velocity of a radar beam that rotates at 150 revolutions per minute?
9. Find the sine, cosine, and tangent of an angle θ in standard position with the point $(-5, 12)$ on its terminal side.
10. Find the secant, cosecant, and cotangent of an angle θ in standard position with the point $(-4, -3)$ on its terminal side.
11. Find the value of $\sin \theta$ if $\cos \theta = 3/4$ and θ is in quadrant IV.
12. Find the value of $\sec \theta$ if $\tan \theta = 1/2$ and θ is in quadrant III.
13. By using a special angle, find $\sin 240°$ and $\cos 240°$.
14. By using a special angle, find $\csc (11\pi/4)$ and $\cot (11\pi/4)$.
15. Express $\sin (9\pi/5)$ and $\cos (9\pi/5)$ in terms of the reference angle.
16. Express $\sec 250°$ and $\cot 250°$ in terms of the reference angle.
17. Find an acute angle θ satisfying $\sin 37° = \cos \theta$.
18. Find an acute angle θ satisfying $\tan (\theta + 45°) = \cot 2\theta$.
19. Find all six trigonometric functions of $-\pi/3$.
20. Find all the trigonometric functions of $7\pi/2$ that are defined.
21. If θ is an acute angle, what is the reference angle for $\theta + \pi$?

22. Express $\cos(\theta + \pi)$ as a trigonometric function of the acute angle θ. (*Hint:* see Exercise 21.)

Exercises 23–29 refer to a right triangle ABC with C = 90°.

23. If $A = 37°$, find angle B.
24. If sides a and b are equal, find angles A and B.
25. If $\cos A = .2$, find $\sin B$.
26. If $\sin A = 2/5$, find $\tan A$.
27. Find angle B if $a = 10$ and $b = 10\sqrt{3}$.
28. Find the radian measure of angle A if $b = 9.9$ and $c = 11$.
29. Find sides b and c if $a = 6.0$ and $B = 35°$.
30. Two sidewalks on a campus intersect. The library is located on one walk 75 yd east of their intersection and the math building is on the other walk 63 yd north of the library. At what angle do the walks intersect?
31. From ground points A and B, 155 ft apart and lying in a vertical plane with a church spire S, the angles of elevation of the spire are observed to be 12°15′ and 28°45′, respectively. How high is S if A and B are on the same side of the spire?
32. Determine the height of S in Exercise 31 if A and B are on opposite sides of the spire.

Exercises 33–35 refer to an oblique triangle in standard position.

33. Find side a of triangle ABC if $\alpha = 45°$, $\beta = 105°$, and $c = 10$.
34. Find angle γ of triangle ABC if $a = 3$, $b = 5$, and $c = 7$.
35. Find side c of triangle ABC if $a = 4$, $b = 6$, and $\gamma = 60°$.

36. In order to find the distance from point A to island B, a third point C is chosen 1,500 ft along the shore from A. If angle ABC is measured as 75° and angle CAB is 60°, find the distance from A to B to the nearest foot.
37. To find the length AB of a proposed tunnel through a mountain, a point C is found from which both A and B are visible. Measurements show that $AC = 500$ yd, $BC = 300$ yd, and angle $ACB = 120°$. What is the length of the tunnel?
38. A triangle is to be made out of three sticks of wood that are 5, 8, and 9 in long, respectively. What angle should the first two sticks make in order that the third one be just long enough to join their ends?
39. How many triangles can be made out of four sticks of length 5 in, 10 in, 15 in, and 20 in, respectively, if three sticks are used at a time?
40. A U.F.O. is sighted at an angle of elevation of 15° from a tracking station. At the same time, another tracking station located 100 mi from the first sights the same U.F.O. at an angle of elevation of 30°. The U.F.O. is located between the two tracking stations, and all three are in a vertical plane. How high is the U.F.O. above the earth?

Graphing Calculator Exercises

In each of the following, verify your answers to the indicated review exercises of this section by using a graphing calculator for the given functions and windows $[x_{min}, x_{max}]$, $[y_{min}, y_{min}]$. *To improve the accuracy of the graphical solutions, use the TRACE and ZOOM features of the calculator.*

41. Exercise 28: $Y_1 = \cos x - 9.9/11$, $[0, \pi/2]$, $[-0.5, 0.5]$
42. Exercise 30: $Y_1 = \tan x - 63/75$, Degree Mode, $[0, 45]$, $[-0.5, 0.5]$
43. Exercise 34: $Y_1 = 7^2 - 3^2 - 5^2 + 30\cos x$, Degree Mode, $[0, 180]$, $[-30, 30]$
44. Exercise 38: $Y_1 = 9^2 - 5^2 - 8^2 + 80\cos x$, Degree Mode, $[0, 180]$, $[-80, 80]$

7 Analytic Trigonometry

As discussed in the last chapter, the study of trigonometry began with angles and triangles, and that part of trigonometry is still applied in such areas as astronomy, surveying, and navigation. However, more recent applications have emphasized the periodic nature of the trigonometric functions in describing such phenomena as sound waves, electromagnetic waves, seismic waves, heartbeats (cardiograms), and business cycles. In these applications, the phenomena occur as functions of *time* rather than as functions of *angles*.

Hence, we now introduce a definition of the trigonometric functions that is independent of angles but equivalent to our earlier definition. This definition leads us to the graphs of the trigonometric functions, which clearly display their periodicity. We then consider simple harmonic motion and its applications. Finally, to prepare for the use of trigonometry in calculus and its many applications, we study inverse trigonometric functions, trigonometric identities, and trigonometric equations.

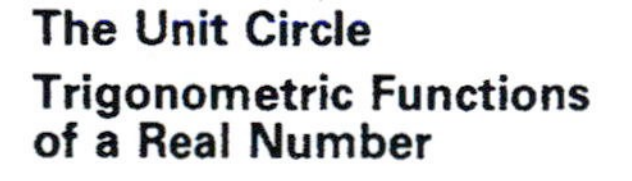

7.1 Trigonometric Functions with Real Domains

We now go from angles to real numbers by way of the unit circle.

The Unit Circle
Trigonometric Functions of a Real Number

The Unit Circle. The circle with center at the origin and radius 1 in a rectangular coordinate system is called the **unit circle.** If A is the fixed point with coordinates $(1, 0)$ and $P(x, y)$ is any point determined by proceeding from A along the unit circle in the *counterclockwise* direction, then there is a unique nonnegative number s equal to the length of the arc AP (Figure 1). If we allow one or more complete revolutions about the unit circle before stopping at an arbitrary point P, the arc length s can equal any nonnegative number. Similarly, by considering *clockwise* rotations from A to points P, we can obtain any negative real number for s.

Since the circumference of the unit circle is 2π, it follows that each point P on the unit circle determines an infinite number of values of s, any two of which differ by an integral multiple of 2π. However, each real number s determines just one point P on the unit circle. For example, $P(0, 1)$ determines $s = \pi/2 + 2n\pi$, where n is any integer, but $s = \pi/2$ determines the unique point $P(0, 1)$.

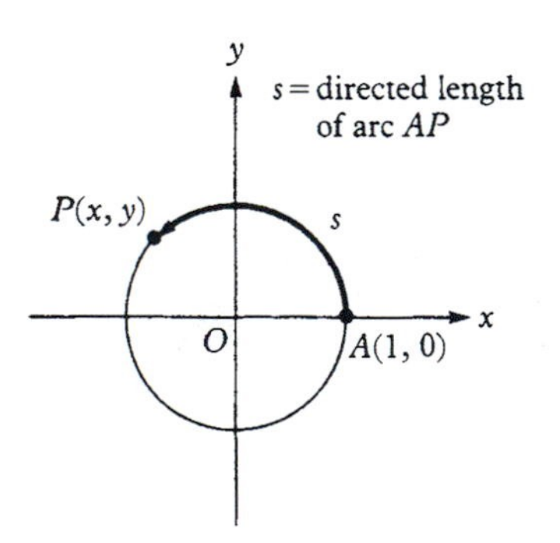

Figure 1 Arc Length Along the Unit Circle

Trigonometric Functions of a Real Number. Now let s be any real number, and let $P(x, y)$ be the unique point on the unit circle for which the directed length of an arc that starts at A and ends at P is s (Figure 2). Then we *define*

$$\sin s = y \quad \text{and} \quad \cos s = x. \tag{1}$$

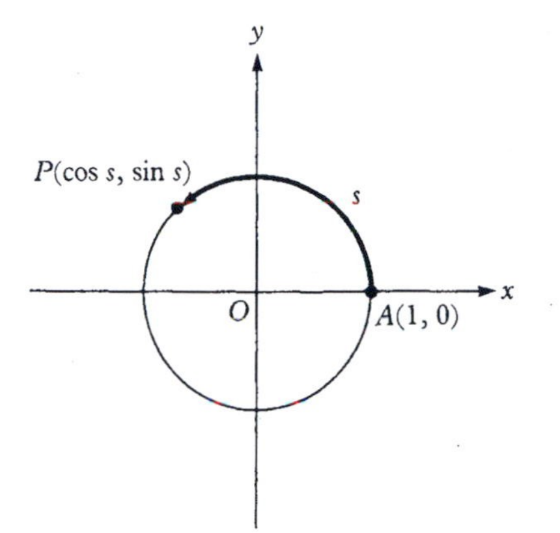

Figure 2 Trigonometric Functions of Arc Length

Thus, for each real number s, we may interpret $\cos s$ and $\sin s$ as the rectangular coordinates of the point P on the unit circle that has s as the directed length of an arc from A to P. In this way the cosine and sine are defined as *functions whose domain is the set of real numbers.*

Since $x^2 + y^2 = 1$ on the unit circle, we obtain the equation

$$(\sin s)^2 + (\cos s)^2 = 1 \tag{2}$$

for any arc length s. Equation (2) is called an **identity** because it is true for all values of s. We will discover other identities in trigonometry, and we call (2) the **fundamental identity of trigonometry.**

The other four trigonometric functions of s are defined as follows:

$$\tan s = \frac{\sin s}{\cos s} \qquad \cot s = \frac{\cos s}{\sin s}$$

$$\sec s = \frac{1}{\cos s} \qquad \csc s = \frac{1}{\sin s}. \tag{3}$$

Example Use definitions (1) and (3) to find the six trigonometric functions for $s = 3\pi/4$.

Solution: The arc length $s = 3\pi/4$, which is 3/4 the length of the unit semicircle, determines a point $P(x, y)$ in quadrant II on the line $y = -x$, as shown in Figure 3. Now the unit circle has the equation

$$x^2 + y^2 = 1.$$

Therefore, the coordinates (x, y) of P satisfy

$$x^2 + (-x)^2 = 1$$
$$2x^2 = 1$$
$$x = -\frac{1}{\sqrt{2}}$$
$$y = \frac{1}{\sqrt{2}}.$$

From definitions (1) and (3), we conclude that

$$\sin\frac{3\pi}{4} = y = \frac{1}{\sqrt{2}} \qquad \cos\frac{3\pi}{4} = x = -\frac{1}{\sqrt{2}}$$

$$\tan\frac{3\pi}{4} = \frac{\sin\frac{3\pi}{4}}{\cos\frac{3\pi}{4}} = -1 \qquad \cot\frac{3\pi}{4} = \frac{\cos\frac{3\pi}{4}}{\sin\frac{3\pi}{4}} = -1$$

$$\sec\frac{3\pi}{4} = \frac{1}{\cos\frac{3\pi}{4}} = -\sqrt{2} \qquad \csc\frac{3\pi}{4} = \frac{1}{\sin\frac{3\pi}{4}} = \sqrt{2}.$$ ■

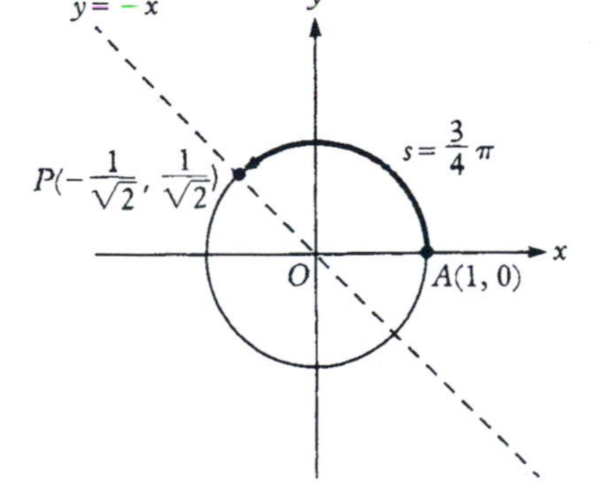

Figure 3

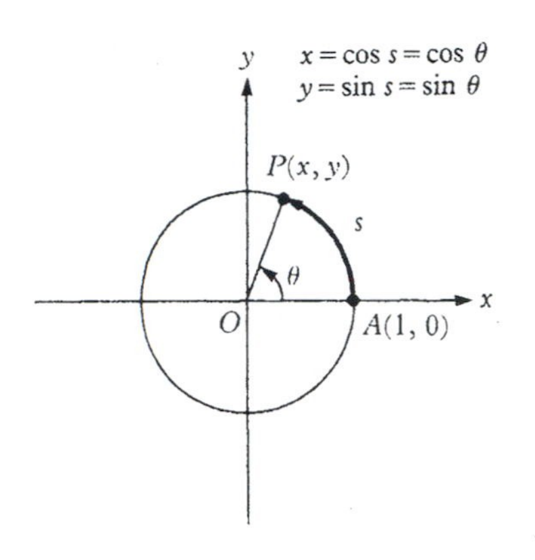

Figure 4

We now compare definitions (1) of sin s and cos s with the definitions of sin θ and cos θ given in Chapter 6, where θ is any angle in standard position. Since the radius r of the unit circle is equal to 1, the radian measure of angle $\theta = AOP$ in Figure 4 is

$$\theta = \frac{s}{r} = \frac{s}{1} = s.$$

Hence, we may interpret s in Figure 4 as either the directed length of arc AP or as the radian measure of angle AOP.

More generally, the relationship

$$\theta = \frac{s}{1}$$

establishes a one-to-one correspondence between the radian measure of central angles and the directed length of arcs along the unit circle. In turn, this one-to-one correspondence establishes the equivalence between the definitions of trigonometric functions of s and trigonometric functions of θ. Both are functions of real numbers, and expressions such as sin $(3\pi/4)$ or tan $(3\pi/4)$ have the same value whether we regard the real number $3\pi/4$ as an arc length on the unit circle or as the radian measure of a central angle. Hence, we can continue to obtain the values of the trigonometric functions from a calculator.

Comment Up to now, we have made a point of distinguishing between an arc along the unit circle and the directed length s of the arc. Now that the point has been made, we will adopt the usual convention of using s to denote both arc and arc length. This is similar to using θ to denote both an angle and the radian or degree measure of the angle.

Exercises 7.1

In all the following exercises, if the starting point of an arc along the unit circle is not explicitly stated, assume that the starting point is (1, 0).

Fill in the blanks to make each statement true.

1. The unit circle in a rectangular coordinate system is the circle with center ________ and radius ________.
2. If s is any real number and $P(x, y)$ is the point on the unit circle for which the directed arc length from A (1, 0) to P is s, then cos s = ________ and sin s = ________.
3. If s_1 and s_2 are directed lengths of arcs on the unit circle with starting point (1, 0) and endpoint (x, y), then $s_2 - s_1 =$ ________.
4. If $P_1(x_1, y_1)$ and $P_2(x_2, y_2)$ are endpoints on the unit circle for arcs s and $s + 2\pi$, respectively, then $x_2 - x_1 =$ ________ and $y_2 - y_1 =$ ________.
5. An arc length s along the unit circle corresponds to a central angle θ whose radian measure is ________.

Write true *or* false *for each statement.*

6. An arc length of $2\pi/3$ units along the unit circle corresponds to a central angle of $2\pi/3$ degrees.
7. sin s and cos s are defined for all real numbers s.
8. If sin s_1 = sin s_2, then $s_1 = s_2$.
9. If $s_2 = s_1 + 2\pi$, then sin s_2 = sin s_1.
10. If $s_1 = s_2 - \pi$, then tan s_1 = tan s_2.

The Unit Circle

For each of the points in Exercises 11–16, determine the smallest corresponding positive value of arc length s on the unit circle and the corresponding negative value of smallest magnitude.

11. $P(0, 1)$ **12.** $P(1/\sqrt{2}, 1/\sqrt{2})$

13. $P(0, -1)$ **14.** $P(1/\sqrt{2}, -1/\sqrt{2})$

15. $P(-1, 0)$ **16.** $P(1, 0)$

For each of the points in Exercises 17–20, determine all *corresponding values of arc length s on the unit circle.*

17. $P(0, 1)$ **18.** $P(1/\sqrt{2}, 1/\sqrt{2})$

19. $P(-1, 0)$ **20.** $P(0, -1)$

For each value of arc length s given in Exercises 21–30, find the unique point $P(x, y)$ determined on the unit circle.

21. $s = 0$ **22.** $s = \frac{\pi}{4}$

23. $s = -\frac{7\pi}{4}$ **24.** $s = \pi$

25. $s = -\pi$ **26.** $s = \frac{5\pi}{4}$

27. $s = \frac{3\pi}{2}$ **28.** $s = 2\pi$

29. $s = 4\pi$ **30.** $s = -4\pi$

Trigonometric Functions of a Real Number

If possible, use definitions (1) and (3) to do Exercises 31–34.

31. Find sin s, cos s, and tan s in Exercises 21, 23, and 25.

32. Find sin s, cos s, and tan s in Exercises 22, 24, and 26.

33. Find sec s, csc s, and cot s in Exercises 27 and 29.

34. Find sec s, csc s, and cot s in Exercises 28 and 30.

Miscellaneous

35. Using definition (1), show that $\sin(-s) = -\sin s$ and $\cos(-s) = \cos s$.

36. Using definition (1), show that $\sin(s + \pi) = -\sin s$ and $\cos(s + \pi) = -\cos s$.

7.2 Graphs of the Trigonometric Functions

The trigonometric functions are similar to rational numbers in that both have repeating patterns.

Graphs of $u = \sin t$ and $u = \cos t$
Graphs of $u = \tan t$ and $u = \cot t$
Graphs of $u = \sec t$ and $u = \csc t$

Whether we consider the trigonometric functions as functions of arc length s or of radian measure θ, they are functions of a *real variable*. That is, for each value of the independent variable in the domain, there is a unique value of the dependent variable in the range. It is this point of view that we will emphasize from now on.

In graphing the trigonometric functions, we will use the letter t to denote the independent variable and u to denote the dependent variable. That is, for our rectangular coordinate system, the horizontal axis will be called the t-axis, and the vertical axis will be called the u-axis. (We use the tu-axes rather than the usual xy-axes in order to avoid confusion between the definitions $x = \cos s$, $y = \sin s$ and the graph of $y = \sin x$.) Here t is simply a real number, but we can interpret t as either radian measure θ or arc length s if doing so is helpful in illustrating a property of a trigonometric function. Later, in applications of the trigonometic functions, we will usually interpet t as *time*.

Graphs of $u = \sin t$ and $u = \cos t$. Before sketching their graphs, we gather some pertinent facts concerning the sine and cosine functions. First, it follows from either the radian definition or the arc length definition that the **domain** of both the sine function and the cosine function is the set of all real numbers. Also, from the arc length definition

$$\sin s = y \qquad \text{and} \qquad \cos s = x,$$

where the points (x, y) range over the entire unit circle

$$x^2 + y^2 = 1,$$

we see that

$$-1 \le \sin s \le 1 \qquad \text{and} \qquad -1 \le \cos s \le 1.$$

That is, the **range** of both the sine function and the cosine function is the interval $[-1, 1]$. The same conclusion can be deduced from the radian definition of sine and cosine.

An important property of the sine and cosine functions is their **periodicity.** We saw in Chapter 6 that coterminal angles have the same values for their trigonometric functions. In particular,

$$\sin(\theta + 2\pi) = \sin\theta \qquad \text{and} \qquad \cos(\theta + 2\pi) = \cos\theta$$

for *any* value of θ. Hence,

$$\sin(t + n \cdot 2\pi) = \sin t \qquad \text{and} \qquad \cos(t + n \cdot 2\pi) = \cos t$$

for any real number t and any integer n. This property means that the graphs of $u = \sin t$ and $u = \cos t$ on the interval $[0, 2\pi]$ are repeated on $[2\pi, 4\pi]$, $[4\pi, 6\pi]$, $[-2\pi, 0]$, $[-4\pi, -2\pi]$, and on any interval of the form $[2n\pi, 2n\pi + 2\pi]$, where n is any integer. Hence, to obtain the entire graphs of $u = \sin t$ and $u = \cos t$, we need only plot points for values of t from 0 to 2π.

In general, a nonconstant function f is called **periodic** if and only if there is a positive real number p for which

$$f(t + p) = f(t)$$

for *all* t in the domain of f. The smallest such value of p is called the **period** of f. It has just been observed that the sine and cosine functions are periodic, and we will see from their graphs that the period for each is 2π. That is, 2π is the *smallest* positive value of p for which

$$\sin(t + p) = \sin t$$

for all values of t, and for which

$$\cos(t + p) = \cos t$$

for all values of t.

To graph $u = \sin t$, we first construct a table of coordinates (t, u) for selected values of t from 0 to 2π. In the table below we have used special values of t and have rounded $u = \sin t$ to one decimal place. In general, to compute $\sin t$, we can use a calculator to construct a coordinate table or, of course, a graphing calculator. After the table is constructed, we can plot the corresponding points and join them by a smooth curve, as shown in Figure 5.

t	u	t	u
0	0		
$\pi/6$	.5	$7\pi/6$	$-.5$
$\pi/4$	.7	$5\pi/4$	$-.7$
$\pi/3$	.9	$4\pi/3$	$-.9$
$\pi/2$	1	$3\pi/2$	-1
$2\pi/3$	.9	$5\pi/3$	$-.9$
$3\pi/4$	.7	$7\pi/4$	$-.7$
$5\pi/6$	.5	$11\pi/6$	$-.5$
π	0	2π	0

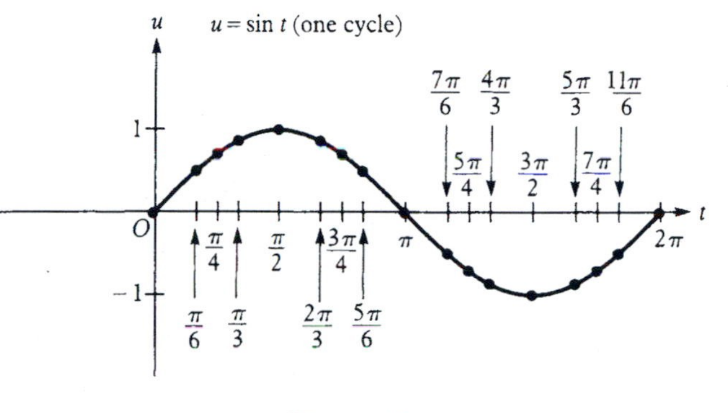

Figure 5

The curve in Figure 5 is one **cycle** of the sine curve. As stated earlier, we can obtain the entire graph of $u = \sin t$ by repeating this cycle on each interval of length 2π to the right and to the left indefinitely. The result is shown in Figure 6.

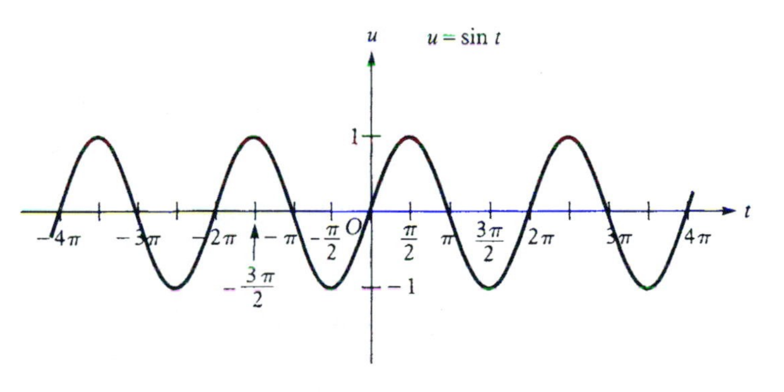

Figure 6

The following five properties are evident from an inspection of the graph in Figure 6.

(i) Vertically, the sine curve oscillates from -1 to 1. The value 1 is called the **amplitude** of $u = \sin t$.

(ii) Horizontally, the sine pattern repeats every 2π units, and 2π is the smallest positive number with this property. That is, $u = \sin t$ has **period** 2π.

(iii) The **t-intercepts** are at $t = n\pi$ for every integer n. That is,

$$\sin(n\pi) = 0 \qquad (n = 0, \pm 1, \pm 2, \ldots). \qquad (1)$$

(iv) The sine curve is **symmetric with respect to the origin.** That is,

$$\sin(-t) = -\sin t \tag{2}$$

for all real numbers t. Equivalently, the sine is an **odd function.**

(v) The sine curve from π to 2π is the negative of the curve from 0 to π, and in general,

$$\sin(t + \pi) = -\sin t \tag{3}$$

for all real numbers t.

Example 1 Given $\sin t = .25$, find each of the following.

(a) $\sin(t + 6\pi)$ (b) $\sin(-t + 2\pi)$ (c) $\sin(t - \pi)$

Solution:

(a) $\sin(t + 6\pi) = \sin(t + 3 \cdot 2\pi)$
$= \sin t$ *Period 2π*
$= .25$

(b) $\sin(-t + 2\pi) = \sin(-t)$ *Period 2π*
$= -\sin t$ *From (2)*
$= -.25$

(c) $\sin(t - \pi) = \sin(t + \pi + (-1)2\pi)$
$= \sin(t + \pi)$ *Period 2π*
$= -\sin t$ *From (3)*
$= -.25$ ■

The cosine also repeats its values every 2π units. Therefore, to draw the entire graph of $u = \cos t$, we first construct one cycle for values of t from 0 to 2π and then repeat that cycle to the left and right indefinitely, as shown in Figure 7.

t	u
0	1
$\pi/4$	.7
$\pi/2$	0
$3\pi/4$	−.7
π	−1
$5\pi/4$	−.7
$3\pi/2$	0
$7\pi/4$	.7
2π	1

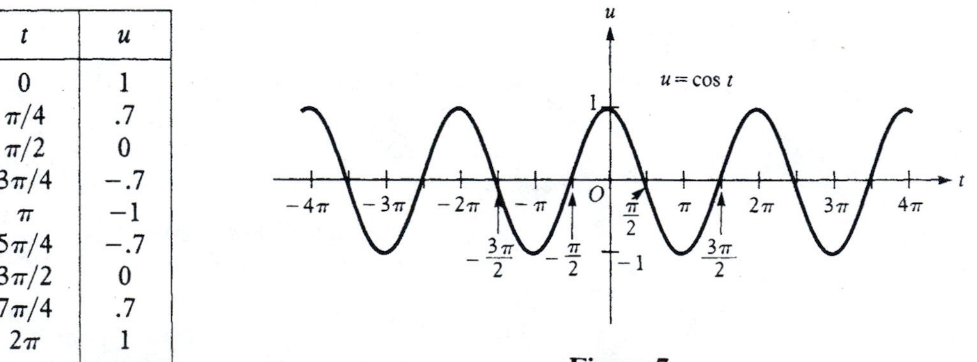

Figure 7

We see from the graph that $u = \cos t$ has **amplitude** 1 and **period** 2π, as does $u = \sin t$. The **t-intercepts** for the cosine curve are at $t = \pi/2 + n\pi$ for every integer n. That is,

$$\cos\left(\frac{\pi}{2} + n\pi\right) = 0 \qquad (n = 0, \pm 1, \pm 2, \ldots). \tag{4}$$

The cosine curve is **symmetric with respect to the u-axis**. That is,

$$\cos(-t) = \cos t \tag{5}$$

for all real numbers t. Equivalently, the cosine is an **even function**.

As is the sine curve, the cosine curve from π to 2π is the negative of the curve from 0 to π. In general,

$$\cos(t + \pi) = -\cos t \tag{6}$$

for all real numbers t.

By comparing Figures 6 and 7, we see that if the sine curve is shifted $\pi/2$ units to the left, the result is the cosine curve. This property can be expressed by the equation

$$\cos t = \sin\left(t + \frac{\pi}{2}\right), \tag{7}$$

which is true for all real numbers t. To see that $u = \sin(t + \pi/2)$ is a shift of $u = \sin t$ by $\pi/2$ units to the left, note that when $t + \pi/2 = 0$, we have $t = -\pi/2$. Hence, the cycle from 0 to 2π for $u = \sin t$ (Figure 5) corresponds to a cycle from $-\pi/2$ to $3\pi/2$ for $u = \sin(t + \pi/2)$.

Example 2 Given $\cos t = .1$, find

(a) $\cos(-t + 4\pi)$ (b) $\cos(-t + 5\pi)$ (c) $\sin\left(t + \frac{3}{2}\pi\right)$.

Solution:

(a)
$$\begin{aligned} \cos(-t + 4\pi) &= \cos(-t + 2 \cdot 2\pi) \\ &= \cos(-t) && \textit{Period } 2\pi \\ &= \cos t && \textit{From (5)} \\ &= .1 \end{aligned}$$

(b)
$$\begin{aligned} \cos(-t + 5\pi) &= \cos((-t + \pi) + 2 \cdot 2\pi) \\ &= \cos(-t + \pi) && \textit{Period } 2\pi \\ &= -\cos(-t) && \textit{From (6)} \\ &= -\cos t && \textit{From (5)} \\ &= -.1 \end{aligned}$$

(c) $\sin\left(t + \frac{3}{2}\pi\right) = \sin\left(t + \frac{\pi}{2} + \pi\right)$

$= -\sin\left(t + \frac{\pi}{2}\right)$ *From (3)*

$= -\cos t$ *From (7)*

$= -.1$ ■

Example 3 By comparing Figures 6 and 7, give a geometric reason for each of the following relationships.

(a) $\sin t = \cos(t - \pi/2)$ (b) $-\sin t = \cos(t + \pi/2)$

Solution:

(a) The curve $u = \cos(t - \pi/2)$ is a shift of $u = \cos t$ by $\pi/2$ units to the right [see the explanation of "shift" following equation (7)]. Hence, the equation $\sin t = \cos(t - \pi/2)$ expresses the property that the sine curve is a shift of the cosine curve by $\pi/2$ units to the right.

(b) The curve $u = \cos(t + \pi/2)$ is a shift of $u = \cos t$ by $\pi/2$ units to the left, and $u = -\sin t$ is a reflection of $u = \sin t$ in the t-axis. Hence, the equation $-\sin t = \cos(t + \pi/2)$ expresses the property that the sine curve reflected in the t-axis is a shift of the cosine curve by $\pi/2$ units to the left. ■

Graphs of $u = \tan t$ and $u = \cot t$. By definition,

$$\tan t = \frac{\sin t}{\cos t},$$

which means that $\tan t$ is undefined whenever $\cos t = 0$. That is, the **domain** of the tangent function is all real numbers t except

$$t = \frac{\pi}{2} + n\pi \qquad (n = 0, \pm 1, \pm 2, \ldots).$$

Also, from equations (3) and (6), we obtain

$$\tan(t + \pi) = \frac{\sin(t + \pi)}{\cos(t + \pi)} = \frac{-\sin t}{-\cos t} = \tan t.$$

Therefore, the tangent repeats its values every π units, and the entire graph can be obtained by repeating the part of its graph for t in the interval $(-\pi/2, \pi/2)$ (Figure 8).

We can deduce the following properties from Figure 8.

(i) Vertically, the tangent curve varies from $-\infty$ to ∞. That is, the **range** of the tangent function is the set of all real numbers.

t	u
0	0
$\pi/4$	1
$9\pi/20$	6.3
$99\pi/200$	63.7
$-\pi/4$	-1
$-9\pi/20$	-6.3
$-99\pi/200$	-63.7

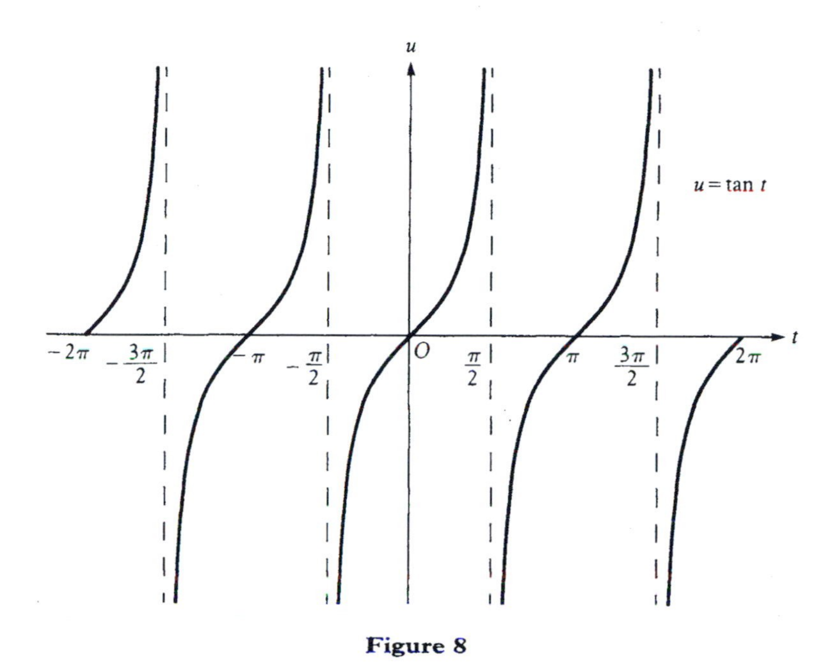

Figure 8

(ii) Horizontally, the tangent pattern repeats every π units, and π is the smallest positive number with this property. That is, $u = \tan t$ has **period** π.

(iii) The **t-intercepts** are $t = n\pi$ for every integer n. That is,

$$\tan n\pi = 0 \qquad (n = 0, \pm 1, \pm 2, \ldots). \tag{8}$$

(iv) The tangent curve is **symmetric with respect to the origin.** That is,

$$\tan(-t) = -\tan t \tag{9}$$

for every real number t in the domain of the tangent. Equivalently, the tangent is an **odd function.**

(v) The vertical lines

$$t = \frac{\pi}{2} + n\pi \qquad (n = 0, \pm 1, \pm 2, \ldots)$$

are **vertical asymptotes** for the tangent curve. More specifically,

$$\tan t \to +\infty \quad \text{as} \quad t \to \left(\frac{\pi}{2} + n\pi\right)^{-},$$

$$\tan t \to -\infty \quad \text{as} \quad t \to \left(\frac{\pi}{2} + n\pi\right)^{+},$$

where $^{+}$ means the approach is from the *right,* and $^{-}$ means the approach is from the *left.*

Example 4 Given $\tan t = 10$, find (a) $\tan(t - 5\pi)$ and (b) $\tan(-t + 8\pi)$; then describe the graph of $\tan t$ (c) as $t \to (5\pi/2)^-$ and (d) as $t \to (-3\pi/2)^+$.

Solution:

(a) $\tan(t - 5\pi) = \tan(t + (-5)\pi)$

$= \tan t$ *Period π*

$= 10$

(b) $\tan(-t + 8\pi) = \tan(-t)$ *Period π*

$= -\tan t$ *From (9)*

$= -10$

(c) $\tan t \to +\infty$ as $t \to \dfrac{5\pi^-}{2}$ $\quad \dfrac{5\pi}{2} = \dfrac{\pi}{2} + 2\pi$

(d) $\tan t \to -\infty$ as $t \to -\dfrac{3\pi^+}{2}$ $\quad -\dfrac{3\pi}{2} = \dfrac{\pi}{2} - 2\pi$ ■

Figure 9 shows the graph of $u = \cot t$, which can be obtained from the tangent curve by using the reciprocal relationship

$$\cot t = \frac{\cos t}{\sin t} = \frac{1}{\tan t}$$

for all values of t for which both $\tan t$ and $\cot t$ are defined. In particular, $\cot t \to 0$ where $\tan t \to \pm\infty$, and $\cot t \to \pm\infty$ where $\tan t \to 0$. The **domain** of the cotangent is all real numbers except

$$t = n\pi \qquad (n = 0, \pm 1, \pm 2, \ldots),$$

and at these points the cotangent has **vertical asymptotes.** The **range** is all real numbers, and $u = \cot t$ has **period** π. The ***t*-intercepts** occur wherever $\cos t = 0$, that is,

$$\cot\left(\frac{\pi}{2} + n\pi\right) = 0 \qquad (n = 0, \pm 1, \pm 2, \ldots).$$

Graphs of $u = \sec t$ and $u = \csc t$. By definition,

$$\sec t = \frac{1}{\cos t} \qquad \text{and} \qquad \csc t = \frac{1}{\sin t}.$$

Therefore, the secant and cosecant are periodic functions with period 2π. In graphing $u = \sec t$, we first observe that $\sec t = \pm 1$ whenever $\cos t = \pm 1$. Also, from the inequality $|\cos t| \leq 1$, we have $|\sec t| \geq 1$ for all t in its domain, and although $\sec t$ is undefined where $\cos t = 0$, our graph must

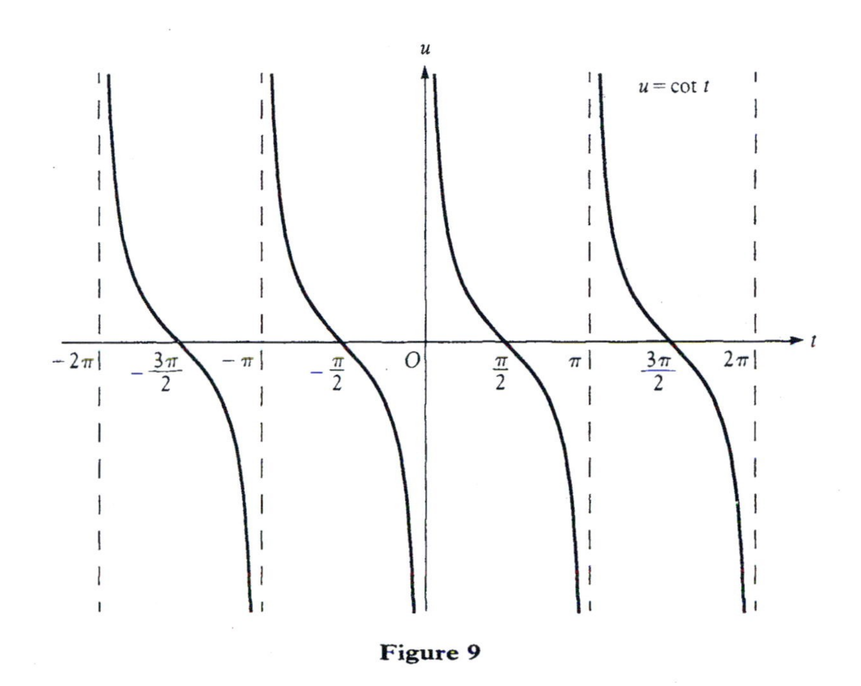

Figure 9

show that sec $t \to +\infty$ where cos $t \to 0^+$, and sec $t \to -\infty$ where cos $t \to 0^-$. Using these facts, we construct the graph of the secant function as shown in Figure 10.

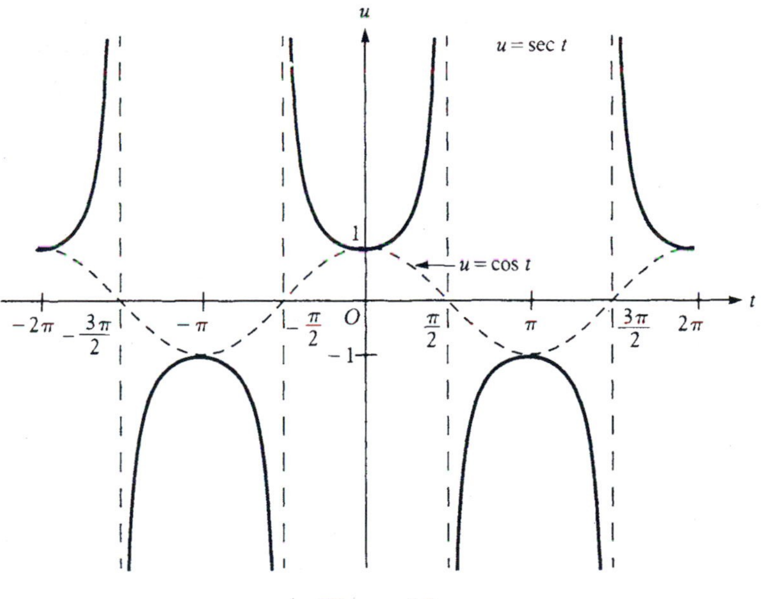

Figure 10

Similarly, the graph of $u = \csc t$ can be constructed from that of $u = \sin t$ as shown in Figure 11.

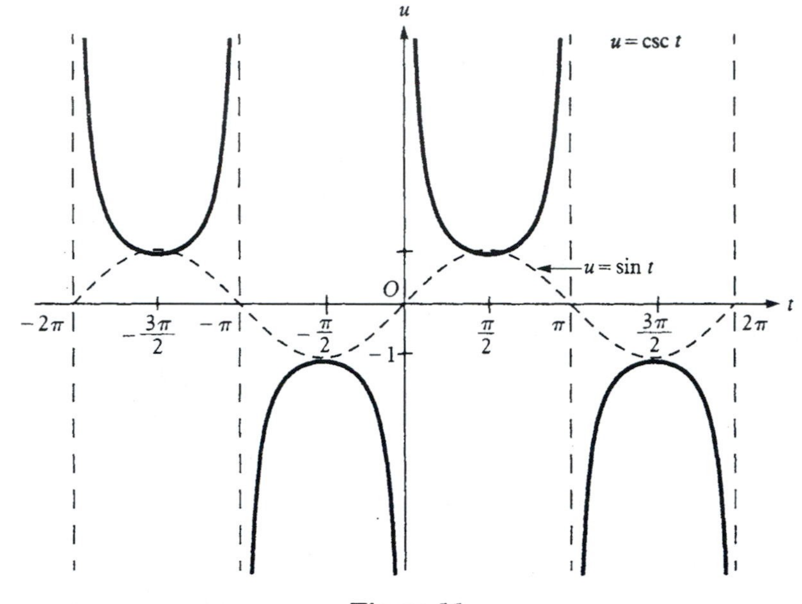

Figure 11

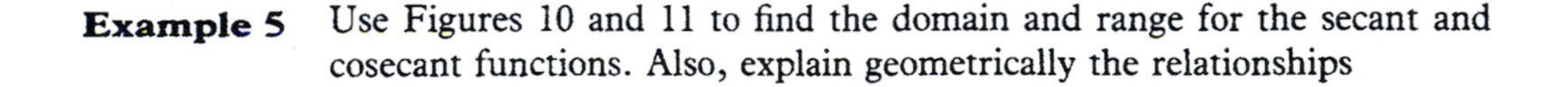

Example 5 Use Figures 10 and 11 to find the domain and range for the secant and cosecant functions. Also, explain geometrically the relationships

$$\sec (t + \pi) = -\sec t$$

and

$$\sec t = \csc \left(t + \frac{\pi}{2}\right).$$

Solution: The domain for the secant is all t except $t = \pi/2 + n\pi$, and the domain for the cosecant is all t except $t = n\pi$, where n is any integer. The range for both functions is $(-\infty, -1] \cup [1, \infty)$.

The graph of $u = \sec t$ on that part of its domain from π to 2π is the negative of the graph from 0 to π. Therefore, by the periodicity of the secant function, we have $\sec (t + \pi) = -\sec t$ for all t in its domain.

If the graph of the cosecant is shifted $\pi/2$ units to the *left,* then the result is the secant curve. Hence, $\sec t = \csc (t + \pi/2)$ whenever both expressions are defined [see the explanation of "shift" following equation (7)]. ■

Exercises 7.2

Fill in the blanks to make each statement true.

1. The basic trigonometric functions with period 2π are ________.
2. The basic trigonometric functions with period π are ________.
3. If the cosine curve is shifted $\pi/2$ units to the ________, the result is the sine curve.
4. If the sine curve is shifted $\pi/2$ units to the ________, the result is the cosine curve.
5. The curve $u = \tan t$ has vertical asymptotes at the lines $t =$ ________.

Write true *or* false *for each statement.*

6. The domain of the sine function is all real numbers.
7. The range of the cosine function is all real numbers.
8. The range of the tangent function is all real numbers.
9. If the tangent curve is shifted $\pi/2$ units to the left, the result is the cotangent curve.
10. If the secant curve is shifted $\pi/2$ units to the right, the result is the cosecant curve.

Use the graphs in Figures 6–11 to do Exercises 11–31. Do not use tables or a calculator.

Sine and Cosine Functions

11. Given $\sin t = .36$, find each of the following.
    (a) $\sin(t + \pi)$ (b) $\sin(t - 3\pi)$ (c) $\sin(t + 4\pi)$
12. Given $\sin t = -.75$, find each of the following.
    (a) $\sin(-t)$ (b) $\sin(-t + 2\pi)$ (c) $\sin(\pi - t)$
13. Given $\cos t = .3$, find each of the following.
    (a) $\cos(t + \pi)$ (b) $\cos(t - \pi)$ (c) $\cos(t - 4\pi)$
14. Given $\cos t = -.4$, find each of the following.
    (a) $\cos(-t)$ (b) $\cos(-t - 2\pi)$ (c) $\cos(\pi - t)$
15. On the same axes, graph $u = \sin t$ for $\pi/2 \le t \le 5\pi/2$ and $u = \cos t$ for $0 \le t \le 2\pi$. Describe the relationship between the two curves in terms of "shift."
16. On the same axes, graph $u = \sin t$ for $0 \le t \le 2\pi$ and $u = \cos t$ for $-\pi/2 \le t \le 3\pi/2$. Describe the relationship between the two curves in terms of "shift."
17. Determine all solutions to the following equations on $[0, 2\pi]$.
    (a) $\sin t = 0$ (b) $\sin t = 1$ (c) $\sin t = -1$
18. Repeat Exercise 17 for the cosine.
19. How many solutions on $[0, 2\pi]$ are there to the equation $\sin t = c$, where $-1 < c < 1$ and $c \ne 0$? If t_0 is one of the solutions, express all solutions in terms of t_0.
20. Repeat Exercise 19 for the cosine.

Tangent and Cotangent Functions

21. On the same axes, graph $u = \tan t$ for $-\pi/2 < t < \pi/2$ and $u = \cot t$ for $0 < t < \pi$. Is one curve a shift of the other?
22. How many solutions are there to the equation $\tan t = c$ on $(-\pi/2, \pi/2)$, where c is any real number?
23. How many solutions are there to the equation $\cot t = c$ on $(0, \pi)$, where c is any real number?
24. What are the vertical asymptotes for $u = \cot t$? Describe the behavior of $\cot t$ as t approaches either side of an asymptote.

Secant and Cosecant Functions

25. On the same axes, graph $u = \csc t$ for $0 < t < \pi$ and $u = \sec t$ for $-\pi/2 < t < \pi/2$. Describe the relationship between the two curves in terms of "shift."
26. On the same axes, graph $u = \csc t$ for $\pi/2 \le t \le 3\pi/2$ ($t \ne \pi$) and $u = \sec t$ for $0 \le t \le \pi$ ($t \ne \pi/2$). Describe the relationship between the two curves in terms of "shift."
27. Locate all vertical asymptotes for $u = \sec t$. Describe the behavior of $\sec t$ as t approaches either side of an asymptote.
28. Repeat Exercise 27 for $u = \csc t$.

Miscellaneous

29. Which of the six basic trigonometric functions are odd functions? Which are even functions? Is every trigonometric function either odd or even?

30. In each of the following, determine which, if any, of the six basic trigonometric functions satisfy the equation.

(a) $f(t+\pi)=f(t)$ (b) $f\left(t+\frac{\pi}{2}\right)=-f(t)$

31. In each of the following, determine which, if any, of the six basic trigonometric functions satisfy the equation.

(a) $f\left(t+\frac{\pi}{2}\right)=f(t)$ (b) $f(t+\pi)=-f(t)$

7.3 Amplitude, Period, and Phase Shift

The sine shifts to the cosine, the secant shifts to the cosecant, but the tangent and cotangent shift only to themselves.

Amplitude
Period
Phase Shift
Sums of Trigonometric Functions

In this section we discuss variations of the basic trigonometic functions. We show how the basic graphs can be stretched or compressed either vertically or horizontally. Other variations include a reflection in either the t-axis or the u-axis and horizontal shifts.

Amplitude. From the basic graph of $u = \sin t$, we can obtain periodic functions of amplitudes other than 1, as shown in the next two examples.

Example 1 Sketch the graph of $u = 4 \sin t$.

Solution: Since $4 \sin(t + 2\pi) = 4 \sin t$, the function $f(t) = 4 \sin t$ has period 2π. Hence, we construct a table of coordinates for $0 \le t \le 2\pi$, plot the points of one cycle, and then extend the graph as indicated in Figure 12. ■

Example 2 Sketch the graph of $u = -4 \sin t$.

Solution: A table for the same values of t as in Example 1 would show that each point (t, u) on the graph of $u = 4 \sin t$ corresponds to a point $(t, -u)$ on the graph of $u = -4 \sin t$. That is, the graph of $u = -4 \sin t$ is a reflection in the t-axis of $u = 4 \sin t$, as Figure 13 shows. ■

Examples 1 and 2 illustrate the general result that for any constant A, the graph of

$$u = A \sin t$$

has **amplitude** $|A|$. That is, the effect of the factor A is to vertically *stretch* ($|A| > 1$) or *compress* ($|A| < 1$) the basic sine curve. Also, if $A < 0$, the basic sine curve is reflected in the t-axis. The same results apply to $u = A \cos t$.

Period. We can also stretch or compress the basic sine curve horizontally. This amounts to changing the period of the sine function.

t	$\sin t$	$4 \sin t$
0	0	0
$\pi/4$	.7	2.8
$\pi/2$	1	4
$3\pi/4$	.7	2.8
π	0	0
$5\pi/4$	−.7	−2.8
$3\pi/2$	−1	−4
$7\pi/4$	−.7	−2.8
2π	0	0

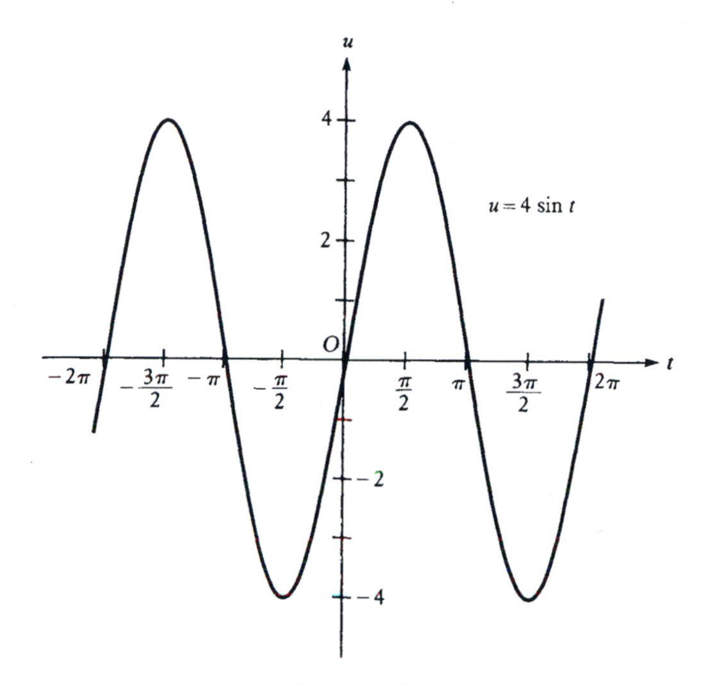

Figure 12

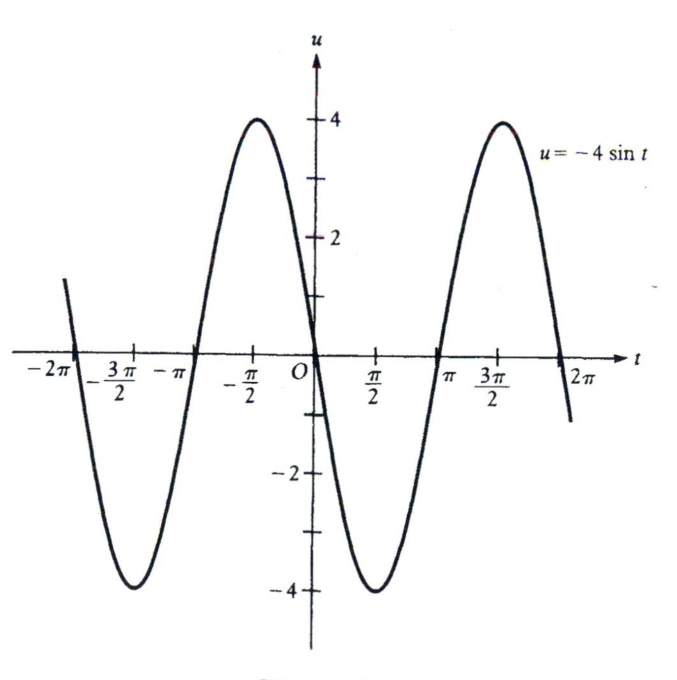

Figure 13

Example 3 Sketch the graph of $u = \sin 3t$.

Solution: A substitution such as $v = 3t$ to obtain $u = \sin v$ shows that the graph is just a modification of the basic sine curve. To proceed, we construct a table of coordinates as shown below.

t	0	$\pi/6$	$\pi/3$	$\pi/2$	$2\pi/3$	$5\pi/6$	π	...
$3t$	0	$\pi/2$	π	$3\pi/2$	2π	$5\pi/2$	3π	...
$u = \sin 3t$	0	1	0	-1	0	1	0	...

Although we could continue the table to $t = 2\pi$, we see that $3t$ has already reached 2π and the values of $\sin 3t$ start to repeat after t reaches $2\pi/3$. This suggests that the period is $p = 2\pi/3$, and we can verify this by letting $f(t) = \sin 3t$ and computing

$$f\left(t + \frac{2\pi}{3}\right) = \sin\left(3\left(t + \frac{2\pi}{3}\right)\right) = \sin(3t + 2\pi) = \sin 3t = f(t).$$

Thus the graph of $u = \sin 3t$ (Figure 14) repeats every $2\pi/3$ units, and we can see from the figure that $2\pi/3$ is the smallest positive number with this property. That is, $u = \sin 3t$ has period $2\pi/3$. ■

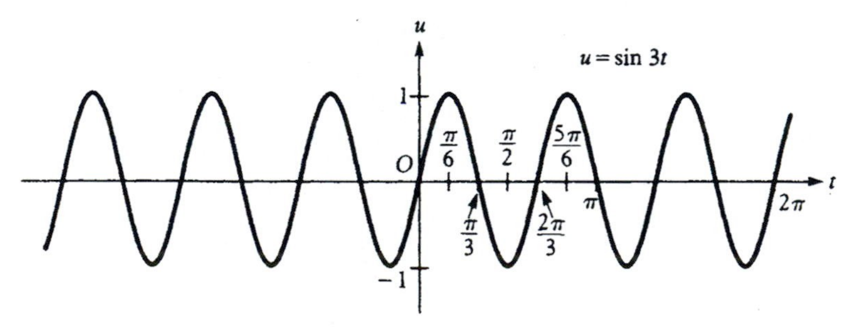

Figure 14

In general, for any positive constant B, the function

$$f(t) = \sin Bt$$

has **period** $p = 2\pi/B$ since $2\pi/B$ is the smallest positive number satisfying

$$f(t + p) = \sin\left(B\left(t + \frac{2\pi}{B}\right)\right) = \sin(Bt + 2\pi) = \sin Bt = f(t).$$

The same result applies to $\cos Bt$, $\sec Bt$, and $\csc Bt$. Similarly, the functions

$$g(t) = \tan Bt \qquad \text{and} \qquad h(t) = \cot Bt$$

have period π/B (*why?*).

Question *Suppose B is negative. For instance, what is the period of $u = \sin(-3t)$?*

Answer: Since $\sin(-3t) = -\sin 3t$, the period is $2\pi/3$. For any nonzero B, positive or negative, the period of $\sin Bt$, $\cos Bt$, $\sec Bt$, and $\csc Bt$ is $2\pi/|B|$, and the period of $\tan Bt$ and $\cot Bt$ is $\pi/|B|$. ■

As shown in Figure 14, the graph of $u = \sin 3t$ has three cycles in an interval of length 2π. We therefore say that the frequency of $\sin 3t$ is $3/(2\pi)$. In general, if $f(t)$ has period p, then $1/p$ is called the **frequency** of f.

Example 4 Determine the period and frequency of each of the following functions.

(a) $f(t) = \sin(t/2)$ (b) $g(t) = \cos \pi t$ (c) $h(t) = \tan(-2\pi t)$

Solution:

(a) Here $B = 1/2$, so the period is $2\pi/(1/2) = 4\pi$, and the frequency is $1/(4\pi)$.

(b) We have $B = \pi$, so the period is $2\pi/\pi = 2$, and the frequency is $1/2$.

(c) First we write $h(t) = -\tan 2\pi t$. Hence, $B = 2\pi$, so the period is $\pi/(2\pi) = 1/2$, and the frequency is 2. ■

Phase Shift. We saw earlier that if the cosine curve is shifted $\pi/2$ units to the right, the result is the sine curve. We now consider the concept of "shifting" a periodic function in more detail.

Example 5 Sketch a graph of $u = \sin(3t + \pi)$.

Solution: To analyze the effect of the factor 3 and the addend π, we make the substitution

$$v = 3t + \pi \quad \text{or} \quad t = \frac{v - \pi}{3}.$$

Now the function $u = \sin v$ has period 2π in v, and a single cycle occurs from $v = 0$ to $v = 2\pi$. The corresponding interval in t is

$$\text{from } t = -\frac{\pi}{3} \quad \text{to} \quad t = \frac{\pi}{3}.$$

This means that a single cycle occurs on the t-interval $[-\pi/3, \pi/3]$, and the length of this interval, namely $2\pi/3$, is the t-period of u. Hence, to graph $u = \sin(3t + \pi)$, we construct a table of coordinates for values of t from $-\pi/3$ to $\pi/3$. From this table, we construct a cycle and then repeat this cycle indefinitely to the right and left (Figure 15). In constructing the table, it is helpful to select convenient values of v from 0 to 2π and convert these into the corresponding values of t. ■

v	t	u
0	$-\pi/3$	0
$\pi/2$	$-\pi/6$	1
π	0	0
$3\pi/2$	$\pi/6$	-1
2π	$\pi/3$	0

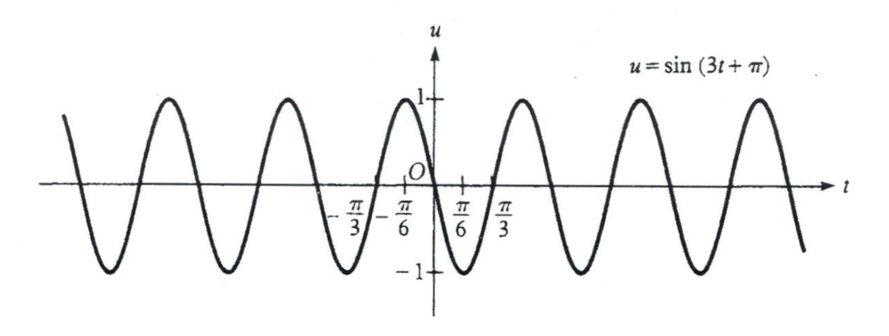

Figure 15

The starting point $-\pi/3$ of the cycle mentioned in Example 5 is called the *phase shift* of the function $u = \sin(3t + \pi)$. In general, if $f(t)$ is periodic and B and C are real numbers with $B > 0$, then the **phase shift** of

$$f(Bt + C)$$

is defined to be $-C/B$, which is the starting point of the cycle obtained by setting $Bt + C = 0$. The shift is to the left if $-C/B$ is negative, and to the right if $-C/B$ is positive. For instance, the equation

$$\cos t = \sin\left(t + \frac{\pi}{2}\right) \qquad \left(B = 1, C = \frac{\pi}{2}\right)$$

says that the sine curve with a shift of $-\pi/2$ units becomes the cosine curve (Figure 16). The shift is to the left because $-\pi/2$ is negative. Or, if we let $v = t + \pi/2$ in the above equation, we get

$$\cos\left(v - \frac{\pi}{2}\right) = \sin v \qquad \left(B = 1, C = -\frac{\pi}{2}\right),$$

which says that the cosine curve with a phase shift of $\pi/2$ units to the right becomes the sine curve.

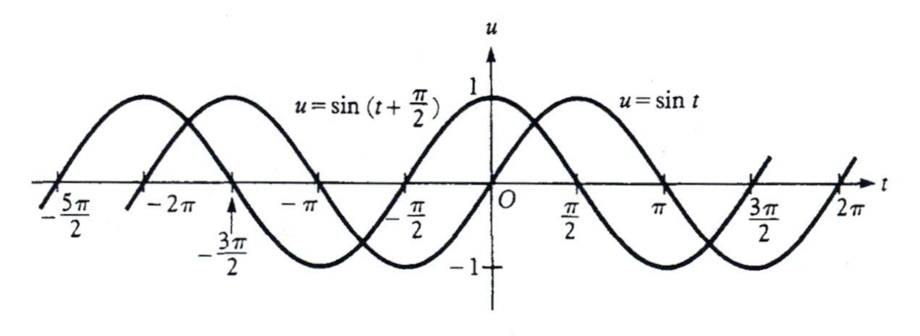

Figure 16

We can apply the reasoning used in the previous examples to conclude that if A, B, and C are real numbers and B is positive, then the function

$$f(t) = A \sin (Bt + C)$$

has **amplitude** $|A|$, **period** $2\pi/B$, and **phase shift** $-C/B$. The same result applies to $A \cos (Bt + C)$. Similarly, $A \sec (Bt + C)$ and $A \csc (Bt + C)$ have period $2\pi/B$ and phase shift $-C/B$, and $A \tan (Bt + C)$ and $A \cot (Bt + C)$ have period π/B and phase shift $-C/B$. (We do not apply the term "amplitude" to the secant, cosecant, tangent, and cotangent because they are unbounded functions.) Also, the condition $B > 0$ can always be met. For example, we can write

$$3 \sin \left(-2t + \frac{\pi}{3}\right) = 3 \sin \left(-\left(2t - \frac{\pi}{3}\right)\right) = -3 \sin \left(2t - \frac{\pi}{3}\right)$$

by using the property $\sin (-\theta) = -\sin \theta$. Similarly,

$$3 \cos \left(-2t + \frac{\pi}{3}\right) = 3 \cos \left(2t - \frac{\pi}{3}\right)$$

by the property $\cos (-\theta) = \cos \theta$.

Example 6 Find the amplitude, period, and phase shift of

$$f(t) = \frac{3}{4} \sin \left(2t + \frac{\pi}{3}\right).$$

Also, graph one cycle of the curve starting at $-\pi/6$.

Solution: Here $A = 3/4, B = 2$, and $C = \pi/3$. Therefore, the amplitude is $3/4$, the period is $2\pi/2 = \pi$, and the phase shift is $-(\pi/3)/2 = -\pi/6$. To graph the desired cycle, we construct a table of coordinates from $t = -\pi/6$ to $-\pi/6 + \pi = 5\pi/6$. As in Example 5, it is helpful to set

$$v = 2t + \frac{\pi}{3} \quad \text{or} \quad t = \frac{v}{2} - \frac{\pi}{6},$$

then select convenient values of v from 0 to 2π, and convert these to the corresponding values of t. The graph is given in Figure 17. ■

v	t	$f(t)$
0	$-\pi/6$	0
$\pi/2$	$\pi/12$	$3/4$
π	$\pi/3$	0
$3\pi/2$	$7\pi/12$	$-3/4$
2π	$5\pi/6$	0

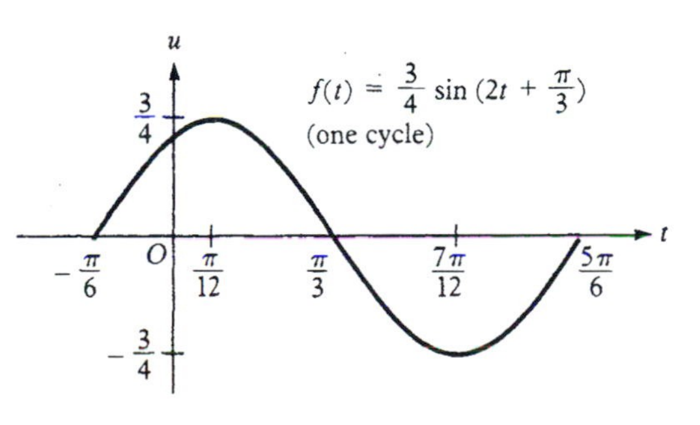

Figure 17

Sums of Trigonometric Functions. By combining trigonometric functions with different amplitudes, periods, and phase shifts, we can generate even more general trigonometric functions. For example, the graph of

$$f(t) = 4 \sin t + \cos 2t$$

is obtained by adding the value of $4 \sin t$ to that of $\cos 2t$ for each value of t. Since $4 \sin t$ has period 2π and $\cos 2t$ has period π, the graph of f will repeat every 2π units. Hence, as in graphing single trigonometric functions, we plot points on an interval of length at least one period, in this case $[0, 2\pi]$, and connect the points by a smooth curve (Figure 18). By repeating this basic cycle, we can get the entire curve.

t	$4 \sin t$	$\cos 2t$	$f(t)$
0	0	1	1
$\pi/4$	2.8	0	2.8
$\pi/2$	4	-1	3
$3\pi/4$	2.8	0	2.8
π	0	1	1
$5\pi/4$	-2.8	0	-2.8
$3\pi/2$	-4	-1	-5
$7\pi/4$	-2.8	0	-2.8
2π	0	1	1

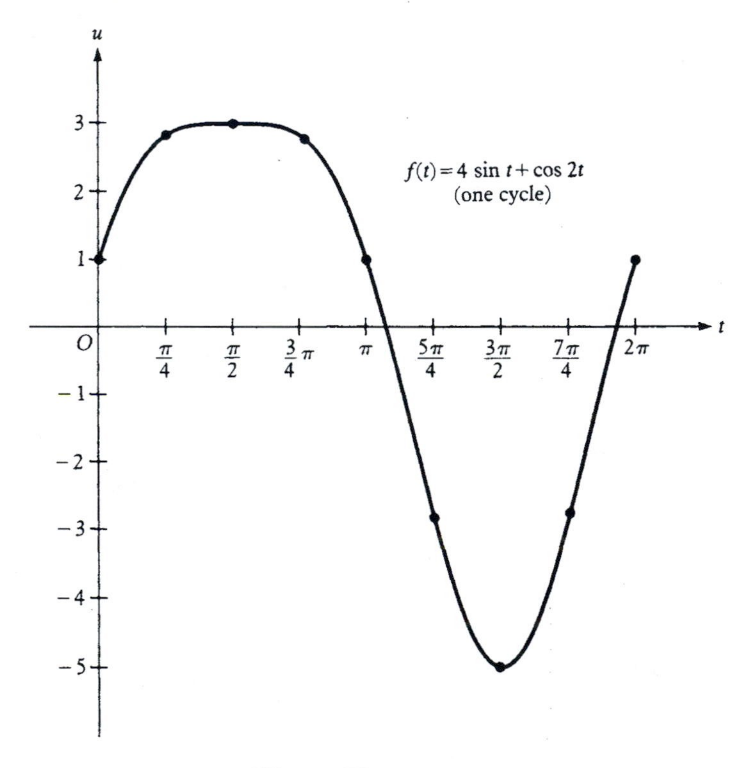

Figure 18

We can also obtain the graph of $f(t) = 4 \sin t + \cos 2t$ given in Figure 18 by first graphing $u_1 = 4 \sin t$ and $u_2 = \cos 2t$ on the same axes and then geometrically adding u_1 and u_2 for selected values of t (Figure 19). In the exercises, you are asked to sketch graphs by both methods.

In general, if $f(t)$ is composed of trigonometric functions with respective periods $p_1, p_2, \ldots, p_n$, then the period of $f(t)$ is at most the least common *integral* multiple of $p_1, p_2, \ldots, p_n$. For example, if $f(t)$ is an

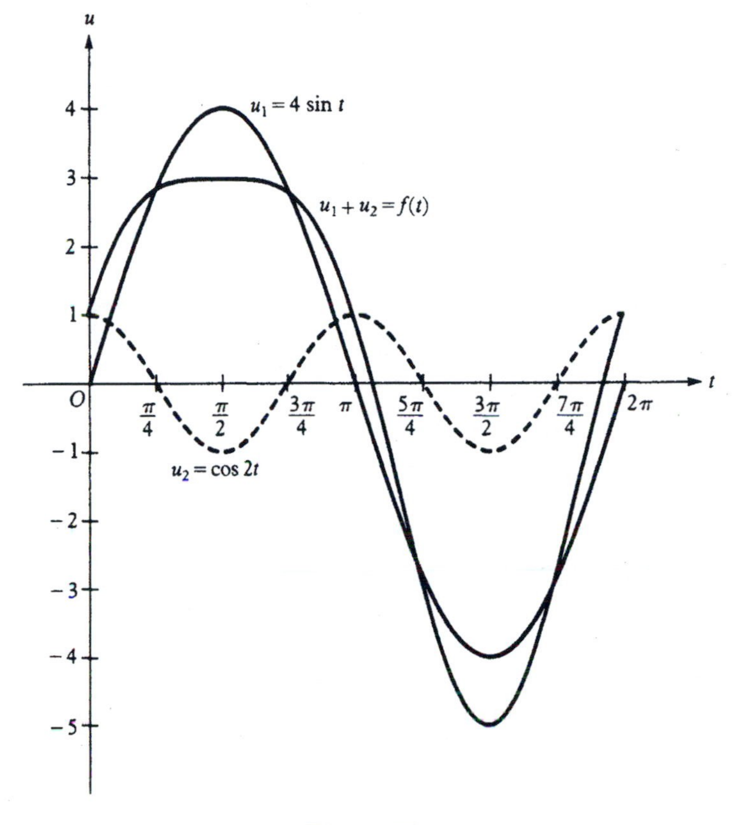

Figure 19

algebraic combination of the functions $\sin t$, $\cos 4t$, and $\cos (2t/3)$ with respective periods 2π, $\pi/2$, and 3π, then $f(t)$ has period at most 6π. Hence, we would graph $f(t)$ on $[0, 6\pi]$ and then repeat this cycle to the right and left in order to obtain the entire graph.

Exercises 7.3

In Exercises 1–10, assume $A \neq 0$, $B \neq 0$, and $C \neq 0$.

Fill in the blanks to make each statement true.

1. If $B > 0$, the function $f(t) = \sin Bt$ has period ________, and if $B < 0$, the period is ________.
2. The function $u = \cos Bt$ has domain ________ and range ________.
3. The function $u = 2 \cos Bt$ has domain ________ and range ________.
4. The phase shift of $u = A \sin (2t + C)$ is ________. The shift is to the left if ________ and to the right if ________.
5. An example of a sine function with amplitude 4, period π, and phase shift $\pi/6$ is ________.

Write true *or* false *for each statement.*

6. If f is a trigonometric function, then $f(Bt)$ and $f(Bt + C)$ have the same period.

7. If f is a trigonometric function, then $f(Bt)$ and $f(Bt + C)$ have the same domain.
8. If f is any basic trigonometric function, then $-f(t)$ has the same range as $f(t)$.
9. The function $f(t) = 3 \sin(-2t + \pi/4)$ has amplitude 3, period π, and phase shift $\pi/8$.
10. The function $f(t) = -3 \cos(2t - \pi/4)$ has amplitude -3, period π, and phase shift $\pi/8$.

Amplitude and Period

Determine the amplitude and period of each of the following, and sketch a graph on the interval $[0, 2\pi]$.

11. $u = 3 \sin t$
12. $u = -5 \cos t$
13. $u = \cos 2t$
14. $u = \sin 2t$
15. $u = -3 \sin 2t$
16. $u = -5 \cos 2t$

In each of the following, determine the amplitude and period, and sketch one cycle.

17. $u = 3 \sin(t/2)$
18. $u = -2 \cos(t/3)$
19. $u = \sin \pi t$
20. $u = \cos(\pi t/2)$
21. $u = -2 \cos(\pi t/4)$
22. $u = 4 \sin(\pi t/3)$

Determine the period of each of the following, and sketch a graph on $[0, \pi]$, *showing all vertical asymptotes.*

23. $u = 2 \tan t$
24. $u = \tan 2t$
25. $u = -\tan 2t$
26. $u = 3 \cot t$
27. $u = \cot 3t$
28. $u = \cot(t/3)$

Determine the period of each of the following, and sketch a graph on $[0, 2\pi]$, *showing all vertical asymptotes.*

29. $u = \sec 2t$
30. $u = \sec 3t$
31. $u = \frac{1}{2} \csc 2t$
32. $u = 2 \csc(t/2)$

Phase Shift

In each of the following, determine the period, amplitude, and phase shift, and graph one cycle.

33. $u = 4 \sin(2t + \pi)$
34. $u = -4 \sin(2t + \pi)$
35. $u = \sin(-2t + \pi)$
36. $u = 3 \cos\left(\frac{t}{2} + \frac{\pi}{2}\right)$
37. $u = -3 \cos\left(-\frac{t}{2} + \frac{\pi}{2}\right)$
38. $u = 2 \sin\left(\pi t + \frac{\pi}{2}\right)$
39. $u = -2 \sin\left(\frac{\pi}{2}t + \frac{\pi}{4}\right)$
40. $u = \cos\left(\pi t - \frac{\pi}{4}\right)$

In each of the following, determine the period and phase shift, and graph one cycle.

41. $u = \tan\left(2t + \frac{\pi}{2}\right)$
42. $u = -\tan\left(2t - \frac{\pi}{2}\right)$
43. $u = -\cot\left(t + \frac{\pi}{4}\right)$
44. $u = -\cot\left(t - \frac{\pi}{2}\right)$
45. $u = 2 \sec\left(t - \frac{\pi}{2}\right)$
46. $u = -\frac{1}{2} \csc\left(t + \frac{\pi}{2}\right)$

Sums of Trigonometric Functions

In each of the following, determine the period and graph one cycle.

47. $f(t) = \sin t + \cos t$
48. $f(t) = \sin t - \cos t$
49. $f(t) = 4 \cos t + \cos 2t$
50. $f(t) = 4 \sin t - \cos 2t$

51. Obtain the graph in Exercise 47 by first graphing $u_1 = \sin t$ and $u_2 = \cos t$ and then adding ordinates for selected values of t.
52. Obtain the graph in Exercise 48 by first graphing $u_1 = \sin t$ and $u_2 = -\cos t$ and then adding ordinates for selected values of t.
53. Obtain the graph in Exercise 49 by first graphing $u_1 = 4 \cos t$ and $u_2 = \cos 2t$ and then adding ordinates.
54. Obtain the graph in Exercise 50 by first graphing $u_1 = 4 \sin t$ and $u_2 = -\cos 2t$ and then adding ordinates.

55. Suppose that $f(t)$ is an algebraic combination of n trigonometric functions whose respective periods are $k_1\pi, k_2\pi, \ldots, k_n\pi$, where the k_i are rational numbers p_i/q_i $(i = 1, 2, \ldots, n)$. Show that the period of $f(t)$ is at most $k\pi$, where k is the least common multiple of $p_1, p_2, \ldots, p_n$ divided by the greatest common divisor of $q_1, q_2, \ldots, q_n$.

56. Apply the result of Exercise 55 to determine the maximum value of the period in each of the following cases.

(a) $k_1 = \frac{1}{3}, \quad k_2 = \frac{5}{6}$

(b) $k_1 = \frac{1}{2}, \quad k_2 = \frac{3}{10}, \quad k_3 = \frac{5}{4}$

(c) $k_1 = \frac{9}{10}, \quad k_2 = \frac{12}{5}, \quad k_3 = \frac{4}{15}$

7.4 Applications of Trigonometric Functions

When it comes to applications of the trigonometric functions, the sine and cosine do all the work.

Motion of a Pendulum
Motion of a Vibrating Spring
Electrical Circuits

Many of the applications of trigonometric functions deal with quantities that oscillate periodically. Some examples are a simple pendulum, a vibrating spring, sound waves, electric circuits, radio waves, tides, and human heartbeats. For all of these, the mathematical model is given by either of the functions

$$u = A \sin Bt \qquad \text{or} \qquad u = A \cos Bt \tag{1}$$

or modifications of the form

$$u = A \sin (Bt + C) \qquad \text{or} \qquad u = A \cos (Bt + C) \tag{2}$$

or sums of such functions. Here, A, B, and C are constants, and t is time. The phenomenon described by any one of the equations in (1) or (2) is called **simple harmonic motion.**

Motion of a Pendulum. A familiar example of simple harmonic motion is a **simple pendulum.** Consider a mass hanging at the end of a string L feet long, and let S be the arc through which the mass travels when displaced from the rest position as indicated in Figure 20. If the mass is displaced an arc of S_0 feet at time $t = 0$ and then released, the pendulum will swing back and forth according to the equation

$$S(t) = S_0 \cos \omega t, \tag{3}$$

where $\omega = \sqrt{g/L}$, and $g = 32$ ft/sec^2 is the acceleration due to gravity.

Equation (3) accurately describes the motion of the pendulum, provided S_0 is not too large in comparison to L. A value of $S_0 \leq L/4$ is permissible. $S(t)$ is the directed length of the arc subtended from the rest position, which is vertical. Counterclockwise arcs are positive and clockwise arcs are negative.

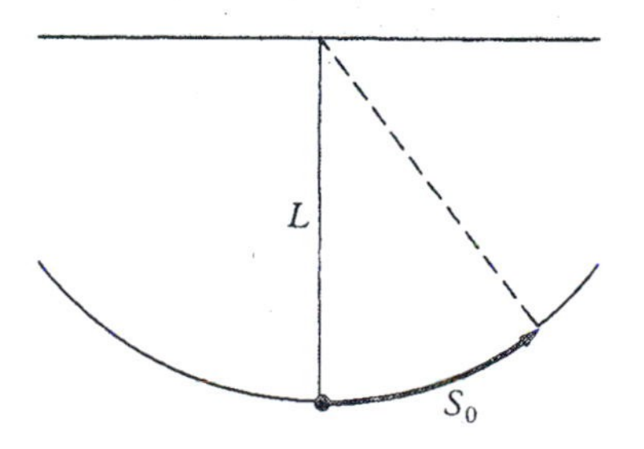

Figure 20 Simple Pendulum

We know from Section 7.3 that the period T of the trigonometric function S given by (3) is

$$T = \frac{2\pi}{\omega} = 2\pi\sqrt{\frac{L}{g}}. \tag{4}$$

Example 1 If $L = 1.5$ feet and $S_0 = 3$ inches in the pendulum in Figure 20, find each of the following.

(a) the period T

(b) the equation for $S(t)$

(c) the first time that the pendulum returns to the vertical position

Solution: Our units for length and time will be feet and seconds, respectively.

(a) From equation (4), we have

$$T = 2\pi\sqrt{\frac{1.5}{32}} \approx 1.36 \text{ (seconds/cycle)}.$$

(b) S_0 is given as 3 inches, but we must express all lengths in terms of feet. Therefore, we write $S_0 = .25$ foot. Also, the coefficient of t in equation (3) for $S(t)$ is

$$\omega = \sqrt{\frac{g}{L}} = \sqrt{\frac{32}{1.5}} \approx 4.6.$$

Therefore, $$S(t) = .25 \cos 4.6t.$$

(c) When the pendulum is in the vertical position, its displacement $S(t)$ is zero. Hence, we solve for the smallest positive value of t that satisfies the equation

$$.25 \cos 4.6t = 0.$$

Since the smallest positive value of x satisfying $\cos x = 0$ is $\pi/2$, we conclude that $4.6t = \pi/2$ or $t = \pi/[2(4.6)] \approx .34$ seconds. ■

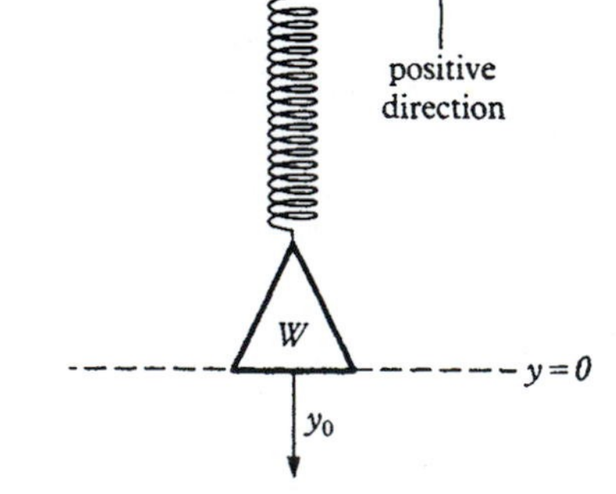

Figure 21 Vibrating Spring

Motion of a Vibrating Spring. If a weight W is suspended from the end of a spring, it will stretch the spring downward to an equilibrium position as indicated in Figure 21. The equilibrium position of W is at 0 on a vertical y-axis, and the positive direction is upward. If W is displaced vertically by y_0 units at time $t = 0$ and then released, it will bob up and down according to the equation

$$y = y_0 \cos \omega t, \tag{5}$$

where

$$\omega = \sqrt{\frac{gk}{W}},$$

g being the acceleration due to gravity and k the so-called "spring constant." The value of k is a measure of the restoring force of the spring. It varies from spring to spring, but each spring has a specific constant associated with it. The derivation of (5) takes into account the force of gravity and the restoring force of the spring, but not air resistance, and therefore is most accurate in a vacuum.

Example 2 A 12-pound weight hangs on the end of a spring whose spring constant is 24 pounds per foot. The spring is stretched 4 inches downward and released.

(a) Find the equation of motion of the weight.

(b) Find the amplitude and period of its motion.

(c) Sketch a graph representing the motion.

Solution: We will express our units in terms of feet, pounds, and seconds.

(a) From the given information, we have $y_0 = -1/3$ feet and

$$\omega = \sqrt{\frac{32 \cdot 24}{12}} = 8.$$

Therefore, the equation of motion is

$$y = -\frac{1}{3}\cos 8t.$$

(b) From Section 7.3, we know that the amplitude is 1/3 foot and the period is

$$T = \frac{2\pi}{8} = \frac{\pi}{4} \text{ (seconds/cycle).}$$

(c) The graph of $y = -1/3 \cos 8t$ is shown in Figure 22. Be sure to interpret the graph properly. The motion of the weight is not *along* the graph; it is up and down. The y-value of a point (t, y) on the graph shows the vertical displacement of the weight from the equilibrium position at time t. Positive values of y correspond to a compression of the spring, and negative values to a stretch of the spring. Note that the graph indicates that the weight will continue to oscillate indefinitely

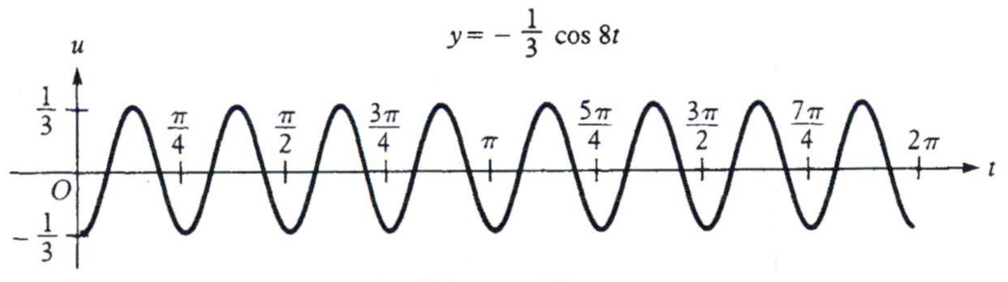

Figure 22

with an amplitude of 1/3 foot; in other words, we have a "perpetual motion machine." The motion of the pendulum in Example 1 leads to a similar graph. We know from experience that the oscillations will actually decrease and eventually die out. If we had taken air resistance into account, then the equation of motion would show decreasing oscillations. ■

If the weight W in Figure 21 is not only displaced vertically by y_0 units but is also given an initial velocity v_0 at time 0, then the equation of motion (again neglecting air resistance) is

$$y = A \cos(\omega t - \phi), \qquad (6)$$

where

$$\omega = \sqrt{\frac{gk}{W}}, \quad A = y_0\sqrt{1 + \left(\frac{v_0}{y_0\omega}\right)^2},$$

and

$$\tan\phi = \frac{v_0}{y_0\omega} \quad \left(-\frac{\pi}{2} < \phi < \frac{\pi}{2}\right).$$

In the terminology of Section 7.3, $|A|$ is the amplitude, $2\pi/\omega$ is the period, and ϕ/ω is the phase shift.

Example 3 Solve Example 2 if the spring is stretched 4 inches downward and also given an initial velocity of 2 ft/sec upward.

Solution:

(a) As in Example 2, we have $\omega = 8$ and $y_0 = -1/3$, but now $v_0 = 2$. (The sign of v_0 is positive because the initial velocity is upward, in the positive direction.) The other constants in (6) are

$$A = -\frac{1}{3}\sqrt{1 + \left[\frac{2(-3)}{8}\right]^2} = -\frac{5}{12},$$

and

$$\tan\phi = \frac{2(-3)}{8} = -\frac{3}{4},$$

which implies $\phi = \approx -.64$ (radians). Hence,

$$y = -\frac{5}{12}\cos(8t + .64).$$

(b) The amplitude of y is 5/12, the period is $2\pi/8 = \pi/4$, and the phase shift is $-.64/8 = -.08$. Since the phase shift is negative, the shift of the cosine curve is to the left.

(c) The graph of y is given in Figure 23. In comparing Figures 22 and 23, we see that in both cases the weight starts at $y = -1/3$ foot and oscillates with a period of $\pi/4$ seconds. But Figure 23 shows that the initial push of 2 feet per second upward results in a larger amplitude. Also, the weight first returns to its equilibrium position faster. That is, in Figure 22 the first return to equilibrium $y = 0$ is in $\pi/16$ seconds, and in Figure 23 the first return to 0 is in $\pi/16 - .08$ seconds. ■

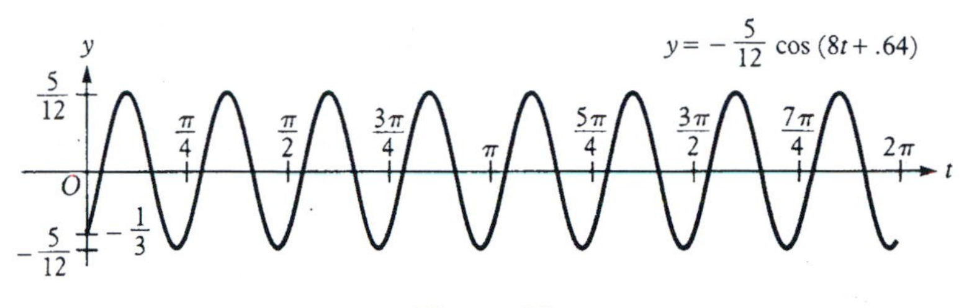

Figure 23

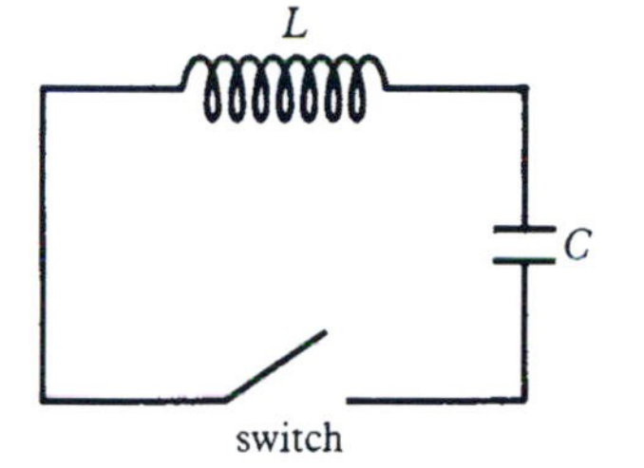

Figure 24 *LC*-Circuit

Electrical Circuits. An LC-circuit consisting of an inductor (coil of wire) with inductance L henries and a capacitor (device that stores an electrical charge) with capacitance C farads is shown in Figure 24. If the charge on the capacitor is Q_0 coulombs at time $t = 0$ when the switch is closed, then the charge $Q(t)$ at time t is

$$Q(t) = Q_0 \cos \omega t, \tag{7}$$

where $\omega = 1/\sqrt{LC}$.

Example 4 Find the charge on a capacitor in an LC-circuit 10 seconds after the switch is closed, given $L = 1$ henry, $C = .0001$ farads, and $Q_0 = 10^{-4}$ coulombs.

Solution: The value of ω in (7) is

$$\omega = 1/\sqrt{1(.0001)} = 100.$$

Therefore, the equation for $Q(t)$ is

$$Q(t) = 10^{-4} \cos 100t,$$

and in 10 seconds we have

$$\begin{aligned} Q(10) &= 10^{-4} \cos 1000 \\ &= 10^{-4}(.5624) \\ &= 5.624 \times 10^{-5} \text{ coulombs.} \end{aligned}$$

For some calculators, 1000 radians is beyond the input range for the cosine. In such cases, 1000 can be replaced by $1000 - 318\pi \approx .9735$ radians. ■

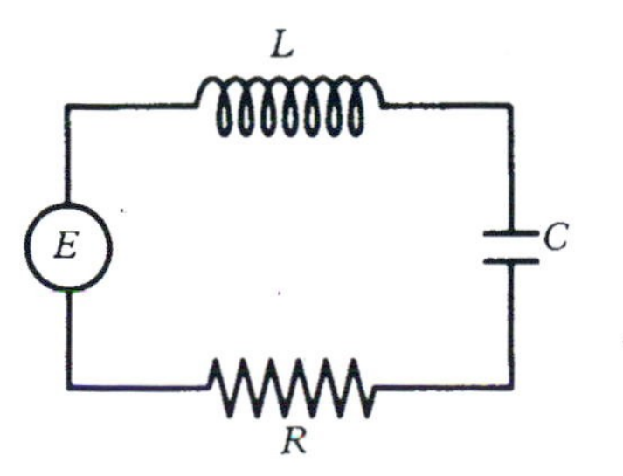

Figure 25 *RLC*-Circuit

The LC-circuit in Example 4 does not take circuit resistance into account, and the result is another "perpetual motion machine" as in Examples 1, 2, and 3. Suppose a resistance of R ohms is placed in the circuit and also a constant electromotive force of E volts, as indicated in Figure 25. We assume that initially the charge on the capacitor is 0 and that the initial current (rate of change of the charge) is also 0. Then at time t a current of $I(t)$ amperes flows in the circuit, and

$$I(t) = Ae^{-bt} \sin \omega t, \tag{8}$$

where $A = E/(L\omega)$, $b = R/(2L)$, and

$$\omega = \frac{\sqrt{\frac{4L}{C} - R^2}}{2L} \qquad \left(\frac{4L}{C} > R^2 \text{ by assumption}\right).$$

Of course, $I(t)$ is not strictly a trigonometric function because of the so called "damping" factor

$$e^{-[R/(2L)]t},$$

whose effect is to compress the amplitude of the sine cycles in the graph of $I(t)$ (Figure 26). The equations in Examples 1, 2, and 3 would be similarly affected if we took into account air resistance.

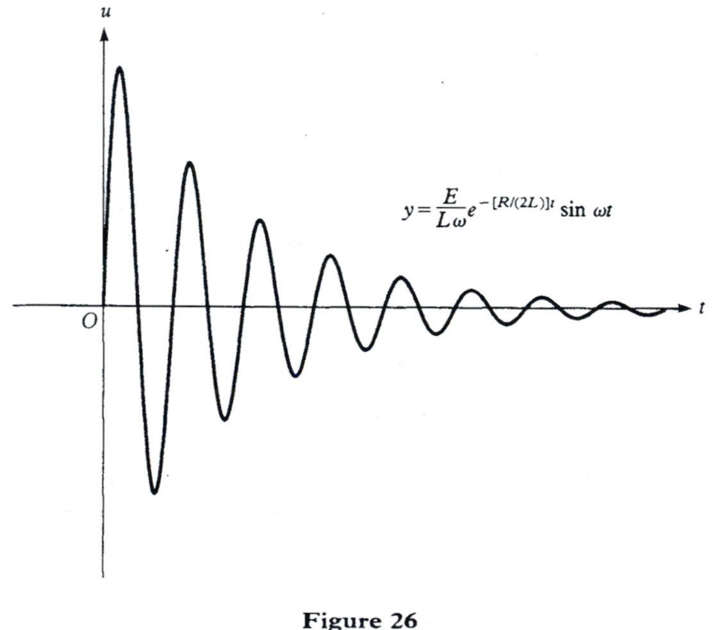

Figure 26

Exercises 7.4

Motion of a Pendulum

If a pendulum has length L feet and initial position S_0 feet as in Figure 20, find its equation of motion and its period in each of the following cases.

1. $L = 1$ ft, $S_0 = .25$ ft
2. $L = 2$ ft, $S_0 = 2$ in
3. $L = .5$ ft, $S_0 = -.125$ ft
4. How must the length of the pendulum in Exercise 2 be changed in order to double the period?
5. The frequency (cycles per second) of a simple pendulum is defined to be $1/T$, where T is the period (seconds per cycle). How must the length of the pendulum be changed in Exercise 1 in order to double the frequency?

Motion of a Vibrating String

A weight W hangs on the end of a spring with spring constant k. The weight is displaced y_0 feet from its equilibrium position $y = 0$ and given an initial velocity of v_0 ft/sec. Find its equaiton of motion, amplitude, and period in each of the following cases. (The positive y direction is upward.)

6. $W = 3$ lb, $k = 6$, $y_0 = .25$ ft, $v_0 = 0$

7. $W = 3$ lb, $k = 6$, $y_0 = .25$ ft, $v_0 = 1$ ft/sec

8. $W = 9$ lb, $k = 54$, $y_0 = .5$ ft, $v_0 = -1$ ft/sec

9. For what value of t in Exercise 7 does the spring first return to the equilibrium position?

10. Graph the equation of motion in Exercise 8.

Electrical Circuits

11. For the LC circuit described by equation (7), graph the curve for $Q(t)$, given $Q_0 = 10^{-2}$ and $\omega = 6$. Construct the graph for t in the interval $[0, 2\pi]$.

12. For the LC circuit described by equation (7), graph the curve for $Q(t)$, given $Q_0 = 10^{-1}$ and $\omega = 1/2$.

13. For the RLC circuit described by equation (8), graph the curve for $I(t)$, given $A = 3$, $b = 1$, and $\omega = 2$.

Miscellaneous

14. If a point $P(x, y)$ moves around a circle of radius r with constant angular velocity ω radians per second, as indicated in the figure, show that after t seconds, the following are true.

(a) $x = r \cos \omega t$ and $y = r \sin \omega t$, given that the point is initially at $(r, 0)$.

(b) $x = r \cos (\omega t + \theta_0)$ and $y = r \sin (\omega t + \theta_0)$, given that the point is initially at $(r \cos \theta_0, r \sin \theta_0)$.

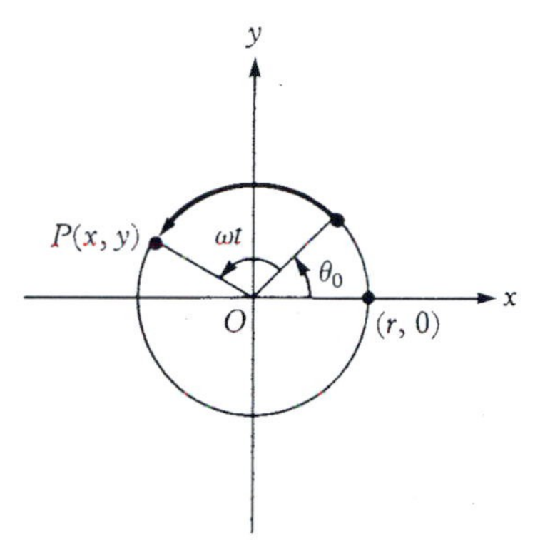

15. A wheel of radius 1 foot is rotating at 2 revolutions per second. If a coordinate system like that in Exercise 14 is used, and the initial location of a point P on the wheel is $\left(\cos \frac{\pi}{4}, \sin \frac{\pi}{4}\right)$, find each of the following.

(a) the coordinates of P at time t

(b) the coordinates of P at time $t = .25$ seconds

(c) all values of t for which the wheel returns to its starting position

7.5 Inverse Trigonometric Functions

The six inverse trigonometric functions are not inverses of the six basic trigonometric functions!

The Inverse Sine Function
The Inverse Tangent Function
Other Inverse Trigonometric Functions

We learned in Section 4.5 that a function has an inverse if and only if the function is one-to-one. Furthermore, a function f and its inverse f^{-1} have graphs that, taken together, are symmetric with respect to the line $y = x$ and also satisfy

$$f^{-1}(f(x)) = x \qquad \text{and} \qquad f(f^{-1}(x)) = x.$$

A function that is not one-to-one may become one-to-one if its domain is restricted to a smaller set. For example, $f(x) = x^2$ is not one-to-one on the domain $(-\infty, \infty)$ and therefore has no inverse. However, if the domain is restricted to $[0, \infty)$, then f is one-to-one and its inverse is $f^{-1}(x) = \sqrt{x}$. We now apply these ideas to the trigonometric functions. We will return to the usual convention of letting x and y, for the most part, stand for the independent and dependent variables, respectively.

The Inverse Sine Function. The function $f(x) = \sin x$ with domain $(-\infty, \infty)$ is not one-to-one. For example, the horizontal line $y = 1/2$ intersects the graph of $y = \sin x$ in an *infinite* number of points (see Figure 27), including points for $x = \pi/6, 5\pi/6$, and $-7\pi/6$.

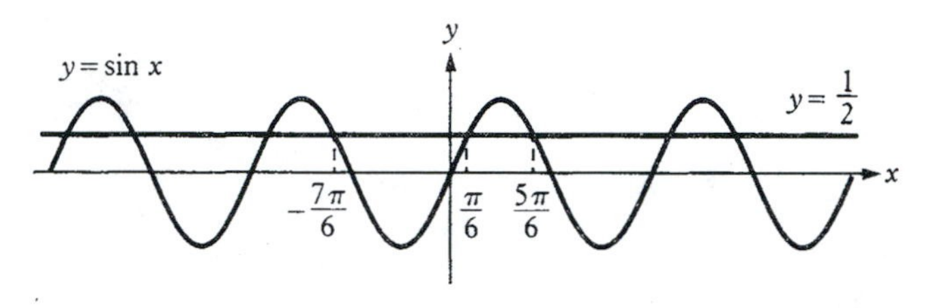

Figure 27

However, $y = \sin x$ with domain $[-\pi/2, \pi/2]$ is an increasing one-to-one function that takes on every value in the range of the sine function, that is, every value from -1 to 1 [Figure 28(a)]. Hence, the **restricted sine function** defined by

$$y = \sin x, \quad \text{with } -\frac{\pi}{2} \le x \le \frac{\pi}{2} \text{ and } -1 \le y \le 1$$

has an **inverse** defined by

$$y = \sin^{-1} x, \quad \text{with } -1 \le x \le 1 \text{ and } -\frac{\pi}{2} \le y \le \frac{\pi}{2},$$

whose graph is the reflection of the restricted $y = \sin x$ in the line $y = x$ [Figure 28(b)].

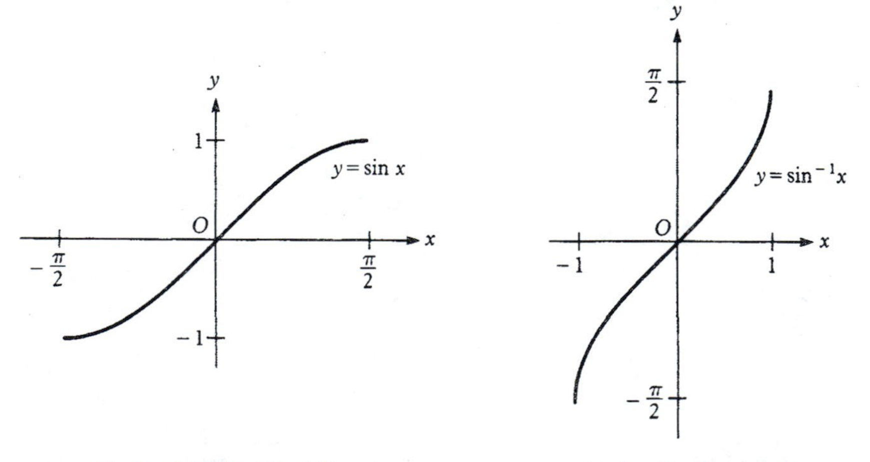

(a) Graph of Restricted Sine

(b) Graph of Inverse Sine

Figure 28

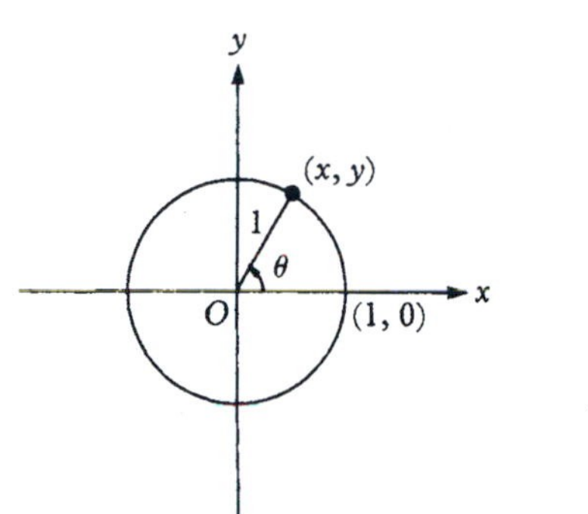

Figure 29

The restricted $y = \sin x$ and its inverse $y = \sin^{-1} x$ satisfy

$$\sin(\sin^{-1} x) = x \text{ for } -1 \leq x \leq 1$$

$$\text{and} \qquad \sin^{-1}(\sin x) = x \text{ for } -\frac{\pi}{2} \leq x \leq \frac{\pi}{2}. \tag{1}$$

Equivalently, in angle notation,

$$\theta = \sin^{-1} y \Leftrightarrow \sin\theta = y, \quad \text{where } -1 \leq y \leq 1 \text{ and } -\frac{\pi}{2} \leq \theta \leq \frac{\pi}{2}. \tag{2}$$

The notation

$$\theta = \text{arc sin } y$$

is also used for the inverse sine function since on the unit circle, θ is the arc or angle in radians whose sine is y (Figure 29).

Example 1 Find the exact value of each of the following.

(a) $\sin^{-1}(1/2)$ (b) arc sin $(-\sqrt{3}/2)$

Solution:

(a) If we let $\theta = \sin^{-1}(1/2)$, then $\sin\theta = 1/2$ with θ in the interval $[-\pi/2, \pi/2]$. The only solution for θ in this interval is $\pi/6$. Hence, $\pi/6 = \sin^{-1}(1/2)$.

(b) The equation $\theta = \text{arc sin}(-\sqrt{3}/2)$ is equivalent to $\sin\theta = -\sqrt{3}/2$ with θ restricted to the interval from $-\pi/2$ to $\pi/2$. The only solution for θ in this interval is $-\pi/3$. Therefore, $-\pi/3 = \text{arc sin}(-\sqrt{3}/2)$. ■

Example 2 Find all values of x that satisfy $\sin x = 1/2$.

Solution: With reference to Figure 27, we see that there are two values of x in the interval $[0, 2\pi]$ satisfying $\sin x = 1/2$, namely

$$x = \frac{\pi}{6} \quad \text{and} \quad x = \frac{5\pi}{6}.$$

Therefore, because $\sin x$ has period 2π, it follows that all solutions of $\sin x = 1/2$ are

$$\frac{\pi}{6} + 2n\pi \quad \text{and} \quad \frac{5\pi}{6} + 2n\pi,$$

where n is any integer. ■

Comment As we have just seen in Example 2, there are an infinite number of x's for which $\sin x = 1/2$. However, only one of these, namely $\pi/6$, is in the interval $[-\pi/2, \pi/2]$. This is the one that is denoted by

$\sin^{-1}(1/2)$ [Example 1(a)], and to distinguish $\pi/6$ from all the others, we call $\pi/6$ the principal value of x satisfying $\sin x = 1/2$. In general, the value of x in the interval $[-\pi/2, \pi/2]$ that satisfies $\sin x = y$ is called the **principal value** solution of the equation $\sin x = y$. The principal value is the one given by a calculator.

Example 3 Compute each of the following.

(a) $\sin\left(\sin^{-1}\frac{\sqrt{2}}{2}\right)$ (b) $\sin^{-1}\left(\sin\frac{\pi}{6}\right)$ (c) $\sin^{-1}\left(\sin\frac{5\pi}{3}\right)$

Solution:

(a) Since $\sqrt{2}/2$ falls between -1 and 1, it follows from (1) that $\sin\left(\sin^{-1}(\sqrt{2}/2)\right) = \sqrt{2}/2$.

(b) Since $\pi/6$ falls between $-\pi/2$ and $\pi/2$, it follows from (1) that $\sin^{-1}\left(\sin(\pi/6)\right) = \pi/6$.

(c) Since $5\pi/3$ does not fall between $-\pi/2$ and $\pi/2$, we cannot apply (1) here. We proceed as follows. First, $\sin(5\pi/3) = -\sqrt{3}/2$. Now, $\sin^{-1}(-\sqrt{3}/2) = -\pi/3$ [Example 1(b)]. Therefore, $\sin^{-1}(\sin(5\pi/3)) = -\pi/3$. ■

Example 4 Use a scientific calculator to approximate each of the following to four decimal places.

(a) $\sin^{-1} .9407$ (b) $\sin^{-1}(-.3393)$

Solution:

As desired, a scientific calculator will give the angle between $-\pi/2$ and $\pi/2$ whose sine has the given value.

(a) First put the calculator in the radian mode. Then press the 2nd or INV key followed by the SIN key, and enter .9407 to obtain $\sin^{-1} .9407 = 1.2247$ to four decimal places.

(b) Following the same procedure as in Part (a), we obtain $\sin^{-1}(-.3393) = -.3462$.

■

The Inverse Tangent Function. The function $f(x) = \tan x$ is not one-to-one on its entire domain. However, if we restrict the domain to $(-\pi/2, \pi/2)$, the result is an increasing one-to-one function that assumes the full range of values of the tangent, namely all real numbers [Figure 30(a)]. Hence, the **restricted tangent function** defined by

$$y = \tan x, \quad \text{with } -\frac{\pi}{2} < x < \frac{\pi}{2} \text{ and } -\infty < y < \infty$$

has an **inverse**

$$y = \tan^{-1} x, \quad \text{with } -\infty < x < \infty \text{ and } -\frac{\pi}{2} < y < \frac{\pi}{2},$$

whose graph is the reflection of the restricted $y = \tan x$ in the line $y = x$ [Figure 30(b)].

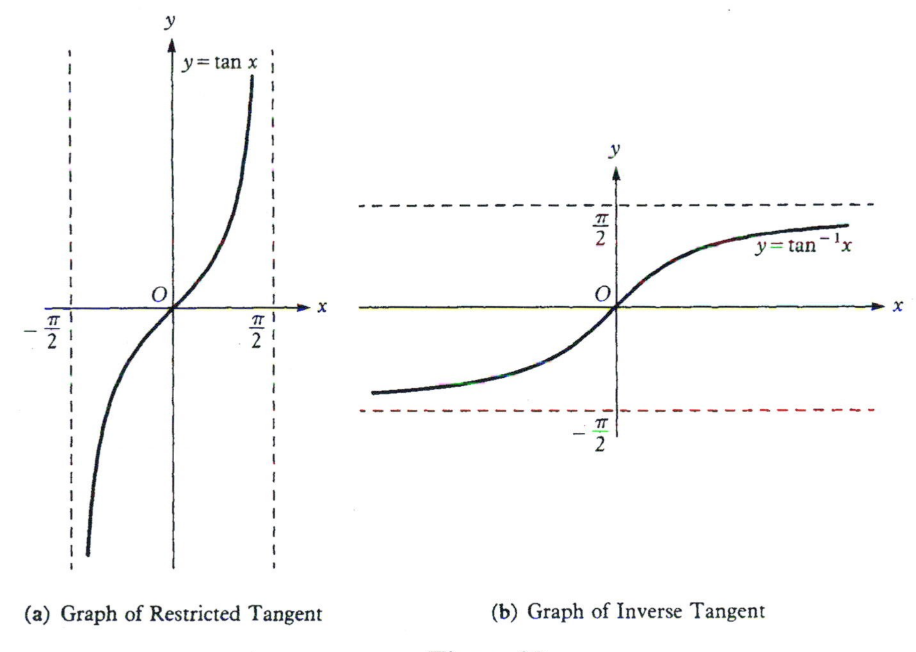

(a) Graph of Restricted Tangent

(b) Graph of Inverse Tangent

Figure 30

We have

$$\tan(\tan^{-1} x) = x \quad \text{for } -\infty < x < \infty$$

and

$$\tan^{-1}(\tan x) = x \quad \text{for } -\frac{\pi}{2} < x < \frac{\pi}{2} \tag{3}$$

which is equivalent to

$$\theta = \tan^{-1} y \Leftrightarrow \tan \theta = y, \quad \text{where } -\infty < y < \infty \text{ and } -\frac{\pi}{2} < \theta < \frac{\pi}{2}. \tag{4}$$

Also, as with the inverse sine, the notation

$$\theta = \text{arc tan } y$$

is sometimes used for the inverse tangent. We read this notation as "θ is the arc, or the angle, whose tangent is y."

Example 5 Find the exact value of each of the following.

(a) $\tan^{-1} \sqrt{3}$ (b) $\tan^{-1}(-1)$

Solution:

(a) By (4), the equation $\theta = \tan^{-1} \sqrt{3}$ is equivalent to $\tan \theta = \sqrt{3}$ with θ between $-\pi/2$ and $\pi/2$. Since $\tan \pi/3 = \sqrt{3}$, we conclude that $\pi/3 = \tan^{-1} \sqrt{3}$.

(b) Since $\tan(-\pi/4) = -1$, it follows from (4) that $-\pi/4 = \tan^{-1}(-1)$. ■

Example 6 Compute each of the following without a calculator.

(a) $\sin\left(\tan^{-1} \frac{3}{4}\right)$ (b) $\tan^{-1}\left(\sin \frac{3\pi}{2}\right)$

Solution:

(a) If we interpret $\tan^{-1}$ as arc tan, then the problem is to compute the sine of the angle whose tangent is 3/4. Let θ denote this angle. Then, since 3/4 is positive, θ is an acute angle in a right triangle as shown in Figure 31. The side opposite θ is 3, the adjacent side is 4, and the hypotenuse is 5 (*why?*). Therefore,

$$\sin \theta = \frac{3}{5} = \sin\left(\tan^{-1} \frac{3}{4}\right).$$

(b) Since $\sin 3\pi/2 = -1$, the problem is to compute $\tan^{-1}(-1)$. As we saw in Example 5, $\tan^{-1}(-1) = -\pi/4$. Hence, $\tan^{-1}(\sin(3\pi/2)) = -\pi/4$. ■

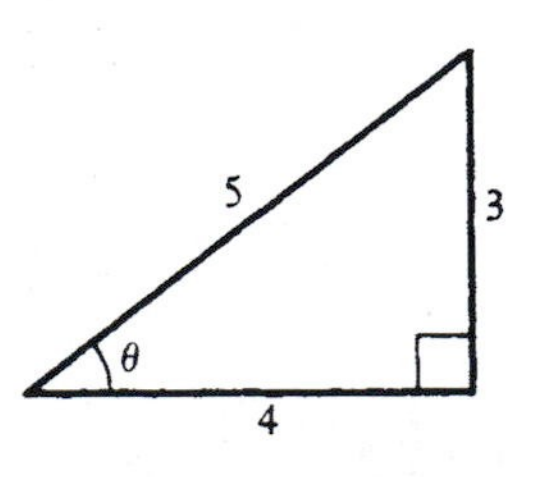

Figure 31

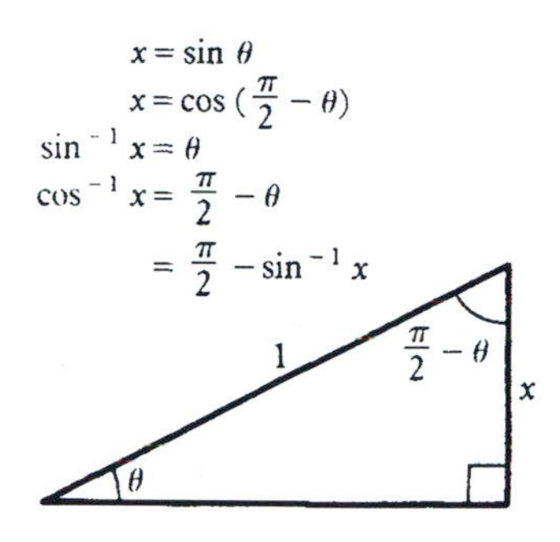

Figure 32

Other Inverse Trigonometric Functions. We now use the cofunction property to define $\cos^{-1} x$ and $\cot^{-1} x$. Figure 32 shows a right triangle in which $\sin \theta = x$ and $\cos(\pi/2 - \theta) = x$. That is, θ is the angle whose sine is x, and $\pi/2 - \theta$ is the angle whose cosine is x.

In general, we define the **inverse cosine** as

$$\cos^{-1} x = \frac{\pi}{2} - \sin^{-1} x.$$

The domain of $\cos^{-1} x$ is the same as the domain of $\sin^{-1} x$, namely $[-1, 1]$. Furthermore, as $\sin^{-1} x$ varies from $-\pi/2$ to $\pi/2$, $\cos^{-1} x$ ranges from 0 to π. Hence, the inverse cosine

$$y = \cos^{-1} x, \text{ with } -1 \leq x \leq 1 \text{ and } 0 \leq y \leq \pi,$$

is in fact the inverse of the **restricted cosine function** defined by

$$y = \cos x, \quad \text{with } 0 \leq x \leq \pi \text{ and } -1 \leq y \leq 1.$$

The corresponding graphs are shown in Figure 33.

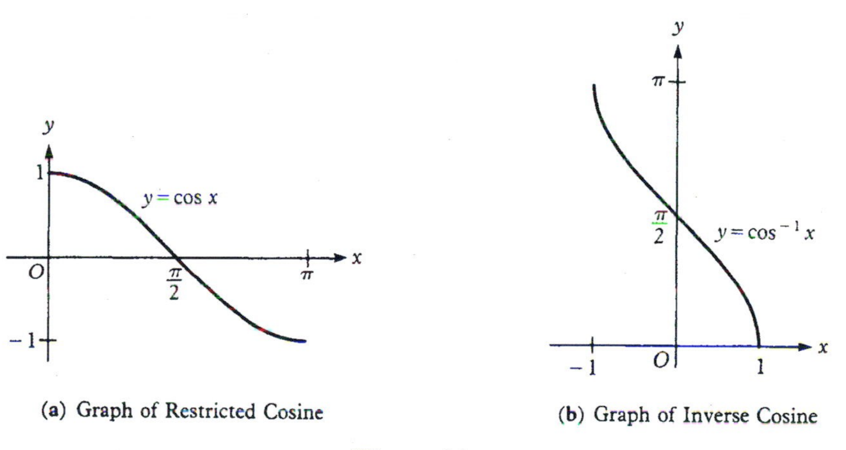

(a) Graph of Restricted Cosine

(b) Graph of Inverse Cosine

Figure 33

Example 7 Find the exact value of each of the following.

(a) $\cos^{-1} 1$

(b) $\cos^{-1}(-1/2)$

Solution:

(a) By definition, $\cos^{-1} 1 = \pi/2 - \sin^{-1} 1 = \pi/2 - \pi/2 = 0$. We can also obtain the result directly from the graph of $y = \cos^{-1} x$ or from the inverse function property $\theta = \cos^{-1} 1 \Leftrightarrow \cos \theta = 1$, where $0 \leq \theta \leq \pi$.

(b) From the inverse function property, $\theta = \cos^{-1}(-1/2)$ means $\cos \theta = -1/2$, where θ is in the interval $[0, \pi]$. The value $\theta = 2\pi/3$ has this property. Therefore, $\cos^{-1}(-1/2) = 2\pi/3$. We could also obtain this result either from the definition of $y = \cos^{-1} x$ or from its graph. ■

Example 8 Find the value of each of the following without a calculator.

(a) $\sin(\cos^{-1}(1/3))$ (b) $\sin(\cos^{-1}(-1/3))$

Solution:

(a) As in Example 6(a), we can use the arc interpretation of the inverse. That is, if $\theta = \cos^{-1}(1/3) = \text{arc cos}(1/3)$, then θ is an acute angle of a right triangle in which the side adjacent to θ is 1 and the hypotenuse is 3 [Figure 34(a)]. Therefore, the side opposite θ is $\sqrt{3^2 - 1^2} = \sqrt{8} = 2\sqrt{2}$, and

$$\sin\theta = \frac{2\sqrt{2}}{3}.$$

That is,

$$\sin\left(\cos^{-1}\frac{1}{3}\right) = \frac{2\sqrt{2}}{3}.$$

(b) If $\theta = \cos^{-1}(-1/3) = \text{arc cos}(-1/3)$, then θ is an angle between $\pi/2$ and π (*why?*) whose cosine is $-1/3$. With θ in standard position, the terminal side of θ lies in the second quadrant. As shown in Figure 34(b), we obtain a right triangle in which $x = -1$, $r = 3$, and $y = \sqrt{3^2 - (-1)^2} = 2\sqrt{2}$. Therefore,

$$\sin\left(\cos^{-1}\left(-\frac{1}{3}\right)\right) = \sin\theta = \frac{2\sqrt{2}}{3}. \quad ∎$$

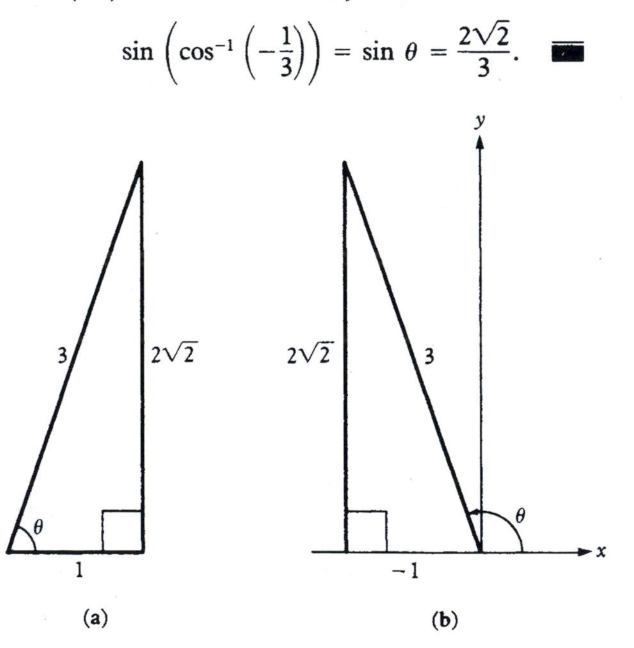

Figure 34

Comment Example 8 can be generalized to

$$\sin(\cos^{-1}x) = \sqrt{1 - x^2}$$

for all real numbers x in the interval $[-1, 1]$ (see Exercise 55). Similarly,

$$\cos(\sin^{-1} x) = \sqrt{1 - x^2}$$

for all x in the same interval (see Exercise 56). These facts are useful in calculus.

Question *Can an inverse cosine function be obtained by restricting the domain of the cosine to $[-\pi/2, \pi/2]$ as we did with the sine?*

Answer: No. As indicated in Figure 7, the cosine with its domain restricted to $[-\pi/2, \pi/2]$ is not one-to-one and its range is only $[0, 1]$. ■

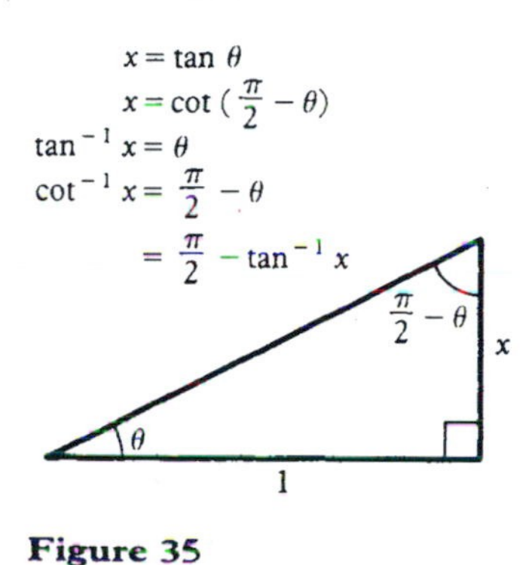

Figure 35

We now define the inverse cotangent. The right triangle in Figure 35 illustrates the cofunction property of the tangent and cotangent. That is, $\tan\theta = x$ and $\cot(\pi/2 - \theta) = x$.

Hence, similar to the inverse cosine, we define the **inverse cotangent** as

$$\cot^{-1} x = \frac{\pi}{2} - \tan^{-1} x.$$

The domain of $\cot^{-1} x$ is $(-\infty, \infty)$, and the range is $(0, \pi)$ (*why?*). The inverse cotangent

$$y = \cot^{-1} x, \quad \text{with } -\infty < x < \infty \text{ and } 0 < y < \pi,$$

is therefore the inverse of the **restricted cotangent function** defined by

$$y = \cot x, \quad \text{with } 0 < x < \pi \text{ and } -\infty < y < \infty,$$

and the corresponding graphs are shown in Figure 36.

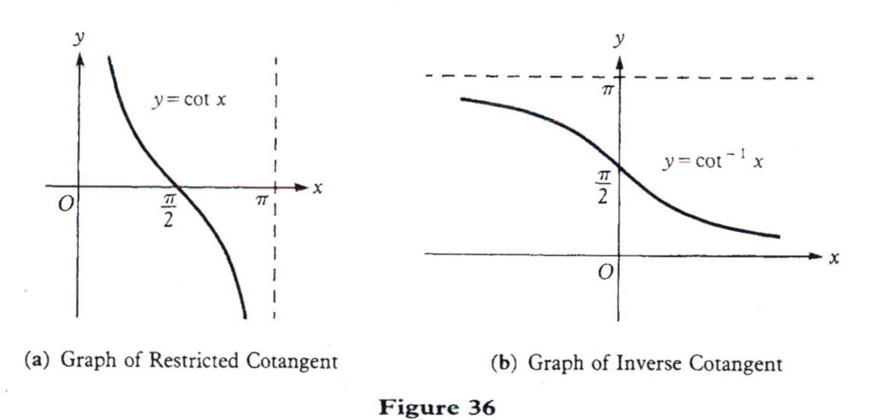

(a) Graph of Restricted Cotangent (b) Graph of Inverse Cotangent

Figure 36

To define $\sec^{-1} x$ and $\csc^{-1} x$, we use the reciprocal relationship between the secant and cosine, and between the cosecant and sine, respectively. That is, the angle whose secant is x is the same angle whose cosine is $1/x$.

Similarly, the angle whose cosecant is x is the same angle whose sine is $1/x$. Hence, we define the **inverse secant** by

$$\sec^{-1} x = \cos^{-1} \frac{1}{x}$$

and the **inverse cosecant** by

$$\csc^{-1} x = \sin^{-1} \frac{1}{x}.$$

Since the domain for both $\cos x$ and $\sin x$ is $|x| \leq 1$, the domain for $\sec^{-1} x$ and $\csc^{-1} x$ is $1/|x| \leq 1$, that is, $|x| \geq 1$. The range for $\sec^{-1} x$ is almost the same as that of $\cos^{-1} x$, but since $1/x$ can never equal zero, $\sec^{-1} x$ can never equal $\pi/2$. Hence, the range for $\sec^{-1} x$ is

$$\left[0, \frac{\pi}{2}\right) \cup \left(\frac{\pi}{2}, \pi\right].$$

Similarly, the range for $\csc^{-1} x$ is

$$\left[-\frac{\pi}{2}, 0\right) \cup \left(0, \frac{\pi}{2}\right].$$

The inverse secant

$$y = \sec^{-1} x, \quad \text{with } |x| \geq 1 \text{ and } y \in \left[0, \frac{\pi}{2}\right) \cup \left(\frac{\pi}{2}, \pi\right]$$

is in fact the inverse of the **restricted secant function** defined by

$$y = \sec x, \quad \text{with } x \in \left[0, \frac{\pi}{2}\right) \cup \left(\frac{\pi}{2}, \pi\right] \text{ and } |y| \geq 1,$$

and the corresponding graphs are shown in Figure 37.

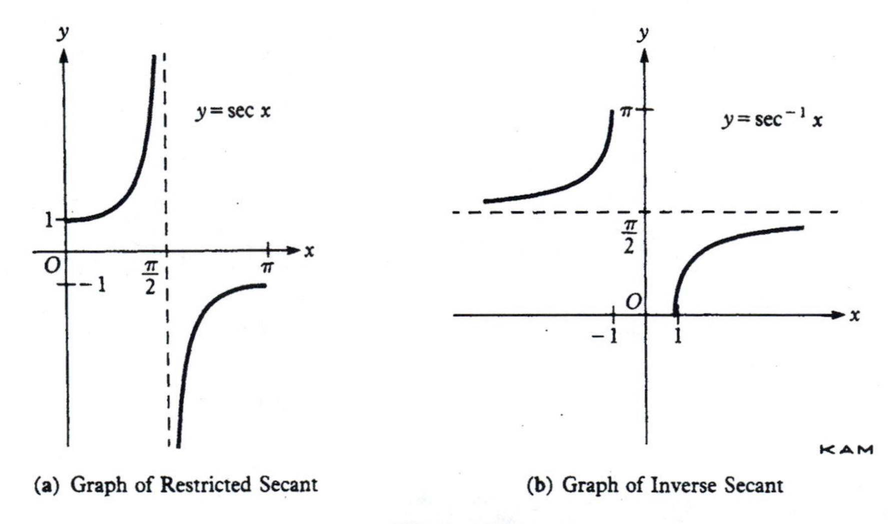

(a) Graph of Restricted Secant

(b) Graph of Inverse Secant

Figure 37

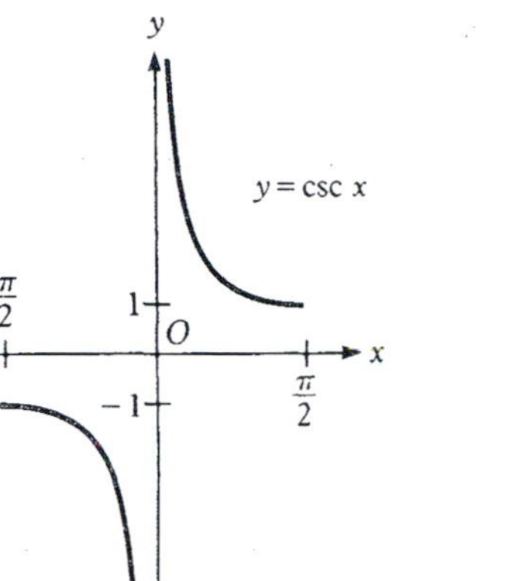

(a) Graph of Restricted Cosecant

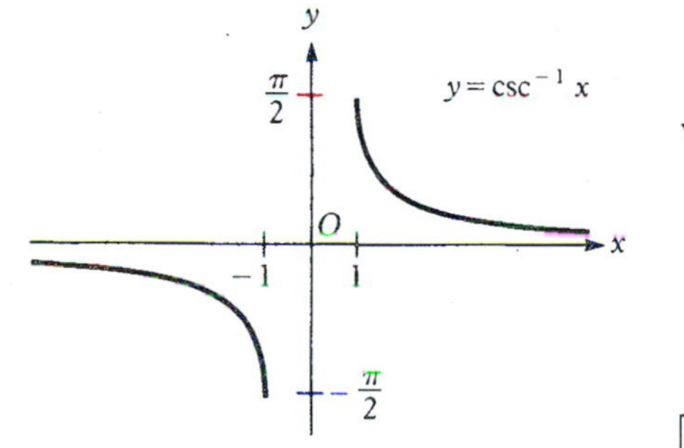

(b) Graph of Inverse Cosecant

Figure 38

Similarly, the inverse cosecant

$$y = \csc^{-1} x, \quad \text{with } |x| \geq 1 \text{ and } y \text{ in } \left[-\frac{\pi}{2}, 0\right) \cup \left(0, \frac{\pi}{2}\right]$$

is the inverse of the **restricted cosecant function** defined by

$$y = \csc x, \quad \text{with } x \text{ in } \left[-\frac{\pi}{2}, 0\right) \cup \left(0, \frac{\pi}{2}\right] \text{ and } |y| \geq 1,$$

and their graphs are shown in Figure 38.

The inverse secant and cosecant also satisfy the cofunction property. That is,

$$\sec^{-1} x = \cos^{-1} \frac{1}{x} = \frac{\pi}{2} - \sin^{-1} \frac{1}{x}$$

and since

$$\csc^{-1} x = \sin^{-1} \frac{1}{x},$$

we obtain

$$\sec^{-1} x + \csc^{-1} x = \frac{\pi}{2}.$$

In summary, we have the following three **cofunction equations:**

$$\sin^{-1} x + \cos^{-1} x = \frac{\pi}{2} \qquad (-1 \leq x \leq 1),$$

$$\tan^{-1} x + \cot^{-1} x = \frac{\pi}{2} \qquad (-\infty < x < \infty),$$

$$\sec^{-1} x + \csc^{-1} x = \frac{\pi}{2} \qquad (|x| \geq 1).$$

For most applications, including those in calculus, $\sin^{-1} x$ and $\tan^{-1} x$ are by far the most important of the six inverse trigonometric functions. This is reasonable since the other four can be defined in terms of these two. Therefore, you should concentrate first on $\sin^{-1} x$ and $\tan^{-1} x$. After these two functions are thoroughly understood, the other four are easy.

Comment The restricted sine, cosine, tangent, and cotangent are standard in mathematics. However, some authors choose the domain for the restricted secant to be $[0, \pi/2) \cup [\pi, 3\pi/2)$, and that for the restricted cosecant to be $(-\pi, -\pi/2] \cup (0, \pi/2]$. These choices result in a simplification of some formulas in calculus at the loss of the reciprocal relationship between the restricted sine and cosecant and between the restricted cosine and secant. It is the reciprocal relationship that enables us to compute the arc secant and arc cosecant on a calculator by using the arc cosine and arc sine keys, respectively. Either convention gives the same answers to problems. The difference is merely in the way we express the answers.

Exercises 7.5

Fill in the blanks to make each statement true.

1. The function $y = \sin x$ with restricted domain $[0, 3\pi/2]$ is not suitable for defining $\sin^{-1} x$ because ________.
2. The function $y = \tan x$ with restricted domain $\left[0, \frac{\pi}{2}\right)$ is not suitable for defining $\tan^{-1} x$ because ________.
3. For any of our restricted trigonometric functions f, if the domain of f is set A and the range is set B, then the domain of f^{-1} is ______ and the range of f^{-1} is ______.
4. The equation $\theta = \sin^{-1} x$ is equivalent to the equation $x =$ ______ with θ restricted to the interval ______.
5. The equality $\sin(\sin^{-1} x) = x$ is true for all real numbers x satisfying ________.

Write true *or* false *for each statement.*

6. The function defined by $y = \sin^{-1} x$ is increasing on $[-1, 1]$, and $y = \cos^{-1} x$ is decreasing on $[-1, 1]$.
7. The equality $\sin^{-1}(\sin x) = x$ is true for all real numbers x.
8. The equality $\tan(\tan^{-1} x) = x$ is true for all real numbers x.
9. If $\tan^{-1} x < 0$, then $x < 0$.
10. The equation $\sec^{-1} x = \cos^{-1}(1/x)$ is equivalent to the equation $\sec^{-1} x = 1/(\cos^{-1} x)$.

The Inverse Sine and Tangent Functions

In Exercises 11-20, find exact values, if possible, without the aid of a calculator.

11. $\sin^{-1} \frac{\sqrt{3}}{2}$
12. $\sin^{-1}\left(-\frac{1}{2}\right)$
13. $\sin^{-1} 1$
14. $\sin^{-1}(-1)$
15. $\arcsin 0$
16. $\arcsin 2$
17. $\tan^{-1} 1$
18. $\arctan \frac{1}{\sqrt{3}}$
19. $\tan^{-1}\left(-\frac{1}{\sqrt{3}}\right)$
20. $\tan^{-1} \sqrt{3}$

In Exercises 21-28, use a calculator to find approximate radian values to four decimal places.

21. $\sin^{-1} .2560$
22. $\sin^{-1}(-.2560)$
23. $\sin^{-1}(-.9580)$
24. $\arcsin .8570$
25. $\tan^{-1} .0175$
26. $\tan^{-1} 5.671$
27. $\tan^{-1}(-19.08)$
28. $\arctan(-114.6)$

Compute each of the following, if possible, without the aid of a calculator.

29. $\sin(\sin^{-1} 1)$
30. $\sin(\sin^{-1}(-1))$
31. $\cos\left(\sin^{-1} \frac{1}{2}\right)$
32. $\cos\left(\sin^{-1}\left(-\frac{1}{2}\right)\right)$
33. $\sin^{-1}(\sin 1.53)$
34. $\sin(\sin^{-1} 1.53)$
35. $\tan^{-1}\left(\tan \frac{3\pi}{4}\right)$
36. $\tan^{-1}\left(\sin \frac{3\pi}{2}\right)$

Other Inverse Functions

In Exercises 37–48, compute exact values, if possible.

37. $\cos^{-1} 0$

38. arc cos (-1)

39. $\cos^{-1} \frac{\sqrt{2}}{2}$

40. $\sec^{-1} 2$

41. $\sec^{-1} (-2)$

42. arc sec 0

43. $\cot^{-1} (-1)$

44. $\cot^{-1} \frac{1}{\sqrt{3}}$

45. arc cot $\sqrt{3}$

46. $\csc^{-1} (-2)$

47. $\csc^{-1} \frac{1}{2}$

48. arc csc (-1)

In Exercises 49–54, use a calculator to find approximate radian values to four decimal places.

49. $\sec^{-1} 1.044$

50. $\sec^{-1} (-1.499)$

51. $\cot^{-1} 3.271$

52. $\cot^{-1} (-3.271)$

53. $\csc^{-1} 2.366$

54. $\csc^{-1} 1.555$

Miscellaneous

55. Draw a right triangle with base $|x|$ and hypotenuse 1 to verify that

$$\sin(\cos^{-1} x) = \sqrt{1 - x^2}$$

for all x in $[-1, 1]$. Consider the case of positive x separately from that of negative x. Although a triangle cannot be drawn when $x = 0$, show that the result does hold for $x = 0$.

56. Proceed as in Exercise 55 to show that

$$\cos(\sin^{-1} x) = \sqrt{1 - x^2}$$

for all x in $[-1, 1]$.

7.6 Basic Identities

Quotient, Reciprocal, and Pythagorean Identities

Identities Involving $-\theta$, $\pi/2 - \theta$, and $\pi \pm \theta$

An identity is an equation that just won't quit.

In algebra, we encountered equations such as

$$(x + 1)(x - 1) = x^2 - 1$$

that are true for *every* number x, and others such as

$$\frac{x^2 - 3x + 2}{x - 1} = x - 2$$

that hold for every value of x for which both sides are defined. These equations are called **identities.** We also solved equations (sometimes called **conditional equations**) such as

$$x^2 - 3x + 2 = 0$$

that are true only for some values of x.[1] Similarly, we derive trigonometric identities in the next several sections and solve conditional trigonometric equations in Section 7.9. Trigonometric identities arise naturally in the context of triangle trigonometry, and we will use the letter θ to denote the independent variable. Of course, θ can be interpreted as an angle, an arc length, or simply a real number.

[1]Every equation, whether an identity or not, is *conditional* in the sense that the equation can hold only if both sides are defined.

Quotient, Reciprocal, and Pythagorean Identities. We have already mentioned some relationships between the trigonometric functions that follow directly from their definitions. These are the **quotient identities**

$$\tan\theta = \frac{\sin\theta}{\cos\theta}, \qquad \cot\theta = \frac{\cos\theta}{\sin\theta} \tag{1}$$

and the **reciprocal identities**

$$\cot\theta = \frac{1}{\tan\theta}, \qquad \sec\theta = \frac{1}{\cos\theta}, \qquad \csc\theta = \frac{1}{\sin\theta}. \tag{2}$$

They hold for all values of θ except where any denominator is zero.

The **fundamental trigonometric identity**

$$(\sin\theta)^2 + (\cos\theta)^2 = 1$$

also was mentioned before. It follows directly from the definitions of sine and cosine and the Pythagorean theorem. The fundamental identity is usually written in the parentheses-saving notation (see comment below)

$$\sin^2\theta + \cos^2\theta = 1,$$

and it holds for every real number θ. If we divide both sides of the fundamental identity by $\cos^2\theta$ we obtain

$$\frac{\sin^2\theta}{\cos^2\theta} + \frac{\cos^2\theta}{\cos^2\theta} = \frac{1}{\cos^2\theta}$$

or

$$\tan^2\theta + 1 = \sec^2\theta,$$

which holds for all values of θ for which both sides are defined—that is, for all numbers except $\pi/2 + n\pi$, where n is any integer. Similarly, if we divide both sides of the fundamental identity by $\sin^2\theta$, we obtain

$$1 + \cot^2\theta = \csc^2\theta,$$

which holds for all values of θ except $n\pi$, where n is any integer.

We combine all three of these identities into one group, called the **Pythagorean identities.**

$$\begin{aligned} \sin^2\theta + \cos^2\theta &= 1 \\ \tan^2\theta + 1 &= \sec^2\theta \\ 1 + \cot^2\theta &= \csc^2\theta \end{aligned} \tag{3}$$

Comment The notation $\sin^n\theta$ is used for $(\sin\theta)^n$ for all positive integers n, and similar notation is used for the other trigonometric functions. Of course, $\sin^{-1}\theta$ denotes the inverse sine function, not $1/\sin\theta$. For exponents other than positive integers, we use parentheses to denote powers of the trignometric functions. For instance,

$$\sqrt[3]{\sin^2\theta} = (\sin\theta)^{2/3} \quad \text{and} \quad \frac{1}{\sin^2\theta} = (\sin\theta)^{-2}.$$

As the following examples show, we can sometimes use identities to simplify trigonometric expressions.

Example 1 Simplify the trigonometric expression

$$\sec\theta - \sin\theta\tan\theta.$$

Solution: We proceed by writing the trigonometric functions in the expression in terms of sines and cosines.

$$\begin{aligned}
\sec\theta - \sin\theta\tan\theta &= \frac{1}{\cos\theta} - \sin\theta\cdot\frac{\sin\theta}{\cos\theta} && \textit{Reciprocal and quotient identities}\\
&= \frac{1-\sin^2\theta}{\cos\theta} && \textit{Algebra}\\
&= \frac{\cos^2\theta}{\cos\theta} && 1-\sin^2\theta = \cos^2\theta \textit{ by the fundamental identity}\\
&= \cos\theta && \textit{Algebra}
\end{aligned}$$

Hence, we have derived the identity

$$\sec\theta - \sin\theta\tan\theta = \cos\theta,$$

which holds for all θ except $\pi/2 + n\pi$, where n is any integer. ■

Example 2 Simplify $\dfrac{\sin\theta + \tan\theta}{1 + \sec\theta}$.

Solution: We again proceed by writing the functions in terms of sines and cosines.

$$\begin{aligned}
\frac{\sin\theta + \tan\theta}{1+\sec\theta} &= \frac{\sin\theta + \dfrac{\sin\theta}{\cos\theta}}{1 + \dfrac{1}{\cos\theta}} && \textit{Quotient and reciprocal identities}\\
&= \frac{\sin\theta\cos\theta + \sin\theta}{\cos\theta + 1} && \textit{Algebra}\\
&= \frac{\sin\theta(\cos\theta + 1)}{\cos\theta + 1} && \textit{Algebra}\\
&= \sin\theta && \textit{Algebra}
\end{aligned}$$

Hence, our original expression is equal to $\sin\theta$, provided $\theta \neq \pi/2 + n\pi$ (so $\tan\theta$ is defined) and $\theta \neq \pi + 2n\pi$ (so $\sec\theta \neq -1$), where n is any integer. ■

In Examples 1 and 2, we actually derived two more identities. The ability to derive and use such identities is crucial in applying trigonometry in advanced mathematics and science. Sometimes the problem is presented as in Examples 3 and 4 below.

Example 3 Prove the identity

$$(\csc\theta + \sec\theta)(\csc\theta - \sec\theta) = \cot^2\theta - \tan^2\theta.$$

Solution: We first multiply on the left side of the equals sign and then use the Pythagorean identities to arrive at the expression on the right.

$$\begin{aligned}(\csc\theta + \sec\theta)(\csc\theta - \sec\theta) &= \csc^2\theta - \sec^2\theta \\ &= (1 + \cot^2\theta) - (\tan^2\theta + 1) \\ &= \cot^2\theta - \tan^2\theta \quad \blacksquare\end{aligned}$$

Example 4 Prove that

$$\sin\theta\tan\theta + \cos\theta\cot\theta = \sec\theta - \cos\theta + \csc\theta - \sin\theta$$

is an identity.

Solution: In a case like this one in which each side of the equation is complicated, we simplify each side *independently* and thereby try to reduce each side to the same expression.

First we work on the left side. Here we express the functions in terms of sines and cosines and then simplify.

$$\begin{aligned}\sin\theta\tan\theta + \cos\theta\cot\theta &= \sin\theta\cdot\frac{\sin\theta}{\cos\theta} + \cos\theta\cdot\frac{\cos\theta}{\sin\theta} \\ &= \frac{\sin^2\theta}{\cos\theta} + \frac{\cos^2\theta}{\sin\theta} \\ &= \frac{\sin^3\theta + \cos^3\theta}{\cos\theta\sin\theta}\end{aligned}$$

This is as far as we can go on the left side. Now we change the right side to sines and cosines and simplify.

$$\begin{aligned}&\sec\theta - \cos\theta + \csc\theta - \sin\theta \\ &= \frac{1}{\cos\theta} - \cos\theta + \frac{1}{\sin\theta} - \sin\theta \\ &= \frac{\sin\theta - \sin\theta\cos^2\theta + \cos\theta - \sin^2\theta\cos\theta}{\cos\theta\sin\theta} \\ &= \frac{\sin\theta(1 - \cos^2\theta) + \cos\theta(1 - \sin^2\theta)}{\cos\theta\sin\theta} \\ &= \frac{\sin\theta\sin^2\theta + \cos\theta\cos^2\theta}{\cos\theta\sin\theta} \\ &= \frac{\sin^3\theta + \cos^3\theta}{\cos\theta\sin\theta}\end{aligned}$$

Since each side reduces to the same expression and all steps are reversible, we conclude that the two sides are identical whenever both are defined. ■

There is no single procedure for proving a trigonometric identity. However, based on the previous examples, we can state the followng general principles.

1. You should first learn the identities (1), (2), and (3). Note that each Pythagorean identity can be expressed in several ways. For instance, $\sin^2\theta + \cos^2\theta = 1$ is equivalent to $\sin^2\theta = 1 - \cos^2\theta$ and to $\cos^2\theta = 1 - \sin^2\theta$.
2. Try to reduce the more complicated side of an identity to the simpler side. If both sides are complicated, then try to reduce each side independently to the same expression.
3. Some expressions suggest the use of a Pythagorean identity (see Example 3). If such is not the case, it is often helpful to write the expression in terms of sines and cosines and simplify.
4. Remember that the usual rules of algebra, such as factoring and adding fractions with the LCD, apply when working with sums, differences, products, and quotients of trigonometric functions.

Identities Involving $-\theta$, $\pi/2 - \theta$, and $\pi \pm \theta$. We now restate as identities some of the relations between trigonometric functions included earlier in the text, and also some related identities.

First, we have the **negative-angle identities**

$$\sin(-\theta) = -\sin\theta \quad \text{and} \quad \cos(-\theta) = \cos\theta, \tag{4}$$

and the **cofunction identities**

$$\sin\left(\frac{\pi}{2} - \theta\right) = \cos\theta \quad \text{and} \quad \cos\left(\frac{\pi}{2} - \theta\right) = \sin\theta. \tag{5}$$

Identities (4) were first deduced in Exercises 7.1 from the definitions of sine and cosine and again in Section 7.2 from the graphs of $u = \sin t$ and $u = \cos t$. Identities (5), for acute angles, follow directly from the definitions of sine and cosine. We will see in Section 7.7 that identities (5) hold for all real numbers θ.

Example 5 Use (4) and (5) to verify each of the following identities.

(a) $\sin(\theta - \pi/2) = -\cos\theta$ (b) $\cos(\theta - \pi/2) = \sin\theta$

Solution:

(a) $\sin\left(\theta - \frac{\pi}{2}\right) = \sin\left(-\left(\frac{\pi}{2} - \theta\right)\right)$

$= -\sin\left(\frac{\pi}{2} - \theta\right)$ *From (4)*

$= -\cos\theta$ *From (5)*

(b) $\cos\left(\theta - \frac{\pi}{2}\right) = \cos\left(-\left(\frac{\pi}{2} - \theta\right)\right)$

$= \cos\left(\frac{\pi}{2} - \theta\right)$ *From (4)*

$= \sin\theta$ *From (5)* ∎

Next, we have the **π-identities**

$$\sin(\pi + \theta) = -\sin\theta$$
$$\cos(\pi + \theta) = -\cos\theta \tag{6}$$

and

$$\sin(\pi - \theta) = \sin\theta$$
$$\cos(\pi - \theta) = -\cos\theta. \tag{7}$$

Identities (6) were deduced from the graphs of $u = \sin t$ and $u = \cos t$ in Section 7.2. We can derive identities (7) for acute angles from the graphs or as indicated in Figure 39. We will see in Section 7.7 that identities (7) hold for all real numbers θ.

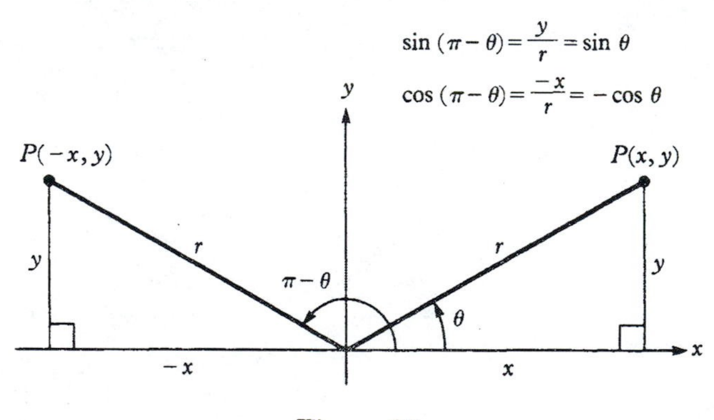

Figure 39

Example 6 Verify each of the following identities.

(a) $\sin(\theta - \pi) = -\sin\theta$ (b) $\cos(\theta - \pi) = -\cos\theta$

Solution:

(a) $\sin(\theta - \pi) = \sin(-(\pi - \theta))$
$= -\sin(\pi - \theta)$ *From (4)*
$= -\sin\theta$ *From (7)*

(b) $\cos(\theta - \pi) = \cos(-(\pi - \theta))$
$= \cos(\pi - \theta)$ *From (4)*
$= -\cos\theta$ *From (7)* ■

As the following examples show, we can use identities for sines and cosines to derive further identities for the other trigonometric functions.

Example 7 Prove the following identities.

(a) $\tan(\pi - \theta) = -\tan\theta$ (b) $\csc(\theta + \pi/2) = \sec\theta$

Solution:

(a) $\tan(\pi - \theta) = \dfrac{\sin(\pi - \theta)}{\cos(\pi - \theta)}$
$= \dfrac{\sin\theta}{-\cos\theta}$ *From (7)*
$= -\tan\theta$

(b) $\csc\left(\theta + \dfrac{\pi}{2}\right) = \dfrac{1}{\sin\left(\theta + \dfrac{\pi}{2}\right)}$

$\sin\left(\theta + \dfrac{\pi}{2}\right) = \sin\left(\theta + \pi - \dfrac{\pi}{2}\right)$
$= -\cos(\theta + \pi)$ *See Example 5(a)*
$= -(-\cos\theta)$ *From (6)*
$= \cos\theta$

Therefore, $\csc(\theta + \pi/2) = \dfrac{1}{\sin\left(\theta + \dfrac{\pi}{2}\right)} = \dfrac{1}{\cos\theta} = \sec\theta,$

which is the desired result. ■

Comment We note that the identities involving π have $\pm$ the same function on both sides, whereas those involving $\pi/2$ equate a function to $\pm$ its cofunction. Also, it can be verified that the sign that is correct when θ is an acute angle is the correct sign for any value of θ for which both sides are defined.

Exercises 7.6

Fill in the blanks to make each statement true.

1. The equality $\sin^2\theta + \cos^2\theta = 1$ is called an identity because ______________.
2. The identity $\tan\theta = \dfrac{\sin\theta}{\cos\theta}$ holds for all θ except ______________.
3. The reciprocal identities are ______________.
4. The Pythagorean identities are ______________.
5. By definition, $\sin^n\theta =$ ________ for all positive integers n, whereas $\sin^{-1}\theta =$ ________, and $(\sin\theta)^{-1} =$ ________.

Write true *or* false *for each statement.*

6. A trigonometric identity in θ is a trigonometric equation that holds for all values of θ for which both sides are defined.
7. For any trigonometric function f, the equation $f(-\theta) = -f(\theta)$ is an identity.
8. For any basic trigonometric function f and its cofunction g, the equation $f\left(\dfrac{\pi}{2} - \theta\right) = g(\theta)$ is an identity.
9. For any basic trigonometric function f and its cofunction g, one of the equations $f(\pi + \theta) = g(\theta)$ and $f(\pi + \theta) = -g(\theta)$ is an identity.
10. Since the equation $\sin(\pi/2 - \theta) = \sin\theta$ is not an identity, it cannot be true for any value of θ.

In the following exercises, it is assumed that all identities are subject to the condition that both sides are defined.

Quotient, Reciprocal, and Pythagorean Identities

11. Verify the fundamental identity $\sin^2\theta + \cos^2\theta = 1$ for the following special cases by substituting exact values of $\sin\theta$ and $\cos\theta$.
    (a) $\theta = 30°$ (b) $\theta = \pi/4$
    (c) $\theta = \pi/3$ (d) $\theta = 90°$
12. Verify the identity $\tan^2\theta + 1 = \sec^2\theta$ for the following angles by substituting exact values of $\tan\theta$ and $\sec\theta$.
    (a) $\theta = 45°$ (b) $\theta = -45°$
    (c) $\theta = 2\pi/3$ (d) $\theta = -2\pi/3$

In Exercises 13 and 14, use a calculator to verify that $\tan\theta = \sin\theta/\cos\theta$ *for the angles given.*

13. $\theta = 37°$
14. $\theta = 2$ (radians)

By using the quotient and reciprocal identities, simplify the trigonometric expressions in Exercises 15–22.

15. $\tan\theta\cos\theta$
16. $\cot\theta\sin\theta$
17. $\sin\theta\sec\theta$
18. $\cos\theta\csc\theta$
19. $\sin\theta\sec\theta\cot\theta$
20. $\cos\theta\csc\theta\tan\theta$
21. $\tan\theta(\cot\theta + \cos\theta)$
22. $\cos\theta(\tan\theta + \sec\theta)$

By using identities, simplify the expressions in Exercises 23–35.

23. $\sec^2\theta(1 - \sin^2\theta)$
24. $\sec^2\theta(1 - \cos^2\theta)$
25. $(1 - \sin\theta)(1 + \sin\theta)$
26. $(1 - \cos\theta)(1 + \cos\theta)$
27. $\cos\theta(\sec\theta - \cos\theta)$
28. $\dfrac{\sin\theta}{1 - \cos^2\theta}$
29. $\dfrac{1 - \sin^2\theta}{\cos\theta}$
30. $\dfrac{\sec^2\theta - 1}{\tan\theta}$
31. $\dfrac{1 - \csc^2\theta}{\cot\theta}$
32. $\dfrac{1 - \sin^2\theta}{1 - \cos^2\theta}$
33. $\dfrac{\sec\theta\csc\theta}{\tan\theta + \cot\theta}$
34. $\dfrac{\tan\theta + \sin\theta}{\sec\theta + 1}$
35. $\dfrac{\sin\theta\sec^2\theta + 1}{\sec\theta + \cot\theta}$

Prove that the following equations are identities.

36. $(\sin\theta + \cos\theta)^2 = 1 + 2\sin\theta\cos\theta$
37. $\dfrac{1 + \sec\theta}{\sin\theta + \tan\theta} = \csc\theta$
38. $\sin\theta + \cos\theta\cot\theta = \csc\theta$
39. $\dfrac{\cot^2\theta - 1}{1 - \tan^2\theta} = \cot^2\theta$
40. $\dfrac{\cos\theta}{\csc\theta + 1} + \dfrac{\cos\theta}{\csc\theta - 1} = 2\tan\theta$
41. $\dfrac{1 - \cos\theta}{\cos\theta} = \dfrac{\tan^2\theta}{\sec\theta + 1}$
42. $\dfrac{\sec\theta - 1}{\sec\theta + 1} = \dfrac{1 - \cos\theta}{1 + \cos\theta}$

43. $\dfrac{\tan\theta - \cot\theta}{\tan\theta + \cot\theta} = 1 - 2\cos^2\theta$

44. $\tan^2\theta - \sin^2\theta = \tan^2\theta\sin^2\theta$

45. $\dfrac{\tan\theta - \sin\theta\cos\theta}{\sin^2\theta} = \tan\theta$

46. $\dfrac{1+\tan\theta}{1-\tan\theta} + \dfrac{1+\cot\theta}{1-\cot\theta} = 0$

47. $\dfrac{1+\cos\theta}{\sin\theta} + \dfrac{\sin\theta}{1+\cos\theta} = 2\csc\theta$

48. $\dfrac{\csc^2\theta - 1}{1-\sin\theta} = \dfrac{1+\sin\theta}{\sin^2\theta}$

49. $\dfrac{\tan\theta + \sec\theta}{\cos\theta + \cot\theta} = \tan\theta\sec\theta$

Identities Involving $-\theta$, $\pi/2 - \theta$, and $\pi \pm \theta$

50. Use the sine and cosine identities for $-\theta$ in this section to derive identities for each of the following.
    (a) $\cot(-\theta)$ (b) $\sec(-\theta)$ (c) $\csc(-\theta)$

51. Verify that identities (4) and (5) hold for quadrantal angles.

52. Verify that identities (6) and (7) hold for quadrantal angles.

53. Use the π-identities for the sine and cosine in this section to derive identities for each of the following.
    (a) $\cot(\theta + \pi)$ (b) $\sec(\theta + \pi)$ (c) $\csc(\theta + \pi)$

54. Use the identities in this section to derive each of the following.
    (a) $\sin\left(\theta + \dfrac{3\pi}{2}\right) = -\cos\theta$
    (b) $\sin\left(\dfrac{3\pi}{2} - \theta\right) = -\cos\theta$

Miscellaneous

55. Show that the following equations are *not* identities.
    (a) $\sin\theta = \sqrt{1-\cos^2\theta}$ (b) $\sec\theta = \sqrt{1+\tan^2\theta}$
    (*Hint:* in each case, find at least one value of θ for which the equation is false.)

56. A student who was poor in algebra concluded from the identity $\sin^2\theta + \cos^2\theta = 1$ that $\sin\theta + \cos\theta = \pm 1$. Why is this a false conclusion?

57. Show that the equation $\sin(\alpha + \beta) = \sin\alpha + \sin\beta$ is *not* an identity.

58. Show that the equation $\sin 2\theta = 2\sin\theta$ is *not* an identity.

59. Although identities are used most often to simplify trigonometric expressions, there are occasions in which we need to replace a simple expression by a more complicated one. For example, for an operation in calculus, it is necessary to replace the function $\sec\theta$ by the function

$$\frac{\sec^2\theta + \sec\theta\tan\theta}{\sec\theta + \tan\theta}.$$

Show that this substitution is valid.

60. Proceed as in Exercise 59 to show that $\csc\theta$ may be replaced by

$$\frac{\csc\theta\cot\theta - \csc^2\theta}{\cot\theta - \csc\theta}.$$

7.7 Sum and Difference Identities

The distance formula and the law of cosines are the basis for the sum and difference identities.

$\cos(\alpha - \beta)$ and $\cos(\alpha + \beta)$
$\sin(\alpha + \beta)$ and $\sin(\alpha - \beta)$
Applications

When using trigonometry in calculus and its applications, the identities from Section 7.6 are very helpful. However, we also need identities involving sums and differences in angles, double- and half-angles, and products of trigonometric functions. These will be derived in this section and the next. We begin by deriving the difference identity for the cosine, from which all the others follow.

cos ($\alpha - \beta$) and cos ($\alpha + \beta$). We apply the law of cosines and the distance formula to triangle OPQ in Figure 40 in order to compute the length of side PQ, denoted by $d(P, Q)$, in terms of the included angle $\alpha - \beta$. As shown by the figure, the coordinates of P are $(\cos \alpha, \sin \alpha)$, and those of Q are $(\cos \beta, \sin \beta)$ since P and Q are on the unit circle.

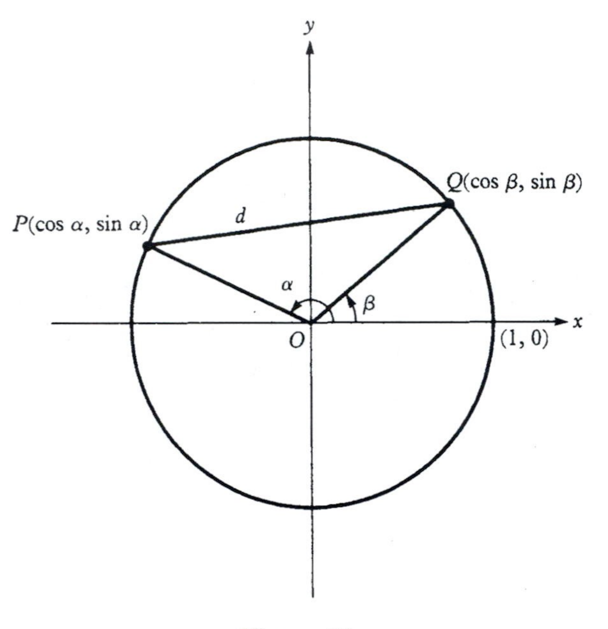

Figure 40

By the law of cosines,

$$\begin{aligned} d^2(P, Q) &= d^2(O, P) + d^2(O, Q) - 2d(O, P)d(O, Q) \cos (\alpha - \beta) \\ &= 1 + 1 - 2 \cdot 1 \cdot 1 \cos (\alpha - \beta) \\ &= 2 - 2 \cos (\alpha - \beta), \end{aligned}$$

where $d(O, P) = 1 = d(O, Q)$ since P and Q are on the unit circle.

By the distance formula,

$$\begin{aligned} d^2(P, Q) &= (\cos \alpha - \cos \beta)^2 + (\sin \alpha - \sin \beta)^2 \\ &= \cos^2 \alpha - 2 \cos \alpha \cos \beta + \cos^2 \beta + \sin^2 \alpha \\ &\quad - 2 \sin \alpha \sin \beta + \sin^2 \beta \\ &= \underbrace{\cos^2 \alpha + \sin^2 \alpha}_{1} + \underbrace{\cos^2 \beta + \sin^2 \beta}_{1} \\ &\quad - 2(\cos \alpha \cos \beta + \sin \alpha \sin \beta) \\ &= 2 - 2(\cos \alpha \cos \beta + \sin \alpha \sin \beta). \end{aligned}$$

Equating the expression for $d^2(P, Q)$ obtained from the law of cosines with that obtained from the distance formula, we get

$$2 - 2\cos(\alpha - \beta) = 2 - 2(\cos\alpha\cos\beta + \sin\alpha\sin\beta).$$

We now subtract 2 on both sides and then divide by -2 to show that

$$\boldsymbol{\cos(\alpha - \beta) = \cos\alpha\cos\beta + \sin\alpha\sin\beta.} \qquad (1)$$

It should be noted that, because of the periodicity of the cosine and also the identity $\cos(-\theta) = \cos\theta$, it follows that equation (1) holds for *all* values of α and β, whether positive or negative and regardless of their relative magnitudes. That is, equation (1) is an identity. Similar identities for $\cos(\alpha + \beta)$ and $\sin(\alpha \pm \beta)$ can now be derived as follows.

If we write $\alpha + \beta = \alpha - (-\beta)$ in (1), then

$$\begin{aligned}\cos(\alpha + \beta) &= \cos(\alpha - (-\beta))\\ &= \cos\alpha\cos(-\beta) + \sin\alpha\sin(-\beta)\\ &= \cos\alpha\cos\beta + \sin\alpha(-\sin\beta)\\ &= \cos\alpha\cos\beta - \sin\alpha\sin\beta.\end{aligned}$$

Therefore, we obtain the identity

$$\boldsymbol{\cos(\alpha + \beta) = \cos\alpha\cos\beta - \sin\alpha\sin\beta.} \qquad (2)$$

In the following example, we use identity (1) to prove that the cofunction identities of Section 7.6, which were derived for acute angles θ, actually hold for *all* angles θ.

Example 1 Show that the following equations hold for all values of θ.

(a) $\cos\left(\frac{\pi}{2} - \theta\right) = \sin\theta$ (b) $\sin\left(\frac{\pi}{2} - \theta\right) = \cos\theta$

Solution:

(a) By substituting $\alpha = \pi/2$ and $\beta = \theta$ in (1), we get

$$\begin{aligned}\cos\left(\frac{\pi}{2} - \theta\right) &= \cos\frac{\pi}{2}\cos\theta + \sin\frac{\pi}{2}\sin\theta\\ &= 0\cdot\cos\theta + 1\cdot\sin\theta\\ &= \sin\theta.\end{aligned}$$

(b) Since part (a) holds for all values of θ, we can use it to derive part (b) as follows. Substitute $\theta = \pi/2 - \alpha$ in (a) to obtain

$$\cos\left(\frac{\pi}{2} - \left(\frac{\pi}{2} - \alpha\right)\right) = \sin\left(\frac{\pi}{2} - \alpha\right),$$

that is,

$$\cos\alpha = \sin\left(\frac{\pi}{2} - \alpha\right),$$

which is equivalent to (b). ■

sin $(\alpha + \beta)$ and sin $(\alpha - \beta)$. By using the cofunction identities proven in Example 1 along with identity (1), we have

$$\begin{aligned}
\sin(\alpha + \beta) &= \cos\left(\frac{\pi}{2} - (\alpha + \beta)\right) && \textit{Example 1(a)} \\
&= \cos\left(\frac{\pi}{2} - \alpha - \beta\right) \\
&= \cos\left(\frac{\pi}{2} - \alpha\right)\cos\beta + \sin\left(\frac{\pi}{2} - \alpha\right)\sin\beta && \textit{Identity (1)} \\
&= \sin\alpha\cos\beta + \cos\alpha\sin\beta. && \textit{Example 1(a) and (b)}
\end{aligned}$$

Hence, we have derived the identity

$$\sin(\alpha + \beta) = \sin\alpha\cos\beta + \cos\alpha\sin\beta. \tag{3}$$

Finally, from (3) and the identities for $-\theta$, we obtain

$$\begin{aligned}
\sin(\alpha - \beta) &= \sin\alpha\cos(-\beta) + \cos\alpha\sin(-\beta) \\
&= \sin\alpha\cos\beta - \cos\alpha\sin\beta,
\end{aligned}$$

giving the identity

$$\sin(\alpha - \beta) = \sin\alpha\cos\beta - \cos\alpha\sin\beta. \tag{4}$$

For easy reference, we gather the preceding identities into one group called the **sum and difference identities.**

$$\begin{aligned}
\sin(\alpha + \beta) &= \sin\alpha\cos\beta + \cos\alpha\sin\beta \\
\sin(\alpha - \beta) &= \sin\alpha\cos\beta - \cos\alpha\sin\beta \\
\cos(\alpha + \beta) &= \cos\alpha\cos\beta - \sin\alpha\sin\beta \\
\cos(\alpha - \beta) &= \cos\alpha\cos\beta + \sin\alpha\sin\beta
\end{aligned}$$

We have written the above identities in the order that seems easiest to remember, not in the order in which they were derived. Remember, the keys to deriving these identities are the law of cosines and the distance formula.

Applications. We now illustrate several applications of the sum and difference identities.

Example 2 Find the exact values of each of the following.

(a) $\cos 15^\circ$ (b) $\sin 75^\circ$

Solution: The idea is to express each of the given angles as a sum or difference of special angles, whose sines and cosines we know exactly.

(a)
$$\begin{aligned}\cos 15^\circ &= \cos(45^\circ - 30^\circ)\\ &= \cos 45^\circ \cos 30^\circ + \sin 45^\circ \sin 30^\circ\\ &= \frac{\sqrt{2}}{2}\cdot\frac{\sqrt{3}}{2} + \frac{\sqrt{2}}{2}\cdot\frac{1}{2}\\ &= \frac{\sqrt{6}+\sqrt{2}}{4}\end{aligned}$$

(b)
$$\begin{aligned}\sin 75^\circ &= \sin(45^\circ + 30^\circ)\\ &= \sin 45^\circ \cos 30^\circ + \cos 45^\circ \sin 30^\circ\\ &= \frac{\sqrt{2}}{2}\cdot\frac{\sqrt{3}}{2} + \frac{\sqrt{2}}{2}\cdot\frac{1}{2}\\ &= \frac{\sqrt{6}+\sqrt{2}}{4}\end{aligned}$$

Observe that $\sin 75^\circ = \cos 15^\circ$, which could also be derived from a cofunction identity since $75^\circ = 90^\circ - 15^\circ$. ■

Example 3 Derive the identity

$$\tan(\alpha + \beta) = \frac{\tan\alpha + \tan\beta}{1 - \tan\alpha\tan\beta}.$$

Solution:
$$\begin{aligned}\tan(\alpha+\beta) &= \frac{\sin(\alpha+\beta)}{\cos(\alpha+\beta)}\\ &= \frac{\sin\alpha\cos\beta + \cos\alpha\sin\beta}{\cos\alpha\cos\beta - \sin\alpha\sin\beta}\end{aligned}$$

To obtain tangents on the right side of the above equation, we divide both the numerator and the denominator by $\cos\alpha\cos\beta$. We then get

$$\begin{aligned}\tan(\alpha+\beta) &= \frac{\dfrac{\sin\alpha\cos\beta + \cos\alpha\sin\beta}{\cos\alpha\cos\beta}}{\dfrac{\cos\alpha\cos\beta - \sin\alpha\sin\beta}{\cos\alpha\cos\beta}}\\ &= \frac{\tan\alpha + \tan\beta}{1 - \tan\alpha\tan\beta}.\end{aligned}$$ ■

Example 4 If $\sin\alpha = -4/5$ with α in quadrant III, and $\sin\beta = 5/13$ with β in quadrant II, find $\sin(\alpha+\beta)$ and the quadrant of $\alpha + \beta$.

Solution: Now

$$\sin(\alpha+\beta) = \sin\alpha\cos\beta + \cos\alpha\sin\beta,$$

and since we are given $\sin\alpha$ and $\sin\beta$, we must determine $\cos\alpha$ and $\cos\beta$.

From the fundamental identity, we have

$$\begin{aligned}\cos\alpha &= \pm\sqrt{1-\sin^2\alpha}\\ &= \pm\sqrt{1-\frac{16}{25}}\\ &= \pm\frac{3}{5}.\end{aligned}$$

Since α is in quadrant III, where the cosine is negative, we conclude that

$$\cos\alpha = -\frac{3}{5}.$$

Similarly, from the fundamental identity and the fact that β is in quadrant II, we can determine that

$$\cos\beta = -\frac{12}{13}.$$

Therefore,

$$\begin{aligned}\sin(\alpha+\beta) &= -\frac{4}{5}\left(-\frac{12}{13}\right) + \left(-\frac{3}{5}\right)\frac{5}{13}\\ &= \frac{33}{65}.\end{aligned}$$

Since $\sin(\alpha+\beta)$ is positive, we can conclude that $\alpha+\beta$ is either in quadrant I or II. To determine which of these is the case, we compute $\cos(\alpha+\beta)$. From its sign, we will know the quadrant of $\alpha+\beta$. Now

$$\begin{aligned}\cos(\alpha+\beta) &= \cos\alpha\cos\beta - \sin\alpha\sin\beta\\ &= \left(-\frac{3}{5}\right)\left(-\frac{12}{13}\right) - \left(-\frac{4}{5}\right)\left(\frac{5}{13}\right)\\ &= \frac{56}{65}.\end{aligned}$$

Since $\cos(\alpha+\beta)$ is also positive, we conclude that $\alpha+\beta$ is in quadrant I. ■

The identities involving $\pi-\theta$ in Section 7.6 were derived for acute angles only. We can now show that they hold for *all* angles θ.

Example 5 Show that each of the following equations is true for all values of θ.

(a) $\sin(\pi-\theta) = \sin\theta$ (b) $\cos(\pi-\theta) = -\cos\theta$

Solution:

(a)
$$\begin{aligned}\sin(\pi-\theta) &= \sin\pi\cos\theta - \cos\pi\sin\theta\\ &= 0\cdot\cos\theta - (-1)\sin\theta\\ &= \sin\theta\end{aligned}$$

$$\begin{aligned}(b)\ \cos(\pi - \theta) &= \cos \pi \cos \theta + \sin \pi \sin \theta \\ &= (-1)\cos \theta + 0 \cdot \sin \theta \\ &= -\cos \theta \quad \blacksquare\end{aligned}$$

Exercises 7.7

Fill in the blanks to make each statement true.

1. All of the trigonometric identities in this section depend upon the difference formula $\cos(\alpha - \beta) =$ ________.
2. The identity for $\cos(\alpha - \beta)$ was derived from the ________ formula and the law of ________.
3. By using the difference identity for $\cos(\alpha - \beta)$, we obtain the cofunction identity $\cos(\pi/2 - \theta) =$ ________.
4. To obtain the exact value of $\sin 15°$, we can let $\alpha =$ ________ and $\beta =$ ________ in the identity for $\sin(\alpha - \beta)$.
5. The quadrant of $\alpha + \beta$ can be determined by computing ________ and ________.

Write true *or* false *for each statement.*

6. For all α and β, $\sin(\alpha + \beta) \neq \sin \alpha + \sin \beta$.
7. The cofunction identity $\sin(\pi/2 - \theta) = \cos \theta$ holds only if θ is acute.
8. The identity for $\sin(\alpha - \beta)$ can be derived from the one for $\sin(\alpha + \beta)$ and the identities for $\sin(-\theta)$ and $\cos(-\theta)$.
9. For any angle α, $\sin \alpha \cos(\pi/2 - \alpha) + \cos \alpha \sin(\pi/2 - \alpha) = 1$.
10. For any angle α, $\cos(\pi/2 - \alpha) \cos(\pi/2 + \alpha) - \sin(\pi/2 - \alpha) \sin(\pi/2 + \alpha) = 0$.

Sum and Difference Identities

Use the trigonometric functions of special angles and the identities in this section to find exact values for each of the functions in Exercises 11–20.

11. $\cos 75°$
12. $\tan 75°$
13. $\sin 15°$
14. $\tan 15°$
15. $\sin 105°$
16. $\cos 105°$
17. $\tan 105°$
18. $\sin 120°$
19. $\cos 120°$
20. $\tan 120°$

Evaluate each of the following without the aid of a calculator.

21. $\sin 10° \cos 80° + \cos 10° \sin 80°$
22. $\cos 100° \cos 10° + \sin 100° \sin 10°$
23. $\sin 70° \cos 40° - \cos 70° \sin 40°$
24. $\cos 25° \cos 65° - \sin 25° \sin 65°$
25. $\dfrac{\tan 20° + \tan 25°}{1 - \tan 20° \tan 25°}$

In Exercises 26–30, find $\sin(\alpha + \beta)$, $\cos(\alpha + \beta)$, *and the quadrant of* $\alpha + \beta$ *from the given information.*

26. $\sin \alpha = 3/5$, $\cos \beta = -3/4$, α and β in quadrant II
27. $\cos \alpha = 12/13$, $\cos \beta = -4/5$, α in quadrant IV, β in quadrant III
28. $\tan \alpha = 3/4$, $\tan \beta = -2$, α in quadrant III, β in quadrant IV
29. $\sin \alpha = -3/5$, α in quadrant IV, $\sin \beta = 4/5$, β in quadrant II
30. $\tan \alpha = -3/4$, α in quadrant IV, $\tan \beta = -2$, β in quadrant II

31. Find $\sin(\alpha - \beta)$ and $\cos(\alpha - \beta)$ in Exercise 27.
32. Find $\sin(\alpha - \beta)$ and $\cos(\alpha - \beta)$ in Exercise 26.

Use the sum and difference identities to derive the following identities.

33. $\sin\left(\theta + \dfrac{\pi}{2}\right) = \cos \theta$

34. $\cos\left(\theta + \frac{\pi}{2}\right) = -\sin\theta$

35. $\sin\left(\theta + \frac{3\pi}{2}\right) = -\cos\theta$

36. $\cos\left(\frac{3\pi}{2} - \theta\right) = -\sin\theta$

37. $\tan\left(\theta + \frac{\pi}{4}\right) = \frac{1 + \tan\theta}{1 - \tan\theta}$

Prove that the following equations are identities.

38. $\sin(\alpha + \beta)\sin(\alpha - \beta) = \sin^2\alpha - \sin^2\beta$

39. $\cos(\alpha + \beta)\cos(\alpha - \beta) = \cos^2\alpha - \sin^2\beta$

40. $\frac{\sin(\alpha + \beta)}{\sin(\alpha - \beta)} = \frac{\tan\alpha + \tan\beta}{\tan\alpha - \tan\beta}$

41. $\frac{\cos(\alpha + \beta)}{\cos(\alpha - \beta)} = \frac{1 - \tan\alpha\tan\beta}{1 + \tan\alpha\tan\beta}$

42. $\tan(\alpha - \beta) = \frac{\tan\alpha - \tan\beta}{1 + \tan\alpha\tan\beta}$

Find the exact value of each of the following.

43. $\sin\left(\sin^{-1}\frac{1}{2} + \tan^{-1} 1\right)$

44. $\cos\left(\sin^{-1}\frac{\sqrt{3}}{2} + \sin^{-1}\left(-\frac{1}{2}\right)\right)$

45. $\tan\left(\tan^{-1}\sqrt{3} - \tan^{-1}\frac{1}{\sqrt{3}}\right)$

46. $\sin\left(\sin^{-1}\frac{4}{5} + \sin^{-1}\left(-\frac{3}{5}\right)\right)$

Miscellaneous

In Exercises 47–50, assume that a and b are not both equal to zero.

47. The function defined by $y = a\sin x + b\cos x$ occurs frequently in applied mathematics, but it is often written in the form $y = \sqrt{a^2 + b^2}\cos(x - \phi)$. Show that ϕ satisfies the equations $\sin\phi = a/\sqrt{a^2 + b^2}$ and $\cos\phi = b/\sqrt{a^2 + b^2}$. [*Hint:* expand $\cos(x - \phi)$.]

48. Use Exercise 47 to explain why the graph of $y = a\sin x + b\cos x$ is a cosine curve, and determine its amplitude, period, and phase shift.

49. Show that $a\sin x + b\cos x = \sqrt{a^2 + b^2}\sin(x + \theta)$, where $\sin\theta = b/\sqrt{a^2 + b^2}$ and $\cos\theta = a/\sqrt{a^2 + b^2}$. [*Hint:* expand $\sin(x + \theta)$.]

50. Use the result of Exercise 47 to graph $y = \cos x + \sin x$ on $[0, 2\pi]$.

51. Use the result of Exercise 49 to graph $y = \sin x - \cos x$ on $[0, 2\pi]$.

52. Use the result of Exercise 47 to graph $y = \cos x - \sqrt{3}\sin x$ on $[0, 2\pi]$.

53. Use the result of Exercise 49 to graph $y = \sin x + \sqrt{3}\cos x$ on $[0, 2\pi]$.

Prove the following two identities, where x and h are real numbers and $h \neq 0$. These identities are used in calculus.

54. $\frac{\sin(x + h) - \sin x}{h}$

$$= \sin x\left(\frac{\cos h - 1}{h}\right) + \cos x\left(\frac{\sin h}{h}\right)$$

55. $\frac{\cos(x + h) - \cos x}{h}$

$$= \cos x\left(\frac{\cos h - 1}{h}\right) - \sin x\left(\frac{\sin h}{h}\right)$$

56. Derive the following result, which is useful in both analytic geometry and calculus: if m_1 is the slope of line L_1 and m_2 is the slope of line L_2 then

$$\tan\theta = \frac{m_2 - m_1}{1 + m_1 m_2},$$

where θ is the angle from L_1 to L_2.

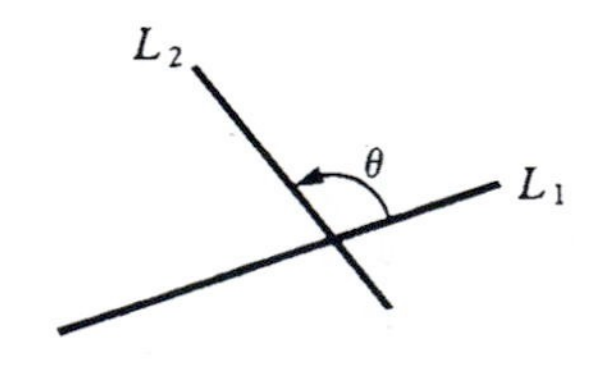

7.8 Multiple-Angle Identities

Multiple-angle identities are spinoffs of the sum and difference identities.

Double- and Half-Angle Identities
Product and Factor Identities

We now use the sum and difference formulas from the last section to derive additional identities that are essential in advanced mathematics.

Double- and Half-Angle Identities. An important application of identities in calculus is to reduce products or powers of trigonometric functions to trigonometric functions of multiple angles. We now derive the identities that accomplish these reductions.

By letting $\alpha = \beta$ in the identity for the sine of a sum, we obtain

$$\begin{aligned}\sin 2\alpha &= \sin(\alpha + \alpha)\\ &= \sin\alpha\cos\alpha + \cos\alpha\sin\alpha\\ &= 2\sin\alpha\cos\alpha.\end{aligned}$$

Hence, the **double-angle identity for the sine** is

$$\boldsymbol{\sin 2\alpha = 2\sin\alpha\cos\alpha} \qquad (1)$$

Similarly, from the identity for the cosine of a sum, we get

$$\begin{aligned}\cos 2\alpha &= \cos(\alpha + \alpha)\\ &= \cos\alpha\cos\alpha - \sin\alpha\sin\alpha\\ &= \cos^2\alpha - \sin^2\alpha.\end{aligned}$$

Also, by writing $\cos^2\alpha$ as $1 - \sin^2\alpha$, we obtain

$$\begin{aligned}\cos 2\alpha &= 1 - \sin^2\alpha - \sin^2\alpha\\ &= 1 - 2\sin^2\alpha,\end{aligned}$$

and by writing $\sin^2\alpha$ as $1 - \cos^2\alpha$, we get

$$\begin{aligned}\cos 2\alpha &= \cos^2\alpha - (1 - \cos^2\alpha)\\ &= 2\cos^2\alpha - 1.\end{aligned}$$

Hence, we have the following three versions of the **double-angle identity for the cosine.**

$$\boldsymbol{\cos 2\alpha = \cos^2\alpha - \sin^2\alpha} \qquad (2)$$

$$\boldsymbol{\cos 2\alpha = 1 - 2\sin^2\alpha} \qquad (3)$$

$$\boldsymbol{\cos 2\alpha = 2\cos^2\alpha - 1} \qquad (4)$$

Example 1 Verify identities (1) and (2) for the special case $\theta = 60°$.

Solution: According to (1) and (2),

$$\begin{aligned}\sin(2\cdot 60°) &= 2\sin 60°\cos 60°\\ &= 2\cdot\frac{\sqrt{3}}{2}\cdot\frac{1}{2}\\ &= \frac{\sqrt{3}}{2},\end{aligned}$$

and

$$\begin{aligned}\cos(2\cdot 60°) &= \cos^2 60° - \sin^2 60°\\ &= \left(\frac{1}{2}\right)^2 - \left(\frac{\sqrt{3}}{2}\right)^2\\ &= -\frac{1}{2}.\end{aligned}$$

Both of these results agree with the computation of sin 120° and cos 120° in Example 5 from Section 6.2. ■

Example 2 Express $\sin 4\alpha$ as a function of α alone.

Solution:

$$\begin{aligned}\sin 4\alpha &= \sin(2\cdot 2\alpha)\\ &= 2\sin 2\alpha\cos 2\alpha && \textit{Identity (1)}\\ &= 2(2\sin\alpha\cos\alpha)(\cos^2\alpha - \sin^2\alpha) && \textit{Identities (1) and (2)}\\ &= 4\sin\alpha\cos^3\alpha - 4\sin^3\alpha\cos\alpha\end{aligned}$$

Here we obtained the desired result in two applications of the double-angle identities, one to reduce 4α to 2α, and another to reduce 2α to α. This technique would work for $\sin 2^n\alpha$ and $\cos 2^n\alpha$, where n is any positive integer. After n applications of the double angle identities, a function of α alone would be obtained. ■

Identities (3) and (4) are equivalent, respectively, to the following:

$$\sin^2\alpha = \frac{1-\cos 2\alpha}{2} \tag{3'}$$

$$\cos^2\alpha = \frac{1+\cos 2\alpha}{2}. \tag{4'}$$

Equations (3′) and (4′) are the ones used in calculus to reduce powers of trigonometric functions to trigonometric functions of multiple angles. If we now replace 2α by β and at the same time take square roots of both sides in (3′) and (4′), we obtain the following **half-angle identities.**

$$\sin\frac{\beta}{2} = \pm\sqrt{\frac{1-\cos\beta}{2}}, \tag{5}$$

$$\cos\frac{\beta}{2} = \pm\sqrt{\frac{1+\cos\beta}{2}}. \tag{6}$$

If the quadrant of $\beta/2$ is known, then the correct sign for the right side of (5) and (6) can be determined.

Example 3 Verify identities (5) and (6) for the special case $\beta = 120°$.

Solution: Since $\beta/2 = 60°$, we take the positive sign in both cases. Then, according to the half-angle identities,

$$\sin 60° = \sqrt{\frac{1 - \cos 120°}{2}} = \sqrt{\frac{1 - \left(-\frac{1}{2}\right)}{2}} = \sqrt{\frac{3}{4}} = \frac{\sqrt{3}}{2},$$

and

$$\cos 60° = \sqrt{\frac{1 + \cos 120°}{2}} = \sqrt{\frac{1 + \left(-\frac{1}{2}\right)}{2}} = \sqrt{\frac{1}{4}} = \frac{1}{2}.$$

Both of these values agree with previous results. ■

In Example 3 from Section 7.7, we derived an identity for $\tan(\alpha + \beta)$ by dividing the identity for $\sin(\alpha + \beta)$ by that for $\cos(\alpha + \beta)$. Similarly, if we divide (5) by (6), we obtain

$$\tan\frac{\beta}{2} = \pm\sqrt{\frac{1 - \cos\beta}{1 + \cos\beta}}.$$

Furthermore, we can transform the above equation into an identity without a ± sign as follows.

$$\begin{aligned}\left|\tan\frac{\beta}{2}\right| &= \sqrt{\frac{1 - \cos\beta}{1 + \cos\beta}} \\ &= \sqrt{\frac{1 - \cos\beta}{1 + \cos\beta} \cdot \frac{1 + \cos\beta}{1 + \cos\beta}} \\ &= \sqrt{\frac{1 - \cos^2\beta}{(1 + \cos\beta)^2}} \\ &= \sqrt{\frac{\sin^2\beta}{(1 + \cos\beta)^2}} \\ &= \frac{|\sin\beta|}{|1 + \cos\beta|}\end{aligned}$$

Now $1 + \cos\beta \geq 0$ for all values of β (*why?*). Therefore, the absolute value signs are not needed in the denominator. Furthermore, it can be shown that whenever $\tan(\beta/2)$ is defined, it has the same sign as $\sin\beta$ (see Exercise 45). Hence, we can remove the absolute value signs from both sides of the above equation to obtain the following **half-angle identity for the tangent.**

$$\tan\frac{\beta}{2} = \frac{\sin\beta}{1 + \cos\beta}. \tag{7}$$

In a similar way (see Exercise 41), the following alternate half-angle identity for the tangent can be proved.

$$\tan \frac{\beta}{2} = \frac{1 - \cos \beta}{\sin \beta}. \qquad (7')$$

Example 4 Find the exact value of $\tan \pi/8$.

Solution: From identity (7), with $\beta = \pi/4$, we have

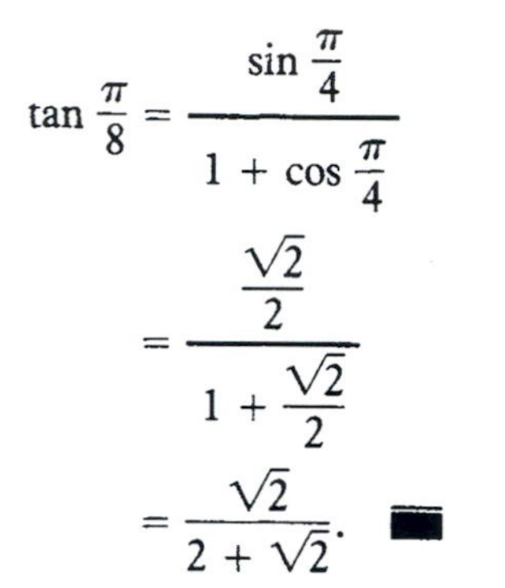

$$\tan \frac{\pi}{8} = \frac{\sin \frac{\pi}{4}}{1 + \cos \frac{\pi}{4}} = \frac{\frac{\sqrt{2}}{2}}{1 + \frac{\sqrt{2}}{2}} = \frac{\sqrt{2}}{2 + \sqrt{2}}. \quad \blacksquare$$

Product and Factor Identities. By adding or subtracting sum and difference identities, we can derive further identities that are useful in calculus. For example, by adding the identity

$$\sin(\alpha + \beta) = \sin \alpha \cos \beta + \cos \alpha \sin \beta$$

to the identity

$$\sin(\alpha - \beta) = \sin \alpha \cos \beta - \cos \alpha \sin \beta,$$

we obtain

$$\sin(\alpha + \beta) + \sin(\alpha - \beta) = 2 \sin \alpha \cos \beta,$$

which is usually written as

$$\sin \alpha \cos \beta = \frac{1}{2}[\sin(\alpha + \beta) + \sin(\alpha - \beta)],$$

and is called a product identity. Other identities of this type can be obtained in a similar manner (see Exercises 65–67). With the one just derived, we list three others in the following group called the **product identities.**

$$\begin{aligned}
\sin \alpha \cos \beta &= \frac{1}{2}[\sin(\alpha + \beta) + \sin(\alpha - \beta)] \\
\cos \alpha \sin \beta &= \frac{1}{2}[\sin(\alpha + \beta) - \sin(\alpha - \beta)] \\
\sin \alpha \sin \beta &= -\frac{1}{2}[\cos(\alpha + \beta) - \cos(\alpha - \beta)] \\
\cos \alpha \cos \beta &= \frac{1}{2}[\cos(\alpha + \beta) + \cos(\alpha - \beta)]
\end{aligned} \qquad (8)$$

Example 5 Express the product $\sin 2\theta \cos 3\theta$ as a sum or difference of trigonometric functions.

Solution: By substituting $\alpha = 2\theta$ and $\beta = 3\theta$ in the product identity (8) for $\sin \alpha \cos \beta$, we obtain

$$\begin{aligned}\sin 2\theta \cos 3\theta &= \frac{1}{2}[\sin (2\theta + 3\theta) + \sin (2\theta - 3\theta)] \\ &= \frac{1}{2}[\sin 5\theta + \sin (-\theta)] \\ &= \frac{1}{2}(\sin 5\theta - \sin \theta).\end{aligned}$$

If we now make the substitutions

$$\theta = \alpha + \beta \qquad \text{and} \qquad \phi = \alpha - \beta,$$

then

$$\frac{\theta + \phi}{2} = \alpha \qquad \text{and} \qquad \frac{\theta - \phi}{2} = \beta,$$

and the product identity for $\sin \alpha \cos \beta$ becomes

$$\sin \left(\frac{\theta + \phi}{2}\right) \cos \left(\frac{\theta - \phi}{2}\right) = \frac{1}{2}(\sin \theta + \sin \phi),$$

which is usually written as

$$\sin \theta + \sin \phi = 2 \sin \left(\frac{\theta + \phi}{2}\right) \cos \left(\frac{\theta - \phi}{2}\right),$$

and is called a factor identity. We list this one along with three others that are similarly derived (see Exercises 68–70) in the following group, called the **factor identities.**

$$\sin \theta + \sin \phi = 2 \sin \left(\frac{\theta + \phi}{2}\right) \cos \left(\frac{\theta - \phi}{2}\right)$$

$$\sin \theta - \sin \phi = 2 \cos \left(\frac{\theta + \phi}{2}\right) \sin \left(\frac{\theta - \phi}{2}\right)$$

$$\cos \theta - \cos \phi = -2 \sin \left(\frac{\theta + \phi}{2}\right) \sin \left(\frac{\theta - \phi}{2}\right)$$

$$\cos \theta + \cos \phi = 2 \cos \left(\frac{\theta + \phi}{2}\right) \cos \left(\frac{\theta - \phi}{2}\right)$$

Comment At this point, you may be wondering where this business of deriving identities is going to end. You may also have come to the conclusion that mastering all of the trigonometric identities is a hopeless task. As for the first point, *we will derive no more identities*. There is also relief on the second point. If you know the values of $\sin\theta$ and $\cos\theta$ at the quadrantal angles and the definitions of $\tan\theta$, $\cot\theta$, $\sec\theta$, and $\csc\theta$ in terms of $\sin\theta$ and $\cos\theta$, then all of the identities that we have derived can be easily obtained from the following *five basic identities:*

1. $\sin^2\theta + \cos^2\theta = 1$
2. $\sin(-\theta) = -\sin\theta$
3. $\cos(-\theta) = \cos\theta$
4. $\sin(\theta + \phi) = \sin\theta\cos\phi + \cos\theta\sin\phi$
5. $\cos(\theta + \phi) = \cos\theta\cos\phi - \sin\theta\sin\phi$.

We suggest that you memorize these five. The remaining identities, such as the double-angle, half-angle, or product identities, can be derived or looked up as the need arises.

Exercises 7.8

Fill in the blanks to make each statement true.

1. The identity for $\sin 2\alpha$ can be derived from the identity for $\sin(\alpha + \beta)$ by substituting ______.
2. Two other versions of the double-angle identity $\cos 2\alpha = \cos^2\alpha - \sin^2\alpha$ are $\cos 2\alpha =$ ______ and $\cos 2\alpha =$ ______.
3. The identity $\sin^2\alpha = (1 - \cos 2\alpha)/2$ is equivalent to the identity $\cos 2\alpha =$ ______.
4. The identity $\sin^2\alpha = (1 - \cos 2\alpha)/2$ is equivalent to the identity $\sin(\beta/2) =$ ______.
5. By adding the identities for $\sin(\alpha + \beta)$ and $\sin(\alpha - \beta)$, we obtain the product identity ______.

Write true *or* false *for each statement.*

6. The equation $\sin 2\alpha = 2\sin\alpha$ is not an identity.
7. The equation $\sin 2\alpha = 2\sin\alpha$ is true for $\alpha = n\pi$, where n is any integer.
8. $\cos(\beta/2) = \sqrt{(1 + \cos\beta)/2}$ if $0 \le \beta \le \pi/2$
9. $\cos(\beta/2) = -\sqrt{(1 + \cos\beta)/2}$ if β is in quadrant II.
10. $\frac{1}{2}\sin 6\theta\cos 6\theta = \sin 3\theta$.

Double and Half-Angle Identities

By using the identities in this section and the trigonometric functions of special angles, find exact values for each of the following.

11. $2\sin 15^\circ\cos 15^\circ$
12. $\sin 75^\circ\cos 75^\circ$
13. $\cos^2 75^\circ - \sin^2 75^\circ$
14. $2\sin^2 165^\circ - 1$
15. $1 - 2\sin^2 15^\circ$
16. $\sin\frac{\pi}{8}\cos\frac{\pi}{8}$
17. $\cos^2\frac{\pi}{8} - \sin^2\frac{\pi}{8}$
18. $\sin 165^\circ$
19. $\cos 105^\circ$
20. $\tan 67.5^\circ$

For each of the following, find $\sin 2\alpha$ *and* $\cos 2\alpha$.

21. $\sin\alpha = \frac{3}{5}$, α in quadrant II
22. $\cos\alpha = -\frac{3}{5}$, α in quadrant III
23. $\tan\alpha = \frac{3}{4}$, α in quadrant III
24. $\tan\alpha = -\frac{5}{12}$, α in quadrant IV
25. $\sin\alpha = -\frac{5}{13}$, α in quadrant III

Find $\sin(\alpha/2)$ *and* $\cos(\alpha/2)$ *for each of the following.*

26. $\sin\alpha = \frac{4}{5},\ \frac{\pi}{2} < \alpha < \pi$
27. $\cos\alpha = -\frac{4}{5},\ \pi < \alpha < \frac{3\pi}{2}$
28. $\tan\alpha = \frac{12}{5},\ \pi < \alpha < \frac{3\pi}{2}$
29. $\cos\alpha = \frac{5}{12},\ -\frac{\pi}{2} < \alpha < 0$
30. $\sin\alpha = -\frac{1}{3},\ -\frac{\pi}{2} < \alpha < 0$

Simplify each of the following.

31. $\sin 2\theta\,(\tan\theta + \cot\theta)$
32. $\dfrac{\cos 2\theta - 1}{\sin 2\theta}$
33. $\dfrac{\sin 2\theta + \sin\theta}{\cos 2\theta + 1 + \cos\theta}$
34. $\cos^4\theta - \sin^4\theta$
35. $\dfrac{1 + \cos 4\theta}{2}$
36. $\dfrac{1 - \cos 6\theta}{2}$
37. $\dfrac{1 + \cos 4\theta}{\sin 4\theta}$
38. $\dfrac{\sin 6\theta}{1 + \cos 6\theta}$

Prove that the following equations are identities.

39. $(\sin\theta + \cos\theta)^2 = \sin 2\theta + 1$
40. $\tan\theta + \cot\theta = 2\csc 2\theta$
41. $\tan\dfrac{\theta}{2} = \dfrac{1 - \cos\theta}{\sin\theta}$
42. $\tan\dfrac{\theta}{2} + \cot\theta = \csc\theta$
43. $\dfrac{\sin\theta + \frac{1}{2}\sin 2\theta}{(1 + \cos\theta)^2} = \tan\dfrac{\theta}{2}$
44. $\dfrac{\sin\theta + \sin 2\theta}{\cos 2\theta + 3\cos\theta + 2} = \tan\dfrac{\theta}{2}$
45. Verify that whenever $\tan(\beta/2)$ is defined, it has the same sign as $\sin\beta$. [*Hint:* consider two cases, namely $\sin\beta > 0$ for $2n\pi < \beta < (2n + 1)\pi$, and $\sin\beta < 0$ for $(2n + 1)\pi < \beta < (2n + 2)\pi$, where n is any positive integer. For negative β, use $\sin(-\theta) = -\sin\theta$ and $\tan(-\theta) = -\tan\theta$.]

Product and Factor Identities

Use the identities in this section and the trigonometric functions of special angles to find the exact value of each of the following.

46. $\frac{1}{2}(\sin 75° + \sin 15°)$
47. $\frac{1}{2}(\cos 75° - \cos 15°)$
48. $\cos 75° + \cos 15°$
49. $\cos 15° \cos 45°$
50. $\sin 75° \cos 15°$

Express each of the following as a sum.

51. $\sin 3\theta \cos 2\theta$
52. $\cos 3\theta \cos 2\theta$
53. $\sin 3\theta \sin 2\theta$
54. $\cos 3\theta \sin 2\theta$
55. $\cos(-3\theta)\sin(-2\theta)$

Express each of the following as a product.

56. $\sin 3\theta + \sin\theta$
57. $\cos 3\theta + \cos\theta$
58. $\sin 3\theta - \sin 5\theta$
59. $\cos 5\theta - \cos 3\theta$
60. $\cos 3\theta - \cos 5\theta$

Prove the following identities.

61. $2\sin 3\theta\cos 4\theta = \sin 7\theta - \sin\theta$
62. $2\cos 3\theta\cos 4\theta = \cos 7\theta + \cos\theta$
63. $\dfrac{\sin 3\theta - \sin\theta}{\cos 3\theta + \cos\theta} = \tan\theta$
64. $\dfrac{\sin 3\theta - \sin\theta}{\cos\theta - \cos 3\theta} = \cot 2\theta$

By adding or subtracting sum and difference identities, derive the following product identities.

65. $\sin\alpha\sin\beta = -\frac{1}{2}[\cos(\alpha + \beta) - \cos(\alpha - \beta)]$
66. $\cos\alpha\cos\beta = \frac{1}{2}[\cos(\alpha + \beta) + \cos(\alpha - \beta)]$

67. $\cos\alpha \sin\beta = \frac{1}{2}[\sin(\alpha + \beta) - \sin(\alpha - \beta)]$

By letting $\alpha + \beta = \theta$ and $\alpha - \beta = \phi$ in the product identities, prove the following factor identities.

68. $\sin\theta - \sin\phi = 2\cos\left(\frac{\theta + \phi}{2}\right)\sin\left(\frac{\theta - \phi}{2}\right)$

69. $\cos\theta - \cos\phi = -2\sin\left(\frac{\theta + \phi}{2}\right)\sin\left(\frac{\theta - \phi}{2}\right)$

70. $\cos\theta + \cos\phi = 2\cos\left(\frac{\theta + \phi}{2}\right)\cos\left(\frac{\theta - \phi}{2}\right)$

Miscellaneous

Derive the following identities.

71. $\sin 3\theta = 3\sin\theta\cos^2\theta - \sin^3\theta$
(*Hint:* write $3\theta = \theta + 2\theta$.)

72. $\cos 3\theta = 4\cos^3\theta - 3\cos\theta$

73. $\sin^4\theta = \frac{3}{8} - \frac{1}{2}\cos 2\theta + \frac{1}{8}\cos 4\theta$
[*Hint:* write $\sin^4\theta = (\sin^2\theta)^2$.]

74. $\cos^4\theta = \frac{3}{8} + \frac{1}{2}\cos 2\theta + \frac{1}{8}\cos 4\theta$

75. $\dfrac{\sin(x + h) - \sin x}{h} = \cos\left(x + \frac{h}{2}\right)\dfrac{\sin\frac{h}{2}}{\frac{h}{2}}$

76. $\dfrac{\cos(x + h) - \cos x}{h} = -\sin\left(x + \frac{h}{2}\right)\dfrac{\sin\frac{h}{2}}{\frac{h}{2}}$

77. In steps (a) through (e), derive the **law of tangents,** which holds for any triangle ABC:

$$\frac{\tan\left(\frac{\alpha - \beta}{2}\right)}{\tan\left(\frac{\alpha + \beta}{2}\right)} = \frac{a - b}{a + b}.$$

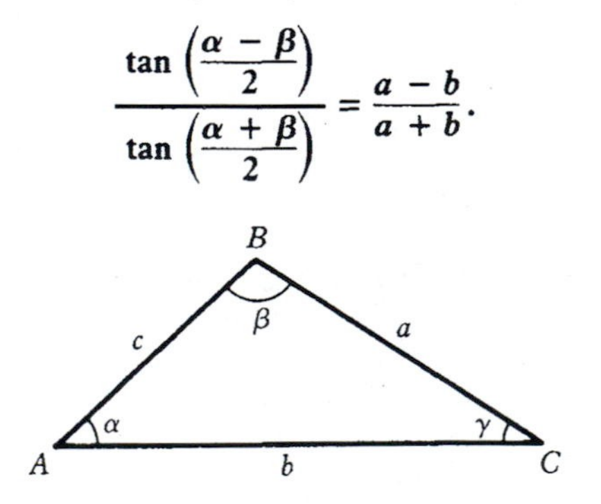

(a) $\frac{a}{b} = \frac{\sin\alpha}{\sin\beta}$ (*why?*)

(b) Show that $\frac{a - b}{b} = \frac{\sin\alpha - \sin\beta}{\sin\beta}$.

(c) Show that $\frac{a + b}{b} = \frac{\sin\alpha + \sin\beta}{\sin\beta}$.

(d) Show that $\frac{a - b}{a + b} = \frac{\sin\alpha - \sin\beta}{\sin\alpha + \sin\beta}$.

(e) Apply factor identities to $\frac{\sin\alpha - \sin\beta}{\sin\alpha + \sin\beta}$ in step (d).

78. **Heron's formula** for the area A of any triangle ABC states that

$$A = \sqrt{s(s - a)(s - b)(s - c)},$$

where $s = (1/2)(a + b + c)$ is one-half the perimeter of the triangle. Derive Heron's formula in the following steps.

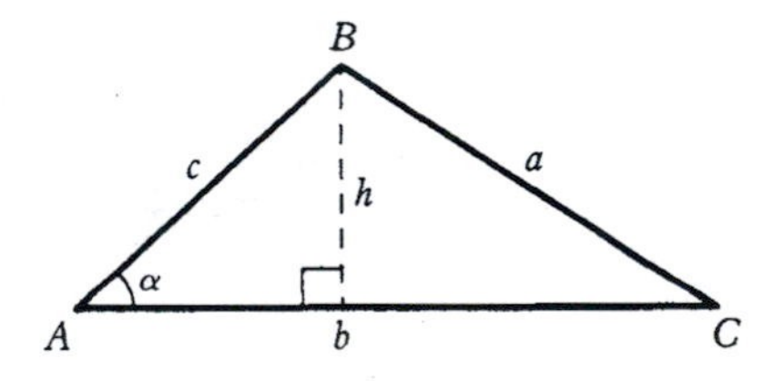

(a) Use the standard formula for the area of a triangle,

$$A = \frac{1}{2}bh,$$

to show that $A = (1/2)bc\sin\alpha$.

(b) Show that $A = bc\sin(\alpha/2)\cos(\alpha/2)$.

(c) Use the law of cosines in the form $\cos\alpha = (b^2 + c^2 - a^2)/(2bc)$ to show that

$$1 + \cos\alpha = \frac{(b + c + a)(b + c - a)}{2bc}$$

and

$$1 - \cos\alpha = \frac{(a + b - c)(a - b + c)}{2bc}.$$

(d) Show that $s - a = (1/2)(b + c - a)$, $s - b = (1/2)(a + c - b)$, and $s - c = (1/2)(a + b - c)$.

(e) Show that $\sin(\alpha/2) = \sqrt{\dfrac{(s-b)(s-c)}{bc}}$

and $\cos(\alpha/2) = \sqrt{\dfrac{s(s-a)}{bc}}$.

(f) Obtain Heron's formula for A by combining the results in parts (b) and (e) above.

(g) Why are only the positive square roots taken in part (e)?

7.9 Trigonometric Equations

If we can find one solution of a trigonometric equation, then we usually can find an infinite number of solutions.

Simple Trigonometric Equations

Trigonometric Equations Reducible to Simple Equations

***Approximate Solutions of Nonreducible Equations**[1]

In solving a right triangle we can say that if $\sin A = 1/2$, then $A = 30°$ since A must be an acute angle. For an oblique triangle, A can range from 0° to 180°, so the same equation $\sin A = 1/2$ could have the two solutions 30° and 150°. More generally, we have observed that the equation

$$\sin x = \frac{1}{2}$$

has an *infinite* number of solutions.

In this section we first consider the problem of finding all solutions to equations of the form

$$f(x) = c,$$

where c is a constant and f is one of the six trigonometric functions. These equations are called **simple trigonometric equations.** We then consider more general equations, such as

$$\cos 2x + \cos x = 0,$$

which can be reduced to simple equations by means of algebraic operations such as factoring or by using trigonometric identities. Finally, we indicate how to find approximate solutions to trigonometric equations that cannot be reduced to simple ones.

Simple Trigonometric Equations. We start with a graphical analysis of the equation

$$\sin x = c,$$

where c is a constant between -1 and 1. Because the sine has period 2π, we can obtain all possible solutions by first finding all solutions in the interval $[0, 2\pi]$ and then adding integral multiples of 2π to these. Also, we would like to express all solutions in terms of the unique solution

$$\sin^{-1} c,$$

[1]The asterisk indicates that this topic may be omitted depending on individual course needs.

which is contained in the interval $[-\pi/2, \pi/2]$, since this is the one that we get from a calculator. Figure 41 is the graph of $y = \sin x$ from $x = -\pi/2$ to 2π.

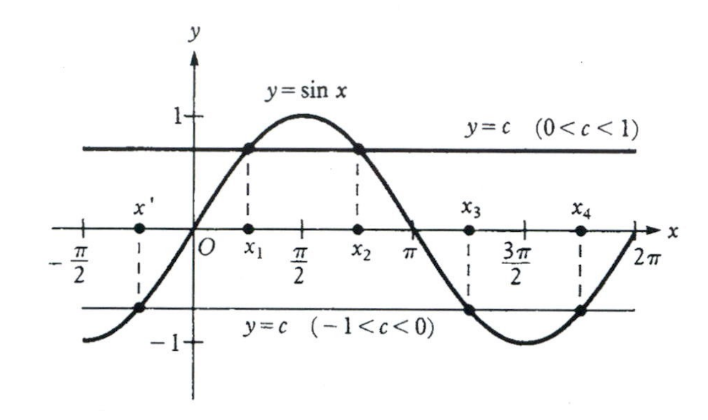

Figure 41

If $0 < c < 1$, then the line $y = c$ cuts the graph for $x = x_1$ and $x = x_2$, where

$$x_1 = \sin^{-1} c \qquad \text{and} \qquad x_2 = \pi - x_1.$$

Hence, all possible solutions are of the form

$$\sin^{-1} c + 2n\pi \qquad \text{and} \qquad (\pi - \sin^{-1} c) + 2n\pi, \tag{1}$$

where n is any integer. Now if $-1 < c < 0$, then the line $y = c$ cuts the graph for x', x_3, and x_4, where

$$x' = \sin^{-1} c, \qquad x_3 = \pi - x', \qquad \text{and} \qquad x_4 = x' + 2\pi.$$

Hence, we can again conclude that all possible solutions are of the form (1). The remaining special cases can also be deduced from the graph. They are

$$\left.\begin{aligned} \sin x = 0 &\Leftrightarrow x = n\pi, \\ \sin x = 1 &\Leftrightarrow x = \frac{\pi}{2} + 2n\pi, \\ \text{and} \qquad \sin x = -1 &\Leftrightarrow x = -\frac{\pi}{2} + 2n\pi, \end{aligned}\right\} \tag{2}$$

where n is any integer. If we examine results (1) and (2) closely, we see that they can be combined into one compact statement as follows.

> The solutions to the equation
>
> $$\sin x = c,$$
>
> where $-1 \le c \le 1$, are all the numbers
>
> $$\sin^{-1} c + 2n\pi \qquad \text{and} \qquad (\pi - \sin^{-1} c) + 2n\pi,$$
>
> where n is any integer. **(3)**

Although (3) is the simplest way to *state* the complete solution of $\sin x = c$, in practice we usually find the principal value of x, namely $\sin^{-1} c$, and then use the graph to find all other solutions.

Example 1 Find all solutions of each of the following.

(a) $\sin x = \sqrt{3}/2$ (b) $\sin x = -.4462$

Solution:

(a) We know that $\sin^{-1}(\sqrt{3}/2) = \pi/3$ since $\pi/3$ is a special angle. Hence, we conclude that the solutions in $[0, 2\pi]$ are $\pi/3$ and $\pi - \pi/3 = 2\pi/3$, as shown in Figure 42(a). Hence, the solutions are

$$\frac{\pi}{3} + 2n\pi \qquad \text{and} \qquad \frac{2\pi}{3} + 2n\pi,$$

where n is any integer.

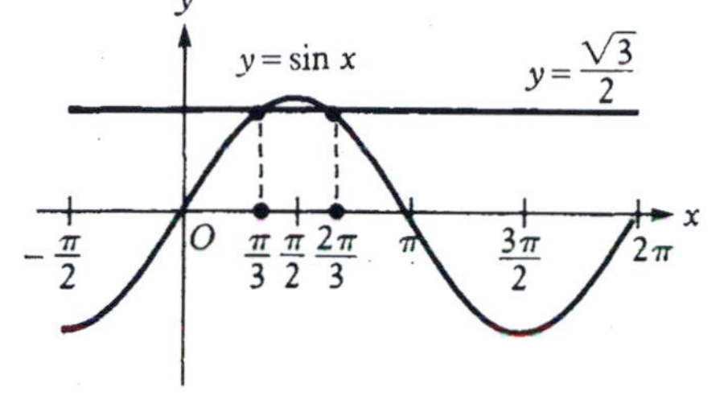

(a)

(b) By means of a calculator, we find that $\sin^{-1}(-.4462) = -.4625$ radians. Hence, we conclude that the solutions are

$$-.4625 + 2n\pi \qquad \text{and} \qquad (\pi + .4625) + 2n\pi,$$

where n is any integer [see Figure 42(b)]. ■

(b)

Figure 42

We now consider the equation

$$\cos x = c,$$

where, as before, c is a constant between -1 and 1. First, we observe the following special cases from the graph of $y = \cos x$ in Figure 43:

$$\left.\begin{aligned} \cos x = 0 &\Leftrightarrow x = \frac{\pi}{2} + n\pi, \\ \cos x = 1 &\Leftrightarrow x = 2n\pi, \\ \text{and} \quad \cos x = -1 &\Leftrightarrow x = \pi + 2n\pi, \end{aligned}\right\} \tag{4}$$

where n is any integer. Now, if $0 < c < 1$ or $-1 < c < 0$, then the line $y = c$ cuts the graph of $y = \cos x$ for

$$x = \cos^{-1} c + 2n\pi \qquad \text{and} \qquad -\cos^{-1} c + 2n\pi, \tag{5}$$

where n is any integer.

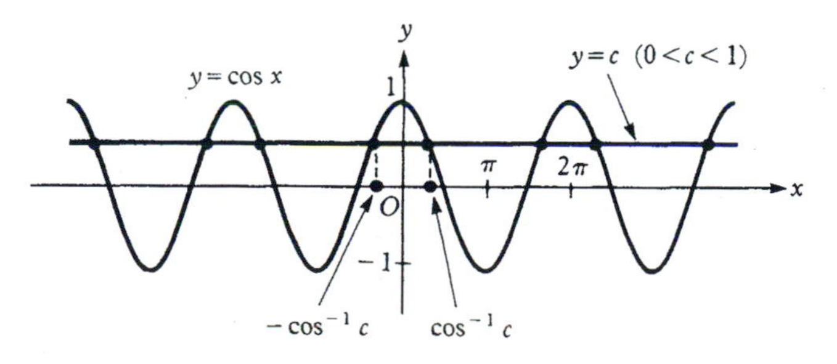

Figure 43

The conclusions in (4) and (5) can be combined into the following single statement.

> The solutions to the equation
>
> $$\cos x = c$$
>
> where $-1 \le c \le 1$, are all numbers
>
> $$\pm\cos^{-1} c + 2n\pi,$$
>
> where n is any integer. **(6)**

Example 2 Find all solutions of each of the following.

(a) $\cos x = -\dfrac{\sqrt{2}}{2}$ (b) $\cos x = .9474$

Solution:

(a)
$$\cos^{-1}\left(-\frac{\sqrt{2}}{2}\right) = \frac{\pi}{2} - \sin^{-1}\left(-\frac{\sqrt{2}}{2}\right)$$
$$= \frac{\pi}{2} - \left(-\frac{\pi}{4}\right)$$
$$= \frac{3\pi}{4}$$

Hence, by (6), the solutions are

$$\pm\frac{3\pi}{4} + 2n\pi,$$

where n is any integer [see Figure 44(a)].

(b) Using a calculator, we find that $\cos^{-1} .9474 = .3258$ radians. Hence, from (6), the solutions are

$$\pm.3258 + 2n\pi,$$

where n is any integer [see Figure 44(b)]. ■

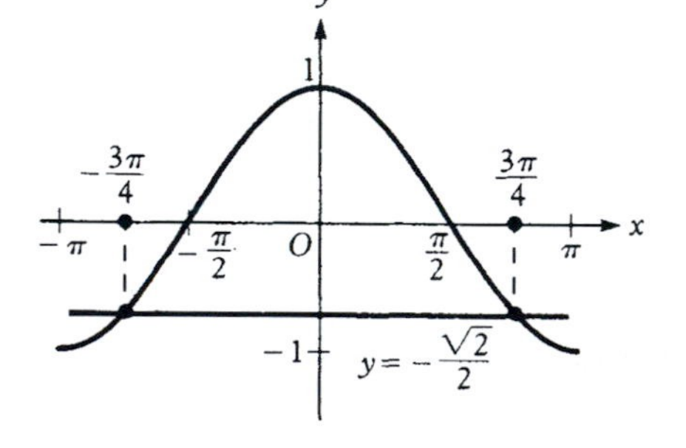

(a)

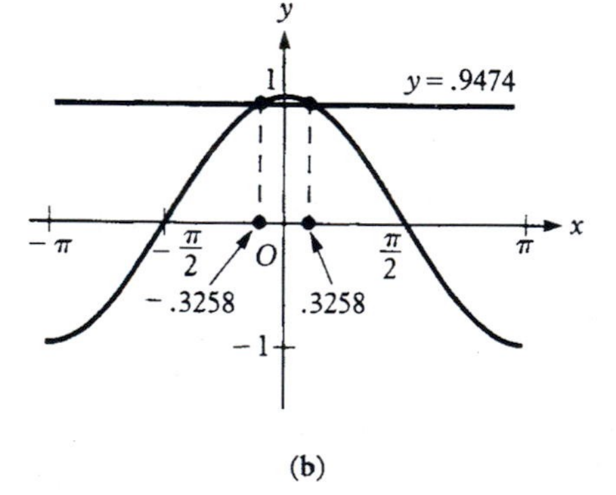

(b)

Figure 44

The simple trigonometric equation

$$\tan x = c$$

lives up to its name since for any real number c, there is exactly one solution, namely

$$\tan^{-1} c$$

in the interval $(-\pi/2, \pi/2)$ (Figure 45). Then, because the tangent has period π, we have the following result.

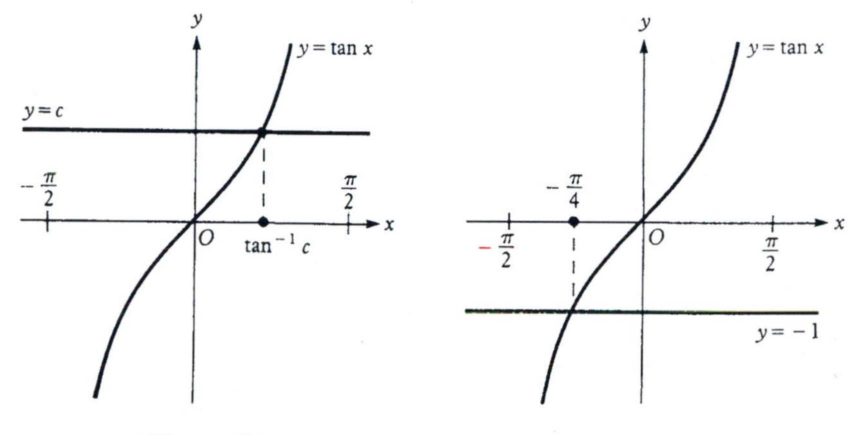

Figure 45 **Figure 46**

> The solutions of the equation
>
> $$\tan x = c,$$
>
> for any real number c, are
>
> $$\tan^{-1} c + n\pi,$$
>
> where n is any real number.

Example 3 Solve $\tan x = -1$.

Solution: Since $\tan^{-1}(-1) = -\pi/4$, the solutions are

$$-\frac{\pi}{4} + n\pi,$$

where n is any integer (Figure 46). ■

Simple trigonometric equations for the secant, cosecant, and cotangent can be solved by using the reciprocal identities.

Example 4 Solve each of the following.

(a) $\sec x = -\dfrac{2}{\sqrt{2}}$ (b) $\csc x = \dfrac{2}{\sqrt{3}}$ (c) $\cot x = -1$

Solution:

(a) The given equation is equivalent to $\cos x = -\sqrt{2}/2$, which was solved in Example 2. Hence, the solutions are

$$\pm\frac{3\pi}{4} + 2n\pi,$$

where n is any integer.

(b) The equation here is equivalent to $\sin x = \sqrt{3}/2$, which was solved in Example 1. Therefore, the solutions are

$$\frac{\pi}{3} + 2n\pi \qquad \text{and} \qquad \frac{2\pi}{3} + 2n\pi,$$

where n is any integer.

(c) This equation is equivalent to $\tan x = -1$, which we solved in Example 3. Hence, the solutions are

$$-\frac{\pi}{4} + n\pi,$$

where n is any integer. ■

Trigonometric Equations Reducible to Simple Equations. If a trigonometric equation involves several functions or several angles, then it may be equivalent to a system of one or more simple equations. There is no general method here, but factoring, substitution, and trigonometric identities are often useful, as illustrated in the following examples.

Example 5 Solve $\sin 2x = 1/2$.

Solution: If we substitute $t = 2x$, then the solutions of the simple trigonometric equation $\sin t = 1/2$ are

$$t = \frac{\pi}{6} + 2n\pi \qquad \text{and} \qquad \frac{5\pi}{6} + 2n\pi,$$

where n is any integer. Since $x = t/2$, the solutions of the original equation are

$$x = \frac{\pi}{12} + n\pi \qquad \text{and} \qquad \frac{5\pi}{12} + n\pi,$$

where n is any integer. ■

Example 6 Solve $3 \sin x - \sin x \sin 2x = 0$.

Solution: By factoring, we obtain

$$\sin x\,(3 - \sin 2x) = 0,$$

which is satisfied if and only if either

$$\sin x = 0$$

or

$$3 - \sin 2x = 0.$$

The solutions of $\sin x = 0$ are $n\pi$, but $3 - \sin 2x = 0$ has no solutions (*why?*). Therefore, the solutions of the given equation are $n\pi$, where n is any integer. ■

Example 7 Solve $3 \sin x - 2 \cos^2 x + 3 = 0$.

Solution: First we use the identity $\cos^2 x = 1 - \sin^2 x$ to replace the given equation by the equivalent one

$$3 \sin x - 2(1 - \sin^2 x) + 3 = 0$$

or

$$2 \sin^2 x + 3 \sin x + 1 = 0.$$

This equation is a quadratic in $\sin x$ that factors into

$$(2 \sin x + 1)(\sin x + 1) = 0,$$

which means that either

$$\text{(a)}\ 2 \sin x + 1 = 0;\ \text{that is, } \sin x = -\frac{1}{2};$$

or

$$\text{(b)}\ \sin x + 1 = 0;\ \text{that is, } \sin x = -1.$$

By our methods for simple trigonometric equations, the solutions to (a) are

$$-\frac{\pi}{6} + 2n\pi \quad \text{and} \quad \left(\pi + \frac{\pi}{6}\right) + 2n\pi,$$

and the solutions to (b) are

$$-\frac{\pi}{2} + 2n\pi,$$

where n is any integer. The set of all values solving either (a) or (b) forms the solution of the original equation. ■

Example 8 Solve $\cos 2x + \cos x = 0$.

Solution: By the double-angle identity $\cos 2x = 2 \cos^2 x - 1$, we can replace the given equation with the equivalent one

$$2 \cos^2 x + \cos x - 1 = 0,$$

which can be factored into

$$(2 \cos x - 1)(\cos x + 1) = 0.$$

This product is zero if and only if either

$$\text{(a)}\ 2 \cos x - 1 = 0;\ \text{that is, } \cos x = \frac{1}{2};$$

or

$$\text{(b)}\ \cos x + 1 = 0;\ \text{that is, } \cos x = -1.$$

The solutions to (a) are

$$\pm\frac{\pi}{3} + 2n\pi,$$

and the solutions to (b) are

$$\pi + 2n\pi,$$

where n is any integer. The set of all these values is the solution to the original equation. ■

Approximate Solutions of Nonreducible Equations. As the previous examples show, solving trigonometric equations is not easy. Although we were able to reduce the equations in Examples 6, 7, and 8 to simple ones, we do not have to look far to find equations that cannot be simplified. For instance, our previous methods are not sufficient to reduce

$$\sin x + 2\cos^3 x = 1$$

to simple equations. Such equations can be solved only by numerical methods that are essentially "trial and error." That is, we substitute values of x, generated in some systematic manner, into the equation until we discover all possible solutions within a given margin of error. Although there are many numerical methods available, we will describe a very basic one to give the general idea. All numerical methods involve lengthy computations and are best done on a calculator or computer.

Suppose we have a trigonometric equation

$$T(x) = 0, \tag{1}$$

where $T(x)$ is an algebraic expression of trigonometric functions of x, for example, $T(x) = \sin x + 2\cos^3 x - 1$. For simplicity, we assume that each trigonometric function in $T(x)$ is a function of x alone, not $2x$ or $x/2$, for example. Then each function in $T(x)$ repeats its values at least every 2π units, and therefore $T(x)$ repeats *its* values at least every 2π units. Hence, we have to determine the solutions of (1) only on the interval $[0, 2\pi]$ in order to find all possible solutions.

The first step is to graph $y = T(x)$ on $[0, 2\pi]$ in order to get a rough estimate of the solutions to $T(x) = 0$. This is done for

$$y = \sin x + 2\cos^3 x - 1$$

in Figure 47 using the values in the table at the top of the facing page.

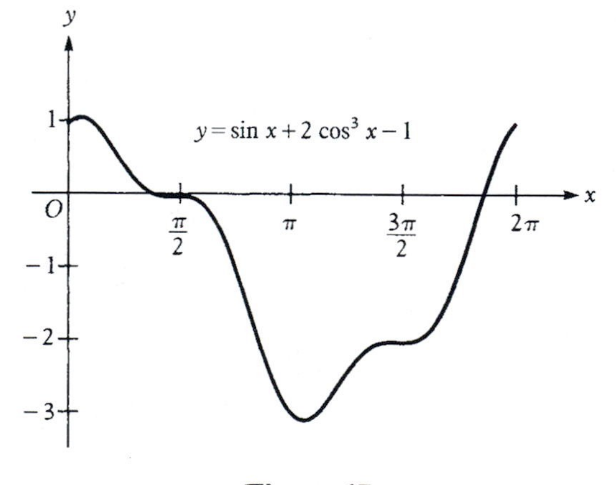

Figure 47

x	0	$\pi/4$	$\pi/2$	$3\pi/4$	π	$5\pi/4$	$3\pi/2$	$7\pi/4$	2π
y	1	.4	0	−1	−3	−2.4	−2	−1	1

From the table and the figure, we see that $\pi/2$ is an exact solution, and there is another solution between $7\pi/4$ and 2π.

The next step is to get a more accurate estimate of the solution between $7\pi/4$ and 2π. Suppose we want our answer to the nearest tenth. Since the length of the interval $[7\pi/4, 2\pi]$ is $\pi/4$, which is less than 1, we can partition the interval $[7\pi/4, 2\pi]$ into 10 equal parts and compute $T(x)$ at each point of the partition. Note that 1/10 of $\pi/4$ is $\pi/40$. Therefore, starting at $x = 7\pi/4 = 70\pi/40$, we add in increments of $\pi/40$ until we get to $x = 80\pi/40 = 2\pi$. The corresponding values of $y = T(x)$ obtained on a calculator are shown in the tables below.

x	$70\pi/40$	$71\pi/40$	$72\pi/40$	$73\pi/40$	$74\pi/40$	$75\pi/40$
y	−1	−.77	−.53	−.28	−.04	.19

x	$76\pi/40$	$77\pi/40$	$78\pi/40$	$79\pi/40$	$80\pi/40$
y	.41	.61	.77	.90	1

Since $74\pi/40$ gives the value of y closest to 0, we take $74\pi/40$ as an approximate solution of $T(x) = 0$. The solutions of the equation

$$\sin x + 2\cos^3 x - 1 = 0$$

are

$$\frac{\pi}{2} + 2n\pi \quad \text{and} \quad \frac{74\pi}{40} + 2n\pi,$$

where n is any integer.

Comment When solving trigonometric equations, even simple ones, we are usually finding approximate values of solutions. In the case of simple equations or equations reducible to simple ones, this is true because calculators and computers provide us answers only to a limited number of decimal places. In the case of equations not reducible to simple ones, the *method of solution* itself leads to approximate values. However, by making enough computations, we can achieve accuracy to any desired degree. For instance, for $T(x) = \sin x + 2\cos^3 x - 1 = 0$, we computed $T(74\pi/40) = -.04$ and $T(75\pi/40) = .19$, and we selected $74\pi/40$ as an approximate value of a solution. We could improve the accuracy by dividing the interval from $74\pi/40$ to $75\pi/40$ into 10 equal parts and repeating the above process. This would lead to another interval that could be subdivided, and so on, until we had achieved the desired accuracy.

Exercises 7.9

Fill in the blanks to make each statement true.

1. The number of solutions of the trigonometric equation $\sin x = c$, where c is any real number, is either ______ or ______.
2. To solve the equation $\sec x = c$, we may use the identity ________.
3. To solve the equation $a \cos^2 x + b \cos x + c = 0$ for x, we first solve the equation for ______.
4. All of the solutions of $\tan x = c$ can be obtained from the solution in any interval of length ______.
5. Approximate solutions of trigonometric equations arise from two sources. For simple equations and equations reducible to simple ones, the source is ____________; for other trigonometric equations, another source is ____________.

Write true *or* false *for each statement.*

6. All solutions of any trigonometric equation can be obtained from the solutions in $[0, 2\pi]$.
7. To solve the equation $\sin 2x = c$, we must use a double-angle identity.
8. If $x_0 = \tan^{-1} c$ is one solution of the equation $\tan x = c$, then all solutions are of the form $x_0 + n\pi$, where n is any integer.
9. If x_0 is a solution of $\sin x = c$, then so is $\pi - x_0$.
10. Every trigonometric equation has an infinite number of solutions.

Simple Trigonometric Equations

In Exercises 11–20, find all solutions in $[0, 2\pi]$.

11. $\sin x = \frac{\sqrt{3}}{2}$
12. $\cos x = -\frac{\sqrt{3}}{2}$
13. $\tan x = -\frac{1}{\sqrt{3}}$
14. $\sin x = 1$
15. $\cos x = 1$
16. $\tan x = 0$
17. $\sin 2x = 1$
18. $\cos 3x = \frac{1}{2}$
19. $\tan \frac{x}{2} = -1$
20. $\sin 4x = 4$

In Exercises 21–26, find all solutions.

21. $\sin x = -\frac{\sqrt{2}}{2}$
22. $\cos x = \frac{1}{\sqrt{2}}$
23. $\tan x = 0$
24. $\cot x = 1$
25. $\sec x = 1$
26. $\csc x = -1$

In Exercises 27–32, find all solutions in $[-\pi, \pi]$.

27. $\sin x = .5760$
28. $\cos x = -.7431$
29. $\tan x = 1.130$
30. $\cot x = -1.130$
31. $\sec x = 1.509$
32. $\csc x = -1.766$

Equations Reducible to Simple Equations

In Exercises 33–38, find all solutions in $[0, 2\pi)$.

33. $\sin x + 1 = 0$
34. $4 \sin^2 x - 1 = 0$
35. $\sqrt{2} - \sec x = 0$
36. $2 \csc x - 1 = 0$
37. $\sin \frac{x}{2} + \frac{1}{2} = 0$
38. $\sin^2 2x - \frac{1}{2} = 0$

In Exercises 39–49, find all solutions in the given intervals.

39. $2 \sin^2 x - \sin x - 1 = 0, \quad [0, 2\pi)$
40. $\sin x \cos x - \cos x = 0, \quad [0, 2\pi)$
41. $\tan x \sec x + \tan x = 0, \quad [-\pi, \pi]$
42. $4 \sin^2 x \tan x - 3 \tan x = 0, \quad [0, 2\pi)$
43. $(\cos 2\theta - 2)(\tan^2 \theta + 1) = 0, \quad [0, 2\pi)$
44. $2 \cos^2 \theta - 3 \cos \theta - 2 = 0, \quad [0°, 360°)$
45. $\tan^2 x + 2 \tan x + 1 = 0, \quad (-\infty, \infty)$
46. $2 \sin^2 x + \sin x - 3 = 0, \quad (-\infty, \infty)$
47. $\sin^2 x - \cos^2 x = 0, \quad [0, 2\pi)$
48. $2 \cos x \tan x + \tan x + 2 \cos x + 1 = 0, \quad (-\infty, \infty)$
49. $2 \sin^2 x - \sqrt{3} \sin x = 0, \quad [0°, 360°)$

In Exercises 50–60, find all solutions in $[0, 2\pi)$.

50. $\cos^2 x - \sin^2 x - 3 \sin x - 2 = 0$
51. $2 \cos^2 x + \sin x = 2$
52. $\cos 2x + \sin^2 2x = 0$
53. $\cos 2x - 3 \sin x - 2 = 0$

54. $\sin^2 x - \cos^2 x - \cos x = 1$

55. $\cos 2x + \cos x = 0$

56. $\sin 2x - \cos x = 0$

57. $\sin^2 \frac{x}{2} - \cos^2 x = 0$

58. $\tan x + \cot x + 2 = 0$

59. $\csc^2 x - \cot x - 1 = 0$

60. $\tan \frac{\theta}{2} + \sin 2\theta - \sin \theta = 0$

In Exercises 61–64, find all solutions for x in $[0, 2\pi)$.

61. $\sin^2 x + \sin x - 1 = 0$
(*Hint:* use the quadratic formula.)

62. $2 \sin^2 x + 2 \sin x - 3 = 0$

63. $\cos x + \cos 3x = 0$
(*Hint:* use a factor identity.)

64. $\sin x - \sin 3x = 0$

Approximate Solutions of Nonreducible Equations

Using a programmable calculator, proceed as in the text to find approximate solutions to each of the following equations $T(x) = 0$. *That is, first graph* $y = T(x)$ *on* $[0, 2\pi]$ *in increments of* $\pi/4$ *to find the interval(s) of length* $\pi/4$ *that contain a root. Subdivide each interval that contains a root into 10 equal parts and compute* $T(x)$ *at each point of the subdivision. Select the point that gives the smallest magnitude of* $T(x)$.

65. $2 \sin^3 x + \cos x + 1 = 0$

66. $3 \sin^2 x + 4 \cos^3 x - 1 = 0$

67. $\cos x + 2 \cos^2 x + 3 \sin x = 0$

68. $\sin^3 x + \cos^3 x - 4 \sin x = 0$

69. $\sin^3 x + 3 \cos x - 2 = 0$

70. $\cos^3 x - 5 \sin x + 4 = 0$

Chapter 7 Review Outline

7.1 Trigonometric Functions with Real Domains

Definitions

The trigonometric functions are defined with real numbers as domain by letting any real number s equal the directed length of an arc from $A(1, 0)$ to $P(x, y)$ on the unit circle $x^2 + y^2 = 1$. Then

$$\sin s = y \qquad \cos s = x$$

$$\tan s = \frac{\sin s}{\cos s} \qquad \cot s = \frac{\cos s}{\sin s}$$

$$\sec s = \frac{1}{\cos s} \qquad \csc s = \frac{1}{\sin s}.$$

7.2 Graphs of the Trigonometric Functions

Periodic Function:
A function f is said to be periodic with period p if p is the smallest positive number for which $f(t + p) = f(t)$ for all t in the domain of f.

Domain, Range, and Periodicity:
The sine and cosine have domain $(-\infty, \infty)$, range $[-1, 1]$, and period 2π.

The tangent and secant have as domain all real numbers except $\pi/2 + n\pi$, where n is any integer. The cotangent and cosecant have as domain all real numbers except $n\pi$, where n is any integer.

The tangent and cotangent have range $(-\infty, \infty)$ and period π.

The secant and cosecant have range $(-\infty, -1] \cup [1, \infty)$ and period 2π.

Asymptotes:
The tangent and secant have vertical asymptotes at the points $\pi/2 + n\pi$, where n is any integer. The cotangent and cosecant have vertical asymptotes at $n\pi$, where n is any integer.

7.3 Amplitude, Period, and Phase Shift

$f(t) = A \sin (Bt + C)$ and $f(t) = A \cos (Bt + C)$, with $B > 0$, have amplitude $|A|$, period $2\pi/B$, and phase shift $-C/B$. The shift is to the left if $-C/B < 0$ and to the right if $-C/B > 0$.

7.4 Applications of Trigonometric Functions

Examples of simple harmonic motion, represented by the equations

$$u = A \sin (Bt + C) \quad \text{or} \quad u = A \cos (Bt + C),$$

include the motion of a pendulum, a vibrating spring, and an electric circuit.

7.5 Inverse Trigonometric Functions

Definitions

$y = \sin^{-1} x$ is the inverse of the restricted sine function

$y = \sin x$, with $-\frac{\pi}{2} \le x \le \frac{\pi}{2}$.

$y = \tan^{-1} x$ is the inverse of the restricted tangent function

$y = \tan x$, with $-\frac{\pi}{2} < x < \frac{\pi}{2}$.

$$\cos^{-1} x = \frac{\pi}{2} - \sin^{-1} x$$

$$\cot^{-1} x = \frac{\pi}{2} - \tan^{-1} x$$

$$\sec^{-1} x = \cos^{-1} \frac{1}{x}$$

$$\csc^{-1} x = \sin^{-1} \frac{1}{x}$$

7.6 Basic Identities

Pythagorean Identities:

$$\sin^2 \theta + \cos^2 \theta = 1$$

$$\tan^2 \theta + 1 = \sec^2 \theta$$

$$\cot^2 \theta + 1 = \csc^2 \theta$$

Quotient and Reciprocal Identities:

$$\tan \theta = \frac{\sin \theta}{\cos \theta} \qquad \cot \theta = \frac{\cos \theta}{\sin \theta}$$

$$\cot \theta = \frac{1}{\tan \theta} \qquad \sec \theta = \frac{1}{\cos \theta}$$

$$\csc \theta = \frac{1}{\sin \theta}$$

Negative Angle Identities:

$\sin (-\theta) = -\sin \theta$ and $\cos (-\theta) = \cos \theta$

Cofunction Identities:

$$\sin \left(\frac{\pi}{2} - \theta\right) = \cos \theta$$

$$\cos \left(\frac{\pi}{2} - \theta\right) = \sin \theta$$

π-Identities:

$$\sin (\pi \pm \theta) = \mp\sin \theta$$

$$\cos (\pi \pm \theta) = -\cos \theta$$

7.7 Sum and Difference Identities

$$\sin (\alpha \pm \beta) = \sin \alpha \cos \beta \pm \cos \alpha \sin \beta$$

$$\cos (\alpha \pm \beta) = \cos \alpha \cos \beta \mp \sin \alpha \sin \beta$$

7.8 Multiple-Angle Identities

Double-Angle Identities:

$$\sin 2\alpha = 2 \sin \alpha \cos \alpha$$

$$\begin{aligned} \cos 2\alpha &= \cos^2 \alpha - \sin^2 \alpha \\ &= 2 \cos^2 \alpha - 1 \\ &= 1 - 2 \sin^2 \alpha \end{aligned}$$

Half-Angle Identities:

$$\sin \frac{\beta}{2} = \pm\sqrt{\frac{1 - \cos \beta}{2}}$$

$$\cos \frac{\beta}{2} = \pm\sqrt{\frac{1 + \cos \beta}{2}}$$

$$\tan \frac{\beta}{2} = \frac{\sin \beta}{1 + \cos \beta} = \frac{1 - \cos \beta}{\sin \beta}$$

Product Identities:

$$\sin \alpha \cos \beta = \frac{1}{2}[\sin (\alpha + \beta) + \sin (\alpha - \beta)]$$

$$\cos \alpha \sin \beta = \frac{1}{2}[\sin (\alpha + \beta) - \sin (\alpha - \beta)]$$

$$\sin \alpha \sin \beta = -\frac{1}{2}[\cos (\alpha + \beta) - \cos (\alpha - \beta)]$$

$$\cos \alpha \cos \beta = \frac{1}{2}[\cos (\alpha + \beta) + \cos (\alpha - \beta)]$$

Factor Identities:

$$\sin \theta + \sin \phi = 2 \sin \left(\frac{\theta + \phi}{2}\right) \cos \left(\frac{\theta - \phi}{2}\right)$$

$$\sin \theta - \sin \phi = 2 \cos \left(\frac{\theta + \phi}{2}\right) \sin \left(\frac{\theta - \phi}{2}\right)$$

$$\cos \theta - \cos \phi = -2 \sin \left(\frac{\theta + \phi}{2}\right) \sin \left(\frac{\theta - \phi}{2}\right)$$

$$\cos \theta + \cos \phi = 2 \cos \left(\frac{\theta + \phi}{2}\right) \cos \left(\frac{\theta - \phi}{2}\right)$$

7.9 Trigonometric Equations

Simple Trigonometric Equations:

Equation	All solutions (n any integer)
$\sin x = c$	$\sin^{-1} c + 2n\pi$ and $\pi - \sin^{-1} c + 2n\pi$
$\cos x = c$	$\pm\cos^{-1} c + 2n\pi$
$\tan x = c$	$\tan^{-1} c + n\pi$

Trigonometric Equations Reducible to Simple Ones:
No general method exists for reduction, but substitution, trigonometric identities, and algebraic techniques such as factoring are often useful.

Nonreducible Equations:
These are solved by a combination of graphical and numerical approximation.

Chapter 7 Review Exercises

1. Give two interpretations of the real number t in $f(t) = \sin t$.
2. For the point $(1/\sqrt{2}, -1/\sqrt{2})$ on the unit circle, determine all corresponding values of the arc length s measured from $(1, 0)$.
3. Find $\sin s$ and $\cos s$ for the values of s determined in Exercise 2.
4. If $\sin t = -.2$, find $\sin (t + 4\pi)$.
5. If $\sin t = .9$, find $\sin (3\pi - t)$.
6. Given $\cos t = .7$, find $\cos (t + 3\pi)$.
7. Given $\cos t = -.7$, find $\sin (t - 3\pi/2)$.

In Exercises 8–12, sketch a graph of the given equation on the given interval.

8. $u = 5 \sin t, \quad [0, 2\pi]$
9. $u = \cos \frac{\pi}{2}t, \quad [0, 4]$
10. $u = 4 \sin \left(\frac{t}{2} - \frac{\pi}{4}\right), \quad \left[\frac{\pi}{2}, \frac{9\pi}{2}\right]$
11. $u = \tan \pi t, \quad [0, 1]$
12. $u = \sec 2t, \quad [0, \pi]$

In Exercises 13–16, evaluate in terms of π.

13. $\sin^{-1}\left(-\frac{1}{\sqrt{2}}\right)$
14. $\tan^{-1}\frac{1}{\sqrt{3}}$
15. $\sec^{-1}(-\sqrt{2})$
16. $\cos^{-1}\frac{\sqrt{3}}{2} + \sin^{-1}\left(-\frac{1}{2}\right)$

In Exercises 17–20, find the exact value.

17. $\tan (\tan^{-1} 1)$
18. $\sin^{-1}\left(\sin \frac{7\pi}{6}\right)$
19. $\cos^{-1}\left(\sin \frac{3\pi}{2}\right)$
20. $\cos (\sin^{-1} (-1))$
21. Simplify $\cos \theta + \sin \theta \tan \theta$.
22. Simplify $\dfrac{\cos \theta + \cot \theta}{1 + \sin \theta}$.

In Exercises 23–26, prove the given identities.

23. $\sec \theta - \cos \theta = \tan \theta \sin \theta$
24. $\cos (\alpha + \beta) \cos (\alpha - \beta) = \cos^2 \alpha + \cos^2 \beta - 1$
25. $\cos^4 \theta - \sin^4 \theta = \cos 2\theta$
26. $\dfrac{\tan \theta - \sin \theta}{2 \tan \theta} = \sin^2 \dfrac{\theta}{2}$
27. Use the sum and difference identities to prove that $\cos (\theta + \pi/6) = -\sin (\theta - \pi/3)$.

In Exercises 28–31, find the exact value of each expression.

28. $2 \sin 75° \cos 75°$
29. $1 - 2 \sin^2 \frac{\pi}{12}$
30. $\cos \frac{5\pi}{8}$
31. $\sin 105° + \sin 15°$
32. Express $\sin 4\theta \sin 3\theta$ as a sum or difference.
33. Express $\sin 5\theta + \sin 3\theta$ as a product.

In Exercises 34–37, find all solutions of the given trigonometric equations.

34. $\sin x = -1/2$

35. $\tan 2x = -1$

36. $2 \sin^2 x + 3 \sin x + 1 = 0$

37. $\sin 2x + \sin x = 0$

38. A weight on the end of a pendulum of length L ft swings through an arc s ft in t sec, where $s(t) = s_0 \cos(\sqrt{32/L}\, t)$. What length L will make the period 1 sec?

39. The electromotive force E (volts) and current I (amperes) in an electrical circuit satisfy $E = 220 \sin(120\pi t)$ and $I = 20 \sin(120\pi t - \pi/4)$.

(a) Find the frequency (cycles/second) for E and I, respectively.

(b) Find the maximum values of E and I, respectively, and the corresponding (positive) values for t.

40. A weight of w lb hanging on the end of a spring, when displaced vertically y_0 ft from equilibrium and then released, has an equation of motion $y(t) = y_0 \cos(\sqrt{kg/w}\, t)$, where $y(t)$ is the vertical displacement from the equilibrium position (positive upward). Here $g = 32$ ft/sec^2 and k is the spring constant. If an 8-lb weight with spring constant 16 lb/ft is pulled downward 3 in below equilibrium and released, find the amplitude and frequency of the resulting motion.

Graphing Calculator Exercises

In each of the following, verify your answers to the indicated review exercises of this section by using a graphing calculator for the given functions and windows $[x_{min}, x_{max}]$, $[y_{min}, y_{max}]$. (If π cannot be entered in a window, use 3.14.)

41. Exercise 5: $Y_1 = \sin x - .9$, $Y_2 = \sin(3\pi - x)$, $[0,2\pi]$, $[-2,1.5]$

42. Exercise 6: $Y_1 = \cos x - .7$, $Y_2 = \cos(x + 3\pi)$, $[0,2\pi]$, $[-2,1.5]$

43. Exercise 7: $Y_1 = \cos x + .7$, $Y_2 = \sin(x - 3\pi/2)$, $[0,2\pi]$, $[-1.5, 2]$

44. Exercise 9: $Y_1 = \cos(\pi x/2)$, $[0,4]$, $[-1.5,1.5]$

45. Exercise 10: $Y_1 = 4\sin(x/2 - \pi/4)$, $[\pi/2,9\pi/2]$, $[-4.5,4.5]$

46. Exercise 11: $Y_1 = \tan(\pi x)$, $[0,.95]$, $[-10,10]$

47. Exercise 12: $Y_1 = 1/\cos(2x)$, $[0,\pi]$, $[-10,10]$

48. Exercise 23: $Y_1 = 1/\cos x - \cos x - (\tan x)(\sin x)$, $[0,2\pi]$, $[-10,10]$

49. Exercise 25: $Y_1 = (\cos x)^4 - (\sin x)^4 - \cos(2x)$, $[0,2\pi]$, $[-2,2]$

50. Exercise 26: $Y_1 = (\tan x - \sin x)/(2\tan x) - (\sin(x/2))^2$, $[0,2\pi]$, $[-10,10]$

51. Exercise 34: $Y_1 = \sin x + 1/2$, $[-2\pi,2\pi]$, $[-1,2]$

52. Exercise 35: $Y_1 = \tan(2x) + 1$, $[-\pi,\pi]$, $[-10,10]$

53. Exercise 36: $Y_1 = 2(\sin x)^2 + 3\sin x + 1$, $[-2\pi,2\pi]$, $[-1.5,6.5]$

54. Exercise 37: $Y_1 = \sin(2x) + \sin x$, $[0,2\pi]$, $[-2.5,2.5]$

8 Trigonometry and Coordinate Geometry

In this chapter we apply trigonometry to magnitude and direction in the plane. We introduce *polar coordinates*, which enable us to graph functions for which the independent variable is an *angle* measured from the positive x-axis, and the dependent variable is *distance* from the origin. We also use polar coordinates to represent complex numbers, and this representation leads to a simple geometric interpretation of multiplication in the complex number system.

Physical quantities that are measured by magnitude and direction, such as force or velocity, can be represented geometrically as directed line segments in the plane, called *vectors*. We show how trigonometry can be used to measure the direction of one vector relative to another and how vectors, in turn, can be used to describe lines in the plane and planes in space.

8.1 Polar Coordinates

Polar coordinates are the coordinates of the stars.

Polar and Rectangular Coordinates
Equivalent Polar Coordinates

Polar and Rectangular Coordinates. To define polar coordinates in the plane, we take as reference a fixed point O, called the **pole,** and a fixed ray from O, called the **polar axis** (Figure 1). Now let P be any point in the plane other than O. If r is the length of the line segment OP, and θ is an angle from the polar axis to OP, then the numbers r and θ are called **polar coordinates** for P and are denoted by the ordered pair $[r, \theta]$[1]. Positive angles θ correspond to counterclockwise rotations and negative angles to clockwise rotations.

Example 1 Graph the following points by using their polar coordinates.

(a) $P[2, \pi/4]$ (b) $Q[1, \pi]$ (c) $R[3/2, -30°]$

Solution: The three points are graphed in parts (a), (b), and (c) of Figure 2. ■

To distinguish them from polar coordinates, our previously defined coordinates (x, y) of P are called the **rectangular coordinates** of P. To work with both coordinate systems in the same context, we position the origin $(0, 0)$ of the rectangular system at the pole O with the positive x-axis

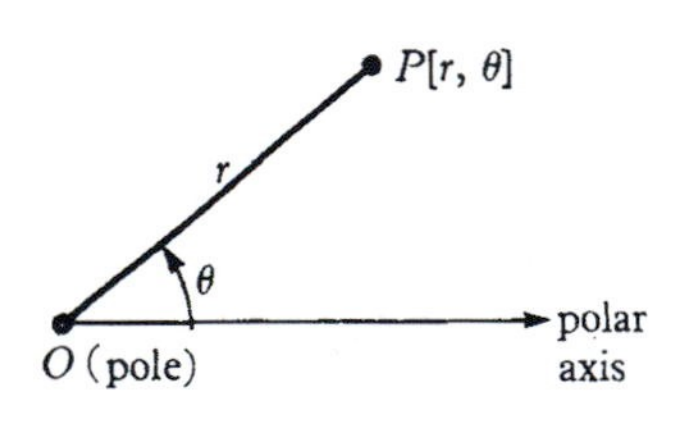

Figure 1 Polar Coordinates

[1]We use brackets [,] for polar coordinates.

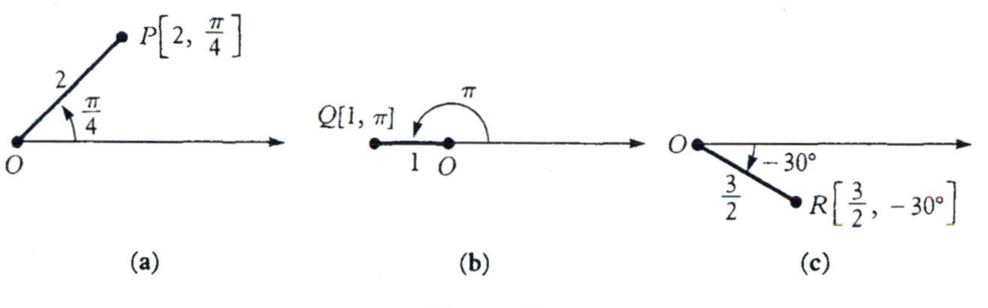

Figure 2

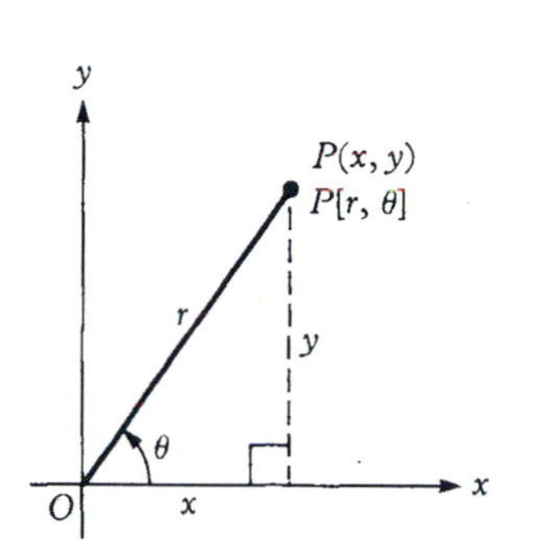

Figure 3 Rectangular and Polar Coordinates

along the polar axis (Figure 3). Polar coordinates $[r, \theta]$ of P can then be changed to rectangular coordinates (x, y) of P by means of equations derived from the definitions of $\sin\theta$ and $\cos\theta$:

$$\mathbf{x = r\cos\theta} \qquad \mathbf{y = r\sin\theta.}$$ **conversion from polar to rectangular coordinates** (1)

Example 2 Find the rectangular coordinates of each of the following points given in polar coordinates and graph each point.

(a) $P[3, \pi/6]$ (b) $Q[2, \pi/2]$ (c) $R[4, -45°]$

Solution: We use equations (1) in each case. The graphs are shown in Figure 4.

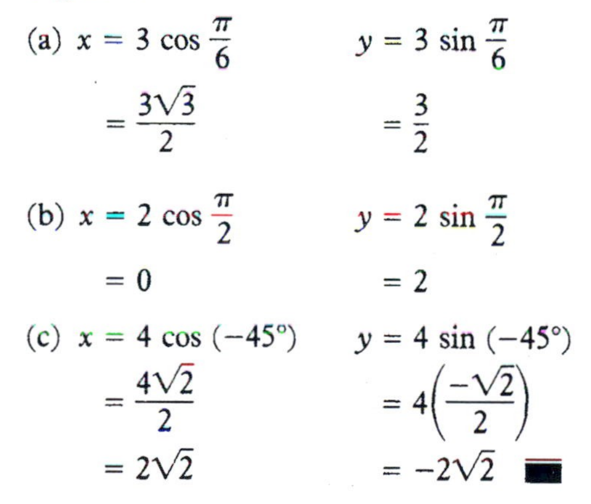

(a) $x = 3\cos\dfrac{\pi}{6} = \dfrac{3\sqrt{3}}{2}$ $\qquad y = 3\sin\dfrac{\pi}{6} = \dfrac{3}{2}$

(b) $x = 2\cos\dfrac{\pi}{2} = 0$ $\qquad y = 2\sin\dfrac{\pi}{2} = 2$

(c) $x = 4\cos(-45°) = \dfrac{4\sqrt{2}}{2} = 2\sqrt{2}$ $\qquad y = 4\sin(-45°) = 4\left(\dfrac{-\sqrt{2}}{2}\right) = -2\sqrt{2}$ ■

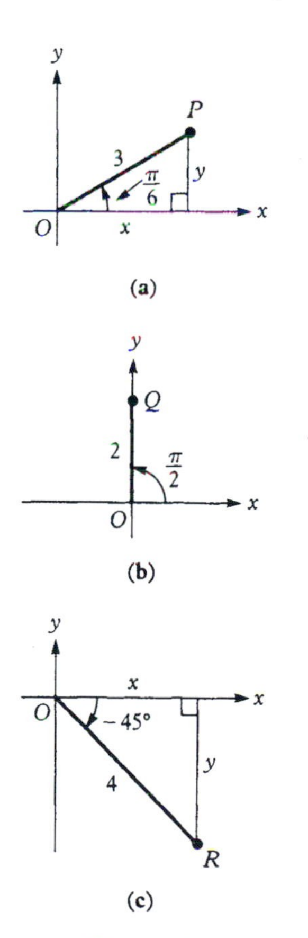

Figure 4

Conversely, if the rectangular coordinates (x, y) of P are given, then polar coordinates $[r, \theta]$ of P can be obtained as follows (see Figure 3).

$$\mathbf{r = \sqrt{x^2 + y^2}} \qquad \mathbf{\tan\theta = \frac{y}{x}}$$ **conversion from rectangular to polar coordinates** (2)

When equations (2) are used, the quadrant for θ is determined by the signs of x and y. If either x or $y = 0$, then θ is a quadrantal angle. In the following example, we choose values of θ between $-\pi$ and π. Later we consider other possible choices for θ.

Example 3 Find polar coordinates for the following points given in rectangular coordinates and graph each point.

(a) $P(2, 2)$ (b) $Q(-2, -2)$ (c) $R(-1, \sqrt{3})$ (d) $S(1, -\sqrt{3})$

Solution: We use equations (2) in each case.

(a) $r = \sqrt{2^2 + 2^2} = \sqrt{8} = 2\sqrt{2}$ $\qquad \tan\theta = \dfrac{2}{2} = 1$, $\quad \theta = 45°$ or $-135°$

The values $x = 2$ and $y = 2$ are both positive. Therefore, P is in quadrant I, and the correct value for θ is 45° [Figure 5(a)]. Polar coordinates for P are $[2\sqrt{2}, 45°]$ or, in radian measure, $[2\sqrt{2}, \pi/4]$.

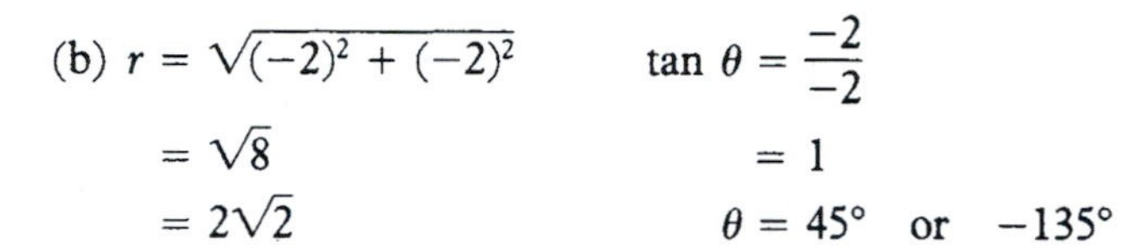

(b) $r = \sqrt{(-2)^2 + (-2)^2} = \sqrt{8} = 2\sqrt{2}$ $\qquad \tan\theta = \dfrac{-2}{-2} = 1$, $\quad \theta = 45°$ or $-135°$

In this case the values $x = -2$ and $y = -2$ are both negative. Therefore, Q is in quadrant III, and the correct value for θ is $-135°$ [Figure 5(b)]. Polar coordinates for Q are $[2\sqrt{2}, -135°]$, or $[2\sqrt{2}, -3\pi/4]$ with θ in radians.

(c) $r = \sqrt{(-1)^2 + (\sqrt{3})^2} = \sqrt{4} = 2$ $\qquad \tan\theta = \dfrac{\sqrt{3}}{-1} = -\sqrt{3}$, $\quad \theta = 120°$ or $-60°$

Since $x = -1$ is negative and $y = \sqrt{3}$ is positive, R is in quadrant II, and $\theta = 120°$ is the correct choice [Figure 5(c)]. Hence, $[2, 120°]$ or $[2, 2\pi/3]$ are polar coordinates for R.

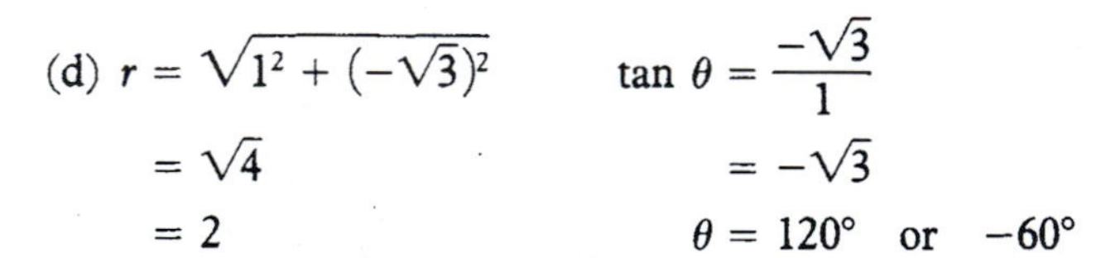

(d) $r = \sqrt{1^2 + (-\sqrt{3})^2} = \sqrt{4} = 2$ $\qquad \tan\theta = \dfrac{-\sqrt{3}}{1} = -\sqrt{3}$, $\quad \theta = 120°$ or $-60°$

Here $x = 1$ is positive and $y = -\sqrt{3}$ is negative, so S is in quadrant IV, and $\theta = -60°$ is the correct choice [Figure 5(d)]. The corresponding polar coordinates for S are $[2, -60°]$ or $[2, -\pi/3]$.

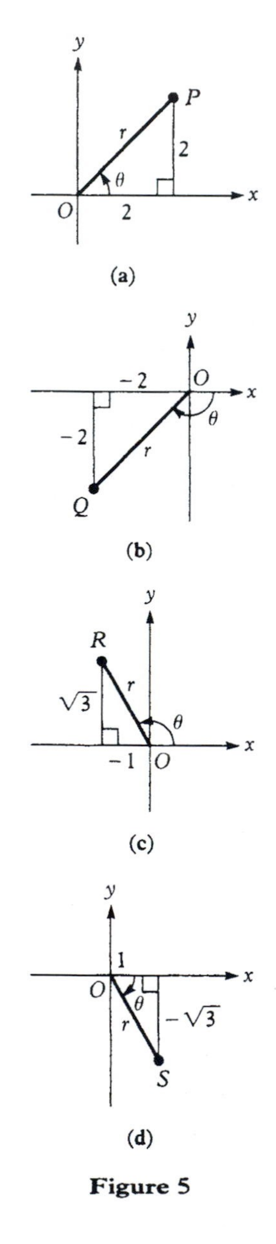

Figure 5

As illustrated in the following example, we can use equations (1) and (2) to transform equations from one coordinate system to the other. ■

Example 4 Transform equations (a) and (b) into polar form and (c) and (d) into rectangular form.

(a) $x^2 + y^2 = 4$ (b) $y = \dfrac{1}{x}$ (c) $r = \cos\theta + \sin\theta$ (d) $r = \dfrac{3}{\sin\theta}$

Solution:

(a) From equations (1),

$$r^2\cos^2\theta + r^2\sin^2\theta = 4$$
$$r^2(\cos^2\theta + \sin^2\theta) = 4$$
$$r^2 = 4.$$

(b) Again, from equations (1),

$$r\sin\theta = \frac{1}{r\cos\theta}$$
$$r^2\sin\theta\cos\theta = 1.$$

(c) First multiply both sides by r to obtain

$$r^2 = r\cos\theta + r\sin\theta,$$

which then becomes

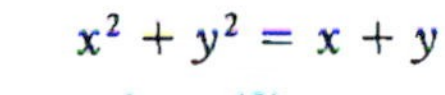

$$x^2 + y^2 = x + y$$

by equations (2).

(d) Here, multiply both sides by $\sin\theta$ to obtain

$$r\sin\theta = 3,$$

which becomes

$$y = 3$$

by equations (2). ■

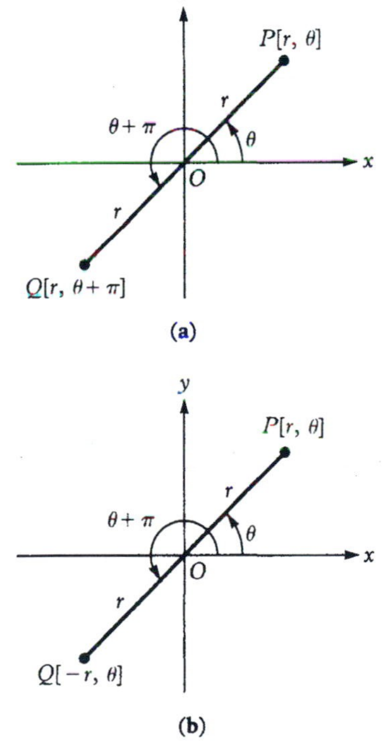

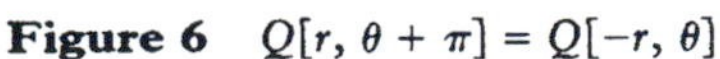

Figure 6 $Q[r, \theta + \pi] = Q[-r, \theta]$

Equivalent Polar Coordinates. Every point in the plane has precisely one ordered pair (x, y) of rectangular coordinates. However, if $[r, \theta]$ are polar coordinates of a point P, then so are

$$[r, \theta + 2n\pi] \qquad (n = \pm 1, \pm 2, \pm 3, \ldots),$$

since angle $\theta + 2n\pi$ has the same terminal side as θ. For example, $[1, \pi/2]$, $[1, \pi/2 + 2\pi]$, and $[1, \pi/2 - 2\pi]$ all represent the same point. Furthermore, the pole, which has the unique rectangular coordinates $(0, 0)$, is assigned the polar coordinates

$$[0, \theta],$$

where θ is *any* angle whatsoever. For example, $[0, 0]$, $[0, \pi/2]$, and $[0, -\pi]$ all represent the pole. Hence, each point in the plane has an *infinite number* of polar coordinates. Now suppose that a point P has polar coordinates $[r, \theta]$; consider the point $Q[r, \theta + \pi]$ [Figure 6(a)]. The rectangular coordinates (x, y) of the point Q satisfy the following equations.

$$x = r\cos(\theta + \pi) \qquad y = r\sin(\theta + \pi)$$
$$= -r\cos\theta \qquad = -r\sin\theta$$

Therefore, we also assign to $Q[r, \theta + \pi]$ the polar coordinates $[-r, \theta]$ [Figure 6(b)]. Similarly, the polar coordinates $[-r, \theta + \pi]$ are assigned to $P[r, \theta]$. We summarize as follows.

1. The polar coordinates of the pole O are $[0, \theta]$, where θ is any angle whatsoever.
2. If a point P, other than the pole, has polar coordinates $[r, \theta]$, then P also has polar coordinates $[-r, \theta + \pi]$, and all possible polar coordinates of P are

$$[r, \theta + 2k\pi] \quad \text{and} \quad [-r, \theta + \pi + 2k\pi],$$

where k is any integer.

We define two pairs of polar coordinates to be **equivalent** if they both represent the same point.

Example 5 Which of the following polar coordinates are equivalent?

$$\text{(a)} \left[3, \frac{\pi}{6}\right] \quad \text{(b)} \left[-3, \frac{\pi}{6}\right] \quad \text{(c)} \left[3, \frac{7\pi}{6}\right] \quad \text{(d)} \left[3, \frac{13\pi}{6}\right]$$

Solution: (a) and (d) are coordinates of the same point in quadrant I and are therefore equivalent. (b) and (c) are coordinates of a single point in quadrant III and are therefore equivalent. ■

Example 6 Determine all possible polar coordinates for the points P and Q given in Example 3.

Solution: We express the angles in radians.

(a) Since $[2\sqrt{2}, \pi/4]$ are polar coordinates for P, so are $[-2\sqrt{2}, \pi/4 + \pi] = [-2\sqrt{2}, 5\pi/4]$. All possible polar coordinates for P are

$$\left[2\sqrt{2}, \frac{\pi}{4} + 2k\pi\right] \quad \text{and} \quad \left[-2\sqrt{2}, \frac{5\pi}{4} + 2k\pi\right],$$

where k is any integer.

(b) Since $[2\sqrt{2}, -3\pi/4]$ are polar coordinates for Q, so are $[-2\sqrt{2}, -3\pi/4 + \pi] = [-2\sqrt{2}, \pi/4]$, and all possible polar coordinates for Q are

$$\left[2\sqrt{2}, \frac{-3\pi}{4} + 2k\pi\right] \quad \text{and} \quad \left[-2\sqrt{2}, \frac{\pi}{4} + 2k\pi\right],$$

where k is any integer. ■

Question *Why don't we restrict the polar coordinates of the pole to $r = 0$, $\theta = 0$? Also, for points other than the pole, why do we bother with negative values of r?*

Answer: Different angles for the pole and negative values for r arise naturally when working with equations and functions in polar coordinates. For example, if we converted the polar equation

$$r = 1 + 2\sin\theta \tag{3}$$

into rectangular coordinates, we would see that the rectangular coordinates (0, 0) and (0, 1) satisfy the equation even though the corresponding polar coordinates [0, 0] and [1, $\pi/2$] do not. However, by our convention, the points (0, 0) and (0, 1) also have polar coordinates [0, $7\pi/6$] and [−1, $3\pi/2$], respectively, and these coordinates do satisfy equation (3). [The graph of equation (3) is shown in the next section.] ■

Exercises 8.1

Fill in the blanks to make each statement true.

1. In polar coordinates the origin is called the ________, and the positive x-axis is called the ________.
2. If P has polar coordinates $[r, \theta]$, then the rectangular coordinates (x, y) of P can be determined by the equations ____________.
3. If P has rectangular coordinates (x, y), then polar coordinates $[r, \theta]$ for P satisfy the equations ____________.
4. If P has polar coordinates $[r, \theta]$, where $r \neq 0$, then all possible polar coordinates for P are ____________.
5. All possible polar coordinates for the origin are ____________.

Write true *or* false *for each statement.*

6. The polar coordinates $[r, \theta]$ and $[-r, \theta + 2\pi]$ are equivalent.
7. If 30° and 60° are each polar angles for P, then P must be the origin.
8. If $0 < \theta < \pi/2$, then $P[r, \theta]$ is in the first quadrant.
9. If $r < 0$, then $P[r, \theta]$ cannot be in the first quadrant.
10. If $r > 0$ and θ is a nonquadrantal angle, then $[-r, \theta]$ is in the quadrant of $\theta + \pi$.

Polar and Rectangular Coordinates

11. Graph each of the following points.
    (a) $P[1, 0°]$ (b) $Q(1, 0)$
    (c) $R\left[-1, \frac{\pi}{2}\right]$ (d) $S\left(-1, \frac{\pi}{2}\right)$
12. Graph each of the following points.
    (a) $P[0, \pi]$ (b) $Q(0, \pi)$
    (c) $R[-5, 0°]$ (d) $S(-5, 0)$
13. Find rectangular coordinates for each of the following.
    (a) $P\left[2, \frac{\pi}{4}\right]$ (b) $Q[3, -30°]$ (c) $R\left[-2, \frac{\pi}{3}\right]$
14. Use a scientific calculator to find rectangular coordinates for each of the following.
    (a) $P[1, 25°]$ (b) $Q\left[-2, \frac{3\pi}{5}\right]$ (c) $R[2, -72°]$
15. Find all possible polar coordinates for each of the following.
    (a) $P\left(\frac{1}{2}, \frac{\sqrt{3}}{2}\right)$ (b) $Q(0, -3)$ (c) $R(-\sqrt{2}, \sqrt{2})$
16. Use a scientific calculator to find all possible polar coordinates for each of the following.
    (a) $P(3, 4)$ (b) $Q(-2, 1)$ (c) $R(-5, -12)$
17. Find a polar equation equivalent to $x^2 + y^2 - 2x + 3y = 1$.
18. Find a polar equation equivalent to
$$\sin^{-1}\frac{1}{\sqrt{x^2+y^2}} + \cos^{-1}\frac{1}{\sqrt{x^2+y^2}} = \frac{\pi}{2}.$$
19. Find a rectangular equation equivalent to
$$r = \frac{1}{\cos\theta} + \frac{1}{\sin\theta} \quad \left(0 < \theta < \frac{\pi}{2}\right).$$
20. Find a rectangular equation equivalent to
$$\ln r + \ln(\sin\theta) - r\cos\theta = 0 \quad (0 < \theta < \pi).$$
21. Find all polar coordinates of the point $P[1, \pi]$ that satisfy the equation $r\cos(\theta/2) = 1$.
22. Determine whether any coordinates of the point $P[-1, \pi/4]$ satisfy the equation $r \sin 2\theta = 1$.

23. Show that the area A of the triangle with vertices O, $P_1[r_1, \theta_1]$, and $P_2[r_2, \theta_2]$ satisfies

$$A = \tfrac{1}{2} r_1 r_2 \sin(\theta_2 - \theta_1).$$

Assume $r_1 > 0$ and $r_2 > 0$, and $0 < \theta_1 < \theta_2 < \pi$. (*Hint:* see the section on the law of sines.)

24. Use the result of Exercise 23 to compute the area of the triangle with vertices O, $P_1[4, \pi/6]$, and $P_2[3, \pi/3]$.

25. Show that if any equivalent coordinates are used for P_1 and P_2, respectively, in Exercise 23, then the formula for area still holds.

Rotation of Axes

26. If the $x'y'$-axes in the figure are obtained by rotating the xy-axes through an angle θ, then every point P in the plane has coordinates (x, y) and (x', y'), where

$$x = r\cos(\theta + \phi), \quad y = r\sin(\theta + \phi),$$

and $\quad x' = r\cos\phi, \quad y' = r\sin\phi.$

Show that

$$x = x'\cos\theta - y'\sin\theta$$
$$y = x'\sin\theta + y'\cos\theta.$$

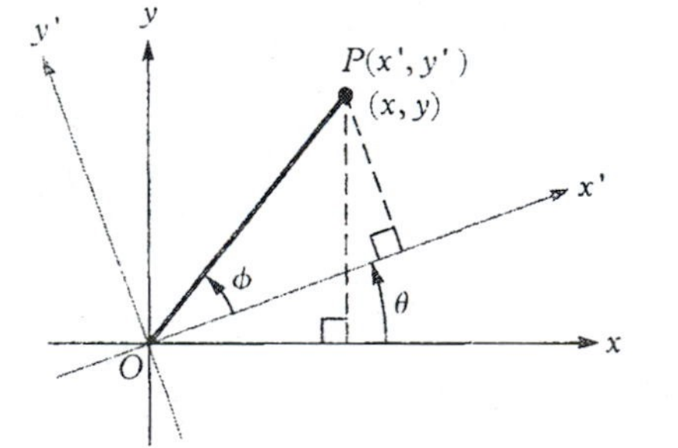

27. Use the result of Exercise 26 to find the $x'y'$-equation of each of the following curves. Draw the $x'y'$-coordinate axes as rotations of the xy-coordinate axes, and graph each curve in $x'y'$-coordinates.

(a) $3x^2 + 2xy + 3y^2 = 16$; $\theta = 45°$ (ellipse)

(b) $xy = 1$; $\theta = 45°$ (hyperbola)

(c) $x^2 + 2xy + y^2 + \sqrt{2}x - \sqrt{2}y = -2$; $\theta = 45°$ (parabola)

28. With reference to Exercise 26, show that if the $x'y'$-axes are a rotation of the xy-axes through angle θ, then the general second-degree equation

$$Ax^2 + By^2 + Cxy + Dx + Ey = F$$

becomes

$$[A\cos^2\theta + B\sin^2\theta + C\sin\theta\cos\theta]x'^2 +$$
$$[A\sin^2\theta + B\cos^2\theta - C\sin\theta\cos\theta]y'^2 +$$
$$[(B - A)\sin 2\theta + C\cos 2\theta]x'y' +$$
$$[D\cos\theta + E\sin\theta]x' + [E\cos\theta - D\sin\theta]y' = F.$$

29. Use the result of Exercise 28 to show that if $C \neq 0$, then the $x'y'$ term can be eliminated from the rotated version of the second-degree equation

$$Ax^2 + By^2 + Cxy + Dx + Ey = F$$

by a rotation through angle θ, where

$$\cot 2\theta = \frac{A - B}{C}.$$

8.2 Graphs in Polar Coordinates

The point $P[r_1, \theta_1]$ may be on the graph of $r = f(\theta)$ even though $r_1 \neq f(\theta_1)$.

Graph of a Polar Equation
Polar and Rectangular Curves

Graph of a Polar Equation. The graph of an equation in r and θ is the set of all points P in the plane for which *at least one pair* $[r, \theta]$ of polar coordinates for P satisfies the equation. For example, the graph of $r = 2$ is a circle with center at the pole and radius 2 [Figure 7(a)], and the graph of $\theta = \pi/4$ is a straight line through the pole that makes an angle of $\pi/4$ radians with the polar axis [Figure 7(b)].

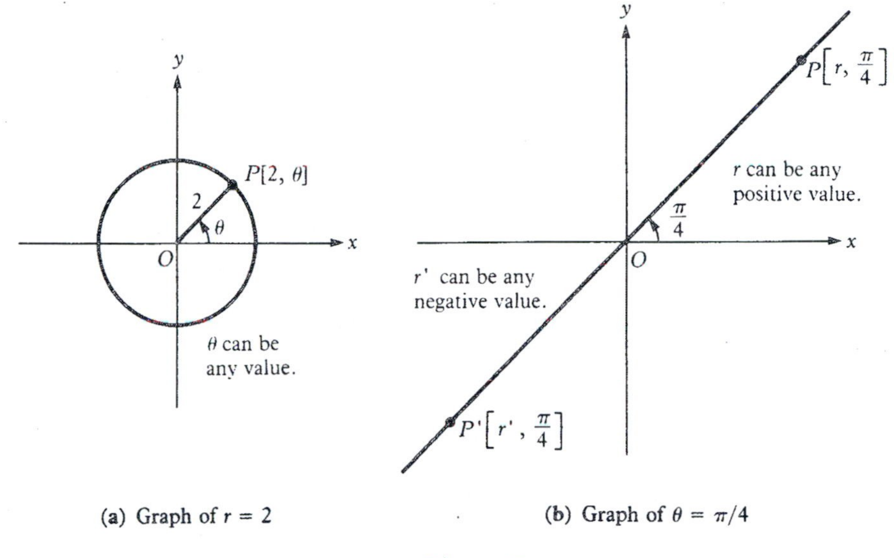

(a) Graph of $r = 2$ (b) Graph of $\theta = \pi/4$

Figure 7

Especially suited to graphing in polar coordinates are equations of the form

$$r = f(\theta),$$

where r is a function of θ.

Example 1 Graph the curve $r = 1 + \cos\theta$ (cardioid).

Solution: First we construct a coordinate table by assigning values to θ in *increasing order* and computing the corresponding values of r. We then plot the points from the table and join them in the order of increasing θ by a smooth curve (Figure 8). ■

θ	r
0	2
$\pi/4$	$1 + \sqrt{2}/2 \approx 1.7$
$\pi/2$	1
$3\pi/4$	$1 - \sqrt{2}/2 \approx .3$
π	0
$5\pi/4$	$1 - \sqrt{2}/2 \approx .3$
$3\pi/2$	1
$7\pi/4$	$1 + \sqrt{2}/2 \approx 1.7$
2π	2

Figure 8 Cardioid: $r = 1 + \cos\theta$

In Example 1, we get the entire curve by considering values of θ from 0 to 2π. This is because $[1 + \cos\theta, \theta]$ and $[1 + \cos(\theta + 2\pi), \theta + 2\pi]$ represent the same point. If a function $f(\theta)$ has period $p\pi$ where p is an integer, then $[f(\theta), \theta]$ and $[f(\theta + 2p\pi), \theta + 2p\pi]$ represent the same point (*why?*). Hence, we can restrict the domain of f to $[0, 2p\pi]$ and still obtain the entire curve. (If p is even, we can restrict the domain to $[0, p\pi]$, and if r assumes negative values in the equation for f, it may be possible to obtain the entire curve on a smaller interval.) A graph of a *nonperiodic function* is given in Example 2.

Example 2 Graph the curve $r = \theta$ (spiral of Archimedes).

Solution: Here the equation is quite simple, but the curve is not at all simple. Starting at the pole $[0, 0]$, as θ increases, the curve spirals outward counterclockwise as shown by the unbroken line in Figure 9. Also, as θ increases in magnitude but through negative values, the curve spirals outward clockwise as shown by the broken line in Figure 9. ■

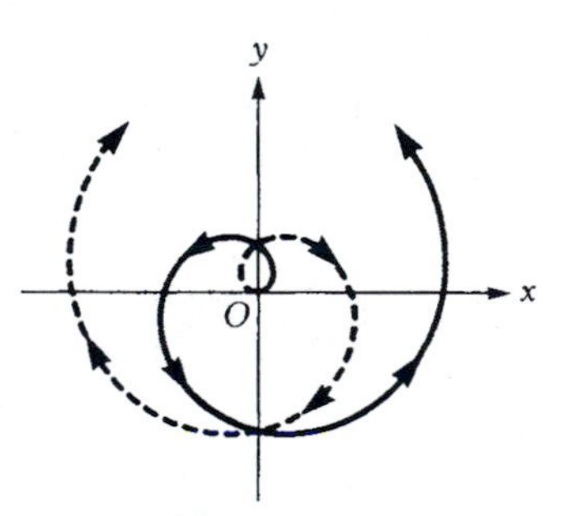

Figure 9 Spiral of Archimedes: $r = \theta$

Example 2 illustrates how coordinates having a negative value for r arise in an algebraic equation. Another equation that produces negative values for r is given in the following example.

Example 3 Graph the curve $r = 1 + 2\sin\theta$ (limaçon).

Solution: Since $\sin\theta$ has period 2π, we need only consider values of θ from 0 to 2π in our coordinate table (see remark after Example 1). Note that $r = 0$ when $\sin\theta = -1/2$, which occurs at $\theta = 7\pi/6$ and at $\theta = 11\pi/6$. We make sure to include these values of θ in our coordinate table. Also, while graphing the curve (Figure 10) keep in mind that *when* $r > 0$, *the point* $[r, \theta]$ *is in the same quadrant as* θ, *but when* $r < 0$, *the point* $[r, \theta]$ *is in the quadrant of* $\theta + \pi$.

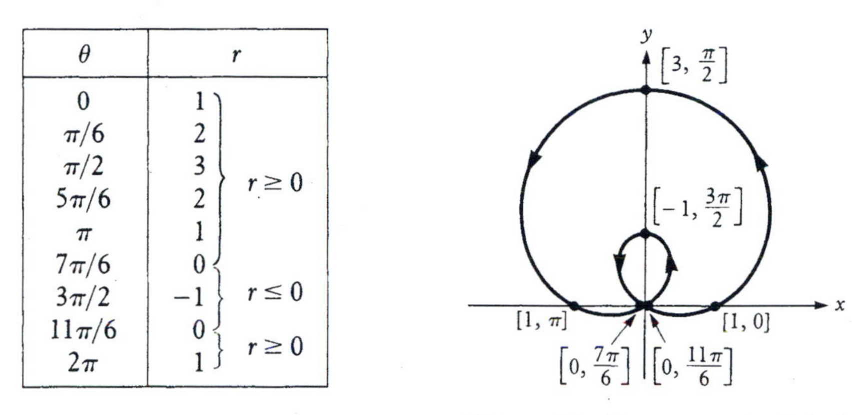

θ	r	
0	1	$r \geq 0$
$\pi/6$	2	
$\pi/2$	3	
$5\pi/6$	2	
π	1	
$7\pi/6$	0	
$3\pi/2$	-1	$r \leq 0$
$11\pi/6$	0	
2π	1	$r \geq 0$

Figure 10 Limaçon: $r = 1 + 2 \sin \theta$

Starting at $\theta = 0$ and the point $[1, 0]$, the curve spirals counterclockwise along the outer loop until we reach $\theta = 7\pi/6$ and the point $[0, 7\pi/6]$. For $7\pi/6 < \theta < 11\pi/6$, the curve spirals counterclockwise along the inner loop (*why?*) until we reach $\theta = 11\pi/6$ and the point $[0, 11\pi/6]$. Here the curve returns to the outer loop and continues there until arriving back at the initial point, this time with coordinates $[1, 2\pi]$. ■

Note that the origin O is on the graph in Examples 1, 2, and 3, but with different coordinates in each case. In general, to determine whether a given point P is on the graph of a polar equation, we must consider all possible polar coordinates for P. If any one of them satisfies the equation, then P is on the graph.

Example 4 Is the point $P[-1, \pi]$ on the curve $r = \sin(\theta/2)$?

Solution: If $\theta = \pi$, then $\sin(\theta/2) = 1$. Therefore, the coordinates $[-1, \pi]$ do not satisfy the given equation. However, P also has polar coordinates $[-1, 3\pi]$, and these do satisfy the given equation. Hence, P is on the curve. ■

Polar and Rectangular Curves. There is a basic difference between the graph of a function $y = f(x)$ in rectangular coordinates and that of a function $r = f(\theta)$ in polar coordinates. A vertical line $x = c$ can intersect the rectangular graph of a function $y = f(x)$ in at most one point. On the other hand, if we are given a polar function $r = f(\theta)$, then $\theta = c$ is a line through the origin that may intersect the graph in more than one point. For instance, the line $\theta = \pi/4$ intersects the limaçon of Example 3 in three points and the spiral of Example 2 in an infinite number of points. The reason is that $\theta = \pi/4$ is the same line as $\theta = \pi/4 + k\pi$, where k is *any* integer. Different points of intersection correspond to different values of k. Hence, we cannot draw the graph of a polar function from a cluster of points unless we know how the points are generated by the function. This is why we always construct the coordinate table for a polar function in terms of *increasing values* of θ, and accordingly draw the graph. We now compare the rectangular and polar versions of the cosine curve.

Example 5 Compare the polar curve $r = \cos 2\theta$ (four-leaved rose) with the rectangular curve $y = \cos 2x$.

Solution: The function $r = \cos 2\theta$ has period π, so we can get the entire polar curve for values of θ from 0 to 2π (see remark after Example 1). Now $\cos 2\theta = 0$ when $\theta = \pi/4, 3\pi/4, 5\pi/4$, or $7\pi/4$, so we include these values in our polar table. Also, since r is computed from 2θ, we include a 2θ-column in the table. Keep in mind, however, that the points are plotted from the θ-column (Figure 11).

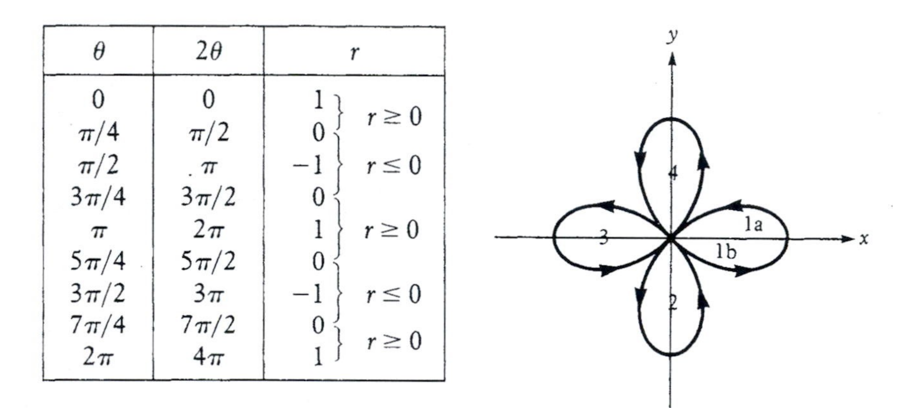

θ	2θ	r	
0	0	1	$r \geq 0$
$\pi/4$	$\pi/2$	0	
$\pi/2$	π	-1	$r \leq 0$
$3\pi/4$	$3\pi/2$	0	
π	2π	1	$r \geq 0$
$5\pi/4$	$5\pi/2$	0	
$3\pi/2$	3π	-1	$r \leq 0$
$7\pi/4$	$7\pi/2$	0	$r \geq 0$
2π	4π	1	

Figure 11 Four-leaved Rose: $r = \cos 2\theta$

From Chapter 7, we know that the graph of $y = \cos 2x$ for x from 0 to 2π is as shown in Figure 12. By examining the rectangular curve, we can see how the leaves of the rose are generated. Note that the x-intercepts of the rectangular curve correspond to the polar crossings of the rose. As y varies from 1 to 0 along the x-interval from 0 to $\pi/4$, r traces out the top half (1a) of leaf 1 in the direction of the arrows in Figure 11. As y varies from 0 to -1 and back to 0 along the x-interval from $\pi/4$ to $3\pi/4$, r traces out leaf 2. Note that r is negative for $\pi/4 < \theta < 3\pi/4$, so leaf 2 is located in the polar sector corresponding to $\pi/4 + \pi < \theta < 3\pi/4 + \pi$. After leaf 2, leaves 3 and 4 are generated in the direction of the arrows, and then the bottom half (1b) of leaf 1 completes the curve. ∎

The two graphs in Example 5 illustrate another basic difference between polar and rectangular curves. The periodic polar curve $r = \cos 2\theta$ traces over itself every 2π radians, whereas the rectangular curve $y = \cos 2x$ repeats its pattern to the right and to the left in intervals of length π. Hence, in this case, the entire polar curve corresponds to two cycles of the rectangular curve.

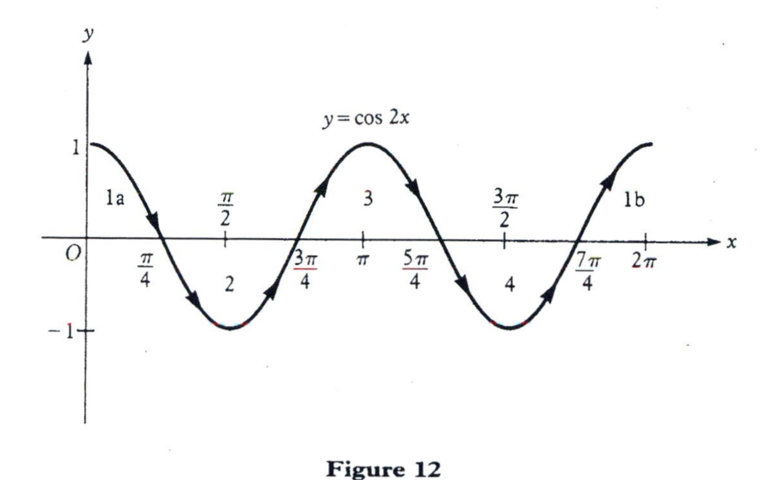

Figure 12

Exercises 8.2

Fill in the blanks to make each statement true.

1. The graph of an equation in r and θ is the set of all points P in the plane for which ____________.
2. The graph of the equation $\theta = \pi/3$ is ____________.
3. The graph of the equation $r = -2$ is ____________.
4. A polar equation of the x-axis is __________.
5. The polar form of the general linear equation $Ax + By = C$ is __________.

Write true *or* false *for each statement.*

6. The graph of a polar equation is the set of all points P for which all the polar coordinates of P satisfy the equation.
7. The point P with polar coordinates $[-2, 4\pi/3]$ is on the graph of the equation $r = 2$.
8. The point with polar coordinates $[-2, 4\pi/3]$ is on the graph of the equation $\theta = \pi/3$.
9. The points with polar coordinates $[\pi/4, \pi/4 + 2\pi]$ and $[\pi/4 + 2\pi, \pi/4]$ are on the graph of $r = \theta$.
10. The line $\theta = c$ can intersect the graph of $r = f(\theta)$ in at most one point.

Graph of a Polar Equation

Graph each of the following polar equations. Assume $a > 0$.

11. $r = 2\theta$ for $\theta \geq 0$ (spiral of Archimedes)
12. $r = a(1 - \sin\theta)$ (cardioid)
13. $r = \sin 2\theta$ (four-leaved rose)
14. $r = a\cos\theta$ (circle)
15. $r = a(1 + \cos\theta)$ (limaçon)
16. $r = e^{\theta}$ (logarithmic spiral[1])
17. $r\sin\theta = -2$ (line)
18. $r = \dfrac{a}{\theta}$ for $\theta > 0$ (hyperbolic spiral)

Polar and Rectangular Curves

Graph and compare each of the following polar and rectangular curves.

19. $r = \sin 3\theta$ (three-leaved rose); $y = \sin 3x$ on the interval $[0, \pi]$
20. $r = 2\cos 3\theta$ (three-leaved rose); $y = 2\cos 3x$ on the interval $[0, \pi]$

[1]The shell of a snail coils outward in a logarithmic spiral of the form $r = ce^{k\theta}$, where c and k are constants. Hence, r measures the outward growth of the shell, which is consistent with the general principle that growth in nature is exponential.

21. $r = 1 + \sin\frac{\theta}{2}$; $y = 1 + \sin\frac{x}{2}$ on $[0, 4\pi]$

22. $r = 2 - \cos\frac{\theta}{2}$; $y = 2 - \cos\frac{x}{2}$ on $[0, 4\pi]$

Symmetry

23. The graph of a polar equation is symmetric with respect to the x-axis if whenever the coordinates $[r, \theta]$ satisfy the equation, *at least one* of the coordinates equivalent to $[r, -\theta]$ also satisfies the equation. Determine which of the curves in Exercises 11 to 18 are symmetric with respect to the x-axis.

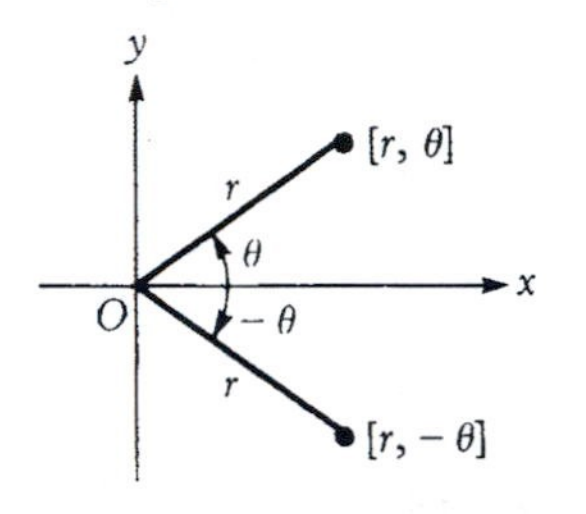

24. The graph of a polar equation is symmetric with respect to the y-axis if whenever the coordinates $[r, \theta]$ satisfy the equation, *at least one* of the coordinates equivalent to $[r, \pi - \theta]$ also satisfies the equation. Determine which of the curves in Exercises 11 to 18 are symmetric with respect to the y-axis.

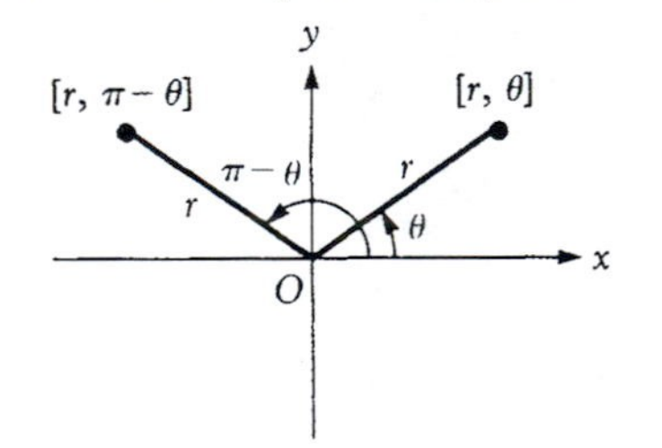

25. The graph of a polar equation is symmetric with respect to the origin if whenever the coordinates $[r, \theta]$ satisfy the equation, *at least one* of the coordinates equivalent to $[r, \theta + \pi]$ also satisfies the equation. Determine which of the curves in Exercises 11–18 are symmetric with respect to the origin.

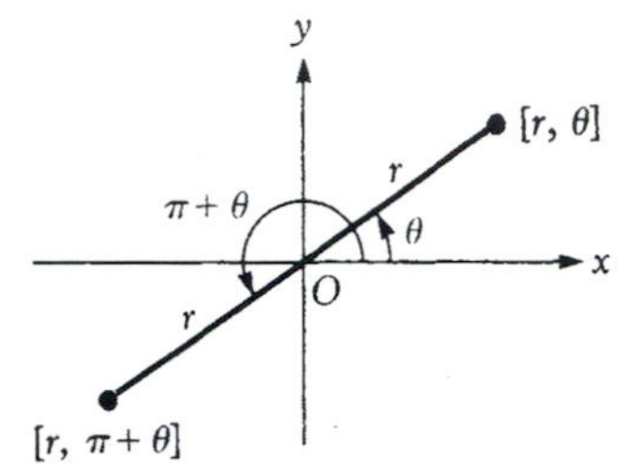

Conic Sections

The graph of

$$r = \frac{k}{1 - e\cos\theta} \qquad (k \neq 0) \qquad (*)$$

is a conic section of ***eccentricity*** *e. That is, the graph is a* circle *if* $e = 0$, *an* ellipse *for* $0 < e < 1$, *a* parabola *for* $e = 1$, *and a* hyperbola *for* $e > 1$.

26. Verify the above statement. [*Hint:* write (*) in the form $r = e(r\cos\theta) + k$, square both sides, and switch to rectangular coordinates.]

27. Graph the ellipse $r = \dfrac{1}{1 - \frac{1}{2}\cos\theta}$.

28. Graph the parabola $r = \dfrac{1}{1 - \cos\theta}$.

29. Graph the hyperbola $r = \dfrac{1}{1 - 2\cos\theta}$.

8.3 Polar Form of Complex Numbers

In polar coordinates, we can see why $i^2 = -1$.

Rectangular and Polar Forms of Complex Numbers
Multiplication of Complex Numbers
Roots of Complex Numbers

Rectangular and Polar Forms of Complex Numbers. The rectangular form of a complex number (see Section 3.2) is

$$z = x + yi \qquad (\text{or } x + iy), \qquad (1)$$

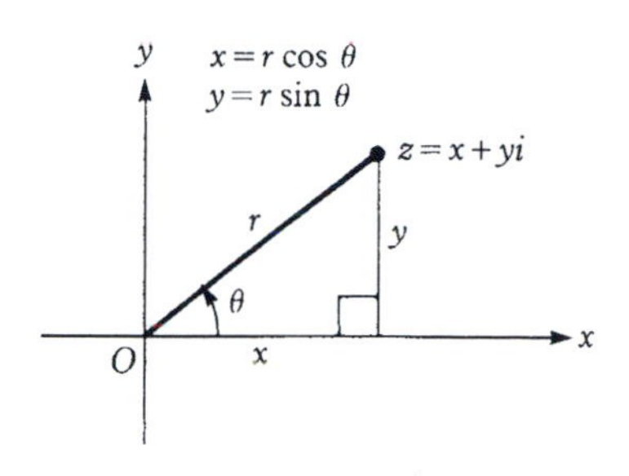

Figure 13 Polar Form of Complex Number $z = x + yi$

where x and y are real numbers and $i = \sqrt{-1}$. With reference to Figure 13, we can replace x by $r \cos \theta$ and y by $r \sin \theta$ in (1), and then factor out r to obtain

$$\mathbf{z = r(\cos \theta + i \sin \theta),} \tag{2}$$

which is called the **polar form** of z. That is, z is a point in the plane with rectangular coordinates (x, y) and corresponding polar coordinates $[r, \theta]$. In Chapter 3 we used arrows to denote complex numbers because arrows illustrate clearly the parallelogram rule for addition. Here we use polar notation for complex numbers and will reserve arrows for vectors.

Example 1 Find the polar form of the complex number $z = 1 + \sqrt{3}i$.

Solution: We use equation (2) of Section 8.1 for changing from rectangular coordinates to polar coordinates. Here $x = 1$ and $y = \sqrt{3}$. Therefore,

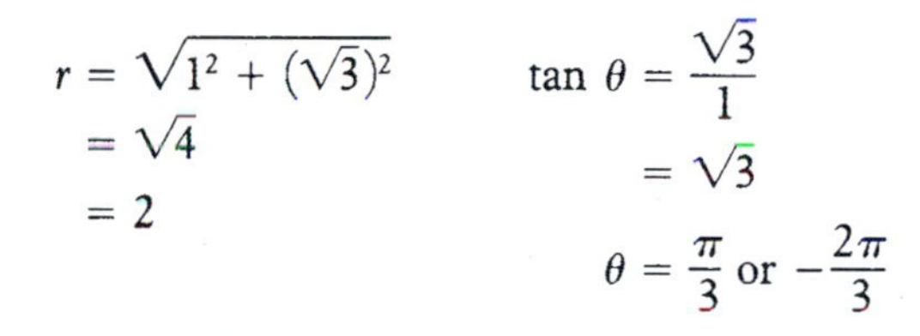

$$r = \sqrt{1^2 + (\sqrt{3})^2} = \sqrt{4} = 2 \qquad \tan \theta = \frac{\sqrt{3}}{1} = \sqrt{3}$$

$$\theta = \frac{\pi}{3} \text{ or } -\frac{2\pi}{3}$$

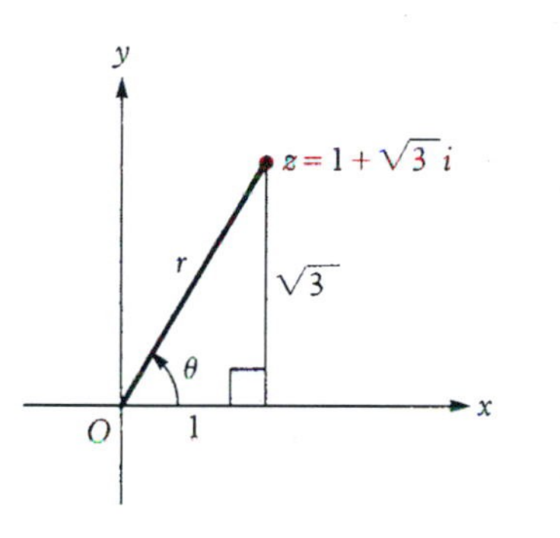

Figure 14

Since $x = 1$ and $y = \sqrt{3}$ are both positive, z is in the first quadrant (Figure 14), and the correct choice for θ is $\pi/3$. Hence, the polar form of z is

$$z = 2\left(\cos \frac{\pi}{3} + i \sin \frac{\pi}{3}\right). \quad ■$$

As usual, the angle θ in the polar form of z can be replaced by $\theta + 2k\pi$ for any integer k. However, we always choose $r \geq 0$ in the polar form of z. We call r the **absolute value** of z and denote it by $|z|$. That is,

$$\mathbf{|z| = \sqrt{x^2 + y^2}.}$$

Note that when $y = 0$, $|z| = \sqrt{x^2} = |x|$, so the absolute value notation $|\ \ |$ for both real and complex numbers is consistent.

Multiplication of Complex Numbers. As indicated in Section 3.2, addition of complex numbers is accomplished geometrically by the parallelogram law. To give a geometric interpretation for multiplication, polar coordinates are more convenient than rectangular ones. To see why, let

$$z_1 = r_1(\cos \theta_1 + i \sin \theta_1) \quad \text{and} \quad z_2 = r_2(\cos \theta_2 + i \sin \theta_2).$$

Then, keeping in mind that $i^2 = -1$, we obtain

$$\begin{aligned} z_1 z_2 = r_1 r_2[&(\cos \theta_1 \cos \theta_2 - \sin \theta_1 \sin \theta_2) \\ &+ i(\sin \theta_1 \cos \theta_2 + \cos \theta_1 \sin \theta_2)]. \end{aligned}$$

We have the identities

$$\cos\theta_1 \cos\theta_2 - \sin\theta_1 \sin\theta_2 = \cos(\theta_1 + \theta_2)$$

and

$$\sin\theta_1 \cos\theta_2 + \cos\theta_1 \sin\theta_2 = \sin(\theta_1 + \theta_2).$$

Therefore,

$$\mathbf{z_1 z_2 = r_1 r_2[\cos(\theta_1 + \theta_2) + i \sin(\theta_1 + \theta_2)].} \qquad \textbf{multiplication} \qquad (3)$$

Equation (3) says that $z_1 z_2$ has polar coordinates $[r_1 r_2, \theta_1 + \theta_2]$. Hence, to multiply two complex numbers, we multiply their absolute values and *add* their polar angles (Figure 15).

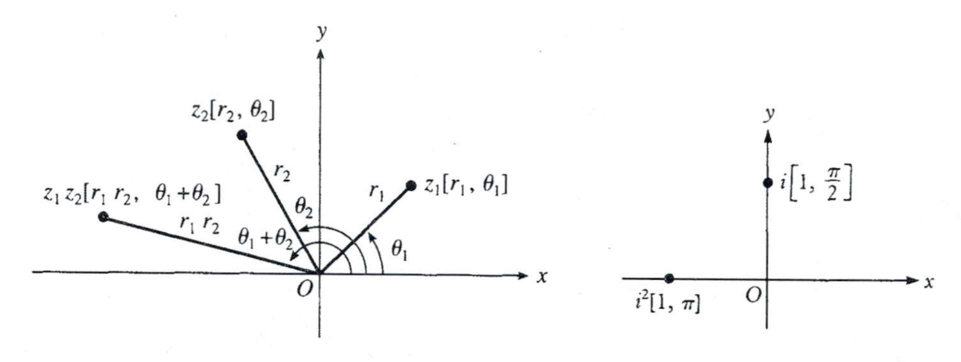

Figure 15 Multiplication of Complex Numbers

Figure 16 $i^2 = -1$

Note in particular that i has polar coordinates $[1, \pi/2]$, and therefore i^2 has polar coordinates $[1 \cdot 1, \pi/2 + \pi/2] = [1, \pi]$ (Figure 16). But $[1, \pi]$ are polar coordinates of the complex number $-1 + 0i = -1$. Hence, we see geometrically why $i^2 = -1$. In general, multiplication of a complex number z by i results in a rotation of z through 90°.

Example 2 Given $z_1 = 1 + i$ and $z_2 = 2\sqrt{3} + 2i$, express z_1, z_2, and $z_1 z_2$ in polar form.

Solution: By the usual equations for converting from rectangular to polar coordinates, we find that z_1 and z_2 have polar coordinates $[\sqrt{2}, \pi/4]$ and $[4, \pi/6]$, respectively (see Figure 17). Therefore,

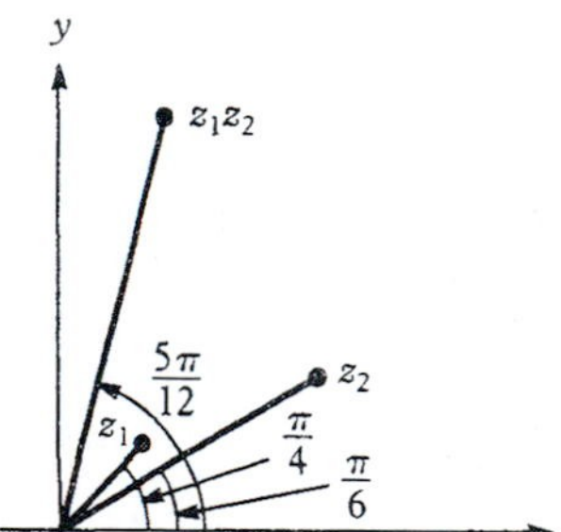

Figure 17

$$z_1 = \sqrt{2}\left(\cos\frac{\pi}{4} + i \sin\frac{\pi}{4}\right)$$

$$z_2 = 4\left(\cos\frac{\pi}{6} + i \sin\frac{\pi}{6}\right)$$

and

$$z_1 z_2 = 4\sqrt{2}\left[\cos\left(\frac{\pi}{4} + \frac{\pi}{6}\right) + i \sin\left(\frac{\pi}{4} + \frac{\pi}{6}\right)\right]$$

$$= 4\sqrt{2}\left(\cos\frac{5\pi}{12} + i \sin\frac{5\pi}{12}\right). \quad ■$$

Example 3 Given $z_1 = r_1(\cos\theta_1 + i\sin\theta_1)$ and $z_2 = r_2(\cos\theta_2 + i\sin\theta_2)$, where $r_2 \neq 0$, show that

$$\frac{z_1}{z_2} = \frac{r_1}{r_2}[\cos(\theta_1 - \theta_2) + i\sin(\theta_1 - \theta_2)]. \qquad \textbf{division} \qquad (4)$$

That is, to divide two complex numbers, we *divide* their absolute values and *subtract* their polar angles.

Solution: Let $z_1/z_2 = z_3$, where the polar coordinates of z_3 are $[r_3, \theta_3]$. Now,

$$z_1 = z_2 z_3$$

and therefore $\quad r_1 = r_2 r_3 \quad$ and $\quad \theta_1 = \theta_2 + \theta_3$.

Hence, $\quad r_3 = \dfrac{r_1}{r_2} \quad$ and $\quad \theta_3 = \theta_1 - \theta_2$,

from which equation (4) follows. ■

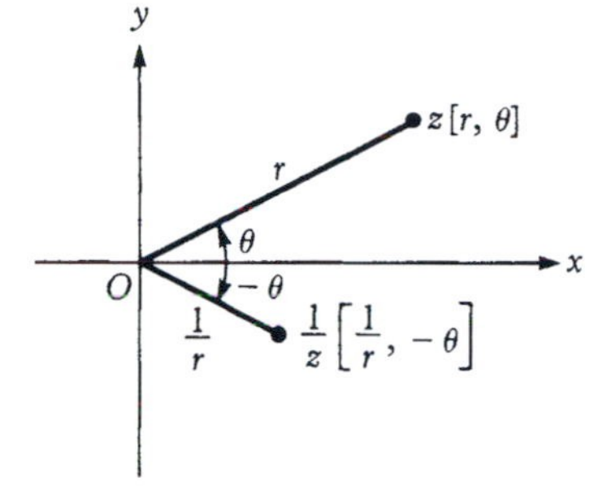

Figure 18 Polar Coordinates of $1/z$

In particular, if z has polar coordinates $[r, \theta]$ and $r \neq 0$, then since 1 has polar coordinates $[1, 0]$, it follows from equation (4) that $1/z$ has polar coordinates $[1/r, -\theta]$ (Figure 18).

We can use equation (3) to derive a very simple formula for computing z^n, where n is any positive integer. For, if $z = r(\cos\theta + i\sin\theta)$, then by (3),

$$\begin{aligned} z^2 = z \cdot z &= r \cdot r[\cos(\theta + \theta) + i\sin(\theta + \theta)] \\ &= r^2(\cos 2\theta + i\sin 2\theta). \end{aligned}$$

Similarly,

$$\begin{aligned} z^3 &= z^2 \cdot z = r^3(\cos 3\theta + i\sin 3\theta), \\ z^4 &= z^3 \cdot z = r^4(\cos 4\theta + i\sin 4\theta), \end{aligned}$$

and in general, for any positive integer n,

$$z^n = r^n(\cos n\theta + i\sin n\theta). \qquad (5)$$

Equation (5) is called **De Moivre's theorem.**[1]

[1] *Abraham De Moivre (1667–1754) was a French-born mathematician who made significant contributions to trigonometry. His actual result was the identity* $(\cos\theta + i\sin\theta)^n = \cos n\theta + i\sin n\theta$.

Example 4 Compute $(1 + i)^6$.

Solution: The polar form of $z = 1 + i$ is

$$z = \sqrt{2}\left(\cos\frac{\pi}{4} + i\sin\frac{\pi}{4}\right).$$

Therefore, by De Moivre's theorem,

$$\begin{aligned} (1 + i)^6 &= (\sqrt{2})^6\left(\cos\frac{6\pi}{4} + i\sin\frac{6\pi}{4}\right) \\ &= 2^3\left(\cos\frac{3\pi}{2} + i\sin\frac{3\pi}{2}\right) \\ &= 8(0 - i) = -8i. \end{aligned}$$ ■

Roots of Complex Numbers. We know that every real number has *one* real nth root if n is an odd positive integer, and every positive real number has *two* real nth roots if n is even. We can now use De Moivre's theorem to derive a most remarkable result concerning roots of complex numbers, namely,

> every nonzero complex number has exactly n distinct nth roots for any positive integer n.

To see why, let $z_0 = r_0(\cos\theta_0 + i\sin\theta_0)$ be an nth root of $z = r(\cos\theta + i\sin\theta)$. That is, $z_0{}^n = z$. Then, by De Moivre's theorem,

$$r_0{}^n(\cos n\theta_0 + i\sin n\theta_0) = r(\cos\theta + i\sin\theta).$$

Therefore, $r_0{}^n$ must equal r, and $n\theta_0$ must equal $\theta + 2k\pi$, for some integer k. That is,

$$r_0 = \sqrt[n]{r} \qquad \text{and} \qquad \theta_0 = \frac{\theta + 2k\pi}{n}.$$

The numbers $k = 0, 1, 2, \ldots, n-1$ give different values for z_0, and for other integers k, the values repeat (see Exercise 48). We summarize as follows.

> The nth roots of $z = r(\cos\theta + i\sin\theta)$ are
>
> $$z_k = \sqrt[n]{r}\left[\cos\left(\frac{\theta + 2k\pi}{n}\right) + i\sin\left(\frac{\theta + 2k\pi}{n}\right)\right], \qquad (6)$$
>
> where $k = 0, 1, 2, \ldots, n-1$.

Example 5 Find the fourth roots of 16.

Solution: In polar form, $16 = 16(\cos 0 + i\sin 0)$. That is, $r = 16$ and $\theta = 0$. Therefore, by (6),

$$z_k = \sqrt[4]{16}\left(\cos\frac{2k\pi}{4} + i\sin\frac{2k\pi}{4}\right),$$

where $k = 0, 1, 2, 3$. That is,

$$z_0 = 2(\cos 0 + i\sin 0) = 2 \qquad z_2 = 2\left(\cos\frac{4\pi}{4} + i\sin\frac{4\pi}{4}\right) = -2$$

$$z_1 = 2\left(\cos\frac{2\pi}{4} + i\sin\frac{2\pi}{4}\right) = 2i \qquad z_3 = 2\left(\cos\frac{6\pi}{4} + i\sin\frac{6\pi}{4}\right) = -2i.$$

Hence, the fourth roots of 16 in the complex number system are $2, 2i, -2$, and $-2i$. ■

Of particular interest are the *nth roots of unity*, by which we mean the nth roots of 1. The polar form of 1 is

$$1 = \cos 0 + i \sin 0,$$

that is, $r = 1$ and $\theta = 0$. Therefore, by (6), the **nth roots of unity** are

$$\mathbf{z_k = \cos \frac{2k\pi}{n} + i \sin \frac{2k\pi}{n}}, \quad \text{where } k = 0, 1, 2, \ldots, n - 1. \tag{7}$$

Starting at $z_0 = 1$, the nth roots of unity are spaced in increments of $2\pi/n$ radians about the unit circle. For instance, if $n = 5$, then the fifth roots of unity are spaced in increments of $2\pi/5 = 72°$ about the unit circle [Figure 19(a)], and if $n = 8$, then the eighth roots of unity are spaced in increments of $2\pi/8 = 45°$ about the unit circle [Figure 19(b)].

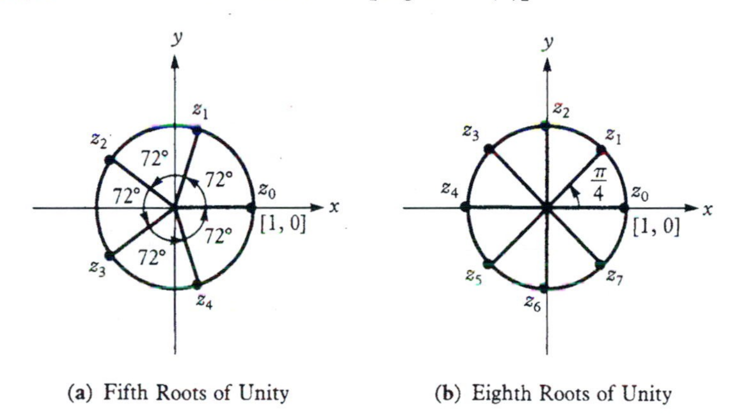

(a) Fifth Roots of Unity (b) Eighth Roots of Unity

Figure 19

Example 6 Find the sixth roots of unity and graph them on the unit circle.

Solution: By (7), the sixth roots of unity are

$$z_0 = \cos 0 + i \sin 0 = 1 \qquad z_3 = \cos \frac{6\pi}{6} + i \sin \frac{6\pi}{6} = -1$$

$$z_1 = \cos \frac{2\pi}{6} + i \sin \frac{2\pi}{6} = \frac{1}{2} + \frac{\sqrt{3}}{2}i \qquad z_4 = \cos \frac{8\pi}{6} + i \sin \frac{8\pi}{6} = -\frac{1}{2} - \frac{\sqrt{3}}{2}i$$

$$z_2 = \cos \frac{4\pi}{6} + i \sin \frac{4\pi}{6} = -\frac{1}{2} + \frac{\sqrt{3}}{2}i \qquad z_5 = \cos \frac{10\pi}{6} + i \sin \frac{10\pi}{6} = \frac{1}{2} - \frac{\sqrt{3}}{2}i$$

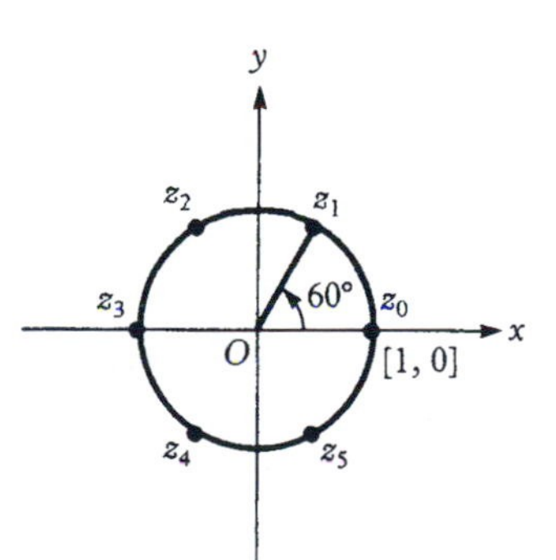

Figure 20

As indicated in Figure 20, starting at the point [1, 0], the roots are spaced every $2\pi/6 = 60°$ around the unit circle. ■

Question *If p and q are nonzero positive integers and z is a nonzero complex number, how many values does the expression $z^{p/q}$ have? [We define $z^{p/q}$ as $(z^{1/q})^p$.]*

Answer: If p and q have no factors other than 1 in common, then it can be shown that $z^{p/q}$ has q values. For example, $16^{3/4}$ has four values in the complex number system, namely 8, -8, $8i$, and $-8i$ (see Example 5). However, if p and q have a factor other than 1 in common, then $z^{p/q}$ will have fewer than q values. For example, $16^{2/4}$ has only two values, namely 4 and -4. For this reason, we always assume that rational numbers p/q are in lowest terms when computing $z^{p/q}$. ■

Comment As stated in Section 3.2, acceptance of complex numbers was a very gradual process. Complex numbers, as solutions to quadratic, cubic (third-degree), and quartic (fourth-degree) equations, began to be taken seriously by some mathematicians in the sixteenth century. Coordinate geometry was introduced in the seventeenth century, but no one at that time thought to represent complex numbers as points in the plane. The first to do so were a Norwegian surveyor named Wessel, in 1797, and a Swiss bookkeeper named Argand, in 1806. However, it was not until 1831, when the German mathematician Gauss interpreted complex addition according to the parallelogram law and complex multiplication by i as a rotation through 90°, that the idea took hold and complex numbers were finally accepted as meaningful numbers. Although Wessel and Argand initiated the idea, it took the authority of Gauss to make others realize its significance.

Exercises 8.3

Fill in the blanks to make each statement true.

1. The absolute value of $z = x + yi$ is equal to __________.
2. If $z_1 = r_1(\cos\theta_1 + i\sin\theta_1)$ and $z_2 = r_2(\cos\theta_2 + i\sin\theta_2)$, then the polar form of z_1z_2 is __________ and, if $r_2 \neq 0$, the polar form of z_1/z_2 is __________.
3. If $z = r(\cos\theta + i\sin\theta)$, then $z^n =$ __________.
4. If $z = r(\cos\theta + i\sin\theta)$, then the nth roots of z are __________.
5. The nth roots of unity are __________.

Write true *or* false *for each statement.*

6. If z has polar coordinates $[r, \theta]$, then iz has polar coordinates $[r, \theta + \pi/2]$.
7. If z has polar coordinates $[r, \theta]$ and $r \neq 0$, $\theta \neq 0$, then $1/z$ has polar coordinates $[1/r, 1/\theta]$.
8. If z has polar coordinates $[r, \theta]$, then z^n has polar coordinates $[nr, n\theta]$.
9. The number -8 has two square roots in the complex number system.
10. If $z = \cos(2\pi/3) + i\sin(2\pi/3)$, then z, z^2, and z^3 are cube roots of unity.

Rectangular and Polar Forms of Complex Numbers

Express each of the following complex numbers in polar form.

11. $1 + i$

12. -5

13. $2i$

14. $2 - 2i$

15. $-1 - \sqrt{3}i$

16. $-\frac{\sqrt{3}}{2} + \frac{1}{2}i$

Multiplication of Complex Numbers

Express the product z_1z_2 in polar form in each of the following cases.

17. $z_1 = 1 + i;\quad z_2 = 2i$

18. $z_1 = 2 - 2i;\quad z_2 = -1 - \sqrt{3}i$

19. $z_1 = -1 - \sqrt{3}i;\quad z_2 = -\frac{\sqrt{3}}{2} + \frac{1}{2}i$

20. $z_1 = -\frac{\sqrt{3}}{2} + \frac{1}{2}i;\quad z_2 = 1 + i$

21. $z_1 = 2\left(\cos\frac{\pi}{4} + i\sin\frac{\pi}{4}\right);$

$z_2 = 2\left(\cos\frac{\pi}{2} + i\sin\frac{\pi}{2}\right)$

22. $z_1 = 2\left(\cos\frac{\pi}{5} + i\sin\frac{\pi}{5}\right);$

$z_2 = \cos\frac{\pi}{10} + i\sin\frac{\pi}{10}$

23. $z_1 = \cos\left(-\frac{\pi}{4}\right) + i\sin\left(-\frac{\pi}{4}\right);$

$z_2 = \sqrt{2}\left(\cos\frac{3\pi}{8} + i\sin\frac{3\pi}{8}\right)$

24. $z_1 = 5(\cos\pi + i\sin\pi);$

$z_2 = \frac{1}{5}[\cos(-\pi) + i\sin(-\pi)]$

Express the quotient z_1/z_2 in polar form in each of the following cases.

25. z_1 and z_2 are as in Exercise 17.

26. z_1 and z_2 are as in Exercise 18.

27. z_1 and z_2 are as in Exercise 19.

28. z_1 and z_2 are as in Exercise 20.

29. z_1 and z_2 are as in Exercise 21.

30. z_1 and z_2 are as in Exercise 22.

31. z_1 and z_2 are as in Exercise 23.

32. z_1 and z_2 are as in Exercise 24.

Express each of the following powers in polar form.

33. $(1 + i)^3$

34. $(-\sqrt{3} + i)^5$

35. $\left(\frac{\sqrt{2}}{2} - \frac{\sqrt{2}}{2}i\right)^{10}$

36. $(1 - i)^8$

37. $\left(\cos\frac{\pi}{5} + i\sin\frac{\pi}{5}\right)^5$

38. $\left(\cos\frac{\pi}{8} + i\sin\frac{\pi}{8}\right)^{12}$

39. $\left[2\left(\cos\frac{4\pi}{5} + i\sin\frac{4\pi}{5}\right)\right]^{10}$

40. $\left[\frac{3}{2}\left(\cos\frac{\pi}{16} + i\sin\frac{\pi}{16}\right)\right]^4$

Roots of Complex Numbers

41. Find the cube roots of -1.

42. Find the fourth roots of unity.

43. Find the fourth roots of -16.

44. Find the cube roots of unity.

45. Find the sixth roots of $32\sqrt{2} - 32\sqrt{2}i$.

46. Find the fourth roots of $-8\sqrt{2} + 8\sqrt{2}i$.

Miscellaneous

47. Given $z = r(\cos\theta + i\sin\theta)$, $r \neq 0$, and n is a positive integer, show that

$$z^{-n} = r^{-n}(\cos n\theta - i\sin n\theta).$$

(*Hint:* by definition, $z^{-n} = 1/z^n$. First find the polar coordinates of z^n and then the polar coordinates of $1/z^n$.)

48. With reference to equation (6) for the nth roots of z, show that the integers $k = 0, 1, 2, \ldots, n - 1$ give different nth roots, and for any other integer k no new roots are obtained. (*Hint:* $\cos\theta_1 + i\sin\theta_1 = \cos\theta_2 + i\sin\theta_2 \Leftrightarrow \theta_2 = \theta_1 + 2j\pi$ for some integer j.)

49. Show that the solution of the equation

$$z^{n-1} + z^{n-2} + \cdots + z^2 + z + 1 = 0$$

consists of all the nth roots of unity except 1. [*Hint:* $z^n - 1$ factors into $(z - 1)(z^{n-1} + z^{n-2} + \cdots + z^2 + z + 1)$.]

8.4 Vectors

Vectors bring geometry to physical science.

Vectors in the Plane. Many physical quantities, including force, velocity, and acceleration, are measured by both *magnitude* and *direction*. These quantities can be represented geometrically in the plane by directed line segments called **vectors** (Figure 21). The length of the vector in Figure 21 corresponds to the magnitude of the velocity, and the arrow points in the direction of the velocity. By definition, any two vectors in the plane that have the same length and direction are **equivalent,** regardless of their starting points (Figure 22). For example, if a wind of 30 miles per hour is blowing across a broad front in the southeast direction, then any vector of magnitude 30 and direction $-45°$ can be used to represent the wind. That is, there is no unique starting point or point of application of the wind.

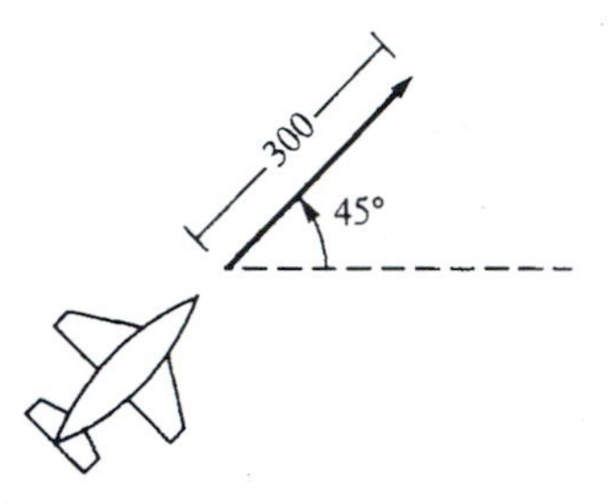

Figure 21 Velocity Vector (The velocity of the plane is 300 mph in the northeast direction.)

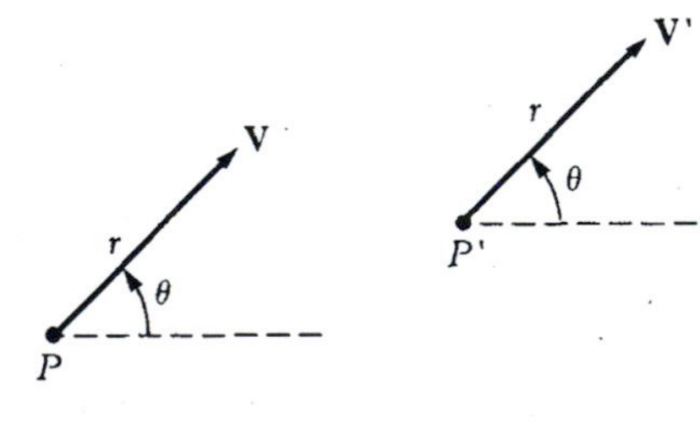

Figure 22 Equivalent Vectors (**V** starts at P and **V**′ at P', but both vectors have the same length r and direction θ.)

Operations with Vectors. If two vectors **V** and **W** have the same starting point P, then their sum **V** + **W** is defined as the directed diagonal, starting at P, of the parallelogram formed by **V** and **W** [Figure 23(a)].[1] Note that if **W** is translated to an equivalent vector **W**′ starting at the endpoint of **V**, then **W**′ and **V** + **W** have the same endpoint [Figure 23(b)]. Similarly, if **V** is translated to an equivalent sector **V**′ starting at the endpoint of **W**, then **V**′ and **V** + **W** have the same endpoint [Figure 23(c)]. It can be seen geometrically (also see Exercise 28) that vector addition is commutative and associative.

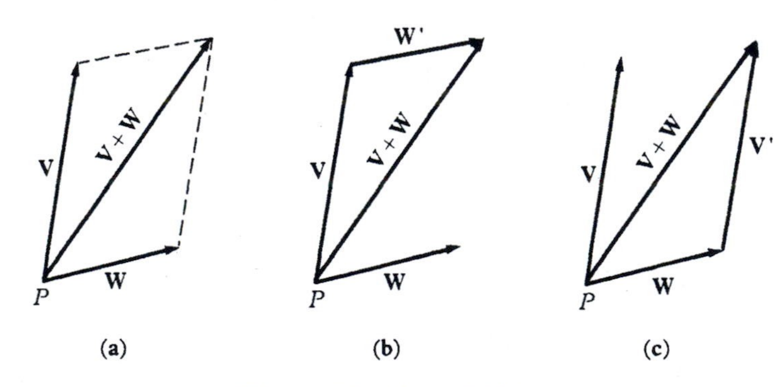

Figure 23 Vector Addition

Example 1 A plane is flying due north at 250 miles per hour when it encounters a wind that is blowing due east at 50 miles per hour. What is the resulting speed and direction of the plane?

[1]There are several conventions for denoting vectors, including capital letters, small boldface letters, and letters with arrows above them. We will use capital letters (toward the end of the alphabet) because they are the easiest to reproduce in notes and on the board.

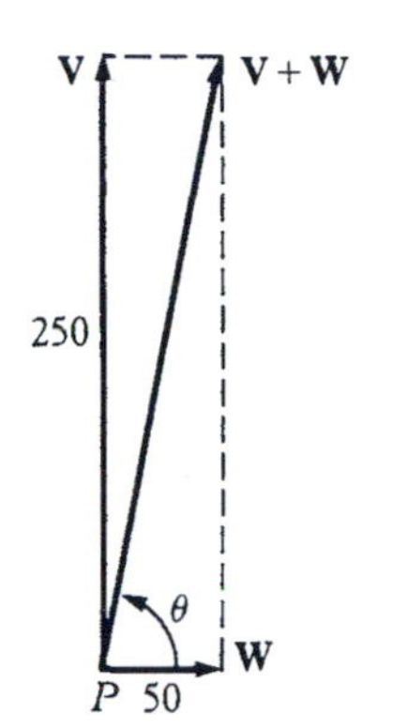

Figure 24

Solution: We represent the velocity of the plane by a vector **V** of magnitude 250 starting at some point P and pointing vertically upward. The wind velocity is a vector **W** of magnitude 50, also starting at P, and pointing horizontally to the right. The resulting velocity of the plane is represented by the vector **V** + **W**. As shown in Figure 24, **V** + **W** is a vector of magnitude approximately 255 miles per hour whose direction θ from the positive horizontal satisfies $\tan\theta = 5$. ■

If **W** is a vector, then by −**W** we mean the vector with the same length and starting point as **W** but in the opposite direction. The difference **V** − **W** is defined to be the sum **V** + (−**W**) (Figure 25(a)]. Note that if −**W** is translated to an equivalent vector −**W′** starting at the endpoint of **V**, then −**W′** and **V** − **W** have the same endpoint [Figure 25(b)].

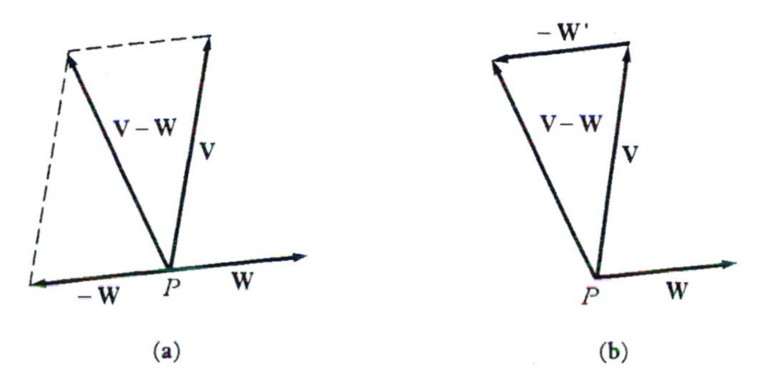

Figure 25 Vector Subtraction

Example 2 A rope is attached to a 100-pound weight and looped over a pulley as shown in Figure 26(a). What force must be exerted by the operator in order to keep the weight suspended? What part of this force is directed away from the wall on the bracket supporting the pulley?

Solution: We can represent the weight as a vector **W** of length 100 directed vertically downward. As shown in Figure 26(b), the force exerted by the operator is a vector **V** that makes an angle of 30° with **W**. The

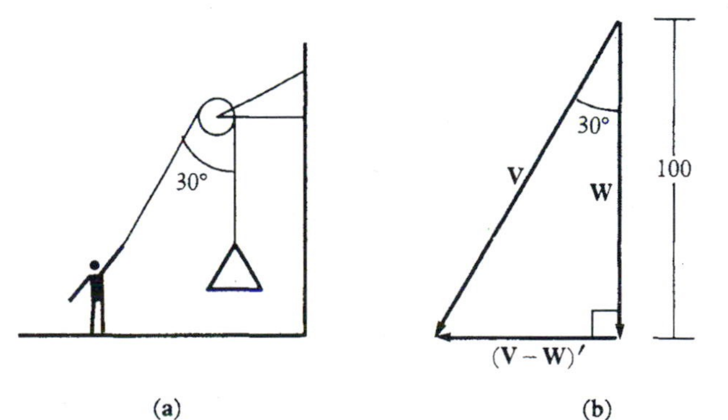

Figure 26

exerted force will keep the weight suspended if the projection of **V** on **W** has magnitude 100, that is, if **V** and **W** form the hypotenuse and height, respectively, of a right triangle. The directed base of the triangle is equivalent to **V** − **W** and represents the part of **V** that is directed horizontally away from the wall. Hence the magnitude v of **V** must satisfy

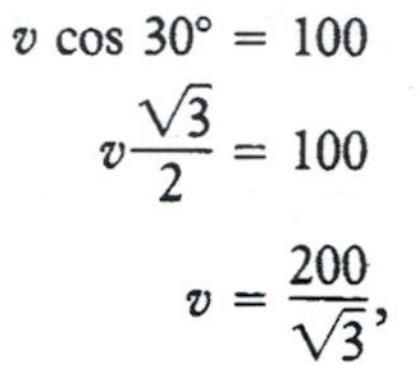

$$v \cos 30^\circ = 100$$
$$v\frac{\sqrt{3}}{2} = 100$$
$$v = \frac{200}{\sqrt{3}},$$

and the magnitude u of **V** − **W** then satisfies

$$u = v \sin 30^\circ$$
$$u = \frac{200}{\sqrt{3}} \cdot \frac{1}{2} = \frac{100}{\sqrt{3}}.$$

Hence, the operator must exert a force of $200/\sqrt{3} \approx 115.5$ pounds, which also results in a force of $100/\sqrt{3} \approx 57.75$ pounds directed away from the wall. ■

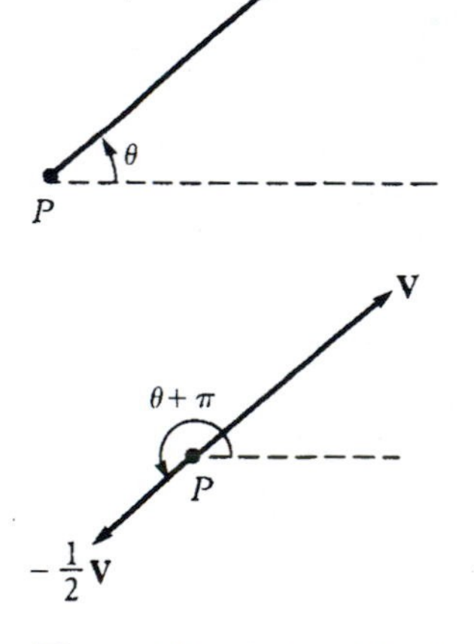

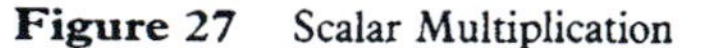

Figure 27 Scalar Multiplication

If k is a real number and **V** is a vector, then by $k\mathbf{V}$ we mean the vector whose length is $|k|$ times the length of **V**, and whose direction is the same as **V**'s if $k > 0$ and the opposite of **V**'s if $k < 0$ (Figure 27). If $k = 0$, then $k\mathbf{V}$ is the **zero vector,** that is, a vector of zero length. In particular,

$$1\mathbf{V} = \mathbf{V} \quad \text{and} \quad (-1)\mathbf{V} = -\mathbf{V}.$$

In the context of vectors, real numbers are called **scalars,** and $k\mathbf{V}$ is called a **scalar multiple** of **V**. It can be seen geometrically (also see Exercise 29) that scalar multiplication satisfies the following properties:

$$k_1(k_2\mathbf{V}) = (k_1k_2)\mathbf{V} \qquad \textbf{associative property}$$

and

$$\left.\begin{aligned} k(\mathbf{V} + \mathbf{W}) &= k\mathbf{V} + k\mathbf{W} \\ (k_1 + k_2)\mathbf{V} &= k_1\mathbf{V} + k_2\mathbf{V}. \end{aligned}\right\} \qquad \textbf{distributive properties}$$

Example 3 Vectors **V** and **W** start at the same point P and end at Q and S, respectively [Figure 28(a)]. Show that the vector

$$t\mathbf{V} + (1 - t)\mathbf{W} \qquad (0 \le t \le 1)$$

ends at a point on the line segment from S to Q.

Solution: Now,

$$\begin{aligned} t\mathbf{V} + (1 - t)\mathbf{W} &= t\mathbf{V} + \mathbf{W} - t\mathbf{W} \\ &= \mathbf{W} + t(\mathbf{V} - \mathbf{W}). \end{aligned}$$

Also, $\mathbf{V} - \mathbf{W}$ is equivalent to a vector $(\mathbf{V} - \mathbf{W})'$ from S to Q. Since $0 \le t \le 1$, $t(\mathbf{V} - \mathbf{W})'$ ends somewhere on the line segment from S to Q. Finally, $t(\mathbf{V} - \mathbf{W})'$ and $t\mathbf{V} + (1 - t)\mathbf{W}$ have the same endpoint, as shown in Figure 28(b). ■

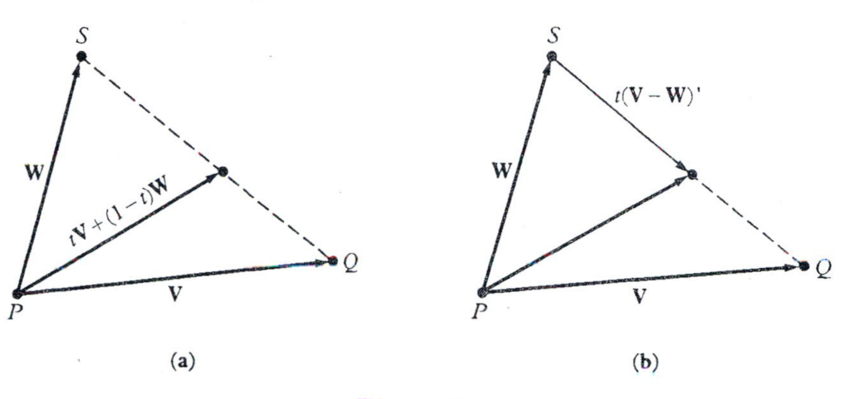

Figure 28

Question *Why don't we define multiplication of two vectors?*

Answer: Vectors represent directed magnitudes such as force or velocity. There is a physical interpretation of the sum of two vectors as the resultant of two forces or two velocities. Similarly, we have an interpretation of scalar multiplication as a magnification, contraction, or reversal of direction. However, there is no physical interpretation for the "product" of two vectors in terms of another *vector* in the plane. Later we will define the "dot product" of two vectors as a *scalar* that has a geometric interpretation in terms of the relative magnitudes and directions of the two vectors. ■

Coordinate Vectors. A vector whose starting point is the origin $O(0, 0)$ is called a **coordinate vector,** and every vector $\mathbf{V}$ in the plane is equivalent to a coordinate vector. In fact, if $\mathbf{V}$ is a vector from point (x_1, y_1) to (x_2, y_2), then $\mathbf{V}$ is equivalent to the coordinate vector $\mathbf{V}'$ with endpoint $(x_2 - x_1, y_2 - y_1)$ (Figure 29).

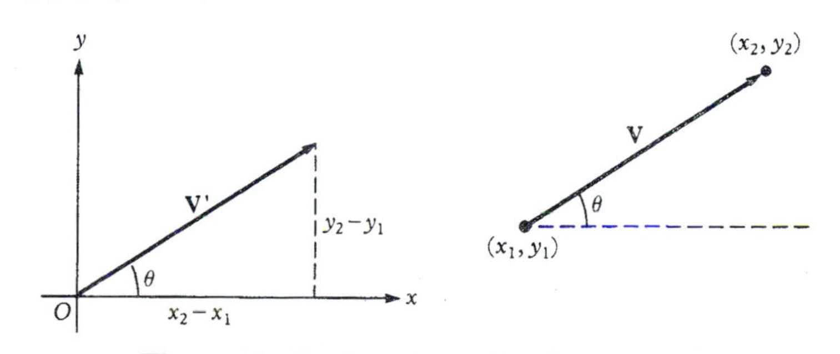

Figure 29 Coordinate Vector $\mathbf{V}'$ Is Equivalent to $\mathbf{V}$

Example 4 For each vector in Figure 30, find an equivalent coordinate vector.

Solution: The vectors **U**, **V**, and **W** are equivalent to the coordinate vectors **U′**, **V′**, and **W′**, respectively, with endpoints as follows.

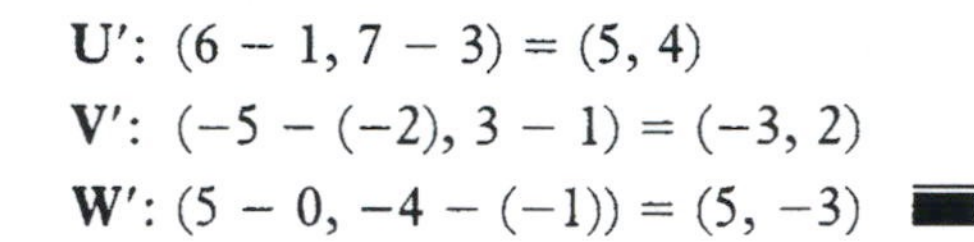

$$\mathbf{U}': (6 - 1, 7 - 3) = (5, 4)$$
$$\mathbf{V}': (-5 - (-2), 3 - 1) = (-3, 2)$$
$$\mathbf{W}': (5 - 0, -4 - (-1)) = (5, -3) \quad ■$$

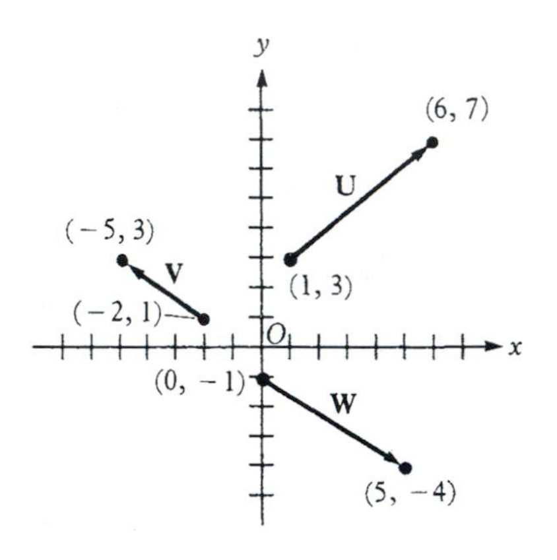

Figure 30

The coordinate vector **V** with endpoint (x, y) will be denoted by $\langle x, y\rangle$, and the numbers x and y are called the **coordinates** or **components** of **V**. There is a natural one-to-one correspondence between the points in the plane and the set of all coordinate vectors. That is, each point in the plane is the endpoint of a unique coordinate vector. Because of this correspondence, we often identify the point (x, y) with the vector $\langle x, y\rangle$. In particular, the origin $(0, 0)$ is identified with the **zero vector** $\langle 0, 0\rangle$.

For coordinate vectors, our geometric rules for **addition** (Figure 31) and **scalar multiplication** (Figure 32) can be expressed algebraically by means of the following equations.

$$\langle x_1, y_1\rangle + \langle x_2, y_2\rangle = \langle x_1 + x_2, y_1 + y_2\rangle \tag{1}$$
$$k\langle x, y\rangle = \langle kx, ky\rangle. \tag{2}$$

Hence, to add two coordinate vectors, we add in each coordinate position, and to multiply a coordinate vector by a scalar k, we multiply each coordinate by k.

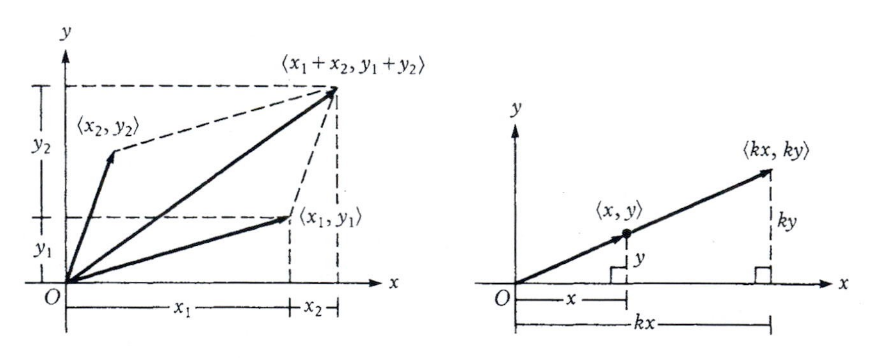

Figure 31 Addition of Coordinate Vectors

Figure 32 Scalar Multiplication for Coordinate Vectors

The rule for addition extends to more than two vectors. For instance,

$$\langle x_1, y_1\rangle + \langle x_2, y_2\rangle + \langle x_3, y_3\rangle = \langle x_1 + x_2 + x_3, y_1 + y_2 + y_3\rangle.$$

Also, since $-\langle x, y\rangle = (-1)\langle x, y\rangle = \langle -x, -y\rangle$, the rule for **subtracting coordinate vectors** is as follows.

$$\langle x_1, y_1\rangle - \langle x_2, y_2\rangle = \langle x_1 - x_2, y_1 - y_2\rangle \tag{3}$$

Example 5 Given $\mathbf{U} = \langle 2, 1\rangle$, $\mathbf{V} = \langle 3, -4\rangle$, and $\mathbf{W} = \langle -5, 6\rangle$, compute each of the following.

(a) $\mathbf{U} + \mathbf{V}$ (b) $\mathbf{V} - \mathbf{W}$ (c) $5\mathbf{U} - 3\mathbf{V} + 2\mathbf{W}$

Solution:

(a) $\langle 2, 1\rangle + \langle 3, -4\rangle = \langle 2 + 3, 1 + (-4)\rangle = \langle 5, -3\rangle$

(b) $\langle 3, -4\rangle - \langle -5, 6\rangle = \langle 3 - (-5), -4 - 6\rangle = \langle 8, -10\rangle$

(c) $$\begin{aligned} 5\langle 2, 1\rangle - 3\langle 3, -4\rangle + 2\langle -5, 6\rangle &= \langle 10, 5\rangle - \langle 9, -12\rangle + \langle -10, 12\rangle \\ &= \langle 10 - 9 - 10, 5 + 12 + 12\rangle \\ &= \langle -9, 29\rangle \end{aligned}$$ ■

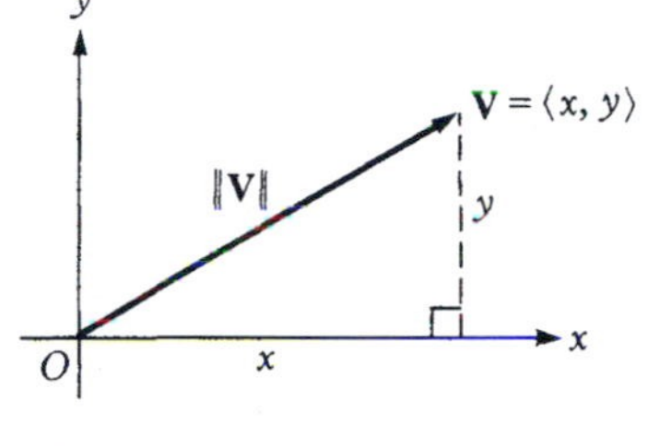

Figure 33 Length of a Coordinate Vector

The **length** or **magnitude of a coordinate vector V** is denoted by $\|\mathbf{V}\|$ (Figure 33) and can be computed by the distance formula (4).

$$\|\mathbf{V}\| = \sqrt{x^2 + y^2} \qquad \textbf{(4)}$$

The quantity $\|\mathbf{V}\|$ is also called the **norm** of **V**. Note that **V** is the zero vector $\langle 0, 0\rangle$ if and only if $\|\mathbf{V}\| = 0$. Also, if $\mathbf{V} \neq \langle 0, 0\rangle$, then $\frac{1}{\|\mathbf{V}\|}\mathbf{V}$ is a vector of length 1 (*why?*) in the same direction as **V** (*why?*). Any vector of length 1 is called a **unit vector.** If **V** is not the zero vector, then the unit vector $\frac{1}{\|\mathbf{V}\|}\mathbf{V}$ is usually written as $\frac{\mathbf{V}}{\|\mathbf{V}\|}$.

Example 6 The velocity of a rocket has a vertical component of 500 feet per second and a horizontal component of 25 feet per second (Figure 34). What is the magnitude of the velocity? Also, find a unit vector in the direction of the velocity.

Solution: If **V** denotes the velocity vector, then

$$\begin{aligned} \|\mathbf{V}\| &= \sqrt{500^2 + 25^2} \\ &= \sqrt{250{,}625} \\ &\approx 500.625 \text{ feet per second.} \end{aligned}$$

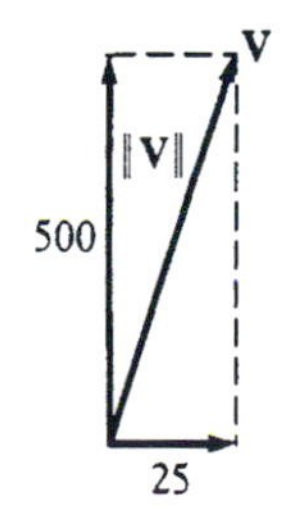

Figure 34

To determine a unit vector in the direction of **V**, we divide each component of **V** by $\|\mathbf{V}\|$:

$$\frac{500}{500.625} \approx .9988 \quad \text{and} \quad \frac{25}{500.625} \approx .0499.$$

Therefore, a unit vector in the direction of **V** is (to 4-place accuracy)

$$\frac{\mathbf{V}}{\|\mathbf{V}\|} = \langle 0.9988, 0.0499\rangle.$$ ■

Example 7 Let $\mathbf{V} = \langle 6, 8 \rangle$ and $\mathbf{W} = \langle -2, 4 \rangle$. Show that $\|3\mathbf{V}\| = 3\|\mathbf{V}\|$ and $\|\mathbf{V} + \mathbf{W}\| < \|\mathbf{V}\| + \|\mathbf{W}\|$.

Solution: First,

$$\|\mathbf{V}\| = \sqrt{6^2 + 8^2} = \sqrt{100} = 10$$

and $$\|3\mathbf{V}\| = \|\langle 18, 24 \rangle\| = \sqrt{18^2 + 24^2} = \sqrt{900} = 30.$$

Therefore, $\|3\mathbf{V}\| = 3\|\mathbf{V}\|$.

Also, $$\|\mathbf{W}\| = \sqrt{(-2)^2 + 4^2} = \sqrt{20} = 2\sqrt{5}$$

and $$\|\mathbf{V} + \mathbf{W}\| = \|\langle 4, 12 \rangle\| = \sqrt{4^2 + 12^2} = \sqrt{160} = 4\sqrt{10}.$$

Now $4\sqrt{10} < 10 + 2\sqrt{5}$ (*why?*). Therefore, $\|\mathbf{V} + \mathbf{W}\| < \|\mathbf{V}\| + \|\mathbf{W}\|$. ■

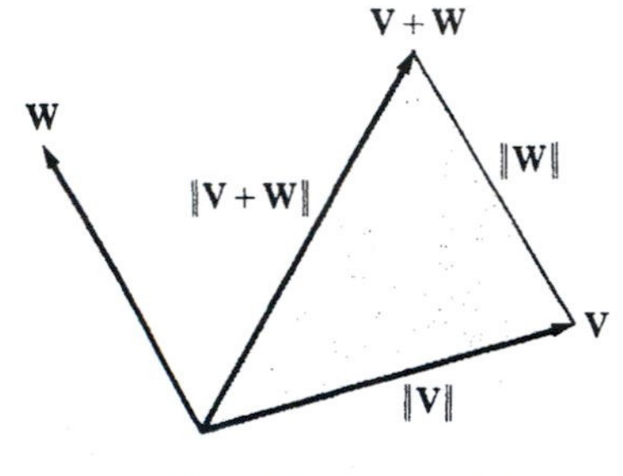

(a) $\|\mathbf{V} + \mathbf{W}\| < \|\mathbf{V}\| + \|\mathbf{W}\|$

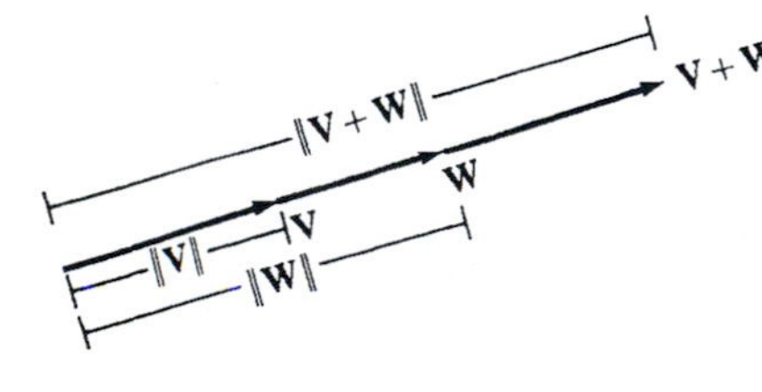

(b) $\|\mathbf{V} + \mathbf{W}\| = \|\mathbf{V}\| + \|\mathbf{W}\|$

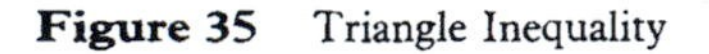

Figure 35 Triangle Inequality

The results in Example 7 illustrate the following general properties that hold for any scalar k and any two vectors $\mathbf{V}$ and $\mathbf{W}$ (see Exercises 39 and 40 in the next section):

$$\|k\mathbf{V}\| = |k|\,\|\mathbf{V}\| \tag{5}$$

and

$$\|\mathbf{V} + \mathbf{W}\| \le \|\mathbf{V}\| + \|\mathbf{W}\|. \quad \textbf{triangle inequality} \tag{6}$$

Comment A triangle inequality for real numbers was stated in Section P.3. However, it is from vectors that the inequality gets its name. That is, any two vectors $\mathbf{V}$ and $\mathbf{W}$ not on the same line determine a triangle [Figure 35(a)], and the inequality states the evident geometric property that the sum of the lengths of any two sides of a triangle is greater than the length of the third side. Equality occurs in the triangle inequality when the two vectors are in the same direction [Figure 35(b)].

Exercises 8.4

Fill in the blanks to make each statement true.

1. A vector quantity is measured by both ________ and ________.
2. In the context of vectors, real numbers are called ________.
3. If $\mathbf{V}$ and $\mathbf{W}$ are vectors starting at P, then $\mathbf{V} + \mathbf{W} =$ ________.
4. A vector whose initial point is at the origin (0, 0) is called a ________.
5. The vector from point (x_1, y_1) to point (x_2, y_2) is equivalent to the coordinate vector ________.

Write true *or* false *for each statement.*

6. If vectors $\mathbf{V}$ and $\mathbf{W}$ have the same magnitude, they are equivalent.
7. If vectors $\mathbf{V}$ and $\mathbf{W}$ are in opposite directions, then $\mathbf{V} + \mathbf{W}$ is perpendicular to both $\mathbf{V}$ and $\mathbf{W}$.
8. Two coordinate vectors are equivalent if and only if they have the same endpoint.

9. If $\mathbf{W} = k\mathbf{V}$, then $\|\mathbf{V} + \mathbf{W}\| = \|\mathbf{V}\| + \|\mathbf{W}\|$.

10. If $\|\mathbf{V} + \mathbf{W}\| = \|\mathbf{V}\| + \|\mathbf{W}\|$, then $\mathbf{W} = k\mathbf{V}$, where $k \geq 0$.

Vectors in the Plane

11. A crate of machine parts weighing 250 lb is resting on a ramp that makes an angle of 10° with the horizontal. How much weight is directed perpendicular to the ramp, and how much is directed parallel to the ramp?

12. A car is traveling 40 mph down a 2-mi hill that makes an angle of 15° with the horizontal. How fast is the car traveling in the horizontal direction and how fast in the vertical direction?

13. The current in a river is 3 ft/sec. Andy wants to maneuver his motorboat straight across the river at 4 ft/sec. At what speed and in what direction must he set the controls on the boat?

14. A hot-air balloon is 1000 ft high, and a 20-mph wind is blowing due east. How fast must the balloon descend in order to touch down 2000 ft east of its present position?

15. A jet plane flying at 400 mph launches a rocket at 200 mph perpendicular to the path of the plane. What are the resulting speed and direction of the rocket?

Coordinate Vectors

Compute and graph $\mathbf{V} + \mathbf{W}$, $\mathbf{V} - \mathbf{W}$, *and* $2\mathbf{V} + 3\mathbf{W}$.

16. $\mathbf{V} = \langle 2, -5 \rangle$; $\mathbf{W} = \langle -5, 2 \rangle$

17. $\mathbf{V} = \langle 1, 6 \rangle$; $\mathbf{W} = \langle -1, 5 \rangle$

Compute each of the following.

18. $2\langle 4, 1 \rangle - 3\langle 0, 5 \rangle + 4\langle -1, 6 \rangle$

19. $5\langle -2, 4 \rangle + 4\langle 1, 1 \rangle - 3\langle 2, 0 \rangle$

Solve for x and y.

20. $\langle x, y \rangle - 5\langle y, x \rangle = \langle 2, 4 \rangle$

21. $\langle 2x, 1 \rangle - 3\langle 4, y \rangle = \langle x, y \rangle + 2\langle 5, -1 \rangle$

Compute $\|\mathbf{V}\|$ *and find a unit vector in the direction of* $\mathbf{V}$.

22. $\mathbf{V} = \langle 2, -2 \rangle$

23. $\mathbf{V} = \langle -3, 4 \rangle$

Solve for x.

24. $\|\langle x, 2 \rangle + \langle 3, 1 \rangle\| = \sqrt{13}$

25. $\|\langle 4, 5 \rangle - \langle x, 3 \rangle\| = 2\sqrt{2}$

Compute each of the following.

26. $\|3\langle -1, 2 \rangle - 2\langle 1, 4 \rangle + \langle 5, 4 \rangle\|$

27. $\|2\langle 4, -5 \rangle + 3\langle 2, 6 \rangle - \langle 4, 1 \rangle\|$

Miscellaneous

28. *Basic Properties for Vector Addition.* Show that the basic properties (1), (2), (6), and (8) of Section P.2 hold for addition of coordinate vectors.

29. *Associative and Distributive Properties for Vector Addition and Scalar Multiplication.* Let $\mathbf{V} = \langle x_1, y_1 \rangle$ and $\mathbf{W} = \langle x_2, y_2 \rangle$ be any two coordinate vectors, and k_1 and k_2 be any two scalars. Verify each of the following.
   (a) $k_1(k_2\mathbf{V}) = (k_1k_2)\mathbf{V}$
   (b) $(k_1 + k_2)\mathbf{V} = k_1\mathbf{V} + k_2\mathbf{V}$
   (c) $k_1(\mathbf{V} + \mathbf{W}) = k_1\mathbf{V} + k_1\mathbf{W}$

8.5 The Dot Product

The dot product combines algebra, geometry, and trigonometry.

The Dot Product of Two Vectors
Vector Interpretation of the General Linear Equation
Angle Between Two Lines

The Dot Product of Two Vectors. We now introduce an important concept linking trigonometry and coordinate geometry. If **V** and **W** are nonzero coordinate vectors and θ is the smaller nonnegative angle between them ($0 \leq \theta \leq \pi$), then the **dot product** of **V** and **W**, denoted by $\mathbf{V} \cdot \mathbf{W}$, is defined as follows:

$$\mathbf{V} \cdot \mathbf{W} = \|\mathbf{V}\|\,\|\mathbf{W}\| \cos \theta. \quad (1)$$

If either $\mathbf{V}$ or $\mathbf{W}$ is the zero vector, then $\mathbf{V} \cdot \mathbf{W} = 0$ by definition. Note that the dot product of two vectors is a *real number* (scalar), not a vector.

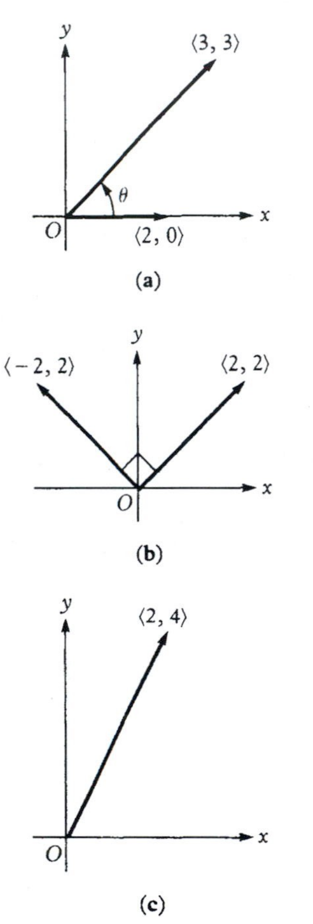

Figure 36

Example 1 Find $\mathbf{V} \cdot \mathbf{W}$ for each of the following cases.

(a) $\mathbf{V} = \langle 2, 0 \rangle$ and $\mathbf{W} = \langle 3, 3 \rangle$

(b) $\mathbf{V} = \langle 2, 2 \rangle$ and $\mathbf{W} = \langle -2, 2 \rangle$

(c) $\mathbf{V} = \langle 2, 4 \rangle$ and $\mathbf{W} = \langle 2, 4 \rangle$

Solution:

(a) By inspecting Figure 36(a), we see that $\theta = \pi/4$. Therefore, by definition (1),

$$\langle 2, 0 \rangle \cdot \langle 3, 3 \rangle = \sqrt{2^2 + 0^2}\,\sqrt{3^2 + 3^2}\,\cos\frac{\pi}{4}$$

$$= 2 \cdot 3\sqrt{2} \cdot \frac{1}{\sqrt{2}} = 6.$$

(b) The angle between $\langle 2, 2 \rangle$ and $\langle -2, 2 \rangle$ in Figure 36(b) is $\pi/2$ (*why?*), and $\cos \pi/2 = 0$. Therefore,

$$\langle 2, 2 \rangle \cdot \langle -2, 2 \rangle = 0.$$

(c) The angle between any nonzero vector and itself is 0 [Figure 36(c)], and $\cos 0 = 1$. Hence,

$$\langle 2, 4 \rangle \cdot \langle 2, 4 \rangle = \sqrt{2^2 + 4^2}\,\sqrt{2^2 + 4^2} \cdot 1$$

$$= 20. \quad \blacksquare$$

As indicated in Figure 37, if neither $\mathbf{V}$ nor $\mathbf{W}$ is the zero vector, then $\|\mathbf{W}\| \cos \theta$ is the real number equal to the directed magnitude of the projection of $\mathbf{W}$ on $\mathbf{V}$. Hence, $\mathbf{V} \cdot \mathbf{W}$ is the magnitude of $\mathbf{V}$ times the directed magnitude of the projection of $\mathbf{W}$ on $\mathbf{V}$.

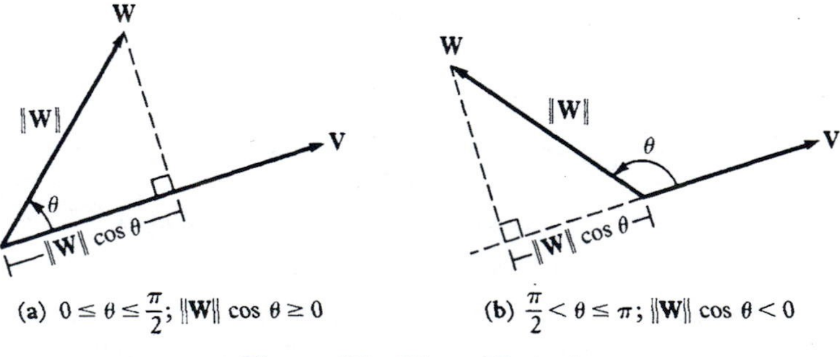

(a) $0 \le \theta \le \frac{\pi}{2}$; $\|\mathbf{W}\| \cos \theta \ge 0$ (b) $\frac{\pi}{2} < \theta \le \pi$; $\|\mathbf{W}\| \cos \theta < 0$

Figure 37 Directed Projection

Example 2 Find the directed magnitude of the projection of $\mathbf{W} = \langle 4, 4 \rangle$ on $\mathbf{V} = \langle \sqrt{3}, 1 \rangle$.

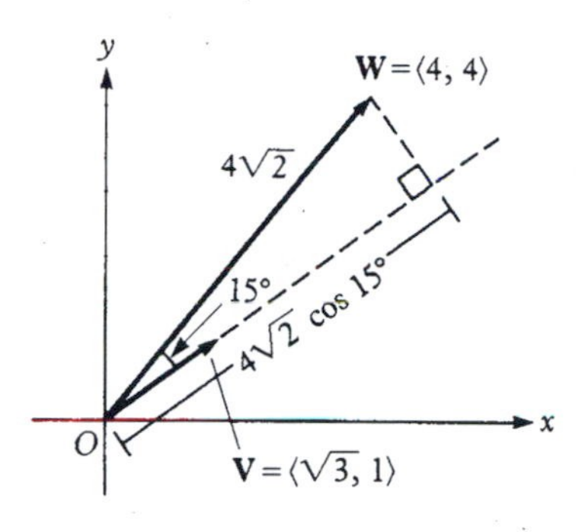

Figure 38

Solution: The vector $\langle 4, 4\rangle$ makes an angle of 45° with the positive x-axis, and $\langle\sqrt{3}, 1\rangle$ makes an angle of 30° with the same axis (*why?*). Therefore, the angle θ from **V** to **W** is 15° (Figure 38), and the projection of **W** on **V** has directed magnitude

$$\begin{aligned}\|\mathbf{W}\| \cos\theta &= \sqrt{4^2 + 4^2}\cos 15^\circ \\ &= 4\sqrt{2}(.9659) \\ &\approx 5.4640.\end{aligned}$$ ■

Since $\cos 90^\circ = 0$, we have the following important result.

Two nonzero vectors V and W are perpendicular ⇔

$$\mathbf{V}\cdot\mathbf{W} = 0. \tag{2}$$

Also, since $\cos 0 = 1$, we obtain

$$\mathbf{V}\cdot\mathbf{V} = \|\mathbf{V}\|^2 \text{ for any vector } \mathbf{V}. \tag{3}$$

Although the dot product is defined in terms of the angle between **V** and **W**, it is possible to compute $\mathbf{V}\cdot\mathbf{W}$ directly from the coordinates of **V** and **W**. That is, (see Exercise 41), if $\mathbf{V} = \langle x_1, y_1\rangle$ and $\mathbf{W} = \langle x_2, y_2\rangle$, then

$$\mathbf{V}\cdot\mathbf{W} = x_1x_2 + y_1y_2. \quad \textbf{algebraic formula for dot product} \tag{4}$$

Example 3 Use formula (4) to compute $\mathbf{V}\cdot\mathbf{W}$ for each of the cases in Example 1.

Solution:

(a) $\langle 2, 0\rangle\cdot\langle 3, 3\rangle = 2\cdot 3 + 0\cdot 3 = 6$

(b) $\langle 2, 2\rangle\cdot\langle -2, 2\rangle = 2(-2) + 2\cdot 2 = 0$

(c) $\langle 2, 4\rangle\cdot\langle 2, 4\rangle = 2\cdot 2 + 4\cdot 4 = 20$ ■

The algebraic formula (4) is often more convenient than the geometric definition (1) for computing $\mathbf{V}\cdot\mathbf{W}$ because in (4) it is not necessary to know the angle θ between **V** and **W**. In fact, we can use (4) to determine $\cos\theta$, as shown in the following example.

Example 4 Find the cosine of the angle between $\mathbf{V} = \langle 4, 3\rangle$ and $\mathbf{W} = \langle 5, 5\rangle$.

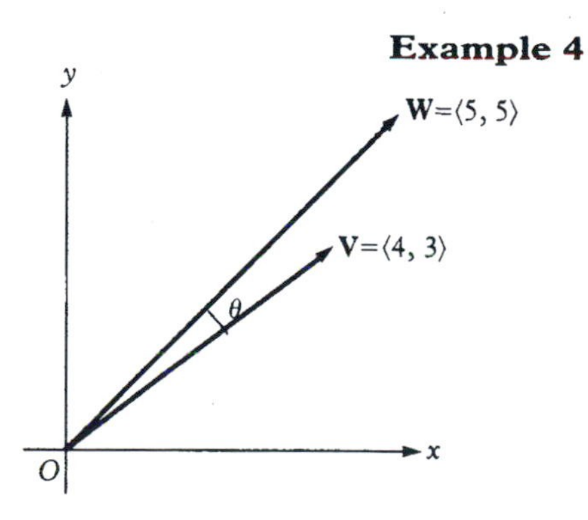

Figure 39

Solution: The vectors **V** and **W** are shown in Figure 39. From definition (1),

$$\begin{aligned}\mathbf{V}\cdot\mathbf{W} &= \|\mathbf{V}\|\,\|\mathbf{W}\|\cos\theta \\ &= \sqrt{4^2 + 3^2}\,\sqrt{5^2 + 5^2}\cos\theta \\ &= 25\sqrt{2}\cos\theta.\end{aligned}$$

From formula (4),

$$\begin{aligned}\mathbf{V}\cdot\mathbf{W} &= 4\cdot 5 + 3\cdot 5 \\ &= 35.\end{aligned}$$

Therefore, $25\sqrt{2}\cos\theta = 35$

or $$\cos\theta = \frac{35}{25\sqrt{2}} = \frac{7\sqrt{2}}{10}.$$ ■

Example 5 Which pairs of the following vectors are perpendicular: $\mathbf{U} = \langle 2, -3\rangle$, $\mathbf{V} = \langle -2, 3\rangle$, and $\mathbf{W} = \langle 6, 4\rangle$?

Solution: According to (2), two nonzero vectors are perpendicular if and only if their dot product is zero. Now, by the formula (4),

$$\mathbf{U}\cdot\mathbf{V} = 2(-2) + (-3)3 = -13$$
$$\mathbf{U}\cdot\mathbf{W} = 2\cdot 6 + (-3)4 = 0$$

and $$\mathbf{V}\cdot\mathbf{W} = (-2)6 + 3\cdot 4 = 0.$$

Therefore, **U** and **W** are perpendicular, and so are **V** and **W**. ■

Vector Interpretation of the General Linear Equation. We now apply the dot product to the equation of a straight line. In Section 2.5 we saw that the graph of the linear equation

$$ax + by = c \qquad (a \text{ or } b \neq 0) \tag{5}$$

is a straight line L in the plane. If (x_0, y_0) is a fixed point on L, then $ax_0 + by_0 = c$, and (5) becomes

$$ax + by = ax_0 + by_0$$

or $$a(x - x_0) + b(y - y_0) = 0. \tag{6}$$

Equation (6) can be interpreted as a dot product. That is,

$$\langle a, b\rangle \cdot \langle x - x_0, y - y_0\rangle = 0, \tag{6'}$$

which says that the vector $\langle a, b\rangle$ is *perpendicular* to the vector $\langle x - x_0, y - y_0\rangle$. But $\langle x - x_0, y - y_0\rangle$ is equivalent to a vector on L (Figure 40).

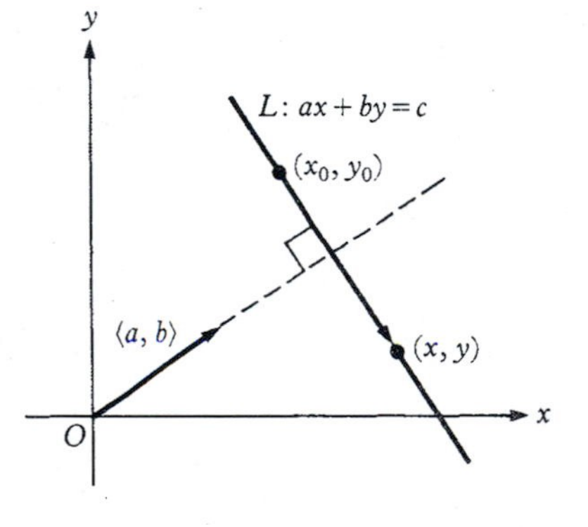

Figure 40 Vectors $\langle x - x_0, y - y_0\rangle$ and $\langle a, b\rangle$ Are Perpendicular

Therefore, we have the following result.

In the general linear equation $L\colon ax + by = c$, the coordinate vector $\langle a, b\rangle$ is perpendicular to the line L.

Example 6 Find an equation of the line L that passes through the point (4, 5) and is perpendicular to the vector $\langle 3, 2\rangle$ (Figure 41).

Solution: From the vector interpretation of the general linear equation, we know that an equation of L is

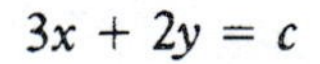

$$3x + 2y = c$$

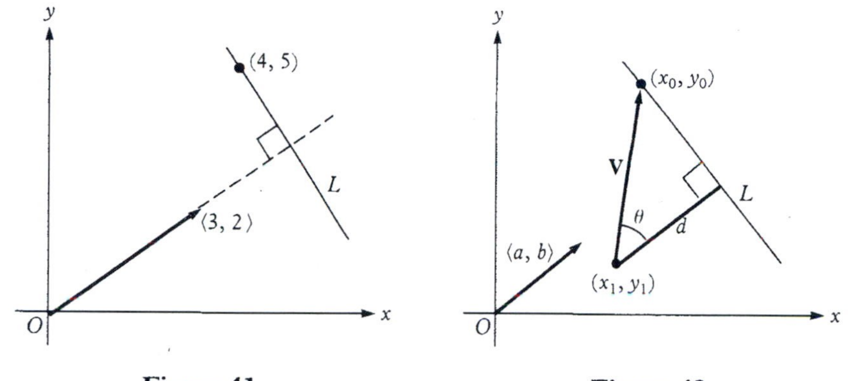

Figure 41

Figure 42

for some value of c. Now when $x = 4$ and $y = 5$, we have

$$3 \cdot 4 + 2 \cdot 5 = c$$
$$22 = c.$$

Therefore, the equation of L is

$$3x + 2y = 22. \blacksquare$$

Example 7 Show that the (shortest) distance from a point (x_1, y_1) to the line $L: ax + by = c$ is

$$d = \frac{|ax_1 + by_1 - c|}{\sqrt{a^2 + b^2}}. \qquad \text{distance from a point to a line} \qquad (7)$$

Solution: (See Section 2.6, Example 11, for a nonvector approach.) Let $\mathbf{V}$ be a vector from (x_1, y_1) to some point (x_0, y_0) on L, where $(x_0, y_0) \neq (x_1, y_1)$. With reference to Figure 42,

$$d = \|\mathbf{V}\| \cos \theta. \qquad \left(0 \leq \theta \leq \frac{\pi}{2}\right)$$

Now since vector $\langle a, b\rangle$ is perpendicular to L, the angle between $\mathbf{V}$ and $\langle a, b\rangle$ is either θ or $\pi - \theta$ (see Exercise 42). Also, $|\cos(\pi - \theta)| = \cos \theta$. Therefore,

$$|\langle a, b\rangle \cdot \mathbf{V}| = \|\langle a, b\rangle\| \, \|\mathbf{V}\| \cos \theta$$
$$|a(x_0 - x_1) + b(y_0 - y_1)| = \sqrt{a^2 + b^2} \cdot d$$
$$d = \frac{|ax_0 + by_0 - ax_1 - by_1|}{\sqrt{a^2 + b^2}}$$
$$= \frac{|c - ax_1 - by_1|}{\sqrt{a^2 + b^2}} = \frac{|ax_1 + by_1 - c|}{\sqrt{a^2 + b^2}},$$

which is the desired result. Note that if (x_1, y_1) lies on L, then $\theta = \pi/2$ and $d = 0$. $\blacksquare$

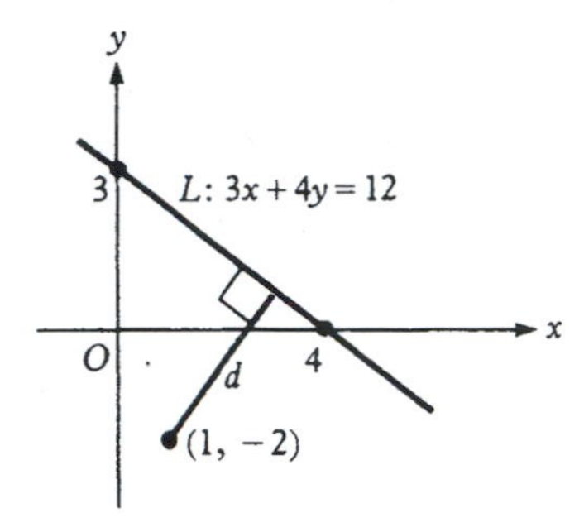

Figure 43

As an application of the result in Example 7, the distance from the point $(1, -2)$ to the line $3x + 4y = 12$ (Figure 43) is

$$d = \frac{|3 \cdot 1 + 4(-2) - 12|}{\sqrt{3^2 + 4^2}}$$

$$= \frac{17}{5}.$$

Comment In Section 2.6, we saw that two lines

$$L_1\colon a_1x + b_1y = c_1$$
$$L_2\colon a_2x + b_2y = c_2$$

are perpendicular if and only if

$$a_1a_2 + b_1b_2 = 0,$$

that is,
$$\langle a_1, b_1\rangle \cdot \langle a_2, b_2\rangle = 0.$$

Either of the last two equations states the evident geometric property that L_1 and L_2 intersect at 90° if and only if their respective perpendicular vectors $\langle a_1, b_1\rangle$ and $\langle a_2, b_2\rangle$ also intersect at 90°.

Angle Between Two Lines. As indicated in Figure 44(a), a pair of nonparallel lines L_1 and L_2 will intersect in two positive angles θ and $\pi - \theta$. If θ is the angle from L_1 to L_2, then $\pi - \theta$ is the angle from L_2 to L_1. Also, if $\mathbf{N}_1$ is a vector perpendicular to L_1, and $\mathbf{N}_2$ is perpendicular to L_2, then the smaller positive angle between $\mathbf{N}_1$ and $\mathbf{N}_2$ is either θ or $\pi - \theta$, depending on the orientation of $\mathbf{N}_1$ and $\mathbf{N}_2$ [Figures 44(b) and 44(c)].

Hence, we can determine the angles of intersection of two lines

$$a_1x + b_1y = c_1$$
$$a_2x + b_2y = c_2$$

by computing the angle between their perpendicular vectors $\langle a_1, b_1\rangle$ and $\langle a_2, b_2\rangle$. By equations (1) and (4), we have the following result.

> The angles of intersection of the nonparallel lines
>
> $$a_1x + b_1y = c_1 \quad \text{and} \quad a_2x + b_2y = c_2$$
>
> are θ and $\pi - \theta$, where
>
> $$a_1a_2 + b_1b_2 = \sqrt{a_1^2 + b_1^2}\,\sqrt{a_2^2 + b_2^2}\cos\theta. \tag{8}$$

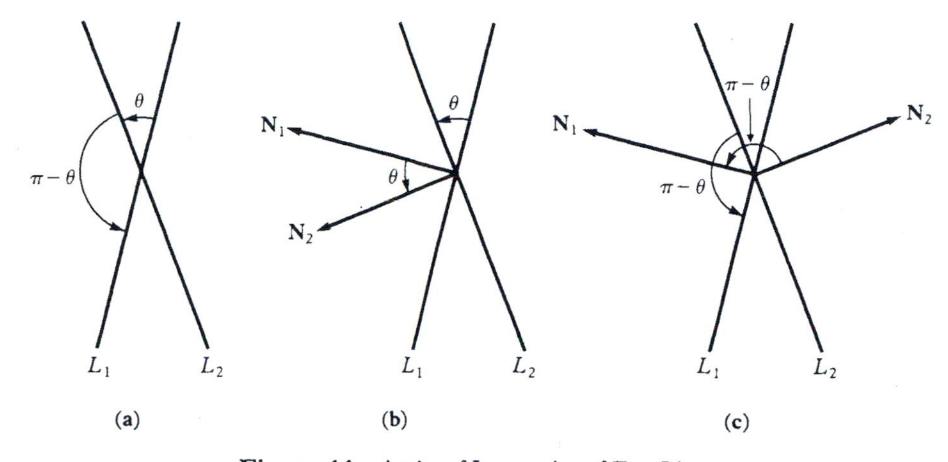

Figure 44 Angles of Intersection of Two Lines

Example 8 Find the angles of intersection of the two lines

$$5x + y = 5 \qquad \text{and} \qquad 2x + 3y = 6.$$

Solution: Here, $a_1 = 5$, $b_1 = 1$, $a_2 = 2$, and $b_2 = 3$. By substituting these values into equation (8) above, we obtain

$$\begin{aligned} 5 \cdot 2 + 1 \cdot 3 &= \sqrt{5^2 + 1^2}\sqrt{2^2 + 3^2}\cos\theta \\ 13 &= \sqrt{26}\sqrt{13}\cos\theta \\ \cos\theta &= \frac{13}{\sqrt{2}\sqrt{13}\sqrt{13}} \\ &= \frac{1}{\sqrt{2}} \\ \theta &= 45°. \end{aligned}$$

Hence, as indicated in Figure 45, the angles are 45° and 135°. ■

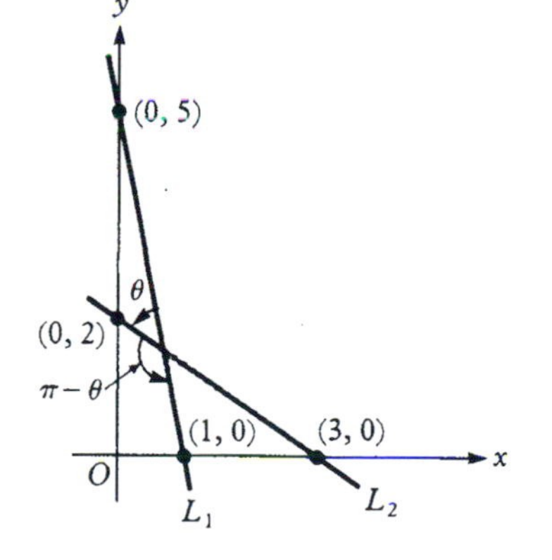

Figure 45

Comment We have used the coordinate plane in three different contexts: first, to graph equations and functions in either rectangular or polar coordinates, second, to represent complex numbers and give a geometric interpretation for complex addition and multiplication, and third, to represent vectors along with vector addition, scalar multiplication, and the dot product. The coordinate plane is also the natural setting for calculus. Hence, the idea to combine algebra and geometry, first conceived by Descartes and Fermat in the seventeenth century, has become an essential part of mathematics.

Exercises 8.5

Fill in the blanks to make each statement true.

1. Two nonzero vectors **V** and **W** are perpendicular if and only if $\mathbf{V} \cdot \mathbf{W} =$ ________.
2. The magnitude $\|\mathbf{V}\|$ of any vector **V** can be expressed in terms of the dot product as ________.
3. In the equation of a line L: $ax + by = c$, the coordinate vector ________ is perpendicular to L.
4. The distance from the point (x_1, y_1) to the line $ax + by = c$ is ________.
5. For nonparallel lines $a_1x + b_1y = c_1$ and $a_2x + b_2y = c_2$, the angles of intersection θ and $\pi - \theta$ can be determined from the equation ________.

Write true *or* false *for each statement.*

6. If (x_1, y_1) and (x_2, y_2) are any two points on the line $ax + by = c$, then $\langle a, b\rangle \cdot \langle x_1, y_1\rangle = \langle a, b\rangle \cdot \langle x_2, y_2\rangle$.
7. If line $a_1x + b_1y = c_1$ is parallel to line $a_2x + b_2y = c_2$, then $\langle a_1, b_1\rangle \cdot \langle a_2, b_2\rangle = 0$.
8. If the smaller nonnegative angle between the coordinate vectors $\langle a_1, b_1\rangle$ and $\langle a_2, b_2\rangle$ is θ, then the nonnegative angles between the lines $a_1x + b_1y = c_1$ and $a_2x + b_2y = c_2$ are θ and $\pi - \theta$.
9. The distance from the origin $(0, 0)$ to the line $ax + by = c$ is $|c|/\sqrt{a^2 + b^2}$.
10. If $\langle a, b\rangle \cdot \langle x_2 - x_1, y_2 - y_1\rangle = 0$, then the points (x_1, y_1) and (x_2, y_2) are on line $ax + by = c$, for some value of c.

The Dot Product of Two Vectors

11. Compute $\mathbf{V} \cdot \mathbf{W}$ for each of the following.
    (a) $\mathbf{V} = \langle 4, 5\rangle$; $\mathbf{W} = \langle 2, 6\rangle$
    (b) $\mathbf{V} = \langle 3, -2\rangle$; $\mathbf{W} = \langle 4, -1\rangle$
12. Given $\mathbf{U} = \langle 3, 2\rangle$, $\mathbf{V} = \langle -4, 1\rangle$, and $\mathbf{W} = \langle 6, 5\rangle$, compute $\mathbf{U} \cdot (\mathbf{V} + \mathbf{W})$.
13. Determine $2\mathbf{U} \cdot (3\mathbf{V} - \mathbf{W})$ for the vectors in Exercise 12.
14. Find the cosine of the angle between **V** and **W** in Exercise 11(a).
15. Find the cosine of the angle between **V** and **W** in Exercise 11(b).
16. Find the directed magnitude of the projection of **W** on **V** in Exercise 11(a).
17. Find the directed magnitude of the projection of **V** on **W** in Exercise 11(b).

In Exercises 18 and 19, determine which pairs of vectors, if any, are perpendicular.

18. $\mathbf{U} = \langle 2, 5\rangle$; $\mathbf{V} = \langle 2, -5\rangle$; $\mathbf{W} = \langle 5, -2\rangle$
19. $\mathbf{U} = \langle 4, 3\rangle$; $\mathbf{V} = \langle 3, 4\rangle$; $\mathbf{W} = \langle -4, 3\rangle$

20. Find a unit vector perpendicular to $\langle 2, 1\rangle$.
21. Find a unit vector perpendicular to $\langle -4, 8\rangle$.
22. Given $\mathbf{V} = \langle 4, 2\rangle$ and $\mathbf{W} = \langle 3, 4\rangle$, do each of the following.
    (a) Find a vector **U** in the direction of **V** whose magnitude is that of the projection of **W** on **V**.
    (b) Show that $\mathbf{V} \cdot (\mathbf{W} - \mathbf{U}) = 0$.
23. Repeat Exercise 22 for $\mathbf{V} = \langle 6, 2\rangle$ and $\mathbf{W} = \langle 4, 4\rangle$.

Vector Interpretation of the General Linear Equation

Illustrate the results in each of the following problems by means of a graph.

24. Find a general linear equation for the line through the point $(2, 4)$ and perpendicular to the vector $\langle 1, 3\rangle$.
25. Find a general linear equation for the line through the origin and perpendicular to the vector $\langle -3, 7\rangle$.
26. For what value of b is the line $3x + by = 2$ perpendicular to the line $2x - 5y = 1$?
27. For what value of a is the line $ax + 4y = 1$ perpendicular to the line $x + ay = 6$?
28. Let $(4, 3)$ be a point on a line L that is perpendicular to the vector $\langle 4, 3\rangle$. Find another point on L.
29. Find a vector perpendicular to the line through the two points $(3, -5)$ and $(-4, 6)$.
30. Find the distance from the point $(2, -3)$ to the line $3x - 4y = 7$.
31. Find the distance from the origin to the line $2x + 6y = 15$.

Angle Between Two Lines

A calculator will be needed for some of the following problems.

32. Find the angles of intersection of the lines $x - y = 2$ and $y = 5$.
33. Find the angles of intersection of the lines $x = 4$ and $x + y = 1$.

34. Find the angles of intersection of the lines $4x - 3y = 1$ and $x + y = 1$.

35. Find the angles of intersection of the lines $3x + 4y = -2$ and $4x + 3y = 2$.

36. Find an equation of the line L through the origin that satisfies each of the following conditions.

(a) The smaller positive angle from the line $y = x$ to L is 30°.

(b) The smaller positive angle from L to the line $y = x$ is 30°.

37. Find an equation of the line $L: ax + by = 1$ through the point $(1, 1/2)$ that satisfies each of the following conditions.

(a) The smaller positive angle θ from the line $x + 2y = 2$ to L satisfies $\cos\theta = 1/\sqrt{5}$.

(b) The smaller positive angle θ from L to the line $x + 2y = 2$ satisfies $\cos\theta = 1/\sqrt{5}$.

Miscellaneous

38. *Distributive Rule for Dot Product and Vector Addition.* Let **U**, **V**, and **W** be any three coordinate vectors. Show that

$$\mathbf{U} \cdot (\mathbf{V} + \mathbf{W}) = \mathbf{U} \cdot \mathbf{V} + \mathbf{U} \cdot \mathbf{W}.$$

Could the dot product equation $(\mathbf{U} \cdot \mathbf{V}) \cdot \mathbf{W} = \mathbf{U} \cdot (\mathbf{V} \cdot \mathbf{W})$ ever be satisfied? Why?

39. Let $\mathbf{V} = \langle x, y \rangle$ be any coordinate vector and k any real number. Show that $\|k\mathbf{V}\| = |k|\,\|\mathbf{V}\|$.

40. *Triangle Inequality.* Let $\mathbf{V} = \langle x_1, y_1 \rangle$ and $\mathbf{W} = \langle x_2, y_2 \rangle$. Show that each of the following statements is true.

(a) $\|\mathbf{V} + \mathbf{W}\|^2 = (\mathbf{V} + \mathbf{W}) \cdot (\mathbf{V} + \mathbf{W})$

(b) $(\mathbf{V} + \mathbf{W}) \cdot (\mathbf{V} + \mathbf{W}) = \|\mathbf{V}\|^2 + 2\mathbf{V} \cdot \mathbf{W} + \|\mathbf{W}\|^2$

(c) $\mathbf{V} \cdot \mathbf{W} \leq \|\mathbf{V}\|\,\|\mathbf{W}\|$ (*Hint:* use the geometric definition of $\mathbf{V} \cdot \mathbf{W}$.)

Use (a), (b), and (c) to conclude that

$$\|\mathbf{V} + \mathbf{W}\| \leq \|\mathbf{V}\| + \|\mathbf{W}\|.$$

41. *Algebraic Formula for the Dot Product.* As indicated in the figure, let **V** and **W** be nonzero vectors with endpoints $P(x_1, y_1)$ and $Q(x_2, y_2)$, respectively. Choose polar coordinates $[\|\mathbf{V}\|, \alpha]$ for P and $[\|\mathbf{W}\|, \beta]$ for Q, where $\theta = \alpha - \beta$ is the angle from **W** to **V**. Use the identity $\cos(\alpha - \beta) = \cos\alpha\cos\beta + \sin\alpha\sin\beta$ to derive the algebraic formula (4) for computing the dot product.

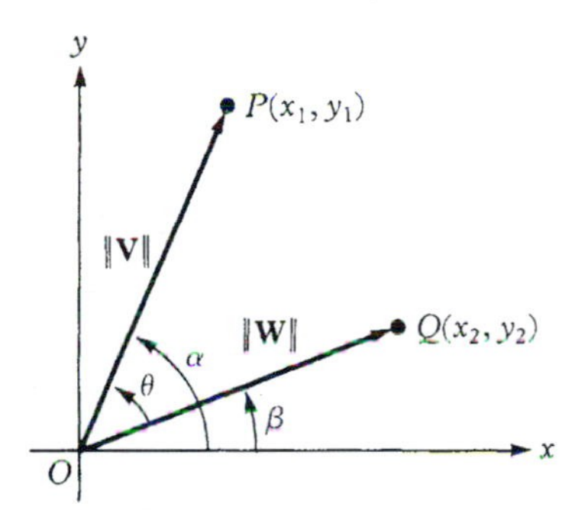

42. With reference to Figure 42, note that the vector **V** from (x_1, y_1) to (x_0, y_0) is equivalent to the coordinate vector $\langle x_0 - x_1, y_0 - y_1 \rangle$. Show graphically that the smaller nonnegative angle between $\langle a, b \rangle$ and $\langle x_0 - x_1, y_0 - y_1 \rangle$ is either θ or $\pi - \theta$.

8.6 Coordinates, Vectors, and Planes in 3-Space

A point is on a line, a line is in a plane, a plane is in space, and space is in . . . .

Coordinates and Distance in 3-Space
Vectors in 3-Space
Planes in 3-Space

Coordinates and Distance in 3-Space. Until now all of our equations have involved one or two variables and have been graphically represented on the real line or in the plane. By adding a third axis to the xy-coordinate system, we can assign coordinates to points in three-dimensional space. As

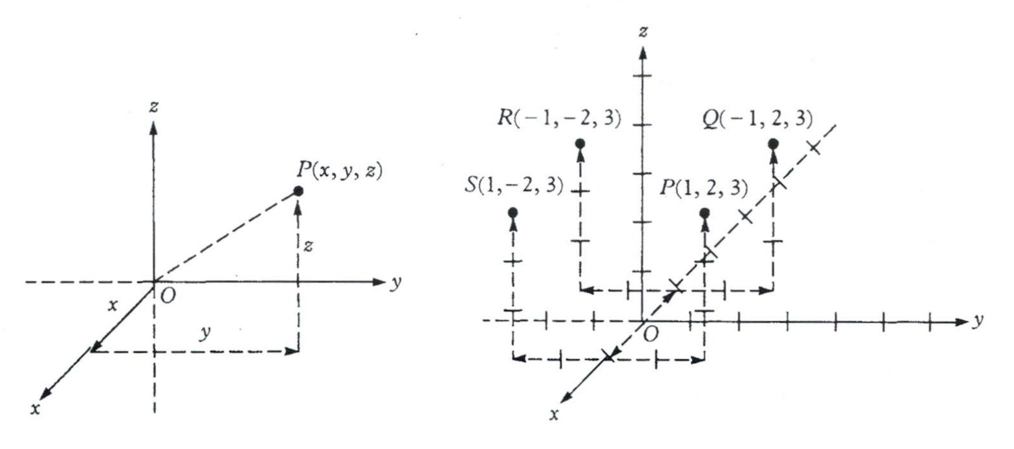

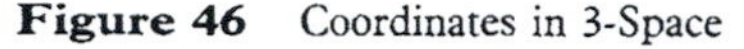

Figure 46 Coordinates in 3-Space

Figure 47 Points in 3-Space

indicated in Figure 46, we position the xy-plane horizontally and construct a z-axis in the vertical direction with its positive direction upward. Every point in space now has associated with it an **ordered triple** (x, y, z) of coordinates. The point of intersection of the three axes is called the **origin** and has coordinates $(0, 0, 0)$. To locate the point P with coordinates (x, y, z), start at the origin, go x directed units along the x-axis, then y directed units parallel to the y-axis, and then z directed units parallel to the z-axis. The terminal point is $P(x, y, z)$ (Figure 46).

Example 1 In a coordinate system, locate each of the following points.

(a) $P(1, 2, 3)$ (b) $Q(-1, 2, 3)$

(c) $R(-1, -2, 3)$ (d) $S(1, -2, 3)$

Solution: The four points are shown in Figure 47. ▬

The three coordinate axes partition 3-space into eight **octants,** four above the xy-plane and four below it. The points in Example 1 all have a positive z-coordinate and are therefore in the upper four octants. In any given octant, the coordinates do not vary in sign. For example, the first octant consists of all sign points $(+, +, +)$. In the remaining seven octants, rotating counterclockwise, the signs are $(-, +, +)$, $(-, -, +)$, $(+, -, +)$, $(+, +, -)$, $(-, +, -)$, $(-, -, -)$, and $(+, -, -)$.

With reference to Figure 48, we can extend the distance formula to 3-space. By applying the Pythagorean theorem to triangle P_1QP_2 in the figure, we get

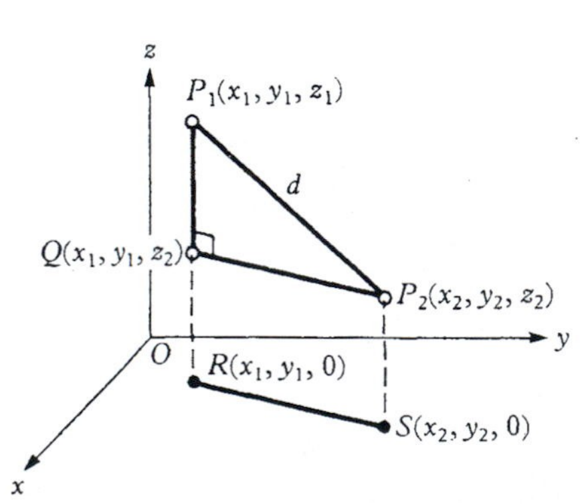

Figure 48 Distance in 3-Space

$$\begin{aligned} d^2(P_1, P_2) &= d^2(Q, P_2) + d^2(Q, P_1) \\ &= d^2(R, S) + d^2(Q, P_1) \\ &= (x_2 - x_1)^2 + (y_2 - y_1)^2 + (z_2 - z_1)^2. \end{aligned}$$

Although the points P_1 and P_2 in Figure 48 are located in the first octant, the above derivation can be verified for any two points in 3-space. We conclude that the distance $d(P_1, P_2)$ between any two points $P_1(x_1, y_1, z_1)$ and $P_2(x_2, y_2, z_2)$ satisfies the equation

$$d(P_1, P_2) = \sqrt{(x_2 - x_1)^2 + (y_2 - y_1)^2 + (z_2 - z_1)^2}. \tag{1}$$

distance formula in 3-space

Example 2 Find the distance between $P_1(1, -2, 3)$ and $P_2(4, 0, 5)$.

Solution: We substitute the given coordinates in (1) to obtain

$$\begin{aligned} d(P_1, P_2) &= \sqrt{(4-1)^2 + (0+2)^2 + (5-3)^2} \\ &= \sqrt{9+4+4} \\ &= \sqrt{17}. \end{aligned}$$

Vectors in 3-Space. In a three-dimensional coordinate system, the directed line segment from the origin (0, 0, 0) to the point (x, y, z) is called a **coordinate vector** and is denoted by $\langle x, y, z\rangle$. **Vector addition** and **scalar multiplication** in three dimensions are defined as follows and illustrated in Figure 49.

$$\langle x_1, y_1, z_1\rangle + \langle x_2, y_2, z_2\rangle = \langle x_1 + x_2, y_1 + y_2, z_1 + z_2\rangle$$
$$k\langle x, y, z\rangle = \langle kx, ky, kz\rangle$$

The **length of a coordinate vector** $\mathbf{V} = \langle x, y, z\rangle$ is denoted by $\|\mathbf{V}\|$. Since $\|\mathbf{V}\|$ is the distance from (0, 0, 0) to the point (x, y, z), we have

$$\|\mathbf{V}\| = \sqrt{x^2 + y^2 + z^2}. \tag{2}$$

As in the case of two-dimensional vectors, we define two vectors in 3-space to be **equivalent** if they have the same length and direction. If $\mathbf{V}$ is a vector from point $P_1(x_1, y_1, z_1)$ to point $P_2(x_2, y_2, z_2)$, then $\mathbf{V}$ is equivalent to the coordinate vector $\mathbf{V}' = \langle x_2 - x_1, y_2 - y_1, z_2 - z_1\rangle$.

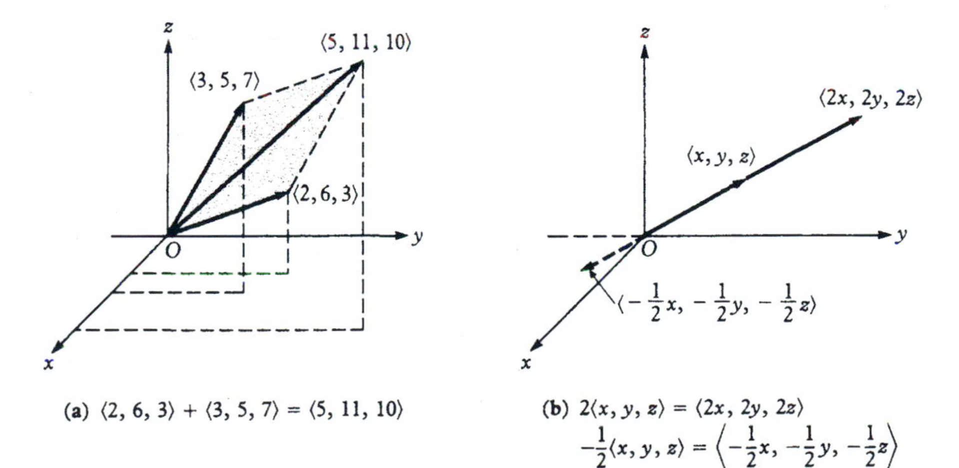

(a) $\langle 2, 6, 3\rangle + \langle 3, 5, 7\rangle = \langle 5, 11, 10\rangle$

(b) $2\langle x, y, z\rangle = \langle 2x, 2y, 2z\rangle$
$-\frac{1}{2}\langle x, y, z\rangle = \left\langle -\frac{1}{2}x, -\frac{1}{2}y, -\frac{1}{2}z\right\rangle$

Figure 49 Vector Addition and Scalar Multiplication

Also as in the case of two dimensions, we define the **dot product** of two nonzero vectors $\mathbf{V} = \langle x_1, y_1, z_1 \rangle$ and $\mathbf{W} = \langle x_2, y_2, z_2 \rangle$ geometrically by the formula

$$\mathbf{V} \cdot \mathbf{W} = \|\mathbf{V}\| \, \|\mathbf{W}\| \cos \theta, \tag{3}$$

where θ $(0 \leq \theta \leq \pi)$ is the smaller nonnegative angle between $\mathbf{V}$ and $\mathbf{W}$. If either $\mathbf{V}$ or $\mathbf{W}$ is the zero vector, then $\mathbf{V} \cdot \mathbf{W} = 0$ by definition. In terms of coordinates, the dot product can be computed by the equation (see Exercise 33)

$$\mathbf{V} \cdot \mathbf{W} = x_1x_2 + y_1y_2 + z_1z_2. \tag{4}$$

Example 3 Compute the dot product of the vectors $\mathbf{V} = \langle 2, 4, 5 \rangle$ and $\mathbf{W} = \langle 4, 0, 8 \rangle$. Also, determine the angle θ between $\mathbf{V}$ and $\mathbf{W}$.

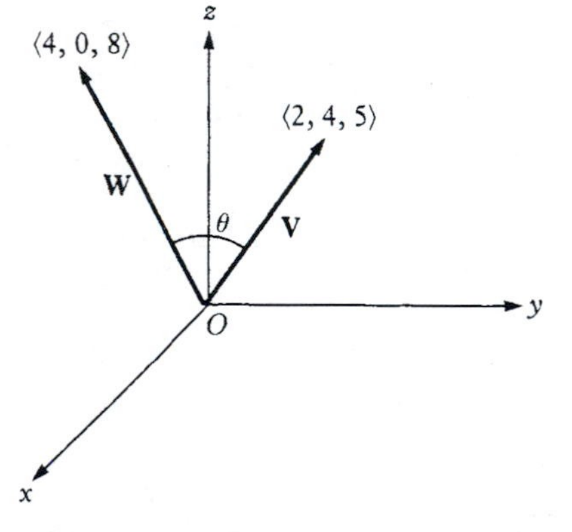

Figure 50

Solution: The vectors $\mathbf{V}$ and $\mathbf{W}$ are shown in Figure 50. By formula (4),

$$\mathbf{V} \cdot \mathbf{W} = 2 \cdot 4 + 4 \cdot 0 + 5 \cdot 8 = 48,$$

and by definition (3),

$$\begin{aligned} \mathbf{V} \cdot \mathbf{W} &= \sqrt{4 + 16 + 25}\,\sqrt{16 + 0 + 64}\cos\theta \\ &= \sqrt{45}\,\sqrt{80}\cos\theta. \end{aligned}$$

Therefore,

$$\begin{aligned} \sqrt{45}\,\sqrt{80}\cos\theta &= 48 \\ 3\sqrt{5}\,4\sqrt{5}\cos\theta &= 48 \\ 60\cos\theta &= 48 \\ \cos\theta &= \frac{4}{5} \\ \theta &\approx 36.87^\circ. \end{aligned}$$ ∎

Example 4 Show that the vectors $\mathbf{V} = \langle 2, -3, 1 \rangle$ and $\mathbf{W} = \langle 4, 1, -5 \rangle$ are perpendicular.

Solution: By formula (4), we have

$$\mathbf{V} \cdot \mathbf{W} = 2 \cdot 4 + (-3)1 + 1(-5) = 0,$$

and by definition (3),

$$\mathbf{V} \cdot \mathbf{W} = \sqrt{14}\,\sqrt{42}\cos\theta.$$

Therefore $\cos\theta = 0$, which means $\theta = 90^\circ$. Hence, $\mathbf{V}$ and $\mathbf{W}$ are perpendicular. ∎

Example 4 illustrates that the condition for perpendicularity in two dimensions extends to 3-space. That is,

Two nonzero vectors V and W are perpendicular ⇔
$$\mathbf{V} \cdot \mathbf{W} = 0. \tag{5}$$

Planes in 3-Space. In three dimensions, the graph of the equation $z = z_0$ is a plane Γ parallel to the xy-plane and z_0 directed units from it. In Figure 51, Γ consists of all points with coordinates (x, y, z_0), where z_0 is fixed, and x and y are any two real numbers.

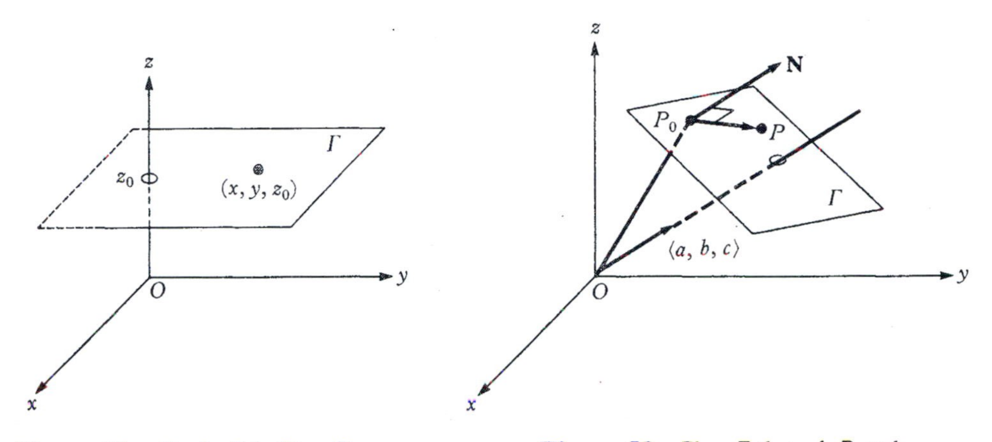

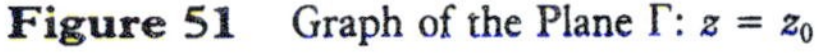

Figure 51 Graph of the Plane Γ: $z = z_0$

Figure 52 Plane Γ through P_0 and Perpendicular to **N**

Similarly, $x = x_0$ is a plane parallel to the yz-plane, and $y = y_0$ is a plane parallel to the xz-plane. More generally, we can use the dot product to show that every plane in 3-space has a *linear* equation of the form

$$ax + by + cz = d, \qquad \textbf{equation of a plane} \tag{6}$$

where a, b, or $c \neq 0$. To see why, let Γ be a plane and $P_0(x_0, y_0, z_0)$ a fixed point on Γ. Also, let **N** be a vector perpendicular to Γ. We can picture **N** as starting at the point P_0, but **N** is equivalent to a coordinate vector $\langle a, b, c\rangle$ (Figure 52). Now if $P(x, y, z)$ is any point in the plane other than P_0, then the vector $\langle x - x_0, y - y_0, z - z_0\rangle$ is perpendicular to **N** and hence to $\langle a, b, c\rangle$. Therefore, by applying (5) and simplifying, we get

$$\left.\begin{aligned} \langle a, b, c\rangle \cdot \langle x - x_0, y - y_0, z - z_0\rangle &= 0 \\ a(x - x_0) + b(y - y_0) + c(z - z_0) &= 0 \\ ax + by + cz &= ax_0 + by_0 + cz_0 \\ ax + by + cz &= d \end{aligned}\right\} \tag{7}$$

where $d = ax_0 + by_0 + cz_0$ is a constant. Furthermore, since $\langle a, b, c\rangle$ is a nonzero vector, at least one of the numbers a, b, and c is not zero. The vector **N** or $\langle a, b, c\rangle$ is called a **normal vector** to the plane Γ. Note that the point (x_0, y_0, z_0) also satisfies equation (7), in any one of its forms.

We have shown that every point $P(x, y, z)$ on Γ satisfies (all forms of) equation (7). Conversely, any point whose coordinates satisfy (7) will lie on the plane Γ.

Example 5 Find the equation of the plane Γ through the point $(2, 6, 3)$ and with normal vector $\langle 5, -1, 4\rangle$.

Solution: We are given $a = 5$, $b = -1$, and $c = 4$. We could start with the first equation in (7) and arrive at the last one as we did above, or we could start with the last equation and solve for the constant d. We illustrate the latter approach. That is, we solve

$$5x - y + 4z = d$$

for d. Now the point $(2, 6, 3)$ lies on Γ. Therefore,

$$5 \cdot 2 - 6 + 4 \cdot 3 = d$$
$$16 = d.$$

Hence, the equation of Γ is

$$5x - y + 4z = 16.$$

Example 6 Given the plane Γ: $3x + 2y + 4z = 12$, do each of the following.

(a) Find a vector normal to Γ. (b) Sketch a graph of Γ.

Solution:

(a) Here we have $a = 3$, $b = 2$, and $c = 4$. Hence, $\langle 3, 2, 4\rangle$ is a normal vector to Γ.

(b) Sketching a plane from its equation is not always a simple task. We try to find three points on the plane that are easy to graph and then draw a plane through these three points. If the plane intersects the coordinate axes for *positive* values of x, y, and z, then the graph is relatively simple. This particular plane has x-intercept $(4, 0, 0)$, which is obtained by setting $y = 0$ and $z = 0$ in the equation for Γ and then solving for x. Similarly, the y-intercept is $(0, 6, 0)$, and the z-intercept is $(0, 0, 3)$. These intercepts are easily plotted, and we obtain a graph of Γ as shown in Figure 53.

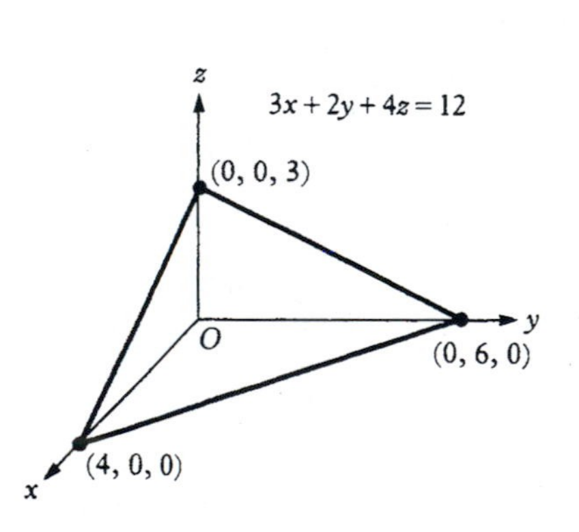

Figure 53

Example 7 Find the equation of the plane Γ that passes through the point $(2, 1, 3)$ and intersects the xy-plane in the straight line $3x - 4y = 8$.

Solution: The equation of Γ is of the form

$$ax + by + cz = d,$$

and we must determine the values of a, b, c, and d. First, the intersection of Γ with the xy-plane is obtained by setting $z = 0$, that is,

$$ax + by = d.$$

Therefore, $a = 3$, $b = -4$, $d = 8$, and the equation of Γ is

$$3x - 4y + cz = 8,$$

with now only c to be determined. Since Γ goes through the point $(2, 1, 3)$, we must have

$$3 \cdot 2 - 4 \cdot 1 + c \cdot 3 = 8.$$

Therefore, $c = 2$, and the equation of Γ is

$$3x - 4y + 2z = 8. \blacksquare$$

Exercises 8.6

Fill in the blanks to make each statement true.

1. In one dimension, the graph of $x = x_0$ is a ________, in two dimensions it is a ________, and in three dimensions it is a ________.
2. In two dimensions the graph of $ax + by = c$, where a or $b \neq 0$, is a ________, and in three dimensions it is a ________.
3. If the line segment from $P_1(x_1, y_1, z_1)$ to $P(x, y, z)$ is parallel to the z-axis, then $x =$ ________ and $y =$ ________.
4. The vector from $(-2, 3, 5)$ to $(4, 0, -1)$ is equivalent to the coordinate vector ________.
5. A vector perpendicular to the plane $3x + 4y + 5z = 2$ and having length 1 is ________.

Write true *or* false *for each statement.*

6. The line through $P_1(x_1, y_1, z_1)$ and $P_2(x_1, y_2, z_1)$ intersects the xz-plane at $P(x_1, 0, z_1)$.
7. The plane $ax + by + cz = d$ intersects the yz-plane in the line $by + cz = d$.
8. For any coordinate vector $\mathbf{V}$, $\|\mathbf{V}\| = \sqrt{\mathbf{V} \cdot \mathbf{V}}$.
9. If (x_1, y_1, z_1) and (x_2, y_2, z_2) are points in the plane $ax + by + cz = d$, then $\langle a, b, c\rangle \cdot \langle x_2 - x_1, y_2 - y_1, z_2 - z_1\rangle = 0$.
10. If $\langle a_1, b_1, c_1\rangle \cdot \langle a_2, b_2, c_2\rangle = 0$, then the plane $a_1x + b_1y + c_1z = d_1$ is perpendicular to the plane $a_2x + b_2y + c_2z = d_2$.

Coordinates and Distance in 3-Space

11. Locate each of the following points in a three-dimensional coordinate system.
    (a) $(2, 3, 4)$ (b) $(5, 0, 3)$
    (c) $(3, 4, 0)$ (d) $3, -2, 1)$
    (e) $(-3, -5, 2)$
12. Find the coordinates of a point $P(x, y, z)$ in each of the eight octants, where $|x| = 2$, $|y| = 3$, and $|z| = 4$.
13. Find two points $P_1(x_1, y_1, z_1)$ and $P_2(x_2, y_2, z_2)$ in the first octant for which P_1, P_2, $(3, 5, 1)$, and $(3, 7, 1)$ are the vertices of a square.
14. Find four points $P_k(x_k, y_k, z_k)$ $(k = 1, 2, 3, 4)$ such that $(2, 2, -1)$, $(2, -2, -1)$, $(-2, -2, -1)$, $(-2, 2, -1)$ and the four P_k are the vertices of a cube.
15. Find the length of a diagonal of the square in Exercise 13.
16. Find the length of a diagonal of the cube in Exercise 14.

Vectors in 3-Space

In Exercises 17–20, compute $2\mathbf{V} - 3\mathbf{W}$, $\|\mathbf{V} + \mathbf{W}\|$, *and* $\mathbf{V} \cdot \mathbf{W}$.

17. $\mathbf{V} = \langle 2, 3, 1\rangle$; $\mathbf{W} = \langle 4, 0, -5\rangle$
18. $\mathbf{V} = \langle 3, -2, 0\rangle$; $\mathbf{W} = \langle -4, 1, 6\rangle$
19. $\mathbf{V} = \mathbf{W} = \langle 2, 1, 7\rangle$
20. $\mathbf{V} = \mathbf{W} = \langle -8, 0, 6\rangle$

In Exercises 21–24, find the cosine of the angle between **V** *and* **W**. *Also, determine whether* **V** *and* **W** *are perpendicular.*

21. $\mathbf{V} = \langle 2, 1, 2\rangle$; $\mathbf{W} = \langle 3, -4, 3\rangle$
22. $\mathbf{V} = \langle -2, 0, -1\rangle$; $\mathbf{W} = \langle 2, 0, 1\rangle$
23. $\mathbf{V} = \langle 5, -1, 1\rangle$; $\mathbf{W} = \langle 2, 12, 2\rangle$
24. $\mathbf{V} = \langle 0, 3, 6\rangle$; $\mathbf{W} = \langle 7, -4, 2\rangle$

Planes in 3-Space

25. Find an equation of the plane parallel to the xy-plane and 5 units above it.
26. Find an equation of the plane perpendicular to the x-axis and 4 units behind the yz-plane.
27. Find an equation of the plane that passes through the point (1, 0, 0) and intersects the yz-plane in the line $10y - 5z = 7$.
28. Find an equation of the plane that is parallel to the z-axis and intersects the xy-plane in the line $3x + 2y = 4$.
29. Find an equation of the plane that passes through the origin and is perpendicular to the vector ⟨1, 1, 1⟩.
30. Find an equation of the plane that passes through the point (4, 2, −5) and is perpendicular to the vector from (1, 3, −2) to (3, 5, 0).
31. Find an equation of the plane that passes through the point (−2, 3, 4) and is parallel to the plane $5x - 3y + 2z = 1$.
32. Find an equation of the plane with intercepts (1, 0, 0), (0, 3, 0), (0, 0, 5).

Miscellaneous

33. *Algebraic Formula for the Dot Product.* Formula (4) of this section can be derived as follows. Let **V** and **W** be coordinate vectors with endpoints (x_1, y_1, z_1) and (x_2, y_2, z_2), respectively. If d is the distance between the endpoints, compute d^2 first by the distance formula and then by the law of cosines (draw a figure). Equate the two expressions for d^2 and simplify to obtain formula (4).
34. *Triangle Inequality.* Apply the reasoning used in Exercise 40 of Section 8.5 to 3-space in order to show that

$$\|\mathbf{V} + \mathbf{W}\| \le \|\mathbf{V}\| + \|\mathbf{W}\|$$

for any two coordinate vectors **V** and **W**.

35. *Distance from a Point to a Plane.* We can apply the reasoning used in Example 7 of Section 8.5 to 3-space in order to show that the distance D from a point (x_1, y_1, z_1) to a plane $ax + by + cz = d$ satisfies

$$D = \frac{|ax_1 + by_1 + cz_1 - d|}{\sqrt{a^2 + b^2 + c^2}}.$$

Use this equation to determine D in the following cases.

(a) $P_1(1, 2, 3)$; Γ: $2x - 3y + z = -1$

(b) $P_1(0, 0, 0)$; Γ: $\frac{x}{2} + \frac{y}{3} - \frac{z}{4} = 1$

36. *Intercept Equation.* The graph of the equation

$$\frac{x}{a} + \frac{y}{b} + \frac{z}{c} = 1 \qquad \text{(intercept equation)}$$

is a plane that intersects the x-axis at $(a, 0, 0)$, the y-axis at $(0, b, 0)$, and the z-axis at $(0, 0, c)$. (Compare this result with the one before Exercises 39–44 in Section 2.5.) Find the intercept equation and graph the plane through the points (2, 0, 0), (0, −3, 0), (0, 0, 5).

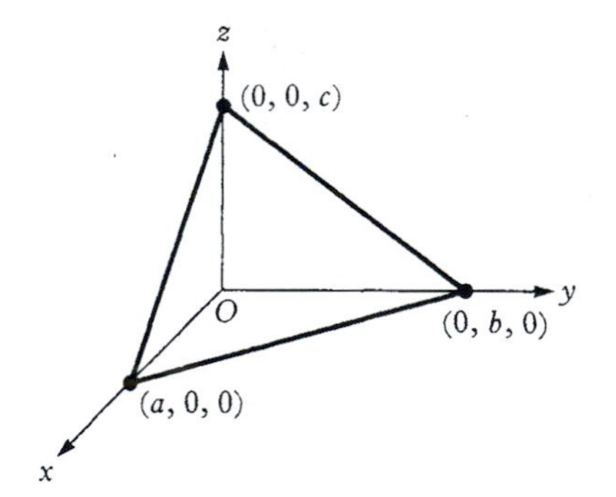

37. *Midpoint Formula.* We can use Figure 48 to show that the midpoint $P(x, y, z)$ of the line segment from $P_1(x_1, y_1, z_1)$ to $P_2(x_2, y_2, z_2)$ has coordinates

$$x = \frac{x_1 + x_2}{2}, \quad y = \frac{y_1 + y_2}{2}, \quad \text{and } z = \frac{z_1 + z_2}{2}.$$

(midpoint coordinates)

(a) Determine the midpoint P of the line segment from $P_1(2, 3, -5)$ to $P_2(6, -1, 5)$.

(b) Use the distance formula to verify that $d(P, P_1) = d(P, P_2)$ in part (a).

38. *Ratio Formula.* We can use Figure 48 to show that if $P(x, y, z)$ is the point on the line segment P_1P_2 for which

$$\frac{d(P_1, P)}{d(P_1, P_2)} = r \qquad (0 \le r \le 1),$$

then

$$\left.\begin{aligned} x &= x_1 + r(x_2 - x_1) \\ y &= y_1 + r(y_2 - y_1) \\ z &= z_1 + r(z_2 - z_1). \end{aligned}\right\} \text{(ratio formula)}$$

(Compare this result with Exercise 46 in Section 2.4.) Determine the coordinates of the point P on the line segment from $P_1(2, -5, 1)$ to $P_2(5, 4, 7)$ for which $d(P_1, P)/d(P_1, P_2) = 1/3$.

39. *Parametric Equations of a Line in 3-Space.* If the parameter r in the ratio formula in Exercise 38 is allowed to vary through *all* real numbers, then the graph of the equations

$$\left.\begin{aligned} x &= x_1 + r(x_2 - x_1) \\ y &= y_1 + r(y_2 - y_1) \\ z &= z_1 + r(z_2 - z_1) \end{aligned}\right\} \text{(parametric equations)}$$

is the entire line through P_1 and P_2. (Compare this result with the one before Exercises 45–50 of Section 2.5.) Find parametric equations of the line through $P_1(3, 6, -2)$ and $P_2(-4, 5, 7)$.

40. *Collinearity.* The three points $P_i(x_i, y_i, z_i)$ $(i = 1, 2, 3)$ are collinear precisely when P_3 is on the line through P_1 and P_2. Therefore, by using the parametric equations in Exercise 39, we can say that the three points are collinear if and only if the equations

$$\left.\begin{aligned} x_3 &= x_1 + r(x_2 - x_1) \\ y_3 &= y_1 + r(y_2 - y_1) \\ z_3 &= z_1 + r(z_2 - z_1) \end{aligned}\right\} \text{(collinearity equations)}$$

are satisfied for some value of r. Show that the points $P_1(3, -2, 5)$, $P_2(4, 1, 0)$, and $P_3(5, 4, -5)$ are collinear by finding a value of r that satisfies the collinearity equations.

41. *Line of Intersection of Two Planes.* Find parametric equations of the line L of intersection of the two planes

$$3x - 2y + z = 8$$
$$x + 2y - 5z = 0.$$

[*Hint:* solve the equations when $z = 0$ to obtain a point $P_1(x_1, y_1, 0)$ on L, and then solve the equations when $z = 1$ to obtain another point $P_2(x_2, y_2, 1)$ on L. Now find parametric equations of L.]

Chapter 8 Review Outline

8.1 Polar Coordinates

Polar Coordinates of a Point:
Given a fixed point O (the pole) and a fixed ray from O (the polar axis) in the plane, then a point P has polar coordinates $[r, \theta]$ if r is the length of OP and θ is an angle from the polar axis to OP.

The coordinates $[r, \theta]$, $[r, \theta + 2k\pi]$, and $[-r, \theta + \pi + 2k\pi]$, where k is any integer, all represent the same point.

The pole O has coordinates $[0, \theta]$, where θ is any angle whatsoever.

Polar and Rectangular Coordinates:
Polar coordinates $[r, \theta]$ of a point P are related to the rectangular coordinates (x, y) of P by the equations

$$x = r\cos\theta, \quad y = r\sin\theta$$

and

$$r = \sqrt{x^2 + y^2}, \quad \tan\theta = \frac{y}{x}.$$

8.2 Graphs in Polar Coordinates

The graph of an equation in r and θ is the set of all points P in the plane for which at least one pair $[r, \theta]$ of polar coordinates for P satisfies the equation.

8.3 Polar Form of Complex Numbers

The complex number $x + yi$ has polar form $r(\cos\theta + i\sin\theta)$, and is represented in the plane by the point with rectangular coordinates (x, y) and polar coordinates $[r, \theta]$.

Products and Quotients in Polar Form:

If $z_1 = r_1(\cos\theta_1 + i\sin\theta_1)$
and $z_2 = r_2(\cos\theta_2 + i\sin\theta_2)$, then

$$z_1 z_2 = r_1 r_2[\cos(\theta_1 + \theta_2) + i\sin(\theta_1 + \theta_2)],$$

and

$$\frac{z_1}{z_2} = \frac{r_1}{r_2}[\cos(\theta_1 - \theta_2) + i\sin(\theta_1 - \theta_2)],$$

where $r_2 \neq 0$.

If $z = r(\cos\theta + i\sin\theta)$ and n is any positive integer, then

$$z^n = r^n(\cos n\theta + i\sin n\theta)$$

(De Moivre's theorem),

and the nth roots of z are

$$z_k = \sqrt[n]{r}\left[\cos\left(\frac{\theta + 2k\pi}{n}\right) + i\sin\left(\frac{\theta + 2k\pi}{n}\right)\right],$$

where $k = 0, 1, 2, \ldots, n - 1$.

8.4 Vectors

A (nonzero) vector is a quantity that has both magnitude and direction and is represented in the plane by a directed line segment.

Vectors with the same magnitude and direction are equivalent.

Coordinate Vector:

A coordinate vector has starting point at the origin and is denoted by $\langle x, y\rangle$ if its endpoint is (x, y).

A vector from (x_1, y_1) to (x_2, y_2) is equivalent to the coordinate vector $\langle x_2 - x_1, y_2 - y_1\rangle$.

Vector Addition and Scalar Multiplication:

Vector addition is defined by the parallelogram rule.

The product of a vector **V** by a scalar k, denoted by $k\mathbf{V}$, is defined as the vector whose magnitude is $|k|$ times the magnitude of **V** and whose direction is the same as **V**'s if $k > 0$ and the opposite of **V**'s if $k < 0$.

If $k = 0$, then $k\mathbf{V}$ is the zero vector which by definition has 0 magnitude and no specific direction. The coordinate vector $\langle 0, 0\rangle$ is the zero vector.

Properties of Coordinate Vectors:

$\langle x_1, y_1\rangle \pm \langle x_2, y_2\rangle = \langle x_1 \pm x_2, y_1 \pm y_2\rangle$

$k\langle x, y\rangle = \langle kx, ky\rangle$

$\|\langle x, y\rangle\| = \sqrt{x^2 + y^2}$ is the magnitude (norm) of $\langle x, y\rangle$

Vector Magnitude:

$\|k\mathbf{V}\| = |k|\,\|\mathbf{V}\|$

$\|\mathbf{V} + \mathbf{W}\| \leq \|\mathbf{V}\| + \|\mathbf{W}\|$
(triangle inequality)

8.5 The Dot Product

Definition

The dot product $\mathbf{V}\cdot\mathbf{W}$ of two nonzero vectors **V** and **W** is $\|\mathbf{V}\|\,\|\mathbf{W}\|\cos\theta$, where θ is the smaller nonnegative angle between **V** and **W** ($0 \leq \theta \leq 180°$).

If either **V** or **W** is the zero vector, then $\mathbf{V}\cdot\mathbf{W} = 0$.

Properties:

Two nonzero vectors **V** and **W** are perpendicular $\Leftrightarrow$ $\mathbf{V}\cdot\mathbf{W} = 0$.

$\mathbf{V}\cdot\mathbf{V} = \|\mathbf{V}\|^2$

$\mathbf{V}\cdot\mathbf{W} = x_1x_2 + y_1y_2$ for coordinate vectors $\mathbf{V} = \langle x_1, y_1\rangle$, $\mathbf{W} = \langle x_2, y_2\rangle$.

The coordinate vector $\langle a, b\rangle$ is perpendicular to line $ax + by = c$.

For two lines $a_1x + b_1y = c_1$ and $a_2x + b_2y = c_2$, the angles of intersection are θ and $\pi - \theta$, where

$$\cos\theta = \frac{\langle a_1, b_1\rangle\cdot\langle a_2, b_2\rangle}{\|\langle a_1, b_1\rangle\|\,\|\langle a_2, b_2\rangle\|}.$$

8.6 Coordinates, Vectors, and Planes in 3-Space

Definitions

A coordinate system in 3-space consists of the xy-axes in a horizontal plane plus a vertical z-axis. Every point in 3-space has coordinates (x, y, z).

The coordinate vector $\langle x, y, z\rangle$ is directed from the origin to the point (x, y, z).

Properties:

$d(P_1, P_2) = \sqrt{(x_2 - x_1)^2 + (y_2 - y_1)^2 + (z_2 - z_1)^2}$

$\langle x_1, y_1, z_1\rangle \pm \langle x_2, y_2, z_2\rangle = \langle x_1 \pm x_2, y_1 \pm y_2, z_1 \pm z_2\rangle$

$k\langle x, y, z\rangle = \langle kx, ky, kz\rangle$

$\|\langle x, y, z\rangle\| = \sqrt{x^2 + y^2 + z^2}$

$\mathbf{V} \cdot \mathbf{W} = \|\mathbf{V}\| \|\mathbf{W}\| \cos \theta$ for any nonzero vectors.

$\mathbf{V} \cdot \mathbf{W} = x_1x_2 + y_1y_2 + z_1z_2$ for coordinate vectors $\mathbf{V} = \langle x_1, y_1, z_1 \rangle$ and $\mathbf{W} = \langle x_2, y_2, z_2 \rangle$.

Nonzero vectors $\mathbf{V}$ and $\mathbf{W}$ are perpendicular $\Leftrightarrow \mathbf{V} \cdot \mathbf{W} = 0$.

Every plane in 3-space has an equation $ax + by + cz = d$, where $\langle a, b, c \rangle$ is a vector perpendicular (normal) to the plane.

Chapter 8 Review Exercises

1. Find all possible polar coordinates for the points with the following rectangular coordinates.
   (a) $(1, 0)$ (b) $(\sqrt{3}, 1)$
2. Find the rectangular coordinates for the points with the following polar coordinates.
   (a) $[2, 3\pi/4]$ (b) $[-2, -\pi/3]$
3. Find a rectangular equation equivalent to the polar equation $r = 2 \cos \theta - 3/\sin \theta$.
4. Find a polar equation equivalent to the rectangular equation $2x^2 + 3y^2 = x - y$.

In Exercises 5–8, sketch the graph of each polar equation by the method of Section 8.2.

5. $r = 1 - \cos \theta$
6. $r = 1 - 2 \cos \theta$
7. $r = 2 \sin \theta$
8. $r = \cos 3\theta$

9. Find the polar form of the complex number $\sqrt{3} - i$.
10. For $z_1 = 2[\cos(\pi/3) + i \sin(\pi/3)]$ and $z_2 = 3[\cos(\pi/6) + i \sin(\pi/6)]$, find z_1z_2 and z_1/z_2 in rectangular form.
11. For $z = 3[\cos(3\pi/4) + i \sin(3\pi/4)]$, find $1/z$ in rectangular form.
12. Find z^{10}, given $z = 1/\sqrt{2} + (1/\sqrt{2})i$.
13. Find the fifth roots of 32.
14. Find the fourth roots of -1.
15. Graph on the unit circle the fourth roots of -1 found in Exercise 14.
16. A parachutist bails out at 5,000 ft above ground and descends at 20 ft/sec. If a wind is blowing at 5 ft/sec due east, how far east from the point of bailing out does the parachutist land? Draw vectors to indicate the downward velocity, the wind velocity, and the resulting velocity along the path of descent.

In Exercises 17–20, let $\mathbf{V} = \langle 5, 3 \rangle$ and $\mathbf{W} = \langle -1, 4 \rangle$. Find and graph the given vector.

17. $\mathbf{V} + \mathbf{W}$
18. $2\mathbf{V}$
19. $-3\mathbf{W}$
20. $3\mathbf{V} - 4\mathbf{W}$

21. Find the magnitude of the vector of Exercise 17, and find a unit vector in the same direction.
22. A vector $\mathbf{V}$ has starting point $(2, 3)$ and is equivalent to the coordinate vector $\langle -5, 7 \rangle$. What is the endpoint of $\mathbf{V}$?
23. Find scalars a and b for which $a\langle 2, 5 \rangle + b\langle -3, 4 \rangle = \langle 1, 7 \rangle$.
24. Use the geometric definition to compute $\mathbf{V} \cdot \mathbf{W}$ for $\mathbf{V} = \langle \sqrt{3}, 1 \rangle$ and $\mathbf{W} = \langle -2\sqrt{3}, 2 \rangle$.
25. Use the algebraic formula to compute $\mathbf{V} \cdot \mathbf{W}$ for the vectors in Exercise 24.
26. For $\mathbf{V} = \langle 4, 2 \rangle$ and $\mathbf{W} = \langle 3, 5 \rangle$, show that $\mathbf{V} \cdot \mathbf{W} < \|\mathbf{V}\| \|\mathbf{W}\|$.
27. For the vectors in Exercise 26, show that $\|\mathbf{V} + \mathbf{W}\| < \|\mathbf{V}\| + \|\mathbf{W}\|$.
28. For the vector $\mathbf{V}$ in Exercise 26, show that $\|-2\mathbf{V}\| = 2\|\mathbf{V}\|$.
29. Find $\cos \theta$, where θ is the angle between the vectors $\mathbf{V}$ and $\mathbf{W}$ in Exercise 26 $(0° \leq \theta \leq 180°)$.
30. Use the dot product to find an equation of the line in the xy-plane through the point $(6, 4)$ and perpendicular to the vector $\langle 1, 3 \rangle$.
31. Find the angles between the lines $2x + 3y = 6$ and $y = x$.
32. Find the distance from the point $(-5, 1)$ to the line $2x + 3y = 6$.

The remaining exercises deal with vectors in 3-space.

33. Graph the points (4, 3, 1) and (2, 6, 7), and compute the distance between them.

34. If $\mathbf{U} = \langle 3, -4, 5\rangle$, $\mathbf{V} = \langle 2, 1, -3\rangle$, and $\mathbf{W} = \langle -2, 6, 2\rangle$, find a unit vector in the direction of $\mathbf{U} + \mathbf{V} + \mathbf{W}$.

***35.** For the vectors, **U**, **V**, and **W** in Exercise 34, compute $\mathbf{U} \cdot (4\mathbf{V} - 7\mathbf{W})$.

***36.** Show that $\langle 2, -3, 4\rangle$ and $\langle 5, 2, -1\rangle$ are perpendicular vectors.

***37.** Find the cosine of an acute angle between the vectors $\mathbf{V} = \langle 4, 5, 1\rangle$ and $\mathbf{W} = \langle 2, 3, 6\rangle$.

***38.** Find an equation of the plane that is parallel to the xz-plane and two units to the left of it.

***39.** Find an equation of the plane through the point (3, 2, 0) and perpendicular to the vector $\langle 2, 1, 5\rangle$.

***40.** Given the plane $x + 3y + 2z = 6$, find a coordinate vector that is normal to it. Next, sketch a graph of the plane and the normal vector.

Graphing Calculator Exercises

The graph of a polar curve $r = f(\theta)$, $\alpha \le \theta \le \beta$, *can be obtained on a graphing calculator such as the TI-81 by using the parametric mode and graphing the functions* $X_{IT} = f(t)\sin t$, $Y_{IT} = f(t)\cos t$, $\alpha \le t \le \beta$. *In each of the following, verify your answers to the indicated review exercises of this section by graphing the functions in the given windows* $[t_{min}, t_{max}]$, $[x_{min}, x_{max}]$, $[y_{min}, y_{max}]$. *(If* π *cannot be entered in a window, use 3.14.)*

41. Exercise 5: $X_{1T} = (1 - \cos t)\cos t$, $Y_{1T} = (1 - \cos t)\sin t$, $[0, 2\pi]$, $[-3, 2]$, $[-2, 2]$

42. Exercise 6: $X_{1T} = (1 - 2\cos t)\cos t$, $Y_{1T} = (1 - 2\cos t)\sin t$, $[0, 2\pi]$, $[-3, 2]$, $[-2, 2]$

43. Exercise 7: $X_{1T} = (2\sin t)\cos t$, $Y_{1T} = (2\sin t)\sin t$, $[0, \pi]$, $[-3, 2]$, $[-2, 2]$

44. Exercise 8: $X_{1T} = (\cos 3t)\cos t$, $Y_{1T} = (\cos 3t)\sin t$, $[0, \pi]$, $[-1, 1]$, $[-1, 1]$

*Asterisks indicate topics that may be omitted, depending on individual course needs.

9 Linear Systems, Determinants and Matrices

In Section 2.6 we solved systems of two linear equations in two unknowns. Geometrically such systems represent two lines in a plane that are parallel, coincide, or intersect in a unique point. To solve the systems, three algebraic methods were used: elimination, substitution, and determinants.

In this chapter we consider linear systems with more than two equations and two unknowns. We begin with three linear equations in three unknowns, whose solutions may be interpreted as intersections of planes in 3-space. The methods of substitution and determinants are applied again here, and we also investigate further properties of determinants.

We shall see that the method of elimination, when employed in a systematic way, leads to the representation of a linear system by its augmented matrix and to the Gaussian elimination method. Gaussian elimination is our most general and efficient way to solve linear systems. Finally, we discuss briefly the algebraic properties of matrices.

9.1 Linear Systems: Reduction Method

We can make a point by intersecting three planes.

Linear Systems in Three Variables
Reduction Method
Geometric Meaning of Solutions
Applications

Linear Systems in Three Variables. Our concern in this section is to solve a system of three linear equations in three unknowns x, y, and z:

$$\begin{aligned}\Gamma_1&: a_1x + b_1y + c_1z = d_1\\ \Gamma_2&: a_2x + b_2y + c_2z = d_2\\ \Gamma_3&: a_3x + b_3y + c_3z = d_3.\end{aligned} \tag{1}$$

Geometrically, the solution of system (1) is the intersection of the three planes Γ_1, Γ_2, and Γ_3 in space.[1] If we assume that the planes are distinct and that no two of them are parallel, then Γ_1 and Γ_2 will intersect in a line L, and Γ_1 and Γ_3 will intersect in a line L'. The lines L and L' are both in the plane Γ_1; hence, if they are not parallel or coincident, they will intersect in a point $P(x_1, y_1, z_1)$ (Figure 1).

We can solve the system algebraically for x_1, y_1, and z_1 by several methods. To facilitate our discussion, we introduce the following terminology. The real numbers a_i, b_i, and c_i $(i = 1, 2, 3)$ in system (1) are called the **coefficients,** x, y, and z are called the **variables or unknowns,** and the d_i $(i = 1, 2, 3)$ are called the **constant terms.**

[1]No knowledge of three-dimensional geometry is required, if it is accepted that a linear equation in x, y, and z represents a plane in 3-space.

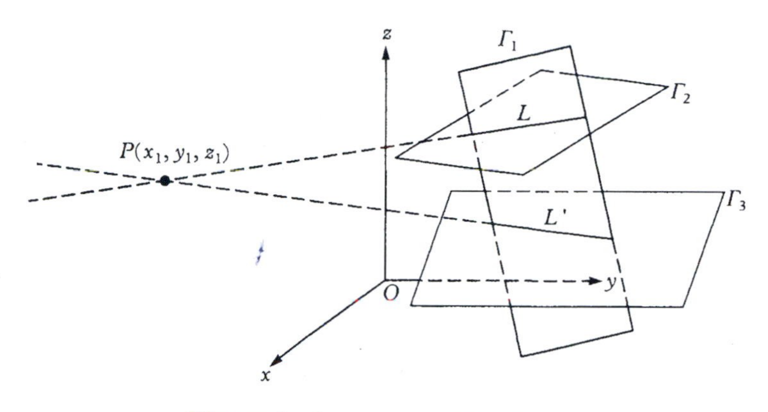

Figure 1 Intersection of Three Planes

Reduction Method. The basic idea in our first method is to express one of the unknowns, for example z, in terms of the remaining ones (x and y) and then by substitution to reduce a system of three equations to a system of two equations. Hence, the procedure is called the method of **reduction by substitution** or simply the **reduction method.**

Example 1 Solve the following system by the method of reduction.

$$\begin{aligned} 3x - 2y + z &= 7 \\ 2x + 5y - 4z &= -3 \\ 4x + y - z &= 6 \end{aligned}$$

Solution: From the first equation,

$$z = -3x + 2y + 7.$$

Substitute this expression for z into the last two equations.

$$\begin{aligned} 2x + 5y - 4(-3x + 2y + 7) &= -3 \\ 4x + y - (-3x + 2y + 7) &= 6 \end{aligned}$$

Equivalently,

$$\begin{aligned} 14x - 3y &= 25 \\ 7x - y &= 13, \end{aligned}$$

whose solution is $x = 2$ and $y = 1$. Therefore, $z = -3 \cdot 2 + 2 \cdot 1 + 7 = 3$, and the solution of the original system is $x = 2$, $y = 1$, and $z = 3$. Geometrically, the given system represents three planes that intersect in a unique point (2, 1, 3). ■

Example 2 Apply the reduction method to the following system.

$$\begin{aligned} x - 2y - 3z &= 2 \\ x - 4y - 13z &= 14 \\ -3x + 5y + 4z &= 2 \end{aligned}$$

Solution: In the first equation, it is easiest to solve for x in terms of y and z. We get

$$x = 2y + 3z + 2.$$

Substituting this expression for x in the last two equations of the given system gives

$$\begin{aligned} (2y + 3z + 2) - 4y - 13z &= 14 \\ -3(2y + 3z + 2) + 5y + \ \ 4z &= 2 \end{aligned}$$

or equivalently

$$\begin{aligned} -y - 5z &= 6 \\ -y - 5z &= 8, \end{aligned}$$

which is a system that has no solution for y and z (*why?*). Hence, the given system has *no* solution for x, y, and z. ■

Example 3 Apply the reduction method to the following system.

$$\begin{aligned} x - 2y - \ \ 3z &= 2 \\ x - 4y - 13z &= 14 \\ -3x + 5y + \ \ 4z &= 0 \end{aligned}$$

Solution: The first equation gives

$$x = 2y + 3z + 2,$$

and substitution in the remaining two equations yields

$$\begin{aligned} (2y + 3z + 2) - 4y - 13z &= 14 \\ -3(2y + 3z + 2) + 5y + \ \ 4z &= 0. \end{aligned}$$

The problem is therefore reduced to solving the system

$$\begin{aligned} -y - 5z &= 6 \\ -y - 5z &= 6, \end{aligned}$$

which says that $y = -5z - 6$, and then by substitution, $x = -7z - 10$. We may write the solution of the system as $x = -7z - 10$, $y = -5z - 6$, and z = any real number. Thus, by varying the value of z, we have an *infinite* number of triples (x, y, z) that satisfy the equation. ■

Geometric Meaning of Solutions. Example 1 shows that the linear system (1) may have a unique solution, which means that (1) represents three planes Γ_1, Γ_2, and Γ_3 intersecting in a unique point as shown in Figure 1. But Examples 2 and 3 illustrate that the linear system (1) may not have a unique solution; geometrically, the planes Γ_1, Γ_2, Γ_3 may not intersect in a unique point. This could happen in any of the following ways.

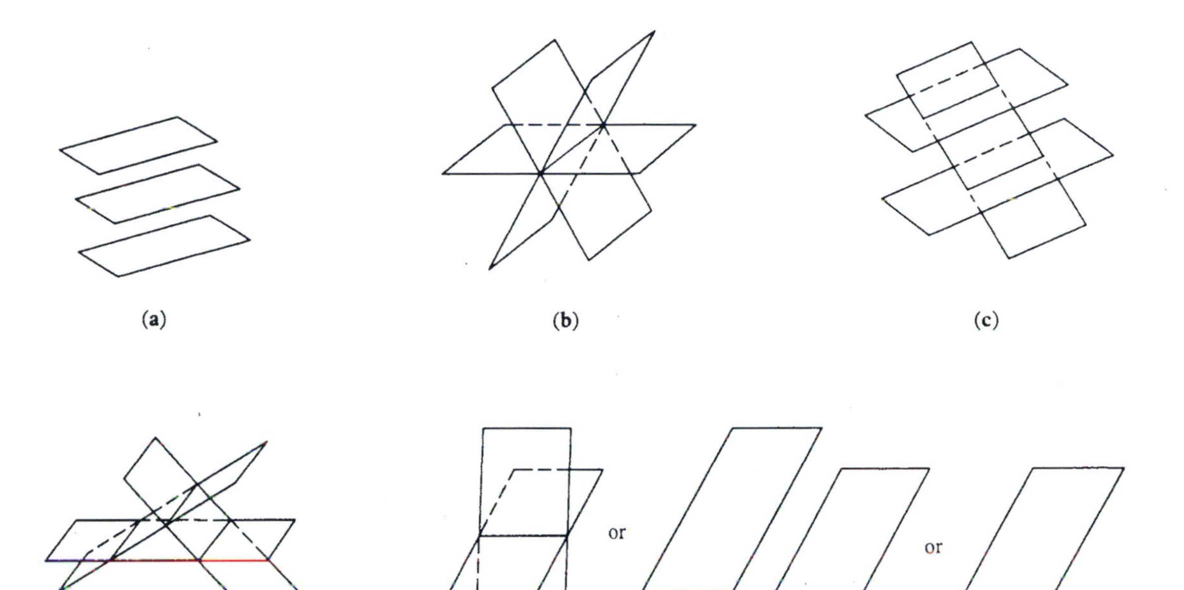

Figure 2

(a) The planes are parallel [Figure 2(a)].

(b) All three planes intersect in a single line [Figure 2(b)].

(c) Two of the planes are parallel, and the third plane intersects each of these in a line, resulting in two parallel lines [Figure 2(c)].

(d) Each pair of planes intersects in a line, resulting in three parallel lines [Figure 2(d)].

(e) Two or all three of the planes are coincident [Figure 2(e)].

In each of these five cases, there is a fourth plane Γ that is perpendicular to Γ_1, Γ_2, and Γ_3, and it can be shown that the converse is also true (see Exercises 31–32). That is, if there is a plane Γ perpendicular to Γ_1, Γ_2, and Γ_3, then the system (1) has no solution [cases (a), (c), (d), and (e)] or an infinite number of solutions [cases (b) and (e)].

Although we have limited our discussion in this section to three equations in three unknowns, the reduction method can be applied to any number of linear equations in any number of unknowns. For example, given four equations in five unknowns, we first solve one of the equations for one of the unknowns in terms of the remaining four. We then substitute this expression into the remaining three equations, and thereby reduce the original system to three equations in four unknowns. We can repeat this process until we arrive at a solution.

Applications. Many of the decisions that must be made in business and in science require the solution of linear systems of many equations in large numbers of unknowns. The most common approach to such problems is to use the method of Gaussian elimination, which will be discussed later in this chapter. Here we limit ourselves to an example involving three linear equations in three unknowns.

Example 4 A zoo dietician must produce 100 pounds of an animal feed that contains 20% protein and 10% fat. Three types of food are available for mixing. Type A is 15% protein and 10% fat, type B is 15% protein and 7.5% fat, and type C is 25% protein and 11% fat. How many pounds of each type should be used to produce the desired feed?

Solution: Let x, y, and z be the number of pounds of food types A, B, and C, respectively, to be mixed. The mixture must be 100 pounds and contain 20 pounds of protein and 10 pounds of fat. Therefore,

Type A	Type B	Type C		
$x +$	$y +$	$z = 100$		*Total pounds*
$.15x +$	$.15y +$	$.25z = 20$		*Pounds of protein*
$.1x +$	$.075y +$	$.11z = 10.$		*Pounds of fat*

This linear system has the unique solution $x = 30$, $y = 20$, and $z = 50$. Hence, the dietician should mix 30 pounds of type A, 20 pounds of type B, and 50 pounds of type C. ■

Exercises 9.1

Fill in the blanks to make each statement true.

1. A system of three linear equations in three unknowns may have ________, ________, or ________ solutions.
2. If a system of three linear equations in three unknowns has a unique solution (x_1, y_1, z_1), then geometrically the system represents three ________ intersecting in a ________.
3. If a system of three linear equations in three unknowns geometrically represents three planes that have a line in common, then algebraically an ________ number of solutions is obtained.
4. The reduction method applied to a linear system in three variables involves substitution for one variable so that the reduced system involves only ________ variables.
5. If a linear system in three variables has the solution $x = y = z = 0$, then all of the ________ terms in the equations are zero.

Write true *or* false *for each statement.*

6. The reduction (substitution) method can be applied to a linear system involving any number of variables.

7. A system of linear equations can be solved only if the number of equations equals the number of variables.
8. A linear system of three equations in three variables represents three planes that either intersect in one point or are parallel.
9. A system of four linear equations in four unknowns may be reduced to a system of three equations in three unknowns by substitution of one variable as a linear function of the others.
10. If a system of three equations in three unknowns has no solution, then at least two of the equations represent parallel planes.

Reduction Method

In Exercises 11–16, solve the linear systems by the reduction method.

11. $\begin{aligned} x - 2y + z &= -2 \\ 2x + y + 5z &= 3 \\ 4x - 5y - 4z &= 6 \end{aligned}$

12. $\begin{aligned} x - y + z &= 2 \\ 2x + y - z &= 1 \\ 2x - y - 3z &= -9 \end{aligned}$

13. $\begin{aligned} x + y - z &= 4 \\ x - 3y + z &= -2 \\ 3x - y - 2z &= 6 \end{aligned}$

14. $\begin{aligned} 2x + 5y - 3z &= 4 \\ 4x - y - 2z &= 2 \\ 5x - 2y + z &= 15 \end{aligned}$

15. $\begin{aligned} x - 2y + 3z &= 5 \\ 2x + 5y - 2z &= -5 \\ 4x + y + z &= -4 \end{aligned}$

16. $\begin{aligned} 3x + 2y - 7z &= 10 \\ 2x - 5y + 3z &= -1 \\ 5x + 6y - z &= 6 \end{aligned}$

17. Show that there is no solution for the system

$$\begin{aligned} x - 2y + 3z &= 2 \\ 2x - 3y + 5z &= 7 \\ 5x + 4y + z &= 1 \end{aligned}$$

by reducing to a system in x and y only.

18. Solve Exercise 17 by reducing to a system involving y and z only.
19. Solve Exercise 17 by reducing to a system involving x and z only.

In Exercises 20–25, use the reduction method to show that the system has infinitely many solutions.

20. $\begin{aligned} 2x - y &= 7 \\ 8x - 4y &= 28 \end{aligned}$

21. $\begin{aligned} x - y + 4z &= 0 \\ 2x + y + 2z &= 0 \\ 3x + 2y + 2z &= 0 \end{aligned}$

22. $\begin{aligned} x + y &= 2 \\ x - z &= 1 \\ 2x + y - z &= 3 \end{aligned}$

23. $\begin{aligned} x + 4y - z &= 12 \\ 3x + 8y - 2z &= 4 \end{aligned}$

24. $\begin{aligned} 2x + y - 4z &= 3 \\ 2x + 3y + 2z &= -1 \\ y + 3z &= -2 \end{aligned}$

25. $\begin{aligned} x + z &= 2 \\ y + z &= 6 \\ y + t &= 0 \\ x + y + z + t &= 2 \end{aligned}$

Solve the systems in Exercises 26–27 by the reduction method.

26. $\begin{aligned} 2x - 3y - z &= 2 \\ 3y - 2z - 4t &= -17 \\ x + y - 2t &= -4 \\ x + z + 2t &= 8 \end{aligned}$

27. $\begin{aligned} x + y + 2z &= 13 \\ 2x - y &= 1 \\ 4x + 3y - t &= 19 \\ x + y + z - t &= 11 \end{aligned}$

Geometric Meaning of Solutions

28. Each of the systems in Exercises 11–16 has a unique solution. What does that mean geometrically?
29. The system of equations given in Exercise 17 represents three distinct planes, no two of which are parallel, but the system has no solution. Explain how this is possible geometrically.
30. The linear system in Exercise 21 represents three planes passing through (0, 0, 0) (*why?*) and has an infinite number of solutions. Given that the three planes are distinct, explain how this is possible geometrically.

Exercises 31 and 32 illustrate that when three planes do not intersect in a unique point, there is a fourth plane perpendicular to all three.

31. Given the linear system

$$\begin{aligned} \Gamma_1&: x - 2y + 3z = 2 \\ \Gamma_2&: 2x - 3y + 5z = 7 \\ \Gamma_3&: 5x + 4y + z = 1, \end{aligned}$$

do each of the following.

(a) Show that the system has no solution.

(b) Find a plane Γ: $Ax + By + Cz = D$ that is perpendicular to Γ_1, Γ_2, and Γ_3. (*Hint:* Γ is perpendicular to Γ_1, Γ_2, and Γ_3 if and only if A, B, and C satisfy the system

$$\begin{aligned} A - 2B + 3C &= 0 \\ 2A - 3B + 5C &= 0 \\ 5A + 4B + C &= 0.) \end{aligned}$$

32. Given the linear system

$$\begin{aligned} \Gamma_1&: x - 2y + 3z = 2 \\ \Gamma_2&: 2x - 3y + 5z = 7 \\ \Gamma_3&: 5x - 8y + 13z = 16, \end{aligned}$$

do each of the following.

(a) Show that the system has an infinite number of solutions.

(b) Find a plane $Ax + By + Cz + D = 0$ that is perpendicular to Γ_1, Γ_2, and Γ_3. (*Hint:* see Exercise 31.)

Applications

Solve Exercises 33–40 by using a system of three linear equations in three unknowns.

33. A couple has $5000 to invest. They decide to divide it into three parts, putting the first part in a certificate of deposit at 4% annual interest, the second part in bonds paying 5% per year, and the third part in a mutual fund paying 6% per year. The third investment equals the sum of the first two. Find the amount of each of the three investments, if at the end of one year they receive $265 income from them.

34. A movie theater charges $5.00 admission for adults, $3.00 for senior citizens, and $2.00 for children. If 75 tickets are sold for a total of $285, with three times as many tickets for senior citizens as for children, how many tickets of each type are sold?

35. A store purchased 100 dresses of three qualities at $45, $30, and $25 each for a total of $3050. The store sold the dresses for $90, $65, and $50, respectively, for a total of $6250. How many of each type were sold?

36. A manufacturer makes three kinds of suits. One week he sold a total of 1050 suits. He sold the first, second, and third kinds to a small retail store at prices of $150, $100, and $80 per suit, respectively, for a total of $22,000. He also sold five times as many of each kind to a large discount store at prices of $125, $90, and $75, respectively, for a total of $94,375. How many suits of each kind were sold to the small retail store?

37. A company packages one-quart cartons of three kinds of blended fruit juices. Each carton of type 1 contains 20 oz of orange juice, 10 oz of grapefruit juice, and 2 oz of pineapple juice. Type 2 has 8 oz of orange, 20 oz of grapefruit, and 4 oz of pineapple per carton, while type 3 has 12 oz of orange and 10 oz each of grapefruit and pineapple per carton. How many cartons of each type are packaged if a total of 14,400 oz of orange, 14,000 oz of grapefruit, and 3,600 oz of pineapple juice are used?

38. A company makes three kinds of calculators. Each calculator of the first kind costs $10 for parts, $20 for labor, and $1 for advertising. For the second kind those costs are $5, $15, and $.50, respectively, and for the third kind, $4, $10, and $.25, respectively. How many of each type are made if the company spends a total of $13,700 on parts, $30,500 on labor, and $1325 on advertising?

39. Find an equation of the circle passing through points $P_1(2, 1)$, $P_2(-1, 3)$, and $P_3(3, -2)$. (*Hint:* a circle has an equation of the form $Ax^2 + Ay^2 + Cz + Dy = E$.)

40. Find an equation of the parabola $y = ax^2 + bx + c$ passing through $P_1(1, 4)$, $P_2(2, 1)$, and $P_3(-1, 8)$.

9.2 Linear Systems and Determinants of Order 3

Determinants of Order 3 and Cramer's Rule

Homogeneous Systems

As a rule, we don't refer to Cramer's result as a theorem.

Determinants of Order 3 and Cramer's Rule. In this section we consider a second method for solving the linear system

$$\begin{aligned} a_1x + b_1y + c_1z &= d_1 \\ a_2x + b_2y + c_2z &= d_2 \\ a_3x + b_3y + c_3z &= d_3. \end{aligned} \tag{1}$$

If $c_1 \neq 0$, the first equation gives

$$z = -\frac{a_1}{c_1}x - \frac{b_1}{c_1}y + \frac{d_1}{c_1}. \tag{2}$$

Substitution of (2) in the last two equations of system (1) yields the reduced system

$$\begin{aligned}(a_2c_1 - a_1c_2)x + (b_2c_1 - b_1c_2)y &= c_1d_2 - c_2d_1\\ (a_3c_1 - a_1c_3)x + (b_3c_1 - b_1c_3)y &= c_1d_3 - c_3d_1.\end{aligned} \tag{3}$$

The solutions of (3) and (2) for x, y, and z are

$$x = \frac{d_1(b_2c_3 - b_3c_2) - b_1(d_2c_3 - d_3c_2) + c_1(d_2b_3 - d_3b_2)}{a_1(b_2c_3 - b_3c_2) - b_1(a_2c_3 - a_3c_2) + c_1(a_2b_3 - a_3b_2)}$$

$$y = \frac{a_1(d_2c_3 - d_3c_2) - d_1(a_2c_3 - a_3c_2) + c_1(a_2d_3 - a_3d_2)}{a_1(b_2c_3 - b_3c_2) - b_1(a_2c_3 - a_3c_2) + c_1(a_2b_3 - a_3b_2)}$$

$$z = \frac{a_1(b_2d_3 - b_3d_2) - b_1(a_2d_3 - a_3d_2) + d_1(a_2b_3 - a_3b_2)}{a_1(b_2c_3 - b_3c_2) - b_1(a_2c_3 - a_3c_2) + c_1(a_2b_3 - a_3b_2)}. \tag{4}$$

Formulas (4) are obtained by first solving (3) for x and y by the methods described in Section 2.6. The resulting expressions for x and y are then substituted in (2) to obtain the expression for z.

The formulas look formidable, but they can be simplified as follows. The expression in the denominator of each of the above equations is

$$a_1(b_2c_3 - b_3c_2) - b_1(a_2c_3 - a_3c_2) + c_1(a_2b_3 - a_3b_2)$$

or

$$a_1\begin{vmatrix} b_2 & c_2 \\ b_3 & c_3 \end{vmatrix} - b_1\begin{vmatrix} a_2 & c_2 \\ a_3 & c_3 \end{vmatrix} + c_1\begin{vmatrix} a_2 & b_2 \\ a_3 & b_3 \end{vmatrix},$$

which is denoted by

$$\begin{vmatrix} a_1 & b_1 & c_1 \\ a_2 & b_2 & c_2 \\ a_3 & b_3 & c_3 \end{vmatrix} \tag{5}$$

and is called a **determinant of order 3.** For example,

$$\begin{aligned}\begin{vmatrix} 1 & 2 & 3 \\ 6 & 5 & 4 \\ 8 & 9 & 7 \end{vmatrix} &= 1\begin{vmatrix} 5 & 4 \\ 9 & 7 \end{vmatrix} - 2\begin{vmatrix} 6 & 4 \\ 8 & 7 \end{vmatrix} + 3\begin{vmatrix} 6 & 5 \\ 8 & 9 \end{vmatrix}\\ &= 1(35 - 36) - 2(42 - 32) + 3(54 - 40)\\ &= 21.\end{aligned}$$

By means of determinants, the solutions (4) can be written as

$$x = \frac{\begin{vmatrix} d_1 & b_1 & c_1 \\ d_2 & b_2 & c_2 \\ d_3 & b_3 & c_3 \end{vmatrix}}{\begin{vmatrix} a_1 & b_1 & c_1 \\ a_2 & b_2 & c_2 \\ a_3 & b_3 & c_3 \end{vmatrix}}, \quad y = \frac{\begin{vmatrix} a_1 & d_1 & c_1 \\ a_2 & d_2 & c_2 \\ a_3 & d_3 & c_3 \end{vmatrix}}{\begin{vmatrix} a_1 & b_1 & c_1 \\ a_2 & b_2 & c_2 \\ a_3 & b_3 & c_3 \end{vmatrix}}, \quad z = \frac{\begin{vmatrix} a_1 & b_1 & d_1 \\ a_2 & b_2 & d_2 \\ a_3 & b_3 & d_3 \end{vmatrix}}{\begin{vmatrix} a_1 & b_1 & c_1 \\ a_2 & b_2 & c_2 \\ a_3 & b_3 & c_3 \end{vmatrix}}. \tag{6}$$

Each a_i in the denominator is replaced by d_i in the numerator — *Each b_i in the denominator is replaced by d_i in the numerator* — *Each c_i in the denominator is replaced by d_i in the numerator*

The determinant (5) is called the **determinant of the coefficients** of system (1). If it is not zero, then (6) gives the unique solution of system (1), and this result is called **Cramer's rule.** In the derivation of Cramer's rule, we assumed $c_1 \neq 0$, but the rule holds whenever the determinant of the coefficients is not zero.

The method of determinants that was used to solve two equations in two unknowns in Section 2.6 is also called Cramer's rule. In fact, determinants can be defined for the coefficients of any linear system in which the number of equations is equal to the number of unknowns. There is a Cramer's rule for any such system.

Example 1 Use Cramer's rule to solve the system of Example 1 in the preceding section.

Solution: The determinant of the coefficients for the system is

$$\Delta = \begin{vmatrix} 3 & -2 & 1 \\ 2 & 5 & -4 \\ 4 & 1 & -1 \end{vmatrix} = 3\begin{vmatrix} 5 & -4 \\ 1 & -1 \end{vmatrix} - (-2)\begin{vmatrix} 2 & -4 \\ 4 & -1 \end{vmatrix} + 1\begin{vmatrix} 2 & 5 \\ 4 & 1 \end{vmatrix} = 7.$$

The numerator for x in equation (6) is

$$\Delta_x = \begin{vmatrix} 7 & -2 & 1 \\ -3 & 5 & -4 \\ 6 & 1 & -1 \end{vmatrix} = 7\begin{vmatrix} 5 & -4 \\ 1 & -1 \end{vmatrix} - (-2)\begin{vmatrix} -3 & -4 \\ 6 & -1 \end{vmatrix} + 1\begin{vmatrix} -3 & 5 \\ 6 & 1 \end{vmatrix} = 14.$$

The numerator for y is

$$\Delta_y = \begin{vmatrix} 3 & 7 & 1 \\ 2 & -3 & -4 \\ 4 & 6 & -1 \end{vmatrix} = 3\begin{vmatrix} -3 & -4 \\ 6 & -1 \end{vmatrix} - 7\begin{vmatrix} 2 & -4 \\ 4 & -1 \end{vmatrix} + 1\begin{vmatrix} 2 & -3 \\ 4 & 6 \end{vmatrix} = 7.$$

Finally, the numerator for z is

$$\Delta_z = \begin{vmatrix} 3 & -2 & 7 \\ 2 & 5 & -3 \\ 4 & 1 & 6 \end{vmatrix} = 3\begin{vmatrix} 5 & -3 \\ 1 & 6 \end{vmatrix} - (-2)\begin{vmatrix} 2 & -3 \\ 4 & 6 \end{vmatrix} + 7\begin{vmatrix} 2 & 5 \\ 4 & 1 \end{vmatrix} = 21.$$

Therefore, the solution of the system is

$$x = \frac{\Delta_x}{\Delta} = \frac{14}{7} = 2, \qquad y = \frac{\Delta_y}{\Delta} = \frac{7}{7} = 1, \qquad z = \frac{\Delta_z}{\Delta} = \frac{21}{7} = 3,$$

which agrees with the values obtained by the method of reduction in the previous section. ∎

Caution Cramer's rule applies only if the determinant of the coefficients in system (1) is not equal to 0, in which case the equations have a unique solution. Geometrically, this means that the three planes represented by system (1) intersect in a unique point.

Example 2 Show that Cramer's rule cannot be applied in Examples 2 and 3 of the previous section.

Solution: In both examples, the determinant of the coefficients is

$$\begin{aligned}\Delta &= \begin{vmatrix} 1 & -2 & -3 \\ 1 & -4 & -13 \\ -3 & 5 & 4 \end{vmatrix} \\ &= 1\begin{vmatrix} -4 & -13 \\ 5 & 4 \end{vmatrix} - (-2)\begin{vmatrix} 1 & -13 \\ -3 & 4 \end{vmatrix} + (-3)\begin{vmatrix} 1 & -4 \\ -3 & 5 \end{vmatrix} \\ &= 49 - 70 + 21 = 0.\end{aligned}$$

Since using Cramer's rule would require $\Delta = 0$ in the denominators of the equations for x, y, and z, the rule cannot be applied to these systems. Furthermore, the solutions obtained in the previous section illustrate that $\Delta = 0$ implies that the corresponding system either has no solution or an infinite number of solutions. ∎

Homogeneous Systems. A linear system

$$\begin{aligned} a_1x + b_1y + c_1z &= 0 \\ a_2x + b_2y + c_2z &= 0 \\ a_3x + b_3y + c_3z &= 0 \end{aligned} \tag{7}$$

in which all of the constant terms are 0 is called a **homogeneous system** and always has the **trivial solution** $x = y = z = 0$ (*why?*). According to Cramer's rule, this is the only solution if and only if determinant (5) is not zero. Hence, *a homogeneous linear system* (7) *has a nontrivial solution if and only if the determinant of the coefficients is equal to zero.*

Example 3 Without actually solving, show that the system

$$\begin{aligned} x - 2y - 3z &= 0 \\ x - 4y - 13z &= 0 \\ -3x + 5y + 4z &= 0 \end{aligned}$$

has nontrivial solutions.

Solution: This homogeneous system has the determinant Δ of Example 2 as its coefficient determinant. Since $\Delta = 0$, we conclude that there are nontrivial solutions. That is, there are solutions other than $x = 0, y = 0$, and $z = 0$. ■

Question *What is the geometric interpretation of a homogeneous system of linear equations?*

Answer: Since every equation in the homogeneous system (7) is satisfied by (0, 0, 0), the system represents three planes that pass through the origin of a three-dimensional coordinate system. ■

Exercises 9.2

Fill in the blanks to make each statement true.

1. Cramer's rule may be used to solve a linear system if and only if the number of __________ and the number of __________ are equal and __________ is not zero.
2. By definition, $\begin{vmatrix} a_1 & b_1 & c_1 \\ a_2 & b_2 & c_2 \\ a_3 & b_3 & c_3 \end{vmatrix} =$ __________.
3. When a linear system of three equations in three unknowns is solvable by Cramer's rule, the solution is __________ and represents geometrically the __________ in which three planes intersect.
4. If the determinant of the coefficients in a linear system of 3 equations in 3 unknowns equals 0, then the number of solutions is ________ or ________.
5. A homogeneous system of 3 linear equations in 3 unknowns has a nontrivial solution if and only if ________________.

Write true *or* false *for each statement.*

6. If the determinant $\begin{vmatrix} a_1 & b_1 & c_1 \\ a_2 & b_2 & c_2 \\ a_3 & b_3 & c_3 \end{vmatrix} \neq 0$, then there is a unique solution for the system $\begin{cases} a_1x + b_1y + c_1z = d_1 \\ a_2x + b_2y + c_2z = d_2 \\ a_3x + b_3y + c_3z = d_3. \end{cases}$
7. For the linear system $\begin{cases} a_1x + b_1y + c_1z + d_1 = 0 \\ a_2x + b_2y + c_2z + d_2 = 0 \\ a_3x + b_3y + c_3z + d_3 = 0 \end{cases}$ with $\Delta = \begin{vmatrix} a_1 & b_1 & c_1 \\ a_2 & b_2 & c_2 \\ a_3 & b_3 & c_3 \end{vmatrix} \neq 0$, Cramer's rule gives the solution $\dfrac{\begin{vmatrix} d_1 & b_1 & c_1 \\ d_2 & b_2 & c_2 \\ d_3 & b_3 & c_3 \end{vmatrix}}{\Delta}$ for x.
8. A linear system of three equations in three unknowns has no solution when the determinant of the coefficients equals 0.
9. In general, a linear system of three equations in two unknowns cannot be solved by Cramer's rule.
10. A homogeneous system of linear equations always has at least one solution.

Determinants of Order 3

Evaluate each of the following determinants.

11. $\begin{vmatrix} 1 & -2 & 3 \\ 4 & 1 & 2 \\ 5 & -1 & 4 \end{vmatrix}$
12. $\begin{vmatrix} 1 & -2 & 3 \\ 4 & 1 & 2 \\ 5 & -1 & 5 \end{vmatrix}$
13. $\begin{vmatrix} 2 & 3 & 0 \\ 1/2 & -6 & 1 \\ 1/4 & 8 & -2 \end{vmatrix}$
14. $\begin{vmatrix} 3 & 1 & -2 \\ 0 & 4 & 7 \\ 0 & 6 & -1 \end{vmatrix}$

Cramer's Rule

15.–20. Use Cramer's rule to solve the linear systems given in Exercises 11–16 of the previous section.

21.–24. Explain why Cramer's rule cannot be used to solve the systems given in Exercises 21–24 of the previous section.

In Exercises 25–28, determine (without solving) whether the given homogeneous system has any nontrivial solutions.

25. $3x - 6y = 0$
$2x - 4y = 0$

26. $x - y + z = 0$
$2x + 3y - 4z = 0$
$3x + 2y - 3z = 0$

27. $2x - 3y + 4z = 0$
$x + y - 5z = 0$
$5x - 5y + 3z = 0$

28. $2x - 3y + 4z = 0$
$x + y - 5z = 0$
$5x - 5y + 2z = 0$

Miscellaneous

Determinant Equation for a Plane. *The three planes corresponding to system (1) determine a unique point if and only if determinant (5) is not zero. The* dual *question is "Under what conditions do the three points $P_1(x_1, y_1, z_1)$, $P_2(x_2, y_2, z_2)$, and $P_3(x_3, y_3, z_3)$ determine a unique plane?" Geometrically, it is evident that* a unique plane is determined precisely when the three points do not lie on the same straight line, i.e., are not collinear. *It can be shown that P_1, P_2, and P_3 are not collinear if and only if* at least one *of the determinants*

$$\begin{vmatrix} 1 & y_1 & z_1 \\ 1 & y_2 & z_2 \\ 1 & y_3 & z_3 \end{vmatrix}, \quad \begin{vmatrix} x_1 & 1 & z_1 \\ x_2 & 1 & z_2 \\ x_3 & 1 & z_3 \end{vmatrix}, \quad \begin{vmatrix} x_1 & y_1 & 1 \\ x_2 & y_2 & 1 \\ x_3 & y_3 & 1 \end{vmatrix} \tag{i}$$

is not equal to zero. Furthermore, when this condition is met, the equation of the plane containing P_1, P_2, and P_3 is

$$\begin{vmatrix} 1 & y_1 & z_1 \\ 1 & y_2 & z_2 \\ 1 & y_3 & z_3 \end{vmatrix} x + \begin{vmatrix} x_1 & 1 & z_1 \\ x_2 & 1 & z_2 \\ x_3 & 1 & z_3 \end{vmatrix} y + \begin{vmatrix} x_1 & y_1 & 1 \\ x_2 & y_2 & 1 \\ x_3 & y_3 & 1 \end{vmatrix} z = \begin{vmatrix} x_1 & y_1 & z_1 \\ x_2 & y_2 & z_2 \\ x_3 & y_3 & z_3 \end{vmatrix}. \tag{ii}$$

In Exercises 29–32, determine whether the given points are collinear. Also, if possible, find an equation for the plane determined by the points.

29. $(2, 0, 0)$, $(0, -3, 0)$, $(0, 0, 5)$

30. $(3, -2, 5)$, $(4, 1, 0)$, $(5, 4, -5)$

31. $(2, 1, -3)$, $(-1, 4, 0)$, $(1, -3, 5)$

32. $(3, 2, 4)$, $(1, 5, 6)$, $(5, -1, 2)$

33. Verify by direct substitution that each of the points $P_1(x_1, y_1, z_1)$, $P_2(x_2, y_2, z_2)$, and $P_3(x_3, y_3, z_3)$ satisfies equation (ii) of the plane determined when condition (i) is satisfied.

34. Show that three points (x_1, y_1), (x_2, y_2), and (x_3, y_3) in the plane are collinear if and only if

$$\begin{vmatrix} x_1 & y_1 & 1 \\ x_2 & y_2 & 1 \\ x_3 & y_3 & 1 \end{vmatrix} = 0.$$

35. Show that an equation of the line through two distinct points (x_1, y_1) and (x_2, y_2) is

$$\begin{vmatrix} x & y & 1 \\ x_1 & y_1 & 1 \\ x_2 & y_2 & 1 \end{vmatrix} = 0.$$

9.3 Properties of Determinants

One property of determinants that is easy to discover is that they take a long time to compute.

Matrices, Determinants, and Minors
Expanding a Determinant by Cofactors
Row and Column Operations

Determinants of order 2 were introduced in Section 2.6 to solve linear systems of two equations in two unknowns. Similarly, in the previous section we used determinants of order 3 to solve three equations in three unknowns. We now review our definition of determinants of order 3 and then generalize to define determinants of order n for any positive integer n. We also introduce properties that simplify their computation.

Matrices, Determinants, and Minors. A determinant of order 3 was introduced as a *number* associated with the coefficients in a linear system of the form

$$\begin{aligned} a_{11}x + a_{12}y + a_{13}z &= b_1 \\ a_{21}x + a_{22}y + a_{23}z &= b_2 \\ a_{31}x + a_{32}y + a_{33}z &= b_3. \end{aligned} \tag{1}$$

The coefficients fall into horizontal rows and vertical columns, forming a square array of numbers called a **3 × 3 matrix** or a **matrix of order 3,** which we denote by the capital letter A. That is,

$$\begin{array}{cccccc} & \textit{Column 1} & \textit{Column 2} & \textit{Column 3} & & \\ A = \Bigg[& a_{11} & a_{12} & a_{13} & \Bigg]. & \textit{Row 1} \\ & a_{21} & a_{22} & a_{23} & & \textit{Row 2} \\ & a_{31} & a_{32} & a_{33} & & \textit{Row 3} \end{array} \tag{2}$$

The numbers a_{ij} ($i = 1, 2, 3; j = 1, 2, 3$) are called the **entries** or **elements** in the matrix A. The first subscript i denotes the row and the second subscript j denotes the column of the entry a_{ij}. For example, a_{23} is in the second row and third column. Similarly, the coefficients of the system

$$\begin{aligned} a_{11}x + a_{12}y &= b_1 \\ a_{21}x + a_{22}y &= b_2 \end{aligned}$$

are represented by the 2×2 matrix

$$\begin{bmatrix} a_{11} & a_{12} \\ a_{21} & a_{22} \end{bmatrix}. \tag{3}$$

In the next two sections of this chapter, we will introduce matrices that are not square. In general, a matrix is a *rectangular* array of numbers in which the number of rows need not equal the number of columns. Also, a matrix need not be associated with a system of equations. Only a *square* matrix has an associated number called its determinant. Note that brackets [ ] represent matrices and vertical bars | | represent determinants.

For the 3×3 matrix A given in (2), its determinant will now be denoted by $|A|$, and as previously given, its definition is

$$|A| = a_{11}\begin{vmatrix} a_{22} & a_{23} \\ a_{32} & a_{33} \end{vmatrix} - a_{12}\begin{vmatrix} a_{21} & a_{23} \\ a_{31} & a_{33} \end{vmatrix} + a_{13}\begin{vmatrix} a_{21} & a_{22} \\ a_{31} & a_{32} \end{vmatrix}. \tag{4}$$

The second-order determinants in definition (4) are called the minors of the elements a_{11}, a_{12}, and a_{13} and will be denoted by M_{11}, M_{12}, and M_{13}, respectively. Hence, we can write (4) in a more compact form as

$$|A| = a_{11}M_{11} - a_{12}M_{12} + a_{13}M_{13}. \tag{5}$$

Similarly, for any element a_{ij} *in a square matrix* A, its **minor** M_{ij} is defined as the determinant of the matrix obtained by deleting the ith row and the jth column of A.

Example 1 For $A = \begin{bmatrix} 2 & -1 & 5 \\ 1 & 2 & -3 \\ 4 & -2 & 6 \end{bmatrix}$, find M_{11}, M_{12}, and M_{13}, and then compute the determinant $|A|$ by using definition (5).

Solution: We have $a_{11} = 2$ and by deleting row 1 and column 1 of A, we get

$$M_{11} = \begin{vmatrix} 2 & -3 \\ -2 & 6 \end{vmatrix} = 6.$$

Similarly, $a_{12} = -1$, and we delete row 1 and column 2 of A to obtain

$$M_{12} = \begin{vmatrix} 1 & -3 \\ 4 & 6 \end{vmatrix} = 18.$$

Finally, $a_{13} = 5$, and when row 1 and column 3 are deleted from A, we get

$$M_{13} = \begin{vmatrix} 1 & 2 \\ 4 & -2 \end{vmatrix} = -10.$$

By equation (5), the determinant of matrix A is

$$\begin{aligned} |A| &= a_{11}M_{11} - a_{12}M_{12} + a_{13}M_{13} \\ &= 2(6) - (-1)18 + 5(-10) = -20. \end{aligned}$$ ■

Expanding a Determinant by Cofactors. To give the most general definition of a determinant in terms of minors, it is first necessary to attach an algebraic sign to each minor. By definition, the **cofactor** A_{ij} of the element in row i and column j of a square matrix is

$$\boldsymbol{A_{ij} = (-1)^{i+j} M_{ij}},$$

where M_{ij} is the minor of the element.

For example, in the 3×3 matrix given in (2), we have

$$\begin{aligned} A_{11} &= (-1)^{1+1}M_{11} = M_{11} \\ A_{12} &= (-1)^{1+2}M_{12} = -M_{12} \\ A_{13} &= (-1)^{1+3}M_{13} = M_{13} \end{aligned}$$

and therefore the definition of the determinant in (5) can be rewritten as

$$\boldsymbol{|A| = a_{11}A_{11} + a_{12}A_{12} + a_{13}A_{13}}. \qquad \textbf{(6)}$$

Example 2 Use equation (6) to compute the determinant of the matrix

$$A = \begin{bmatrix} 2 & -1 & 5 \\ 1 & 2 & -3 \\ 4 & -2 & 6 \end{bmatrix}.$$

Solution: From Example 1, we have $M_{11} = 6$, $M_{12} = 18$, and $M_{13} = -10$. The corresponding cofactors are

$$A_{11} = (-1)^{1+1}M_{11} = +M_{11} = 6$$
$$A_{12} = (-1)^{1+2}M_{12} = -M_{12} = -18$$
$$A_{13} = (-1)^{1+3}M_{13} = +M_{13} = -10.$$

Therefore, by equation (6),

$$\begin{aligned}|A| &= a_{11}A_{11} + a_{12}A_{12} + a_{13}A_{13}\\ &= 2(6) + (-1)(-18) + 5(-10) = -20. \quad \blacksquare\end{aligned}$$

It can be shown that the same value of the determinant $|A|$ is obtained by using the *cofactors of the elements of any row or of any column of* A. We now illustrate this for the determinant of Examples 1 and 2.

Example 3 Compute the determinant of the matrix

$$A = \begin{bmatrix} 2 & -1 & 5 \\ 1 & 2 & -3 \\ 4 & -2 & 6 \end{bmatrix}$$

by using the cofactors of (a) the elements in the second row and (b) the elements in the third column.

Solution:
(a) Using the cofactors of the second row, we get

$$\begin{aligned}|A| = \begin{vmatrix} 2 & -1 & 5 \\ 1 & 2 & -3 \\ 4 & -2 & 6 \end{vmatrix} &= a_{21}A_{21} + a_{22}A_{22} + a_{23}A_{23}\\
&= 1(-1)^{2+1}\begin{vmatrix} -1 & 5 \\ -2 & 6 \end{vmatrix} + 2(-1)^{2+2}\begin{vmatrix} 2 & 5 \\ 4 & 6 \end{vmatrix}\\
&\quad + (-3)(-1)^{2+3}\begin{vmatrix} 2 & -1 \\ 4 & -2 \end{vmatrix}\\
&= 1(-1)(4) + 2(1)(-8) + (-3)(-1)(0) = -20.\end{aligned}$$

(b) If we use the cofactors of the third column, we obtain

$$\begin{aligned}\begin{vmatrix} 2 & -1 & 5 \\ 1 & 2 & -3 \\ 4 & -2 & 6 \end{vmatrix} &= a_{13}A_{13} + a_{23}A_{23} + a_{33}A_{33}\\
&= 5(-1)^{1+3}\begin{vmatrix} 1 & 2 \\ 4 & -2 \end{vmatrix} + (-3)(-1)^{2+3}\begin{vmatrix} 2 & -1 \\ 4 & -2 \end{vmatrix}\\
&\quad + 6(-1)^{3+3}\begin{vmatrix} 2 & -1 \\ 1 & 2 \end{vmatrix}\\
&= 5(1)(-10) + (-3)(-1)(0) + 6(1)5 = -20.\end{aligned}$$

As expected, the results in (a) and (b) are the same, and they agree with the results in Examples 1 and 2. ■

Our examples can be generalized to define the determinant of any square matrix. By a **matrix A of order n**, we mean an array of n rows and n columns of elements. The **determinant of A** is equal to the sum of n terms, each of which is the product of an element in a fixed row (or a fixed column) by its cofactor. The determinant of A is denoted by $|A|$ and is called a determinant of order n. We illustrate with a determinant of order 4.

Example 4 Evaluate the determinant of the matrix

$$A = \begin{bmatrix} 1 & 3 & -2 & 4 \\ 2 & 0 & 0 & -3 \\ -1 & 2 & 1 & 0 \\ 5 & 0 & -1 & 2 \end{bmatrix}.$$

Solution: We expand by the cofactors of the second row, since the zeros in that row will simplify the computation. Thus,

$$|A| = \begin{vmatrix} 1 & 3 & -2 & 4 \\ 2 & 0 & 0 & -3 \\ -1 & 2 & 1 & 0 \\ 5 & 0 & -1 & 2 \end{vmatrix}$$

$$= 2(-1)^{2+1}\begin{vmatrix} 3 & -2 & 4 \\ 2 & 1 & 0 \\ 0 & -1 & 2 \end{vmatrix} + 0 + 0 + (-3)(-1)^{2+4}\begin{vmatrix} 1 & 3 & -2 \\ -1 & 2 & 1 \\ 5 & 0 & -1 \end{vmatrix}$$

$$= \underbrace{-2\left(3\begin{vmatrix} 1 & 0 \\ -1 & 2 \end{vmatrix} - 2\begin{vmatrix} -2 & 4 \\ -1 & 2 \end{vmatrix}\right)}_{\textit{Expanding by first column}} \underbrace{- 3\left(5\begin{vmatrix} 3 & -2 \\ 2 & 1 \end{vmatrix} + (-1)\begin{vmatrix} 1 & 3 \\ -1 & 2 \end{vmatrix}\right)}_{\textit{Expanding by third row}}$$

$$= -2(3 \cdot 2 - 2 \cdot 0) \qquad - 3(5 \cdot 7 - 1 \cdot 5)$$

$$= -102.$$ ■

Row and Column Operations. A determinant of order $n > 3$ requires lengthy computations unless a row or column of its matrix contains some zeros, as in Example 4. If zeros do not exist in a given matrix, we can perform row and column operations to produce a matrix that does contain zeros and has the same determinant. To produce such a matrix, we use the following properties.

(1) If every element of any row (or column) is multiplied by a number k, then the determinant is multiplied by k.
(2) If two rows (or two columns) are interchanged, then the determinant is multiplied by -1.
(3) If every element of one row (or one column) is multiplied by a constant and then added to the corresponding element of another row (or column), then the value of the determinant is not changed.
(4) If two rows (or columns) are equal, then the determinant is equal to zero.

We leave the general proof of these properties to more advanced courses, but include verification for determinants of order 3 in the exercises. We also note that property (4) follows directly from (2), as shown in Example 8 later in this section.

We now illustrate the use of these properties in simplifying the computation of a determinant. *The idea is to replace all but one of the elements of a row (or column) by zeros.*

Example 5 Simplify the computation of the determinant of the matrix A in Example 1 by using properties (1) through (4) above.

Solution: Observing that 2 is a factor of each element in row 3, we use property (1) to write

$$|A| = \begin{vmatrix} 2 & -1 & 5 \\ 1 & 2 & -3 \\ 4 & -2 & 6 \end{vmatrix} = 2\begin{vmatrix} 2 & -1 & 5 \\ 1 & 2 & -3 \\ 2 & -1 & 3 \end{vmatrix}.$$

Next we introduce two zeros in column 2 by using property (3).

$$|A| = 2\begin{vmatrix} 2 & -1 & 5 \\ 1 & 2 & -3 \\ 2 & -1 & 3 \end{vmatrix}$$

$$= 2\begin{vmatrix} 2 & -1 & 5 \\ 5 & 0 & 7 \\ 0 & 0 & -2 \end{vmatrix}$$

2 times each element in row 1 was added to the corresponding element below it in row 2, and

-1 times each element in row 1 was added to the corresponding element below it in row 3

$$= 2(-1)(-1)^{1+2}\begin{vmatrix} 5 & 7 \\ 0 & -2 \end{vmatrix}$$

Expanding by cofactors of column 2

$$= 2(-1)(-1)(-10) = -20$$ ■

Example 6 Repeat Example 5 by introducing zeros in row 1.

Solution: We keep the entry -1 and use it to replace the 2 and the 5 in the first row by 0s.

$$|A| = \begin{vmatrix} 2 & -1 & 5 \\ 1 & 2 & -3 \\ 4 & -2 & 6 \end{vmatrix}$$

$$= \begin{vmatrix} 0 & -1 & 0 \\ 5 & 2 & 7 \\ 0 & -2 & -4 \end{vmatrix}$$

2 times each element of column 2 was added to the corresponding element to the left in column 1, and

5 times each element of column 2 was added to the corresponding element to the right in column 3

$$= (-1)(-1)^{1+2}\begin{vmatrix} 5 & 7 \\ 0 & -4 \end{vmatrix}$$

Expanding by the cofactors of row 1

$$= -20 \quad ■$$

Examples 5 and 6 made use of properties (1) and (3) above. We now illustrate properties (2) and (4).

Example 7 Compute $|A|$, where $A = \begin{bmatrix} -1 & 5 & 2 \\ 2 & -3 & 1 \\ -2 & 6 & 4 \end{bmatrix}$.

Solution: Here, the matrix A has the same columns as the matrix in Example 6 but not in the same order. Using property (2), we change the sign of the determinant each time we interchange two columns. Thus,

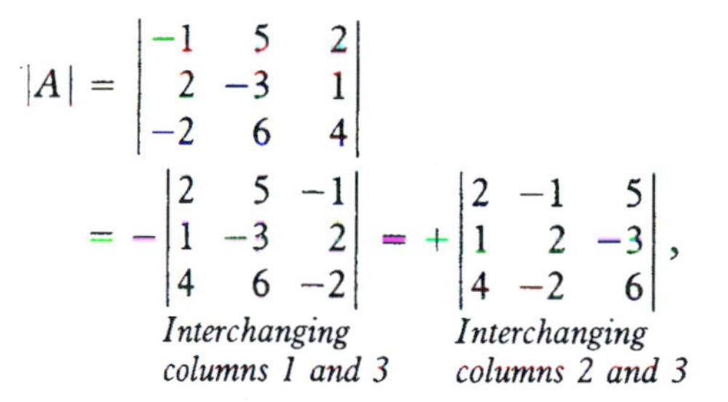

$$|A| = \begin{vmatrix} -1 & 5 & 2 \\ 2 & -3 & 1 \\ -2 & 6 & 4 \end{vmatrix} = -\begin{vmatrix} 2 & 5 & -1 \\ 1 & -3 & 2 \\ 4 & 6 & -2 \end{vmatrix} = +\begin{vmatrix} 2 & -1 & 5 \\ 1 & 2 & -3 \\ 4 & -2 & 6 \end{vmatrix},$$

which is the determinant of Example 6 and has the value -20. ■

Example 8 Illustrate property (4) for the matrix

$$A = \begin{bmatrix} a_{11} & a_{12} & a_{11} \\ a_{21} & a_{22} & a_{21} \\ a_{31} & a_{32} & a_{31} \end{bmatrix}.$$

Solution 1: If we interchange the first and third columns of A, then we still have A, but according to property (2), $|A|$ is changed to $-|A|$. Hence, $|A| = -|A|$, which means that $|A| = 0$.

Solution 2: By adding -1 times column 1 to column 3, we get

$$|A| = \begin{vmatrix} a_{11} & a_{12} & 0 \\ a_{21} & a_{22} & 0 \\ a_{31} & a_{32} & 0 \end{vmatrix}.$$

Then, expanding by the cofactors of the third column, we have $|A| = 0$. Thus, we see why a determinant equals zero if it contains two equal columns (or rows), as stated above in property (4). ■

Exercises 9.3

Fill in the blanks to make each statement true.

1. In a determinant of order 3, each minor is a determinant of order ________.
2. The minor of an element in a square matrix is the determinant of the matrix obtained by deleting the ________ and the ________ containing the element.
3. The cofactor of an element a_{ij} in a square matrix is the minor of a_{ij} multiplied by $(-1)^p$, where $p =$ ________.
4. To compute a determinant, we may multiply each element of one row by its ________ and then ________.
5. The value of a determinant $|A|$ is unchanged if the row operation performed on A is ________.

Write true *or* false *for each statement.*

6. The determinant of a square matrix is unchanged if two rows of the matrix are interchanged.
7. If two columns of a square matrix are equal, then its determinant equals 0.
8. If a square matrix contains a row of zeros, then its determinant is zero.
9. By definition, a determinant of order n is the sum of n determinants of order $n - 1$.
10. If every element of a matrix of order n is multiplied by a constant k, then its determinant is multiplied by k^n.

Matrices, Determinants, and Minors

Exercises 11–20 refer to the following matrices A and B.

$$A = \begin{bmatrix} 5 & -1 & 2 \\ 3 & 4 & -7 \\ 2 & 6 & -3 \end{bmatrix}$$

$$B = \begin{bmatrix} 1 & 0 & -1 & 2 \\ 0 & 2 & 0 & -1 \\ -2 & 5 & 6 & -2 \\ 3 & -4 & 0 & 7 \end{bmatrix}$$

11. For A, determine the following.
    (a) a_{11} and M_{11} (b) a_{12} and M_{12} (c) a_{13} and M_{13}
12. For A, determine the following.
    (a) a_{21} and M_{21} (b) a_{22} and M_{22} (c) a_{23} and M_{23}
13. For A, determine the following.
    (a) a_{13} and M_{13} (b) a_{23} and M_{23} (c) a_{33} and M_{33}
14. For B, determine the following.
    (a) b_{22} and M_{22} (b) b_{24} and M_{24}
15. For B, determine the following.
    (a) b_{13} and M_{13} (b) b_{33} and M_{33}

Expanding a Determinant by Cofactors

16. Compute $|A|$ by expanding by cofactors along the second row.
17. Compute $|A|$ by expanding by cofactors along the first row.

18. Compute $|B|$ by expanding by cofactors along the second row.

19. Compute $|A|$ by expanding by cofactors along the third column.

20. Compute $|B|$ by expanding by cofactors along the third column.

Use expansion by cofactors to evaluate the determinants in Exercises 21–23.

21. $\begin{vmatrix} 0 & 3 & 0 \\ -1 & 4 & 7 \\ 2 & -2 & 1 \end{vmatrix}$

22. $\begin{vmatrix} 2 & 0 & 4 \\ 1 & -2/3 & 3 \\ -5/6 & 0 & 6 \end{vmatrix}$

23. $\begin{vmatrix} 3 & 4 & -1 & 2 \\ 0 & -1 & 0 & 0 \\ 5 & 0 & -3 & 1 \\ -2 & 1 & 2 & 1 \end{vmatrix}$

24. Show that a determinant of order 3 may be expanded by the cofactors of the second row. [*Hint:* show that $a_{21}A_{21} + a_{22}A_{22} + a_{23}A_{23}$ yields the same six terms as definition (4).]

25. Show that a determinant of order 3 may be expanded by the cofactors of the third column. [*Hint:* compare this expansion with equation (4).]

Row and Column Operations

Justify each of the equalities in Exercises 26–31 by one of the properties (1) through (4) of determinants.

26. $\begin{vmatrix} 2 & -4 \\ 1 & 3 \end{vmatrix} = 2\begin{vmatrix} 1 & -2 \\ 1 & 3 \end{vmatrix}$

27. $\begin{vmatrix} 2 & -4 \\ 1 & 3 \end{vmatrix} = \begin{vmatrix} 0 & -10 \\ 1 & 3 \end{vmatrix}$

28. $\begin{vmatrix} 2 & -1 & 5 \\ 6 & 7 & 4 \\ 2 & -1 & 5 \end{vmatrix} = 0$

29. $\begin{vmatrix} 2 & -3 \\ 1 & 5 \end{vmatrix} = -\begin{vmatrix} -3 & 2 \\ 5 & 1 \end{vmatrix}$

30. $\begin{vmatrix} 2 & -3 \\ 1 & 5 \end{vmatrix} = -\begin{vmatrix} 1 & 5 \\ 2 & -3 \end{vmatrix}$

31. $\begin{vmatrix} 2 & 1 & 3 \\ 4 & -7 & 5 \\ 0 & 6 & 0 \end{vmatrix} = -\begin{vmatrix} 2 & 1 & 3 \\ 4 & -7 & 5 \\ 0 & -6 & 0 \end{vmatrix}$

In Exercises 32–35, first find a determinant equal to the given one in which a row or a column contains only one nonzero element, and then evaluate.

32. $\begin{vmatrix} 2 & -1 & 5 \\ 3 & 8 & -6 \\ 1 & -4 & 7 \end{vmatrix}$

33. $\begin{vmatrix} 1 & .5 & -4 \\ 2 & 1 & 1 \\ 4 & -.5 & 6 \end{vmatrix}$

34. $\begin{vmatrix} -3 & 1 & 0 & 2 \\ 6 & -2 & 5 & -1 \\ 9 & 3 & 7 & 3 \\ -12 & 4 & 1 & 4 \end{vmatrix}$

35. $\begin{vmatrix} 1/2 & 1/4 & -3/8 & 3/4 \\ 0 & 1 & 3 & 2 \\ -1 & 0 & -2 & 3 \\ 5 & 4 & -6 & 1 \end{vmatrix}$

36. Give a partial proof for order 3 of property (1) of determinants by showing that

$$\begin{vmatrix} ka_{11} & ka_{12} & ka_{13} \\ a_{21} & a_{22} & a_{23} \\ a_{31} & a_{32} & a_{33} \end{vmatrix} = k\begin{vmatrix} a_{11} & a_{12} & a_{13} \\ a_{21} & a_{22} & a_{23} \\ a_{31} & a_{32} & a_{33} \end{vmatrix}.$$

37. Give a partial proof for order 3 of property (2) of determinants by showing that

$$\begin{vmatrix} a_{21} & a_{22} & a_{23} \\ a_{11} & a_{12} & a_{13} \\ a_{31} & a_{32} & a_{33} \end{vmatrix} = -\begin{vmatrix} a_{11} & a_{12} & a_{13} \\ a_{21} & a_{22} & a_{23} \\ a_{31} & a_{32} & a_{33} \end{vmatrix}.$$

(*Hint:* expand the first determinant by cofactors of row 2.)

38. Give a partial proof for order 3 of property (3) of determinants by showing that

$$\begin{vmatrix} a_{11} + ka_{21} & a_{12} + ka_{22} & a_{13} + ka_{23} \\ a_{21} & a_{22} & a_{23} \\ a_{31} & a_{32} & a_{33} \end{vmatrix} = \begin{vmatrix} a_{11} & a_{12} & a_{13} \\ a_{21} & a_{22} & a_{23} \\ a_{31} & a_{32} & a_{33} \end{vmatrix}.$$

39. Show that

$$\begin{vmatrix} a_{11} & a_{12} & a_{13} & a_{14} \\ 0 & a_{22} & a_{23} & a_{24} \\ 0 & 0 & a_{33} & a_{34} \\ 0 & 0 & 0 & a_{44} \end{vmatrix} = a_{11}a_{22}a_{33}a_{44}.$$

40. Show that

$$\begin{vmatrix} a_{11} & a_{12} & 0 & 0 \\ a_{21} & a_{22} & 0 & 0 \\ 0 & 0 & a_{33} & a_{34} \\ 0 & 0 & a_{43} & a_{44} \end{vmatrix} = \begin{vmatrix} a_{11} & a_{12} \\ a_{21} & a_{22} \end{vmatrix}\begin{vmatrix} a_{33} & a_{34} \\ a_{43} & a_{44} \end{vmatrix}.$$

41. (a) Show that $\begin{vmatrix} x & y & 1 \\ x_1 & y_1 & 1 \\ x_2 & y_2 & 1 \end{vmatrix} = 0$ is the equation of a straight line in the plane provided $x_1 \neq x_2$ or $y_1 \neq y_2$. (*Hint:* show it is an equation of the form $Ax + By + C = 0$ with A or $B \neq 0$.)

(b) Using one of the determinant properties (1) through (4), show that the line in part (a) passes through (x_1, y_1) and (x_2, y_2).

9.4 Linear Systems: Gaussian Elimination

First we eliminate variables, and then we eliminate equations.

Elimination Methods
Augmented Matrix of a Linear System
Gaussian Elimination Method
Gauss-Jordan Elimination Method

The third method we use for solving linear systems is *elimination*. Although we begin by solving a linear system in three variables by an elimination procedure similar to that used in Section 2.6 for two variables, our aim is to develop a more systematic procedure that generalizes to linear systems in many variables and can easily be implemented both by hand computation and on a computer.

Elimination Methods. A linear system in three variables can be solved by reducing it to one with two variables. At each step, one variable is eliminated from two equations by adding a multiple of one of the equations to the other.

Example 1 Solve the following system by elimination.

$$2x + 5y + 4z = 4 \quad (1)$$
$$x + 4y + 3z = 1 \quad (2)$$
$$x - 3y - 2z = 5 \quad (3)$$

Solution: By multiplying equation (2) by -2 and adding the result to (1), we eliminate x from the first two equations and obtain equation (4) below. Similarly, if (3) is multiplied by -1 and added to (2), the result is (5).

$$\begin{array}{r} 2x + 5y + 4z = 4 \\ -2x - 8y - 6z = -2 \\ \hline -3y - 2z = 2 \end{array} \quad (4) \qquad \begin{array}{r} x + 4y + 3z = 1 \\ -x + 3y + 2z = -5 \\ \hline 7y + 5z = -4 \end{array} \quad (5)$$

The reduced system

$$-3y - 2z = 2 \quad (4)$$
$$7y + 5z = -4 \quad (5)$$

has the solution $y = -2$ and $z = 2$. Substituting these values in one of the original equations, say (2), we have $x + 4(-2) + 3 \cdot 2 = 1$ or $x = 3$. Hence, the solution of the given system is $x = 3$, $y = -2$, and $z = 2$. ■

The method used in Example 1 is satisfactory for a linear system in three variables with a unique solution. However, we prefer a **general elimination procedure** that can be used conveniently on any linear system, no matter how many equations and variables are involved and regardless of the number of solutions. Our aim is to reduce a given system to **triangular form.** We illustrate the general method by again solving the system in Example 1.

Solution:

Step 1: Make the coefficient of x equal to 1 in the first equation.
In Example 1, this is easily done by interchanging the first two equations, giving the equivalent system

$$\begin{aligned} x + 4y + 3z &= 1 \\ 2x + 5y + 4z &= 4 \\ x - 3y - 2z &= 5. \end{aligned}$$

Step 2: Eliminate x in all equations following the first one.

$$\begin{aligned} x + 4y + 3z &= 1 \\ -3y - 2z &= 2 \qquad \textit{-2 times the first equation added to the second} \\ -7y - 5z &= 4 \qquad \textit{-1 times the first equation added to the third} \end{aligned}$$

Step 3: Make the coefficient of y equal to 1 in the second equation.
In this example, we can do this by multiplying the second equation above by $-1/3$. Our system now is

$$\begin{aligned} x + 4y + 3z &= 1 && (1') \\ y + \frac{2}{3}z &= -\frac{2}{3} && (2') \\ -7y - 5z &= 4. && (3') \end{aligned}$$

Step 4: Eliminate y in all equations following the second one.

$$\begin{aligned} x + 4y + 3z &= 1 \\ y + \frac{2}{3}z &= -\frac{2}{3} \\ -\frac{1}{3}z &= -\frac{2}{3} \qquad \textit{7 times (2') was added to (3')} \end{aligned}$$

Step 5: Make the coefficient of z equal to 1 in the third equation by multiplying both sides by -3.

$$\begin{aligned} x + 4y + 3z &= 1 && (1'') \\ y + \frac{2}{3}z &= -\frac{2}{3} && (2'') \\ z &= 2. && (3'') \end{aligned}$$

Since there are no equations following the third one in this system, the elimination process stops here. We now proceed to find the solution by **back-substitution**.

Step 6: We substitute $z = 2$ from (3″) into (2″) to get $y = -2$. Then both $z = 2$ and $y = -2$ are substituted in (1″) to give $x = 3$. Thus, we have $x = 3$, $y = -2$, and $z = 2$, as in our first method for solving Example 1. ■

We note that each step in the general elimination procedure used above involves one of the following operations:

(1) interchanging two equations,

(2) multiplying an equation by a constant $k \neq 0$,

(3) adding a constant multiple of one equation to another equation.

Before solving any more examples by this method, we introduce a more efficient way to carry out these operations.

Augmented Matrix of a Linear System. The general elimination procedure used above to solve a linear system can be simplified if we observe that only the *positions* of the variables x, y, and z are important, not the symbols used to denote them. That is, if we keep the variables in the same order in each equation, then a given linear system is completely determined by the array of coefficients of the variables and the constant terms. For instance, the system in Example 1,

$$\begin{aligned} 2x + 5y + 4z &= 4 \\ x + 4y + 3z &= 1 \\ x - 3y - 2z &= 5, \end{aligned}$$

can be represented by the rectangular array of numbers

$$\left[\begin{array}{rrr|r} 2 & 5 & 4 & 4 \\ 1 & 4 & 3 & 1 \\ 1 & -3 & -2 & 5 \end{array}\right],$$

which is called the **augmented matrix** of the linear system. The adjective "augmented" is used here to distinguish that matrix from the **coefficient matrix** of the system, namely

$$\begin{bmatrix} 2 & 5 & 4 \\ 1 & 4 & 3 \\ 1 & -3 & -2 \end{bmatrix}.$$

The broken vertical line in the augmented matrix is optional, but it is a convenient device to separate the coefficient matrix from the column of constant terms.

Each horizontal row of the augmented matrix represents one equation, and to solve the linear system, we simply perform on the *rows* of the augmented matrix the operations that correspond exactly to the operations performed on the *equations* in the general elimination procedure. These three operations, which are called **elementary row operations,** are:

(1) interchanging two rows,

(2) multiplying each entry of a row by a constant $k \neq 0$,

(3) adding a constant times each entry in a row to the corresponding entry in another row.

We repeat the solution to Example 1 below with the reduction of the equations to triangular form on the left and the corresponding row operations applied to the augmented matrix on the right.

Solution:

Linear System	Augmented Matrix	Operations
$\begin{aligned} 2x + 5y + 4z &= 4 \\ x + 4y + 3z &= 1 \\ x - 3y - 2z &= 5 \end{aligned}$	$\left[\begin{array}{ccc\|c} 2 & 5 & 4 & 4 \\ 1 & 4 & 3 & 1 \\ 1 & -3 & -2 & 5 \end{array}\right]$	

Step 1: Make the coefficient of the first variable in the first equation (first entry in first row) equal to 1.

Linear System	Augmented Matrix	Operations
$\begin{aligned} x + 4y + 3z &= 1 \\ 2x + 5y + 4z &= 4 \\ x - 3y - 2z &= 5 \end{aligned}$	$\left[\begin{array}{ccc\|c} 1 & 4 & 3 & 1 \\ 2 & 5 & 4 & 4 \\ 1 & -3 & -2 & 5 \end{array}\right]$	*Equations (rows) 1 and 2 were interchanged*

Step 2: Make the coefficient of the first variable (entry in the first column) equal to 0 in all equations (rows) below the first.

Linear System	Augmented Matrix	Operations
$\begin{aligned} x + 4y + 3z &= 1 \\ -3y - 2z &= 2 \\ -7y - 5z &= 4 \end{aligned}$	$\left[\begin{array}{ccc\|c} 1 & 4 & 3 & 1 \\ 0 & -3 & -2 & 2 \\ 0 & -7 & -5 & 4 \end{array}\right]$	*-2 times equation (row) 1 was added to equation (row) 2, and -1 times equation (row) 1 was added to equation (row) 3*

Step 3: Make the coefficient of the second variable in the second equation (second entry in second row) equal to 1.

Linear System	Augmented Matrix	Operations
$\begin{aligned} x + 4y + 3z &= 1 \\ y + \frac{2}{3}z &= -\frac{2}{3} \\ -7y - 5z &= 4 \end{aligned}$	$\left[\begin{array}{ccc\|c} 1 & 4 & 3 & 1 \\ 0 & 1 & 2/3 & -2/3 \\ 0 & -7 & -5 & 4 \end{array}\right]$	*Equation (row) 2 was multiplied by $-1/3$*

Step 4: Make the coefficient of the second variable (entry in second column) equal to 0 in all equations (rows) below the second.

$$\begin{aligned} x + 4y + 3z &= 1 \\ y + \frac{2}{3}z &= -\frac{2}{3} \\ -\frac{1}{3}z &= -\frac{2}{3} \end{aligned} \qquad \left[\begin{array}{ccc|c} 1 & 4 & 3 & 1 \\ 0 & 1 & 2/3 & -2/3 \\ 0 & 0 & -1/3 & -2/3 \end{array}\right]$$

7 times equation (row) 2 was added to equation (row) 3

Step 5: Make the coefficient of the third variable in the third equation (third entry in third row) equal to 1.

$$\begin{aligned} x + 4y + 3z &= 1 \\ y + \frac{2}{3}z &= -\frac{2}{3} \\ z &= 2 \end{aligned} \qquad \left[\begin{array}{ccc|c} 1 & 4 & 3 & 1 \\ 0 & 1 & 2/3 & -2/3 \\ 0 & 0 & 1 & 2 \end{array}\right]$$

Equation (row) 3 was multiplied by −3

Step 6: Use back-substitution of $z = 2$ to obtain $y = -2$ from equation (row) 2 and then use $z = 2$ and $y = -2$ to obtain $x = 3$ from equation (row) 1. ■

Gaussian Elimination Method. The matrix in step 5 above corresponding to the triangular form of the linear system is said to be in **row-echelon form,** and the general elimination method, when applied to the augmented matrix of the system, is called *Gaussian elimination.*[1] As mentioned previously, Gaussian elimination is our most general and efficient way to solve linear systems. We now apply this method to the augmented matrix of several linear systems to illustrate that it works equally well if the system has one solution, none, or an infinite number of solutions and regardless of the number of variables or equations.

[1] *Carl Friedrich Gauss (1777–1855) is considered to be one of the greatest mathematicians of all time. See his fundamental theorem in the chapter on roots of polynomials.*

Example 2 Solve the following system by Gaussian elimination.

$$\begin{aligned} x + y + z &= 4 \\ 2x + y + z &= 3 \\ 3x + 2y - z &= 1 \end{aligned}$$

Solution: The augmented matrix of the system is

$$\left[\begin{array}{ccc|c} 1 & 1 & 1 & 4 \\ 2 & 1 & 1 & 3 \\ 3 & 2 & -1 & 1 \end{array}\right],$$

which can be reduced to row-echelon form as follows.

We begin by noting that the entry in the first row and first column is 1. Therefore, we proceed to introduce 0's in the rest of column 1. To do this, row 1 is multiplied by -2 and the result is added to row 2. Similarly, row 1 is multiplied by -3 and added to row 3, with the following result:

$$\left[\begin{array}{rrr|r} 1 & 1 & 1 & 4 \\ 0 & -1 & -1 & -5 \\ 0 & -1 & -4 & -11 \end{array}\right].$$

Next, the entry in row 2, column 2 is made equal to 1 by multiplying the second row by -1:

$$\left[\begin{array}{rrr|r} 1 & 1 & 1 & 4 \\ 0 & 1 & 1 & 5 \\ 0 & -1 & -4 & -11 \end{array}\right].$$

To introduce 0 as the entry in row 3 and column 2, we now add row 2 to row 3:

$$\left[\begin{array}{rrr|r} 1 & 1 & 1 & 4 \\ 0 & 1 & 1 & 5 \\ 0 & 0 & -3 & -6 \end{array}\right].$$

Finally, row 3 is multiplied by $-1/3$ to make its first nonzero entry equal to 1:

$$\left[\begin{array}{rrr|r} 1 & 1 & 1 & 4 \\ 0 & 1 & 1 & 5 \\ 0 & 0 & 1 & 2 \end{array}\right].$$

The last matrix represents the system

$$\begin{aligned} x + y + z &= 4 \\ y + z &= 5 \\ z &= 2. \end{aligned}$$

Back-substitution yields $y = 3$ and then $x = -1$. Therefore, the solution is $x = -1$, $y = 3$, and $z = 2$. ■

In the following examples, we will simplify explanations and use R_1, R_2, . . . to denote row 1, row 2, and so on.

Example 3 Use Gaussian elimination to solve

$$\begin{aligned} x - 2y - 3z &= 2 \\ x - 4y - 13z &= 14 \\ -3x + 5y + 4z &= 2. \end{aligned}$$

Solution: We first write the augmented matrix and then perform the indicated elementary row operations. (Read the + sign in the steps as "added to.")

$$\left[\begin{array}{rrr|r} 1 & -2 & -3 & 2 \\ 1 & -4 & -13 & 14 \\ -3 & 5 & 4 & 2 \end{array}\right]$$

1. $(-1)R_1 + R_2$,
   $3R_1 + R_3$:
$$\left[\begin{array}{rrr|r} 1 & -2 & -3 & 2 \\ 0 & -2 & -10 & 12 \\ 0 & -1 & -5 & 8 \end{array}\right].$$

2. $(-1/2)R_2$:
$$\left[\begin{array}{rrr|r} 1 & -2 & -3 & 2 \\ 0 & 1 & 5 & -6 \\ 0 & -1 & -5 & 8 \end{array}\right].$$

3. $R_2 + R_3$:
$$\left[\begin{array}{rrr|r} 1 & -2 & -3 & 2 \\ 0 & 1 & 5 & -6 \\ 0 & 0 & 0 & 2 \end{array}\right].$$

The last row of the reduced matrix represents the equation

$$0x + 0y + 0z = 2,$$

which has no solution. Therefore, the given system has no solution. ■

When a system of equations has no solution, as in Example 3, we say that the system is **inconsistent.**

Example 4 By reducing the augmented matrix to row-echelon form, solve

$$\begin{aligned} x + y + 2z + 3t &= 13 \\ 3x + y + z - t &= 1 \\ x - 2y + z + t &= 8. \end{aligned}$$

Solution: The augmented matrix is

$$\left[\begin{array}{rrrr|r} 1 & 1 & 2 & 3 & 13 \\ 3 & 1 & 1 & -1 & 1 \\ 1 & -2 & 1 & 1 & 8 \end{array}\right],$$

which is reduced as follows. (Read the + sign in the steps as "added to.")

1. $(-3)R_1 + R_2$,
   $-R_1 + R_3$:
$$\left[\begin{array}{rrrr|r} 1 & 1 & 2 & 3 & 13 \\ 0 & -2 & -5 & -10 & -38 \\ 0 & -3 & -1 & -2 & -5 \end{array}\right].$$

2. $(-1/2)R_2$: $\left[\begin{array}{cccc|c} 1 & 1 & 2 & 3 & 13 \\ 0 & 1 & 5/2 & 5 & 19 \\ 0 & -3 & -1 & -2 & -5 \end{array}\right].$

3. $3R_2 + R_3$: $\left[\begin{array}{cccc|c} 1 & 1 & 2 & 3 & 13 \\ 0 & 1 & 5/2 & 5 & 19 \\ 0 & 0 & 13/2 & 13 & 52 \end{array}\right].$

4. $(2/13)R_3$: $\left[\begin{array}{cccc|c} 1 & 1 & 2 & 3 & 13 \\ 0 & 1 & 5/2 & 5 & 19 \\ 0 & 0 & 1 & 2 & 8 \end{array}\right].$

The last matrix is in row-echelon form and corresponds to the system of equations

$$\begin{aligned} x + y + 2z + 3t &= 13 \\ y + \frac{5}{2}z + 5t &= 19 \\ z + 2t &= 8. \end{aligned}$$

By back-substitution, we first get

$$z = 8 - 2t,$$

where t can be *any real number*. If this expression for z is substituted into the second equation, we obtain

$$y = -1,$$

and if $y = -1$ and the expression for z are substituted in the first equation, we get

$$x = -2 + t.$$

Hence, all solutions can be written as $x = -2 + t$, $y = -1$, $z = 8 - 2t$, and t, where t is any real number. In other words, the system has *infinitely many* solutions. ■

When reducing the augmented matrix to row-echelon form, it is possible to obtain an entire row of 0's. If this occurs, we simply move the row of 0's to the bottom row and continue with the reduction process.

Example 5 Solve

$$\begin{aligned} y + z &= 2 \\ x + y + z &= 5 \\ x + 2y + 2z &= 7 \\ 2x + y - z &= 4. \end{aligned}$$

Solution: The augmented matrix is

$$\left[\begin{array}{ccc|c} 0 & 1 & 1 & 2 \\ 1 & 1 & 1 & 5 \\ 1 & 2 & 2 & 7 \\ 2 & 1 & -1 & 4 \end{array}\right],$$

which we reduce to row-echelon form in the following steps:

1. $R_1 \leftrightarrow R_2$: $\left[\begin{array}{ccc|c} 1 & 1 & 1 & 5 \\ 0 & 1 & 1 & 2 \\ 1 & 2 & 2 & 7 \\ 2 & 1 & -1 & 4 \end{array}\right].$ *Row 1 interchanged with Row 2*

2. $-R_1 + R_3$, $-2R_1 + R_4$: $\left[\begin{array}{ccc|c} 1 & 1 & 1 & 5 \\ 0 & 1 & 1 & 2 \\ 0 & 1 & 1 & 2 \\ 0 & -1 & -3 & -6 \end{array}\right].$

3. $-R_2 + R_3$, $R_2 + R_4$: $\left[\begin{array}{ccc|c} 1 & 1 & 1 & 5 \\ 0 & 1 & 1 & 2 \\ 0 & 0 & 0 & 0 \\ 0 & 0 & -2 & -4 \end{array}\right].$

4. $R_3 \leftrightarrow R_4$: $\left[\begin{array}{ccc|c} 1 & 1 & 1 & 5 \\ 0 & 1 & 1 & 2 \\ 0 & 0 & -2 & -4 \\ 0 & 0 & 0 & 0 \end{array}\right].$

5. $(-1/2)R_3$: $\left[\begin{array}{ccc|c} 1 & 1 & 1 & 5 \\ 0 & 1 & 1 & 2 \\ 0 & 0 & 1 & 2 \\ 0 & 0 & 0 & 0 \end{array}\right].$

The last matrix is in row-echelon form and represents the system

$$\begin{aligned} x + y + z &= 5 \\ y + z &= 2 \\ z &= 2, \end{aligned}$$

whose solution is $x = 3$, $y = 0$, and $z = 2$. ■

Gauss-Jordan Elimination Method. As an alternative to the back-substitution used in the preceding examples, we can continue row operations on the row-echelon form of the augmented matrix in order to introduce 0's *above* the leading 1's in each row. For instance, in Example 2 above, the row-echelon form of the augmented matrix was determined to be

$$\left[\begin{array}{ccc|c} 1 & 1 & 1 & 4 \\ 0 & 1 & 1 & 5 \\ 0 & 0 & 1 & 2 \end{array}\right].$$

We now work from the last row upward to introduce 0's *above* the leading 1's as follows:

1. $-R_3 + R_2$, $-R_3 + R_1$: $\left[\begin{array}{ccc|c} 1 & 1 & 0 & 2 \\ 0 & 1 & 0 & 3 \\ 0 & 0 & 1 & 2 \end{array}\right]$.

2. $-R_2 + R_1$: $\left[\begin{array}{ccc|c} 1 & 0 & 0 & -1 \\ 0 & 1 & 0 & 3 \\ 0 & 0 & 1 & 2 \end{array}\right]$

This matrix represents the solution

$$x = -1$$
$$y = 3$$
$$z = 2.$$

The matrix in step 2 above is said to be in **reduced row-echelon form,** and the procedure used to obtain it is called *Gauss-Jordan*[2] *elimination.* We now summarize the previous two matrix methods.

[2]*Wilhelm Jordan (1842–1899) used the method to solve systems of equations that occur in the subject of geodesy.*

To solve a linear system by the method of **Gaussian elimination,** we first perform the following steps on the augmented matrix by means of elementary row operations.

1. Make the element in the first row and first column equal to 1, if possible. This may require interchanging rows.
2. Make all elements below the leading 1 of step 1 equal to 0.
3. Place any row consisting entirely of 0's under all nonzero rows.
4. Repeat steps 1, 2, and 3 on the smaller matrix remaining after the first row and column are deleted.
5. Repeat step 4 until the process automatically stops. The augmented matrix is now in *row-echelon form.*

We then solve for the variable represented by the leading 1 in the last nonzero row and back-substitute to obtain the complete solution.

To solve a linear system by the method of **Gauss-Jordan elimination,** we perform steps 1 through 5 above as well as the following step.

6. Make all elements above the leading 1's in each nonzero row of the row-echelon form equal to 0. The matrix is now in *reduced row-echelon form* and represents the simplest form of the original system of equations.

Question *Why is Gaussian elimination more efficient than Cramer's rule for solving n linear equations in n unknowns?*

Answer: First we note that Cramer's rule can be used only if the determinant of the coefficients is not zero. Now to solve n equations in n unknowns by Cramer's rule, we must evaluate the $n \times n$ coefficient determinant and one $n \times n$ determinant for each of the n unknowns. To evaluate an nth order determinant by means of cofactors (without using elementary row operations to introduce 0's) takes $n!$ multiplications. Hence, if we were solving a system of 10 equations in 10 unknowns and could perform one multiplication per second, it would take $10!/3600 = 1008$ hours just to compute the coefficient determinant. On the other hand, to reduce an $n \times (n + 1)$ augmented coefficient matrix to row-echelon form, as required by Gaussian elimination, takes approximately n^3 multiplications. Hence, the above system of 10 equations in 10 unknowns would take about $10^3/3600 \approx .3$ hours or 18 minutes to solve by Gaussian elimination. ■

Exercises 9.4

Fill in the blanks to make each statement true.

1. A linear system of three equations in three unknowns may be solved by using elimination to reduce the system to ________ equations in ________ unknowns.
2. When solving a linear system by the general elimination method, the aim is to reduce the system to ________ form.
3. In order to represent a linear system by its augmented matrix, we write the system with the ________ in the same order in each equation and with the ________ on the right side of each equation.
4. If an augmented matrix represents a system of three equations in three variables, then its first three columns constitute its ________ matrix, and its last column contains the ________ terms.
5. In the Gaussian elimination method, the three types of row operations that may be performed on the augmented matrix are ________, ________ and ________.

Write true *or* false *for each statement.*

6. The general elimination method can be used for solving linear systems in any number of variables.
7. The determinant method is restricted to solving linear systems with a unique solution, but the elimination method is not.
8. In the Gaussian elimination method, a row of the augmented matrix may be multiplied by any constant.
9. Interchanging two rows of the augmented matrix of a linear system corresponds to interchanging two equations in the system.
10. If the last column of an augmented matrix contains all 0's, then that column is not changed in the Gaussian elimination method.

Elimination Methods

In Exercises 11–14, solve the system by reducing it to two equations in two unknowns.

11. $\begin{aligned} x + y - z &= 4 \\ 2x + 2y - 3z &= 3 \\ 3x + y - z &= 2 \end{aligned}$

12. $\begin{aligned} 2x - y + 2z &= 9 \\ x + y + z &= 13 \\ x - 5y + 2z &= 12 \end{aligned}$

13. $\begin{aligned} x + y - 3z &= 1 \\ 2x - y - 3z &= -1 \\ x + 3y - 5z &= 3 \end{aligned}$

14. $\begin{aligned} x + y + z &= 3 \\ 2x + y - z &= 0 \\ 3x + 2y + z &= 7 \end{aligned}$

15–18. Solve Exercises 11–14 by the general elimination procedure.

Use the general elimination procedure to solve the systems in Exercises 19–20.

19. $x + y - 3z = 1$
$x + 3y - 5z = 2$

20. $x + y + 2z + 2t = 7$
$x - y - z - t = 0$
$2x + 3y + z + t = 5$
$2x + y - z - 2t = -5$

Augmented Matrix of a Linear System

In Exercises 21–25, write a linear system represented by the given augmented matrix.

21. $\left[\begin{array}{cc|c} 2 & 1 & -3 \\ 4 & -5 & 6 \end{array}\right]$

22. $\left[\begin{array}{ccc|c} 1 & -3 & 2 & 5 \\ 6 & 1 & -2 & 7 \\ 2 & 5 & -1 & 3 \end{array}\right]$

23. $\left[\begin{array}{cccc|c} 1 & 2 & -3 & 4 & 5 \\ 2 & -1 & 0 & 7 & 6 \end{array}\right]$

24. $\left[\begin{array}{ccc|c} 1 & -2 & 0 & 3 \\ 2 & 0 & 1 & 5 \\ 3 & 1 & 1 & 1 \\ 4 & -5 & 2 & 3 \end{array}\right]$

25. $\left[\begin{array}{ccc|c} 1 & 0 & 0 & 2 \\ 0 & 1 & 0 & -3 \\ 0 & 0 & 1 & 4 \end{array}\right]$

In Exercises 26–30, write the augmented matrix that represents the given linear system.

26. $x - 2y = 5$
$2x + 3y = 1$

27. $x - y - 6 = 0$
$y + 2x - 3 = 0$

28. $x - y + z = 5$
$2x + y - 3z = 7$
$4x + y - 2z = 6$

29. $2x - y + 3z - 8 = 0$
$y + 7x - z = 4$
$z - x + 2y = 3$

30. $x - y + z = 8$
$x + 2y + 3t = 5$
$x - z + 2t = 6$
$x + y + 3z = 0$

Gaussian Elimination Method

In Exercises 31–35, the matrix given is the reduced augmented matrix obtained by Gaussian elimination. Write the solution of the corresponding system of equations.

31. $\left[\begin{array}{cc|c} 1 & 0 & 2 \\ 0 & 1 & 3 \end{array}\right]$

32. $\left[\begin{array}{cc|c} 1 & 1 & 2 \\ 0 & 0 & 0 \end{array}\right]$

33. $\left[\begin{array}{ccc|c} 1 & 0 & 0 & 3 \\ 0 & 1 & 0 & -2 \\ 0 & 0 & 1 & 1 \end{array}\right]$

34. $\left[\begin{array}{ccc|c} 1 & 0 & 1 & 3 \\ 0 & 1 & -2 & 1 \\ 0 & 0 & 1 & 2 \end{array}\right]$

35. $\left[\begin{array}{ccc|c} 1 & 0 & 1 & 3 \\ 0 & 1 & -2 & 1 \\ 0 & 0 & 0 & 0 \end{array}\right]$

In Exercises 36–40, solve by Gaussian elimination.

36. $x - 2y = 5$
$2x + 3y = -4$

37. $x - y + z = 6$
$2x + y + z = 3$
$x + y + z = 2$

38. $2x + 5y - 4z = -3$
$4x + y - z = 6$
$3x - 2y + z = 7$

39. $x - 2y - 3z = 2$
$x - 4y - 13z = 14$
$-3x + 5y + 4z = 2$

40. $x - 2y - 3z = 2$
$x - 4y - 13z = 14$
$-3x + 5y + 4z = 0$
$y + 5z = -6$

Gauss-Jordan Elimination Method

41–45. Solve Exercises 36–40 by Gauss-Jordan elimination.

Use the Gauss-Jordan elimination method to solve the systems in Exercises 46–50.

46. $x + y = 9$
$2x - \frac{1}{2}y = 8$

47. $x - 3y = 5$
$4x - 12y = 20$

48. $x + y + z = 0$
$2x - y + 4z = 1$
$x + 2y - z = -3$

49. $x + y + z = 0$
$2x - y + 4z = 1$
$x + 2y - z = -3$
$-x - y + 2z = 6$

50. $x + y + z + 2t = 0$
$3x + y + 3z + 5t = 1$
$x - y - z + 3t = -1$
$2x - 3y + z + t = 0$

9.5 Matrix Algebra

In matrix algebra, $(A + B)^2 \neq A^2 + 2AB + B^2$.

Matrices were used in the previous section to simplify the solution of linear systems. They have other uses as well, and their study has been intensified because operations with matrices can be programmed easily on a computer. With certain restrictions, matrices can be added, subtracted, and multiplied, and although we do not call it division for matrices, the equivalent operation of multiplication by an inverse is possible for one class of matrices. This section gives an introduction to the algebra of matrices.

Matrices: Size and Notation. A matrix has already been described as a **rectangular array** of real numbers. It may have any number of rows and any number of columns, and these two numbers specify its size. For example, in the following matrices,

$$\begin{bmatrix} 1 \\ 2 \\ 3 \end{bmatrix}, \quad \begin{bmatrix} 1 & -2 \\ 4 & 1 \\ 0 & 3 \end{bmatrix}, \quad \begin{bmatrix} 1 & 4 & 0 \\ -2 & 1 & 3 \end{bmatrix},$$

the first has 3 rows and 1 column (size 3×1), the second has 3 rows and 2 columns (3×2), and the third has 2 rows and 3 columns (2×3). A 1×1 matrix contains only one real number, and in this case we sometimes drop the bracket notation. It is customary to denote matrices by capital letters and to represent each real number in the array by a lowercase letter. In this context, a real number is called a **scalar** and the symbol a_{ij} is used to denote the entry in row i and column j of matrix A.

As stated earlier, a matrix with n rows and n columns is called **a square matrix of order n.** The entries a_{ii} $(i = 1, 2, \ldots, n)$ in an $n \times n$ square matrix A are said to form the **main diagonal** of A.

Two matrices are **equal** if and only if they are identical; that is, if they have the same size and all corresponding entries are the same. In symbols, for two $m \times n$ matrices A and B,

$A = B$ if and only if $a_{ij} = b_{ij}$ for $i = 1, 2, \ldots, m$ and $j = 1, 2, \ldots, n$.

For instance,
$$\begin{bmatrix} 1 & 2 \\ 3 & 4 \end{bmatrix} = \begin{bmatrix} 1 & 2 \\ 3 & 4 \end{bmatrix},$$

but
$$\begin{bmatrix} 1 & 2 \\ 3 & 4 \end{bmatrix} \neq \begin{bmatrix} 2 & 3 \\ 1 & 4 \end{bmatrix}, \quad \text{and} \quad \begin{bmatrix} 1 & 2 \\ 3 & 4 \end{bmatrix} \neq \begin{bmatrix} 1 & 2 & 0 \\ 3 & 4 & 0 \end{bmatrix}.$$

Matrix Addition and Multiplication by a Scalar. If two matrices have the same size, their sum is obtained by adding the corresponding entries.

Thus, if A and B are both of size $m \times n$, then

$$A + B = [a_{ij}] + [b_{ij}] = [a_{ij} + b_{ij}]$$
$$\text{for } i = 1, 2, \ldots, m \quad \text{and} \quad j = 1, 2, \ldots, n.$$

addition of matrices

Similarly,

$$A - B = [a_{ij}] - [b_{ij}] = [a_{ij} - b_{ij}]$$
$$\text{for } i = 1, 2, \ldots, m \quad \text{and} \quad j = 1, 2, \ldots, n.$$

subtraction of matrices

Caution If A and B are not the same size, then $A \pm B$ are not defined.

Example 1 Suppose an appliance dealer sells 500 radios, 450 TVs and 225 VCRs in 1986, and 575 radios, 380 TVs and 500 VCRs in 1987. Use matrices to represent the sales each year and the sales for the two years together.

Solution: If we let

$$A = \begin{bmatrix} 500 \\ 450 \\ 225 \end{bmatrix} \quad \text{and} \quad B = \begin{bmatrix} 575 \\ 380 \\ 500 \end{bmatrix}$$

represent the sales of the three appliances for 1986 and 1987, respectively, then the total sales of these items for the two years are represented by the matrix sum

$$A + B = \begin{bmatrix} 500 + 575 \\ 450 + 380 \\ 225 + 500 \end{bmatrix} = \begin{bmatrix} 1075 \\ 830 \\ 725 \end{bmatrix} \begin{matrix} \textit{Radios} \\ \textit{TVs} \\ \textit{VCRs} \end{matrix}$$

Thus, the dealer sold 1075 radios, 830 TVs, and 725 VCRs in the two years. ■

To multiply a matrix by a real number (scalar) c means to multiply every entry by c. That is,

$$\text{If } A = [a_{ij}], \text{ then } cA = [ca_{ij}].$$

multiplication of a matrix by a scalar

Example 2 If in 1987 the appliance dealer in Example 1 had doubled the 1986 sales of the three items, what matrix would represent that result?

Solution: The sales for 1986 were given by

$$A = \begin{bmatrix} 500 \\ 450 \\ 225 \end{bmatrix},$$

so doubling those sales in 1987 would be indicated by the matrix

$$2A = \begin{bmatrix} 2 \cdot 500 \\ 2 \cdot 450 \\ 2 \cdot 225 \end{bmatrix} = \begin{bmatrix} 1000 \\ 900 \\ 450 \end{bmatrix}. \quad ■$$

Since matrix addition and multiplication by a scalar involve ordinary addition and multiplication of real numbers, it is not surprising that these operations on matrices have many of the properties discussed earlier for the real number system. We list these properties here and leave their verification to the exercises. It is understood that matrices to be added must be of the same size.

Matrix Addition

$A + B = B + A$	**commutative propety of addition**
$(A + B) + C = A + (B + C)$	**associative property of addition**
If O is a matrix whose entries are all zeros, then $A + O = O + A = A$.	**identity matrix for addition**
If $-A$ denotes $(-1)A$, then $A + (-A) = (-A) + A = O$.	**additive inverse of a matrix**

Multiplication by a Scalar (c and k are scalars)

$c(kA) = (ck)A$	**associative property**
$(c + k)A = cA + kA$	**distributive property**
$c(A + B) = cA + cB$	**distributive property**

Matrix Multiplication. We might expect that multiplication of matrices would involve multiplying corresponding entries, but this definition turns out not to be useful. Instead, we define multiplication of matrices only if *the number of columns of the first matrix equals the number of rows of the second matrix,* and we give the definition in two steps. Examples will illustrate the usefulness of our definition. (See also Exercises 37 and 44.)

We first consider the case in which A has size $1 \times n$ and B has size $n \times 1$. Here, A is called a **row matrix** and B is called a **column matrix,** and we define AB as follows.

$$[a_{11}\ a_{12}\ \ldots\ a_{1n}]\begin{bmatrix} b_{11} \\ b_{21} \\ \vdots \\ b_{n1} \end{bmatrix} = [a_{11}b_{11} + a_{12}b_{21} + \cdots + a_{1n}b_{n1}].$$

That is, if A is a $1 \times n$ matrix and B is an $n \times 1$ matrix, then AB is a 1×1 matrix.[1]

Example 3 If in 1986 the appliance dealer in Example 1 made profits of \$5, \$30, and \$50 on each radio, TV, and VCR, respectively, indicate the total profit as a matrix multiplication.

[1]The n-tuples $(a_{11}, a_{12}, \ldots, a_{1n})$ and $(b_{11}, b_{21}, \ldots, b_{n1})$ are n-dimensional vectors, and the real number $a_{11}b_{11} + a_{12}b_{21} + \cdots + a_{1n}b_{n1}$ is called their dot product.

Solution: We let the profits on the three items be represented by the matrix

$$P = [5 \quad 30 \quad 50].$$

To compute the total profit, we multiply the profit on each item by the number of that item sold and then add these three products. Hence, the total profit is equal to the matrix product

$$PA = [5 \quad 30 \quad 50]\begin{bmatrix}500\\450\\225\end{bmatrix} = [5\cdot 500 + 30\cdot 450 + 50\cdot 225]$$
$$= [27{,}250]. \quad ■$$

We are now ready to define the product AB of matrices A and B in which the number of columns of A equals the number of rows of B. The entry in row i and column j of AB is the product of row i of A and column j of B, as defined above. That is, if A is an $m \times n$ matrix and B is an $n \times p$ matrix, then, by definition,

$$\begin{bmatrix} a_{11} & a_{12} & \cdots & a_{1n} \\ \vdots & \vdots & & \vdots \\ a_{i1} & a_{i2} & \cdots & a_{in} \\ \vdots & \vdots & & \vdots \\ a_{m1} & a_{m2} & \cdots & a_{mn} \end{bmatrix} \begin{bmatrix} b_{11} & \cdots & b_{1j} & \cdots & b_{1p} \\ b_{21} & \cdots & b_{2j} & \cdots & b_{2p} \\ \vdots & & \vdots & & \vdots \\ b_{n1} & \cdots & b_{nj} & \cdots & b_{np} \end{bmatrix} = \begin{bmatrix} c_{11} & \cdots & c_{1j} & \cdots & c_{1p} \\ \vdots & & \vdots & & \vdots \\ c_{i1} & \cdots & c_{ij} & \cdots & c_{ip} \\ \vdots & & \vdots & & \vdots \\ c_{m1} & \cdots & c_{mj} & \cdots & c_{mp} \end{bmatrix},$$

where

$$c_{ij} = a_{i1}b_{1j} + a_{i2}b_{2j} + \cdots + a_{1n}b_{nj}$$
$$\text{for } i = 1, 2, \ldots, m \quad \text{and} \quad j = 1, 2, \ldots, p.$$

matrix multiplication

Hence, the product AB of an $m \times n$ matrix A and an $n \times p$ matrix B is an $m \times p$ matrix C. *If the number of columns of A is not equal to the number of rows of B, then AB is not defined.*

Example 4 Find AB for $A = \begin{bmatrix}1 & 2 & 3\\4 & 5 & 6\end{bmatrix}$ and $B = \begin{bmatrix}2 & -1 & 3\\1 & 2 & -1\\4 & 0 & -4\end{bmatrix}$.

Solution: Since A is 2×3 and B is 3×3, we know AB is 2×3. We start by multiplying row 1 by column 1:

$$\begin{bmatrix}1 & 2 & 3\\ \_ & \_ & \_\end{bmatrix}\begin{bmatrix}2 & \_ & \_\\1 & \_ & \_\\4 & \_ & \_\end{bmatrix}$$

$$= \begin{bmatrix}1\cdot 2 + 2\cdot 1 + 3\cdot 4 & \_ & \_\\ \_ & \_ & \_\end{bmatrix} = \begin{bmatrix}16 & \_ & \_\\ \_ & \_ & \_\end{bmatrix}.$$

Similarly, we obtain the remaining entries in the first row of AB by multiplying row 1 by column 2 and then row 1 by column 3, giving

$$\begin{bmatrix} 1 & 2 & 3 \\ \text{—} & \text{—} & \text{—} \end{bmatrix}\begin{bmatrix} \text{—} & -1 & 3 \\ \text{—} & 2 & -1 \\ \text{—} & 0 & -4 \end{bmatrix}$$

$$= \begin{bmatrix} 16 & 1(-1) + 2\cdot 2 + 3\cdot 0 & 1\cdot 3 + 2(-1) + 3(-4) \\ \text{—} & \text{—} & \text{—} \end{bmatrix}$$

$$= \begin{bmatrix} 16 & 3 & -11 \\ \text{—} & \text{—} & \text{—} \end{bmatrix}.$$

Finally, multiplying row 2 of A by each of the columns of B, we get

$$\begin{bmatrix} \text{—} & \text{—} & \text{—} \\ 4 & 5 & 6 \end{bmatrix}\begin{bmatrix} 2 & -1 & 3 \\ 1 & 2 & -1 \\ 4 & 0 & -4 \end{bmatrix}$$

$$= \begin{bmatrix} \text{—} & \text{—} & \text{—} \\ 4\cdot 2 + 5\cdot 1 + 6\cdot 4 & 4(-1) + 5\cdot 2 + 6\cdot 0 & 4\cdot 3 + 5(-1) + 6(-4) \end{bmatrix}$$

$$= \begin{bmatrix} \text{—} & \text{—} & \text{—} \\ 37 & 6 & -17 \end{bmatrix}.$$

Therefore, $AB = \begin{bmatrix} 16 & 3 & -11 \\ 37 & 6 & -17 \end{bmatrix}$. ■

Example 5 Suppose the appliance dealer in Example 1 purchased from each of two suppliers 550 radios, 500 TVs, and 300 VCRs in 1986 and 600 radios, 450 TVs, and 500 VCRs in 1987. If the prices were the same in both years and supplier 1 charges \$25 per radio, \$200 per TV, and \$395 per VCR, while supplier 2 charges \$20 per radio, \$185 per TV, and \$425 per VCR, use matrix multiplication to display total costs for the two years from the two suppliers.

Solution: The matrix product

Numbers of items (Radio, TV, VCR) · *Cost per item from supplier* (1, 2) = *Total cost from supplier* (1, 2)

$$\begin{matrix} (1986) \\ (1987) \end{matrix}\overset{\textbf{Radio}\quad\textbf{TV}\quad\textbf{VCR}}{\begin{bmatrix} 550 & 500 & 300 \\ 600 & 450 & 500 \end{bmatrix}}\overset{\textbf{1}\qquad\textbf{2}}{\begin{bmatrix} 25 & 20 \\ 200 & 185 \\ 395 & 425 \end{bmatrix}} = \overset{\textbf{1}\qquad\qquad\textbf{2}}{\begin{bmatrix} 232{,}250 & 231{,}000 \\ 302{,}500 & 307{,}750 \end{bmatrix}}\begin{matrix} (1986) \\ (1987) \end{matrix}$$

shows that in 1986 the cost from supplier 1 is \$232,250 and from supplier 2 is \$231,000, while in 1987, the cost from supplier 1 is \$302,500 and from 2 is \$307,750. ■

Example 6 Show how a linear system of m equations in n unknowns can be represented in terms of matrices.

Solution: By the definitions of matrix multiplication and equality of matrices, the system

$$\begin{array}{l} a_{11}x_1 + a_{12}x_2 + \ldots + a_{1n}x_n = b_1 \\ a_{21}x_1 + a_{22}x_2 + \ldots + a_{2n}x_n = b_2 \\ \vdots \\ a_{m1}x_1 + a_{m2}x_2 + \ldots + a_{mn}x_n = b_m \end{array}$$

can be written

$$\begin{bmatrix} a_{11} & a_{12} & \ldots & a_{1n} \\ a_{21} & a_{22} & \ldots & a_{2n} \\ \vdots & \vdots & & \vdots \\ a_{m1} & a_{m2} & \ldots & a_{mn} \end{bmatrix} \begin{bmatrix} x_1 \\ x_2 \\ \vdots \\ x_n \end{bmatrix} = \begin{bmatrix} b_1 \\ b_2 \\ \vdots \\ b_m \end{bmatrix}$$

or more briefly $AX = B$, where A is the coefficient matrix and X and B are both column matrices. ■

In the algebra of matrices, we have some additional properties similar to those for real numbers, namely

$(AB)C = A(BC)$	**associative property of multiplication**
$A(B + C) = AB + AC$	**left distributive property**
$(A + B)C = AC + BC.$	**right distributive property**

Caution One major difference between real numbers and matrices is that *matrix multiplication (even when both AB and BA are defined) is not a commutative operation*. For example, if $A = \begin{bmatrix} 1 & -2 \\ 4 & 3 \end{bmatrix}$ and $B = \begin{bmatrix} 2 & 5 \\ -1 & 3 \end{bmatrix}$, then $AB = \begin{bmatrix} 4 & -1 \\ 5 & 29 \end{bmatrix}$ but $BA = \begin{bmatrix} 22 & 11 \\ 11 & 11 \end{bmatrix}$. Also, for matrices *we may have* $AB = 0$ *with* $A \neq 0$ *and* $B \neq 0$, and $AB = AC$ *with* $A \neq 0$ *and* $B \neq C$ (see Exercise 36).

Inverse of a Square Matrix. For square matrices of order n, there exists an **identity matrix** I for multiplication. I has 1's on its main diagonal and 0's elsewhere. For example, if $n = 3$, then

$$I = \begin{bmatrix} 1 & 0 & 0 \\ 0 & 1 & 0 \\ 0 & 0 & 1 \end{bmatrix}.$$

It is easy to verify by matrix multiplication that

$$\begin{bmatrix} a_{11} & a_{12} & a_{13} \\ a_{21} & a_{22} & a_{23} \\ a_{31} & a_{32} & a_{33} \end{bmatrix} \begin{bmatrix} 1 & 0 & 0 \\ 0 & 1 & 0 \\ 0 & 0 & 1 \end{bmatrix} = \begin{bmatrix} a_{11} & a_{12} & a_{13} \\ a_{21} & a_{22} & a_{23} \\ a_{31} & a_{32} & a_{33} \end{bmatrix};$$

that is, $AI = A$, and similarly, $IA = A$.

An $n \times n$ matrix B is called the **multiplicative inverse** of an $n \times n$ matrix A if

$$\boldsymbol{AB = BA = I}.$$ **inverse of a square matrix**

If such an inverse matrix B exists for A, then B is denoted by A^{-1}, and A is called an **invertible matrix.** It can be shown that if there is a matrix B satisfying $AB = I$, then B also satisfies $BA = I$, and B is unique. Hence, to determine whether a square matrix A is invertible, we determine whether the single equation $AB = I$ has a solution for B. If a solution exists, then $B = A^{-1}$.

Example 7 Find A^{-1}, if possible, for

$$A = \begin{bmatrix} 2 & 3 \\ 4 & 5 \end{bmatrix}.$$

Solution: The matrix equation $AB = I$, or

$$\begin{bmatrix} 2 & 3 \\ 4 & 5 \end{bmatrix} \begin{bmatrix} x & u \\ y & v \end{bmatrix} = \begin{bmatrix} 1 & 0 \\ 0 & 1 \end{bmatrix}$$

requires $\begin{cases} 2x + 3y = 1 \\ 4x + 5y = 0 \end{cases}$ and $\begin{cases} 2u + 3v = 0 \\ 4u + 5v = 1 \end{cases}$, for which the solutions are

$$\begin{matrix} x = -5/2 \\ y = 2 \end{matrix} \quad \text{and} \quad \begin{matrix} u = 3/2 \\ v = -1. \end{matrix}$$

Hence,

$$A^{-1} = \begin{bmatrix} -5/2 & 3/2 \\ 2 & -1 \end{bmatrix}.$$

This result is verified by showing that

$$AA^{-1} = \begin{bmatrix} 2 & 3 \\ 4 & 5 \end{bmatrix} \begin{bmatrix} -5/2 & 3/2 \\ 2 & -1 \end{bmatrix} = \begin{bmatrix} 1 & 0 \\ 0 & 1 \end{bmatrix},$$

and

$$A^{-1}A = \begin{bmatrix} -5/2 & 3/2 \\ 2 & -1 \end{bmatrix} \begin{bmatrix} 2 & 3 \\ 4 & 5 \end{bmatrix} = \begin{bmatrix} 1 & 0 \\ 0 & 1 \end{bmatrix}.$$ ■

Example 8 Find A^{-1}, if possible, for $A = \begin{bmatrix} 1 & 0 \\ 0 & 0 \end{bmatrix}$.

Solution: Consider the matrix equation $AB = I$, or

$$\begin{bmatrix} 1 & 0 \\ 0 & 0 \end{bmatrix}\begin{bmatrix} x & u \\ y & v \end{bmatrix} = \begin{bmatrix} 1 & 0 \\ 0 & 1 \end{bmatrix}$$

$$\begin{bmatrix} x & u \\ 0 & 0 \end{bmatrix} = \begin{bmatrix} 1 & 0 \\ 0 & 1 \end{bmatrix},$$

which implies

$$\begin{matrix} x = 1 \\ 0 = 0 \end{matrix} \quad \text{and} \quad \begin{matrix} u = 0 \\ 0 = 1 \end{matrix}.$$

Since $0 \neq 1$, we have shown that there is *no* solution for the matrix B. Hence, A is *not* invertible, which means A^{-1} does not exist. ■

Although solving the equation $AB = I$ for B is easy enough when A is of order 2, as in Example 7, for matrices of order 3, 4, 5, . . . , or n, the above procedure involves the solution of 9, 16, 25, . . . , or n^2 equations, respectively. Hence, we want a more efficient method for finding the inverse of a matrix. We now describe such a method for finding A^{-1}, if it exists, when A is of order 2, and this method can be generalized to matrices of higher order. Given matrix $A = \begin{bmatrix} a & b \\ c & d \end{bmatrix}$, we seek a matrix $B = \begin{bmatrix} x & u \\ y & v \end{bmatrix}$ such that $AB = I$:

$$\begin{bmatrix} a & b \\ c & d \end{bmatrix}\begin{bmatrix} x & u \\ y & v \end{bmatrix} = \begin{bmatrix} 1 & 0 \\ 0 & 1 \end{bmatrix}.$$

Hence, we must try to solve the systems

$$\begin{cases} ax + by = 1 \\ cx + dy = 0 \end{cases} \quad \text{and} \quad \begin{cases} au + bv = 0 \\ cu + dv = 1. \end{cases} \tag{1}$$

The corresponding augmented matrices are

$$\left[\begin{array}{cc|c} a & b & 1 \\ c & d & 0 \end{array}\right] \quad \text{and} \quad \left[\begin{array}{cc|c} a & b & 0 \\ c & d & 1 \end{array}\right],$$

and if there is a unique solution by Gauss-Jordan elimination, these matrices must row-reduce to

$$\left[\begin{array}{cc|c} 1 & 0 & x \\ 0 & 1 & y \end{array}\right] = \left[\begin{array}{c|c} I & \begin{matrix} x \\ y \end{matrix} \end{array}\right] \quad \text{and} \quad \left[\begin{array}{cc|c} 1 & 0 & u \\ 0 & 1 & v \end{array}\right] = \left[\begin{array}{c|c} I & \begin{matrix} u \\ v \end{matrix} \end{array}\right],$$

respectively. In both cases, this means that the matrix A is reduced to I, and when this is possible, the other two columns $\begin{bmatrix} 1 \\ 0 \end{bmatrix}$ and $\begin{bmatrix} 0 \\ 1 \end{bmatrix}$, which are the columns of I, will be replaced by $\begin{bmatrix} x \\ y \end{bmatrix}$ and $\begin{bmatrix} u \\ v \end{bmatrix}$, which are the columns of A^{-1}. To organize this row-reduction method in compact form, we represent both systems (1) by one matrix, namely,

$$\left[\begin{array}{cc|cc} a & b & 1 & 0 \\ c & d & 0 & 1 \end{array}\right] = [A \mid I]. \tag{2}$$

If it is possible to row-reduce this matrix in such a way that A is replaced by I, then at the same time I will be replaced by

$$\begin{bmatrix} x & u \\ y & v \end{bmatrix} = A^{-1}.$$

That is, matrix (2) will be replaced by

$$\left[\begin{array}{cc|cc} 1 & 0 & x & u \\ 0 & 1 & y & v \end{array}\right] = [I \mid A^{-1}].$$

If we find that A cannot be row-reduced to I, then A^{-1} does not exist, and A is not invertible.

Example 9 Repeat Example 7 by the row-reduction method.

Solution: Since $A = \begin{bmatrix} 2 & 3 \\ 4 & 5 \end{bmatrix}$, we row-reduce the matrix $\left[\begin{array}{cc|cc} 2 & 3 & 1 & 0 \\ 4 & 5 & 0 & 1 \end{array}\right]$ as follows.

1. $\frac{1}{2}R_1$: $\left[\begin{array}{cc|cc} 1 & 3/2 & 1/2 & 0 \\ 4 & 5 & 0 & 1 \end{array}\right]$.

2. $-4R_1 + R_2$: $\left[\begin{array}{cc|cc} 1 & 3/2 & 1/2 & 0 \\ 0 & -1 & -2 & 1 \end{array}\right]$.

3. $-1R_2$: $\left[\begin{array}{cc|cc} 1 & 3/2 & 1/2 & 0 \\ 0 & 1 & 2 & -1 \end{array}\right]$.

4. $-\frac{3}{2}R_2 + R_1$: $\left[\begin{array}{cc|cc} 1 & 0 & -5/2 & 3/2 \\ 0 & 1 & 2 & -1 \end{array}\right]$.

We conclude that A is invertible and

$$A^{-1} = \begin{bmatrix} -5/2 & 3/2 \\ 2 & -1 \end{bmatrix},$$

as in Example 7. ■

The procedure explained above for finding A^{-1} is applicable if A is of order $n > 2$, as is illustrated in the following example.

Example 10 Find A^{-1}, if possible, for

$$A = \begin{bmatrix} 1 & 2 & 3 \\ 3 & 2 & 1 \\ 1 & 2 & 1 \end{bmatrix}.$$

Solution: Use elementary row operations on $[A \mid I]$ as follows.

1. $-3R_1 + R_2$, $-R_1 + R_3$:
$$\left[\begin{array}{ccc|ccc} 1 & 2 & 3 & 1 & 0 & 0 \\ 0 & -4 & -8 & -3 & 1 & 0 \\ 0 & 0 & -2 & -1 & 0 & 1 \end{array}\right].$$

2. $-\frac{1}{4}R_2$:
$$\left[\begin{array}{ccc|ccc} 1 & 2 & 3 & 1 & 0 & 0 \\ 0 & 1 & 2 & 3/4 & -1/4 & 0 \\ 0 & 0 & -2 & -1 & 0 & 1 \end{array}\right].$$

3. $-\frac{1}{2}R_3$:
$$\left[\begin{array}{ccc|ccc} 1 & 2 & 3 & 1 & 0 & 0 \\ 0 & 1 & 2 & 3/4 & -1/4 & 0 \\ 0 & 0 & 1 & 1/2 & 0 & -1/2 \end{array}\right].$$

4. $-2R_3 + R_2$, $-3R_3 + R_1$:
$$\left[\begin{array}{ccc|ccc} 1 & 2 & 0 & -1/2 & 0 & 3/2 \\ 0 & 1 & 0 & -1/4 & -1/4 & 1 \\ 0 & 0 & 1 & 1/2 & 0 & -1/2 \end{array}\right].$$

5. $-2R_2 + R_1$:
$$\left[\begin{array}{ccc|ccc} 1 & 0 & 0 & 0 & 1/2 & -1/2 \\ 0 & 1 & 0 & -1/4 & -1/4 & 1 \\ 0 & 0 & 1 & 1/2 & 0 & -1/2 \end{array}\right].$$

Therefore,

$$A^{-1} = \begin{bmatrix} 0 & 1/2 & -1/2 \\ -1/4 & -1/4 & 1 \\ 1/2 & 0 & -1/2 \end{bmatrix}.$$

It is a good idea to check that $AA^{-1} = I$. ■

Example 11 Show that the invertible matrix A in Example 10 has a determinant that is not equal to 0.

Solution:
$$\begin{vmatrix} 1 & 2 & 3 \\ 3 & 2 & 1 \\ 1 & 2 & 1 \end{vmatrix} = \begin{vmatrix} 1 & 2 & 3 \\ 0 & -4 & -8 \\ 0 & 0 & -2 \end{vmatrix} = 8 \qquad (why?)$$ ■

The result in Example 11 can be generalized to show that *a square matrix A is invertible if and only if its determinant is not* 0.

A system of n equations in n unknowns may be written in matrix form (Example 6) as

$$AX = B,$$

where A is $n \times n$, and X and B are $n \times 1$. Now if A is invertible, then

$$A^{-1}AX = A^{-1}B$$
$$IX = A^{-1}B$$
$$X = A^{-1}B.$$

Hence, *a linear system $AX = B$, in which A is invertible, has the matrix solution $X = A^{-1}B$.*

Example 12 Solve the system $\begin{cases} 2x + 3y = 1 \\ 4x + 5y = 6 \end{cases}$ by inverting the coefficient matrix.

Solution: The matrix representation of the given system is

$$\begin{bmatrix} 2 & 3 \\ 4 & 5 \end{bmatrix}\begin{bmatrix} x \\ y \end{bmatrix} = \begin{bmatrix} 1 \\ 6 \end{bmatrix}.$$

The coefficient matrix is $A = \begin{bmatrix} 2 & 3 \\ 4 & 5 \end{bmatrix}$, and by Example 9 we have

$$A^{-1} = \begin{bmatrix} -5/2 & 3/2 \\ 2 & -1 \end{bmatrix}.$$

Therefore, the solution of the system can be given by

$$X = A^{-1}B$$

or

$$\begin{bmatrix} x \\ y \end{bmatrix} = \begin{bmatrix} -5/2 & 3/2 \\ 2 & -1 \end{bmatrix}\begin{bmatrix} 1 \\ 6 \end{bmatrix} = \begin{bmatrix} 13/2 \\ -4 \end{bmatrix}.$$

That is, $x = 13/2$ and $y = -4$ is the solution. ■

Exercises 9.5

Fill in the blanks to make each statement true.

1. A matrix has size 2×5 if it has 2 __________ and 5 __________.
2. Matrices A and B can be added only if they have the __________.
3. If every entry in matrix A is multiplied by a constant c, then the resulting matrix is denoted by __________.
4. The product AB of matrices A and B is defined if and only if the number of __________ of A equals the number of __________ of B.
5. A square matrix A has a multiplicative inverse B if and only if $AB = BA =$ __________, and then A is said to be __________.

Write true *or* false *for each statement.*

6. If A and B are matrices of different size, then $A + B = 0$.
7. To add two matrices A and B of the same size, we add corresponding entries a_{ij} and b_{ij}.
8. We can multiply two matrices if they have the same size.
9. Multiplication of matrices is not an associative operation.
10. A square matrix has a multiplicative inverse if and only if its determinant is not 0.

Matrices: Size and Notation

11. State the size of each of the following matrices.

(a) $\begin{bmatrix} 1 \\ 3 \\ 2 \end{bmatrix}$ (b) $\begin{bmatrix} 1 & 3 & 2 \end{bmatrix}$

(c) $\begin{bmatrix} 1 & -2 & 3 \\ 6 & 5 & 7 \end{bmatrix}$ (d) $\begin{bmatrix} 2 & 1 & 3 \\ 4 & 6 & 8 \\ 1 & 5 & 7 \end{bmatrix}$

(e) $\begin{bmatrix} 2 & 1 & 3 & 9 & 7 \\ 4 & 6 & 8 & 1 & 1 \\ 1 & 5 & 7 & 2 & 3 \end{bmatrix}$

12. For the matrix $A = \begin{bmatrix} 2 & 1 & 3 \\ 4 & 7 & -1 \\ 3 & -5 & 6 \end{bmatrix}$, find each of the following.
 (a) a_{32} (b) a_{13}
 (c) a_{22}
 (d) the entries on the main diagonal

13. Find x and y for each equation.
 (a) $\begin{bmatrix} 2 & 1 \\ -3 & 4 \end{bmatrix} = \begin{bmatrix} x & 1 \\ -3 & y \end{bmatrix}$
 (b) $\begin{bmatrix} 1 & 0 & 0 \\ 0 & 2x & 0 \\ 0 & 0 & 2+y \end{bmatrix} = \begin{bmatrix} 1 & 0 & 0 \\ 0 & x-3 & 0 \\ 0 & 0 & 3y \end{bmatrix}$

Matrix Addition and Multiplication by a Scalar

In Exercises 14–19, add the matrices A and B or explain why the addition is not possible.

14. $A = \begin{bmatrix} 1 & 0 \\ 3 & -4 \end{bmatrix}$; $B = \begin{bmatrix} 2 & 1 \\ 2 & 1 \end{bmatrix}$

15. $A = \begin{bmatrix} 1 & 2 & 3 \\ 4 & 5 & 6 \\ 7 & 8 & 9 \end{bmatrix}$; $B = \begin{bmatrix} 1 & -1 & 1 \\ 2 & -5 & -2 \\ 3 & 2 & 1 \end{bmatrix}$

16. $A = \begin{bmatrix} 1 & 0 \\ 3 & -4 \end{bmatrix}$; $B = \begin{bmatrix} 2 & 1 & 3 \\ 2 & 1 & 4 \end{bmatrix}$

17. $A = \begin{bmatrix} 1 \\ 2 \\ 3 \end{bmatrix}$; $B = \begin{bmatrix} -1 \\ -2 \\ -3 \end{bmatrix}$

18. $A = \begin{bmatrix} 1 \\ 2 \\ 3 \end{bmatrix}$; $B = \begin{bmatrix} 3 & -1 & 0 \end{bmatrix}$

19. $A = \begin{bmatrix} 1 & 2 & 3 \\ 4 & 5 & 6 \end{bmatrix}$; $B = \begin{bmatrix} 1 & 4 \\ 2 & 5 \\ 3 & 6 \end{bmatrix}$

In Exercises 20–23, find the indicated matrix.

20. $2\begin{bmatrix} 1 & 2 \\ 4 & 3 \end{bmatrix}$

21. $-3\begin{bmatrix} 1 & 0 & -1 \\ 2 & -2 & 1 \\ 3 & 4 & 0 \end{bmatrix}$

22. $0\begin{bmatrix} 1 & 2 \\ 4 & -3 \end{bmatrix}$

23. $\frac{1}{2}\begin{bmatrix} 2 & -4 & 6 \\ 3 & 5 & 7 \end{bmatrix}$

24. For the matrices

$$A = \begin{bmatrix} 1 & 3 & -2 \\ 0 & -4 & 3 \end{bmatrix}, \quad B = \begin{bmatrix} 2 & -1 & 1 \\ 3 & 0 & 5 \end{bmatrix},$$
$$C = \begin{bmatrix} 0 & -3 & 1 \\ 2 & 3 & -1 \end{bmatrix},$$

verify each of the following equations.
 (a) $A + B = B + A$
 (b) $(A + B) + C = A + (B + C)$
 (c) $A + O = A$
 (d) $B + (-B) = O$

25. For the matrices given in Exercise 24, verify each of the following.
 (a) $2(3A) = 6A$
 (b) $(2 + 3)B = 2B + 3B$
 (c) $-2(A + C) = -2A - 2C$

26. For the matrices

$$A = \begin{bmatrix} a_{11} & a_{12} \\ a_{21} & a_{22} \end{bmatrix}, \quad B = \begin{bmatrix} b_{11} & b_{12} \\ b_{21} & b_{22} \end{bmatrix}, \quad C = \begin{bmatrix} c_{11} & c_{12} \\ c_{21} & c_{22} \end{bmatrix},$$

verify each of the following.
 (a) $A + B = B + A$
 (b) $(A + B) + C = A + (B + C)$
 (c) $c(kA) = (ck)A$
 (d) $k(B + C) = kB + kC$

Matrix Multiplication

In Exercises 27–34, find the matrix AB or explain why AB is not defined.

27. $A = \begin{bmatrix} 3 & 4 \end{bmatrix}$; $B = \begin{bmatrix} 1 \\ 2 \end{bmatrix}$

28. $A = \begin{bmatrix} 1 \\ 2 \end{bmatrix}$; $B = \begin{bmatrix} 3 & 4 \end{bmatrix}$

29. $A = \begin{bmatrix} 3 & -4 \\ 1 & 2 \end{bmatrix}$; $B = \begin{bmatrix} 5 & -1 \\ 7 & 3 \end{bmatrix}$

30. $A = \begin{bmatrix} -1 & 2 & 0 \\ 3 & 1 & 4 \end{bmatrix}$; $B = \begin{bmatrix} 2 & -1 & 5 \\ 3 & 0 & 4 \\ 1 & 6 & 2 \end{bmatrix}$

31. $A = \begin{bmatrix} 2 & -3 & 5 \\ 1 & -1 & 0 \\ 0 & 4 & 6 \end{bmatrix}$; $B = \begin{bmatrix} 1 & -2 & 4 \\ 1 & 3 & -5 \\ 1 & 2 & 4 \end{bmatrix}$

32. $A = \begin{bmatrix} 1 & 0 & -1 \\ 0 & 2 & 3 \\ -4 & 5 & 2 \end{bmatrix}$; $B = \begin{bmatrix} 1 \\ 2 \\ -3 \end{bmatrix}$

33. $A = \begin{bmatrix} 1 & 2 \\ 2 & 0 \\ 3 & 5 \\ -4 & 6 \end{bmatrix}$; $B = \begin{bmatrix} 2 & 1 & 0 \\ -1 & 1 & 3 \end{bmatrix}$

34. $A = \begin{bmatrix} 1 & 2 \\ 2 & 0 \\ 3 & 5 \\ 4 & 6 \end{bmatrix}$; $B = \begin{bmatrix} 1 & 2 \\ 1 & 3 \\ 0 & 4 \\ 1 & -1 \end{bmatrix}$

35. For the matrices

$$A = \begin{bmatrix} 1 & -1 \\ 2 & 0 \\ 3 & 1 \end{bmatrix}, \quad B = \begin{bmatrix} 3 & 1 \\ 2 & -4 \end{bmatrix}, \quad C = \begin{bmatrix} -5 & -2 \\ 1 & 6 \end{bmatrix},$$

verify each of the following.

(a) $(AB)C = A(BC)$

(b) $A(B + C) = AB + AC$

36. Let $A = \begin{bmatrix} 2 & -1 \\ 6 & -3 \end{bmatrix}$, $B = \begin{bmatrix} 1 & 4 \\ 2 & 8 \end{bmatrix}$, and $C = \begin{bmatrix} 2 & 3 \\ 4 & 6 \end{bmatrix}$.

(a) Show that $A \neq 0$ and $B \neq 0$ but $AB = 0$.

(b) Show that $AB = AC$, but $B \neq C$.

37. Let $f(x) = ax + b$ and $g(x) = cx + d$. Also, let $\begin{bmatrix} a & b \\ 0 & 1 \end{bmatrix}$ correspond to f and $\begin{bmatrix} c & d \\ 0 & 1 \end{bmatrix}$ correspond to g. Show that $\begin{bmatrix} a & b \\ 0 & 1 \end{bmatrix}\begin{bmatrix} c & d \\ 0 & 1 \end{bmatrix}$ corresponds to $f \circ g$.

Inverse of a Square Matrix

In Exercises 38–43, find A^{-1} or show that A is not invertible.

38. $A = \begin{bmatrix} 2/5 & -3/5 \\ -4 & 6 \end{bmatrix}$

39. $A = \begin{bmatrix} 1 & 2 \\ -1 & 3 \end{bmatrix}$

40. $A = \begin{bmatrix} 1 & 0 & -1 \\ 0 & 2 & 3 \\ 1 & 2 & 0 \end{bmatrix}$

41. $A = \begin{bmatrix} 1 & 0 & -1 \\ 2 & 2 & 1 \\ 0 & 2 & 3 \end{bmatrix}$

42. $A = \begin{bmatrix} 1 & -1 & 1 \\ 3 & 4 & 5 \end{bmatrix}$

43. $A = \begin{bmatrix} 2 & 1 & 0 \\ -1 & 1 & 3 \\ 3 & -1 & 5 \end{bmatrix}$

44. Let $f(x) = ax + b$ correspond to $\begin{bmatrix} a & b \\ 0 & 1 \end{bmatrix}$ as in Exercise 37. Assume that $a \neq 0$ and show that $f^{-1}(x)$ corresponds to $\begin{bmatrix} a & b \\ 0 & 1 \end{bmatrix}^{-1}$.

In Exercises 45–47, write the given system in the matrix form $AX = B$, and then solve, if possible, by finding $X = A^{-1}B$.

45. $\begin{aligned} 2x - y &= 4 \\ 8x + y &= 1 \end{aligned}$

46. $\begin{aligned} x + 2y &= 0 \\ y + 3z &= 1 \\ x - z &= 3 \end{aligned}$

47. $\begin{aligned} 3x + 6y + z &= 0 \\ 3x - 3y + 2z &= 2 \\ 6x + 9y + 2z &= 1 \end{aligned}$

In Exercises 48–49, show that all parts can be put in the matrix form $AX = B$ with the same A but different B's. Find A^{-1} and then solve for $X = A^{-1}B$.

48. Henry Rich invests \$10,000, some in a money market fund paying 3% per year and the rest in a mutual fund paying 5% per year. How much is each investment if his yearly income from both amounts to the following?

(a) \$400 (b) \$440 (c) \$460

49. A manufacturer produces two clocks, type I and type II. The cost of parts is \$25 for type I and \$40 for type II. The type I clock requires 5 hr of labor and type II requires 4 hr of labor. If 320 hr of labor are used, how many clocks of each type are produced if the amount the manufacturer spends on parts is as follows?

(a) \$1700 (b) \$2000 (c) \$2500

Chapter 9 Review Outline

9.1 Linear Systems: Reduction Method

To solve a system of three linear equations in three unknowns by the reduction method, one equation is solved for one variable in terms of the other two; then, by substitution, the remaining two equations involve only two variables and are solved by the methods in Chapter 2. A linear system of three equations in three unknowns represents three planes in space. It has a unique solution if the three planes intersect in one point, but such a system may have no solutions or infinitely many solutions.

9.2 Linear Systems and Determinants of Order 3

A determinant of order 3, denoted by

$$\Delta = \begin{vmatrix} a_1 & b_1 & c_1 \\ a_2 & b_2 & c_2 \\ a_3 & b_3 & c_3 \end{vmatrix},$$

is the number

$$a_1 \begin{vmatrix} b_2 & c_2 \\ b_3 & c_3 \end{vmatrix} - b_1 \begin{vmatrix} a_2 & c_2 \\ a_3 & c_3 \end{vmatrix} + c_1 \begin{vmatrix} a_2 & b_2 \\ a_3 & b_3 \end{vmatrix}.$$

Cramer's Rule:

The linear system

$$a_i x + b_i y + c_i z = d_i \quad (i = 1, 2, 3),$$

with determinant of coefficients $\Delta \neq 0$, has the unique solution

$$x = \frac{\Delta_x}{\Delta}, \quad y = \frac{\Delta_y}{\Delta}, \quad z = \frac{\Delta_z}{\Delta},$$

where

Δ_x has the a's in Δ replaced by d's,

Δ_y has the b's in Δ replaced by d's,

Δ_z has the c's in Δ replaced by d's.

A homogeneous system ($d_1 = d_2 = d_3 = 0$) always has the trivial solution $x = y = z = 0$, and has nontrivial solutions if and only if $\Delta = 0$.

9.3 Properties of Determinants

Matrix:

A matrix is a rectangular array of numbers. A matrix with the same number of rows and columns is called a square matrix and has an associated number called its determinant. We use the notation

$$A = \begin{bmatrix} a_{11} & a_{12} \\ a_{21} & a_{22} \end{bmatrix} \text{ and } A = \begin{bmatrix} a_{11} & a_{12} & a_{13} \\ a_{21} & a_{22} & a_{23} \\ a_{31} & a_{32} & a_{33} \end{bmatrix}$$

for square matrices of orders 2 and 3, respectively, and $|A|$ for their corresponding determinants.

Minors and Cofactors:

For a square matrix A, the minor M_{ij} of an element a_{ij} is the determinant of the matrix obtained by deleting row i and column j of A. The quantity $A_{ij} = (-1)^{i+j} M_{ij}$ is called the cofactor of a_{ij}.

Computation of $|A|$ by Cofactors:

If A is a matrix of order n, then

$$\begin{aligned} |A| &= a_{i1}A_{i1} + a_{i2}A_{i2} + \cdots + a_{in}A_{in} \\ &= a_{1j}A_{1j} + a_{2j}A_{2j} + \cdots + a_{nj}A_{nj} \end{aligned}$$

for any $i = 1, 2, \ldots, n$ and any $j = 1, 2, \ldots, n$.

To simplify the computation of a determinant, we make all but one element of a column (row) equal to 0 by using the following elementary row (column) operations.

(1) Multiply each element of a row (column) by a nonzero constant k, which has the effect of multiplying the determinant by k.

(2) Interchange two rows (columns), which multiplies the determinant by -1.

(3) Multiply each element of a row (column) by a constant and add it to the corresponding element of another row (column), which has no effect on the determinant.

9.4 Linear Systems: Gaussian Elimination

Solving Linear Systems:

Three general elimination procedures for solving a linear system are as follows.

(1) Reduce the given system to a triangular system of equations by means of elementary (row) operations; then back-substitute to obtain the solution.

(2) (Gaussian elimination) Reduce the augmented matrix of the system to row-echelon form by means of elementary row operations. The corresponding linear system is now in triangular form and can be solved by back-substitution.

(3) (Gauss-Jordan elimination) Further reduce the row-echelon form of the augmented matrix to reduced row-echelon form, which represents the simplest form of the original system.

9.5 Matrix Algebra

Definitions

A matrix A with m rows and n columns is called an $m \times n$ matrix.

Two matrices are equal if their entries are identical.

If A is $m \times n$ and B is $m \times n$, then

$$A \pm B = C,$$

where

$$c_{ij} = a_{ij} \pm b_{ij} \quad (i = 1, \ldots, m; j = 1, \ldots, n).$$

If A is $m \times n$ and B is $n \times p$, then

$$AB = C,$$

where

$$c_{ij} = a_{i1}b_{1j} + a_{i2}b_{2j} + \cdots + a_{in}b_{nj}$$
$$(i = 1, 2, \ldots, m; j = 1, 2, \ldots, p).$$

The $n \times n$ multiplicative identity matrix I has 1's on its main diagonal and 0's elsewhere.

A square matrix A of order n is invertible if there is a matrix B of order n satisfying $AB = BA = I$. If such a B exists, it is denoted by A^{-1}.

Properties

A square matrix A is invertible $\Leftrightarrow$ the matrix $[A \mid I]$ can be row-reduced to $[I \mid B]$, and then $B = A^{-1}$.

A square matrix A is invertible $\Leftrightarrow |A| \neq 0$.

A linear system of n equations in n unknowns can be written in matrix form as

$$AX = B,$$

where A is the $n \times n$ coefficient matrix, X is the $n \times 1$ matrix of unknowns, and B is the $n \times 1$ matrix of constants. If A is invertible, then the solution is

$$X = A^{-1}B.$$

Chapter 9 Review Exercises

In Exercises 1–5, solve the following system by the indicated method.

$$\begin{aligned} x + y + z &= 3 \\ x - 2y + 2z &= 1 \\ 2x + 3y + z &= 4 \end{aligned}$$

1. Reduction by substitution
2. Cramer's rule
3. Elimination that reduces the system to triangular form
4. Gaussian elimination
5. Gauss-Jordan elimination

6. Show that the following linear system cannot be solved by Cramer's rule.

$$\begin{aligned} x + y - 3z &= 1 \\ x - y + z &= 1 \\ 3x + y - 5z &= 3 \end{aligned}$$

7. Solve the system in Exercise 6 by using elimination to reduce the system to two equations in two unknowns.
8. Solve the system in Exercise 6 by Gaussian elimination.
9. Show that the following system has no solutions by reducing it to a system in x and y only.

$$\begin{aligned} x - y + z &= 5 \\ 2x + 3y + z &= 7 \\ 3x + 2y + 2z &= 4 \end{aligned}$$

10. Apply Gaussian elimination to the system in Exercise 9.
11. Show that the following system has nontrivial solutions by computing the determinant of its coefficients.

$$\begin{aligned} x + 2y - z &= 0 \\ 2x - y + z &= 0 \\ 5x + 5y - 2z &= 0 \end{aligned}$$

12. Find all solutions of the system in Exercise 11 by Gaussian elimination.
13. Solve the following system by Gauss-Jordan elimination.

$$\begin{aligned} x + y + z + t &= 4 \\ x - y - z \phantom{{}+t} &= 0 \\ y + 2z - t &= 0 \\ 2y + 3t &= 2 \end{aligned}$$

Compute by the definition the determinant of each matrix.

14. $\begin{bmatrix} 2 & 3 \\ -1 & 5 \end{bmatrix}$

15. $\begin{bmatrix} 1 & 2 & 3 \\ 0 & -2 & 4 \\ 0 & 1 & 5 \end{bmatrix}$

16. $\begin{bmatrix} 1 & -2 & 3 \\ 4 & 1 & 5 \\ 8 & -1 & 2 \end{bmatrix}$

Compute each determinant by cofactors.

17. $\begin{bmatrix} 1 & -2 \\ 3 & -6 \end{bmatrix}$

18. $\begin{bmatrix} 7 & 1 & -2 \\ 0 & 0 & 0 \\ 1 & 4 & -5 \end{bmatrix}$

19. $\begin{bmatrix} 3 & -4 & 1 & 2 \\ 0 & 0 & 3 & 0 \\ 1 & -1 & 2 & 1 \\ 2 & 1 & -3 & 1 \end{bmatrix}$

Explain why the indicated operations cannot be performed.

20. $\begin{bmatrix} 2 & 1 & 0 \\ 3 & 4 & 0 \end{bmatrix} + \begin{bmatrix} 3 & -2 \\ 5 & 4 \end{bmatrix}$

21. $\begin{bmatrix} 2 & 1 & 0 \\ 3 & 4 & 0 \end{bmatrix}\begin{bmatrix} 3 & -2 \\ 5 & 4 \end{bmatrix}$

22. $\begin{bmatrix} 3 & -2 \\ -6 & 4 \end{bmatrix}^{-1}\begin{bmatrix} 2 & 1 \\ 3 & 4 \end{bmatrix}$

23. $\begin{vmatrix} 2 & 1 & 3 \\ 4 & 6 & 5 \end{vmatrix}$

Perform the indicated operations in Exercises 24–29.

24. $\begin{bmatrix} 2 & -3 \\ 1 & 4 \end{bmatrix} + 3\begin{bmatrix} 0 & 5 \\ 4 & 3 \end{bmatrix}$

25. $\begin{bmatrix} 2 & -3 \\ 1 & 4 \end{bmatrix}\begin{bmatrix} 0 \\ 4 \end{bmatrix}$

26. $\begin{bmatrix} 1 & 2 & -3 \\ 3 & -1 & 2 \\ 4 & 5 & 1 \end{bmatrix}\begin{bmatrix} 1 & 3 \\ 1 & -2 \\ 2 & 1 \end{bmatrix}$

27. $\begin{bmatrix} 2 & -3 \\ 1 & 4 \end{bmatrix}\begin{bmatrix} 1 & -3 \\ -2 & 5 \end{bmatrix}$

28. $\begin{vmatrix} 2 & -3 \\ 1 & 4 \end{vmatrix}\begin{vmatrix} 1 & -3 \\ -2 & 5 \end{vmatrix}$

29. $\begin{bmatrix} 2 & -3 \\ 1 & 4 \end{bmatrix}\begin{bmatrix} 2 & -3 \\ 1 & 4 \end{bmatrix}^{-1}$

30. Find the multiplicative inverse of

$$A = \begin{bmatrix} 1 & 1/2 \\ -2 & -3 \end{bmatrix}$$

by solving $AB = I$ for B.

31. Find A^{-1} for matrix A in Exercise 30 by row reduction of the matrix $[A \mid I]$.

32. Find the inverse of $\begin{bmatrix} 1 & -1 & 1 \\ 0 & 2 & -2 \\ 1 & 3 & -1 \end{bmatrix}$.

33. The equation of any plane in 3-space has the form $ax + by + cz = d$. Find an equation for the plane that passes through the points $(1, -1, 1)$, $(2, 0, 2)$, and $(0, -2, 1)$.

34. Find the point of intersection of the planes

$$\begin{aligned} 2x - 5y - 2z &= 9 \\ x - y + z &= 2 \\ x - 3y - 2z &= 6. \end{aligned}$$

35. Write the system

$$\begin{aligned} 2x - y &= 7 \\ x + 3y &= 0 \end{aligned}$$

in the matrix form $AX = B$, find A^{-1}, and use it to solve for $\begin{bmatrix} x \\ y \end{bmatrix}$.

36. Write the following system in the form $AX = B$. Then find A^{-1} and use it to solve for $\begin{bmatrix} x_1 \\ x_2 \\ x_3 \end{bmatrix}$.

$$\begin{aligned} x_1 + x_2 + 3x_3 &= 2 \\ -x_2 + x_3 &= -4 \\ x_1 \quad\quad + x_3 &= 1 \end{aligned}$$

37. Let $A = \begin{bmatrix} a_{11} & a_{12} \\ a_{21} & a_{22} \end{bmatrix}$. Show that if $|A| \neq 0$, then A is invertible. (*Hint:* show that $AB = I$ can be solved for B and find all entries in $B = A^{-1}$.)

38. Use expansion by cofactors to prove that

$$\begin{vmatrix} a_{11} & ka_{12} & a_{13} \\ a_{21} & ka_{22} & a_{23} \\ a_{31} & ka_{32} & a_{33} \end{vmatrix} = k\begin{vmatrix} a_{11} & a_{12} & a_{13} \\ a_{21} & a_{22} & a_{23} \\ a_{31} & a_{32} & a_{33} \end{vmatrix}.$$

39. For any matrix A, its transpose A^t is obtained by interchanging the rows and columns of A. For example,

$$\text{if } A = \begin{bmatrix} a & b \\ c & d \end{bmatrix},$$

$$\text{then } A^t = \begin{bmatrix} a & c \\ b & d \end{bmatrix}.$$

Show that

$$\text{if } B = \begin{bmatrix} a' & b' \\ c' & d' \end{bmatrix},$$

$$\text{then } (AB)^t = B^t A^t.$$

40. A company makes three kinds of dresses, and each type requires a certain amount of time for cutting, sewing, and finishing, as indicated in the following matrix.

	Type A	Type B	Type C
Cutting	15 min	10 min	20 min
Sewing	20 min	15 min	30 min
Finishing	30 min	20 min	20 min

If the company uses a total of 8 1/3, 11 2/3, and 15 hr labor for cutting, sewing, and finishing, respectively, how many dresses of each type are produced?

Graphing Calculator Exercises

The following exercises, written in the notation of a TI-81, are intended to provide practice in using the matrix capabilities of a graphing calculator.

41. Verify the values for Δ, Δ_x, Δ_y, and Δ_z, respectively, in Example 1 of Section 9.2 by a calculator computation of the determinants of the 3 × 3 matrices. Also make a calculator computation of the linear combinations $a \cdot det[A] + b \cdot det[B] + c \cdot det[C]$ on the right side of the equations.

42. Use the calculator's elementary row operations to carry out the Gaussian elimination process illustrated in Examples 3, 4, and 5 of Section 9.4.

43. Use the calculator to check your answers for the matrix sums $[A] + [B]$ in Exercises 15–19 of Section 9.5.

44. Use the calculator to check your answers for the matrix products $[A][B]$ in Exercises 27–33 of Section 9.5.

45. Use the calculator to check your answers for the matrix inverses $[A]^{-1}$ in Exercises 38–43 of Section 9.5.

46. Check your answers to Review Exercises 6, 9, and 11 by performing Gaussian elimination on the calculator.

47. Check your answer to Review Exercise 13 by performing Gauss-Jordan elimination on the calculator.

48. Check your answers to Review Exercises 17–19 by computing det$[A]$ on the calculator.

49. Check your answers to Review Exercises 24–29 by performing the indicated algebraic operations on the calculator.

50. Check your answers to Review Exercises 35 and 36 by computing $[A]^{-1}[B]$ on the calculator.

10 Roots of Polynomial Equations

The traditional goal of algebra has been to solve equations, and we are concerned here with polynomial equations of degree n. That is, we wish to solve equations of the form

$$a_n x^n + a_{n-1} x^{n-1} + \cdots + a_1 x + a_0 = 0,$$

where the coefficients a_j ($j = 0, 1, 2, \ldots, n$) are real numbers and $a_n \neq 0$. So far we have learned how to solve linear equations

$$ax + b = 0$$

and quadratic equations

$$ax^2 + bx + c = 0.$$

It is also possible to solve cubic equations

$$ax^3 + bx^2 + cx + d = 0$$

and quartic equations

$$ax^4 + bx^3 + cx^2 + dx + e = 0$$

by algebraic methods. However, the procedures for solving a given cubic or quartic equation involve lengthy computations with complex numbers and therefore are not very practical. For polynomial equations of degree $\geq$ 5, the problem is more than one of practicality because of the following **insolvability theorem.**[1]

> The general polynomial equation of degree ≥ 5 cannot be solved by algebraic methods.

More specifically, the insolvability theorem says that for a general polynomial $P(x)$ of degree ≥ 5, the roots of $P(x) = 0$ cannot be expressed in terms of sums, differences, products, quotients, and radicals involving the coefficients of $P(x)$. This "can't do" theorem is in sharp contrast to the following result, which is called the **fundamental theorem of algebra.**[2]

> Every polynomial equation of degree n has n roots in the complex number system.

Hence, the fundamental theorem tells us that the roots are "out there," but the insolvability theorem says that we can't find them by algebra! As a sort of compromise to these two great opposing theorems, we are able to *approximate* roots of polynomial equations by numerical methods. In place of formulas, computers can be programmed to search for roots by a systematic trial-and-error procedure. Although answers obtained in this manner are not exact, they can be made to approximate exact roots to just about any desired degree of accuracy. Hence, today we can plug in our computers and let the electrons do the work.

[1] *Niels Henrik Abel (1802–1829) proved the insolvability theorem for $n = 5$ around 1828. The result for $n > 5$ follows from the work of Evariste Galois (1811–1832).*

[2] *Carl Friedrich Gauss (1777–1855) proved the fundamental theorem for his Ph.D. dissertation in 1799.*

In this chapter we first study general theory that applies to polynomial equations of all orders. Although the general theory does not provide us with all the roots of a given equation, it does tell us what *rational numbers* can possibly be roots when the equation has integer coefficients, *how many* real roots there may be, and the largest possible *magnitude* of the real roots. We then digress to the subject of partial fractions, a topic that could have been considered earlier but does depend on the fundamental theorem. Finally, we investigate two numerical methods for approximating the real roots of a polynomial equation. To implement these methods, a programmable calculator or a computer will be especially helpful.

10.1 Remainder Theorem, Synthetic Division, and Factor Theorem

Theory of Equations: Part I

Remainder Theorem
Synthetic Division
Factor Theorem
Complex and Real Factored Forms

In this section we obtain results that apply to polynomials of every degree. As usual, we consider polynomials $P(x)$ with real coefficients, but the roots of $P(x) = 0$ must be considered in the complex number system.

Remainder Theorem. We saw in Section 1.1 that if a polynomial $P(x)$ of degree $n \geq 1$ is divided by a linear expression $x - a$, then

$$\boldsymbol{P(x) = (x - a)Q(x) + R}, \qquad \textbf{division formula} \tag{1}$$

where the quotient $Q(x)$ is a polynomial of degree $n - 1$, and the remainder R is a constant. For instance, if $2x^3 + 7x^2 + 8x + 9$ is divided by $x + 2$, then

$$2x^3 + 7x^2 + 8x + 9 = (x + 2)(2x^2 + 3x + 2) + 5. \tag{2}$$

Here $a = -2$ and $R = 5$. The division formula (1) holds for any real or complex value of a. If we let $x = a$ in (1), then the term $(x - a)Q(x)$ drops out, and the equation becomes

$$P(a) = R. \tag{3}$$

For instance, when $x = -2$, equation (2) becomes

$$2(-2)^3 + 7(-2)^2 + 8(-2) + 9 = 5.$$

Hence, by combining equations (1) and (3), we obtain the following result, which is called the **remainder theorem.**

If a polynomial $P(x)$ of degree ≥ 1 is divided by $x - a$, then the remainder R is equal to $P(a)$.

Example 1 Find the remainder when $3x^5 - 2x^4 + x^3 + 7x^2 - 5x + 4$ is divided by $x + 1$.

Solution: Here $a = -1$, and by the remainder theorem,

$$\begin{aligned} R &= P(-1) \\ &= 3(-1)^5 - 2(-1)^4 + (-1)^3 + 7(-1)^2 - 5(-1) + 4 \\ &= 10. \quad ■ \end{aligned}$$

Example 2 Verify the remainder theorem for the polynomial $P(x) = x^3 - 2x^2 - 5x + 6$ by actually dividing $P(x)$ by $x - a$.

Solution:

$$\begin{array}{r l} & x^2 + (a-2)x + (a^2 - 2a - 5) \\ x - a \overline{)} & x^3 - 2x^2 - 5x + 6 \\ & \underline{x^3 - ax^2} \\ & (a-2)x^2 - 5x + 6 \\ & \underline{(a-2)x^2 - (a-2)ax} \\ & \underbrace{-5x + 6 + (a-2)ax} \\ & (a^2 - 2a - 5)x + 6 \\ & \underline{(a^2 - 2a - 5)x - a(a^2 - 2a - 5)} \\ & \underbrace{6 + a(a^2 - 2a - 5)} \\ & a^3 - 2a^2 - 5a + 6 \\ & \textit{Remainder} \end{array}$$

Hence, the quotient is $x^2 + (a - 2)x + (a^2 - 2a - 5)$ and, as stated in the remainder theorem, $R = a^3 - 2a^2 - 5a + 6 = P(a)$. ■

Synthetic Division. Synthetic division is a shortcut method to divide a polynomial $P(x)$ by a linear expression $x - a$. However, its importance from a computational point of view is that, whether we are computing by hand or by a calculator or computer, synthetic division enables us to rapidly compute values of $P(x)$ for given values of x. To explain, suppose we wish to determine $P(a)$. The remainder theorem tells us that $P(a)$ is the remainder when $P(x)$ is divided by $x - a$. Therefore, if we have a quick way to divide $P(x)$ by $x - a$, then we have a quick way to determine $P(a)$.

We illustrate the method of synthetic division by means of the following example, in which $P(x) = 3x^4 - 2x^3 - 10x^2 + 15$ is divided by $x - 2$. In the usual division process, we have the following.

$$
\begin{array}{lrl}
 & 3x^3 + 4x^2 - 2x - 4 & \textit{Quotient} \\
\textit{Divisor} \quad x - 2 & \overline{)3x^4 - 2x^3 - 10x^2 + 0x + 15} & \textit{Dividend} \\
 & 3x^4 - 6x^3 \qquad\qquad\qquad\qquad & \\
 & \overline{4x^3 - 10x^2 \qquad\qquad} & \\
 & 4x^3 - 8x^2 \qquad\qquad & \\
 & \overline{- 2x^2 + 0x \qquad} & \\
 & - 2x^2 + 4x \qquad & \\
 & \overline{- 4x + 15} & \\
 & - 4x + 8 & \\
 & \overline{7} & \textit{Remainder}
\end{array}
$$

Only the shaded terms below the dividend are essential to the process; the others are repetitions. We delete the repeated terms and move the essential ones upward as follows.

$$
\begin{array}{rl}
 & 3x^3 + 4x^2 - 2x - 4 \\
x - 2 & \overline{)3x^4 - 2x^3 - 10x^2 + 0x + 15} \\
 & -6x^3 - 8x^2 + 4x + 8 \\
 & \overline{4x^3 - 2x^2 - 4x + 7}
\end{array}
$$

Next we drop all plus signs and all x's, which results in the following.

$$
\begin{array}{lrrrrrrl}
 & & 3 & 4 & -2 & -4 & & \textit{Coefficients of quotient} \\
\textit{Coefficients} & 1-2) & 3 & -2 & -10 & 0 & 15 & \textit{Coefficients of dividend} \\
\textit{of divisor} & & & -6 & -8 & 4 & 8 & \\
 & & & 4 & -2 & -4 & 7 & \textit{Remainder line}
\end{array}
$$

The coefficient of x in the divisor is always 1 and may be omitted. Also, the coefficients in the quotient line may be deleted since they are repeated in the remainder line, provided we include the first coefficient 3 as follows.

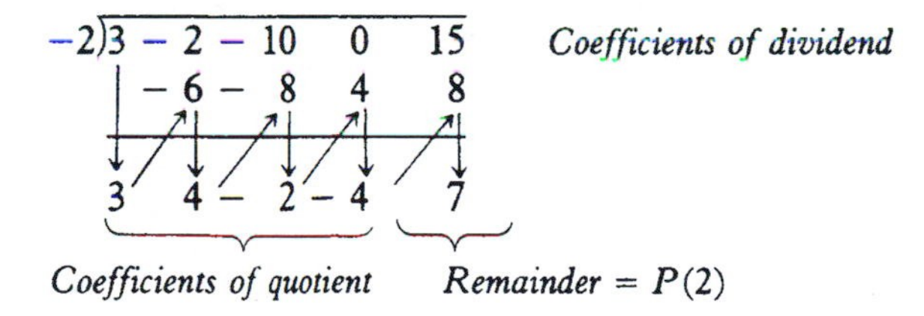

In the display above, the vertical arrows ↓ indicate subtraction, and the diagonal arrows ↗ indicate multiplication by -2. Finally, it is customary to replace the subtraction by addition by changing -2 in the divisor to 2 and letting the vertical arrows represent addition. Hence, for the **synthetic division** of $3x^4 - 2x^3 - 10x^2 + 15$ by $x - 2$, we obtain

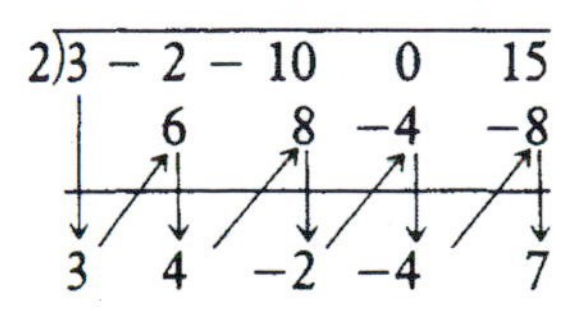

where ↓ means *addition* and ↗ means *multiplication* by 2 (in general, by a). Since $P(x) = 3x^4 - 2x^3 - 10x^2 + 15$, the result can be stated as

$$P(x) = (x - 2)(3x^3 + 4x^2 - 2x - 4) + 7,$$

and therefore $P(2) = 7$.

Example 3 Find $P(-3)$ for $P(x) = 3x^4 - 2x^3 - 10x^2 + 15$.

Solution: We use synthetic division to divide $P(x)$ by $x - (-3)$. We get

$$\begin{array}{r|rrrrr} -3 & 3 & -2 & -10 & 0 & 15 \\ & & -9 & 33 & -69 & 207 \\ \hline & 3 & -11 & 23 & -69 & 222 \end{array}$$

Each ↓ means addition, and each ↗ means multiplication by -3

and conclude that $P(-3) = 222$. ■

To compute $P(a)$ directly, we must compute *powers* of a, whereas in synthetic division, we compute simple *multiples* of a. Powers, especially high powers of large numbers, take a longer time to compute. Another method for computing $P(a)$ quickly, called **Horner's method,** is covered in the exercises.

Factor Theorem. From equation (1), we see that if $R = 0$, then $x - a$ is a *factor* of $P(x)$. But $R = P(a)$ by the remainder theorem, and therefore the condition $R = 0$ is equivalent to $P(a) = 0$, or a is a *root* of $P(x) = 0$. Hence, we obtain the following result, which is called the **factor theorem.**[1]

If a is a root of $P(x) = 0$,
then $x - a$ is a factor of $P(x)$.

For instance, we saw in Chapter 3 that if a_1 and a_2 are the roots of a quadratic equation $ax^2 + bx + c = 0$, then

$$ax^2 + bx + c = a(x - a_1)(x - a_2). \tag{4}$$

Here, a_1 and a_2 can be fractions, irrational numbers, or even complex conjugates, so equation (4) goes beyond the method of factoring covered in Chapter 1.

Example 4 Factor $x^2 - 2x + 5$.

Solution: By the quadratic formula, the roots of $x^2 - 2x + 5 = 0$ are $1 + 2i$ and $1 - 2i$. Therefore, $[x - (1 + 2i)]$ and $[x - (1 - 2i)]$ are factors of $x^2 - 2x + 5$. That is,

$$x^2 - 2x + 5 = [x - (1 + 2i)][x - (1 - 2i)].$$ ■

Example 5 Factor $P(x) = 8x^3 - 12x^2 - 10x + 15$, given that $3/2$ is a root of $P(x) = 0$.

[1] *The factor theorem appeared in the work* La Géométrie *by René Descartes in 1637.*

Solution: Since 3/2 is a root of $P(x) = 0$, $x - 3/2$ is a factor of $P(x)$. Now if $8x^3 - 12x^2 - 10x + 15$ is divided by $x - 3/2$, we obtain the quotient $8x^2 - 10$. That is,

$$8x^3 - 12x^2 - 10x + 15 = \left(x - \frac{3}{2}\right)(8x^2 - 10).$$

The roots of $8x^2 - 10 = 0$ are $\sqrt{5}/2$ and $-\sqrt{5}/2$. Therefore,

$$8x^2 - 10 = 8\left(x - \frac{\sqrt{5}}{2}\right)\left(x + \frac{\sqrt{5}}{2}\right),$$

and
$$\begin{aligned} 8x^3 - 12x^2 - 10x + 15 &= \left(x - \frac{3}{2}\right)8\left(x - \frac{\sqrt{5}}{2}\right)\left(x + \frac{\sqrt{5}}{2}\right) \\ &= 8\left(x - \frac{3}{2}\right)\left(x - \frac{\sqrt{5}}{2}\right)\left(x + \frac{\sqrt{5}}{2}\right). \end{aligned}$$ ■

Complex and Real Factored Forms. Let

$$P(x) = a_nx^n + a_{n-1}x^{n-1} + \cdots + a_1x + a_0 \tag{5}$$

be a polynomial of degree n, where the coefficients $a_0, a_1, \ldots, a_{n-1}, a_n$ are real numbers and $a_n \neq 0$. By the fundamental theorem of algebra, $P(x)$ has n zeros $R_1, R_2, \ldots, R_n$ (not necessarily distinct), and by the factor theorem we can deduce (see Exercise 51) that

$$P(x) = a_n(x - R_1)(x - R_2) \ldots (x - R_n). \tag{6}$$

The roots of $P(x) = 0$ are considered in the complex number system, and it is understood that the **order** of each root is taken into account. For instance, the polynomial

$$\begin{aligned} P(x) &= (x - 7)(x + 5)^2(x - 4)^3(x - i)(x + i) \\ &= (x - 7)(x + 5)(x + 5)(x - 4)(x - 4)(x - 4)(x - i)(x + i) \end{aligned}$$

is of degree 8, and the eight roots of $P(x) = 0$ are

$$7, -5, -5, 4, 4, 4, i, -i.$$

That is, 7 is a root of order 1, as are i and $-i$, -5 is a root of order 2, and 4 is a root of order 3.

The factored form of $P(x)$ given by (6) tells us that the coefficients of $P(x)$ are determined, up to a constant multiple, by its zeros. That is, if

$$Q(x) = b_nx^n + b_{n-1}x^{n-1} + \cdots + b_1x + b_0$$

is another polynomial that also has $R_1, R_2, \ldots, R_n$ for its zeros, then

$$Q(x) = b_n(x - R_1)(x - R_2) \ldots (x - R_n),$$

which means that

$$Q(x) = \frac{b_n}{a_n}P(x).$$

Hence, the coefficients b_k of $Q(x)$ are related to the a_k of $P(x)$ by

$$b_k = ca_k, \quad k = 0, 1, 2, \ldots, n,$$

where c is the constant b_n/a_n.

Example 6 Find all polynomials whose zeros are 1, 2, 5, 5.

Solution: Any polynomial $P(x)$ with the given zeros has the factored form

$$P(x) = c(x-1)(x-2)(x-5)^2,$$

where c is a nonzero constant. Therefore,

$$\begin{aligned} P(x) &= c(x^2 - 3x + 2)(x^2 - 10x + 25) \\ &= c(x^4 - 13x^3 + 57x^2 - 95x + 50). \quad \blacksquare \end{aligned}$$

We now derive two more specific versions of (6). First, we know from Chapter 3 that if a complex number $z = \alpha + \beta i$ is a root of a quadratic equation $ax^2 + bx + c = 0$, where a, b, and c are real numbers, then so is the **complex conjugate** $\bar{z} = \alpha - \beta i$. We obtained this result from the quadratic formula, but we could also obtain it as follows.

Let $P(z) = az^2 + bz + c$, where a, b, and c are real numbers. Then

$$\begin{aligned} P(\bar{z}) &= a\bar{z}^2 + b\bar{z} + c \\ &= a\overline{z^2} + b\bar{z} + c && \text{$\bar{z}^2 = \overline{z^2}$ because the product of complex conjugates equals the conjugate of the product (see Exercise 62)} \\ &= \bar{a}\overline{z^2} + \bar{b}\bar{z} + \bar{c} && \text{The conjugate of a real number is equal to itself} \\ &= \overline{az^2} + \overline{bz} + \bar{c} && \text{The conjugate of a product equals the product of the conjugates} \\ &= \overline{az^2 + bz + c} && \text{The conjugate of a sum equals the sum of the conjugates (see Exercise 61)} \\ &= \overline{P(z)}. \end{aligned}$$

Therefore, if z is a root of $P(x) = 0$, then

$$P(\bar{z}) = \overline{P(z)} = \bar{0} = 0,$$

which means that $\bar{z}$ is also a root.

The above method can be applied to a polynomial $P(x)$ of any degree, provided $P(x)$ has real coefficients. That is, *if $z = \alpha + \beta i$ is a root of $P(x) = 0$, where $P(x)$ is any polynomial with real coefficients, then $\bar{z} = \alpha - \beta i$ is also a root.*[2]

Now suppose that the n roots of $P(x) = 0$ separate into k distinct real roots $r_1, r_2, \ldots, r_k$ of orders $m_1, m_2, \ldots, m_k$, respectively, and j distinct

[2]*This result was proved by Sir Isaac Newton in his work* Arithmetica Universalis, *which was published in 1707.*

pairs of complex conjugate roots $c_1, \bar{c}_1, c_2, \bar{c}_2, \ldots, c_j, \bar{c}_j$ of orders $n_1, n_2, \ldots, n_j$, respectively. Then the factoring (6) of $P(x)$ becomes

$$P(x) = a_n(x - r_1)^{m_1}(x - r_2)^{m_2} \ldots (x - r_k)^{m_k}(x - c_1)^{n_1}(x - \bar{c}_1)^{n_1} \cdot (x - c_2)^{n_2}(x - \bar{c}_2)^{n_2} \ldots (x - c_j)^{n_j}(x - \bar{c}_j)^{n_j}, \tag{7}$$

which is called the **complex factored form** of $P(x)$. The expression in (7) looks complicated, but it is just equation (6) with a little more detail. Now for any complex number $c = \alpha + \beta i$, we have

$$\begin{aligned}[x - (\alpha + \beta i)]^n[x - (\alpha - \beta i)]^n &= \{[x - (\alpha + \beta i)][x - (\alpha - \beta i)]\}^n \\ &= [x^2 - 2\alpha x + (\alpha^2 + \beta^2)]^n \\ &= (x^2 + Ax + B)^n,\end{aligned}$$

where $A = -2\alpha$ and $B = \alpha^2 + \beta^2$ are real numbers. Therefore, (7) can be written as

$$P(x) = a_n(x - r_1)^{m_1}(x - r_2)^{m_2} \ldots (x - r_k)^{m_k}(x^2 + A_1x + B_1)^{n_1} \cdot (x^2 + A_2x + B_2)^{n_2} \ldots (x^2 + A_jx + B_j)^{n_j}, \tag{8}$$

which is called the **real factored form** of $P(x)$. In this form, $P(x)$ factors into powers of linear and quadratic factors with real coefficients. The quadratic factors are **irreducible** because they cannot be factored in the real number system. Note that the coefficient a_n appears as a constant factor in both the complex and the real factored forms.

Example 7 Find the complex and real factored forms of a polynomial $P(x)$ whose real zeros are 2 and -3 of orders 1 and 2, respectively, and whose complex zeros are $3 \pm 4i$ and $1 \pm 2i$ of orders 1 and 3, respectively. What is the degree of $P(x)$?

Solution: The complex factored form of $P(x)$ is

$$P(x) = c(x - 2)(x + 3)^2[x - (3 + 4i)][x - (3 - 4i)] \cdot [x - (1 + 2i)]^3[x - (1 - 2i)]^3,$$

where c can be any nonzero real number. Now

$$[x - (3 + 4i)][x - (3 - 4i)] = x^2 - 6x + 25,$$

and

$$[x - (1 + 2i)][x - (1 - 2i)] = x^2 - 2x + 5.$$

Therefore, the real factored form of $P(x)$ is

$$P(x) = c(x - 2)(x + 3)^2(x^2 - 6x + 25)(x^2 - 2x + 5)^3.$$

The degree of $P(x)$ is obtained by adding the degrees of its factors in either form. From the real factored form we get

$$\text{degree } P(x) = 1 + 2 + 2 + 3 \cdot 2 = 11. \quad ■$$

Question *Can we find the real or complex form for any polynomial $P(x)$?*

Answer: Here we again see the opposition between the fundamental theorem of algebra and the insolvability theorem mentioned at the beginning of this section. The existence of the real and complex factored forms of $P(x)$ follows directly from the fundamental theorem of algebra, but to actually perform the factorization, we must know the roots of $P(x) = 0$, and the insolvability theorem says that we cannot find the roots by algebra for general polynomials of degree ≥ 5. Hence, we cannot write down the factored forms for a general polynomial $P(x)$ of degree ≥ 5 even though we know they exist!

We can state the situation in other words as follows: the fundamental theorem, by way of the factored forms, tells us that the n roots of a polynomial equation $P(x) = 0$ of degree n determine its coefficients, up to a constant multiple. Conversely, we also know that the coefficients of $P(x) = 0$ somehow determine its roots. But the insolvability theorem tells us that, for $n \geq 5$, the "somehow" goes beyond addition, subtraction, multiplication, division, and radicals. ■

Exercises 10.1

Fill in the blanks to make each statement true.

1. If 5 is a root of the polynomial equation $P(x) = 0$, then ________ is a factor of $P(x)$.
2. If $x + 3$ is a factor of $P(x)$, then ________ is a root of $P(x) = 0$.
3. The roots of $64x^3 +$ ________ $x^2 +$ ________ $x +$ ________ $= 0$ are 1/2, 1/4, and 1/8.
4. The polynomial $x^2 - 2x + 2$ factors into ________.
5. The notation $4\overline{)6 \quad -5 \quad 1 \quad 0 \quad 3}$ in synthetic division indicates that the polynomial ________ is to be divided by ________.

Write true *or* false *for each statement.*

6. If a is a root of a polynomial equation $P(x) = 0$, then $x + a$ is a factor of $P(x)$.
7. If $x + a$ is a factor of a polynomial $P(x)$, then $-a$ is a root of $P(x) = 0$.
8. The roots of $x^7 - 4x^6 - 14x^5 + 56x^4 + 49x^3 - 196x^2 - 36x + 144 = 0$ are $1, -1, 2, -2, 3, -3, 4$ and -4.
9. We can find the factors of any polynomial with real coefficients by algebraic methods.
10. Synthetic division provides us with a rapid method for evaluating polynomials.

Remainder Theorem

In Exercises 11–16, determine the remainder when the given polynomial is divided by **(a)** $x - 1$ *and* **(b)** $x + 2$.

11. $2x^3 - 4x^2 + 3x + 9$
12. $5x^3 - 7x^2 - 10x + 8$
13. $2x^4 + x^3 - 7x + 4$
14. $x^4 - 2x^3 + x^2 - 4$
15. $x^5 - 3x^4 - 5x^3 + 15x^2 + 4x - 12$
16. $2x^5 - x^4 + 3x^2 - x + 5$

As in Example 2, verify the remainder theorem in Exercises 17–20 by actually dividing $P(x)$ by $x - a$.

17. $P(x) = x^2 + 2x - 3$

18. $P(x) = 2x^2 - 3x + 4$
19. $P(x) = 2x^3 + x^2 - 4x + 5$
20. $P(x) = x^3 - 6x^2 + 3x - 2$

Synthetic Division

In Exercises 21–24, perform the indicated division by the method of synthetic division. State the quotient and the remainder.

21. $x - 2\overline{)x^3 - 3x^2 + 2x - 5}$
22. $x - 1\overline{)x^3 + x^2 + x + 1}$
23. $x + 1\overline{)2x^5 + 5x^3 - 3x + 4}$
24. $x + 2\overline{)3x^5 - 4x^4 + 2x^2 - 1}$

Use synthetic division to compute $P(5)$ and $P(-5)$ in each of the following.

25. $P(x) = x^4 + 2x^3 - x + 4$
26. $P(x) = 4x^5 + 2x^4 - 3x^3 + 1$
27. $P(x) = 6x^6 - 2x^5 + 4x^3 - x^2 + 3$
28. $P(x) = 5x^6 + 6x^5 - 3x^4 - x^3 + 7$

Use synthetic division to compute $P(3/2)$ and $P(-2.5)$ for the following.

29. $P(x) = x^3 + 3x^2 - 2x + 1$
30. $P(x) = x^3 - x^2 + x - 1$

Factor Theorem

Use the factor theorem to verify each of the following.

31. $x + 1$ is a factor of $P(x) = 4x^2 + 12x + 8$.
32. $x - 1$ is a factor of $P(x) = 4x^3 - 8x^2 + 3x + 1$.
33. $x - 2$ is a factor of $P(x) = 3x^5 - 8x^4 + 5x^3 - 7x^2 + 11x - 2$.
34. $x + 2$ is a factor of $P(x) = x^4 - x^3 - 2x^2 + 3x - 10$.
35. $x - \frac{1}{2}$ is factor of $P(x) = 2x^3 - x^2 - 6x + 3$.
36. $x - \sqrt{2}$ is a factor of $P(x) = x^4 + x^3 - x^2 - 2x - 2$.

Use the factor theorem in each of the following to find all factors $x - a$, where a is an integer *from -3 to 3. State the order of each factor.*

37. $x^3 - 7x + 6$
38. $x^3 - 3x + 2$
39. $x^3 + x^2 - 8x - 12$
40. $x^3 + 2x^2 - 5x - 6$

Complex and Real Factored Forms

In each of the following, give the complex and real factored forms of a polynomial $P(x)$ of lowest degree with real coefficients and the given zeros.

41. $a_1 = 2,\quad a_2 = -2,\quad a_3 = 1$
42. $a_1 = \frac{1}{2},\quad a_2 = 2,\quad a_3 = \frac{1}{3}$
43. $a_1 = -1,\quad a_2 = 1,\quad a_3 = -2,\quad a_4 = 2$
44. $a_1 = 2,\quad a_2 = -1,\quad a_3 = i,\quad a_4 = -i$
45. $a_1 = -2,\quad a_2 = 1 + i,\quad a_3 = 1 - i$
46. $a_1 = 1,\quad a_2 = -2,\quad a_3 = 3,\quad a_4 = -4$
47. $a_1 = 1,\quad a_2 = -\frac{1}{2},\quad a_3 = \frac{1}{3},\quad a_4 = -\frac{1}{4}$
48. $a_1 = \frac{1}{\sqrt{2}},\quad a_2 = -\frac{1}{\sqrt{2}},\quad a_3 = \frac{1}{\sqrt{3}},\quad a_4 = -\frac{1}{\sqrt{3}}$
49. $a_1 = \frac{1}{3},\quad a_2 = \frac{1}{3},\quad a_3 = 3 - 2i$
50. $a_1 = \sqrt{2},\quad a_2 = -\sqrt{2},\quad a_3 = 2i,\quad a_4 = 2 - i$

Miscellaneous

51. Let $P(x)$ be a polynomial of degree n with *distinct* zeros $a_1, a_2, \ldots, a_n$. Fill in the blanks for each of the following steps, which lead to the conclusion that $P(x) = c(x - a_1)(x - a_2)\ldots(x - a_n)$.

    $P(x) = (x - a_1)Q_1(x)$ by the ________ theorem, where $Q_1(x)$ is a polynomial of degree ________.

    $Q_1(x) = (x - a_2)Q_2(x)$, where $Q_2(x)$ is a polynomial of degree ________.

    ⋮

    $Q_{n-1}(x) = (x - a_n)Q_n(x)$, where $Q_n(x)$ is a polynomial of degree ________.

    Therefore $Q_n(x)$ is a ________, and $P(x) = c(x - a_1)(x - a_2)\ldots(x - a_n)$.

52. Given that n is a positive integer, show that:
    (a) $(x - a)$ is a factor of $x^n - a^n$;
    (b) $(x + a)$ is a factor of $x^n - a^n$, if n is even;
    (c) $(x + a)$ is a factor of $x^n + a^n$, if n is odd.

53. Why is it true that every polynomial with real coefficients of degree n, where n is *odd*, has at least one real zero?

54. Let $P(x)$ and $Q(x)$ be polynomials.
   (a) Show that if $P(x) = 0$ for all real x, then all the coefficients of $P(x)$ are zero.
   (b) Show that if $P(x) = Q(x)$ for all real x, then $P(x)$ and $Q(x)$ have the same coefficients.

Horner's Method

The standard computation of $P(x) = 2x^3 + 4x^2 + 5x + 3$ *for a given value of* x *requires six multiplications and three additions. However, by successively factoring out* x*, we can obtain*

$$\begin{aligned} P(x) &= (2x^2 + 4x + 5)x + 3 \\ &= [(2x + 4)x + 5]x + 3. \end{aligned}$$

We call the last expression ***Horner's form*** *of* $P(x)$. *In this form only three multiplications and three additions are needed to evaluate* $P(x)$. *For example,*

$P(11) = [(2 \cdot 11 + 4)11 + 5]11 + 3$	Horner's form
$= [(22 + 4)11 + 5]11 + 3$	Multiplication
$= [26 \cdot 11 + 5]11 + 3$	Addition
$= [286 + 5]11 + 3$	Multiplication
$= 291 \cdot 11 + 3$	Addition
$= 3201 + 3$	Multiplication
$= 3204.$	Addition

This method of successively factoring out x *in order to obtain Horner's form for rapid computation can be applied to any polynomial and is known as* ***Horner's method.***

Apply Horner's method in Exercises 55–58.

55. $P(x) = 2x^3 - 4x^2 + 5x + 7,\ x = 6$
56. $P(x) = 4x^3 + 3x^2 - 8x + 10,\ x = 5$
57. $P(x) = x^4 + x^3 + 12x^2 - 4x - 15,\ x = 7$
58. $P(x) = 2x^4 - 5x^3 - 15x^2 + 12x + 3,\ x = 8$

59. Use synthetic division to evaluate the polynomials in Exercises 55 and 57.
60. Use synthetic division to evaluate the polynomials in Exercises 56 and 58.

Complex Conjugates

61. Show that $\overline{(3 + 2i) + (4 + 5i)} = \overline{(3 + 2i)} + \overline{(4 + 5i)}$, and in general, $\overline{(a + bi) + (c + di)} = \overline{(a + bi)} + \overline{(c + di)}$ (the conjugate of a sum equals the sum of the conjugates).
62. Show that $\overline{(3 + 2i)(4 + 5i)} = \overline{(3 + 2i)}\,\overline{(4 + 5i)}$, and in general, $\overline{(a + bi)(c + di)} = \overline{(a + bi)}\,\overline{(c + di)}$ (the conjugate of a product equals the product of the conjugates).

10.2 Rational Roots, Bounds, and Descartes' Rule of Signs

Theory of Equations: Part II

Rational Roots
Upper and Lower Bounds
Descartes' Rule of Signs

Rational Roots. In Example 5 of the previous section we were given that $3/2$ is a root of $8x^3 - 12x^2 - 10x + 15 = 0$. Note that 3 divides the constant term 15 and 2 divides the coefficient 8 of x^3. More generally, suppose that a rational number p/q is a root of the polynomial equation

$$a_3x^3 + a_2x^2 + a_1x + a_0 = 0, \tag{1}$$

where the coefficients a_3, a_2, a_1, and a_0 are *integers*, $a_3 \neq 0$, and $a_0 \neq 0$. As usual, we assume that p and q are integers with no common factors except ± 1, and of course, $q \neq 0$. Then (1) becomes

$$a_3\left(\frac{p}{q}\right)^3 + a_2\left(\frac{p}{q}\right)^2 + a_1\left(\frac{p}{q}\right) + a_0 = 0. \tag{2}$$

If we multiply both sides of (2) by q^3, we get

$$a_3p^3 + a_2p^2q + a_1pq^2 + a_0q^3 = 0. \qquad (3)$$

Equation (3) is equivalent to

$$a_3p^3 + a_2p^2q + a_1pq^2 = -a_0q^3. \qquad (4)$$

The left side of (4) is divisible by p, and therefore the right side must also be divisible by p. But, by assumption, p and q have no common factors. It can then be shown that p and q^3 have no common factors, and we may conclude that *a_0 is divisible by p.*

Similarly, equation (3) is equivalent to

$$a_3p^3 = -a_2p^2q - a_1pq^2 - a_0q^3. \qquad (5)$$

The right side of (5) is divisible by q and therefore the left side must also be divisible by q. Since p and q have no common factors, it can be shown that p^3 and q have no common factors. Hence, we may conclude that *a_3 is divisible by q.*

The above procedure can be applied to any polynomial equation with integer coefficients and leads to the following general result, which is called the **rational root theorem.**[1]

[1]*Descartes utilized the idea of the rational root theorem in his work* La Géométrie *(1637).*

> If the rational number p/q is a root of the equation
>
> $$a_nx^n + a_{n-1}x^{n-1} + \cdots + a_1x + a_0 = 0,$$
>
> where the coefficients a_k are integers, $a_n \neq 0$, and $a_0 \neq 0$, then a_0 is divisible by p, and a_n is divisible by q.

Question 1 *In the rational root theorem, why do we assume that $a_0 \neq 0$?*

Answer: $a_0 = 0$ if and only if 0 is a root of $P(x) = 0$, but a_0 is not divisible by 0. However, if 0 is a root of order k of a polynomial equation $P(x) = 0$, then by the factor theorem, $P(x) = x^kQ(x)$, where the constant term of $Q(x)$ is not zero. We can then apply the rational root theorem to the polynomial $Q(x)$. ■

As illustrated in the following examples, we can use the rational root theorem to determine all the rational roots of a given polynomial equation with integral *or* rational coefficients. However, even small values of a_0 and a_n can generate many possible rational numbers p/q to be tested. Therefore, a rapid method for computing $P(p/q)$ is needed. We can use synthetic division or a calculator to help with the computations. For large values of a_0 and a_n that may have many factors, it is helpful to use a programmable calculator or a computer to test all the possibilities.

Example 1 Find all rational roots of the equation

$$15x^4 + 8x^3 + 6x^2 - 9x - 2 = 0.$$

Solution: Here, $a_0 = -2$ and $a_4 = 15$. The factors of -2 are ± 1 and ± 2, and those of 15 are ± 1, ± 3, ± 5, and ± 15. Therefore, by the rational root theorem, the *possible* rational roots are

$$\pm 1, \pm\frac{1}{3}, \pm\frac{1}{5}, \pm\frac{1}{15}, \pm 2, \pm\frac{2}{3}, \pm\frac{2}{5}, \pm\frac{2}{15}.$$

If each of these numbers is tested in the given equation, we find that

$$\frac{2}{3} \quad \text{and} \quad -\frac{1}{5}$$

are the only rational roots. ■

Example 2 Find all rational roots, including order, of the equation

$$x^4 - \frac{3}{2}x^3 + \frac{41}{16}x^2 - 3x + \frac{9}{8} = 0.$$

Solution: We first clear the fractions by multiplying both sides of the equation by 16. We get the equivalent equation

$$16x^4 - 24x^3 + 41x^2 - 48x + 18 = 0.$$

Here, $a_0 = 18$ has the positive factors 1, 2, 3, 6, 9, and 18, and $a_4 = 16$ has the positive factors 1, 2, 4, 8, and 16. The possible rational roots are $\pm$ the following.

$$1, 2, 3, 6, 9, 18, \frac{1}{2}, \frac{3}{2}, \frac{9}{2}, \frac{1}{4}, \frac{3}{4}, \frac{9}{4}, \frac{1}{8}, \frac{3}{8}, \frac{9}{8}, \frac{1}{16}, \frac{3}{16}, \frac{9}{16}.$$

Since there are 36 numbers to be tested, a programmable calculator would be helpful. We discover that 3/4 is the only rational root. Division by $x - 3/4$ gives

$$\begin{aligned} &16x^4 - 24x^3 + 41x^2 - 48x + 18 \\ &\quad = \left(x - \frac{3}{4}\right)(16x^3 - 12x^2 + 32x - 24). \end{aligned}$$

We find that 3/4 is also a zero of the quotient $16x^3 - 12x^2 + 32x - 24$, and when the quotient is divided by $x - 3/4$, we get

$$\begin{aligned} 16x^3 - 12x^2 + 32x - 24 &= \left(x - \frac{3}{4}\right)(16x^2 + 32). \\ &= 16\left(x - \frac{3}{4}\right)(x^2 + 2). \end{aligned}$$

Therefore,

$$16x^4 - 24x^3 + 41x^2 - 48x + 18 = 16\left(x - \frac{3}{4}\right)^2(x^2 + 2).$$

Hence, 3/4 is a root of order 2 of the original equation. Also, the quadratic $x^2 + 2$ is irreducible over the real number system (*why?*), and therefore the right side of the above equation is the real factored form of the polynomial on the left side. ■

Caution The rational root theorem does not say that a polynomial with integer coefficients *must* have a rational zero. For example, $x^2 - 2 = 0$ has no rational roots.

Question 2 *What can be said about the rational zeros of a polynomial with integer coefficients if the leading coefficient a_n is 1?*

Answer: Since any rational root p/q must have the property that q divides a_n, it follows that p/q must be an integer if $a_n = 1$. ■

Upper and Lower Bounds. We can sometimes reduce the number of rational numbers that must be tested in a given polynomial equation by obtaining upper and lower bounds for the real roots of the equation. First, recall that if a polynomial $P(x)$ of degree $n \geq 1$ is divided by $x - a$, then the division formula says that

$$P(x) = (x - a)Q(x) + R, \tag{6}$$

where $Q(x)$ is a polynomial of degree $n - 1$ and R is a real number. Suppose R and the coefficients of $Q(x)$ are ≥ 0. Then, if $b > a$, we must have $P(b) > 0$ (*why?*). Hence, all real roots of $P(x) = 0$ are less than or equal to a, in which case we call a an **upper bound** for the real roots of $P(x) = 0$. We now have the following **upper bound test.**

> If a polynomial $P(x)$ of degree ≥ 1 is divided by $x - a$ and the remainder R as well as the coefficients of the quotient $Q(x)$ are ≥ 0, then a is an upper bound for the real roots of $P(x) = 0$.

Caution We apply the upper bound test to polynomials $P(x)$ whose leading coefficient a_n is *positive*. Otherwise, the leading coefficient of $Q(x)$ in equation (6) would always be negative. Hence, if a_n is negative, first replace $P(x)$ by $-P(x)$ and then apply the upper bound test to $-P(x)$. The roots of $-P(x) = 0$ are the same as those of $P(x) = 0$.

We can use synthetic division to divide $P(x)$ by $x - a$ for $a = 1, 2, 3, \ldots$ until we reach a positive integer a that gives a nonnegative R and nonnegative coefficients for $Q(x)$. Such an a is an upper bound for the real roots of $P(x) = 0$. If $R = 0$, then a is also a root of $P(x) = 0$. If $R > 0$, then all real roots of $P(x) = 0$ are strictly less than a.

Example 3 Use the upper bound test to find an upper bound for the real roots of the polynomial equation $16x^4 - 24x^3 + 41x^2 - 48x + 18 = 0$ in Example 2.

Solution: We first use synthetic division to divide the given polynomial by $x - 1$.

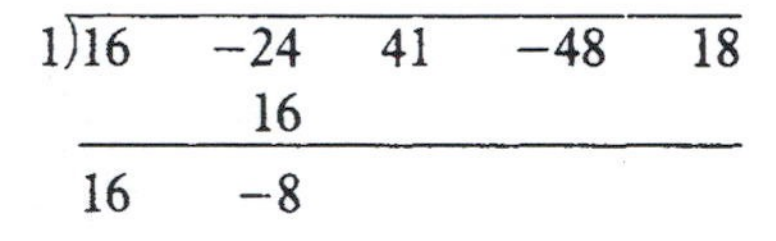

1)	16	−24	41	−48	18
		16			
	16	−8			

Because the negative term −8 appears in the quotient, we stop dividing and proceed to $x - 2$.

2)	16	−24	41	−48	18
		32	16	114	132
	16	8	57	66	150

Since the quotient's coefficients and the remainder are nonnegative, we conclude that 2 is an upper bound. ■

If we had performed the upper bound test before doing Example 2, we could have eliminated the rational numbers 2, 3, 6, 9, 18, 9/2, and 9/4 from consideration (*why is 2 eliminated?*)

A number $-a$ is called a **lower bound** for the real roots of $P(x) = 0$ if all the real roots are greater than or equal to $-a$. The upper bound test can also be used to find a lower bound. We first replace $P(x)$ by $P(-x)$. Now r is a positive root of $P(-x) = 0$ if and only if $-r$ is a negative root of $P(x) = 0$. We therefore obtain the following **lower bound test.**

If a is a positive upper bound for the real roots of $P(-x) = 0$, then $-a$ is a negative lower bound for the real roots of $P(x) = 0$.

Example 4 Use the lower bound test to find a lower bound for the real roots of the polynomial equation $16x^4 - 24x^3 + 41x^2 - 48x + 18 = 0$ in Example 2.

Solution: We first replace $P(x) = 16x^4 - 24x^3 + 41x^2 - 48x + 18$ by $P(-x) = 16x^4 + 24x^3 + 41x^2 + 48x + 18$. We then use synthetic division to divide $P(-x)$ by $x - 1$.

1)	16	24	41	48	18
		16	40	81	129
	16	40	81	129	147

Since the quotient's coefficients and the remainder are nonnegative, we can conclude that 1 is an upper bound for the real roots of $P(-x) = 0$. Hence, -1 is a lower bound for the real roots of $P(x) = 0$. Also, since $R = 147 \neq 0$, -1 is not a root of $P(x) = 0$ and all of the real roots are strictly greater than -1. ■

The upper bound test enabled us to eliminate the rational numbers 2, 3, 6, 9, 18, 9/2, and 9/4 from consideration as roots of $P(x) = 0$ in Example 2. The lower bound test now allows us to eliminate also the rational numbers -1, -2, -3, -6, -9, -18, $-3/2$, $-9/2$, $-9/4$, and $-9/8$. Hence, of the original 36 possible rational roots, we can eliminate 17 of them by our upper and lower bound tests. Furthermore, all of the real roots are contained in the interval $(-1, 2)$.

Descartes' Rule of Signs. If all the coefficients of a polynomial equation are positive (negative), then the equation cannot have a positive root (*why?*). Hence, if such an equation does have a positive root, then there must be at least one sign change among the coefficients. The following rule, which is known as **Descartes' rule of signs,**[2] gives us more precise information concerning the positive roots. We omit the proof.

[2] *The rule of signs appeared in Descartes' work* La Géométrie *(1637).*

> The number of positive roots of the polynomial equation
>
> $$a_n x^n + a_{n-1} x^{n-1} + \cdots + a_1 x + a_0 = 0 \quad (a_n \neq 0)$$
>
> is at most equal to the number of sign changes in the sequence of coefficients
>
> $$a_n, a_{n-1}, \ldots, a_1, a_0.$$

For instance, in the polynomial equation

$$3x^4 - 2x^3 - x^2 + 5x + 4 = 0,$$

the coefficients

$$3, -2, -1, 5, 4$$

have two sign changes, namely from 3 to -2 and from -1 to 5. Hence, according to Descartes' rule of signs, the equation can have at most two positive roots.

In counting the number of sign changes for a polynomial, we adopt the convention that a zero coefficient has the same sign as the coefficient of its preceding higher-power term. For example, the polynomial

$$P(x) = x^5 - 3x^2 + 2$$

has coefficients

$$1, 0, 0, -3, 0, 2$$

whose signs are

$$+, +, +, -, -, +,$$

and there are two sign changes.

Question 3 *Does Descartes' rule of signs take into account the order of a positive root?*

Answer: Yes. A positive root of order k counts as k roots in Descartes' rule of signs. For example, $P(x) = (x - 1)^3 = x^3 - 3x^2 + 3x - 1$ has three sign changes, and the root 1 of order 3 counts as three positive roots. Hence, Descartes' rule says that the number of positive roots, *including order*, is at most equal to the number of sign changes in the coefficients. ■

We can also use Descartes' rule of signs to determine the maximum number of negative roots of a polynomial. We use the property that $-r$ is a negative root of $P(x) = 0$ if and only if r is a positive root of $P(-x) = 0$. Hence, the number of negative roots of $P(x) = 0$ is equal to the number of positive roots of $P(-x) = 0$. Therefore, we can state the following version of Descartes' rule of signs.

> The number of negative roots of a polynomial equation $P(x) = 0$ is at most equal to the number of sign changes in the coefficients of $P(-x)$.

Finally, by using Descartes' rule of signs for both the positive and the negative roots, and by testing 0 separately, we can determine the maximum number of *real roots*, including order, of a polynomial equation.

Example 5 Given $P(x) = x^5 - 3x^2 + 2$, find each of the following.

(a) the maximum number of real roots of $P(x) = 0$

(b) upper and lower bounds for the real roots of $P(x) = 0$

(c) all rational roots of $P(x) = 0$

Solution:

(a) Since $P(x)$ has two sign changes, there are at most two positive roots. Now

$$\begin{aligned} P(-x) &= (-x)^5 - 3(-x)^2 + 2 \\ &= -x^5 - 3x^2 + 2. \end{aligned}$$

Hence, $P(-x)$ has one sign change, so there is at most one negative root. Finally, $P(0) = 2$, so 0 is not a root. Therefore, $P(x) = 0$ has at most three real roots, including order. (Since $P(x)$ is of degree 5, it follows that $P(x) = 0$ has at least two complex roots.)

(b) To find an upper bound for the real roots, we use synthetic division to divide $P(x)$ by $x - a$, for $a = 1, 2, 3, \ldots$. When we divide by $x - 1$, there are negative coefficients in the quotient, but when the divisor is $x - 2$, we get the following.

$$\begin{array}{r|rrrrrr} 2 & 1 & 0 & 0 & -3 & 0 & 2 \\ & & 2 & 4 & 8 & 10 & 20 \\ \hline & 1 & 2 & 4 & 5 & 10 & 22 \end{array}$$

Since the remainder and the coefficients of the quotient are all nonnegative, 2 is an upper bound for the real roots of $P(x) = 0$. Also, since $R = 22 \neq 0$, 2 is not a root and all real roots are strictly less than 2.

To find a lower bound, we apply the upper bound test to $P(-x)$. However, $P(-x) = -x^5 - 3x^2 + 2$ has a negative leading coefficient, so we must replace $P(-x)$ by $-P(-x) = x^5 + 3x^2 - 2$ (see Caution following the statement of the upper bound test). When $x^5 + 3x^2 - 2$ is divided by $x - 1$, we get

$$\begin{array}{r|rrrrrr} 1 & 1 & 0 & 0 & 3 & 0 & -2 \\ & & 1 & 1 & 1 & 4 & 4 \\ \hline & 1 & 1 & 1 & 4 & 4 & 2. \end{array}$$

Since the remainder and the coefficients of the quotient are all nonnegative, -1 is a lower bound for the real roots of $P(x) = 0$. Also, since $R = 2 \neq 0$, -1 is not a root and all real roots are strictly greater than -1. We have found that the real roots must be in the interval $(-1, 2)$.

(c) By the rational root theorem, the only possible rational roots are ± 1 and ± 2. However, -1 and ± 2 were eliminated in part (b). This leaves 1, which is in fact a root. ■

Example 5 illustrates the general approach to polynomial equations. We use Descartes' rule of signs to determine the maximum number of real roots, the bound tests to find upper and lower bounds for the real roots, and the rational root theorem to find all rational roots. To implement the bound tests, we use synthetic division. We can also use synthetic division to implement the rational root theorem if there are a large number of rational numbers to be tested. Later in this chapter we describe two numerical methods that can be used to find approximations to the real roots of a polynomial equation.

Exercises 10.2

Fill in the blanks to make each statement true.

1. According to the rational root theorem, the possible rational roots of $2x^2 + bx + 3 = 0$, where b is an integer, are ______.
2. If $x^3 + bx + c = 0$ has 5 as a root, where b and c are integers and $c \neq 0$, then the possible values of c are ______.
3. If $ax^4 + bx^3 + cx + 4 = 0$, where a, b, and c are integers, has 2/3 as a root, then the possible nonzero values of a are ______.
4. The polynomial $ax^4 + bx^3 - cx + d = 0$, where a, b, c, and d are positive integers, has at most ______ positive root(s).

5. The polynomial $ax^3 - bx^2 + cx + d = 0$, where a, b, c, and d are positive integers, has at most ________ negative root(s).

Write true *or* false *for each statement.*

6. If the coefficients of $P(x) = a_nx^n + a_{n-1}x^{n-1} + \cdots + a_1x + a_0$ are integers and p/q is a rational number for which a_0 is divisible by p and a_n is divisible by q, then $P(p/q) = 0$.
7. The equation $7x^3 + bx^2 + cx - 11 = 0$, where b and c are integers, could have 7 as a root.
8. The equation $6x^3 + bx^2 + cx + 6 = 0$, where b and c are integers, has only ± 1 as its possible rational roots.
9. The equation $x^4 + x^3 + x^2 + x + 1 = 0$ has no rational roots.
10. The equation $x^6 + 2x^4 + 5x^2 + 3 = 0$ has no real roots.

Rational Roots

Find all rational roots of the following equations.

11. $x^3 - 2x^2 - x + 2 = 0$
12. $x^3 - 6x^2 + 11x - 6 = 0$
13. $x^4 - 5x^3 + 20x - 16 = 0$
14. $2x^3 - 7x^2 - 2x + 12 = 0$
15. $4x^4 + 7x^3 + 2x^2 + 7x - 2 = 0$
16. $x^4 + 2x^3 - x - 2 = 0$

For the polynomials in Exercises 17–24, perform each of the following steps.

(a) *Find all rational roots of $P(x) = 0$.*

(b) *Use the factor theorem to find any remaining real roots and the real factored form of $P(x)$.*

(c) *Find all complex roots of $P(x) = 0$.*

17. $P(x) = x^3 - x^2 - 4x + 4$
18. $P(x) = x^3 - x^2 - 4x - 2$
19. $P(x) = 2x^3 - x^2 + 2x - 1$
20. $P(x) = 2x^3 + x^2 + 2x + 1$
21. $P(x) = x^4 + x^3 - 5x^2 - 3x + 6$
22. $P(x) = x^4 + 2x^3 - 25x^2 - 50x$
23. $P(x) = x^4 + 2x^3 + x^2 - 2x - 2$
24. $P(x) = x^4 + 2x^3 - 3x^2 - 2x + 2$

Upper and Lower Bounds

Find integers a and b that are lower and upper bounds, respectively, for the real roots of each of the following polynomial equations.

25. $2x^3 - 4x^2 + 5x + 7 = 0$
26. $4x^3 + 3x^2 - 8x + 10 = 0$
27. $x^4 + 7x^3 + 12x^2 - 4x - 16 = 0$
28. $2x^4 - 5x^3 - 15x^2 + 12x + 3 = 0$

Descartes' Rule of Signs

Use Descartes' rule of signs to determine the maximum number of real roots of each of the following equations.

29. $2x^3 - x^2 - 2x - 1 = 0$
30. $x^4 - x^3 + x^2 - 4x - 5 = 0$
31. $5x^5 + 4x^3 - x + 2 = 0$
32. $x^4 - 2x^2 + 7 = 0$
33. $3x^6 + x^5 + 2x^3 + x = 0$
34. $2x^6 - 4x^4 + x^2 + x = 0$

For the polynomials in Exercises 35–40, peform each of the following steps.

(a) *Find the maximum number of positive and negative roots of $P(x) = 0$.*

(b) *Find integers a and b for which $[a, b]$ contains the real roots of $P(x) = 0$.*

35. $P(x) = x^3 + 3x^2 + 3x - 7$
36. $P(x) = x^3 - 3x^2 + 3x + 7$
37. $P(x) = x^3 - 3x^2 - 21x + 55$
38. $P(x) = x^3 + 9x^2 + 21x + 13$
39. $P(x) = x^4 + 8x^3 + 6x^2 - 4x - 2$
40. $P(x) = x^4 - 8x^3 + 30x^2 - 56x + 43$

Miscellaneous

41. Use the equation $x^2 - 2 = 0$ to conclude that $\sqrt{2}$ is an irrational number.
42. Use the equation $x^2 + 1 = 0$ to conclude that $\sqrt{-1}$ is not a real number.

10.3 Partial Fractions

Partial Fraction Decomposition
Decomposition of Proper Rational Expressions
Decomposition of Improper Rational Expressions

In partial fraction problems, we are given the answer, and the object is to find the question.

Partial Fraction Decomposition. In Section 1.3 we added two or more rational expressions by means of a common denominator in order to obtain a single rational expression. Here we reverse the process. That is, we decompose a given rational expression into a sum of fractional expressions whose denominators are factors of the given denominator. The fractional expressions in the sum are called **partial fractions.** For example, each of the decompositions

$$\frac{x-2}{x(x-1)^2} = \frac{-2}{x} + \frac{2}{x-1} + \frac{-1}{(x-1)^2}$$

and

$$\frac{x^2-x+13}{(x+1)(x^2+4)} = \frac{3}{x+1} + \frac{-2x+1}{x^2+4}$$

can be verified by adding the fractions on the right to obtain the rational expressions on the left. Partial fraction decomposition is an important operation in calculus, and it is made possible by the property that every polynomial with real coefficients has a real factored form as defined in the first section of this chapter.

Decomposition of Proper Rational Expressions. Let $P(x)/Q(x)$ be a given rational expression, where $P(x)$ and $Q(x)$ are polynomials in x. We consider first the case where $P(x)/Q(x)$ is a **proper rational expression,** which means that the degree of the numerator $P(x)$ is *less than* the degree of the denominator $Q(x)$. Then $P(x)/Q(x)$ can be decomposed into partial fractions in the following three steps.

1. Factor $Q(x)$ into its real factored form of linear and irreducible quadratic factors:

$$Q(x) = a \cdot \ldots \cdot (x-r)^m \cdot \ldots \cdot (x^2+bx+c)^n \cdot \ldots .$$

If $m > 1$, the factor $(x-r)^m$ is called a **repeated linear factor,** and if $n > 1$, $(x^2+bx+c)^n$ is called a **repeated quadratic factor.**

2. Write

$$\begin{aligned}\frac{P(x)}{Q(x)} = \cdots &+ \frac{A_1}{x-r} + \frac{A_2}{(x-r)^2} + \cdots + \frac{A_m}{(x-r)^m} + \cdots \\ \cdots &+ \frac{B_1x+C_1}{x^2+bx+c} + \frac{B_2x+C_2}{(x^2+bx+c)^2} + \cdots \\ \cdots &+ \frac{B_nx+C_n}{(x^2+bx+c)^n} + \cdots,\end{aligned}$$

where the A's, B's, and C's are constant real numbers to be determined. Note that there are m terms corresponding to each factor $(x - r)^m$ and n terms corresponding to each factor $(x^2 + bx + c)^n$.

3. Solve for the constant A's, B's, and C's.

It can be shown by advanced algebra that every proper rational expression has a decomposition into partial fractions as indicated in step 2. The most general method for finding the constants is illustrated in the following examples. Another method, which requires less computation in some cases, is described in Exercises 35 and 36. Also, we note in step 1 that for a general polynomial $Q(x)$ of degree ≥ 5, its real factored form must be given (*why?*).

Example 1 Decompose the proper rational function $1/(x^2 + x - 2)$ into partial fractions.

Solution: First factor the denominator into $(x - 1)(x + 2)$. These factors are linear and nonrepeating. Now let

$$\frac{1}{x^2 + x - 2} = \frac{A}{x - 1} + \frac{B}{x + 2}$$

in accordance with step 2. We now solve for A and B. By adding on the right side with the LCD, we obtain

$$\frac{1}{x^2 + x - 2} = \frac{A(x + 2)}{(x - 1)(x + 2)} + \frac{B(x - 1)}{(x + 2)(x - 1)}$$

or

$$\frac{1}{x^2 + x - 2} = \frac{(A + B)x + (2A - B)}{(x - 1)(x + 2)}. \quad (1)$$

Since the denominators in equation (1) are equal, the numerators must also be equal. Therefore,

$$1 = (A + B)x + (2A - B), \quad (2)$$

which implies the system of equations

$$\begin{aligned} A + B &= 0 \\ 2A - B &= 1, \end{aligned}$$

{*In order for equation (2) to hold for all values of x, the coefficient $A + B$ of x must equal 0, and the constant term $2A - B$ must equal 1*

whose solution is $A = 1/3$ and $B = -1/3$. Therefore,

$$\frac{1}{x^2 + x - 2} = \frac{\frac{1}{3}}{x - 1} + \frac{-\frac{1}{3}}{x + 2}$$

$$= \frac{1}{3}\left(\frac{1}{x - 1}\right) - \frac{1}{3}\left(\frac{1}{x + 2}\right). \quad ■$$

Example 2 Decompose $(2 - x)/[x(x - 1)^2]$ into partial fractions.

Solution: The numerator has degree 1 and the denominator has degree 3, so the given rational expression is proper. The factors of the denominator are linear, but $(x - 1)^2$ is a repeated linear factor. Therefore, let

$$\frac{2-x}{x(x-1)^2} = \frac{A}{x} + \frac{B}{x-1} + \frac{C}{(x-1)^2}$$

in accordance with step 2. We now find A, B, and C. The partial fractions on the right have LCD $= x(x-1)^2$. Therefore,

$$\frac{2-x}{x(x-1)^2} = \frac{A(x-1)^2 + Bx(x-1) + Cx}{x(x-1)^2}$$

or

$$\frac{2-x}{x(x-1)^2} = \frac{(A+B)x^2 + (-2A - B + C)x + A}{x(x-1)^2}. \tag{3}$$

By equating the numerators of equation (3), we obtain

$$2 - x = (A+B)x^2 + (-2A - B + C)x + A, \tag{4}$$

which is equivalent to the system of equations

$$\begin{aligned} A + B &= 0 \\ -2A - B + C &= -1 \\ A &= 2, \end{aligned}$$

In order for equation (4) to hold for all values of x, the coefficient of each power of x on the right must equal the coefficient of the same power of x on the left

whose solution is $A = 2$, $B = -2$, $C = 1$. Therefore,

$$\frac{2-x}{x(x-1)^2} = \frac{2}{x} - \frac{2}{x-1} + \frac{1}{(x-1)^2}.$$ ■

Example 3 Decompose the proper rational expression $x/(x+1)^2$ into partial fractions.

Solution: As in step 2,

$$\begin{aligned} \frac{x}{(x+1)^2} &= \frac{A}{x+1} + \frac{B}{(x+1)^2} \\ &= \frac{A(x+1) + B}{(x+1)^2} \\ &= \frac{Ax + (A+B)}{(x+1)^2}. \end{aligned}$$

By equating numerators, we obtain the equations

$$\begin{aligned} A &= 1 \\ A + B &= 0. \end{aligned}$$

Therefore, $A = 1$, $B = -1$, and

$$\frac{x}{(x+1)^2} = \frac{1}{x+1} - \frac{1}{(x+1)^2}.$$ ■

Example 4 Decompose the proper rational expression $(x - 5)/[(x + 1)(x^2 + 1)]$ into partial fractions.

Solution: The denominator is factored into one linear and one irreducible quadratic factor, both nonrepeating. Hence, by step 2,

$$\frac{x - 5}{(x + 1)(x^2 + 1)} = \frac{A}{x + 1} + \frac{Bx + C}{x^2 + 1}.$$

As in the preceding examples, we add the partial fractions with their LCD, thereby obtaining

$$\begin{aligned}\frac{x - 5}{(x + 1)(x^2 + 1)} &= \frac{A}{x + 1} + \frac{Bx + C}{x^2 + 1}\\ &= \frac{A(x^2 + 1) + (Bx + C)(x + 1)}{(x + 1)(x^2 + 1)}\\ &= \frac{(A + B)x^2 + (B + C)x + (A + C)}{(x + 1)(x^2 + 1)}.\end{aligned}$$

Therefore,

$$x - 5 = (A + B)x^2 + (B + C)x + (A + C),$$

which implies

$$\begin{aligned}A + B &= 0\\ B + C &= 1\\ A + C &= -5.\end{aligned}$$

The solution of this system is $A = -3$, $B = 3$, and $C = -2$. Hence,

$$\frac{x - 5}{(x + 1)(x^2 + 1)} = \frac{-3}{x + 1} + \frac{3x - 2}{x^2 + 1}.$$ ■

Example 5 Decompose the proper rational expression

$$\frac{x^4 + 3x^3 + 3x^2 + 4x + 1}{x(x^2 + 1)^2}$$

into partial fractions.

Solution: The denominator is factored into one nonrepeating linear factor and one repeating, irreducible quadratic factor. Therefore, by step 2,

$$\begin{aligned}&\frac{x^4 + 3x^3 + 3x^2 + 4x + 1}{x(x^2 + 1)^2}\\ &= \frac{A}{x} + \frac{Bx + C}{x^2 + 1} + \frac{Dx + E}{(x^2 + 1)^2}\\ &= \frac{A(x^2 + 1)^2 + (Bx + C)x(x^2 + 1) + (Dx + E)x}{x(x^2 + 1)^2}\\ &= \frac{(A + B)x^4 + Cx^3 + (2A + B + D)x^2 + (C + E)x + A}{x(x^2 + 1)^2}.\end{aligned}$$

By equating numerators as in the previous examples, we arrive at the system of equations

$$\begin{aligned} A + B &= 1 \\ C &= 3 \\ 2A + B + D &= 3, \\ C + E &= 4 \\ A &= 1, \end{aligned}$$

whose solution is $A = 1, B = 0, C = 3, D = 1$, and $E = 1$. Thus, the partial fraction decomposition is

$$\frac{x^4 + 3x^3 + 3x^2 + 4x + 1}{x(x^2 + 1)^2} = \frac{1}{x} + \frac{3}{x^2 + 1} + \frac{x + 1}{(x^2 + 1)^2}. \quad ■$$

Comment We note that in all cases the number of constants to be determined is equal to the degree of the given denominator. In Example 1, the degree of the denominator $x^2 + x - 2$ is 2, and we found two constants A and B. Similarly, in Example 5, in which the denominator $x(x^2 + 1)^2 = x^5 + 2x^3 + x$ has degree 5, we solved for five constants.

Decomposition of Improper Rational Expressions. Suppose that $P(x)/Q(x)$ is an improper rational expression, which means that degree $P(x) \geq$ degree $Q(x)$. We then divide $Q(x)$ into $P(x)$ to obtain

$$\frac{P(x)}{Q(x)} = q(x) + \frac{R(x)}{Q(x)},$$

where the quotient $q(x)$ is a polynomial, and the remainder $R(x)$ is a polynomial whose degree is *less than* the degree of $Q(x)$. The partial fraction decomposition may now be applied to the *proper* rational expression $R(x)/Q(x)$.

Example 6 Decompose $(3x^2 + 3x - 5)/(x^2 + x - 2)$ into partial fractions.

Solution:

$$\frac{3x^2 + 3x - 5}{x^2 + x - 2} = 3 + \frac{1}{x^2 + x - 2} \qquad \textit{By division}$$

$$= 3 + \frac{1}{3}\left(\frac{1}{x - 1}\right) - \frac{1}{3}\left(\frac{1}{x + 2}\right) \qquad \textit{By Example 1} \quad ■$$

Example 7 Decompose $(3x^3 + 7x^2 + 3x + 1)/(x^2 + x)$ into partial fractions.

Solution:

$$\frac{3x^3 + 7x^2 + 3x + 1}{x^2 + x} = 3x + 4 + \frac{-x + 1}{x^2 + x} \qquad \textit{By division}$$

Now

$$\frac{-x + 1}{x^2 + x} = \frac{-x + 1}{x(x + 1)} = \frac{A}{x} + \frac{B}{x + 1}, \qquad \textit{By step 2}$$

where we obtain, by the methods of the previous examples, $A = 1$ and $B = -2$. Therefore,

$$\frac{3x^3 + 7x^2 + 3x + 1}{x^2 + x} = 3x + 4 + \frac{1}{x} - \frac{2}{x + 1}.$$ ■

Caution If the denominator of a proper rational expression is a *nonrepeating* linear or irreducible quadratic polynomial, then no further partial fraction decomposition is possible. For example,

$$\frac{x^4 + 5x^2 + 6x + 11}{x^2 + 4} = x^2 + 1 + \frac{6x + 7}{x^2 + 4},$$

and the proper rational expression $(6x + 7)/(x^2 + 4)$ is already in partial fraction form.

Exercises 10.3

Fill in the blanks to make each statement true.

1. Partial fraction decomposition is applied to ________ rational expressions.
2. In a proper rational expression, the numerator has ________ less than that of the denominator.
3. If the degree of the numerator is ________ or ________ the degree of the denominator, the first step needed before a partial fractional decomposition is ________.
4. The type of partial fraction decomposition of a proper rational expression depends upon the ________ of the denominator.
5. For each factor $(x - r)^n$ of the denominator, there are ________ constants to be determined; for each irreducible factor $(x^2 + bx + c)^m$, there are ________ constants to be determined.

Write true *or* false *for each statement.*

6. For some constants A and B, $\dfrac{3x^2 + 3x - 5}{x^2 + x - 2} = \dfrac{A}{x - 1} + \dfrac{B}{x + 2}$.
7. If a proper rational expression has denominator $(x + 3)(x - 3)^2$, its decomposition is of the form $\dfrac{A}{x + 3} + \dfrac{B}{(x - 3)^2}$.
8. If a proper rational expression has denominator $x^2 + 3x + 2$, its decomposition is of the form $\dfrac{A}{x + 1} + \dfrac{B}{x + 2}$.
9. If a proper rational expression has denominator $(x + 2)(x^2 + x + 2)$, its decomposition is of the form $\dfrac{A}{x + 2} + \dfrac{Bx + C}{x^2 + x + 2}$.
10. If a proper rational expression has denominator $x^2(x + 1)^3$, its decomposition is of the form $\dfrac{A}{x} + \dfrac{B}{x^2} + \dfrac{C}{x + 1} + \dfrac{D}{(x + 1)^2} + \dfrac{E}{(x + 1)^3}$.

Partial Fraction Decomposition of Rational Expressions

Decompose, if possible, each of the following rational expressions into partial fractions. If necessary, divide first.

11. $\dfrac{x - 13}{x^2 - x - 6}$
12. $\dfrac{2x + 1}{x^2 + 5x + 6}$
13. $\dfrac{5x - 4}{x^2 - 4x}$
14. $\dfrac{2x^2}{x^2 - 3x + 2}$
15. $\dfrac{2x^2 + 3}{x(x - 1)^2}$
16. $\dfrac{2x + 4}{x^2(x - 2)}$
17. $\dfrac{x^2 - 2x - 8}{x(x - 2)^3}$
18. $\dfrac{x - 1}{(x + 1)^2}$

19. $\frac{1}{(x-1)^3}$

20. $\frac{x}{(x-1)^3}$

21. $\frac{x^2}{(x-1)^3}$

22. $\frac{x^3}{(x-1)^3}$

23. $\frac{x^3+2x^2-2x+2}{x^2-3x+2}$

24. $\frac{x^2}{x^2-3x+2}$

25. $\frac{x}{x^2+1}$

26. $\frac{x}{x^2+x+1}$

27. $\frac{1}{x(x^2+x+1)}$

28. $\frac{3x^2+x+13}{(x-1)(x^2+16)}$

29. $\frac{4}{(x^2-4)(x^2+4)}$

30. $\frac{3x^3+3x^2+3x+6}{(x^2+1)(x^2+4)}$

31. $\frac{1}{(x^2+1)^2}$

32. $\frac{x^2}{(x^2+1)^2}$

33. $\frac{x^2}{(x^2-1)^2}$

34. $\frac{x^3}{(x^2+1)^2}$

35. The general method used in this section for finding the constants in a partial fraction decomposition requires the solution of a system of equations. An alternate (and sometimes easier) method is now illustrated.

The decomposition in Example 1 of the text is

$$\frac{1}{x^2+x-2}=\frac{A}{x-1}+\frac{B}{x+2}=\frac{A(x+2)+B(x-1)}{(x-1)(x+2)},$$

and by setting numerators equal, we have

$$1 = A(x+2)+B(x-1). \qquad \text{(i)}$$

Substitution of $x = 1$ in (i) gives $1 = A(3)$ or $A = 1/3$. Substitution of $x = -2$ in (i) gives $1 = B(-3)$ or $B = -1/3$. Thus,

$$\frac{1}{x^2+x-2}=\frac{\frac{1}{3}}{x-1}-\frac{\frac{1}{3}}{x+1}.$$

Apply this substitution method to solving Exercises 11–14 on the preceding page. Note that these examples involve only nonrepeated linear factors.

36. The substitution method in Exercise 35 may be applied with some modification to cases in which the factors are not all linear and nonrepeated. We solve Example 2 by this method as follows. We have

$$\frac{2-x}{x(x-1)^2}=\frac{A}{x}+\frac{B}{x-1}+\frac{C}{(x-1)^2}=\frac{A(x-1)^2+Bx(x-1)+Cx}{x(x-1)^2}$$

and by setting numerators equal,

$$2-x = A(x-1)^2+Bx(x-1)+Cx. \qquad \text{(ii)}$$

Substituting $x = 0$ in (ii) gives $2 = A$, and substituting $x = 1$ in (ii) gives $1 = C$, but this method will not work for B (*why?*). However, if we substitute $A = 2$, $C = 1$, and any value of $x(\neq 0 \text{ or } 1)$ in (ii), we will get an equation for B. For instance, if $x = 2$, then (ii) becomes

$$0 = 2(1) + B(2)(1) + 1(2)$$

so that $-4 = 2B$ or $B = -2$. Therefore,

$$\frac{2-x}{x(x-1)^2}=\frac{2}{x}-\frac{2}{x-1}+\frac{1}{(x-1)^2}.$$

Solve Exercises 15, 16, 28, and 29 by this method of substitution.

10.4 Numerical Methods for Approximating Solutions

The bisection method is to Newton's method as the tortoise is to the hare.

Rough Estimates
The Bisection Method
Newton's Method

We know from the fundamental theorem of algebra that a polynomial equation of degree n has n roots, including order, in the complex number system. In this section we investigate two methods for obtaining approximate values for the *real* roots of a polynomial equation. The idea in both cases is to first do a point graph of the polynomial in order to get rough estimates of the real roots and then to use a numerical procedure to obtain closer approximations.

Most numerical methods for approximating a root of $P(x) = 0$ are *iterative* in nature. That is, one approximation of the root is used to generate another approximation that is better than the first, and so on until we achieve the desired degree of accuracy.

Accuracy can be measured in two ways. One way, of course, is to find a value of x for which $|P(x)|$ is very small in magnitude. Another way is to find an interval $[a, b]$ of very small length that is known to contain an exact root of $P(x) = 0$. We will use both ways.

One final note about approximations: because most numerical procedures are iterative and may require many repetitions to achieve acceptable results, we usually rely on calculators or computers to do the computations. Our results therefore depend on the level of precision of the computing tool used. Most calculators display 8 or 10 digits, and microcomputers display 7 digits in single precision and 16 in double precision.

Rough Estimates. As illustrated in the following example, we can obtain rough estimates for the real roots of a polynomial equation $P(x) = 0$ by graphing $y = P(x)$ on an interval $[a, b]$, where a is a lower bound and b is an upper bound for the real roots.

Example 1 Find rough estimates for the real roots of $P(x) = 0$, where

$$P(x) = 5x^4 - 10x^3 - 3x^2 - 25.$$

Solution: We use synthetic division to divide $P(x)$ by $x - a$ for $a = 1$, 2, 3, . . . and find that for $x - 3$, the remainder and the coefficients of the quotient are positive. Therefore, 3 is an upper bound for the real roots of $P(x) = 0$. Similarly, by dividing $P(-x)$ by $x - a$ for $a = 1, 2, 3, \ldots$, we find that -2 is a lower bound. Hence, all of the real roots of $P(x) = 0$ are contained in the interval $[-2, 3]$. We then plot points to get a rough sketch of the graph of $y = P(x)$ on the interval $[-2, 3]$ (Figure 1). As the graph shows, $P(x)$ changes sign as x goes from -2 to -1 and from 2 to 3. Therefore, we conclude that $P(x) = 0$ has a real root in the interval $[-2, -1]$ and another in $[2, 3]$. ■

We now investigate two methods, one algebraic and the other geometric, for closely approximating the real roots of the polynomial equation in Example 1.

The Bisection Method. If $P(a)$ and $P(b)$ have opposite signs, where $a < b$, then $P(x) = 0$ has a root in the interval $[a, b]$. For example, suppose $P(a) < 0$ and $P(b) > 0$, as in Figure 2. We then let $c = (a + b)/2$, the *midpoint* of the interval $[a, b]$, and compute $P(c)$. If $P(c) < 0$, we know

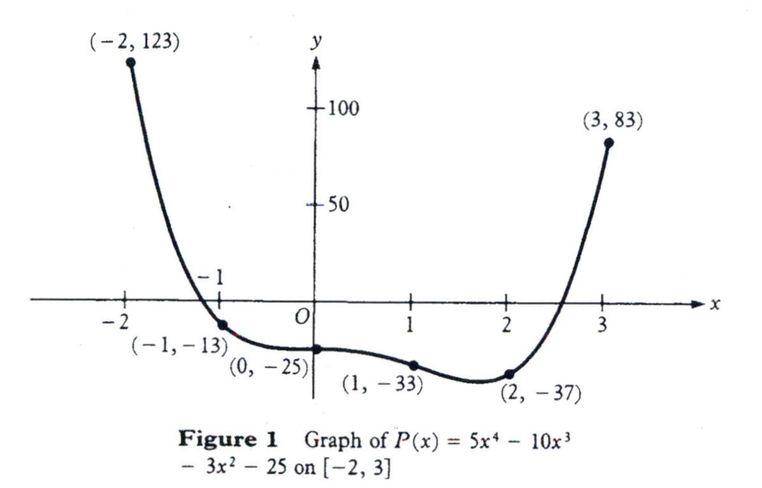

Figure 1 Graph of $P(x) = 5x^4 - 10x^3 - 3x^2 - 25$ on $[-2, 3]$

there is a root in the interval $[c, b]$, and if $P(c) > 0$, we know there is a root in the interval $[a, c]$. (In Figure 2, $c = (2 + 3)/2 = 2.5$ and $P(2.5) < 0$, so there is a root in the interval $[2.5, 3]$.) Whichever is the case, we now have an interval that contains a root of $P(x) = 0$ and is one half the length of the original interval. By repeating this bisection process, we can generate a sequence of intervals containing a root of $P(x) = 0$, and each interval is one half the length of the one before. Of course, if $P(c) = 0$ for one of the midpoints c, then we have found an *exact* root. Otherwise, we continue the bisection process until an interval is reached whose length is less than or equal to some preassigned small positive value, say δ. Then any value of x in this last interval is within δ of an exact root.

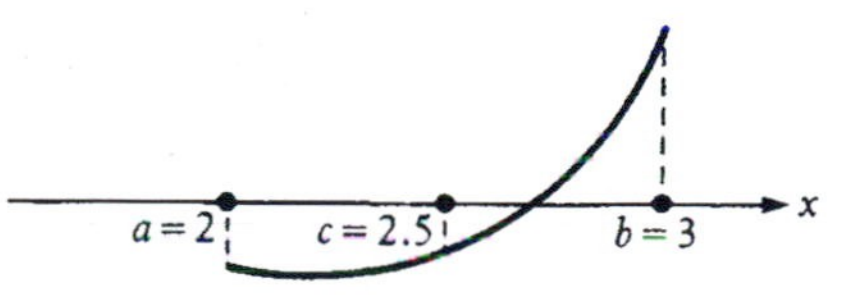

Figure 2 The Bisection Method

Note that the length of the original interval is $b - a$. After the first bisection, the length is $(b - a)/2$; after the second, it is $(b - a)/2^2$, then $(b - a)/2^3$, and so on. To reach an interval whose length is less than or equal to δ will require n bisections, where

$$\frac{b - a}{2^n} \leq \delta.$$

Example 2 For the polynomial

$$P(x) = 5x^4 - 10x^3 - 3x^2 - 25$$

in Example 1, find a subinterval of [2, 3] of length $\leq .001$ that contains a root of $P(x) = 0$.

Solution: The interval [2, 3] has length $3 - 2 = 1$, and

$$\frac{1}{2^{10}} \approx .00098 < .001.$$

Therefore, in 10 bisections, we will obtain an interval of length $\leq .001$ that contains a root of $P(x) = 0$. The steps are listed in Table 1, where all numbers are computed with five digits and rounded to four decimal places.

Table 1 Bisection Method for $P(x) = 5x^4 - 10x^3 - 3x^2 - 25$ on [2, 3]

Step	a	Sign of $P(a)$	b	Sign of $P(b)$	$b - a$	$c = \frac{a+b}{2}$	Sign of $P(c)$
1	2.0000	−	3.0000	+	1.0000	2.5000	−
2	2.5000	−	3.0000	+	.5000	2.7500	+
3	2.5000	−	2.7500	+	.2500	2.6250	+
4	2.5000	−	2.6250	+	.1250	2.5625	+
5	2.5000	−	2.5625	+	.0625	2.5313	−
6	2.5313	−	2.5625	+	.0312	2.5469	+
7	2.5313	−	2.5469	+	.0156	2.5391	−
8	2.5391	−	2.5469	+	.0078	2.5430	+
9	2.5391	−	2.5430	+	.0039	2.5410	+
10	2.5391	−	2.5410	+	.0019	2.5401	−
Final Interval	2.5401	−	2.5410	+	.0009		

Hence, after ten steps, we arrive at the interval [2.5401, 2.5410] of length less than .001 that contains a root of $P(x) = 0$. ■

The numbers in Table 1 are rounded to four decimal places. If the computations were performed on a calculator with 8 or 10 digits, then the calculator numbers might differ slightly from those in the table. Naturally, greater accuracy is achieved if the computations are made with more than five digits. Also, the number of steps will increase as δ decreases. If a very close approximation of a root is desired, say $\delta = .000001$, then the computations are best performed on a computer or a programmable calculator. For this purpose, we provide the flow chart and BASIC computer program in Figure 3. When the program was run on a microcomputer, the following results were obtained for the $P(x)$ in Example 2.

For $\delta = .000001$, 20 iterations resulted in

$$[a, b] = [2.5409212, 2.5409222],$$

and for $\delta = .00000001$, 27 iterations gave

$$[a, b] = [2.540921621, 2.540921628].$$

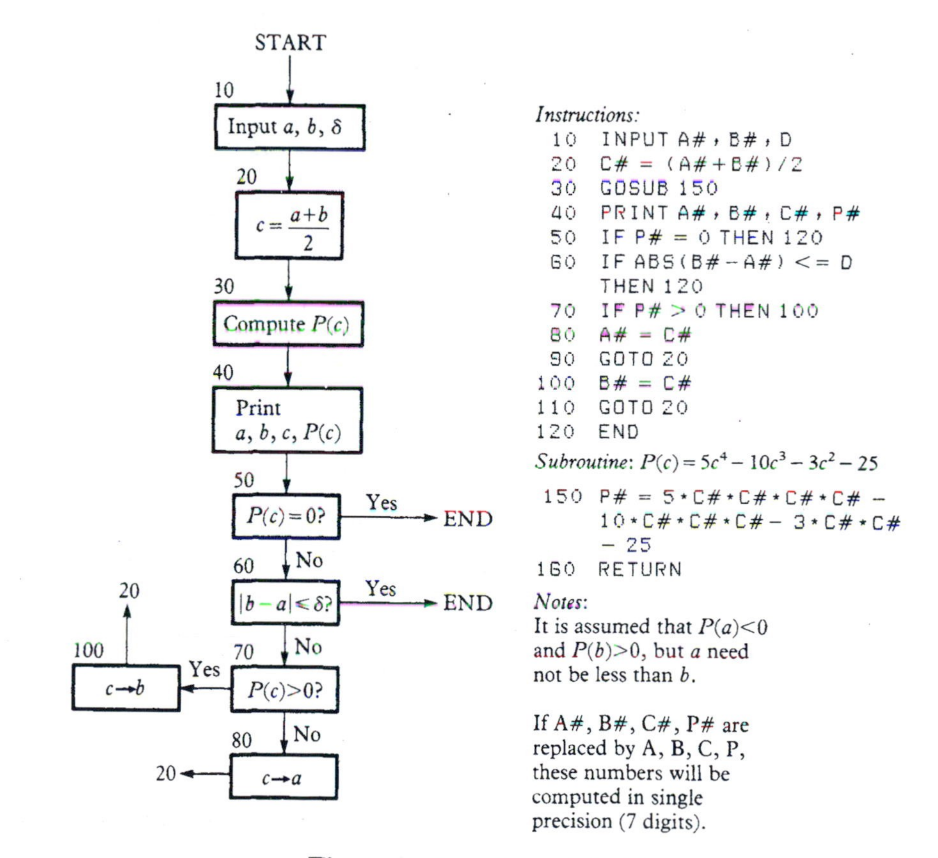

Figure 3 Flow Chart and BASIC Program in Double Precision for the Bisection Method

(The calculations were made in double precision with 16 digits. The numbers in the above intervals are rounded off to 8 and 10 digits, respectively.)

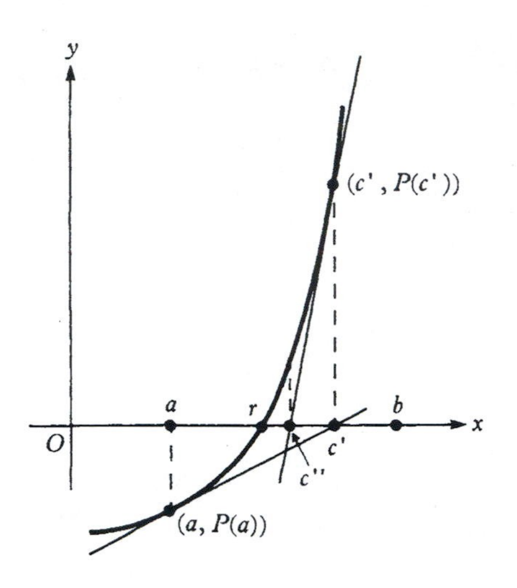

Figure 4 Newton's Method

Newton's Method. Whereas the bisection process is *algebraic*, Newton's method is a *geometric* procedure for finding an approximate root of a polynomial equation. If we know that $P(x) = 0$ has a root in the interval $[a, b]$, we can start the procedure at either a or b. In Figure 4, starting at a, we draw the tangent line to the curve $y = P(x)$ at the point $(a, P(a))$.

Now let c' be the value of x for which the tangent line intersects the x-axis. We then draw another tangent line at $(c', P(c'))$ that intersects the x-axis at some point c'', and repeat the process at c'', and so on. As indicated in Figure 4,

$$a, c', c'', \ldots$$

form a sequence of numbers that approach the root r of $P(x) = 0$. We continue to draw tangent lines until we reach a point c in the sequence for which

$$|P(c)| \leq \epsilon,$$

where ϵ is some preassigned small positive number.

In carrying out Newton's method, we don't actually *draw* the tangent lines. We can *compute* their point of intersection with the x-axis by using the following formula, which is derived in calculus.

> Let
>
> $$P(x) = a_n x^n + a_{n-1}x^{n-1} + \cdots + a_2x^2 + a_1x + a_0.$$
>
> Then the function
>
> $$P'(x) = na_nx^{n-1} + (n-1)a_{n-1}x^{n-2} + \cdots + 2a_2x + a_1 \quad \textbf{(1)}$$
>
> gives the slope of the tangent line to the curve $y = P(x)$. That is, at the point $(c, P(c))$ on the curve, the slope of the tangent line is $P'(c)$.

For instance, in Example 2 we have

$$P(x) = 5x^4 - 10x^3 - 3x^2 - 25,$$

and therefore

$$P'(x) = 20x^3 - 30x^2 - 6x.$$

Now, for any $P(x)$, a point-slope equation of the tangent line to $y = P(x)$ at the point $(c, P(c))$ is (see Section 2.5)

$$y - P(c) = P'(c)(x - c),$$

where $P'(c)$ is the slope, according to (1). If $P'(c) \neq 0$, this line intersects the x-axis at the point

$$x = c - \frac{P(c)}{P'(c)},$$

which is obtained by setting $y = 0$ in the point-slope equation. Hence, if c is one of the points of intersection of the tangent line with the x-axis in Newton's method, then the next point of intersection is

$$c' = c - \frac{P(c)}{P'(c)}. \qquad \textit{Points generated by Newton's method} \qquad (2)$$

After ϵ has been specified, we start Newton's method at either a or b and use (2) to generate c', c'', . . . until a c is reached for which $|P(c)| \leq \epsilon$.

Example 3 For the polynomial

$$P(x) = 5x^4 - 10x^3 - 3x^2 - 25$$

in Examples 1 and 2, use Newton's method, starting at 3, to find a value c for which $|P(c)| \leq .01$.

Solution: From (1) we know that the slope of the tangent line at $(c, P(c))$ is

$$P'(c) = 20c^3 - 30c^2 - 6c.$$

Starting at the point $c = 3$ on the x-axis, we use (2) to determine the next point in Newton's method. The results, computed with a ten-digit calculator and rounded off, are listed in Table 2.

Table 2 Newton's Method for $P(x) = 5x^4 - 10x^3 - 3x^2 - 25$ on [2, 3]

Step	c	$P(c)$	$P'(c)$	$c - \frac{P(c)}{P'(c)}$
1	3.0000000	83.000000	252.00000	2.6706349
2	2.6706349	17.473539	150.96236	2.5548872
3	2.5548872	1.6866317	122.38512	2.5411059
4	2.5411059	.0219631	119.20634	2.5409217
5	2.5409217	.0000092		

In step 5 of the table we have $P(2.5409217) = .0000092$, and since $|.0000092| \leq .01$, we stop the process. Note that Newton's method has given an approximation in five steps comparable to the approximation that the bisection method gave after twenty steps. ■

The flow chart and BASIC program in Figure 5 can be used for Newton's method.

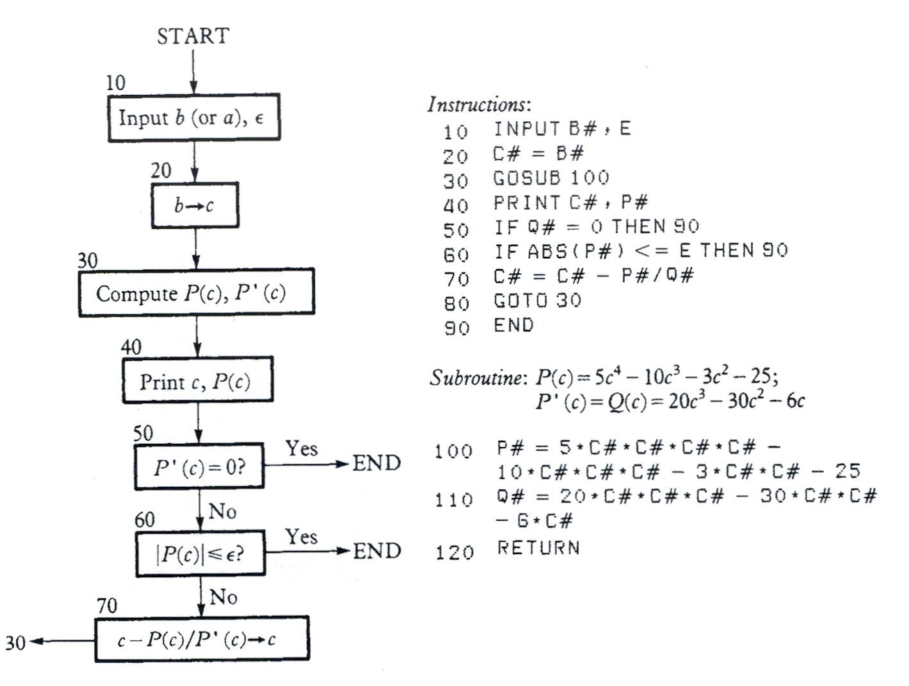

Figure 5 Flow Chart and BASIC Program in Double Precision for Newton's Method

When the program was run on a microcomputer, starting at $c = 3$, the following results were obtained for the $P(x)$ of Example 3.

Iterations	c	$P(c)$
5	2.54092165...	$3.88... \times 10^{-6}$
6	2.54092162...	$1.19... \times 10^{-13}$

Note that the value .00000388 for $P(c)$ obtained at the fifth iteration on the computer differs slightly from the .0000092 obtained at the fifth iteration in Example 3. This is because the computer was working with 16 digits while Example 3 was done on a calculator with 10 digits.

For seven or more iterations, the results on the computer remained the same, namely

$$c = 2.540921622692491 \quad \text{and} \quad P(c) = 4.44... \times 10^{-16}.$$

Hence, by only the seventh iteration, Newton's method obtained an approximate root c for which the decimal representation of $P(c)$ has 15 leading zeros! Furthermore, this is the best possible result within the limits of double precision accuracy of the computer.

Comparing the bisection method with Newton's method is like comparing "the tortoise and the hare." When it works, Newton's method usually arrives at a close approximation of a root much faster (in fewer iterations) than the bisection method. However, we have to look out for the following possibilities in Newton's method.

1. If $P'(c') = 0$ for some c' in the sequence generated, then Newton's method will stop at c'. Geometrically, this is because the tangent line at $(c', P(c'))$ is horizontal and does not intersect the x-axis (Figure 6).

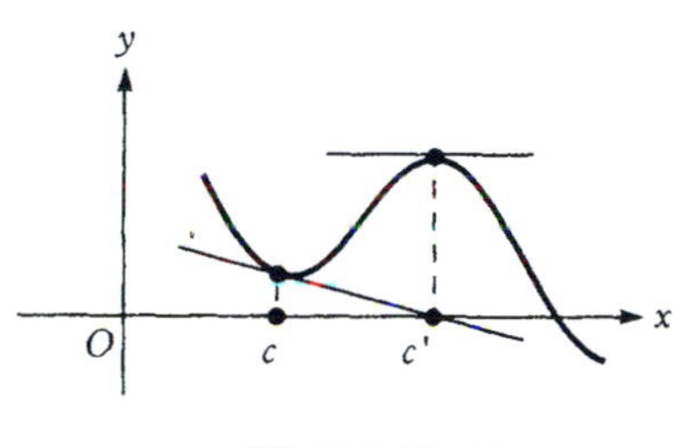

Figure 6

2. The sequence of c's may not approach a unique point. As indicated in Figure 7, if Newton's method for $P(x) = 3x^5 - 5x^3$ reaches the point $c' = 5/(3\sqrt{3}) \approx .96$, then from that point on, the sequence oscillates between c' and $c'' = -c'$.

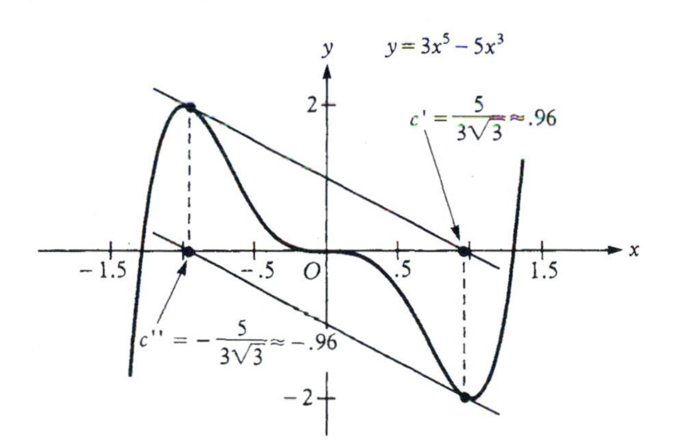

Figure 7

3. The sequence of c's could approach a root outside the interval $[a, b]$. As indicated in Figure 8, if Newton's method is started at $c = b$, the sequence generated approaches a root r that is not contained in $[a, b]$.

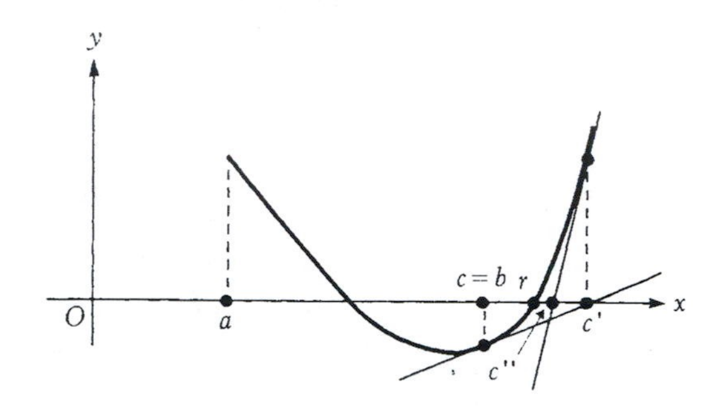

Figure 8

4. Suppose we are computing with n digits and $|P(c)|$ is greater than some value ϵ_0 for every n-digit number in $[a, b]$. If an ϵ less than ϵ_0 is chosen as a measure of accuracy, then Newton's method will arrive at some value of c and then repeat the same value over and over in an endless search to find one satisfying $|P(c)| \leq \epsilon$.

Situations 1, 2, and 3 above are due to the geometric nature of Newton's method. They can be rectified by choosing a different value of c in the interval $[a, b]$ and starting over. Situation 4 is due to the physical limitations of the computing equipment. It can be rectified by choosing a larger value of ϵ and settling for less accuracy.

There are no limitations on the bisection method comparable to the above four. If we are working with n digits, then even if a value of 0 is chosen for δ, the method will eventually produce an interval $[a, b]$, where $a = b$ in the first n digits. The exact root will agree with a and b in the first n digits. However, as we have seen, the bisection method can be much slower than Newton's method.

Comment In the case of a quadratic equation, we have a formula for finding the roots exactly. However, even for quadratic equations, we may have to sacrifice perfect accuracy when obtaining a numerical solution. For example, suppose we wish to solve the quadratic equation

$$x^2 - \sqrt{2}x + 23.1752194 = 0$$

using an 8-place calculator. Since we cannot enter exact values of the coefficients $\sqrt{2}$ and 23.1752194 in the calculator, we would have to replace the given equation by

$$x^2 - (1.4142136)x + 23.175219 = 0.$$

Hence, even before applying the quadratic formula, we are dealing with an approximation to the original equation. Solving the second equation by the quadratic formula will introduce further approximations. In any given applied problem, we must first determine the amount of accuracy that is required and then proceed accordingly.

We still have not touched on two points.

1. How can approximations to the *complex* roots of a polynomial equation be obtained?
2. How do we know when we have obtained *all* the real roots, including order, of a given polynomial equation?

These questions are best answered with the help of calculus. We therefore leave them for another day.

Exercises 10.4

Use a calculator (or computer) to perform the following steps for each of the polynomials $P(x)$ in Exercises 1–10.

(a) *Find integers a and b for which all of the real roots of $P(x) = 0$ are contained in the interval $[a, b]$.*

(b) *Graph $y = P(x)$ on $[a, b]$ by computing $P(x)$ for each integer value of x from a to b.*

(c) *Use (b) to find intervals $[d, d + 1]$ that contain a real root of $P(x) = 0$.*

(d) *On each such interval $[d, d + 1]$, use the bisection method to find an interval of length $\leq \delta = .01$ that contains a root of $P(x) = 0$. If you have access to a computer, set $\delta = .000001$ and use the BASIC program for the bisection process.*

(e) *Use Newton's method to find a value c in $[d, d + 1]$ for which $|P(c)| < \delta$, for δ of part (d).*

1. $P(x) = x^3 + x^2 - x - 3$
2. $P(x) = x^3 - x^2 - 1$
3. $P(x) = 4x^3 - 2x^2 + 3x + 1$
4. $P(x) = 4x^3 + 8x^2 + 5x + 2$
5. $P(x) = x^3 - 4x + 1$
6. $P(x) = x^3 - 2x^2 - x + 1$
7. $P(x) = x^4 - x^3 - 2x^2 - 1$
8. $P(x) = x^4 - 4x^3 + 3x^2 + 1$
9. $P(x) = x^4 - 5x^3 + 6x^2 - 1$
10. $P(x) = x^4 + x^3 - 6x^2 + 1$

Chapter 10 Review Outline

Insolvability Theorem:

The general polynomial equation of degree ≥ 5 cannot be solved by algebraic methods (Abel, Galois).

Fundamental Theorem of Algebra:

Every polynomial equation of degree n has n roots in the complex number system (Gauss).

10.1 Remainder Theorem, Synthetic Division, and Factor Theorem

Remainder Theorem:

If a polynomial $P(x)$ of degree ≥ 1 is divided by $x - a$, the remainder is $P(a)$.

Synthetic division provides a rapid method to evaluate polynomials.

Factor Theorem:

If a is a root of the polynomial equation $P(x) = 0$, then $x - a$ is a factor of $P(x)$.

Complex Conjugate Roots:

If $P(x)$ has real coefficients and the complex number $z = \alpha + \beta i$ is a root of $P(x) = 0$, then the complex conjugate $\bar{z} = \alpha - \beta i$ is also a root.

Complex and Real Factored Forms:

If the real roots of $P(x) = 0$ are $r_1, \ldots, r_k$, of orders $m_1, \ldots, m_k$, respectively, and the complex roots are $c_1, \bar{c}_1, \ldots, c_j, \bar{c}_j$ of orders $n_1, \ldots, n_j$, respectively, then

$$P(x) = a_n(x - r_1)^{m_1} \ldots (x - r_k)^{m_k}(x - c_1)^{n_1} \cdot (x - \bar{c}_1)^{n_1} \ldots (x - c_j)^{n_j}(x - \bar{c}_j)^{n_j}.$$

By combining the factors of complex conjugate roots into real irreducible quadratics $x^2 + Ax + B$,

$$P(x) = a_n(x - r_1)^{m_1} \ldots (x - r_k)^{m_k} \cdot (x^2 + A_1x + B_1)^{n_1} \ldots (x^2 + A_jx + B_j)^{n_j}.$$

10.2 Rational Roots, Bounds, and Descartes' Rule of Signs

Rational Root Theorem:

If the rational number p/q is a root of $a_nx^n + a_{n-1}x^{n-1} + \cdots + a_1x + a_0 = 0$, where the a_k are integers, $a_n \neq 0$, and $a_0 \neq 0$, then p divides a_0 and q divides a_n.

Upper and Lower Bounds:

If the remainder and all the coefficients of the quotient are nonnegative when $P(x)$ is divided by $x - a$, then a is an upper bound for the real roots of $P(x) = 0$.

If a is a positive upper bound for the real roots of $P(-x) = 0$, then $-a$ is a lower bound for the real roots of $P(x) = 0$.

Descartes' Rule of Signs:

The number of positive roots of the polynomial equation $P(x) = 0$ is at most equal to the number of sign changes in the coefficients of $P(x)$.

The number of negative roots of $P(x) = 0$ is at most equal to the number of sign changes in the coefficients of $P(-x)$.

10.3 Partial Fractions

Definition

A rational expression $P(x)/Q(x)$ is called a proper rational expression if the degree of $P(x)$ is less than the degree of $Q(x)$.

Partial Fraction Decomposition:

A proper rational expression $P(x)/Q(x)$ can be decomposed into partial fractions by using the following steps.

(1) Factor $Q(x)$ into its real factored form.

(2) For each linear factor of order m, $(x - r)^m$, there are m terms

$$\frac{A_j}{(x - r)^j} \quad (j = 1, 2, \ldots, m),$$

and for each irreducible quadratic factor of order n, $(x^2 + bx + c)^n$, there are n terms

$$\frac{B_jx + C_j}{(x^2 + bx + c)^n} \quad (j = 1, 2, \ldots, n).$$

(3) By setting $P(x)/Q(x)$ equal to its partial fraction decomposition and equating coefficients of like powers of x, a system of linear equations in the unknowns A_j, B_j, and C_j is determined. The system can be solved by the methods in the chapter on linear systems.

If degree $P(x) \geq$ degree $Q(x)$, divide $Q(x)$ into $P(x)$ to obtain

$$\frac{P(x)}{Q(x)} = q(x) + \frac{R(x)}{Q(x)},$$

where $q(x)$ is a polynomial and $R(x)/Q(x)$ is a proper rational expression. Partial fraction decomposition can be applied to $R(x)/Q(x)$.

10.4 Numerical Methods for Approximating Solutions

Real roots of a polynomial equation $P(x) = 0$ can be approximated by using the following steps.

(1) Determine an integral upper bound b and an integral lower bound a for the real roots of $P(x) = 0$.

(2) Sketch a graph of $y = P(x)$ on $[a, b]$ by computing $P(x)$ for each integer x from a to b.

(3) Use the graph in (2) to determine intervals $[d, d + 1]$ that contain a root of $P(x) = 0$.

(4) Apply the bisection method or Newton's method to each interval from step 3.

Roots can be determined only within the accuracy of the computing equipment.

Chapter 10 Review Exercises

Follow the directions or fill in the blanks in Exercises 1–28, as required.

1. State the fundamental theorem of algebra.
2. State the insolvability theorem.
3. Without dividing, find the remainder when $P(x) = x^4 - 3x^3 + 2x^2 - 5x + 1$ is divided by $x - 1$.
4. Without dividing, find the remainder when $P(x) = 2x^3 + 3x^2 - 4x + 4$ is divided by $x + 2$.
5. Use synthetic division to evaluate the polynomial in Exercise 3 at $x = 5$.
6. Use synthetic division to evaluate the polynomial in Exercise 4 at $x = -5$.
7. Without dividing, verify that $x - \sqrt{2}$ is a factor of $P(x) = x^5 + 3x^4 - 4x^3 - x^2 + 4x - 10$.
8. Without dividing, verify that $x - i$ is a factor of $P(x) = 3x^4 - 2x^3 + 4x^2 - 2x + 1$.
9. Find all factors $x - a$, where a is an integer, of the polynomial $P(x) = x^5 + x^4 - x^3 + x^2 - 2x$.
10. Find a polynomial $P(x)$ with real coefficients of lowest degree that has $1/2$ and i among the roots of $P(x) = 0$.
11. Find the complex factored form of the polynomial $P(x)$, if $-2, 3, 3, \sqrt{2}, 1 + 2i$, and $1 - 2i$ are the roots of $P(x) = 0$.
12. Find the real factored form of the polynomial in Exercise 11.
13. The numbers $4, 5/2, -\sqrt{3}, 2 + i, 2 - i$, and $3i$ cannot be the roots of $P(x) = 0$, where $P(x)$ is a polynomial of degree 6 with real coefficients, because __________.
14. Find all rational roots of $2x^4 - x^3 - 5x^2 + 2x + 2 = 0$.
15. Show that the polynomial equation $x^5 - 5x^4 + 7x^3 - 2x^2 + 3x + 2 = 0$ has no rational roots.
16. Find upper and lower bounds for the real roots in Exercise 14.
17. Find upper and lower bounds for the real roots in Exercise 15.
18. Find all rational roots, including order, for $x^4 - 2x^3 + 2x^2 - 2x + 1 = 0$.
19. Find the real factored form of the polynomial in Exercise 18. (*Hint:* use the factor theorem.)
20. If $2x^3 + bx^2 + cx + d = 0$, where b, c, and d are integers, has $3/2$ for a root, then the possible nonzero values for d are __________.
21. The possible rational roots of $x^3 + bx^2 + x = 0$, where b is a nonzero integer, are __________.
22. According to Descartes' rule of signs, the polynomial equation $3x^4 - 2x^3 + 6x^2 - 3x - 5 = 0$ can have at most ______ positive root(s).
23. By Descartes' rule of signs, the polynomial equation in Exercise 22 can have at most ______ negative root(s).
24. Use Descartes' rule of signs to determine the maximum number of real roots of $x^6 + 2x^5 - 4x^4 - 7x^2 - 5x + 5 = 0$.
25. The equation $ax^5 + bx^3 + cx^2 - dx + e = 0$ where a, b, c, d, and e are positive integers, can have at most ______ real root(s).
26. Use Descartes' rule of signs to show that the equation $3x^6 + 5x^4 + 2x^2 + 8 = 0$ has no real roots.
27. Find all *positive* rational roots of the polynomial equation $10x^6 + 3x^5 + 2x^4 + 3x^3 + x^2 + 2x + 2 = 0$.
28. Find all real roots of $x^4 - 3x^3 + 6x - 4 = 0$. (*Hint:* first find all rational roots and then apply the factor theorem.)

Decompose the rational expressions in Exercises 29–31 into partial fractions. Do not *solve for the constants in the decomposition.*

*29. $\dfrac{x}{(x - 1)^4}$

*30. $\dfrac{x^2}{(x^2 + 2x + 5)^2}$

*Asterisks indicate topics that may be omitted, depending on individual course needs.

*31. $\dfrac{2x^2 - 3x + 1}{(x + 1)(x - 2)^3(x^2 + x + 5)(x^2 + 3x + 4)^2}$

Decompose the rational expressions in Exercises 32–35 into partial fractions, and solve for the constants.

*32. $\dfrac{6x^2 + x - 6}{x^3 + x^2 - 6x}$ *33. $\dfrac{1}{x^4 - x^2}$

*34. $\dfrac{2x^3 - x^2 + 2x - 7}{x^2 - x - 2}$ *35. $\dfrac{x^5 + x^2 + 2x - 1}{x^3 + x}$

36. Find an interval $[a, b]$ that contains all real roots of $x^3 - 2x^2 - 5x + 5 = 0$.

37. Graph $y = x^3 - 2x^2 - 5x + 5$ on the interval $[a, b]$ obtained in Exercise 36.

38. Use the bisection method, with $\delta = .1$, to approximate the real roots of $x^3 - 2x^2 - 5x + 5 = 0$.

39. Apply Newton's method, with $\epsilon = .05$, to the polynomial in Exercise 38.

40. If a computer is available, use the BASIC programs in this section to approximate the real roots of $x^3 - 2x^2 - 5x + 5 = 0$ by the bisection method with $\delta = .00001$, and by Newton's method with $\epsilon = .00001$.

Graphing Calculator Exercises

In each of the following, verify your answers to the indicated review exercises of this section by using a graphing calculator for the given functions and windows $[x_{min}, x_{max}]$, $[y_{min}, y_{max}]$. Also, use the TRACE and ZOOM features to estimate all real zeros to three decimal places.

41. Exercise 14: $Y_1 = 2x^4 - x^3 - 5x^2 + 2x + 2$, $[-2.25, 2.5]$, $[-5, 5]$. Answer: -1.414, -0.5, 1, 1.414

42. Exercise 15: $Y_1 = x^5 - 5x^4 + 7x^3 - 2x^2 + 3x + 2$, $[-4.5, 5]$, $[-10, 10]$. Answer: -0.388 is the only real zero.

43. Exercise 18: $Y_1 = x^4 - 2x^3 + 2x^2 - 2x + 1$, $[-2.25, 2.5]$, $[-2, 5]$. Answer: 1 (order 2) is the only real zero.

44. Exercise 22: $Y_1 = 3x^4 - 2x^3 + 6x^2 - 3x - 5$, $[-2.25, 2.5]$, $[-10, 10]$. Answer: -0.614 and 1.062 are the only real zeros.

45. Exercise 24: $Y_1 = x^6 + 2x^5 - 4x^4 - 7x^2 - 5x + 5$, $[-4.5, 5]$, $[-150, 100]$. Answer: -3.335, -0.924, .543, and 1.765 are the only real roots.

46. Exercise 26: $Y_1 = 3x^6 + 5x^4 + 2x^2 + 8$, $[-2.25, 2.5]$, $[0, 100]$

47. Exercise 27: $Y_1 = 10x^6 + 3x^5 + 2x^4 + 3x^3 + x^2 + 2x + 2$, $[-2.25, 2.5]$, $[0, 10]$. Answer: No real zeros

48. Exercise 28: $Y_1 = x^4 - 3x^3 + 6x - 4$, $[-2.25, 2.5]$, $[-10, 5]$. Answer: -1.414, 1, 2, 1.414

49. Exercise 37: $Y_1 = x^3 - 2x^2 - 5x + 5$, $[-4.5, 5]$, $[-10, 10]$. Answer: -1.931, .837, 3.094

*Asterisks indicate topics that may be omitted, depending on individual course needs.

11 Sequences, Series and Probability

We started with the counting numbers in Chapter P, and now in the last chapter we return to the subject of counting. Here we are not concerned with counting individual objects, but collections of them. We begin with the principle of mathematical induction, which enables us to prove the binomial theorem and other equations that hold for the entire set of positive integers. We then consider sequences of numbers and their sums. We finish with permutations and combinations and their application to probability.

11.1 Mathematical Induction

The Principle of Mathematical Induction
The Binomial Coefficients

We reason by induction and deduction, and mathematical induction is a form of deduction.

The Principle of Mathematical Induction. Induction is a process by which we arrive at *probable* general conclusions based on evidence contained in particular cases. For example, by observing the cases

$$1 = 1^2$$
$$1 + 3 = 2^2$$
$$1 + 3 + 5 = 3^2$$
$$1 + 3 + 5 + 7 = 4^2$$
$$1 + 3 + 5 + 7 + 9 = 5^2,$$

we would arrive at the general conclusion that the sum of the first n odd integers is probably n^2, or

$$1 + 3 + 5 + 7 + \cdots + (2n - 1) = n^2. \quad (1)$$

Although we feel confident that formula (1) is correct, we must admit that our feeling is based on the *assumption* that the pattern established for the first five positive integers will continue for all positive integers. Mathematical induction is a method by which we can actually *prove* that equation (1) is true for all positive integers n. Since it leads to conclusions that are certain, rather than probable, mathematical induction is not induction in the usual sense. The **principle of mathematical induction**[1] is as follows.

[1] *Mathematical induction has been used explicitly by mathematicians since the late sixteenth century. Francesco Maurolycus used it to prove (1) in 1575.*

> Let $E(n)$ be an equation in n. If
>
> (i) $E(1)$ is true, and
>
> (ii) whenever $E(k)$ is true for some integer k, $E(k + 1)$ is also true,
>
> then $E(n)$ is true for all positive integers n.

We can think of mathematical induction as a process similar to a loop in computer programming. We start by proving that $E(1)$ is a true equation [condition (i) is satisfied]. Then, after proving that $E(k + 1)$ is true whenever $E(k)$ is true [condition (ii) is satisfied], we can logically loop through all the positive integers:

$E(1)$ is true,
$E(2)$ is true because $E(1)$ is true,
$E(3)$ is true because $E(2)$ is true,
$E(4)$ is true because $E(3)$ is true,

and so on.

Example 1 Use mathematical induction to prove that equation (1) is true for all positive integers n.

Solution:

1. When $n = 1$, equation (1) says $1 = 1^2$, which is true. Hence, $E(1)$ is true.
2. Suppose $E(k)$ is true for some integer k. That is, suppose

$$1 + 3 + 5 + \cdots + 2k - 1 = k^2.$$

The next odd integer after $2k - 1$ is $2k + 1$, and

$$\begin{aligned} 1 + 3 + 5 + \cdots + 2k + 1 &= \underbrace{1 + 3 + 5 + \cdots + 2k - 1}_{k^2 \textit{ by assumption}} + 2k + 1 \\ &= k^2 + 2k + 1 \\ &= (k + 1)^2. \end{aligned}$$

Hence, $E(k + 1)$ is true whenever $E(k)$ is true.

Steps 1 and 2 imply that conditions (i) and (ii) of the principle of mathematical induction are satisfied. Hence, equation (1) is true for every positive integer n. ■

Question 1 *Why is condition (i) necessary in the principle of mathematical induction? That is, if we can prove that condition (ii) is satisfied, why can't we conclude that $E(k + 1)$ is true?*

Answer: Condition (ii) does not say that $E(k + 1)$ is true. It does say that *if* $E(k)$ is true, then $E(k + 1)$ is true. The emphasis is on the word "if." For instance, we cannot conclude that $E(3)$ is true unless we know that $E(2)$ is true, and we cannot conclude that $E(2)$ is true unless we know that $E(1)$ is true. Therefore, a proof that $E(1)$ is true is needed in order to get the process started. ■

Question 2 *Suppose we can prove that condition (ii) is satisfied and can also prove that $E(5)$ is true. What can we conclude?*

Answer: We can conclude that $E(n)$ is true for all integers $n \geq 5$. In general, if we can prove that (ii) is satisfied and also that $E(n_1)$ is true for some integer n_1 (positive, negative, or 0), then we can conclude that $E(n)$ is true for all integers $n \geq n_1$. ■

Now that formula (1) has been established, we can use it for any positive integer n. For instance, the sum of the first 100 odd integers is

$$\begin{aligned} 1 + 3 + 5 + 7 + \cdots + 199 &= 100^2 \\ &= 10{,}000. \end{aligned}$$

The one-hundredth odd integer is $2 \cdot 100 - 1 = 199$

Example 2 Use mathematical induction to prove that

$$\mathbf{1 + 2 + 3 + 4 + \cdots + n = \frac{n(n+1)}{2}} \tag{2}$$

for all positive integers n.

Solution:

1. Since $1 = \frac{1 \cdot 2}{2}$, equation (2) is true for $n = 1$ or condition (i) is satisfied.
2. Now suppose that equation (2) holds when n is equal to some integer k. Then

$$\begin{aligned} 1 + 2 + 3 + \cdots + k + 1 &= \underbrace{1 + 2 + 3 + \cdots + k}_{\frac{k(k+1)}{2} \text{ by assumption}} + k + 1 \\ &= \frac{k(k+1)}{2} + k + 1 \\ &= \frac{k(k+1) + 2(k+1)}{2} \\ &= \frac{(k+1)(k+2)}{2}, \end{aligned}$$

which is equation (2) for the case $n = k + 1$. That is, equation (2) is true for $k + 1$ whenever it is true for k, or condition (ii) is satisfied. Hence, by the principle of mathematical induction, equation (2) is true for all positive integers n. ■

Equation (2) gives us a formula for finding the sum of the first n positive integers. For instance,

$$\begin{aligned} 1 + 2 + 3 + \cdots + 100 &= \frac{100 \cdot 101}{2} \\ &= 5050. \end{aligned}$$

Also, we can use equation (2) and a little algebra to determine the sum of the first n even integers as follows.

$$\begin{aligned} 2 + 4 + 6 + 8 + \cdots + 2n &= 2(1 + 2 + 3 + 4 + \cdots + n) \\ &= 2\frac{n(n+1)}{2} \\ &= n(n+1) \end{aligned}$$

For instance, for $n = 50$,

$$\begin{aligned} 2 + 4 + 6 + 8 + \cdots + 100 &= 50 \cdot 51 \\ &= 2550. \end{aligned}$$

Example 3 Use mathematical induction to show that for all positive integers n,

$$1^2 + 2^2 + 3^2 + \cdots + n^2 = \frac{n(n+1)(2n+1)}{6}. \tag{3}$$

Solution:

1. Since $1^2 = (1 \cdot 2 \cdot 3)/6$, equation (3) is true for $n = 1$.
2. Now suppose that (3) holds for some integer k. Then

$$\begin{aligned} 1^2 + 2^2 + 3^2 + \cdots + (k+1)^2 &= 1^2 + 2^2 + 3^2 + \cdots + k^2 + (k+1)^2 \\ &= \frac{k(k+1)(2k+1)}{6} + (k+1)^2 \\ &= \frac{k(k+1)(2k+1)}{6} + \frac{6(k+1)^2}{6} \\ &= \frac{(k+1)[k(2k+1) + 6(k+1)]}{6} \\ &= \frac{(k+1)(2k^2 + 7k + 6)}{6} \\ &= \frac{(k+1)(k+2)(2k+3)}{6} \\ &= \frac{(k+1)[(k+1)+1][2(k+1)+1]}{6}, \end{aligned}$$

which is equation (3) for $k + 1$. Hence, equation (3) holds for $k + 1$ whenever it holds for k. Therefore, by the principle of mathematical induction, equation (3) is true for all positive integers n. ■

The Binomial Coefficients. In Chapter 1 we let ${}_nC_r$ denote the coefficient of $a^{n-r}b^r$ in the binomial expansion of $(a + b)^n$. We said that

$${}_nC_r = \frac{n!}{(n-r)!r!}, \tag{4}$$

where n and r are integers satisfying $0 \leq r \leq n$, and, by definition, $k! = k(k-1)(k-2)\ldots 1$ if k is any positive integer and $0! = 1$. For example, the coefficient of a^3b^4 in $(a+b)^7$ is

$$\begin{aligned} {}_7C_4 &= \frac{7!}{3!4!} \\ &= \frac{7 \cdot 6 \cdot 5 \cdot 4 \cdot 3 \cdot 2 \cdot 1}{3 \cdot 2 \cdot 1 \cdot 4 \cdot 3 \cdot 2 \cdot 1} \\ &= 35, \end{aligned}$$

and the coefficient of a^7 is

$${}_7C_0 = \frac{7!}{7!0!} = 1.$$

We stated in Chapter 1 that formula (4) could be deduced from the properties

$$\text{(a)}\ {}_nC_0 = 1 = {}_nC_n \quad (n \geq 0)$$

and $$\text{(b)}\ {}_nC_r = {}_{n-1}C_r + {}_{n-1}C_{r-1} \quad (1 \leq r \leq n-1),$$

which we arrived at by means of Pascal's triangle. Property (a) says that the coefficient of a^n and that of b^n is 1 in the expansion of $(a+b)^n$. Property (b) also can be obtained by writing

$$(a+b)^n = (a+b)^{n-1}(a+b)$$

and then expanding the right side to express the coefficient of $a^{n-r}b^r$ in $(a+b)^n$ in terms of corresponding coefficients in $(a+b)^{n-1}$. For instance, in the case ${}_7C_4$, we write

$$\begin{aligned} (a+b)^7 &= (a+b)^6(a+b) \\ &= a^6 \cdot a + \cdots + {}_6C_4a^2b^4 \cdot a + \cdots + {}_6C_3a^3b^3 \cdot b + \cdots + b^6 \cdot b \\ &= a^7 + \cdots + {}_6C_4a^3b^4 + {}_6C_3a^3b^4 + \cdots + b^7 \\ &= a^7 + \cdots + [{}_6C_4 + {}_6C_3]a^3b^4 + \cdots + b^7. \end{aligned}$$

Therefore, since ${}_7C_4$ denotes the coefficient of a^3b^4 in $(a+b)^7$, we must have

$${}_7C_4 = {}_6C_4 + {}_6C_3,$$

which is property (b).

In the following example, we show how properties (a) and (b) can be used to prove formula (4).

Example 4 Given properties (a) and (b) above, use mathematical induction to prove formula (4).

Solution: This is an advanced application of mathematical induction and may take several readings to understand. The main difficulty is keeping track of both n and r. Also, the algebra toward the end gets a little tricky. We proceed as follows.

1. By property (a), formula (4) holds for $n = 1$ ($r = 0, 1$):

$$_1C_0 = 1 \text{ by (a), and } 1 = \frac{1!}{1!0!} \quad \text{[formula (4) for } n = 1, r = 0]$$

$$_1C_1 = 1 \text{ by (a), and } 1 = \frac{1!}{0!1!} \quad \text{[formula (4) for } n = 1, r = 1].$$

2. Now suppose that formula (4) holds for some integer k and $0 \le r \le k$. Then, by property (b), with $n = k + 1$,

$$\begin{aligned} {}_{k+1}C_r &= {}_kC_r + {}_kC_{r-1} \quad (1 \le r \le k) \\ &= \frac{k!}{(k-r)!r!} + \frac{k!}{[k-(r-1)]!(r-1)!}. \end{aligned}$$

Now we multiply the numerator and denominator of the first expression on the right by $(k + 1)(k + 1 - r)$ and the numerator and denominator of the second expression by $(k + 1)r$. On the right side we get

$$\begin{aligned} &\frac{(k+1)(k+1-r)\cdot k!}{(k+1)(k+1-r)\cdot(k-r)!r!} + \frac{(k+1)r\cdot k!}{(k+1)r\cdot(k+1-r)!(r-1)!} \\ &= \frac{(k+1-r)(k+1)!}{(k+1)(k+1-r)!r!} + \frac{r(k+1)!}{(k+1)(k+1-r)!r!} \\ &= \frac{[(k+1-r)+r](k+1)!}{(k+1)(k+1-r)!r!} \\ &= \frac{(k+1)(k+1)!}{(k+1)(k+1-r)!r!} \\ &= \frac{(k+1)!}{(k+1-r)!r!}, \end{aligned}$$

which is the right side of (4) for $k + 1$. Hence, when formula (4) is true for k $(0 \le r \le k)$, it is also true for $k + 1$ $(1 \le r \le k)$. We want to show that it is in fact true for $k + 1$ $(0 \le r \le k + 1)$, and therefore we must consider the cases $r = 0$ and $r = k + 1$ separately. But, by property (a), formula (4) always holds for $r = 0$ and $r = k + 1$ when $n = k + 1$. Hence, when formula (4) is true for k $(0 \le r \le k)$, it is also true for $k + 1$ $(0 \le r \le k + 1)$. Therefore, by the principle of mathematical induction, formula (4) is true for all positive integers n. ■

Comment Although mathematical induction provides us with a method of proving that a given equation $E(n)$ is true for all positive integers n, it does not help us in the original formulation of the equation $E(n)$. We have to somehow produce the formula $E(n)$, and *then* mathematical induction will help us to determine whether it is true for all positive integers n. As in our original equation (1), we often arrive at a possible $E(n)$ by observing patterns for several choices of n. There may be several $E(n)$'s that appear plausible, and we then use mathematical induction to determine which, if any, is correct.

There are variations of the statement of the principle of mathematical induction that enable us to prove equations $E(n)$ for sets of integers other than the positive ones. For example, if $E(-1)$ is true, and if $E(k-1)$ is true whenever $E(k)$ is true, then $E(n)$ is true for all *negative* integers n. Also, the following version of the principle of mathematical induction is equivalent to our original one.

> Let $E(n)$ be an equation in n. If
>
> (i) $E(1)$ is true, and
>
> (ii) whenever $E(k)$ is true for all integers k less than some integer n, $E(n)$ is also true,
>
> then $E(n)$ is true for all positive integers n.

This statement is sometimes called the **second principle of mathematical induction.** We leave it as an exercise in logic for the reader to prove that the two statements of the principle of mathematical induction are equivalent.

Exercises 11.1

Fill in the blanks to make each statement true.

1. By the usual induction process, we arrive at ________ conclusions.
2. By mathematical induction, we arrive at ________ conclusions.
3. If we can show that an equation $E(k+1)$ is true whenever $E(k)$ is true, then we __________ (can, cannot) conclude that $E(n)$ is true for all positive integers n.
4. If we can show that an equation $E(1)$ is true and that $E(k+2)$ is true whenever $E(k)$ is true, then we can conclude that ________________.
5. If we want to prove that an equation $E(n)$ is true for all positive integral multiples of 5, it is sufficient to prove that $E(5)$ is true, and that __________________ is true whenever $E(k)$ is true.

Write true *or* false *for each statement.*

6. Mathematical induction is just induction in the usual sense.
7. If we can prove that an equation is true for all integers from 1 to 100, then the equation must be true for all positive integers.
8. If an equation $E(k+1)$ is true whenever $E(k)$ is true, then it may be possible that $E(n)$ is false for all integers n.
9. If $E(0)$ is true, and $E(k+2)$ is true whenever $E(k)$ is true, then $E(n)$ is true for all nonnegative even integers.
10. If $E(0)$ is true, and $E(k+1)$ and $E(k-1)$ are both true whenever $E(k)$ is true, then $E(n)$ is true for *all* integers n.

In Exercises 11–24, use mathematical induction to prove the given equations.

11. $1 + 4 + 7 + 10 + \cdots + 3n - 2 = \dfrac{n(3n - 1)}{2}$

12. $1 + 5 + 9 + 13 + \cdots + 4n - 3 = \dfrac{n(4n - 2)}{2}$

13. $1 + 6 + 11 + 16 + \cdots + 5n - 4 = \dfrac{n(5n - 3)}{2}$

14. $1 + 7 + 13 + 19 + \cdots + 6n - 5 = \dfrac{n(6n - 4)}{2}$

15. $1 + 2 + 2^2 + 2^3 + \cdots + 2^{n-1} = 2^n - 1$

16. $1 + 3 + 3^2 + 3^3 + \cdots + 3^{n-1} = \dfrac{3^n - 1}{2}$

17. $1 + 4 + 4^2 + 4^3 + \cdots + 4^{n-1} = \dfrac{4^n - 1}{3}$

18. $1 + 5 + 5^2 + 5^3 + \cdots + 5^{n-1} = \dfrac{5^n - 1}{4}$

19. $\dfrac{1}{1 \cdot 2} + \dfrac{1}{2 \cdot 3} + \dfrac{1}{3 \cdot 4} + \cdots + \dfrac{1}{n(n + 1)} = \dfrac{n}{n + 1}$

20. $\dfrac{1}{1 \cdot 3} + \dfrac{1}{3 \cdot 5} + \dfrac{1}{5 \cdot 7} + \cdots + \dfrac{1}{(2n - 1)(2n + 1)} = \dfrac{n}{2n + 1}$

21. $1 \cdot 2 + 2 \cdot 3 + 3 \cdot 4 + \cdots + n(n + 1) = \dfrac{n(n + 1)(n + 2)}{3}$

22. $1 \cdot 2 \cdot 3 + 2 \cdot 3 \cdot 4 + 3 \cdot 4 \cdot 5 + \cdots + n(n + 1)(n + 2) = \dfrac{n(n + 1)(n + 2)(n + 3)}{4}$

23. $1^3 + 2^3 + 3^3 + \cdots + n^3 = \dfrac{n^2(n + 1)^2}{4}$

24. $1^4 + 2^4 + 3^4 + \cdots + n^4 = \dfrac{n(n + 1)(2n + 1)(3n^2 + 3n - 1)}{30}$

In Exercises 25–28, use mathematical induction to prove that the equation is true for all positive integers n. Assume that m is a fixed integer and that x and y are positive real numbers.

25. $x^m x^n = x^{m+n}$

26. $(xy)^n = x^n y^n$

27. $\left(\dfrac{x}{y}\right)^n = \dfrac{x^n}{y^n}$

28. $(x^m)^n = x^{mn}$

In Exercises 29 and 30, show that the equation in part (a) is not true but has the property that $E(k + 1)$ is true whenever $E(k)$ is true. Use mathematical induction to show that the equation in part (b) is true.

29. **(a)** $1 + \dfrac{1}{2} + \dfrac{1}{2^2} + \dfrac{1}{2^3} + \cdots + \dfrac{1}{2^n} = 1 - \dfrac{1}{2^n}$

(b) $\dfrac{1}{2} + \dfrac{1}{2^2} + \dfrac{1}{2^3} + \cdots + \dfrac{1}{2^n} = 1 - \dfrac{1}{2^n}$

30. **(a)** $2 + 4 + 6 + \cdots + 2(n - 2) = n^2 - 3n$

(b) $-2 + 0 + 2 + 4 + 6 + \cdots + 2(n - 2) = n^2 - 3n$

In Exercises 31–34, prove that the equation $E(n)$ is true for all integers $n \geq n_1$ by showing (1) $E(n_1)$ is true, and (2) $E(k + 1)$ is true whenever $E(k)$ is true.

31. $5 + 6 + 7 + \cdots + n = \dfrac{n^2 + n - 20}{2}, \quad n_1 = 5$

32. $8 + 12 + 16 + \cdots + 4n = 2n^2 + 2n - 4, \quad n_1 = 2$

33. $\dfrac{1}{2^2} + \dfrac{1}{2} + 1 + 2 + 2^2 + 2^3 + \cdots + 2^n = \dfrac{2^{n+3} - 1}{4}$, $n_1 = -2$

34. $3^{-2} + 3^{-1} + 1 + 3 + 3^2 + 3^3 + \cdots + 3^n = \dfrac{3^{n+3} - 1}{18}$, $n_1 = -2$

35. By observing the equations in Exercises 11–14, complete the following equation. Prove your answer by mathematical induction.

$$1 + 8 + 15 + 22 + \cdots + 7n - 6 = \_\_\_\_\_\_\_\_$$

36. By observing the equations in Exercises 15–18, complete the following equation. Prove your answer by mathematical induction.

$$1 + 6 + 6^2 + 6^3 + \cdots + 6^{n-1} = \_\_\_\_\_\_\_\_$$

37. By observing Example 2 and Exercises 21 and 22, complete the following equation. Prove your answer by mathematical induction.

$$1 \cdot 2 \cdot 3 \cdot 4 + 2 \cdot 3 \cdot 4 \cdot 5 + 3 \cdot 4 \cdot 5 \cdot 6 + \cdots + n(n + 1)(n + 2)(n + 3) = \_\_\_\_\_\_\_\_$$

38. The result

$$_nC_0 + {_nC_1} + {_nC_2} + \cdots + {_nC_n} = 2^n \qquad (*)$$

follows directly from the binomial expansion for $(1 + x)^n$ if we let $x = 1$. Use the property

$$_{n+1}C_r = {_nC_r} + {_nC_{r-1}} \quad (1 \le r \le n)$$

to prove (*) by mathematical induction. (*Hint:* when showing that $E(k + 1)$ is true whenever $E(k)$ is true, write

$$_{k+1}C_0 + {_{k+1}C_1} + {_{k+1}C_2} + \cdots + {_{k+1}C_k} + {_{k+1}C_{k+1}}$$

as $$1 + {_{k+1}C_1} + {_{k+1}C_2} + \cdots + {_{k+1}C_k} + 1$$

before applying the above property.)

39. Use mathematical induction to show that for every positive integer n, the integer $n(n + 1)(n + 2)$ is divisible by 6. That is, show that

$$n(n + 1)(n + 2) = 6j,$$

where j is an integer depending on n.

40. Use mathematical induction to show that if x is a fixed positive real number, then $(1 + x)^n > 1 + nx$ for every positive integer $n \ge 2$. That is, show that

$$(1 + x)^n - (1 + nx) = p,$$

where p is a positive real number depending on n.

11.2 Arithmetic and Geometric Sequences

The math professor announced: "We will begin a lecture series on sequences and series with series following sequences in the series. Is that clear?"

Finite and Infinite Sequences
Arithmetic and Geometric Sequences
Graphs of Sequences

Finite and Infinite Sequences. We all have taken tests in which terms from some sequence of numbers are given and we are asked to find some missing entries. For example,

$$1, 3, 5, 7, \_\_ \qquad \text{(answer: 9)}$$

or $$1, 4, 2, 4, 3, 4, \_\_, \_\_, 5, 4, \ldots \qquad \text{(answers: 4, 4).}$$

In any such sequence, there is a first number a_1, a second number a_2, a third number a_3, and so on. If the sequence ends at the nth term a_n, then it is called a **finite sequence.** The first sequence above is a finite sequence of five terms. If the sequence continues for all positive integers, as in the second sequence above, then it is called an **infinite sequence.** From a mathematical point of view, a sequence is a function f whose domain is either the integers from 1 to n (finite sequence) or the set of all positive integers (infinite sequence). We call $f(k)$ $(k = 1, 2, 3, \ldots)$ the kth **term** or kth **entry** of the sequence f. However, as indicated above, we usually write a_k in place of $f(k)$. Also, we usually denote the entire sequence by

$$a_1, a_2, a_3, \ldots, a_n \qquad \text{or} \qquad \{a_k\}_{k=1}^n$$

if it is finite, and by

$$a_1, a_2, a_3, \ldots \qquad \text{or} \qquad \{a_k\}_{k=1}^\infty$$

if it is infinite. We also use the notation

$$\{a_k\}$$

to denote *any* sequence, finite or infinite.

Example 1 Find a_7, a_8, and the general term a_k $(k = 1, 2, 3, \ldots)$ of the sequence

$$-2,\ 4,\ -6,\ 8,\ -10,\ 12,\ \ldots .$$

Solution: The terms given are a_1 through a_6. The next two terms are

$$a_7 = -14 \qquad \text{and} \qquad a_8 = 16.$$

To find an expression for the general term a_k we note that the magnitude of a_k is $2k$ $(k = 1, 2, 3, \ldots)$, and the sign of $a_1, a_3, a_5, \ldots$ is $-$, while that of $a_2, a_4, a_6, \ldots$ is $+$. That is, *odd* subscripts correspond to $-$ and *even* subscripts to $+$. Now $(-1)^k$ is negative when k is odd, and positive when k is even. Therefore,

$$a_k = (-1)^k 2k, \quad k = 1, 2, 3, \ldots . \quad ■$$

Example 2 Find the general term a_k $(k = 1, 2, 3, \ldots)$ for each of the following sequences.

(a) $\frac{1}{1\cdot 2}, \frac{1}{2\cdot 3}, \frac{1}{3\cdot 4}, \frac{1}{4\cdot 5}, \ldots$

(b) $1 - \frac{1}{2}, 1 - \frac{1}{4}, 1 - \frac{1}{8}, 1 - \frac{1}{16}, \ldots$

(c) $1, -\frac{1}{2}, \frac{1}{3}, -\frac{1}{4}, \frac{1}{5}, -\frac{1}{6}, \ldots$

Solution:

(a) Here, $a_1 = 1/(1\cdot 2)$, $a_2 = 1/(2\cdot 3)$, $a_3 = 1/(3\cdot 4)$, . . . , and the general term a_k is $1/[k(k+1)]$.

(b) The terms are $1 - 1/2$, $1 - 1/2^2$, $1 - 1/2^3$, $1 - 1/2^4$, ... The general term a_k is $1 - 1/2^k$.

(c) The terms $a_1, a_3, a_5, \ldots$ with *odd* subscripts are positive, and the terms $a_2, a_4, a_6, \ldots$ with *even* subscripts are negative. The expression $(-1)^{k+1}$ is positive when k is odd and negative when k is even. The general term a_k is $(-1)^{k+1}(1/k)$. ■

Arithmetic and Geometric Sequences. The two most basic sequences are arithmetic sequences and geometric sequences. In an **arithmetic sequence,** the *difference* of any two successive terms is a fixed constant d. For example,

$$1,\ 4,\ 7,\ 10,\ 13,\ 16,\ \ldots$$

is an arithmetic sequence in which the constant difference d is 3. The terms of this sequence can be written as follows.

$$\begin{aligned}
a_1 &= 1\\
a_2 &= 1 + 3\\
a_3 &= 1 + 2 \cdot 3\\
a_4 &= 1 + 3 \cdot 3\\
a_5 &= 1 + 4 \cdot 3\\
&\vdots\\
a_k &= 1 + (k - 1)3\\
&\vdots
\end{aligned}$$

In general, if $\{a_k\}$ is an arithmetic sequence whose first term is a and whose constant difference is d, then the terms of $\{a_k\}$ are

$$a,\ a + d,\ a + 2d,\ a + 3d,\ \ldots,\ a + (k - 1)d,\ \ldots.$$

That is,

$$\{a_k\} = \{a + (k - 1)d\}, \quad k = 1, 2, 3, \ldots. \qquad \textbf{arithmetic sequence}$$

Example 3 Write out the terms of the following arithmetic sequences.

(a) $\{a_k\}_{k=1}^{4}$, where $a = 3$ and $d = 5$

(b) $\{a_k\}_{k=1}^{5}$, where $a = 1$ and $d = \frac{2}{3}$

(c) $\{a_k\}_{k=1}^{6}$, where $a = 0$ and $d = -2$

Solution:

(a) $a_1 = 3,\ a_2 = 8,\ a_3 = 13,\ a_4 = 18$

(b) $a_1 = 1,\ a_2 = \frac{5}{3},\ a_3 = \frac{7}{3},\ a_4 = \frac{9}{3},\ a_5 = \frac{11}{3}$

(c) $a_1 = 0,\ a_2 = -2,\ a_3 = -4,\ a_4 = -6,\ a_5 = -8,\ a_6 = -10$ ■

Example 4 Which of the following sequences are arithmetic? For those that are arithmetic sequences, find the next two terms and an expression for a_k $(k = 1, 2, 3, \ldots)$.

(a) $1, \frac{3}{2}, 2, \frac{5}{2}, 3, \frac{7}{2}, 4, \ldots$

(b) $2, 5, 2, 5, 2, 5, 2, 5, \ldots$

(c) $6, 4, 2, 0, -2, -4, -6, \ldots$

Solution:

(a) The sequence is arithmetic with constant difference $d = 1/2$. The next two terms are $9/2$ and 5, and the general term is $a_k = 1 + (k - 1)1/2$, which can be written as $(k + 1)/2$.

(b) The difference of any two successive terms is either 3 or -3. Since the difference is not constant, the sequence is not an arithmetic one.

(c) Here the difference of any two successive terms is -2. The next two terms are -8 and -10, and the general term is $a_k = 6 + (k - 1)(-2)$, which simplifies to $8 - 2k$. ■

In a **geometric sequence,** the *ratio* of any two successive terms is a constant $r \neq 0$. For example,

$$3, 6, 12, 24, 48, 96, \ldots$$

is a geometric sequence in which the constant ratio is $r = 2$. The terms of this sequence can be written as follows.

$$\begin{aligned} a_1 &= 3 \\ a_2 &= 3 \cdot 2 \\ a_3 &= 3 \cdot 2^2 \\ a_4 &= 3 \cdot 2^3 \\ a_5 &= 3 \cdot 2^4 \\ &\vdots \\ a_k &= 3 \cdot 2^{k-1} \\ &\vdots \end{aligned}$$

In general, if $\{a_k\}$ is a geometric sequence whose first term is a and whose constant ratio is r, then the terms of $\{a_k\}$ are

$$a,\ a \cdot r,\ a \cdot r^2,\ a \cdot r^3,\ a \cdot r^4,\ \ldots\ a \cdot r^{k-1},\ \ldots.$$

That is,

$$\{a_k\} = \{a \cdot r^{k-1}\}, \quad k = 1, 2, 3, \ldots. \qquad \textbf{geometric sequence}$$

Example 5 Write out the terms of each of the following geometric sequences.

(a) $\{a_k\}_{k=1}^{4}$, where $a = 2$ and $r = 3$

(b) $\{a_k\}_{k=1}^{5}$, where $a = 1$ and $r = \dfrac{1}{2}$

(c) $\{a_k\}_{k=1}^{6}$, where $a = 1$ and $r = -1$

Solution:

(a) $2,\ 2 \cdot 3,\ 2 \cdot 3^2,\ 2 \cdot 3^3$ or $2, 6, 18, 54$

(b) $1, \dfrac{1}{2}, \left(\dfrac{1}{2}\right)^2, \left(\dfrac{1}{2}\right)^3, \left(\dfrac{1}{2}\right)^4$ or $1, \dfrac{1}{2}, \dfrac{1}{4}, \dfrac{1}{8}, \dfrac{1}{16}$

(c) $1, -1, (-1)^2, (-1)^3, (-1)^4, (-1)^5$ or $1, -1, 1, -1, 1, -1$ ■

Example 6 Determine which of the following sequences are geometric. For the geometric ones, find the kth term $(k = 1, 2, 3, \ldots)$.

(a) $1, \frac{3}{4}, \frac{9}{16}, \frac{27}{64}, \frac{81}{256}, \ldots$

(b) $2, -1, \frac{1}{2}, -\frac{1}{4}, \frac{1}{8}, -\frac{1}{16}, \ldots$

(c) $1, 2, -4, -8, 16, 32, -64, -128, \ldots$

Solution:

(a) The sequence is geometric with $a = 1$ and $r = 3/4$. The kth term is $a_k = (3/4)^{k-1}$ $(k = 1, 2, 3, \ldots)$.

(b) The sequence here is geometric with $a = 2$ and $r = -1/2$. The kth term is $2(-1/2)^{k-1}$ $(k = 1, 2, 3, \ldots)$.

(c) If a sequence is geometric and $r > 0$, then the signs of all of the terms are the same; if $r < 0$, the signs alternate from one term to the next (*why?*). In the given sequence, the signs are neither the same nor do they alternate from one term to the next. Hence, the sequence is not geometric. ■

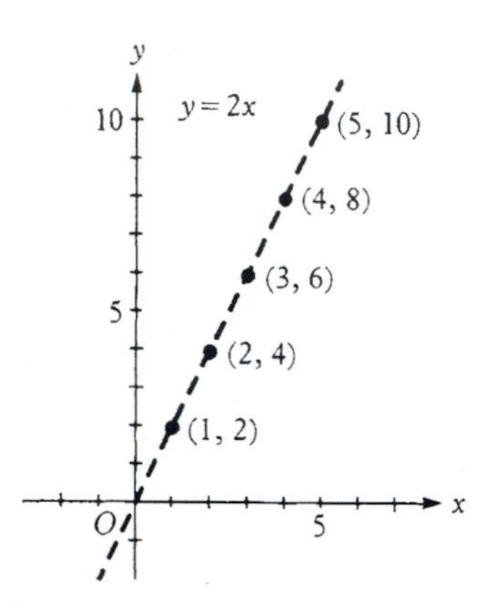

Figure 1 Graph of the Arithmetic Sequence $\{2k\}$

Graphs of Sequences. Since a sequence $\{a_k\}$ is a function, we can graph the sequence by plotting the points (k, a_k) in the plane. For example, the sequence

$$2, 4, 6, 8, 10, \ldots,$$

where $a_k = 2k$, has for its graph all points $(k, 2k)$, for $k = 1, 2, 3, \ldots$. As indicated in Figure 1, these points all lie on the line $y = 2x$.

In general, the points of an arithmetic sequence

$$a_k = a + (k - 1)d, \quad k = 1, 2, 3, \ldots,$$

lie along the line

$$y = a + (x - 1)d$$
$$= dx + (a - d),$$

which has slope d and y-intercept $a - d$. In this sense, an arithmetic sequence could be called a *linear sequence.*

On the other hand, the points of the geometric sequence

$$2, 2^2, 2^3, 2^4, \ldots,$$

in which $a_k = 2^k$ $(k = 1, 2, 3, \ldots)$, lie on the curve $y = 2^x$ (Figure 2). In general, if $0 < r < 1$ or $r > 1$, then the points of a geometric sequence

$$a_k = ar^{k-1}, \quad k = 1, 2, 3, \ldots,$$

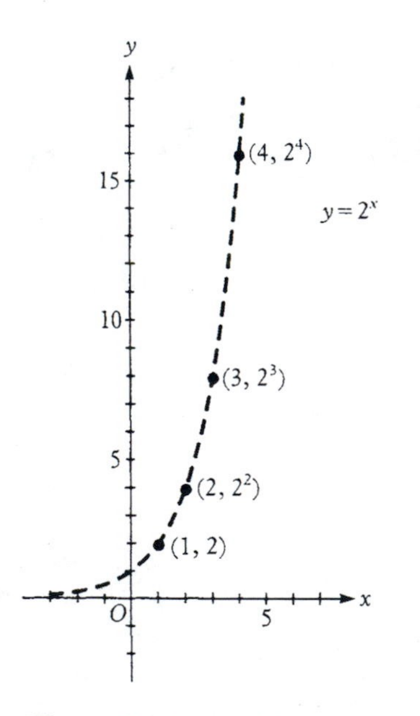

Figure 2 Graph of the Geometric Sequence $\{2^k\}$

lie on the exponential curve

$$y = ar^{x-1} = \left(\frac{a}{r}\right)r^x,$$

and hence a geometric sequence for $0 < r < 1$ or $r > 1$ could be called an *exponential sequence.*

Example 7 Graph the following sequences for $k = 1, 2, 3, 4, 5, 6$.

(a) $a_k = -2 + \frac{1}{2}(k - 1)$ (b) $a_k = \frac{1}{2}\left(\frac{3}{2}\right)^{k-1}$

Solution:

(a) As shown in Figure 3(a), the points of the sequence lie on the line

$$y = -2 + \frac{1}{2}(x - 1) = \frac{1}{2}x - \frac{5}{2}.$$

(b) As shown in Figure 3(b), the points of this sequence lie on the exponential curve

$$y = \frac{1}{2}\left(\frac{3}{2}\right)^{x-1} = \frac{1}{2}\left(\frac{3}{2}\right)^{-1}\left(\frac{3}{2}\right)^x = \frac{1}{3}\left(\frac{3}{2}\right)^x. \blacksquare$$

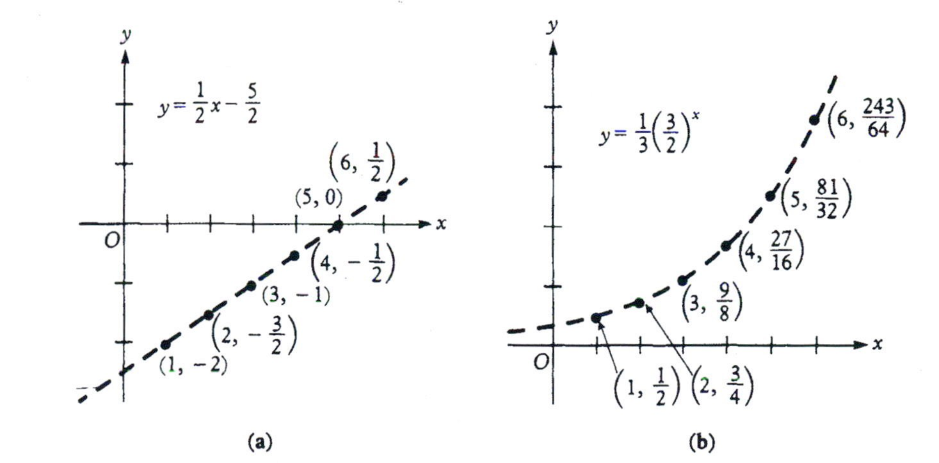

Figure 3

Example 8 Graph each of the following sequences.

(a) $a_k = \frac{1}{k}$, $k = 1, 2, 3, \ldots$

(b) $a_k = 1 + (-1)^k$, $k = 1, 2, 3, \ldots$

Solution:

(a) The points lie on the curve

$$y = \frac{1}{x}.$$

As k increases, the points approach the x-axis asymptotically [Figure 4(a)].

(b) Here $a_k = 0$ if k is odd, and $a_k = 2$ if k is even. Since we have not defined the exponential function $y = (-1)^x$ for the negative base -1, there is no corresponding curve $y = f(x)$ in this case [see Figure 4(b)]. ■

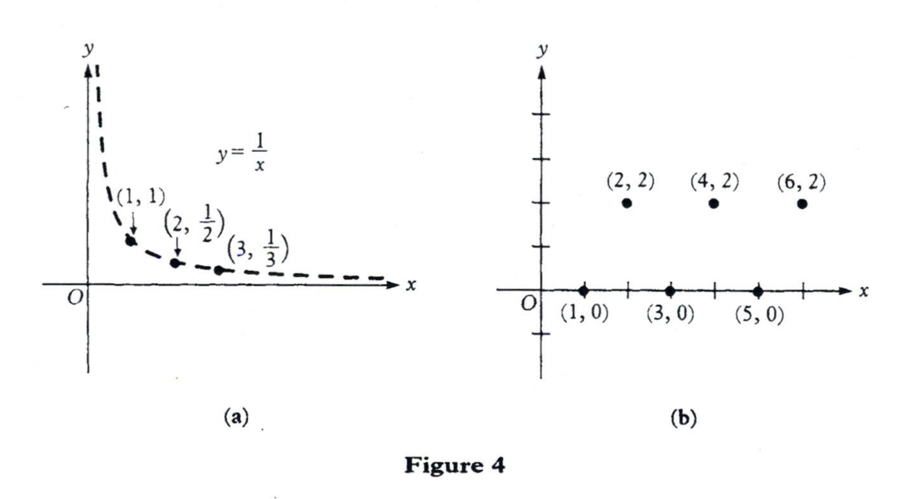

Figure 4

Exercises 11.2

Fill in the blanks to make each statement true.

1. A finite sequence of n terms is a function whose domain is ________, and an infinite sequence is a function with domain ________.
2. A sequence is an arithmetic sequence if ________.
3. A sequence is a geometric sequence if ________.
4. The graph of the sequence $a_k = 2 - (k - 1) \cdot 3$ lies on the line ________.
5. The graph of the sequence $a_k = 3 \cdot (3/4)^{k-1}$ lies on the exponential curve ________.

Write true *or* false *for each statement.*

6. Any function whose domain is either the integers from 1 to n or the set of all positive integers and whose range is a subset of the real numbers is a sequence.
7. Every sequence is either arithmetic or geometric.
8. In a geometric sequence, either all of the terms have the same sign or the signs alternate from one term to the next.
9. The terms of an arithmetic sequence all have the same sign.
10. In an arithmetic sequence, let a be the first term, a_k the kth term, and d the constant difference. If any two of the three numbers a, a_k, and d are known, then the third one can be determined.

Finite and Infinite Sequences

In Exercises 11–16, write out the terms of the given sequences.

11. $\{2k - 1\}_{k=1}^{6}$
12. $\left\{k + \frac{1}{k}\right\}_{k=1}^{7}$
13. $\left\{\left(1 + \frac{1}{k}\right)^k\right\}_{k=1}^{4}$
14. $\left\{\frac{1}{k^2 + 1}\right\}_{k=1}^{5}$
15. $\{1 + 2 + 3 + \cdots + k\}_{k=1}^{6}$
16. $\left\{1 + \frac{1}{2} + \frac{1}{3} + \cdots + \frac{1}{k}\right\}_{k=1}^{5}$

In Exercises 17–22, find the next two terms and the general term a_k $(k = 1, 2, 3, \ldots)$ of each sequence.

17. $\frac{1}{2}, \frac{2}{3}, \frac{3}{4}, \frac{4}{5}, \ldots$
18. $1 - \frac{1}{2}, 1 - \frac{1}{4}, 1 - \frac{1}{6}, 1 - \frac{1}{8}, \ldots$
19. $-\frac{1}{2}, \frac{1}{4}, -\frac{1}{8}, \frac{1}{16}, -\frac{1}{32}, \frac{1}{64}, \ldots$
20. $\frac{2}{3}, -\frac{4}{5}, \frac{6}{7}, -\frac{8}{9}, \frac{10}{11}, -\frac{12}{13}, \ldots$
21. $\frac{1}{1 \cdot 3}, \frac{1}{3 \cdot 5}, \frac{1}{5 \cdot 7}, \frac{1}{7 \cdot 9}, \ldots$
22. $\frac{1}{2 \cdot 4}, \frac{1}{4 \cdot 6}, \frac{1}{6 \cdot 8}, \frac{1}{8 \cdot 10}, \ldots$

Arithmetic and Geometric Sequences

In Exercises 23–30, write out the terms of the sequence and determine if the sequence is either arithmetic or geometric.

23. $\left\{\frac{k + 2}{3}\right\}_{k=1}^{6}$
24. $\left\{\frac{2^k}{3^k}\right\}_{k=1}^{5}$
25. $\left\{\frac{3k - 5}{6}\right\}_{k=1}^{5}$
26. $\left\{\frac{k^2}{2^k}\right\}_{k=1}^{6}$
27. $\{(k + 1)^2 - k^2\}_{k=1}^{6}$
28. $\{2^{k+1} - 2^k\}_{k=1}^{5}$
29. $\left\{\frac{3 \cdot 4^{k+1}}{2^k}\right\}_{k=1}^{5}$
30. $\left\{\frac{(-3)^k}{2^{2k-1}}\right\}_{k=1}^{4}$
31. If $\{a_k\}$ is an arithmetic sequence with $a = 2$ and $d = 3$, find a_{30}.
32. If $\{a_k\}$ is an arithmetic sequence with $a = 2$ and $a_{35} = 104$, find d.
33. If $\{a_k\}$ is an arithmetic sequence with $d = .5$ and $a_{45} = 30$, find a.
34. If $\{a_k\}$ is an arithmetic sequence with $a_1 = 5$ and $a_{16} = -25$, find a_7.
35. If $\{a_k\}$ is a geometric sequence with $a = 3$ and $r = 2$, find a_8.
36. If $\{a_k\}$ is a geometric sequence with $r = -1/2$ and $a_6 = 3/16$, find a.
37. If $\{a_k\}$ is a geometric sequence with $a = 8$ and $a_4 = 512$, find r.
38. If $\{a_k\}$ is a geometric sequence with $a_{12} = 64$ and $a_7 = -2$, find a_3.

Graphs of Sequences

Graph each of the following sequences in the plane, and if possible, identify the curve $y = f(x)$ on which the graph of the sequence lies.

39. $a_k = -1 + \frac{3}{4}(k - 1), \quad k = 1, 2, 3, 4, 5$
40. $a_k = 2 - \frac{1}{2}(k - 1), \quad k = 1, 2, 3, 4, 5$
41. $a_k = \left(\frac{1}{2}\right)^{k-1}, \quad k = 1, 2, 3, 4$
42. $a_k = (-2)^{k-1}, \quad k = 1, 2, 3, 4$

43. $a_k = \frac{1}{1 \cdot 2} + \frac{1}{2 \cdot 3} + \frac{1}{3 \cdot 4} + \cdots + \frac{1}{k(k+1)}$,
$k = 1, 2, 3, 4, 5$ (*Hint:* show $a_k = \frac{k}{k+1}$.)

44. $a_k = \frac{5}{2 \cdot 4} + \frac{5}{4 \cdot 6} + \frac{5}{6 \cdot 8} + \cdots + \frac{5}{2k(2k+2)}$,
$k = 1, 2, 3, 4, 5$

Miscellaneous

45. An object dropped from above the ground falls $16t^2$ ft in t sec. Let a_k equal the distance fallen from $t = k - 1$ to $t = k$ sec for $k = 1, 2, 3, \ldots$. Show that $\{a_k\}_{k=1}^{n}$ is an arithmetic sequence for any integer $n \geq 2$.

11.3 Arithmetic and Geometric Series

A student responded: "Professor, you said that series would follow sequences in the lecture series, but since a series is a sum of terms of a sequence, doesn't it follow that wherever our study of series leads, sequences will follow?"

Finite Series and Summation Notation. A **finite series** is a sum of a finite number of **terms.** For example, the series

$$5 \cdot 1 + 5 \cdot 2 + 5 \cdot 3 + 5 \cdot 4 + 5 \cdot 5 + 5 \cdot 6$$

has for its terms the first six numbers of the sequence $5k \quad (k = 1, 2, \ldots)$. By adding the terms, we obtain 105 for the **value** of the series. When working with series, it is convenient to introduce a compact notation. For instance, the series above is written in compact form as

$$\sum_{k=1}^{6} 5 \cdot k,$$

read "the summation of $5 \cdot k$ from $k = 1$ to $k = 6$." The Greek letter Σ is called the **summation symbol,** k is called the **summation index,** and $5 \cdot k$ is called the **general term** of the series. The starting value of the index k is written below the summation symbol and the end value of k is written above. It is understood that k increases by 1 from its starting value to each succeeding value until it reaches its end value.

In general, a series whose terms are $a_1, a_2, \ldots, a_n$ is written as

$$\sum_{k=1}^{n} a_k.$$

Note that the summation notation does not provide us with the *value* of the series, but merely permits us to write it in compact form.

Example 1 Write out all the terms of each of the following series.

(a) $\sum_{k=1}^{5} (2k - 1)$ (b) $\sum_{k=2}^{7} \frac{1}{k}$ (c) $\sum_{k=0}^{4} (-1)^k 3^k$

Solution:

(a) $\sum_{k=1}^{5}(2k-1) = (2\cdot 1 - 1) + (2\cdot 2 - 1) + (2\cdot 3 - 1) + (2\cdot 4 - 1) + (2\cdot 5 - 1)$
$= 1 + 3 + 5 + 7 + 9$

(b) $\sum_{k=2}^{7}\frac{1}{k} = \frac{1}{2} + \frac{1}{3} + \frac{1}{4} + \frac{1}{5} + \frac{1}{6} + \frac{1}{7}$

(c) $\sum_{k=0}^{4}(-1)^k 3^k = (-1)^0 3^0 + (-1)^1 3^1 + (-1)^2 3^2 + (-1)^3 3^3 + (-1)^4 3^4$
$= 1 - 3 + 3^2 - 3^3 + 3^4$ ■

Example 2 Write each of the following series in compact form.

(a) $2 + 4 + 6 + 8 + 10 + 12 + \cdots + 20$

(b) $\frac{1}{2} + \frac{1}{2^2} + \frac{1}{2^3} + \frac{1}{2^4} + \frac{1}{2^5} + \cdots + \frac{1}{2^{10}}$

(c) $\frac{5}{2(-3)+1} + \frac{5}{2(-2)+1} + \frac{5}{2(-1)+1} + 5 + \frac{5}{2\cdot 1+1} + \frac{5}{2\cdot 2+1} + \frac{5}{2\cdot 3+1}$

Solution:

(a) $\sum_{k=1}^{10} 2k$ (b) $\sum_{k=1}^{10}\frac{1}{2^k}$ (c) $\sum_{k=-3}^{3}\frac{5}{2k+1}$ ■

Arithmetic Series. In the first section of this chapter, we used the principle of mathematical induction to prove the formula

$$1 + 2 + 3 + \cdots + n = \frac{n(n+1)}{2}. \tag{1}$$

The numbers $1, 2, 3, \ldots, n$ on the left side of the above equation form an arithmetic sequence of n terms whose first term is 1 and whose last term is n. In general, a series whose terms form an arithmetic sequence is called an **arithmetic series,** and the value of the series is

$$a + (a + d) + (a + 2d) + \cdots + [a + (n-1)d] = \frac{n(a+l)}{2}, \tag{2}$$

where

$$l = a + (n-1)d \tag{3}$$

is the last term of the series and n is the number of terms. For example, in the following arithmetic series, $a = 3$, $l = 15$, $n = 7$, and

$$3 + 5 + 7 + 9 + 11 + 13 + 15 = \frac{7(3+15)}{2} = 63.$$

You are asked to complete a proof of equation (2) by mathematical induction in the exercises. Here we give a proof that does not use mathematical induction.

Proof of equation (2): Let

$$S = a + (a + d) + (a + 2d) + \cdots + l, \tag{4}$$

where $l = a + (n - 1)d$. The term before the last one in (4) is $l - d$, and the one before that is $l - 2d$, and so on, down to the first term a. Hence, by writing the terms of S *from the last to the first*, we get

$$S = l + (l - d) + (l - 2d) + \cdots + a. \tag{5}$$

We now add equations (4) and (5) to obtain

$$\begin{aligned} 2S &= \underbrace{(a + l) + (a + l) + (a + l) + \cdots + (a + l)}_{n \text{ terms}} \quad \textit{All the d's cancel} \\ &= n(a + l) \\ S &= \frac{n(a + l)}{2}, \end{aligned}$$

which completes the proof.[1]

[1] *It is said that the mathematician Gauss discovered this proof when he was a ten-year-old schoolboy. Given an assignment to add the terms of an arithmetic sequence, he instantly hit upon the idea of adding the terms from the last to the first along with those from first to last, thus arriving at equation (2).*

Example 3 Find the value of the arithmetic series

$$5 + 10 + 15 + 20 + 25 + \cdots + 250.$$

Solution: We can do this sum in two ways. First, we can use $a = 5$, $d = 5$, and $l = 250$ in equation (3) to get n. That is, $250 = 5 + (n - 1)5$, which means that $n = 50$. Therefore, by formula (2),

$$S = \frac{50(5 + 250)}{2} = 6375.$$

A second way uses equation (1) as follows.

$$\begin{aligned} S &= 5 + 10 + 15 + 20 + 25 + \cdots + 250 && \textit{Given} \\ &= 5(1 + 2 + 3 + 4 + 5 + \cdots + 50) && \textit{Factor out 5} \\ &= 5 \cdot \frac{50 \cdot 51}{2} && \textit{By (1)} \\ &= 6375 \ \blacksquare \end{aligned}$$

The general term a_k of an arithmetic series is $a + (k - 1)d$, and formula (2) can be written in compact form as

$$\sum_{k=1}^{n} [a + (k - 1)d] = \frac{n(a + l)}{2}. \tag{2'}$$

In particular, when $a = 1$ and $d = 1$, we have $l = n$, and

$$\sum_{k=1}^{n} k = \frac{n(n+1)}{2}, \qquad (1')$$

which is equation (1) in compact form.

Geometric Series. A series whose terms form a geometric sequence is called a **geometric series.** For example, the terms $2, 2 \cdot 4, 2 \cdot 4^2, 2 \cdot 4^3, \ldots, 2 \cdot 4^{10}$ define the geometric series

$$2 + 2 \cdot 4 + 2 \cdot 4^2 + 2 \cdot 4^3 + \cdots + 2 \cdot 4^{10},$$

which is written in compact form as

$$\sum_{k=1}^{11} 2 \cdot 4^{k-1}.$$

We can find the value of the series as follows.

$$\text{Let} \quad S = 2 + 2 \cdot 4 + 2 \cdot 4^2 + 2 \cdot 4^3 + \cdots + 2 \cdot 4^{10}. \qquad (6)$$

We multiply both sides of (6) by the constant ratio 4 to obtain

$$4S = 2 \cdot 4 + 2 \cdot 4^2 + 2 \cdot 4^3 + 2 \cdot 4^4 + \cdots + 2 \cdot 4^{11}. \qquad (7)$$

If we now subtract (7) from (6), all of the terms after 2 in (6) cancel with all of the terms before $2 \cdot 4^{11}$ in (7), resulting in

$$S - 4S = 2 - 2 \cdot 4^{11}$$
$$S(1-4) = 2(1-4^{11})$$
$$S = \frac{2(1-4^{11})}{1-4}.$$

The above procedure can be applied to any **geometric series** in which $r \neq 1$ (*why not* $r = 1$?) to obtain the following formula.

$$\textbf{If } r \neq 1, \textbf{ then } a + a \cdot r + a \cdot r^2 + \cdots + a \cdot r^{n-1} = \frac{a(1-r^n)}{1-r}. \qquad (8)$$

If $r = 1$, each term on the left side of (8) is equal to a and there are n terms in all, so the sum equals $n \cdot a$ when $r = 1$. (In the exercises, you are asked to give a proof of formula (8) by mathematical induction.)

Example 4 Find the value of the geometric series

$$1 + \frac{1}{2} + \frac{1}{4} + \frac{1}{8} + \frac{1}{16} + \frac{1}{32} + \frac{1}{64} + \frac{1}{128}.$$

Solution: We have $a = 1, r = 1/2$, and $n = 8$. Therefore, by formula (8),

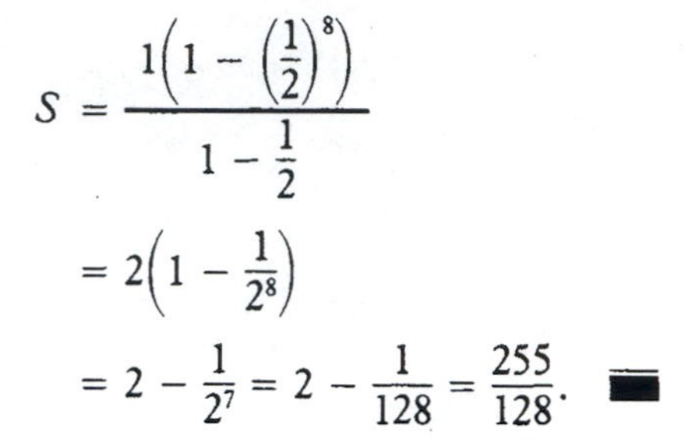

$$S = \frac{1\left(1 - \left(\frac{1}{2}\right)^8\right)}{1 - \frac{1}{2}}$$
$$= 2\left(1 - \frac{1}{2^8}\right)$$
$$= 2 - \frac{1}{2^7} = 2 - \frac{1}{128} = \frac{255}{128}.$$ ■

In compact notation, formula (8) can be written as

$$\sum_{k=1}^{n} a \cdot r^{k-1} = \frac{a(1 - r^n)}{1 - r}, \quad r \neq 1. \tag{8'}$$

Example 5 There is a story about a king and a subject that involves a geometric series. The subject had done a favor for the king, and in return the king offered to grant any reasonable request. The subject produced a chess board and asked that one grain of wheat be given for the first square on the board, 2 grains for the second square, 4 grains for the third, 8 for the fourth, and so on, doubling at each stage, until all 64 squares were accounted for. The king, thinking the request a modest one, immediately agreed. How many grains were due the subject?

Solution: The total number of grains is

$$S = 1 + 2 + 2^2 + \cdots + 2^{63} = \sum_{k=1}^{64} 2^{k-1},$$

which is a geometric series with $a = 1$, $r = 2$, and $n = 64$. Therefore

$$S = \frac{1 - 2^{64}}{1 - 2}$$
$$= 2^{64} - 1$$
$$\approx 1.84 \times 10^{19}$$
$$= 18{,}400{,}000{,}000{,}000{,}000{,}000 \text{ grains.}$$

It seems the king knew little about geometric sequences and series! ■

Even if n is very large, the value of a geometric series does not have to be large as in Example 5. For example, if $a = 1/2$ and $r = 1/2$, then by formula (8),

$$\frac{1}{2} + \frac{1}{2^2} + \frac{1}{2^3} + \cdots + \frac{1}{2^n} = \frac{\frac{1}{2}\left[1 - \left(\frac{1}{2}\right)^n\right]}{1 - \frac{1}{2}} = 1 - \left(\frac{1}{2}\right)^n. \tag{9}$$

As n gets large, $(1/2)^n$ gets close to 0 and the value of the geometric series (9) approaches 1. More generally, if $|r| < 1$, then $r^n \to 0$ as $n \to \infty$ and by (8′),

$$\sum_{k=1}^{n} ar^{k-1} \to \frac{a}{1-r}, \quad |r| < 1.$$

Formally, we write

$$\sum_{k=1}^{\infty} ar^{k-1} = \frac{a}{1-r}, \quad |r| < 1. \tag{10}$$

That is, the symbol ∞ in (10) indicates the value approached by the sum from $k = 1$ to $k = n$ as n increases without bound. Hence, for the geometric series (9), we write

$$\sum_{k=1}^{\infty} \left(\frac{1}{2}\right)^k = 1.$$

Example 6 Find the value approached by the geometric series

$$1 - \frac{1}{3} + \left(\frac{1}{3}\right)^2 - \left(\frac{1}{3}\right)^3 + \cdots + (-1)^{n-1}\left(\frac{1}{3}\right)^{n-1}$$

as $n \to \infty$. Express the answer in terms of equation (10).

Solution: Here $a = 1$ and $r = -1/3$. Since $|-1/3| < 1$, the given series approaches the value $1/[1 - (-1/3)] = 3/4$. In terms of equation (10) we write

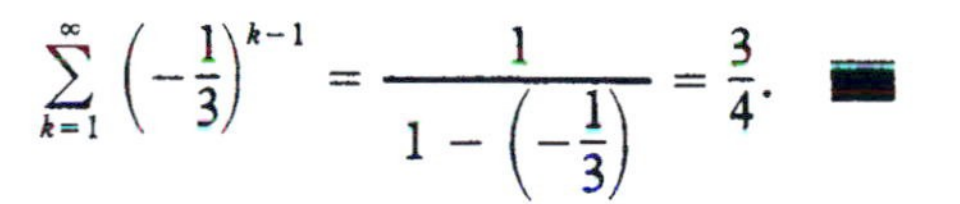

$$\sum_{k=1}^{\infty} \left(-\frac{1}{3}\right)^{k-1} = \frac{1}{1 - \left(-\frac{1}{3}\right)} = \frac{3}{4}. \quad \blacksquare$$

Question *What value does a geometric series approach as $n \to \infty$ if $|r| \geq 1$?*

Answer: If $|r| > 1$, then the magnitude of r^n increases without bound, and therefore the series (8) does not approach any finite value. If $r = 1$, the series

$$a + a + a + a + \cdots + a = na$$

also does not approach any finite value (we assume $a \neq 0$). Finally, if $r = -1$, the series

$$a - a + a - a + \cdots + (-1)^{n-1}a$$

has the value 0 if n is even and the value a if n is odd. Hence, in none of the cases corresponding to $|r| \geq 1$ does a geometric series approach any specific finite value. $\blacksquare$

Comment The previous discussion of the value approached by a geometric series as $n \to \infty$ leads to a general investigation of series that have an infinite number of terms. However, a precise treatment of this concept goes beyond the fundamentals of algebra and is covered in calculus.

Exercises 11.3

Fill in the blanks to make each statement true.

1. A finite series is ______________.
2. An arithmetic series is ______________.
3. A geometric series is ______________.
4. The sum of the n terms of an arithmetic sequence from a to l is __________.
5. The sum of the n terms of a geometric sequence from a to $a \cdot r^{n-1}$ is __________.

Write true *or* false *for each statement.*

6. Every finite series is either an arithmetic series or a geometric series.
7. A geometric series in which $r = 1$ is also an arithmetic series.
8. An arithmetic series in which $d = 1$ is also a geometric series.
9. The value of $\sum_{k=1}^{n} a_k$ is equal to the value of $\sum_{k=0}^{n-1} a_{k+1}$.
10. $\sum_{k=1}^{n} a_k - \sum_{k=1}^{n-1} a_k = a_n$

Finite Series and Summation Notation

Write out the terms of each of the following series.

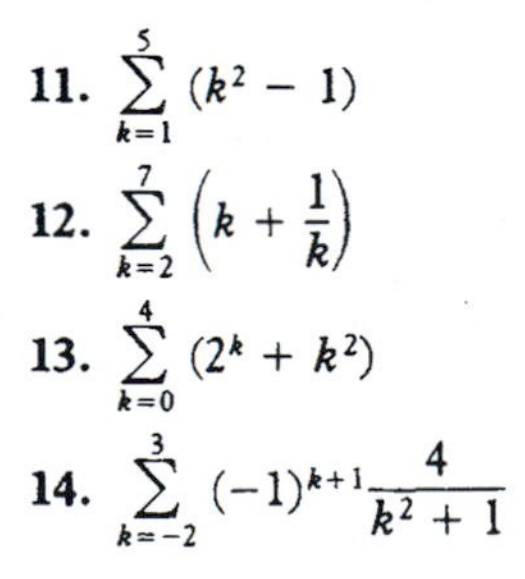

11. $\sum_{k=1}^{5} (k^2 - 1)$
12. $\sum_{k=2}^{7} \left(k + \frac{1}{k}\right)$
13. $\sum_{k=0}^{4} (2^k + k^2)$
14. $\sum_{k=-2}^{3} (-1)^{k+1} \frac{4}{k^2 + 1}$

Write each of the following series in compact form.

15. $1^2 + 3^2 + 5^2 + 7^2 + 9^2$
16. ${}_nC_0 + {}_nC_1 + {}_nC_2 + \cdots + {}_nC_n$
17. $\frac{1}{2} - \frac{2}{3} + \frac{3}{4} - \frac{4}{5} + \frac{5}{6} - \frac{6}{7} + \frac{7}{8}$
18. $\frac{10!}{0!} + \frac{9!}{1!} + \frac{8!}{2!} + \cdots + \frac{0!}{10!}$

Find the value of each of the following series.

19. $\sum_{k=1}^{20} 3k$
20. $\sum_{k=0}^{5} 2^k$
21. $\sum_{k=1}^{10} 4$
22. $\sum_{k=1}^{4} k^k$
23. Show that formula (2′) can be written as $\sum_{k=0}^{n-1} (a + kd) = \frac{n(a + l)}{2}$.
24. Write formula (8′) with the index k starting at 0. (*Hint:* see the previous exercise.)

Arithmetic Series

In Exercises 25–28, find the value of the arithmetic series whose terms are those of the given arithmetic sequence.

25. $\{a_k\}_{k=1}^{50}, \quad a = -1, \quad d = 2$
26. $\{a_k\}_{k=1}^{100}, \quad a = 30, \quad d = 2.5$
27. $\{3 + .25(k - 1)\}_{k=1}^{40}$
28. $a = -5, \quad a_2 = 0, \quad n = 25$

Find the value of each of the arithmetic series in Exercises 29–32.

29. $\sum_{k=1}^{25} [1 + 4(k - 1)]$

30. $\sum_{k=1}^{40} [-3 + 2(k - 1)]$

31. $\sum_{k=0}^{50} (2k + 3)$ (*Hint:* see Exercise 23.)

32. $\sum_{k=0}^{30} (-k + 2)$

33. The value of an arithmetic series is 150. If $a = -5$ and $n = 20$, find l.

34. In an arithmetic series of 150 terms, the value is 5 and the last term is 1. Find d.

35. The sum of the terms of an arithmetic sequence from -3 to 25 is 176. How many terms are there?

36. The value of an arithmetic series of 100 terms is 25. If $d = .5$, what is a?

Geometric Series

Find the value of each geometric series.

37. $1 + \frac{1}{2} + \frac{1}{2^2} + \frac{1}{2^3} + \cdots + \frac{1}{2^8}$

38. $1 - 2 + 4 - 8 \pm \cdots - 128$

39. $\frac{3}{10} - \frac{3}{10^2} + \frac{3}{10^3} - \frac{3}{10^4} + \frac{3}{10^5} - \frac{3}{10^6}$

40. $\left(\frac{4}{5}\right)^2 + \left(\frac{4}{5}\right)^3 + \left(\frac{4}{5}\right)^4 + \cdots + \left(\frac{4}{5}\right)^7$

41. $\sum_{k=1}^{6} 5(-3)^{k-1}$

42. $\sum_{k=1}^{7} -4\left(\frac{2}{3}\right)^{k-1}$

43. $\sum_{k=0}^{8} 3(-2)^k$

44. $\sum_{k=0}^{4} 10(.5)^k$

45. The value of a geometric series that has three terms and starts at 1 is .75. Find r.

46. The value of a geometric series is 6.24992. If $a = 5$ and $r = .2$, find n.

47. The value of a geometric series is 117. If $r = 5$ and $n = 4$, what is a?

48. The value of a geometric series of 4 terms is 180 and the first term is 12. What is r?

If possible, find the value approached by each of the following geometric series as the number of terms increases without bound.

49. The series in Exercise 37.

50. The series in Exercise 38.

51. The series in Exercise 39.

52. The series in Exercise 40.

53. The series in Exercise 41.

54. The series in Exercise 42.

Miscellaneous

(Some of the following exercises require a calculator.) Exercises 55–57 involve geometric sequences or series and are based on the result that if P dollars are invested at R% interest, compounded annually, then the amount accumulated after n years is

$$P\left(1 + \frac{R}{100}\right)^n = Pr^n, \text{ where } r = 1 + \frac{R}{100}.$$

55. If one of your ancestors invested \$1 for you 200 years ago at 10% interest compounded annually, what would the accumulated amount be now? If you decide to leave half of the amount in the bank at the same terms for another 200 years for one of your heirs, what will the heir receive?

56. If \$2000 is deposited at the beginning of each year in an IRA account at 10% compounded annually, how much is accumulated at the end of 35 years? (*Hint:* express the amount accumulated as a geometric series.)

57. Parents of a newborn child would like to have \$20,000 available for college at the end of 18 years. If they plan on making 18 equal deposits, one at the beginning of each year, how much should each deposit be? Assume that interest is at 8%, compounded annually. (*Hint:* treat each deposit as a term of a geometric series.)

58. Use mathematical induction to prove the equation

$$a + (a + d) + (a + 2d) + \cdots + [a + (n - 1)d] = \frac{n[2a + (n - 1)d]}{2}$$

for an arithmetic series.

59. Use mathematical induction to prove the equation

$$a + a \cdot r + a \cdot r^2 + \cdots + a \cdot r^{n-1} = \frac{a(1 - r^n)}{1 - r}, \quad r \neq 1,$$

for a geometric series.

11.4 Permutations and Combinations

For every combination of n distinct objects, there are n! permutations.

Fundamental Counting Principle
Permutations
Combinations

In a state-run lottery, a daily number consisting of 3 digits from 0 to 9 is chosen. To win \$500 on a \$1 ticket in a "straight" play, the player must get all three digits *in the correct order.* However, a player can choose to "box" the number, which means that if the three digits appear *in any order* on the ticket, the player wins \$80. For example, if the daily number is 123, then, on a straight play, only the number 123 wins. On a box play, the winning numbers are

123
132
213
231
312
321.

In mathematical terminology, the straight play, consisting of three digits in a certain order, is called a **permutation,** and the box play, consisting of three digits in any order, is called a **combination.** In this section, we develop systematic ways for counting the number of permutations or combinations associated with the possible outcomes of a given event. In the next section, we will apply these counting procedures to determine the **probability** of a specific outcome of an event.

Fundamental Counting Principle. The basis for all formulas on permutations and combinations is the following **fundamental counting principle.**

> If an event A has n possible outcomes, and if for each one of these, a second event B has m possible outcomes, then the event AB consisting of A followed by B has exactly $n \cdot m$ possible outcomes. **(1)**

Example 1 There are two roads from Newtown to Bay City, and three roads from Bay City to Mountain Valley. How many routes are there from Newtown to Mountain Valley that go through Bay City?

Solution: According to the fundamental counting principle, there are $2 \cdot 3 = 6$ such routes. Specifically, if the roads from Newtown to Bay City are A_1 and A_2, and those from Bay City to Mountain Valley are B_1, B_2, and B_3, then the six routes are A_1B_1, A_1B_2, A_1B_3, A_2B_1, A_2B_2, and A_2B_3. ■

Although the fundamental counting principle is stated for two events, it can be extended to more than two events, as shown in the following example.

Example 2 A restaurant menu offers 3 appetizers, 4 main courses, and 2 desserts. How many different complete meals can be ordered?

Solution: According to the fundamental principle, the appetizer and main course can be chosen in $3 \cdot 4 = 12$ ways. With each of these, there are 2 choices for dessert. Hence, by another application of the counting principle, we get $12 \cdot 2 = 24$ choices in all. That is, the total number of choices is the product $3 \cdot 4 \cdot 2$ of all three choices. If we denote the appetizers by A_1, A_2, and A_3, the main courses by M_1, M_2, M_3, and M_4, and the desserts by D_1 and D_2, then the 24 complete meals are as follows.

$A_1M_1D_1$	$A_1M_4D_1$	$A_2M_3D_1$	$A_3M_2D_1$
$A_1M_1D_2$	$A_1M_4D_2$	$A_2M_3D_2$	$A_3M_2D_2$
$A_1M_2D_1$	$A_2M_1D_1$	$A_2M_4D_1$	$A_3M_3D_1$
$A_1M_2D_2$	$A_2M_1D_2$	$A_2M_4D_2$	$A_3M_3D_2$
$A_1M_3D_1$	$A_2M_2D_1$	$A_3M_1D_1$	$A_3M_4D_1$
$A_1M_3D_2$	$A_2M_2D_2$	$A_3M_1D_2$	$A_3M_4D_2$ ■

In general, if there are k events $A_1, A_2, \ldots, A_k$ with $n_1, n_2, \ldots, n_k$ possible compatible outcomes, then the event $A_1A_2 \ldots A_k$ has

$$n_1 \cdot n_2 \cdot \ldots \cdot n_k$$

possible outcomes.

You should note the pattern of enumeration on the subscripts in Example 2. The enumeration is ordinary counting given that you can use only 1 or 2 in the units place, 1, 2, 3, or 4 in the tens place, and 1, 2, or 3 in the hundreds place. For events in which the number of outcomes is not very large, as in Examples 1 and 2, we can list all the possible outcomes. However, if a restaurant had six appetizers, a dozen main courses, and five desserts, then there would be $6 \cdot 12 \cdot 5 = 360$ complete meals in all, which would be too many to list.

Example 3 How many different five-digit numbers are there?

Solution: Any such number can be obtained by filling in the blanks

$$\underline{\quad}\ \underline{\quad}\ \underline{\quad}\ \underline{\quad}\ \underline{\quad}$$

with numbers from 0 to 9, but only 1 to 9 can be used in the leftmost position. Hence, we have nine choices for the first blank and ten choices for each of the others. That is, there are

$$\underline{\ 9\ } \cdot \underline{\ 10\ } \cdot \underline{\ 10\ } \cdot \underline{\ 10\ } \cdot \underline{\ 10\ } = 90{,}000$$

five-digit numbers. This result should come as no surprise since we have merely counted the numbers from 10,000 to 99,999. ■

Example 4 How many ways can a ten-question true-false test be answered?

Solution: Here we fill in the blanks

$$\underline{\quad}_1\ \underline{\quad}_2\ \underline{\quad}_3\ \underline{\quad}_4\ \underline{\quad}_5\ \underline{\quad}_6\ \underline{\quad}_7\ \underline{\quad}_8\ \underline{\quad}_9\ \underline{\quad}_{10}$$

with either T or F. Since there are two choices for each blank, we have

$$\underline{2}_1 \cdot \underline{2}_2 \cdot \underline{2}_3 \cdot \underline{2}_4 \cdot \underline{2}_5 \cdot \underline{2}_6 \cdot \underline{2}_7 \cdot \underline{2}_8 \cdot \underline{2}_9 \cdot \underline{2}_{10} = 2^{10} = 1024$$

possible ways to fill in the blanks with T or F. ■

Complex events, such as championship playoffs in sports, must be broken down into simpler events before the fundamental counting principle can be applied. In the following example, we count the possible outcomes of the Super Bowl playoffs.

Example 5 In 1985, the results of the Super Bowl playoffs were as shown in the following diagram.

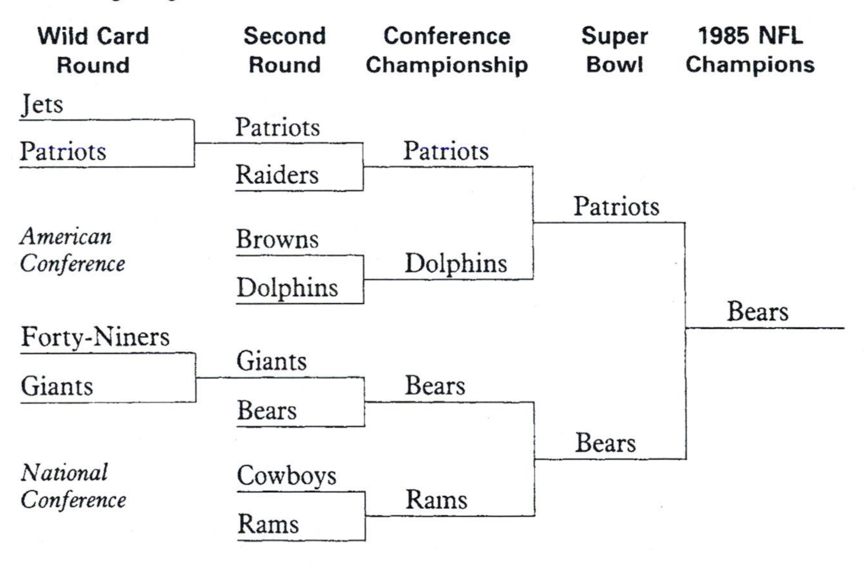

Use the diagram and the fundamental counting principle to determine the total number of possible outcomes in the Super Bowl playoffs.

Solution: As indicated in the diagram, the event called the Super Bowl playoffs can be broken down into three simpler events, the American Football Conference (AFC) playoffs, the National Football Conference (NFC) playoffs, and the Super Bowl. Five teams participate in the AFC playoffs, which consist of four games and result in one wild card winner, two second-round winners, and a conference champion. The NFC playoffs are similar. The winner of the Super Bowl game is the National Football League (NFL) champion.

Now for the counting. In each conference, there are two possibilities for the wild card winner, two possibilities for each second-round winner, and two possibilities for the conference champion. Hence, the number of possible outcomes in the AFC playoffs is

$$\underset{\textit{wild card winner}}{\underline{\quad 2 \quad}} \cdot \underset{\textit{second-round winner}}{\underline{\quad 2 \quad}} \cdot \underset{\textit{second-round winner}}{\underline{\quad 2 \quad}} \cdot \underset{\textit{conference champion}}{\underline{\quad 2 \quad}} = 16.$$

With each of these, there are also 16 possible outcomes in the NFC playoffs. Therefore, there are

$$16 \cdot 16 = 256$$

possible outcomes in the two conference playoffs, taken together. With each of these, there are two possible winners of the Super Bowl game. Hence, the total number of possible outcomes in the Super Bowl playoffs is

$$256 \cdot 2 = 512. \quad \blacksquare$$

Permutations. The word *permutation* means change or alteration. In mathematics, a **permutation** of a set of objects or elements is an *arrangement* or an *ordering* of the elements. For example, as stated before, the digits in the number 123 can be arranged in six ways, namely

$$123 \qquad 132 \qquad 213 \qquad 231 \qquad 312 \qquad 321.$$

Technically, a permutation of a set of elements is a one-to-one function from the set onto itself. In the above example, the set is $\{1, 2, 3\}$, and there are six one-to-one functions with $\{1, 2, 3\}$ as both domain and range. They are

1.	2.	3.	4.	5.	6.
$1\to1$	$1\to1$	$1\to2$	$1\to2$	$1\to3$	$1\to3$
$2\to2$	$2\to3$	$2\to1$	$2\to3$	$2\to1$	$2\to2$
$3\to3$	$3\to2$	$3\to3$	$3\to1$	$3\to2$	$3\to1$.

In the function representation, the range elements fill the positions indicated by the domain elements.

If a set S has n distinct elements, then the number of permutations of S is the number of ways in which we can fill the blanks

$$\underline{\quad}_1 \; \underline{\quad}_2 \; \underline{\quad}_3 \cdots \underline{\quad}_n$$

with elements of S. In any one arrangement, no element of S can be in more than one blank. Since there are n elements in S, we have n choices for filling the first blank, then $n - 1$ choices for the second blank (no repetitions), $n - 2$ choices for the third blank, and so on, with one less choice for each new blank until we get to exactly 1 choice for the nth blank. Hence, there are

$$\underline{\quad n \quad}_1 \cdot \underline{\ n-1\ }_2 \cdot \underline{\ n-2\ }_3 \cdot \cdots \cdot \underline{\quad 1 \quad}_n = n!$$

permutations of S.

Example 6 List all permutations of the set $S = \{a, b, c, d\}$.

Solution: Here S has 4 elements, and according to the above result, there are $4! = 24$ permutations of S. They are listed below.

abcd	bacd	cabd	dabc
abdc	badc	cadb	dacb
acbd	bcad	cbad	dbac
acdb	bcda	cbda	dbca
adbc	bdac	cdab	dcab
adcb	bdca	cdba	dcba ■

Question *What is the method by which the 24 permutations of* {a, b, c, d} *are listed in the above example?*

Answer: Each arrangement of the letters a, b, c, and d can be thought of as a word, and the 24 words are listed *alphabetically* in each column. Although the particular method for listing the permutations is not in itself important, what is important is that *some systematic method* be used. If you just write down the arrangements at random, you are sure to miss some of them. Try it! Note that we also had 24 arrangements in Example 2. There we used a counting method to list them. ■

Now let us count the number of arrangements that can be made with a fixed number of objects drawn from a set of distinct objects. For example, the number of five-letter words (not necessarily meaningful) that can be made from the word GEOPHYSICAL is

$$11 \cdot 10 \cdot 9 \cdot 8 \cdot 7 = 55{,}440.$$

In general, if r objects are drawn from a set of n distinct objects, then we have r blanks,

$$\underline{\quad\quad}_1 \ \underline{\quad\quad}_2 \ \underline{\quad\quad}_3 \ \cdots \ \underline{\quad\quad}_r,$$

to fill, and the number of choices is

$$\underline{\quad n \quad}_1 \cdot \underline{\ n-1\ }_2 \cdot \underline{\ n-2\ }_3 \cdot \cdots \cdot \underline{n-(r-1)}_r. \qquad \textit{Why } r-1?$$

We use ${}_nP_r$ to denote the **number of permutations of n distinct objects taken r at a time** $(1 \le r \le n)$, and the above result says that

$$ {}_nP_r = n(n-1)(n-2)\ldots[n-(r-1)]. \tag{2}$$

Note that when $r = n$, that is, when we permute all n objects, then $n - (r-1) = 1$, and (2) becomes ${}_nP_n = n!$, which agrees with our previous result for the total number of permutations of a set of n (distinct) elements. Note also that

$$\begin{aligned} n(n-1)(n-2)\ldots[n-(r-1)] \\ &= \frac{n(n-1)(n-2)\ldots[n-(r-1)]\cdot(n-r)!}{(n-r)!} \\ &= \frac{n!}{(n-r)!}. \end{aligned}$$

Therefore,

$$_nP_r = \frac{n!}{(n-r)!}, \tag{2'}$$

which may be a little easier to remember than (2). Also, although (2′) was derived under the assumption that $r \geq 1$, it actually has a meaningful interpretation for $r = 0$. That is, when $r = 0$, equation (2′) reads

$$_nP_0 = 1,$$

which we can interpret as saying that there is 1 way to arrange 0 objects from a set of n distinct objects: just leave all the objects in the original set. On the other hand, there is no need to memorize either (2) or (2′) for computational purposes. We can always obtain the value of $_nP_r$ by using r blanks and forming the product $n(n-1)(n-2)\ldots$ until we get to the last blank. The product will automatically stop at $n-(r-1)$.

Example 7 The main presidential candidates in the 1984 Democratic primaries were Alan Cranston, John Glenn, Gary Hart, Jesse Jackson, and Walter Mondale. List all possible tickets containing a presidential and vice-presidential candidate chosen from among these candidates.

Solution: The number of ways of arranging two people from a group of five is $5 \cdot 4 = 20$. The possible selections, with presidential candidate first, are listed below.

Cranston, Glenn	Glenn, Cranston
Cranston, Hart	Hart, Cranston
Cranston, Jackson	Jackson, Cranston
Cranston, Mondale	Mondale, Cranston
Glenn, Hart	Hart, Glenn
Glenn, Jackson	Jackson, Glenn
Glenn, Mondale	Mondale, Glenn
Hart, Jackson	Jackson, Hart
Hart, Mondale	Mondale, Hart
Jackson, Mondale	Mondale, Jackson

Note that each pair of names appears on two different tickets, depending on which one is the presidential candidate. ■

Example 8 The six teams in the Patrick Division of the National Hockey League are the Capitals, Devils, Flyers, Islanders, Penguins, and Rangers. The teams that finish the regular season in first, second, third, or fourth place make it to the Stanley Cup Playoffs. How many possible first- to fourth-place finishes are there?

Solution: The number of permutations of four objects from a set of six distinct objects is

$$6 \cdot 5 \cdot 4 \cdot 3 = 360.$$

Hence, there are 360 different first- to fourth-place finishes possible in the Patrick Division. ∎

Combinations. In Example 7, we determined that there were twenty possible tickets consisting of a presidential and vice-presidential candidate chosen from the five main presidential candidates in the 1984 Democratic Party primaries. If we are concerned only with the two people chosen and not with their respective place on the ticket, then there are only ten possible tickets as indicated in the first column of Example 7. A selection of objects, without regard to their order or arrangement, is called a **combination** of the objects. Hence, there are ten possible combinations of the five candidates taken two at a time.

In general, when counting the number of permutations of n distinct objects taken r at a time, we can first select the r objects and then permute them in $r!$ ways. That is, each combination of r objects results in $r!$ permutations of them. Therefore, if ${}_nC_r$ denotes the total **number of combinations of n distinct objects taken r at a time,** we have the relationship

$${}_nP_r = {}_nC_r \cdot r!$$

or

$${}_nC_r = \frac{{}_nP_r}{r!}, \tag{3}$$

which, by equation (2), becomes

$${}_nC_r = \frac{n(n-1)(n-2)\ldots[n-(r-1)]}{r!} \tag{4}$$

In Example 8, we determined that there were 360 first- to fourth-place finishes for the six teams in the Patrick Division of the National Hockey League. That is,

$${}_6P_4 = 360.$$

If we are concerned only with the first four teams and not with their respective places, then by formula (4) the total number is

$${}_6C_4 = \frac{6 \cdot 5 \cdot 4 \cdot 3}{4 \cdot 3 \cdot 2 \cdot 1} = 15.$$

Hence, there are fifteen ways in which four out of six teams can qualify for the Stanley Cup Playoffs.

Example 9 A poker hand consists of five cards drawn from a deck of 52 (different) cards. The order is not important. How many poker hands are there?

Solution: The number is

$$_{52}C_5 = \frac{52 \cdot 51 \cdot 50 \cdot 49 \cdot 48}{5 \cdot 4 \cdot 3 \cdot 2 \cdot 1} = 2{,}598{,}960. \blacksquare$$

Example 10 A play has five male roles and seven female roles. If eight men and eleven women audition for the parts, how many casts can be selected?

Solution: The number of selections for the male roles is

$$_8C_5 = \frac{8 \cdot 7 \cdot 6 \cdot 5 \cdot 4}{5 \cdot 4 \cdot 3 \cdot 2 \cdot 1} = 56,$$

and the number of selections for the female roles is

$$_{11}C_7 = \frac{11 \cdot 10 \cdot 9 \cdot 8 \cdot 7 \cdot 6 \cdot 5}{7 \cdot 6 \cdot 5 \cdot 4 \cdot 3 \cdot 2 \cdot 1} = 330.$$

Therefore, by the fundamental counting principle, there are

$$56 \cdot 330 = 18{,}480$$

possible casts. Note that we are counting only the people selected, not the individual roles they are playing. $\blacksquare$

Just as $_nP_r$ can be written according to equation (2) or (2′), we can also write $_nC_r$ as in (4) or as

$$_nC_r = \frac{n!}{(n-r)!r!}, \tag{4'}$$

which follows from equations (3) and (2′). Equation (4′) is easier to remember than (4), but (4) is more useful for computation than (4′). Also, from equation (4′), we have

$$_nC_0 = 1,$$

which can be interpreted as saying that there is 1 way to select 0 objects from n distinct objects, that is, select the empty set.

You have probably noticed that $_nC_r$ is also the coefficient of $a^{n-r}b^r$ in the binomial expansion of $(a+b)^n$. To see why the binomial coefficients are related to combinations,[1] consider the following expansion of $(a+b)^3$.

$$(a+b)^3 = (a_1+b_1)(a_2+b_2)(a_3+b_3)$$ *We number the a's and b's to keep track of them in the expansion*

$$= a_1a_2a_3 + a_1a_2b_3 + a_1a_3b_2 + a_2a_3b_1 + a_1b_2b_3 + a_2b_1b_3 + a_3b_1b_2 + b_1b_2b_3$$

$$= {}_3C_0a^3 + {}_3C_1a^2b + {}_3C_2ab^2 + {}_3C_3b^3$$

$_3C_0a^3$	$_3C_1a^2b$	$_3C_2ab^2$	$_3C_3b^3$
number of ways 0 b's can be selected from the 3 factors	*number of ways 1 b can be selected from the 3 factors*	*number of ways 2 b's can be selected from the 3 factors*	*number of ways 3 b's can be selected from the 3 factors*

[1]*The relationship between combinations and binomial coefficients was noted by Blaise Pascal (1623–1662), one of the founders of probability theory. James Bernoulli (1654–1705), another founder of probability theory, used the relationship to prove the binomial theorem.*

Hence, in general, we can interpret ${}_nC_r$ as the number of ways in which r b's can be selected from n factors $(a + b)$.

From equation (4′) we obtain the result

$$ {}_nC_r = {}_nC_{n-r}. \tag{5}$$

In terms of the binomial coefficients, equation (5) says that in the expansion of $(a + b)^n$, the coefficient of $a^{n-r}b^r$ [left side of (5)] is the same as that of a^rb^{n-r} [right side of (5)]. In terms of combinations, equation (5) says that for each selection of r objects from n distinct objects, there corresponds a selection of $n - r$ objects that are left after the first r objects are selected.

Exercises 11.4

Fill in the blanks to make each statement true.

1. The fundamental counting principle states that ________.
2. If events $A_1, A_2, \ldots, A_k$ have $n_1, n_2, \ldots, n_k$ possible compatible outcomes, then the event $A_1A_2 \ldots A_k$ has ________ possible outcomes.
3. There are ________ arrangements of n distinct objects taken r at a time $(0 \le r \le n)$.
4. There are ________ ways to select r objects from n distinct objects $(0 \le r \le n)$.
5. A ________ involves both selection and order, whereas a ________ involves selection only.

Write true *or* false *for each statement.*

6. If an event A has n possible outcomes, and an event B has possible outcomes, then the event AB has $n + m$ possible outcomes.
7. A permutation of a set of elements is a one-to-one function from the set onto itself.
8. ${}_nP_r = {}_nP_{n-r}$
9. ${}_nC_r = {}_nC_{n-r}$
10. ${}_nP_r \ge {}_nC_r$

Fundamental Counting Principle

11. Gus can go by train or plane from Philadelphia to Indianapolis and by car or bus from Indianapolis to Terre Haute. How many different ways can he travel from Philadelphia to Terre Haute?
12. Buses from Martinville arrive in Allentown at 10 and 11 A.M., 12 noon, and 1 P.M. daily. Departures from Allentown for Gibbsburg are at 10:30 and 11:30 A.M. and 12:30 and 1:30 P.M. If Lynne is planning a day trip from Martinville to Gibbsburg, how many choices does she have?
13. A new car has three transmissions to choose from, with or without power steering and with or without air conditioning. How many choices involving these three items can be made?
14. On sale at Stereo Sounds are four turntables, five amplifiers, two tuners, three cassette decks, two compact disc players, and four speaker systems. All the items are of different makes. How many systems, consisting of one of each type of item, are possible?
15. Four soccer teams, A, B, C, and D, compete in a league. There are two games involving all four teams scheduled each week, and each team plays each of the other three teams exactly once. How many possible

three-week schedules are there? (*Hint:* each week's schedule is determined by the team that A competes against.)

16. The World Series is a best-out-of-seven series. In how many ways can a series between A and N end in n games, where $n = 4, 5, 6$, and 7?

17. How many ways can a ten-question multiple-choice test be answered if each question has four choices?

18. How many ways can a ten-question test be answered if the first five questions are true–false and the last five are multiple-choice with four choices each?

19. How many numbers are there from 1 to 1000 that contain exactly two 5s?

20. How many numbers are there from 1 to 1000 that contain at least two 5s?

Permutations

21. Evaluate each of the following.

(a) ${}_8P_5$ (b) ${}_7P_1$ (c) ${}_6P_0$ (d) ${}_5P_5$

22. How many ways can four people line up side by side for a group picture?

23. How many possible finishes are there in a race with six horses if no two horses tie?

24. A multiple-choice test has five questions, and each question has five choices, A, B, C, D, or E. How many ways can the test be answered with a different choice for each question?

25. How many five-digit numbers can be made from the digits 1, 2, 3, 4, and 5 if no digit is used more than once?

26. How many five-digit numbers can be made from the digits 1 through 9 if no digit is used more than once?

27. How many win, place, and show finishes are possible in a race with seven horses if there are no ties?

28. How many ways can a president, vice-president, and secretary be chosen from a high school senior class of 150 students?

29. The waiting room in a dentist's office has eight chairs. How many ways can five people be seated in the room?

30. How many license plate codes are there consisting of three letters followed by three digits, with no repetitions?

Combinations

31. Evaluate each of the following.

(a) ${}_{10}C_5$ (b) ${}_{10}C_0$ (c) ${}_{10}C_{10}$

32. Any three noncollinear points in 3-space determine a plane. How many planes are determined by twelve points, no three of which are collinear?

33. Nine players try out for the five positions on a basketball team. How many ways can five players be chosen?

34. Based on past experience with a certain professor, a student decides to answer six questions T and four questions F in a ten-question true-false test. How many ways can the student complete the test?

35. A senior must take eight independent courses over the next two semesters. If four courses are taken each semester, how many combinations are possible?

36. How many ways can a baseball manager choose eight pitchers and two catchers from twelve pitchers and four catchers?

37. A traveler decides to take two books and four magazines on a trip. If there are seven books and ten magazines to choose from, how many choices are there?

38. A bridge hand consists of 13 cards from a 52-card deck. If four people play bridge, how many possibilities are there for the four hands?

39. How many 3-digit combinations (order not important) can be made from the digits 1, 2, 3, and 4 if

(a) no two digits are the same,

(b) exactly two digits are the same,

(c) exactly three digits are the same.

40. A student has ten hours to prepare for three tests—math, language, and history. If each test will take at least two hours of preparation, in how many ways can the preparation time be allotted?

11.5 Basic Probability

Probability gives us a precise measure of our uncertainty.

Equally Likely Outcomes
The Rules of Probability

Probability is a branch of mathematics that was originally developed to bring mathematical precision to games of chance. It is used to determine the odds in casino gambling and the payoffs in state lotteries. But in addition to recreational gambling, probability has important applications in science and business. In particular, it is used to explain atomic structure[1] and to establish insurance premiums. In fact, since probability attaches itself to both deductive and inductive reasoning, there is hardly a human activity that does not involve probability.

[1] *Albert Einstein was the most notable holdout in applying probability to certain parts of atomic theory. His famous quote on the subject is "I shall never believe that God plays at dice with the universe."*

Equally Likely Outcomes. A die is a cube with its sides numbered 1 to 6. When a balanced or "fair" die is tossed, then the side numbered 4 is as likely to land face up as any other side. Since there are six sides, the likelihood of getting a 4 is one in six. We therefore say that the probability of getting a 4 in one toss of a die is $1/6$. To determine the probability of getting an even number, we reason that there are three outcomes that result in an even number, namely 2, 4, and 6. Therefore, the likelihood of getting an even number is three in six, and we say the probability of an even number is $3/6$ or $1/2$. In the same way, we would say that the probability of getting a number greater than 1 is $5/6$ since there are five outcomes, 2, 3, 4, 5, and 6, that result in a number greater than 1.

In probability language, the toss of the die is called an **experiment,** and, as stated before, the possible results 1, 2, 3, 4, 5, and 6 are called the **outcomes** of the experiment. The set S of outcomes is called the **sample space,** and any subset of S is called an **event.** For example, the subset $\{2, 4, 6\}$ is the event consisting of all even outcomes, and the subset $\{2, 3, 4, 5, 6\}$ is the event consisting of all outcomes greater than 1. We can consider an outcome as an element of S or as a one-element subset (event) of S. Also, the empty set $\varnothing$ and the entire set S are included as events. For example, the event of getting both an even and an odd number in one toss of a die is the empty set, and the event of getting at least one number between 1 and 6 is all of S.

More generally, if a sample space S has n outcomes $a_1, a_2, \ldots, a_n$, and if the outcomes are **equally likely,** then the **probability** of each outcome is defined as $1/n$, and we write

$$P(a_i) = \frac{1}{n}, \quad i = 1, 2, \ldots, n. \tag{1}$$

If A is an event consisting of k outcomes $(0 \le k \le n)$, then, by definition,

$$P(A) = \frac{k}{n}. \tag{2}$$

Note that with this definition, the probability of the empty set $\varnothing$ is 0, and the probability of the entire sample space S is 1. That is,

$$P(\varnothing) = 0 \qquad \text{and} \qquad P(S) = 1.$$

We think of events that define the empty set as **impossible** events and those that define the entire sample space as **certain** events.

Example 1 An experiment consists of tossing two fair dice one time. Describe the sample space, and determine the probability of each of the following events.

(a) The sum of the two dice is 3.

(b) The sum of the two dice is 7.

(c) The sum is greater than 8.

Solution: We label the dice as first and second, and then an outcome can be described by a pair of numbers (a, b) where a is the number face up on the first die and b is that on the second. There are six possible choices for a, and with each of these, there are six choices for b. Hence, by the fundamental counting principle, there are 36 outcomes in the sample space.

(a) The outcomes that result in a sum of 3 are (1, 2) and (2, 1). Therefore,

$$P(\text{sum} = 3) = \frac{2}{36} = \frac{1}{18}.$$

(b) The outcomes that result in a sum of 7 are (1, 6), (2, 5), (3, 4), (4, 3), (5, 2), and (6, 1). Therefore,

$$P(\text{sum} = 7) = \frac{6}{36} = \frac{1}{6}.$$

(c) The outcomes that result in a sum greater than 8 are those whose sums are 9, 10, 11, or 12. They are

9: (3, 6), (4, 5), (5, 4), (6, 3)
10: (4, 6), (5, 5), (6, 4)
11: (5, 6), (6, 5)
12: (6, 6).

Hence, there are 10 outcomes that result in a sum greater than 8, and

$$P(\text{sum} > 8) = \frac{10}{36} = \frac{5}{18}. \quad ■$$

Definitions (1) and (2) define probabilities of outcomes and events when the outcomes are equally likely. If k outcomes define a certain event A, then we can say that there are k outcomes **favorable** to A, and we can rewrite (2) as

$$\boldsymbol{P(A) = \frac{\textbf{the number of outcomes favorable to } A}{\textbf{the total number of outcomes}}}. \qquad (2')$$

Example 2 An experiment consists of tossing a fair coin three times. Describe the sample space, and compute the corresponding probabilities.

(a) A is the event that there are three heads.

(b) B is the event that there are at least two heads.

(c) C is the event that the first and last tosses are tails.

Solution: If H denotes heads and T denotes tails, then the sample space consists of the eight outcomes

$$HHH, \quad HHT, \quad HTH, \quad HTT, \quad THH, \quad THT, \quad TTH, \quad \text{and} \quad TTT.$$

(a) Only the first listed outcome is favorable to A. Therefore,

$$P(A) = \frac{1}{8}.$$

(b) By inspecting the above list, we see that there are four outcomes favorable to B. Hence,

$$P(B) = \frac{4}{8} = \frac{1}{2}.$$

(c) In event C, the middle toss can be H or T, so there are two outcomes favorable to C, and

$$P(C) = \frac{2}{8} = \frac{1}{4}. \quad ■$$

The Rules of Probability. Not all outcomes of an experiment need be equally likely. For example, suppose a letter is chosen at random from the letters in the word

INTERNATIONAL.

There are thirteen letters in the word, and we assume that each one is as likely to be chosen as any other. However, there are only eight *different* letters in the word, and the possible outcomes for the different letters are

I, N, T, E, R, A, O, L.

Since I appears twice in the word, there are two chances in thirteen of selecting I. Hence, we define

$$P(\text{I}) = \frac{2}{13}.$$

Proceeding in this way for the other letters, we define

$$P(\text{N}) = \frac{3}{13}, \quad P(\text{T}) = \frac{2}{13} = P(\text{A}), \quad P(\text{E}) = \frac{1}{13} = P(\text{R}) = P(\text{O}) = P(\text{L}).$$

To define the probability of an event in this sample space, we merely add the probabilities of the outcomes that make up the event. For example, the probability of selecting a vowel is equal to

$$P(\mathrm{A}) + P(\mathrm{E}) + P(\mathrm{I}) + P(\mathrm{O}) = \frac{2}{13} + \frac{1}{13} + \frac{2}{13} + \frac{1}{13} = \frac{6}{13},$$

and the probability of selecting a letter that rhymes with *sea* is

$$P(\mathrm{T}) + P(\mathrm{E}) = \frac{2}{13} + \frac{1}{13} = \frac{3}{13}.$$

In general, any set S of n elements $a_1, a_2, \ldots, a_n$ can be a sample space. We assign probabilities $P(a_i)$ $(i = 1, 2, \ldots, n)$ to the elements, subject to the following **rules of probability:**

$$0 \le P(a_i) \le 1 \quad \text{and} \quad P(a_1) + P(a_2) + \cdots + P(a_n) = 1. \tag{3}$$

By definition, the **probability of any subset (event) of S** is the sum of the probabilities of the elements (outcomes) that make up the subset. Also, the **probability of the empty set** is defined to be 0. These ideas also can be extended to infinite sets, but doing so involves concepts from calculus and therefore will not be covered here.

The question that naturally arises is "How do we assign reasonable values to the $P(a_i)$?" In the case of a "fair" die or coin or in the random choice of a letter in the word INTERNATIONAL, we arrived at the probabilities of the outcomes *deductively*. In the case of a "loaded" coin or die, or in situations in which the number of factors determining an outcome are too complicated to analyze completely, we can arrive at reasonable probabilities *inductively*, that is, from data generated by actual experiments or from statistical records. For instance, if a coin is unbalanced (and every coin is unbalanced to some extent), we can assign a probability of getting a head on a single toss by actually tossing the coin a large number of times, say 1000 times, and dividing the number of heads obtained by the number of tosses. That is, we define the probability based on the frequency of success in a large number of trials. If we wanted to arrive at a probability that a person chosen at random in the city of San Diego will live to be 75, about the best we could do is go to the city records and divide the number of people aged 75 or older by the total number of people in the city. Of course, these numbers will change, perhaps significantly, over the course of years, and the probability will change accordingly.

There are many questions of a philosophical nature concerned with the subject of probability, but this is not the place to go into them. As mathematicians, we are concerned that a sample space has been defined and probabilities assigned to the outcomes according to the two rules given in (3). We can then determine the probability of any event associated with the sample space. The significance of our results depends on the reasonableness of the probabilities assigned to the individual outcomes.

Exercises 11.5

Fill in the blanks to make each statement true.

1. A fair coin is tossed three times. The sample space is $\{HHH, HHT,$ ________$\}$.
2. In Exercise 1, the event of getting at least two heads corresponds to the subset $\{$________$\}$.
3. In Exercise 1, the subset $\{HTH, THT\}$ corresponds to the event ________.
4. If a sample space consists of outcomes $a_1, a_2, \ldots, a_n$, with corresponding probabilities $P(a_1), P(a_2), \ldots, P(a_n)$, then each $P(a_i)$ must satisfy ________, and the sum of the $P(a_i)$'s must equal ________.
5. If an event A consists of outcomes $a_1, a_2, \ldots, a_n$, then the probability of A is defined as ________.

Write true *or* false *for each statement.*

6. The probability of any outcome is a number in the interval $[0, 1]$.
7. The rules of probability require that all outcomes in a sample space be equally likely.
8. If the probability of one outcome in a sample space is 1, then the probability of every other outcome in the space is 0.
9. If the probability of one event in a sample space is 1, then the probability of every other event in the space is 0.
10. If A and B are events in a sample space and $P(A) \leq P(B)$, then A is a subset of B.

Equally Likely Outcomes

11. A fair die is tossed.
    (a) Find $P(3)$. (b) Find $P(\text{odd number})$.
12. A fair coin is tossed three times.
    (a) Find $P(HHT)$. (b) Find $P(\text{two heads})$.
13. One ball is drawn from an urn containing 4 red and 6 black balls.
    (a) Find $P(\text{red})$. (b) Find $P(\text{black})$.
14. A letter is chosen at random from the word PROBABILITY.
    (a) Find $P(A)$. (b) Find $P(B)$.
15. Two different numbers from 1 to 4 are selected.
    (a) Find $P(\text{even sum})$.
    (b) Find $P(\text{odd sum})$.
16. Two numbers (not necessarily different) from 1 to 4 are selected.
    (a) Find $P(\text{even sum})$.
    (b) Find $P(\text{odd sum})$.
17. Two fair dice are tossed.
    (a) Find $P(\text{sum} = 6)$.
    (b) Find $P(\text{sum} = 5 \text{ or } 7)$.
18. Two fair dice are tossed.
    (a) Find $P(\text{at least one die is } 4)$.
    (b) Find $P(\text{exactly one die is } 4)$.
19. One card is drawn from a standard 52-card deck, replaced, and then a second card is drawn.
    (a) Find $P(2 \text{ aces})$. (b) Find $P(2 \text{ hearts})$.
20. Two cards are drawn (without replacement) from a standard 52-card deck.
    (a) Find $P(2 \text{ aces})$. (b) Find $P(2 \text{ hearts})$.
21. Two balls are drawn (without replacement) from an urn containing three red and two black balls.
    (a) Find $P(2 \text{ red})$. (b) Find $P(2 \text{ black})$.
22. Each of two urns contains three red and two black balls. A ball is drawn from each urn.
    (a) Find $P(2 \text{ red})$. (b) Find $P(2 \text{ black})$.
23. A three-digit number is formed from the digits 1, 2, 3, 4, 5, 6, and 7. No digit can be used more than once.
    (a) Find $P(\text{number is less than } 500)$.
    (b) Find $P(\text{number is divisible by } 2)$.
24. A three-digit number is formed from the digits 1, 2, 3, 4, 5, 6, and 7. A digit can be used more than once.
    (a) Find $P(\text{number is less than } 500)$.
    (b) Find $P(\text{number is divisible by } 2)$.
25. Two dice are tossed.
    (a) Find $P(\text{odd sum})$.
    (b) Find $P(\text{at least one even number})$.

26. Three cards are chosen (without replacement) from a deck of 52 cards.
    (a) Find P(exactly 2 kings).
    (b) Find P(at most two diamonds).
27. Three cards are chosen (without replacement) from a deck of 52 cards.
    (a) Find P(a spade followed by a heart followed by a diamond).
    (b) Find P(1 spade, 1 heart, and 1 diamond in any order).
28. Do Exercise 27 if the cards are chosen with replacement.

In Exercises 29–31, E denotes an even digit and O denotes an odd digit.

29. Ten cards numbered 1 to 10 are shuffled and placed in a row.
    (a) Find $P(EEEEEOOOOO)$.
    (b) Find $P(EOEOEOEOEO)$.
30. If only four of the ten cards in Exercise 29 are placed in a row, find $P(EOEO$ or $OEOE)$.
31. Four of seven cards marked 1 to 7 are placed in a row.
    (a) Find $P(EOEO)$. (b) Find $P(OEOE)$.
32. Five cards from a 52-card deck are turned up in succession.
    (a) Find P(ace, king, queen, jack, and 10 in that order).
    (b) Find P(ace, king, queen, jack, and 10 in any order).
33. A poker hand consists of five cards drawn from a 52-card deck. A full house is three of one face value and two of another, that is, three of a kind and two of a kind.
    (a) Find P(3 aces and 2 kings).
    (b) Find P(a full house).
34. A five-card poker hand is drawn.
    (a) Find P(4 aces). (b) Find P(4 of a kind).
35. A straight in poker consists of five cards that can be arranged in order from 1 (ace) to 5, or from 2 to 6, and so on, up to from 10 to ace.
    (a) Find P(10, jack, queen, king, ace straight).
    (b) Find P(straight).
36. A flush in poker consists of five cards of the same suit.
    (a) Find P(5 hearts). (b) Find P(a flush).

The Rules of Probability

Use the rules of probability to do each of the following exercises.

37. If A and B are events in a sample space and A is a subset of B, show that $P(A) \leq P(B)$.
38. If A and B are disjoint subsets of a sample space, show that $P(A \cup B) = P(A) + P(B)$, where $A \cup B$ is the event consisting of all outcomes in either A or B.
39. A is an event in a sample space. $\overline{A}$ denotes the set of all outcomes not in A. Show that $P(\overline{A}) = 1 - P(A)$.
40. If A and B are events in a sample space, then $A \cup B$ is the event consisting of all outcomes in either A or B, and $A \cap B$ consists of all outcomes in both A and B. Show that
$$P(A \cup B) = P(A) + P(B) - P(A \cap B).$$

11.6 Conditional Probability and Mathematical Expectation

The probability of an event can change as information about the event changes.

Conditional Probability
Mathematical Expectation

Conditional Probability. In many applications of probability, we are concerned with the probability of one event, given that another event has taken place. For example, suppose we toss two dice, one red and the other white. What is the probability that the sum is 5, given that the red die turns up 2? The answer is 1/6, since the white die must turn up 3 and the probability of doing so is 1/6. We denote this result as

$$P(\text{sum} = 5 \mid \text{red die} = 2) = \frac{1}{6}, \tag{1}$$

where the event after the vertical bar is the given event. Now let's change the question. What is the probability that the sum is 5, given that exactly one of the dice turns up 2? The outcomes in which we get exactly one 2 are

$$(2, 1),\ (2, 3),\ (2, 4),\ (2, 5),\ (2, 6),$$
$$(1, 2),\ (3, 2),\ (4, 2),\ (5, 2),\ (6, 2),$$

where the first entry of each die indicates the outcome of the red die and the second entry indicates that of the white die. There are ten outcomes with exactly one 2, and these form our new sample space. Among these, there are two outcomes that result in 5, namely (2, 3) and (3, 2). Hence,

$$P(\text{sum} = 5 \mid \text{exactly one die} = 2) = \frac{2}{10} = \frac{1}{5}. \tag{2}$$

If we now replace the given event by the event that *at least* one die turns up 2, then the new sample space contains the outcome (2, 2) in addition to the ten outcomes that contain exactly one 2. Hence, we now have eleven outcomes in the sample space, and

$$P(\text{sum} = 5 \mid \text{at least one die} = 2) = \frac{2}{11}. \tag{3}$$

Results (1), (2), and (3) are examples of **conditional probability.** In each case the event of getting a sum of 5 was the same, but the sample space changed depending on the event that was given. There is a way to compute a conditional probability without changing the sample space. For simplification, let us denote the events involved in the previous example as follows.

A = "the sum is 5"
B = "the red die turns up 2"
C = "exactly one of the dice turns up 2"
D = "at least one of the dice turns up 2"

The sample space for the experiment of tossing both dice has 36 outcomes (a, b), where a is the number turned up on the red die and b is that on the white die. For this sample space, we obtain the probabilities listed in the following table.

Event	Favorable Outcomes	Probability
A	(1, 4), (2, 3), (4, 1), (3, 2)	$4/36 = 1/9$
B	(2, 1), (2, 2), (2, 3), (2, 4), (2, 5), (2,6)	$6/36 = 1/6$
C	(2, 1), (2, 3), (2, 4), (2, 5), (2, 6), (1, 2), (3, 2), (4, 2), (5, 2), (6, 2)	$10/36 = 5/18$
D	(2, 1), (2, 2), (2, 3), (2, 4), (2, 5), (2, 6), (1, 2), (3, 2), (4, 2), (5, 2), (6, 2)	$11/36$
$A \cap B$	(2, 3)	$1/36$
$A \cap C$	(2, 3), (3, 2)	$2/36 = 1/18$
$A \cap D$	(2, 3), (3, 2)	$2/36 = 1/18$

In the table, $A \cap B$ denotes the event "A and B," which consists of all outcomes that are in both A and B. $A \cap C$ and $A \cap D$ are similar. By comparing the conditional probabilities (1), (2), and (3) with the probabilities in the table, we obtain the following results:

$$P(A|B) = \frac{P(A \cap B)}{P(B)}, \quad P(A|C) = \frac{P(A \cap C)}{P(C)}, \quad P(A|D) = \frac{P(A \cap D)}{P(D)},$$

which lead to the definition below.

> Let A and B be any two events for the same sample space, where $P(B) > 0$. Then
>
> $$P(A|B) = \frac{P(A \cap B)}{P(B)} \tag{4}$$
>
> is called the **conditional probability of A given B**, or, simply, the probability of A given B.

Example 1 The daily number in a state lottery is a three-digit number from 000 to 999. To win in a "straight" play, a player must get all three digits in the correct order. For a single lottery ticket, compute each of the following.

(a) the probability of winning

(b) the probability of winning given that the first two digits are correct

(c) the probability of winning given that at least two digits are correct

Solution: Let

$$A = \text{"the ticket is a winner,"}$$
$$B = \text{"the first two digits are correct,"}$$

and $$C = \text{"at least two digits are correct."}$$

Then $$A \cap B = A \quad \text{and} \quad A \cap C = A \ (why?).$$

(a) There are 1000 three-digit numbers from 000 to 999, and only one of these is a winner. Therefore,

$$P(A) = \frac{1}{1000} = .001.$$

(b) There are 10 three-digit numbers with the correct first two digits (*why?*). Therefore,

$$P(B) = \frac{10}{1000} = .01.$$

Also, $$P(A \cap B) = P(A) = .001.$$

Hence, by definition (4),

$$P(A|B) = \frac{P(A \cap B)}{P(B)} = \frac{.001}{.01} = .1.$$

(c) We count the numbers with at least two correct digits as follows:

first and second digits correct, but not the third:	9 numbers;
first and third digits correct, but not the second:	9 numbers;
second and third digits correct, but not the first:	9 numbers;
all three digits correct:	1 number;
total:	28 numbers.

Therefore, $$P(C) = \frac{28}{1000} = .028,$$

and, by definition (4),

$$P(A|C) = \frac{P(A \cap C)}{P(C)} = \frac{.001}{.028} \approx .036. \blacksquare$$

Example 2 A fair coin is tossed three times. Let A be the event that a head is obtained on the third toss, and let B be the event that the first two tosses are heads. Find $P(A)$ and $P(A|B)$.

Solution: The sample space (see Example 2 in the previous section) is

$$HHH,\ HHT,\ HTH,\ HTT,\ THH,\ THT,\ TTH,\ TTT,$$

and the outcomes for the events in question are as follows:

$$\begin{aligned} A&: HHH, HTH, THH, TTH, \\ B&: HHT, HHH, \\ A \cap B&: HHH. \end{aligned}$$

Therefore,
$$P(A) = \frac{4}{8} = .5.$$

Also, $P(B) = \frac{2}{8} = .25$ and $P(A \cap B) = \frac{1}{8} = .125$,

which means that

$$P(A|B) = \frac{.125}{.25} = .5.$$

Hence, the probability of getting a head on the third toss, given that heads were obtained on the first two tosses, is the same as the probability of getting a head on the third toss regardless of the first two. This should come as no surprise, since the outcome of a toss of a *fair* coin should not be affected by previous outcomes. ■

Mathematical Expectation. We use mathematical expectation to measure the fairness of a game of chance. For instance, as in Example 1, let us say that the daily number in a state lottery is a three-digit number from 000 to 999, and to win in a "straight" play, a lottery ticket must have all three digits in the correct order. The net payoff is \$499 on a \$1 winning ticket. The probability of winning is 1/1000 = .001, and the probability of losing is 999/1000 = .999. The sum

$$499(.001) + (-1)(.999) = -.5$$

is called the **mathematical expectation** for a \$1 lottery ticket. The factor 499 represents the net amount collected from the state for a winning ticket, and the factor −1 represents the dollar paid to the state on a losing ticket. We can think of the −1 as *negative winnings*. The quantity −.5 is the amount a player can expect to "win" on the average for each dollar played. That is, if a player buys a lottery ticket five days a week (260 days a year), then in the long run the player can expect to *lose* an average of

$$.5(\$260) = \$130$$

per year.

In general, if a game has n outcomes $a_1, a_2, \ldots, a_n$ with corresponding probabilities $p_1, p_2, \ldots, p_n$ $(p_1 + p_2 + \cdots + p_n = 1)$, and if the payoff to the player on outcome a_i is w_i (positive for a win, negative for a loss), then the quantity

$$E = w_1p_1 + w_2p_2 + \cdots + w_np_n$$

is called the **mathematical expectation** for the player. A positive w_i is what a player wins if outcome a_i occurs; a negative w_i is what a player pays if a_i occurs. E is the amount that the player can expect to "win" on the average for each time the game is played. If E is positive, then the game is in the player's favor. If E is negative, then the game is biased against the player. That is, negative expected winnings represent losses. Of course, when playing a lottery or gambling in a casino, E is always negative.

Example 3 A player tosses two dice. If the dice come up 7 or 11, the player wins $7, and for any other number, the player loses $2. Determine the player's mathematical expectation.

Solution: Eight outcomes out of a possible 36 will result in 7 or 11, namely

$$(1, 6), (2, 5), (3, 4), (4, 3), (5, 2), (6, 1), (5, 6), (6, 5).$$

Hence, $P(7 \text{ or } 11) = \frac{8}{36}$ and $P(\text{not } 7 \text{ or } 11) = \frac{28}{36}$.

Therefore, $$E = 7 \cdot \frac{8}{36} + (-2)\frac{28}{36} = 0.$$

Hence, the player should break even over the long run. ■

When the mathematical expectation is 0, as in Example 3, the game is called a **fair game.** As stated in the example, the player's winnings and losings should be about equal if a fair game is played a large number of times.

Example 4 In a "box" play of the three-digit daily number, a lottery ticket wins if it has the correct digits in *any* order. The net payoff is $79 on a $1 ticket. What is a player's mathematical expectation? What should the payoff be in order for the game to be fair?

Solution: Here we are counting three-digit combinations, not three-digit numbers. Also, some digits may be repeated. The situation is similar to having three boxes, each containing ten cards numbered 0 to 9. A card is selected from each box, and the order is not important. We note that any combination of three digits is as likely to be drawn as any other combination.

Now we can begin counting possible combinations.

There are ${}_{10}C_3 = 120$ combinations of *three different digits.*

There are 10 three-digit combinations with *all three digits the same.*

There are 90 three-digit combinations with *exactly two digits the same,* namely, 9 three-digit combinations with two 0's, 9 three-digit combinations with two 1's, and so on, to 9 three-digit combinations with two 9's.

Therefore, there are a total of

$$120 + 10 + 90 = 220$$

three-digit combinations in all, and as already mentioned, the combinations are equally likely. Hence, the probability of winning is 1/220, that of losing is 219/220, and a player's mathematical expectation is

$$E = 79 \cdot \frac{1}{220} + (-1)\frac{219}{220}$$
$$\approx -.64.$$

(We note that a player can expect to lose more in a "box" play than in a "straight" play whose mathematical expectation is $-.5$.)

For the game to be fair, the payoff should be x dollars, where

$$x \cdot \frac{1}{220} + (-1)\frac{219}{220} = 0$$
$$\frac{x - 219}{220} = 0$$
$$x = 219. \quad \blacksquare$$

Of course, a zero mathematical expectation as a measure of fairness in Example 4 does not take into account overhead costs for running the lottery, or the entertainment value of the game. These considerations would result in a more realistic definition of fairness.

This concludes our brief treatment of probability. Although we were able to touch on only a few of its many topics, we have attempted to provide you with enough tools to solve some meaningful problems, and to show you enough of the theory to give you some feel for the subject.

Exercises 11.6

Fill in the blanks to make each statement true.

1. We denote the probability of event A given that event B has occurred by __________, which is called the __________ probability of A given B.
2. To compute $P(A|B)$ we can either change the sample space to B or use the formula $P(A|B) =$ __________, where $P(B) > 0$.
3. If $P(A \cap B) = P(A)P(B)$ and $P(B) > 0$, then $P(A|B) =$ __________.
4. If a game has outcomes a_i with corresponding probabilities p_i and payoffs w_i $(i = 1, 2, \ldots, n)$, then a player's mathematical expectation is $E =$ __________.
5. A game with mathematical expectation E is in the player's favor if __________, is biased against the player if __________, and is fair if __________.

Write true *or* false *for each statement.*

6. If $P(B) = 0$, then $P(A|B)$ is not defined.

7. $P(A|B) = P(B|A)$
8. If $P(B) > 0$, then $P(A|B) \geq P(A \cap B)$.
9. If a player's mathematical expectation in a game is positive, then the player will win every time the game is played.
10. If a player's mathematical expectation in a game is 0, then in the long run, the player should about break even.

Conditional Probability

11. A and B are events in the same sample space, where $P(A) = .7$, $P(B) = .4$, and $P(A \cap B) = .2$.
    (a) Find $P(A|B)$. (b) Find $P(B|A)$.
12. A and B are events in the same sample space, where $P(A) = .6$, $P(A|B) = .5$, and $P(B|A) = .8$.
    (a) Find $P(A \cap B)$. (b) Find $P(B)$.
13. Two fair dice are tossed.
    (a) Find $P(7 \mid \text{sum is odd})$.
    (b) Find $P(6 \mid \text{one die is } 5)$.
14. A fair coin is tossed three times.
    (a) Find $P(3 \text{ heads} \mid \text{at least 1 head})$.
    (b) Find $P(2 \text{ heads} \mid \text{at least 1 tail})$.
15. Two different integers from 1 to 4 are selected.
    (a) Find $P(\text{exactly one is even} \mid \text{at least one is odd})$.
    (b) Find $P(\text{sum is odd} \mid \text{at least one is even})$.
16. Two integers (not necessarily different) from 1 to 4 are selected.
    (a) Find $P(\text{one is even} \mid \text{at least one is odd})$.
    (b) Find $P(\text{sum is odd} \mid \text{at least one is even})$.
17. Three cards are drawn in succession (without replacement) from a 52-card deck.
    (a) Find $P(3 \text{ aces} \mid \text{first card is an ace})$.
    (b) Find $P(3 \text{ aces} \mid \text{first two cards are aces})$.
18. Three cards are drawn (without replacement) from a 52-card deck. Order is not important.
    (a) Find $P(3 \text{ kings} \mid \text{at least one card is a king})$.
    (b) Find $P(3 \text{ kings} \mid \text{at least two cards are kings})$.
19. What is the probability of guessing all answers correctly in a five-question true-false quiz given that exactly four answers are T?
20. What is the probability of guessing all answers correctly in a five-question true-false quiz given that exactly three answers are F?
21. What is the probability of winning the three-digit lottery in a "straight" play given that the first digit on the ticket is correct?
22. What is the probability of winning the three-digit lottery in a "straight" play given that at least one of the digits on the ticket is correct?
23. Two balls are selected without replacement from an urn containing two red and three black balls.
    (a) Find $P(2 \text{ black} \mid \text{first is black})$.
    (b) Find $P(2 \text{ red} \mid \text{second is red})$.
24. Two balls are selected with replacement from an urn containing two red and three black balls.
    (a) Find $P(2 \text{ black} \mid \text{first is black})$.
    (b) Find $P(2 \text{ red} \mid \text{second is red})$.

Use the poker hands described in Exercises 33–36 of the previous section for the following.

25. Find $P(\text{straight} \mid \text{jack of diamonds and ace of hearts})$.
26. Find $P(\text{flush} \mid 5 \text{ of spades and queen of spades})$.
27. Find $P(4 \text{ aces} \mid \text{at least 3 aces})$.
28. Find $P(\text{full house} \mid \text{exactly 3 aces})$.

Mathematical Expectation

29. A game consists of tossing a fair coin twice. To win, a player must get exactly 1 head. The payoff is \$1 for a win and $-\$1$ for a loss. What is a player's mathematical expectation?
30. A game consists of tossing a fair coin four times. To win, a player must get at least two heads. The payoff is \$3 for a win and $-\$4$ for a loss. What is a player's mathematical expectation?
31. A game consists of tossing two fair dice until a sum of either 5 or 7 appears. A player wins with a 5 and loses with a 7. The payoff for a loss is $-\$2$. What should the winning payoff be if the game is fair? (*Hint:* the probability of getting a 5 before a 7 is $2/5$.)
32. A game consists of tossing two fair dice until a sum of 2, 3, 12, 7, or 11 appears. The player wins with 7 and 11 and loses with 2, 3, and 12. If the winning payoff is \$1, what should be the losing payoff for the game to

be fair? (*Hint:* the probability of getting 7 or 11 before 2, 3, or 12 is 2/3.)

33. A game consists of tossing two fair dice. A player wins if the sum is even and loses if the sum is odd. A winning payoff is the amount of the sum, and a losing payoff is the negative of the amount of the sum. What is a player's mathematical expectation?

34. A game consists of drawing two cards (without replacement) from a 52-card deck. For a player to win, the cards must have the same face value or be of the same suit. The winning payoffs are \$5 for the same face value and \$10 for the same suit. The losing payoff is −\$1. What is a player's mathematical expectation?

Chapter 11 Review Outline

11.1 Mathematical Induction

Principle of Mathematical Induction
Let $E(n)$ be an equation in n. If
(i) $E(1)$ is true, and
(ii) whenever $E(k)$ is true for some integer k, $E(k+1)$ is also true,
then $E(n)$ is true for all positive integers n.

11.2 Arithmetic and Geometric Sequences

Definitions

A sequence of real numbers is a function f whose domain is either the integers from 1 to n (finite sequence) or the set of all positive integers (infinite sequence). The range is a subset of the set of real numbers.

$f(k)$ is denoted by a_k, and the entire sequence by $\{a_k\}$.

Arithmetic Sequences:
A sequence $\{a_k\}$ is arithmetic if $a_{k+1} - a_k = d$, a constant, for any two successive members of the sequence.

The general term of an arithmetic sequence is $a_k = a + (k-1)d$, where a is the first term.

Geometric Sequences:
A sequence $\{a_k\}$ is geometric if $a_{k+1}/a_k = r$, a nonzero constant, for any two successive members of the sequence.

The general term of a geometric sequence is $a_k = a \cdot r^{k-1}$, where a is the first term.

Graphs of Sequences

The points (k, a_k) of an arithmetic sequence lie on the line

$$y = a + (x-1)d = dx + (a-d).$$

The points (k, a_k) of a geometric sequence, with $0 < r < 1$ or $r > 1$, lie on the exponential curve

$$y = a \cdot r^{x-1} = (a/r)r^x.$$

11.3 Arithmetic and Geometric Series

Definitions

A finite series is a sum of a finite number of terms.

The notation $\sum_{k=1}^{n} a_k$ denotes the series $a_1 + a_2 + \cdots + a_n$.

A series is arithmetic if its terms form an arithmetic sequence.

A series is geometric if its terms form a geometric sequence.

Arithmetic Series:

$\sum_{k=1}^{n} [a + (k-1)d] = \dfrac{n(a+l)}{2}$, where $l = a + (n-1)d$ is the last term.

Geometric Series:

$$\sum_{k=1}^{n} a \cdot r^{k-1} = \frac{a(1-r^n)}{1-r}, \quad r \neq 1.$$

$$\sum_{k=1}^{\infty} a \cdot r^{k-1} = \frac{a}{1-r}, \quad |r| < 1.$$

11.4 Permutations and Combinations

Fundamental Counting Principle:
If one event has n possible outcomes, and if for each one of these a second event has m possible outcomes, then the first event followed by the second has $n \cdot m$ possible outcomes.

Permutations:
A permutation of a set of n distinct objects is an arrangement or an ordering of the objects.

More precisely, a permutation of a set is a one-to-one function from the set onto itself.

${}_nP_r$ denotes the number of permutations of n distinct objects taken r at a time.

$${}_nP_r = \frac{n!}{(n-r)!}, \quad 0 \le r \le n$$

Combinations:
A selection of r objects, without regard to order, from n distinct objects is a combination of r objects.

${}_nC_r$ denotes the number of combinations of n distinct objects taken r at a time.

$${}_nC_r = \frac{n!}{(n-r)!r!}, \quad 0 \le r \le n$$

11.5 Basic Probability

Definitions

A sample space is a set. The elements of the sample space are called outcomes, and the subsets of the sample space are called events.

The Rules of Probability:
If $a_1, a_2, \ldots, a_n$ are the elements of a sample space, then a probability on the sample space is an assignment of a real number $P(a_i)$ to each element a_i, subject to the following rules:

1. $0 \le P(a_i) \le 1, \quad i = 1, 2, \ldots, n$
2. $\sum_{i=1}^{n} P(a_i) = 1.$

$P(a_i)$ is called the probability of the outcome a_i. The probability of an event A, denoted $P(A)$, is the sum of the probabilities of the outcomes that make up A.

The probability of the empty set is 0.

Equally Likely Outcomes:
If $P(a_i) = 1/n$ for each a_i in the sample space, the outcomes a_i are called equally likely.

11.6 Conditional Probability and Mathematical Expectation

Conditional Probability:

$$P(A|B) = \frac{P(A \cap B)}{P(B)}, \quad P(B) > 0$$

is called the conditional probability of event A given event B.

Mathematical Expectation:
If a game has outcomes a_i with corresponding probabilities p_i and payoffs w_i $(i = 1, 2, \ldots, n)$, then a player's mathematical expectation is

$$E = w_1p_1 + w_2p_2 + \cdots + w_np_n.$$

A positive w_i is what a player wins if outcome a_i occurs; a negative w_i is what a player pays if a_i occurs.

The game is called fair if $E = 0$.

Chapter 11 Review Exercises

In Exercises 1–3, use mathematical induction to prove the given equation.

1. $2 + 5 + 8 + 11 + \cdots + 3n - 1 = \dfrac{n(3n+1)}{2}$

2. $3 + 3^2 + 3^3 + 3^4 + \cdots + 3^n = \dfrac{3(3^n - 1)}{2}$

3. $\dfrac{1}{1 \cdot 4} + \dfrac{1}{4 \cdot 7} + \dfrac{1}{7 \cdot 10} + \cdots + \dfrac{1}{(3n-2)(3n+1)} = \dfrac{n}{3n+1}$

4. Show that the equation
$$\frac{1}{1 \cdot 2} + \frac{1}{2 \cdot 3} + \frac{1}{3 \cdot 4} + \cdots + \frac{1}{n(n+1)} = \frac{2n+1}{n+1}$$
is *not* true but has the property that $E(k+1)$ is true whenever $E(k)$ is.

5. Show that the equation
$$1 + 2 + 3 + 4 + \cdots + n = \frac{n(n+1)}{2}$$

has the property that $E(n + 2)$ is true whenever $E(n)$ is true.

6. Write out all the terms of the sequence $\left\{(-1)^{k-1}\frac{k}{k+1}\right\}_{k=1}^{6}$.

7. Find the general term a_k $(k = 1, 2, 3, \ldots)$ for the sequence

$$1 - \frac{1}{2},\ 2 - \frac{1}{4},\ 3 - \frac{1}{8},\ 4 - \frac{1}{16}, \ldots.$$

8. Find the general term a_k $(k = 1, 2, 3, \ldots)$ of the arithmetic sequence

$$1,\ 1\frac{3}{4},\ 2\frac{1}{2},\ 3\frac{1}{4}, \ldots.$$

9. Given that $\{a_k\}$ is an arithmetic sequence with $a_1 = -2$ and $a_{25} = 2$, find d.

10. Find the general term a_k $(k = 1, 2, 3, \ldots)$ of the geometric sequence

$$-4,\ \frac{8}{3},\ \frac{-16}{9},\ \frac{32}{27}, \ldots.$$

11. Given that $\{a_k\}$ is a geometric sequence, where $a_1 = 1/25$ and $a_9 = 25$, find r.

12. Graph in the plane the arithmetic sequence

$$a_k = 1 + \frac{3}{4}(k - 1) \quad (k = 1, 2, 3, 4, 5).$$

13. Graph in the plane the geometric sequence

$$a_k = \left(\frac{3}{2}\right)^{k-1} \quad (k = 1, 2, 3, 4).$$

14. Write out the terms of the series $\sum_{k=1}^{6} (-1)^{k+1}2k$.

15. Write the series $1/2 + 2/3 + 3/4 + \cdots + 99/100$ in compact form.

16. Find the sum of the first fifteen terms of the arithmetic sequence in Exercise 8.

17. Find the sum of the first six terms of the geometric series in Exercise 10.

Evaluate each series in Exercises 18–21.

18. $\sum_{k=1}^{30} [-3 + (k - 1) \cdot 5]$

19. $\sum_{k=1}^{7} \left(\frac{2}{3}\right)^{k-1}$

20. $\sum_{k=1}^{\infty} \left(\frac{2}{3}\right)^{k-1}$

21. $\sum_{k=1}^{10} \frac{1}{k(k+1)}$

(*Hint:* write $\frac{1}{k(k+1)} = \frac{1}{k} - \frac{1}{k+1}$.)

Compute each value in Exercises 22–23.

22. (a) ${}_8P_3$ (b) ${}_5P_0$ (c) ${}_{10}P_{10}$

23. (a) ${}_{10}C_2$ (b) ${}_{25}C_0$ (c) ${}_{25}C_{25}$

24. If there are five morning flights from New York to Chicago and four afternoon flights from Chicago to Los Angeles, how many flights from New York to Los Angeles can be made using these connections?

25. A house builder offers six different paint colors, five different choices of wallpaper, and three different types of rugs. How many selections of those three items can be made?

26. A committee of ten, consisting of four seniors, three juniors, two sophomores, and one freshman, must be formed from eight seniors, ten juniors, twelve sophomores, and twenty freshmen. How many choices are there?

27. How many five-digit numbers can be formed from the digits 1, 2, 3, 4, 5, and 6 in each of the following cases?

(a) No digit can be used more than once.

(b) Any digit can be used any number of times.

(c) Digits can be used more than once, but the number must be even.

28. In an audition of twenty equally talented people, three are to be chosen for a part. What is the probability that any one person will be chosen?

29. A fair coin is tossed four times.

(a) Find P(exactly 2 heads).

(b) Find P(at least 2 heads).

30. Two fair dice are tossed.

(a) Find P(sum is even).

(b) Find P(sum is 7 or 11).

31. Three cards are drawn (without replacement) from a 52-card deck.

(a) Find P(3 hearts). (b) Find P(3 of a kind).

32. An integer between 1 and 100 is chosen at random. Find the probability that the number is divisible by 3.

***33.** In Exercise 32, find the probability that the number is divisible by 3 given that an even number is drawn.

34. An integer from 10 to 99 is chosen at random. What is the probability that the sum of the digits is 7?

***35.** Determine the probability in Exercise 34 given that each of the two digits is greater than 2.

36. Two evenly matched teams A and B play a three-game series. Find the probability that team A will win

(**a**) exactly 2 games; (**b**) at least two games.

***37.** Find the probabilities in Exercise 36, given that A wins the first game.

***38.** Two fair dice are tossed. Find the probability that the sum is 7, given that the sum is odd.

***39.** A math professor gives an "extra-credit" problem on a test. If it is done correctly, three points are added to the test score, and if it is done only partially correctly, one point is added; otherwise, one point is subtracted. If a student's probability of getting the problem completely right is 1/4 and only partially correct is 1/2, what is the student's mathematical expectation for extra credit?

***40.** A game consists of tossing two fair dice until a sum of 3, 4, or 7 is obtained. A player wins with 3 or 4 and loses with 7. If the winning payoff is $3, what should be the losing payoff for the game to be fair? (*Hint:* the probability of getting a 3 or 4 before a 7 is 5/11.)

Graphing Calculator Exercises

In Exercises 41–44, verify your answers to the indicated review exercises of this section by using a graphing calculator and the TRACE feature for the given functions and windows $[x_{min}, x_{max}]$, $[y_{min}, y_{max}]$.

41. Exercise 6: $Y_1 = (-1)^{(x-1)}x/(x+1)$, Dot Mode, $[0,9.5]$, $[-1.5,1.5]$

42. Exercise 10: $Y_1 = -4(-2/3)^{(x-1)}$, Dot Mode, $[0,19]$, $[-4,4]$

43. Exercise 12: $Y_1 = 1 + .75(x - 1)$, $[0,9.5]$, $[0,8]$

44. Exercise 13: $Y_1 = (1.5)^{(x-1)}$, $[0,9.5]$, $[0,32]$

In Exercises 45–50, verify your answers to the indicated review exercises of this section by using the following TI-81 program SUM for computing $\sum_{K=1}^{N} a_K$.

```
Program: SUM
Disp "N="
Input N
1→K
0→S
Lbl 1
S+Y1→S
1+K→K
If K>N
Goto 2
Goto 1
Lbl 2
Disp "SUM="
Disp S
End
```

45. Exercise 16: $a_k = Y_1 = 1 + (K - 1)3/4$, $N = 15$

46. Exercise 17: $a_K = Y_1 = -4(-2/3)^{(K-1)}$, $N = 6$

47. Exercise 18: $a_K = Y_1 = -3 + 5(K - 1)$, $N = 30$

48. Exercise 19: $a_K = Y_1 = (2/3)^{(K-1)}$, $N = 7$

49. Exercise 20: $a_K = Y_1 = (2/3)^{(K-1)}$, $N = 25, 50, 75$

50. Exercise 21: $a_K = Y_1 = 1/(K(K + 1))$, $N = 10$

*Asterisks indicate topics that may be omitted, depending on individual course needs.

Appendices

Appendix A: Significant Digits

When performing computations, we often work with approximate values of numbers. This may be due to the numbers themselves or to the way in which the values are obtained. For example, to express $1/3$ or $\sqrt{2}$ as a decimal, we can use only a finite number of decimal places. Also, if values are obtained by measurements, then the numbers can be only as accurate as the measuring devices used. As a general rule,

> any number computed by means of approximate values
> is only as accurate as the least accurate number used.

For example, suppose we measure the diameter of a circle as 11.5 centimeters using a tape measure graded in tenths of a centimeter. If we compute the circumference C of the circle on an 8-place calculator that has a key for π, we get

$$C = \pi d = 36.128316 \text{ centimeters.}$$

However, the measured value 11.5 of d has only three significant digits. Therefore, according to the above rule, our answer should be expressed with three significant digits. That is, we should take for our answer

$$C = 36.1 \text{ centimeters.}$$

We now define precisely what is meant by "significant digit" as well as the rules for performing computations with approximate values. First note that numbers in a table are usually rounded off according to the following convention. If, for example, 12.73 is an approximation of the number x to two decimal places, then it is understood that the difference between 12.73 and x is at most .005 in magnitude. That is,

$$12.725 \leq x \leq 12.735.$$

Similarly, an approximation 12.730 for x means that the difference is at most .0005 in magnitude, or

$$12.7295 \leq x \leq 12.7305.$$

If the whole number 1273 is an approximation to x, then we assume

$$1272.5 \leq x \leq 1273.5.$$

We call 3 the round-off digit in 12.73 and 1273, and 0 the round-off digit in 12.730. In general, the last digit of any decimal whatsoever or of any whole number not ending in 0 is by definition the **round-off digit.** If the last digit of a whole number is 0, then the round-off digit must be indicated in some manner. Here are several examples in which the round-off digit is indicated by underlining in the first column and by scientific notation in the third column.

Approximate Value of *x*	True Value of *x*	Scientific Notation
$\underline{2}0$	$15 \le x \le 25$	2×10
$2\underline{0}$	$19.5 \le x \le 20.5$	2.0×10
$\underline{2}00$	$150 \le x \le 250$	2×10^2
$2\underline{0}0$	$195 \le x \le 205$	2.0×10^2
$20\underline{0}$	$199.5 \le x \le 200.5$	2.00×10^2

The number of **significant digits** in a decimal or whole number is by definition the number of digits, counting from the first nonzero digit on the left to the round-off digit. We give several examples below.

Number	Number of Significant Digits
0.0215	3
4.0215	5
0.003	1
0.0030	2
405	3
$40\underline{5}0$	3
40.50	4
4.2×10^6	2
4.20×10^{-6}	3

We note that every nonzero digit in a number is significant, and 0 is significant unless its only purpose is to fix the location of the decimal point.

If we wish to round off a given number *x* to fewer significant digits, we proceed as illustrated in the following examples.

x	Rounded-Off Approximation	Significant Digits
2.7148674	2.715	4
2.7148674	2.71	3
2.7148674	2.7	2
2.7148674	3	1

That is, the approximation differs from *x* by at most 1/2 unit in the round-off digit. When the last significant digit of *x* is 5 and we wish to round off to one less significant digit, then our convention is to round off to an *even* digit. Here are some examples.

x	Rounded-Off Value
.0135	.014
.0125	.012
2715	2720
2.715	2.72

Note that if we first round off 2.7148674 to four digits, we get 2.715, and if 2.715 is then rounded off to three digits, we get 2.72. However, if 2.7148674 is rounded off in one step to three digits, we get 2.71. Hence, successive round-offs are not necessarily equivalent to a single one.

Our convention for **addition** and **subtraction** with approximate values depends more on the number of decimal places (digits after the decimal point) than on the number of significant digits of the numbers. The convention is as follows.

> Add or subtract the numbers as given, and round off the result to the decimal place of the number(s) with the fewest decimal places.

For example, when adding approximate values 23.714, .0002, and 5.0129, we first obtain 28.7271, which is then rounded off to 28.727.

The number of significant digits is the important factor in our convention for **multiplication** and **division**, which is given below.

> Multiply or divide the numbers as given, and round off the result to have as many significant digits as the number(s) with the fewest significant digits.

For example, as indicated earlier, the product $\pi \cdot 11.5$, when performed on an 8-place calculator, gives $3.1415927 \cdot 11.5 = 36.128316$, which is rounded off to 36.1.

The above conventions for performing operations with approximate values are merely general guidelines, and they do not guarantee the accuracy of the final result. A precise analysis of error buildup when computing with approximate values requires methods of advanced mathematics. Error analysis is especially important these days when problems requiring thousands or even millions of calculations are performed on computers. The computational examples and exercises in this text do not require a large number of calculations, and our conventions will give accurate results.

When working with a calculator, we should keep in mind that all numbers entered and computed have the same number of significant digits. Rounding off occurs in the last place only. If our numbers actually have fewer significant digits than those of the calculator, we must do our own rounding off, according to the above conventions.

Answers to Selected Exercises

To the Student

A *Study and Solutions Manual* is available to help you study and review the course material. It includes step-by-step solutions to the odd-numbered exercises in the text and a set of practice chapter tests, also with complete solutions.

Chapter P

Section P.1 (page 4)

1. rational, irrational **2.** rational **3.** neither terminates nor has an indefinitely repeating pattern **4.** 0 **5.** addition, subtraction, multiplication **6.** true **7.** false **8.** false **9.** true **10.** true **11.** 7/10, 3/10 **13.** .34999 . . . **15.** .666999 . . . **17.** .333 . . . **19.** 123/1000 **21.** 5/9 **23.** One possible answer is .50005.

25. $-3.5 < -\sqrt{9.1} < -3 < 1.41 < \sqrt{2} < 3.14 < \pi < 22/7$ **27.**

29. $100 < 110 < 135 < 140 < 150$; median = 135 lb **31.** $16 < 17 < 18 < 19 < 20 < 21 < 22$; median = 19 yr **33.** $125 < 250 < 325 < 350 < 475 < 500 < 625 < 750$; median = \$412.50 **35.** $5.25 < 5.5 < 8.5 < 9.75 < 10 < 10.5$; median = 9.125% **37.** $\cdot, \div$ **39.** $+, -, \cdot$ **41.** One possible answer is the set of odd integers. **43.** \$0

Section P.2 (page 13)

1. commutative property, associative property **2.** $a \cdot (b + c) = a \cdot b + a \cdot c$, $(a + b) \cdot c = a \cdot c + b \cdot c$ **3.** 3, 3 **4.** 0, 0, 1, 1 **5.** 1, 0, 1/0 is not a real number **6.** false **7.** true **8.** false **9.** true **10.** false **11.** (a) commutative property of multiplication (b) associative property of addition (c) left distributive property **13.** associative property of addition **15.** right distributive property **17.** all three **19.** 2, −2 **21.** ±1, ±2 **23.** \$25(1.06) = \$26.50 **25.** \$2662.50/1.065 = \$2,500 **27.** The solution of the equation is $x = -2$, and therefore $x + 2 = 0$. Hence, multiplying by $(x + 2)^{-1}$ is equivalent to dividing by 0. **33.** additive inverse property, 16, 32 **35.** (a) 24 (b) 152 (c) 216 (d) 153 **37.** (a) 34.607477 (b) 16.080997 (c) 34.607477 (d) 24.700685 **39.** extended associative property of addition **41.** extended commutative and associative properties of addition, right distributive property **43.** −E = E; −O = O **45.** (a) 4 (b) 8 (c) 6.5 (d) 6 (e) 12 (f) 12 **47.** (a) yes (b) no (c) yes (d) yes **49.** 6M(2A12) = 7, but (6M2)A(6M12) = 9

Section P.3 (page 23)

1. Subtract the smaller absolute value from the larger and precede the result with the sign of the term with the larger absolute value. **2.** changing the sign **3.** $a + (-b)$ **4.** $a \cdot b^{-1}$ or $a \cdot (1/b)$ **5.** positive, negative **6.** true **7.** false **8.** true **9.** true **10.** false **11.** −9 **13.** 13 **15.** 15 **17.** −2 **19.** −64 **21.** 6 **23.** −28 **25.** −1/3 **27.** $x - 7$ **29.** $-6 + y$ **31.** $-x - y$ **33.** $-4b$ **35.** $-5y$ **37.** 1 **39.** −5 **41.** ±2, −3 **43.** positive **45.** 6 **47.** < 3 **49.** algebraic definition of greater than, algebraic property of equality, associative and commutative properties of addition, algebraic definition of greater than **51.** algebraic definition of greater than, algebraic definition of greater than, substitution for b, associative property of addition, positive numbers are closed under addition, algebraic definition of greater than **63.** preservation property for addition, transitive property of greater than

Section P.4 (page 31)

1. 0, 0 **2.** b, 0 **3.** $(a + c)/b$, $(ac + ab)/(bc)$ **4.** numerators, denominators **5.** invert, multiplication **6.** false **7.** false **8.** false **9.** true **10.** true **11.** 3 **13.** 9 **15.** 2 **17.** $b/3$; $a \neq 0$ **19.** $(1 + 2b)/(2b)$; $a \neq 0$, $b \neq 0$

21. $(b-c)/(b+c)$; $a \neq 0$, $b+c \neq 0$ **23.** $1/(2a)$; $a \neq 0$, $2x-3y+4z \neq 0$ **25.** $a+b$; $x \neq y$ **27.** $4/(x+1)$; $x \neq -1$
29. $2x/(2x-1)$; $x \neq 1/2$ **31.** 1; $x \neq 1/3$ **33.** $(2x+6)/(3x+1)$; $x \neq -1/3$ **35.** $3/x$; $x \neq 0$, $x \neq -1$ **37.** $8x/15$
39. $(x+3)/6$ **41.** $(x+1)/x$; $x \neq 0$ **43.** $(7x-2)/[(x+2)(x-2)]$; $x \neq \pm 2$ **45.** $(7x-3)/5$ **47.** 0; $x \neq a$
49. $(2ax+2bx)/[(2x+a)(2x-b)]$; $x \neq -a/2$, $x \neq b/2$ **51.** $1/4$; $a \neq 0$ **53.** $3a/2$ **55.** $1/4$; $x \neq -3/2$ **57.** $5/4$; $x \neq 1/2$
59. $1/4$; $x \neq 1/2$, $x \neq 3$ **61.** $1/4$; $a \neq 0$ **63.** $27ad/(8bc)$; $b \neq 0$, $c \neq 0$, $d \neq 0$ **65.** 1; $a \neq 0$ **67.** $2/(2x+3)$; $x \neq -3/2$
69. $1/(2x+2)$; $x \neq -1$ **71.** $(x-1)/(x+2)$; $x \neq \pm 2$ **73.** $8/3$; $x \neq -1$, $x \neq 0$ **75.** $8/3$; $x \neq -1/2$, $x \neq 2$
77. 1; $a \neq \pm b$ **79.** a/b; $a \neq 0$, $b \neq 0$, $ab+1 \neq 0$ **81.** $2x+7$; $x \neq -3$, $x \neq -4$ **83.** $(x+3)(x+1)/[(x+2)(2x+3)]$; $x \neq -1$, $x \neq -2$, $x \neq -3/2$ **85.** 1; $x \neq 1$, $x \neq -1$ **87.** $bc/[a(b+c)]$; $a \neq 0$, $b \neq 0$, $c \neq 0$, $(b+c) \neq 0$
89. $c(ab+1)/b$; $b \neq 0$, $c \neq 0$ **91.** $(a+b)/(a-b)$; $a \neq 0$, $b \neq 0$, $a \neq b$ **93.** a/c; $a \neq 0$, $b \neq 0$, $c \neq 0$, $abc \neq -1$
95. $(2x+3)(x+3)/[(x+1)(x+2)]$; $x \neq -1$, $x \neq -2$, $x \neq -3$ **97.** $3-2x$; $x \neq 1$, $x \neq 2$

Section P.5 (page 40)

1. (a) $4^3 \cdot 4^5 = 4^8$ (b) $(3^2)^4 = 3^8$ (c) $(5 \cdot y)^5 = 5^5 \cdot y^5$ **2.** x^n **3.** $2^6 = 64$, $y^n = x$ **4.** x, $|x|$ **5.** two, no; one, one
6. true **7.** false **8.** false **9.** true **10.** true **11.** (a) $(-2)^3$ (b) $(-2)^4$ (c) $(-2)^5$ (d) $(-2)^6$ (e) $(-2)^{11}$
13. (a) 2^5 (b) 2^8 (c) 2^{18} **15.** (a) $2^4 \cdot 3^3$ (b) $2^9 \cdot 3^2$ (c) $2^{11} \cdot 3^7$ **17.** x^{15} **19.** $(-3)^{17}$ or -3^{17} **21.** a^{12}
23. $(-2)^{15}$ or -2^{15} **25.** x^{10n} **27.** 12^9 **29.** $a^{12}b^{12} = (ab)^{12}$ **31.** $3^3 \cdot 2^3 = 6^3$ **33.** x^4y^5; $x \neq 0$, $y \neq 0$ **35.** 3^4
37. 2^3 **39.** 8^2 **41.** a; $a \neq 0$ **43.** xy^4; $x \neq 0$, $y \neq 0$ **45.** a^2b^{16}; $a \neq 0$, $b \neq 0$ **47.** a^5/b^{10}; $a \neq 0$, $b \neq 0$ **49.** 1,680
51. $-6ab^2$ **53.** $2a^2/(5b)$; $b \neq 0$ **55.** -15 **57.** -15 **59.** $8\sqrt{2}$ **61.** $2abc^2\sqrt[3]{2ab^2}$ **63.** $|ab|c^2d^2\sqrt{bd}$
65. $(a^2/b)\sqrt[3]{a/b^2}$; $b \neq 0$ **67.** 5.28×10^3 **69.** 1.76×10^2 **71.** 2.5×10^{-4} **73.** 1.76×10^5 **75.** 2.5×10^7
77. 1.0×10^{-7} **79.** 1.32×10^{12} **81.** 6.25×10^6 **83.** 1.0×10^{-7} **85.** 2.25×10^{-4} **87.** 3.0×10^{-11}
89. 2.99776×10^{10} cm/sec **91.** 36,000,000 mi **93.** 92,956,000 mi **95.** 483,400,000 mi **97.** 1,782,000,000 mi
99. 3,664,000,000 mi **101.** 6.624×10^{-27} **103.** .000 000 000 000 000 000 000 01328 grams
105. .000 000 000 000 000 000 000 327 grams **107.** (a) \$260.96 (b) \$205.79 (c) \$617.17 (d) \$482.77

Chapter P Review Exercises (page 45)

1. All except $\sqrt{-8}$ and 4/0 are real numbers. Of the real numbers, $\sqrt{5}$ is irrational, and the rest are rational.
2. $75/99 = 25/33$ **3.** 0.571428571428 . . . **4.** Let $x = .9999 \ldots$. Then $10x = 9.9999 \ldots$, $9x = 9$, $x = 1$.
5. $3/8 = (3 \cdot 11)/(8 \cdot 11) = 33/88$; $3/8 = (3 \cdot 111)/(8 \cdot 111) = 333/888$ **6.** (a) commutative property of addition (b) associative property of multiplication (c) left distributive property (d) additive identity

7. (a)
$$\begin{aligned} 7+a &= 7+b \\ -7+(7+a) &= -7+(7+b) \\ (-7+7)+a &= (-7+7)+b \\ 0+a &= 0+b \\ a &= b \end{aligned}$$
(b)
$$\begin{aligned} 5 \cdot a &= 5 \cdot b \\ 5^{-1} \cdot (5 \cdot a) &= 5^{-1} \cdot (5 \cdot b) \\ (5^{-1} \cdot 5) \cdot a &= (5^{-1} \cdot 5) \cdot b \\ 1 \cdot a &= 1 \cdot b \\ a &= b \end{aligned}$$

8. (a) $1/2 = 1/3 + 1/6$, and $1/6$ is positive (b) $-1/3 = -1/2 + 1/6$, and $1/6$ is positive

9.
$$\begin{aligned} 2x - 18 &> 5x - 3 \\ 2x - 18 + 18 &> 5x - 3 + 18 \\ 2x &> 5x + 15 \\ 2x - 5x &> 5x + 15 - 5x \\ -3x &> 15 \\ x &< -5 \end{aligned}$$

10. (a) true (b) false (unless $c \neq 0$) (c) true (d) true (e) true (f) true (g) false (unless $a \geq 0$) (h) false (i) true (j) true **11.** (a) 23/20 (b) 5/9 (c) 36 (d) 1 (e) 12/5 (f) 5 **12.** (a) $10a + 10b + 12c$ (b) $12a + 12b + 66$
13. If $a \neq b$ $(b - a \neq 0)$, then $(a-b)/(b-a) = -(b-a)/(b-a) = -1$. **14.** If $x \neq -3$, then $(x+3)/(x^2-9) = (x+3)/[(x+3)(x-3)] = 1/(x-3) \neq 0$. If $x = -3$, then $(x+3)/(x^2-9) = 0/0$ is not defined.
15. (a) 1, if $x \neq 3$ (b) $(x-3)/6$ (c) $(x+4)/(2x)$, if $x \neq 0$ **16.** $2^{13} \cdot 3^9$ **17.** (a) $2/x$ (b) $5x^2$ (c) $4x^{10}$ (d) a^2/b^2 (e) a/b **18.** (a) 2 (b) -2 (c) 1 **19.** (a) $5|a|b\sqrt{2b}$ (b) $2ab\sqrt[3]{2a^2}$ (c) $|x|$ (d) $-x$ **20.** (a) 5/2 (b) 20 (c) $7\sqrt{2}$ **21.** 1.414213562; 4 **22.** 3, 6, 7, or 9; 3, 6, or 9; 6 **23.** (a) 7962624 (b) .03125 (c) .0022675737 (d) 3.16227766 (e) 2.924017738 **24.** (a) = (b) = (e) = 0.50505051; (c) = (d) 10.22727273
25. (a) 9.722222222 (b) 12.9 (c) 68.25 **26.** (a) 2.99776E10 (b) 6.023E23 (c) 1.6E-19 (d) 6.624E-27
27. (a) 92,956,000 (b) 141,600,000 (c) 1.673E-24 (d) 1.328E-23 **28.** (a) 1.805550848E34 (b) 24,154,589.37 (c) 1.31625696E16 (d) .1259789157 **29.** (a) \$12,833.59 (b) \$16,470.09 (c) \$27,126.40 **30.** (a) \$313.36 (b) \$244. (c) \$202.76

Chapter 1

Section 1.1 (page 57)

1. addition, subtraction, multiplication **2.** $-3x^2 + 5x + 2$ **3.** multiplying each term of P by each term of Q **4.** quotient, remainder **5.** root, zero **6.** true **7.** true **8.** false **9.** false **10.** false **11.** $2x^2 + 5xy + 4y$ **13.** $-7x + 5$ **15.** $3x^3 - 2x^2 + x - 8$ **17.** $3ax^2 + 4bx + 5c$ **19.** $(1 + a)x^3 + (1 + b)x^2 + (1 + c)x + (1 + d)$ **21.** $6x^2 + 8x$ **23.** $2x^3y + 2xy$ **25.** $x^2 + 3x + 2$ **27.** $x^2 - x - 6$ **29.** $x^3y^3 + 2x^2y^2 + xy$ **31.** $25x^2 + 40x + 16$ **33.** $4x^4 + 12x^2 + 9$ **35.** $4x^4 - 25$ **37.** $x^3 + 6x^2 + 12x + 8$ **39.** $x^4 + 8x^3 + 24x^2 + 32x + 16$ **41.** $6x^3 - 12x^2 - 90x$ **43.** $15x^5 + 20x^4 + 15x^3 + 20x^2$ **45.** $24x^3 - 98x^2 + 133x - 60$ **47.** $x^4 - 5x^2 + 4$ **49.** $x^5 + 10x^4 + 35x^3 + 50x^2 + 24x$ **51.** $x^2 + x - 1$ **53.** $x^3 + 2x^2 + 3x + 4 + 5/(x - 1)$ **55.** $x^4 + 4x^3 + 11x^2 + 44x + 178 + 716/(x - 4)$ **57.** $2 + 12/(2x - 3)$ **59.** $3x^2 + 4x - 2$ **61.** $3x^2 - 4x + 11 + (-36x + 16)/(x^2 + 2x - 1)$ **63.** $x^3 - x^2 + x + 1/(x^2 + x + 1)$ **65.** $x^6 - 5x^4 + 3x^2 - 6 + 5/(x^2 + 1)$ **67.** $5 + (10x + 12)/(x^2 - 2x - 1)$ **69.** $x - 1 + 1/(x^3 + x^2 + x + 1)$ **71.** $5x - 2, 4$ **73.** $x^6 + x^5 + x^4 + x^3 + x^2 + x + 1, 0$ **75.** $2, 7x + 3$ **77.** $x + 1, -3x + 2$ **79.** $x^4 - x^3 + x^2 - x + 1, 0$ **81.** (a) -1 (b) 1 (c) 4 **83.** (a) 15 (b) 39 (c) 0 **85.** (a) $-2 - 2\sqrt{2}$ (b) -2 (c) -2 **87.** $-2/3, 5/3, -2, 7$ **89.** $\sqrt{5}, -\sqrt{5}$ **91.** $2, 3, -2, -3$ **93.** (a) $(20 - 1)^2 = 400 - 40 + 1 = 361$ (b) $(20 - 2)^2 = 400 - 80 + 4 = 324$ (c) $(30 - 1)^2 = 900 - 60 + 1 = 841$ (d) $(30 - 2)^2 = 900 - 120 + 4 = 784$

Section 1.2 (page 64)

1. $x^2 + 4x - 5$ **2.** $(x + 6)(x - a)$ **3.** $-, a - b, x - y$ **4.** $x^k + m, x^k + n$ **5.** B, AC, grouping **6.** true **7.** true **8.** false **9.** false **10.** false **11.** $(x + 2)^2$ **13.** not a perfect square **15.** not a perfect square **17.** $(2x - 5y)^2$ **19.** $(x^2 + 3)^2$ **21.** $(x - 8)(x + 8)$ **23.** $(x - y)(x + y)(x^2 + y^2)$ **25.** $4ax$ or $(2a)(2x)$ **27.** $3(x - 2)(x + 2)$ **29.** $2(x + 2)^2$ **31.** $(x + 4)(x + 1)$ **33.** $3(x + 3)(x - 4)$ **35.** $x(x - 9)(x + 1)$ **37.** $(x - 1)(x - 2)$ **39.** $(x + 1)(x - 2)$ **41.** not factorable by integers **43.** $3(x + 1)(x - 2)$ **45.** $(x - 1)(2x - 3)$ **47.** $(x - 8)(x - 10)$ **49.** $(x + 4)(x + 25)$ **51.** $(x + 11)(x + 12)$ **53.** $3(x + 1)(x + 4)$ **55.** $2(x + 1)(x - 9)$ **57.** $(2x + 1)(x + 4)$ **59.** $(5x + 1)(x + 4)$ **61.** $(5x + 7)(x + 1)$ **63.** $(2x - 1)(2x - 5)$ **65.** $(2x - 5)(5x + 2)$ **67.** $(x + 2)(x^2 - 2x + 4)$ **69.** $(2x - 3y)(4x^2 + 6xy + 9y^2)$ **71.** $2(2a^2 - 3b)(4a^4 + 6a^2b + 9b^2)$ **73.** $(x - b)(x - 4b)$ **75.** $(x - a)(x - b)$ **77.** $(ax + 2)(ax + 3)$ **79.** $(a - b)(x - 1)(x - 2)$ **81.** $x(x + 2a)$ **83.** $(x - a + 2)(x + a + 4)$ **85.** $(a + 2)(x - y)(x + y)$ **87.** $2a(x + y)(x - y)$ **89.** $a(2x + 3)(3x + 2)$ **91.** $x(x + 4)(5x - 2)$ **93.** $(3x^2 + 5)(x^2 + 1)$ **95.** $(2x^2 + 1)(x^2 + a^2)$ **97.** $(x - \sqrt{2})(x + \sqrt{2})$ **99.** $4(x - \sqrt{3})(x + \sqrt{3})$ **101.** $(x + 1/2)(x + 3/2)$

Section 1.3 (page 71)

1. least, factor **2.** LCM **3.** multiple **4.** $2(x - 6)(x + 6)^2$ **5.** rational numbers **6.** true **7.** false **8.** true **9.** true **10.** false **11.** 126 **13.** $(x^2 - 1)(x^2 - 9)$ **15.** $3x^3y^2$ **17.** $2(x + 1)(x^2 - 4)$ **19.** $-2/63$ **21.** $-h/[x(x + h)]$ **23.** $(1 - 2x)/x^3$ **25.** $(x^2 + 5x - 10)/[2(x - 1)(x - 3)]$ **27.** $(-6x^2 + 5x - 6)/[2(2x - 1)(2x + 1)]$ **29.** $(x^2 + 5x + 8)/[(x + 1)(x + 2)(x + 4)]$ **31.** $(-x^2 + 4x + 2)/[x(x - 3)(2x + 1)]$ **33.** $3/(x^2 - 4)$ **35.** $(2x^2 - 4x + 1)/[(x - 2)^2(x - 3)]$ **37.** $(4x^2 - 15x + 18)/[x(x - 2)^2]$ **39.** $(x^4 + 3x^3 - 6x^2 + 9)/[x^2(x + 3)^2]$ **41.** $(-3x^2 - 11x - 8)/[(x + 1)(x + 2)(x + 3)]$ **43.** $(x + a - b)/[(x + a)(x + b)]$ **45.** $(x + a)^2/(x^3 - a^3)$ **47.** $4a^2y^2/(3bx)$ **49.** $(5x - 2)(2x - 1)/[(3x - 2)(2x - 5)]$ **51.** $5bc^2yz/(4x)$ **53.** $5(x + 3)$ **55.** $3/(x^2 - 1)$ **57.** $x/(x^2 - 1)$ **59.** $3x + 2$

Section 1.4 (page 77)

1. $\sqrt[n]{x}$, even **2.** positive nth **3.** $(x^{1/n})^m$, $1/x^{m/n}$, 0 **4.** $(x^{1/n})^m$, $(x^m)^{1/n}$ **5.** positive **6.** true **7.** false **8.** false **9.** false **10.** true **11.** (a) 2 (b) not a real number (c) -4 **13.** (a) 729 (b) 1/729 (c) not a real number **15.** $(9^{1/2})^3 = 3^3 = 27$, and $(9^3)^{1/2} = \sqrt{729} = 27$ **17.** $[(-32)^{1/5}]^2 = [-2]^2 = 4$, and $[(-32)^2]^{1/5} = \sqrt[5]{1024} = 4$ **19.** $(1250^{1/5})^2 \approx (4.162766)^2 \approx 17.3286$, and $(1250^2)^{1/5} = \sqrt[5]{1{,}562{,}500} \approx 17.3286$ **21.** (a) $64^{1/6} = 2$ (b) $64^{1/2} = 8$ (c) -1 **23.** (a) $8^4 = 4096$ (b) $1/27^3 = 1/19{,}683$ (c) $2^3 = 8$ **25.** $2^{1/3} \cdot 3^{3/4}$ **27.** (a) x^4 (b) x^9 (c) $1/x^{1/4}$ **29.** 1/2 **31.** $5^{2/3}/(7^4x^{4/3}y^{12})$ **33.** any negative x **35.** $y^{2/3}(x^{2/3} + y)$ **37.** $x^2(y^3 + 1)/y^2$ **39.** 0

Section 1.5 (page 83)

1. $\sqrt{x}/x$ **2.** $x^{2/3}/x$ or $\sqrt[3]{x^2}/x$ **3.** $x^{3/4}/x^2$ or $\sqrt[4]{x^3}/x^2$ **4.** $\sqrt{a} + \sqrt{b}$ **5.** $(x + 2\sqrt{a}\sqrt{x} + a)/(x - a)$ **6.** true **7.** true **8.** false **9.** false **10.** true **11.** $(x^2 + 1)^{1/2}$ **13.** $(x^2 + 3)^{3/4}$ **15.** $(x + x^{1/2})^{1/2}$ **17.** $(x + 2)^{3/4}$ **19.** $\sqrt[6]{(x^2 + 1)^5}$ **21.** $(x + 1)\sqrt[6]{x + 1}$ **23.** $(x^2 - 1)\sqrt{x^2 - 1}$ **25.** $x - 2$ **27.** 1 **29.** $6\sqrt[3]{x^2 - 1}$ **31.** $(2a + 3b)\sqrt{x^2 + x + 1}$ **33.** $(x + 2)\sqrt[3]{x - 4}$ **35.** $\sqrt{x - 3}/(x - 3)$ **37.** $x\sqrt{x}$ **39.** $\sqrt[3]{x + 1}/(x + 1)$ **41.** $(\sqrt{x + 1} - \sqrt{x - 1})/2$ **43.** $(1 + x)\sqrt{x}/x$ **45.** $(a\sqrt{a} - b\sqrt{ab})/(ab)$ **47.** $(a + \sqrt{a} + a\sqrt{a})/a^2$ **49.** $\sqrt{x} + x\sqrt[4]{x^3} + x^3\sqrt[6]{x^5}$ **51.** $(1 + 2a)\sqrt{a^2 - b^2}/(a^2 - b^2)$ **53.** $2/(x - 1)$ **55.** $(2x^2 + 1)/\sqrt{x^2 + 1}$ or $(2x^2 + 1)\sqrt{x^2 + 1}/(x^2 + 1)$ **57.** $-(x + 10)/(2x^2\sqrt{x + 5})$ or $-(x + 10)\sqrt{x + 5}/[2x^2(x + 5)]$ **59.** $(5x^2 - 3)/(3\sqrt[3]{(x^2 - 1)^2})$ or $(5x^2 - 3)\sqrt[3]{x^2 - 1}/[3(x^2 - 1)]$ **61.** $-2/(x^2\sqrt{x^2 + 2})$ or $-2\sqrt{x^2 + 2}/[x^2(x^2 + 2)]$

Section 1.6 (page 90)

1. $n + 1$ **2.** n **3.** $n!/[(n - r)!r!]$ **4.** a^4b^8 **5.** $12!/(6!6!)$ **6.** false **7.** true **8.** false **9.** false **10.** false **11.** $(x^2 - 2x + 1)(x - 1) = x^3 - 3x^2 + 3x - 1$ **13.** $(x^2 + 4x + 4)(x + 2)(x + 2) = (x^3 + 6x^2 + 12x + 8)(x + 2) = x^4 + 8x^3 + 24x^2 + 32x + 16$ **15.** $(x^2 - 2x + 1)(x - 1)(x - 1)(x - 1) = (x^3 - 3x^2 + 3x - 1)(x - 1)(x - 1) = (x^4 - 4x^3 + 6x^2 - 4x + 1)(x - 1)$ $= x^5 - 5x^4 + 10x^3 - 10x^2 + 5x - 1$ **17.** $x^5, x^4, x^3, x^2, x, 1$ **19.** $a^4x^4, a^3x^3y, a^2x^2y^2, axy^3, y^4$ **21.** $x^4 + 8x^3 + 24x^2 + 32x + 16$ **23.** $16x^4 + 96x^3 + 216x^2 + 216x + 81$ **25.** $1 + 6x + 15x^2 + 20x^3 + 15x^4 + 6x^5 + x^6$ **27.** 1120 **29.** (a) 64 (b) 2^n **31.** (a) 24 (b) 120 (c) 5040 **33.** (a) $(n + 2)(n + 1)$ (b) $n(n + 1)$ (c) $(2n + 2)(2n + 1)$ **35.** (a) ${}_5C_3 = 10$, ${}_4C_3 + {}_4C_2 = 4 + 6 = 10$ (b) ${}_6C_1 = 6$, ${}_5C_1 + {}_5C_0 = 5 + 1 = 6$ **37.** $x^4 + 4x^3y + 6x^2y^2 + 4xy^3 + y^4$ **39.** $16x^4 - 96x^3y + 216x^2y^2 - 216xy^3 + 81y^4$ **41.** $x^6 + 12x^5 + 60x^4 + 160x^3 + 240x^2 + 192x + 64$ **43.** $32x^5 + 60x^4 + 45x^3 + (135/8)x^2 + (405/128)x + (243/1024)$ **45.** 28 **47.** 56 **49.** 60 **51.** $(1.1)^5 = 1 + 5(.1) + 10(.1)^2 + 10(.1)^3 + 5(.1)^4 + (.1)^5 = 1.61051$ **53.** $(1.01)^4 = 1 + 4(.01) + 6(.01)^2 + 4(.01)^3 + (.01)^4 = 1.04060401$ **55.** $(9.8)^3 = (10 - .2)^3 = 10^3 + 3 \cdot 10^2(-.2) + 3 \cdot 10(-.2)^2 + (-.2)^3 = 941.192$ **57.** $2^n = (1 + 1)^n = {}_nC_0 + {}_nC_1 + {}_nC_2 + \cdots + {}_nC_n$ **59.** $(x + y)^3 + 3(x + y)^2z + 3(x + y)z^2 + z^3 = x^3 + y^3 + z^3 + 3x^2y + 3xy^2 + 3x^2z + 3xz^2 + 3y^2z + 3yz^2 + 6xyz$ **61.** $(1 + .01)^4 \approx \$1.04$ **63.** $100(1.08)^3 \approx \$125.97$ **65.** ${}_{30}C_3 = 4060$ ways **67.** ${}_{10}C_8 = 45$ choices **69.** ${}_{100}C_2 = 4950$ ways

Chapter 1 Review Exercises (page 93)

1. $8ab - 4a + 1$ **2.** $3x^9 - 48x$ **3.** $(x^3 - 3x^2 + 5)/(x + 3) = x^2 - 6x + 18 - 49/(x + 3)$ or $x^3 - 3x^2 + 5 = (x^2 - 6x + 18)(x + 3) - 49$ **4.** quotient $= x^3 + 3x$; remainder $= 4x - 3$ **5.** $(2x + 1)^2 = 4x^2 + 4x + 1$; $(2x - 1)^3 = 8x^3 - 12x^2 + 6x - 1$ **6.** $P(-2) = -11$; $P(\sqrt{2}) = 1 + 2\sqrt{2}$ **7.** $(\sqrt{2} - 2)^2 + 4(\sqrt{2} - 2) + 2 = 2 - 4\sqrt{2} + 4 + 4\sqrt{2} - 8 + 2 = 0$ **8.** $P(x) \geq 1$ for all real x. **9.** $(4x - 7y)(4x + 7y)$ **10.** $5(x - 3)^2$ **11.** $(3x - 2)(2x + 5)$ **12.** $2(x - 2y)(x^2 + 2xy + 4y^2)$ **13.** $(5x + 2)(x - 6)$ **14.** $(a^2 + b^2 + 1)(a^2 - b^2 + 1)$ **15.** $(x^5 - 32)/(x - 2) = x^4 + 2x^3 + 4x^2 + 8x + 16$; therefore, $x^5 - 32 = (x^4 + 2x^3 + 4x^2 + 8x + 16)(x - 2)$ **16.** $2(x - \sqrt{3})(x + \sqrt{3})$ **17.** $(x + 2)(x + .1)$ **18.** $24(x - 1)(x + 1)^2$ **19.** $(2x + 3)/[(x - 1)(x + 2)(x + 3)]$ **20.** $(2x^2 + 11x - 4)/(x^2 - 4)$ **21.** $(2x - 1)/2$ **22.** $(2 - x)/[3(x - 1)]$ **23.** (a) 64 (b) 1/64 (c) undefined (d) 1 (e) undefined **24.** $2 \cdot 5 = 10$ **25.** $64^{-2/3} = 1/64^{2/3} = 1/16$; $64^{-2/3} = (64^{1/3})^{-2} = 4^{-2} = 1/16$; $64^{-2/3} = (64^{-2})^{1/3} = \sqrt[3]{1/4096} = 1/16$ **26.** The base must be positive in the power-of-a-rule **27.** $4x^{1/2}y$ **28.** $x^{2/3}$ **29.** $9a^2$ **30.** $(x - y)^2/(x + y)$ **31.** $(x + 2 + \sqrt{2x + 4})/x$ **32.** $3/(x - 1)$

33.

1
1 1
1 2 1
1 3 3 1
1 4 6 4 1
1 5 10 10 5 1

$(a+b)^2 = a^2 + 2ab + b^2$
$(a+b)^3 = a^3 + 3a^2b + 3ab^2 + b^3$
$(a+b)^4 = a^4 + 4a^3b + 6a^2b^2 + 4ab^3 + b^4$
$(a+b)^5 = a^5 + 5a^4b + 10a^3b^2 + 10a^2b^3 + 5ab^4 + b^5$

34. **(a)** 21 **(b)** 1 **(c)** 120 **35.** ${}_8C_4 = 70$; ${}_7C_4 + {}_7C_3 = 35 + 35 = 70$ **36.** **(a)** $x^5 + 10x^4 + 40x^3 + 80x^2 + 80x + 32$ **(b)** $x^4 - 12x^3 + 54x^2 - 108x + 81$ **37.** $[(x-y)+1]^4 = (x-y)^4 + 4(x-y)^3 + 6(x-y)^2 + 4(x-y) + 1 = x^4 - 4x^3y + 6x^2y^2 - 4xy^3 + y^4 + 4x^3 - 12x^2y + 12xy^2 - 4y^3 + 6x^2 - 12xy + 6y^2 + 4x - 4y + 1$

Chapter 2

Section 2.1 (page 103)

1. equation **2.** equivalent **3.** linear **4.** $-b/a$ **5.** the set of all real numbers, the empty set **6.** true **7.** false **8.** true **9.** false **10.** false **11.** 2 **13.** 10 **15.** $-57/2$ **17.** 2.6 **19.** $-122/43$ **21.** all real numbers **23.** no solution **25.** $(\sqrt{3}+8)/(\sqrt{3}-3)$ **27.** $(3-\sqrt{5})/3$ **29.** $-51/13$ **31.** $12/(2-c)$ **33.** If $c = 5$, $x =$ any real number; if $c \neq \pm 5$, $x = 1/(c+5)$. **35.** $-(by+c)/a$ **37.** $-2ab/(a+b)$ **39.** $-a^2y/b^2$ **41.** -1 **43.** $7/2$ **45.** $-1/5$ **47.** 5 **49.** no solution **51.** -4 **53.** 1 **55.** all real numbers except ± 2 **57.** $-9/2$ **59.** no solution **61.** 3 **63.** $5/4$ **65.** -3 **67.** all real numbers except -1 and -2 **69.** $-8/3$ **71.** $y/(1-y); y \neq 1$ **73.** $y/(y-1); y \neq 0, y \neq 1$ **75.** $(a-b)/c$ **77.** $a/2$ **79.** $1/2$ if $y \neq 0$, any nonzero real number if $y = 0$ **81.** $15/2$ **83.** no solution **85.** -4 **87.** $2/3$ **89.** $3/2$ **91.** -2 **93.** no solution **95.** 0 **97.** $-3/4$ **99.** $-3/4$ **101.** all real numbers except 0 **103.** $-1, 5$ **105.** $4, 2/3$ **107.** no solution **109.** $-37/4, -47/4$ **111.** $21/13, 19/17$ **113.** $1/2, 3/2$ **115.** $5, 1/7$ **117.** $5, -3/5$ **119.** $16/7, -4/7$ **121.** $8/3, -2/7$

Section 2.2 (page 113)

1. inequality **2.** $>, \geq, <, \leq$ **3.** $<, >$ **4.** $M > c, M < -c$ **5.** $M < c, M > -c$ **6.** true **7.** false **8.** true **9.** false **10.** true

11. $x > 2$, $(2, \infty)$

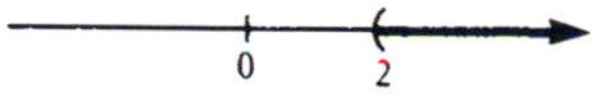

13. $x \leq 7/2$, $(-\infty, 7/2]$

15. $x < 7/2$, $(-\infty, 7/2)$

17. $x \geq -1/3$, $[-1/3, \infty)$

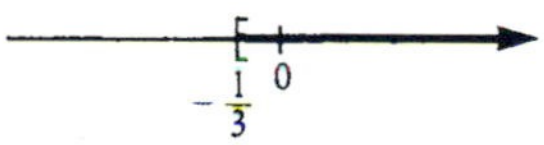

19. $x > 3/2$, $(3/2, \infty)$

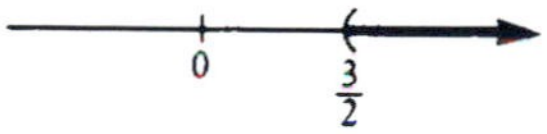

21. $x \leq 14$, $(-\infty, 14]$

23. $x > -5$, $(-5, \infty)$

25. all x, $(-\infty, \infty)$

27. $x < -b/a$

29. $x < a(b-y)/b$

31. $x > (d-b)/(a-c)$

33. $(-1/4, \infty)$

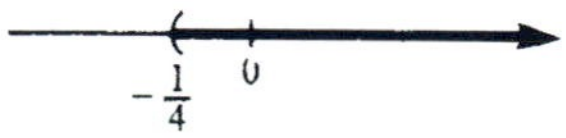

35. $[0, 1)$

37. $(-3, 1/2)$

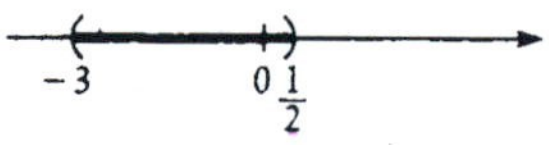

39. $(1, 3/2]$

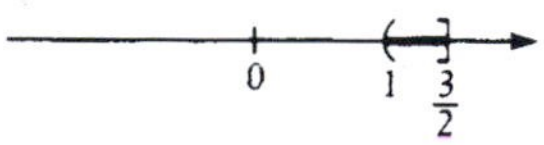

41. $(-\infty, -3) \cup (-1, \infty)$

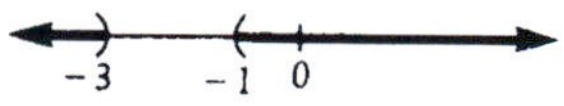

43. $(1, 3/2]$

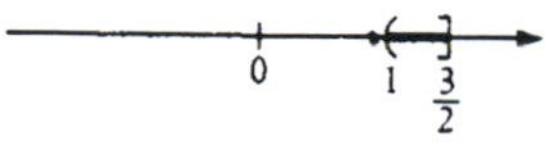

45. $(-\infty, -3)$

47. $[-13/4, -3)$

49. $(a, a+1)$

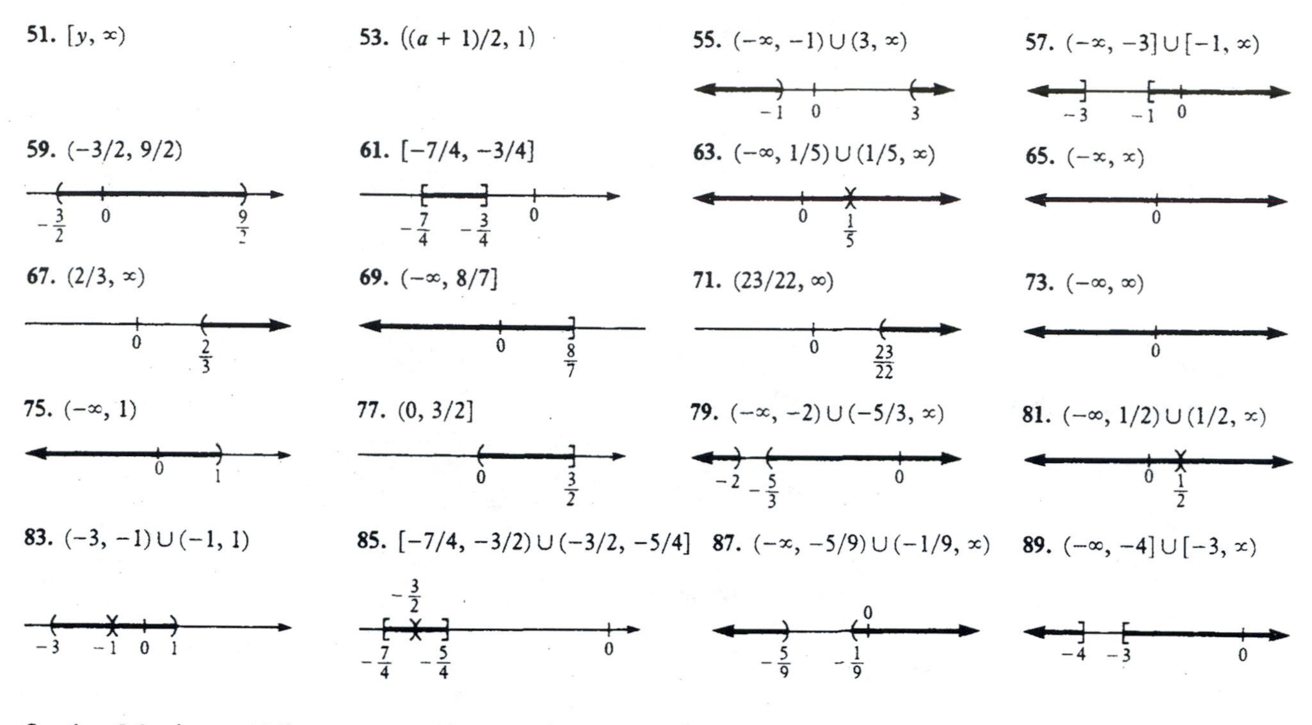

Section 2.3 (page 124)

1. \$19.96 3. 25% 5. $33\frac{1}{3}$% 7. decrease of 1.6% 9. \$236 11. about 38.46 mph 13. They meet 48 mi from the starting point of the car traveling at 40 mph, at 11:12 A.M. 15. 3:33 P.M. 17. $33\frac{1}{3}$ meters 19. Patrick rows at 12.5 mph, and the current is 2.5 mph. 21. 38% 23. 2.5 liters 25. 1 lb 27. 34% 29. 90.8 31. 1/3 hr, or 20 min 33. 2.4 hr 35. The two machines are 1.5 times as fast as A and 3 times as fast as B. 37. 1.5 min 41. 92 45. .5 and 5.5 47. about 60 micrograms 49. 6

Section 2.4 (page 135)

1. abscissa, ordinate 2. 0 3. III, I 4. IV, IV 5. Pythagorean 6. false 7. false 8. false 9. true 10. false

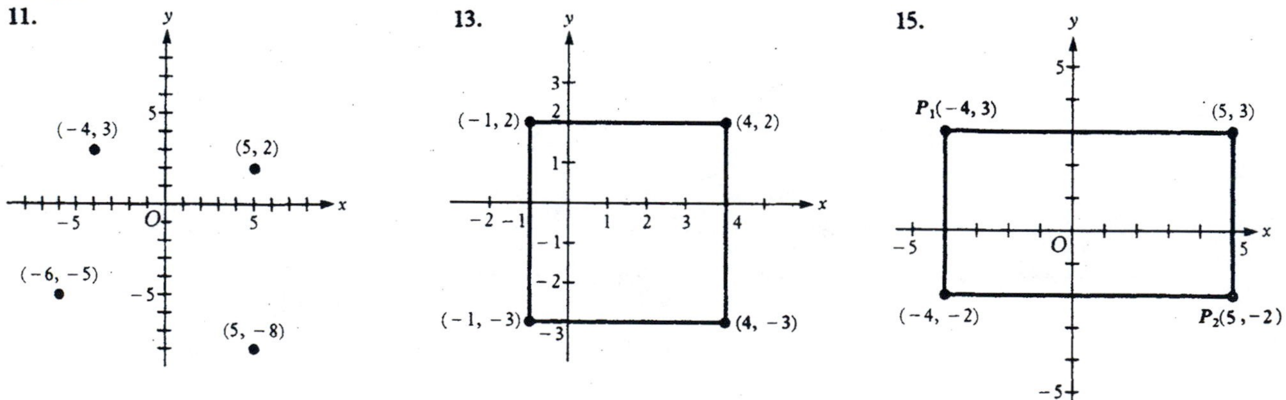

17.

19. One possibility is shown below.

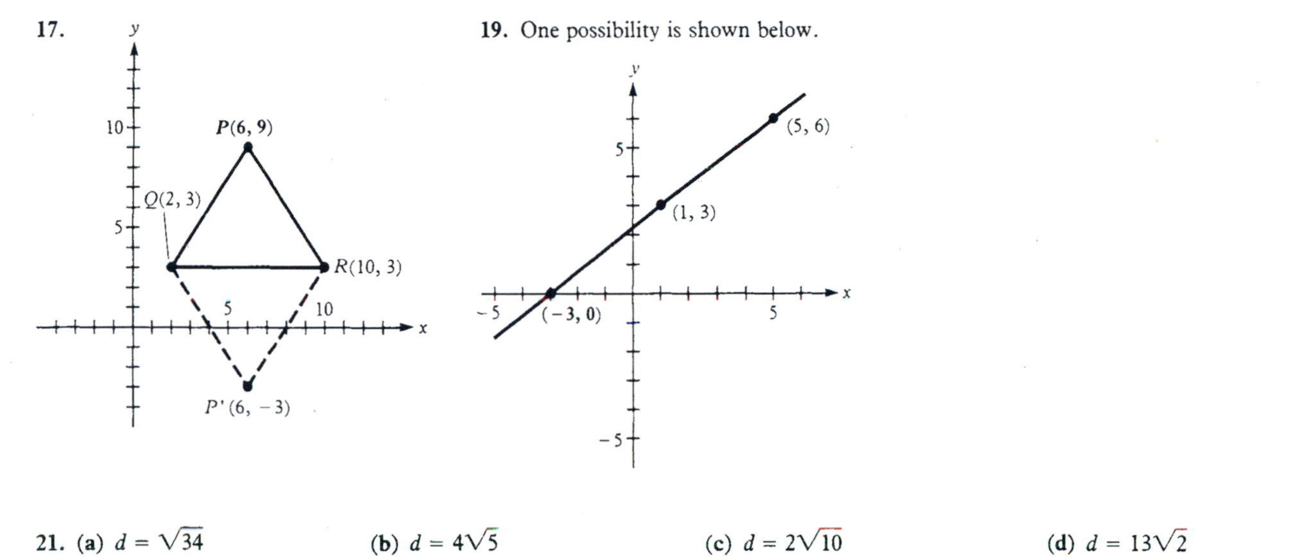

21. **(a)** $d = \sqrt{34}$

(−3, 4)
(2, 1)

(b) $d = 4\sqrt{5}$

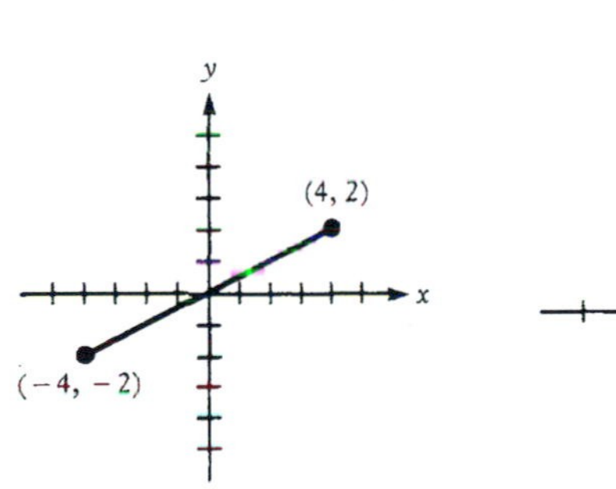

(c) $d = 2\sqrt{10}$

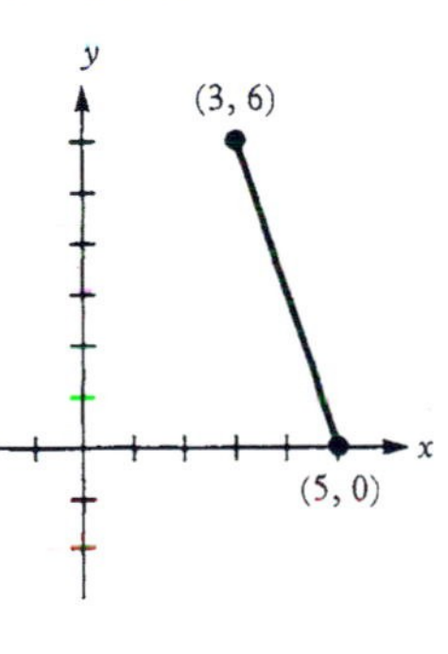

(d) $d = 13\sqrt{2}$

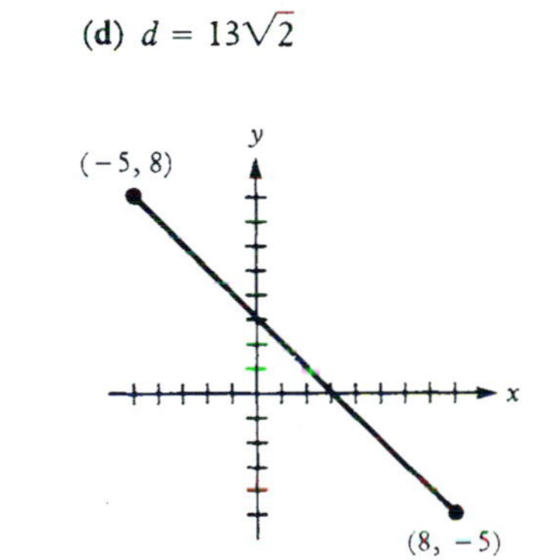

23. $\sqrt{61}$ and $\sqrt{41}$

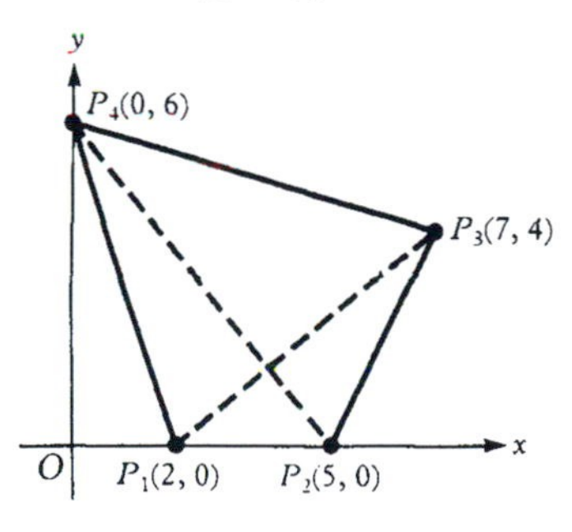

25. $y = 7$

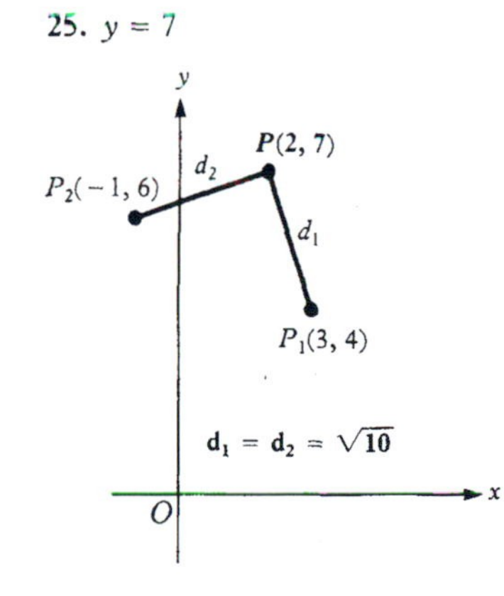

27. $d(P_1, P_2) = d(P_2, P_3) = d(P_1, P_3)$

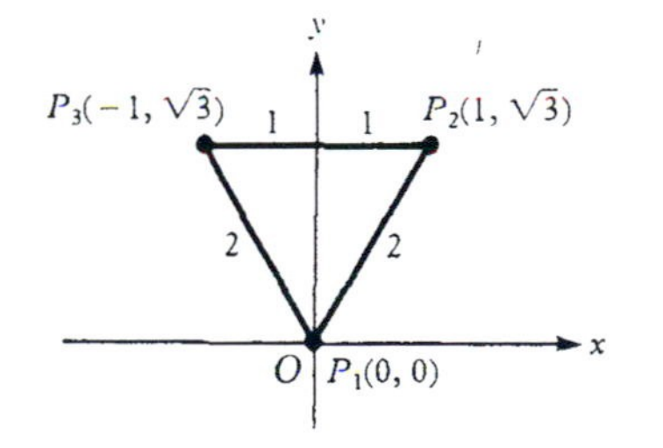

29. no; $d(P_1, P_3) \neq d(P_2, P_4)$

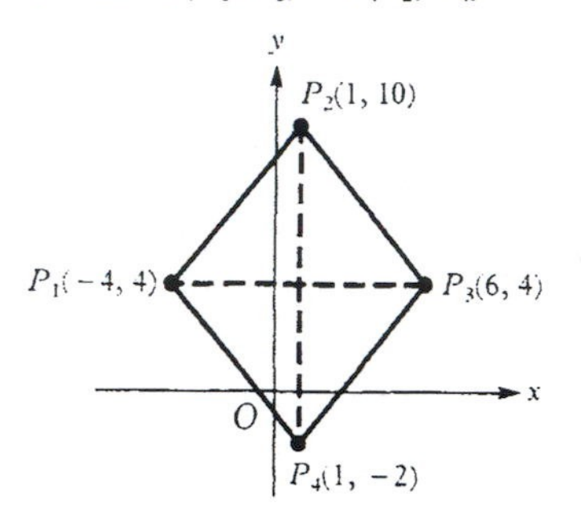

31. yes; center of circle at (9/2, 5/2) and radius = $\sqrt{130}/2$

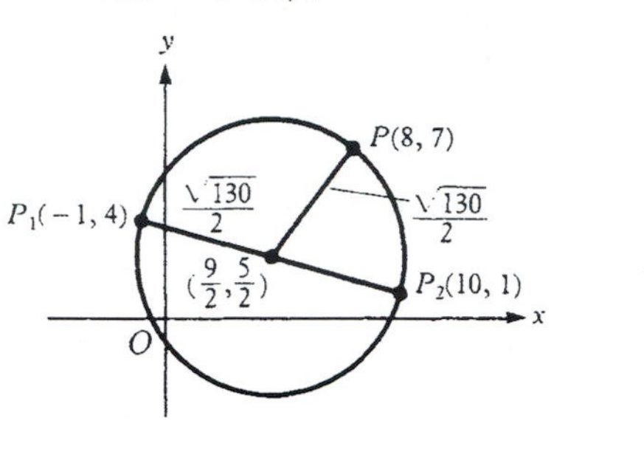

33. $d(P_1, P_2) + d(P_2, P_3) = d(P_1, P_3)$

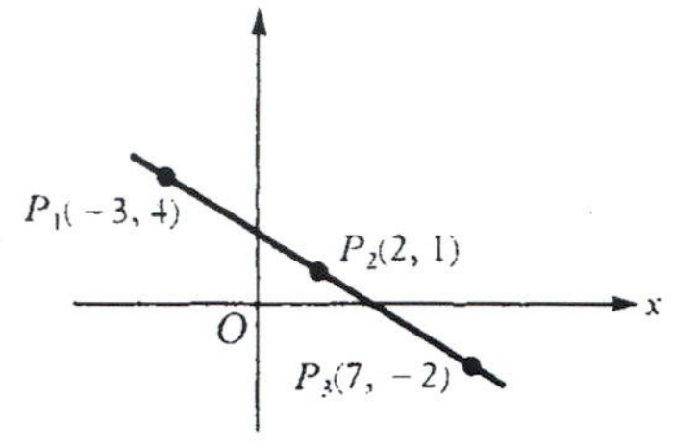

35. $4 + 2\sqrt{2}$ mi

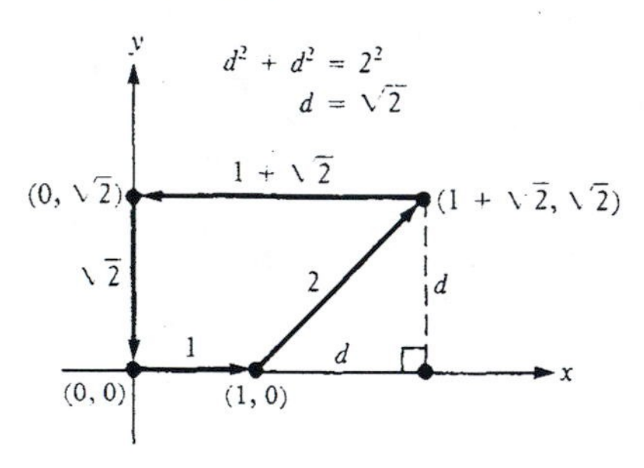

37.

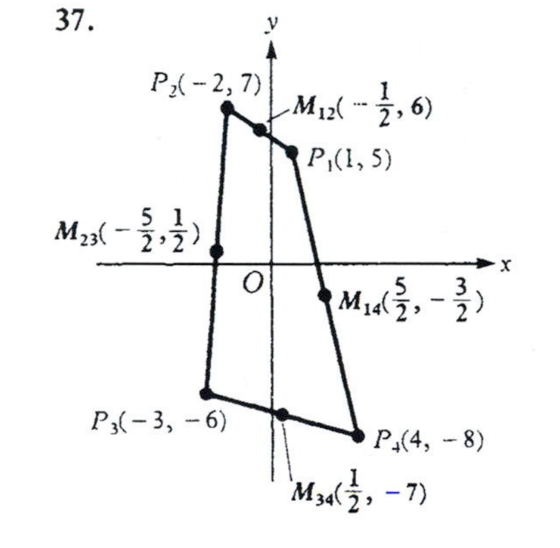

39. $x = -5/2$; $y_2 = 3$

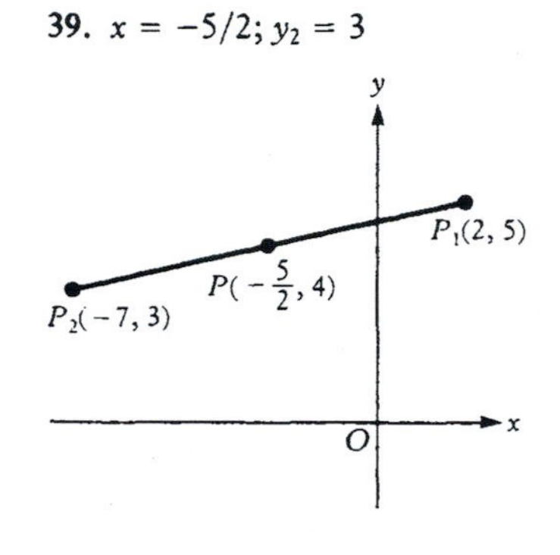

41.

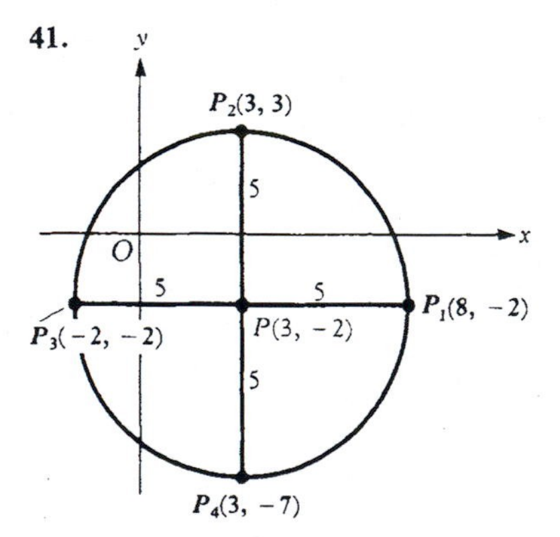

43. $P_3(5, 6)$, $P_4(-1, 4)$ (shown) and $P_3'(17, 6)$, $P_4'(11, 4)$

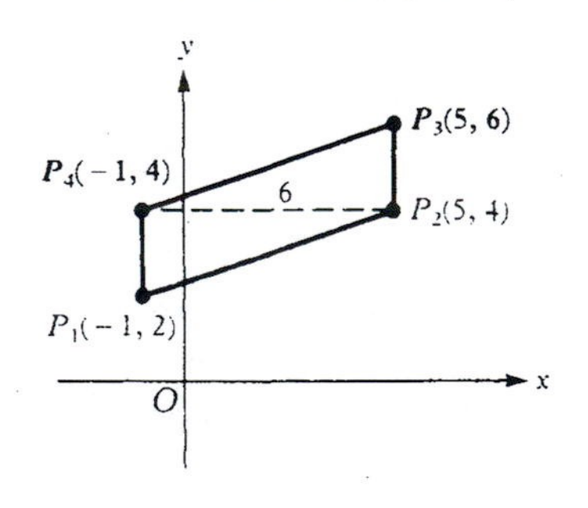

Section 2.5 (page 146)

1. graph **2.** slope **3.** (3, 5), (1, −3) **4.** point-slope, slope-intercept, general linear **5.** $y = -4$, $x = 3$ **6.** true **7.** true **8.** false **9.** true **10.** true **11.** −1 **13.** −1/4 **15.** undefined slope **17.** $y = 7$ **19.** $x = -5/2$

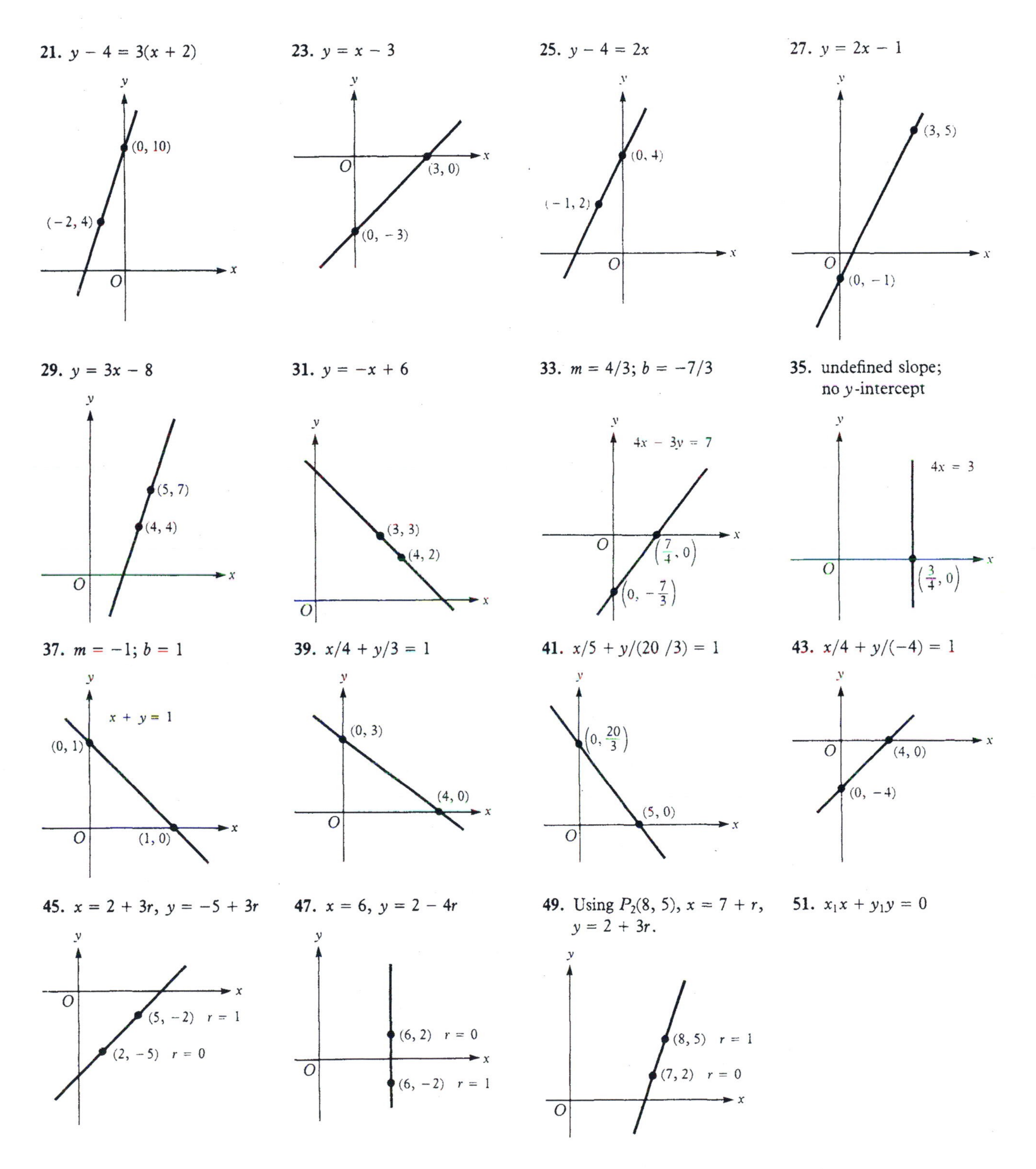
21. y − 4 = 3(x + 2)
(0, 10)
(−2, 4)
23. y = x − 3
(3, 0)
(0, −3)
25. y − 4 = 2x
(0, 4)
(−1, 2)
27. y = 2x − 1
(3, 5)
(0, −1)
29. y = 3x − 8
(5, 7)
(4, 4)
31. y = −x + 6
(3, 3)
(4, 2)
33. m = 4/3; b = −7/3
4x − 3y = 7
(7/4, 0)
(0, −7/3)
35. undefined slope;
no y-intercept
4x = 3
(3/4, 0)
37. m = −1; b = 1
x + y = 1
(0, 1)
(1, 0)
39. x/4 + y/3 = 1
(0, 3)
(4, 0)
41. x/5 + y/(20 /3) = 1
(0, 20/3)
(5, 0)
43. x/4 + y/(−4) = 1
(4, 0)
(0, −4)
45. x = 2 + 3r, y = −5 + 3r
(5, −2) r = 1
(2, −5) r = 0
47. x = 6, y = 2 − 4r
(6, 2) r = 0
(6, −2) r = 1
49. Using P₂(8, 5), x = 7 + r,
y = 2 + 3r.
(8, 5) r = 1
(7, 2) r = 0
51. x₁x + y₁y = 0
x
y
O

53. (a) 5 mph (b) .5 mi (c) $y = 5x + .5$

55. (a) \$151 (b) 1% (c) $P = 1.51t + 151;\ t \geq 0$

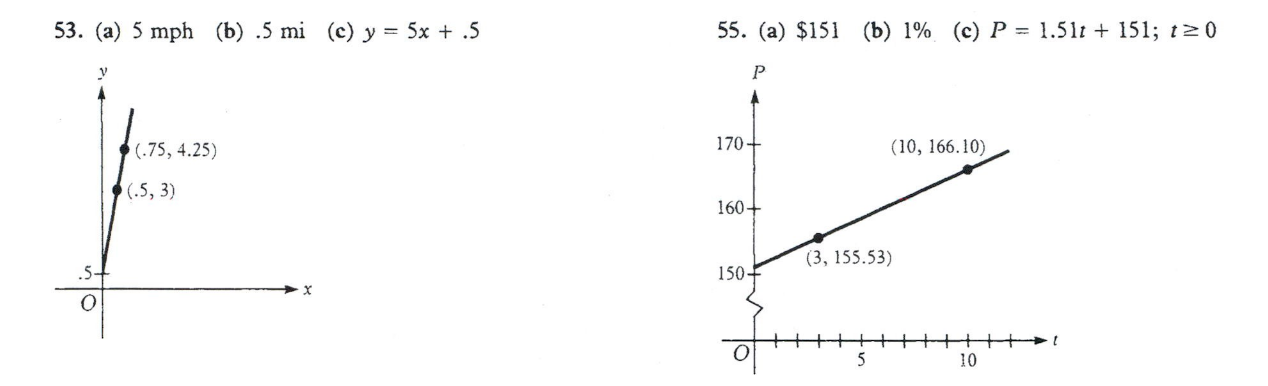

Section 2.6 (page 157)

1. $m_1 = m_2$ **2.** $A_1B_2 - A_2B_1 = 0$ **3.** $m_1m_2 = -1$ **4.** $A_1A_2 + B_1B_2 = 0$ **5.** $A_1B_2 - A_2B_1 \neq 0$ **6.** true **7.** false **8.** true **9.** true **10.** true **11.** parallel **13.** perpendicular **15.** perpendicular **17.** coincident **19.** parallel **21.** perpendicular **23.** $4x - 7y = -23$ **27.** $m_1 = 4/3$ and $m_2 = -3/4$ **33.** $x = 28/13, y = 4/13$ **35.** $x = 5, y = 2$ **37.** $x = 25, y = 32$ **39.** $x = 1/2, y = 4$ **41.** $x = 38, y = -29$ **43.** $x = 9, y = 13$ **45.** $x = -23/6, y = 11/3$ **47.** $x = 9/16, y = -37/32$ **49.** $x = 7/2, y = 13/3$ **51.** 12 ml of 20% acid and 6 ml of 50% acid **53.** \$27.90 fixed charge and \$1.00 for each word after the fifteenth **55.** 2000 of the first style and 1500 of the second **57.** 6

Section 2.7 (page 169)

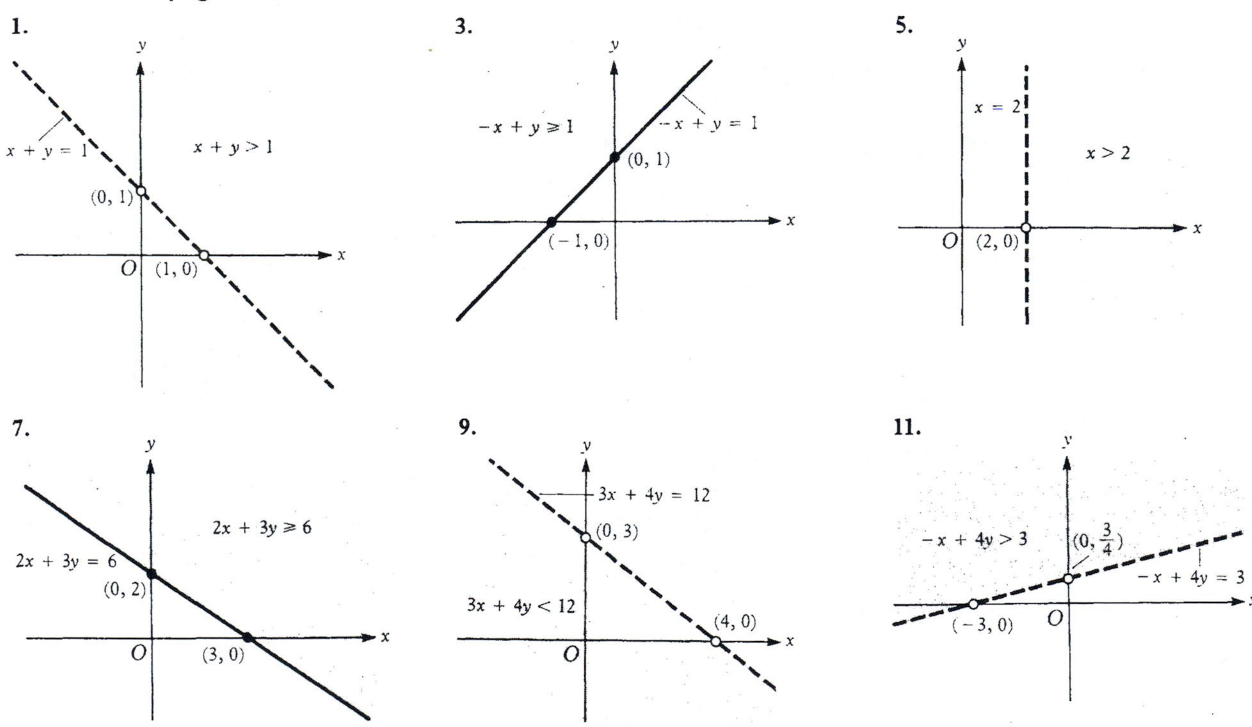

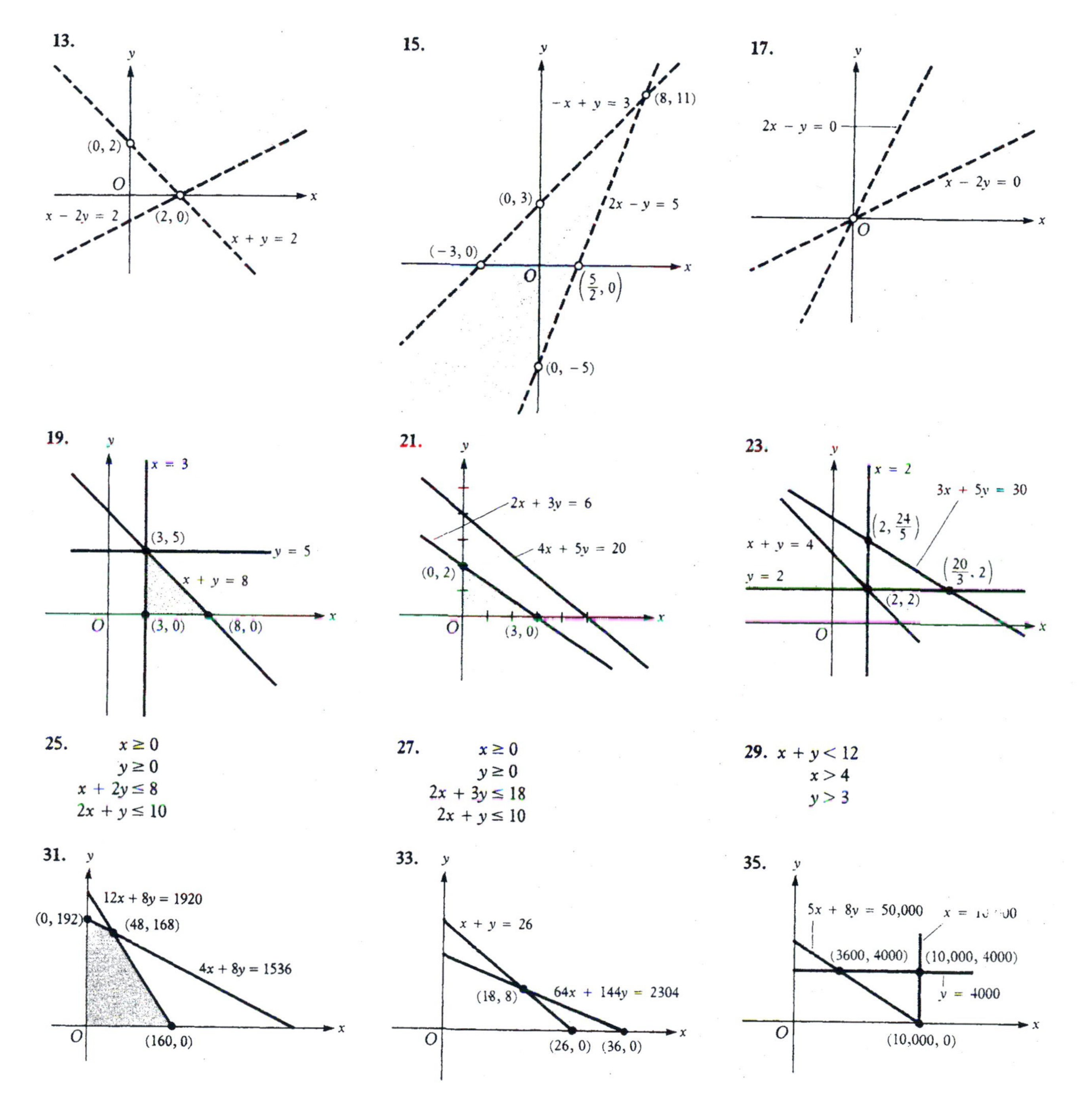
13.
(0, 2)
O
x − 2y = 2
(2, 0)
x + y = 2
15.
−x + y = 3
(8, 11)
(0, 3)
2x − y = 5
(−3, 0)
O
(5/2, 0)
(0, −5)
17.
2x − y = 0
x − 2y = 0
O
19.
x = 3
(3, 5)
y = 5
x + y = 8
O
(3, 0)
(8, 0)
21.
2x + 3y = 6
4x + 5y = 20
(0, 2)
O
(3, 0)
23.
x = 2
3x + 5y = 30
(2, 24/5)
x + y = 4
(20/3, 2)
y = 2
(2, 2)
O
25. x ≥ 0
y ≥ 0
x + 2y ≤ 8
2x + y ≤ 10
27. x ≥ 0
y ≥ 0
2x + 3y ≤ 18
2x + y ≤ 10
29. x + y < 12
x > 4
y > 3
31.
12x + 8y = 1920
(0, 192)
(48, 168)
4x + 8y = 1536
O
(160, 0)
33.
x + y = 26
(18, 8)
64x + 144y = 2304
O
(26, 0)
(36, 0)
35.
5x + 8y = 50,000
(3600, 4000)
(10,000, 4000)
y = 4000
O
(10,000, 0)

37. **39.**

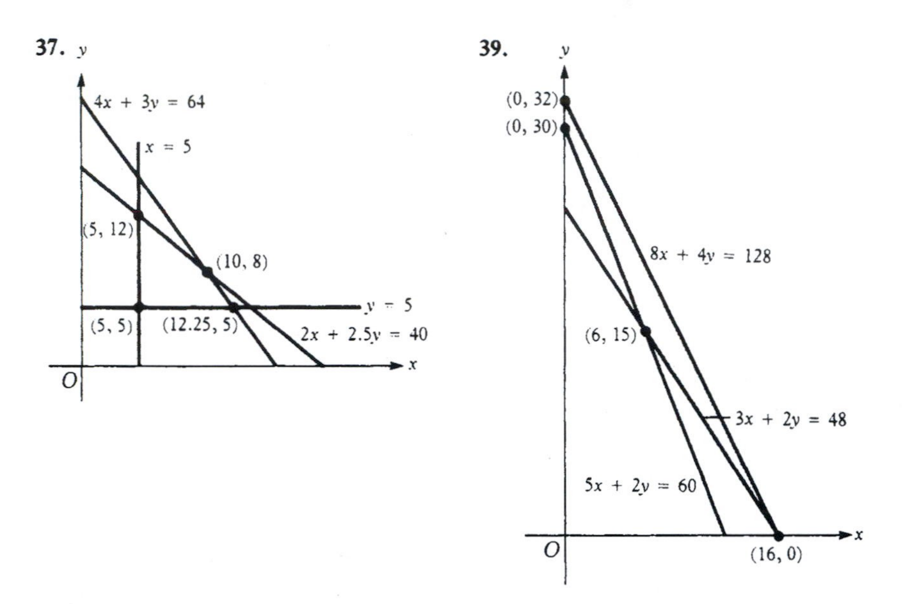

41. (a) \$204 (b) \$160 (c) \$174
43. (a) \$70 (b) \$52 (c) \$52
45. 7600 tickets
47. (a) \$252 (b) \$230 (c) \$246
49. (a) \$90 (b) \$96 (c) \$102

Chapter 2 Review Exercises (page 173)

1. $x = -7/2$ **2.** $x = 725/300$ **3.** $x > -7/2$ **4.** $x \geq 1$ **5.** $x = 7$ **6.** no solution **7.** $x = 0$ **8.** $x = 9$ **9.** no solution **10.** $-6 < x < 2$ **11.** $x \leq 1/3$ or $x \geq 3$ **12.** (a) all x (b) no x **13.** $x = 3$ or $x = 5$ **14.** $x = \pm 2$ **15.** If $a = -4$, $x =$ any real number; if $a \neq \pm 4$, $x = 1/(a-4)$. **16.** $x = (5 - 3a)/5$; $a \neq 5/3$, $a \neq -5/2$ **17.** $-3 < x < -1/2$ **18.** $(-\infty, -3] \cup (-1/2, \infty)$ **19.** $-a < x < 0$ **20.** $M = -1/2$ and $d = 13$ **21.** $M = (-1/2, 1/2)$ and $d = \sqrt{194}$ **22.** (a) $d^2(P_1, P_3) + d^2(P_2, P_3) = d^2(P_1, P_2)$; that is, $41 + 41 = 82$ (b) $m(P_2P_3) = -5/4$, and $m(P_1P_3) = 4/5$ **23.** $-3/10$ **24.** $x = 7, y = 3$ **25.** $6x - 8y = -11$ **26.** $y = 4, x = 6$ **27.** $y = (-3/5)x - 1/5$ **28.** $y = (3/2)x - 2$ **29.** $6x - 8y = -11$ **30.** $2x - 3y = -17$ **31.** $x = 4, y = -3$ **32.** $x = 4, y = -3$ **33.** $x = 4, y = -3$ **34.** $x = 4, y = 2$

35. **36.**

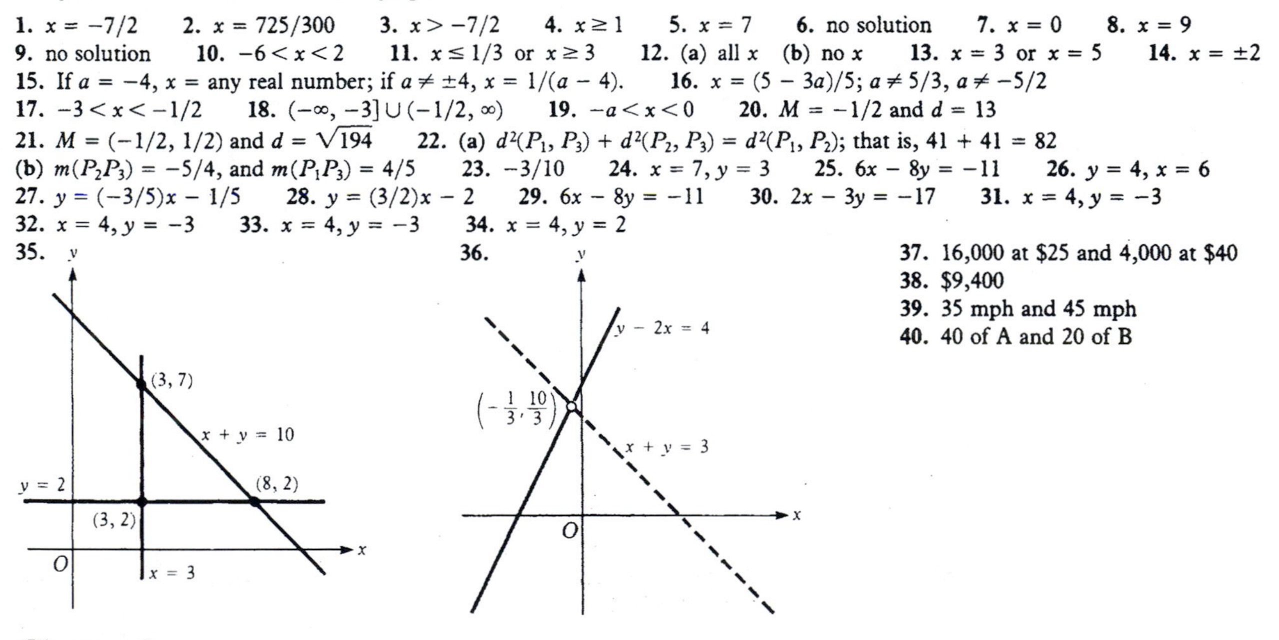

37. 16,000 at \$25 and 4,000 at \$40
38. \$9,400
39. 35 mph and 45 mph
40. 40 of A and 20 of B

Chapter 3

Section 3.1 (page 184)

1. $ax^2 + bx + c = 0$, constant real numbers, a **2.** two, one **3.** factoring, completing the square, quadratic formula **4.** $(-b \pm \sqrt{b^2 - 4ac})/(2a)$ **5.** $b^2 - 4ac > 0$, $b^2 - 4ac = 0$, $b^2 - 4ac < 0$ **6.** true **7.** true **8.** true **9.** true **10.** true **11.** 3, 4 **13.** $-2, 8$ **15.** 4 **17.** $\pm 3/2$ **19.** $-1, -2/3$ **21.** $-a, -b$ **23.** $x^2 - 9 = 0$

25. $6x^2 - 17x + 12 = 0$ **27.** $4x^2 - 28x + 49 = 0$ **29.** $x^2 - 6x + 7 = 0$ **31.** $(1 \pm \sqrt{5})/2$ **33.** $-2 \pm \sqrt{2}$
35. $(3 \pm \sqrt{3})/2$ **37.** $(5 \pm \sqrt{13})/3$ **39.** $(5 \pm \sqrt{41})/4$ **41.** a, b **43.** 0, 6 **45.** $-1, -1/2$ **47.** $(1 \pm \sqrt{3})/2$
49. $1/2, -1/3$ **51.** 2 **53.** 1 **55.** 0 **57.** 2 **59.** 1 **61.** $2(x - 5/2)(x - 4)$ **63.** $[x - (5 + \sqrt{3})][x - (5 - \sqrt{3})]$
65. $4(x - 3/2)^2$ **67.** irreducible **69.** $4[x - (1 + \sqrt{2})/2][x - (1 - \sqrt{2})/2]$ **71.** $(5 \pm \sqrt{5})/2$ **73.** $(-2 \pm \sqrt{10})/2$
75. $(-1 \pm \sqrt{2})/2$ **77.** $1, -1/3$ **79.** $2(x + 3)^2 - 15$ **81.** $4(x + 1/4)^2 - 21/4$ **83.** $1(x + 1/2)^2 + 3/4$
87. Exercise 51: $a_1 + a_2 = 100$, $a_1a_2 = 275$; Exercise 53: $a_1 + a_2 = -24$, $a_1a_2 = 144$; Exercise 57: $a_1 + a_2 = -75/107$, $a_1a_2 = -240/107$; Exercise 59: $a_1 + a_2 = 17/3$, $a_1a_2 = 289/36$

Section 3.2 (page 192)

1. real numbers, $\sqrt{-1}$ **2.** $a - bi$ **3.** 0 **4.** pure imaginary **5.** diagonal **6.** false **7.** true **8.** true
9. true **10.** false **11.** $7 - 4i$ **13.** $3 + 6i$ **15.** $14 - 5i$ **17.** 25 **19.** $3 + 3i$ **21.** $(5/2) - (1/2)i$
23. $(-2/29) + (5/29)i$ **25.** $4 - 7i$ **27.** (a) i (b) -1 (c) $-i$ (d) 1 **29.** $\pm 3i$ **31.** $2 \pm i$ **33.** $-4 \pm 3i$
35. $(1 \pm i)/2$ **37.** $(2 \pm \sqrt{2}i)/2$ **39.** $(x - 3i)(x + 3i)$ **41.** $[x - (-4 + 3i)][x - (-4 - 3i)]$
43. $[x - (-5 + i)][x - (-5 - i)]$ **45.** $2[x - (-1 + \sqrt{2}i/2)][x - (-1 - \sqrt{2}i/2)]$
47. $z_1 + z_2 = 5 + 5i$ **49.** $z_1 + z_2 = -2 + 6i$ 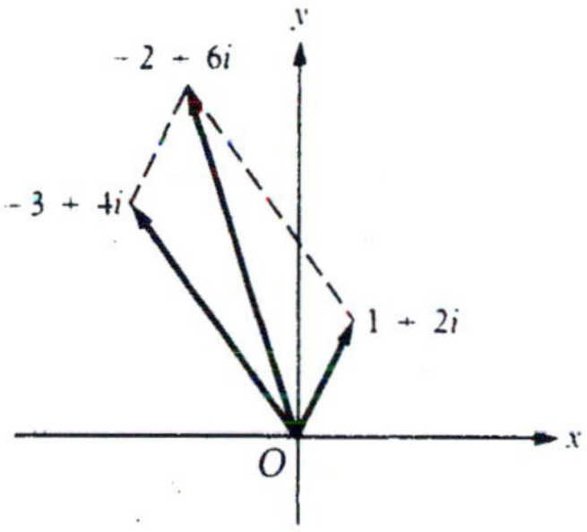 **51.** $z_1 + z_2 = 4i$

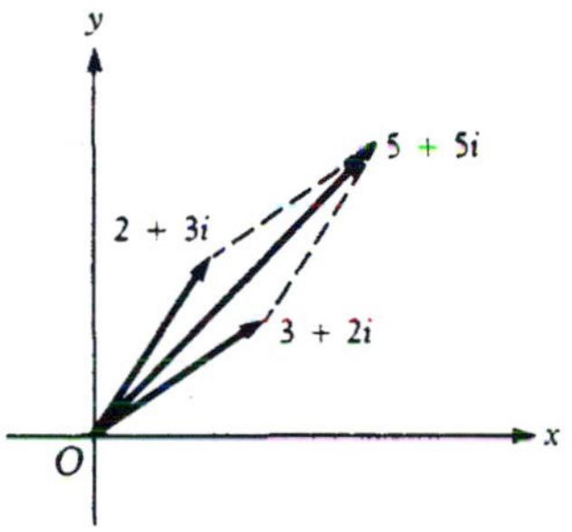

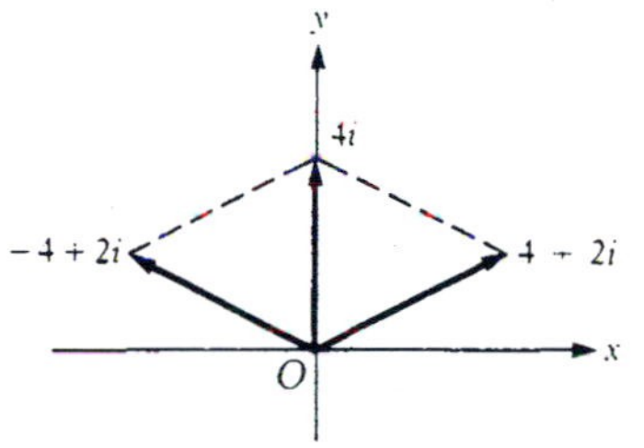

53. $z_1 + z_2 = 2 + 2i$ 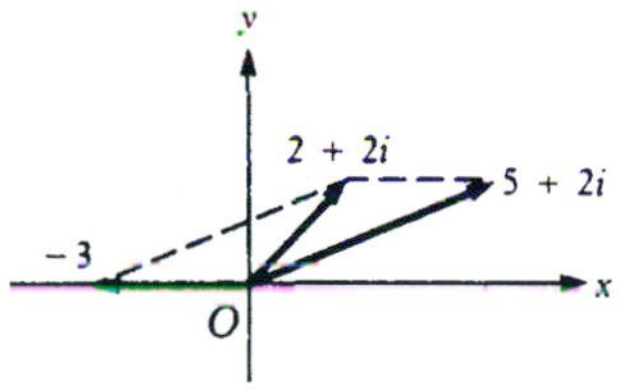 **57.** (a) $-1 + 7i$ 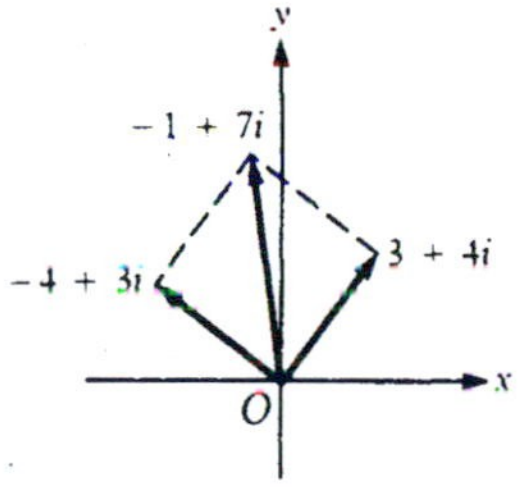 (b) $7 + i$ 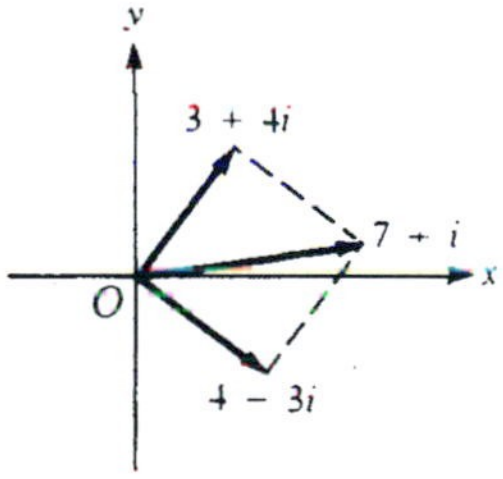 (c) $-7 - i$

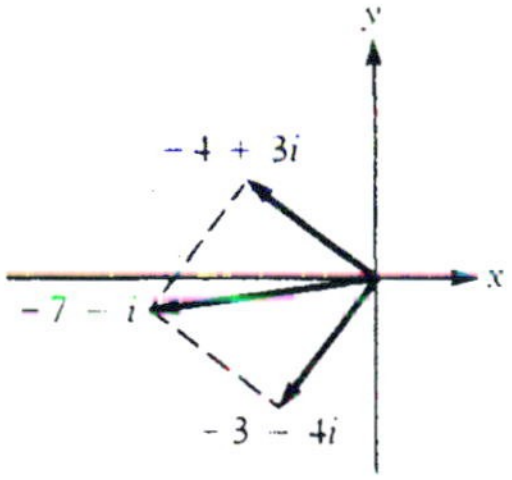

Section 3.3 (page 197)

1. substitution **2.** 0 **3.** the original equation **4.** nonnegative **5.** $(x + 1)^{3/2}$ **6.** false **7.** false **8.** false
9. true **10.** false **11.** $\pm 1, \pm\sqrt{2}$ **13.** 81/16 **15.** $(3 \pm \sqrt{5})/2$ **17.** $\pm\sqrt{5}$ **19.** $-8/5, -13/8$ **21.** 2, 3
23. 1, 4 **25.** -3 **27.** $2, -1$ **29.** $(11 \pm \sqrt{73})/4$ **31.** 4, 5 **33.** 1/3 **35.** 12 **37.** 5 **39.** $-1, -5/2$

Section 3.4 (page 202)

1. a, roots **2.** less than **3.** $(-\infty, \infty)$, $\varnothing$ **4.** $a < 0$ **5.** $(-\infty, a_1] \cup [a_2, \infty)$ **6.** false **7.** false **8.** true
9. false **10.** false
11. $(-\infty, 1] \cup [3, \infty)$ **13.** $(-4, 2)$ **15.** $(-\infty, 1/2] \cup [3, \infty)$ **17.** $[-2/3, 4]$

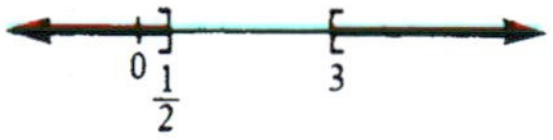

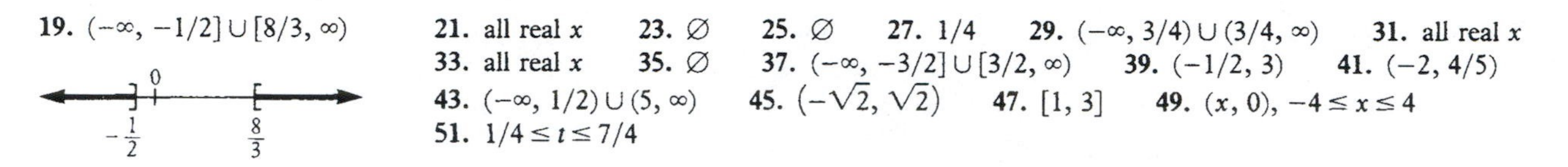

19. $(-\infty, -1/2] \cup [8/3, \infty)$ **21.** all real x **23.** $\varnothing$ **25.** $\varnothing$ **27.** $1/4$ **29.** $(-\infty, 3/4) \cup (3/4, \infty)$ **31.** all real x **33.** all real x **35.** $\varnothing$ **37.** $(-\infty, -3/2] \cup [3/2, \infty)$ **39.** $(-1/2, 3)$ **41.** $(-2, 4/5)$ **43.** $(-\infty, 1/2) \cup (5, \infty)$ **45.** $(-\sqrt{2}, \sqrt{2})$ **47.** $[1, 3]$ **49.** $(x, 0), -4 \le x \le 4$ **51.** $1/4 \le t \le 7/4$

Section 3.5 (page 210)

1. 5 **3.** 40 **5.** 25% the first year and 50% the second **7.** 7 **9.** Jean: 10 mph; Margo: 15 mph **11.** 300 mph **13.** 1 **15.** (a) 4.5 sec after firing (b) 324 ft (c) 9 sec after firing **17.** yes **19.** $16\sqrt{6}$ ft/sec **21.** one carrier: 2 hr; other carrier: 3 hr **23.** A: 10 min; B: 15 min **25.** A: 4.8 min; B: 8 min; C: 12 min **27.** 2 and 7 **29.** 75 yd by 100 yd **31.** 1 in

Section 3.6 (page 221)

1. A, B **2.** completing the square **3.** A, B **4.** axis of symmetry **5.** $a > 0, a < 0$ **6.** true **7.** false **8.** true **9.** true **10.** true **11.** $(x-2)^2 + (y-1)^2 = 9$ **13.** $(x-2)^2 + (y+2)^2 = 8$ **15.** $(x+1)^2 + y^2 = 1$ **17.** $x^2 + y^2 = 1/2$ **19.** $(x-2)^2 + (y-1)^2 = 20$

21. circle

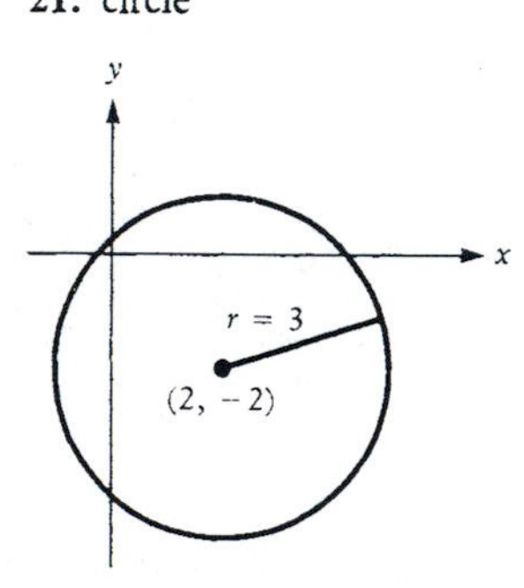

23. circle

25. point

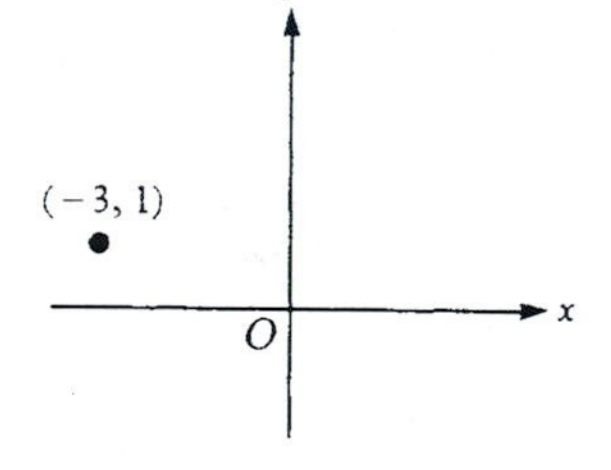

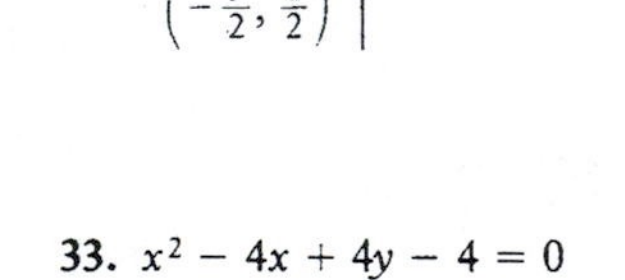

27. $\varnothing$ **29.** $\varnothing$ **31.** $x^2 + 2y - 3 = 0$ **33.** $x^2 - 4x + 4y - 4 = 0$

35. vertex: (0, 0);
axis of symmetry: y-axis

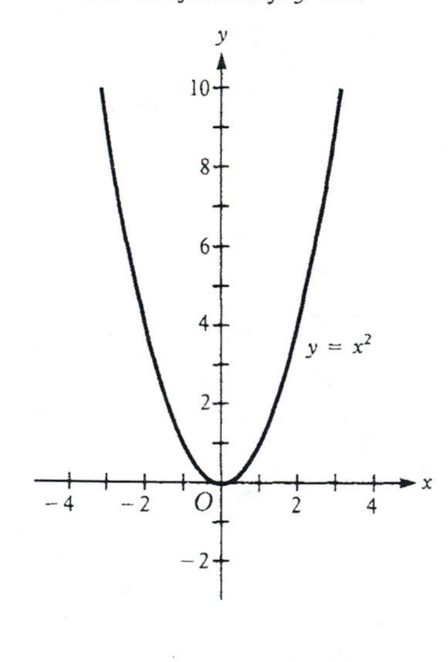

37. vertex: (0, 2);
axis of symmetry: y-axis

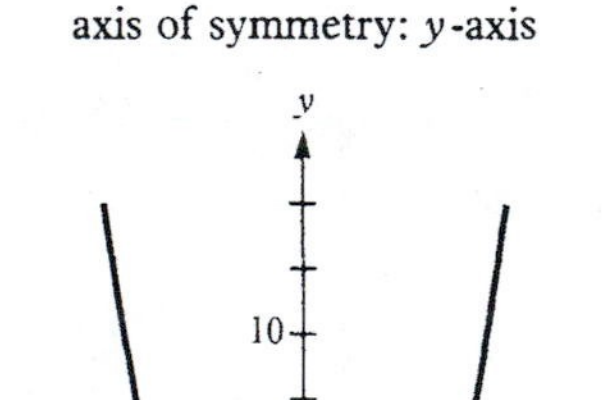

39. vertex: (−3, 1);
axis of symmetry: $x = -3$

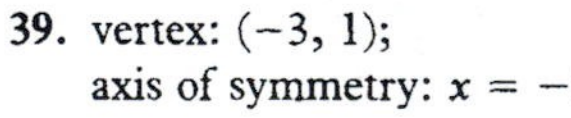

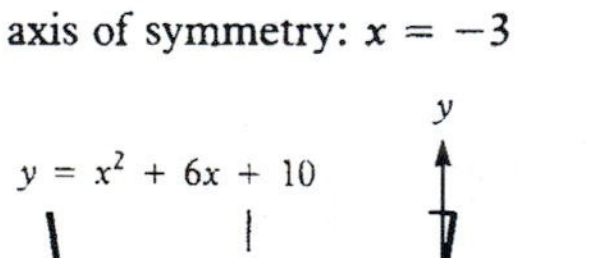

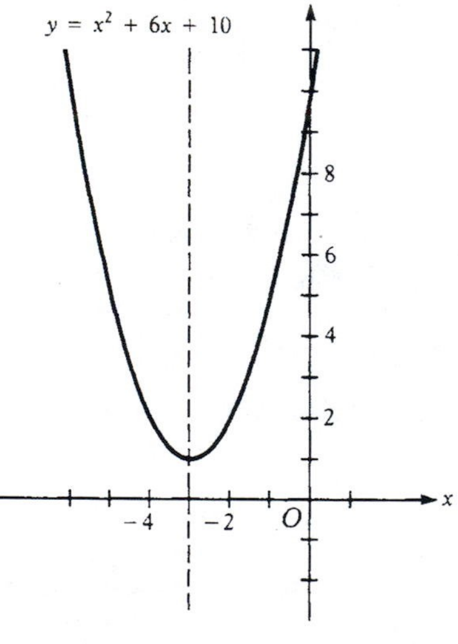

41. vertex: $(2, -4)$; axis of symmetry: $x = 2$

43. vertex: $(-2, -1)$; axis of symmetry: $x = -2$

45. vertex: $(0, 0)$; axis of symmetry: x-axis

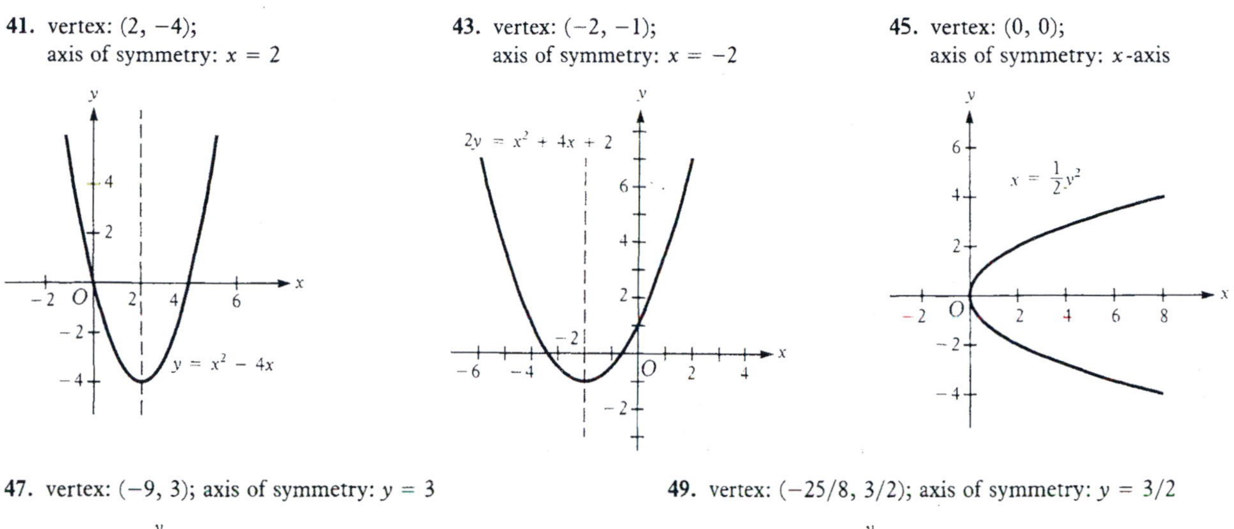

47. vertex: $(-9, 3)$; axis of symmetry: $y = 3$

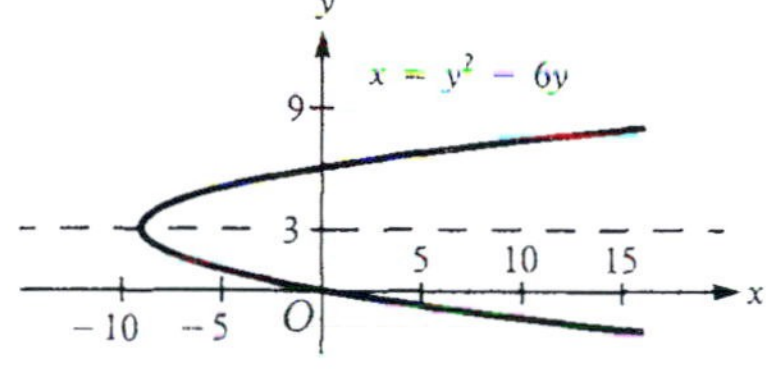

49. vertex: $(-25/8, 3/2)$; axis of symmetry: $y = 3/2$

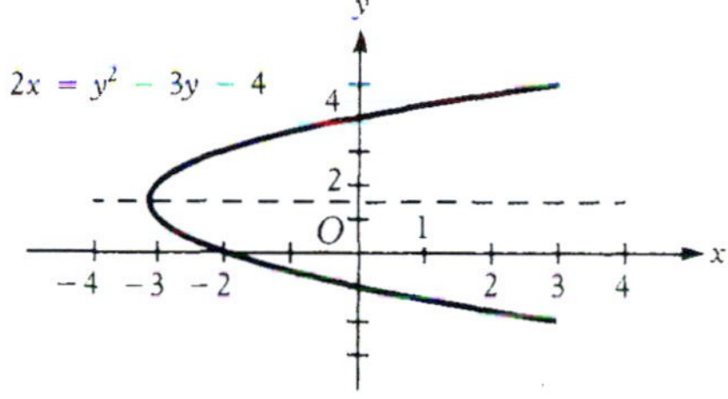

51. length: 500 ft; width: 250 ft **53.** 2 weeks **55.** (a) 8 sec (b) 512 ft (c) 256 ft (d) $y = -x^2/256 + 2x$

Section 3.7 (page 229)

1. $x' = x - x_0$ and $y' = y - y_0$ **2.** $(x')^2 + (y')^2 = r^2$ **3.** $x^2 + y^2 = r^2$ **4.** completing the square **5.** four **6.** true **7.** true **8.** true **9.** false **10.** true **11.** (a) $(1, -7)$ (b) $(-2, -5)$ (c) $(0, 0)$ **13.** $x' = x + 2; y' = y - 4$ **15.** $x' = x + 1; y' = y - 2$ **17.** $x' = x + 1; y' = y - 2$ **19.** $x' = x; y' = y + 2$

21. $2x' + 3y' = 8$

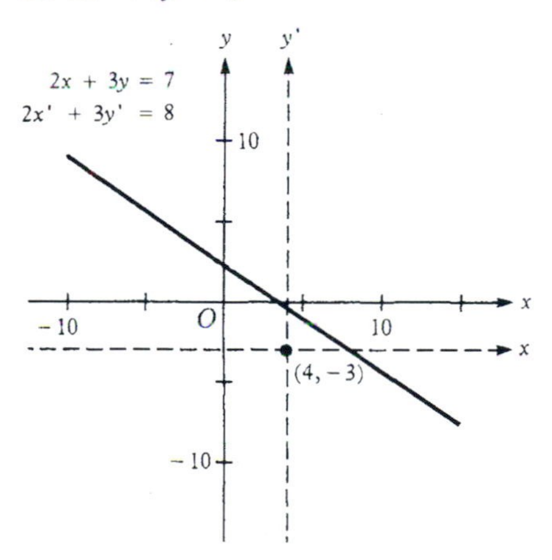

23. $(x' + 2)^2 + (y' - 4)^2 = 5$

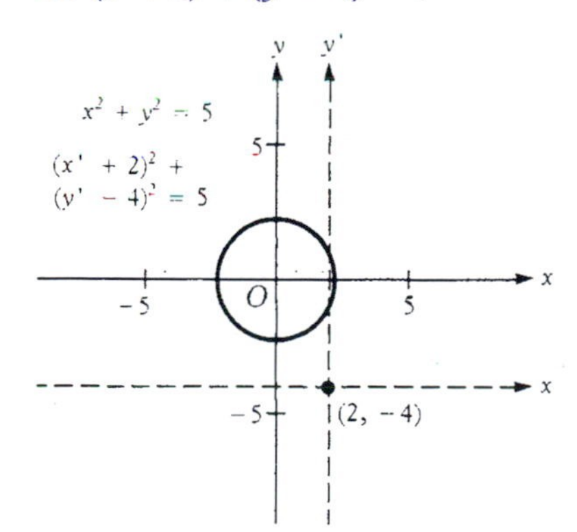

25. $y' = (x')^2 - 5$

27. $x^2 + y^2 = 25$

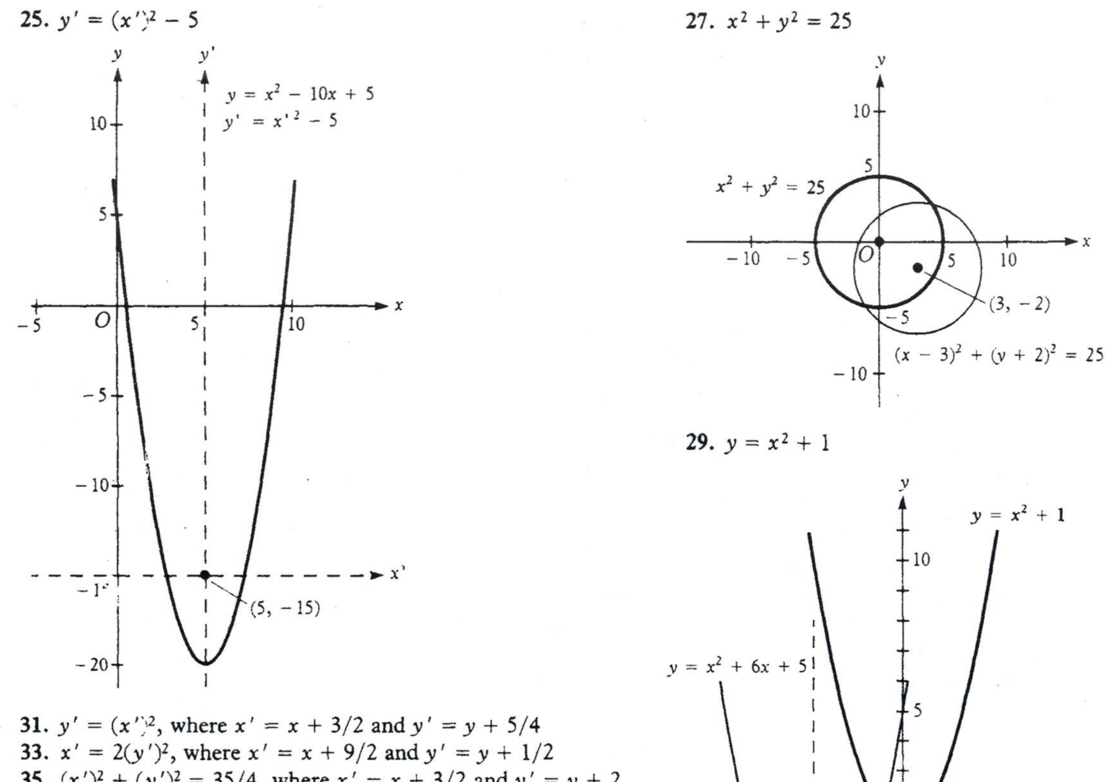

29. $y = x^2 + 1$

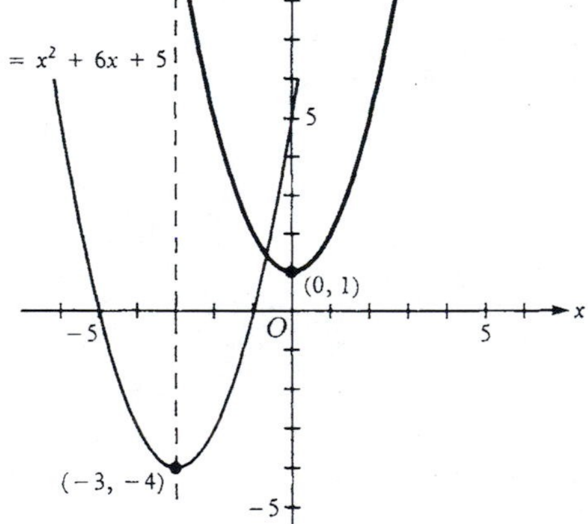

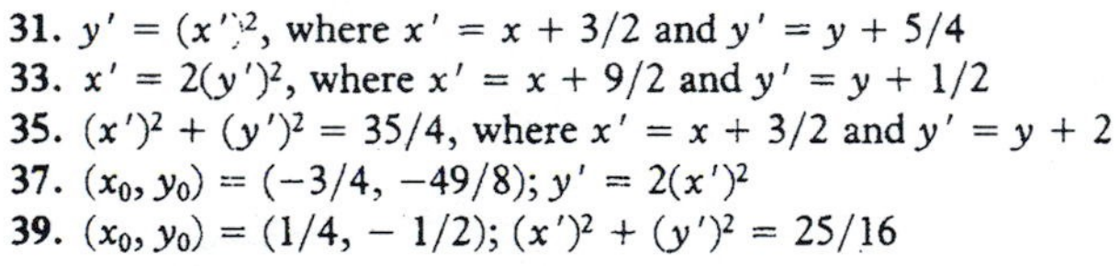

31. $y' = (x')^2$, where $x' = x + 3/2$ and $y' = y + 5/4$
33. $x' = 2(y')^2$, where $x' = x + 9/2$ and $y' = y + 1/2$
35. $(x')^2 + (y')^2 = 35/4$, where $x' = x + 3/2$ and $y' = y + 2$
37. $(x_0, y_0) = (-3/4, -49/8)$; $y' = 2(x')^2$
39. $(x_0, y_0) = (1/4, -1/2)$; $(x')^2 + (y')^2 = 25/16$

41. $(-1, 3)$ and $(3, -5)$

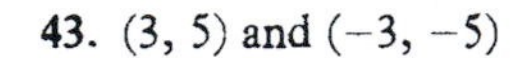

43. $(3, 5)$ and $(-3, -5)$

45. $(\sqrt{2}, 1)$ and $(-\sqrt{2}, 1)$

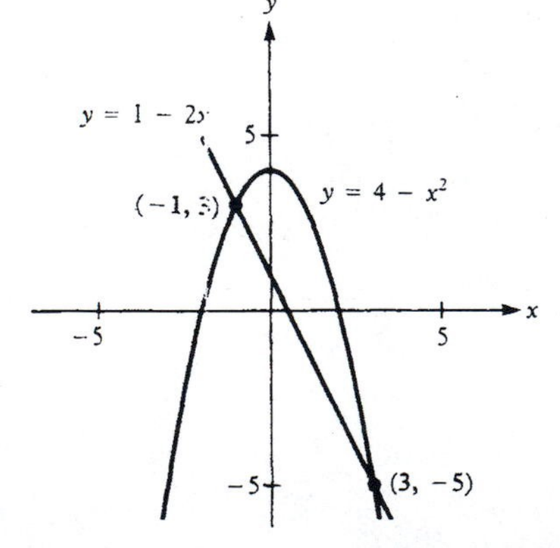

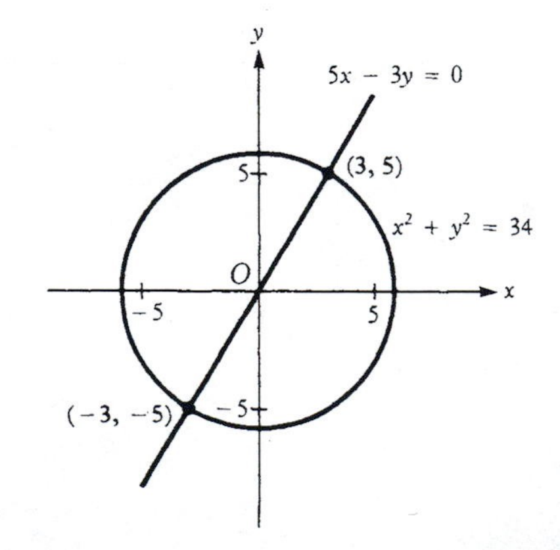

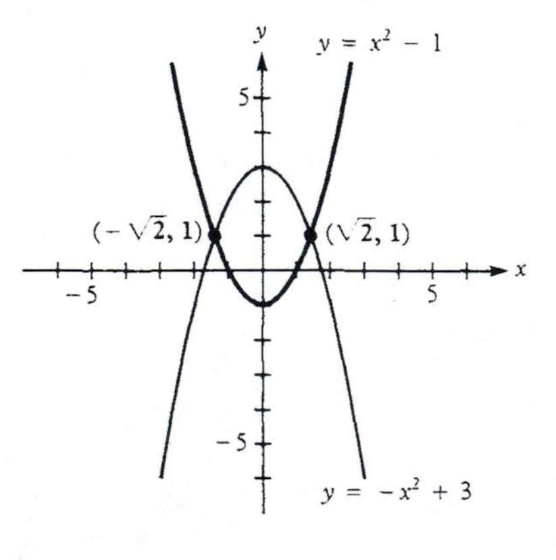

47. (3, 1) and (3, 3)

49. $(\sqrt{2}, \sqrt{2})$ and $(-\sqrt{2}, -\sqrt{2})$

51.

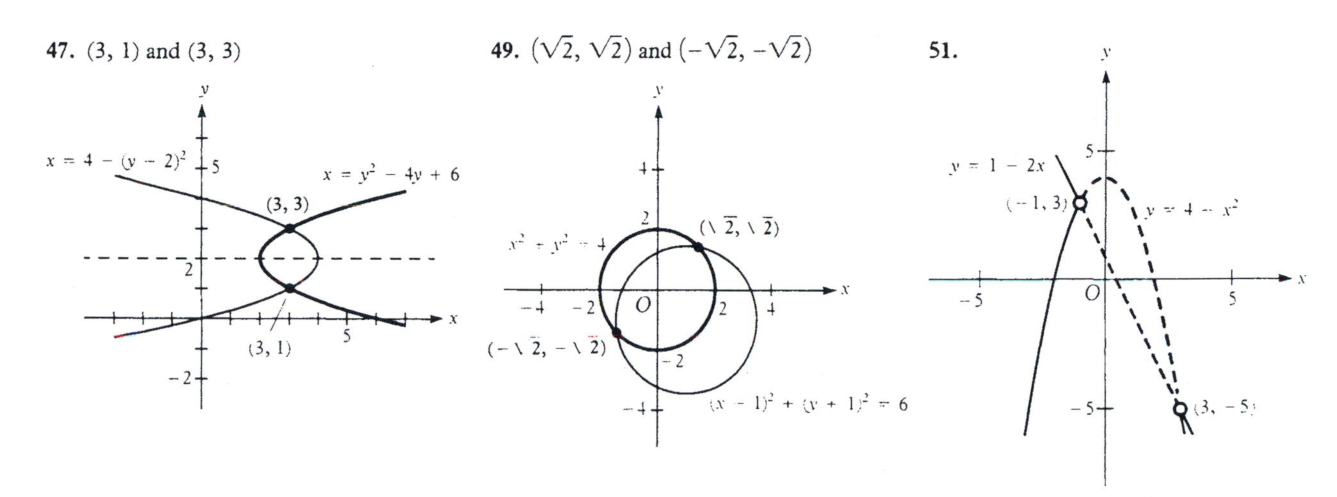

53.

55.

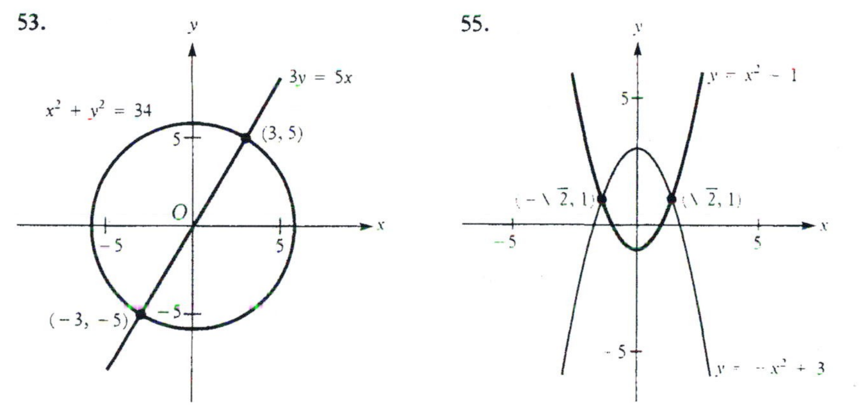

Section 3.8 (page 241)

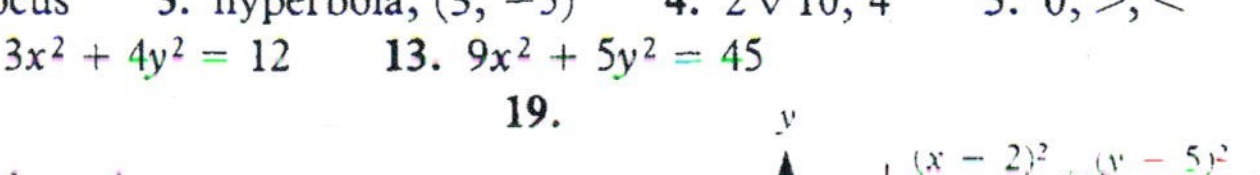

1. $d(P_1, P) + d(P_2, P) = c$, $d(P_2, P) - d(P_1, P) = \pm c$ **2.** focus **3.** hyperbola, $(3, -5)$ **4.** $2\sqrt{10}$, 4 **5.** 0, >, <
6. false **7.** false **8.** true **9.** false **10.** false **11.** $3x^2 + 4y^2 = 12$ **13.** $9x^2 + 5y^2 = 45$

15.

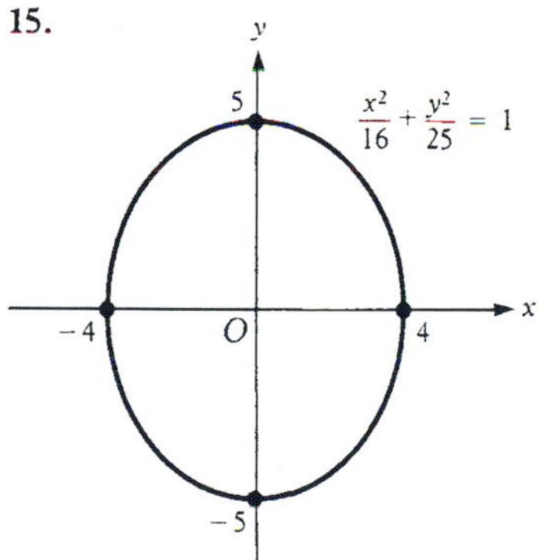

17.

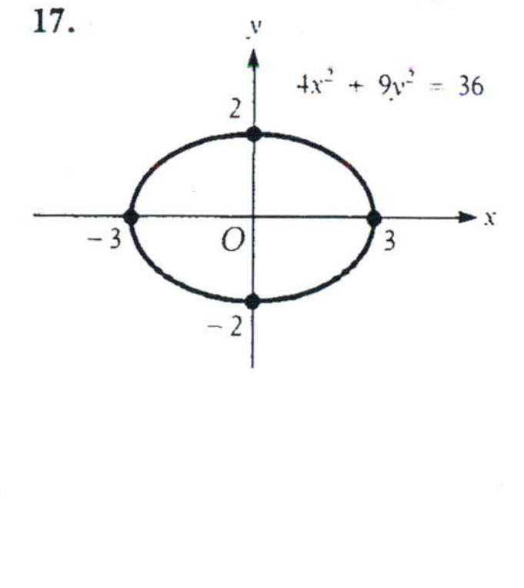

19.

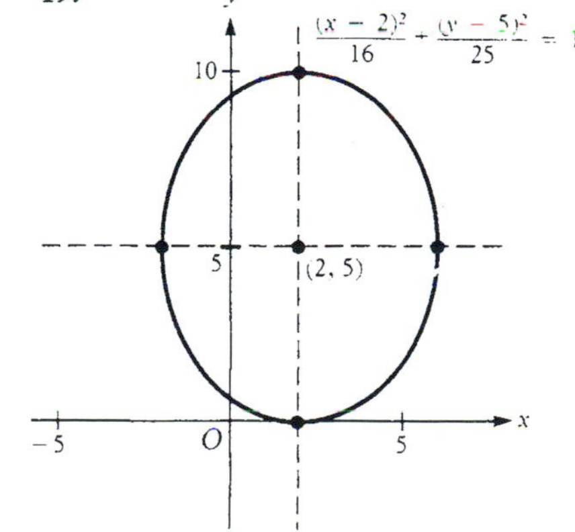

23. ellipse $(x + 2)^2 + 4(y - 2)^2 = 100$
25. point $(-1, -2)$
27. ellipse $4(x + 2)^2 + 9(y - 1)^2 = 36$
31. $12x^2 - 4y^2 = 3$
33. $3y^2 - x^2 = 3$
35. two lines $y = 2x, y = -2x - 1$
37. hyperbola $5(y + 1)^2 - 2(x - 1)^2 = 5$
39. hyperbola $8(y - 1/2)^2 - x^2 = 1$

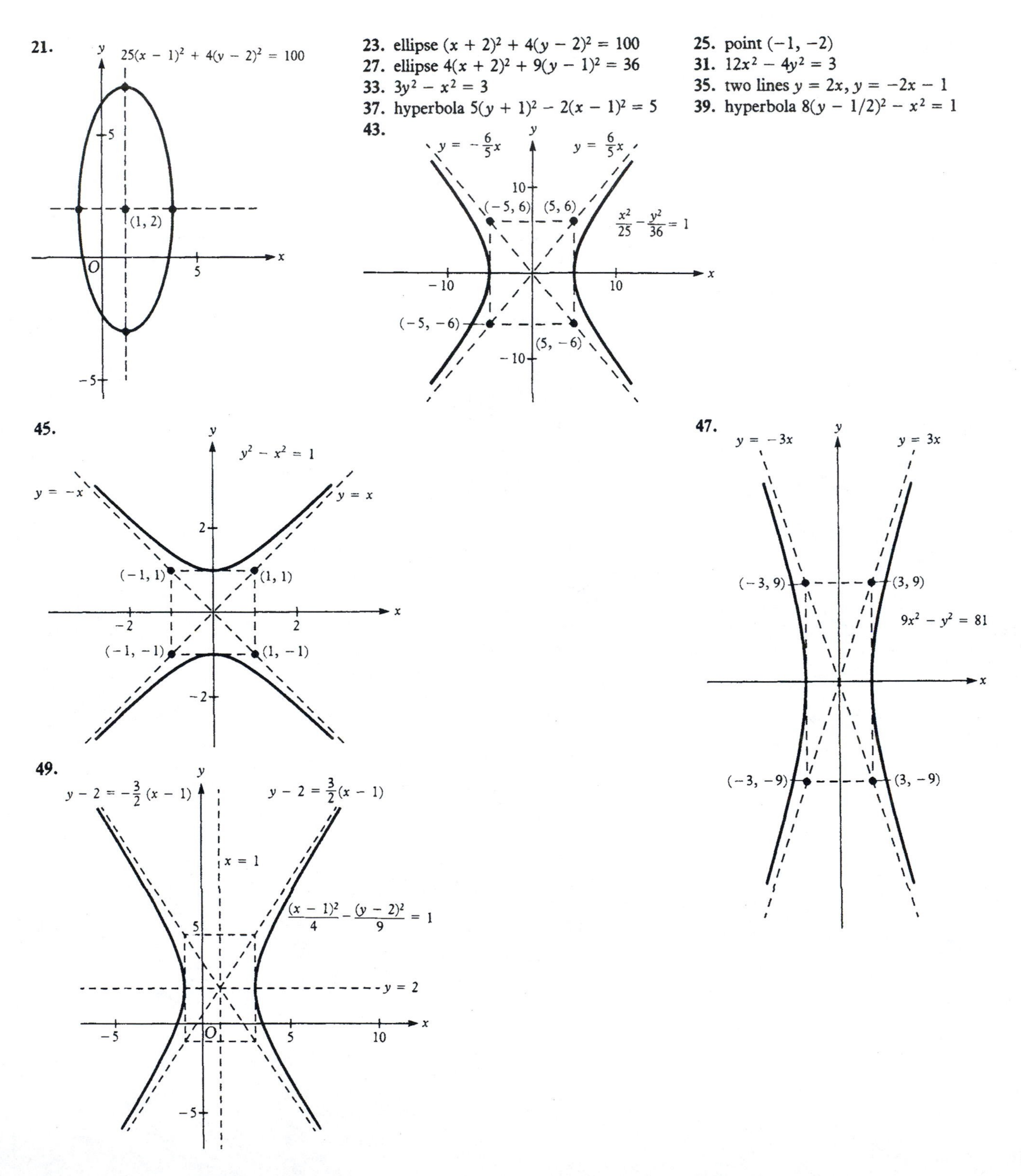

Chapter 3 Review Exercises (page 245)

1. $-5, 10$ **2.** $-2/3, 4$ **3.** -3 **4.** $(-1 \pm \sqrt{5}i)/3$ **5.** $(-1 \pm \sqrt{7})/3$ **6.** $(5 \pm \sqrt{23}i)/12$
7. $8(x + 3/4)(x - 1/2)$ **8.** $x^2 - 6x + 11$ has no real factors because its discriminant is equal to -8, which is less than 0.
9. $12x^2 + x - 6 = 0$ **10.** $\pm\sqrt{3}$ **11.** 4 **12.** $4, -4/5$ **13.** $8 - i$

14.

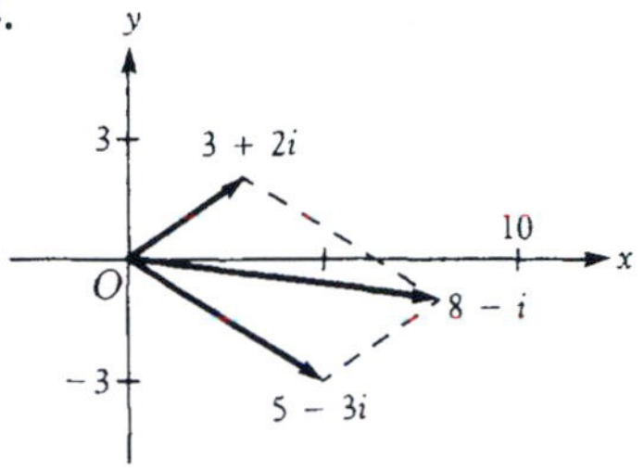

15. $11 + 2i, 1 + 2i$ **16.** $(-\infty, 1] \cup [3, \infty)$ **17.** $(-\infty, -4) \cup (2/3, \infty)$
18. $5/2$ **19.** $\varnothing$

20.

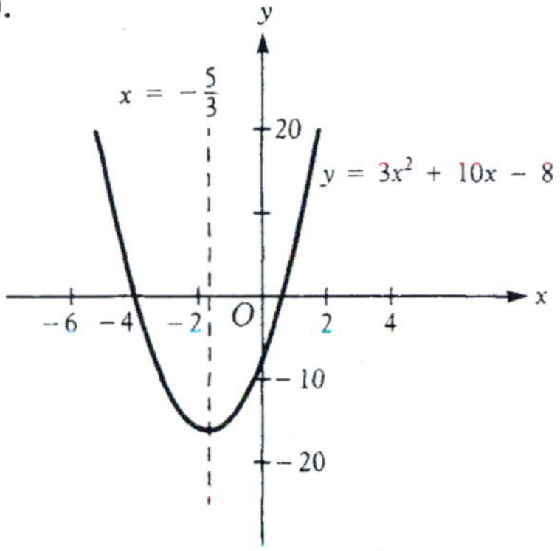

21.

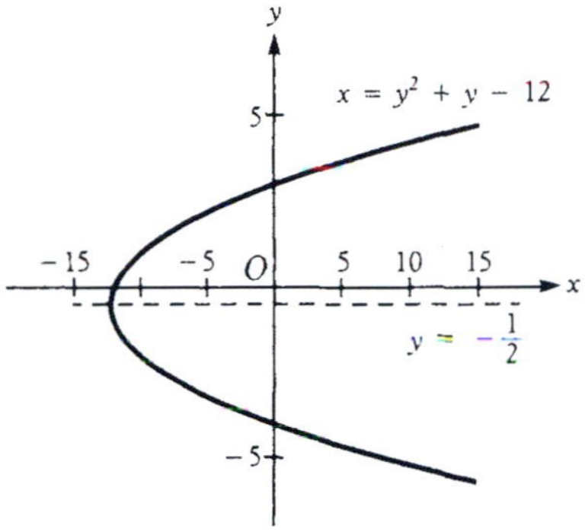

22. $(x - 4)^2 + (y - 2)^2 = 37$

23.

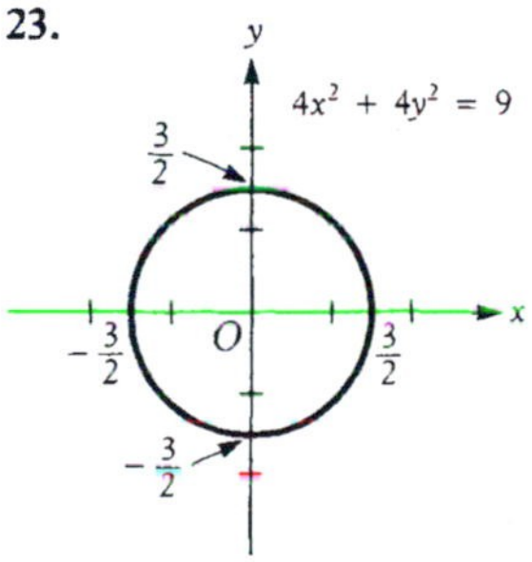

24.

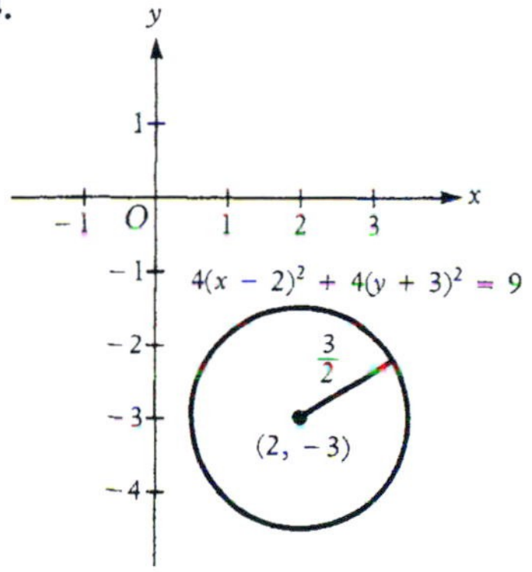

25.

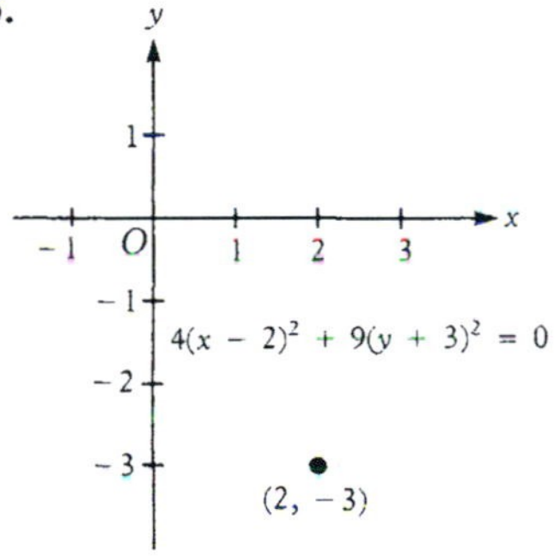

26.

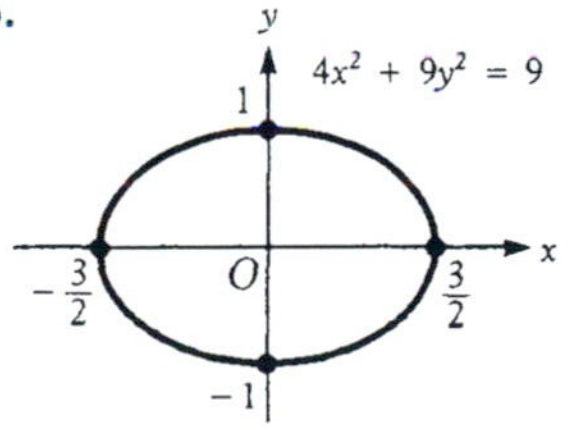

27.

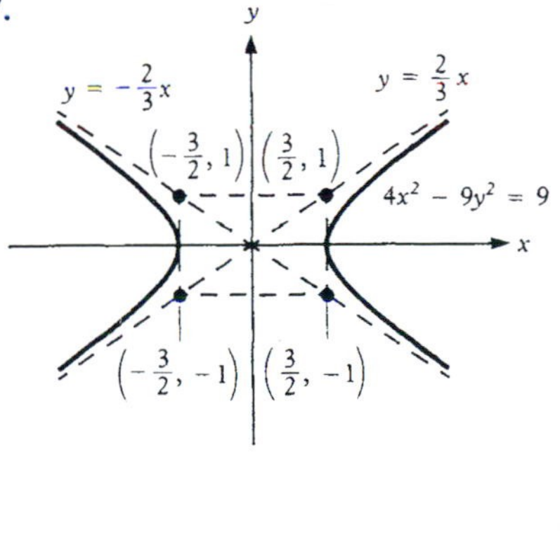

28.

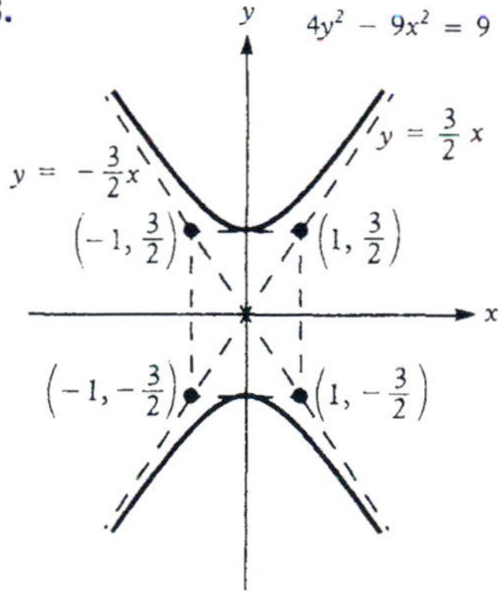

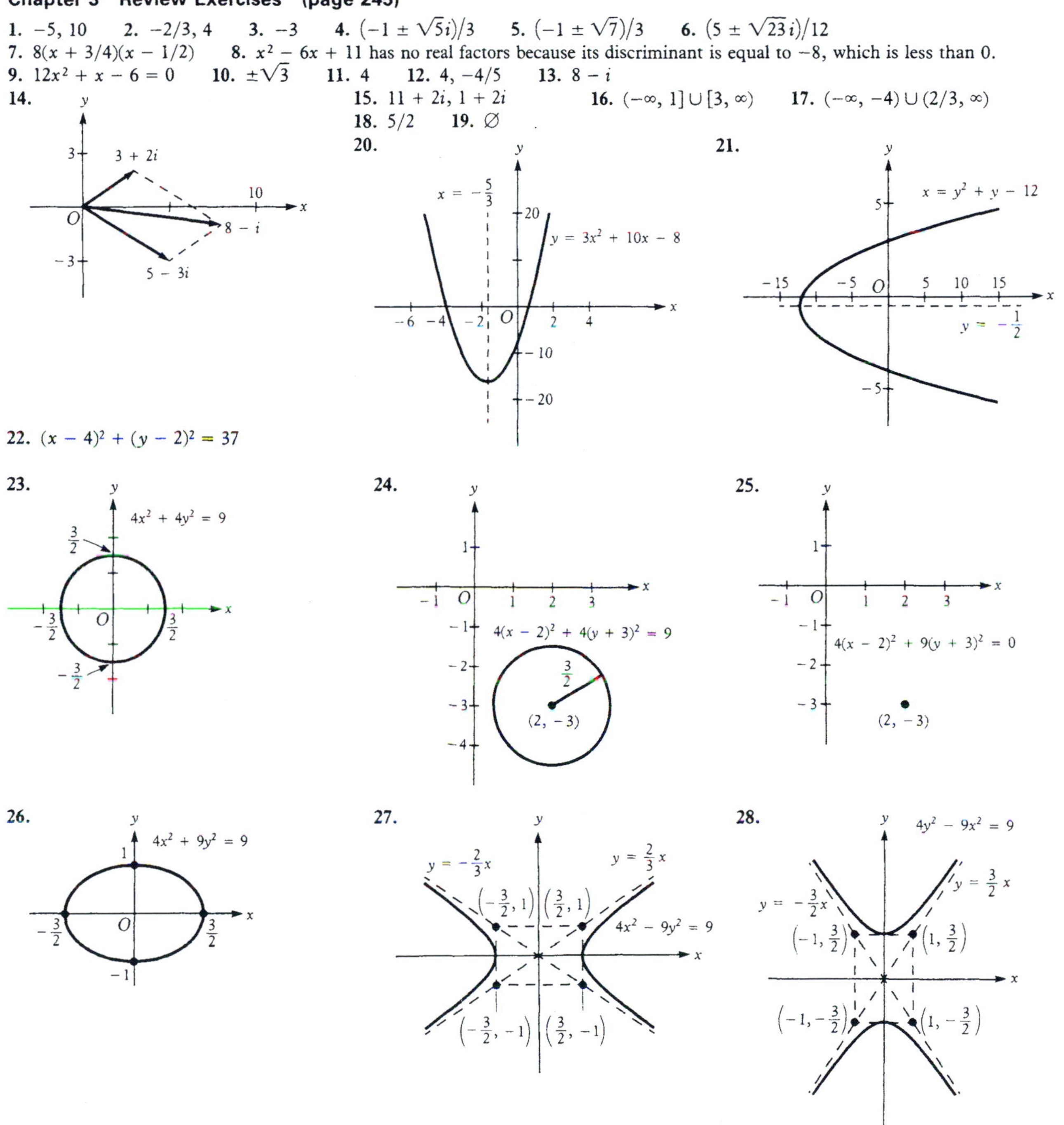

29. center: (1, 6); circle **30.** center: (−5, 4); hyperbola **31.** $(x')^2 + (y')^2 = 5$ **32.** $4(x')^2 - (y')^2 = 35$
33. $y' = 2(x')^2$ **34.** (5, 0), (−4, −3) **35.** (3, 0), (−1/2, −7/4) **36.**
37. 3 (3% or .03) **38.** Dick: 8 mph; Frank: 7.5 mph
39. Janet: 15 min; Dan: 10 min **40.** .5 (.5% or .005)

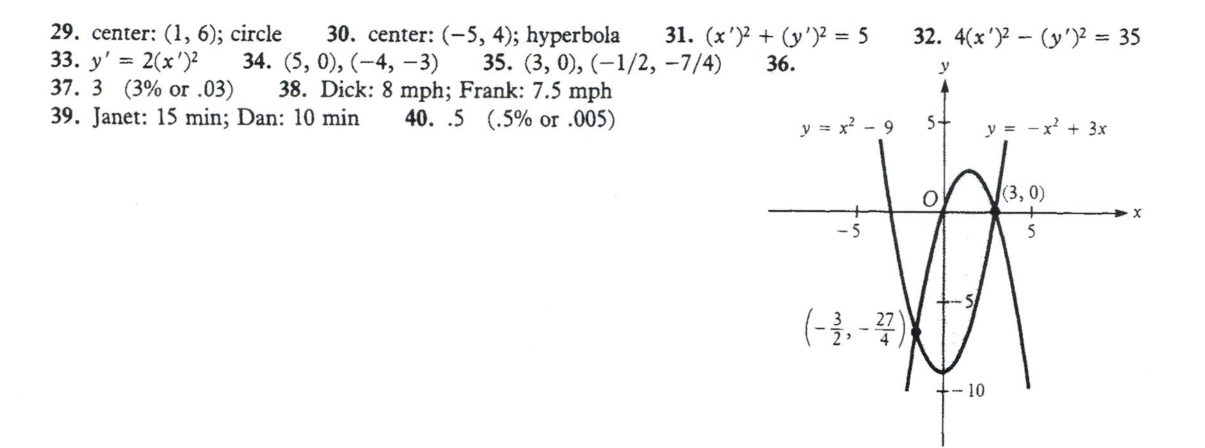

Chapter 4

Section 4.1 (page 255)

1. exactly one element **2.** domain **3.** 8, −8, range **4.** in both D_f and D_g **5.** f, $f(x)$ **6.** false **7.** false
8. true **9.** false **10.** false **11.** (a) 1 (b) 9 (c) 9 (d) 5 (e) 5 (f) $2\pi^2 + 1$ **13.** (a) 1 (b) 1 (c) 1 (d) −1
(e) −1 (f) 1 **15.** $(-\infty, \infty)$ **17.** $[5/2, \infty)$ **19.** $(-\infty, 5/2) \cup (5/2, \infty)$ **21.** $(-\infty, -1] \cup [1, \infty)$ **23.** $(-\infty, 0) \cup (0, \infty)$
25. $(-\infty, 1] \cup [2, \infty)$ **27.** ±2 **29.** 2 **31.** ∅ **33.** (a) $\frac{2x - 1}{2x + 1}$ (b) $\frac{x^2 - 1}{x^2 + 1}$ (c) $\frac{x}{x + 2}$
(d) $\frac{1 - x}{1 + x}$ if $x \neq 0$, undefined if $x = 0$ (e) $\frac{x + h - 1}{x + h + 1}$ (f) $\frac{x - h - 1}{x - h + 1}$ **35.** not a function of x **37.** $f(x) = \sqrt[3]{x}$
39. $f(x) = x^2/(x - 1)$
41. the graph of a function of x

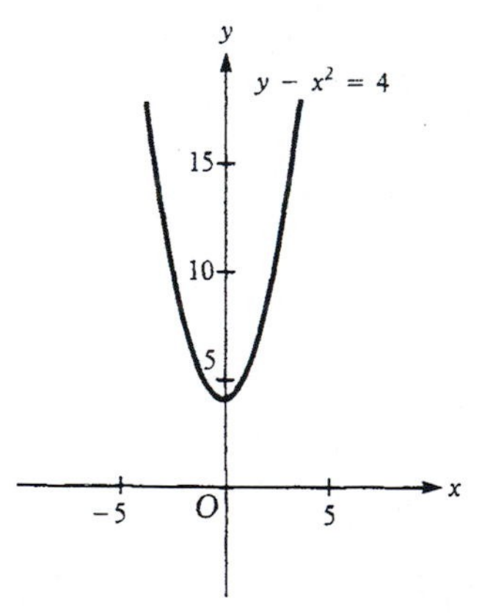

43. not the graph of a function of x

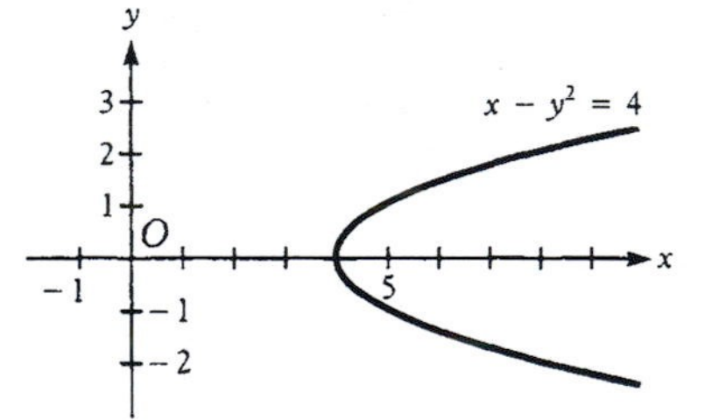

45. not the graph of a function of x **47.** the graph of a function of x
49. $(f + g)(x) = 5x - 7$, $(f - g)(x) = -x + 7$, $(fg)(x) = 6x^2 - 14x$, $(f/g)(x) = 2x/(3x - 7)$; $D_{f+g} = D_{f-g} = D_{fg} = (-\infty, \infty)$, $D_{f/g} = (-\infty, 7/3) \cup (7/3, \infty)$
51. $(f + g)(x) = x^2 + \sqrt{x}$, $(f - g)(x) = x^2 - \sqrt{x}$, $(fg)(x) = x^2\sqrt{x}$, $(f/g)(x) = x^2/\sqrt{x}$; $D_{f+g} = D_{f-g} = D_{fg} = [0, \infty)$, $D_{f/g} = (0, \infty)$
53. $(f + g)(x) = x^3 + 1/x^3$, $(f - g)(x) = x^3 - 1/x^3$, $(fg)(x) = 1$ if $x \neq 0$, undefined if $x = 0$, $(f/g)(x) = x^6$ if $x \neq 0$, undefined if $x = 0$; $D_{f+g} = D_{f-g} = D_{fg} = D_{f/g} = (-\infty, 0) \cup (0, \infty)$
55. $(f + g)(x) = x/(x^2 + 1) + x^2 + 1$, $(f - g)(x) = x/(x^2 + 1) - (x^2 + 1)$, $(fg)(x) = x$, $(f/g)(x) = x/(x^2 + 1)^2$; $D_{f+g} = D_{f-g} = D_{fg} = D_{f/g} = (-\infty, \infty)$

57. $(g \circ f)(x) = 4x^2$, $D_{g\circ f} = (-\infty, \infty)$ **59.** $(g \circ f)(x) = x$, $D_{g\circ f} = (-\infty, \infty)$ **61.** $(g \circ f)(x) = |x|$, $D_{g\circ f} = (-\infty, \infty)$
63. $(g \circ f)(x) = \sqrt{x+1}$, $D_{g\circ f} = [-1, \infty)$
65. $(g \circ f)(x) = 1/(x^2 - 4)$, $D_{g\circ f} = (-\infty, -2) \cup (-2, 2) \cup (2, \infty)$; $(f \circ g)(x) = 1/x^2 - 4$, $D_{f\circ g} = (-\infty, 0) \cup (0, \infty)$
69. $A(r) = \pi r^2$ **71.** $V(s) = s^3$ **73.** $C(x) = 5000 + 1500x$ **75.** $S(x) = 10x^2$ **77.** $L(x) = 2x + 10{,}000/x$
79. **(a)** 80 ft/sec **(b)** $(32 + 16\Delta t)$ ft/sec

Section 4.2 (page 265)

1. $a_4 \neq 0$ **2.** the set of all real numbers **3.** even, symmetric with respect to the y-axis **4.** 0, −2, 3 **5.** 4 **6.** true
7. false **8.** true **9.** true **10.** false **11.** **(a)** and **(b)** are polynomial functions.
13. The graphs of **(a)**, **(c)**, **(d)**, and **(e)** are symmetric with respect to the y-axis. **15.** **(a)**, **(c)**, and **(d)** are odd functions.
17. **(a)** $x = -1, y = 1$ **(b)** $x = 1, y = 1$ **(c)** no x-intercept, $y = 7$ **(d)** no x-intercept, $y = -1$ **(e)** $x = \pm 2, y = -4$
(f) no x-intercept, $y = 4$

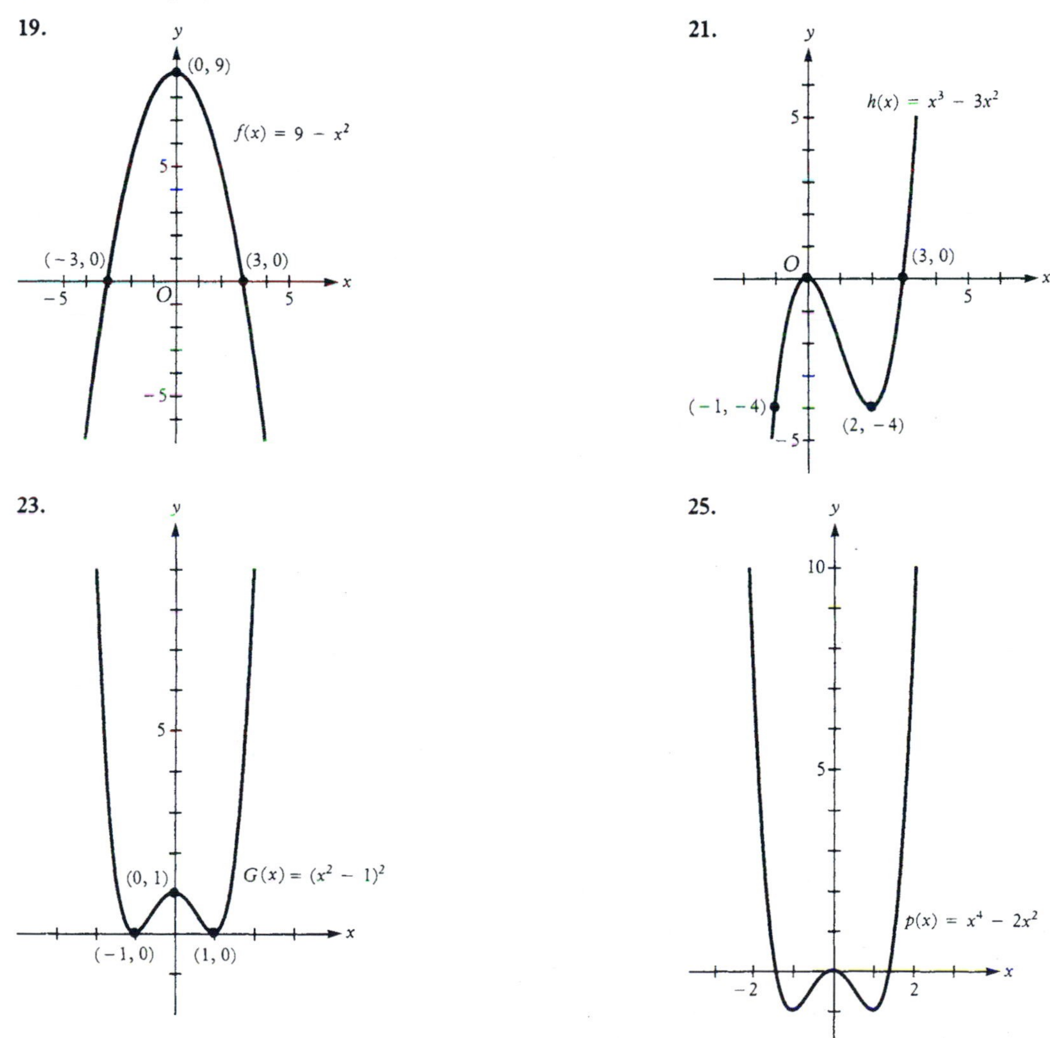

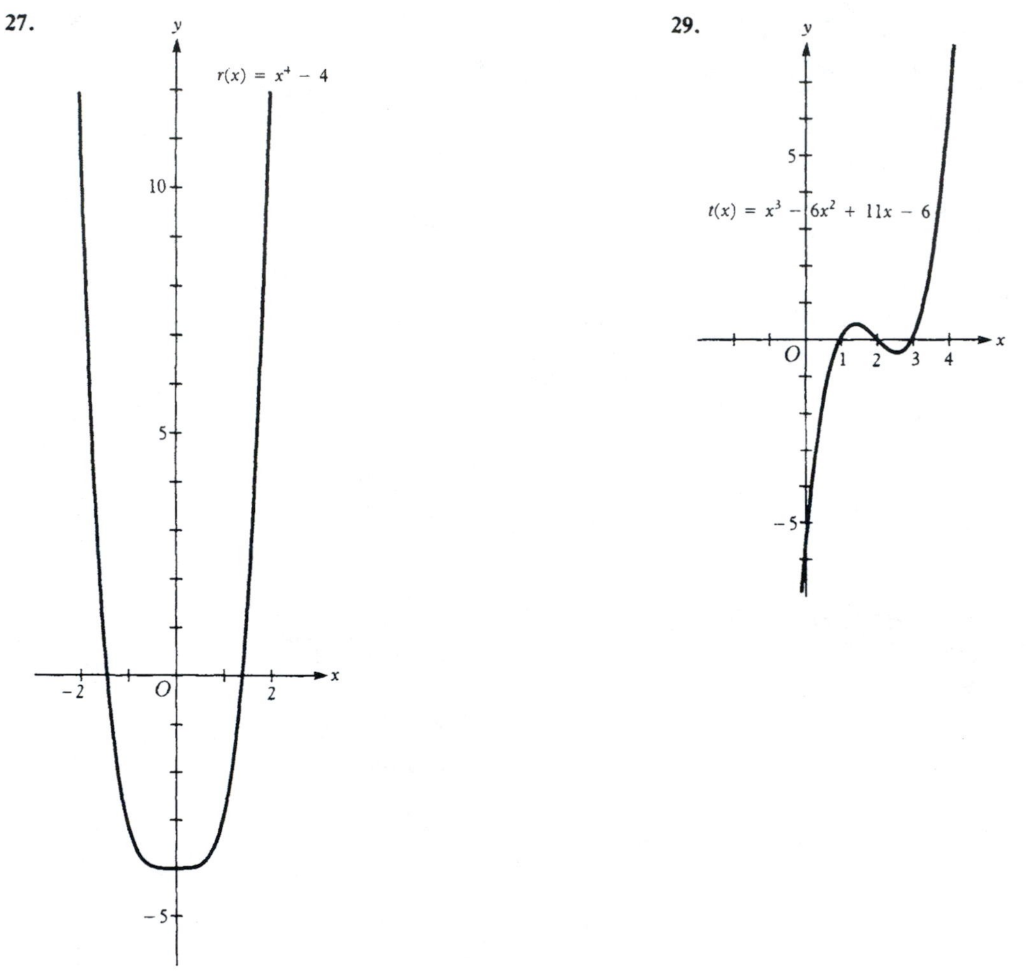

Section 4.3 (page 276)

1. two polynomial functions **2.** $Q(a) = 0$ **3.** 3/2 **4.** the degree of the numerator is less than that of the denominator **5.** the degree of the numerator is greater than that of the denominator **6.** true **7.** true **8.** false **9.** false **10.** true **11.** (a), (b), (c), (d), and (f) are rational functions. **13.** $x = -1$ **15.** $x = 1, x = -1$ **17.** $x = -1, x = -3$ **19.** $x = 1$ **21.** $y = 2$ **23.** $y = -3/2$ **25.** no horizontal asymptote **27.** no horizontal asymptote

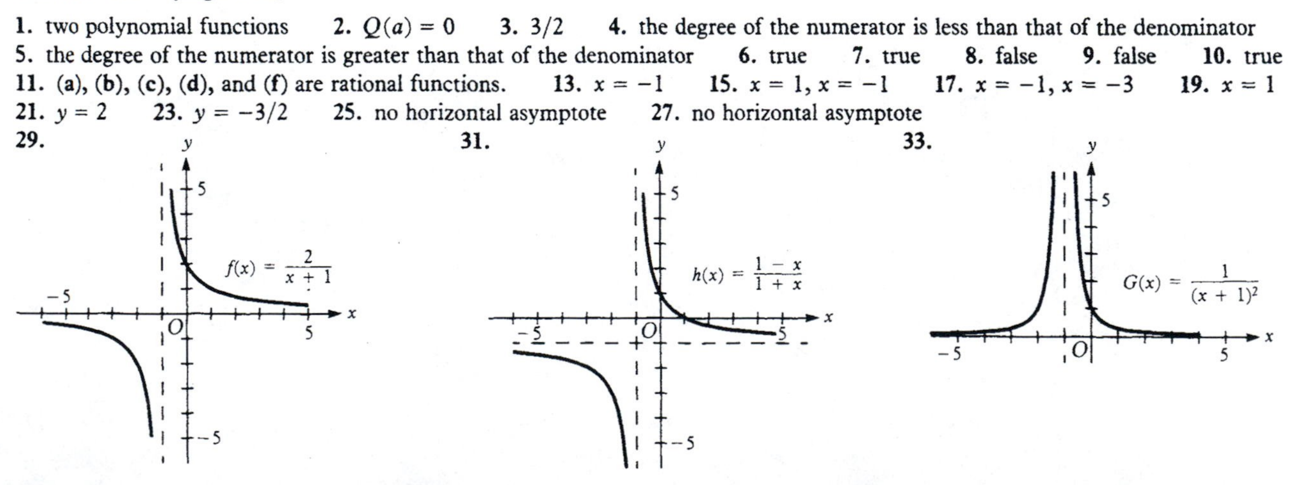

35. **37.** **39.**

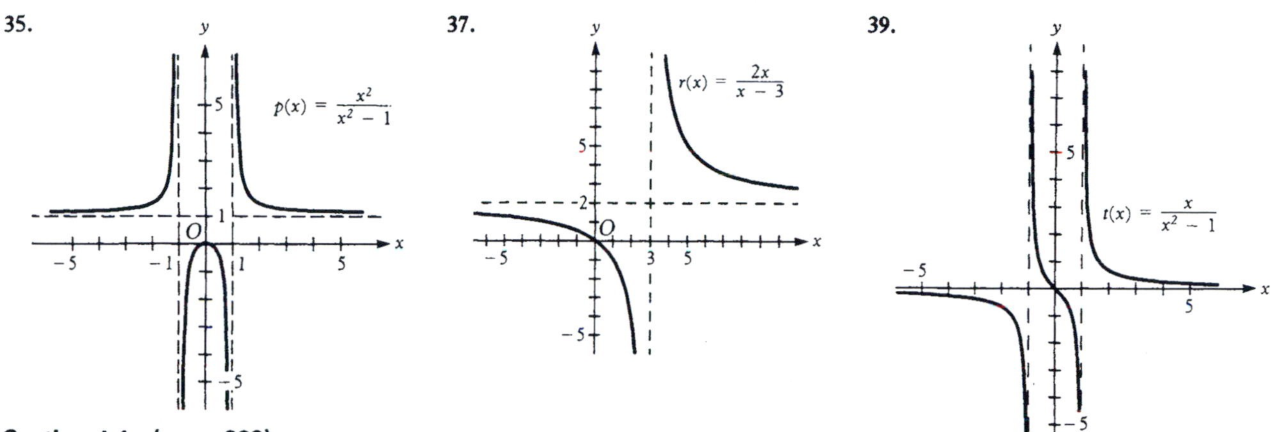

Section 4.4 (page 283)

1. addition, subtraction, multiplication, division, extraction of roots **2.** the radicand is negative **3.** $(-\infty, \infty)$
4. $\geq 1, \leq 1$ **5.** algebraic, polynomial **6.** false **7.** true **8.** true **9.** true **10.** false
11. (b), (c), (e), and (f) are not rational functions. **13.** **(a)** $(-\infty, 1) \cup (1, \infty)$ **(b)** $(-\infty, 1) \cup (1, \infty)$ **(c)** $(-\infty, -2] \cup (1, \infty)$
(d) $(-\infty, 1) \cup (1, \infty)$ **(e)** $(-\infty, -1] \cup [1, \infty)$ **(f)** $(-\infty, -2) \cup [-1, 1] \cup (2, \infty)$

15.

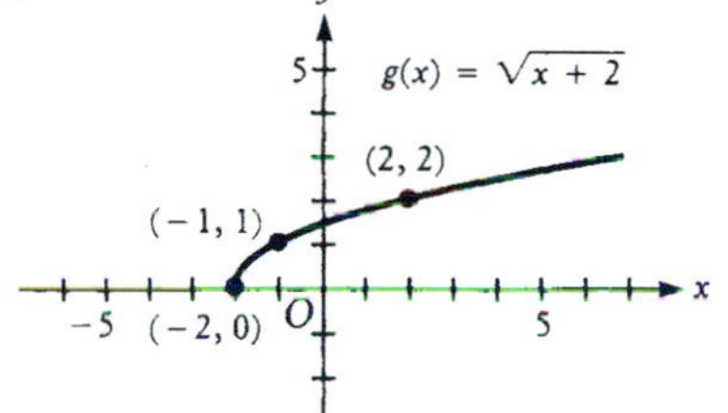

17.

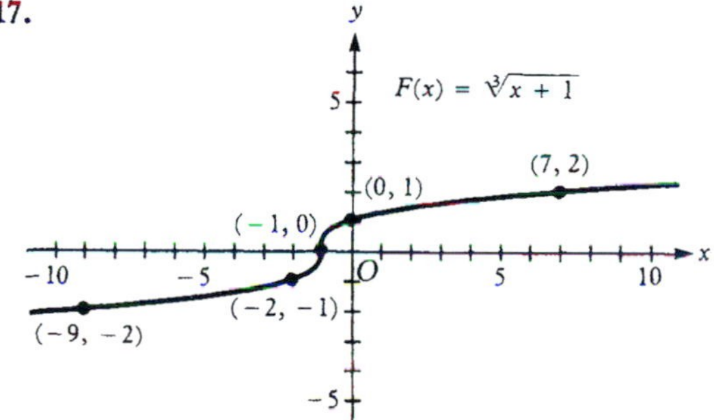

19.

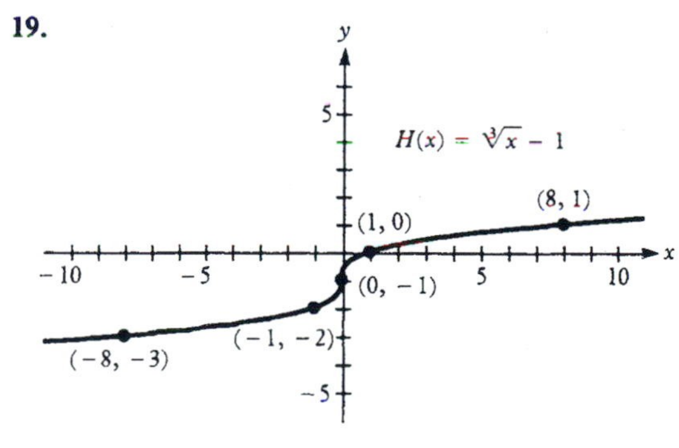

21.

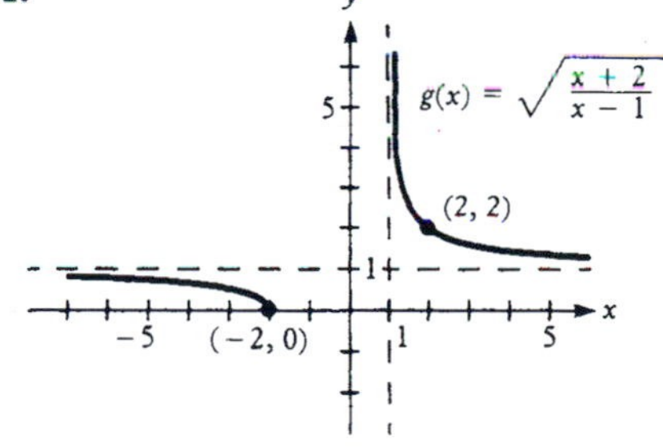

23.

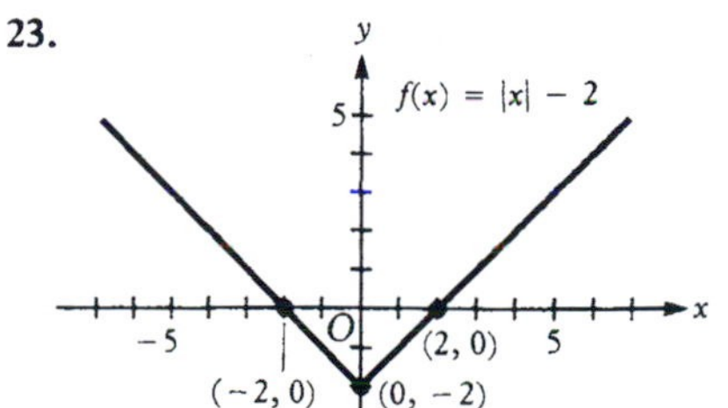

25.

$F(x) = |x - 2|$ (0, 2) (4, 2) (2, 0)

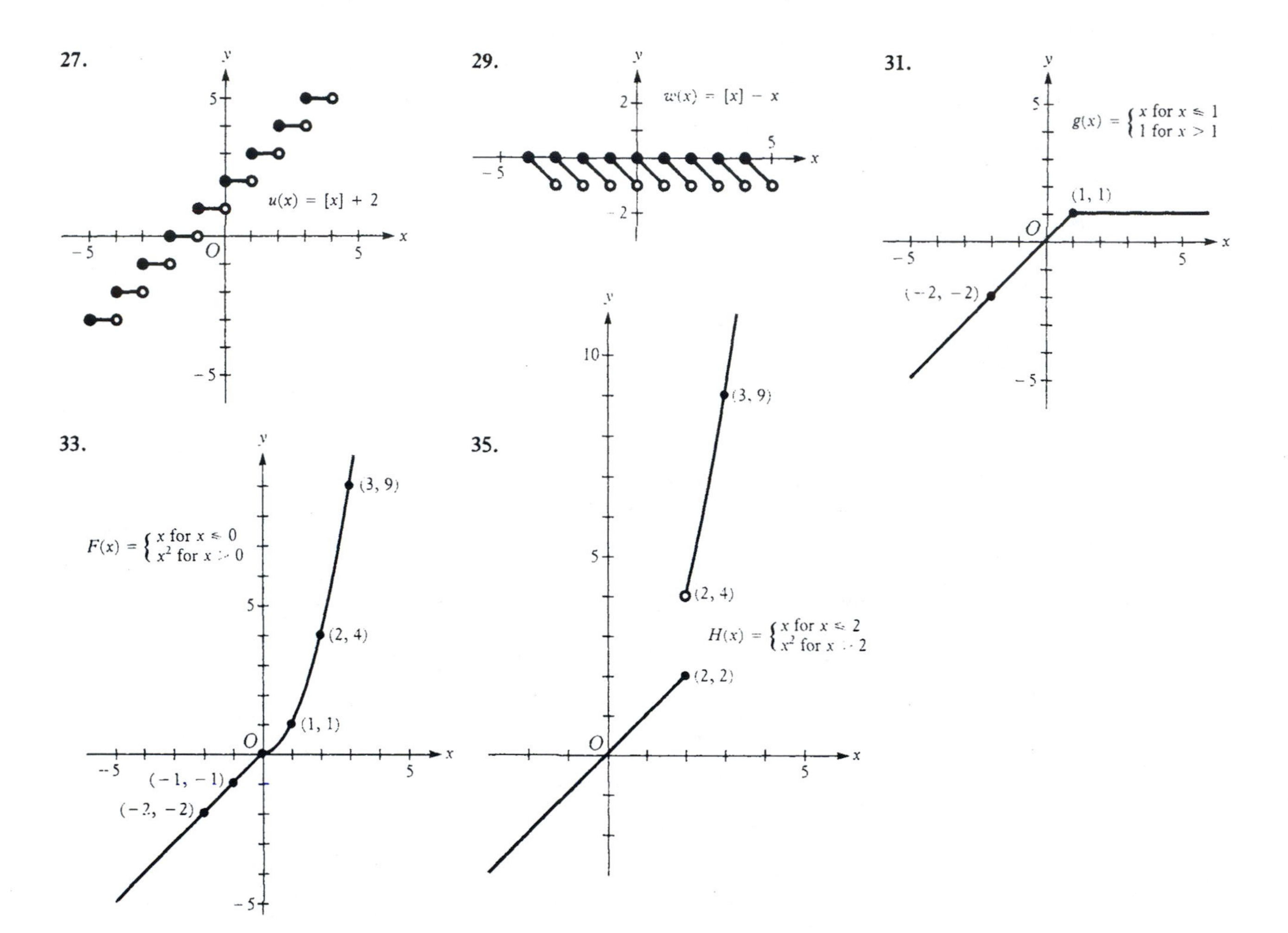

Section 4.5 (page 294)

1. x, y **2.** y, x **3.** at most one point **4.** $(g \circ f)(x)$, $(f \circ g)(x)$ **5.** the range, the domain **6.** false
7. true **8.** true **9.** false **10.** true **11.** (a), (b), (e), and (f) are one-to-one.
13. (a), (b), and (d) are graphs of one-to-one functions. **15.** $f^{-1}(x) = \sqrt[3]{x+1}$; $D_{f^{-1}} = R_{f^{-1}} = (-\infty, \infty)$
17. $f^{-1}(x) = x^2 - 2$; $D_{f^{-1}} = [0, \infty)$; $R_{f^{-1}} = [-2, \infty)$ **19.** $f^{-1}(x) = (x-2)^3$; $D_{f^{-1}} = R_{f^{-1}} = (-\infty, \infty)$
21.

(Exercise 15) $f(x) = x^3 - 1$; $f^{-1}(x) = \sqrt[3]{x+1}$

(Exercise 17) $f^{-1}(x) = x^2 - 2$; $f(x) = \sqrt{x+2}$; (2, 2); (−2, 0); (0, −2)

(Exercise 19) $f^{-1}(x) = (x-2)^3$; $f(x) = \sqrt[3]{x} + 2$

23. $g^{-1}(x) = -\sqrt{x}$; $D_{g^{-1}} = [1, \infty)$; $R_{g^{-1}} = (-\infty, -1]$ 25. $g^{-1}(x) = -\sqrt{x+1}$; $D_{g^{-1}} = [0, \infty)$; $R_{g^{-1}} = (-\infty, -1]$
27. For Exercise 23, $(g^{-1} \circ g)(x) = g^{-1}(g(x)) = g^{-1}(x^2) = -\sqrt{x^2} = -|x| = -(-x) = x$ (since x is in D_g, x is negative and $|x| = -x$); $(g \circ g^{-1})(x) = g(g^{-1}(x)) = g(-\sqrt{x}) = (-\sqrt{x})^2 = x$ (since x is in $D_{g^{-1}}$, x is positive).
For Exercise 25, $(g^{-1} \circ g)(x) = g^{-1}(g(x)) = g^{-1}(x^2 - 1) = -\sqrt{x^2 - 1 + 1} = -\sqrt{x^2} = -|x| = -(-x) = x$ (since x is in D_g, x is negative and $|x| = -x$); $(g \circ g^{-1})(x) = g(g^{-1}(x)) = g(-\sqrt{x+1}) = (-\sqrt{x+1})^2 - 1 = x + 1 - 1 = x$ (since x is in $D_{g^{-1}}$, $x + 1$ is positive).

Section 4.6 (page 299)

1. a nonzero constant 2. inversely, x^2 3. directly, inversely 4. jointly as the base and the height, constant of variation
5. $v = kx^2y^3/z$ $(k \neq 0)$ 6. true 7. false 8. true 9. true 10. false 11. $d = kt$ 13. $I = kR$
15. $T = k\sqrt{L}$ 17. The force exerted by flowing water varies directly as the sixth power of its velocity.
19. The time to travel between two cities varies inversely as the average rate of speed. 21. 108 23. \$112
25. 59.16 mph 31. $z = kxy^3$ 33. $I = kV/R$ 35. $P = kn/\sqrt{c}$
37. h varies directly as V and inversely as the square of r. 39. W varies jointly with w and the square of d and inversely with l.
41. 37 1/7 cubic inches 43. 3 2/3 ohms

Chapter 4 Review Exercises (page 302)

1. (a) 30 (b) 48 (c) 4 2. (a) $(6x + 4)/(4x^2 + 5)$ (b) $(3\sqrt{x} + 4)/(x + 5)$ (c) $(3x + 1)/(x^2 - 2x + 6)$
3. (a) $D_f = R_f = (-\infty, \infty)$ (b) $D_g = (-\infty, \infty)$, $R_g = [1, \infty)$ (c) $D_h = (-\infty, \infty)$, $R_h = [0, \infty)$ 4. (a) $(-\infty, 4/3) \cup (4/3, \infty)$
(b) $(4/3, \infty)$ (c) $(-\infty, -2) \cup (-2, 2) \cup (2, \infty)$ 5. (a) $x^2 + 2 + \sqrt{x - 2}$ (b) $(x^2 + 2)\sqrt{x - 2}$ (c) $(x^2 + 2)/\sqrt{x - 2}$
6. (a) $27 + \sqrt{3}$ (b) $(4x^2 + 2)\sqrt{2x - 2}$ (c) $(x^4 + 2)/\sqrt{x^2 - 2}$ 7. $D_{f+g} = [2, \infty)$, $D_{fg} = [2, \infty)$, $D_{f/g} = (2, \infty)$
8. (a) $\sqrt{1 + x}$ (b) $1 + \sqrt{x}$ 9. (a) $\sqrt{6}$ (b) $1 + \sqrt{2x}$ 10. $D_{g \circ f} = [-1, \infty)$, $D_{f \circ g} = [0, \infty)$
11. (a) and (b) are polynomial functions. 12. (b) and (c) are rational functions. 13. (a), (b), and (c) are even functions.
14. (b) and (c) are odd functions. 15. (a), (b), and (c) have graphs that are symmetric with respect to the y-axis.
16. (b) and (c) have graphs that are symmetric with respect to the origin. 17. (a) $x = 5/2$ (b) $x = 2$, $x = -2$
(c) $x = 1$, $x = 3$ 18. (a) $y = 1/2$ (b) $y = 0$ (c) no horizontal asymptote
19. vertical asymptote $x = 2$, horizontal asymptotes $y = 1$ and $y = -1$

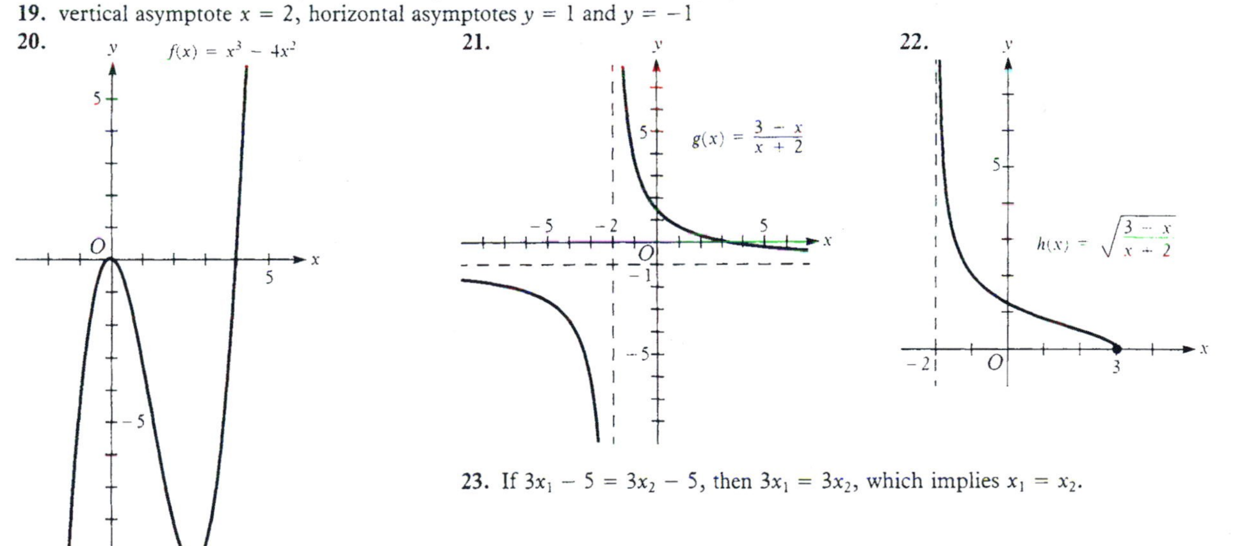

23. If $3x_1 - 5 = 3x_2 - 5$, then $3x_1 = 3x_2$, which implies $x_1 = x_2$.

24. The graph of $g(x) = x^3 + 1$ is intersected in exactly one point by any horizontal line $y = c$.

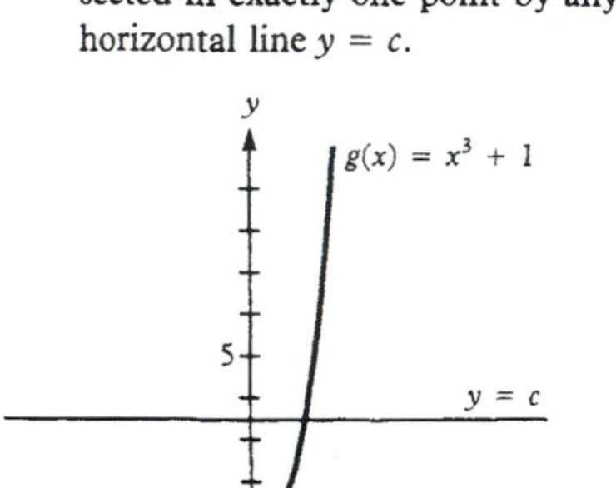

25. The x-axis intersects the graph in the two points (0, 0) and (4, 0).

26.

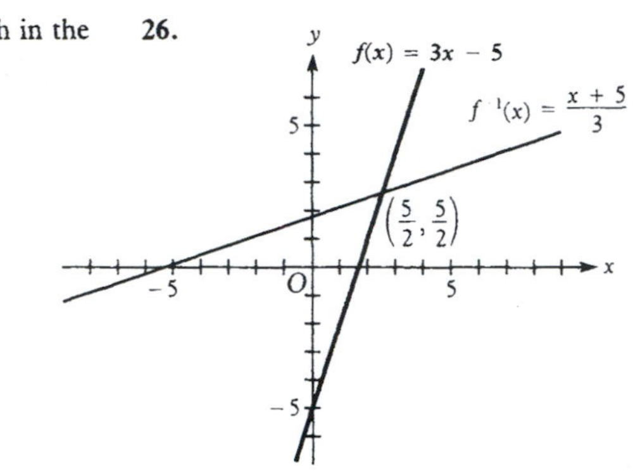

27.

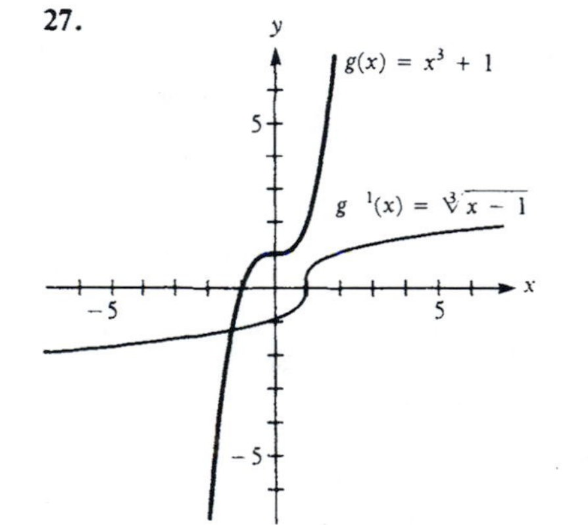

28. (c) is decreasing **29.** (c) is one-to-one

30. Suppose $x_1 > x_2$; then $3x_1 > 3x_2$ and $3x_1 + 5 > 3x_2 + 5$. Therefore, $f(x_1) > f(x_2)$.

31. Suppose $x_1 > x_2$; then $-3x_1 < -3x_2$ and $-3x_1 + 5 < -3x_2 + 5$. Therefore, $g(x_1) < g(x_2)$.

32.

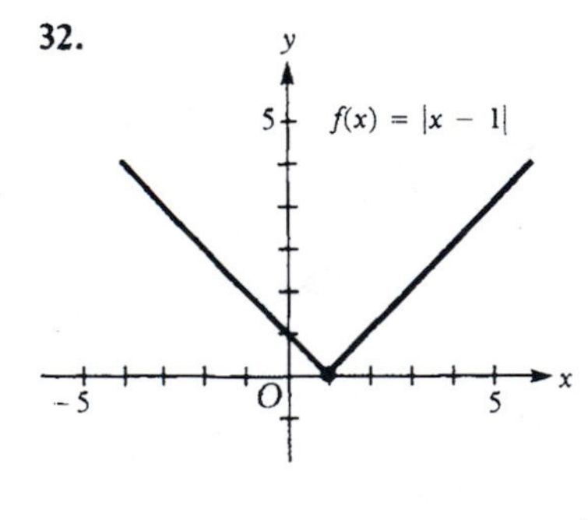

33.

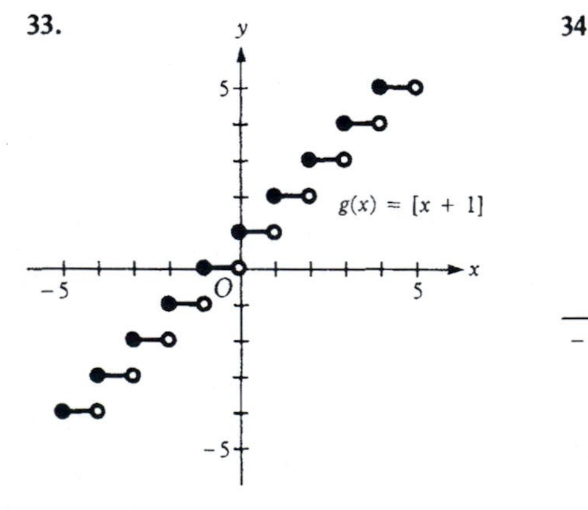

34.

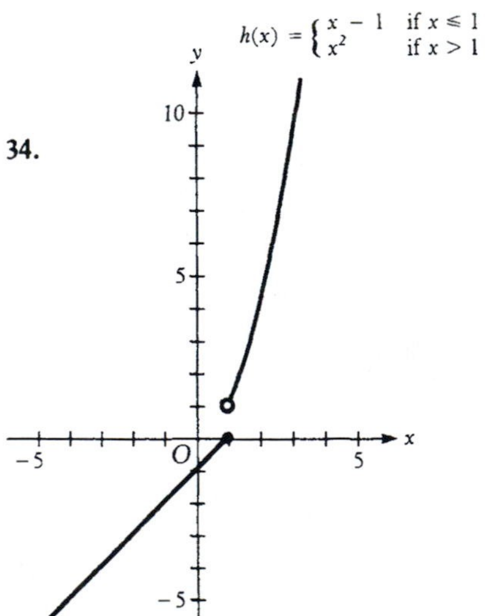

35. $D_g = (-\infty, \infty)$, R_g = the set of all integers **36.** $D_h = (-\infty, \infty)$, $R_h = (-\infty, 0] \cup (1, \infty)$ **37.** $y = kuv/w$
38. L varies jointly as w and the square of d and inversely as r. **39.** 80 **40.** about 1.1×10^{-12}

Chapter 5

Section 5.1 (page 312)

1. positive **2.** $(-\infty, \infty)$, $(0, \infty)$ **3.** > 1, < 1 (and > 0) **4.** $1/a$ **5.** one-to-one **6.** true **7.** true **8.** false
9. true **10.** true

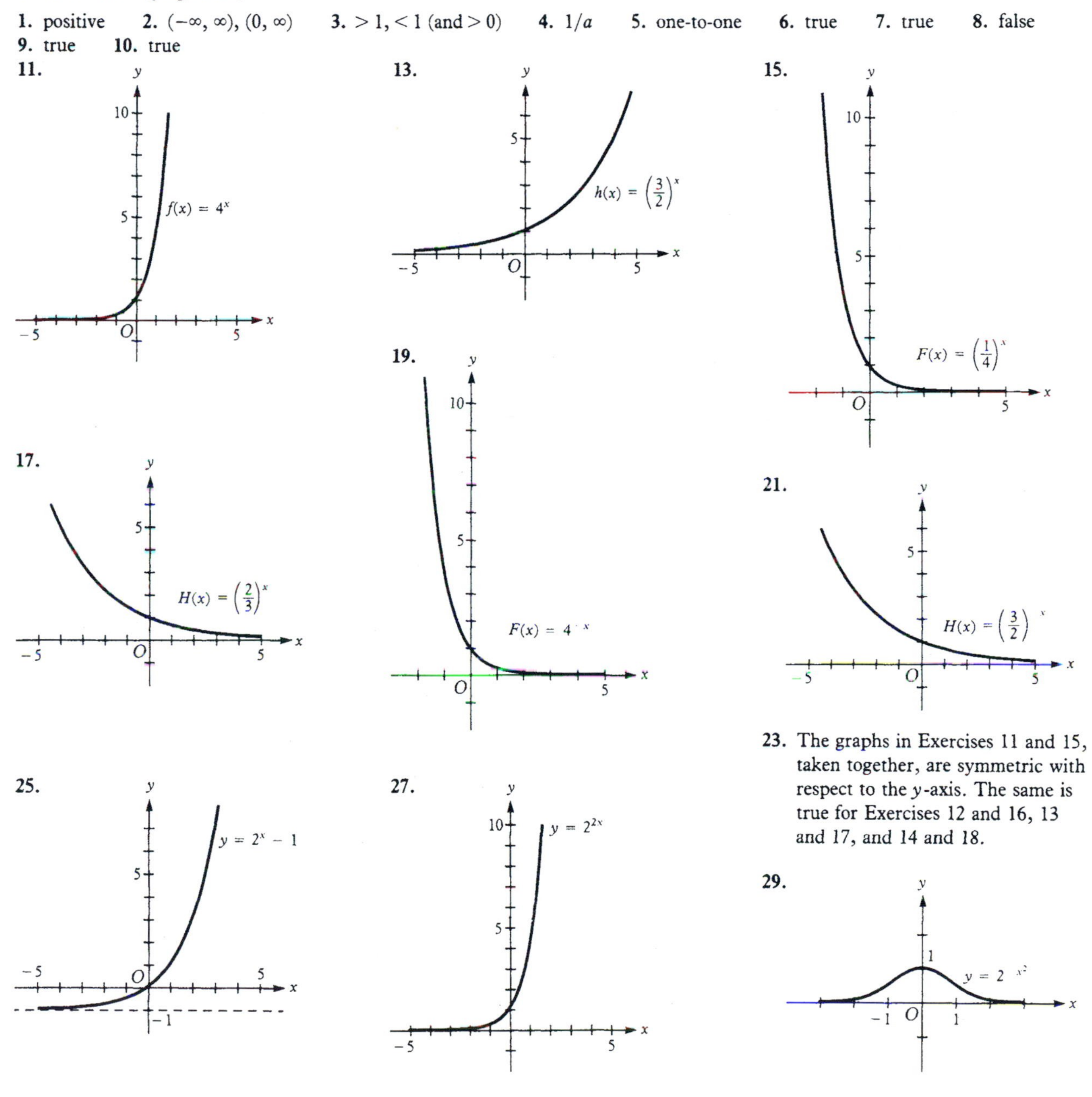

23. The graphs in Exercises 11 and 15, taken together, are symmetric with respect to the y-axis. The same is true for Exercises 12 and 16, 13 and 17, and 14 and 18.

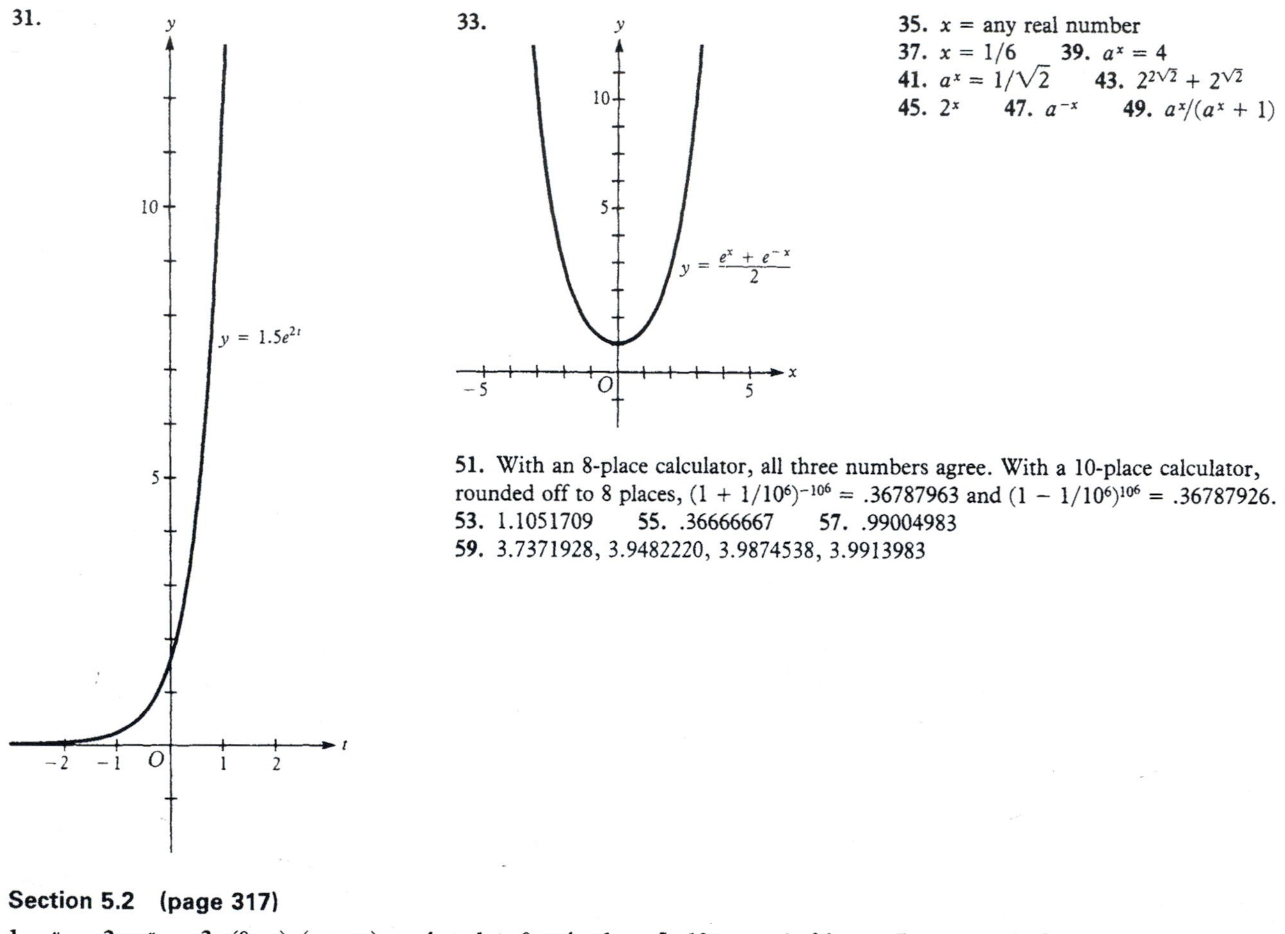

35. x = any real number
37. $x = 1/6$ **39.** $a^x = 4$
41. $a^x = 1/\sqrt{2}$ **43.** $2^{2\sqrt{2}} + 2^{\sqrt{2}}$
45. 2^x **47.** a^{-x} **49.** $a^x/(a^x + 1)$

51. With an 8-place calculator, all three numbers agree. With a 10-place calculator, rounded off to 8 places, $(1 + 1/10^6)^{-10^6} = .36787963$ and $(1 - 1/10^6)^{10^6} = .36787926$.
53. 1.1051709 **55.** .36666667 **57.** .99004983
59. 3.7371928, 3.9482220, 3.9874538, 3.9913983

Section 5.2 (page 317)

1. a^y **2.** a^x **3.** $(0, \infty)$, $(-\infty, \infty)$ **4.** > 1, > 0 and < 1 **5.** 10, e **6.** false **7.** true **8.** false **9.** true
10. true **11.** $\log_2 16 = 4$ **13.** $\log_{10} \sqrt[3]{100} = 2/3$ **15.** $\log_{16} 64 = 3/2$ **17.** $10^2 = 100$ **19.** $3^{-3} = 1/27$
21. $a^0 = 1$ **23.** 4 **25.** 2 **27.** $-19/9$, $-17/9$ **29.** 100 **31.** -1000 **33.** 5 **35.** 0 **37.** -4

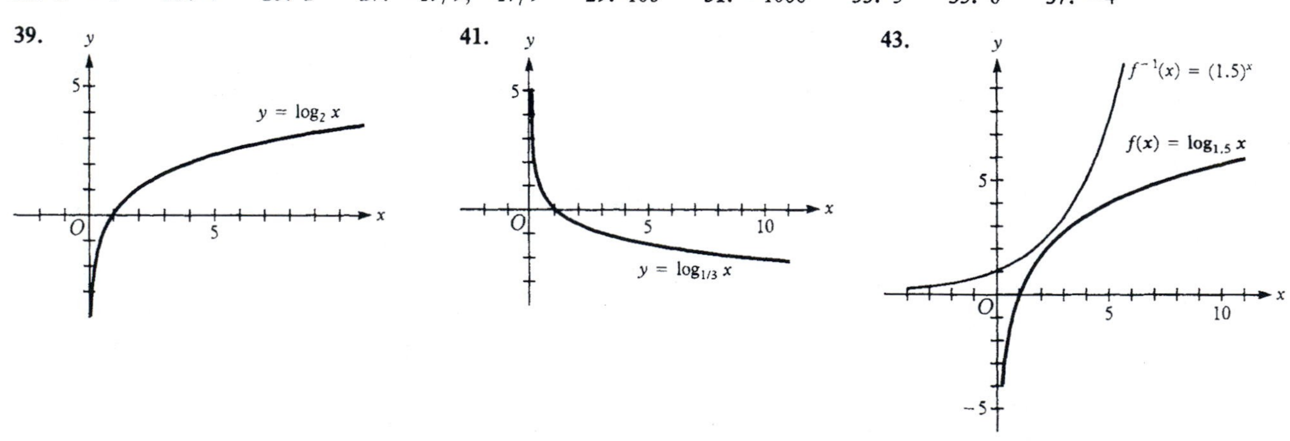

45.

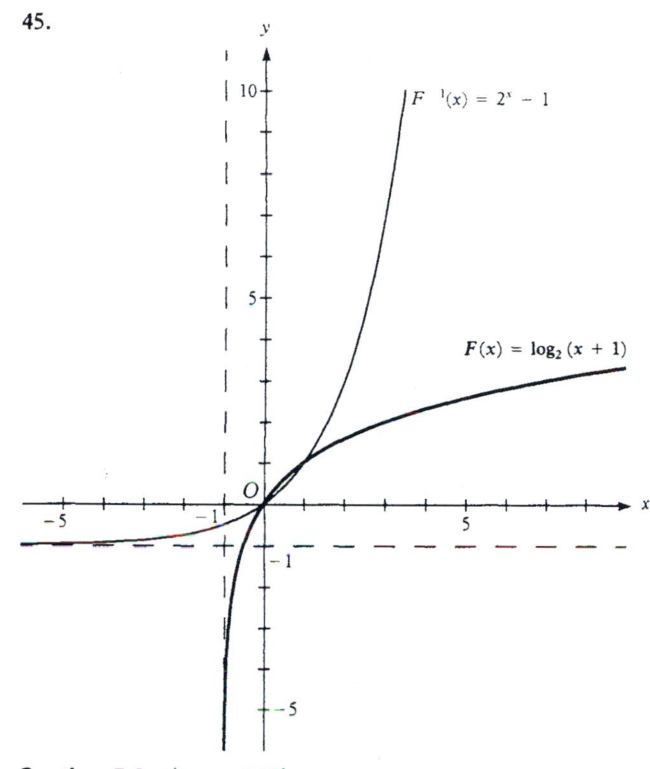

49. $\log_2 100 = (\log_{10} 100)/\log_{10} 2 \approx 6.6439$
51. $\log_6 .25 = (\log_{10} .25)/\log_{10} 6 \approx -.7737$
53. $\log_{17} 4981 = (\log_{10} 4981)/(\log_{10} 17) \approx 3.0049$
61. .795 **63.** 1.47, 7.86

Section 5.3 (page 325)

1. x **2.** x^3 **3.** 2^x **4.** $x^{3/2}$ **5.** 1, a^2 **6.** false **7.** false **8.** true **9.** true **10.** true **11.** 1.5441 **13.** −3.4559 **15.** −.6990 **17.** $\log_a[(x-1)^{1/2}/x^7]$ **19.** $\log[(x+2)/(x-2)]$ **21.** $2\log b + \log c - 4\log d$ **23.** $(1/3)\log b + (1/6)\log c - \log d$ **25.** 7 **27.** 100 **29.** 2 **31.** 2 **35.** 2.3802 **37.** 6.1673 **39.** 5.4807 **41.** 14.2211 **43.** 6.12 **45.** .612
47. $Y = \log_a y = \log_a(ca^{kx}) = \log_a c + \log_a(a^{kx}) = \log_a c + kx = kx + \log_a c = mx + b$, where $m = k$ and $b = \log_a c$
49. .470 **51.** 1.26, 14.8

Section 5.4 (page 336)

1. (a) \$141.76 (b) \$141.90 (c) \$141.91 **3.** \$1000 at 7.95% compounded daily yields \$23.13 more. **5.** yes **7.** \$318.41 **9.** 40,960 bacteria **11.** 640,000 bacteria **13.** about 287 million people **15.** about 45 days **17.** 15,601 years **21.** (a) $c \approx 1144$ deer (b) about 276 deer **23.** (a) 20 decibels (b) 100 decibels **25.** about 60.4 decibels **27.** 10^{2k+16} watts **29.** (a) approximately 50,119 joules (b) approximately 1.585×10^{12} joules **31.** $10^{2k-4.7}$ joules **35.** about 106°F **37.** (a) 6.4 (b) 2.2
39.

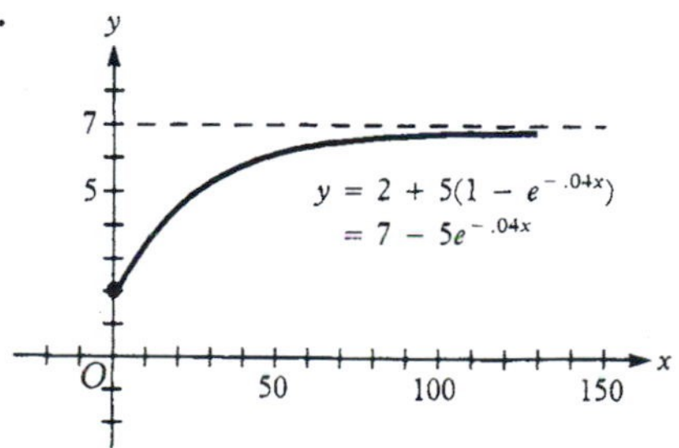

Chapter 5 Review Exercises (page 339)

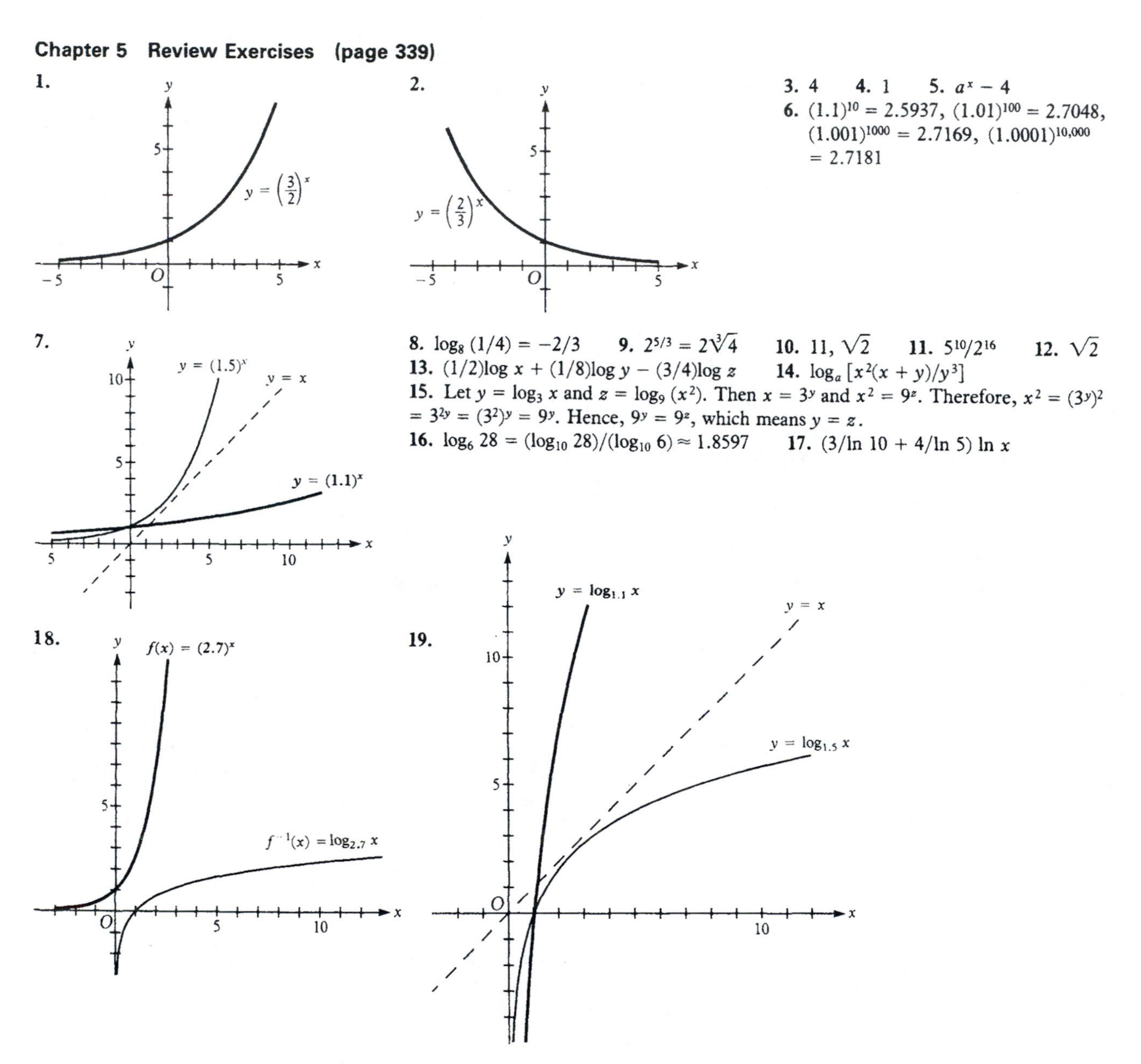

3. 4 **4.** 1 **5.** $a^x - 4$
6. $(1.1)^{10} = 2.5937$, $(1.01)^{100} = 2.7048$, $(1.001)^{1000} = 2.7169$, $(1.0001)^{10{,}000} = 2.7181$

8. $\log_8 (1/4) = -2/3$ **9.** $2^{5/3} = 2\sqrt[3]{4}$ **10.** 11, $\sqrt{2}$ **11.** $5^{10}/2^{16}$ **12.** $\sqrt{2}$
13. $(1/2)\log x + (1/8)\log y - (3/4)\log z$ **14.** $\log_a [x^2(x + y)/y^3]$
15. Let $y = \log_3 x$ and $z = \log_9 (x^2)$. Then $x = 3^y$ and $x^2 = 9^z$. Therefore, $x^2 = (3^y)^2 = 3^{2y} = (3^2)^y = 9^y$. Hence, $9^y = 9^z$, which means $y = z$.
16. $\log_6 28 = (\log_{10} 28)/(\log_{10} 6) \approx 1.8597$ **17.** $(3/\ln 10 + 4/\ln 5) \ln x$

20. $x = e^{-2}$ **21.** $\ln x = -2$ **22.** 1/8 **23.** $(\log 5)/(2 \log 3 - \log 2)$ **24.** approximately 550
25. approximately .0550 **26.** approximately 9.0 **27.** approximately 10.3189 **28.** $4 \log 5 - \frac{1}{2} \log 7 = 2.37333$
29. $\log 1.4 = \log (7/5) = \log 7 - \log 5 = .14613$ **30.** $\log_a x = [\log_{a^n} (x)]/[\log_{a^n} (a)] = [\log_{a^n} (x)]/[1/n] = n \log_{a^n} (x)$
31. $4/(5 \log 1.02) \approx 93$ **32.** approximately 872.04 **33.** $r \approx .104$ **34.** \$3483.94 **35.** \$3499.08
36. \$6376.64 **37.** about 14 years **38.** about 830 years **39.** $D_2 = 10 \log (10^{16} I_2) = 10 \log (10^{16} \cdot 10 I_1) = 10 \log (10^{16} I_1 \cdot 10) = 10[\log (10^{16} I_1) + \log 10] = 10 \log (10^{16} I_1) + 10 \log 10 = D_1 + 10$ **40.** 1.0×10^{14} joules

Chapter 6

Section 6.1 (page 346)

1. rotation **2.** 180° **3.** 2π **4.** its initial side is along the positive x-axis **5.** its terminal side lies in the third quadrant **6.** true **7.** false **8.** true **9.** true **10.** true **11.** (a) 90° (b) 1260° **13.** 45°, 405°, and −315°; 135° and −225° **15.** $\pi/6$ in I, $5\pi/6$ in II, $-7\pi/6$ in II, 4 in III

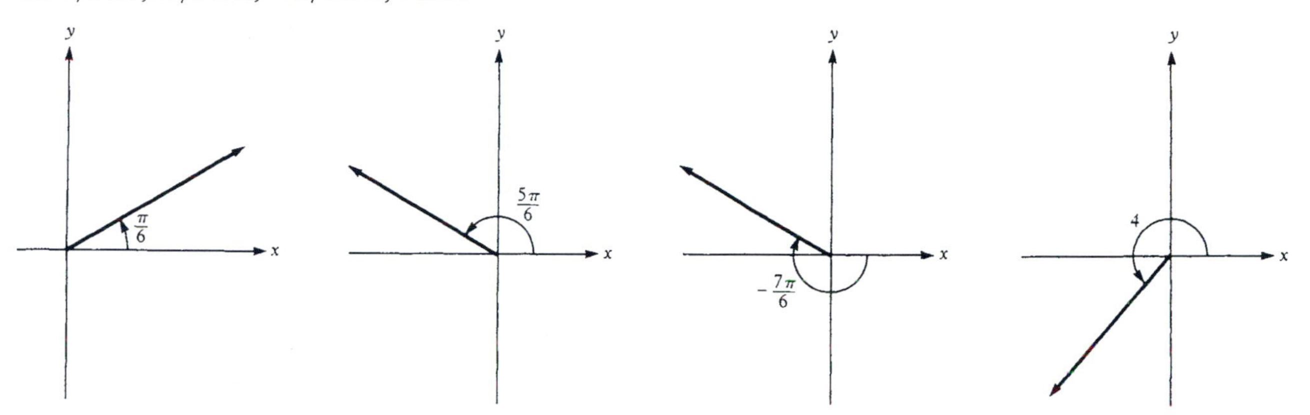

17. $\pi/4$, $3\pi/2$, $-7\pi/4$ **19.** 47.2625° **21.** (a) 1 (b) .01 **23.** about 720.3 rotations **25.** (a) about 2793 mi (b) about 4538 mi **27.** 200π radians per min **29.** 900π radians, about 707 ft **31.** $A = r^2\theta/2$

Section 6.2 (page 359)

1. standard position **2.** ratios **3.** acute **4.** isosceles, equilateral **5.** reference angle, quadrant **6.** true **7.** false **8.** false **9.** false **10.** true **11.** $\sin\theta = 4/5$, $\cos\theta = 3/5$, $\tan\theta = 4/3$, $\cot\theta = 3/4$, $\sec\theta = 5/3$, $\csc\theta = 5/4$

13. $\sin\theta = 3/5$, $\cos\theta = -4/5$, $\tan\theta = -3/4$, $\cot\theta = -4/3$, $\sec\theta = -5/4$, $\csc\theta = 5/3$

15. $\cos\theta = 4/5$, $\tan\theta = 3/4$, $\cot\theta = 4/3$, $\sec\theta = 5/4$, $\csc\theta = 5/3$

17. $\sin\theta = -5/13$, $\cos\theta = -12/13$, $\cot\theta = 12/5$, $\sec\theta = -13/12$, $\csc\theta = -13/5$

19. $\sin 0° = 0$, $\cos 0° = 1$, $\tan 0° = 0$, $\sec 0° = 1$, $\cot 0°$ and $\csc 0°$ undefined

21. $\sin 270° = -1$, $\cos 270° = 0$, $\cot 270° = 0$, $\csc 270° = -1$, $\tan 270°$ and $\sec 270°$ undefined

23. $\sin(-630°) = 1$, $\cos(-630°) = 0$, $\cot(-630°) = 0$, $\csc(-630°) = 1$, $\tan(-630°)$ and $\sec(-630°)$ undefined

25. $\sin(-\pi) = 0$, $\cos(-\pi) = -1$, $\tan(-\pi) = 0$, $\sec(-\pi) = -1$, $\cot(-\pi)$ and $\csc(-\pi)$ undefined

27. $\sin(-5\pi) = 0$, $\cos(-5\pi) = -1$, $\tan(-5\pi) = 0$, $\sec(-5\pi) = -1$, $\cot(-5\pi)$ and $\csc(-5\pi)$ undefined

29. $\sin 225° = -1/\sqrt{2}$, $\cos 225° = -1/\sqrt{2}$, $\tan 225° = 1$, $\cot 225° = 1$, $\sec 225° = -\sqrt{2}$, $\csc 225° = -\sqrt{2}$

31. $\sin(-60°) = -\sqrt{3}/2$, $\cos(-60°) = 1/2$, $\tan(-60°) = -\sqrt{3}$, $\cot(-60°) = -1/\sqrt{3}$, $\sec(-60°) = 2$, $\csc(-60°) = -2/\sqrt{3}$

33. $\sin 420° = \sqrt{3}/2$, $\cos 420° = 1/2$, $\tan 420° = \sqrt{3}$, $\cot 420° = 1/\sqrt{3}$, $\sec 420° = 2$, $\csc 420° = 2/\sqrt{3}$

35. $\sin(-5\pi/6) = -1/2$, $\cos(-5\pi/6) = -\sqrt{3}/2$, $\tan(-5\pi/6) = 1/\sqrt{3}$, $\cot(-5\pi/6) = \sqrt{3}$, $\sec(-5\pi/6) = -2/\sqrt{3}$, $\csc(-5\pi/6) = -2$

37. $\sin(5\pi/6) = 1/2$, $\cos(5\pi/6) = -\sqrt{3}/2$, $\tan(5\pi/6) = -1/\sqrt{3}$, $\cot(5\pi/6) = -\sqrt{3}$, $\sec(5\pi/6) = -2/\sqrt{3}$, $\csc(5\pi/6) = 2$

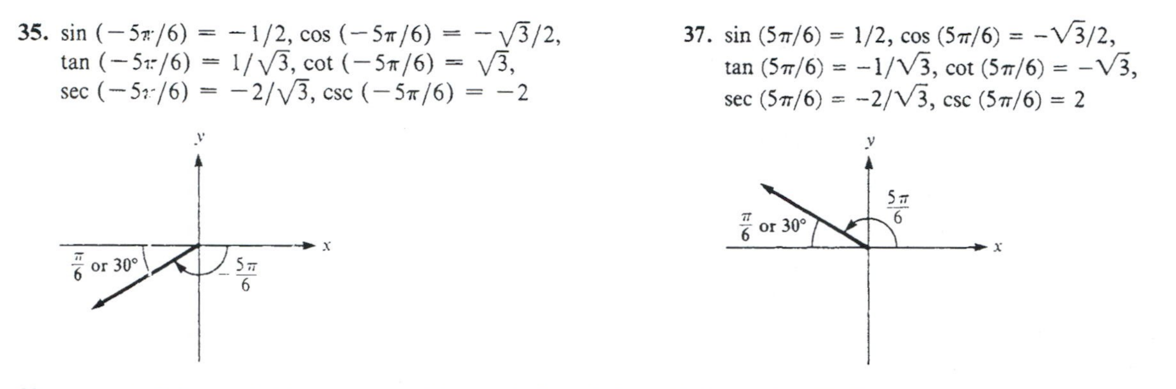

39. $n = 1, 3, 5, 7, \ldots$ (odd integers) **41.** $n = 1, 2, 4, 5, 7, 8, 10, 11, 13, 14, 16, 17, \ldots$ **43.** $n = 1, 5, 7, 11, 13, 17, 19, 23, 25, 29, 31, \ldots$ **45.** .4540 **47.** 1.540 **49.** −.8910 **51.** .9435 **53.** 1.236 **55.** .6361 **57.** .5972 **59.** 6.392

Section 6.3 (page 366)

1. 180, 90 **2.** two sides, one side and one acute angle **3.** tangent (or cotangent) **4.** Pythagorean **5.** horizontal **6.** true **7.** true **8.** false **9.** true **10.** true **11.** $B = 58°$, $b \approx 32$, $c \approx 38$ **13.** $A = 75°$, $a \approx 75$, $c \approx 77$ **15.** $B = 17°30'$, $a \approx 19.1$, $b \approx 6.01$ **17.** $A \approx 40°$, $B \approx 50°$, $c \approx 12$ **19.** about 456 ft **21.** about 77,300 ft (14.6 mi) **23.** about 1160 ft **25.** about 70.5 ft **27.** about 7.10 mi **29.** S 50°12′ E

Section 6.4 (page 375)

1. three, side **2.** the law of sines **3.** two sides and an opposite angle **4.** zero, one, two **5.** $\frac{1}{2}bc \sin \alpha$ **6.** true **7.** false **8.** false **9.** true **10.** false **11.** $\gamma = 78°$, $a \approx 30$, $b \approx 44$ **13.** $\beta = 47°$, $a \approx 8.7$, $c \approx 20$ **15.** $\alpha \approx 117°$, $\gamma \approx 23°$, $a \approx 6.9$ **17.** no triangle **19.** yes **21.** about 148 m **23.** $a = 10$ in and $b = 20$ in **25.** about 140 yd

Section 6.5 (page 381)

1. Pythagorean **2.** two sides and their included angle, three sides **3.** the sum of any two is greater than the remaining one **4.** one **5.** greater **6.** false **7.** true **8.** true **9.** true **10.** true **11.** $c = 5\sqrt{3}$, $\alpha = 90°$, $\beta = 30°$ **13.** $b \approx 21$, $\alpha \approx 10°$, $\gamma \approx 148°$ **15.** $\alpha \approx 29°$, $\beta \approx 47°$, $\gamma \approx 104°$ **17.** no triangle **19.** about 119 mi or 162 mi **21.** about 624 mi, about 87° **23.** about 320 nautical mi, about 326° **25.** $a = 9$, $b = 5$, $c = 6$

Chapter 6 Review Exercises (page 383)

1. $\pi/8$ or $.125\pi$ **2.** 20° **3.** 35.2° **4.** 27° 21′ **5.** (a) IV (b) I (c) III (d) IV **6.** 650° and −70° **7.** $\pi/5$ ft **8.** 300π radians per min **9.** $\sin\theta = 12/13$, $\cos\theta = -5/13$, $\tan\theta = -12/5$ **10.** $\sec\theta = -5/4$, $\csc\theta = -5/3$, $\cot\theta = 4/3$ **11.** −√[illegible]/4 **12.** $-\sqrt{5}/2$ **13.** $\sin 240° = -\sqrt{3}/2$, $\cos 240° = -1/2$ **14.** $\csc(11\pi/4) = \sqrt{2}$, $\cot(11\pi/4) = -1$ **15.** $\sin(9\pi/5) = -\sin(\pi/5)$, $\cos(9\pi/5) = \cos(\pi/5)$ **16.** $\sec 250° = -\sec 70°$, $\cot 250° = \cot 70°$ **17.** 53° **18.** 15° **19.** $\sin(-\pi/3) = -\sqrt{3}/2$, $\cos(-\pi/3) = 1/2$, $\tan(-\pi/3) = -\sqrt{3}$, $\cot(-\pi/3) = -1/\sqrt{3}$, $\sec(-\pi/3) = 2$, $\csc(-\pi/3) = -2/\sqrt{3}$ **20.** $\sin(7\pi/2) = \csc(7\pi/2) = -1$, $\cos(7\pi/2) = \cot(7\pi/2) = 0$ **21.** θ **22.** $-\cos\theta$ **23.** 53° **24.** $A = B = 45°$ **25.** .2 **26.** $2/\sqrt{21}$ **27.** 60° **28.** about .45 radians **29.** $b \approx 4.2$, $c \approx 7.3$ **30.** about 40° **31.** about 55.7 ft **32.** about 24.1 ft **33.** $10\sqrt{2}$ **34.** 120° **35.** $2\sqrt{7}$ **36.** 2049 ft **37.** 700 yd **38.** about 84° **39.** one (10 in, 15 in, 20 in) **40.** about 18 mi

Chapter 7

Section 7.1 (page 388)

1. (0, 0), 1 **2.** x, y **3.** $2n\pi$ for any integer n **4.** 0, 0 **5.** s **6.** false **7.** true **8.** false **9.** true **10.** true **11.** $\pi/2, -3\pi/2$ **13.** $3\pi/2, -\pi/2$ **15.** $\pi, -\pi$ **17.** $\pi/2 + 2n\pi$ for any integer n **19.** $\pi + 2n\pi$ for any integer n **21.** (1, 0) **23.** $(1/\sqrt{2}, 1/\sqrt{2})$ **25.** (−1, 0) **27.** (0, −1) **29.** (1, 0)

31. $\sin 0 = 0$, $\cos 0 = 1$, $\tan 0 = 0$; $\sin(-7\pi/4) = 1/\sqrt{2} = \cos(-7\pi/4)$, $\tan(-7\pi/4) = 1$; $\sin(-\pi) = 0$, $\cos(-\pi) = -1$, $\tan(-\pi) = 0$ **33.** $\sec(3\pi/2)$ is undefined, $\csc(3\pi/2) = -1$, $\cot(3\pi/2) = 0$; $\sec 4\pi = 1$, $\csc 4\pi$ is undefined, $\cot 4\pi$ is undefined

Section 7.2 (page 399)

1. sine, cosine, secant, and cosecant **2.** tangent and cotangent **3.** right **4.** left **5.** $\pi/2 + n\pi$ for any integer n **6.** true **7.** false **8.** true **9.** false **10.** true **11.** (a) $-.36$ (b) $-.36$ (c) $.36$ **13.** (a) $-.3$ (b) $-.3$ (c) $.3$ **15.** The cosine curve is the sine curve shifted $\pi/2$ units to the left. **17.** (a) $0, \pi, 2\pi$ (b) $\pi/2$ (c) $3\pi/2$

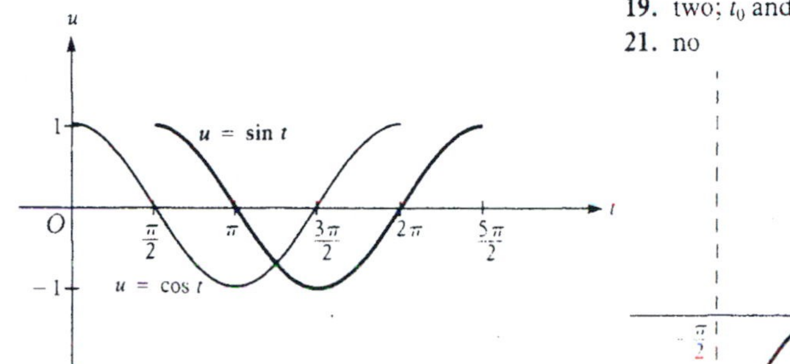

19. two; t_0 and $\pi - t_0$ if $0 < c < 1$, t_0 and $3\pi - t_0$ if $-1 < c < 0$
21. no **23.** one

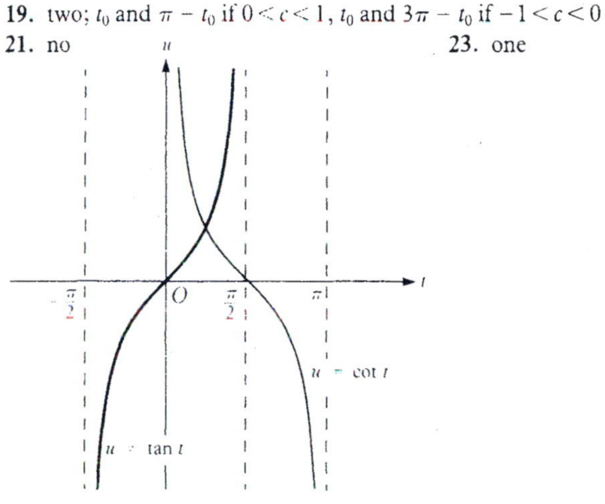

25. The secant curve on $(-\pi/2, \pi/2)$ is the cosecant curve on $(0, \pi)$ shifted $\pi/2$ units to the left.

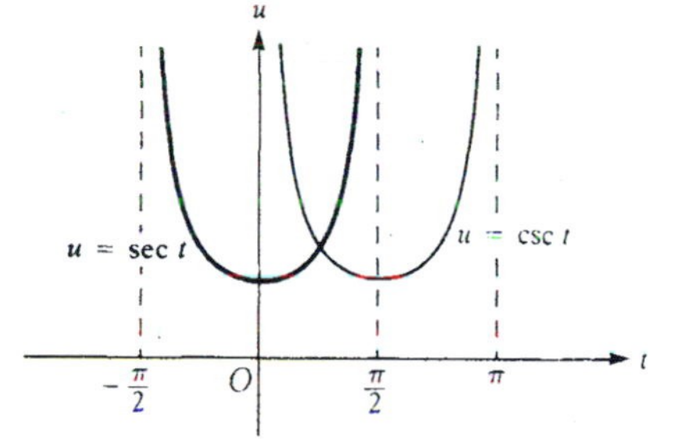

27. $t = \pi/2 + n\pi$ for any integer n; $\sec t \to +\infty$ as $t \to (\pi/2 + 2n\pi)^-$ and as $t \to (-\pi/2 + 2n\pi)^+$; $\sec t \to -\infty$ as $t \to (\pi/2 + 2n\pi)^+$ and as $t \to (-\pi/2 + 2n\pi)^-$ **29.** sine, tangent, cosecant, and cotangent are odd functions; cosine and secant are even functions; no
31. (a) none (b) sine, cosecant, cosine, secant

Section 7.3 (page 407)

1. $2\pi/B$, $2\pi/(-B)$ **2.** $(-\infty, \infty)$, $[-1, 1]$ **3.** $(-\infty, \infty)$, $[-2, 2]$ **4.** $-C/2$, $-C/2$ is negative, $-C/2$ is positive **5.** $u = 4 \sin(2t - \pi/3)$ **6.** true **7.** false **8.** true **9.** true **10.** false
11. amplitude $= 3$, period $= 2\pi$

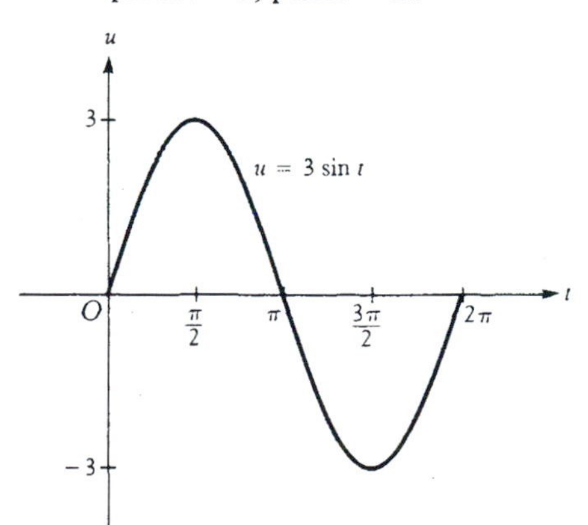

13. amplitude $= 1$, period $= \pi$

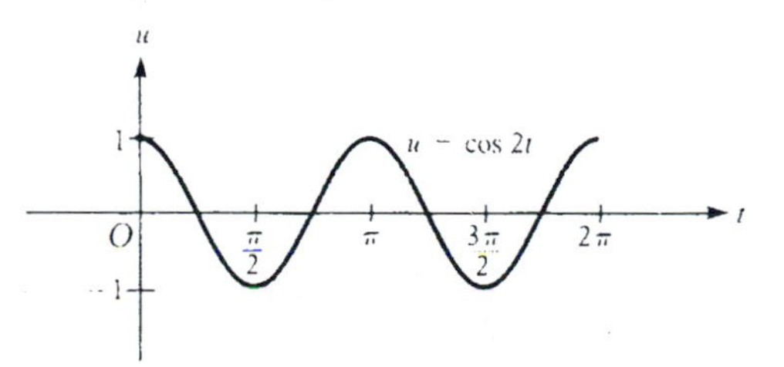

15. amplitude = 3, period = π

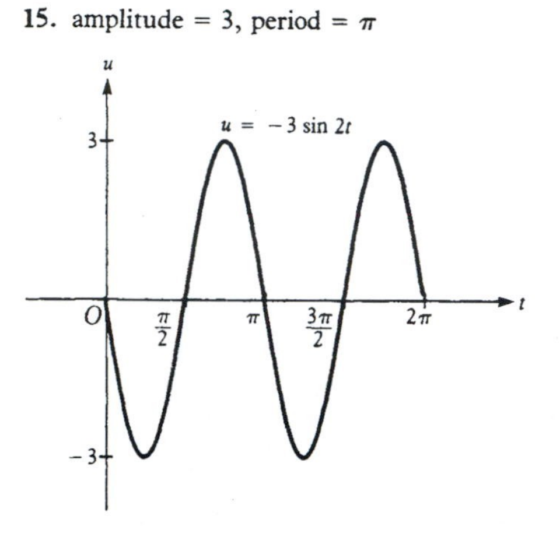

17. amplitude = 3, period = 4π

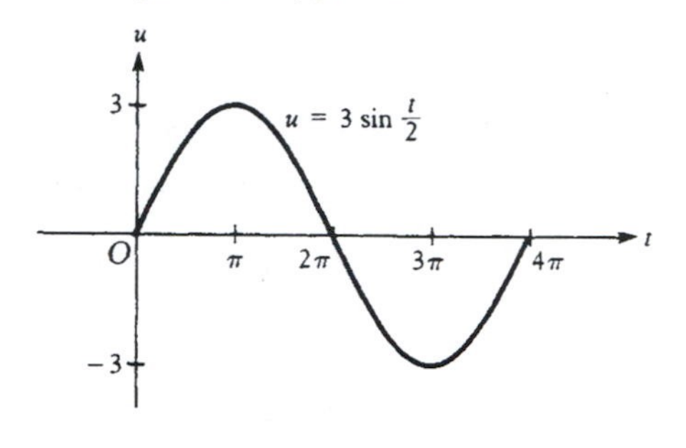

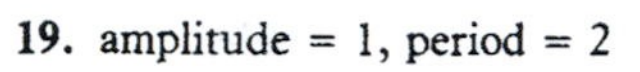

19. amplitude = 1, period = 2

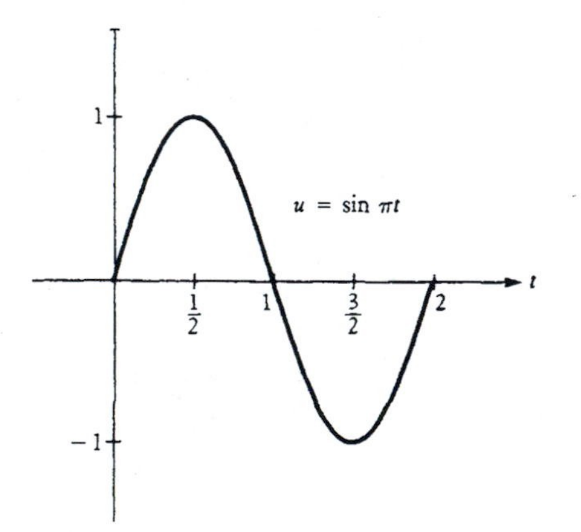

21. amplitude = 2, period = 8

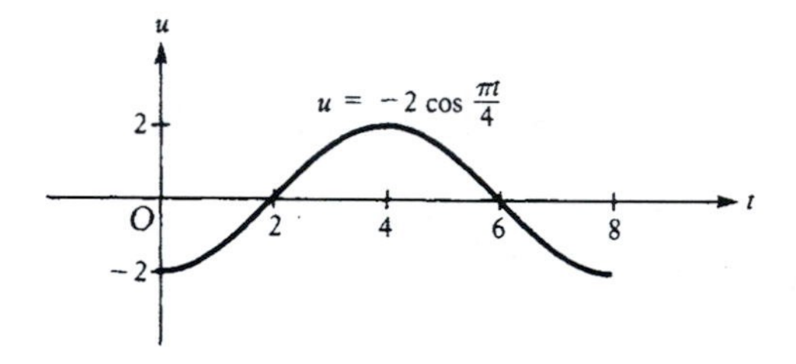

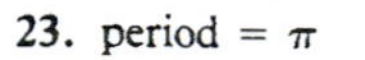

23. period = π

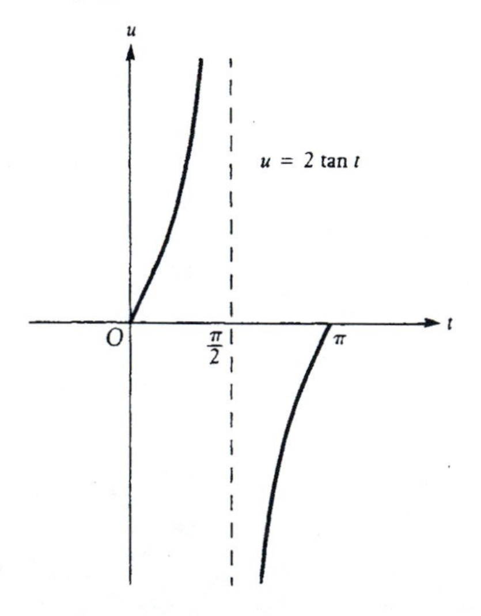

25. period = $\pi/2$

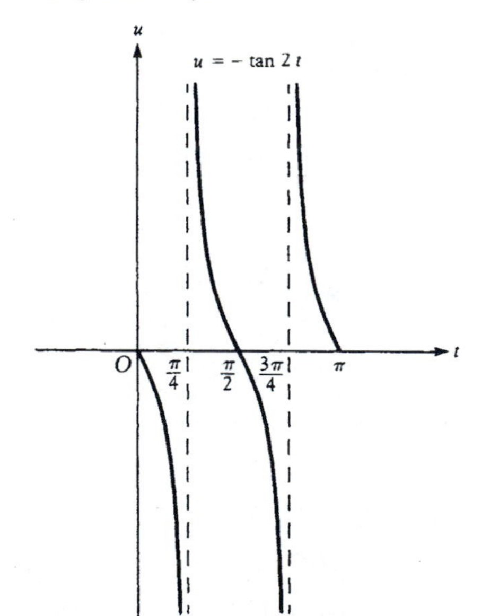

27. period = $\pi/3$

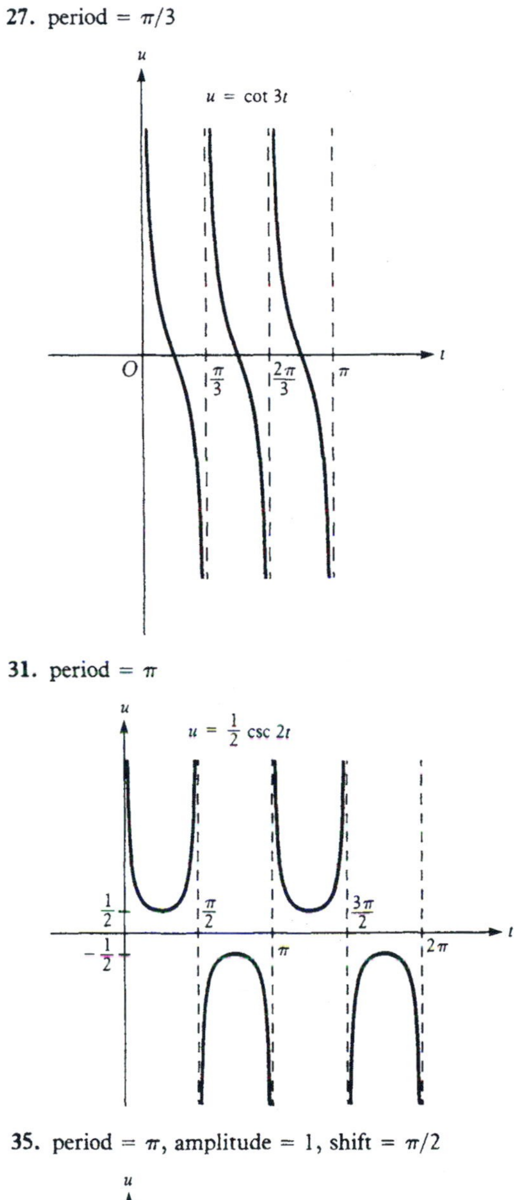

29. period = π

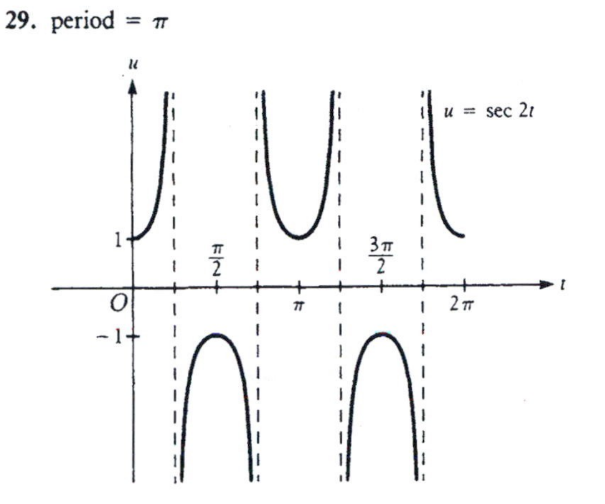

31. period = π

u = ½ csc 2t

33. period = π, amplitude = 4, shift = $-\pi/2$

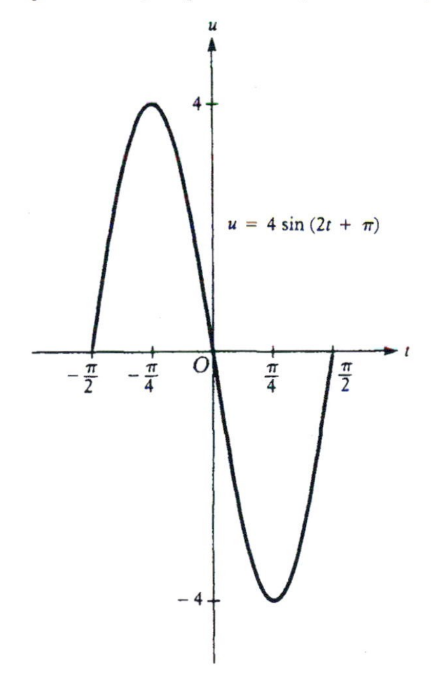

35. period = π, amplitude = 1, shift = $\pi/2$

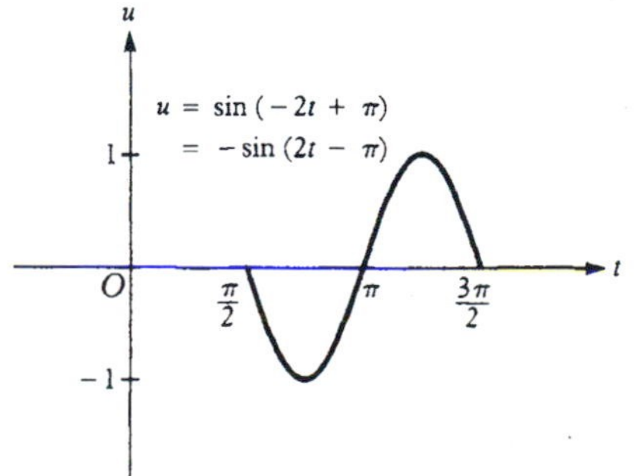

37. period $= 4\pi$, amplitude $= 3$, shift $= \pi$

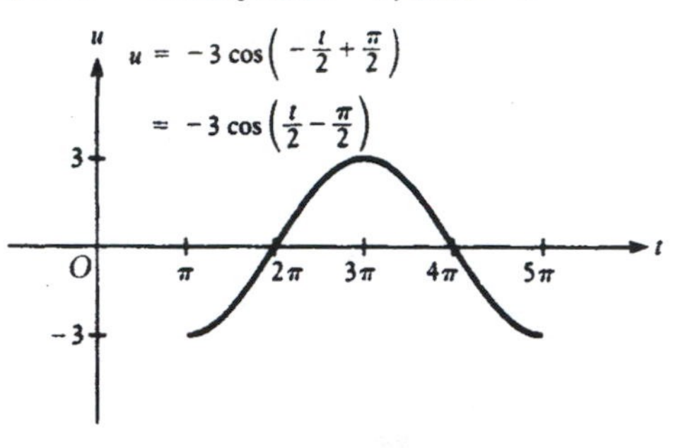

39. period $= 4$, amplitude $= 2$, shift $= -1/2$

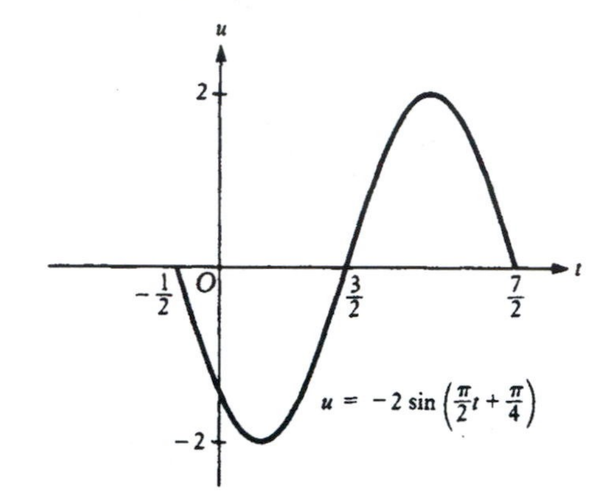

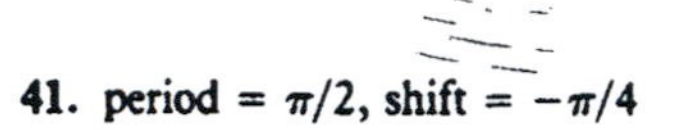

41. period $= \pi/2$, shift $= -\pi/4$

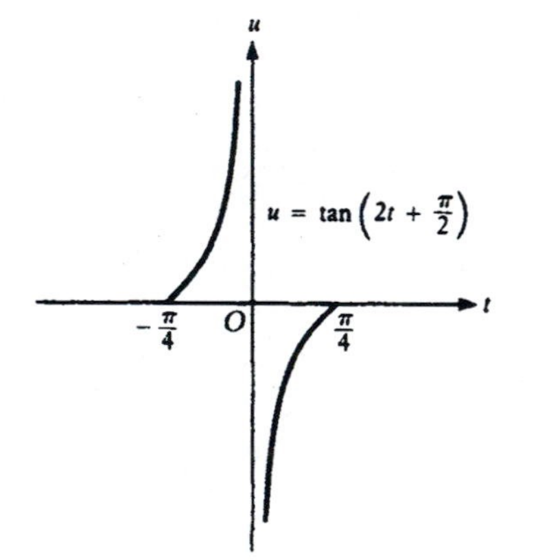

43. period $= \pi$, shift $= -\pi/4$

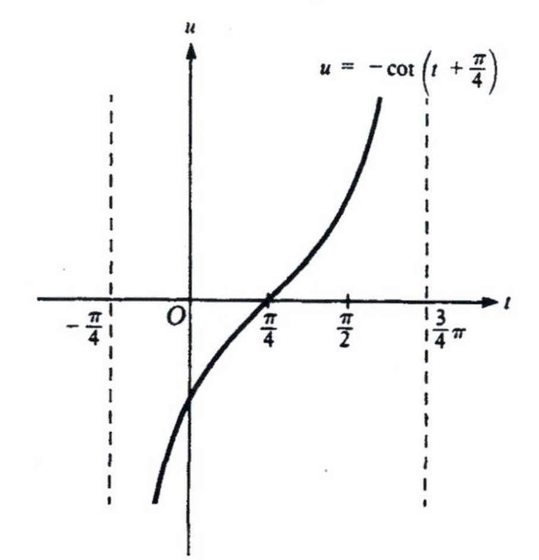

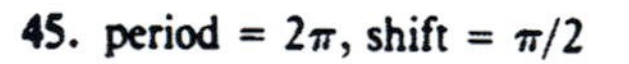

45. period $= 2\pi$, shift $= \pi/2$

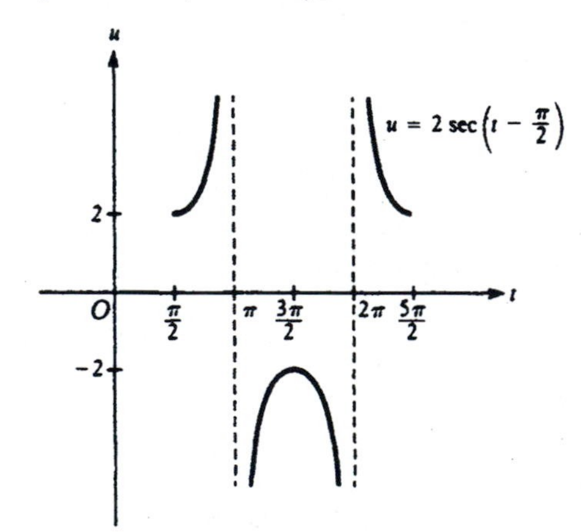

47. period $= 2\pi$

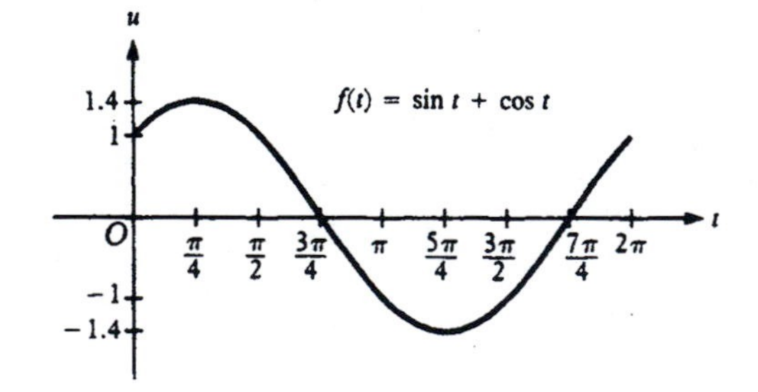

49. period $= 2\pi$

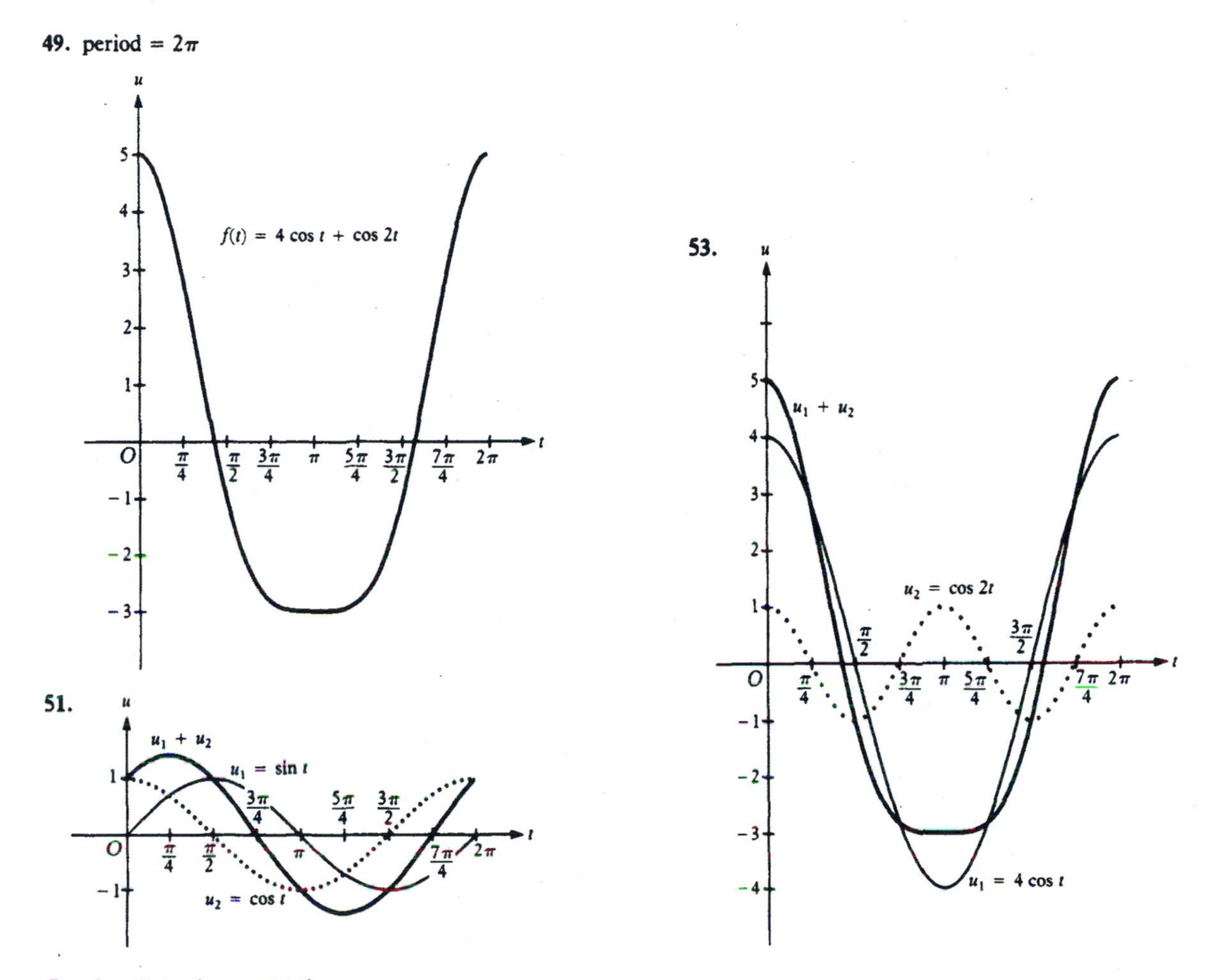

Section 7.4 (page 414)

1. $s = .25\cos(4\sqrt{2}\,t)$, period: about 1.11 sec/cycle **3.** $s = -.125\cos 8t$, period: $\pi/4$ sec/cycle
5. L must be shortened to .25 ft **7.** $y = (\sqrt{5}/8)\cos(8t - .46)$, $|A| = \sqrt{5}/8$ ft, $T = \pi/4$ sec/cycle **9.** $t \approx .25$ sec
11. **13.**

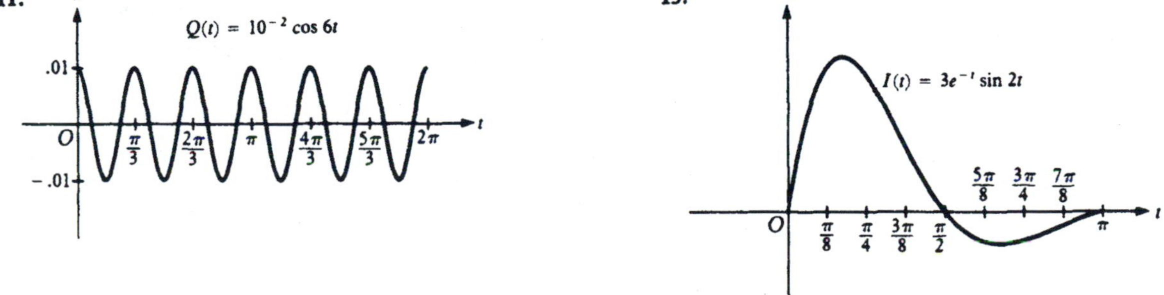

15. **(a)** $x = \cos(4\pi t + \pi/4)$, $y = \sin(4\pi t + \pi/4)$ **(b)** $x = -.7$, $y = -.7$ **(c)** $t = n/2$ sec for any positive integer n

Section 7.5 (page 426)

1. $\sin x$ is not one-to-one on $[0, 3\pi/2]$ **2.** the range of $\tan x$ on $[0, \pi/2)$ is not $(-\infty, \infty)$ **3.** B, A **4.** $\sin\theta, [-\pi/2, \pi/2]$
5. $-1 \le x \le 1$ **6.** true **7.** false **8.** true **9.** true **10.** false **11.** $\pi/3$ **13.** $\pi/2$ **15.** 0 **17.** $\pi/4$
19. $-\pi/6$ **21.** .2589 **23.** -1.2799 **25.** .0175 **27.** -1.5184 **29.** 1 **31.** $\sqrt{3}/2$ **33.** 1.53 **35.** $-\pi/4$
37. $\pi/2$ **39.** $\pi/4$ **41.** $2\pi/3$ **43.** $3\pi/4$ **45.** $\pi/6$ **47.** undefined **49.** .2914 **51.** .2967 **53.** .4364

Section 7.6 (page 434)

1. it is true for all values of θ **2.** $\pi/2 + n\pi$ for any integer n **3.** $\csc\theta = 1/\sin\theta$, $\sec\theta = 1/\cos\theta$, and $\cot\theta = 1/\tan\theta$
4. $\sin^2\theta + \cos^2\theta = 1$, $\tan^2\theta + 1 = \sec^2\theta$, $1 + \cot^2\theta = \csc^2\theta$ **5.** $(\sin\theta)^n$, arc $\sin\theta$, $1/\sin\theta$ **6.** true **7.** false
8. true **9.** false **10.** false **15.** $\sin\theta$ **17.** $\tan\theta$ **19.** 1 **21.** $1 + \sin\theta$ **23.** 1 **25.** $\cos^2\theta$ **27.** $\sin^2\theta$
29. $\cos\theta$ **31.** $-\cot\theta$ **33.** 1 **35.** $\tan\theta$ **53 (a)** $\cot\theta$ **(b)** $-\sec\theta$ **(c)** $-\csc\theta$

Section 7.7 (page 441)

1. $\cos\alpha\cos\beta + \sin\alpha\sin\beta$ **2.** distance, cosines **3.** $\sin\theta$ **4.** 45°, 30° **5.** $\sin(\alpha+\beta)$, $\cos(\alpha+\beta)$ **6.** false
7. false **8.** true **9.** true **10.** false **11.** $(\sqrt{6}-\sqrt{2})/4$ **13.** $(\sqrt{6}-\sqrt{2})/4$ **15.** $(\sqrt{6}+\sqrt{2})/4$
17. $-(2+\sqrt{3})$ **19.** $-1/2$ **21.** 1 **23.** 1/2 **25.** 1
27. $\sin(\alpha+\beta) = -16/65$, $\cos(\alpha+\beta) = -63/65$, third quadrant
29. $\sin(\alpha+\beta) = 1$, $\cos(\alpha+\beta) = 0$, quadrantal angle coterminal with 90°
31. $\sin(\alpha-\beta) = 56/65$, $\cos(\alpha-\beta) = -33/65$ **43.** $(\sqrt{6}+\sqrt{2})/4$ **45.** $1/\sqrt{3}$
51. **53.**

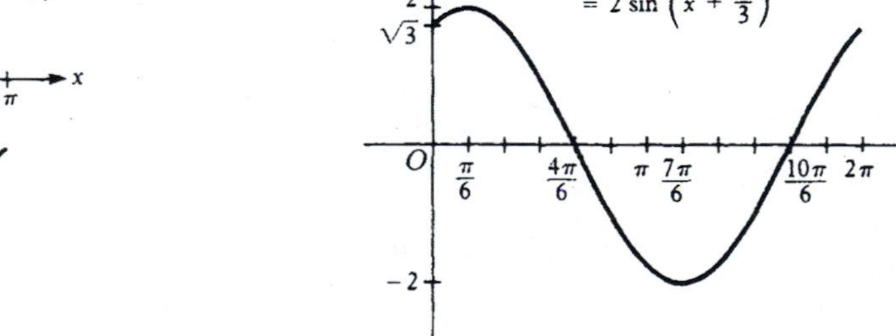

Section 7.8 (page 448)

1. $\beta = \alpha$ **2.** $2\cos^2\alpha - 1$, $1 - 2\sin^2\alpha$ **3.** $1 - 2\sin^2\alpha$ **4.** $\pm\sqrt{(1-\cos\beta)/2}$
5. $\sin\alpha\cos\beta = \frac{1}{2}[\sin(\alpha+\beta) + \sin(\alpha-\beta)]$ **6.** true **7.** true **8.** true **9.** false **10.** false **11.** 1/2
13. $-\sqrt{3}/2$ **15.** $\sqrt{3}/2$ **17.** $\sqrt{2}/2$ **19.** $-(\sqrt{2-\sqrt{3}})/2$ **21.** $\sin 2\alpha = -24/25$, $\cos 2\alpha = 7/25$
23. $\sin 2\alpha = 24/25$, $\cos 2\alpha = 7/25$ **25.** $\sin 2\alpha = 120/169$, $\cos 2\alpha = 119/169$ **27.** $\sin(\alpha/2) = 3/\sqrt{10}$, $\cos(\alpha/2) = -1/\sqrt{10}$
29. $\sin(\alpha/2) = -\sqrt{42}/12$, $\cos(\alpha/2) = \sqrt{102}/12$ **31.** 2 **33.** $\tan\theta$ **35.** $\cos^2 2\theta$ **37.** $\cot 2\theta$ **47.** $-\sqrt{2}/4$
49. $(1+\sqrt{3})/4$ **51.** $(1/2)\sin 5\theta + (1/2)\sin\theta$ **53.** $(-1/2)\cos 5\theta + (1/2)\cos\theta$ **55.** $(1/2)\sin\theta - (1/2)\sin 5\theta$
57. $2\cos 2\theta\cos\theta$ **59.** $-2\sin 4\theta\sin\theta$

Section 7.9 (page 460)

1. 0, an infinite number **2.** $\sec x = 1/\cos x$ **3.** $\cos x$ **4.** π
5. approximate values of trigonometric functions, approximation methods **6.** false **7.** false **8.** true **9.** true
10. false **11.** $\pi/3, 2\pi/3$ **13.** $5\pi/6, 11\pi/6$ **15.** $0, 2\pi$ **17.** $\pi/4, 5\pi/4$ **19.** $3\pi/2$
21. $-\pi/4 + 2n\pi$, $5\pi/4 + 2n\pi$ for any integer n **23.** $n\pi$ for any integer n **25.** $2n\pi$ for any integer n **27.** .6138, 2.5278
29. .8464, -2.2952 **31.** .8464, $-.8464$ **33.** $3\pi/2$ **35.** $\pi/4, 7\pi/4$ **37.** no solution in $[0, 2\pi)$
39. $\pi/2, 7\pi/6, 11\pi/6$ **41.** $0, \pi, -\pi$ **43.** no solution **45.** $-\pi/4 + n\pi$ for any integer n **47.** $\pi/4, 3\pi/4, 5\pi/4, 7\pi/4$
49. 240°, 300° **51.** $0, \pi, \pi/6, 5\pi/6$ **53.** $7\pi/6, 11\pi/6, 3\pi/2$ **55.** $\pi/3, 5\pi/3, \pi$ **57.** $\pi/3, 5\pi/3, \pi$
59. $\pi/2, 3\pi/2, \pi/4, 5\pi/4$ **61.** .6662, 2.4754 **63.** $\pi/2, 3\pi/2, \pi/4, 3\pi/4, 5\pi/4, 7\pi/4$ **65.** $\pi, 66\pi/40$
67. $44\pi/40, 71\pi/40$ **69.** $15\pi/40, 71\pi/40$

Chapter 7 Review Exercises (page 463)

1. radian measure of an angle, arc length on unit circle 2. $-\pi/4 + 2n\pi$ for any integer n 3. $\sin s = -1/\sqrt{2}$, $\cos s = 1/\sqrt{2}$
4. $-.2$ 5. $.9$ 6. $-.7$ 7. $-.7$
8. 9. 10. 11. 12.

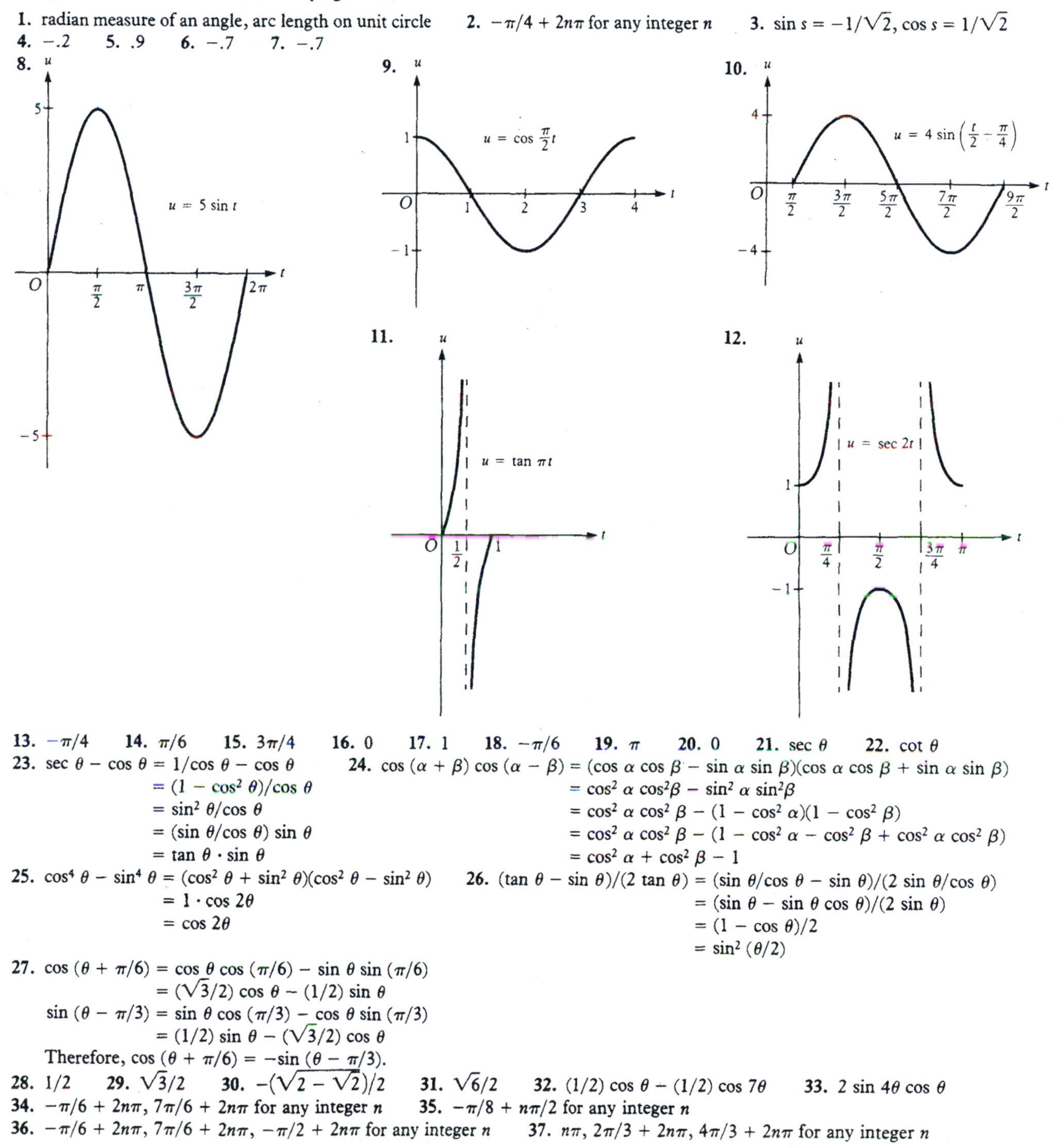

13. $-\pi/4$ 14. $\pi/6$ 15. $3\pi/4$ 16. 0 17. 1 18. $-\pi/6$ 19. π 20. 0 21. $\sec\theta$ 22. $\cot\theta$

23. $\sec\theta - \cos\theta = 1/\cos\theta - \cos\theta$
$= (1-\cos^2\theta)/\cos\theta$
$= \sin^2\theta/\cos\theta$
$= (\sin\theta/\cos\theta)\sin\theta$
$= \tan\theta\cdot\sin\theta$

24. $\cos(\alpha+\beta)\cos(\alpha-\beta) = (\cos\alpha\cos\beta - \sin\alpha\sin\beta)(\cos\alpha\cos\beta + \sin\alpha\sin\beta)$
$= \cos^2\alpha\cos^2\beta - \sin^2\alpha\sin^2\beta$
$= \cos^2\alpha\cos^2\beta - (1-\cos^2\alpha)(1-\cos^2\beta)$
$= \cos^2\alpha\cos^2\beta - (1-\cos^2\alpha-\cos^2\beta+\cos^2\alpha\cos^2\beta)$
$= \cos^2\alpha + \cos^2\beta - 1$

25. $\cos^4\theta - \sin^4\theta = (\cos^2\theta+\sin^2\theta)(\cos^2\theta-\sin^2\theta)$
$= 1\cdot\cos 2\theta$
$= \cos 2\theta$

26. $(\tan\theta - \sin\theta)/(2\tan\theta) = (\sin\theta/\cos\theta - \sin\theta)/(2\sin\theta/\cos\theta)$
$= (\sin\theta - \sin\theta\cos\theta)/(2\sin\theta)$
$= (1-\cos\theta)/2$
$= \sin^2(\theta/2)$

27. $\cos(\theta+\pi/6) = \cos\theta\cos(\pi/6) - \sin\theta\sin(\pi/6)$
$= (\sqrt{3}/2)\cos\theta - (1/2)\sin\theta$
$\sin(\theta-\pi/3) = \sin\theta\cos(\pi/3) - \cos\theta\sin(\pi/3)$
$= (1/2)\sin\theta - (\sqrt{3}/2)\cos\theta$
Therefore, $\cos(\theta+\pi/6) = -\sin(\theta-\pi/3)$.

28. $1/2$ 29. $\sqrt{3}/2$ 30. $-(\sqrt{2-\sqrt{2}})/2$ 31. $\sqrt{6}/2$ 32. $(1/2)\cos\theta - (1/2)\cos 7\theta$ 33. $2\sin 4\theta\cos\theta$
34. $-\pi/6 + 2n\pi$, $7\pi/6 + 2n\pi$ for any integer n 35. $-\pi/8 + n\pi/2$ for any integer n
36. $-\pi/6 + 2n\pi$, $7\pi/6 + 2n\pi$, $-\pi/2 + 2n\pi$ for any integer n 37. $n\pi$, $2\pi/3 + 2n\pi$, $4\pi/3 + 2n\pi$ for any integer n

38. $8/\pi^2$ ft **39.** (a) 60 cycles/sec for both (b) max. $E = 220$ volts when $t = (1/240 + n/60)$ sec, and max. $I = 20$ amperes when $t = (1/160 + n/60)$ sec, where n is a nonnegative integer **40.** 1/4 ft, $4/\pi$ cycles/sec

Chapter 8

Section 8.1 (page 471)

1. pole, polar axis **2.** $x = r\cos\theta$ and $y = r\sin\theta$ **3.** $r = \sqrt{x^2 + y^2}$ and $\tan\theta = y/x$
4. $[r, \theta + 2k\pi]$ and $[-r, \theta + \pi + 2k\pi]$ for any integer k **5.** $[0, \theta]$ for any angle θ **6.** false **7.** true **8.** false
9. false **10.** true
11.

$S\left(-1, \frac{\pi}{2}\right)$ $Q(1, 0)$ $P[1, 0°]$ $R\left[-1, \frac{\pi}{2}\right]$

13. (a) $(\sqrt{2}, \sqrt{2})$ (b) $(3\sqrt{3}/2), -3/2)$ (c) $(-1, -\sqrt{3})$
15. (a) $[1, \pi/3 + 2n\pi]$, $[-1, 4\pi/3 + 2n\pi]$
(b) $[3, 3\pi/2 + 2n\pi]$, $[-3, \pi/2 + 2n\pi]$ (c) $[2, 3\pi/4 + 2n\pi]$, $[-2, 7\pi/4 + 2n\pi]$ for any integer n
17. $r^2 - 2r\cos\theta + 3r\sin\theta = 1$ **19.** $xy = y + x$
21. $[-1, 2\pi + 4n\pi]$ for any integer n

27. (a) $2(x')^2 + (y')^2 = 8$ (b) $(x')^2 - (y')^2 = 2$ (c) $y' = (x')^2 + 1$

Section 8.2 (page 477)

1. at least one pair of polar coordinates $[r, \theta]$ of P satisfies the equation
2. a straight line through the pole at an angle of $\pi/3$ with the polar axis **3.** a circle with center at the pole and radius 2
4. $\theta = 0$ **5.** $Ar\cos\theta + Br\sin\theta = C$ **6.** false **7.** true **8.** true **9.** true **10.** false
11.

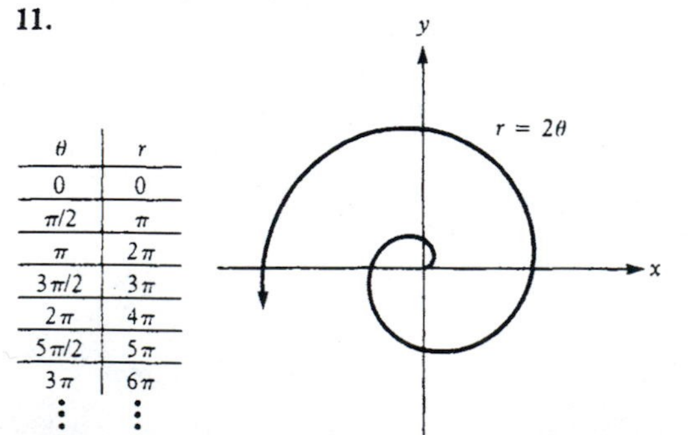

θ	r
0	0
$\pi/2$	π
π	2π
$3\pi/2$	3π
2π	4π
$5\pi/2$	5π
3π	6π
⋮	⋮

13.

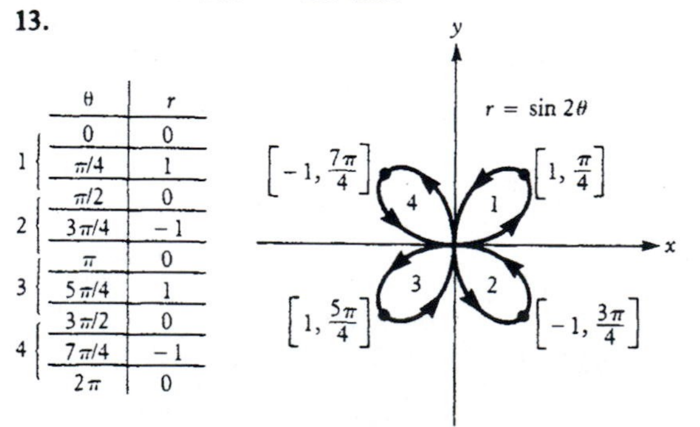

	θ	r
	0	0
1	$\pi/4$	1
	$\pi/2$	0
2	$3\pi/4$	−1
	π	0
3	$5\pi/4$	1
	$3\pi/2$	0
4	$7\pi/4$	−1
	2π	0

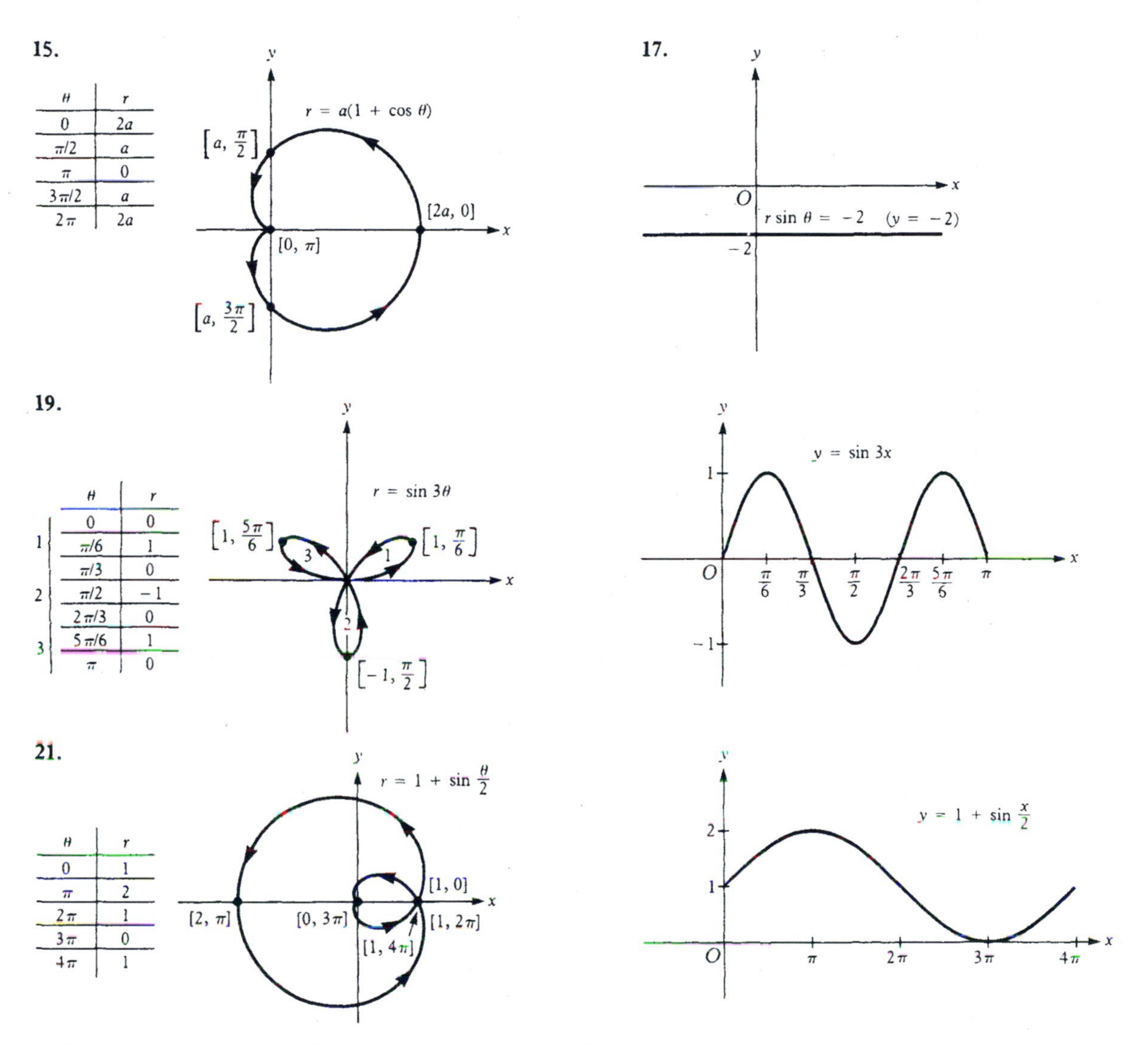

23. The curves for Exercises 13, 14, and 15 are symmetric with respect to the x-axis.

25. The curve for Exercise 13 is symmetric with respect to the origin.

27.

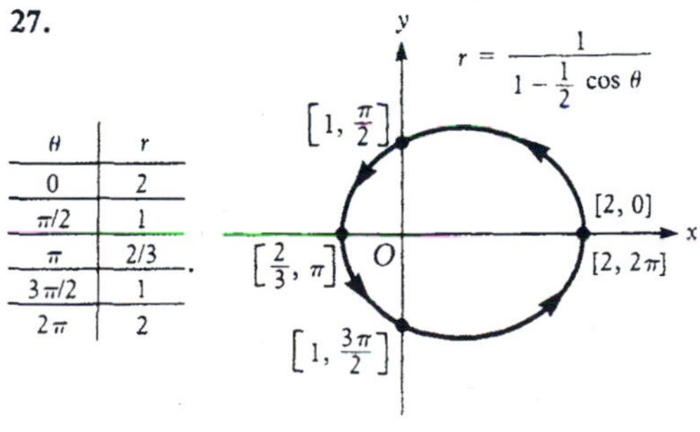

29.

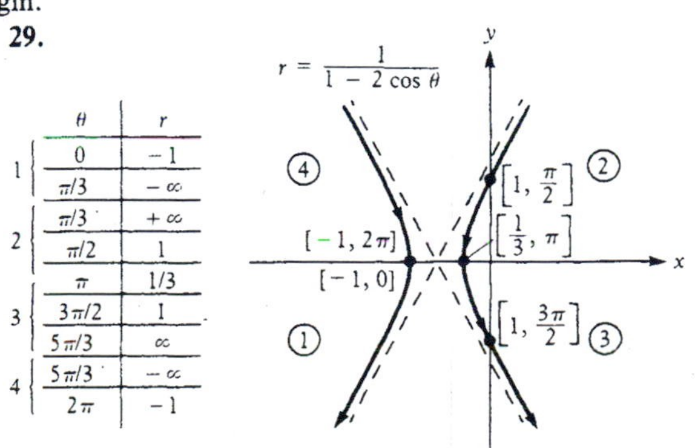

Section 8.3 (page 484)

1. $\sqrt{x^2 + y^2}$ **2.** $r_1r_2[\cos(\theta_1 + \theta_2) + i \sin(\theta_1 + \theta_2)]$, $(r_1/r_2)[\cos(\theta_1 - \theta_2) + i \sin(\theta_1 - \theta_2)]$ **3.** $r^n(\cos n\theta + i \sin n\theta)$
4. $z_k = \sqrt[n]{r}\{\cos[(\theta + 2k\pi)/n] + i \sin[(\theta + 2k\pi)/n]\}$, $k = 0, 1, 2, \ldots, n - 1$
5. $z_k = \cos(2k\pi/n) + i \sin(2k\pi/n)$, $k = 0, 1, 2, \ldots, n - 1$ **6.** true **7.** false **8.** false **9.** true **10.** true
11. $\sqrt{2}[\cos(\pi/4) + i \sin(\pi/4)]$ **13.** $2[\cos(\pi/2) + i \sin(\pi/2)]$ **15.** $2[\cos(4\pi/3) + i \sin(4\pi/3)]$
17. $2\sqrt{2}[\cos(3\pi/4) + i \sin(3\pi/4)]$ **19.** $2[\cos(\pi/6) + i \sin(\pi/6)]$ **21.** $4[\cos(3\pi/4) + i \sin(3\pi/4)]$
23. $\sqrt{2}[\cos(\pi/8) + i \sin(\pi/8)]$ **25.** $(\sqrt{2}/2)[\cos(-\pi/4) + i \sin(-\pi/4)]$ **27.** $2[\cos(\pi/2) + i \sin(\pi/2)]$
29. $\cos(-\pi/4) + i \sin(-\pi/4)$ **31.** $(1/\sqrt{2})[\cos(-5\pi/8) + i \sin(-5\pi/8)]$ **33.** $2\sqrt{2}[\cos(3\pi/4) + i \sin(3\pi/4)]$
35. $\cos(3\pi/2) + i \sin(3\pi/2)$ **37.** $\cos \pi + i \sin \pi$ **39.** $1024[\cos 0 + i \sin 0]$ **41.** $(1/2) + (\sqrt{3}/2)i$, -1, $(1/2) - (\sqrt{3}/2)i$
43. $\sqrt{2} + \sqrt{2}i$, $-\sqrt{2} + \sqrt{2}i$, $-\sqrt{2} - \sqrt{2}i$, $\sqrt{2} - \sqrt{2}i$ **45.** $z_0 = 2[\cos(7\pi/24) + i \sin(7\pi/24)]$,
$z_1 = 2[\cos(5\pi/8) + i \sin(5\pi/8)]$, $z_2 = 2[\cos(23\pi/24) + i \sin(23\pi/24)]$, $z_3 = 2[\cos(31\pi/24) + i \sin(31\pi/24)]$,
$z_4 = 2[\cos(13\pi/8) + i \sin(13\pi/8)]$, $z_5 = 2[\cos(47\pi/24) + i \sin(47\pi/24)]$

Section 8.4 (page 492)

1. magnitude, direction **2.** scalars **3.** the directed diagonal, starting at P, of the parallelogram formed by **V** and **W**
4. coordinate vector **5.** $\langle x_2 - x_1, y_2 - y_1\rangle$ **6.** false **7.** false **8.** true **9.** false **10.** true
11. 246.2 lb directed perpendicular to the ramp and 43.4 lb parallel to the ramp
13. 5 ft/sec at an angle of approximately 53° with the river bank
15. about 447 mph, at an angle of approximately 26.5° with the path of the plane

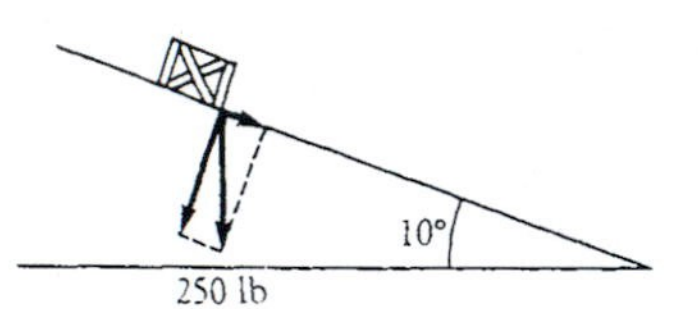

17. $\langle 0, 11\rangle, \langle 2, 1\rangle, \langle -1, 27\rangle$

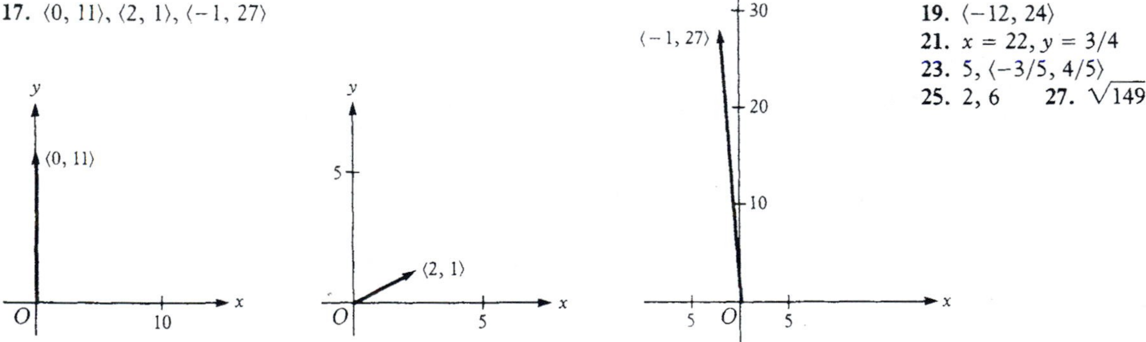

19. $\langle -12, 24\rangle$
21. $x = 22, y = 3/4$
23. $5, \langle -3/5, 4/5\rangle$
25. 2, 6 **27.** $\sqrt{149}$

Section 8.5 (page 500)

1. 0 **2.** $\sqrt{\mathbf{V}\cdot\mathbf{V}}$ **3.** $\langle a, b\rangle$ **4.** $|ax_1 + by_1 - c|/\sqrt{a^2 + b^2}$ **5.** $a_1a_2 + b_1b_2 = \sqrt{a_1^2 + b_1^2}\sqrt{a_2^2 + b_2^2}\cos\theta$ **6.** true
7. false **8.** true **9.** true **10.** true **11.** (a) 38 (b) 14 **13.** −116 **15.** $14/(\sqrt{13}\sqrt{17})$ **17.** $14/\sqrt{17}$
19. **V** and **W** **21.** $\langle 2/\sqrt{5}, 1/\sqrt{5}\rangle$ **23.** (a) $\mathbf{U} = \langle 24/5, 8/5\rangle$ (b) $\mathbf{V}\cdot(\mathbf{W} - \mathbf{U}) = \langle 6, 2\rangle \cdot \langle -4/5, 12/5\rangle = 0$
25. $-3x + 7y = 0$ **27.** 0 **29.** any nonzero scalar multiple of $\langle 11, 7\rangle$ **31.** $15/(2\sqrt{10})$ **33.** $\pi/4$ and $3\pi/4$
35. approximately 16° and 164° **37.** (a) $3x - 4y = 1$ (b) $x = 1$

Section 8.6 (page 507)

1. point, line, plane **2.** line, plane **3.** x_1, y_1 **4.** $\langle 6, -3, -6\rangle$ **5.** $\langle 3/(5\sqrt{2}), 4/(5\sqrt{2}), 1/\sqrt{2}\rangle$ **6.** true **7.** true
8. true **9.** true **10.** true

11.

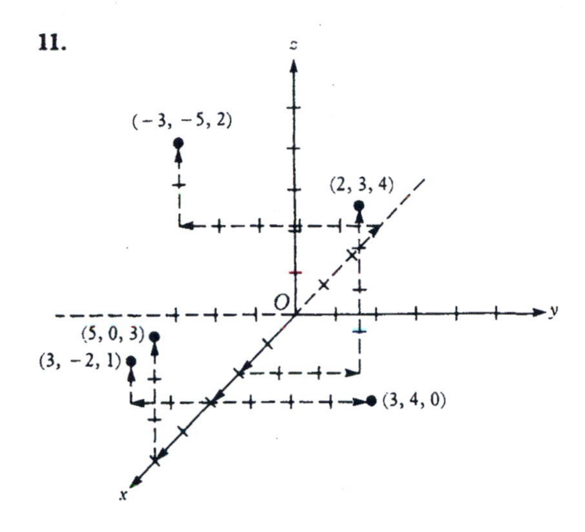

13. There are an infinite number of solutions, including the points (3, 5, 3) and (3, 7, 3), the points (5, 5, 1) and (5, 7, 1), and the points (1, 5, 1) and (1, 7, 1). **15.** $2\sqrt{2}$
17. $2\mathbf{V} - 3\mathbf{W} = \langle -8, 6, 17\rangle$, $\|\mathbf{V} + \mathbf{W}\| = \sqrt{61}$, $\mathbf{V}\cdot\mathbf{W} = 3$
19. $2\mathbf{V} - 3\mathbf{W} = \langle -2, -1, -7\rangle$, $\|\mathbf{V} + \mathbf{W}\| = 6\sqrt{6}$, $\mathbf{V}\cdot\mathbf{W} = 54$
21. $8/(3\sqrt{34})$, not perpendicular **23.** 0, perpendicular
25. $z = 5$ **27.** $7x + 10y - 5z = 7$ **29.** $x + y + z = 0$
31. $5x - 3y + 2z = -11$ **35.** (a) 0 (b) $12/\sqrt{61}$
37. (a) (4, 1, 0) (b) $d(P, P_1) = d(P, P_2) = \sqrt{33}$
39. $x = 3 - 7r, y = 6 - r, z = -2 + 9r$
41. $x = 2 + r, y = -1 + 2r, z = r$

Chapter 8 Review Exercises (page 511)

1. (a) $[1, 2n\pi]$ and $[-1, \pi + 2n\pi]$ for any integer n (b) $[2, \pi/6 + 2n\pi]$ and $[-2, 7\pi/6 + 2n\pi]$ for any integer n
2. (a) $(-\sqrt{2}, \sqrt{2})$ (b) $(-1, \sqrt{3})$ **3.** $(y + 3)(x^2 + y^2) = 2xy$ **4.** $r(2\cos^2\theta + 3\sin^2\theta) = \cos\theta - \sin\theta$

5.

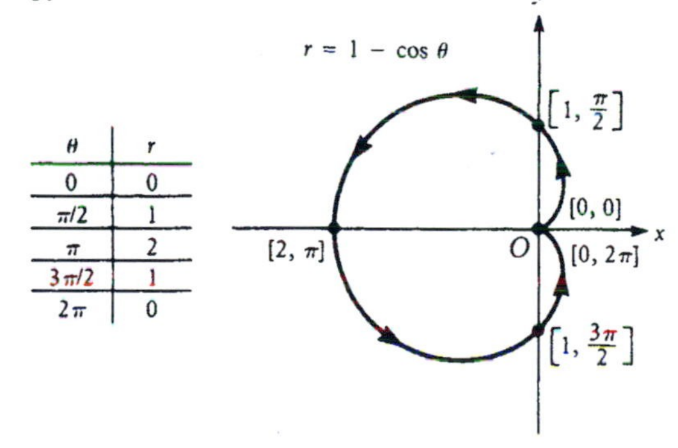

θ	r
0	0
$\pi/2$	1
π	2
$3\pi/2$	1
2π	0

6.

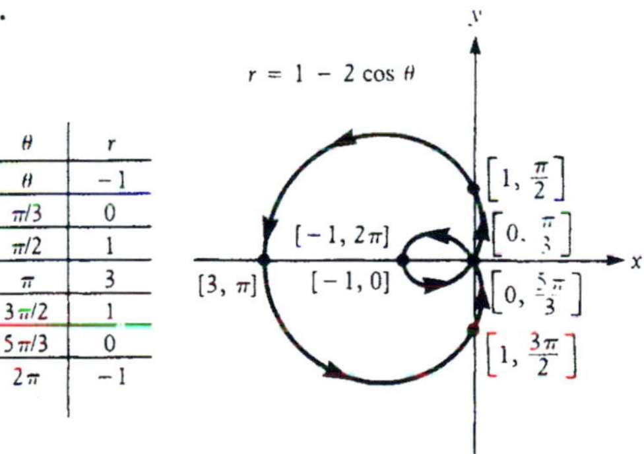

θ	r
0	−1
$\pi/3$	0
$\pi/2$	1
π	3
$3\pi/2$	1
$5\pi/3$	0
2π	−1

7. **8.**

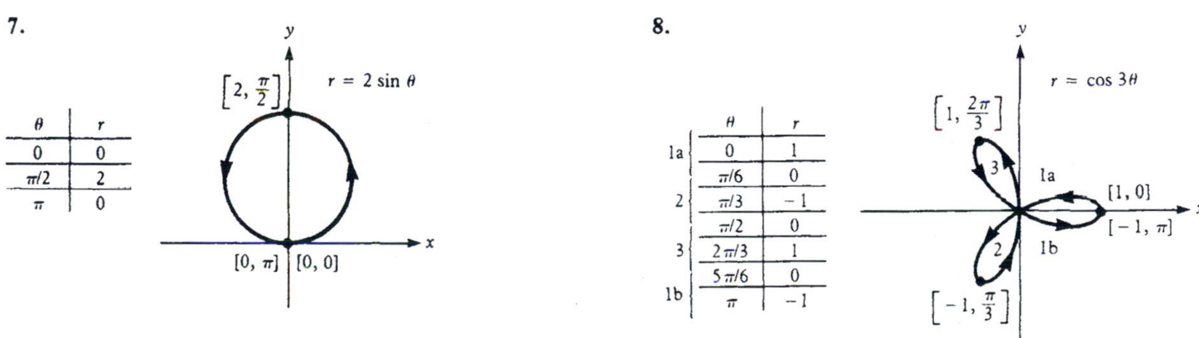

θ	r
0	0
$\pi/2$	2
π	0

	θ	r
1a	0	1
	$\pi/6$	0
2	$\pi/3$	−1
	$\pi/2$	0
3	$2\pi/3$	1
	$5\pi/6$	0
1b	π	−1

9. $2[\cos(11\pi/6) + i\sin(11\pi/6)]$ **10.** $z_1z_2 = 6i$, $z_1/z_2 = \sqrt{3}/3 + i/3$ **11.** $-\sqrt{2}/6 - i\sqrt{2}/6$ **12.** i
13. 2, $2[\cos(2\pi/5) + i\sin(2\pi/5)]$, $2[\cos(4\pi/5) + i\sin(4\pi/5)]$, $2[\cos(6\pi/5) + i\sin(6\pi/5)]$, $2[\cos(8\pi/5) + i\sin(8\pi/5)]$
14. $\sqrt{2}/2 + i\sqrt{2}/2$, $-\sqrt{2}/2 + i\sqrt{2}/2$, $-\sqrt{2}/2 - i\sqrt{2}/2$, $\sqrt{2}/2 - i\sqrt{2}/2$

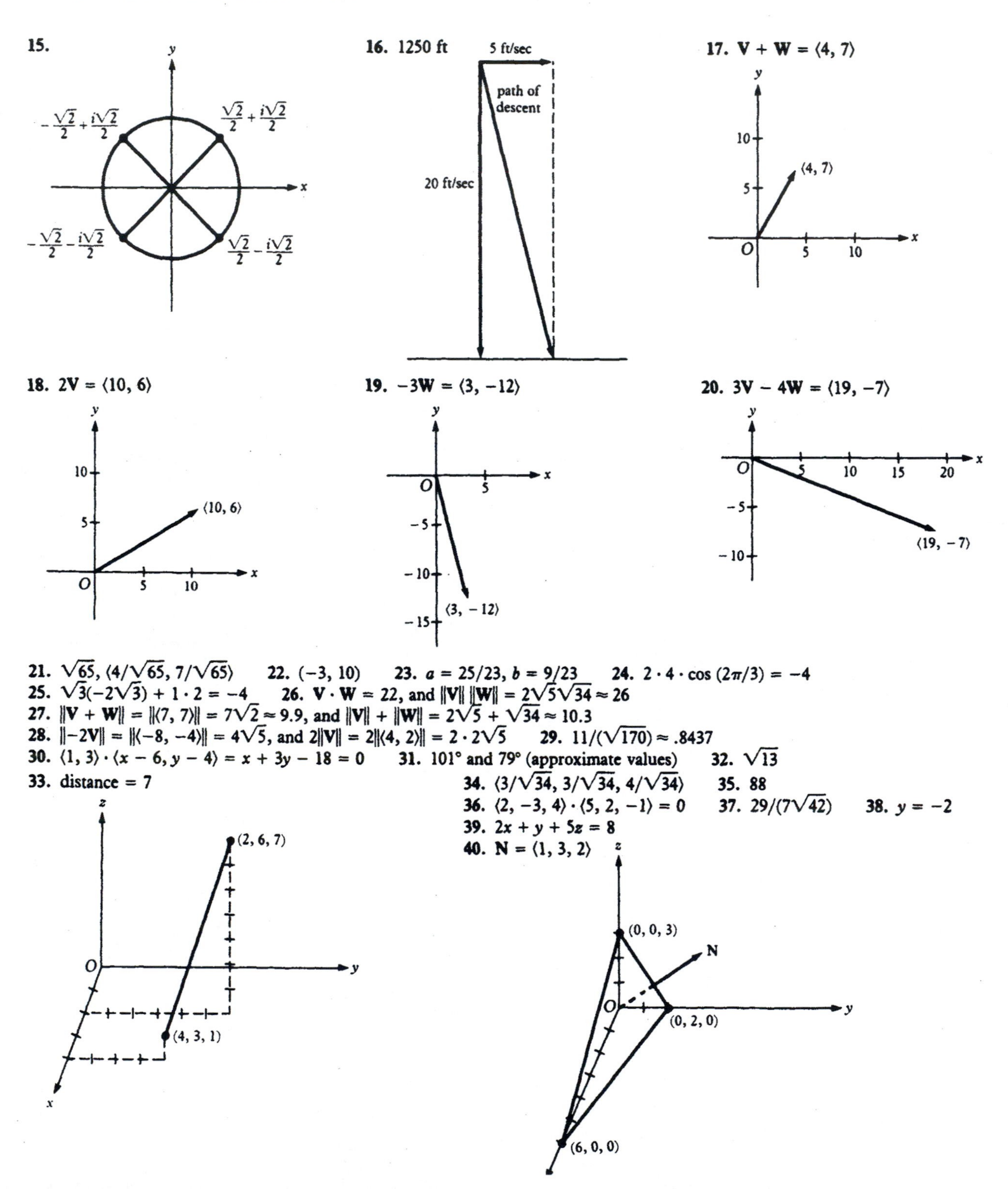

21. $\sqrt{65}$, $\langle 4/\sqrt{65}, 7/\sqrt{65}\rangle$ **22.** $(-3, 10)$ **23.** $a = 25/23$, $b = 9/23$ **24.** $2 \cdot 4 \cdot \cos(2\pi/3) = -4$
25. $\sqrt{3}(-2\sqrt{3}) + 1 \cdot 2 = -4$ **26.** $\mathbf{V} \cdot \mathbf{W} = 22$, and $\|\mathbf{V}\|\,\|\mathbf{W}\| = 2\sqrt{5}\sqrt{34} \approx 26$
27. $\|\mathbf{V} + \mathbf{W}\| = \|\langle 7, 7\rangle\| = 7\sqrt{2} \approx 9.9$, and $\|\mathbf{V}\| + \|\mathbf{W}\| = 2\sqrt{5} + \sqrt{34} \approx 10.3$
28. $\|-2\mathbf{V}\| = \|\langle -8, -4\rangle\| = 4\sqrt{5}$, and $2\|\mathbf{V}\| = 2\|\langle 4, 2\rangle\| = 2 \cdot 2\sqrt{5}$ **29.** $11/(\sqrt{170}) \approx .8437$
30. $\langle 1, 3\rangle \cdot \langle x - 6, y - 4\rangle = x + 3y - 18 = 0$ **31.** 101° and 79° (approximate values) **32.** $\sqrt{13}$
33. distance = 7
34. $\langle 3/\sqrt{34}, 3/\sqrt{34}, 4/\sqrt{34}\rangle$ **35.** 88
36. $\langle 2, -3, 4\rangle \cdot \langle 5, 2, -1\rangle = 0$ **37.** $29/(7\sqrt{42})$ **38.** $y = -2$
39. $2x + y + 5z = 8$
40. $\mathbf{N} = \langle 1, 3, 2\rangle$

Chapter 9

Section 9.1 (page 518)

1. one, none, infinitely many **2.** planes, point **3.** infinite **4.** two **5.** constant **6.** true **7.** false **8.** false **9.** true **10.** false **11.** $x = 3, y = 2, z = -1$ **13.** $x = 5/2, y = 3/2, z = 0$ **15.** $x = -2, y = 1, z = 3$
17. The reduced system is of the form $x + y = c_1, x + y = c_2$, where $c_1 \neq c_2$.
19. The reduced system is of the form $x + z = c_1, x + z = c_2$, where $c_1 \neq c_2$. **21.** $x = -2z, y = 2z, z =$ any real number
23. $x = -20, y =$ any real number, $z = 4y - 32$ **25.** $x = -4 - t, y = -t, z = 6 + t, t =$ any real number
27. $x = 2, y = 3, z = 4, t = -2$ **29.** Each pair of planes intersects in a line, resulting in three parallel lines.
31. (a) The reduced system is of the form $y - z = c_1, y - z = c_2$, where $c_1 \neq c_2$. (b) $x - y - z = D$, where D is any real number
33. \$1000 in certificates, \$1500 in bonds, and \$2500 in mutual funds **35.** 20 valued at \$45, 30 at \$30, and 50 at \$25
37. 500 of type 1, 400 of type 2, and 100 of type 3 **39.** $7x^2 + 7y^2 + 31x + 29y = 126$

Section 9.2 (page 524)

1. equations, unknowns, determinant of the coefficients **2.** $a_1\begin{vmatrix} b_2 & c_2 \\ b_3 & c_3 \end{vmatrix} - b_1\begin{vmatrix} a_2 & c_2 \\ a_3 & c_3 \end{vmatrix} + c_1\begin{vmatrix} a_2 & b_2 \\ a_3 & b_3 \end{vmatrix}$ **3.** unique, point
4. none, infinitely many **5.** the determinant of the coefficients is zero **6.** true **7.** false **8.** false **9.** true
10. true **11.** -9 **13.** $47/4$ **21.** The determinant of the coefficients equals zero.
23. The number of equations does not equal the number of unknowns. **25.** The system has a nontrivial solution.
27. The system has a nontrivial solution. **29.** not collinear; $15x - 10y + 6z = 30$ **31.** not collinear; $12x + 7y + 5z = 16$

Section 9.3 (page 532)

1. 2 **2.** row, column **3.** $i + j$ **4.** cofactor, add the resulting products
5. adding a constant multiple of each element of one row to the corresponding element of another row **6.** false **7.** true
8. true **9.** false **10.** true **11.** (a) $a_{11} = 5$ and $M_{11} = \begin{vmatrix} 4 & -7 \\ 6 & -3 \end{vmatrix} = 30$ (b) $a_{12} = -1$ and $M_{12} = \begin{vmatrix} 3 & -7 \\ 2 & -3 \end{vmatrix} = 5$
(c) $a_{13} = 2$ and $M_{13} = \begin{vmatrix} 3 & 4 \\ 2 & 6 \end{vmatrix} = 10$ **13.** (a) $a_{13} = 2$ and $M_{13} = \begin{vmatrix} 3 & 4 \\ 2 & 6 \end{vmatrix} = 10$ (b) $a_{23} = -7$ and $M_{23} = \begin{vmatrix} 5 & -1 \\ 2 & 6 \end{vmatrix} = 32$
(c) $a_{33} = -3$ and $M_{33} = \begin{vmatrix} 5 & -1 \\ 3 & 4 \end{vmatrix} = 23$ **15.** (a) $b_{13} = -1$ and $M_{13} = \begin{vmatrix} 0 & 2 & -1 \\ -2 & 5 & -2 \\ 3 & -4 & 7 \end{vmatrix} = 23$
(b) $b_{33} = 6$ and $M_{33} = \begin{vmatrix} 1 & 0 & 2 \\ 0 & 2 & -1 \\ 3 & -4 & 7 \end{vmatrix} = -2$ **17.** $|A| = 5 \cdot 30 + (-1)(-5) + 2 \cdot 10 = 175$
19. $|A| = 2 \cdot 10 + (-7)(-32) + (-3)23 = 175$ **21.** 45 **23.** 0
27. -2 times row 2 was added to row 1, which does not change the determinant.
29. Interchanging the two columns changes the sign of the determinant.
31. Since every element of row 3 is multiplied by -1, the determinant is multiplied by -1.
33. By introducing 0's in the first column, we obtain $\begin{vmatrix} 1 & .5 & -4 \\ 0 & 0 & 9 \\ 0 & -2.5 & 22 \end{vmatrix} = 22.5$.
35. By introducing 0's in the second row, we obtain $\begin{vmatrix} 1/2 & 1/4 & -9/8 & 1/4 \\ 0 & 1 & 0 & 0 \\ -1 & 0 & -2 & 3 \\ 5 & 4 & -18 & -7 \end{vmatrix} = 32$.

Section 9.4 (page 544)

1. 2, 2 **2.** triangular **3.** variables, constant term **4.** coefficient, constant
5. interchanging two rows, multiplying each entry of a row by a nonzero constant, adding a constant times each entry of one (fixed) row to the corresponding entry of another (fixed) row
6. true **7.** true **8.** false **9.** true **10.** true **11.** $x = -1, y = 10, z = 5$

13. $x = 2z, y = z + 1, z =$ any real number **19.** $x = 2z + 1/2, y = z + 1/2, z =$ any real number **21.** $\begin{aligned} 2x + y &= -3 \\ 4x - 5y &= 6 \end{aligned}$

23. $\begin{aligned} x + 2y - 3z + 4t &= 5 \\ 2x - y + 7t &= 6 \end{aligned}$ **25.** $x = 2, y = -3, z = 4$ **27.** $\left[\begin{array}{cc|c} 1 & -1 & 6 \\ 2 & 1 & 3 \end{array}\right]$ **29.** $\left[\begin{array}{ccc|c} 2 & -1 & 3 & 8 \\ 7 & 1 & -1 & 4 \\ -1 & 2 & 1 & 3 \end{array}\right]$ **31.** $x = 2, y = 3$

33. $x = 3, y = -2, z = 1$ **35.** $x = 3 - z, y = 1 + 2z, z =$ any real number **37.** $x = 1, y = -2, z = 3$
39. no solution **41.** $x = 1, y = -2$ **43.** $x = 2, y = 1, z = 3$ **45.** $x = -7z - 10, y = -5z - 6, z =$ any real number
47. $x = 3y + 5, y =$ any real number **49.** $x = -3, y = 1, z = 2$

Section 9.5 (page 556)

1. rows, columns **2.** same size **3.** cA **4.** columns, rows **5.** I, invertible **6.** false **7.** true **8.** false
9. false **10.** true **11.** (a) 3×1 (b) 1×3 (c) 2×3 (d) 3×3 (e) 3×5 **13.** (a) $x = 2, y = 4$

(b) $x = -3, y = 1$ **15.** $\begin{bmatrix} 2 & 1 & 4 \\ 6 & 0 & 4 \\ 10 & 10 & 10 \end{bmatrix}$ **17.** $\begin{bmatrix} 0 \\ 0 \\ 0 \end{bmatrix}$ **19.** A and B cannot be added because they are not the same size.

21. $\begin{bmatrix} -3 & 0 & 3 \\ -6 & 6 & -3 \\ -9 & -12 & 0 \end{bmatrix}$ **23.** $\begin{bmatrix} 1 & -2 & 3 \\ 3/2 & 5/2 & 7/2 \end{bmatrix}$ **25.** (a) $2(3A) = 2\begin{bmatrix} 3 & 9 & -6 \\ 0 & -12 & 9 \end{bmatrix} = \begin{bmatrix} 6 & 18 & -12 \\ 0 & -24 & 18 \end{bmatrix} = 6A$

(b) $2B + 3B = \begin{bmatrix} 4 & -2 & 2 \\ 6 & 0 & 10 \end{bmatrix} + \begin{bmatrix} 6 & -3 & 3 \\ 9 & 0 & 15 \end{bmatrix} = \begin{bmatrix} 10 & -5 & 5 \\ 15 & 0 & 25 \end{bmatrix} = 5\begin{bmatrix} 2 & -1 & 1 \\ 3 & 0 & 5 \end{bmatrix} = 5B = (2 + 3)B$

(c) $-2(A + C) = -2\begin{bmatrix} 1 & 0 & -1 \\ 2 & -1 & 2 \end{bmatrix} = \begin{bmatrix} -2 & 0 & 2 \\ -4 & 2 & -4 \end{bmatrix}$; $-2A - 2C = \begin{bmatrix} -2 & -6 & 4 \\ 0 & 8 & -6 \end{bmatrix} - \begin{bmatrix} 0 & -6 & 2 \\ 4 & 6 & -2 \end{bmatrix} = \begin{bmatrix} -2 & 0 & 2 \\ -4 & 2 & -4 \end{bmatrix}$

27. $[11]$ **29.** $\begin{bmatrix} -13 & -15 \\ 19 & 5 \end{bmatrix}$ **31.** $\begin{bmatrix} 4 & -3 & 43 \\ 0 & -5 & 9 \\ 10 & 24 & 4 \end{bmatrix}$ **33.** $\begin{bmatrix} 0 & 3 & 6 \\ 4 & 2 & 0 \\ 1 & 8 & 15 \\ -14 & 2 & 18 \end{bmatrix}$

35. (a) $(AB)C = \begin{bmatrix} 1 & 5 \\ 6 & 2 \\ 11 & -1 \end{bmatrix}\begin{bmatrix} -5 & -2 \\ 1 & 6 \end{bmatrix} = \begin{bmatrix} 0 & 28 \\ -28 & 0 \\ -56 & -28 \end{bmatrix}$; $A(BC) = \begin{bmatrix} 1 & -1 \\ 2 & 0 \\ 3 & 1 \end{bmatrix}\begin{bmatrix} -14 & 0 \\ -14 & -28 \end{bmatrix} = \begin{bmatrix} 0 & 28 \\ -28 & 0 \\ -56 & -28 \end{bmatrix}$

(b) $A(B + C) = \begin{bmatrix} 1 & -1 \\ 2 & 0 \\ 3 & 1 \end{bmatrix}\begin{bmatrix} -2 & -1 \\ 3 & 2 \end{bmatrix} = \begin{bmatrix} -5 & -3 \\ -4 & -2 \\ -3 & -1 \end{bmatrix}$; $AB + AC = \begin{bmatrix} 1 & 5 \\ 6 & 2 \\ 11 & -1 \end{bmatrix} + \begin{bmatrix} -6 & -8 \\ -10 & -4 \\ -14 & 0 \end{bmatrix} = \begin{bmatrix} -5 & -3 \\ -4 & -2 \\ -3 & -1 \end{bmatrix}$ **39.** $\begin{bmatrix} 3/5 & -2/5 \\ 1/5 & 1/5 \end{bmatrix}$

41. A is not invertible because $|A| = 0$. **43.** $A^{-1} = \begin{bmatrix} 4/15 & -1/6 & 1/10 \\ 7/15 & 2/6 & -2/10 \\ -1/15 & 1/6 & 1/10 \end{bmatrix}$

45. $\begin{bmatrix} 2 & -1 \\ 8 & 1 \end{bmatrix}\begin{bmatrix} x \\ y \end{bmatrix} = \begin{bmatrix} 4 \\ 1 \end{bmatrix}$; $\begin{bmatrix} x \\ y \end{bmatrix} = \begin{bmatrix} 1/10 & 1/10 \\ -4/5 & 1/5 \end{bmatrix}\begin{bmatrix} 4 \\ 1 \end{bmatrix} = \begin{bmatrix} 1/2 \\ -3 \end{bmatrix}$

47. $\begin{bmatrix} 3 & 6 & 1 \\ 3 & -3 & 2 \\ 6 & 9 & 2 \end{bmatrix}\begin{bmatrix} x \\ y \\ z \end{bmatrix} = \begin{bmatrix} 0 \\ 2 \\ 1 \end{bmatrix}$; $\begin{bmatrix} x \\ y \\ z \end{bmatrix} = \begin{bmatrix} -8/3 & -1/3 & 5/3 \\ 2/3 & 0 & -1/3 \\ 5 & 1 & -3 \end{bmatrix}\begin{bmatrix} 0 \\ 2 \\ 1 \end{bmatrix} = \begin{bmatrix} 1 \\ -1/3 \\ -1 \end{bmatrix}$

49. (a) 60 type I clocks and 5 type II clocks (b) 48 type I clocks and 20 type II clocks
(c) 28 type I clocks and 45 type II clocks

Chapter 9 Review Exercises (page 560)

1.–5. $x = -3, y = 2, z = 4$ **6.** The determinant of the coefficients equals zero.
7.–8. $x = z + 1, y = 2z, z =$ any real number **9.** The reduced system is of the form $x + 4y = c_1$, $x + 4y = c_2$, where $c_1 \neq c_2$.

10. $\left[\begin{array}{ccc|c} 1 & -1 & 1 & 5 \\ 0 & 5 & -1 & -3 \\ 0 & 0 & 0 & -8 \end{array}\right]$ is in row-echelon form, and the last row corresponds to the equation $0 = -8$.

11. The determinant of the coefficients equals zero. **12.** $x = -z/5, y = 3z/5, z =$ any real number
13. $x = -1, y = -8, z = 7, t = 6$ **14.** 13 **15.** -14 **16.** -93 **17.** 0 **18.** 0 **19.** 12
20. Matrices of different sizes cannot be added.
21. The number of columns of the first matrix is not equal to the number of rows of the second.
22. The inverse of the first matrix does not exist. **23.** A determinant is defined only for a square matrix.

24. $\begin{bmatrix} 2 & 12 \\ 13 & 13 \end{bmatrix}$ **25.** $\begin{bmatrix} -12 \\ 16 \end{bmatrix}$ **26.** $\begin{bmatrix} -3 & -4 \\ 6 & 13 \\ 11 & 3 \end{bmatrix}$ **27.** $\begin{bmatrix} 8 & -21 \\ -7 & 17 \end{bmatrix}$ **28.** -11 **29.** $\begin{bmatrix} 1 & 0 \\ 0 & 1 \end{bmatrix}$ **30.–31.** $\begin{bmatrix} 3/2 & 1/4 \\ -1 & -1/2 \end{bmatrix}$

32. $\begin{bmatrix} 1 & 1/2 & 0 \\ -1/2 & -1/2 & 1/2 \\ -1/2 & -1 & 1/2 \end{bmatrix}$ **33.** $x - y = 2$ **34.** $(5, 1, -2)$ **35.** $\begin{bmatrix} 2 & -1 \\ 1 & 3 \end{bmatrix}\begin{bmatrix} x \\ y \end{bmatrix} = \begin{bmatrix} 7 \\ 0 \end{bmatrix}$; $A^{-1} = \begin{bmatrix} 3/7 & 1/7 \\ -1/7 & 2/7 \end{bmatrix}$; $x = 3, y = -1$

36. $\begin{bmatrix} 1 & 1 & 3 \\ 0 & -1 & 1 \\ 1 & 0 & 1 \end{bmatrix}\begin{bmatrix} x_1 \\ x_2 \\ x_3 \end{bmatrix} = \begin{bmatrix} 2 \\ -4 \\ 1 \end{bmatrix}$; $A^{-1} = \begin{bmatrix} -1/3 & -1/3 & 4/3 \\ 1/3 & -2/3 & -1/3 \\ 1/3 & 1/3 & -1/3 \end{bmatrix}$; $x_1 = 2, x_2 = 3, x_3 = -1$

37. The matrix equation $AB = I$ is equivalent to the two systems of equations $\begin{cases} a_{11}b_{11} + a_{12}b_{21} = 1 \\ a_{21}b_{11} + a_{22}b_{21} = 0 \end{cases}$ and $\begin{cases} a_{11}b_{12} + a_{12}b_{22} = 0 \\ a_{21}b_{12} + a_{22}b_{22} = 1 \end{cases}$.
If $|A| \neq 0$, the solutions for the entries in B are $b_{11} = a_{22}/|A|$, $b_{21} = -a_{21}/|A|$, $b_{12} = -a_{12}/|A|$, and $b_{22} = a_{11}/|A|$.
38. By expanding the determinant on the left by the cofactors of column 2, we obtain $ka_{12}A_{12} + ka_{22}A_{22} + ka_{32}A_{32} = k(a_{12}A_{12} + a_{22}A_{22} + a_{32}A_{32}) = k|A|$.

39. $B^tA^t = \begin{bmatrix} a'a + c'b & a'c + c'd \\ b'a + d'b & b'c + d'd \end{bmatrix}$; $(AB)^t = \begin{bmatrix} aa' + bc' & ab' + bd' \\ ca' + dc' & cb' + dd' \end{bmatrix}^t = \begin{bmatrix} aa' + bc' & ca' + dc' \\ ab' + bd' & cb' + dd' \end{bmatrix}$

40. 20 of type A, 10 of type B, 5 of type C

Chapter 10

Section 10.1 (page 572)

1. $x - 5$ **2.** -3 **3.** $-56, 14, -1$ **4.** $[x - (1 + i)][x - (1 - i)]$ **5.** $6x^4 - 5x^3 + x^2 + 3, x - 4$ **6.** false
7. true **8.** false **9.** false **10.** true **11.** (a) 10 (b) -29 **13.** (a) 0 (b) 42 **15.** (a) 0 (b) 0

17.
$$\begin{array}{r|l} & x + (2 + a) \\ \hline x - a & x^2 + 2x - 3 \\ & x^2 - ax \\ \hline & (2 + a)x - 3 \\ & (2 + a)x - a(2 + a) \\ \hline & a^2 + 2a - 3 = P(a) \end{array}$$

19.
$$\begin{array}{r|l} & 2x^2 + (2a + 1)x + (2a^2 + a - 4) \\ \hline x - a & 2x^3 + x^2 - 4x + 5 \\ & 2x^3 - 2ax^2 \\ \hline & (2a + 1)x^2 - 4x \\ & (2a + 1)x^2 - a(2a + 1)x \\ \hline & (2a^2 + a - 4)x + 5 \\ & (2a^2 + a - 4)x - a(2a^2 + a - 4) \\ \hline & 2a^3 + a^2 - 4a + 5 = P(a) \end{array}$$

21. quotient $= x^2 - x$; remainder $= -5$ **23.** quotient $= 2x^4 - 2x^3 + 7x^2 - 7x + 4$; remainder $= 0$
25. $P(5) = 874$; $P(-5) = 384$ **27.** $P(5) = 87{,}978$; $P(-5) = 99{,}478$ **29.** $P(3/2) = 8.125$; $P(-2.5) = 9.125$
31. $P(-1) = 0$ **33.** $P(2) = 0$ **35.** $P(1/2) = 0$ **37.** $x - 1$, $x - 2$, and $x + 3$, all of order 1
39. $x + 2$ of order 2 and $x - 3$ of order 1 **41.** The complex and real factored form of $P(x)$ is $(x - 2)(x + 2)(x - 1)$.
43. The complex and real factored form of $P(x)$ is $(x + 1)(x - 1)(x + 2)(x - 2)$.
45. The complex factored form of $P(x)$ is $(x + 2)[x - (1 + i)][x - (1 - i)]$, and the real factored form is $(x + 2)(x^2 - 2x + 2)$.
47. The complex and real factored form of $P(x)$ is $(x - 1)(x + 1/2)(x - 1/3)(x + 1/4)$.
49. The complex factored form of $P(x)$ is $(x - 1/3)^2[x - (3 - 2i)][x - (3 + 2i)]$, and the real factored form is $(x - 1/3)^2(x^2 - 6x + 13)$.
53. Complex zeros appear in conjugate *pairs*. **55.** $P(x) = [(2x - 4)x + 5]x + 7$; $P(6) = [(2 \cdot 6 - 4)6 + 5]6 + 7 = 325$
57. $P(x) = \{[(x + 1)x + 12]x - 4\}x - 15$; $P(7) = \{[(7 + 1)7 + 12]7 - 4\}7 - 15 = 3289$

59. Exercise 55: $6\overline{)2 \quad -4 \quad 5 \quad 7}$

$6)$	2	−4	5	7
		12	48	318
	2	8	53	$325 = P(6)$

Exercise 57:

$7)$	1	1	12	−4	−15
		7	56	476	3304
	1	8	68	472	$3289 = P(7)$

Section 10.2 (page 581)

1. $\pm 1, \pm 3, \pm 1/2, \pm 3/2$ **2.** $5n$, where n is a nonzero integer **3.** $3n$, where n is a nonzero integer **4.** two **5.** one **6.** false **7.** false **8.** false **9.** true **10.** true **11.** $1, -1, 2$ **13.** $1, 2, -2, 4$ **15.** $-2, 1/4$
17. (a) $1, \pm 2$ (b) $P(x) = (x - 1)(x - 2)(x + 2)$ (c) none **19.** (a) $1/2$ (b) $P(x) = 2(x - 1/2)(x^2 + 1)$ (c) $\pm i$
21. (a) $1, -2$ (b) $\pm\sqrt{3}$, $P(x) = (x - 1)(x + 2)(x - \sqrt{3})(x + \sqrt{3})$ (c) none
23. (a) ± 1 (b) $P(x) = (x - 1)(x + 1)(x^2 + 2x + 2)$ (c) $-1 + i, -1 - i$ **25.** $a = -1, b = 2$ **27.** $a = -7, b = 1$
29. three **31.** three **33.** two (including 0) **35.** (a) one positive, two negative (b) $a = -3, b = 1$
37. (a) two positive, one negative (b) $a = -5, b = 7$ **39.** (a) one positive, three negative (b) $a = -8, b = 1$

Section 10.3 (page 588)

1. proper **2.** degree **3.** greater than, equal to, division **4.** real factored form **5.** $n, 2m$ **6.** false **7.** false
8. true **9.** true **10.** true **11.** $\frac{3}{x+2} - \frac{2}{x-3}$ **13.** $\frac{1}{x} + \frac{4}{x-4}$ **15.** $\frac{3}{x} - \frac{1}{x-1} + \frac{5}{(x-1)^2}$
17. $\frac{1}{x} - \frac{1}{x-2} + \frac{3}{(x-2)^2} - \frac{4}{(x-2)^3}$ **19.** $\frac{1}{(x-1)^3}$ **21.** $\frac{1}{x-1} + \frac{2}{(x-1)^2} + \frac{1}{(x-1)^3}$ **23.** $x + 5 - \frac{3}{x-1} + \frac{14}{x-2}$
25. $\frac{x}{x^2+1}$ **27.** $\frac{1}{x} - \frac{x+1}{x^2+x+1}$ **29.** $\frac{1}{8(x-2)} - \frac{1}{8(x+2)} - \frac{1}{2(x^2+4)}$ **31.** $\frac{1}{(x^2+1)^2}$
33. $\frac{1}{4(x-1)} + \frac{1}{4(x-1)^2} - \frac{1}{4(x+1)} + \frac{1}{4(x+1)^2}$

Section 10.4 (page 599)

In each of the following, for $\delta = .01$, computations were made on a 10-place programmable calculator and rounded off; for $\delta = .000001$, computations were made on a microcomputer in double precision (16 places) and rounded off.

1. (a) $a = -2, b = 2$
(b)

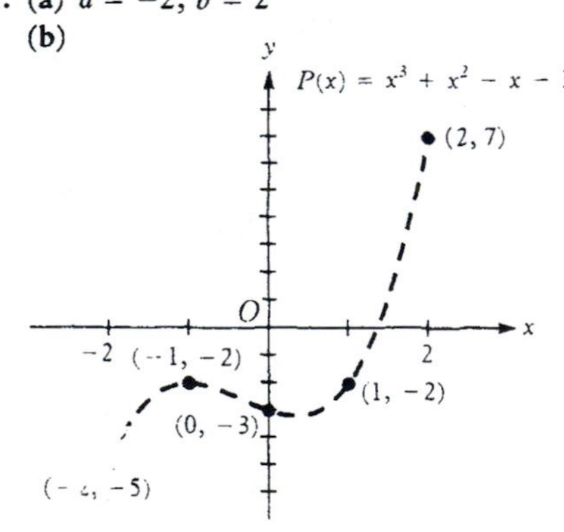

(c) $[1, 2]$
(d) $\delta = .01$, $[1.3516, 1.3594]$;
$\delta = .000001$, $[1.35930347, 1.35930443]$
(e) $\delta = .01$, $c = 1.3595$; $\delta = .000001$, $c = 1.35930412$

3. (a) $a = -1, b = 1$
(b)

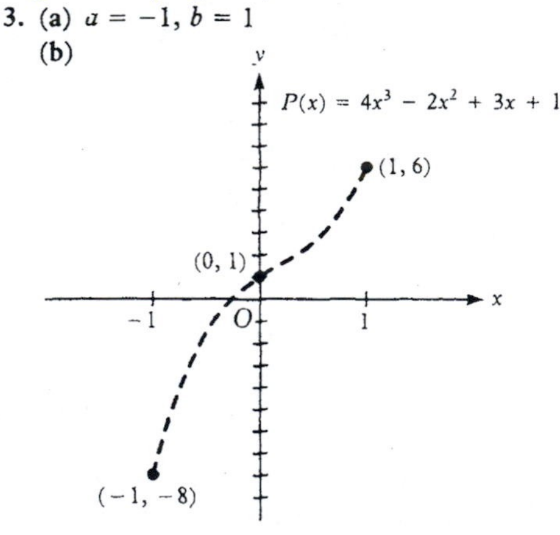

(c) $[-1, 0]$
(d) $\delta = .01$, $[-.2656, -.2578]$;
$\delta = .000001$, $[-.26297951, -.26297855]$
(e) $\delta = .01$, $c = -.2630$; $\delta = .000001$, $c = -.26297874$

5. **(a)** $a = -3$, $b = 2$

(b)

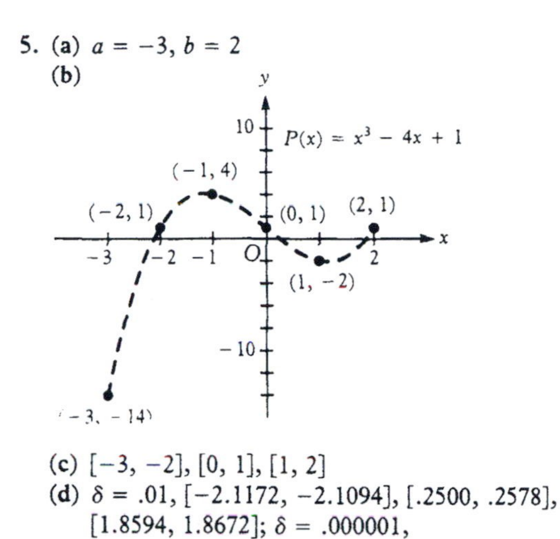

(c) $[-3, -2]$, $[0, 1]$, $[1, 2]$

(d) $\delta = .01$, $[-2.1172, -2.1094]$, $[.2500, .2578]$, $[1.8594, 1.8672]$; $\delta = .000001$, $[-2.11490822, -2.11490726]$, $[.25410080, .25410175]$, $[1.86080551, 1.86080647]$

(e) $\delta = .01$, $c = -2.1150$, $c = .2541$, $c = 1.8610$; $\delta = .000001$, $c = -2.11490754$, $c = .25410169$, $c = 1.86080588$

7. **(a)** $a = -2$, $b = 3$

(b)

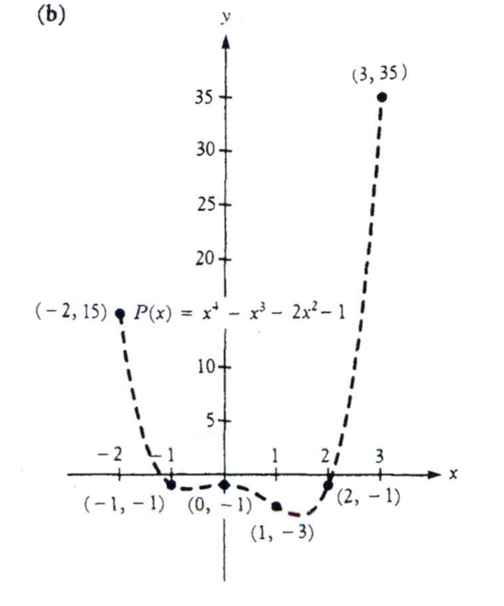

(c) $[-2, -1]$, $[2, 3]$

(d) $\delta = .01$, $[-1.2188, -1.2109]$, $[2.0703, 2.0781]$; $\delta = .000001$, $[-1.21195984, -1.21195888]$, $[2.07548237, 2.07548332]$

(e) $\delta = .01$, $c = -1.2125$, $c = 2.0758$; $\delta = .000001$, $c = -1.21195979$, $c = 2.07548287$

9. **(a)** $a = -1$, $b = 5$

(b)

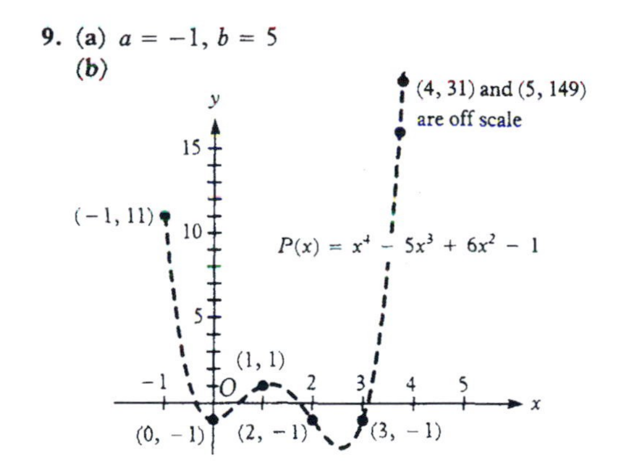

(c) $[-1, 0]$, $[0, 1]$, $[1, 2]$, $[3, 4]$ **(d)** $\delta = .01$, $[-.3594, -.3516]$, $[.5156, .5234]$, $[1.7344, 1.7422]$, $[3.0938, 3.1016]$; $\delta = .000001$, $[-.35567474, -.35567379]$, $[.52273941, .52274036]$, $[1.73763943, 1.73764038]$, $[3.09529305, 3.09529400]$

(e) $\delta = .01$, $c = -.3560$, $c = .5227^\dagger$, $c = 1.7377$, $c = 3.0959$; $\delta = .000001$, $c = -.35567444$, $c = .52274000^\dagger$, $c = 1.73764031$, $c = 3.09529399$

†When Newton's method was started at either 0 or 1, the process stopped before completion because a point c was reached for which $P'(c) = 0$. The process was then started at .5 and reached completion.

Chapter 10 Review Exercises (page 601)

1. Every polynomial of degree n has n roots in the complex number system.
2. The general polynomial equation of degree ≥ 5 cannot be solved by algebraic methods. **3.** remainder $= P(1) = -4$
4. remainder $= P(-2) = 8$ **5.** $P(5) = 276$ **6.** $P(-5) = -151$ **7.** $P(\sqrt{2}) = 0$ **8.** $P(i) = 0$ **9.** $x, x - 1, x + 2$
10. $P(x) = 2x^3 - x^2 + 2x - 1$ **11.** $P(x) = (x + 2)(x - 3)^2(x - \sqrt{2})[x - (1 + 2i)][x - (1 - 2i)]$
12. $P(x) = (x + 2)(x - 3)^2(x - \sqrt{2})(x^2 - 2x + 5)$ **13.** if $3i$ is a root, the complex conjugate $-3i$ is also a root
14. $1, -1/2$ **15.** The only possible rational roots are ± 1 and ± 2, but $P(1) = 6$, $P(-1) = -16$, $P(2) = 8$, and $P(-2) = -180$.
16. 2 is an upper bound; -2 is a lower bound. **17.** 5 is an upper bound; -1 is a lower bound. **18.** 1 is a root of order 2.
19. $(x - 1)^2(x^2 + 1)$ **20.** $3n$, where n is any nonzero integer **21.** 0 and ± 1 **22.** 3 **23.** 1 **24.** 4 **25.** 3
26. By Descartes' rule of signs, there are no positive roots and no negative roots. Also, 0 is not a root. **27.** none
28. $1, 2, \sqrt{2}, -\sqrt{2}$ **29.** $\dfrac{A}{x-1} + \dfrac{B}{(x-1)^2} + \dfrac{C}{(x-1)^3} + \dfrac{D}{(x-1)^4}$ **30.** $\dfrac{Ax+B}{x^2+2x+5} + \dfrac{Cx+D}{(x^2+2x+5)^2}$
31. $\dfrac{A}{x+1} + \dfrac{B}{x-2} + \dfrac{C}{(x-2)^2} + \dfrac{D}{(x-2)^3} + \dfrac{Ex+F}{x^2+x+5} + \dfrac{Gx+H}{x^2+3x+4} + \dfrac{Ix+J}{(x^2+3x+4)^2}$ **32.** $\dfrac{1}{x} + \dfrac{2}{x-2} + \dfrac{3}{x+3}$
33. $\dfrac{-1}{x^2} + \dfrac{1}{2(x-1)} - \dfrac{1}{2(x+1)}$ **34.** $2x + 1 + \dfrac{4}{x+1} + \dfrac{3}{x-2}$ **35.** $x^2 - 1 - \dfrac{1}{x} + \dfrac{2x+3}{x^2+1}$ **36.** $[-2, 4]$

37.

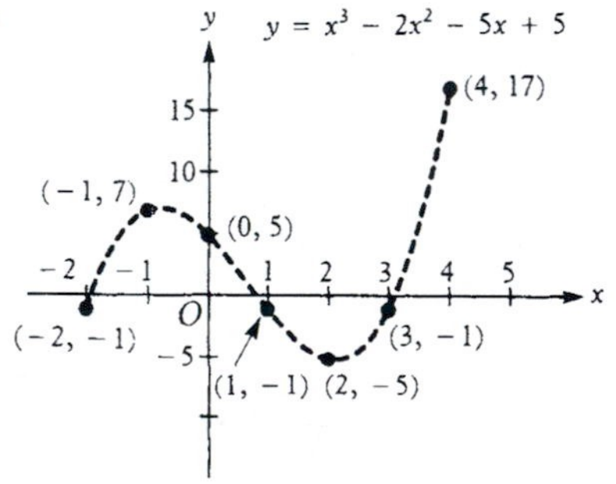

y
$y = x^3 - 2x^2 - 5x + 5$
(4, 17)
15
10
(−1, 7)
(0, 5)
−2 −1 1 2 3 4 5 x
O
(−2, −1)
−5
(3, −1)
(1, −1) (2, −5)

38. $x^3 - 2x^2 - 5x + 5 = 0$ has roots in the intervals $[-2, -1]$, $[0, 1]$, and $[3, 4]$. The bisection method with $\delta = .1$ reduces these intervals to $[-1.9375, -1.875]$, $[.8125, .875]$, and $[3.0625, 3.125]$, respectively.
39. Using Newton's method with $\epsilon = .05$ and starting at $b = -1$, 1, and 4 gives the approximate roots -1.9327, $.8333$, and 3.0947, respectively.
40. The bisection method with $\delta = .00001$ reduces the intervals $[-2, -1]$, $[0, 1]$, and $[3, 4]$ of Exercise 38 to $[-1.93080902, -1.93080139]$, $[.83704376, .83703613]$, and $[3.09375763, 3.09376526]$, respectively. The corresponding approximate roots obtained by Newton's method, with $\epsilon = .00001$, are -1.93080160, $.83703704$, and 3.09376399, respectively.

Chapter 11

Section 11.1 (page 610)

1. probable **2.** certain **3.** cannot **4.** $E(n)$ is true for all positive odd integers **5.** $E(k + 5)$ **6.** false **7.** false
8. true **9.** true **10.** true **35.** $n(7n - 5)/2$ **37.** $n(n + 1)(n + 2)(n + 3)(n + 4)/5$

Section 11.2 (page 618)

1. the set of integers from 1 to n, the set of all positive integers
2. the difference of any two successive terms is a fixed constant
3. the ratio of any two successive terms is a fixed nonzero constant **4.** $y = 2 - 3(x - 1)$ **5.** $y = 3(3/4)^{x-1}$ **6.** true
7. false **8.** true **9.** false **10.** true **11.** 1, 3, 5, 7, 9, 11 **13.** $2, (3/2)^2, (4/3)^3, (5/4)^4$ **15.** 1, 3, 6, 10, 15, 21
17. $5/6, 6/7, k/(k + 1)$ **19.** $-1/128, 1/256, (-1)^k/2^k$ **21.** $1/(9 \cdot 11), 1/(11 \cdot 13), 1/[(2k - 1)(2k + 1)]$
23. 3/3, 4/3, 5/3, 6/3, 7/3, 8/3; arithmetic **25.** $-2/6$, 1/6, 4/6, 7/6, 10/6; arithmetic
27. 3, 5, 7 9, 11, 13; arithmetic **29.** $3 \cdot 2^3, 3 \cdot 2^4, 3 \cdot 2^5, 3 \cdot 2^6, 3 \cdot 2^7$; geometric **31.** 89 **33.** 8 **35.** 384 **37.** 4

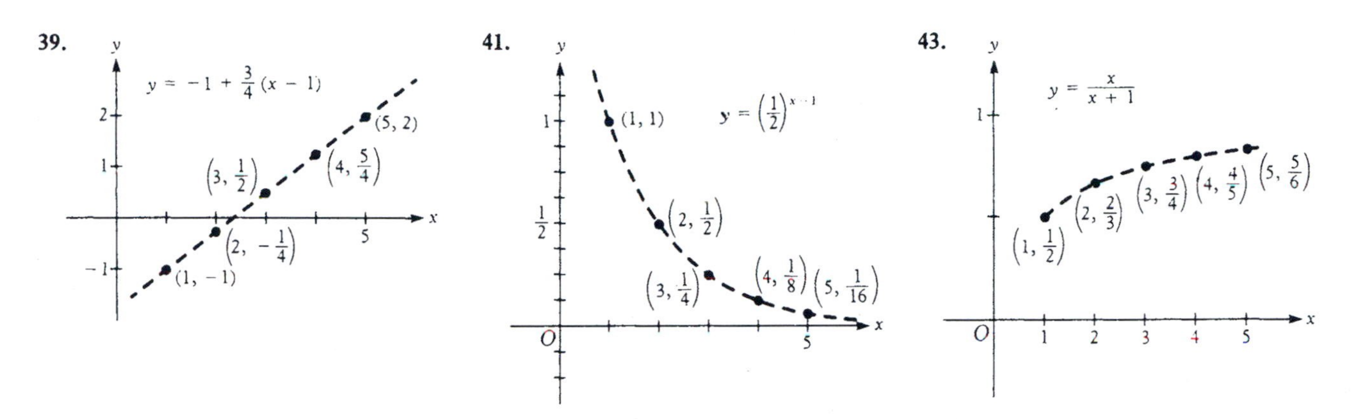

45. $a_k = 16 + 32(k - 1)$, $k = 1, 2, \ldots, n$, is an arithmetic sequence with $a = 16$ and $d = 32$.

Section 11.3 (page 626)

1. a sum of a finite number of terms **2.** a series whose terms form an arithmetic sequence **3.** a series whose terms form a geometric sequence **4.** $n(a + l)/2$ **5.** $a(1 - r^n)/(1 - r)$ if $r \neq 1$, na if $r = 1$ **6.** false **7.** true **8.** false **9.** true **10.** true **11.** $0 + 3 + 8 + 15 + 24$ **13.** $1 + 3 + 8 + 17 + 32$ **15.** $\sum_{k=1}^{5}(2k - 1)^2$ **17.** $\sum_{k=1}^{7}(-1)^{k+1}\frac{k}{k + 1}$ **19.** 630 **21.** 40

23. If the substitution $j = k - 1$ is made, then $\sum_{k=1}^{n}[a + (k - 1)d] = \sum_{j=0}^{n-1}[a + jd]$, which equals $\sum_{k=0}^{n-1}(a + kd)$. **25.** 2400 **27.** 315 **29.** 1225 **31.** 2703 **33.** 20 **35.** 16 **37.** $(2^9 - 1)/2^8 = 511/256$ **39.** .272727 **41.** −910 **43.** 513 **45.** −.5 **47.** .75 **49.** 2 **51.** 3/11 **53.** not possible because $r = -3$ and $|-3| > 1$ **55.** approximately \$189,910,000; approximately \$18,032,000,000,000,000 **57.** \$494.48

Section 11.4 (page 636)

1. if one event has n possible outcomes, and if for each one of these, a second event has m possible outcomes, then the first event followed by the second has $n \cdot m$ possible outcomes **2.** $n_1 n_2 \ldots n_k$ **3.** $n!/(n - r)!$ **4.** $n!/[(n - r)!r!]$ **5.** permutation, combination **6.** false **7.** true **8.** false **9.** true **10.** true **11.** 4 ways **13.** 12 choices **15.** 6 schedules **17.** 1,048,576 ways **19.** 27 numbers **21.** (a) 6720 (b) 7 (c) 1 (d) 120 **23.** 720 finishes **25.** 120 numbers **27.** 210 finishes **29.** 6720 ways **31.** (a) 252 (b) 1 (c) 1 **33.** 126 ways **35.** 70 combinations **37.** 4410 choices **39.** (a) 4 (b) 12 (c) 4

Section 11.5 (page 642)

1. $HTH, HTT, THH, THT, TTH, TTT$ **2.** HHH, HHT, HTH, THH **3.** of not getting two consecutive heads or tails **4.** $0 \leq P(a_i) \leq 1$, 1 **5.** $P(a_1) + P(a_2) + \ldots + P(a_n)$ **6.** true **7.** false **8.** true **9.** false **10.** false **11.** (a) 1/6 (b) 1/2 **13.** (a) 2/5 (b) 3/5 **15.** (a) 1/3 (b) 2/3 **17.** (a) 5/36 (b) 5/18 **19.** (a) 1/169 (b) 1/16 **21.** (a) 3/10 (b) 1/10 **23.** (a) 4/7 (b) 3/7 **25.** (a) 1/2 (b) 3/4 **27.** (a) $13^3/(52 \cdot 51 \cdot 50) \approx .017$ (b) $13^3 \cdot 3!/(52 \cdot 51 \cdot 50) \approx .1$ **29.** (a) $5!5!/10! \approx .004$ (b) $5 \cdot 5 \cdot 4 \cdot 4 \cdot 3 \cdot 3 \cdot 2 \cdot 2 \cdot 1 \cdot 1/10! \approx .004$ **31.** (a) $3 \cdot 4 \cdot 2 \cdot 3/(7 \cdot 6 \cdot 5 \cdot 4) = 3/35$ (b) $4 \cdot 3 \cdot 3 \cdot 2/(7 \cdot 6 \cdot 5 \cdot 4) = 3/35$ **33.** (a) ${}_4C_3 \cdot {}_4C_2/{}_{52}C_5 \approx .000009$ (b) $13{}_4C_3 \cdot 12{}_4C_2/{}_{52}C_5 \approx .0014$ **35.** (a) $4 \cdot 4 \cdot 4 \cdot 4 \cdot 4/{}_{52}C_5 \approx .0004$ (b) $4 \cdot 4 \cdot 4 \cdot 4 \cdot 4 \cdot 10/{}_{52}C_5 \approx .004$

Section 11.6 (page 649)

1. $P(A|B)$, conditional **2.** $P(A \cap B)/P(B)$ **3.** $P(A)$ **4.** $w_1p_1 + w_2p_2 + \ldots + w_np_n$ **5.** $E > 0, E < 0, E = 0$ **6.** true **7.** false **8.** true **9.** false **10.** true **11.** (a) 1/2 (b) 2/7 **13.** (a) 1/3 (b) 2/11 **15.** (a) 4/5 (b) 4/5 **17.** (a) ${}_3C_2/{}_{51}C_2 \approx .0024$ (b) $1/25 = .04$ **19.** 1/5 **21.** 1/100 **23.** (a) 1/2 (b) 1/4 **25.** $4 \cdot 4 \cdot 4/{}_{50}C_3 \approx .0033$ **27.** $48/(48 + {}_4C_3 \cdot {}_{48}C_2) \approx .01$ **29.** 0 **31.** \$3 **33.** 0

Chapter 11 Review Exercises (page 652)

1. (1) The given equation is true for $n = 1$ since $2 = 1(3 \cdot 1 + 1)/2$.

(2) Now suppose the equation holds when $n = k$ for some positive integer k. Then

$$\begin{aligned} 2 + 5 + 8 + \dots + 3k - 1 + [3(k+1) - 1] &= 2 + 5 + 8 + \dots + 3k - 1 + 3k + 2 \\ &= \frac{k(3k+1)}{2} + 3k + 2 \\ &= \frac{k(3k+1) + 2(3k+2)}{2} \\ &= \frac{3k^2 + 7k + 4}{2} \\ &= \frac{(k+1)[3(k+1)+1]}{2}, \end{aligned}$$

which is the given equation for $n = k + 1$.

2. (1) The given equation is true for $n = 1$ since $3 = 3(3^1 - 1)/2$.

(2) Now suppose that the equation holds when $n = k$ for some positive integer k. Then

$$\begin{aligned} 3 + 3^2 + \dots + 3^k + 3^{k+1} &= \frac{3(3^k - 1)}{2} + 3^{k+1} \\ &= \frac{3^{k+1} - 3 + 2 \cdot 3^{k+1}}{2} \\ &= \frac{3 \cdot 3^{k+1} - 3}{2} \\ &= \frac{3(3^{k+1} - 1)}{2}, \end{aligned}$$

which is the given equation for $n = k + 1$.

3. (1) The given equation is true for $n = 1$ since $1/(1 \cdot 4) = 1/(3 \cdot 1 + 1)$.

(2) Suppose the equation is true for $n = k$. Then

$$\begin{aligned} &\frac{1}{1 \cdot 4} + \frac{1}{4 \cdot 7} + \dots + \frac{1}{(3k-2)(3k+1)} + \frac{1}{[3(k+1)-2][3(k+1)+1]} \\ &= \frac{k}{3k+1} + \frac{1}{(3k+1)(3k+4)} \\ &= \frac{k(3k+4)+1}{(3k+1)(3k+4)} \\ &= \frac{3k^2+4k+1}{(3k+1)(3k+4)} \\ &= \frac{(3k+1)(k+1)}{(3k+1)(3k+4)} \\ &= \frac{k+1}{3k+4} \\ &= \frac{k+1}{3(k+1)+1}, \end{aligned}$$

which is the given equation for $n = k + 1$.

4. The left side is equal to $n/(n+1)$ for all positive integers n (see Exercise 19 in the first section of this chapter). However, *if* the equation were true for some integer k, then we would have

$$\frac{1}{1\cdot 2}+\frac{1}{2\cdot 3}+\ldots+\frac{1}{k(k+1)}+\frac{1}{(k+1)(k+2)}$$
$$=\frac{2k+1}{k+1}+\frac{1}{(k+1)(k+2)}$$
$$=\frac{(2k+1)(k+2)+1}{(k+1)(k+2)}$$
$$=\frac{2k^2+5k+3}{(k+1)(k+2)}$$
$$=\frac{(2k+3)(k+1)}{(k+1)(k+2)}$$
$$=\frac{2k+3}{k+2}$$
$$=\frac{2(k+1)+1}{(k+1)+1},$$

which is the given equation for $k+1$.

5. $$1+2+\ldots+n+n+1+n+2=\frac{n(n+1)}{2}+2n+3$$
$$=\frac{n(n+1)+2(2n+3)}{2}$$
$$=\frac{n^2+n+4n+6}{2}$$
$$=\frac{n^2+5n+6}{2}$$
$$=\frac{(n+2)(n+3)}{2},$$

which is the given equation for $n+2$.

6. 1/2, −2/3, 3/4, −4/5, 5/6, −6/7 **7.** $a_k = k - 1/2^k$ **8.** $a_k = 1 + (k-1)3/4$ or $a_k = (3k+1)/4$ **9.** 1/6
10. $a_k = -4(-2/3)^{k-1}$ or $a_k = 4(-1)^k(2/3)^{k-1}$ **11.** $\pm\sqrt{5}$

12.

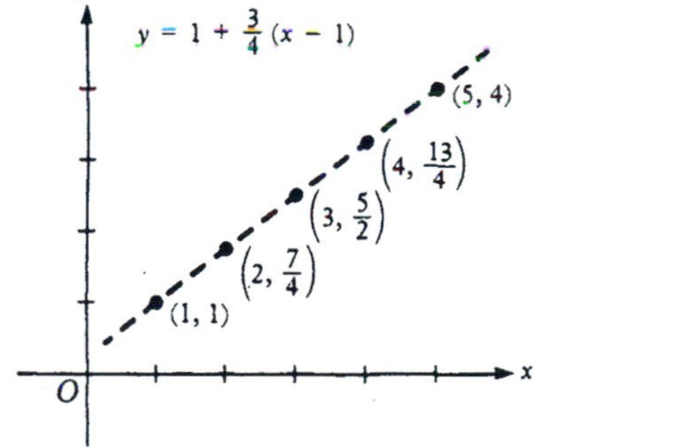

13.

y
$y = \left(\frac{3}{2}\right)^{x-1}$
(4, 27/8)
(3, 9/4)
(2, 3/2)
(1, 1)
O 1 2 3 4 5 x

14. 2 − 4 + 6 − 8 + 10 − 12 **15.** $\sum_{k=1}^{99}\frac{k}{k+1}$ **16.** 375/4 **17.** −532/243 **18.** 2085 **19.** $(3^7 - 2^7)/3^6 = 2059/729$
20. 3 **21.** 10/11 **22.** (a) 336 (b) 1 (c) 10! = 3, 628, 800 **23.** (a) 45 (b) 1 (c) 1 **24.** 20 flights
25. 90 selections **26.** 11,088,000 choices **27.** (a) 720 (b) 7776 (c) 3888 **28.** 3/20 **29.** (a) 3/8 (b) 11/16
30. (a) 1/2 (b) 2/9 **31.** (a) ${}_{13}C_3/{}_{52}C_3 \approx .013$ (b) $13 \cdot {}_4C_3/{}_{52}C_3 \approx .0024$ **32.** 33/100 **33.** 8/25 **34.** 7/90
35. 2/49 **36.** (a) 3/8 (b) 1/2 **37.** (a) 1/2 (b) 3/4 **38.** 1/3 **39.** 1 point **40.** −$2.50

Index

What is Plane Trigonometry?

Trigonometry grew out of the application of geometry to astronomy. The ancient Babylonians divided the circumference of a circle into 360 degrees, and, using their base-60 number system, divided each degree into 60 minutes and each minute into 60 seconds.

The Greek Hipparchus is called the father of trigonometry. In about 140 B.C. he constructed the first trigonometric table, a table of chord lengths corresponding to central angles of a circle. The first table of sines (half-chords) appeared around 400 A.D., and about 100 years later mathematicians began to use the cosine. The tangent and cotangent came into use around 900 A.D., but it was not until 1550 that the trigonometric functions were defined in terms of right triangles as opposed to central angles. Also in the sixteenth century, trigonometry came into its own as a mathematical discipline independent of astronomy. There began an analytic approach to trigonometry similar to our modern functional approach.

Trigonometry is linked to the complex number system by the relationship

$$e^{ix} = \cos x + i \sin x, \quad i^2 = -1,$$

which was discovered in the eighteenth century. From this formula and the algebraic properties of the exponential, all trigonometric identities can be derived.

Today, numerical trigonometry is still applied to astronomy as well as surveying and navigation, and analytic or functional trigonometry is used in mathematics and science wherever periodic phenomena appear.

Trigonometric Identities

Quotient and Reciprocal Identities

$$\tan \theta = \frac{\sin \theta}{\cos \theta} \qquad \cot \theta = \frac{\cos \theta}{\sin \theta}$$

$$\csc \theta = \frac{1}{\sin \theta} \qquad \sec \theta = \frac{1}{\cos \theta}$$

$$\cot \theta = \frac{1}{\tan \theta}$$

Negative Identities

$\sin(-\theta) = -\sin \theta \qquad \cos(-\theta) = \cos \theta$

$\tan(-\theta) = -\tan \theta$

Pythagorean Identities

$\sin^2 \theta + \cos^2 \theta = 1$

$1 + \tan^2 \theta = \sec^2 \theta$

$1 + \cot^2 \theta = \csc^2 \theta$

Cofunction Identities

$\sin(\pi/2 - \theta) = \cos \theta$

$\cos(\pi/2 - \theta) = \sin \theta$

$\tan(\pi/2 - \theta) = \cot \theta$

Sum and Difference Identities

$$\sin(\alpha + \beta) = \sin\alpha\cos\beta + \cos\alpha\sin\beta$$
$$\cos(\alpha + \beta) = \cos\alpha\cos\beta - \sin\alpha\sin\beta$$
$$\tan(\alpha + \beta) = \frac{\tan\alpha + \tan\beta}{1 - \tan\alpha\tan\beta}$$
$$\sin(\alpha - \beta) = \sin\alpha\cos\beta - \cos\alpha\sin\beta$$
$$\cos(\alpha - \beta) = \cos\alpha\cos\beta + \sin\alpha\sin\beta$$
$$\tan(\alpha - \beta) = \frac{\tan\alpha - \tan\beta}{1 + \tan\alpha\tan\beta}$$

Product Identities

$$\sin\alpha\cos\beta = \frac{1}{2}[\sin(\alpha + \beta) + \sin(\alpha - \beta)]$$
$$\cos\alpha\sin\beta = \frac{1}{2}[\sin(\alpha + \beta) - \sin(\alpha - \beta)]$$
$$\cos\alpha\cos\beta = \frac{1}{2}[\cos(\alpha + \beta) + \cos(\alpha - \beta)]$$
$$\sin\alpha\sin\beta = -\frac{1}{2}[\cos(\alpha + \beta) - \cos(\alpha - \beta)]$$

Double-Angle Identities

$$\sin 2\alpha = 2\sin\alpha\cos\alpha$$
$$\cos 2\alpha = \cos^2\alpha - \sin^2\alpha = 2\cos^2\alpha - 1 = 1 - 2\sin^2\alpha$$
$$\tan 2\alpha = \frac{2\tan\alpha}{1 - \tan^2\alpha}$$

Factor Identities

$$\sin\alpha + \sin\beta = 2\sin\left[\frac{1}{2}(\alpha + \beta)\right]\cos\left[\frac{1}{2}(\alpha - \beta)\right]$$
$$\sin\alpha - \sin\beta = 2\cos\left[\frac{1}{2}(\alpha + \beta)\right]\sin\left[\frac{1}{2}(\alpha - \beta)\right]$$
$$\cos\alpha + \cos\beta = 2\cos\left[\frac{1}{2}(\alpha + \beta)\right]\cos\left[\frac{1}{2}(\alpha - \beta)\right]$$
$$\cos\alpha - \cos\beta = -2\sin\left[\frac{1}{2}(\alpha + \beta)\right]\sin\left[\frac{1}{2}(\alpha - \beta)\right]$$

Half-Angle Identities

$$\sin(\alpha/2) = \pm\sqrt{\frac{1 - \cos\alpha}{2}} \qquad \cos(\alpha/2) = \pm\sqrt{\frac{1 + \cos\alpha}{2}}$$
$$\tan(\alpha/2) = \frac{\sin\alpha}{1 + \cos\alpha} = \frac{1 - \cos\alpha}{\sin\alpha}$$

Oblique Triangles

Law of Sines: $\dfrac{\sin\alpha}{a} = \dfrac{\sin\beta}{b} = \dfrac{\sin\gamma}{c}$

Law of Cosines:

$a^2 = b^2 + c^2 - 2bc\cos\alpha$
$b^2 = a^2 + c^2 - 2ac\cos\beta$
$c^2 = a^2 + b^2 - 2ab\cos\gamma$

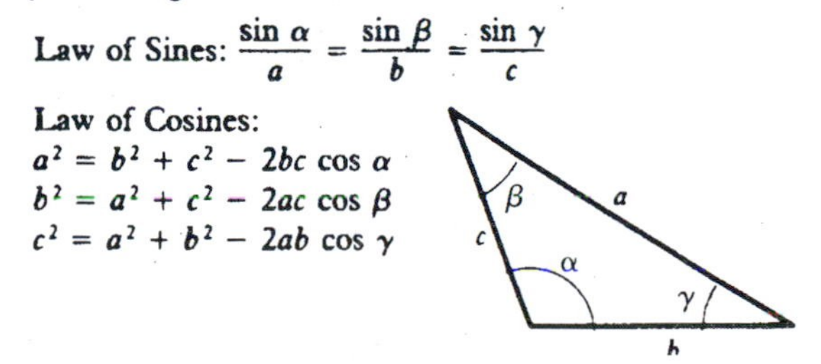

SIENA, FLORENCE AND PADUA:
ART, SOCIETY AND RELIGION 1280–1400